# ADVANCED MATERIALS TECHNOLOGY '87

SOCIETY FOR THE ADVANCEMENT OF
MATERIAL AND PROCESS ENGINEERING

32ND INTERNATIONAL SAMPE SYMPOSIUM
AND EXHIBITION

VOLUME 32

# ADVANCED MATERIALS TECHNOLOGY '87

EDITED BY
Ralph Carson
Martin Burg
Kendall J. Kjoller
Frank J. Riel

Anaheim Convention Center
Anaheim, California
April 6-9, 1987

Additional copies of this publication may be obtained from
Society for the Advancement of Material and Process Engineering

SAMPE International Business Office
P.O. Box 2459
Covina, California 91722

ISBN 0-938994-34-4

32nd INTERNATIONAL SAMPE SYMPOSIUM AND EXHIBITION

## INTERNATIONAL OFFICERS

| | |
|---|---|
| President | Charles J. Hurley |
| Past President | Charles J. Weizenecker |
| 1st Vice President | Gary W. Valentine |
| 2nd Vice President | G. B. Wadsworth |
| Secretary | Brian A. Wilson |
| Treasurer | Susan Lockerby |

## CHAPTER OFFICERS
### (San Diego Chapter)

| | |
|---|---|
| Chairman | Donald G. Harmston |
| Vice Chairman | Edward H. Hanson |
| Secretary | Tseng-Hua Tsiang |
| Treasurer | N. R. Adsit |
| Sr. International Director | Robert H. Greer |
| Jr. International Director | Mark T. Beckerson |

PUBLICITY CHAIRMAN

DON HARMSTON

REGISTRATION CHAIRMAN

MARK BECKERSON

VENDOR REGISTRATION

SUSAN PERRY

FINANCE CHAIRMAN

BARBARA STEVINSON

Other Committee Members

Arrangements Chairman
Tseng-Hua Tsiang

Audio/Visual Chairman
Ed Hanson

SESSIONS AND SESSION CHAIRMEN

NEW BISMALEIMIDE RESIN TECHNOLOGY

 Charles Segal, OMNIA

INTRODUCTION TO ADVANCED COMPOSITES TECHNOLOGY

 Gail DiSalvo, Ciba-Geigy Corporation

MANUFACTURING TECHNOLOGY

 Art James, Lockheed-California Company

AFML FORECAST II

 H. M. Burte, Air Force Materials Laboratory

 Charles E. Browning, Air Force Materials Laboratory

 Charles Y. Lee, Air Force Materials Laboratory

AIRCRAFT APPLICATIONS

 James Rengon, Cessna Aircraft Company

HIGH PERFORMANCE THERMOPLASTIC MATERIALS

 Martin Harvey, I.C.I. Americas

NEW FIBER TECHNOLOGY

 Dan O'Neil, Georgia Technical Research Institute

 Dave Gibson, HITCO

ADHESIVE MATERIALS AND PROCESSES

 Weldon Scardino, Air Force Materials Laboratory

PROCESSING/MANUFACTURING SCIENCE

 Toby Cordell, Air Force Materials Laboratory

DESIGN AND ANALYSIS OF COMPOSITE STRUCTURES

 Leslie J. Cohen, McDonnell Douglas

 Sam J. Dastin, Grumman Aerospace

AIRCRAFT ENGINE APPLICATIONS

 Robert Miller, Pratt & Whitney Aircraft

THERMOPLASTIC COMPOSITE APPLICATIONS

 Mike Fortson, Air Force Materials Laboratory

CERAMIC COMPOSITES

 William Long, Babcock & Wilcox

ADHESIVE MATERIALS AND PROCESSES

 Raymond B. Krieger, Jr., American Cyanamid

MANUFACTURING IN SPACE

 Curtis R. Davies, McDonnell Douglas

SIMPLIFIED DESIGN AND FRACTURE MECHANICS

 Stephen Tsai, Air Force Materials Laboratory

AUTOMOTIVE APPLICATIONS

 Peter Beardmore, Ford Motor Company

 Carl Johnson, Ford Motor Company

SESSIONS AND SESSION CHAIRMEN

HIGH TEMPERATURE THERMOSET RESINS

Robert Jones, TRW, Inc.

CERAMIC COMPOSITES

Willard Hauth, Dow Corning Corporation

PRESSURE SENSITIVE ADHESIVES

Yehuda Ozari, Avery International

ADVANCED FILAMENT WINDING TECHNOLOGY

Brian Wilson, Aerojet Strategic Propulsion

SPACECRAFT MATERIALS APPLICATIONS

Jerry Bauer, Jet Propulsion Laboratory

METAL MATRIX COMPOSITES

James Cornie, Massachusetts Institute of Technology

IMPACT DAMAGE TOLERANCE AND CONTROL IN FILAMENT WOUND STRUCTURES

Brian Wilson, Aerojet Strategic Propulsion

COMPOSITE DESIGN TECHNOLOGY

Donald Humphrey, Brunswick Corporation

SPACECRAFT MATERIALS APPLICATIONS

Robert A. Dunaetz, Hughes Aircraft Company

HIGH TEMPERATURE THERMOSET RESINS

C. H. Sheppard, Boeing Aerospace Company

EPOXY RESIN TECHNOLOGY

Jerry M. Hartman, American Cyanamid

AUTOMATED PROCESSING EQUIPMENT

Paul Pirrung, Air Force Materials LaboratoryASBESTOS REPLACEMENT FIBERS

ASBESTOS REPLACEMENT FIBERS

Novis Smith, R. K. Textiles

THERMALLY HARDENED MATERIALS FOR ELECTRONICS

Allan Goldberg, Harry Diamond Laboratories

CARBON/CARBON COMPOSITES

Charles Rowe, NSWC

EPOXY RESIN TECHNOLOGY

Larry Hopper, Narmco Materials

PULTRUSION TECHNOLOGY

Brian Jones, Compositek Engineering

ASBESTOS REPLACEMENT FIBERS

Roger Barker, North Carolina State University

COMMERCIAL APPLICATIONS

Clyde Yates, HITCO

# PREFACE

The San Diego Chapter of SAMPE is pleased to host the 32nd International SAMPE Symposium & Exhibition at the Anaheim Convention Center. The steady growth of SAMPE becomes more evident each year, and this will be the first SAMPE Symposium in fifteen years to be held in this large, first-class facility. The meeting rooms are bigger and there is a record number of exhibits.

The theme for this symposium is "Advanced Materials Technology '87" and this expresses succinctly the primary goals of the program co-chairmen. Session chairmen were chosen because of their broad knowledge of the subjects, and they provided invaluable assistance to the committee responsible for selecting the technical papers. Among the general subjects to be covered in the technical sessions are high-performance thermoplastic and thermosetting resins, including adhesives; several types of advanced fibers; design, processing, and manufacturing technology for advanced composites; and various applications of high-performance composites. The sessions are arranged so that attendees with a particular area of interest -- such as composite design, will not have to choose between concurrent sessions on the same subject.

Not included in this symposium are papers on metallic materials, since these will be presented at another SAMPE technical conference later this year. Most papers on electronic materials will also be given at a separate SAMPE conference. The organizing committee received many helpful suggestions from Henry Brown before his fatal accident last September. Henry insisted (and we agreed) that program chairmen should strive for quality rather than quantity, so session subjects had to be both important and timely. Session chairmen were individually contacted rather than waiting for volunteers, and only those papers which met certain criteria were accepted.

Our sincere thanks to all of the speakers, co-authors, and session chairmen. Also deserving much credit is a sizable team of volunteers from the San Diego and other Southern California chapters who willingly contributed their time and expertise. As always, special thanks are due to Marge Smith and Doris Weaver at the International Business Office for their experience, dedication, and long hours of effort to make each annual symposium a smooth-running, successful, and memorable event.

Ralph Carson  
General Chairman

Ken Kjoller  
Program Chairman

Frank Riel  
Program Chairman

Marty Burg  
Program Chairman

CONTENTS

PAPERS AND AUTHORS

PAPERS AND AUTHORS

PAPERS AND AUTHORS

PAPERS AND AUTHORS

PAPERS AND AUTHORS

PAPERS AND AUTHORS

PAPERS AND AUTHORS

PAPERS AND AUTHORS

BISMALEIMIDE FORMULATIONS FOR RESIN TRANSFER MOULDING

Kevin D Potter
BP Research Centre, Sunbury on Thames
Middx, UK

Frank C Robertson
BP Advanced Composites Ltd, Avonmouth
Bristol, UK

# 1. INTRODUCTION

## 1.1 Chemistry of Bismaleimides

Abstract

Bismaleimide (BMI) resins are being considered for a number of primary and secondary aerospace components. BMIs offer ease of processing, high temperature capability and hot/wet environmental resistance.

Considerable work has been carried out on the development of resin transfer moulding (RTM) as a viable alternative to conventional compression and autoclave moulding. RTM has the capability of producing high quality components of complex shape with tight tolerances. Little work however, has been reported on combining BMI technology with RTM.

The paper will present work in three areas,

1) screening of resins suitable for RTM
2) preparation of laminates using RTM
3) manufacture of a complex demonstrator component using BMI resin and RTM.

Bismaleimide (BMI) resins are being considered for a number of aerospace applications. These resins have been developed to bridge the gap between epoxies which are easy to process but have limited retention of properties above $130^{\circ}C$ (wet) and polyimides which have excellent retention of mechanical properties at high temperature but are generally difficult to process.

The most commonly used building block in BMI chemistry is bis (4-maleimidodiphenyl) methane. When homopolymerised, a high cross-link density and high backbone aromatic content results (Figure 1) producing a very thermally stable matrix. The cured material, however, is brittle. Considerable formulation work has been carried out to improve fracture toughness and processability while retaining high temperature properties. To achieve these aims BMIs have been blended and co-reacted with eg., other BMIs, dinucleophiles (1, 2), allyl/vinyl monomers (3, 4), epoxies (5), thermoplastics (6) and reactive rubbers (7).

1

## 1.2 Potential of RTM/BMI

BP has demonstrated the capability to manufacture complex components by RTM with the commonly used fibre/matrix combinations (8). A comparison of the advantages of RTM with component requirements in various sectors of the aerospace industry indicated that aeroengines are a good market for the utilisation of the technology (Table 1). A prerequisite for the development of aeroengine components is that matrix systems with acceptable properties at temperatures in excess of $200^{o}C$ are required. The realisation of this fact has led directly to the development of techniques to handle BMIs by RTM, building on the broad base of RTM technology within BP.

There are opportunities for BMI/RTM outside the aeroengine sector, e.g., gearbox casings (due to their retention of properties at high temperature) and aircraft interiors (due to their excellent fire, smoke and toxicity behaviour). This illustrates some of the technical drives that inspired BP to progress with RTM of BMIs.

## 2. RESIN FORMULATION

The resin formulations developed are required to fulfil a number of functions:-

Processable, in terms of, viscosity, pot life, injection time and cure time.

Properties, a use temperature of at least $200^{o}C$, with greater than 50% retention of room temperature resin dominated properties.

## 3. TESTING PROGRAMME

### 3.1 Resin

#### 3.1.1 Viscosity Profile

Time-temperature-viscosity curves were generated for each formulation developed using a Viscometers UK Ltd LV-8 viscometer fitted with a TCU-2A temperature controller.

#### 3.1.2 Differential Scanning Calorimetry (DSC).

Resin formulations were assessed using a Mettler instrument. Onset temperature, peak exotherm temperature and polymersation energy was determined.

#### 3.1.3 Neat Resin

Resin formulations were thoroughly degassed and cast in a metal mould, cured for a standard time of 4 hours at $180^{o}C$, demoulded and post-cured for 12 hours at $240^{o}C$. The flexural strength/modulus ($20^{o}C$ and $200^{o}C$), using ASTM D 790, was determined for each formulation.

### 3.2 Laminate Mould Design

The mould used consists of a base plate and side walls, (Figure 2).

Outside the mould area four pillars carry the top plate which is positioned hydraulically. Dimensional control is by internal or external mould stops. Both mould faces have integral heaters which can operate up to $250^{o}C$. The mould cavity can be varied to allow the use of insert plates, to enable investigation of gating options.

### 3.3 Laminate Properties

The mechanical properties of cured laminates were determined before and after post-curing. The flexural strength/modulus ($22^{\circ}$ and $250^{\circ}$C), ILSS and $V_f/V_v$ was determined for each laminate.

4. RESULTS AND DISCUSSION

4.1 Chemistry

For the purpose of this study a commercially available blend of two aromatic BMIs was used as the base resin. The maleimide double bond is known to react via a number of routes. Several have been reported in the literature e.g. Michael Addition, free radical co-polymerisation, Diels-Alder reaction and "Ene" reaction. These reaction pathways offer methods of converting base BMI resins to materials capable of being processed by RTM.

The modifying agents used in this study are shown schematically in Figure 3. The vinyl terminated aromatic compounds (structures A and B) are postulated to co-polymerise with maleimides by a free radical mechanism. The linear chain extension is expected to produce a reduction in the matrix cross-link density.

The reaction of a maleimide group with allyl terminated molecules is thought to proceed via an "Ene" reaction followed by a Diels-Alder reaction. This yields a cyclic structure and chain extension Figure 4. Stenzenberger (4) and Chaudhari (9) both describe such a reaction, producing cured resins with high fracture toughness.

Allylic monomers e.g. o-diallylphthalate and triallycyanurate are capable of reacting with BMIs by a high temperature free radical reaction. Alternatively DAP could react via an "Ene" reaction with BMI in a similar

way to BMI with allyletherified ortho cresol novolac as proposed by Nakamura (10). DSC suggests that BMI resin homopolymerises at lower temperatures and co-polymerises with allylic monomers at high temperature. The method of modification is not favoured since little increase in fracture toughness is thought to occur.

4.2 Resin Properties

4.2.1 Modifiers A and B

The addition of modifier A produced acceptable mechanical properties at an addition level of 20 parts as shown in Table 2. However, at this level of addition the viscosity of the resin melt at feasible injection temperatures was too high and unacceptable for RTM.

As shown in Figure 3 modifier B has the same terminal groups as modifier A but is essentially the monomeric form. The viscosity of modifier B is less than 2 poise at temperatures greater than $40^{\circ}$C and therefore is ideal for reducing the viscosity of base resins. As the level of modifier B was increased a reduction in flexural strength at $23^{\circ}$C and $200^{\circ}$C was noted (Table 3). The corresponding reduction in modulus was less apparent. The Tg of the cured resin remained high even when 35 parts of modifier were added. This is surprising since Tg of the cured modifier is only $135^{\circ}$C. Laminates prepared from wet prepreg exhibited microcracking after post-curing at $240^{\circ}$C. Therefore no attempt to carryout RTM experiments was made.

4.2.2 Modifiers C and D

Modifier C is a viscous semi-solid at ambient temperature, but at RTM

3

injection temperature the viscosity is reduced to an acceptable level of 1-2 poise. The addition of C significantly reduced the viscosity of the base resin (Figure 5) by a factor of 2-4. At each addition level investigated the viscosity of the resin remained stable at $125^\circ$C for at least one hour. Despite the long pot-life at $125^\circ$C the formulated material begins to gel rapidly at $175^\circ$C. As the concentration of modifier was raised the reactivity of the blend was observed to increase.

DSC was used to gain information on the curing of the modified resins. As the level of modifier was increased, polymerisation onset temperature decreased (Figure 7). A shoulder in the DSC trace at approximately $310^\circ$C became more pronounced as the level of modifier was increased. The associated chemical reaction has not been identified. The modifier, however, does not thermally homopolymerise.
Polymerisation energy was noted to decrease from 320 J/g to 260 J/g as the level of modifier was raised from 15 parts to 35 parts. The reduction in exotherm is expected to reduce the thermal stresses generated during laminate curing.

The addition of allyl terminated modifier C had little effect on room temperature mechanical prop-erties (Table 4). The retention of strength and modulus at $200^\circ$C was also acceptable and superior to formulations modified with D (see below). This is probably related to the higher aromatic backbone content of modifier C compared to D which contains flexible isopropylidene groups.

Modifier D behaves in a similar manner to C in terms of viscosity profile (Figure 8). Modifier D, however, is of higher viscosity and is less reactive. The mechanical properties of base resin modified with D are shown in Table 5. The trends are similar to those observed with Modifier C.

A general trend towards lower modulus at high temperature as modifier level is increased (C and D) was observed. This is expected due to the higher backbone flexibility of the modifiers compared to the relatively stiff BMI. From the work of Chaudhari (9) and Stenzenberger (4) the reaction of BMI with allyl terminated moieties, is expected to lead to increased resin fracture toughness. The resistance to microcracking in carbon fibre reinforced laminates is expected to be improved by this modification.

4.3 Design of injection equipment and the development of injection parameters

Standard polyester/epoxy RTM machines are unsuitable for BMIs. The requirements for BMI production equipment was considered as a separate issue to the process development per se, and the simplest possible equipment was chosen for the latter work.

The equipment employed consisted of a heated pressure pot connected through heated pipework to a heated mould. The pressure pot temperature was held slightly above the resin melting point. The resin was heated further to the injection temperature in the pipework. Pressure was varied during the injection cycle.

Injection parameters varied with resin type but some general conclusion can be made.

1. The total thermal input to the resin should be minimised.

2. The mould should be maintained at a slightly elevated temperature in relation to the resin feed.

3. The resin temperature should be minimised.

Following injection, the mould temperature is raised to approximately 200°C, followed by a dwell of 3 to 4 hours to allow resin cure.

4.4 Laminate Properties

The mechanical properties of fabric laminates prepared from modifier D (15 parts) show good retention of ILSS up to 200°C when non post-cured. At 250°C ILSS had dropped to 28 MPa. The non post cured values are comparable to some commercially available BMIs.

Increasing the level of Modifier D to 35 parts and further modification with allylic monomer dramatically reduced the ILSS values at 250°C (Tables 7 and 8). This is due to the high level of flexible modifier.

The formulation containing 15 parts modifier D exhibited a 38% loss and 9% loss in room temperature flexural strength and modulus when tested at 250°C. The loss at high temperature became more significant as the level of modifier was increased to 35 parts. Further modification using allylic monomers led to an improvement in non-post cured high temperature flexural properties.

Laminate samples were post-cured for 5 hours at 240°C to improve the high temperature properties. Tables 6 - 8 show that improvements in high temperature flexural properties could be achieved. However this was at the expense of a slight reduction in room temperature ILSS. Overall the properties achieved after the above post cure are acceptable. Further work however will be necessary to optimise the curing cycle and hence final properties.

4.5 Case Study - The Manufacture of an Article of Complex Geometry by RTM

The component chosen to demonstrate the feasibility of BMI/RTM is illustrated in Figure 9.

a) Mould Design

The external parts consisted of an upper and lower mould plate. The internal parts had to be multipiece to accommodate the undercuts in the component. An aluminium annular split along its diameter and a silicone elastomer ring of variable thickness completed the internal components.

b) Lay-up

To minimise the lay-up time, use was made of the known deformation behaviour of cloth reinforcements (11). Sheets of carbon cloth were taken, sheared and starched to shape. Strips at +60° to the fibre angle were cut from the sheared cloth and wound on a former, with a powder binder used to hold the plies together. The internal mould parts were clamped around this layup. The layup was deformed to cover the internal mould parts using a light solvent spray to release the binder. This produced a handleable preform.

Various other techniques have been developed by BP for the rapid transformation of reinforcement stock into preforming for moulding.

## c) Injection/Cure

Injection of molten resin into the demonstrator was complete after 10 minutes. Resin gelation was noted after approximately 30 minutes and curing carried out at 205°C for 2 hours. The disassembly of the mould after cooling was simple with no damage to the elastomeric ring. Moulding quality was excellent with radii faithfully followed and surface finish comparable to that of the mould face. The demonstrator component was considered to have been fully successful in meeting the set objectives.

## 5. CONCLUSIONS

1) BMI resin systems have been developed which are suitable for resin transfer moulding applications.

2) High quality laminates with low void content and acceptable mechanical properties have been prepared using RTM.

3) A demonstration component of complex geometry with tight corner radii and undercuts has successfully been manufactured.

4) The programme of work has demonstrated that BMI and RTM technology can successfully be combined to produce a component of aerospace quality.

## 6. REFERENCES

1. GB Patent 1,190,718 (1970)
2. US Patent 4,211,860 (1980)
3. H.C. Nash et al ACS Organic Coatings and Plast Chem 46, 877 (1979)
4. H.D. Stenzenberger et al, 29th Nat. SAMPE, 1043 (1985)
5. GB Patent 2,139,628A (1984)
6. Y. Yamamoto et al, 30th Nat. SAMPE, 903, (1985)
7. S.J. Shaw and A.J. Kinloch, Int.J. of Adhesion and Adhesives 5, 123, (1985).
8. Advances in the Resin Injection Process for the reliable production of complex structural components.
   KD Potter in Progress in Advanced Materials & Process, Durability, Reliability and Quality Control. Ed G. Bartelds & R.J. Schliekelmann (1985).

9. M.A. Chaudhari and J.J. King SME Conference, MF85-503, (1985).
10. H. Nakamura et al 18th SAMPE Technical Conference, 694, (1986).
11. Deformation Machining of Fibre Reinforcements and their Influences on the Lubrication of Complex Structural Parts. KD Potter - Proc ICCM3 Paris 1980.

## 7. BIOGRAPHY

Kevin Potter graduated from Imperial College, London with an Honours Degree in Material Science in 1974. He is currently involved in the development of composite components using RTM.

Frank Robertson graduated from Heriot-Watt University, Edinburgh with an Honours Degree in Chemistry in 1977 and with a Ph.D in Polymer Chemistry in 1980. He is currently responsible for the development of high temperature resins and prepregs for aerospace applications.

Table 1 - Advantages of RTM Processing

| OBJECTIVE | EXAMPLE/COMMENT |
|---|---|
| Tight tolerances can be obtained on all surfaces and a surface finish identical to that of the mould. | Flywheel hubs made by RTM can be spin tested to destruction without any balancing. |
| Components can be complex. | Casing components made by RTM have ribs, flanges, 'O' ring seals and accurate bores for bearing housings - all moulded in. |
| Simple cure cycles. | Single stage cures, with no dwells for all matrices used to date. |
| A wide variety of resins can be used without prepregging. | All quality limiting factors are in the hands of the fabricator. |
| A wide variety of reinforcements can be used. | Unidirectional, bidirectional, and three dimensional reinforcements can be used, as well as a wide variety of knitted goods. |
| Fibre contents are controllable and predictable. | The fibre content of a moulding can be verified independent of resin, thus $V_f$ and hence mechanical properties can be held consistently. |
| Voidage can be eliminated. | Hubs for the flywheel system are 7mm thick with glass reinforcement. These components are almost transparent. |
| Quality Control | Samples of both laminate and unreinforced resin can be produced in an RTM component mould for QC purposes. |

TABLE 2        Resin + Modifier A

| Property | Test Temperature ($^\circ$C) | Addition Level (parts) | | |
|---|---|---|---|---|
| | | 0 | 20 | 35 |
| Flexural Strength (MPa) | 23 | 108 | 114 | 10.8 |
| Flexural Strength (MPa) | 200 | 61 | 48 | 31 |
| Flexural Modulus  (GPa) | 23 | 3.8 | 3.7 | 3.6 |
| Flexural Modulus  (GPa) | 200 | 2.4 | 1.8 | 1.2 |

TABLE 3        Resin + Modifier B

| Property | Temperature ($^{o}$C) | Addition Level (parts) | | | |
|---|---|---|---|---|---|
| | | 0 | 15 | 25 | 35 |
| Flexural Strength (MPa) | 23 | 108 | 106 | 84 | 76 |
| Flexural Strength (MPa) | 200 | 61 | 54 | 29 | 31 |
| Flexural Modulus  (GPa) | 23 | 3.8 | 3.5 | 3.5 | 3.5 |
| Flexural Modulus  (GPa) | 200 | 2.4 | 2.4 | 1.9 | 1.7 |
| Tg ($^{o}$C) | - | | 354$^{o}$ | 348$^{o}$ | 341$^{o}$ |

TABLE 4        Resin + Modifier C

| Property | Temperature ($^{o}$C) | Addition Level | | | |
|---|---|---|---|---|---|
| | | 0 | 15 | 25 | 35 |
| Flexural Strength (MPa) | 23 | 108 | 104 | 120 | 118 |
| Flexural Strength (MPa) | 200 | 61 | 62 | 65 | 63 |
| Flexural Modulus  (GPa) | 23 | 3.8 | 3.6 | 3.7 | 3.6 |
| Flexural Modulus  (GPa) | 200 | 2.4 | 2.9 | 2.0 | 2.0 |

TABLE 5        Resin + Modifier D

| Property | Temperature ($^{o}$C) | Addition Level (parts) | | | |
|---|---|---|---|---|---|
| | | 0 | 15 | 25 | 35 |
| Flexural Strength (MPa) | 23 | 108 | 105 | 134 | 112 |
| Flexural Strength (MPa) | 200 | 61 | 50 | 53 | 53 |
| Flexural Modulus  (GPa) | 23 | 3.8 | 3.5 | 3.8 | 3.4 |
| Flexural Modulus  (GPa) | 200 | 2.4 | 2.6 | 2.0 | 1.7 |

TABLE 6     Laminate Properties

Base Resin – Modifier D (15 parts)

TORAY T300B 3K – 5HS

| Property | Test Temperature ($^{o}$C) | Post Cure (hrs @ 240$^{o}$C) | |
|---|---|---|---|
| | | 0 | 5 |
| ILSS | 23 | 44 MPa | 44 |
| ILSS | 170 | 39 MPa | – |
| ILSS | 200 | 39 MPa | – |
| ILSS | 250 | 28 MPa | 34 |
| Flexural Strength (Warp) | 23 | 902 MPa | 890 MPa |
| Flexural Strength (Warp) | 250 | 557 MPa | 630 MPa |
| Flexural Modulus (Warp) | 23 | 54 GPa | 52 GPa |
| Flexural Modulus (Warp) | 250 | 49 GPa | 50 GPa |
| $V_f$ | – | 52.7% | – |
| $V_v$ | – | 1.57% | – |
| Microcracking | – | None | – |

TABLE 7     Laminate Properties

Base Resin  Modifier D (35 parts)

TORAY T300B 3K – 5HS

| Property | Test Temperature ($^{o}$C) | Post Cure (hrs @ 240$^{o}$C) | |
|---|---|---|---|
| | | 0 | 5 |
| ILSS | 23 | 62 MPa | 48 |
| ILSS | 170 | 39 MPa | – |
| ILSS | 200 | 31 MPa | – |
| ILSS | 250 | 12 MPa | 34 |
| Flexural Strength (Warp) | 23 | 734 MPa | 715 MPa |
| Flexural Strength (Warp) | 250 | 294 MPa | 580 MPa |
| Flexural Modulus (Warp) | 23 | 47 GPa | 48 GPa |
| Flexural Modulus (Warp) | 250 | 36 GPa | 43 GPa |
| $V_f$ | – | 53.0% | – |
| $V_v$ | – | 0.1% | – |
| Microcracking | – | None | – |

9

TABLE 8    Laminate Properties

Base Resin + Modifier D (35 parts) - reactive diluent (5 parts)

| Property | Test Temperature (°C) | Post Cure (hours @ 240°C) | |
|---|---|---|---|
| | | 0 | 5 |
| ILSS | 23 | 49 MPa | 46 MPa |
| ILSS | 170 | 38 MPa | - |
| ILSS | 200 | 32 MPa | - |
| ILSS | 250 | 14 MPa | 28 MPa |
| Flexural Strength (Warp) | 23 | 747 MPa | 765 MPa |
| Flexural Strength (Warp) | 250 | 550 MPa | 630 MPa |
| Flexural Modulus (Warp) | 23 | 52 GPa | 51 GPa |
| Flexural Modulus (Warp) | 250 | 45 GPa | 40 GPa |
| $V_f$ | - | 57.7% | 57.7% |
| $V_v$ | - | 0.3% | 0.3% |
| Microcracking | - | None | - |

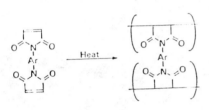

Fig. 1 Bismaleimide
Homopolymerisation

$$CH_2=CH-\overset{\overset{O}{\|}}{C}-O\left[R\right]_{n=3-5}O-\overset{\overset{O}{\|}}{C}-\overset{CH_3}{\underset{}{CH}}=CH_2 \qquad A$$

$$CH_2=\overset{CH_3}{\underset{}{C}}-\overset{\overset{O}{\|}}{C}-O\left[R\right]_{n=1}O-\overset{\overset{O}{\|}}{C}-\overset{CH_3}{\underset{}{C}}=CH_2 \qquad B$$

$$CH_2=CH-CH_2\left[Ar\right]CH_2-CH=CH_2 \qquad C$$

$$CH_2=CH-CH_2\left[R-C(CH_3)-R\right]CH_2-CH=CH_2 \qquad D$$

$$\bigcirc(\overset{\overset{O}{\|}}{C}-O-CH_2CH=CH_2)_{2\,or\,3} \qquad E$$

Fig. 3    RTM Modifiers

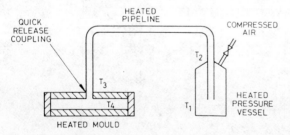

IN GENERAL :

$$T_1 < T_2 < T_3 < T_4$$

MOULD EQUIPED WITH -:

INTEGRAL HEATERS
HYDRAULIC CLAMPING
EJECTOR PINS
INTERNAL & EXTERNAL MOULD STOPS
INTERCHANGEABLE MOULD FACES

Fig. 2 Schematic diagram of
Resin Transfer Equipment
and Flat Plate Mould

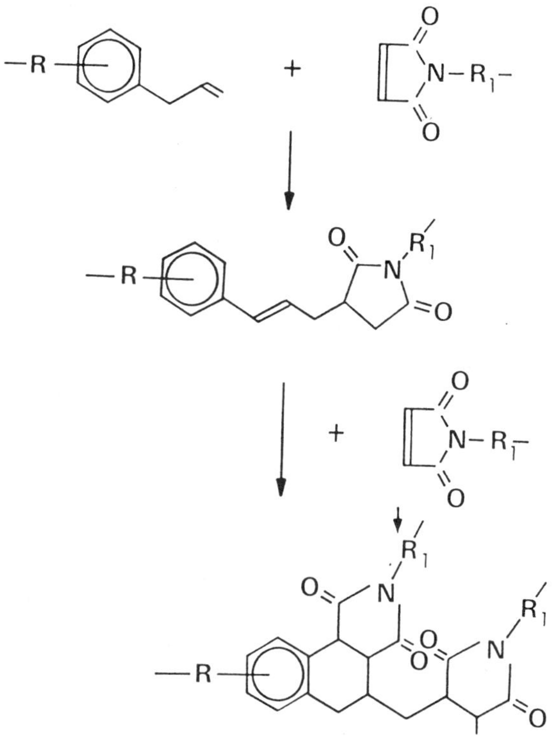

Fig. 4 Proposed Reaction of BMI with
Allyl-Phenyl Compounds

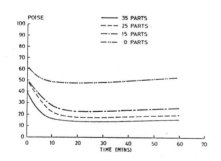

Fig. 5
Modifier C Viscosity
at 125C

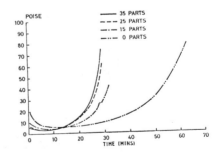

Fig. 6
Modifier C Viscosity
at 175C

11

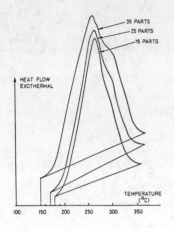

Fig. 7 DSC Traces –
Base Resin +
Modifier C

Fig. 8  Comparison Viscosity
Modifier C + D

ID  =  140mm
OD  =  260mm

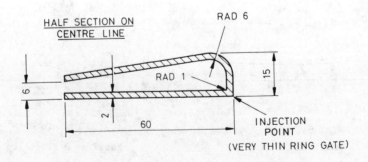

HALF SECTION ON
CENTRE LINE

RAD 6

RAD 1

15

6

2

60

INJECTION
POINT
(VERY THIN RING GATE)

Fig. 9  Demonstrator Component Showing
Tight Radii

32nd International SAMPE Symposium
April 6-9, 1987

SOLVENT AND AMINE-FREE POLYIMIDE MATRICES:
A NEW CLASS OF KERIMID RESINS FOR STRUCTURAL ADVANCED COMPOSITES

Neale A. Mc Garry, George Pouzols and Patrick R. Lopez
Rhone Poulenc Inc.
P.O. Box 125, Monmouth Junction, 08852 New Jersey

## Abstract

Development polyimide systems representing second generation hot melt thermosetting matrices based on bismaleimide chemistry are presented in this paper, namely KERIMID FE 70006 AND KERIMID FE 70011. They are a first answer to meet very precise requirements which were not met by state-of-the art products such as KERIMID 601, particularly in the area of processing. The physical evaluation of model end-products allows optimistic conclusions for future developments.

## 1. INTRODUCTION

KINEL and KERIMID are respectively the registered trade-names of polyimide compounds and impregnation resins marketed by RHONE POULENC. The basic chemistry is typically of the addition polyaminobismaleimide type (Figure 1). They offer excellent thermomechanical properties, withstanding high stresses at temperatures at which phenolics, epoxies and most high-quality engineering plastics are no longer satisfactory.

However, in recent years new industrial needs have caused a strong emphasis on research concerning resins with very special characteristics such as tack and drape or improved interlaminar toughness, particularly for the aerospace sector.
On the ground of our experience in this area (Figures 2 to 6) and thanks to a common collaboration with worldwide industrial leaders, RHONE POULENC is now proposing new products to meet this technological challenge.

## 2. EXPERIMENTAL

All samples which were processed or examined in this paper were extracted from standard productions, as they would be sampled or sold to customers.
Fiber reinforcements were T300 woven carbon fibers, 3K-PW Toray equivalent, Filkar, provided by ELF AQUITAINE of France (HERCULES AS4 fibers showed similar final results). T300 fibers were our reference, although their thermostability or their usage for KERIMID matrices may be discussed.

13

Thermal analysis investigations were conducted on a Dupont 1090 thermal analyser driving all usual Dupont peripherals. All scans were performed at a speed of 10°C/min, except for DMA, at 3°C/min.
Mechanical evaluation was based on ASTM and ISO methods since no discrepancy was observed in our case.

## 3. PROCESSING CONDITIONS

### 3.1 Prepregging

Standard pilot plant carbon prepregs of KERIMID 601 were used. KERIMID FE 70006 AND FE 70011 were impregnated at temperatures in the range of 80-100°C (175-210°F) in a standard hot-melt impregnation line. These are acceptable conditions since the pot life is superior to 10 hours at 100°C (210°F). Wettability was correct, but tackiness was considered as insufficient for FE 70011. Work has been engaged in this area and results can be obtained without any sacrifice in terms of properties or processing. Prepregs from the same campaign showed no evolution when stored for three months in a cold chamber, at room temperature or in a humid environment. Studies of prepreg storing ability were discontinued at 3 months.
All prepreg systems presented here incorporated 60 to 65% by weight of carbon fibers.

### 3.2 Curing

The direct, addition-type, low-activated reaction chemistry of all KERIMID systems allows very soft and forgiving processing conditions. A typical cure cycle, represented in Figure 7, illustrates this feature. It consists of a 1 to 2-hour plateau at 150°C (300°F). A short application of vacuum is proposed, not so much for extracting volatiles, but rather for eliminating trapped air pockets between plies. A standard 5-bar (80 Psi) pressure, achievable by most autoclaves is sufficient to attain proper compaction. This profile can easily be cutomized to meet particular needs such as filament winding or curing of large structure. A free-standing postcure of at least 15 hours at 250°C is absolutely necessary to provide best final mechanical properties. In some instances, such as long term cycling around or above 300°C (600°F), even higher postcures may be advised.
After postcure, composite and pressureless neat resin plaques were cut and examined by optical and electron microscopy for detection of cracks and voids and morphological examinations. Figure 8 shows the excellent quality of the samples. Flat bidirectional bars were then cut out for mechanical testing.

## 4. PHYSICAL EVALUATION

### 4.1 Thermal analysis

Thermogravimetric analysis performed on KERIMID 601 and KERIMID FE 70006 showed a high thermo-oxydative stability (Figure 9). DMA (Figure 10) revealed a glass transition over 360°C (670°F) for FE 70006 by comparison to 290°C for KERIMID 601 (550°F). In conjunction to TGA, DMA results make new KERIMIDS serious candidates for long term high temperature usage.

### 4.2 Mechanical properties

Figures 11 and 12 show a good correlation between the properties of neat resins and those of the resulting composites. Although there is some controversy concerning the translation of toughness from neat resin to laminates, the improvement of the polymer matrix is generally benefitial to the composite for common thermosets. Characterization of cross-ply laminates has now been initiated to determine which level of toughness is realistic to attain in order to get commercially viable laminates, and if these resins meet the requirements.

## 5. CONCLUSION

The field of engineering plastics R & D requires a strong commitment for long term development. The area of matrices for advanced structural composites is very fast moving in itself. Thanks to heavy basic molecular research and adequate development methodology RHONE POULENC is able to propose state-of-the art polymers and is now willing to improve the present physical properties of these new materials beyond present limits. KERIMID FE products are representative members of this promising generation of high-tech plastics.

## 6. REFERENCES

1. Pouzols, G., and Rakoutz, M., Rhone Poulenc's polyaminobismaleimides Kerimid and Kinel: Further Developments Through Solventless Matrix, 30th National Symposium, 1985, pp 606-609.

2. Lopez, P.R., Characterization and Processing of Polyimide Matrix Composites, Master of Science Thesis, University of Washington, 1986.

3. Woo, E.M., Lopez, P.R., and Seferis, J.C., Processing and Characterization of PMR-15 Polyimide Matrices, to be submitted to Polym. Comp., 1987.

4. Scola, D.A., Addition Polyimides, Chemistry and Properties, Proceedings of the Interdisciplinary Workshop on the Chemistry and Properties of Polyimides, sponsored by the Division of Polymer Chemistry, American Chemical Society, Reno, Nevada, 1985.

5. Mittal, K.L., Polyimides, Plenum Press, New York, 1984.

6. Halpin, J.C., et al, Processing Science: an Approach for Prepreg Composite Systems, Pure and Applied Chem., 55, 5, 893, 1983.

7. Tough Composites Materials, NASA Conference Publication 2334, Langley Research Center, Hampton, Virginia, May 1983.

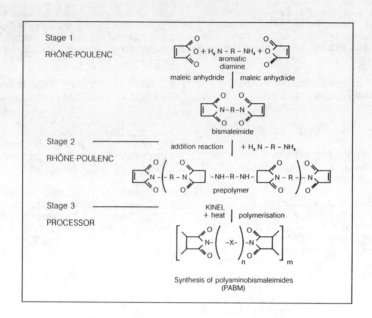

Fig.1   Synthesis of polyaminobismaleimides.

Fig.2   General Electric/SNECMA 50-50 joint venture CFM 56 civil engine equipping most BOEING and AIRBUS modern jets: nose cone out of KINEL 5504 superseded lightweight steel.

16

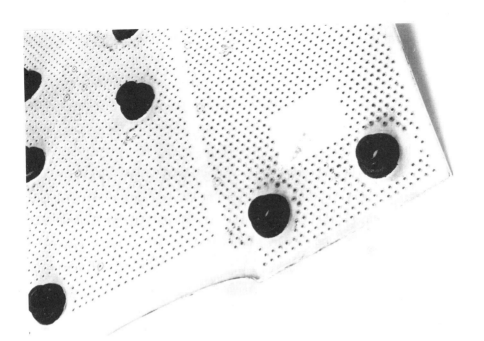

3. Acoustic panel spacer (KINEL 4525).

Fig.4 SNECMA "LARZAC" military aircraft engine motor case
(KERIMID 601 + glass cloth).

Fig.5  Air/oil seal for CFM 56 jet engine (KERIMID 601 + glass
cloth and KINEL 5504).

Fig.6 Printed board for electronics (KERIMID 601).

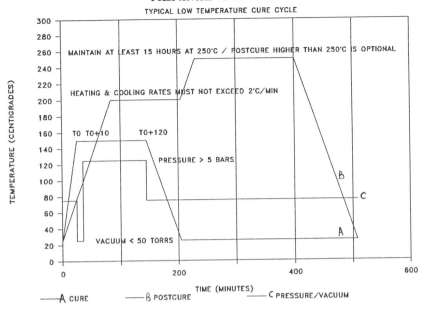

# KERIMID FE 70006

### TYPICAL LOW TEMPERATURE CURE CYCLE

MAINTAIN AT LEAST 15 HOURS AT 250°C / POSTCURE HIGHER THAN 250°C IS OPTIONAL

HEATING & COOLING RATES MUST NOT EXCEED 2°C/MIN

TO TO+10    TO+120

PRESSURE > 5 BARS

VACUUM < 50 TORRS

TEMPERATURE (CENTIGRADES)

TIME (MINUTES)

—— A CURE      —— B POSTCURE      —— C PRESSURE/VACUUM

Fig.7 A typical cure cycle for hot-melt, addition KERIMID FE resins.

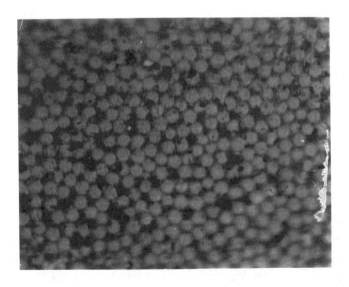

Fig.8 Optical microscopy examination of a Kerimid FE 70006 composite sample showing the validity of the cure cycle.

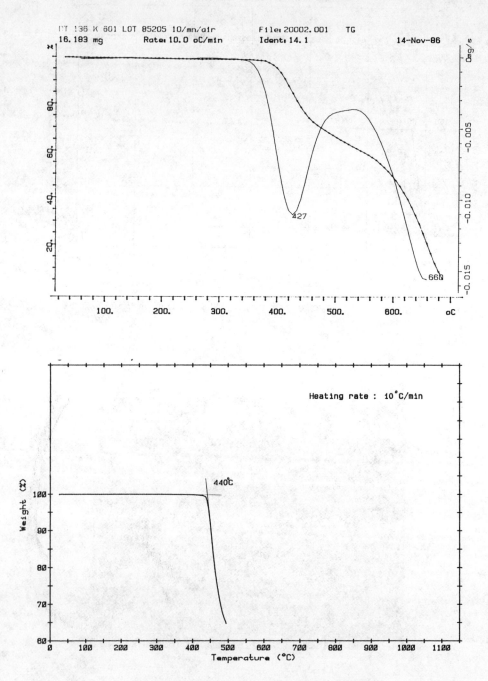

Fig.9 TGA for KERIMID 601 and FE 70006.

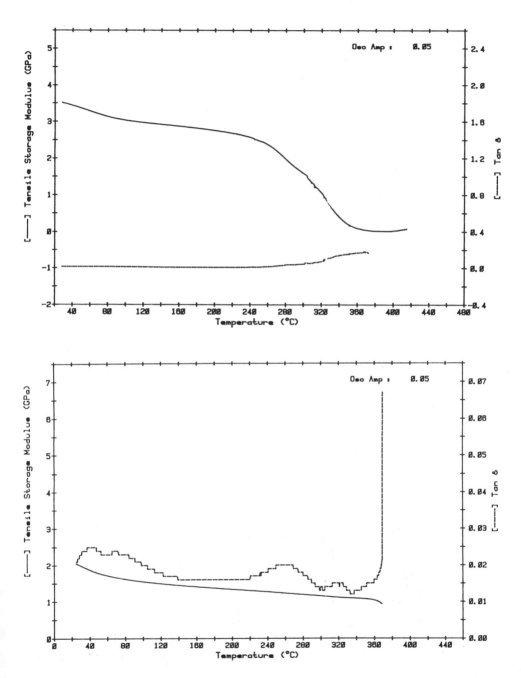

Fig.10 DMA for KERIMID 601 and FE 70006.

# FLEXURAL PROPERTIES OF NEAT KERIMID

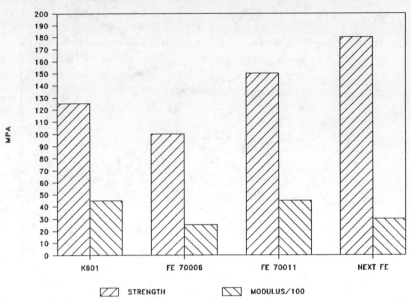

Fig.11 Flexural Properties of KERIMID neat resin systems.

# FLEX. PROPERTIES OF KERIMID COMPOSITES

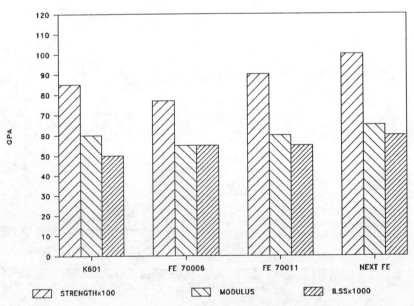

Fig.12 Flexural properties and toughness of KERIMID woven carbon fiber composites.

| PROPERTY | UNIT | K 601 | FE 70006 | FE 70011 | NEXT FEs |
|---|---|---|---|---|---|
| TG | °C | 290 | 370 | 360 | >350 |
| FLEXURAL MODULUS (NEAT RESIN) | GPa | 4.5 | 2.5 | 4.5 | >3.0 |
| FLEXURAL STRENGTH (NEAT RESIN) | MPa | 125 | 100 | 150 | >180 |
| FLEXURAL MODULUS (WOVEN COMPOSITE) | GPa | 60 | 55 | 60 | >60 |
| FLEXURAL STRENGTH (WOVEN COMPOSITE) | MPa | 850 | 770 | 900 | >900 |
| ILSS (WOVEN COMPOSITE) | MPa | 50 | 55 | 55 | >60 |

TABLE OF PHYSICAL PROPERTIES OF KERIMIDS

A NEW BISMALEIMIDE MATRIX RESIN
FOR
HIGH PERFORMANCE ADVANCED COMPOSITES

M. Chaudhari, J. King and B. Lee
CIBA-GEIGY Corporation
Ardsley, New York 10502

## ABSTRACT

For the past few years there has been an increased effort in the development of high temperature, high performance resin systems for composite applications. This paper reports the preliminary characterization data on a new bismaleimide resin developed by CIBA-GEIGY Corporation. This bismaleimide resin has been designated RD85-101 and is based on a proprietary aromatic diamine. Due to the structural nature of this amine, the resultant BMI resin is a unique isomeric blend with low melting temperature, 90-100°C. The lower melting point and enhanced solubility of this resin is common organic solvents, leads to improved processing characteristics compared to most of the current commercially available bismaleimides. When cured with traditional cocuring agents for BMI's such as allylphenols or aromatic amines, the resultant polymer has outstanding thermal/hygrothermal and mechanical properties.

## 1. INTRODUCTION

Fiber reinforced polymeric composites have been used in numerous structural applications over the past few decades. More recently, the increased concern for fuel conservation have placed greater emphasis on the production of light weight parts and less energy intensive manufacturing techniques.

Traditionally, epoxy resins in conjunction with high strength, high modulus fibers, have been utilized in high performance structural composites. These materials have been meeting the past and current requirements but more recently greater demands are being put on the composite structures to perform under extremely hostile environments beyond the capability of the traditional epoxy based matrices. Some of these demands include performance at elevated temperatures and in rapidly changing environments.

To meet these new demands, researchers have been focusing their attention on other classes of materials which can perform at temperatures beyond the reach of current epoxy chemistries. Most sought after class of materials in this category has been bismaleimides (BMI's) and polyimides (PI's). BMI's and PI's are known for their temperature resistance but many traditional BMI's and PI's have serious drawbacks of being brittle and require demanding processing conditions. For bismaleimides or polyimides to gain wider acceptance as a matrix resin, the material has to demonstrate easy processability, preferably as similar to epoxy matrices, as possible.

CIBA-GEIGY Corporation has been very active in the high performance resin technology and has in the past introduced state of-the-art epoxy resin systems and more recently has also responded to the emerging needs of the composite industry by introducing bismaleimide and polyimide systems.

As a result of our on going research programs in high temperature matrix resins, we have now developed another new bismaleimide resin. This paper discusses some of the salient characteristics of this new bismaleimide resin which is based on a proprietary aromatic diamine. This new resin has extremely desirable processing features such as:

- Low melt temperature, 90-100°C.

- Readily soluble in common organic solvents like MEK, acetone, methylene chloride, etc.

- Stable in solution form.

- Stable in melt condition.

- Extremely good reactivity with aromatic amines and reaction takes place at lower temperatures compared to the other commercial bismaleimides.

When cured with traditional curing agents for BMI's such as allylphenols (Matrimid 5292B), the resultant polymer has extremely good thermal/ mechanical properties and good moisture resistance.

## 2. EXPERIMENTAL METHODS AND RESULTS

### 2.1 PREPARATION OF RD85-101

The experimental material, RD85-101 was synthesized from diaminophenylindane and maleic anhydride via the well established synthetic procedures (1,2,3) for the bismaleimide materials.

Typical chemical structure is shown below:

## 2.2 Materials

The materials used in this study are CIBA-GEIGY's experimental material RD85-101 shown above and CIBA-GEIGY's Matrimid™ 5292 A&B (4&5) represented by:

A

B

Methylene dianiline (MDA), commercially available as CIBA-GEIGY's HT972

HT 972

and two other commercial BMI resins identified here as I & II.

Resin I is a sort of prepolymer of MDA based BMI with less than stoichiometric amount of free MDA and resin II is an eutectic mixture of aromatic and aliphatic amine based BMI's.

### 2.3 FORMULATION/PROCESSING

The new BMI (RD85-101) can be formulated and cured with allylphenols, amines and variety of other coreactants. The cured neat resin mechanical properties presented in this paper are based on the formulation when stoichiometric ratios of diallyl bisphenol A (Matrimid 5292 B) are used as a coreactant with RD85-101.

## 2.4 PREPOLYMER

The term prepolymer used in this article refers to a clear homogeneous liquid obtained after mixing and heating RD85-101/ diallylbisphenol A (1:1 M) components for 10-25 minutes at 100-125$^{\circ}$C.

## 2.5 SOLUTION STABILITY

RD85-101 is readily soluble in high concentration in common organic solvents e.g. acetone, MEK and methylene chloride, etc.

Solution stability of the resin and the prepolymer (RD85-101/ Matrimid 5292B) was checked in acetone and MEK.

A 50 w/o solutions of RD85-101 and prepolymer in MEK and acetone were prepared. The stability was monitored by checking the change in viscosity with time, using Cannon-Fenske technique. The solutions were stored at R.T. for periods of up to 6 months and periodic viscosity determinations were made.

No change in viscosity was observed and no materials settled out of the solution, for both RD85-101 alone and the prepolymer.

## 2.6 REACTIVITY OF RD85-101

Differential scanning calorimetric (DSC) technique using DuPont's 910 DSC module, was used to check the reactivity of RD85-101 and other commercial BMI's. 10$^{\circ}$C/minute heat rate was used for this study,

## 2.7 HOMOPOLYMERIZATION

Figures 1,2,&3 are DSC scans for RD85-101, Matrimid 5292A and resin II homopolymerization, respectively.

RD85-101 has the lowest initiation and peak exotherm temperatures compared to the other BMI resins which indicates the possibility of lower curing temperature.

## 2.8 REACTIVITY WITH AROMATIC AMINES

RD85-101 has extremely good reactivity with aromatic amines. Methylene dianiline (commercially available from CIBA-GEIGY as HT972) reacts with RD85-101 at much lower temperatures compared to the other commercial BMI's.

Figures 4,5&6 are DSC scans of RD85-101/HT972, resin I and resin II/HT972, prepolymers, respectively.

RD85-101/HT972 system has the lowest reaction temperature, TI 67$^{\circ}$C and TPK 115$^{\circ}$C compared to the other BMI's. This increased reactivity of RD85-101 with aromatic amines can be very useful in lower curing temperature systems.

## 2.9 RD85-101 MELT STORAGE

RD85-101 MELTS AT 90-100$^{\circ}$C and it has been demonstrated that it can stay in melt form at 100-125$^{\circ}$C for a considerably long time before any appreciable change in melt viscosity or reactivity occurs.

Figure 7 shows DSC scans for RD85-101 showing the reactivity of the melt stored RD85-101 at 125$^{\circ}$C for 5 hours. This is desirable because in case of high melting BMI's melt temperature and homopolymerization temperatures are so close that there is a very narrow processing window.

## 3.0 CURED NEAT RESIN PROPERTIES

For neat resin properties shown in table 1, the prepolymer using RD85-101/Matrimid 5292B (1:1 M) was prepared as described under prepolymer section. The following cure cycle was used:

1 Hr/180°C + 2 Hrs/200°C + 6 Hrs/250°C

No attempt was made at this time to optimize the stoichiometry or the cure cycle. Excellent mechanical properties were obtained.

Figure 8 shows a dynamic mechanical analysis of cured neat resin. Excellent modulus retention is seen.

### 3.1 CONCLUSIONS

A new bismaleimide (RD85-101) resin with outstanding performance characteristics has been developed. This new BMI is based on an aromatic diamine and due to the structural nature of this amine, the resultant BMI is a unique isomeric blend with low melt temperature, 90-100°C and good solubility characteristics in common organic solvents. RD85-101 in melt as well as in solution form has good storage stability. RD85-101 compared to other commercial BMI's has good reactivity with aromatic diamines and reaction takes place at comparatively lower temperatures. The low melt temperature and good solubility in common organic solvents provides processing advantages and makes it a very desirable candidate for advanced composite structures as well as printed wire board applications. RD85-101 when formulated with stoichiometric amounts of diallylbisphenol A (Matrimid 5292B) and properly cured, provides excellent thermal/mechanical properties with good toughness characteristics.

REFERENCES:

1. U.S. Patent 3,127,414 (March 4, 1964)

2. N.E. Searle, U.S. Patent 2,444,536

3. Byung Lee, Mohammand A. Chaudhari and Thomas Galvin "Bismaleimides - Synthesis and applications" 17th National SAMPE Technical Conference, 1985, pp 172-178.

4. J. King, M. Chaudhari and S. Zahir, "A New Bismaleimide System for High Performance composite Applications", 29th National SAMPE Symp., 1984, pp 392-408.

5. M. Chaudhari, T. Galvin and J. King" Characterization of Bismaleimide system, XU292" 30th National SAMPE Symp., 1985, pp 735-746.

Acknowledgements: The authors greatly appreciate the contribution of the entire Matrix R&D staff for their help in generating the data presented in this paper.

TABLE 1

CURED NEAT RESIN MECHANICAL PROPERTIES

RD85-101/Matrimid 5292B (1:1M)

Cure Cycle:    1 Hr./180°C + 2 Hrs./200°C + 6 Hrs./250°C

| PROPERTY | TEST TEMPERATURE | | |
| --- | --- | --- | --- |
| | 25°C | 177°C | 232°C |
| Flexural Strength, KSI | 17.5 | 15.9 | 13.2 |
| Flexural Modulus, KSI | 531 | 428 | 348 |
| *Hot/Wet: Flexural Strength, KSI | -- | 8.8 | -- |
| Flexural Modulus, KSI | -- | 269 | -- |
| Tensile Strength, KSI | 11.5 | -- | -- |
| Tensile Modulus, KSI | 539 | -- | -- |
| Tensile Elongation, % | 1.84 | -- | -- |

Tg (TMA Pen)  °C   297
Tg (DMA       °C   285

*Specimens were immersed in 160°F water for 48 hours prior to testing

## DSC THERMOGRAM OF RD85-101 HOMOPOLYMERIZATION

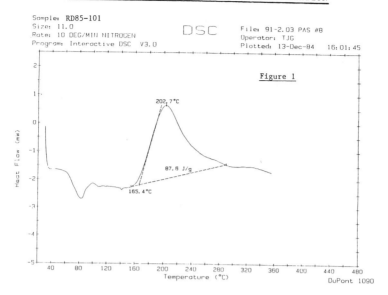

Sample: RD85-101
Size: 11.0
Rate: 10 DEG/MIN NITROGEN
Program: Interactive DSC  V3.0

DSC

File: 91-2.03 PAS #8
Operator: TJG
Plotted: 13-Dec-84   16:01:45

Figure 1

202.7°C

87.6 J/g

165.4°C

DuPont 1090

## DSC THERMOGRAM OF MATRIMID 5292A HOMOPOLYMERIZATION

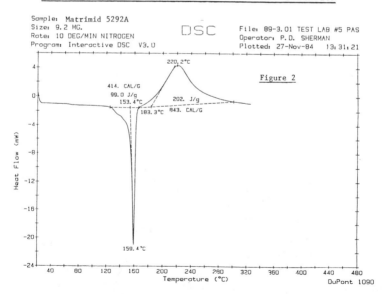

Sample: Matrimid 5292A
Size: 9.2 MG.
Rate: 10 DEG/MIN NITROGEN
Program: Interactive DSC  V3.0

DSC

File: 89-3.01 TEST LAB #5 PAS
Operator: P.D. SHERMAN
Plotted: 27-Nov-84   13:31:21

220.2°C

414. CAL/G
99.0 J/g
153.4°C
183.3°C
202. J/g
843. CAL/G

Figure 2

159.4°C

DuPont 1090

29

## DSC THERMOGRAM OF RESIN II HOMOPOLYMERIZATION

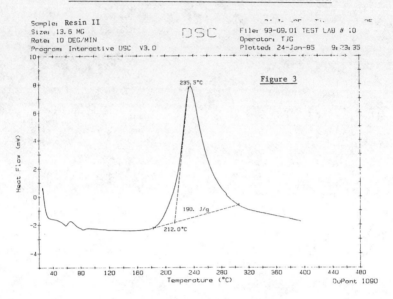

Sample: Resin II
Size: 13.6 MG
Rate: 10 DEG/MIN
Program: Interactive DSC V3.0

DSC

File: 93-09.01 TEST LAB # 10
Operator: TJG
Plotted: 24-Jan-85    9:23:35

Figure 3

235.5°C

190. J/g

212.0°C

DuPont 1090

## DSC THERMOGRAM OF RD85-101/HT972

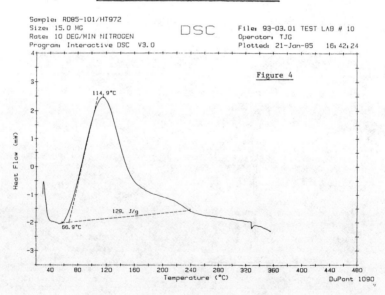

Sample: RD85-101/HT972
Size: 15.0 MG
Rate: 10 DEG/MIN NITROGEN
Program: Interactive DSC V3.0

DSC

File: 93-03.01 TEST LAB # 10
Operator: TJG
Plotted: 21-Jan-85    16:42:24

Figure 4

114.9°C

129. J/g

66.9°C

DuPont 1090

30

## DSC THERMOGRAM OF RESIN I

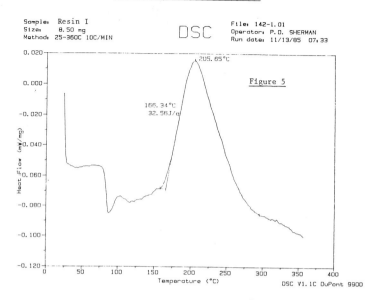

Sample: Resin I
Size: 8.50 mg
Method: 25-360C 10C/MIN

DSC

File: 142-1.01
Operator: P.D. SHERMAN
Run date: 11/13/85 07:33

Figure 5

205.65°C

166.34°C
32.56J/g

Heat Flow (mW/mg)

Temperature (°C)

DSC V1.1C DuPont 9900

## DSC THERMOGRAM OF RESIN II/HT972

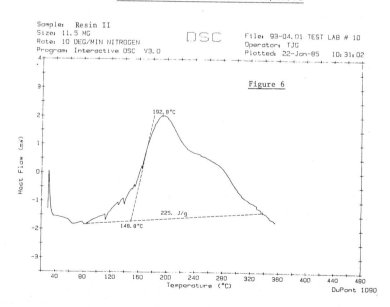

Sample: Resin II
Size: 11.5 MG
Rate: 10 DEG/MIN NITROGEN
Program: Interactive DSC V3.0

DSC

File: 93-04.01 TEST LAB # 10
Operator: TJG
Plotted: 22-Jan-85 10:31:02

Figure 6

192.8°C

225. J/g

148.8°C

Heat Flow (mW)

Temperature (°C)

DuPont 1090

31

## DSC THERMOGRAM SHOWING MELT STABILITY OF RD85-101

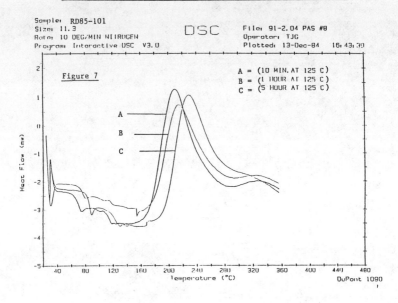

Sample: RD85-101
Size: 11.3
Rate: 10 DEG/MIN NITROGEN
Program: Interactive DSC V3.0

DSC

File: 91-2.04 PAS #8
Operator: TJG
Plotted: 13-Dec-84   16:43:39

Figure 7

A = (10 MIN. AT 125 C)
B = (1 HOUR AT 125 C)
C = (5 HOUR AT 125 C)

DuPont 1090

## DMA SCAN OF CURED NEAT RESIN RD85-101/MATRIMID 5292B

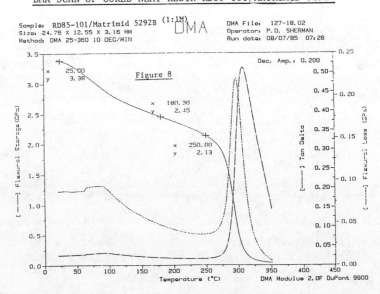

Sample: RD85-101/Matrimid 5292B (1:1M)
Size: 24.76 X 12.55 X 3.16 MM
Method: DMA 25-360 10 DEG/MIN

DMA

DMA File:  127-18.02
Operator: P.D. SHERMAN
Run date: 08/07/85  07:26

Osc. Amp.: 0.200

Figure 8

x  25.00
y  3.38

x  180.38
y  2.45

x  250.00
y  2.13

DMA Modulus 2.0F DuPont 9900

32

SYNTHESIS AND PROPERTIES OF NEW POLYFUNCTIONAL MALEIMIDE

Y. Ohnuma, T. Sugimoto, Y. Nemoto and K. Kanayama

Mitsubishi Petrochemical Co., Ltd.

8-3-1, Ami, Inashiki, Ibaraki 300-03, Japan

## Abstract

A new polyfunctional maleimide of
aromatic polyamine precursor
which consisting of $\alpha$, $\alpha,\alpha'$, $\alpha,'-$
tetrakis-4-aminophenyl-p-xylene
and its oligomer was prepared.
This polyfunctional maleimide
showed excellent solubility, more
than 50% soluble in common solvent
such as acetone, 2-butanone,
cellosolve, carbitol and other
thermoset resins, i.e. epoxy resin,
phenol resin, acrylate resin. This
maleimide was able to cure with
aromatic amines at moderate temperature and produced a heat stable
cross-linked product.
Heat stability was investigated by
programmed and isothermal thermogravimetric analysis (TGA) in air.
Decomposition temperature of this
cured product showed more than 400°C
and weight loss at 400°C/20 min. was
less than 7%. These values were 40°C
and 18% more stable than other commercially available bismaleimide
cured products. Thermomechanical
analysis was measured by the DuPont
982 DMA.
No outstanding decline of flexural
modulus was detected below 400°C.
The curing system of epoxy modified
polyfunctional maleimide was also
examined. This curing system could
cure under the same condition as
epoxy resin (moderate temperature,
short time). It was found that this
modified system showed good heat
resistance, mechanical properties
and excellant formulation latitude.

## 1. INTRODUCTION

Epoxy resins are widely used in the
electrical, electronic and
aerospace industries because of
their excellent properties and
processability. However epoxy
resins exhibit insufficient heat
stability, so their use is limited
in elevated temperature
applications. For the past few
years, extensive work has been
done in the area of high temperature performance materials to
improve tractability and heat

stability.[1]~[6] Bismaleimides are well known heatstable materials and their polymerization has been extensively studied. Homo-polymerization of simple bismaleimide (Fig.1) gives a heat-stable but brittle product.

$$\xrightarrow{\Delta}\ \text{brittle cured product}$$

( Fig.1)

Increased physical property has been achieved by mixing non-stoichiometric amounts of diamine with bismaleimide. Two types of reaction are possible in this system. First, Michael addition to produce liner polymers and secondly, opening of the double bond followed by a free radical cure leading to cross-linking (Fig.2).[7]

$$\xrightarrow{\Delta}\ \text{Cross Linked Product}$$

( Fig. 2)

This system is now commercially produced by Rhone-Poulenc under the trade name "Kinel,Kerimid". Many other curing systems and modification of bismaleimide have been proposed. CIBA-GEIGY introduced ismaleimide/o,o'-Diallylbisphenol A resin system, XU-292 for advanced composite materials.[8] The Boots Company produced new resin system of bismaleimide/amino acid hydrazides under the trade name "COMPIMIDES".[9] There are however problems with these bismaleimide resin systems, such as poor solubility in common solvents, and high temperature often required for curing, a short shelflife and insufficient heat stability. The Mitsubishi Petro-chemical Corporation recently developed a new polyfunctional maleimide to overcome these problems.[10]

2. EXPERIMENTAL METHODS

Preparation of precursor polyamine
Benzenedialdehyde, excess aromatic amine and a catalytic amount of HCl were charged to a three necked flask equipped with a thrmometer, condenser and stirrer. The mixture was heated to 100°C for 3hrs then diluted with MIBK(methylisobuty-ketone) and neutralized with NaOH aq. solution. The organic layer was separated after washing three times with distilled water. MIBK and excess aromatic amine were removed under reduced pressure. A pale reddish, glassy product was obtained.

Synthesis of polymaleimide
Preparation of polymaleimide was the same as for the conventional method of bismaleimide.[11] Precursor

34

amine was allowed to dissolve in
DMF and the solution then cooled
to about 0°C. Maleic anhydride
(1.05eq.) was added to this
solution in drops at a rate to
keep the temperature below 20°C.
After 1hr at 10-25°C, a catalytic
amount of AcONi, Et$_3$N and acetic
anhydride (1.20eq.) were added and
reaction carried out at 60°C for
2hrs. The obtained polymaleimide
solution was coagulated from 5
parts of water to 1 part of
reaction mixture. The pale brown
coagulate which formed was
collected by vacuum filtration.
The product was repeatedly washed
on the filter until the filtrate
was neutral and then dried in a
vacuum oven at 70°C for 48 hrs.
Fig. 3 showes this reaction.

Thermogravimetric Analysis(TGA)
SEIKO I&E SSC/580 thermal
controller equipped with a TGA
module, was used for TGA studies.
Samples were cured in a Al sample
pan and TGA was then measured.

Dynamic Mechanical Analysis(DMA)
A DuPont 9900 thermal analyzer
equipped with a 982 dynamic
mechanical analyzer module, was
used for DMA studies.

Gel Permiation Chromatographic
Analysis(GPC)
A Waters Associates GPC unit
equipped with a series of Shodex
KF columns, was used to determine
molecular distribution.

3. RESULTS AND DISCUSSION
Table 1 shows the effect of
starting materials to the softening

( Fig. 3 )

point (SP) of the amine.  It is clear that mixing aniline and o-toluidine is effective in reducing precursor amine SP. The same effect is also observed in the case of using 1,3-benzenedialdehyde (isopthalaldehyde) and incorporating a small amount of formaldehyde into 1,4-benzenedialdehyde (therephthalaldehyde).

## Polymaleimide Characterization

Table 2 shows the physical properties of Polyfunctional maleimide "MP-2000" & "MP-2030" from low SP precursor amines). "MP-2000" & "MP-2030 both show excellent solubility in common solvents such as acetone, methylethylketone, cellosolves and carbitols (Table 3).  Fig.3 shows their GPC.

## Curing Systems

Table 4 shows the "MP-2030" & "MP-2000" curing system.  "MP-2000" /epoxy system can select any type of curing agent and catalyst for epoxy resins.

## Properties of "MP-2030"/Diamine System

Fig.4 shows the solution stability of the "MP-2030"/diamine system. This system exhibits excellent solution stability compared to the conventional bismaleimide/diamine system.  Fig.5 shows the DMA of this system's cured product.  This system's Tg is 420 C, 100 C higher than conventional bismaleimide. It was clear that steric hindrance strongly affected the solution stability and polyfunctionality

caused a rigid, heat stable polymer network.

## Properties of "MP-2000"/Epoxy system

This epoxy modified system can cure with various curing agents and catalysts. The epoxy improves adhesive property, toughness, processability and imide increases heatstability.  Fig.6 and Fig.7 shows heat aging data and TGA of the epoxy modified system cured with various curing agents. It is surprising that this system produces heatstable polymer net work like as bismaleimide, even it contains only small amount of "MP-2000".  When phenol is selected for the curing agent, 17% imide content is enough to show the same heatstability of bismaleimide and more the "MP-2000" is added to the system, more heatstability is promoted.  The reaction of phenol with epoxide produces most heatstable ether linkage and catalytic amount of imidazole initiates heatresistant homopolymerization of maleimide.  Table 5 presents formulations of "MP-2000"/epoxy systems for various use.

## 4. CONCLUSIONS

New polyfunctional maleimide "MP-2000" & "MP-2030" have outstanding performance properties, both showing excellent solubility and processabilty, similar to epoxy resin.  Heat-stability is surprisingly high and remarkable solution stability was observed.

These systems are viable candidates for advanced composite materials, insulators and IC and printed circuit board coating.

5. REFERENCE

1) H.C.Nash, C.F.Paranski Jr. and R.I.Ting, ACT. Org. Coatings and Plastic Chemistry Vol.40, 877 (1979).
2) S.Street, V378A. A New Modified Bismaleimide Matrix Resin For High Modulus Graphite, SAMPE, Vol.25 366-375 (1980).
3) S.Street, D.Beckley, U.S.Pat. 4,351,932 (1982).
4) A.J.Kinloch, S.J.Shae, ACS, Polymer Material Science and Engineering, Vol.49, 307 (1983).
5) S.Ming-ta, J.A.Parker, T.S.Chen and H.Heimbuch, 29th National Sampe Symposium, 29, 1034-1042 (1984).
6) H.D.Stenzenberger, M.Herzog, W. Romer, R.Scheiblich, N.J.Reeves and S.Pierce, ibid, 29,1043-1059 (1984).
7) Fr.Pat.114,135
8) M.Chaudhari, T.Galvin, J.King, 30th National SAMPE Symposium march 735-746 (1985).
9) H.D.Stenzenberger, M.Herzog, W.Romer, and R.Scheiblich, ibid, 1568-1584 (1984).
10) Y.Ohnuma and K.Kanayama, U.S. Pat. 4,579,957, E.p.C.99,268.
11) G.T.Kwiatkowski, L.M.Robeson, G.L.Brode and A.W.Bedwin, Jurnal of Polymer Sci. Polymer Chem. Ed. Vol.13,961-972 (1975).

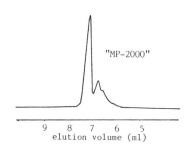

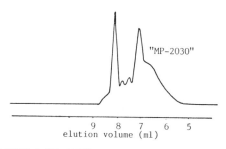

Fig. 3 GPC of "MP-2000" & "MP-2030"
conditions : column Shodex KF-802x1
eluent THF 1.0 ml/min.
temp. 40°C

366-375 (1980).

3) S.Street, D.Beckley, U.S.Pat. 4,351,932 (1982).

4) A.J.Kinloch, S.J.Shae, ACS, Polymer Material Science and Engineerig, Vol.49, 307 (1983).

5) S.Ming-ta, J.A.Parker, T.S.Chen and H.Heimbuch, 29th National Sampe Symposium, 29, 1034-1042 (1984).

6) H.D.Stenzenberger, M.Herzog, W. Romer, R.Scheiblich, N.J.Reeves and S.Pierce, ibid, 29,1043-1059 (1984).

7) Fr.Pat.114,135

8) M.Chaudhari, T.Galvin, J.King, 30th National SAMPE Symposium march 735-746 (1985).

9) H.D.Stenzenberger, M.Herzog, W.Romer, and R.Scheiblich, ibid, 1568-1584 (1984).

10) Y.Ohnuma and K.Kanayama, U.S. Pat. 4,579,957, E.p.C.99,268.

11) G.T.Kwiatkowski, L.M.Robeson, G.L.Brode and A.W.Bedwin, Jurnal of Polymer Sci. Polymer Chem. Ed. Vol.13,961-972 (1975).

Table 1   Softening Point of Precursor Amines

| No. | Aldehyde(mole) | | | Amine(mole) | | softening point (°C ) | solubility |
|-----|-----|-----|-----|-----|-----|-----|-----|
| | TPA | IPA | FA | AN | OT | | |
| 1 | 1 | – | – | 8 | – | 240 – 250 | poor |
| 2 | 1 | – | – | – | 8 | 215 – 225 | " |
| 3 | 1 | – | – | 2 | 6 | 118 – 124 | excellent |
| 4 | 1 | – | – | 4 | 4 | 122 – 128 | " |
| 5 | 1 | – | – | 6 | 2 | 125 – 130 | " |
| 6 | – | 1 | – | 8 | – | 115 – 130 | " |
| 7 | 0.3 | – | 0.7 | 8 | – | 70 – 85 | " |

Reactions were carried out at 100 C/3hrs.

TPA:Terephthalaldehyde, IPA:Isophthalaldehyde, FA:Formaldehyde,

AN:Aniline, OT:o-Toluidine

Table 2  Physical Properties of "MP-2000" & "MP-2030"

|  | "MP-2000" | "MP-2030" |
|---|---|---|
| Mn | 880 | 740 |
| Softening Point( C) | 170 - 180 | 120 - 130 |
| Weight per Imide moiety (g/imide) | 210 | 200 |

Table 4  Curing System of "MP-2000" & "MP-2030"

| Resin | Curing Agent |
|---|---|
| "MP-2030" | 3,3'-Dimethyl-4,4'-diaminodiphenylmethane |
| "MP-2000" / Epoxy | Aromatic amine, Phenol, Acid Anhydride, Imidazole, Dicyanodiamide |
| "MP-2000" / Acrylate | Photochemical initiator, Peroxide |

Table 5  Formulations of "MP-2000"

| Modifier | Curing Agent | Application |
|---|---|---|
| Epoxy | Dicy/imidazole | pr nted circui   boa |
|  | DDS/BF$_3$complex | CFRP |
|  | phenol,imidazole | for high temp. use |
|  | DDM/BF$_3$complex | for low temp. & rapid cure |
| Epoxy/rubber | DDS/Dicy | Adhesive |
| Acrylate | BEE | UV curing |

Dicy:Dicyanodiamide, DDS:Diaminodiphenylsulfone,
DDM:Diaminodiphenylmethane, BEE:Benzoin Ethyl Ether

Table 3 Solubility of "MP-2000" & "MP-2030"

| soluvent | "MP-2000" | "MP-2030" | BMI[1] |
|---|---|---|---|
| Methyl Cellosolve | >50 (wt%) | >50 (wt%) | <10(wt%) |
| Buthyl Cellosolve | 10 – 20 | <10 | " |
| Ethyl Cellosolve A[2] | >50 | >50 | " |
| Carbitol | " | " | " |
| Carbitol A | " | " | " |
| Acetone | " | " | " |
| MEK[3] | " | " | " |
| Toluene | <10 | <10 | " |
| Ethyl Acetate | " | " | " |
| Methylene Dichloride | >50 | >50 | " |
| Chloroform | " | " | " |
| Tetrahydrofurane | " | " | " |
| 1,4-Dioxane | " | " | " |
| DMF[4] | " | " | 10 – 20 |
| DMA[5] | " | " | " |
| NMP[6] | " | " | 20 – 30 |

at room temp.

1) 4,4'-Diaminodiphenylmethane Bis Maleimide, 2) Acetate,

3) Methyl Ethyl Ketone, 4) N,N-Dimethylformamide,

5) N,N-Dimethylacetoamide, 6)N-Methyl-2-Pyrrolidone

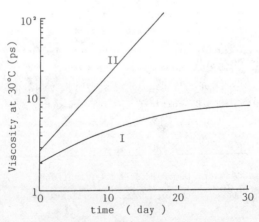

Fig.4 Solution stability of maleimide
I:"MP-2030"/3,3'-dimethyl-DDM
II: DDM-Bismaleimide/DDM
conditions; 50% NMP solution,stored
at 30°C, measured at 30°C by E-type
viscometer

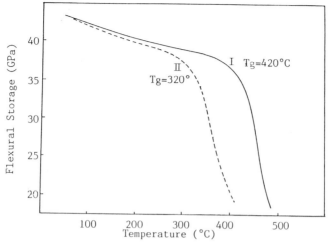

Fig 5  DMA of Maleimide
I : "MP-2030" Dimethyl-DDM curing,   II : DDM-Bis
maleimide/DDM curing  (laminate)

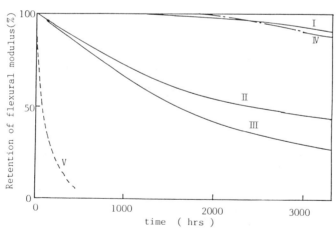

Fig. 6  Heat aging data at 230°C (neat resin)

I :"MP-2000"/Epikote 828 =30/70(wt%) imidazole curing
II : DDM curing of I,   III : Acid Anhydride curing of I
IV : DDM-Bismaleimide/DDM curing  V : DDM curing of Epikote 828

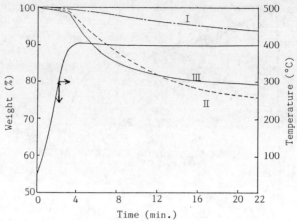

Fig. 7 Isothermal Dagradation of maleimide at 400°C in air

I: "MP-2030"/Dimethyl-DDM curing, II: "MP-2000"(25%)/E 828(75%) phenol curing, III: DDM-Bismaleimide/DDM curing

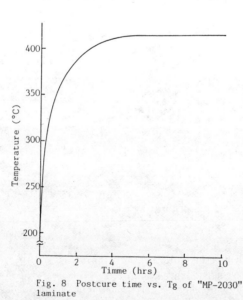

Fig. 8 Postcure time vs. Tg of "MP-2030" laminate

conditions; cured with dimethyl-DDM at 180°C x 1h, postcured at 230°C, measured by DMA

## 6. BIOGRAPHIES

KAORU KANAYAMA received his M.S.
in Applied Chemistry from Hokkaido
University, Sapporo, Japan, in
1976. He joined the Central
Reserch Laboratory of Mitsubishi
Petrochemical Co., Ltd. Japan.
He has five years experience in
heat resistant epoxy compound
synthesis and his research work has
recently concentrated in the
preparation of processable
polyimides. In 1986, he joined
the Advanced Material Laboratory
of the Mitsubishi Petrochemical Co., Ltd.

YHOSUI NEMOTO received his M.S.
degree in Chemical Engineering
from Tokyo Institute of Technology,
Japan, in 1975. He joined the
Advanced Material Laboratory of
the Mitsubishi Petrochemical Co., Ltd.
His lately reserch
work has concentrated in fiber
reinforced plastics and
application of new thermoset resins.

TOSHIO SUGIMOTO is a senior chemist
in the Advanced Material Laboratory
of the Mitsubishi Petrochemical Co.,
Ltd. He has a M.S. Degree in
Industrial Chemistry from University
of Tokyo, Japan. Since joining the
Mitsubishi Petrochemical Co., Ltd.
in 1968, he has concentrated in
photopolymerization and application
of new thermoset resins.

YOSHINOBU OHNUMA joined the
Central Reserch Laboratory of the
Mitsubishi Petrochemical Co., Ltd.
after graduated the Yamagata
Technical school in 1975. He has a
eleven years experience in
synthesis of thermoset resins and
his reserch work has recently
concentrated in the application of
maleimide resins. In 1986 he joined
the Advance Material Laboratory of the
Mitsubishi Petrochemical Co., Ltd.

BISMALEIMIDE RESINS
IMPROVED NOVEL TOUGHENING MODIFIERS FOR BMI RESINS

H.D. Stenzenberger, P. König, M. Herzog, W. Römer
Technochemie GmbH-Verfahrenstechnik, Dossenheim
S. Pierce, M.S. Canning
The Boots Company PLC, Nottingham

## Abstract

Polybismaleimides are a class of
thermosetting resins with high temperature stability and superior hot
wet environmental resistance compared to epoxies. However, improvements in high fracture toughness are
still required.
A new family of alkenyl terminated
arylene-ether-ketones (TM123) has
been synthesised. These are useful
comonomers for bismaleimide resins.
The copolymerization takes place at
temperatures between 150-210$^{\circ}$C, providing tough cured resins. The ratio
between bismaleimide resin and the
alkenyl terminated arylene-ether-
ketone can be widely varied - BMI:
TM123 = 85:15 to 55:45. However,
only small quantities (20% by weight)
of toughening modifier are necessary
to significantly improve the neat
resin fracture toughness. Neat resin
mechanical data are provided for a
series of Compimide 796 bismaleimide
resin/TM123 blends. The optimized
composition has been evaluated as a
matrix resin for carbon fibre composites. Basic laminate properties
for standard T300/6000 and Grafil
XAS carbon fibres are presented.

## 1. INTRODUCTION

Bismaleimide resins are being increasingly used as matrix resins for
advanced carbon fibre composites.
This is due to their improved temperature stability, environmental
resistance (in particular in hot
wet conditions) and better fire
smoke toxicity properties compared
to epoxy resins. Other thermosetting
imide resins, such as acetylene terminated imides or PMR resins, although
better in temperature performance,
are still difficult to process, and
therefore a less attractive prospect.
Bismaleimide resins and prepregs
with epoxy-like processing are
available from several sources.
Good tack and drape characteristics
provide easy handling and autoclave
techniques are used to cure at
temperatures of around 180$^{\circ}$C
(1,2,3,4).
To fully develop the high temperature

performance of bismaleimide resins, postcure at temperatures of up to 250°C is necessary. Recently, bismaleimide resin products for wet filament winding and resin transfer moulding have also become commercially available (5). This provides the possibility for cost effective production of continuous fibre reinforced advanced composites.

The more extensive use of composites for load carrying primary aircraft structures demands improved fracture toughness and impact resistance. Bismaleimide resins and laminates are still considered "brittle", however, several approaches have been tried to overcome this deficiency. One such approach is to increase the molecular weight between cross-links by using long-chain totally aromatic bismaleimides as building blocks. However, this conflicts with processing properties such as flow (low melt viscosity) and low temperature cure. The use of low melting comonomers is applicable as these copolymerize with the BMIs via a linear chain extension reaction to provide a lower cross-link density in the finally cured material (6,7). Recently, allylphenyl type comonomers have been described which copolymerize with bismaleimides via an "ene" type linear chain extension reaction (8). These comonomers provide tough modified bismaleimides with no loss in processability. The use of 4,4'-bis(o-propenylphenoxy)diphenyl sulfone as coreactant for bismaleimides is particularly attractive. This is because pro-

penylphenyl compounds react at low temperatures with bismaleimides via a linear Diels-Alder reaction to provide tough networks (9). Second phase toughening of BMIs has also been tried employing reactive elastomers or thermoplastics as modifiers (10). This paper describes the synthesis and properties of a family of new propenyl terminated arylene-ether-ketones which copolymerize with bismaleimides via a Diels-Alder reaction, and also their use as carbon fibre matrix resins.

## 2. SYNTHESIS OF PROPENYL TERMINATED ARYLENE-ETHER-KETONES

The classical synthesis route of aromatic arylene-ether-ketones involves the reaction of 4,4'-difluorobenzophenone with diphenols such as hydroquinone and/or bisphenol-A. The reaction of o-allylphenol with 4,4'-difluorobenzophenone is very straight forward, using N-methylpyrrolidone as solvent and potassium carbonate as catalyst at 160°C to provide 4,4'-bis(o-propenylphenoxy)benzophenone (Fig.1). The H-NMR spectrum of this compound clearly shows that the allyl group of the o-allylphenol is almost quantitatively converted into the propenyl group (Fig.2). The H-NMR spectrum also indicates the presence of 2 stereo isomeric forms of the propenyl group (83% E- and 17% Z-isomers).

The reaction temperature must be very carefully controlled because the propenyl groups formed can easily undergo a self-polymerization at the synthesis temperature. Instead of o-allylphenol, eugenol was used as a p-substituted allylphenol and 4,4'-bis-(o-methoxy-p-propenylphenoxy)benzophenone was obtained in almost quantitative yield. This product is coded TM123-1 throughout this work. Another propenyl compound was synthesised by reacting 1 mole of 2,4'-difluorobenzophenone with 2 moles of allylphenol. This provided 2,4'-bis-(o-propenylphenoxy)benzophenone, coded TM123-2.

These low molecular weight propenyl compounds are of particular interest for the formulation of low-melting, low-viscosity blends with BMIs. However, we also synthesised a series of oligomeric arylene-ether-ketones by using hydroquinone and bisphenol-A as comonomers, as outlined in Fig.3. These higher molecular weight oligomers are of interest because they can be used as second phase tougheners. We will report at a later stage on the use of high molecular weight propenyl-terminated AEK as second phase tougheners for BMI resins. (AEK = arylene-ether-ketones.)

## 2.1 Properties of propenyl-terminated arylene-ether-ketones

4,4'-bis(o-propenylphenoxy)benzophenone (TM123) and 2,4'-bis(o-propenylphenoxy)benzophenone (TM123-2) are glassy solids at room temperature, and low viscosity fluids at $80^{\circ}C$ (Fig.4). 4,4'-bis(o-methoxy-p-propenylphenoxy)benzophenone (TM123-1) shows a much higher melt viscosity

compared with TM123 and TM123-2. All these compounds however, can easily be melt-blended with low-melting eutectic bismaleimide resins such as Compimide 353 and Compimide 796. Higher molecular weight propenyl-terminated arylene-ether-ketones show significantly higher melt viscosities and can be blended with bismaleimide resin via solution techniques. Bis(propenylphenoxy)benzophenones can undergo a thermally induced self-polymerization at temperatures of around $200^{\circ}C$; however at temperatures of $\sim$ $80-120^{\circ}C$, the compounds show almost no advancement, even over long periods of time.

## 3. BISMALEIMIDE /BIS(PROPENYLPHENOXY) BENZOPHENONE RESINS

In a recent paper (9), we reported on the copolymerization of 4,4'-bis(o-propenylphenoxy)diphenylsulfone with bismaleimide. This proceeds via a Diels-Alder reaction similar to the reaction between maleic acid anhydride and styrene (11). The new propenyl-terminated arylene-ether-ketones react with bismaleimide in the same way, as outlined in Fig.5.

## 3.1 Compimide 796/TM123 (TM123-1, TM123-2) resins

Of major interest is the influence of the new propenyl-terminated ether-ketones on the fracture toughness of the resulting copolymer when used in combination with one of our proprietary low-melting eutectic bismaleimide resins. Compimide 796 was used as the bismaleimide and melt-blended with TM123-1 modifier at temperatures of 110-130°C, cast into a steel mould and cured for 4 hours at 140°C, plus 6 hours at 210°C under a pressure of 4 bars. Neat resin castings were postcured at 240°C for 5 hours and basic mechanical properties, including fracture toughness, were measured. The method for evaluating neat resin $G_{IC}$ has already been described in detail (12). Compimide 796/TM123 blends were cured for 2 hours at 170°C plus 2 hours at 190°C, plus 4 hours at 210°C, and then postcured for 5 hours at 240°C.

The following properties were of particular interest :

△ Fracture toughness and mechanical properties as a function of the toughener (comonomer) concentration.

△ Influence of the toughener structure on mechanical properties.

△ Contribution of the toughener structure and concentration to moisture absorption.

△ Glass transition temperature of toughened bismaleimides.

Before casting neat resin samples, the cure behaviour of BMI/TM123 blends was studied by means of differential scanning calorimetry. The DSC traces for Compimide 796/TM123 and Compimide 796/TM123-1 blends are given in Fig.6. The unmodified Compimide 796 provides a single DSC peak, maximizing at 250°C. The addition of increasing amounts of 4,4'-bis(o-propenylphenoxy)benzophenone finally provides a bimodal DSC trace with a main peak at 225°C and well-developed

shoulder at 289°C. With sufficient amounts of toughener (TM123) a second peak at 289°C is observed.

Different behaviour is obtained with Compimide 796/TM123-1 [4,4'-bis(o-methoxy-p-propenylphenoxy)benzophenone] blends. Small additions of toughener (20% by weight) lead to a bimodal DSC trace with two well-developed peaks at 207°C and 264°C respectively. Further increase in TM123-1 concentration (40% by weight) leads eventually to a single DSC peak with a maximum at 195°C.

The difference in copolymerization behaviour of the o- and p-propenyl compounds can only be explained by considering steric hindrance as preventing o-propenyl phenoxy benzophenone from undergoing complete copolymerization with the bismaleimide. Interestingly, the cure peak temperatures for optimized mixtures occur well below 250°C, which means that the resins should be well cured when exposed to 250°C for postcure.

## 3.2.  Mechanical properties of neat resin castings

The mechanical properties of cured Compimide 796/TM123 resins [4,4'-bis(o-propenylphenoxy)benzophenone] are compiled in Table 1. The unmodified Compimide 796 resin shows low strength properties but high elastic moduli at both room temperature and 250°C. The fracture toughness of this basic resin is however very low (63 Joules/m$^2$). The addition of 20,30 or 40% of TM123 significantly improves the flexural

47

strength at low and high temperatures and is optimum at 30% toughener concentration. As was expected, the flexural moduli values are lower for the toughened resins. What is required, is a high modulus retention at 250°C. This is obtained with toughener concentrations up to 35% by weight. The fracture toughness values increase significantly with increasing comonomer (toughener TM-123) concentration, being in the range of 400-430 Joules/m$^2$ for the balanced system, achieved with the composition consisting of 65% Compimide 796 and 35% TM123.

4,4'-bis(o-methoxy-p-propenylphenoxy) benzophenone was evaluated in the same way. Qualitatively, the same general trend is observed. Again, the optimum toughener concentration is around 35% by weight. The highest fracture toughness (545 Joules/m$^2$) was found for the 70:30 Compimide 796:TM123-1 composition. It is of interest to note that compared with the o-propenyl compound (TM123), the p-propenyl compound provides better high temperature elastic moduli (Table 2).

2,4'-bis(o-propenylphenoxy)benzophenone (TM123-2) was initially of particular interest because of its low-melt viscosity. Mechanical evaluation of neat resins showed however, no improvement on TM123 and TM123-1 (Table 3). Nevertheless, the same trend is seen for all three modifiers. Increased toughener concentration up to a level of 35% in BMI (Compimide

796) provides :

△ increased strength properties at room temperature and at 250°C.

△ reduction in elastic moduli at both room temperature and 250°C.

△ significantly increased fracture toughness (7-9 times higher than Compimide 796).

△ balanced mechanical properties at toughener levels of 35% by weight.

3.3.  Glass transition temperature

Glass transition temperatures were measured employing TMA analysis (expansion mode). The individual results can be found in Tables 1-3. The unmodified Compimide 796 shows a Tg well in excess of 300°C. The dry Tgs for the balanced toughened systems are as follows :

△ Compimide 796/TM123     (70:30)   265°C
△ Compimide 796/TM123-1   (65:35)   260°C
△ Compimide 796/TM123-2   (65:35)   246°C

3.4.  Water absorption

Highly cross-linked thermosetting bismaleimide resins normally show high moisture absorption. Therefore, their wet Tg is sometimes significantly reduced when saturated with moisture. That new and improved thermosetting resins should absorb less moisture is now a target. The moisture absorption was studied for all new resins contained in Tables 1,2 and 3. Neat resin samples were aged at 70°C in 94% relative humidity over a period of 1000 hours, and the weight increase then measured. After 500-600 hours, equilibrium was almost obtained. The following observations were made :

△ The unmodified Compimide 796 resin shows a moisture absorption of 4.3% by weight.

△ Increased toughener concentrations lead to a decreased moisture uptake (Fig.7).

△ The optimized toughener concentration of 35% provided ~3% equilibrium uptake for all three new tougheners.

The reduced water uptake is presumably the result of a lower free volume in the cured modified networks.

## 4. PROCESSING OF COMPIMIDE RESIN/ TM123 BLENDS

Prepregs are normally produced via hot melt techniques for tape and via solvent/solution techniques for fabric. Ideal resins are therefore soluble in low boiling solvents and show melting at reasonably low temperatures. In addition, a prepreg resin must have a Tg of ~10-15$^{o}$C to provide tack and drape properties for optimum handling.

In this work, a Compimide 796/TM123 [4,4'-bis(o-propenylphenoxy)benzophenone] 60:40 blend was selected for further evaluation in carbon fibre laminates. Of particular interest is gaining insight into correlations between neat resin properties and laminate performance, mainly whether or not improved neat resin fracture energy provides better edgewise delamination strength.

Unidirectional prepregs were prepared via a drum-winding process. Release paper is attached to the mandrel and the tape is fabricated in a filament winding process. The impregnation bath contained an ~50% by weight solution of the resin. The prepreg was dried on the mandrel to strip off the methylene chloride solvent. After removal from the drum, the prepregs were B-staged in a vacuum bag at 70$^{o}$C. The prepregs obtained showed limited drapeability and no tack. The fibres used were Torayca T300-6000-50B and Grafil XAS-6000 (standard size). At this point, a comment must be made about the possibilities of formulating drape and tack into a prepreg system. It is most important that low viscosity reactive diluents, which are normally used as tackifiers, have to be compatible with the basic resin system. Compatibility means that the diluent does not adversely affect toughness, moisture absorption and the glass transition temperature when cured into the system. DAP (diallylphthalate) and TAC (triallylcyanurate) were tried with the Compimide 796/ TM123 system and a negative influence on the fracture toughness was experienced. Even small quantities ( 5-10%) showed a significant reduction in the fracture toughness of the neat resin. Work is underway on testing liquid diluents as "tackifiers", including allyl, vinyl, acrylate and epoxy compounds. The results will be reported at a later stage.

### 4.1 Cure cycle

The cure cycle used throughout this work is given in Fig.8. A restricted bleed lay-up was used to bag the

49

laminates. A cure temperature of 180 -190°C was applied for 3 hours. The temperature profile underwent a dwell at 140°C during which period the full lamination pressure was applied. Postcure was normally performed for 2 hours at 210°C, plus 5 hours at 250°C.

## 5. LAMINATE PROPERTIES

Typical laminate properties measured for T300-6000-50B and Grafil XAS fibres are given in Table 4. For T300-6000-50B lamiantes, two different postcure cycles were used. Cure cycle A consisted of a postcure of 2 hours at 210°C, plus 5 hours at 250°C, whilst cycle B consisted of 12 hours at 210°C. Both postcure cycles provide very similar properties (i.e. flexural strength and short beam shear) in unidirectional laminates. The short beam shear properties in both unidirectional and 0±45° laminates are a little higher for the 250°C postcured laminates. Also, the 90° flexural strength in UD laminates is higher for the 250°C postcured material. Short beam shear strength for the 0±45° laminate is surprisingly high. In our experience, such high values can only be achieved with tough resins. Encouraging EDL first failure strength values were gained for the laminates postcured at 250°C. Surprisingly, the higher postcure temperature provided the better edgewise delamination resistance. The mechanical properties for Grafil XAS fibres with the

same matrix are very similar to those of T300/50B. Only the 90° flexural strength is somewhat lower and the first failure edgewise delamination strength is lower when compared with T300 fibres. At this stage, it should be noted that fibre sizes have not yet been optimized for any of the bismaleimide resins. Therefore, there are slight differences in laminate properties for T300 and Grafil XAS. After optimization of the prepreg manufacturing technique however, both fibres seem to be almost interchangeable.

## 6. SUMMARY

4,4'-bis(propenylphenoxy)benzophenones (TM123 series) are a new class of comonomers for bismaleimide resins. High neat resin fracture energies are obtained for optimized Compimide 796/ TM123 ratios (BMI:TM123 = 65:35). Outstanding EDL strength properites are achieved for laminates containing T300-6000-50B fibres when postcured at 250°C for 4-5 hours. Torayca T300-6000-50B and Grafil XAS carbon fibres provide almost the same mechanical properties in laminates.

## 7.REFERENCES

1) S. Street, in Polyimides, Ed. K.
L. Mittal, Plenum Press, New York
and London, 1984, p.77.

2) V. Ho. 31st International SAMPE
Symposium. Vol. 31, p.1362 (1986).

3) J.D. Boyd, T.F. Biermann. 31st
International SAMPE Symposium, Vol.
31, p.977 (1986).

4) P.A. Steiner, J.H. Browne, M.T.
Blair, J. McKillen. 18th International SAMPE Technical Conference, Vol.
18, p. 193 (1986).

5) N.N. Boots-Technochemie Data
sheet for COMPIMIDE 65FWR.

6) M. Bergain, A. Combet, P. Grosjean, British Patent 1190780 (1968)

7) H.D. Stenzenberger, U.S. Patent
4,211,861 (1980)

8) H.D. Stenzenberger, W. Römer, M.
Herzog, S. Pierce, M.S. Canning, K.
Fear. 31st International SAMPE
Symposium Vol. 31 p.920 (1986).

9) H.D. Stenzenberger, P. König, M.
Herzog, W. Römer, M.S. Canning, S.
Pierce. 18th International SAMPE
Technical Conference, p.500 (1986)

10) A.J. Kinloch, S.J. Shaw, ACS.
Polym. Mat. Sci. and Eng. Vol 49,
p.307 (1985)

11) V. Brückner, Ber. Deutsch Chem.
Ges. 75 (1942) 2043

12) H.D. Stenzenberger, P. König, W.
Römer, E. Haberbosch, S. Pierce, M.S.
Canning, 7th International SAMPE
Conference, Europ. Chapter Proceedings p.141 (1986)

Acknowledgement :
Part of this work was performed under
contract 03M1003E5 from the West
German Ministry of Research and
Technology. Their support is gratefully appreciated.

TABLE 1 : MECHANICAL PROPERTIES OF TOUGHENED BISMALEIMIDE NEAT RESINS.
MODIFIER TM123 : 4,4'-bis(o-propenylphenoxy)benzophenone.

| Composition % C796  % TM123 | | Test temp. ($^{o}$C) | Mechanical properties | | | | | | |
|---|---|---|---|---|---|---|---|---|---|
| | | | FS (MPa) | FM (GPa) | FE (%) | $G_{IC}$ (J/m$^2$) | $K_{IC}$ (KN/m$^{-3/2}$) | WA (%) | Tg ($^{o}$C) |
| 100 | – | 23 | 76 | 4.64 | 1.70 | 63 | 526 | 4.30 | >300 |
| | | 250 | 31 | 3.03 | 1.03 | – | – | | |
| 80 | 20 | 23 | 106 | 3.96 | 2.34 | 191 | 847 | 3.66 | 266 |
| | | 250 | 65 | 2.66 | 2.52 | – | – | | |
| 70 | 30 | 23 | 132 | 3.61 | 3.88 | 397 | 1188 | 3.37 | 265 |
| | | 250 | 90 | 2.47 | 5.00 | – | – | | |
| 60 | 40 | 23 | 132 | 3.70 | 3.75 | 439 | 1209 | 2.59 | 249 |
| | | 250 | 56 | 1.71 | 4.86 | – | – | | |

Abbreviations :
FS  : flexural strength
FM  : flexural modulus
FE  : flexural strain
$G_{IC}$ : fracture energy
$K_{IC}$ : stress intensity factor

WA  : water absorption, 1000h at 70$^{o}$C, 94% rel. humidity
Tg  : glass transition temperature

C796   : COMPIMIDE 796
TM123 : 4,4'-bis(o-propenylphenoxy)benzophenone


TABLE 2 : MECHANICAL PROPERTIES OF TOUGHENED BISMALEIMIDE NEAT RESINS.
MODIFIER TM123-1 : 4,4'-bis(o-methoxy-p-propenylphenoxy)-benzophenone.

| Composition % C796  % TM123-1 | | Test temp. ($^{o}$C) | Mechanical properties | | | | | | |
|---|---|---|---|---|---|---|---|---|---|
| | | | FS (MPa) | FM (GPa) | FE (%) | $G_{IC}$ (J/m$^2$) | $K_{IC}$ (KN/m$^{-3/2}$) | WA (%) | Tg ($^{o}$C) |
| 100 | – | 23 | 76 | 4.64 | 1.70 | 63 | 526 | 4.30 | >300 |
| | | 250 | 31 | 3.03 | 1.03 | – | – | | |
| 80 | 20 | 23 | 114 | 4.17 | 2.87 | 247 | 1001 | 3.46 | 252 |
| | | 250 | 78 | 2.47 | 3.73 | – | – | | |
| 70 | 30 | 23 | 119 | 3.79 | 3.20 | 545 | 1435 | 3.26 | 258 |
| | | 250 | 85 | 2.55 | 4.42 | – | – | | |
| 60 | 40 | 23 | 122 | 3.59 | 3.44 | 466 | 1264 | 2.90 | 260 |
| | | 250 | 81 | 2.44 | 4.52 | – | – | | |

Abbreviations :
FS  : flexural strength
FM  : flexural modulus
FE  : flexural strain
$G_{IC}$ : fracture energy
$K_{IC}$ : stress intensity factor

WA  : water absorption, 1000h at 70$^{o}$C, 94% rel. humidity
Tg  : glass transition temperature

C796   : COMPIMIDE 796
TM123-1 : 4,4'-bis(o-methoxy-p-propenylphenoxy)-benzophenone

TABLE 3 : MECHANICAL PROPERTIES OF TOUGHENED BISMALEIMIDE NEAT RESINS.
MODIFIER TM123-2 : 2,4-bis(o-propenylphenoxy)benzophenone.

| Composition | | Test temp. | Mechanical properties | | | | | | |
|---|---|---|---|---|---|---|---|---|---|
| % C796 | % TM123-2 | (°C) | FS (MPa) | FM (GPa) | FE (%) | $G_{IC}$ (J/m$^2$) | $K_{IC}$ (KN/m$^{3/2}$) | WA (%) | Tg (°C) |
| 100 | – | 23 | 76 | 4.64 | 1.70 | 63 | 526 | 4.30 | >300 |
|  |  | 250 | 31 | 3.03 | 1.03 | – | – |  |  |
| 80 | 20 | 23 | 104 | 3.99 | 2.51 | 293 | 1076 | 3.50 | 249 |
|  |  | 250 | 66 | 2.51 | 2.81 | – | – |  |  |
| 70 | 30 | 23 | 111 | 3.87 | 2.90 | 323 | 1090 | 3.16 | 252 |
|  |  | 250 | 65 | 2.50 | 3.30 | – | – |  |  |
| 60 | 40 | 23 | 82 | 3.82 | 2.17 | 467 | 1306 | 2.71 | 241 |
|  |  | 250 | 50 | 1.74 | 4.48 | – | – |  |  |

Abbreviations :
FS  : flexural strength
FM  : flexural modulus
FE  : flexural strain
$G_{IC}$ : fracture energy
$K_{IC}$ : stress intensity factor

WA  : water absorption, 1000h at 70°C, 94% rel. humidity
Tg  : glass transition temperature

C796    : COMPIMIDE 796
TM123-2 : 2,4-bis(o-propenylphenoxy)benzophenone

TABLE 4 :  PROPERTIES OF COMPIMIDE 796/TM123 LAMINATES

| Property | | Fibre type | | |
|---|---|---|---|---|
|  |  | T300/6000/50B | | Grafil XAS |
|  |  | A | B |  |
| Fibre content | % by vol. | 67 | 66 | 63 |
| 0° Flexural strength | (MPa) |  |  |  |
|  | 23°C | 2036 | 2106 | 2096 |
|  | 250°C | 1486 | 1393 | 1430 |
| 0° Flexural modulus | (GPa) |  |  |  |
|  | 23°C | 139 | 137 | 133 |
|  | 250°C | 140 | 140 | 139 |
| 90° Flexural strength | (MPa) |  |  |  |
|  | 23°C | 121 | 110 | 112 |
|  | 250°C | 89 | 78 | 74 |
| 90° Flexural modulus | (GPa) |  |  |  |
|  | 23°C | 9.8 | 9.9 | 9.7 |
|  | 250°C | 7.1 | 6.9 | 6.4 |
| 0° short beam shear strength | (MPa) |  |  |  |
|  | 23°C | 111 | 108 | 103 |
|  | 175°C wet | 54 | 52 | 42 |
|  | 250°C | 54 | 55 | 55 |
| 0/±45 short beam shear strength | (MPa) |  |  |  |
|  | 23°C | 87 | 81 | 85 |
|  | 250°C | 53 | 58 | 47 |
| $G_{IC}$ (DCB test) | (J/m$^2$) | 210 | – | 297 |
| (±25)$_2$90$_s$ edge delamination | (MPa) |  |  |  |
|  | first failure | 307 | 259 | 215 |
|  | ultimate | 461 | 471 | 524 |

A :  Postcure 2h at 210°C, plus 5h at 250°C.        B :  Postcure 12 h at 210°C.

Figure 1 : Synthesis of 4,4'-bis(o-propenylphenoxy)benzophenone

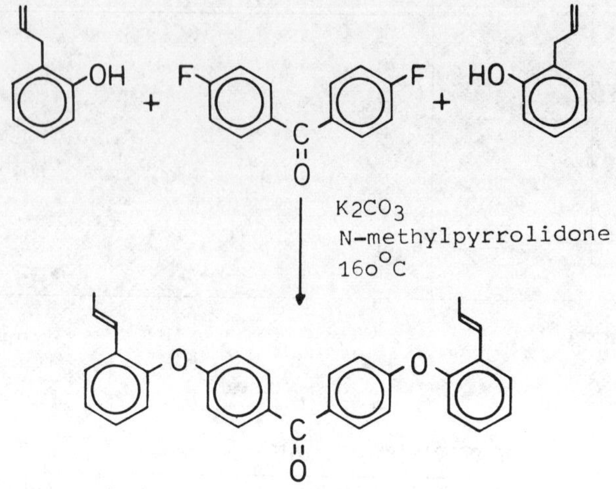

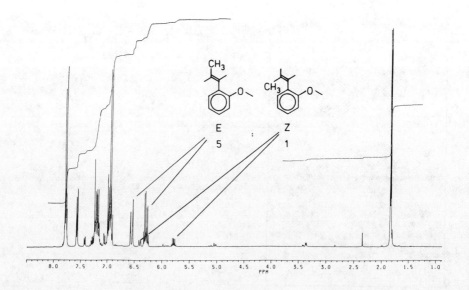

Figure 2 : H-NMR spectrum of 4,4'-bis(o-propenylphenoxy)benzophenone (100 MHz)

Figure 3 : Synthesis of propenyl-terminated arylene-ether-ketone oligomers.

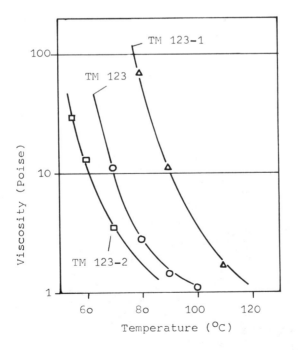

Figure 4 : Viscosity profiles of TM123 toughening modifiers.

Figure 5 : Proposed reaction route for the copolymerization of bismaleimide with propenyl-terminated arylene-ether-ketones.

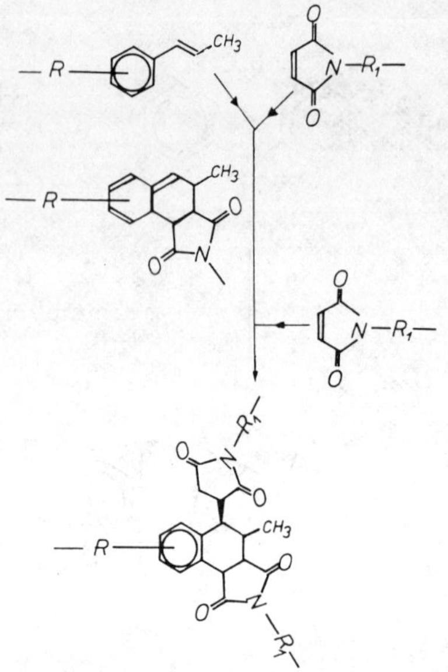

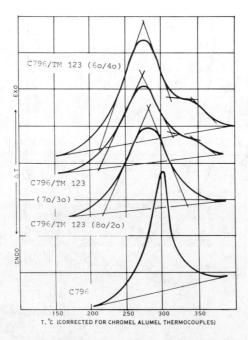

Figure 6a : DSC traces of Compimide 796/TM123 resins.

Figure 6b : DSC traces of Compimide 796/TM123-1 resins.

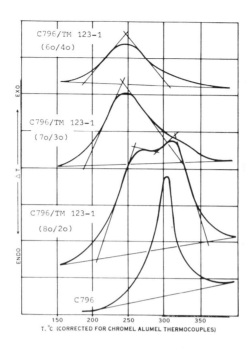

Figure 7 : Moisture absorption of C796/TM123 blends as a function of toughener concentration

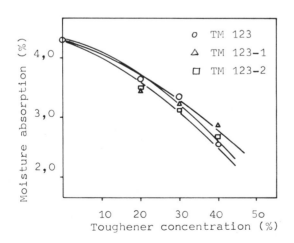

Figure 8 : Cure cycle for COMPIMIDE 796/TM123 prepreg. (restricted bleed lay-up)

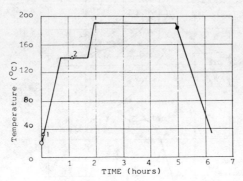

o   pull vacuum

△ 1  apply pressure of 1 bar

△ 2  apply pressure of 4 bars

●   vent vacuum and release pressure

THERMOSETTING RESINS FOR HIGH
PERFORMANCE COMPOSITES
A REVIEW

J. KING, M. CHAUDHARI, G. DISALVO
CIBA-GEIGY CORPORATION
ARDSLEY, NEW YORK 10502

Fiber-reinforced resin composites have been used in many applications over the past few decades. More recently, there has been greater emphasis on these lightweight materials due to various tactical reasons.

The need for structural materials that meet the exacting requirements of aerospace applications and provide weight savings over metals initiated the development of a new class of materials. Since these needs could not be met by any single material, the concept of combining two materials in an integral mixture was explored and found to provide unique advantages over single-phase materials.

Particularly useful material systems were derived from high-strength, high-modulus fibers embedded in a polymeric matrix phase in desired orientation. Unidirectional orientation of the fibers was found to maximize the fiber strength and stiffness for loads along the fiber axis. This resulted in material systems with high specific strength and stiffness. These material systems, now commonly called advanced composites, could be manufactured in the form of tapes by impregnating fibers with resin and precisely collimating the impregnated material onto removable backing paper. These tapes, known in the industry as prepregs, could be formed into a structure by stacking them in selected orientations to match loading conditions and then consolidating the

structure with heat and pressure. The progression of advanced composites from conception to application depended on the availability of reinforcing fibers and matrix resins that provide the necessary performance and processability. Several types of fibers were available for reinforcing members; graphite fibers had greatest utility because of their extremely high strength and modulus. A polymeric material was needed for the matrix phase that could offer necessary composite integrity under rigorous conditions, and be formed into prepreg and consolidated at a temperature of $350^\circ F$ ($177^\circ C$) and a pressure of 100 psi or below.

Amine or anhydride cured systems based on bisphenol A or novolac type epoxies have been used for composite applications for some time now. But these systems do not provide thermal/mechanical properties required for advanced structural applications. Tables 1 and 2 list the high performance thermoset resin and trends and the types of most popular thermoset resins available today, respectively.

ADVANCED COMPOSITE SYSTEMS

CIBA-GEIGY developed an epoxy/amine system consisting of tetraglycidyl methylene diani-line (MY720) and diaminodiphenylsulfone (HT976) that gives much better mechanical properties than basic epoxy/amine systems up to $350^\circ F$ ($177^\circ C$). In addition, MY720/HT976 provides easy prepreg forming through melt impregnation of fibers and easy part consolidation at 350°F (177°C) and 100 psi pressure. MY720/HT976 reinforced with graphite formed the first composite materials to meet the performance demands of aerospace structural applications and have gained acceptance in the industry. Since the introduction of MY720 resin, continuous efforts have been made to improve the consistency of this product in terms of viscosity and other properties. Table 3 shows the chronological history of this product to date.

## MY720/HT976

MY720 is prepared from the precisely controlled reaction of 4,4'-methylenedianiline and epichlorohydrin. The major constituent of MY720 is N,N,N',N'-tetraglycidyl-4,4'-methylenedianiline as:

$$CH_2-CH-CH_2 \quad \quad CH_2-CH-CH_2$$
$$\overset{N}{\underset{CH_2-CH-CH_2}{}} \!\!-\!\!\bigcirc\!\!-CH_2-\bigcirc\!\!-\overset{N}{\underset{CH_2-CH-CH_2}{}}$$

.   Translucent brown semi liquid
.   117-134 g/epoxide equivalent
.   8,000 to 18,000 cp at 50oC

HT976 is 4,4' diaminodiphenylsulfone, a high-performance amine hardener, shown below:

$$H_2N-\bigcirc-\overset{O}{\underset{O}{S}}-\bigcirc-NH_2$$

.   Tan powder
.   172°C melting point.

Formulations of MY720/HT976 constitute the matrix resins of the most widely used aerospace composites. Data on MY720/HT976 in Table 5 is used as a standard against which the developmental resin systems can be judged and are typically needed in the preliminary stages of development for an advanced composite matrix candidate. They indicate the conditions needed to develop the system's mechanical properties and shows how the systems mechanical properties respond to simulated service environments.

MY720/HT976 has progressed well beyond the initial stages of development and into application. Work done internally at CIBA-GEIGY has been concentrating on the characterization and monitoring of the resin system's physical properties, chemical composition, and reactivity (2-6). Many other studies of MY720/HT976 also contributed to its success as the advanced composite matrix resin of choice. Design engineers can benefit from the engineering database on these composites that was developed under funding by the U.S. Air Force (7). The Aircraft Energy Efficiency (8) program coordinated by NASA (National Aeronautics and Space Agency) with the participation of the major airframe manufacturers was also important in promoting utilization of MY720/HT976-based composites in aerospace applications.

The collected data on MY720/HT976 and its composites completely assess the system's performance potential. These data, crucial to the success of MY720/HT976 composites in many rigorous applications, will help maximize the use of its many outstanding properties in the future. The data have also defined two limits on MY720/HT976 performance: a limited resistance to impact damage and a maximum use temperature of 350°F (177°C).

## ADVANCED COMPOSITE SYSTEMS

Many applications require material systems that not only have better resistance to impact damage (toughness) and better thermal stability than MY720/HT 976 composites but retain the same conditions for prepreg forming and part consolidation. Graphite fibers and the newer aramid fibers provide the necessary reinforcement for these applications. As with the first application of composites, therefore, the development of matrix resin systems with the required properties has become the critical element for bringing higher performance composites into use.

Toughened Epoxy Resins. To improve resistance to impact damage, many attempts were made to modify MY720/HT976, usually by the addition of rubber-based compounds. The rubber in these systems comes out of the epoxy during cure resulting in a dispersion of discrete rubber particles throughout the matrix. These rubber particles do help limit impact damage; however, they also degrade the system's mechanical properties, especially in humid and elevated temperature conditions. The degradation of mechanical properties with the addition of rubber modifiers could not be overcome, and this limited their use in advanced composite matrices. We also developed a matrix resin system, XU276, which is based on a new epoxy resin. It offers improved toughness over MY720/HT976 without rubber modifiers and their resultant degradation of mechanical properties.

XU276

XU276 has been developed specifically to extend the use of advanced composite materials in applications requiring greater resistance to impact damage than is provided by MY720/HT976 composites (9,10). These applications, as typified by commercial primary aircraft structures, require a significant increase in resin toughness

without sacrifice of other mechanical and thermal properties and processing. XU276 chemistry is the subject of a pending U.S. patent. The formulation and cure cycle used for the data are shown in Table 4. The typical properties of the resin are:

.  Appearance: translucent red-brown thixotropic liquid
.  Viscosity: 800 to 1,700 cps (0.8 to 1.7 Pa·s) at 100°C
.  Equivalent Weight: 200-220 g/epoxide

The typical mechanical and thermal properties of XU276/HT976 are compared with MY720/HT976 in Table 5. The results verify the enhanced toughness of XU276 over MY720, as shown by the fourfold increase in fracture toughness value. In addition, the results show that XU276 exhibits outstanding room temperature mechanical properties compared to MY720, and only slightly lower glass transition temperature and elevated temperature mechanical properties. Based on these comparisons XU276/HT976 shows definite promise as a matrix resin system for advanced composites with improved resistance to impact damage.

High-Temperature Thermosetting Resin. Much of the effort to develop matrix resin systems with improved thermal stability focused on aromatic polyimides. These resins offer definite improvements in high-temperature performance over MY720/HT976 but are difficult to form into prepreg and their major curing mechanism involves imidization that releases water and causes voids in cured structures.

Several preimidized systems with terminal reactive end groups were developed to eliminate the problem of condensation curing reaction, but these have had limited success because they are very difficult to process. A two-component bismaleimide system, recently developed and commercially available as Matrimid 5292[R] from CIBA-GEIGY, offers significantly better thermal stability than MY720/HT976 while retaining its desirable processability. It also offers improved toughness over MY720/HT 976

## MATRIMID 5292 [R]

Matrimid 5292 is a new resin system exhibiting improvements in elevated temperature properties and toughness over both MY720/HT976 and XU276 systems. It is based on a new, patented technology involving the reaction of a bismaleimide with an allylphenol. The resulting ease of processing and remarkable thermal and mechanical properties place this product in a class by itself. Matrimid 5292 is supplied as a two-component system. The Matrimid 5292 component A consists primarily of 4,4'-bismaleimidodiphenyl methane methane shown here:

. Yellow crystalline powder
. Maleimide double-bond content: 85% of theoretical
. 150-160°C melting point

Matrimid 5292 B is an allylphenol of the type:

. Amber liquid
. Viscosity @ 25°C: 12,000-20,000 cps (12-20 Pa·s).

The formulation and cure cycle used for the data (11) are given in Table 4. The mechanical properties and glass transition temperature are presented in Table 5. Matrimid 5292 shows significantly better properties than MY720/HT976 in all respects. The room temperature and elevated temperature mechanical properties, fracture toughness properties, and glass transition temperature are all much higher for Matrimid 5292. In addition, the test results show that Matrimid 5292 retains good properties up to 230°C. The versatility provided by this unique combination of high-temperature stability and resistance to impact damage provides the basis for new composite material systems which will meet the demands of a wide range of applications.

More recently many newer bismaleimide (BMI) materials are being introduced to the composite industry. The major thrust has been in improving the processability aspect of the high temperature materials. CIBA-GEIGY has also developed a second generation BMI material (RD85-101) which is at experimental stages at this time and would soon be available for customer sampling. This newer BMI solves some of the processing difficulties associated with the current commercial BMI's.

SUMMARY

Advanced composite materials have become a viable class of materials for use in aerospace and other demanding applications. The timely development of MY720/HT976 as a matrix system and the quality and consistency improvements made with time have helped the advancement of polymeric based composites and their acceptance as structural materials. The newly developed BMI's e.g. Matrimid 5292 with their superior performance characteristics are envisioned to be the materials of choice for the future military and commercial aerospace application.

References:

1. Knott, I., and B. Met, "Fundamentals of Fracture Mechanics," John Wiley and Sons (1973).
2. King, J., R. Castonguay, and J. Zizzi, "HPLC Evaluation of MY 720," Nat. SAMPE Tech. Conf. p. 53 (1981).
3. King, J., R. Castonguay, and J. Zizzi, "HPLC Evaluation of MY 720 II," Nat. SAMPE Symp., p. 163 (1982).
4. Cobuzzi, C., J. King, and R. Castonguay, "HPLC Evaluation of MY 720 III," Nat. SAMPE Symp., p. 877 (1983).
5. Chaudhari, M., and J. King, "Characterization of MY 720 IV," Nat. SAMPE Tech. Conf., p. 676 (1983).
6. Cobuzzi, C., J. King, and M. Chaudhari, "Characterization of MY 720 V," Nat. SAMPE Symp., p. 1261 (1984)
7. Hofer, K.E. et al, "Development of Engineering Data on the Mechanical and physical properties of Advanced Composite Materials," AFML TR-74-266 Feb. 1, 1975).
8. "A look at NASA's Aircraft Energy Efficiency Program" Report to the U.S. Congress, NTIS:PB 81-137523 (July 28, 1980).
9. King, J., R. Castonguay, and R. Sellers, "A New Generation of High performance, Rubber Free Toughened Epoxy Resin Systems," Nat. SAMPE Symp., p. 359 (1983).
10. Cobuzzi, C., J. King, and R. Castonguay, " New Generation of High Performance, Rubber Free toughened Epoxy Resin Systems," Nat. SAMPE Symp., p. 359 (1983).
11. King, J., M. Chaudhari, and S. Zahir, "A New Bismaleimide System for High Performance Applications," Nat. SAMPE Symp., p. 392 (1984).

TABLE 1

## HIGH PERFORMANCE THERMOSET RESIN
## NEED AND TRENDS

. Tougher without loss of other
  properties.
. More consistent
. Higher purity
. Higher temperature capability
. More economic processing

TABLE 2

## HIGH PERFORMANCE THERMOSET RESINS

| RESIN | USE TEMPERATURE | MECHANICAL CHARACTERISTICS |
|-------|------------------|----------------------------|
| Epoxy | 350°F | Modulus, Brittle |
| BMI | 400°F | High Modulus, Brittle |
| Polyimide | 550°F | High Modulus, Brittle |

TABLE 3

MORE CONSISTENT HIGH PERFORMANCE EPOXY RESINS

| YEAR | VISCOSITY OF STANDARD EPOXY RESIN * cps @ 50°C |
|------|------|
| 1970 | Up to 25,000 |
| 1980 | 8 - 20,000 |
| 1982 | 12 $\pm$ 2,000 |
|      | 16 $\pm$ 2,000 |
| 1985 | 8,000 + 1,000 |
|      | 12,000 $\pm$ 1,000 |
|      | 14,000 $\pm$ 1,000 |
|      | 18,000 $\pm$ 1,000 |
| 1986 ** | 5,000 $\pm$ 1,000 |

*ARALDITE MY 720: CIBA-GEIGY CORPORATION

** MY 721

TABLE 4

FORMULATIONS

(PARTS BY WEIGHT)

|  | MY 720/HT 976 | XU 276/HT 976 | XU 292 |
|---|---|---|---|
| MY 720 | 100 | --- | |
| XU 275 | --- | 100 | |
| HT976 | 44 | 27 | |
| MATRIMID 5292 | --- | --- | 100 |
| MATRIMID 5292 | --- | --- | 75 |

---

| STOICHIOMETRY | 1:0.85 | 1:0.90 | 1:0.87 |
|---|---|---|---|
| CURE CYCLE | A | A | B |
| | 3 h @ 80°C | | 1 h @ 180°C |
| | 2 h @ 100°C | | 2 h @ 200°C |
| | 5 h @ 175°C | | 6 h @ 250°C |
| | 7 h @ 200°C | | |

66

TABLE 5   MECHANICAL AND THERMAL PROPERTIES

|  | MY 720/HT976 | XU 276/HT 976 | MATRIMID 5292 |
|---|---|---|---|
| **R.T. TENSILE** | | | |
| Strength, (kpsi) | 8.5 | 11.9 | 13.6 |
| Modulus, (kpsi) | 542 | 427 | 564 |
| Elongation,% | 1.8 | 4.3 | 3.0 |
| **R.T. Flexural** | | | |
| Strength, (kpsi) | 13.3 | 19.6 | 26.8 |
| Modulus, (kpsi | 499 | 400 | 580 |
| **R.T. Compression** | | | |
| Yield Strength, (kpsi) | 29.2 | 20.5 | 30.4 |
| Modulus, (kpsi) | 284 | 280 | 360 |
| **150°C Tensile** | | | |
| Strength, (kpsi) | 6.5 | 6.0 | 10.1 |
| Modulus, (kpsi) | 378 | 243 | 412 |
| Elongation,% | 1.9 | 9.5 | 3.1 |
| **200°C Tensile** | | | |
| Strength, (kpsi) | -- | -- | 10.4 |
| Modulus, (kpsi) | -- | -- | 394 |
| Elongation,% | -- | -- | 4.6 |
| **100°C /Wet* Tensile** | | | |
| Strength, (kpsi) | 6.8 | 5.8 | -- |
| Modulus, (kpsi) | 392 | 333 | -- |
| Elongation,% | 1.9 | 5.5 | -- |
| % $H_2O$ Pickup* | 4.3 | 2.8 | 2.5 |
| **150°C/Wet* Tensile** | | | |
| Strength, (kpsi) | 31. | -- | 6.3 |
| Modulus, (kpsi) | 236 | -- | 304 |
| Elongation,% | 1.5 | -- | 3.4 |
| **Composite Short-Beam Shear** | | | |
| R.T. Strength, (kpsi) | 103.4(15.0) | 104.8(15.2) | 17.8 |
| 230°C Strength, (kpsi) | -- | -- | 11.4 |
| **Glass Transition** | | | |
| Temperature, °C (°F) | 238(460) | 195(383) | 285(545) |
| Fracture Toughness | | | |
| $G_{IC}$ Value, | | | |
| (in. $\cdot$ lb/in.$^2$) | 0.3 | 1.2 | 1.4 |

* Stored 2 weeks at 70°C and∿95% Humidity.

# HONEYCOMB MATERIALS AND APPLICATIONS

John L. Corden
Thomas N. Bitzer
Hexcel Corporation
Dublin, CA

## Abstract

An overview is given of the materials used in honeycomb products. Many new types of honeycomb products have been fabricated as a result of composite material developments. Graphite/Phenolic and Kevlar/Epoxy honeycomb products are typical examples of composite honeycomb products.

The second part of the paper summarizes the various applications in which honeycomb products are well suited. Structural sandwich panels represent the main use of honeycomb. A sandwich structure provides one of the highest strength-to-weight and stiffness-to-weight structures possible. Other applications include energy absorption, thermal radiators and barriers, acoustics, RF shielding and flow directionalization.

## 1. INTRODUCTION

Honeycomb products consist of thin sheets attached to form connected cells - See Figure 1.

HONEYCOMB

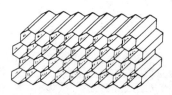

FIGURE 1

The majority of honeycomb used today is fabricated by adhesively bonding the sheets together. Other types of honeycomb products are currently being produced. The most common being cores utilizing resistance welded or brazed attachments of the sheets. Honeycomb products have been used for structural applications since 1940, although nonstructural uses have dated back 2,000 years.

## 2. CORE TYPES

### 2.1 Materials

Honeycomb can be manufactured from virtually any thin sheet material. Common metallic core materials are aluminum, corrosion resistant steel, titanium and nickel based alloys. The most common non-metallic core materials are NOMEX, fiberglass and kraft paper. Non-metallic core is normally dipped in liquid phenolic, polyester or polyimide resin to achieve the final density, although other resin systems can be used. Ideally the resin content should be around 50 percent. Therefore, a range of ribbon thicknesses must be available to make cores of various densities, while maintaining the resin content as close to 50 percent as possible.

To date, over 500 different types of honeycomb have been produced with the most recent types being graphite, kevlar and ceramic.

### 2.2 Cell Configuration

There are three basic cell configurations; hexagonal, overexpanded and Flexcore.

These and some of the less common configurations (square and reinforced) are shown in Figure 2.

CELL CONFIGURATIONS

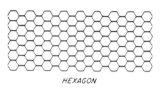
*HEXAGON*

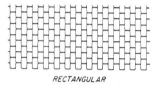
*RECTANGULAR*

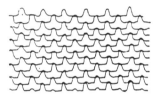
*FLEX-CORE*

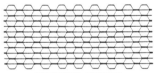
*REINFORCED HEXAGON*

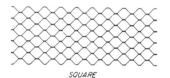
*SQUARE*

FIGURE 2

NOMEX is a registered trademark of E.I. DuPont

69

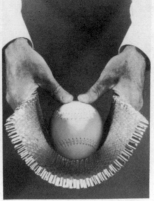

FIGURE 3

The overexpanded cell configuration is obtained by expanding the hexagonal cell until
it forms a rectangle. The
primary advantage of this configuration is that it is easily formed in the "L" direction; the hexagonal cell configuration can only be formed
in this direciton by roll
forming or heat forming. In
addition, the "W" shear properties tend to increase while
the "L" shear properties decrease slightly.

Flex-core offers added formability over other types of
honeycomb. It forms easily
in both the "L" and "W" directions, hence it is ideally
suited for applications that
require the core to be formed
into compound curvatures.
Flex-core can be formed into
compound curvatures without

buckling the cell walls and
exhibits greatly reduced anti-
clastic curvature. When
formed in tight radii, Flex-core
has higher shear strengths
than comparable hexagonal cell
core of equivalent density.
Figure 3 shows a typical example of the formability of
Flex-core.

The square cell configuration
is used predominantly in
welded corrugated core. It
has a very narrow node and the
free walls often have corruga-
tions in them to increase the
cell wall resistance to buckling.

Reinforced core has a flat
sheet placed in-between the
nodes. This results in both
increased densities and mechanical properties. Aluminum corrugated core has been made in

70

this manner with a density as high as 55 pcf.

Tube-core is manufactured by spirally wrapping a corrugated sheet with adhesive applied to the nodes and a flat sheet around a mandrel. The inside and outside diameters are variable. This type of core is used exclusively for energy absorption applications.

2.3 Standard Core Types

Standard honeycomb core types available are listed below.

5052 Aluminum Alloy - Specification grade 5052 H39 aluminum alloy with a corrosion resistant coating applied. For general purpose applications, particularly in aerospace.

5056 Aluminum Alloy - Specification grade 5056 H39 aluminum alloy with a corrosion resistant coating applied. Offers slightly higher strength over 5052 alloy honeycomb.

2024 Aluminum Alloy - Heat treatable 2024 aluminum alloy combines high room temperature properties with good strength retention at elevated temperatures. A corrosion resistant coating is applied in the same manner as the other aluminum alloy honeycombs.

Commercial Grade Aluminum - A low cost aluminum honey-comb (usually 3000 series aluminum alloy) with a corrosion resistant coating applied. For use in non military specification requirements.

Naval Grade Aluminum - A commercial grade 5052 alloy honeycomb with a corrosion resistant coating applied. Primarily for use in Naval applications (bulkheads, decks and joiner panels).

Note that all aluminum honeycomb products come with a corrosion resistant coating. The coating effectively eliminates corrosion of the core as verified by the ASTM B117 salt spray corrosion test.

Fiberglass - A glass fabric reinforced plastic honeycomb in which heat resistant phenolic resin is used for both initial impregnation and dip coats. Varieties are available in either a 0 degree, 90 degree fiber orientation or a +45 degree, -45 degree fiber orientation. The +45 degree, -45 degree (bias) fiber orientation offers increased shear properties over the 0 degree, 90 degree orientation. In addition to the phenolic resin system, fiberglass honeycomb is also available with a polyimide node adhesive, impregnation and dip. This product exhibits excellent elevated temperature properties and low dielectric constants.

NOMEX - This product consists of NOMEX aramid fiber paper dipped in phenolic resin. NOMEX honeycomb has high strength and toughness, yet low density. NOMEX honeycomb is also available with a polyimide resin dip rather than the phenolic. This product has excellent dielectric and loss tangent properties.

Kraft Paper - A kraft paper core dipped in phenolic resin. Offers low cost and high strength. Used extensively in Air Transportable Shelters for the military.

Graphite - A graphite fiber reinforced plastic honeycomb available in a variety of resin systems. Highest strength non-metallic core available, strength and shear moduli approaching 5052 aluminum alloy.

Kevlar - A kevlar fabric reinforced honeycomb with epoxy resin used for initial impregnation and dip coats. Used primarily for space type applications that require core with extremely low coefficient of thermal expansion properties.

2.4 Special Honeycomb

The above honeycomb products are available with either fiberglass or foam filled cells.

Fiberglass filled cells are used to improve the sound absorption characteristics of honeycomb. Side benefits include lower smoke produced in the N.B.S. smoke chamber test for aramid honeycombs and reduced thermal conductivity of the core.

Foam used for cell filling is typically a closed cell, hydrophobic polyimide based material. Polyimide foam filled honeycomb offers several advantages over unfilled core. The polyimide foam precludes the intrusion and build up of water in honeycomb core sandwich panels. The foam provides a high degree of thermal insulation, particularly in non-metallic cores. The foam also increases the mechanical properties of the honeycomb, although, not on a specific basis. Table 1 characterizes the properties of a typical foam in this category, Hexcel Corporation's Airlite$^{TM}$ 600.

## AIRLITE 600 POLYIMIDE FOAM CHARACTERISTICS

. 500°F CONTINUOUS SERVICE TEMPERATURE
. 600°F+ SHORT TERM EXPOSURE
. CLOSED CELL, HYDROPHOBIC MATERIAL
. LESS THAN 3% MOISURE PICKUP BY WEIGHT
  AT 140°F, 98% RELATIVE HUMIDITY
. GOOD THERMAL INSULATION PROPERTIES,
  $k = 0.652$ AT 375°F, $k = 0.742$ AT 500°F
. NONFLAMMABLE, VERY LOW SMOKE
  $D_{mc} = 2$ NBS SMOKE TEST
. LOW DIELECTRIC CONSTANT, 1.2
. DENSITY RANGE 2 PCF TO 15 PCF

TABLE 1

## 3. APPLICATIONS

### 3.1 General

Honeycomb, as mentioned previously, can be manufactured from almost any flat sheet material, and the cell sizes can range from 1/16" to 1" or larger with densities from 1 to 55 pcf. Because of this vast collection of cores available honeycomb can be used advantageously in many diverse applications. In the following paragraphs some of these uses are discussed; such as structural sandwich panels, energy absorption, thermal versatility, acoustics, radio frequency shielding and flow directionalization.

### 3.2 Structural

The main use of honeycomb is in structural sandwich panels, and bonded honeycomb sandwich construction has been a basic structural concept in the aero-

space industry since the early 1940's. Virtually every aircraft flying today uses some type of honeycomb panel.

A typical sandwich consists of two thin, dense facings bonded to a thick lightweight core (See Figure 4). Each component by itself is relatively weak and flexible but when combined in the sandwich panel they provide one of the highest strength-to-weight and stiffness-to-weight structures possible.

Figure 5 shows very clearly how efficient a honeycomb sandwich can be. Here a solid 0.064" thick aluminum beam is compared to two honeycomb beams made with 0.032" aluminum skins. The thicker beam is seven times stronger and thirty-seven times stiffer than the solid beam with only a small weight increase of approximately ten percent.

Other strong advantages of honeycomb sandwiches are its ability to prevent thin skins from buckling and its outstanding sonic fatigue resistance. In sheet and stringer air foils, thin skins may buckle under aerodynamic loading thus causing increased drag while the thin skins continuously supported by the honeycomb remain smooth in a sandwich structure.

# SANDWICH CONSTRUCTION

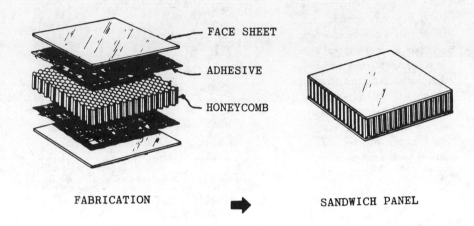

FACE SHEET

ADHESIVE

HONEYCOMB

FABRICATION ➡ SANDWICH PANEL

FIGURE 4

## HONEYCOMB SANDWICH EFFICIENCY

| | T | 2T | 4T |
|---|---|---|---|
| RELATIVE STIFFNESS | 1 | 7 | 37 |
| RELATIVE STRENGTH | 1 | 3 | 7 |
| RELATIVE WEIGHT | 1 | 1.05 | 1.09 |

FIGURE 5

Also, since the facings are continuously bonded to the core the sonic fatigue life is greatly increased - no stress risers are present such as rivets.

Honeycomb sandwich structures are almost unbeatable whenever performance (stiffness, strength and low weight) is the chief design criteria because the honeycomb allows the thin skins to work at their ultimate compressive stresses. The basic concept of a sandwich panel is that the facings take the bending load (one facing in compression and the other in tension), and the core takes the shear load.

One of the drawbacks of sheet and stringer, extrusions and integral stiffened structures is that unless the skins are quite thick the entire compression skin will not be capable of reaching the material's ultimate strength due to buckling. Only the facing area around the web will be able to reach the maximum stress - See Figure 6. To get the entire skin at the ultimate stress requires a very thick laminate or very closely spaced stiffeners. Both of these solutions require the structure to be heavier than a honeycomb sandwich. So whenever the structure must be as light as possible, it is extremely difficult to beat a honeycomb panel.

### COMPRESSION FACING

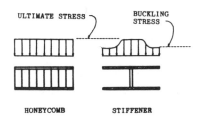

FIGURE 6

## 3.3 Energy Absorption

A few years after the introduction of honeycomb as a sandwich core material, it was discovered that this material also possessed an unusual ability to absorb energy at a constant rate.

Figure 7 shows a typical load-deformation curve. If a piece of core is tested, the load goes up to a peak value (bare compressive strength) and then starts to crush at a uniform load (about 50% of the peak load) until it "bottoms-out" (can no longer crush). Usually the peak load spike is not desired so the core is "precrushed". This entails slightly prefailing the core by bending over the cell edges for aluminum core or using a razor blade to cut the surfaces of fiberglass or NOMEX honeycomb. The "stroke" or distance the honeycomb crushes is around 70 to 80% of it's initial thickness, and this depends on the core type and density. The area under the curve is the amount of energy the honeycomb absorbed. Figure 8 shows a crushed aluminum specimen while Figure 9 is a close-up of the crushed cells.

## TYPICAL HONEYCOMB CRUSH CURVE

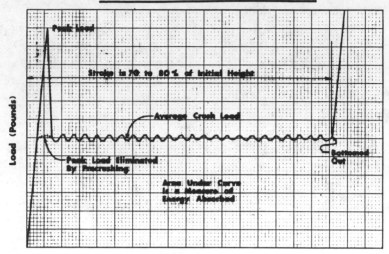

Load (Pounds)

Peak load

Stroke is 70 to 80% of Initial Height

Average Crush Load

Peak Load Eliminated By Precrushing

Area Under Curve Is a Measure of Energy Absorbed

Bottomed Out

Deformation (Inches)

FIGURE 7

FIGURE 8

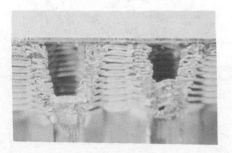

FIGURE 9

The primary advantages in using honeycomb as an energy absorber are that it crushes at a very uniform and predictable level, high crush strength-to-weight ratio, many crush strengths available, has a very long crush stroke and is lightweight. These properties have allowed many different types of honey-comb energy absorbers to be designed. Some of them are the following: Apollo LEM Landing structs, Military Air-Drop Pallets, Alaska Pipeline Restraints and Caltran Attenu-ators.

3.4 Thermal Versatility

Honeycomb sandwich panels can be made to transmit heat from one face to the other or to act as an insulative barrier. If, for example, it is required to keep both skins near the same temperature to minimize thermal curvature, metallic honeycombs can be used to transmit the heat from one side to the other. Aluminum honeycomb has a very high thermal conductivity being from 25 to 100 Btu-in/hr-ft$^2$-°F depending on the core's density.

The heat is transferred from one skin to the other by conduction through the cell walls, air convection currents in the cell and radiation - See Figure 10.

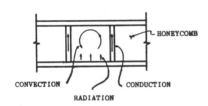

**HEAT TRANSMISSION**

HONEYCOMB

CONVECTION  CONDUCTION
RADIATION

FIGURE 10

If an insulative type panel is desired, a non-metallic core should be used and the cells filled with an insulative material such as fiberglass or foam. In general, most non-metallic honeycombs (fiberglass, NOMEX and kraft paper) have a "k" value of between 0.6 to 0.8 Btu-in/hr-ft$^2$-°F with unfilled cells and about 0.3 to 0.4 with the cells filled. This combination gives good insulative properties along with excellent structural sandwich strength and stiffness.

## 3.5 Acoustics

Sandwich panels can be designed to act as sound absorbing systems or as sounding boards. Aircraft interiors are one example of the former. Here, NOMEX honeycomb is filled with a fiberglass batting, and a thin porous Tedlar skin is used as the interior facing. The sound from the cabin goes through the small facing holes and is absorbed by the fiberglass in the honeycomb cells - See Figure 11. This greatly reduces the noise level inside the cabin. Also, to quiet jet aircraft engines some nacelles are made with honeycomb and a perforated inside skin. This sandwich can be tuned (cell size, core thickness, hole area) to absorb certain frequency ranges by the Helm-Holtz phenomenon.

## SOUND ABSORPTION

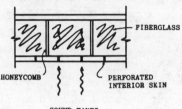

FIGURE 11

### 3.6 Radio Frequency Shielding

Honeycomb offers the designer a unique material which will shield openings used for heating, ventilating and lighting from the passage of radio frequency (R.F.) interference signals. It can be compared to a myriad of wave-guides which if properly designed as to cell size and core thickness will attenuate a required db level through a wide frequency range - See Figure 12.

### RADIO FREQUENCY SHIELDING

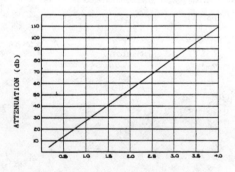

FIGURE 12

Aluminum honeycomb is usually used; however, when high temperature or low frequency attenuation is required, steel core may be required.

### 3.7 Flow Directionalization

Honeycomb has been used successfully as a means of altering turbulent flow into laminar flow or changing the direction of moving fluids. The thin cell walls provide an array of straight passages with a maximum percent open area (95 to 99%) which results in very low pressure drops. The major resistance is related to the skin frictional drag of the cell walls. As would be expected, the larger cells and thinner core slices have the least pressure drops. At the same time, honeycomb will have more stability than other methods, and since the pressure drop is less the overall efficiency is far superior.

Some of the more common uses for these properties are duct fans, frozen food display cases, air conditioners, heaters, wind tunnels and grilles or registers. Corrosion protective coatings for aluminum honeycomb have helped to overcome some of the concerns of environmental exposure condition; also, NOMEX core types are used to minimize impact damage.

## CHANGING FLOW TYPE

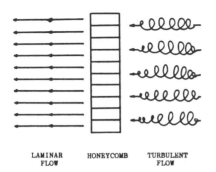

LAMINAR    HONEYCOMB    TURBULENT
FLOW                     FLOW

FIGURE 13

## CHANGING FLOW DIRECTION

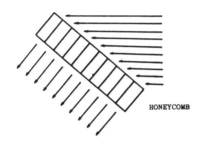

HONEYCOMB

FIGURE 14

4.  REFERENCES

1.  Bitzer, Thomas N.; Honey-
    comb Sandwich Design and
    Testing.
2.  Hexcel Corporation; TSB
    102 - Honeycomb in Air
    Directionalizing Applica-
    tions.
3.  Hexcel Corporation; TSB
    113 - Radio Frequency
    Shielding Properties of
    Hexcel Metallic Honeycomb.
4.  Hexcel Corporation; TSB
    116 - Aluminum Honeycomb
    in Tooling Applications.
5.  Hexcel Corporation; TSB
    120 - Mechanical Properties
    of Hexcel Honeycomb
    Materials.
6.  Hexcel Corporation; TSB
    122 - Design Data for the
    Preliminary Selection of
    Honeycomb Energy Absorp-
    tion Systems.
7.  Hexcel Corporation; TSB
    124 - The Basics on Bonded
    Sandwich Construction.

5.  BIOGRAPHY

Mr. John Corden is a Honeycomb
Product Manager at Hexcel Cor-
poration.  Prior to joining
Hexcel he was employed by the
Northrop Corporation, Hawthorne,
CA, as a Senior Engineer in
the Materials and Processes
Applications Group.  Mr. Corden
received a B.S. degree in
Materials Science Engineering
from Rensselaer Polytechnic
Institute and is a member of
SAMPE.

Mr. Thomas Bitzer is the Manager of Hexcel's R & D Test
Laboratory where he has been
involved with honeycomb for
about twenty years.  He has a
Ph.D.  from Century University
and a M.S. degree in Structural
Engineering from the University
of California at Berkeley and
is a registered Civil Engineer.

FIBER ORIENTATION CONTROL BY THE LOADING OF MAGNETIC MOMENT

S. Yamashita, H. Hatta

T. Sugano, K. Murayama

Mitsubishi Electric Corporation

1-1-57 Miyashimo, Sagamihara, Kanagawa, 229 Japan

## Abstract

An attempt was made to clarify the possibility of fiber orientation control in a short fiber composite by loading  magnetic moment on a fiber coated with ferromagnetic substance. Firstly, effect of various parameters, say coating thickness, aspect ratio of the coated fiber, and intensity of applied magnetic field, on torque and rotation time was examined analitically. Then, the maximum value of the fiber volume fraction $V_{fc}$ at which the orientation control can be made is examined experimentally as a function of the aspect ratio. Finally, in order to force to increase $V_{fc}$, ultrasonic vibration together with the magnetic moment was loaded at high concentration of the  fibers to loosen entanglement of the fibers.

The ultrasonic vibration gave remarkable improvement resulted in increase of $V_{fc}$.

## 1 INTRODUCTION

Short fiber composites are used extensively in many fields due to cheapness of fabrication cost and ease of processing into complex shapes. One of difficulties to use the short fiber composites is lack of stability in material properties. The instability of the material properties is usually originated from anomalous fiber orientation induced by flow in a die. The weld line [1] and the waving pattern [2] are such typical defects and ,for instance,it is reported that the strength of the short fiber composite is reduced to 30 - 50 % of the original material  by the existence of the such defects[2].

Another difficulty encountered in the short fiber composites is narrowness of tailorable domain of material properties. Usually ,the short fibers in the composites are considered to have 2- or 3-dimensional random orientation distribution and we treat it as if control of the orientation distribution of the fibers in the composite were impossible. Accordingly, it becomes rather difficult task to realize optimum material design of the short fiber composite compared with that of the lamination type composite in which the optimum design can be easily made by arranging orientation of long fibers in each layer to be laminated .

To overcome the before-mentioned difficulties, some measure to control the orientation of the short fiber composite must be established. In this paper, we picked up magnetic moment as a tool for the fiber orientation control. The reason for this selection is two-fold.Firstly, by use of the method, the fiber orientation control becomes possible to almost all the reinforcements provided a ferromagnetic sabstance is coated on the reinforcing fibers. Secondary, high torque is expected to load on the coated fiber by means of the magnetic moment compared with the electric loading[3].

For exploring the possibility of the orientation control of the short fiber in a composite, an analitical study is conducted firstly under the condition of dilute fiber volume fraction. The formulations to predict the torque applied on a coated fiber and time required to rotate the coated fiber are presented in section 2. To demonstrate effects of various parameters, numerical results based on the present formulation are also shown in section 2. In section 3, experimental study was carried out to examine effect of volume fraction of fibers $V_f$ on the rotation of the coated fiber. The critical volume fraction of fibers $V_{fc}$ that is defined as the maximum value of $V_f$ at which the orientation control of the fiber is possible ,is examined as a function of aspect ratio. Then, a method to force to increase $V_{fc}$ is explored to extend present method so as to be applicable to actual composite. Finally, the conclusion is stated in section 4.

## 2 A Dilute composite
## 2.1 Formulation

In order to clarify the possibility of the fiber orientation control by the loading of the magnetic moment, an analitical study under the assumption of dilute fiber concentration was conducted. A model for the analysis is shown in Fig.1, where a fiber coated with ferromagnetic substance is embedded in an infinite matrix. The coated fiber inclined at an angle $\theta$ with coordinate axis $x_3$ has placed in uniform magnetic field $H^0$. The permeabilities of fiber, coating, and matrix are denoted by $\mu_f$, $\mu_c$, and $\mu_m$, respectively, and we assume they are constant. The major and miner semiaxis of the coated fiber are denoted by c and a, respectively, and the thickness of the coating by w. The major fiber axis is set along the $x'_3$ axis.

In order to determine magnetic field around the coated fiber, we paied attention to the fact that the solution of steady state heat conduction is applicable to that of transport phenomena such as electric conduction, steady state diffusion [4],[5], by only replacement of parameters and variables. An analysis of magnetic field is also the phenomenon for which the correspondence is effective. For this case the correspondence is as follows.

Thermal conductivity

↓

Permeability

Temperature gradient

↓

Magnetic field intensity

Temperature

↓

Magnetic potential

On the other hand, we analysed temperature field around the coated fiber in the previous paper [6]. Thus, we can use the previous formulation to obtain the magnetic field around the coated fiber. In order to calculate torque apllied to the coated fiber due to the magnetic moment, we used the following well known relation.

$$T = \partial W/\partial \theta$$

$$W = 0.5 \int \underline{B} \cdot \underline{H} dv \qquad (1)$$

where W denotes the magnetic energy, $\underline{H}$ the intensity of the magnetic field and $\underline{B}$ the magnetic flux density. Substituting the solution of the magnetic field into eq.(1), we can obtain formulation to predict the torque due to the magnetic moment loaded on the coated fiber $T^M$ as

$$T^M = \mu_m (H^0)^2 \sin 2\theta \left\{ V_1 (1 - \mu_f/\mu_m)(^1H_1^U - {}^1H_3^U) + V_2 (1 - \mu_c/\mu_m)(<{}^2H_1^U> - <{}^2H_3^U>) \right\}$$

$$= C_1 \sin 2\theta \qquad (2)$$

where $V_1$ and $V_2$ are the volume of the fiber and coating, respectively, the square bracket denotes the average value over the coating region, and $C_1$ is constant. In the derivation of eq.(2), we have introduced ,for convienience , new value of magnetic field intensity that stands for ones under unit boundary condition, namely, the magnetic field intensity in the fiber and coating under the boundary

82

condition of $H^0=1$ and $\theta=0°$, as $^1H_3^{\parallel}$ and $^2H_3^{\parallel}$, and of $H^0=1$ and $\theta=90°$, as $^1H_1^{\parallel}$ and $^2H_1^{\parallel}$.

To predict time t required to force to rorate the coated fiber from some referrence angle $\theta_0$ to arbitrary angle $\theta$, we can use the method adopted by Demetriades[7,8]. According to Jeffery[9], hydrodynamical torque due to rotation of a fiber is given by

$$T^H = -C_2\omega \qquad (3)$$

where $\omega$ is angular velocity of the coated fiber and the constant $C_2$ is given in references [7-9]. Then, from the condition that the sum of torques $T^M$ and $T^H$ vanishes, we get

$$\omega = \partial\theta/\partial t = (C_2/C_1)\sin2\theta \qquad (4)$$

The integration of eq.(4) yields

$$t = (C_2/C_1)$$
$$\times\ln(\tan\theta/\tan\theta_0) \qquad (5)$$

Up to here, we considered the case where the ferromagnetic region, the coating, did not attain the level of the saturation magnetization. Next, let us consider the magnetic field around the coated fiber after the saturation. For this case, the magnetic field H may be devided into two parts, namely;

$$\underline{H} = \underline{H}_b + \underline{H}_a \qquad (6)$$

where $\underline{H}_b$ denotes the magnetic field

intensity formed when all the ferromagnetic region (coating) attains the saturation magnetization $I_s$ and $\underline{H}_a$ denotes uniform magnetic field given by

$$\underline{H}_a = \underline{H}^0 - \underline{H}_b^0 \qquad (7)$$

Here $H_b^0$ denotes the applied magnetic field when the magnetic field intesity in the coating becomes $H_b$. $H_b^0$ can be calculated from the solution of the previous paper [7] under the assumption that the average of magnetization in the coating is equal to $I_s$. Thus,

$$\left|\underline{H}_b^0\right| = I_s/\{(\mu_c-\mu_m)$$
$$\times(\langle^2H_1\rangle\sin^2\theta$$
$$+\langle^2H_3\rangle\cos^2\theta)^{1/2}\} \qquad (8)$$

Substituting eqs. (6)-(8) into eq.(1), $T^M$ loaded on the coated fiber after the saturation is obtained as

$$T^M = \mu_m\mu_c I_s H^0\sin2\theta$$
$$\times\{v_1(^1H_1^{\parallel}+^1H_3^{\parallel})+v_2(\langle^2H_1^{\parallel}\rangle- \langle^2H_3^{\parallel}\rangle)\}/\{2($$
$$\mu_c-\mu_m)(\langle^2H_1^{\parallel}\rangle\sin^2\theta+\langle^2H_3^{\parallel}\rangle\cos^2\theta )^{1/2}\}$$
$$= C_3/(\langle^2H_1^{\parallel}\rangle\sin^2\theta$$
$$+\langle^2H_3^{\parallel}\rangle\cos^2\theta )^{1/2}\sin2\theta \qquad (9)$$

where we introduced constant $C_3$. Combining eqs.(3) and (9), the relation between $\theta$ and t is obtained as

$$\partial\theta/\partial t = C_3 \sin 2\theta$$

$$\Big/ \Big\{ C_1 (\langle {}^2H_1^{\parallel} \rangle \sin^2\theta + \langle {}^2H_3^{\parallel} \rangle \cos^2\theta )^{1/2} \Big\}$$

(10)

This is the ordinary differencial equation to obtain explicit relation between t and $\theta$. Eq.(10) can be integrated numerically.

## 2.2 Numerical Results

The formulation described in the previous section is used to obtain the numerical values of the torque and time required to make a coated short fiber rotated in magnetic field. In the computation, the following data for the costituents which simulate a polymer matrix composite with nickel coated graphite short fibers are used, though some of the data are made round numbers.

$$A_s = 100$$

$$W/a = 0.07$$

$$\mu_f / \mu_m = 1$$

$$\mu_c / \mu_m = 20 \qquad (11)$$

$$I_s = 8.8 \ H/m$$

$$\eta = 0.705 \ Pa \cdot s$$

where $A_s$ denotes the aspect ratio of the coated fiber and $\eta$ is viscosity of the matrix.

To examine the effect of various parameters on the torque $T^M$ and time t of the coated fiber, we have conducted a parametric study. $T^M$ and t as a function of applied magnetic field are shown in Fig. 2. The magnetization in the coating has dependency on the angle. Hence, the point of saturation begins at the angle of $\theta = 0°$ ($H^0 = H_1$) and the saturation completes by the saturation of $\theta = 90°$ ($H^0 = H_2$). As the present formulation is not effective between $H_1$ and $H_2$, we draw the region with dashed line connecting the two points smoothly. In the figure, $T^M$ is proportional to $(H^0)^2$ up to $H_1$ and is proportional to $H^0$ above $H_2$ as it can be deduced from eqs. (2) and (9). The rotation time t in the figure stands for a time in which the coated fiber rotates from $\theta = 1°$ to 89°. t reduces rapidly to small value at the rather small value of $H^0$ as shown in Fig.3. Hence we can consider that the orientation cotrol of the nickel coated short fiber is rather easy by the loading of the magnetic moment.

The rotation time t is also dependent upon geometory of the coated fiber. In Fig.4, effect of the coating thickness w on the rotation time t is shown. It follows from the figure that for the case of thin coating (possibly a coated fiber actually used) the reduction of t is rapidly even though corresponding increase of w is small.

In Fig.5, relation between t and the aspect ratio of the coated fiber $A_s$ is shown. It is obvious from the figure that t increases rapidly with increase of $A_s$. Accordingly, small $A_s$ is advantageous for the sake of easy rotation of the fiber. However, to retain high level of strength that reinforcing fiber has, high aspect ratio is required. Thus,in order to fabricate high quality composites by controling the fiber orientation, it is important task to find the optimum condition making the two requirements compromised.

Next, effect of the coating permeability $\mu_c$ on the rotation time t is examined and we confirmed that $\mu_c$ has not a remarkable effct on t in the region of high coating permeability. Finally, the dimensional effect was examined to find that t does not depend on the absolute dimensions of the coated fiber. It is due to the fact that both of $T^H$ and $T^M$ are proportional to the volume of the coated fiber as verified by the dimensional analysis.

## 3 A CONCENTRATED COMPOSITE

In the previous section, it was shown under the condition of dilute concentration of the coated fiber that the fiber orientation control is possible by the loading of the magnetic moment. In following this section, we discuss about a composite with concentrated fiber volume fraction. When the fiber volume fraction $V_f$ becomes high, the fibers interact to make a structure by which movement of the fibers are restrained. Then, at some $V_f$, the fibers fall into almost complete restraint. Thus, from this $V_f$ the fiber orientation control becomes impossible. Let us call this point critical volume fraction $V_{fc}$. The structure the fibers make is considered to be the phenomenon in which fiber geometory plays a decisive role. Thus, in this section, we devoted to establish relationship between $V_{fc}$ and the aspect ratio of the fiber $A_s$. As an analysis of a concentrated composite is difficult, the discussion of this section is based on experimental data.

### 3.1 Experimental procedure

Three kinds of nickel coated graphite fibers which have different aspect ratio were used as the magnetic fibers. The geometory of the coated short fibers are shown in Table 1. The coated fiber mixed in unsaturated polyestel (UPE) was stirred sufficiently in a transparent container (30mm x50mm x60mm) to set the mixture in the condition of the random fiber orientation. Then, the sample in the container was placed between a pair of magnetic coiles as shown in Fig.6. After that magnetic field was loaded. The maximum magnetic field the magnet yields is $4 \times 10^5$ A/m, viscosity of UPE 0.705 Pa·s, and Gel time 30 min. Thus, in view of the data in the previous section, we can consider that there is enough time to make the coated fiber rotated completely until the matrix is solidified.

To apply present method to actual composites, it is preferable to increase $V_{fc}$. Thus, we examined effect of ultrasonic vibration loaded on the fiber simultaneously with magnetic moment to loosen entangled fibers. For this case, we added the magnetic fibers little by little into the container so as to be able to disperse high concentration of the magnetic fibers. The frequency, electrical load, and amplitude between peek to peek of the vibration were 16(KHz), 1(KW) and 30 ($\mu$m), respectively.

## 3.2 Evaluation of fiber orientation distribution

For the fiber with rather large dimension, we can confirm visually the alignment of the coated fiber as shown in Fig.7. However, it is difficult for small fiber to identify when the alignment is performed. According to the analysis [6], the susceptibility of the coated short fiber composite is sensitive to fiber orientation distribution. Hence, as a measure of fiber alignment, we decided to use the ratio $r = X_0/X_{90}$, where $X_0$ and $X_{90}$ are the susceptibilities in the

direction of the magnetic field and that perpendicular to it, respectively.

From the solidified composite, the cylindrical test pieces, diameter 5mm and length 6 mm, were cut out to measure the susceptibilities. The manetic field intensity H - magnetization I curves were measured using vibrating sample magnetmeter. The measurements were carried out in the two directions. A typical result of H-I curevs is shown in Fig.8. In the figure, the initial magnetization curve is not linear. Thus we use the maximum susceptibility , the slope of line 0-B , as the susceptibility for the evaluation of the fiber orientation.

## 3.3 Results

The ratio of the maximum susceptibilities for the fibers with three kinds of the aspect ratio is shown in Fig.9 as a function of volume fraction of fiber $V_f$. In the figure, the plots are the mean values of 5 test pieces and chain and dotted lines are the predicted values for the aligned coated short fiber composites. As $r$ should be 1 for composite with random fiber orientation, the critical volume fraction $V_{fc}$ can be identified to be the point that $r$ abruptly approaches to unity. $V_{fc}$ drastically increases with decrease of the aspect ratio As. Thus in order to increase $V_{fc}$, $A_s$ should be lowered so as not to sacrifice greatly the strength of the composite. Fig.10 shows effect of the ultrasonic vibration. For this case, $V_{fc}$ can not be determined ,as shown by the open traiangles. However, we can conclude at least that the addition of the ultrasonic vibration gives remarkable effect to increase $V_{fc}$.

## 4 Conclusion

The possibility of fiber orientation control by loading of magnetic moment is explored by the use of the reinfocing fiber coated with ferromagnetic substance. The following conclusions was obtained,

1. The control of the fiber orientation in composites is possible by the loading of the magnetic moment.

2. The effect of parameters on the time until fiber aligned was made clear analitically for the case of a composite with dilute fiber concentration.

3.The maximum value of volume fraction at which the fiber orientation control is possible, decrease rapidly with increase of the aspect ratio of the fiber.

4. Loading of ultrasonic vibration together with the magnetic moment is effective to increase the maximum value of the volume fraction we can control the orientation of the fiber.

## Acknowledgement

This work was performed under management of Research and Development Institute of Metals and Composites for Future Industries sponsored by Agency of Industrial Science and Technology, MITI.

## References

(1) Caston, J. J., Reitz, J. A., Polym. Plast. Technol. Eng.,15(2), 1980, 97-114.

(2) Ishizuka, Y., Koyama, H., Kanayama, T., Kyoka Plastics (in Japanese), 29(2), 1983,16-20.

(3) Toy, A., Proc. of the Fall Meeting of the Metallurgical Society of AIME, Cleveland, Ohio, 15-19, ·1967-October.

(4) Ashton, J. E., Halpin, J. C., Petit, P.H.,"Primer on Composite Materials: Analysis", Technomic, Westport, CT, 1969.

(5) Hatta, H., Taya, M., Int. J. Eng. Sci. 24(7), 1986, 1159-1172.

(6) Hatta, H., Taya, M.,J. Appl. Phys., 59(6), 1986, 1851-1860.

(7) Demetriades, S. T., J. Chem. Phys., 29, 1958, 1054-1063.

(8) Goldsmith, H., Mason, S. G., in "Rheology:Theory and Applications", ed. by Eirich, Academic Press, 1967 85-230.

(9) Jeffery, G. B., Proc. Roy. Soc. London Λ102, 1922, 161-179.

## Biographies

Shu Yamashita is a mechanical engineer in charge of evaluation of mechanical properties of composite materials and design of composie structures used in the space at Mitsubishi Electric Corporation (MELCO). He recieved B.S. in 1982 from department of Aeronautical Engineering, Faculty of Engineering, Unversity of Kyushu.

Hiroshi Hatta is a assistant maneger in charge of material design of fiber reinforced plastics in MELCO. He recieved B.S. and D. of Eng. from Keio Univ. in 1973 and 1980, respectively.

Toshiyuki Sugano is a leader of composite material group in Materials and Electronic Devices Laboratory at MELCO. He mainly engaged in development of polymer matrix from 1980. He recieved B.S. from department of Engineering Science, Faculty of Science and Engineering, Ritsumeikan University. Kunihiko Murayama is a project maneger for R & D of advanced composite materials at MELCO. He recieved B.S. from department of Applied Chemistry, Faculty of Engineering, University of Osaka Prefecture. He is a professional member of the Japan chapiter of SAMPE.

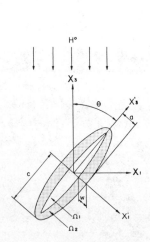

Fig.1 Analitical model

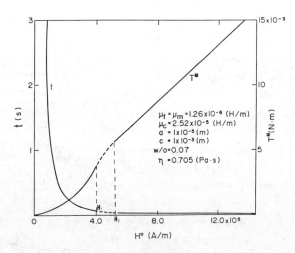

Fig.2 Torque due to magnetic moment $T^M$ and rotation time t as a function of applied magnetic field $H^O$.

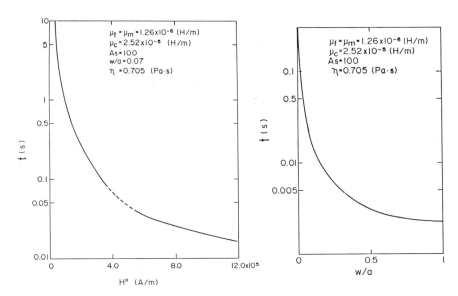

Fig.3 Relation between t and H⁰.

Fig.4 Effect of coating thickness w on t.

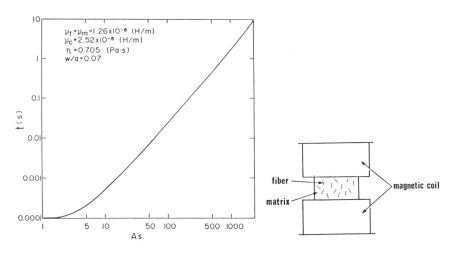

Fig.5 Relation between t and aspect ratio $A_s$.

Fig.6 Explanatory diagram for a method of the fiber orientation control.

(a) before the loading of the magnetic moment

(b) after the loading of the magnetic moment

Fig.7 Effect of magnetic moment on the alignment of the coated fibers. $H^0$:4.0x10$^5$A/m, $V_f$:0.1%, Fiber length:3mm

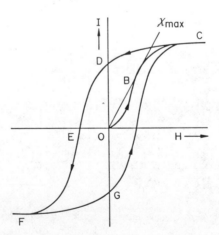

Fig.8 Magnetization curve.

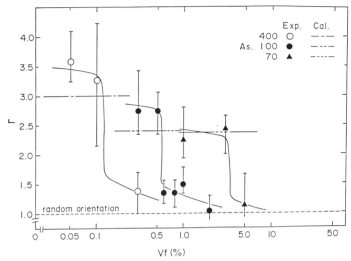

Fig.9 Relation between the ratio of the maximum susceptibility $r = \chi_0 / \chi_{90}$ and fiber volume fraction $V_f$.

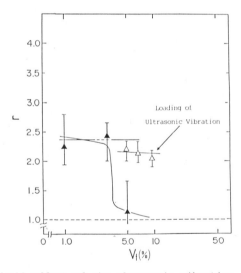

Fig.10 Effect of the ultrasonic vibration.

Table 1 The geometry of the coated short fibers.

| Fiber | | Fiber length(μm) | Fiber diameter(μm) | Aspect ratio | Coating thickness(μm) |
|---|---|---|---|---|---|
| Ni-coated CF | TypeI | 3000 | 7.5 | 400 | 0.25 |
| | TypeII | 1000 | 7.2 | 100 | 0.10 |
| | | 500 | | 70 | |

A NEW 3D BRAID FOR INTEGRATED PARTS MANUFACTURE AND IMPROVED
DELAMINATION RESISTANCE - THE 2-STEP PROCESS

Peter Popper
Ronald McConnell
Engineering Technology Laboratory
E. I. Du Pont de Nemours & Co., Delaware

## Abstract

A new method was developed for forming 3D braids that can form a wide range of structural shapes. The microstructure of this braid consists of layers of parallel yarns interconnected with braiding yarns that pass completely through the fabric in straight, diagonally oriented paths. The braiding pattern which forms this fabric structure is surprisingly simple. It requires an unusually small number of braid carriers and can therefore be automated with relative ease. We are now developing fully automated machines. A number of composite structures were made with Kevlar® and graphite fibers and both thermosetting and thermoplastic matrices. Several complex shapes were formed: I-beams, T-beams, beams with structural voids, and bifurcations. Test results show that the braiding yarns reinforce the structure and minimize delamination crack growth.

## 1. INTRODUCTION

Three-dimensional fabric structures that include though-thickness fibers offer two key advantages for reinforcing composites: they provide a means of forming complex geometric shapes, and they resist delamination. (1,2) A number of such fabrics have been developed, but most of them cannot be made by high-speed automated processes.

Almost every conventional textile process has been modified or enhanced to produce 3D fabrics. A brief description of each follows.

Weaving has been used in several ways. Conventional looms weave multilayer fabrics in which some yarns connect the layers. In addition, several new types of looms have been developed to weave complex shapes. (3). Knitting has also been used. Warp knitting machines stitch-bond multilayers of yarns oriented at different angles. And entirely new machinery has been developed to knit thick-walled contoured cylindrical shapes.(4). Conventional sewing machines have been used to interstitch multilayer fabrics. (5) A number of 3D shapes have also been made by hand-layup processes.

Braiding offers several possible process alternatives. Conventional circular machines braid multilayer, contoured structures, which are usually formed over a shaped mandrel. Primarily hand-operated, 3D equipment has been used to form highly complex shapes. (6,7)

Of the various fabric formation routes, we believe that braiding offers a number of advantages: Complex shapes can be formed because the pattern of relative yarn motion can be varied extensively. Furthermore, this process feeds yarns "on demand" and thereby provides the distribution of yarn length needed for shaped parts. Another key feature of braiding is that the process can incorporate non-yarn elements that remain in the fabric or create structural voids.

Since great difficulty has been encountered in automating 3D braiding, we have concentrated our work on braiding patterns that simplify the required machine actions. In this paper we describe a new type of 3D braid, the 2-Step Braid, which can form a range of shapes and can be automated.

## 2. GENERAL DESCRIPTION OF 2-STEP BRAIDING

The 2-Step Braid consists of two principal sets of yarn - axials and braiders. The axials are placed in the longitudinal or formation direction of the fabric and arranged in the pattern of the desired cross-section. They can be angled relative to each other if desired. The braiders move between the axials in a special geometric pattern that cinches the axials and stabilizes the shape of the braid.

Fig. 1 shows an unimpregnated T-section shaped fabric made by this technique. The structure forms a stable fabric that retains the desired shape. It can be used in this form as a textile product or else combined with a matrix material to form a composite. The matrix material can be impregnated into the braided structure or combined with the yarns before braiding. The composite can then be consolidated by applying the appropriate temperature and pressure.

We can describe the pattern by which the braiders move by referring to Fig. 2, which shows the relative position of all yarns at one location along the length of the braid. The axials are indicated by closed circles and the braiders by open circles. The braiders are distributed around the perimeter of the axial array and occupy alternate positions as shown. The basic process consists of two steps. In each one, the braiders are moved by the 2-Step pattern shown in Fig. 2. The arrows indicate the direction and extent of motion required.

As the two machine steps are repeated, the braiders completely intercinch the structure after several cycles. The braiding yarns follow a helix-like pattern that often reverses direction.

A surprising feature of this relatively simple sequence of braiding moves is its ability to produce a wide range of shapes. In fact, with relatively few technical restrictions, this braiding pattern can form any cross-sectional shape. This feature distinguishes the process from other 3D processes, which generally require a specialized pattern for each shape.

Several factors characterize this type of braid. First, the braiding yarns pass straight through the structure in diagonal paths and, after each step, are completely outside the axial array. This arrangement offers a number of advantages: For example, during processing, the braid can be pulled tight by yarn tension alone, which cannot readily be done in other 3D braiding processes since they require that the yarns be "beat-up" as in weaving. Having the braiders simultaneously outside the array also creates the possibility of adding various inserts to the structure and rearranging the axial array geometry.

Another important characteristic of the braid pattern shown in Fig. 2 relates to the direction of braider motion. The braiders alternate their direction in every row. In the case of complex shapes, several braiders may be positioned on the same motion line in different regions of the part. Even though each of these braiders moves a different amount, the alternating motion pattern from line to line remains unchanged.

## 3. STRUCTURAL SHAPES

The 2-Step braiding process facilitates the direct formation of many regular and irregular cross-section beams and tubes; and it can be extended to form other complex shapes. Figs. 3,4 show the braiding pattern for forming a number of very different examples. In each case, the axials fill the required shape and the braiders are positioned at alternating locations around the perimeter. The braiders are moved simultaneously in the two steps described previously. Note that the tubular structure shown is made by the same procedure as the others. However, for that shape, the axials are arranged in a circular rather than an orthogonal array.

Any required shape can be formed by several alternative routes. First, the scale of the yarn structure can be adjusted by changing the number of axial positions in each direction of the axial array. The geometry of this array automatically determines the required number and location of braiders.

Another structural variation can be achieved for a particular shape by varying the shape and number of independent braiding yarn paths. For example, a T-beam can be braided with all braiders following the identical path around the axials. Changing the total number of axials in each direction by one or two units makes it possible to braid the same shape with multiple paths. In that case, the braiders will follow one of a number of paths through the structure. This

structural variation is particularly useful if a hybrid system is desired and different braiding yarns need to be restricted to one part of the structure. The simplest method of changing the number of paths is to change the count of axials in each direction from odd to even or vice versa.

The microstructure of a particular shape can be changed by varying a number of parameters, including the relative weight of axials and braiders, the braiding angle, and the aspect ratio used in the final consolidation.

Structural shapes braided with the 2-Step pattern demonstrate some of the geometric possibilities. A brief description of each follows.

BEAMS OF ARBITRARY CROSS-SECTION — A T-beam made of Kevlar® axials and braiders and impregnated with an epoxy matrix is shown in Fig. 5. I-beams have been made with a similar technique.

BENT BEAMS — The bent T-beam in Fig. 6 shows the results of selectively bending the output fabric at the point of braid formation. Both axials and braiders automatically feed at a greater rate when they fall at the outside of the bend. The braid structure locks in this bent geometry, and buckling is avoided during consolidation.

BEAMS WITH STRUCTURAL VOIDS — Several braiding techniques create large internal voids, which lighten the structure and improve properties per weight. For example, several of the axial yarns can be replaced by flexible, nonfibrous elements such as rubber rods. These rods are braided into the fabric and later removed to create voids. One example of this type, a graphite/Kevlar®/epoxy I-beam is shown in Fig. 7. Another example, a cellular beam, is shown in Fig. 8. In this case, the braid was made of graphite and Kevlar® tows coated with a thermoplastic resin. This beam consists of approximately 70% voids. It is internally supported by braiding yarns that form a truss-like support.

STRUCTURES WITH LENGTHWISE VARIATIONS - Any of the above-mentioned structural parameters can be varied along the braid length. In the simplest case, the braiding angle is varied. In more complex circumstances, the braiding pattern itself is changed. One example is the bifurcation shown in Fig. 9. In this part, a rectangular pattern is braided for a fixed length and then the pattern is changed to braid two rectangles.

## 4. MICROSTRUCTURE

The spatial orientation of yarns in a 2-Step braid can by determined from the braid pattern, the rate of takeoff, and the consolidation configuration. Fig. 10 shows a computer generated image of the braiders/axials microstructure. As each braider follows its prescribed pattern, it traces out a three-dimensional path. The end-view projection of this path can be obtained by following a particular yarn through a number of braid cycles. This end-view pattern is attenuated in the formation direction by the takeoff motion, and it may be modified as the product is consolidated.

Fig. 11 shows the path followed by the braiders, which are moved by the pattern of Fig. 2 to form a T-beam. For this example, it can be shown that the braiders all follow one of two paths, depending on their starting positions. In some patterns, all braiders follow the same path and in others many different ones. When the braiders follow the same path, they are offset from each other as shown in Fig. 10.

In actual structures the idealized braiding yarn paths are modified near the surface and tend to flatten as they cross the axials on the sides and corners of the structure. Fig. 12 shows a the cross section of a rectangular braid made of graphite axials and Kevlar®

braiders. The diagonal path taken by the braiders in that section can readily be distinguished from the axial array.

The spatial angles of the braiders can be described by the angle of their projections in two planes. Typical orientations for braiders in a flat slab are 45° in the plane perpendicular to braid axis and 60° in the plane of braid.

## 5. PROPERTIES

Improvements in delamination resistance of through-thickness reinforced composites have been reported by a number of investigators. (1,2,6) We have found that small amounts of through-thickness reinforcement improve delamination resistance but have a small effect on in-plane properties.

We conducted experiments in Mode I and II delamination of graphite/Kevlar® 49/epoxy 2-Step braids. In Mode I tests, we were unable to propagate cracks along a length of a braided rectangular slab. Control samples without braiding yarns delaminated readily.

In another experiment, we created a Mode II failure by taking test slabs deliberately damaged in the central plane by small drill holes. These samples were then subjected to repeated loading with a short beam shear test configuration. The ends were painted white to show any developing cracks. The total length of cracks were reported vs. the imposed number of cycles. The results for slabs with and without braiding yarns are shown in Fig. 13.

The results show that the samples without braiding yarns failed completely within 500 cycles in a single crack. The braided samples failed at a slower rate. The cracks grew in a number of locations at angles to the plane of the braid. Rather than form a single crack, the braids were able to spread the damage across the section.

These results indicate that the through-thickness reinforcement provided by the 2-Step braid increases delamination resistance.

## 6. AUTOMATION OF 2-STEP BRAIDING

Although automatic braiding machines have been been in use for many years, none is available for 3D braiding of shapes, probably because of the complexity of the required equipment. The design requirements can be simplified greatly by limiting the number of braiding yarns, since each of them must be moved independently during operation. The 2-Step process has been selected for study because it requires a smaller number of braiders than other 3D processes.

In 2-Step braids most of the structural weight comes from the axial yarns. These yarns simply flow through the process, and their supply packages do not have to be moved. A limited number of braiders following special paths can move throughout the structure and interlock the axials.

The number of braiders required for a particular axial array is shown in Fig. 14. This relation applies to a simple rectangular section, but similar results apply to other shapes. The results show that a relatively small number of braiders are needed for each group of axial yarns, and this number drops as the number of layers or the aspect ratio increases. For example, an 11-layer structure with an aspect ratio of 5:1 requires only 0.1 braider per axial. (Each braider connects 10 axial groups). In other types of braid, circular or 3D, the minimum number of braiders per axial is typically between 1 and 2.

Another simplifying factor in this process is the relatively large braider motions occurring at each step; only two steps are needed for a full cycle. In each step the braiders are moved completely through the axial array. In other 3D processes, each braider is moved across one yarn position at each step.

## 7. SUMMARY

The 2-Step braid offers a number of features for making composite parts - particularly beams of varying sections and beams with large structural voids. Its microstructure consists of a "straight-through," diagonal reinforcement by braiding yarns that can be oriented at relatively high angles to the principal braid directions. This process can be fully automated with less complexity than other 3D braiding operations. We have proto-typic equipment available and we are developing automated machines.

## 8. ACKNOWLEDGMENT

We would like to acknowledge the valuable contributions of Irvin Carroll, Ulrich Rosenbaum, and William C. Walker.

## 9. REFERENCES

1. S-S Yar, T. N. Chow, F. K. Ko, "Flexural and Axial Compressive Failures of Three-Dimensionally Braided Composite I-Beams", Composites Vol. 17, No. 3, July 1986, p. 227-232.

2. Macander, A. B., Crane, R.M., Camponeschi, E.T., "Fabrication and Mechanical Properties of Multi-dimensionally (XD) Braided Composite Materials", Composite Materials: Testing and Design (7th Conference) ASTM STP893, American Society for Testing & Materials, Philadelphia, 1986, p. 422-443.

3. Fukuta, et al, "Three-Dimensionally Latticed Flexible-Structure Composite," U.S. Patent 4,336,296, June 1982.

4. Cahuzac, G.J.J., "Hollow Reinforcements of Revolution Made By Three-Dimensional Weaving Method and Machine for Fabricating Such Reinforcements", U.S. Patent 4,492,096, January 1985.

5.  N. C. Olsen, "Advanced Pressure
Molding (Autocomp) and Fiber Form
Manufacturing Technology for Com-
posite Aircraft/Aerospace Com-
ponents", Proceedings Composites
Group of Society of Manufacturing
Engineers, Los Angeles, Jan. 1986,
EM 86-104.

6.  Brown, R.T., "Through-the-
Thickness Braiding Technology",
30th National SAMPE Symposium,
March 1985.

7.  Florentine, R.A., "Apparatus
for Weaving a Three-Dimensional
Article", U.S. Patent 4,312,261,
January 1982.

## 10.  BIOGRAPHY

Peter Popper is a Senior Research
Associate in Du Pont's Engineering
Technology Laboratory located at
the Experimental Station,
Wilmington Delaware.  He has over
20 years experience in research on
fibrous processes and structures.
He holds a BS in Textile
Engineering from the University of
Lowell and the SM, MechE and ScD in
Mechanical Engineering from MIT.

Ronald McConnell is an Engineering
Associate in Du Pont's Engineering
Development Laboratory located in
Wilmington Delaware.  He has over
30 years experience in process and
equipment development.  He holds a
BSME from Columbia University.

FIG. 1. BRAIDED 2-STEP T-BEAM WITHOUT MATRIX

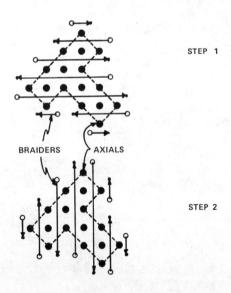

STEP 1

BRAIDERS    AXIALS

STEP 2

FIG. 2. BASIC 2-STEP BRAIDING PATTERN

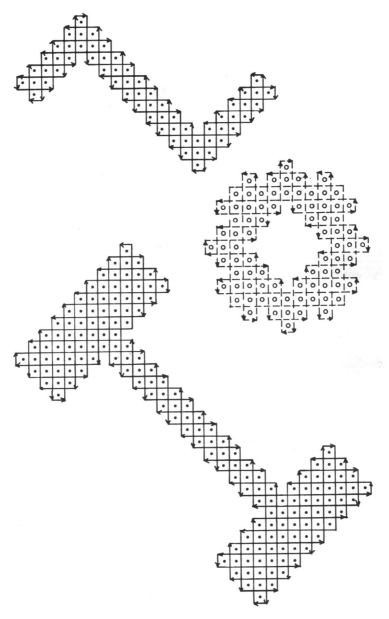

FIG. 3. BRAID PATTERNS FOR VARIOUS SHAPES

FIG. 4. TUBULAR 2—STEP BRAID PATTERN

FIG. 5. CONSOLIDATED T—BEAM
KEVLAR® AXIALS AND BRAIDERS

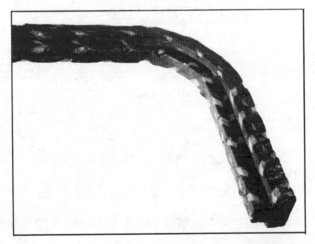

FIG. 6. T—BEAM BRAIDED IN BENT CONFIGURATION

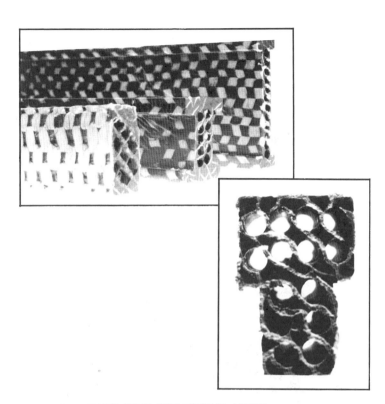

FIG. 7. BEAMS WITH STRUCTURAL VOIDS

FIG. 8. CELLULAR BOX BEAM

FIG. 9. BRAIDED BIFURCATION

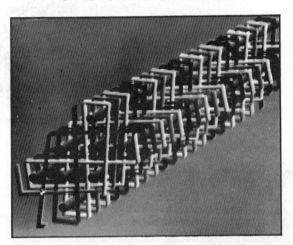

FIG. 10. COMPUTER GENERATED MODEL OF 2—STEP BRAID

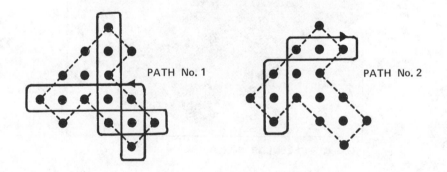

PATH No. 1      PATH No. 2

FIG. 11. PATH OF BRAIDING YARNS

FIG. 12. CROSS SECTION OF 2—STEP BRAID

FIG. 13. MODE II FATIGUE OF PREDAMAGED SLABS

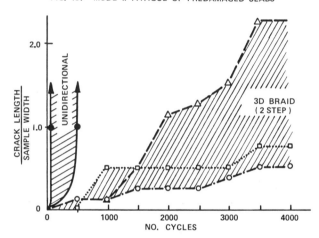

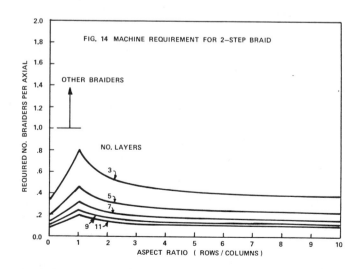

# PROCESS VARIABLES EVALUATION OF PEEK APC-2 THERMOPLASTIC MATRIX COMPOSITE

JOHN M. IACONIS
LTV AIRCRAFT PRODUCTS GROUP

## ABSTRACT

Temperature, pressure, and time were varied in processing of AS4/PEEK graphite/thermo- plastic unidirectional tape composite laminates. The effect of processing cycle on both physical and mechanical properties of laminates was evaluated in two separate stages of testing. Physical properties of laminates included surface finish, NDI results, per ply thickness and other criteria. Mechanical properties consisted of flexural and short beam shear test results.

A qualitative ranking of overall laminate physical properties allowed for some recommendation for forming limits. Comparison of flexural and short beam shear test results enabled the pinpointing of the highest strength forming method.

A laminate physical characteristics evaluation method is presented, forming limit conclusions for AS4/PEEK tape are discussed, flexural and short beam shear test methods for thermoplastic are examined, and results of mechanical testing are reported. Reforming of unacceptable laminates is demonstrated, and directions for further study are explored.

## 1. INTRODUCTION

Graphite fiber reinforced thermoplastic matrix composites offer advantages in handling, processing, design and performance for aerospace applications. LTV Aerospace began working with AS4/PEEK unidirectional composite tape in early 1985. We have subsequently evaluated design allowables for the material, have experimented with different fabrication techniques, and are investigating application to full-scale demonstration items. This report addresses fabrication process evaluation studies intended to develop optimum appearance and strength qualities for the composite.

Fabrication process studies are divided into two parts, physical characteristics evaluation and mechanical properties comparison. The forming method was the same for both physical and mechanical properties testing: 1) A heated iron was used to tack prepreg plies together; and 2) Laminates were compression molded using matched mold steel die tooling. Compression molding began with placing prepreg tape and tooling into a press heated to melt temperature. A five-minute soak period allowed tool to heat and prepreg to melt fully, then specified pressure was applied to the prepreg for the specified forming time. The hot tool was wrapped in a fiberglass blanket and quickly moved to a cold press for quenching, although transfer between presses was found to cause poor surface finish and fiber buckling. The cold press was cooled by ambient water (75°F) channeled through the platens. Forming temperatures were monitored by reading a thermocouple placed between the tool and part. All laminates were quenched from 700°F to 390°F within two minutes as specified by ICI Industries. Tooling and

laminates are shown in Figures 1, 2 and 3.

2. PHYSICAL CHARACTERISTICS TESTING

Fourteen processing cycles were used, as listed in Table 1, for physical characteristics testing. Forming times, temperatures, and pressures were varied to note effects on laminates. Generally one parameter was varied per laminate. Processes in group A in Table 1 varied temperature; processes in group B varied pressure, and processes in group C varied time in fabrication. The process in group D varied all parameters.

Laminates were visually inspected, measured for thickness at nine points on the edges and the center, and inspected by ultra sound NDI (non-destructive inspection) for void areas. Evaluation of laminates considered matrix flow, interply integrity, thickness variation, surface finish, per ply thickness, and NDI results. Laminate rankings are listed in Table 2.

Laminates 1, 5, and 10 were reformed in an attempt to improve their scores in qualitative rankings. PEEK film was bonded to the surfaces of laminate 1, and 'Kapton' film was sandwiched between the tool and laminate in an attempt to improve laminates 5 and 10. Scores of reformed laminates were all improved, as listed in Table 3. In each instance reforming eliminated voids, improved interply integrity, and improved surface finish. Laminate 1 was made thicker and the surface finish was smoother when the PEEK film was bonded on. Laminates 5 and 10 had shinier and smoother surface finishes, and surface resin-poor areas and through thickness voids were eliminated after having been reformed with Kapton.

### 3. MECHANICAL PROPERTIES EVALUATION

The effect of temperature, pressure, and time processing variations on flexural and short beam shear strengths of zero degree layups was evaluated. Eight processes were used as described in Table 5. All formed laminates passed NDI inspection. Flexural specimens were fabricated and tested according to Methods I and II of ASTM D790 using a support span of 32 to 34 times sample thickness. Short beam shear specimens were fabricated and tested according to ASTM D2344 using a span of 4 times thickness. Crosshead travel rate was set at 0.05 in/min., and all testing was done at room temperature.

Tensile failure was expected in flexural test specimens. The incidence of compressive failure in initial specimens prompted exploration of alternative flexural test methods. Loading nose diameter was increased from 0.25 to 0.75 inches to restrict load concentration at the pin in Method I three-point loading, Method II four-point loading was also tried as a way to reduce load concentration. Despite these efforts, flexural specimens continued to break at the compression surface, and it was concluded that compression was probably the flexural failure mode. Consequently, testing using 0.25" diameter loading nose in 3-point bend configuration was resumed, and all specimens failed in compression in the center of the gage length. A typical plot of load and deflection for the flexural tests is illustrated in Figure 4. Numerical averages for flexural testing data are given in Table 6. Typical failures are shown in Figures 7 and 8. Figure 8 shows that compressive failure area in the cross section was larger than tensile failure area; initial compressive failure on the top surface moved the neutral axis below the center of the flexural specimen.

Short beam shear specimens failed by plastic deformation. Due to the non-catastrophic failure of short beam shear specimens, an offset method was used that defined failure strength at a deflection of 0.002 inches. The 0.002 offset measurement was chosen as the smallest easily-measured and accurate deformation. Bottom-side deflection was measured by a deflectometer. Span length was about 0.32 inches; deflection at failure, consequently, was about 0.6% of span length. A typical plot for load and deflection of short beam shear tests is illustrated in Figure 6.

## 4.0 CONCLUSION

Process evaluation studies indicated that AS4/PEEK laminate physical and mechanical properties could be optimized through proper processing. Some conclusions were drawn as to the optimum processing range. Physical characteristic and mechanical properties evaluation indicated that laminates should be formed at temperatures above 675 degrees F, at forming times at melt of five minutes or longer, and at 50 to 600 psi pressure. Additionally, strengths were significantly increased when an anneal or slow cool step was included in processing. Laminate per ply thickness between 5 and 5.5 mils, and a non-transfer forming process, are recommended.

| | | TEMPERATURE | PRESSURE | TIME | QUENCH |
|---|---|---|---|---|---|
| Mfr.<br>Recommended | | 700-730°F | 200psi heat up<br>300psi cool down | 5 min @<br>pressure | Cool to 390°F<br>within 2 min. |
| A. Vary | #1 | 650°F | 200/300 | 5 | same |
| Temperature | #2 | 675°F | 200/300 | 5 | same |
| | #4 | *740°F | 200/300 | 5 | same |
| B. Vary | #5 | 700°F | **50/50 | 5 | same |
| Pressure | #6 | 700°F | **50/50 | 5 | same |
| | #7 | 700°F | 100/100 | 5 | same |
| | #8 | 700°F | 600/600 | 5 | same |
| | #9 | 700°F | 300/300 | 5 | same |
| C. Vary | #10 | 700°F | 100/100 | 1 | same |
| Time | #11 | 700°F | 100/100 | 2 | same |
| | #12 | 700°F | 100/100 | 3 | same |
| | #13 | 700°F | 100/100 | 4 | same |
| | #14 | 700°F | 100/100 | 10 | same |
| D. Vary<br>All | # 3 | 730°F | 300/300 | 2 | same |

\*   Maximum temperature of press

\*\* 50 psi pressure was minimum for press

TABLE 2
RANKING OF LAMINATES
FOR PHYSICAL TESTING

| LAMINATE NUMBER | AVERAGE THICKNESS (IN) | MATRIX FLOW | INTERPLY INTEGRITY | THICKNESS VARIATION (IN) | SURFACE FINISH | PER PLY THICKNESS (IN) | NDI RESULTS | SUM |
|---|---|---|---|---|---|---|---|---|
| Optimum Laminate | 0.040 | 5 | 5 | 0.000 | 5 | 0.0050 | 5 | 20 |
| 9 | 0.040 | 5 | 5 | 0.004 | 4 | 0.0050 | 5 | 19 |
| 8 | 0.040 | 4 | 4.5 | 0.005 | 5 | 0.0050 | 5 | 18.5 |
| 4 | 0.041 | 5 | 4.5 | 0.004 | 4 | 0.0051 | 4.5 | 18 |
| 7 | 0.040 | 4.5 | 4 | 0.003 | 4 | 0.0050 | 4.5 | 17 |
| 12 | 0.048 | 5 | 3 | 0.008 | 5 | 0.0060 | 4 | 17 |
| 6 | 0.042 | 5 | 4 | 0.006 | 3 | 0.0052 | 4.75 | 16.75 |
| 3 | 0.040 | 3 | 5 | 0.003 | 4 | 0.0050 | 4.5 | 16.5 |
| 2 | 0.042 | 3 | 4 | 0.005 | 5 | 0.0052 | 4 | 16 |
| 11 | 0.042 | 4.5 | 3 | 0.005 | 3 | 0.0052 | 4 | 14.5 |
| *5 | 0.042 | 5 | 4 | 0.004 | 1 | 0.0052 | 4 | 14 |
| *13 | 0.041 | 4.5 | 4 | 0.006 | 1 | 0.0051 | 4.5 | 14 |
| *14 | 0.042 | 5 | 3 | 0.005 | 1 | 0.0052 | 4 | 13 |
| *10 | 0.042 | 4 | 2 | 0.006 | 3 | 0.0052 | 3.5 | 12.5 |
| *1 | 0.045 | 5 | 1 | 0.003 | 1 | 0.0056 | 3 | 10 |

*Unacceptable laminate.

## TABLE 3
## RANKING OF REFORMED LAMINATES

| LAMINATE NUMBER | AVERAGE THICKNESS (IN) | MATRIX FLOW | INTERPLY INTEGRITY | THICKNESS VARIATION (IN) | SURFACE FINISH | PER PLY THICKNESS (IN) | NDI RESULTS |
|---|---|---|---|---|---|---|---|
| 1 | * 0.061 | 3 | 5 | 0.008 | 5 | ***0.0052 | 5 |
|  | **0.058 |  |  |  |  |  |  |
| 5 | * 0.040 | 3 | 5 | 0.003 | 5 | 0.0049 | 5 |
|  | **0.039 |  |  |  |  |  |  |
| 10 | * 0.040 | 3 | 5 | 0.004 | 5 | 0.0048 | 5 |
|  | **0.038 |  |  |  |  |  |  |

\* EXPECTED

\*\* ACTUAL

\*\*\*When 0.016" of Stabar K PEEK film is subtracted and result is divided by 8 plies.

TABLE 4
MECHANICAL TEST RESULTS

| PANEL SHEAR ID | TEMP/PRES/PRES/TIME °F PSI PSI MIN | FLEX STRENGTH KSI | FLEX MODULUS MPSI | SBS STRENGTH KSI |
|---|---|---|---|---|
| 1 (b) | (+ anneal @ 335°C for 45 min) | 290 | 17.8 | 14.3 |
| 6 | 700/200/20 slow cool in press | 298 | 17.8 | 14.2 |
| 1 (a) | 700/600/600/5 | 264 | 17.6 | 13.2 |
| 5 | 700/vacuum/20 slow cool in oven | 258 | 17.4 | 13.9 |
| 3 (b) | (reform) 700/200/300/10 | 262 | 17.3 | 13.5 |
| 3 (a) | 675/150/150/5 | 254 | 17.0 | 14.1 |
| | Manufacturer's Recommended 700/200/300/5 | 238 | 17.8 | 13.6 |
| 4 | 740/50/50/2 | 226 | 17.0 | 12.1 |
| 2 | 700/100/100/1 | 224 | 17.2 | 11.8 |

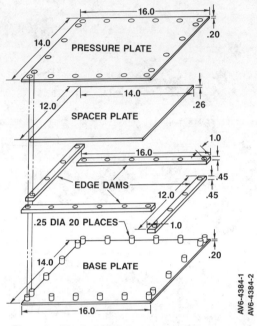

Figure 1. Matched Mold Forming Tool

Figure 2. The Compression Molding Package

112

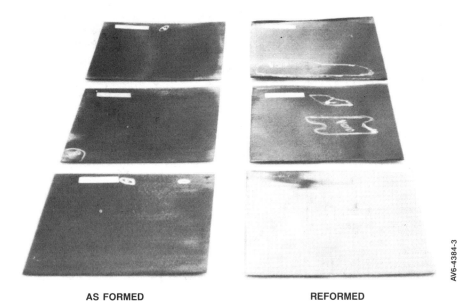

AS FORMED                                    REFORMED

Figure 3.  Laminates from Physical Characteristics Testing

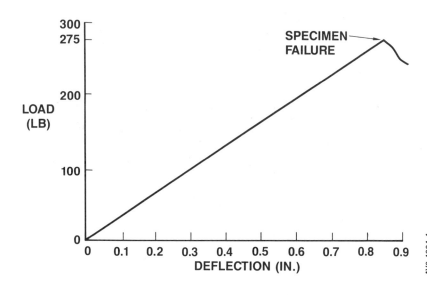

Figure 4  Typical Flexural Plot

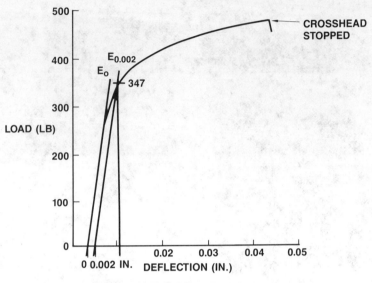

Figure 5   Typical Short Bean Shear Plot

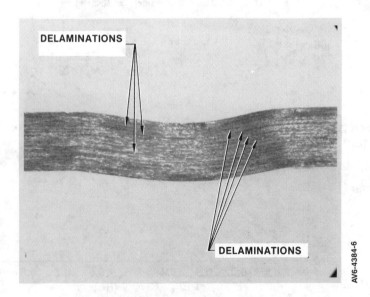

Figure 6   Typical Short Beam Shear Failure

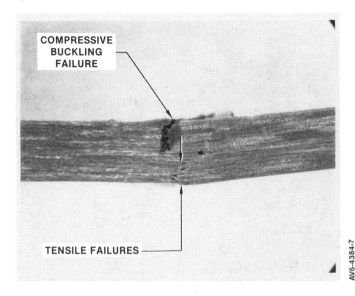

COMPRESSIVE BUCKLING FAILURE

TENSILE FAILURES

AV6-4384-7

Figure 7 Typical Flexural Failure

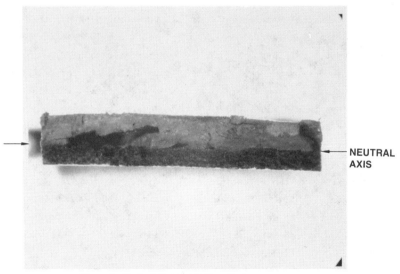

NEUTRAL AXIS

Figure 8 Typical Flexural Failure Surface. Initial compression failure, on upper half, depressed the neutral axis.

FIBER VOLUME OF RESIN MATRIX COMPOSITES
BY DENSITY MEASUREMENT
Sindee Simon and Louis Strunk
Beech Aircraft Corporation
9709 East Central
Wichita, Kansas  67201

### Abstract

In fabrication of aircraft parts
using composites, it is necessary
to be able to assess the quality
of material simply and quickly,
both for design purposes and as a
means of ensuring satisfactory
manufacturing techniques.  One
such parameter is fiber volume.
The use of density for the
determination of fiber volume in
composite parts was developed in
order to reduce the time and cost
of quality assurance while
maintaining reliability.  Density
measurements may also be used in
conjunction with NDI techniques
for void content to determine the
fiber volume of specific parts
through the use of theoretical
fiber volume versus density
graphs.

### 1.  INTRODUCTION

The properties of fiber-
reinforced composite parts are
largely determined by fiber
volume and void content.  This
program was studied to replace
the present method at Beech
Aircraft in determining fiber
volume and void content.[1]  The
presently used method is very
time-consuming and takes a very
small sample size into account
for determining fiber volume and
void content, thus reducing its
confidence level.  Since resin
density and fiber density of material lots are known, the composite
density could be used for fiber
volume[2] and void content by NDI.

Fiber volume and void content are
directly related to the composite
density by the following
equation:[2] [3]

$$c = \frac{(\rho_f - \rho_r) \, F + (100 - V) \, \rho_r}{100}$$

where  $c$ = composite density

$f$ = fiber density

$\rho_r$ = cured resin density

F = fiber volume, %

V = void content, %

The graph of this equation yields
a set of parallel lines, each
line corresponding to a specific
void content. The slope of the
lines is $(\rho_f - \rho_r)/100$. The
smaller the difference between
fiber and resin densities, the
more horizontal the lines will
be. Reference Figures 1, 2 and
3. Graphs can be made for any
resin/fiber combination as well
as for mixtures of different
combinations. The reliability of
such a graph for specific batches
of prepreg or wet lay-up
materials is only dependent on
the accuracy and variability of
the resin and fiber densities
supplied by the material
manufacturer for the batches
involved.

Maximum and minimum acceptable
values for composite part density
may be determined from the fiber
volume versus density graph for
screening purposes. For example,
if the fiber volume requirement
is 49% - 62% and void content is
3% maximum, the minimum
acceptable composite density will
be at 49% fiber volume and 3%
void content. In Figure 1, the
minimum acceptable density would
be 1.472 g/cc. Likewise, the
maximum acceptable composite
density will fall at 62% fiber
volume and 0% void content or

1.583 g/cc in Figure 1. All
parts not falling in the range
1.472 to 1.583 should be rejected
and those parts falling in the
range should be further
evaluated.

The specific fiber volume of a
part may be found by using the
fiber volume versus density graph
in conjunction with NDI
techniques such as A-scan or C-
scan. The void content is
determined by NDI and the fiber
volume is read off of the graph
or calculated by solving the
initial equation for F:

$$F = \frac{100(\rho_c - \rho_r) + \rho_r V}{\rho_f - \rho_r}$$

The accuracy of the fiber volume
value obtained in this manner is
dependent on the reliability of
the measured composite density,
the fiber and resin densities
supplied by the prepregger, and
the void content obtained by NDI.
We have found A-scan void
contents to be conservative when
compared to acid digestion
values. This results in lower
and more conservative values for
fiber volume.

2. EXPERIMENTAL

2,083 graphite/epoxy fabric
laminates were evaluated. Fiber
volume results obtained by resin
digestion were compared to values
determined from the theoretical
fiber volume versus density

117

graph, Figure 1. A SAS [4] computer program was written to determine the density range in which parts met fiber volume and void content requirements. The confidence level reported for Level II parts is the percentage of parts in the density range evaluated which meet the requirement of 49-62% fiber volume and less than 3% void content. For Level III parts, it is the percentage having a fiber volume of 49-62% with less than 2% void content. Density ranges were determined by optimizing the confidence levels with and without NDI.

Nineteen (19) carbon/graphite epoxy fabric co-cure honeycomb panels were also evaluated. These panels were A-scanned and acid digested at two labs. The fiber volumes and void contents thus obtained for each panel were compared.

### 3. RESULTS

From the 2083 fabric laminates, the number of observations that fell outside the density range of 1.472 to 1.583 in Figure 1 was 10 observations. The number of parts that failed to meet fiber volume requirement by digestion was 237 of 2083 observations (see Figure 4). A density range of 1.511-1.565 g/cc was established by use of the SAS [4] computer program. In this range, 98.7% of the Level II and 96.8% of the Level III parts meet the specified fiber volume/void content requirements. When NDI is used in conjunction with the density test, a confidence level of 99.1% is obtained. 74% of the 2,083 parts fell into this range, thus, reducing resin digestion requirements by this amount. The void content obtained for the nineteen carbon/graphite epoxy fabric honeycomb panels are displayed in Figure 5. NDI in most cases indicates a higher void content then by resin digestion.

The density ranges 1.511 to 1.565 selected for the 2083 parts were compared to fiber volume and density of the nineteen honeycomb sandwich panels. 84% of these panels fell into this range (see Figure 6).

### 4. CONCLUSION

Density can be used in conjunction with NDI to determine whether parts meet fiber volume and void content requirements. Errors associated with sample size and edge effects can also be reduced by increasing the size of the density specimen. Thus, the density methods outlined offer substantial time and cost savings as well as reliability. Additional work needs to be done to increase the confidence level over the total density range from

118

the theoretical chart, Figure 1.

### 5. REFERENCES

1) ASTM D3171, "Fiber Content of Resin - Matrix Composites by Matrix Digestion"

2) ASTM D792, "Specific Gravity and Density of Plastics by Displacement"

3) ASTM D2734, "Void Content of Reinforced Plastics"

4) SAS - Statistical Analysis System Software by IBM

### 6. BIOGRAPHY

Sindee Simon: Materials and Process Engineer, Composite Development at Beech 1983 through 1986. Received her B.S. in Chemical Engineering at Yale. Sindee is currently attending Princeton working on her PhD.

Louis Strunk: Lead Engineer M&P Composite R&D. Has been employed at Beech Aircraft for 24 years. Worked as Process Analyst to supervisor of Metal Bond Lab prior to his assignment to Beech Starship Program. Attended Wichita State University 1958 through 1961 in School of Engineering. Member of ASNT, SAE and SAMPE.

FIGURE 1          21-JUL-86 9:53 Page 1

DEN1

E7K8 AS4 FABRIC: FIBER VOLUME VS. DENSITY

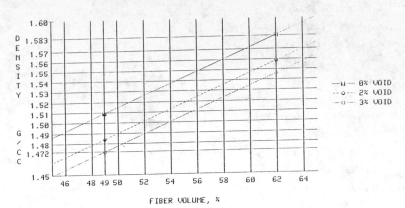

FIBER DENSITY= 1.80 G/CC
RESIN DENSITY= 1.23 G/CC

DEN3                    FIGURE 2          23-MAY-86 10:34 Page 1

E7K8 AS4 TAPE: FIBER VOLUME VS. DENSITY

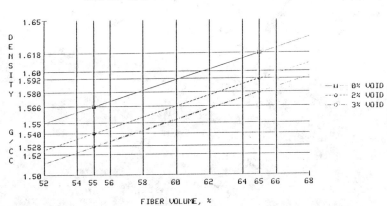

FIBER DENSITY= 1.80 G/CC
RESIN DENSITY= 1.28 G/CC

E7K8 KEVLAR FABRIC: FIBER VOLUME VS. DENSITY

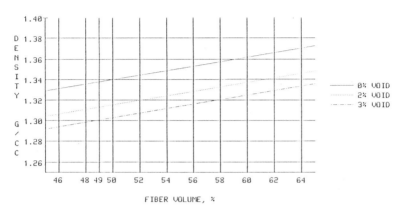

FIBER DENSITY = 1.45 G/CC
RESIN DENSITY = 1.23 G/CC


FIGURE 4

GRAPHITE/EPOXY ALL-FABRIC LAMINATES
2083 PRODUCTION PARTS, 1/1/86 - 6/19/86

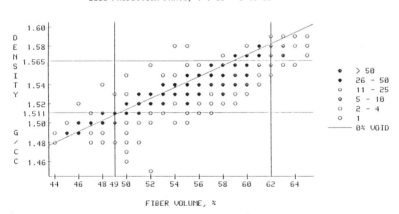

PROPOSED ACCEPTABLE DENSITY RANGE = 1.511 - 1.565

0% VOID LINE BASED ON 1.80 FIBER DENSITY AND 1.23 RESIN DENSITY

## FIGURE 5

COMPARISON OF VOID CONTENTS FOR 19 HONEYCOMB SANDWICH PANEL SPECIMEN
AUTOCLAVE CURE: 30 MIN HOLD AT 200F, 120-140 MIN AT 290-320F AND 45 PSI
MARCH 1986

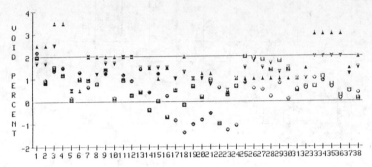

FACESHEET NUMBER

- ▣ PLANT 1 ACID DIGESTION
- ◔ PLANT 4 ACID DIGESTION
- ▲ PLANT 3 A-SCAN
- ▼ PLANT 4 A-SCAN

## FIGURE 6

COMPARISON OF DENSITY/ACID DIGESTION DATA
19 ALL-FABRIC HONEYCOMB SANDWICH PANEL SPECIMEN
CURE: 30 MIN HOLD AT 200F, 140-160 MIN AT 290-320F, 45 PSI

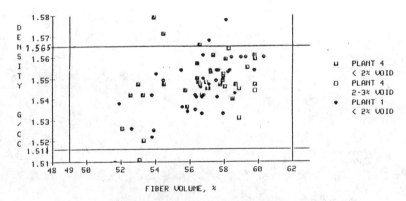

FIBER VOLUME, %

- ⊔ PLANT 4 < 2% VOID
- ◻ PLANT 4 2-3% VOID
- ● PLANT 1 < 2% VOID

PROPOSED ACCEPTABLE DENSITY RANGE = 1.516 - 1.565
84% OF THE SPECIMEN FALL IN THE PROPOSED RANGE
CONFIDENCE = 100% FOR BOTH PLANT 1 AND 4 DATA

FIBER FORM/PRESSURE MOLDING PROCESSING
FOR AIRCRAFT FUSELAGE SUBSTRUCTURE
Gary Van Schooneveld
XERKON Inc.
Minneapolis, Minnesota

## ABSTRACT

As the application of composite
materials continues to expand into
aircraft primary structures, new
manufacturing technologies are
required to ensure repeatability
and cost effectivity of the structures. Fuselage frames represent
a major challenge due to their
complex curvature and variable
cross-section. The fiber form/
Autocomp® pressure molding manufacturing approach has been demonstrated on a number of fuselage
frame sections. Savings up to 70%
have been projected for this integrated composite manufacturing
approach as compared to conventional hand layup/autoclave processing.

The fiber form process uses textile equipment modified for compatibility with high modulus
fibers such as carbon and aramid.
The reinforcing fiber is knitted
into unidirectional and multiple
ply fabrics without bending the
reinforcing fiber and assembled
via stitching to the final configuration of the molded part. The
dry fiber form is impregnated by
means of a resin infusion process.
Final curing of the frame is
achieved with the patented Auto-
comp$^R$ pressure molding process.
The process utilizes integrally
heated, matched surface tooling in
conjunction with vacuum and
pressure. Precise computer control of the process provides high
manufacturing throughput and part-
to-part consistency.

## 1. INTRODUCTION

Application of composite materials
in aircraft structures to date has
generally been limited to secondary structures such as control
and fairing surfaces. Even in
these types of structures, the
application of composite material
is often limited to the skins.

As the capability and reliability
of composite materials have been
demonstrated, the applications
have expanded into primary
structures such as wing and
fuselage. Applications have also
expanded to include not only skins
but substructures also. This
trend is seen in more recent aircraft, such as the AV8B and the F-
18. The expanding use of composite materials will continue as
performance requirements dictate
lighter airframes. It has been
projected that the V-22 Osprey
will be approximately 60% composite materials by structural
weight, while the Advanced Tactical Fighter will possess 40-60%
composite material by structural
weight.

In order to realize these levels
of composite applications in a
cost effective manner, manufacturing methods must be developed
which utilize lower cost material
forms and reduce the labor required to convert the raw materials to
finished structure. This paper
will review the application of
fiber form technology and Auto-
comp® pressure molding as it is
applied to the fabrication of
composite fuselage substructures.

## 2. TACTICAL OBJECTIVES

With the strategic objective established as the reduction of procured article costs to the end user without compromising structural performance, a number of tactical objectives must be established to realize this general strategic goal. Such tactical objectives include:

- The reduction or elimination of fasteners, sealant and shims through parts consolidation and improved dimensional tolerance.

- Effective utilization of composite materials through innovative composite design versus simply substituting material.

- Tailored areal weight laminate construction versus standard ply building blocks.

- Utilization of through-the-thickness reinforcement to enhance areas of high out-of-plane loads. Such areas would include ply drop-offs, cut-outs, corners and fillets.

- Use of material forms which are conformable to simple and complex contours yet maximize the load carrying efficiency of the reinforcing fiber by maintaining fiber straightness.

- Use of material forms which minimize material waste and/or generate waste early in the manufacturing process before a significant level of value is added to the intermediate product.

- Integration of part manufacturing by utilizing materials in their lowest cost form and directly convert them to the finished structural configuration.

- Use of automated and/or computer controlled processing equipment to maximize quality and minimize "touch-labor".

- Designing materials for automation versus designing automation for materials.

- Reducing or eliminating ancillary materials typically associated with "conventional" composite processing such as bleeder and bagging systems.

While this list is obviously a panacea and no composite material and manufacturing processing combination can satisfy all these objectives for all types of structures, the fiber form/pressure molding process does address a large number of these objectives for substructural applications, particularly for composite fuselage frames.

In order to describe the fiber form/pressure molding process, the generic methodology used to fabricate the frames for the NASA/Lockheed acoustic test fuselage section will be described (Figure 1). In this program, 33 constant radius "J" section frames were fabricated using Union Carbide T300 carbon fiber and Hercules 3501-6 resin. Shear clips used to tie the frames, longitudinal stringers and fuselage skins together were also fabricated via this process but will not be discussed in this paper.

## 3. KNITTED FABRIC

The starting point for the XERKON fiber form manufacturing process is the reinforcing fiber on the spool. The fiber is then converted to a unidirectional knitted fabric (Figure 2). The fabric consists of tows of reinforcing fibers held straight and parallel by a fine knitting thread. In the case of most aircraft applications, the reinforcing fiber is carbon fiber such as T-300, AS-4, C6000, IM-6, etc. The knitting fiber can be made of a variety of materials including poly(ethylene terephthalate) or PET, polycarbonate, polyetheretherketone (PEEK), etc. The predominant knitting material used currently is PET.

The unidirectional fabric is produced on specially modified textile processing equipment which allows for the use of high modulus fibers such as carbon, aramid and ceramic fibers (Figure 3). The equipment is designed to provide highly controlled and predictable areal weight which is necessary to control the fiber/resin ratio in the finished parts.

Individual unidirectional plies can be used as manufactured for conversion into a fiber form or can run through a secondary assembly process to produce a multi-ply fabric. Up to 6 individual layers can be assembled in one operation with ply orientations of 0 and 90 degrees and bias orientations ranging from 30 to 90 degrees. Individual fabrics of various areal weights can be used to provide additional design flexibility.

A key feature of the knitted fabric is its ability to conform to changes in contour or curvature. This ability is depicted in Figure 4. For a single layer fabric the flexibility of the knitting fiber allows for local variations in areal weight as the material transitions the radius. A similar capability is found with the knitted bias fabric as the fibers are allowed to fan as the material goes around a radius. This feature allows for curved frames which do not require the darting and splicing typical of prepreg layups. This results in a significant reduction in labor required in the layup of the frame fiber form and reduces the amount of material waste as a narrow width of fabric can be used to produce the entire section. In addition to reducing labor requirements, it is anticipated that improved weight and structural performance will also result from the elimination of overlaps and splices.

## 4. FIBER FORM ASSEMBLY

The second operation in the fiber form manufacturing is the fiber form assembly. This procedure entails the conversion of the dry knitted fabric to the configuration of the finished molded part (Figure 5). Depending on the design of the specific frame, the web is assembled using various layers of bias and 90 degree fabric. 0 degree fibers (those running parallel to the arc of the frame section) are interleafed and inserted at the proper ply locations or are added as a previously manufactured fabric assembly.

The entire assembly is stitched together to lock the fibers into position and stabilize the assembly. The stitch fiber and stitch spacing and pitch are selected based on whether the stitching serves solely as a manufacturing aid or is used as a though-the-thickness reinforcement. Typical stitch fibers used by XERKON are PET, polycarbonate, PEEK, fiber glass, aramid and carbon. Completed fuselage frame fiber forms are seen in Figure 6.

## 5. RESIN INFUSION

Once the fiber form has been completed, the next operation entails impregnation of the fiber form with an appropriate resin system. The technique currently utilized by XERKON is resin film infusion (RFI). This procedure consists of positioning filmed resin on the fiber form and heating the resin and fiber form in a vacuum chamber (Figures 7 and 8). The fiber form is impregnated via gravity and capillary wetting. Vacuum is maintained throughout infusion to remove entrained air and volatiles from the resin. Precise computer control of the process is used to ensure the proper level of advancement of the resin prior to molding.

While procedures such as resin transfer molding can also be used to impregnate fiber forms, the

125

resin film infusion process is more amenable to currently certified hot melt resin systems including epoxies and bismaleimides. Resin transfer molding is generally restricted to resin systems which are liquid at room temperature. The RFI process is also suited to resin systems with high solvent contents as the solvents are removed during the heating and vacuum cycle.

## 6. AUTOCOMP PRESSURE MOLDING

The final operation in the process being described is the Autocomp pressure molding process. This operation establishes the final dimensions and cure condition of the molded part.

The manufacturing method combines aspects of compression molding and autoclave molding into one process. Like compression molding, pressure molding utilizes matched surface tooling (Figure 9). The tools are integrally heated via resistance heaters, and thermocouples are located in the tool to provide constant feedback to the computer controller.

The matched surface tooling provides the capability for extremely tight dimensional tolerances and repeatability. Thickness tolerances of less than +- .005" are not unusual. The matched surface tooling also ensures uniformity in critical structural areas such as radii and mating surfaces.

While the tooling provides the heat to cure the parts, the Autocomp vessel provides the vacuum and pressure required for final consolidation. Vacuum is provided by means of a reusable vacuum bag which is integral to the Autocomp vessel. The tool is inserted into the vessel and is effectively enveloped by the integral vacuum bag. Vacuum is drawn on the part at the beginning of the cycle and is maintained until pressure is applied by the Autocomp vessel. At the appropriate point of the cure cycle, pressure is applied,

typically 85-100 psi, to close the tool to stops. Due to the precise direct computer control of the individual part versus control of an autoclave, an accelerated cure cycle is generally used. This allows for more effective utilization of the tool and Autocomp unit. For example, during the peak production period for the frames described above, up to 5 parts were fabricated during an 8 hour shift on a single tool. Assuming 2 shifts per day and 20 working days per month, up to 400 parts/month/tool is quite feasible. Figure 10 shows a frame section immediately following Autocomp molding.

The direct computer control also provides the opportunity to stop the cure at any point within the cycle. This allows for the fabrication of preconsolidated, partially cured components which can be co-cured into a larger component such as a fuselage skin. Local zone control of the tool can also be achieved to provide parts with different levels of cure advancement throughout the part, for example, a fuselage stringer can be fabricated with a fully gelled cap and web and a base which is only partially cured. This detail could then be co-cured to a skin without the need for further consolidation in the cap or web.

## 7. APPLICATIONS

In addition to the NASA/Lockheed program, the fiber form/pressure molding process has been successfully demonstrated on a number of other fuselage programs.

In Figure 11, a representative section of body frame for a commercial transport is seen. This frame section is one of three sections which mate to form a closed 12.3 foot diameter fuselage frame. The fiber utilized in this component is Hercules AS4-6K and the resin is Hercules 3501-6 epoxy. Cost studies performed on the fiber form/pressure molded frame indicated a potential cost reduction of 69% in recurring costs as compared to a comparable frame fabricated with prepreg using hand layup and autoclave molding. Full-scale static testing of the frame demonstrated equivalency to the autoclave component for the applied loading condition.

The final frame application to be discussed is seen in Figure 12. This frame section is a generic representation of a complex frame which might be found in a small fuselage, inlet duct or engine nacelle. This "J" section frame incorporates features such as jogs, continuously changing contour, cap build-ups and varying cap-to-web angle. The frame section is fabricated using Hercules IM6-12K carbon fiber and Hercules 3501-6 epoxy resin. A key feature demonstrated by this component is the conformability of the knitted fabric. The entire component is fabricated without a single splice or overlap. This unique feature of the fabric significantly reduces the labor required to fabricate the component as well as reducing the scrap normally involved with this type of configuration.

## 8. SUMMARY

The fiber form/Autocomp pressure molding combination has demonstrated a significant potential for reducing the costs of composite fuselage structure. These savings are realized by utilizing materials in their lowest cost forms; fiber on the spool and bulk resin. The fiber is then converted to fabric by means of high speed textile equipment. Taking advantage of the inherent conformability of the knitted fabric, fiber forms are assembled to near-net shapes with minimium waste. The fiber form is impregnated by a hot-melt procedure which does not disturb the orientation of the fibers. Final part dimensions are obtained in an accelerated manner through precise computer control of matched surface tooling and the application of vacuum, pressure and heat in the curing vessel. By using this integrated composite manufacturing approach, the recurring costs for frame structures can be reduced by up to 70% as compared to hand layup/autoclave molding techniques.

## BIOGRAPHY

Mr. Van Schooneveld is Manager, Autocomp® Manufacturing Research and Development for XERKON Inc., Minneapolis, Minnesota. He has over nine years experience in the area of manufacturing development and materials science for composite materials.

Prior to joining XERKON in 1984, Mr. Van Schooneveld held a number of technical positions within General Dynamics/Fort Worth Division's Manufacturing Development organization. He participated in the development of resin transfer molding, compression molding, injection molding, thermoplastic fabrication and low observable materials technology. He was Program Manager on the Air Force contract entitled "Advanced Materials/Processing Science".

Mr. Van Schooneveld holds a Bachelor of Science in Materials Engineering and a Master of Science in Materials Engineering from Rensselaer Polytechnic Institute in Troy, New York, and a Master's of Business Administration from the University of Texas at Arlington.

Figure 1. Acoustic Test Fuselage Section
(Courtesy of Lockheed-Georgia Company)

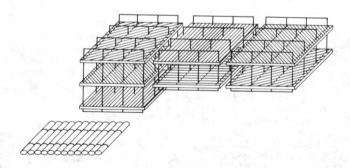

Figure 2. Unidirectional and Multi-directional
Knitted Fabric

Figure 3. Knitted Fabric Processing Equipment

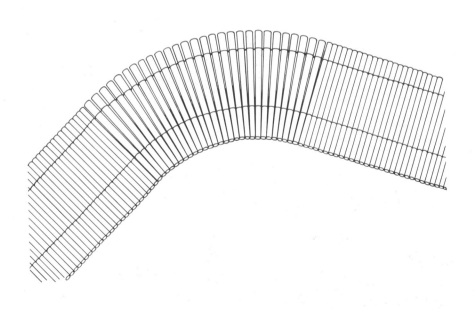

Figure 4. Conformability of Knitted
Unidirectional Fabric

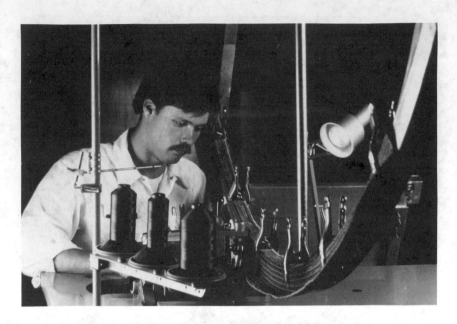

Figure 5. Frame Fiber Form Assembly

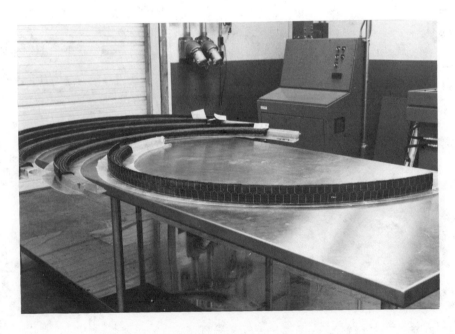

Figure 6. Complete Fuselage Frame Fiber Forms

Figure 7. Resin Film Application

Figure 8. Resin Film Infusion (RFI)

131

Figure 9. Fuselage Frame Autocomp® Molding Tool

Figure 10. Molded Fuselage Frame

Figure 11. Representative Commercial
Transport Fuselage Frame

Figure 12. Complex Frame Section

THE DELAMINATION BEHAVIOR OF CARBON FIBER REINFORCED PPS

Davies P., Benzeggagh M.L., de Charentenay F.X.
Division Polymères et Composites
Université de Technologie de Compiègne
BP 233,    60206    Compiègne    France

## Abstract

In this paper a study of the delamination behavior of unidirectional carbon fiber reinforced poly(phenylene sulfide), (PPS), is presented and a comparison made with two carbon/epoxy systems. The influence of a number of parameters on this behavior are evaluated. These include (a) the type of loading, Mode I and Mode II, (b) the annealing of specimens, (c) the specimen geometry, and (d) the temperature. Results are discussed both in terms of fracture mechanics parameters and failure mechanisms, and the limitations of the test methods are examined.

## 1.  INTRODUCTION

The development of a range of composites based on thermoplastic matrices has stimulated widespread interest in recent years. However, despite the general acceptance of the potential of these systems the influence of parameters such as the semi-crystalline nature of PPS and PEEK matrices on composites properties, and in particular on delamination resistance, has not been established.

The use of fracture mechanics tests to determine toughness values has been shown to be a useful approach for the relatively brittle first generation carbon/epoxy composites (1). While not yet standardized, the Double Cantilever Beam (DCB) specimen is widely used to determine Mode I (opening mode) delamination resistance, though data reduction methods vary (2,3). Mode II (shear) testing is more controversial however and a range of tests has been applied to composites (4-6).

In previously published work a value of Interlaminar Fracture Toughness, $G_{IC}$, of 1350 $J/m^2$ (7.8 in-1b/$in^2$) was obtained for film-stacked C/PPS (7) and more recently a value of 820 $J/m^2$ (4.7 in-1b/$in^2$) for continuously produced composite based on AS4 fibres (8). In both cases the Areas method (2) was used and the value of $G_{IC}$ therefore characterizes the propagation of a delamination. This point is emphasized as in the present work both initiation and propagation are characterized.

In this paper the delamination resistance of C/PPS under different types of loading is examined. In addition, the fabrication of the unidirectional panels of this system involves a rapid cooling step so the initial degree of crystallinity, $\Phi$, is low (9) ; this has allowed us to anneal selected specimens in order to examine the influence of $\Phi$ on the delamination behavior under both Mode I and Mode II loading. Mode I tests were performed on specimens of 2 thicknesses, thin (3 to 5 mm) and thick (20 mm), to assess the effect of specimen geometry on $G_{IIC}$ values. The effect of temperature was also studied. Two different Mode II tests were used for the C/PPS ; the ENF (End Notch Flexure) for the thinner specimens and the ELS (End Loaded Split) for the thick specimens. Results are compared with values obtained for two C/epoxy systems, T300/914 and IM6/6376.

## 2. EXPERIMENTAL

The carbon/PPS was supplied in the form of 3 and 20 mm thick unidirectional panels by Phillips Petroleum Company. The fibers were of the AS4 type and nominal fiber volume fraction, $Vf = 56$ %. The carbon/epoxies, T300/914 in 5 and 20 mm thick and IM6/6376 in 4 mm thick panels were of nominal fiber volume fraction 60 %. All panels contained starter cracks of Kapton or Vak-Pak film.

- Dynamic Mechanical Analysis of the C/PPS was performed on Polymer Laboratories DMTA equipment in the bending mode.

- X-ray diffraction was used to determine the degree of crystallinity, following the method described in reference (9).

- Mode I. Specimen dimensions are given in Figure 1. Tests were performed under displacement control at 2 and 0.8 mm/min. for the thin and thick specimens respectively.

- Thin specimens ; 6 as-received and 6 treated (150°C, 1 hr.) specimens of C/PPS were tested and values of $G_{IC}$ determined for the propagation.

- Thick specimens ; at 25°C, 4 as-received and 4 treated (200°C, 2 hr.) specimens with starter crack lengths from 40 to 80 mm were tested. A compliance calibration enabled values of $G_{IC}$ to be determined at initiation, indicated by acoustic emission, and during propagation by R-curves and the Areas method. Elevated temperature tests, at 80°C and 120°C were carried out inside an Instron temperature chamber ; compliance calibrations were obtained at each temperature.

- Mode II. Specimen dimensions are also given in Figure 1. Crosshead speed was 0.5 mm/min. for the ENF test and 1 mm/min. for the ELS. For the ENF test $G_{IIC}$ values were obtained using the shear-corrected beam theory solution (10) and the maximum load. For the ELS the method proposed by

Vanderkley was used (5).

## 3.  RESULTS AND DISCUSSION

In order to examine the effect of
annealing, DMTA scans were first
performed on specimens taken from
thin panels. Figure 2a shows the Tg of
the as-received material at 90 °C followed by a crystallization peak.
Figure 2b, the scan of a different
specimen annealed at 150 °C for 1 hour,
shows a reduced damping peak and
the Tg shifted to 115 °C.

### 3.1. Mode I

The results obtained for thick and
thin, annealed and non-annealed C/PPS
specimens are presented in Table 1
below, together with results for the
carbon/epoxy systems.

### 3.1.1. The Influence of Specimen Thickness

The most striking feature of these
results is the disparity between initiation and propagation values for all
materials except T300/914. This is
caused at least in part by extensive
fiber bridging during propagation.

Thin specimens can open a sufficient
distance to allow these bridging fibers
to break, and an equilibrium length of
tied zone is rapidly established. Thus
a constant value of $G_{IC}$ is obtained
from the Areas method which will
include any toughening due to matrix
deformation as well as the fiber
bridging contribution.

For thick specimens however the small
displacements involved require a longer
crack propagation before equilibrium
is reached and the Areas method
values rise throughout the test, until
reaching a value similar to that found
for the thin specimens. The mean
values presented in Table 1 are thus
somewhat misleading.

In contrast, no bridging was seen in
the T300/914 specimens although other

|  |  |  |  | INITIATION | PROPAGATION |
| --- | --- | --- | --- | --- | --- |
| SPECIMEN TYPE |  | $\Phi$ (%) | Compliance (ref.3) | Areas (ref.2) |
| C/PPS | Thin (3mm) | As rec[d] | 5 | – | 918 (89) |
| C/PPS | Thin | Annealed | 30 | – | 799 (74) |
| C/Epoxy | Thin (4mm) | T300/914 | – | 124(10) | 137 (9) |
| C/Epoxy | Thin | IM6/6376 | – | 182(17) | 633 (35) |
| C/PPS | Thick (20mm) | As rec[d] | 22 | 196(12) | 818 (273) |
| C/PPS | Thick | Annealed | 31 | 152(35) | 756 (236) |
| C/Epoxy | Thick | T300/914 | – | 185(22) | 151 (13) |

Table 1 : Mode I results, values of $G_{IC}$ $(J/m^2)$

authors have seen bridging in this system (11). Evidently the nesting of fibers is sensitive to forming conditions which determine fiber mobility, and to fiber volume fraction. In addition, bridges will be more likely to form in tough composites for which the plastic zone extends beyond the interply layer.

These observations raise serious doubts about the applicability of the data reduction methods to tests in which extensive fiber bridging takes place.

### 3.1.2. The Validity of Data Reduction Methods

The problems posed by fiber bridging in thick specimens are twofold and are illustrated in Figure 3 ; first, the compliance values recorded during propagation include a significant contribution from these fibers after the crack front has passed, and second there is a large residual load when the visually-observed crack reaches the end of the specimen.

The values of $G_{IC}$ calculated for initiation will be unaffected by these phenomena and these values indicate the resistance of a composite to the first damage. It may be seen that for the T300/914 this value characterizes the material, but for the tougher systems its use will be conservative ; for these materials toughening mechanisms act to resist macroscopic propagation and a characterization of propagation resistance is required. The two approaches considered here both present difficulties.

From the experimental compliance calibration for C/PPS an effective crack length is calculated which is considerably shorter than the observed crack. (This is in contrast to the case of T300/914, for which the calculated crack length is slightly longer than that observed, due to the concealed damage zone at the crack tip). The values presented on the R-curve will include contributions from both bridging and other toughening mechanisms, so an intrinsic toughness parameter characterizing the C/PPS system is very hard to extract. In addition, the increasing non-linearity of the unloading cycles observed in Figure 3 suggests that the Areas method, which takes account of non-linear elastic behavior, might be better suited to these materials.

The Areas method involves the direct determination of the energy necessary to propagate a crack a certain distance. However the total energy necessary to advance the visually-observed crack the full length of the specimen is considerably less than that required to separate the two specimen halves, as shown by the residual load at the end of the test. It might therefore be argued that this method underestimates the resistance of the material.

A means of avoiding this problem of data reduction might be to modify the specimen as suggested by Johnson & Mangalgiri (12), by placing the specimen halves at a small angle to each other during molding, thus avoiding fiber nesting and enabling a lower bound value to be obtained. However for tough composites the relatively large deformation zone at the crack tip may still promote bridging.

### 3.1.3. Effect of Temperature

When the test temperature is raised the maximum propagation values increase for thick C/PPS specimens with identical starter crack lengths, Figure 4. Given the discussion above, there is some doubt over the absolute values but increased ductility of the matrix at elevated temperature, which has been observed in other thermoplastic composites (13), is illustrated in Figure 8. The initiation value will correspond to the first damage, probably at the fiber-matrix interface, but subsequent drawing of the matrix away from the fibers will lead to an increase in $G_R$.

### 3.1.4. Effect of Annealing

Annealing will not only affect the matrix, by increasing the degree of crystallinity as shown by the X-ray measurements, but may also promote changes at the fiber-matrix interface.

Changes to the stress state at the interface may result from matrix shrinkage during crystallization or stress relief. The net effect is a slight decrease in initiation and propagation values of strain energy release rate. Other authors have noted a decrease in ductility of unreinforced thermoplastics on increasing the degree of crystallinity (14) but the differences recorded here are relatively small.

### 3.2. Mode II

The results from the Mode II ENF tests on thin specimens are presented in Table 2.

The effect of annealing is again to reduce the toughness values slightly. The $G_{IIC}$ values represent initiation from a short Mode I precrack, to avoid the influence of the starter crack.

The comparison of Mode I and Mode II values requires propagation from similar defects to be considered. Thus on a macroscopic scale, the ratio of $G_{IIC}$ to $G_{IC}$ for propagation from a defect linked to an equilibrium tied zone, i.e for thin specimens, is approximately 1 for IM6/6376 and for

| SPECIMEN TYPE | | $\Phi$ (%) | $G_{IIC}$ (J/m$^2$) |
|---|---|---|---|
| C/PPS | As rec$^{d.}$ | 5 | 933 (49) |
| C/PPS | Annealed | 30 | 802 (30) |
| C/Epoxy | T300/914 | – | 518 (27) |
| C/Epoxy | IM6/6376 | – | 658 (44) |

Table 2 : Results from ENF tests.

138

C/PPS independent of whether the specimen has been annealed. However, the Mode I values depend heavily on the bridging effect. For T300/914 Mode II values are considerably higher than Mode I values ; Bradley & Jordan found similar results, Mode II values roughly 3 times Mode I values, for other relatively low ductility epoxy based systems (15).

Slight non-linearities were noted in the load-displacement behavior of C/PPS and IM6/6376, but no macroscopic propagation was observed before the load drop. The development of a reliable Mode II test for tough composites requires an un-ambiguous initiation criterion and this criterion has not yet been established ; the maximum load criterion ignores any subcritical crack propagation and acoustic emission detection is not straightforward as there are several sources of noise.

Results from a limited number of ELS tests on C/PPS are not presented in the table as the precracking procedure led to unacceptable scatter in the results. It proved extremely difficult to determine the defect length either by direct measurement or compliance calibration. Tests on 5 specimens without precracks gave values from 900 to 1400 $J/m^2$ using the expression proposed by Vanderkley (5). However, the specimens tested were very thick, so a correction for shear deformation should also be included.

### 3.3. Fracture Surface Examination, S.E.M., Figures 5-10

Under both Mode I and Mode II loading, failure of C/PPS specimens occurred at the fiber/matrix interface following a small amount of matrix deformation, Figure 5. No differences were observed between fracture surfaces of as-received and annealed specimens, but surfaces from elevated temperature tests showed increased matrix ductility, Figure 8.

In contrast, for the C/Epoxy systems failure takes place in the matrix a short distance from the fiber in Mode I, Figures 6-7. Widespread phase separation is also evident in these systems, on a finer scale in the IM6/6376 than in the T300/914.

Under Mode II loading, numerous cracks transverse to the propagation direction are observed in T300/914, Figure 9. In IM6/6376, Figure 10, full hackle formation is observed and some clean fibers are also seen, where the tougher matrix is more resistant than the interface.

These observations help to explain both the relatively low values of $G_{IC}$ and $G_{IIC}$ for C/PPS and the similarity between these values ; under both normal and shear loading failure initiates at the interface. In the epoxy systems two very different mechanisms act according to the type of loading imposed.

## 4. CONCLUSIONS

- Delamination tests on unidirectional composites characterize mechanisms particular to this lay-up, such as fibre bridging, and both initiation and propagation values should therefore be presented. Testing of a specimen designed to avoid fiber bridging would yield a more useful value of $G_{IC}$.

- The delamination behavior of the C/PPS system is relatively insensitive to changes induced by annealing. Increased ductility of the PPS matrix is observed when Mode I tests are performed at elevated temperature.

- Mode II testing yields higher values for the PPS based than for the epoxy based systems ; the mechanisms acting differ considerably.

## 5. ACKNOWLEDGEMENTS

The authors thank Drs Alex Lou and Andy Stirling of Phillips Petroleum Company for supplying C/PPS panels, Ciba-Geigy for C/Epoxy panels and colleagues in the Polymer group, Cranfield Inst. of Technology for the use of DMTA equipment.

## 6. REFERENCES

1) Wilkins D.J., Eisenmann J.R., Camin R.A., Margolis W.S., Benson R.A., ASTM STP775, p. 168-83, 1982.
2) Whitney J.M., Browning C.E., Hoogsteden W., J. Reinf. Plast. & Comp. 1, p. 297, 1982.
3) Benzeggagh M.L., Prel Y., De Charentenay F.X., Proc. ICCM-V, San Diego, 1985.
4) Russell A.J., Street K.N., Proc. ICCM-IV, Tokyo, p. 279, 1982.
5) Vanderkley P.S., MSc Thesis, Texas A&M University, Dec. 1981.
6) Benzeggagh M.L., Prel Y., De Charentenay F.X., Proc. ECCM-1, Bordeaux, p. 291, Sept. 1985.
7) Martin Ma C.C., O'Connor J.E., Lou A.Y., Proc. 29th SAMPE Symp., p. 753-64, April 1984.
8) O'Connor J.E., Lou A.Y., Beever W.H. Proc. ICCM-V, San Diego 1985, p.963-70.
9) Johnson T.W., Ryan C.L., Proc. 31st SAMPE Symp., April 1986.
10) Carlsson L.A., Gillespie J.W., Pipes R.B., to appear in J. Comp. Materials.
11) Marais C., Merienne M.C., Sigety P., Proc. 5th French Composites Conference, JNC-5, p. 3, Sept. 1986.
12) Johnson W.S., Mangalgiri P.D., NASA Tech. Memorandum 87716, April 1986.
13) Davies P., Benzeggagh M.L., De Charentenay F.X., Proc. 5th French Composites Conf., JNC-5, p. 17, Sept. 1986.
14) Bessell T.J., Hull D., Shortall J.B., J. Mat. Sci. 10, p. 1127-36, 1975.
15) Bradley W.L., Jordan W.M., Proc. ISCMS, Beijing, June 1986, p. 445-50.

## 7. BIOGRAPHY

Peter Davies worked on material forming problems at TI Research, England for 3 years after graduating. He then completed an MSc in Polymer Engineering and a Diplôme d'Ingénieur through a Cranfield/Compiègne exchange program. Following this he worked on

rubber toughening of thermoplastics
before joining the composites group
in Compiègne in 1984 to study delamination and viscoelastic behavior of
thermoplastic composites.

Malk Benzeggagh, lecturer in the
Mechanical Engineering Department
at the Université de Technologie de
Compiègne graduated from the
Université de Paris VII in 1976. He
completed a doctorate on the application of fracture mechanics to
composite materials in 1980. His
current research interests are damage
mechanics and mechanisms during
delamination under static and cyclic
loading.

François-Xavier de Charentenay,
Docteur es Sciences, Professor,
graduated from the Ecole Supérieure
de Physique et Chimie in 1961. He
completed a PhD in 1966 on semi-
conducting polymers at the Université
de Paris. After post-doctoral research
at Imperial College, London he spent
six years as research engineer in
Material Science at the Institut
Français du Pétrole. Since 1973, he
has been Professor in the Mechanical
Engineering department at the
Université de Technologie de Compiègne
where he is in charge of the Polymers
and Composites research division.

a) <u>DCB thin</u>

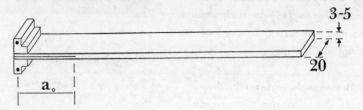

3-5

20

a₀

b) <u>DCB thick</u>

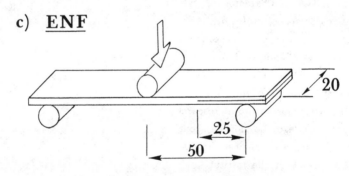

20
20

a₀

c) <u>ENF</u>

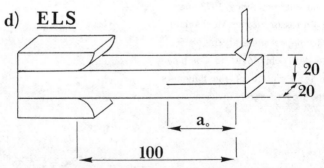

20

25

50

d) <u>ELS</u>

20
20

a₀

100

Figure 1 : Specimen geometries.

142

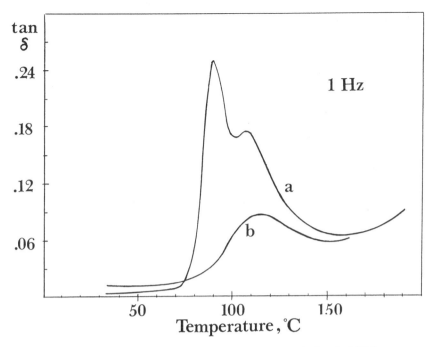

Figure 2 : Dynamic Mechanical Thermal Analysis of C/PPS
a) As received
b) After annealing at 150°C, 1 hr.

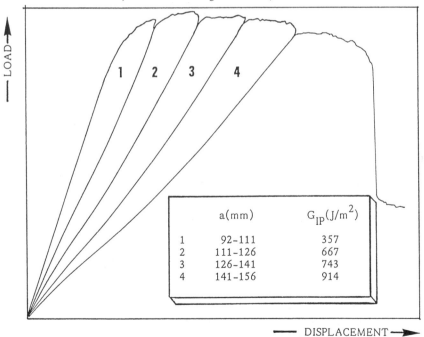

| | a(mm) | $G_{IP}(J/m^2)$ |
|---|---|---|
| 1 | 92–111 | 357 |
| 2 | 111–126 | 667 |
| 3 | 126–141 | 743 |
| 4 | 141–156 | 914 |

Figure 3 : Areas method load-unload cycles
As received C/PPS, thick specimen ($a_o$ = 71 mm)

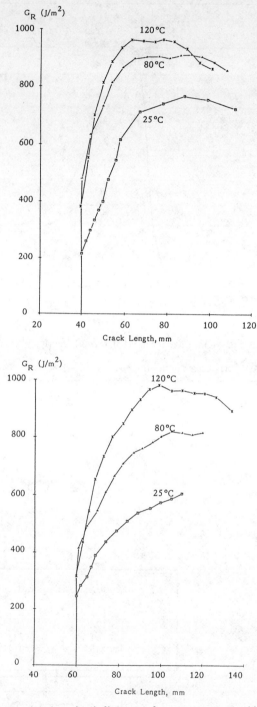

<u>Figure 4</u> : R curves showing the influence of temperature on Mode I behavior of annealed C/PPS, thick specimens, for 2 starter crack lengths.

Figure 5
Mode I C/PPS, 25 °C
10 μ m
⊢———⊣

Figure 6
Mode I T300/914
10 μm
⊢———⊣

Figure 7
Mode I IM6/6376
10 μm
⊢————————⊣

Figure 8
Mode I C/PPS, 120°C
10 μ m

Figure 9
Mode II T300/914
10 μ m

Figure 10
Mode II IM6/6376
10 μ m

## PAS-2 HIGH PERFORMANCE PREPREG AND COMPOSITES

Merlin R. Lindstrom and Robert W. Campbell
Phillips Petroleum Company
Phillips Research Center
Bartlesville, OK 74004

## ABSTRACT

A family of polyarylene sulfide polymers is being developed by Phillips Petroleum as engineering thermoplastic resins. One potential use for these resins is as a matrix for high performance composites. Ryton® polyphenylene sulfide is the first member of this family. It has been reinforced with chopped fiber, long random fibers, woven fabric and continuous unidirectional fibers for use in composites. A new member of the polyarylene sulfide family, PAS-2, has been reinforced with carbon fibers to form a unidirectional prepreg tape. This tape has been molded into composite laminates of excellent quality. As with any thermoplastic, the prepreg has the advantages of having unlimited shelf life and the molding cycle is very short. The composite laminates have excellent mechanical properties, good retention of these properties at 350°F and good chemical resistance.

## 1.0 INTRODUCTION

In 1983, developmental polyarylene sulfide composites were introduced into the marketplace. The matrix for these thermoplastic composites was the semicrystalline polymer, polyphenylene sulfide (PPS). The PPS based-composites exhibit excellent chemical and solvent resistance and good thermal stability. However, PPS unidirectional composites retain only about 30% of their room temperature strength at 350°F.[1] Therefore, use of PPS composites as structural materials is limited.

In 1985, J. E. O'Connor reported on the development of a carbon fiber composite based on the semi-crystalline polymer, PAS-1.[2] More recently, a high temperature, amorphous polyarylene sulfide polymer, PAS-2, has been combined with unidirectional carbon fibers to form a high temperature composite. Properties of neat PAS-2 and PPS resins are compared in Table I. As can be seen, the glass transition temperature, Tg, of PAS-2 (i.e. 215°C) is about 125°C greater than that of PPS. It accounts for the major difference in the elevated temperature properties of these two materials.

## 1.1 PAS-2 Polymer

PAS-2 is a high Tg resin with exceptional chemical resistance for an amorphous polymer as shown in Table II. With the exception of some aromatic oxygenated hydrocarbons and amines, PAS-2 shows no decrease in tensile properties after 24 hours of exposure to a variety of solvents. The reason for this exceptional solvent resistance, when compared to other amorphous resins such as polysulfone, is believed to be due to "pseudocrystallinity" or strong association of the PAS-2 molecules.

In addition to having good chemical resistance, PAS-2 resin is also characterized as having good toughness and impact resistance. Good toughness is indicated by the presence of a tensile yield coupled with good elongation. Impact resistance, as indicated by unnotched Izod, is over twice that of PPS. The combination of a high Tg, good chemical resistance, good toughness and high elongation make PAS-2 an excellent matrix resin for high temperature, high performance composites.

## 1.2 Prepreg Production

Continuous carbon fiber prepreg is prepared using a proprietary prepregging process which has been shown to be effective for prepregging a variety of resins.[1,2] The prepreg produced from this process results is a composite material with good fiber impregnation and dispersion. To date, PAS-2 has been prepregged with AS-4 carbon fibers from Hercules. Future plans include prepreg production using other continuous fibers such as glass, Kevlar and ceramic fibers as well as preparing a fabric prepreg.

As with any thermoplastic, PAS-2 prepreg has the advantage of having unlimited shelf life with no special storage conditions. Further, laminates can be readily prepared from the prepreg by spot welding several plies together. Thus various lay-ups can be accomplished even though the prepreg exhibits no tack.

## 1.3 Laminate Production

PAS-2/carbon fiber laminates having various fiber orientations have been prepared by placing the appropriate stacked laminates between the platens of an electrically heated press. Laminates have been prepared using either picture frame or positive pressure molds or by simply pressing the plies together between two flat metal plates. Once fully consolidated, the laminates can be cooled slowly in the press, or they can be transferred to a cold press and cooled rapidly.

Preliminary evaluation of PAS-2 processing parameters show that the preferred molding conditions are 625°F and 300 psi pressure. Since PAS-2 is an amorphous resin, significantly reduced consolidation pressures at higher temperatures are expected to be sufficient. The effect of temperature on the processability of PAS-2 composites is presently being evaluated, and the results will be reported elsewhere.

## 1.4 Composite Properties

Mechanical properties of PAS-2/carbon fiber laminates were determined using either standard ASTM procedures or procedures supplied by various end users. Table III shows a comparison of the room temperature properties of PAS-2/carbon fiber composites with PPS/carbon fiber composites. As can be seen, properties of the two composites are quite comparable.

Table IV compares the 350°F properties of PAS-2/carbon fiber composites with PPS/carbon fiber composites. At this temperature, PAS-2 composites retain 80-90% of their strength and essentially 100% of their modulus. PPS composites, on the other hand, retain only about 30% of their room temperature strength at 350°F. Consequently, PAS-2/carbon fiber composites have utility in structural applications at 350°F where PPS composites do not. This significant increase in elevated temperature property retention is due to the 135°C higher Tg of PAS-2 as compared to PPS.

As mentioned previously, PAS-2 has exceptional chemical resistance for an amorphous polymer (Table II). This chemical resistance translates to good solvent and fuel resistance for the composite laminate toward materials encountered in aerospace applications. Table V shows that PAS-2 composites retain 80-90% of ultimate tensile and flexural strengths and essentially 100% of the respective moduli after solvent exposure. It should be noted that the tests were run on unstressed test specimens. Work is in progress to determine the effect of loading on the chemical resistance.

## 2.0 CONCLUSIONS

A new polyarylene sulfide resin, PAS-2, has been prepregged into a unidirectional carbon fiber tape and used to prepare composite laminates. These laminates have been found to have ambient temperature properties similar to those observed for PPS composites. However, the 350°F composite properties of the PAS-2 composite are significantly greater than those of the PPS composites indicating that the PAS-2 composites can be used for structural applications at elevated temperatures. In addition, it has been shown that the PAS-2 resin has exceptional chemical resistance for an amorphous polymer. This added chemical resistance results in good solvent and fuel resistance for PAS-2 composites for aerospace applications.

## 3.0 REFERENCES

1. J. E. O'Connor, W. H. Beever and J. F. Geibel, "A New Polyarylene Sulfide Polymer Prepreg and High Performance Composite", International SAMPE Symposium Proceedings, 31, 1313 (1986).

2. J. E. O'Connor, A. Y. Lou and D. G. Brady, "Polyarylene Sulfide Composites", Proceedings of the American Society for Composites, 1, 21 (1986).

## 4.0 BIOGRAPHIES

**Merlin R. Lindstrom** is Supervisor of the New Advanced Polymers and Composites Section with Phillips Petroleum Company. He received his Ph.D. in Organic Chemistry from North Dakota State University in 1978. He is responsible for the development of processes, products and markets for high performance thermoplastic composite materials.

**Robert W. Campbell** is Manager of the Engineering Materials Branch with Phillips Petroleum Company. He received his Ph.D. in Organic Chemistry from Purdue University in 1966. He is responsible for new product and process development in the areas of engineering plastics and advanced materials.

150

## TABLE I: COMPARISON OF NEAT RESIN PROPERTIES OF PPS AND PAS-2

| Property | PPS | PAS-2 |
|---|---|---|
| Density, g/cc | 1.36 | 1.40 |
| Tensile Strength, psi | 11,400 | 14,500 |
| Elongation, % | 2-20* | 8 |
| Flexural Strength, psi | 21,300 | 25,700 |
| Flexural Modulus, psi | 490,000 | 460,000 |
| Izod Impact, ft lb/in | | |
| Notched | 0.4 | 0.8 |
| Unnotched | 10.8 | 25.2 |
| Oxygen Index, % | 44 | 46 |
| Glass Transition Temp., °C | 85 | 215 |
| Crystalline Melting Pt., °C | 285 | none |

*Depending upon crystallinity.

## TABLE II: CHEMICAL RESISTANCE OF AMORPHOUS RESINS

(% Property Retention After 24 Hours at 93°C)

| Solvent | PAS-2 | Polysulfone | Polycarbonate |
|---|---|---|---|
| Conc. HCl | 90 | 100 | 0 |
| 30% NaOH | 102 | 100 | 7 |
| 10% FeCl$_3$ | 100 | 100 | 100 |
| Water | 97 | 100 | 100 |
| Acetic Acid | 102 | 91 | 67 |
| n-Butyl Alcohol | 130 | 100 | 94 |
| 2-Ethoxy Ethanol | 123 | 0 | 78 |
| Pyridine | 19 | 0 | 0 |
| n-Butyl Amine | 96 | 0 | 0 |
| Methyl Ethyl Ketone | 45 | 0 | 0 |
| Ethyl Acetate | 116 | 0 | 0 |
| Tetraheydrofuran | 38 | 0 | 0 |
| Cyclohexane | 112 | 99 | 75 |
| Toluene | 101 | 0 | 0 |
| m-Cresol | 0 | 0 | 0 |

**TABLE III: PAS-2 AND PPS ROOM TEMPERATURE COMPOSITE PROPERTIES**

Unidirectional Laminates/60 Wt % Carbon Fiber

| Property | PAS-2 | PPS |
|---|---|---|
| Long. Tensile Strength (KSI) | 172 | 182 |
| Long. Tensile Modulus (MSI) | 19 | 17 |
| Long. Flexural Strength (KSI) | 153 | 198 |
| Long. Flexural Modulus (MSI) | 15 | 16 |
| Long. Compressive Strength (KSI) | 75 | 92 |
| Tran. Tensile Strength (KSI) | 5.6 | 5.2 |

**TABLE IV: PAS-2 AND PPS 350°F COMPOSITE PROPERTIES**

Unidirectional Laminates/60 Wt % Carbon Fibers

| Property | PAS-2 | | PPS | |
|---|---|---|---|---|
| | Value | % Ret. | Value | % Ret. |
| Long. Ten. Str. (KSI) | 146 | 85 | 68 | 37 |
| Long. Ten. Mod. (MSI) | 19 | 100 | 16 | 94 |
| Long. Flex. Str. (KSI) | 119 | 78 | 51 | 26 |
| Long. Flex. Mod. (MSI) | 16 | 107 | 13 | 81 |

152

## TABLE V: PAS-2 COMPOSITE CHEMICAL RESISTANCE[1]

|  | JP-4 Jet Fuel[2] (14 days) | 5606A Hydraulic Fluid[2] (14 days) | 2% NaOH[3] (24 hrs) | Methyl[3] Ethyl Ketone (24 hrs) | Methylene[3] Chloride/ Phenol (24 hrs) |
|---|---|---|---|---|---|
| **Tensile Strength, KSI** |  |  |  |  |  |
| Control | 25 | 25 | 25 | 25 | 25 |
| Exposed | 22 | 20 | 24 | 19 | 24 |
| % Retention | 88 | 80 | 96 | 76 | 96 |
| **Tensile Modulus, MSI** |  |  |  |  |  |
| Control | 1.8 | 1.8 | 1.8 | 1.8 | 1.8 |
| Exposed | 1.8 | 2.1 | 1.9 | 1.8 | 1.9 |
| % Retention | 100 | 117 | 108 | 100 | 106 |

(1) 60% AS-4 Carbon fibers, 12 ply ± 45 laminates

(2) Immersion at 180°F

(3) Saturated cloth at RT

AN EVALUATION OF A HIGH TEMPERATURE
THERMOPLASTIC POLYIMIDE COMPOSITE

J. Timothy Hartness
University of Dayton
Research Institute
300 College Park Dr.
Dayton, OH 45469

## Abstract

A limited evaluation has been carried out on a unique linear aromatic polyimide used as a matrix for graphite composites, LARC-TPI. This material was developed by NASA Langley Research Center. Initial evaluations in which graphite fibers were impregnated from the amic acid solution in diglyme resulted in poor consolidation due to inadequate melt flow behavior. During this evaluation a molding powder version of the material was received from Mitsui Toatsu of Japan. This powder contained a high degree of imide and was thus not still soluable in diglyme. It was observed that this powder possessed a large endothern upon thermal analysis using Differential Scanning Calorimetry (DSC). It was also observed that the powder exhibited considerable flow as compared to the film or imidized from the amic acid solution. Rheological characterization of the powder indicates that it has a low melt viscosity that could be taken advantage of if powder and graphite fiber could be combined into a useful prepreg form. At least five lots of LARC-TPI powder were investigated. It was observed that the various lots exhibited considerable differences in melt viscosity and residual volatiles. Composites were fabricated from prepreg using a proprietary process. Considerable differences were observed in the quality and mechanical performance of the composites fabricated from various lots of powder.

## 1. INTRODUCTION

As part of the continuing effort to evaluate and characterized composite materials that may offer improvements in properties as well as lower cost manufacturing, the University of Dayton Research Institute has been involved in the research and development of advanced thermoplastic and pseudothermoplastic composites. It is the purpose of this paper to report on the initial findings on the evaluation and characterization of a unique form of a linear aromatic polyimide matrix, LARC-TPI. Various lots of material were investigated. The effect on mechanical properties, such as flexural and shear, due to observed and measured differences between the various lots is the major scope of this paper.

## 2. EXPERIMENTAL

### 2.1 Polymer

The LARC-TPI powders that were characterized were prepared and supplied by Mitsui Toatsu Company, Tokyo, Japan. The lots are as follows: 72-501, 72-502, 92-702, 92-712, and 92-713. Analysis on two separate lots assayed by Mitsui Toatsu are as follows: 72-501, 91.9% imide, 3.1% isoimide, 5.8% amic acid, 1.7% weight loss on drying; Lot 92-713, was reported as 91.13% imide, 2.0% imide, 2.0% isomide, 6.87% amic acid, trace percent remaining solvent. Other lots were not reported. The synthesis and evaluation of LARC-TPI based on the polymerization of 3,3',4,4' - benzophenone tetracarboxylic dianhydride (BTDA) and 3,3' - diaminobenzophenone (3,3'-DABP) has been reported[1]. The molecular structure is shown in Figure 1. A characterization of LARC-TPI powder Lot 72-501 was reported[2] by H. Burks, Hou, and St. Clair. Characterization of LARC-TPI composites from the polyamide acid has been reported by N. Johnston and T. St. Clair[3]. Observations in their work indicated that poor melt flow occurred from this process with poor resulting properties.

### 2.2 Thermal Properties

Thermal analysis was carried out on as received lots of LARC-TPI powder. No drying was done prior to analysis. Glass transition temperatures (Tg), melt onset, and peak melt temperatures (Tm) were determined calorimetrically using a DuPont Model 910 Thermal Analyzer/Differential Scanning Calorimeter (DSC) in air at 10°C/min. Thermal stability and weight loss as a function of temperature was determined using a DuPont Model 951 Thermogravimetric Analyzer (TGA) at a heating rate of 10°C/min under air. Table 1 lists the various lots versus Tm and onset of Tm. Several observations were made of the various lots of powder. Relatively broad melt endotherms were observed for all lots except Lot 92-702 which had the higher onset of melt temperature. It was also observed that the material possessed a much higher melt viscosity although no actual viscosity measurements were made on Lot 92-702, DSC plots are shown for each lot (Figures 2-6). As can be observed in some lots a rather pronounced endotherm occurs just prior to the melt. A less pronounced endotherm occurs prior to this.

Lot 92-702 shows no indication of the first two endotherms. It was observed that Lot 72-501 and 72-502 both had distinct odors of acetic acid upon heating in the open air. This could be the results of residual acetic anhydride used to imidized the amic acid solution. Thermogravimetric analysis of various lots indicate a correlation between weight loss and the observed endotherm prior to melt. Lot 92-702 had a weight loss of 1.1% when run to 400°C in air. Lot 72-501 had a weight loss of 4.4% when run to 400°C in air. In a separate study on Lot 92-713 a DSC sample was heated from room temperature to 300°C and held for 30 minutes followed by rapid cooling. The sample was then rerun to 400°C. The rerun DSC indicates a transition at 215°C considered to be the glass transition and a new higher peak melt at 338°C (See Figures 6 and 7). This is the exact effect observed by Burks et al. in their study of the powder. This information is important for the understanding of the processing of the material be it either neat resin or composite. As was pointed out in the above study, thermal holds of the material below 320°C result in a semi-crystalline material; on the other hand a purely amorphous material results when samples are held at temperatures above 330°C. As was observed in our study, both the glass transition temperature (Tg) and the melting temperature (Tm) increased with longer hold times. The possibility of having a semi-crystalline material with a Tg at or near its maximum and also still possessing a Tm opens up some questions as to effect on mechanical properties, improved chemical resistance and possible advantages in a post-forming operation. The increase in chain extension and melt viscosity in the amorphous regions of the polymer may offset advantages from the melt. No investigations have been made in this area as of yet.

2.3 Rheological Properties

Rheological profiles are very necessary for successful processing of composite materials. Issues such as determining correct times for application of pressure to the composite is one of the most important ones. Rheological measurements were made on incoming lots of LARC-TPI powder. Variations were observed in minimum viscosity between lots of powder. Additional study is necessary in order to correlate the differences. Two separate runs were completed on Lot 92-706. A Rheometrics RDS-II 7700 equipped with a parallel-plate test fixture was used for rheological measurements. A powder sample, 0.5 gm, is spread on 25mm diameter plates and adjusted as the powder starts to soften with temperature. The top plate was driven to oscillate at a frequency of 100 Hz. The bottom plate was driven to oscillate at a frequency of 100 Hz. The bottom plate is attached to a torque transducer which measures force. The plates are enclosed in a heated chamber and purged with nitrogen gas. Measurements were made in the first case (Figure 8) between 300°C and 400°C with a heatup rate of 2°C/min. Strain levels were set at 1% throughout the run. The second case (Figure 9) was run isothermally at 316°C for 30 minutes. Differences in the two runs were similar to that reported by Burks, et al. In the first case, in which the powder is heated above 320°C, storage module G' increased less rapidly when compared to the second case in which the heat is kept below 320°C. As was discussed earlier the thermal analysis made by DSC indicates that a higher crystalline melt occurs when the sample is held at 300°C. It was observed by Burks, et al.[2] that "The modulus level that can be achieved by processing the TPI powders is several orders of magnitude higher for semi-crystalline materials than for materials which are purely amorphous." What significance this may have in the composite has yet to be determined. Thermal mechanical measurements were made on the processed composite. A torson bar specimen approximately 1.25 cm x 6.25 cm x 0.15 cm was run using the Rheometrics RDS-II 7700. The specimen was heated in a nitrogen atmosphere from 40°C to 377°C at 2°C/min.

Figure 10 shows the thermal profile. Storage modulus starts to deviate from linearily at approximately 250°C. This temperature corresponds to the Tg as measured by other techniques such as DSC or TMA, others prefer to use the peak of G" or tan Δ as the Tg of the material.

### 3. COMPOSITES

Composites were fabricated from prepreg manufactured by a proprietary process. Graphite fiber was Celion 12K-30. Unidirectional tape 12 inches wide was cut and layed up in a traditional manner. Consolidation was completed in a matched die metal mold. Processing cycles were developed to account for the differences between the different lots of LARC-TPI powder. Process development had to consider the trade-off between residual volatiles observed in some lots of powder and viscosity variations. Volatiles were measured in all lots of powder. Lot 72-501 and 72-502 had a distinct odor of acetic acid. Thermal gravemeteric analysis (TGA) indicated that most residual solvent was loss prior to 260°C. Processing composites fabricated from these lots had to be accomplished with volatile removal and related void formation in mind. One also had to be aware of changes in viscosity which could also affect both of the above concerns as well as affecting the wetting of the fiber. Basically two different processing cycles were used in this early screening procedure. Process cycle #1 (see Table 2) was employed to remove any residual volatiles prior to the application of pressure. This cycle seemed to work well with little void formation, less than 2%, yet good properties with a maximum in glass transition (Tg) of 249°C. It was observed that this cycle worked well where residual volatiles were high and melt viscosity was low. Panels #1-6 fit this category. Panels made from lot 92-712 and 92-713 differed in the amount of residual solvent and melt viscosity (see Table 4). It was observed that powder from these lots possessed lower volatile content but also higher melt viscosity. When cycle #1 was used on these materials the void content approached zero but wetting of the fiber was poor as observed in scanning electron micrographs (SEM's) of the fractured surface. Mechanical properties were also lower (see Panel #9). Processing cycle #2 (see Table 3) was used to improve wetting with the hope that void formation would not be to detrimental to composite properties. Dramatic results were observed in mechanical properties especially in four point flexure where the shear component is doubled that in three point. As can be seen in Panel #19 the maximum in flexural properties is being reached with two of the three specimens recording 2411 Kpa (350 Ksi). The fracture surface of

Panel #19 also indicated excellent adhesion. SEM's were taken of three representative specimens to (Figs. 11-14) determined fiber wetting as well as fracture morphology. As can be observed in all the photos there is considerable plastic deformation. Only in Panel #19 is there the quality of bonding to the fiber which results in almost total cohesive failure with almost all fibers coated with polymer. Again a very rough surface has resulted indicating that significant plastic deformation is occurring. Very limited mechanical characterization was completed due to time constraints and changes in powder from lot to lot. Mode I and Mode II fracture toughness was measured on the neat resin (lot 92-713) using aluminum adherens 1.25 cm x 1.25 cm x 30 cm. This technique was developed by Dr. Herzl Chai[4] and is used to simulate the resin layer between the plies without the influence of fiber and multiple crack fronts. The Mode I and Mode II values recorded were 430 (2.5) and 875 (5.0) $J/m^2$ (in. lbs) respectively. These tests will also be generated on the composite along with other mechanical properties in the near future. Data generated by Chai on three matrix materials, Hercules 3502, American Cyamide BP-907, and Imperial Chemical Industries PEEK indicates that LARC-TPI is very similar to BP-907 in Mode I fracture energy. The measured fracture energies in Mode I for 3502, BP-907, and PEEK were 79 (0.45), 420 (2.4) and 1137 (6.5) $J/m^2$ (in./lbs) respectively. Mode I fracture energy generated on composites by Chai indicates a one-to-one agreement when comparing the two tests. In the composite case one must use data generated in the early portion of the beam before fully developed damage occurs which results in fiber bridging and often multiple crack fronts.

## 4. CONCLUSIONS

Initial characterization of LARC-TPI composites indicates that this is a very unique material that possesses pseudothermoplastic behavior and possible true thermoplastic character due to its transient crystallinity. Its ability to be used as a matrix resin is due to its ability to be prepregged using a proprietary technique as well as its low melt viscosity resulting in excellent fiber wetting, bonding and composite consolidation. This limited characterization and evaluation indicates that the material has excellent potential as a matrix resin.

## 5. ACKNOWLEDGMENTS

The author gratefully acknowledges Dr. Norman J. Johnston, NASA Langley Research Center for useful and important discussion and encouragement and Dr. Herzl Chai, University of Dayton Research Institute, for neat resin Mode I and Mode II discussion and testing.

## 6. REFERENCES

1. A.K. St. Clair and T.L. St. Clair, NASA Langley Research Center, SAMPE Quarterly, October (1981).

2. H. Burks and T. St. Clair, NASA Langley Research Center, SAMPE Quarterly, Vol. 13, No. 1, October (1986).

3. N.J. Johnston and T. St. Clair, NASA Langley Research Center Preprint, 18th National SAMPE Tech. Conf. (1986).

4. H. CHai, University of Dayton Research Institute, Engineering Fracture Mechanics, Vol. 24, No. 3, pp. 413-431, (1986).

## 7. BIOGRAPHY

J. Timothy Hartness is an Associate Research Chemist and group supervisor of the Structural Composites Group at the University of Dayton Research Institute. Mr. Hartness has been with the Institute in the composite field since 1968. The received his Bachelor of Technology Degree from the University of Dayton in 1975.

## TABLE I

### MELT ONSET AND PEAK MELT VERSUS LOT NUMBER

| Lot No. | Melt Onset | PEEK Melt |
|---------|-----------|-----------|
| 72-501 | 254.2°C | 274.7°C |
| 72-502 | 250.2°C | 267.7°C |
| 92-702 | 273.8°C | 286.1°C |
| 92-712 | 261.1°C | 283.0°C |
| 92-713 | 265.3°C | 286.0°C |

## TABLE II

### PROCESS CYCLE #1

1. RT → 260°C (500°F) as fast as possible using no pressure

2. Hold at temperature 1/2 hour

3. Heat 260°C (500°F) → 343°C (650°F) as fast as possible using no pressure

4. Hold at temperature 15 minutes

5. Apply 1378 kpa (200 psi)

6. Hold 45 minutes

7. Cool under pressure to below 93°C (200°F)

## TABLE III

### PROCESS CYCLE #2

1. RT → 260°C (500°F) as fast as possible using no pressure

2. Hold at temperature 1/2 hour

3. Apply 1378 kap (200 psi)

4. Heat 260°C (500°F) → 343°C (650°F) as fast as possible

5. Hold 1 hour

6. Cool under pressure to below 93°C (200°F)

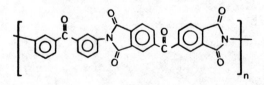

Figure 1. Molecular Structure of LARC-TPI.

TABLE IV

MECHANICAL PROPERTIES

0°

| Panel # | Temperature °C (°F) | 3-Point Flex | | 4-Point Flex | | 4-Point Shear MPa (ksi) | SBS MPa (ksi) | FV % | Void % |
|---|---|---|---|---|---|---|---|---|---|
| | | Strength MPa (ksi) | Mod. GPa (msi) | Strength MPa (ksi) | Mod. GPa (msi) | | | | |
| 3 | RT | 1946 (282.4) | 121 (17.6) | | | | | 49.7 | 2.47 |
| 3 | 177 (350) | 1288 (187.0) | 126 (18.3) | | | | | | |
| 5 | 177 (350) | 1609 (233.6) | 126 (18.3) | | | | | | |
| 6 | -- | -- | -- | | | | | | |
| 9 | RT | 1631 (236.7) | 141 (20.5) | | | 40 (5.8) | | 57.10 | 2.00 |
| 13 | RT | 1617 (234.7) | 116 (16.9) | 983 (142.7) | 92 (13.4) | | | | |
| 16 | RT | 1414 (205.3) | -- | | | | | | |
| 17 | RT | 1962 (284.7) | 113 (16.4) | 951 (138.1) | | | | | |
| 18 | RT | -- | -- | -- | -- | | | | |
| 19 | RT | 2309 (335.1) | 149 (21.6) | 1753 (254.4) | 112 (16.3) | 54 (7.8) | 63 (9.1) | 58.05 | 4.11 |

Note: All specimens are an average of three.

3 Pt Flexural    Span-to-Depth Ratio 32-1, ASTM-D-790

4 Pt Flexural    Span-to-Depth Ratio 32-1, ASTM-D-790

4 Pt Shear       Span-to-Depth Ratio 16-1

SBS              Span-to-Depth Ratio 4-1

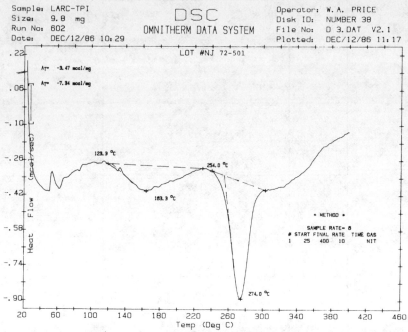

Figure 2. DSC of LARC-TPI, Lot #72-501,
10°C/min in Nitrogen.

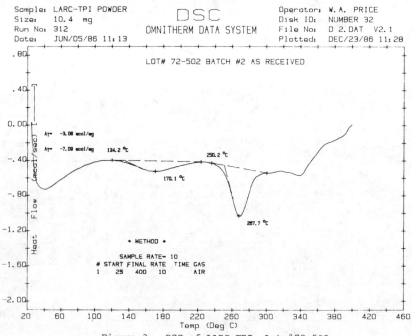

Figure 3. DSC of LARC-TPI, Lot #72-502,
10°C/min in air.

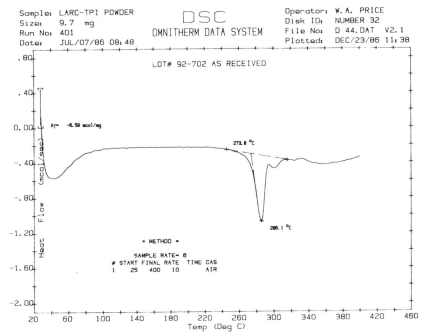

Figure 4. DSC run of LARC-TPI, Lot #92-702,
10°C/min in air.

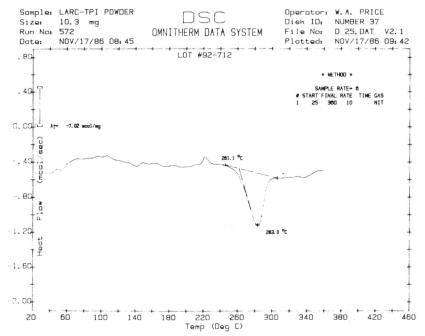

Figure 5. DSC of LARC-TPI, Lot #92-712,
10°C/min in Nitrogen.

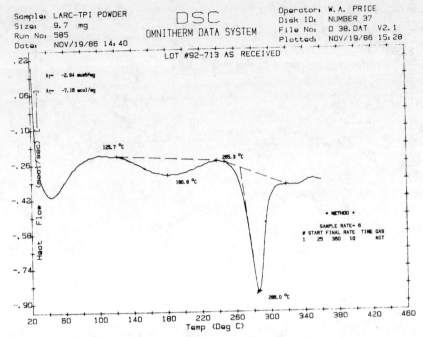

Figure 6.   DSC of LARC-TPI, Lot #92-713,
10°C/min in Nitrogen.

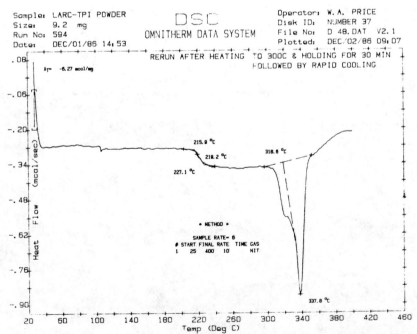

Figure 7.   Rerun DSC of LARC-TPI, Lot #92-713, after 300°C
hold Followed by Rapid Cooling.

LARC-TPI 25mm PARALLEL PLATE 2C/MIN

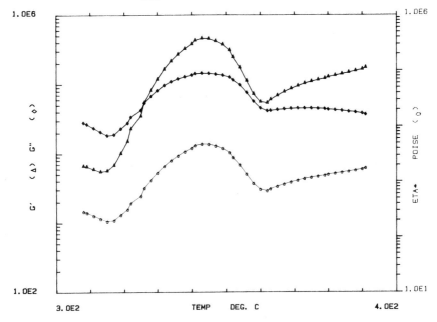

Figure 8.   Parallel Plate Viscosity LARC-TPI
316°-390°C, 2°C/min

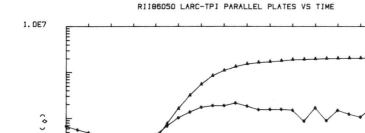

Figure 9.   Parallel Plate Viscosity LARC-TPI
30 minutes at 316°C.

165

RII86051 LARC-TPI TORSION BAR 2C/MIN 100HZ 0.01% STRAIN

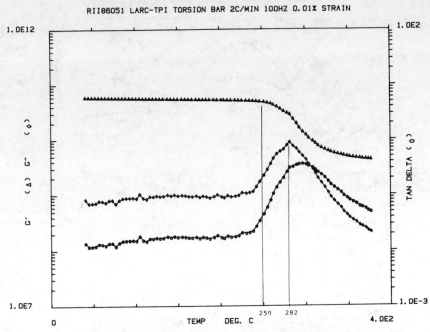

Figure 10.   Torson Bar LARC-TPI/C-12K Composite
             G' (dynes/cm$^2$) Storage Modulus Versus
             Temperature 40°C-377°C, 2°C/min.

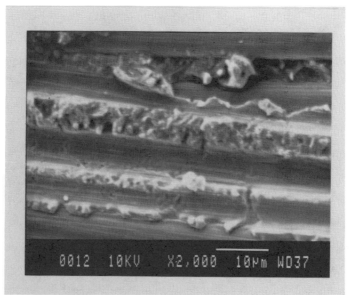

Figure 11. SEM of Fracture Surface of Panel #16 2000X.

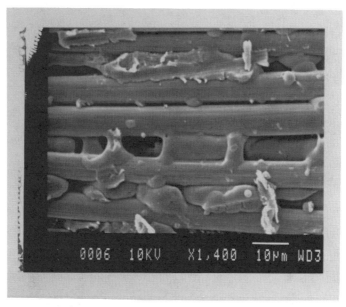

Figure 12. SEM of Fracture Surface of Panel #17 1400X.

167

Figure 13.   SEM of Fracture Surface of Panel #19
            230X.

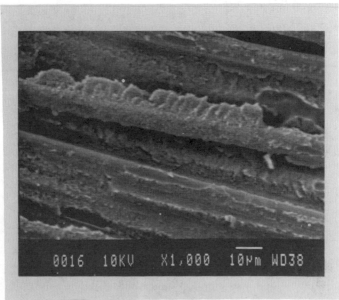

Figure 14.   SEM of Fracture Surface of Panel #19
            1000X.

PROCESSING, PROPERTIES AND APPLICATIONS
OF K-POLYMER COMPOSITE MATERIALS
BASED ON AVIMID® K-III PREPREGS

R. J. Boyce, T. P. Gannett, H. H. Gibbs
and A. R. Wedgewood
Textile Fibers Department
E. I. du Pont de Nemours & Co., Inc.
Wilmington, Delaware 19898

## Abstract

The processing characteristics, mechanical properties, environmental resistance, and potential applications of Avimid® K-III composite materials are described. Control of melt viscosity allows large, thick and variable thickness laminates of excellent quality (<0.5% voids) to be made by autoclave molding. Initial results suggest that there is a broad processing window. The excellent strength, toughness and environmental resistance of the K-III matrix resin, fully equivalent to those previously reported for K-II, are retained in fiber reinforced laminates. As a result, such laminate structures are highly damage resistant and tolerant. With an upper end-use temperature potentially as high as 232°C (450°F) Avimid® K-III composite materials are being evaluated for applications in the primary structures of fixed wing aircraft and missiles.

## 1. INTRODUCTION

To achieve their full potential in the aerospace industry there is a need to develop for advanced composites new matrix resins which significantly out-perform the 350°F epoxies[1]. Not only must such matrix resins be functional at higher temperatures, they must also show increased strain capability, improved toughness and good environmental resistance. To be immediately useful, laminate processing with these resins must be reliable and cost effective using currently available technology, i.e. vacuum bag, autoclave molding.

The desired matrix resin properties were largely achieved with the Avimid® K-II polyimide material[2]. With a resin fracture toughness of 1.7 kJ/m$^2$, a value an order of magnitude higher than that of 350°F epoxies, carbon fiber laminates of Avimid® K-II were both highly resistant to and tolerant of impact damage. These laminates showed good resistance when exposed to fluids which are important in aerospace applications. With a glass transition temperature (Tg) of 250°-275°C these laminates also showed excellent retention of physical properties to 232°C. This material's shortcoming lay in its difficult processing.

Although K-II resin is formulated as a linear polyimide, polymer viscosity increases rapidly during the high temperature portion of the autoclave molding cycle. Management of resin viscosity, especially in laminates with variable cross section, is problematic making it difficult to consistently prepare high quality laminates.

In Avimid® K-III the K-II resin system has been modified to control excessive viscosity build-up. As a result, it is possible to reliably prepare large, thick low void (<0.5%) laminates. In a single operation low void laminates from 8 to 48 plies thick (1.14 to 6.86mm) have been made. A wide range of processing conditions, where heat-up rates, consolidation temperature and hold times have varied, have been successfully used to prepare high quality laminates. Processing directly against a metal tool surface has also been demonstrated. The strength, stiffness and environmental resistance of K-II are fully retained in the K-III resin. Damage resistance and tolerance of Avimid® K-III

composite laminates are equivalent to Avimid® K-II[2]. With K-III the dry Tg is 250° ± 5°C and hot-wet property testing suggests that the water considered upper end use temperature should be greater than 204°C (400°F).

The purpose of this paper is to report on the processing characteristics, properties and potential applications of Avimid® K-III carbon fiber reinforced composites.

## 2. NEAT RESIN PROPERTIES

### 2.1 Polyimide Chemistry

Avimid® K polymers are amorphous linear condensation polyimides produced from monomer solutions by the reaction of an aromatic diethyl ester diacid with an aromatic diamine in N-methyl pyrrolidone (NMP) solvent. The reaction proceeds with loss of water, ethanol and solvent to form the imide ring. The K-III polyimide is structurally identical

$$\overset{O}{\underset{\overset{\|}{C}-OH}{\overset{\|}{C}}}-OCH_2CH_3 \quad + H_2N - + NMP$$

Ester Acid     Heat     Amine     Solvent

$$\downarrow - CH_3 CH_2 \overset{- NMP}{\underset{H_2O}{OH}}$$

$$\overset{O}{\underset{\overset{\|}{C}}{\overset{\|}{C}}} N -$$

Imide ring

to K-II polyimide except it is slightly modified to prevent excessive melt viscosity build up at elevated temperatures.

## 2.2 Neat Resin Properties

The modification which differentiates Avimid® K-III from K-II produces enhanced melt flow. We have assessed the relative melt viscosity of K-II and K-III neat resins prepared by completely polymerizing and devolatilizing the monomer solution. Measurements[3] for a range of identical resin cure conditions show K-III resin to have 15-20 times greater melt flow than K-II. It is this improvement in flow which produces the improved processing characteristics of Avimid® K-III.

The physical properties and environmental resistance of K-III were determined from test specimens taken from compression molded plaques. Table 1 summarizes K-III resin physical properties. For practical purposes these values are identical to those previously reported for K-II[2]. The Tg for K-III resin is independent of processing conditions, unlike K-II resin which has a process dependent Tg. Table 2 summarizes K-III resin environmental resistance. Once again these results are comparable to those obtained for K-II resin[4]. From these results we conclude that the physical and chemical properties of K-II and K-III resin derive from the basic chemical structure common to both.

## 3. VACUUM BAG/AUTOCLAVE MOLDING OF LAMINATES

### 3.1 Starting Material

Autoclave processible Avimid® K-III is provided as a tacky, drapable prepreg tape or fabric. The prepreg is a combination of K-III monomer solution, referred to as binder, and continuous reinforcing fiber. Laminates referred to in this paper were prepared from either "Magnamite" AS-4 or "Magnamite" IM-6 carbon fiber based prepreg tape. The tape used was 12 inches wide and, on a fully cured basis, contained 57 ±2 vol% reinforcing fiber. The volatile content of the tape, which included both condensation polymerization by-products and solvent was 15-17 wt %.

### 3.2 Layup

Laminates were fabricated by vacuum bag/autoclave molding using the following layup sequence:
- Vacuum bag
- Glass fabric bleeder/ breather plies
- Peel plies ("Bleederlease" E or "Emfab" TX-1040)
- Prepreg part
- Prepared Tool

This layup is an example of true one-sided breathing. The part, prepared by stacking individual prepreg plies in the appropriate fiber orientation and sequence, is placed directly against the tool. To fabricate a laminate with a high gloss blister free surface, which accurately replicates the tool, it is necessary to properly prepare the tool surface. This can be accomplished by applying three coats of release agents such as "Freekote" 44 or "Mono-coat" E100 with air drying (15 minutes) between coatings. The tool is then baked at the maximum process temperature to insure complete devolatilization of the release layer.

## 3.3 Cure cycle

Chart 1 presents two cure cycles which have been used to prepare high quality Avimid® K-III laminates. The cycle represented by the dashed line was one developed for Avimid® K-II (see Appendix A for complete details of the K-II cycle) except that application of consolidation pressure was delayed until 343°C. Late pressurization allows complete devolatilization prior to consolidation. Separation of these operations greatly simplifies the fabrication of low void laminates. For K-II a combination of holds and variable heat up rates were necessary to prepare high quality laminates, and the particular combination was dependent on laminate thickness and size. However, with this cycle excellent K-III laminates varying in thickness from 8 to at least 48 plies have been prepared in a single run.

For Avimid® K-III the cure cycle need not be so complicated. The solid line cure cycle in Chart[1] also produces excellent 8 to 48 ply laminates, but in a significantly shorter time. In general, adjusting heat-up rate achieves a shorter cycle time than using holds. However, holds are useful for two purposes. First to insure that the entire laminate is within the 343° - 360°C consolidation temperature range prior to pressure application, and secondly to allow sufficient time for consolidation. By slowing the heat-up rate to 0.5°C/min, 14" x 14" quasi-isotropic laminates up to 96 plies thick, with a void content below 0.5%, have been prepared.

## 3.4 Laminate Quality

Laminate quality, as defined in this paper, is assessed by determining void volume and distribution. Void volume can be determined by measuring laminate density and comparing it to the calculated theoretical density using the following expression:

% void volume $= (1 - \rho_{plaque}/\rho_{theo}) \times 100$

Microscopy and C-scanning can be used to determine void distribution and size.

Over 100 high quality laminates of Avimid® K-III ranging in dimension from 4" x 4" to 2' x 7' and in thickness from 8 to 96 plies have been fabricated by autoclave molding from prepreg tape. The C-scan histograms in Charts 2 and 3 and the micrographs in Figures 1 and 2 are characteristic of the high quality routinely achieved. The C-scans show uniformly low energy loss and the micrographs show that voids are widely distributed and small (<0.5 vol %). The laminates represented in these charts and figures, one 8 ply and the other 48 plies, were made in the same autoclave run, demonstrating an important processing capability for Avimid® K-III.

## 4. LAMINATE PROPERTIES

### 4.1 Laminate Physical Properties

The final test of laminate quality and utility is in the physical properties it possesses. The excellent properties of neat K-III resin are retained in the laminate allowing full realization of the reinforcing fiber's strength and stiffness.

Flexural strength and stiffness, and short beam shear values over a range of temperatures are given in Table 3. The high matrix resin Tg results in the retention of useful property levels at least up to 232°C. Results under hot-wet conditions are not complete at this time, but preliminary results indicate that they are retained at useful levels in excess of 204°C. The high toughness and elongation of K-III matrix resin also translates into useful laminate properties. Table 4 presents the interlaminar fracture toughness that was measured for several different laminates. The measured values of 1.3 - 1.8 KJ/M$^2$ are 5 to 10 times greater than that of untoughened 350°F epoxy. As will be shown, this results in superior damage resistance and tolerance for Avimid® K-III laminates. Table 5 contains additional test results for compressive, transverse tensile, shear and edge delamination properties.

### 4.2 Environmental Resistance

The excellent environmental resistance of the matrix resin is also retained in the laminates. Table 6 lists weight changes for laminate samples exposed to a variety of aerospace fluids. Laminates were virtually unaffected even in the more aggressive fluids like methylene chloride and alkaline cleaner. The weight change of samples under stress was equivalent to that found for unstressed samples. No evidence of stress cracking was found in either stressed or unstressed specimens. Table 7 further indicates that the flexural strength and stiffness of laminates are not effected by exposure.

### 4.3 Damage Resistance and Tolerance

As anticipated the high toughness and elongation exhibited by the K-III matrix resin results in laminates with a high degree of damage resistance and tolerance. This is demonstrated by the retention of a high level of compressive stress and strain after impact, measured for Avimid® K-III plaques made with either "Magnamite" AS4 or IM6 fiber. The results of these tests are presented in Charts 4 and 5. As can be seen, penetration occurs only at very high impact energies and that the loss of compressive properties is gradual over the entire range of impact

173

energies. Details of the impact test conditions are summarized in Appendix B.

Damage areas were determined by analysis of C-scans for plaques at all impact energies tested. Regardless of impact energy, the damage zone was 7 ±2cm$^2$ in area and circular in shape. This corresponds to a diameter about twice that of the impactor. The matrix resin toughness restricts the damage to the impact area and prevents the catastrophic loss of compressive properties.

## 5. APPLICATIONS

The unique combination of properties offered by Avimid® K-III composites should be useful in a broad range of aerospace applications. With at least a 204°C service temperature, applications are possible in certain engine parts and for some aerodynamically heated surfaces. Its toughness and high specific strength and stiffness makes Avimid® K-III a candidate for structural members like wingboxes and spars. The ability to lay up directly against a tool surface means that aerodynamic surfaces, like wing skins and nose cones, can be made directly.

As a demonstration we have made 2' x 7', 32 ply quasi-isotropic laminates. In one run the part was layed up directly against a stainless steel tool. This part was creased in spots due to thermal mismatch between the tool and laminate, causing localized compressive buckling. Nevertheless, the wrinkled areas showed only a 6-7 dB energy loss which was attributed to off axis reflection of the sound waves by the non planar laminate plies and not by voids. A second large laminate was prepared against a composite slip sheet. Lay up against the composite slip sheet greatly reduced the extent and severity of the creasing. Otherwise both laminates had near perfect density and C-scan energy loss of 2-3 dB. Microscopy showed the void level was well below 0.5 vol %, even in the creased areas. A third large laminate was made against the same composite slip sheet using an envelope vacuum bag to isolate the slip sheet from the metal tool. This laminate gave perfect replication of the tool surface with no apparent creasing. This demonstration illustrates that use of a high temperature composite tool should eliminate creasing in large structures.

The object of the above work was to address the problems associated with scale up to large surface areas. This, in combination with a demonstrated ability to simultaneously process variable thickness parts, should allow fabrication of large, complex, variable thickness structures such as wings skins and fuselages.

## 6. SUMMARY

Avimid® K-III prepreg and composite materials have been shown to possess the necessary processing and property characteristics to make them broadly useful in aerospace applications. Laminates made with Avimid® K-III display high specific strength and stiffness, which are retained at temperatures above 232°C. Preliminary hot-wet measurements indicate that useful property levels are retained above 204°C. Laminates were highly resistant to aerospace fluids even in a stressed state. The high neat resin fracture toughness translates into laminate structures having excellent interlaminar fracture toughness and high damage resistance and tolerance.

Of equal importance, Avimid® K-III laminates can be fabricated using a wide range of processing conditions. Thus, large area laminates with variable thickness can be made. Additionally, lay-up directly against a tool for the preparation of high gloss surfaces was easily accommodated.

## 7. APPENDICES

### 7.1 Appendix A

The Avimid® K-II processing cycle was as follows:

1. Apply 5" Hg vacuum
2. Heat to 100° at 1°C/min.
3. Hold 30 min. Apply full vacuum.
4. Heat to 177°C at 1°C/min.
5. Hold 60 min.
6. Apply 200 psi.
7. Heat to 280°C at 1°C/min.
8. Heat to 343°C at 0.3°C/min.
9. Heat to 360°C at 0.3°C/min.
10. Hold one hour.
11. Cool under pressure.

### 7.2 Appendix B

The conditions for the impact testing of Avimid® K-III laminates were:
- ORIENTATION: $(+45, 0, -45, 90)_{4S}$
- PLIES: 32
- SAMPLE DIMENSIONS: 4 IN X 6 IN X 0.18 IN
- VOLUME % FIBERS: 57 ± 2
- TUP DIAMETER: 1.58cm (UP TO 13,500 J/M) 1.24cm (18,000 J/M - THROUGH PENETRATION)

## 8. REFERENCES

1 B. A. Byres, NASA Contractor Report 159290, August, 1980.

2 H. H. GIBBS, "K-Polymer: A New Experimental Thermoplastic Matrix Resin for Advanced Structural Aerospace Composites", pgs. 1073-1084, 29th Annual SAMPE Symposium and Exhibition, Reno, Nevada, April 3-5, 1984.

3 Melt Flow was measured in a melt index apparatus after 5 min. at 380°C.

4 a) H. H. GIBBS, "K-Polymer Composite Materials - A New Approach to Damage Tolerant Aerospace Structures", pgs. 81-91, SAE Aerospace Congress, Long Beach, California, October 15-19, 1984;
b) H. H. GIBBS, "The Processing and Properties of Damage Tolerant composites Based on K-Polymer Materials, pgs. 971-993, Fifth Annual Conference on Composite Materials (ICCM-V), San Diego, California, July 29 - August 1, 1985.

## TABLE 1

### PROPERTIES OF NEAT "AVIMID" K-III

| PROPERTY | VALUE |
|---|---|
| TENSILE STRENGTH | 102 MPa |
| TENSILE MODULUS | 3764 MPa |
| ELONGATION | 14% |
| GIC - BLUNT NOTCH | 10 kJ/m$^2$ |
| SHARP NOTCH | 1.9 kJ/m$^2$ |
| SHEAR MODULUS | 1365 Mpa |
| POISSON'S RATIO | 0.365 |
| DENSITY | 1.31 g/cc |
| $T_g$ (TMA) | 251°C |

## TABLE 2

### ENVIRONMENTAL RESISTANCE OF NEAT "AVIMID" K-III

|  | TEMPERATURE, °C | TIME, HOURS | % WEIGHT GAIN |
|---|---|---|---|
| WATER | 71 | 336 | 2.2 |
| ALKALINE CLEANER[1] | 71 | 336 | 1.9 |
| HYDRAULIC FLUID [2] | 71 | 336 | 0.23 |
| METHYLENE CHLORIDE | 30 | 24 | 5.8 |

(1)  PACE B-82 Alkaline Cleaner - Diluted with 2 parts water to one part cleaner

(2)  Monsanto's Low Density Aviation Hydraulic Test Fluid

Chart 1. "AVIMID" K-III AUTOCLAVE MOLDING CYCLE

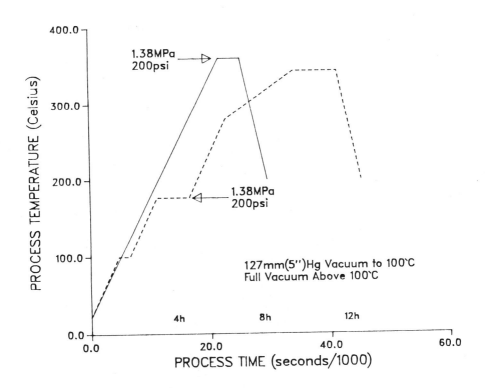

## TABLE 3

### MECHANICAL PROPERTIES OF "AVIMID" K-III COMPOSITES

REINFORCING FIBER:   "MAGNAMITE" AS-4
ORIENTATION:        QUASI-ISOTROPIC
VOLUME OF FIBERS:      55
THICKNESS:         90 - 100 MILS

| TEST | TEMPERATURE, °C | VALUE |
|------|-----------------|-------|
| FLEX STRENGTH | 52 | 793 MPa |
|  | 149 | 725 |
|  | 177 | 648 |
|  | 204 | 662 |
|  | 232 | 614 |
| FLEX MODULUS | 52 | 36 GPa |
|  | 149 | 32 |
|  | 177 | 34 |
|  | 204 | 28 |
|  | 232 | 29 |
| SHORT BEAM SHEAR STRENGTH | 52 | 63 MPa |
|  | 149 | 53 |
|  | 177 | 48 |
|  | 204 | 45 |
|  | 232 | 36 |

## TABLE 4

### INTERLAMINAR FRACTURE TOUGHNESS
(Mode I on 32 Ply 0° Laminate)

| REINFORCING FIBER | CONSOLIDATION TEMP., °C | FINAL MOLDING CONDITIONS °C (HOURS) | FRACTURE TOUGHNESS, $KJ/m^2$ |
|-------------------|-------------------------|-------------------------------------|------------------------------|
| AS-4 | 177 | 343 (2) | 1.8 |
| IM-6 | 177 | 343 (2) | 1.7 |
| IM-6 | 310 | 343 (2) | 1.3 |
| IM-6 | 310 | 360 (1) | 1.5 |

## TABLE 5

### MISCELLANEOUS MECHANICAL PROPERTIES OF "AVIMID" K-III LAMINATES

| TEST | REINFORCING FIBER | PLIES | ORIENTATION | CONSOLIDATION TEMP., °C. | FINAL MOLDING CONDITIONS, °C. (HOURS) | VALUE |
|---|---|---|---|---|---|---|
| INCIPIENT EDGE DELAMINATION STRESS | AS-4 | 10 | $|(\pm25)_2 90|_s$ | 177 | 343 (2) | 224 MPa |
| | IM-6 | " | " | 310 | 360 (1) | 262 MPa |
| IN-PLANE SHEAR STRENGTH MODULUS | AS-4 | 8 | ±45 | 177 | 343 (2) | 119 MPa 4.1 GPa |
| TRANSVERSE TENSILE STRENGTH MODULUS POISSON'S RATIO | AS-4 | 8 | 0 | 177 | 343 (2) | 37 - 48 MPa 8.3 - 10.3 GPa 0.020 - 0.0027 |
| IITRI COMPRESSIVE STRENGTH MODULUS | AS-4 | 32 | 0 | 177 | 343 (2) | 1.0 GPa 110 GPa |
| IITRI COMPRESSIVE STRENGTH MODULUS | IM-6 | 32 | 0 | 177 | 343 (2) | 1.0 GPa 103 GPa |

## TABLE 6

## ENVIRONMENTAL RESISTANCE OF
## "AVIMID" K-III/IM-6 LAMINATES

| FLUID | ORIENTATION | PLIES | % STRAIN[5] | DAYS | °C | % WEIGHT GAIN |
|---|---|---|---|---|---|---|
| Hydraulic A[1] | ±45 | 8 | 0.6 | 14 | 82 | 0.053 |
| " | " | " | 0 | " | " | 0.076 |
| JP-4 | " | " | 0.6 | " | " | 0.14 |
| " | " | " | 0 | " | " | 0.10 |
| Hydraulic A[1] | 0° | 16 | 0 | " | " | -0.004 |
| JP-4 | " | " | " | " | " | -0.003 |
| Hydraulic B[2] | " | " | " | " | 23 | 0.14 |
| MEK | " | " | " | " | " | 0.070 |
| Deicing[3] | " | " | " | " | " | 0.16 |
| $CH_2Cl_2$ | " | " | " | 1 | " | 0.21 |
| $H_2O$ | " | " | " | 14 | 82 | 0.66 |
| Alk. Cleaner[4] | " | " | " | " | 23 | 0.33 |

(1)  Mil Spec 5606 Hydraulic Fluid

(2)  Monsanto's Low Density Aviation Hydraulic Test Fluid

(3)  Dow Deicing Fluid 146AR

(4)  PACE B-82 Alkaline Cleaner - Diluted with 2 parts water to one part cleaner

(5)  No signs of stress cracking observed on stressed laminates

## TABLE 7

## ENVIVONMENTAL TESTING OF "AVIMID" K-III/IM-6 LAMINATES*
### (12 Plies, 0°, 70 Mils Thick, 58 Volume % Fibers)

| Fluid | Exp. Temp., °C. | Exp. Time, Days | % Strain | % Wt. Gain | Flex Strength MPa | Flex Modulus GPa |
|---|---|---|---|---|---|---|
| None - Control | | | | | 1,340 | 132 |
| Mil Spec 5606 Hydraulic Fluid | 82 | 14 | 0.6 | 0.008 | 1,430 | 143 |
| | 82 | 14 | 0 | -0.004 | 1,400 | 139 |
| JP-4 | 82 | 14 | 0.6 | 0.041 | 1,370 | 137 |
| | 82 | 14 | 0 | -0.004 | 1,370 | 139 |
| Monsanto Low Density Hydraulic Fluid | 23 | 14 | 0 | 0.089 | 1,350 | 135 |
| $CH_2Cl_2$ | 23 | 1 | 0 | 0.43 | | |
| | | 14 | | 2.6 | 1,340 | 138 |
| MEK | 23 | 14 | 0 | 0.10 | 1,300 | 134 |
| Dow Deicing Fluid 146 AR | 23 | 14 | 0 | 0.091 | 1,360 | 138 |
| PACE B-82 Alkaline Cleaner | 23 | 14 | 0 | 0.32 | 1,360 | 136 |

* Consolidated at 343°C., molded 1 hour at 360°C.

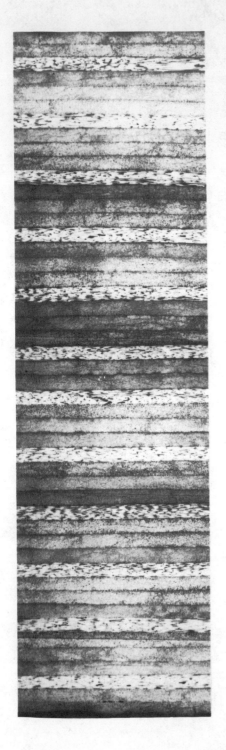

Figure 1 (left)  Photomicrograph of a polished cross-section of a 48 ply Quasi-isotropic Laminate (50X Magnification).

Figure 2  Photomicrograph of a polished cross-section of an 8 ply Uni-directional Laminate (50X Magnification).

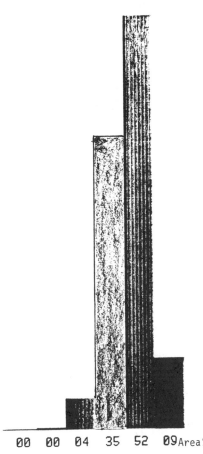

```
00    00    04    35    52    09 Area%

06    05    04    03    02    01   dB
```

Chart 2.   C-SCAN HISTOGRAM FOR
           THE 48 PLY QUASI-
           ISOTROPIC LAMINATE

```
00    01    06    62    30  Area%

05    04    03    02    01   dB
```

Chart 3.   C-SCAN HISTOGRAM FOR
           THE 8 PLY UNI-
           DIRECTIONAL LAMINATE

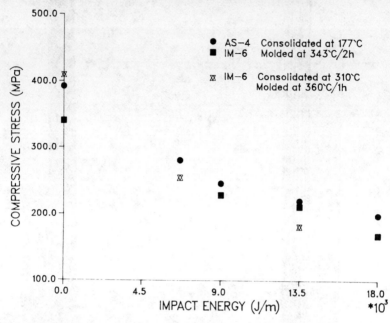

Chart 5.  EFFECT OF IMPACT ENERGY ON COMPRESSIVE STRESS
OF "AVIMID" K-III QUASI-ISOTROPIC LAMINATES

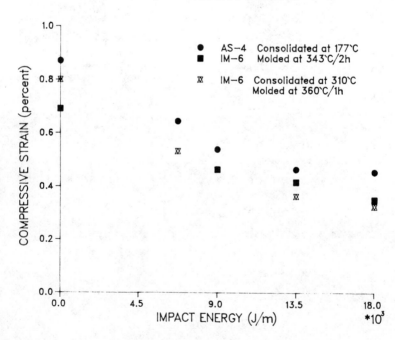

Chart 4.  EFFECT OF IMPACT ENERGY ON THE COMPRESSIVE STRAIN
OF "AVIMID" K-III QUASI-ISOTROPIC LAMINATES

STATUS OF FASTENERS OF VECTRA LIQUID CRYSTAL POLYMER
FOR CARBON FIBER REINFORCED PANELS

Frank C. Jaarsma
Celanese Corporation
Vectra Business Unit
86 Morris Avenue, Summit, NJ   07901

## Abstract

Celanese Vectra liquid crystal polymers are a new class of materials composed of stiff, linear molecules. Orientation of these molecules results in a high level of anisotropic mechanical properties.

In the aerospace industry, rivets of Vectra offer carbon fiber composite panel users a new opportunity. The rivets can fasten these panels without the expense and weight of titanium, or the corrosion problems associated with aluminum. A further advantage is that the head forming technique, which uses ultrasonic energy, requires only light pressure. This greatly reduces the risk of damage to the panels.

Rivets of Vectra have a specific gravity of only 1.4 to 1.8, depending on the particular grade of material used. This results in mechanical properties which compare favorably with metals on a strength/weight basis. Other advantages include that it is unaffected by most corrosive environments such as gasoline, hydraulic fluid, cleaning solutions, chlorinated solvents and paint strippers. Furthermore, Vectra is not electrically conductive. Galvanic corrosion will therefore not occur when joining dissimilar materials.

## 1. Introduction

This report summarizes the status of a program to develop a new type of polymeric rivet. The report is intended to inform persons with an interest in fasteners of this type of the design possibilities becoming available. This will allow designers to consider the incorporation of these rivet concepts into their designs at an early stage. Their feedback will, in turn, assist in directing further research. The program is on-going with additional data being continuously generated. All data presented is therefore subject to change and should not be used for design purposes until verified by the end user. The polymer being used is commercially available but the fabricated rivets presently remain developmental. Prototype quantities of the rivets are available.

Vectra thermotropic liquid crystal polymers (LCP's) are a new family of materials based on aromatic polyester chemistry. Although these are classified as polyesters, the structure and morphology are very different from other polyester polymers. This results in dramatically different properties and processing characteristics.

The polymer has a wholly aromatic structure which results in a highly oriented, fibrous molecular morphology. During flow, the molecules align to a very high degree. It is this highly ordered structure, in the molten state, which gives rise to the generic term "liquid crystal polymer." Although these molecular chains are very highly aligned, they do not form a three dimensional crystalline structure as do conventional crystalline polymers. The structure may be visualized as analogous to a stack of pencils which allows the individual units to slide easily one over the other while still retaining the same ordered orientation. This orientation results in a high level of anisotropic mechanical properties. In a rivet, where longitudinal tensile strength and transverse shear strength are of primary interest, this is a considerable advantage. A detailed description of the structure of liquid crystal polymers has been made by Calundann and Jaffe, see reference 1.

## Applications

Vectra liquid crystal polymer offers an attractive alternative to metal for fasteners in certain applications. The polymer is strong, light weight, corrosion resistant and easily formed.

In the aerospace industry, rivets of Vectra offer carbon fiber-composite panel users a particularly interesting opportunity. The rivets have been shown to securely fasten these panels without the cost and weight of titanium, or the corrosion problems associated with aluminum. A further advantage is that the head forming technique, which uses ultrasonic energy, requires only light pressure. This greatly reduces the risk of damage to the panels.

## Material Properties

Although rivets of Vectra are not as strong, on a size basis, as aluminum or titanium rivets, they do offer other significant advantages. Vectra is about 30% lower density than aluminum and only about one third the density of titanium alloy. When the value for absolute shear strength is divided by specific gravity the resulting calculated specific shear strength indicates Vectra to be in the same range as aluminum and somewhat superior to titanium alloy. (Table 1) Since titanium is generally used in carbon fiber panels for its corrosion resistance, rather than its strength, a rivet of Vectra could be a more weight effective method to join the panels.

Both Vectra B130 and B150 are non-electrically conductive. Galvanic corrosion will therefore not occur when joining dissimilar materials. The volume resistivity for these materials exceeds $1 \times 10^{15}$ ohm-cm as measured by ASTM D257. (Other grades which are electrically conductive are available if desired.)

When materials are considered for interior uses in passenger aircraft, flammability, smoke density and toxic products of combustion are of prime concern. Vectra resins are inherently resistant to burning and require no additives to reduce flammability. They easily pass the flammability requirements for Underwriters Laboratories 94 V-0, the highest rating available. When a 1/16 thick panel of the polymer, without glass fibers, was evaluated for smoke density in the NBS Smoke Density Chamber, a Ds at 4.0 minutes value of 5 for flaming and 0 for the smoldering condition was obtained. (Table 2) A Dragen Tube was also connected to the NBS chamber to test

for toxic gases. The results
indicate only low levels of carbon
dioxide, carbon monoxide and some
hydrocarbons. (Table 3) All values
for smoke density and products of
combustion are among the lowest of
any polymeric material available
today.

The unusual resistance of Vectra
resins to solvents and chemicals is,
in part, a result of the dense,
tight molecular packing which is
difficult for solvents to penetrate.
This polymer is highly resistant to
swelling, softening and loss of
properties when exposed to organic
solvents, even when the material is
subjected to mechanical stress
and/or elevated temperature.

Test bars of Vectra have been
stressed to 90% of breaking load and
subjected to a variety of solvents
including acetone, methanol,
methylene chloride (paint stripper),
Skydrol, JP4 fuel, and gasoline with
no evidence of stress cracking or
other detrimental effects.

If the final assembly requires a
coating, good adhesion has been
obtained with both epoxy and
urethane aircraft paint systems.

## Mechanical Properties

Developmental evaluations have been
directed towards three popular types
of rivets: 1/4 inch shank diameter
with a flush head, 1/4 inch shank
diameter with a protruding head and
a 3/16 inch shank diameter with a
flush head (Figure 1). Vectra B130
has been found to offer an
attractive combination of properties
and has been the most extensively
tested. A summary of properties is
shown in Table 4. Data is also
available on other grades for
particular requirements.

Because rivets are primarily used in
applications where shear strength,
rather than tensile strength is
required, most of the laboratory
efforts to date have been directed
towards maximizing this value.
Research in forming technology is
currently underway which promises to
improve the formed head strength of
the rivets.

## Rivet Design

Prototype rivets are presently being
made by injection molding with one
head molded on. Virtually any head
configuration is possible on the
molded end. The tooling presently
available produces head designs
shown in Figure 1. As interest for
specific applications warrants,
additional styles can be developed.

To install a rivet, the shank is cut to the desired length, machined, inserted through the panels and formed.

Most interest has been expressed in forming flush heads, as opposed to protruding, and therefore this technology development has been emphasized. When a flush head is to be formed, the shank is cut to length, and a recess is machined on the end to be formed. A typical design found to work well is shown in Figure 2.

## Forming The Head

As noted earlier, Vectra develops high mechanical properties as a result of favorable molecular orientation. Careful control of the molding process can further enhance this orientation in the rivets. For optimum strength, it is imperative that this orientation not be lost during the forming process.

Ultrasonic energy has been found to be the preferred method to achieve maximum formed head strength. This equipment is readily available from a number of domestic sources. While a 20 KHz frequency has been used, 40 KHz will produce stronger heads. Maximum strengths have been obtained using a frequency of 40 khz, an amplitude of 2-3 mils and light to moderate pressure. Typical times are: forming, 0.5 to 1.0 seconds and cooling hold, 0.8 to 2.0 seconds. The total time to form the rivet head is therefore less than 3 seconds. The steps for this process are diagrammed in figure 3.

If desired, careful control of the ultrasonic energy can also be used to bulge the rivet shank slightly. This will fill any clearance resulting from a size difference between the hole diameter and the rivet shank diameter. Although this technique has not been fully defined, no loss of shear strength Has been observed when this is done.

## Summary

Rivets made of liquid crystal polymer hold the potential to fasten carbon fiber reinforced panels without the cost and weight of titanium or the galvanic corrosion of aluminum. They have a comparable specific strength to metal fasteners and are easily formed with low stress using ultrasonic energy. The polymer is resistant to all chemicals and solvents commonly found in an aircraft environment.

## BIOGRAPHY

The author is a graduate of the University of Lowell where he received a M.S. degree in plastics engineering in 1970. He has previously held positions at Stevens Molded Products Co. and Servtex Corporation where he was involved in the design of industrial devices using high-performance engineering thermoplastics.

Presently, the author is a staff engineer for Celanese Corporation where he joined the technical organization in 1977. Since that time he has worked extensively with acetal, pbt, nylon and other high-performance thermoplastic materials with the industrial, electronic and aerospace industries. In 1981 he became involved with engineering application development of Vectra liquid crystal polymer (LCP).

## References

1.  Anisotropic Polymers, Their Synthesis and Properties Gordon W. Calundann and Michael Jaffe chapter 7, pg.247-291 Proceedings of the Robert A. Welch Conferences on Polymer Research xxvi Synthetic Polymers, November 15-17, 1982, Houston, TX.

2.  A Processing Guide to Liquid Crystal Polymers Matthew H. Naitove Plastics Technology, April, 1985

3.  Liquid Crystal Polymers: Now They Are Melt-Processable A. Stuart Wood Modern Plastics, April 1985, pg 78.

4.  New Liquid cRYSTAL polymer for High Performance Electronic Components John C. Chen, Frank C. Jaarsma Eighteenth Annual Connectors and Interconnection Technology Symposium Proceedings, November 18 - 20, 1985 Electronic Connector Study Group Proceedings of Symposium, pg. 298.

Vectra® is a trademark of the Celanese Corporation.

Table 1


Strength and Specific Gravity

of Selected Vectra Compounds


| Vectra Grade | Specific Gravity | Shear Strength ASTM B 565-76 GPA (KPSI) | Specific Shear Strength (Shear Str/sp. gr.) GPA (KPSI) |
|---|---|---|---|
| B130 | 1.6 | 0.13 (19.2) | 0.08 (12.0) |
| B150 | 1.8 | 0.14 (21.0) | 0.08 (11.7) |
| Metals | | | |
| Aluminum | 2.7 | 0.19 - 0.31 (28.0 - 45.0) | 0.07 - 0.12 (10.4 - 16.7) |
| Titanium Alloy | 6.2 | 0.30 - 0.37 (44.0 - 54.0) | 0.05 - 0.06 (7.1 - 8.7) |

## Table 2

### SMOKE DENSITY*
### VECTRA® A950

|  | 1/16 Inch (1.6mm) | | 1/8 Inch (3.2mm) | |
|---|---|---|---|---|
|  | Flaming | Smoldering | Flaming | Smoldering |
| Ds @1.5 min | 0 | 0 | 0 | 0 |
| Ds @4.0 min | 7 | 0 | 3 | 0 |
| Time To 90% Dm, (min) | 16.5 | 20.0 | 17.0 | 18.7 |
| Ds | 95 | 2 | 94 | 1 |
| Dmc | 94 | 2 | 92 | 0 |

### VECTRA® B950

|  | 1/16 Inch (1.6mm) | | 1/8 Inch (3.2mm) | |
|---|---|---|---|---|
|  | Flaming | Smoldering | Flaming | Smoldering |
| Ds @ 1.5 min | 0 | 0 | 0 | 0 |
| Ds @4.0 min | 5 | 0 | 1 | 0 |
| Time To 90% Dm, (min) | 14.5 | 20.0 | 17.5 | 19.0 |
| Dm | 70 | 2 | 80 | 1 |
| Dmc | 69 | 2 | 70 | 0 |

*NBS Smoke Density Chamber
ASTM E-662,NFP 258
Above are typical values, not to be used for specification purposes.

## Table 3
## VECTRA® PRODUCTS OF COMBUSTION (PPM)

| GAS | A950 SMOLDERING | A950 FLAMING | B950 SMOLDERING | B950 FLAMING |
|---|---|---|---|---|
| CHLORINE | 0 | 0 | 0 | 0 |
| PHOSGENE | 0 | 0 | 0 | 0 |
| HYDROGEN CHLORIDE | 0 | 0 | 0 | 0 |
| HYDROGEN FLUORIDE | 0 | 0 | 0 | 0 |
| FORMALDEHYDE | 0 | <2 | 0 | <2 |
| AMMONIA | 0 | 0 | 0 | 0 |
| CARBON MONOXIDE | <10 | 320 | <10 | 300 |
| CARBON DIOXIDE | 600 | 8000 | 6000 | 7000 |
| NITROGEN OXIDES | <2 | 5 | <2 | 12 |
| HYDROGEN CYANIDE | 0 | <2 | 0 | <2 |
| SULFUR DIOXIDE | 0 | 0 | 0 | 0 |
| HYDROCARBONS, AS N-OCTANE | 0 | 250 | 0 | 300 |

THE ABOVE DATA WERE GENERATED ON 3 INCH X 3 INCH X 1/8 INCH (76.2mm X 76.2mm X 3.2mm) THICK PLAQUES. THE GASES WERE MEASURED USING DRAGEN TUBES ATTACHED TO A SAMPLING POST AND CIRCULATING SYSTEM ADDED TO THE NBS SMOKE DENSITY CHAMBER.

## Table 4

Typical Properties Of Rivets Of Vectra B130

| | 1/4 inch Flush | 1/4 inch Protruding | 3/16 inch Flush |
|---|---|---|---|
| Specific Gravity | 1.6 | 1.6 | 1.6 |
| Shear Strength ASTM B565-76 GPa (KPSI) | 0.13 (19.2) | 0.13 (19.2) | 0.13 (19.2) |
| Molded Head Load Bearing Capacity Newton (LBS) | 1960 (400) | 3110 (700) | 1020 (230) |
| Formed Head Load Bearing Capacity Newton (LBS) | 890* (200*) | ND** | 490 - 580 (110 - 130) |

* -- Estimated
** - Not Determined

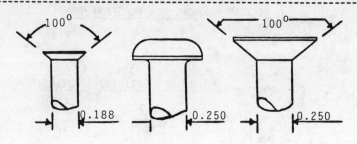

Figure 1
Available Rivet Configurations

Figure 2
Rivet For Forming Flush Head

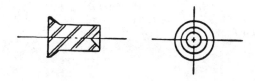

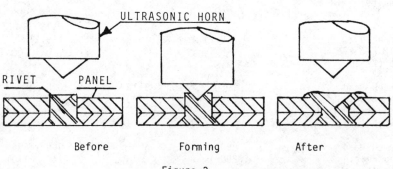

Before          Forming          After

Figure 3
Forming The Rivet Head

194

AUTOWEAVE™ – A UNIQUE AUTOMATED 3D WEAVING TECHNOLOGY
Paul G. Rolincik, Jr.*
Avco Specialty Materials
Subsidiary of Textron Inc.
Lowell, Massachusetts 01851

## Abstract

The AUTOWEAVE™ technology transfer of three-directional (3D) automated weaving from France to the United States is described. This unique technology was developed by Brochier S.A. (BSA) starting in 1972 and was financed by the Commissariat a l'Energie Atomique (CEA). The process was licensed from BSA/CEA in 1983 for the exclusive use of Avco/Textron in the United States. In 1986 the AUTOWEAVE™ equipment designated as BR900 and BR2000 were installed for operation in the Avco Specialty Materials/Textron (ASMT) facility in Lowell, Massachusetts. The BR900 system has a 3D preform fabrication capability up to 900mm (36 inches) diameter and the BR2000 system has a capability up to 2100mm (84 inches) diameter.

The AUTOWEAVE™ technique, which uses low cost foam mandrels for fabricating 3D preforms, requires minimal manual labor. The radial reinforcement is a unique screw-like rod, which is produced as continuous stock, and is simultaneously cut and inserted into the foam mandrel at a rate of one per second. The axial and circumferential reinforcements are placed in the radial corridors at a rate of 381mm per second (15 inches per second) and one revolution per second, respectively. Various types of contoured near net-shape 3D preforms have been produced and reproduced to date.

The AUTOWEAVE™ process has shown to provide a 1/3 reduction in preform production cost and a 1/2 reduction in preform production time compared with state of the art 3D fabrication techniques.

## 1. INTRODUCTION

The transfer of automated 3D weaving preform technology into the United States resulted from the

---

recognition that materials technology is rapidly growing outside of the United States. Studies by Battelle Memorial Institute indicated that in the 1960's more than 75% of the world's new materials technology was generated in this country. Today, that figure has dropped to 50% and it is expected to drop to 35% by 1995. In 1980-81 the Air Force Wright Aeronautical Laboratory (AFWAL) performed a study of 3D preform fabrication capability in the United States and Europe. It was concluded that France and specifically Brochier S.A. (BSA) had a demonstrated automated 3D fabrication technology base. In 1982, in response to an AFWAL Request for Proposal, Avco teamed with BSA of Lyon, the leading European 3D materials developer, to compete against a Hercules/Aerospatiale team. The Avco team won the competition and was awarded an AFWAL/-Materials Laboratory Manufacturing Technology contract (F33615-82-C-5032) for Automated 3D Preform Fabrication. The AFWAL/ML contract percipitated the technology transfer into the United States of the 3D automated preform fabrication process called AUTOWEAVE$^{TM}$. Avco, BSA and CEA signed an exclusive license agreement in 1983 for Avco to use the technology with the right to sublicense it in the United States. In 1986, two AUTO-WEAVE$^{TM}$ systems for fabricating

3D contoured, cylindrical and conical near-net-shape preforms were installed in the Avco Specialty Materials/Textron (ASMT) facility. The BR900 system, which was financed by Avco/Textron, was installed in ASMT in February 1986. It has the capability to fabricate near-net-shape contoured preforms having diameters up to 900mm (36 inches). The BR2000 system, which was financed by AFWAL/ML, was installed in ASMT in September 1986. It has the capability to fabricate near-net-shape contoured preforms having diameters up to 2100mm (84 inches). This paper describes the AUTO-WEAVE$^{TM}$ process and various rocket nozzle component applications. With the transfer of this automated 3D preform technology into the United States, 3D composites are becoming a viable material for applications to advanced propulsion systems and aerospace structures.

2. PREFORM DEFINITION
The preform refers to the skeleton 3D reinforced composite less the matrix. A matrix is used to fill the voids and hold the composite together, for instance using either carbon deposits or cured resins. The AUTOWEAVE$^{TM}$ preform construction is shown in a simple cylindrical configuration on Figure 1. The foundation for building the preform is a phenolic foam mandrel which is machined to conform to the inside diametrical

shape of the final product. Screw-like round radial rods, composed of cured reinforced phenolic, are implanted into the foam mandrel forming distinct helical tapered corridors. Axial and circumferential yarns, either dry or prepreg, are then placed precisely into these corridors to form the 3D preform. It must be noted that the AUTOWEAVE$^{TM}$ process was the first 3D technique to use screw-like round radial rods which improves the fiber/matrix bond in a 3D composite. On Figure 2 is shown a schematic of the AUTO-WEAVE$^{TM}$ 3D material which shows distinctly the screw-like round radial rods.

A computer program is used to prepare a radial rod insertion plan and a weave plan. The program defines the foam mandrel shape, raw materials required, process controls, and characteristics of the woven preform (Figure 3). The preform characteristics include the radial, axial and circumferential yarn spacings and the number of helical spirals, radial rods and axial corridors. These parameters are used to calculate preform density and fiber percentages in the 3 directions as well as the total fiber volume.

From the computer program output is developed a radial rod insertion plan as shown on Figure 4. The plan provides the number of corridors, rod diameter(s), and spacing for the axial and circumferential yarns. The length of the rods is specified by the required preform thickness which can change along the length of the preform. A typical weave plan is shown on Figure 5 which provides the number and type of axial and circumferential yarns to be placed in the corridors. The number of layers (axial and circ yarns) are specified by the required wall thickness.

3. FABRICATION PROCEDURE

The AUTOWEAVE$^{TM}$ fabrication procedure consists of four operations. Firstly, a machined foam mandrel provides the inside diametrical configuration; secondly, the continuous pre-stiffened radial rod material is provided on a large diameter spool; thirdly, is the automatic cutting and simultaneous implantation of radial rods into the foam mandrel; and fourthly, is the automatic weaving/winding of axial/-circumferential yarns in the corridors formed by the radial rods.

The foam mandrel is an expendable part of the tooling, whereas the metal end-plates and shaft are reusable (see Figure 6). The radial rod material is produced on a continuous basis using a separate automated process as shown on Figure 7. This provides twisted yarns plied together to form a rod structure of a desired diameter. The cured pre-stiffened continuous rod material, which is simultaneously wound on a large diameter spool, is directly placed on the insertion system (Figure 8). The rod material is fed continuously from the spool through the insertion head. The insertion head accurately cuts the radial rods to the desired length, prepares the foam mandrel with pre-holes and implants the radial rods in the pre-holes. These three functions are preformed in a continuous automated motion at a nominal rate of one rod per second. As shown on Figure 9, the insertion system has a continuous helical screw shaft motion to provide the required circumferential spacing and the stepping motor indexing to provide the axial spacing. The vertical position of the insertion head is controlled by a template which follows the contour of the foam mandrel. In addition, the insertion head cuts the radial rods to lengths required for the preform wall thickness.

Following completion of the radial rod insertion, the foam mandrel/radial rod assembly is relocated on the automated axial weaving/circumferential winding system shown on Figure 10. The axial yarns are fed into the axial corridors by a micro-processor controlled shuttle as shown on Figure 11. Each pass of the axial yarn is followed by a stepped rotation of the preform to align the next axial corridor with the shuttle. The shuttle loops the axial yarn around the crown at each end of the mandrel before passing through the next corridor.

The shuttle motion is controlled by a template which follows the shape of the contour of the foam mandrel. The axial weaving operation is performed as a continuous automated motion providing a nominal axial yarn placement of 381mm per second (15 inches per second). The circumferential yarns are tensioned and fed into the helical corridors by a micro-processor controlled shuttle as shown on Figure 12. The circumferential yarns anchor the previously placed axial yarns to the prescribed preform shape. The winding operation is performed as a continuous automated motion providing a nominal winding speed of one revolution per second. The alternating axial yarn and circumferential yarn placement produces a layer, which is repeated to build up the preform wall thickness. Either dry or prepreg yarns are used for the axial and circumferential reinforcements as shown on Figure 13.

The characteristics of a typical woven preform is presented on Figure 14. This shows the overall distribution of layer thicknesses, range of fiber distributions in 3 directions, and the packing efficiency or fiber volume for various stations along the preform length.

4.   AUTOWEAVE[TM] EQUIPMENT

The equipment installed at ASMT in Lowell, MA includes the Avco BR900 and Air Force BR2000 automated 3D preform weaving systems. These systems are supported by two radial rod manufacturing production lines. The rod line shown on Figure 15 includes the assembly apparatus in the foreground and the continuous curing oven in the background where the cured rod material is wound on a large diameter spool. The quality of the rod material is monitored through the various operations and the rod diameter is inspected on an audit basis with the data recorded in a preform fabrication log book.

The BR900 system was installed at ASMT in February 1986. On Figure 16 is shown the BR900 insertion system which includes the micro-processor in the foreground and the inserter in the background with the rod material spool attached to the rear of the machine. As shown, the inserter can place long radial rods precisely and uniformly into the foam mandrel. Following radial rod insertion, the mandrel is installed on the winder/weaver as shown on Figure 17. In the foreground is the micro-processor and the axial weaving shuttle for placing a yarn in a corridor is shown on the left of the preform. Note the end plate crown pins around which the axial yarn is looped prior to entering the next corridor. The axial yarn, either dry or prepreg, is fed from the top of the machine whereas the circumferential yarn, either dry of prepreg, is fed from the rear of the machine.

The BR2000 system was installed at ASMT in September 1986. For the BR2000 system, both the insertion and winding/weaving operations are performed on the same equipment. The sketch on Figure 18 illustrates the various functions performed on right and left sides of the equipment which are controlled by a mobile micro-processor control panel. Note that the radial rod insertion operation is performed horizontally whereas on the BR900 it is performed vertically. On Figure 19 is shown the BR2000 system with its electronic and computer control cabinets in the background and the moveable rod insertion and winding table on the left side of the machine. The mobile micro-processor control panel is partially hidden in the upper right hand side.

The chart on Figure 20 compares the capabilities of the BR900 and BR2000 systems. The primary differences are the number of machines utilized for each system, the maximum diameter capability and the conical half angle capability.

5. PREFORM CONFIGURATIONS

To date, over 100 3D carbon preforms have been fabricated using the AUTOWEAVE$^{TM}$ process. The 3D preforms have been produced using carbon yarn reinforcement because of their primary application to rocket motors, specifically integral throat entrances (ITE's), and exit cones. The configurations fabricated have included cylinders for ITE's, contoured near-net-shape ITE's, and contoured exit cones. There has not been any rejects of 3D preforms produced either in France or in the United States. Nearly all the preforms have been densified into carbon-carbon composites using a variety of densification processes. Nine of these carbon-carbon composites have been rocket motor tested as ITE's and exit cones in several different industry and government test facilities.

Typical 3D carbon preforms fabricated using the AUTOWEAVE$^{TM}$ equipment are as follows: long cylinder/small diameter (Figure 21); short cylinders/mid-size diameter (Figure 22); large diameter cylinder (Figure 23); near-net-shape contoured preform (Figure 24) for mid-sized ITE application (Figure 25); back-to-back contoured ITE's (Figures 26 and 27); and small, medium and large sized exit cones (Figures 28-30).

A hybrid cylinder was also fabricated using different fibers (Figure 31). The fibers from ID to OD were carbon, quartz, Kevlar (duPont), Nextel (3M alumina boro-silicate), Nicalon (Nippon silicon carbide) and glass. The radial rods were carbon reinforced. This hybrid cylinder demonstrated the AUTOWEAVE$^{TM}$ equipment capability to handle various types of fibers from the very stiff to the more flexible by adjusting the tension devices.

The potential capability for the AUTOWEAVE$^{TM}$ process includes: hybrids, 4D and 5D reinforcements, integral flanges, elliptic cross-sections and asymmetric cross-sections. This potential capability including the use of either dry or prepreg yarns provides AUTOWEAVE$^{TM}$ with the flexibility to produce 3D preforms for various future aerospace structure applications.

6. SUMMARY

AUTOWEAVE$^{TM}$ is firmly established in the United States. Under the exclusive license agreement from BSA/CEA and with the BR900 and BR2000 systems installed at ASMT in Lowell, MA, automated 3D preform production has begun in the U.S. The AUTOWEAVE$^{TM}$ process has been demonstrated for the fabrication of various contoured near net-shape preforms for rocket motor ITE's and exit cones. Successful rocket motor test firings of AUTOWEAVE$^{TM}$ carbon-carbon components has shown the viability of these 3D composites.

The production of 3D preforms has shown a 1/3 reduction in preform cost compared with current state-of-the-art 3D fabrication techniques. In addition, the automation has nearly eliminated the manual labor associated with state of the art 3D fabrication techniques and has reduced preform fabrication time by 50%.

The reproducibility of the AUTO-WEAVE[TM] 3D preforms has been proven for numerous identical preforms. This is certainly expected considering the minimal manual labor involved and the computer controlled radial rod insertion and winding/weaving operations. It should also be noted that costs associated with rework and rejection are minimal which has an appreciable impact on reducing quality control costs. The important factor is that the AUTO-WEAVE[TM] equipment has machine parts designed for manufacturability. This provides the basis for the repetitive AUTOWEAVE[TM] operations to be performed consistently.

*Paul Rolincik is Program Director of AUTOWEAVE[TM] Applications at Avco Specialty Materials/Textron. Paper prepared for 32nd International SAMPE Symposium/Exhibition, 6-9 April 1987, Anaheim, CA

Acknowledgements

The author wishes to acknowledge the AFWAL/ML Project Engineer, Mr. Walter Gloor, and the support of the Brochier S.A. team under the technical guidance of Mr. Bruno Bompard, BSA Plant Manager.

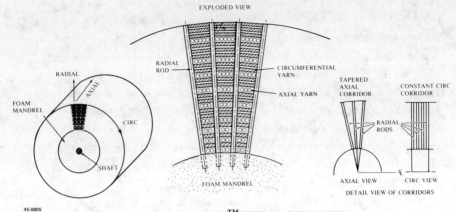

45-0805

Figure 1. AVCO AUTOWEAVE™ PREFORM CONSTRUCTION

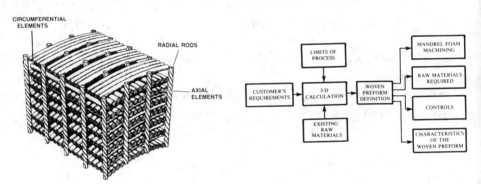

Figure 2. SCHEMATIC OF AUTOWEAVE™ 3D MATERIAL

Figure 3. SCHEMATIC OF 3D PREFORM CALCULATIONS

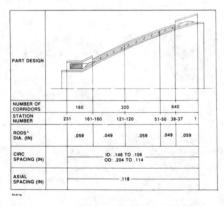

| PART DESIGN | | | | | | |
|---|---|---|---|---|---|---|
| NUMBER OF CORRIDORS | | 160 | 320 | | 640 | |
| STATION NUMBER | 231 | 161-160 | 121-120 | 51-50 | 38-37 | 1 |
| RODS* DIA. (IN) | | .059 | .049 | .059 | .049 | .059 |
| CIRC SPACING (IN) | | ID: .146 TO .106 OD: .204 TO .114 | | | | |
| AXIAL SPACING (IN) | | .118 | | | | |

45-4719

Figure 4. TYPICAL RODS INSERTION PLAN

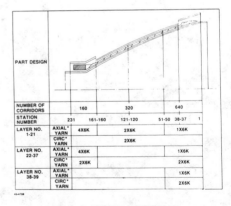

| PART DESIGN | | | | | | | |
|---|---|---|---|---|---|---|---|
| NUMBER OF CORRIDORS | | | 160 | 320 | | 640 | |
| STATION NUMBER | | 231 | 161-160 | 121-120 | 51-50 | 38-37 | 1 |
| LAYER NO. 1-21 | AXIAL* YARN | | 4X6K | 2X6K | | 1X6K | |
| | CIRC* YARN | | | 2X6K | | | |
| LAYER NO. 22-37 | AXIAL* YARN | | 4X6K | | | 1X6K | |
| | CIRC* YARN | | 2X6K | | | 2X6K | |
| LAYER NO. 38-39 | AXIAL* YARN | | | | | 1X6K | |
| | CIRC* YARN | | | | | 2X6K | |

45-4708

Figure 5. TYPICAL WEAVE PLAN

202

Figure 6. TYPICAL FOAM BLOCK AND MANDRELS WITH
END-PLATES

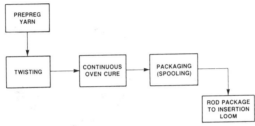

Figure 7. AUTOMATED RADIAL ROD MANUFACTURE

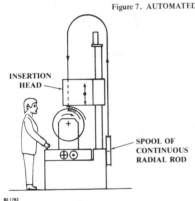

82-1763

Figure 8. SIDE VIEW OF RADIAL INSERTION SYSTEM

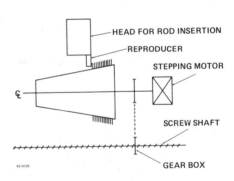

43-5026

Figure 9. RADIAL ROD PLACEMENT

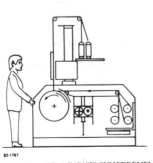

82-1767

Figure 10. AXIAL WEAVING/CIRCUMFERENTIAL
WINDING SYSTEM

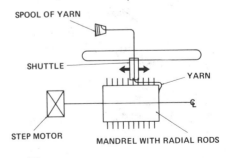

43-5027

Figure 11. AUTOMATED WEAVING OF AXIAL YARNS

203

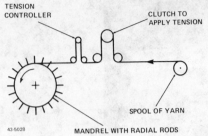

TENSION
CONTROLLER

CLUTCH TO
APPLY TENSION

SPOOL OF YARN

MANDREL WITH RADIAL RODS

43-5028

Figure 12.  AUTOMATED WINDING OF
CIRCUMFERENTIAL YARNS

Figure 13.  DRY AND PREPREG YARN
REINFORCEMENTS

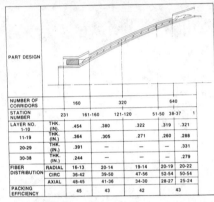

| PART DESIGN | | | | | | |
|---|---|---|---|---|---|---|
| NUMBER OF CORRIDORS | | 160 | 320 | | 640 | |
| STATION NUMBER | 231 | 161-160 | 121-120 | 51-50 | 38-37 | 1 |
| LAYER NO. 1-10 | THK. (IN.) | .454 | .380 | .322 | .319 | .321 |
| 11-19 | THK. (IN.) | .364 | .305 | .271 | .260 | .288 |
| 20-29 | THK. (IN.) | .391 | — | — | — | .331 |
| 30-38 | THK. (IN.) | .244 | — | — | — | .279 |
| FIBER DISTRIBUTION | RADIAL | 16-13 | 20-14 | 19-14 | 20-19 | 20-22 |
| | CIRC | 36-42 | 39-50 | 47-56 | 52-54 | 50-54 |
| | AXIAL | 48-45 | 41-36 | 34-30 | 28-27 | 25-24 |
| PACKING EFFICIENCY | | 45 | 43 | 42 | 43 | |

Figure 14.  TYPICAL PREFORM CHARACTERISTICS

Figure 15.  RADIAL ROD MANUFACTURING SYSTEM

Figure 16.  BR900 INSERTION SYSTEM

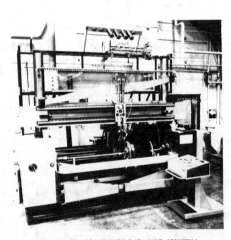

Figure 17.  BR900 WINDER/WEAVER SYSTEM

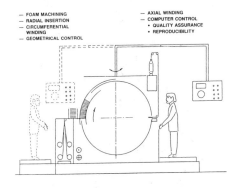

— FOAM MACHINING
— RADIAL INSERTION
— CIRCUMFERENTIAL
  WINDING
— GEOMETRICAL CONTROL

— AXIAL WINDING
— COMPUTER CONTROL
  • QUALITY ASSURANCE
  • REPRODUCIBILITY

Figure 18.  BR2000 SCALE-UP EQUIPMENT OVERVIEW

Figure 19.  BR2000 SYSTEM

| | BR 900 | BR 2000 |
|---|---|---|
| • NO. OF MACHINES IN SYSTEM | TWO — RADIAL INSERTER AND CIRC/AXIAL WINDER/WEAVER | ONE — RADIAL INSERTION AND CIRC/AXIAL WINDING/WEAVING |
| • PREFORM FABRICATION CAPABILITY | | |
| LENGTH | 59 IN. (1500mm) | 59 IN. (1500mm) |
| MAXIMUM DIAMETER (O.D.) | 36 IN. (900mm) | 84 IN. (2150mm) |
| MINIMUM DIAMETER (I.D.) | 2 IN. (50mm) | 3 IN. (75mm) |
| WALL THICKNESS | 0.24 TO 6.7 IN. (6 TO 170mm) | 0.24 TO 8.3 IN. (6 TO 210mm) |
| AXIAL SPACING | 0.10 TO 0.16 IN. (2.5 TO 4mm) | SAME |
| CIRCUMFERENTIAL | 0.08 TO 0.20 (2.5 TO 5mm) | SAME |
| RADIAL SPACING | 0.016 IN. MIN. (0.4mm MIN.) | SAME |
| FIBER VOLUME | 35 TO 55 PERCENT | SAME |
| CONE HALF ANGLE | 22 DEGREES | 35 DEGREES |
| SHAPES | BOTH SYSTEMS WILL PRODUCE CYLINDRICAL, CONICAL, CONTOURED, ASYMMETRICAL, FLANGED (INTERNAL AND EXTERNAL), VARIABLE THICKNESS AND COMBINATIONS | |

46-1872

Figure 20.  COMPARISON BETWEEN BR900 AND BR2000

Figure 21.  LONG CYLINDER/SMALL DIAMETER PREFORM

34/83.052A

Figure 22.  SHORT CYLINDERS/MID-SIZE DIAMETER PREFORM

Figure 23. LARGE DIAMETER CYLINDER PREFORM

Figure 24. CONTOURED PREFORM FOR ITE

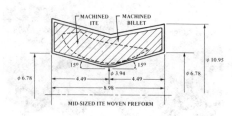

NOTE: DIMENSIONS IN INCHES

Figure 25. AUTOWEAVE$^{TM}$ MID-SIZED CONTOURED ITE PREFORM AND MACHINED ITE SHAPE

Figure 26. BACK-TO-BACK CONTOURED ITE PREFORM – TYPE 1

206

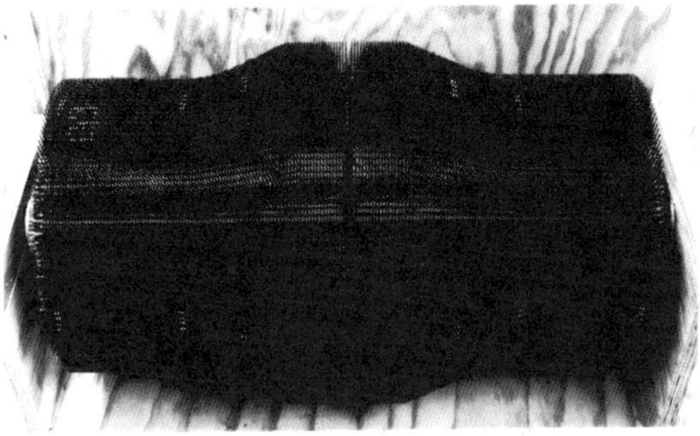

Figure 27. BACK-TO-BACK CONTOURED ITE PREFORM – TYPE 2

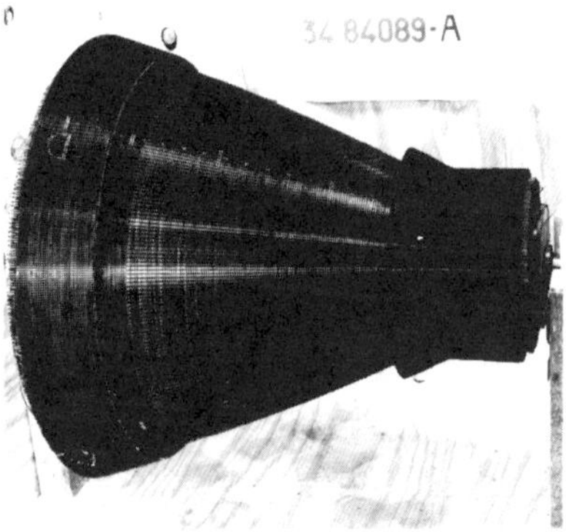


Figure 28. SMALL EXIT CONE PREFORM

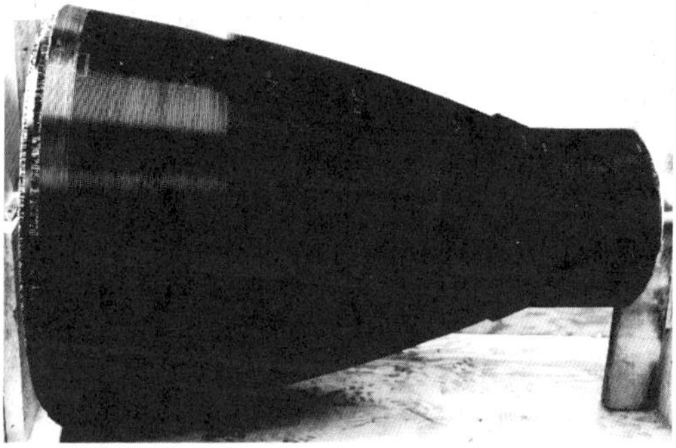

Figure 29. MEDIUM SIZED EXIT CONE PREFORM

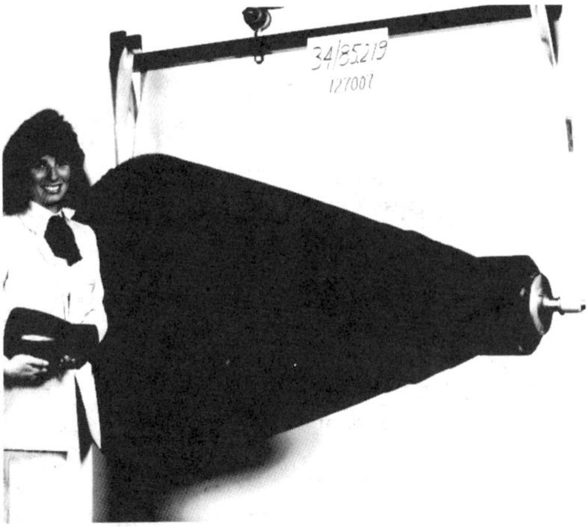

Figure 30. LARGE SIZED EXIT CONE PREFORM

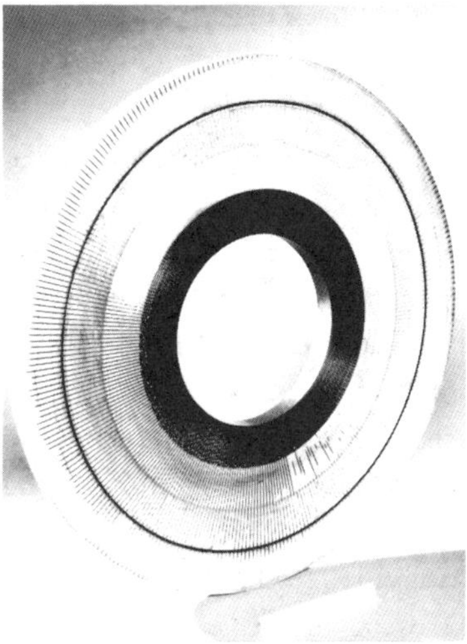

Figure 31. HYBRID FIBER CYLINDER

MODIFICATION OF POLYACRYLONITRILE FIBERS WITH POTASSIUM
DICHROMATE TO MAKE HIGH PERFORMANCE CARBON FIBERS

Tse-Hao Ko, Meng-Liang Tsai
*Hsing-Yie Ting and Chung-Hua Lin
Department of Textile Engineering
*Graduate Institute of Chemical Engineering
Feng Chia University
Taichung, Taiwan 40724
Republic of China

## Abstract

A special variety of PAN fibers, Courtelle fiber, was used in this study. The precursor was modified with potassium dichromate, the modulus and tensile strength of the resulting carbon fibers are above 270 GPa and 4.5 GPa, respectively, the latter figure is 20-40% higher than that made of the unmodified precursors.

The pretreatment increases the crystallite size and the preferred orientation of the PAN fibers. Formation of the ladder polymer in the modified PAN fiber is slightly slower but more gradually than that of the untreated, during the stabilization, and this is one of the reasons that gave better mechanical properties to the carbon fibers. The cross-section of the partially stabilized PAN fibers displays a two-zone morphology, and the time required for full stabilization could be reduced by one hour when stabilized at the same temperature.

## 1.  INTRODUCTION

Since Shindo and Watt et al. published their researchs (1-3), polyacrylonitrile fibers (PAN) have been found to be the most suitable precursor for making high-modulus high-strength carbon fibers.  Recent trends in research and development of PAN-based carbon fibers may be classified as follows (4): (a) high strength and high strain,  to increase the allowable design strain and impact resistance of Carbon Fiber Reinforced Plastics (CFRP); (b) high modulus, mainly requested for space applications; (c) high modulus combined with high strength. To increase strength is one of several methods in increasing the strain of carbon fibers. In the past few years, various methods of modifying PAN fibers to improve the mechanical properties of carbon fibers have been tried and reported (5-9). Recently Bahl and coworkers used CuCl as catalyst to modify PAN fibers to obtain high-quality carbon fibers (10). This report presents our work in the study of another catalyst, potassium dichromate, in modifying PAN fibers to produce high-strength carbon fibers. The mechanical and physical properties and also the microstructure of modified PAN, stabilized fibers and carbon fibers were studied in detail.

## 2. EXPERIMENTAL

A special grade polyacrylonitrile fiber tows, Courtelle, containing 6000 filaments of 1.1 denier were used in the experiments reported here. The PAN fibers were modified with a wound frame. A fixed length of precursor was first wound under constant tension and refluxed in a bath containing hot potassium dichromate solution for few minutes, and then washed with distilled water and dried to a constant weight in oven.

Stabilization of the unpretreated fibers as well as the pretreated precursors was carried out in a constant temperature horizontal furnace with fixed length method at 250 °C and 270 °C from 1 hr. to 10 hrs., under purified air atmosphere. The stabilized fiber tows were carbonized to 1300 °C in a ceramic reaction tube under oxygen free argon atmosphere. A heating rate of 250 °C per hour was programmed to reach 1000 °C and was continued at the rate of 60 °C per hour until reaching 1300 °C and the specimens were cooled down immediately.

Tensile strength and modulus were measured according to JIS R 7601-1980 testing methods with Instron 1122 tensile-testing machine with cross-head speed of 0.5mm/min and load cell of 20g for carbon fibers and load cell of 10g for PAN and stabilized fibers, For each sample at least 25 filaments were tested and mean values determined.

The diameter of the fibers, was measured under an Olympus BHT microscope with closed circuit television camera and a mean value of at least 30 filaments were taken.

A Rigaku X-ray diffractometer were carried out on a tow of fibers, using Ni-filtered Cu $K_\alpha$ radiation. Step-scan method was used to determine the d spacing and stacking size (Lc), the step interval was set at 0.02 degree. The orientation of the fibers was determined with a Fiber Specimen Attachment.

The IR spectra were recorded on a Nicolet model 5-SX FTIR spectrometer. Each spectrum was recorded at a resolution of 2 $cm^{-1}$ utilizing 128 scans.

A Perkin-Elmer 240 C elemental analyser was used to determine the content of carbon, hydrogen and nitrogen of stabilized fibers and carbon fibers. For elemental analy-

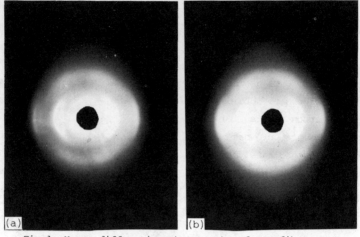

Fig.1. X-ray diffraction photographs of PAN fibers :
(a) original PAN fibers, (b) modified PAN fibers.

ser runs, about 3 mg of the sample was studied at 950 °C for 3 minutes in purified oxygen atmosphere and the sample were burned down completely. Each specimen was run several times and the average value was reported.

3.   RESULTS AND DISCUSSION
3.1 Structure and X-ray Diffraction of PAN
X-Ray diffraction analysis on PAN fibers had been used by many authors. But, none of the results could successfully establish a structure unit. Stefani and his coworkers (11,12) assigned the orthorhombic structure (a=10.2 $\overset{\circ}{A}$, b=6.1 $\overset{\circ}{A}$, c=5.1 $\overset{\circ}{A}$) to PAN fiber and others(13) suggested a hexagonal cell (a=6 $\overset{\circ}{A}$, c=5.1 $\overset{\circ}{A}$). Bahl et al.(14) studied three types of PAN fibers made under three different sets of spinning conditions, they concluded that the crystal struc-

tures are dependent  on the spinning conditions, the composition and the stretch ratio etc. We found that the structure of Courtelle fibers are hexagonal, according to Colvin and Storr's method(16). The wide angle X-ray diffraction photographs of PAN and  modified PAN fibers are given in Fig. 1. The first diffraction maximum with d spacing of 5.4 $\overset{\circ}{A}$ was the (100) reflection. This intense reflection is an indication of existence of good lateral order in both precursors, the original fiber and the modified fiber. The elongated equatorial arc in the pattern represents the orientation of molecular chains in the samples, and the line broadening should give a rough estimate on the size of lamellar crystals. In Fig. 1, since the modified PAN fiber has smaller arc length and narrower line broadening than the original PAN fiber it should have better orientation and larger crystallite size. The wide angle X-ray diffraction patterns of both fibers has a sharp intense peak at $2\theta=16.73$ and $2\theta=16.52$, respectively. The peak positions were measured several times under the step-scan method, and d spacing of the lattice was calculated.

According to the Scherrer formula the crystallite size t which perpendicular to the (100) plane is related to B (in radians) and Bragg angle $\theta$ for this plane :

$$t = K \lambda / B \cos\theta$$

where K is the Scherrer parameter,

can be considered as the factor by which the apparent size must be multiplied to give the true dimension of crystallites, and K is generally given the value of 0.89 for half width or 1.0 for integral breadth. As is found in Table 1, the structural parameters of original PAN fibers have changed after modification. The orientation and size of crystallites for the modified PAN fiber are higher than those of the original PAN samples. Bahl and his coworkers suggested that for making good quality carbon fibers one should look for the PAN precursor having better crystalline orientation(15).

Table 1.  Structural parameters of PAN fibers

| ——— | original PAN | modified PAN |
|---|---|---|
| d spacing (Å) | 5.00±0.01 | 5.36±0.01 |
| crystallite size ( Å ) | 44.3 | 44.8 |
| orientation ( % ) | 82.6 | 84.3 |

3-2. Thermal Analysis of Original and Modified PAN

Fitzer and Muller(17) have studied the kinetics of cyclization and stabilization of PAN fibers during the thermal treatment using DTA technique with varied heating rate. They suggested that the rate of cyclization reaction is a first order with an activation energy of about 34 Kcal/mole in air. Poporske et al.(18) suggested that the acti-

vation energy of any chemical reaction can be calculated from the Arrhenius equation:

$$K = K_0 \exp(-Ea/RT)$$

where $K_0(s^{-1})$ is the reaction rate constant, R is the gas constant and Ea (KJ/mole) is the activation energy, in the present work a TC-10 microprocessor coupled to the DSC unit was used to obtain the final

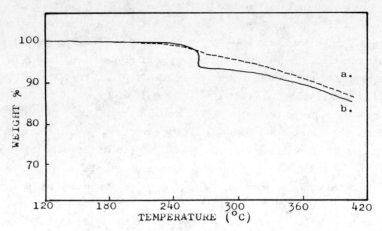

Fig.2. TGA curves of PAN fibers : (a) modified PAN
fiber, (b) original PAN fibers.

data of Ea. The activation energy (Ea) gives the minimum amount of energy required to raise the molecules to the activated state to start the chemical reaction. A lower value of Ea suggests that the chemical reaction could be initiated at low temperature, and also reflects in the reduced time that was required for cyclization reaction at constant temperature during thermal treatment. In general, the time of stabilization required for completion should be shorter for the modified PAN, with Ea equal to 116 KJ/mole, than that of the original PAN, its Ea is 120 KJ/mole. The time required could be reduced by around one hour for the modified PAN, based on the comparison of examinations on the cross-sections of both fibers having morphology changes from two zone, sheath-core structure, to single-unit structure, no sheath-core structure.

From thermogravimetry (TG) analysis, Fig. 2, weight loss starts at a lower temperature (188°C) for the modified PAN and the rate of weight loss also decreases slower than that for the original PAN during thermal treatment in air. This phenomenon indicates that during the stabilization stage the cyclization and dehydrogenation reactions, in which the conversion of C≡N triple bonds to C=N double bonds, starts steadily at lower temperature. This suggests that the stabilization reaction could prevent excessive bond rupture and chain scission in the precursor during thermal treatment. This is one of the reasons why carbon fibers developed from modified PAN has better modulus.

3-3. Tensile Strength of Stabilized
      and Carbon Fibers
The reactions occurred during ther-

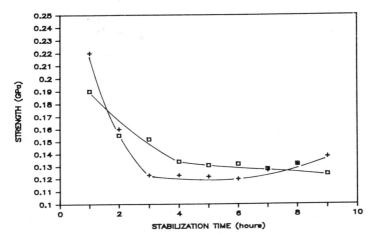

Fig.3. Variation of tensile strength of stabilized fibers with different treatment time at 250 °C : (+) developed from original PAN fibers, (□) developed from modified PAN fibers.

mal treatment of the PAN fibers are the cyclization reaction, the dehydrogenation and the oxygen uptake reaction(17). Cyclization of the molecular chain enhances the carbon fibers of good mechanical properties. However, the tensile strength of the PAN fibers was expected to decrease during the stabilization process, because of the conversion of the C≡N triple bonds to the C=N double bonds at cyclization. During bond conversion process, the cohesive energy between molecular chains in PAN dropped appreciably which reduces the tensile strength of the stabilized fibers(19). The decrease in tensile strength with heat stabilization time, of the original and the modified PAN fibers, are  shown in Fig. 3, the samples were heat-treated from 1 to 9 hours at 250°C. The strength of both fibers decreased rapidly at

early stage and then decreased gradually at latter stages indicated that the attack of oxygen on the molecular chain took place faster at first and then slow down gradually. However, it has been observed that PAN fibers modified with potassium dichromate initially has more activation centers that enhances the cyclization reaction in the early stage (the first hour) of stabilization, After 1 hr, the decrease in strength of the modified PAN fiber was slower than that of the original PAN fiber suggested that the cyclization reaction in the former is slower, the effect was to increase the length of the ring sequence, and enhance the extent of ladder polymer. Therefore, the carbon fibers developed from the modified PAN fiber have better modulus.(Table 2)  From Fig.3, the strength of the original

PAN fiber increased after 7 hours during the stabilization process suggested that intermolecular crosslinking reaction may have occurred in the stabilized fiber. The oxygen content in the fibers decreased suddenly after 7 hours during the stabilization process, the analysis was taken with a Perkin Elmer 240 C Elemental Analyser, we found the ratio of the oxygen atoms to the hydrogen atoms in the stabilized fiber equals 1/2, an indication that evolution of water at stabilization stage did occur that promotes crosslinking of molecular chains in the fiber. Crosslinking between the adjacent chains suppressed chain scission and contributed to make the carbon fibers having improved mechanical properties.

3-4 X-ray Analysis of Carbon Fibers
Moreton et al.(20) reported that the Young's modulus of the carbon fiber increased with heat treatment during carbonization. Another method to increase Young's modulus of carbon fibers is to increase the crystallinity in the precursor fibers (21). In this work, each stabilized sample was carbonized at the same set of heating condition. We would like to point out that the average amount of the preferred alignment of basal layers in the fibrillar along the fiber axis will affect the mechanical properties of carbon fiber, consequently, the factor of preferred orientation is not the main reason to affect those

properties of carbon fiber. Cooper and Mayer (22) showed the effects of boron doping and neutron irradiation on the PAN-based carbon fibers will increase the stiffness of the fiber, this increased both the preferred orientation and the shear stiffness of the graphite. The content of chromium in the carbon fibers developed from the modified PAN fiber increased 30 time than that of the original PAN fibers, analyzed with an IL-257 Atomic Emission Spectrometer. Chromium metal powder, has been used for the catalytic graphitization of carbons, that gave large graphitic crystal flakes at early stage of the reaction, has been studied in the fields of graphite made from resin (23). Carbon fibers developed from modified PAN fibers have larger $L_c$, the stacking size, more basal layer planes, better mechanical properties, poorer preferred orientation than those developed from the original PAN fibers (24) (in Table 2); on the other hand, we propose that the chromium atoms acts like the boron atoms, it performed as a solid-solution hardener which increases both the tensile strength and tensile modulus of carbon fibers. And also the chromium promoted the progress of graphite crystallite plane, that led to larger stacking size $L_c$ (see Table 2, 3 ), but decreased the crystallite orientation. In our view, the chromium atoms dope among crystallite planes which cause the dislo-

Table 2. Mechanical properties and structure parameters of
carbon fibers stabilization temperature: 270 °C.

| carbon fibers | developed from the original PAN fibers | | | | developed from the modified PAN fibers | | | |
|---|---|---|---|---|---|---|---|---|
| Stabilization time (hr) | 2 | 4 | 6 | 8 | 2 | 4 | 6 | 8 |
| Lc (Å) | 16.2 | 16.3 | 16.6 | 16.2 | 16.8 | 16.9 | 17.4 | 17.8 |
| Lc/d | 4.7 | 4.7 | 4.8 | 4.7 | 4.8 | 4.9 | 5.0 | 5.1 |
| orientation(%) | 78.6 | 78.7 | 78.8 | 79.0 | 78.4 | 78.2 | 78.4 | 78.4 |
| strength (GPa) | - | 2.8 | 3.6 | 3.7 | 3.2 | 4.0 | 4.1 | 4.6 |
| modulus (GPa) | - | 182 | 265 | 239 | 189 | 196 | 207 | 277 |

Table 3. Elementary analyses of carbon fibers.
stabilization: 3hr, 250 °C

| carbon fiber developed from Element (wt%) | original PAN fibers | | modified PAN fibers | |
|---|---|---|---|---|
| | C | N | C | N |
| 1000 °C | 80.61 | 11.19 | 80.73 | 11.03 |
| 1300 °C | 96.09 | 3.51 | 97.30 | 2.76 |

cation to propagate from one crystallite plane to the other during crystallite growth in the carbonization process. This process gives the reason why carbon fibers developed from the modified PAN fiber have poorer preferred orientation.

4. CONCLUSIONS

In our study on modifying the PAN fibers with potassium dichromate to make high performance carbon fibers, the applicable conclusions are:

(1) The modified PAN fibers has better crystallite size and molecular orientation. This is one of the reasons that PAN was used for making high performance carbon fibers.

(2) Based on the TG analysis, the occurrence of cyclization of the PAN molecular chains starts at lower temperature (188 °C) for the modified PAN fibers, which decrease the required time for stabilization. The molecular chain of the stabilized fibers were formed slowly and steadily which improves modulus in the carbon fibers.

(3) The evolution of water during stabilization was attributed to the crosslinking reaction occurred among molecular chains which improves mechanical properties of the carbon fiber developed from the original PAN fibers.

(4) Chromium atom acts as catalyst, it promotes the growth of crystallite planes during carbonization, but it also causes dislocation in the crystal that leads to poor preferred orientation in the carbon fiber.

(5) The stabilization process was found to be much faster and the strength of the carbon fiber so obtained is around 20-40 % higher than that prepared from the untreated original PAN fibers.

ACKNOWLEDGEMENTS

We would like to thank Prof. Chen-Chi Ma(National Tsing Hwa Univ.) for his interest in the work and Mr. H. Y. Lin (Materials Research Laboratory, ITRI) for his valuable technical assistance in DSC. Thanks are also to Mr. J.C. Chen and his co-workers (National Taiwan University) for TEM studies. We are also grateful to Mr. W. L. Chou for his helps and Miss J.T. Wang for typing ping this paper. The financial support of this work was provided by The National Science Council, Contract No. NSC75-0405-E035-01.

REFERENCES

1. Shindo, A., Rep. Govt. Ind. Res. Inst., Osaka, No.317 (1961)

2. Johnson, W., Philips, L.N., and Watt, W., Br. Pat. 1,110,791 (1968)

3. Bahl, O.P., and Manocha, L.M., Carbon, 13, 297 (1974)

4. Matsui, J., Matsuda H.S., and Maeda, K., J. Ind. Fabrics, 3, 43 (1985)

5. Watt, W., and Johnson, W., Applied Polymer Symposia, 9, 215 (1969)

6. Raskovic, V., and Marinkovic, S., Carbon, 16, 351 (1978)

7. Gupta, A.K., Fibre Sci. Tech., 12, 159 (1979)

8. Mathur, R.B., et al., Fibre Sci. Tech., 20, 227 (1984)

9. Bahl, O.P., Mathur, R.B., and Dhami, T.L., Materials Sci. Eng., 73, 105 (1985)

10. Mathur, R.B., Bahl, O.P., and Kundra, K.D., J. Mater. Sci. Lett., 5, 757 (1986)

11. Stefani, R., et al., Comp. Rend., 248, 2006 (1959)

12. Stefani, R., Comp. Rend., 251, 2174 (1960)

13. Urbarczyk, G.W., Przeglad Wlokienizy, 15, 216 (1961)

14. Bahl, O.P., Mathur, R.B., and Kundra, K.D., Fibre Sci. Tech. 15, 147 (1981)

15. Colvin, B.G., and Storr, P., Eur. Polym. J., 10, 337 (1974)

16. Alexander, L.E., X-Ray Diffraction Method in Polymer Science, Interscience, New York, 1969.

17. Fitzer, E., and Muller, D.J., Carbon, 13, 63 (1975)

18. Poporske, N., and Mladenov, I., Carbon, 21, 33 (1983)

19. Manocha, L.M., and Bahl, O.P., Fibre Sci. Tech., 13, 199 (1980)

20. Moreton, R., Watt, W., and Johnson, W., Nature, 213, 690 (1967)

21. Moreton, R., Proc. 3rd Conf. on Industrial Carbons and Graphite (Soc. Chem. Ind., London) (1970) p.472

22. Cooper, G.A., and Mayer, R.M., J. Mater. Sci., 6, 60 (1971)

23. Oya, A., and Otani, S., Carbon, 17, 131 (1979)

24. Johnson, D.J., Chem. Indu., 18, 692 (1982)

## BIOGRAPHIES

Tse-Hao Ko is a Ph. D. program student of the Graduate of Textile Engineering, Feng Chia University. He received a B.S. in 1979 and a M.S. in Textile Engineering from Feng Chia University in 1985. He is working in the Polymer Physics Laboratory, Department of Textile Engineering, Feng Chia University as teaching assistant. He has more than 2700 working hours experience for SEM and 2500 working hours experience for X-ray diffractometer, and has co-published more than 15 research papers, and received the Ministry of Education's Youth Research and Invention Awards in 1980 and 1985. At present, he is studying the carbon fiber process from PAN fiber, microstructure and physical properties of carbon fi-bers and stabilized fibers, and the surface treatment of carbon fibers.

Meng-Liang Tsai has received a B.S. in 1984 and an M.S. in 1986 from the Department of Textile Enginee-ring of Feng Chia University. He has co-published many papers on the microstructure and properties of carbon fibers developed from polya-crylonitrile fibers pretreated with potassium dichromate.

Hsing-Yie Ting is presently profes-sor and director of the Institute of Chemical Engineering, Feng Chia University. He received a B. S. degree in Chemistry from National Chung Hsing University in 1965 and a Ph. D. in Physical Chemistry from Texas Christian University in 1971. Professor Ting has been teaching at Feng Chia University since 1972, his research interests are in the field of microstructure analysis of fibers and composite materials using X-rays, simulation and separation processes in Chemical Engineering and the formulation of warp yarn sizes.

Chung-Hua Lin is presently profes-sor of Textile Engineering and the dean of Graduate School at the Feng Chia University. He graduated in 1959 with a B.S. from Gunma Univer-sity and in 1961 with an M.S. from Tokyo Institute of Technology, Ja-pan. He has spent more than 30 years in the field of fiber techno-logies and engineerings. He is the

president of the Chinese Institute
of Textile Engineers. Professor
Lin's present research interests
include the carbon fiber process,
composites and polymer processing.

DYNAMIC MECHANICAL BEHAVIOUR OF THE
MODIFIED POLYACRYLONITRILE FIBERS

Tse-Hao Ko, **P. Chiranairadul,
*Hsing-Yie Ting and Chung-Hua Lin
Department of Textile Engineering
*Graduate Institute of Chemical Engineering
**Graduate Institute of Textile Engineering
Feng Chia University
Taichung, Taiwan 40724
Republic of China

## Abstract

Polyacrylonitrile (PAN) fibers pre-treated with potassium permanganate and with potassium dichromate to make precursors that gave good quality carbon fibers was studied. After the pretreatment, the amount of manganese and chromium in the PAN fibers increased by 60 times and 30 times respectively than that of the untreated samples. Manganese and chromium are highly effective catalysts in the stabilization and carbonization of the precursors, both metals reduced the time required for stabilization and also improved the mechanical properties of the carbon fibers thus produced. In this work, an attempt was made to establish the relationship between the mechanism of formation of the ladder polymers and of the dynamic mechanical absorption phenomena of PAN fibers observed in the temperature range between 30 and 280 °C at a frequency of 35 Hz. Raw and pretreated PAN fibers were measured on dynamic modulus, loss modulus and dynamic loss tangent in air at a heating rate of 2 °C/min. The loss tangent curves for the PAN fibers were characterized by two strong absorption peaks at about 125 °C (beta) and about 254 °C (alpha). The alpha peak was attributed to cyclization of the acrylonitrile units in the polymer, and the formation of the ladder polymer in the structure contributes the increase of dynamic modulus of the stabilized PAN fibers. The viscoelastic technique is another characterization method in the determination of the formation of ladder polymer during the stabilization of the PAN fibers.

## 1. INTRODUCTION

For many years the pyrolysis of PAN fibers, stabilization process was found to be necessary to obtain high quality carbon fibers. Chemists and textile scientists centered their interests almost in how chemical reactions occurred during stabilization. However, important changes in physical characteristics of the fiber during stabilization process, such as the viscoelastic properties, which relate to molecular motion in the fiber were not covered. In this work, we try to discuss the stabilization effects from the point of view of the latter.

Gupta et al. (1-3) reported that the thermal treatment of the polyacrylonitrile (PAN) in air produced considerable changes in dielectric relaxation in the glass transition region of the fiber. Other workers also reported two transition peaks on heating the fiber at temperatures near 80-100 °C and 140-160 °C (4-6). All of the authors had tried to discuss and explain the reasons for these transitions. Andrews et al. (4) proposed the concept of a heterobonded solid-state structure to explain these transitions and suggested that transition at the lower temperature was the result of chain mobility caused by weakening of the van der Waals forces, while the transition at the higher temperature resulted from intermolecular dipole-dipole dissociation of the nitrile groups in more localized sections of the chain. Okajima et

al.(7) attributed the lower transition peak to the molecular motion in the amorphous phase, and the higher transition to the molecular motion in the crystalline region. Ferguson et al.(8) found that the loss tangent curves for the polyacrylonitrile fibers were characterized by a strong peak at 110 °C (beta) and 280 °C (alpha). They suggested that the alpha peak was attributed to cyclization of the acrylonitrile units in the polymers which was enhanced by the acid comonomers. In this paper, we found that loss tangent curves for PAN fibers have two observable peaks at about 125 °C (beta) and around 254 ° C (alpha). The dynamic viscoelasticity study, which combining with known thermal properties, chemical reaction, and mechanical properties of the fiber during thermal treatment will give a better understanding on the nature of stabilization at the molecular level in the manufacture of carbon fibers process.

2. EXPERIMENTAL

A special grade polyacrylonitrile fiber tows, Courtelle, containing 6000 filaments of 1.1 denier were used in the experiments reported here. The PAN fibers were modified with hot potassium permanganate solution and potassium dichromate solution for some minutes, and then washed with distilled water and dried to a constant weight in oven respectively.

Temperature dependence of dynamic

modulus E', loss modulus E" and tangent delta were measured by using a Rheovibron DDV-IIC (manufactured by the Toyo-Boldwin Co., Ltd., Japan), under the following conditions. Frequency f was 35 Hz and average heating rate was 2° K/min. For increasing the measured temperature, the heater was modified to cover the range to 350 °C, and was controlled with a digital PID programmed controller with a built-in microprocessor.

The stress-strain properties of the fibers were determined using Instron 1122 tensile tester with cross-head speed of 0.5 mm/min, load cell of 10g and 2cm of testing gauge. Results were averaged from at least 25 specimens in each case. A Rigaku X-ray diffractometer was carried out on a tow of fibers, using Ni-filtered Cu $K_\alpha$ radiation. The step-scan method was used to determine the d spacing and crystallite size L(hkl), the step-interval was set at 0.02 degree. The orientation of the fibers O(hkl) was determined with a Fiber Specimen Attachment. After fixing the preselected diffraction angle $2\theta$, the X-ray diffraction intensity along with azimuthal direction was recorded by rotating the specimen holder in the (100) plane of the PAN fiber which was supposed to have the hexagonal structure. L(hkl) and O(hkl) were calculated by using eq. 1 and 2, respectively:

$$L(hkl)(\text{in } \overset{\circ}{A} \text{ unit}) = K\lambda \, / \, B\cos\theta \quad (1)$$

$$O(hkl)(\text{in } \%)=[(360-H)/360]*100 \quad (2)$$

in which $\lambda = 1.542 \overset{\circ}{A}$, K the apparatus constant (=1.0), B the half width value in radian of the X-ray diffraction intensity (I) vs.(100) curve and H the half width value in degree of the curve of I vs. azimuthal angle. The preferred orientation, O(hkl), has a value of 0, if the specimen is completely unoriented. While if the crystallites are all arranged perfectly, parallel to one another, it is equal to 100.

## 3. RESULTS AND DISCUSSION
### 3-1 Dynamic Viscoelastic Properties from 30 to 200 °C

The loss tangent curves shown in Fig.1 has two well defined peaks at around 125 °C (beta) and about 254° C (alpha). The beta peak was attributed to molecular motion. The alpha peak in the range of 200 to 280 °C is believed to be due to some type of chemical reaction: cyclization, crosslinking and degradation, during the stabilization process. The beta peak at around 150 °C has an additional shoulder absorption. This means that it can be resolved further into two different absorption peaks: beta(1) and beta(2). In other words, the peak temperatures of tan$\delta$ can be evaluated by seperating the absorption into beta(1) and beta(2) absorptions.

Ferguson and Ray (8) assumed that the beta(1) peak in pure and in copolymer PAN is due to the ordered

221

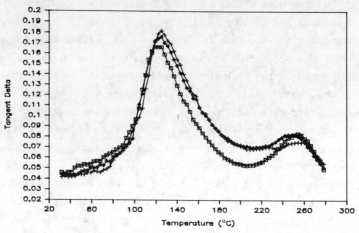

Fig.1 Temperature dependence of loss tangent for PAN fibers :
(□) sample A, (+) sample B, (◇) sample C

regions and the beta(2) peak intensity increases as the concentration of acrylic acid comonomer increased in the polymer, and it can be assumed that beta(2) peak represents the disordered regions. Okajima et al. (7) also gave similar conclusion. S. Minami (9) suggested that the beta(1) peak should be attributed to the molecular motion in the amorphous phase, and the beta(2) peak to the molecular motion in the crystalline phase of PAN. For discussing the phenomena, sample A was carried out in Rheovibron under the same condition of thermal treatment to 280 °C, and then cooled down to room temperature. The preheated sample A was measured again with Rheovibron. The loss tangent curves of preheated sample A was shown in Fig.2. We found the intensity of peak beta was very small and broad, and peak beta(1) disappeared. There are two

different structures in sample A after the thermal treatment of the Rheovibron analysis. One has the original PAN structure, AN unit; and the other has the ladder polymer structure, which was converted from AN units. Each of these different structures has two phases, the ordered region and the amorphous region. After PAN fiber was heat treated, a majority of the AN units converted into the ladder polymer. Such transformation could be (a) initially from AN units at the amorphous phase, and/or (b) then from the interrupted AN sequences and/or at the boundaries of the crystallites. That implies that the beta peak absorption (from 60 to 160 °C temperature range) arised from the molecular motions in the AN structure during thermal treatment. Since the ordered phase of the ladder polymer in the stabilized fiber, has a more stable

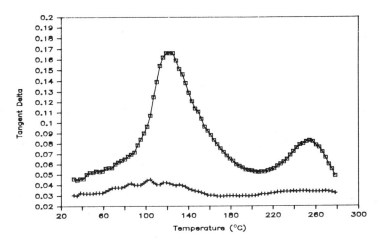

Fig.2 Temperature dependence of loss tangent for PAN and pretreated
PAN fibers : (□) sample A, (+) pretreated PAN fibers

structure, even heated to 300 °C, the structure remains solid state, hence no observable peak was found for the pretreated smaple A. And we suggest that the alpha peak for the pretreated sample A was attributed either to the chemical reaction occurred in the AN units that remain in the preheat-treated fiber and/or due to molecular motion of the transformed ladder polymer, in the amorphous phase, repacking to form the ordered phase.

The mechanical properties of three kinds of PAN fibers, the original PAN and two pretreated PAN fibers, which were modified with potassium dichromate and potassium permanganate solution respectively, are shown in Table 1. And according to the Scherrer formula the crystallite size, L(100), and preferred orientation, O(100), were shown in Table 2. The color of fibers changed from white to tan and buff after been modified with potassium permanganate and potassium dichromate solution respectively. The

Table 1. Mechanical properties of PAN fibers

| Sample | Pretreatment solution | Young's modulus (GPa) | Broken modulus (GPa) | Tensile strength (GPa) | Elongation at break (%) |
|--------|----------------------|------|------|------|------|
| A | none | 10.62 | 3.97 | 0.66 | 17 |
| B | KMnO$_4$ | 7.92 | 3.34 | 0.60 | 18 |
| C | K$_2$Cr$_2$O$_7$ | 7.75 | 2.67 | 0.54 | 20 |

Table 2. Structural parameters of PAN fiber

| Sample | Modified solution | d spacing (Å) | Crystallite size (Å) | Preferred orientation (%) |
|--------|-------------------|---------------|----------------------|---------------------------|
| A | none | 5.00±0.01 | 44.3 | 82.6 |
| B | $KMnO_4$ | 5.40±0.04 | 40.2 | 80.7 |
| C | $K_2Cr_2O_7$ | 5.36±0.01 | 44.8 | 84.3 |

true reason why the chemical reaction occurred during the pretreatment process is not explained well. However the content of Mn and Cr of in each kind of carbon fibers developed from the modified PAN fibers had increased 60 and 30 times than that of original PAN fibers by atomic emission spectrometer analysis, respectively. In other words, the manganese and chromium atoms had been dopped in the PAN fiber during the modification process.

The IR spectrum of PAN fibers has two prominent peaks at 2930 cm$^{-1}$ and 2240 cm$^{-1}$ due to stretch vibration of the methylene ($CH_2$) and the nitrile (C≡N) groups. For sample B, modified with potassium permanganate solution, gives the new peaks at 2340 cm$^{-1}$ (due to −C=N conjugation ) and has a shoulder at 1600 cm$^{-1}$(due to C=C and C=N groups) (9). But, no new peaks were found for sample C, which was modified with potassium dichromate solution. The shoulder at 1600 cm$^{-1}$ for sample B implies that potassium permanganate, acts as catalyst, attacks AN units to initiate the cyclization. Because some AN units participated

the initial cyclization reaction for sample B, the number of intermolecular bonded nitrile groups will decrease and has a lower dipole moment (owing to larger number of free nitrile groups) than those of sample A.

From Table 2, sample B has larger Bragg spacing, smaller crystallite size and poorer preferred orientation than those of sample A. The larger Bragg spacing implies that the crystal lattice of sample B contain interrupted AN sequences in the crystallite phase after the modification process. And that imperfections in the crystalline lattice, either at the boundaries of the crystallites or inside the crystallite phase, may give rise to redistribution of the Bragg spacing (10), which decrease both in the size of crystallites, and in the orientation of molecules. The lower amount of intermolecular nitrile group bonding in the ordered region and the higher free nitrile groups in the amorphous phase for sample B will increase the segmental mobility. Hence the loss tangent curves for sample B should have higher

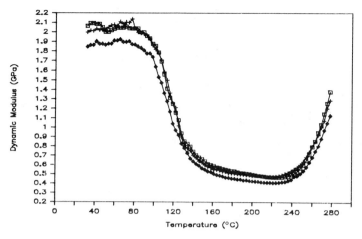

Fig.3 Effect of temperature on E' for PAN fibers :
(□) sample A, (+) sample B, (◇) sample C

intensity and broader half width than those of sample A at the beta peak position.

In Fig.1, the beta(1) peak at 146° C for the original PAN fiber decrease in intensity, and will shift upwards about 12 degrees when sample A was modified with either potassium permanganate solution or potassium dichromate solution. It is possible that since the crystallite size of sample A is larger that the amount of dipole moment increases that reduces the molecular motion and the loss tangent intensity. The temperature shifted to 158 °C for samples B and C could be attributed to the crosslinking, between the metal atom ions and the C≡N groups, derived from the modification process. And the higher peak intensity for both samples could be related to molecular motions of the interrupted AN sequences in the ordered phase.

For sample C, the phenomena was very strange, it has larger crystallite size and higher preferred orientation, but has the worst mechanical properties. Currently, no satisfactory explainations were given. We suspect that some amorphous phase in the fiber has been solubilized during the modification process and that led to have poor mechanical properties for sample C. Fig.3 and Fig.4 shows the temperature dependence of the dynamic storage modulus (E') and the loss modulus (E"). The storage modulus fell sharply by almost two decades over the temperature range in which the loss peaks beta(1) and beta(2) occurred. The existence of the dynamic storage modulus were due to molecular motion before 160 °C. From Table 1, sample A has the highest broken modulus, Young's modulus and tensile strength, and sample C has the worst mechanical

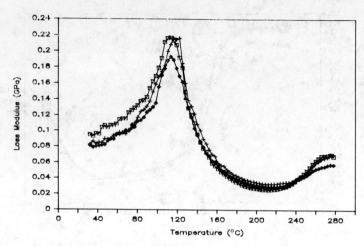

Fig.4 Effect of temperature on E" for PAN fibers :
(□) sample A, (+) sample B, (◇) sample C

properties. It means that sample A has the highest intensity for dynamic storage modulus before chemical reactions occurred. In the loss modulus (E") curves, Fig.4, similar patterns exists for the samples A, B and C. The temperature of the maximum is 113 °C for sample A and C, and is 116 °C for sample B which leads to the conclusion that sample B has more free AN units in the amorphous phase.

3-2 Dynamic Viscoelastic Properties
    from 200 to 280 °C
The dynamic viscoelastic properties over the temperature range from about 200 to 280 °C is believed to

be due to some type of chemical reaction: cyclization, crosslinking, dehydrogenation and degradation (11-13). Houtz (11) discussed thermal stability at low temperature and found that the color of fibers change from white to brown at about 200 °C. We have studied the coloration reaction of PAN film under IR and found a series of color change during stabilization process as shown in Table 3(10): The color variation indicates that chemical reactions did occur during stabilization process. The acidic constituent in the PAN fiber acts as an initiator for ladder polymer formation and this occurs at 160 °C

Table 3. Color changes for PAN films

| Temperature(°C) | 25 | 160 | 190 | 210 | 230 | 250 | 260 |
|---|---|---|---|---|---|---|---|
| Sample A | white | buff | deep-yellow | coco | brown | sorrel | black |
| Sample B | tan | tan | tan | brown | sorrel | black | black |

## Table 4. A.I. values for stabilized fibers

| Sample No. | sample A | sample B | sample C |
|------------|----------|----------|----------|
| A.I.(%)    | 54.5     | 50.4     | 50.7     |

for sample A. From the variation of colors in the films, we found that to obtain the brown color, sample A should be heated to 230 °C and to 210 °C for sample B, this in an additional evidence that by treating sample A with potassium permanganate solution the time required for the completion of cyclization was reduced.

From Fig. 3 the dynamic storage modulus (E') curves decreased with increasing temperature, and then increased with increasing temperature for sample A at 224 °C, for sample B at 218 °C and for sample C at 221 °C. The initial transition temperature for samples B and C shifted to lower temperature indicated that ladder polymer could be formed at lower temperature. But, ladder polymers in samples B and C were formed gradually than those of sample A during stabilization process. Also from Fig.3, the E' curves for sample A has the highest intensity than that of samples B and C after 263 °C. Which indicated that more ladder polymer were formed and the rate of cyclization reaction also increased faster after 263 °C for sample A. These were the reasons why samples B and C can be developed into higher quality carbon fibers (14).

As shown in Fig.1, the peak alpha of sample A has both the highest intensity and broadest half width which indicated that it has more AN units took part in the cyclization, dehydrogenation, crosslinking and degradation reactions during the stabilization process. The designation of "aromatization index", A.I. value, has contributed a useful method in estimating the progress of the oligomerization of the nitrile groups and the subsequent oxidation and dehydrogenation processes (15). Sample A has the highest A.I. value, indicated that sample A has largest amount of ladder polymers among the stabilized fibers, (see Table 4).

The same group of prepared samples were studied further by wide angle X-ray diffraction after the Rheovibron analysis. The X-ray diffraction patterre shown in Fig.5. The original PAN structures ($2\theta=17°$) still remain in the stabilized fibers, and the turbostratic carbon, ladder polymer structure ($2\theta=25°$), were formed. These results in the presence of a two-phase structure for the stabilized fibers, in which one phase is the original PAN structure and the other phase is a new phase which was attributed to ladder polymers

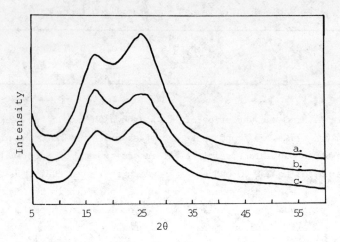

Fig.5 X-ray diffraction patterns of PAN fibers after dynamic
mechanical analysis : (a) sample A, (b) sample B, (c) sample C

converted from the AN units. These
suggested that the alpha peak was
resulted from both chemical reactions, and molecular motions occurred in the formation of the ordered
region of the stabilized fibers.

4.  SUMMARY

The following conclusions can be
drawn.

(a). The loss tangent curves at 125°
C, the beta(1) peak, was  attributed to molecular motion in the
disordered regions. And at about
254 °C, the alpha peak was attributed to the chemical reactions and
molecular motion occurred in the
formation of the crystalline region
of the stabilized fibers.

(b). Modified PAN fibers has more
free AN units in the amorphous
phase, hence the loss tangent curves at the beta peak   has higher
intensity and broader half width.
On the  contrary, the alpha peak

has lower absorption for modified
PAN fibers, because it has less
amount of the ladder polymers formed.

(c). From the dynamic storage analysis, modified PAN fibers have a
lower initial transition temperature, and the ladder polymers were
formed gradually and steadily.

ACKNOWLEDGEMENTS

We would like to thank Dr. Ru-Yu Wu
(Materials  Research Laboratory,
ITRI) for his encouragement in this
work and Prof. Chiou-Guang Leu for
his helpful discussions.  We are
also thank Mr. W. L. Chou for his
helps in the lab, and  Miss J. T.
Wang for  typing  this  paper.
Financial  support  of this work was
provided by the National  Science
Council of the Republic of China,
contract No. NSC76-0405-E035-01.

REFERENCES

1. Gupta, A.K., Singhal, R.P., and
   Agarwal, V.K., J. Appl. Polym.
   Sci., $\underline{26}$, 3599 (1981)
2. Gupta, A.K., and Maiti, A.K.,
   J. Appl. Polym. Sci., $\underline{27}$,
   2409 (1982)
3. Gupta, A.K., Singhal, R.P., and
   Maiti, A.K., J. Appl. Polym.
   Sci., $\underline{27}$, 4101 (1982)
4. Andrews, R.D., and Kimmel,
   R.M., J. Polym. Sci., 3B, 167
   (1965)
5. Minami, S., Appl. Polym.
   Sympo., $\underline{25}$, 145 (1974)
6. Hayakawa, R., et al., J. Polym.
   Sci., A-2, $\underline{5}$, 165 (1967)
7. Okajima, S., Ikeda, M., and
   Takeuchi, A., J. Polym. Sci.,
   A-1, $\underline{6}$, 1925 (1968)
8. Ferguson, A., and Ray, N.G.,
   Fibre Sci. Tech., $\underline{15}$, 99 (1981)
9. Ko, Tse-Hao, Master Thesis
10. Gupta A.K., and Singhal, R.P.,
    J. Polym. Sci., Polym. Phys.
    Edi., $\underline{21}$, 2243 (1983)
11. Houtz, R.C., Text. Res. J., $\underline{20}$,
    786 (1950)
12. Clarke, A., and Bailey, J.,
    Nature, $\underline{234}$, 529 (1971)
13. Potter, W.D., and Scott, G.,
    Nature, $\underline{236}$, 30 (1972)
14. Ko, Tse-Hao, et al., Pro. of
    the 2nd China & Japan Sem. on
    Carbon Fibers, Taichung, Taiwan
    (1985) p.59
15. Ushida, T., et al., Pro. of
    the 10th Bienn. Conf. on Carbon,
    Bethlehem, Pennsylvania (1971)
    p.31

BIOGRAPHIES

Tse-Hao Ko is a Ph. D. program
student of the Graduate of Textile
Engineering, Feng Chia University.
He received a B.S. in 1979 and a
M.S. in Textile Engineering from
Feng Chia University in 1985. He is
working in the Polymer Physics
Laboratory, Department of Textile
Engineering, Feng Chia University
as teaching assistant. He has more
than 2700 working hours experience
for SEM and 2500 working hours
experience for X-ray diffractometer, and has co-published more than
15 research papers, and received
the Ministry of Education's Youth
Research and Invention Awards in
1980 and 1985. At present, he is
studying the carbon fiber process
from PAN fiber, microstructure and
physical properties of carbon fibers and stabilized fibers, and the
surface treatment of carbon fiber.

Phaichit Chiranairadul, a graduate
student of Textile Engineering of
Feng Chia University, received a
B.S. in 1985. He works as research
assistant in the Textile Physics
Laboratory for many years and is
currently studying the manufacture
process of PAN-based carbon fibers
and its physical properties analysis.

Hsing-Yie Ting is presently professor and director of the Institute
of Chemical Engineering, Feng Chia
University. He received a B. S.
degree in Chemistry from National

Chung Hsing University in 1965 and
a Ph. D. in Physical Chemistry from
Texas Christian University in 1971.
Professor Ting has been teaching at
Feng Chia University since 1972,
his research interests are in the
field of microstructure analysis of
fibers and composite materials u-
sing X-rays, simulation and separation processes in Chemical Engineering and the formulation of warp
yarn sizes.

Chung-Hua Lin is presently professor of Textile Engineering and the
dean of Graduate School at the Feng
Chia University. He graduated in
1959 with a B.S. from Gunma University and in 1961 with an M.S. from
Tokyo Institute of Technology, Japan. He has spent more than 30
years in the field of fiber technologies and engineerings. He is the
president of the Chinese Institute
of Textile Engineers. Professor
Lin's present research interests
include the carbon fiber process,
composites and polymer processing.

VIBRATION DAMPING CHARACTERISTICS OF HIGHLY ORIENTED
POLYETHYLENE FIBER REINFORCED EPOXY COMPOSITES

Ronald F. Gibson, Vidish S. Rao and Raju Mantena
Mechanical Engineering Department
University of Idaho
Moscow, Idaho

Abstract

A new high strength, high modulus
fiber of highly oriented
polyethylene (SPECTRA 900) has been
developed by Allied Corporation, and
this paper reports the results of
the first phase of a program to
investigate the vibration damping
characteristics of epoxy matrix
composites reinforced with this
fiber. An impulse-frequency
response technique developed
previously at the University of
Idaho was used to measure the
complex modulus (dynamic stiffness
and damping) of SPECTRA 900/EPON 826
composites in unidirectional and
woven configurations at different
fiber orientations and vibration
frequencies. Results are compared
with similar results from previous
tests of graphite and aramid fiber
reinforced composites.

## 1. INTRODUCTION

An extended chain polyethylene
reinforcing fiber, designated
SPECTRA 900, has been developed
recently by Allied Corporation for
use in polymer matrix composite
materials. Previous publications
have reported the results of
research on static mechanical
properties [1], electrical
properties [2], and impact
performance [3] of SPECTRA 900
composites. Desirable
characteristics such as high
specific strength and high specific
stiffness make SPECTRA composites
attractive for applications in a
variety of structures and machines.
Many of these applications (e.g.,
helicopter panels, aircraft
interiors, tennis rackets, skis and
golf clubs) involve dynamic loading,
and this means that good vibration
damping characteristics are needed
for noise and vibration control,
improved fatigue performance and
rapid dynamic response. Thus,
vibration damping is an important
design parameter in many potential
applications of SPECTRA composites,
but no previous research had been
done on damping in these materials.
It was believed, however, that the
viscoelastic nature of the fiber
would result in good damping
characteristics.

An experimental study of damping
characteristics of SPECTRA
composites was initiated at the
University of Idaho during 1986, and
the results of the first phase of
this study are summarized in this
paper. The purposes of the first
phase of the research program were
(a) to develop a data base for
vibration damping characteristics of
unidirectional and woven SPECTRA
900/EPON 826 composites under
standard laboratory conditions, and
(b) to compare this data with
existing data for other advanced
composites such as graphite/epoxy
and aramid/epoxy having similar
fiber volume fractions.

## 2. DESCRIPTION OF EXPERIMENTS

Two variations of the University of Idaho computer-aided impulse-frequency response technique were used to measure the storage modulus and the loss factor of SPECTRA 900/EPON 826 unidirectional and woven composites at various frequencies and fiber orientations under standard laboratory conditions. As described in Refs. [4-7], the technique involves excitation of flexural (Fig. 1) or extensional (Fig. 2) modes of vibration in the specimen with an electromagnetic impulse hammer which has a piezoelectric force transducer in its tip. The specimen response is measured by a non-contacting eddy current proximity detector in the flexural test and by a miniature piezoelectric accelerometer in the extensional test. Input and response signals are fed into a dual channel frequency spectrum analyzer and the desired frequency response function on the desired frequency span is displayed on the analyzer. A desktop computer reads the digital data from the analyzer and does the calculations required to obtain the storage modulus and loss factor for each mode desired.

Dynamic stiffness and damping are conveniently expressed in terms of the complex modulus

$$E^* = E' + iE'' = E'(1 + i\eta)$$

where $E'$ = storage modulus
$\quad E''$ = loss modulus
$\quad \eta$ = loss factor
$\quad i$ = imaginary operator = $\sqrt{-1}$

Since the specimens in these tests were either unidirectional or woven, the lamina properties did not vary through the thickness and the extensional modulus was the same as the flexural modulus. Thus, both extensional and flexural properties are presented on the same figures later in this paper. As shown in Refs. [4-7], the storage modulus is found from the resonant frequency and the appropriate eigenvalue equation, while the loss factor is found from the half-power bandwidth of the resonant peak for a given mode of vibration. Maximum vibration amplitudes were less than 5% of the specimen thickness, so that the resulting aerodynamic damping was much smaller than the material damping.

The flexural vibration technique was used for unidirectional specimens and woven specimens of various lengths at low frequencies, while the extensional vibration apparatus was used for those same specimens at high frequencies and for all off-axis unidirectional specimens. The use of the extensional technique for off-axis specimens eliminated the bending-twisting coupling that would have resulted if the flexural vibration technique had been used for those tests [4-7]. All data was taken from the peak corresponding to the first mode in the frequency response function. Resonant frequencies were controlled by varying both the length and the end masses in the extensional tests.

Unidirectional panels of SPECTRA 900/EPON 826 and EPON 826 neat resin samples were supplied by the Composite Materials Research Group at the University of Wyoming, while the woven 900/826 panels were supplied by Allied Corporation. All test specimens were machined from these panels at the University of Idaho. Data on Fiberite T-300/934 graphite/epoxy, Fiberite Kevlar 49/934 aramid epoxy, and Fiberite 934 neat resin from previous experiments at the University of Idaho [6,7] were used for comparison. Woven T-300/826 graphite/epoxy panels were also supplied by Allied for comparison with the woven 900/826. Fiber volume fractions and specimen dimensions were kept as close as possible in order to make the comparisons valid. Descriptions of materials and ranges of dimensions and properties are given in Table 1.

## 3. RESULTS

The 934 resin cures at 177C (350F) and would therefore not be compatible with the polyethylene

SPECTRA 900 fiber. Thus, the 826 resin, which cures at 120C (250F) was used. The differences between the damping and stiffness of the two resins are shown in Figs. 3 and 4, where the nonlinear regression curve-fits to the data are plotted. The corresponding equations from the regression analyses are given in Table 2. The fact that both damping and stiffness of the 826 resin are lower than the corresponding values for the 934 resin will be taken into account in later comparisons of the composite properties.

As shown in Figs. 5 and 6, the loss factor for unidirectional SPECTRA 900/826 decreases rapidly in the low frequency range, while the modulus shows a corresponding rapid increase. Since these trends in the composite properties are opposite from those shown by the 826 resin in Figs. 3 and 4, the fiber properties must be dominant over the matrix properties. Both properties show little change above 500 Hz, however. Good agreement between flexural vibration and extensional vibration results is also seen in Figs. 5 and 6. As a check on the modulus data at low frequencies, static tests of the same specimens were conducted in an MTS testing machine. Three-point flexure tests yielded an average modulus of 6 Mpsi (41.37 GPa), while static tensile tests gave an average modulus of 7.1 Mpsi (48.95 GPa). Testing machine crosshead speeds were 1 inch/min. (2.54 cm/min.) for flexure and 0.1 inch/min. (0.254 cm/min.) for tension. The static test data supports the conclusion that the material exhibits a significant strain rate sensitivity in the range 0-100 Hz, and that there is a need for more data in the range 0-50 Hz where the changes occur most rapidly. In order to cover this range, a different technique such as the hysteresis loop method [8] would be needed, however.

Using nonlinear regression analysis of the data, best-fit curves for loss factor and storage modulus of SPECTRA 900/826 are compared with corresponding curves for T-300/934 graphite/epoxy and Kevlar 49/934 aramid/epoxy in Figs. 7 and 8.

Equations from the regression analyses are given in Table 2. SPECTRA 900 in the unidirectional [0] configuration appears to have properties which are intermediate between those of graphite and aramid, while showing more rate sensitivity than either of the other materials. The properties of the polyethylene composite are even more attractive when the material densities Table 1) and the resin data (Figs. 3 and 4) are taken into account. This is because of the low density of the polyethylene and the low damping and stiffness of the 826 resin compared with the 934 resin.

All off-axis tests to determine fiber orientation effects were conducted in the extensional vibration apparatus. In order to maintain approximately the same resonant frequencies at different fiber angles, both the specimen length and the end mass were adjusted as required. As shown in Figs. 5 and 6, the properties were essentially independent of frequency in the extensional vibration frequency range, so any observed changes were a result of fiber orientation. Figs. 9 and 10 show that the damping increases and the stiffness decreases as the fiber angle is increased from 0 to 90 degrees. This behavior is similar to that of graphite and aramid composites as shown by the regression analysis curves in Figs. 11 and 12, but polyethylene is clearly superior in damping for most off-axis orientations. Polyethylene and aramid composites show similar behavior in off-axis stiffness, well below that of graphite. Again, normalization to the material specific gravities and the 826 resin makes SPECTRA 900 even more superior in damping and more attractive from the stiffness point of view.

Finally, comparisons of woven SPECTRA 900/826 and T-300/826 graphite/epoxy materials are given in Figs. 13 and 14. Woven aramid material was not available for these tests. Tests were conducted at orientations of 0, 45 and 90 degrees orientation from the fill direction, but only 0 and 45 degree data are presented. The 0 and 90 degree data shows insignificant differences as expected. SPECTRA 900 shows superior damping characteristics in all cases, but the graphite has greater stiffness. It is interesting to speculate on the potential for hybrid polyethylene/-graphite composites combining the high damping and low density of polyethylene with the high stiffness and strength of graphite. The static and impact properties of such hybrids have been studied by Adams and Zimmerman [3].

## 4.  CONCLUDING REMARKS

The complex moduli of high performance polyethylene fiber-reinforced epoxy composite materials have been characterized as functions of vibration frequency and fiber orientation and the resulting data has been compared with corresponding data for graphite/epoxy and aramid/-epoxy composites having similar fiber volume fractions and specimen dimensions. In the unidirectional configuration, the polyethylene composite has damping and stiffness intermediate between that of aramid and graphite, while exhibiting more strain rate sensitivity than the other materials at low frequencies. More data is needed in the range 0-50 Hz, where the properties appear to be particularly sensitive to frequency. The polyethylene composite has much better damping than the other materials for most off-axis fiber orientations, while both polyethylene and aramid stiffness is below that of graphite. When the material densities and the differences between the epoxy matrix resins for the three composites are taken into account, the polyethylene composite looks even more attractive. In the woven configuration, polyethylene again proved to be far superior to graphite in terms of damping, but graphite has the greatest stiffness.

Hybridization of polyethylene with graphite would appear to offer the potential of a stiff, light, highly damped composite material that should have numerous practical applications.

## 5.  ACKNOWLEDGEMENTS

The authors gratefully acknowledge the financial support of Allied Corporation, Fibers Division. We also wish to thank the University of Wyoming Composite Materials Research Group for supplying cured composite and neat resin panels for the tests. Special thanks go to Darrel Brown for machining the specimens, and to Valerie Smith for careful typing of the manuscript.

## 6. REFERENCES

[1]  Adams, D.F. and Zimmerman, R.S., "Properties of a Polymer Matrix Composite Incorporating Allied A-900 Polyethylene Fiber," Adv. Tech. in Materials and Processes, Proc. 30th National SAMPE Symposium, Anaheim, CA, 1985, pp. 280-89.

[2]  Chang, H.W., Lin, L.C. and Bhatnagar, A., "Properties and Applications of Composites Made of Polyethylene Fibers," Materials Sciences for the Future, Proc. 31st National SAMPE Symposium, Anaheim, CA, 1986, pp. 859-66.

[3]  Adams, D.F. and Zimmerman, R.S., "Static and Impact Performance of Polyethylene Fiber/Graphite Fiber Hybrid Composites," Materials Sciences for the Future, Proc. 31st National SAMPE Symposium, Anaheim, CA, 1986, pp. 1456-68.

[4]  Suarez, S.A., Gibson, R.F. and Deobald, L.R., "Random and Impulse Techniques for Measurement of Damping in Composite Materials," Experimental Techniques, $8$, No. 10, 1984, pp. 19-24.

[5]  Suarez, S.A. and Gibson, R.F., "Computer-Aided Dynamic Testing of Composite Materials," Proc. 1984 SEM Fall Conference, Milwaukee, WI, pp. 118-23.

[6]  Suarez, S.A., Gibson, R.F. Sun, C.T. and Chaturvedi, S.K., "The Influence of Fiber Length and Fiber Orientation on Damping and Stiffness of Polymer Composite Materials," Experimental Mechanics, $26$, No. 2, 1986, pp. 175-84.

[7] Suarez, S.A. and Gibson, R.F., "Improved Impulse-Frequency Response Techniques for measurement of Dynamic Mechanical Properties of Composite Materials," J. Testing and Evaluation, March 1987, in print.

[8] Gibson, R.F., "Vibration Damping Characteristics of Graphite/Epoxy Composites for Large Space Structures," Large Space Systems Technology, NASA Conference Pub. 2215, Part 1, 1981, pp. 123-32.

## 7.  BIOGRAPHIES

Ronald F. Gibson received his BS degree in Mechanical Engineering from the University of Florida in 1965, and was employed as a Development Engineer by the Nuclear Division of Union Carbide Corporation. He received a MSME degree from the University of Tennessee in 1971, and a PhD degree in Mechanics from the University of Minnesota in 1975. Since that time, he has served on the faculties of Iowa State University, the University of Idaho, and the University of Florida. He is currently a Professor of Mechanical Engineering at the University of Idaho, where he teaches courses in applied mechanics and materials, and conducts research in dynamic behavior of advanced materials.

Born in Karnatoka, India, Vidish S. Rao completed his BS degree in Mechanical Engineering at the Bangalore University, India. He is currently working on a Masters degree in Mechanical Engineering at the University of Idaho. Vidish works as a graduate assistant at the Department of Mechanical Engineering and does his research in the optimization of dynamic properties of advanced materials.

Raju Mantena received a BS degree in Mechanical Engineering from Andhra University in 1973, and a Post Graduate Diploma in Design Engineering from the Indian Institute of Technology, Delhi in 1975. He was employed as a Development Engineer for eight years by Bharat Heavy Electricals Ltd. where he was responsible for the establishment of the company's Corporate Research and Development Laboratory Complex at Hyderabad, India. Raju received a MSME degree from the University of Idaho in 1985, and is currently pursuing research in the area of non-destructive evaluation of advanced materials and adhesive joints while working on a PhD degree at the University of Idaho.

235

Table 1

Specimen Dimensions and Properties

| Material | Length(mm) | Width(mm) | Thickness(mm) | Average Density (g/cm$^3$) | Average Fiber Volume Fraction |
|---|---|---|---|---|---|
| SPECTRA 900/826 Unidirectional | 76.2-228.6 | 19.02-19.28 | 1.42-1.47 | 1.062 | 0.67 |
| SPECTRA 900/826 Woven | 101.6-203.2 | 19.02-19.20 | 1.22-1.27 | 1.063 | 0.64 |
| T-300/934[*] Unidirectional | 101.6-228.6 | 18.49-19.30 | 1.49-1.59 | 1.58 | 0.65 |
| Kevlar 49/934[*] Unidirectional | 101.6-228.6 | 18.62-19.35 | 1.32-1.45 | 1.36 | 0.67 |
| T-300/826 Woven | 101.6-203.2 | 18.95-19.07 | 1.47-1.55 | 1.41 | 0.64 |
| EPON 826 Neat Resin | 76.2-228.6 | 19.05-19.18 | 2.56-2.62 | 1.18 | - |
| Fiberite 934[*] Neat Resin | 22.2-211.9 | 18.92-18.95 | 3.15-3.20 | 1.30 | - |

Note: Temperature range was 17.8C-22.2C and relative humidity range was 49-55%.
      For 90% of the tests, the temperature range was 21.1C-22.2C and the
      relative humidity range was 49-51%.
*See references 6 and 7.

Table 2

Equations From Regression Analyses

| Material | Description of Equation | Equation |
|---|---|---|
| SPECTRA 900/826 Unidirectional | $\eta$ vs. f for 0° | $\eta = 0.014512f^{-0.145}$ |
| | E' vs. f for 0° | $E' = 8.878f^{0.0261}$ |
| | $\eta$ vs. $\theta$ for 0°-90° | $\eta = 0.001792 + 0.00208\theta - 3.18(10^{-5})\theta^2 + 1.628(10^{-8})\theta^3$ |
| | E' vs. $\theta$ for 0°-20° | $E' = 10.647 - 0.666\theta + 0.0024\theta^2 - 0.00039\theta^3$ |
| | E' vs. $\theta$ for 20°-90° | $E' = 2.738 - 0.694\theta + 0.000468\theta^2$ |
| T-300/934* Unidirectional | $\eta$ vs. f for 0° | $\eta = 0.00183 + 4.15(10^{-8})f$ |
| | E' vs. f for 0° | $E' = 15.407 + 0.0010987f$ |
| | $\eta$ vs. $\theta$ for 0°-90° | $\eta = 0.01116 + 0.00118\theta - 2.125(10^{-5})\theta^2 + 1.33(10^{-7})\theta^3$ |
| | E' vs. $\theta$ for 0°-20° | $E' = 18.4117 - 0.41735\theta - 0.03042\theta^2 + 0.001095\theta^3$ |
| | E' vs. $\theta$ for 20°-90° | $E' = 11.199 - 0.27183\theta + 0.001858\theta^2$ |
| Kevlar 49/934* Unidirectional | $\eta$ vs. f for 0° | $\eta = 0.01344 - 1.286(10^{-7})f$ |
| | E' vs. f for 0° | $E' = 9.445 + 0.00705f$ |
| | $\eta$ vs. $\theta$ for 0°-90° | $\eta = 0.001531 + 0.001124\theta - 1.818(10^{-5})\theta^2 + 8.595(10^{-8})\theta^3$ |
| | E' vs. $\theta$ for 0°-20° | $E' = 11.584 - 0.60786\theta - 0.004392\theta^2 + 0.000022\theta^3$ |
| | E' vs. $\theta$ for 20°-90° | $E' = 4.978 - 0.12795\theta + 0.0008758\theta^2$ |
| SPECTRA 900/826 Woven | $\eta$ vs. f for 0° | $\eta = 0.01806f^{-0.02714}$ |
| | $\eta$ vs. f for 45° | $\eta = 0.02302f^{-0.01726}$ |
| | E' vs. f for 0° | $E' = 1.828f^{0.006105}$ |
| | E' vs. f for 45° | $E' = 0.3093f^{0.10127}$ |
| T-300/826 Woven | $\eta$ vs. f for 0° | $\eta = 0.002109 + 5.9974(10^{-8})f$ |
| | $\eta$ vs. f for 45° | $\eta = 0.005155 + 1.0944(10^{-6})f$ |
| | E' vs. f for 0° | $E' = 5.367 + 4.5845(10^{-4})f$ |
| | E' vs. f for 45° | $E' = 1.641 + 7.36(10^{-7})f$ |
| EPON 826 Neat Resin | $\eta$ vs. f | $\eta = 0.00144 + 0.00577(\log f)$ |
| | E' vs. f | $E' = 0.411 + 0.0000673(f)$ |
| Fiberite 934* Neat Resin | $\eta$ vs. f | $\eta = 0.004 + 0.006(\log f)$ |
| | E' vs. f | $E' = 0.5712 + 0.00005565(f)$ |

Note: In all equations, f = frequency (Hz), $\theta$ = fiber angle (degrees),
$\eta$ = loss factor (dimensionless), and E' = storage modulus (MPsi).
*See references 6 and 7.

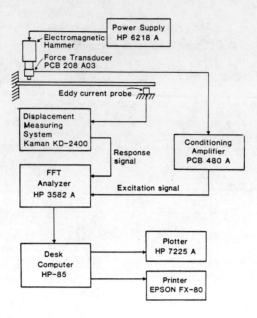

Figure 1.   Flexural vibration apparatus.

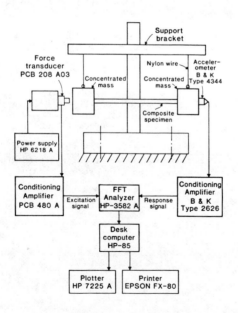

Figure 2.   Extensional vibration apparatus.

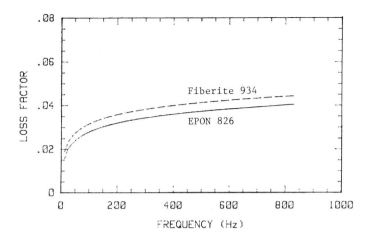

Figure 3.  Variation of loss factor with frequency
for 826 and 934 neat resin specimens.

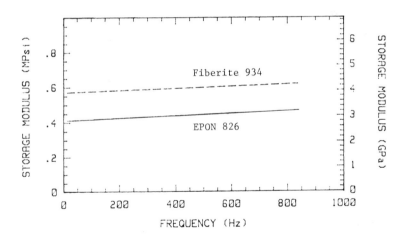

Figure 4.  Variation of storage modulus with frequency
for 826 and 934 neat resin specimens.

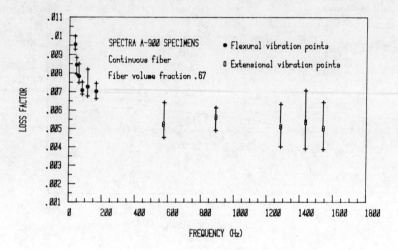

Figure 5.   Variation of loss factor with frequency for
SPECTRA 900/826 unidirectional [0] composite.

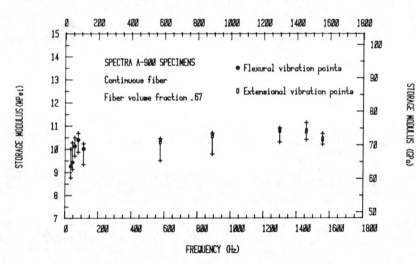

Figure 6.  Variation of storage modulus with
frequency for SPECTRA 900/826
unidirectional composite.

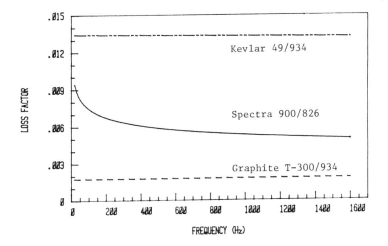

Figure 7.    Comparison of loss factor vs. frequency
             for three unidirectional [0] composites.

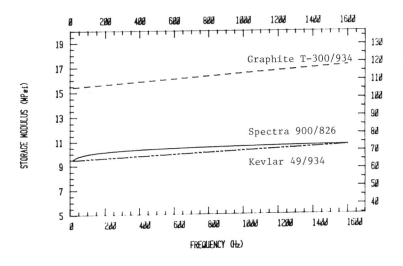

Figure 8.    Comparison of storage modulus vs. frequency
             for three unidirectional [0] composites.

241

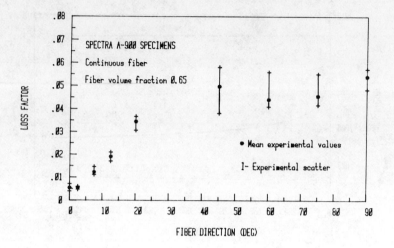

Figure 9. Variation of loss factor with fiber orientation for SPECTRA 900/826 unidirectional composite.

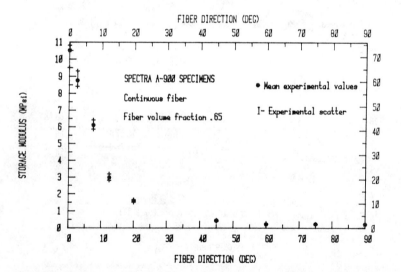

Figure 10. Variation of storage modulus with fiber orientation for SPECTRA 900/826 unidirectional composite.

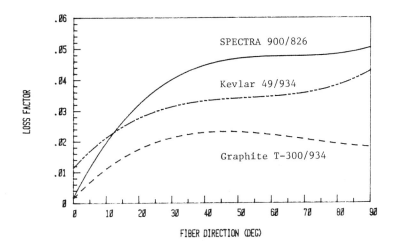

Figure 11.  Comparison of loss factor vs. fiber orientation
for three unidirectional composites.

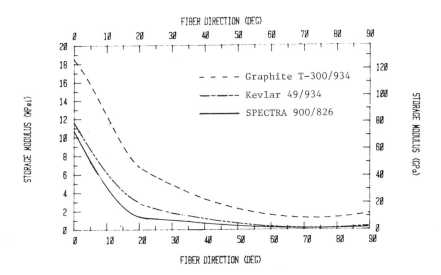

Figure 12.  Comparison of storage modulus vs. fiber orientation
for three unidirectional composites.

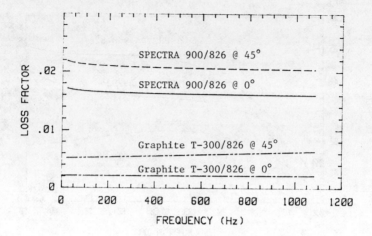

Figure 13.  Comparison of loss factor vs. frequency
for woven composites at two orientations.

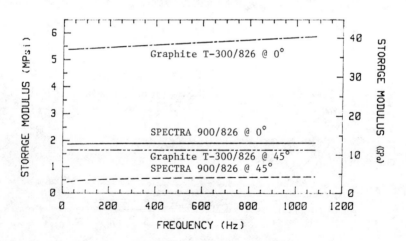

Figure 14.  Comparison of storage modulus vs. frequency
for woven composites at two orientations.

244

# HIGH TEMPERATURE PROPERTIES
# OF THREE NEXTEL CERAMIC FIBERS

Allan R. Holtz and Michael F. Grether
Ceramic Materials Department
3M Company
3M Center, St. Paul, Minnesota

## ABSTRACT

For several years Nextel 312 ceramic fibers have been commercially available from 3M for applications requiring textiles useful above 1000°C. Recently two new ceramic fibers, Nextel 440 and Nextel 480, have been introduced by 3M. This paper describes these three fibers and compares their mechanical properties measured at room temperature before and after exposure up to 1400°C in air and vacuum, and 1200°C in hydrogen. Tensile strength, tensile modulus, and creep results of the three fibers at test temperatures up to 1300°C are also compared. Measurement techniques are described. The results suggest these ceramic fibers are suitable high temperature structural reinforcements, especially Nextel 480.

## 1.  INTRODUCTION

Continuous ceramic fibers, due to their high melting temperature, high modulus, high strength, and good resistance to chemical attack, are potential reinforcements of metal and ceramic matrix composites. Variables such as fiber chemistry, purity, crystal size, glassy phases, fiber diameter, and coatings affect the translation of that potential into useful composites. To help define the limits for the fabrication and use of a composite, information on the properties of the reinforcement measured both after thermal exposure and at elevated temperature is needed.

This paper presents four types of mechanical property behavior measured on three continuous, microcrystalline oxide fibers: Nextel 312, 440 and 480. Creep rate was measured to define structural stability under load. Modulus and tensile strength were measured to compare the sensitivity of fiber stiffness and strength to temperature. These tests help define temperature limits for a composite reinforcement. For evaluating composite fabrication limits, room-temperature properties before and after exposure to high temperature in oxidizing and reducing environments were compared.

2. MATERIAL DESCRIPTION

Table I lists typical physical properties for the three fibers. In general, Nextel 480 fiber is the most crystalline and Nextel 312 fiber is the least crystalline. All three fibers have a 3 to 2 mole ratio of alumina to silica. Nextel 312 fiber contains 14 weight percent boria, while Nextel 440 and 480 fibers contain 2 weight percent boria. All three fibers have geometric surface area. The crystal size is small resulting in good strength and flexibility. The fiber shape is oval in each case, with the major diameter up to two times the minor diameter.

One spool each of sized and unsized Nextel 312, 440, and 480 fiber was used for all tests. These three fibers are available with appropriate sizings and finishes for textile handling and composite bonding. Rovings are produced with

Table I.  Typical Physical Properties of Nextel Fibers

| Property | Units | Nextel 312 | Nextel 440 | Nextel 480 |
|---|---|---|---|---|
| Filament diameter | μm | 10-12 | 10-12 | 10-12 |
| Filament count | | 740-780 | 740-780 | 740-780 |
| Yarn denier | g/9000m | 1800 | 2000 | 2000 |
| Crystal size | nm | <500 | <500 | <500 |
| Crystal type | | $9Al_2O_3:2B_2O_3$ + amorph. $SiO_2$ | gamma $Al_2O_3$ + mullite + amorph. $SiO_2$ | mullite |
| Density | g/cc | 2.7-2.9 | 3.05 | 3.05 |
| Filament tensile strength (51 mm gauge) | MPa | 1700 | 2000 | 1900 |
| Filament tensile modulus | GPa | 150 | 190 | 220 |
| Surface area | $m^2/g$ | <.2 | <.2 | <.2 |
| Fiber chemistry | wt % | 62 $Al_2O_3$ 24 $SiO_2$ 14 $B_2O_3$ | 70 $Al_2O_3$ 28 $SiO_2$ 2 $B_2O_3$ | 70 $Al_2O_3$ 28 $SiO_2$ 2 $B_2O_3$ |

filament counts of approximately 130, 400, and 780 fibers. In addition fibers are supplied in an 8-9 or a 10-12 μm diameter range. For this study 780 filament rovings of 10-12 μm fibers were selected. For creep and all room temperature testing of heat treated fiber, samples were selected from fiber with sizing 170. This is an organic sizing that can be removed cleanly in air at 500-1000°C. Unsized fiber was used for all hot testing of single fibers to facilitate fiber removal from the roving.

## 3. EXPERIMENTAL PROCEDURES

### 3.1 Fiber Properties after Heat Treatment

An overview of the mechanical testing program is given in Table II. This section covers the effect of heat treatment temperature (700 to 1400°C) and atmosphere (air, vacuum, and hydrogen) on the properties of single fibers measured subsequently at room temperature.

To prepare these samples, hanks containing approximately 16 m of 780-filament sized roving were first heat cleaned in air at 700°C for five minutes to remove the 170 sizing. The samples were then weighed for later comparison. All heat treaments consisted of a four hour hold at the chosen peak temperature, with heating and cooling rates typically being 200°C per hour. In each case, the samples were placed on high-purity alumina substrates in the furnace.

Resistance-heated box furnaces were used for the runs in air. Vacuum heat treatments were done in a system having a graphite furnace and a typical operating pressure of 2.7 Pa. Heat treatments in hydrogen were done in a tube furnace using purified process gas at ambient pressure. Since prior work has shown some degree of interaction between the two fiber chemistries (Nextel 312 and Nextel 440/480),

Table II. Fiber Testing Program

| Test | Sample Source | Sample Type | Tested Length | Tests Per Combination |
|------|---------------|-------------|---------------|------------------------|
| Strength and modulus after heat treatment | Roving with sizing 170 | Single filament | 51 mm | 10 |
| Hot strength | Unsized roving | Single filament | 250 mm total; 38 mm heated | 5-10 |
| Hot modulus | Unsized roving | Single-filament loop | 53 mm circumference | 2-8 |
| High-temperature creep | Roving with sizing 170 | Roving | 480 mm between grips; 100 mm heated | 1 |

separate furnace runs were used in air and a physical separation was maintained in vacuum and hydrogen. The resulting test samples were reweighed to reveal any weight change.

Single filaments at least 80 mm in length were separated from the rovings and used in tensile testing. A standard Instron test frame with a 500-gram load cell and self-aligning, spring-loaded grips was used. The strain rate on the 51 mm gauge length was 0.025/min. For each filament tested, the major and minor diameters were measured using a Supramess micrometer readable to 0.25 µm. The slope and height of the chart record was then interpreted to yield tensile strength and elastic modulus, with suitable corrections for test frame compliance, grip effects, and filament shape. In addition, SEM photos were taken of selected samples.

## 3.2  Hot Strength

The tensile strength of single filaments was measured in air at temperatures up to 1300°C. For this work, a slotted box furnace similar to that used by Sawko and Tran[1] was built using platinum wire as the heating element. The furnace was then mounted on sliding supports within the Instron test frame.

These strength measurements were made using the same room-temperature filament grips already described. To run a test, a filament at least 300 mm long was loaded into the grips with the furnace retracted. The furnace was then pulled forward to surround the filament, and the slot in the furnace was closed. After one to two minutes for temperature stabilization, the filament was pulled to failure at a strain rate of 0.020/minute. Diameter measurements and strength calculations were then made in the same manner as described earlier.

## 3.3  Hot Modulus

The cold-grip testing described above for hot strength is not, unfortunately, suitable for obtaining values of hot elastic modulus. This is due to the small proportion of the total gauge length actually at the chosen test temperature. Instead, a hot-gripping technique was developed which involved bonding measured lengths of filament into closed loops. This was accomplished by positioning the ends so as to give an approximate 2 mm overlap and applying a very small amount of diluted fiber-spinning material of the same ceramic composition. These loops were then fired (800°C for the Nextel 312 fibers; 1000°C for the Nextel 440 and 480 fibers) to yield a strong bond.

Hot grips for pull testing these loops were hooks fabricated from strips of mullite ceramic approximately 100 mm long. The working surfaces of these hooks were radiused and polished to prevent damage to the filament. Each test was run by positioning a loop over these hooks in the test frame and moving the crosshead to give a slight preload. The slotted box furnace was then positioned around the sample and closed. The loops were pulled to failure at a strain rate of 0.020/minute, with the slope of the stress-strain curve used to determine the tensile modulus.

## 3.4 High-Temperature Creep

Data on the creep rate of the three fibers was obtained using a laboratory-fabricated tester. The horizontal tube furnace in this unit was constructed to give a long (100 mm) constant-temperature hot zone with abrupt transitions at the furnace ends. This kept the total furnace length small (160 mm), as well as minimizing the amount of roving under test which was at intermediate temperatures. Both dead-weight loading and measurement of displacement were achieved using a pivoted linkage at one end of the furnace. The roving was secured to an adjustable support at the other end, which allowed the initial position of the magnified scale pointer on the linkage to be set to zero at the start of a test. The duration of each test was fifty hours, or until the sample failed.

## 4. RESULTS AND DISCUSSION

### 4.1 Fiber Properties after Heat Treatment

Figure 1 plots the tensile strength of the three fibers versus temperature for three atmospheres. Ninety-five percent confidence

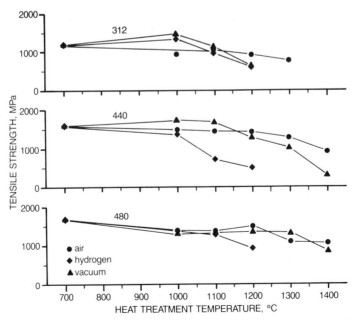

Figure 1.  Single-filament tensile strength after heat treatment in three atmospheres for three Nextel fibers.

intervals were typically ± 20% of plotted averages. All three fibers retained at least 75% of their heat cleaned (HC) strength through 1000°C. They showed a slight change in air up to 1200°C, beyond which Nextel 312 fibers showed a 67% retention through 1300°C and Nextel 440 and 480 fibers a 60% retention after 1400°C.

Some trends in vacuum and hydrogen differed from those in air. Nextel 312 fibers were stronger after 1000°C in both hydrogen and vacuum than after 700°C.

Above 1000°C Nextel 312 fibers showed steady loss of strength similar for both vacuum and hydrogen, reaching 50% of the HC strength after 1200°C. The Nextel 312 fibers were too brittle to measure strength after 1300°C in vacuum.

The strength of Nextel 440 fibers in vacuum improved slightly through 1100°C after which it decreased to 20% by 1400°C. In hydrogen Nextel 440 fibers retained their strength through 1000°C with a loss to 30% by 1200°C.

Nextel 480 fibers retained the most strength after all temperature treatments. In vacuum they retained 80% of their initial strength to 1300°C with a total retention of 50% after 1400°C. In hydrogen Nextel 480 fibers retained 80% of initial strength through 1100°C and 55% through 1200°C.

The elastic modulus of heat treated fibers increased 7-17% over the entire temperature range for all fiber and atmosphere types. 95% confidence intervals were typically ± 4% of the mean for each condition. The modulus of Nextel 312 fiber was initially 140 GPa. It increased 10% by 1000°C and 5% between 1200 and 1300°C. Nextel 440 fiber had an initial modulus of 186 GPa, which was constant through 1100°C and then increased 17% by 1400°C in air. The modulus of Nextel 480 fiber was a constant 225 GPa through 1200°C with a 7% increase by 1400°C.

Table III shows that Nextel 312 fiber had significant weight loss at and above 1100°C. Previous chemical analysis in our laboratory suggests most of this loss is boria, with some silica loss in vacuum at 1300°C. Nextel 440 fiber showed a slight weight loss in $H_2$ at 1200°C and both Nextel 440 and 480 fiber a small weight loss at 1400°C in vacuum, which from ICP analysis is felt to be mostly silica.

Figure 2 shows some microstructural differences among the three fibers before and after heat treatment. As made, fracture surfaces of Nextel 312 and 440 fibers have hackle lines indictative of glassy or amorphous structure. They are missing in the more crystalline Nextel 480 fibers. After four hours at 1300°C in air the fracture surfaces show a crystalline pattern for all three fibers. The SEM negatives at

Table III.  Percent Weight Loss During Heat
Treatment for Three Nextel Fibers

| Atmosphere | Temperature °C | Fiber Tested | | |
| | | Nextel 312 | Nextel 440 | Nextel 480 |
|---|---|---|---|---|
| Air | 1000 | .1 | 0 | 0 |
| | 1100 | .4 | 0 | 0 |
| | 1200 | 2.2 | 0 | .1 |
| | 1300 | 5.9 | 0 | 0 |
| | 1400 | | .1 | .1 |
| Hydrogen | 1000 | .1 | 0 | 0 |
| | 1100 | .3 | 0 | 0 |
| | 1200 | 1.6 | .3 | .1 |
| Vacuum | 1000 | .1 | 0 | 0 |
| | 1100 | .4 | 0 | 0 |
| | 1200 | 2.7 | 0 | 0 |
| | 1300 | 6.7 | .1 | .1 |
| | 1400 | | 1.0 | .8 |

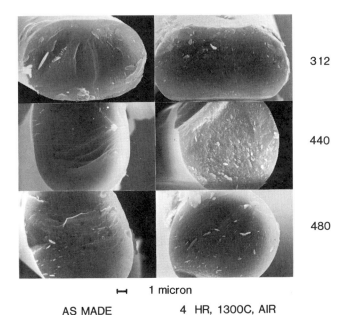

312

440

480

⊢⊣  1 micron

AS MADE          4 HR, 1300C, AIR

Figure 2.  Fracture surfaces of Nextel fibers before
and after heat treatment.

20,000X reveal a fine grain structure. Most grains are less than .2 μm initially and .25 μm after 1300°C. A few grains are as large as .5 μm before and after heat treatment with Nextel 312 fibers having more large grains.

## 4.2 Hot Strength

Figure 3 is a graph of the tensile strength of the three Nextel fiber systems at test temperatures up to 1300°C. The error bars indicate 95% confidence intervals for the mean. For Nextel 312 fibers, strength remained essentially constant through 800°C, beyond which the values decreased in a nearly linear fashion to its test limit of 1200°C. At 1000°C, as the strength fell below 60% of its initial level, the stress-strain curves began to show significant plastic deformation.

The strengths of the Nextel 440 and 480 fibers were similar through perhaps 1100°C, with the values for the former marginally higher at room temperature and 1000°C. Above 1100°C, the strength of the Nextel 440 fibers fell to 32% of its initial level at 1150°C and plastic deformation began to be seen. For the Nextel 480 fiber, the tensile strength decreased less quickly with temperature. The average value at 1250°C was still 531 MPa, and essentially elastic behavior was seen through the 1300°C testing limit.

## 4.3 Hot Modulus

The results of elastic modulus measurements at test temperatures up to 1350°C are presented in Figure 4, again with error bars to show the 95% confidence intervals for the mean. A small scaling factor was applied to each of the three curves

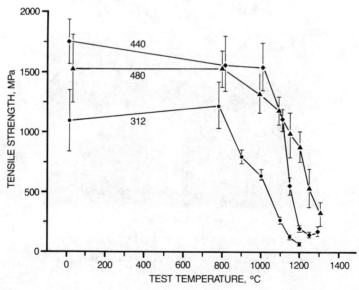

Figure 3. Hot tensile strength of Nextel single filaments in air.

so that the average modulus obtained at room temperature would correspond with reference data from normal, cold-grip tests.

Data taken on the Nextel 312 fibers showed a small (16%) decrease in modulus between room temperature and 800°C. Above this point modulus decreased more quickly, reaching 32% of its initial value at 1100°C.

For Nextel 440 fibers, the modulus at 900°C had fallen about 23% from its starting point. The rate of change above 900°C was more rapid, with data difficult to obtain above 1200°C.

The elastic modulus of Nextel 480 fibers did not appear to change through 900°C and decreased only a small amount through 1100°C. Data was obtained on this fiber through 1350°C, at which point the modulus

reached 70 MPa, or about 30% of its initial value.

## 4.4 High-Temperature Creep
The effects of temperature and stress on the creep rate of each fiber are shown in Figure 5. Each data point was obtained from the region of the elongation curve showing a constant strain rate. As indicated, the rovings were tested under two levels of stress: 69 MPa and 138 MPa.

These results showed large differences in the creep behavior of the three fiber types, with Nextel 480 having the best creep resistance and Nextel 312 the poorest. For values lower than about $10^{-5}$ per hour, creep rates for the Nexel 480 fibers departed from the expected, nearly linear trend. This was taken simply as an indication of the lower

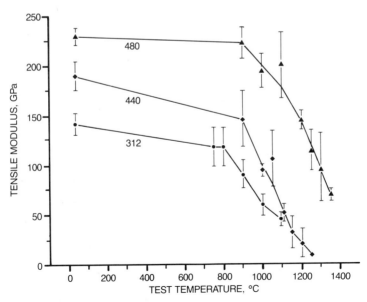

Figure 4. Hot elastic modulus of Nextel single filaments in air.

limit of resolution for this test.

Multiple linear regression was applied to the data above this limit to obtain values for activation energy and stress exponent (n) for each fiber type. These results were as follows:

Nextel 312:   67 KJ/mol, n = 2.2
Nextel 440:  210 KJ/mol, n = 2.8
Nextel 480:  472 KJ/mol, n = 5.4

The values for the Nextel 480 fiber are consistent with creep controlled by dislocation motion, whereas the other fibers may include a contribution from viscous or diffusion processes.[2]

The level of creep which is acceptable in a fiber will vary with the application. If a limit of $10^{-4}$ per hour (1% in 100 hours) under a load of 69 MPa was chosen as a benchmark, temperature limits for the Nextel 440 and 480 fibers would be 1010 and 1190°C respectively. An extrapolated value for the Nextel 312 fiber would appear to be no higher than 800°C.

## 5. CONCLUSIONS

The results presented above cover the physical properties of three oxide fibers for a variety of heat treatment conditions and test temperatures. To simplify comparisons among the fibers, Table IV shows the temperature for each fiber which corresponds with a particular level of the tested property. For tensile strength and elastic modulus, retention of 50% of

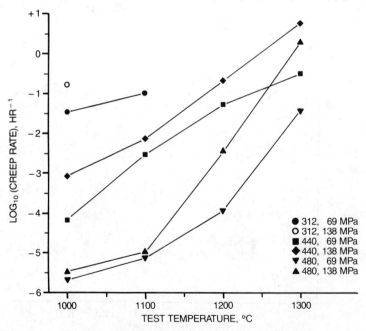

Figure 5.  Steady-state creep rate of Nextel fibers as a function of temperature and stress.

the initial level was chosen. A strain rate of $10^{-4}$/hour was used for comparisons on creep.

In almost every case, these temperatures follow a consistent ranking, with Nextel 312 fibers the lowest and Nextel 480 fibers the highest. As might be expected, properties are generally retained to higher temperatures for testing after heat treatment than for testing at temperature. It should also be noted that the range of temperatures drawn from the various tests is substantially narrower for the Nextel 480 fiber.

These trends are seen as a reflection of the fiber micro-structures described earlier. The Nextel 312 fibers are the least crystalline of the three as made, resulting in the greatest

sensitivity to testing at temperature. Property retention after heat treatment is much improved, however, by the concomitant change to a more crystalline form beginning at 900°C. The Nextel 480 fiber, conversely, is fully crystalline as made. This corresponds with a large improvement in stability during hot testing, as well as a decreased sensitivity to the atmosphere used during heat treatment.

The above test results suggest these fibers, especially Nextel 480, retain sufficient integrity to withstand ceramic composite fabrication temperatures up to 1300°C. They also suggest Nextel 480 fibers should perform structurally in a composite up to 1200°C, assuming compatibility with matrix materials.

Table IV.   Summary of Property-Temperature Relationships for Three Nextel Fibers

| Property | Benchmark | Comparative Temperature, °C | | |
| --- | --- | --- | --- | --- |
| | | Nextel 312 | Nextel 440 | Nextel 480 |
| Strength after heat treatment | 50% of initial: | | | |
| | in air | >1300 | 1400 | >1400 |
| | in hydrogen | 1200 | 1100 | 1200 |
| | in vacuum | 1200 | 1350 | 1400 |
| Hot strength | 50% of initial | 1000 | 1120 | 1200 |
| Hot modulus | 50% of initial | 950 | 1000 | 1250 |
| Creep rate | $10^{-4}$/hr at 69 MPa | 800 | 1010 | 1190 |

## 6. ACKNOWLEDGEMENTS

The preparation of SEMS by Dr. Chris Goodbrake and Dr. Elizabeth Richards, XRD analysis by Dr. Walt Thatcher and consultations with Dr. Vincent Nehring on inter-pretation of creep behavior is gratefully acknowledged.

## 7. REFERENCES

1. P.M. Sawko and H.K. Tran, "Strength and Flexibility Properties of Advanced Ceramic Fabrics", SAMPE Quarterly, Vol. 17, No. 1, October, 1985, pp 7-13.

2. W.R. Cannon and O.D. Sherby, "Third-Power Stress Dependence in Creep of Polycrystalline Nonmetals," J. Amer. Ceram. Soc., Vol. 56, No. 3, March, 1973, pp 157-160.

## 8. BIOGRAPHIES

Mr. Holtz is a Senior Ceramic Process Development Engineer of the Ceramic Materials Department at 3M. He received a B.S. degree in Ceramic Engineering from Iowa State University in 1972. In the past, he developed the process for making 3M's Nextel Ultrafiber and conducted studies to improve process under-standing of various Nextel ceramic fibers. His current responsibili-ties include development of fibers for composites.

Mr. Grether received B.S. (1977) and M.S. (1979) degrees in Ceramic Engineering from Clemson University. He then joined 3M, where he is currently a Senior Product Development Engineer in the Ceramic Materials Department. Areas of involvement since then have included compatibility studies for thick-film circuit materials, compositional development of ceramics for computer memory applications, and various hot pressing and vacuum sintering techniques. Mr. Grether's current responsibilities include characterization of the high-temperature behavior of ceramic fibers for composites.

IMPACT BEHAVIORS OF HYBRID FIBER
REINFORCED THERMOPLASTIC COMPOSITES

CHEN-CHI M. MA AND BIING-YNG SHIEH

DEPARTMENT OF CHEMICAL ENGINEERING
NATIONAL TSING HUA UNIVERSITY
HSIN-CHU, TAIWAN, REPUBLIC OF CHINA

ABSTRACT

Impact behaviors of hybrid fiber
reinforced polystyrene have been
investigated. Fiber reinforcements
used include chopped strand mats of
glass, carbon, interply glass/carbon
hybrid fiber and sandwich type
glass/carbon hybrid fiber.

Both initiation and propagation
impact energies of composites were
measured by instrumented impact
tester. The "Ductility Index (DI)",
an indication of the brittleness of
composite material, was calculated.
Results show that the impact strength and the DI value increase with
the increasing of fiber content in
the hybrid systems. Effects of the
lamination sequence, the type of
hybridization (i.e. glass/carbon
ratio) and core/shell arrangement
on impact behaviors are reported.
A comparison of the instrumented
impact test results with Izod impact
test data is also presented.

1.0 INTRODUCTION

Carbon fiber reinforced composites
posses the characteristics of high
strength, high modulus and low
density. In advanced materials,
particularly those used in the aircraft and aerospace industry, the
high strength-to-weight ratio character is very attractive. However,
low impact strength is one of the
disadvanges of the carbon fiber
reinforced composites. Combining of
expensive and brittle fiber, such as
carbon fiber, with cheaper and
ductile fiber, such as glass fiber,
could improve composite properties
and reduce some undesirable qualit-
ites and cost as well. By using
hybrid fiber it is possible to meet
specific design requirements and to
broaden the composite materials
applications.

Toughness is one of the important properties of engineering materials, especially in aircraft structures. For example, strikes of hail or bird sometimes lead to a catastrophic failure of the materials. As a result, research on the improvement of impact strength has been extensive. One of the impact modified composites is the long fiber reinforced thermoplastic composites. Researches have been conducted and demonstrated the potential of using these materials to substitute some metallic material in near future (1,2).

In order to develop tough composites, one must investigate the impact behavior of materials first. Unfortunately, most of the traditional impact testing methods fall short to some degrees in providing data to evaluate and predict impact performance of testing materials (3). For instance, a high strength brittle material may have the same total impact energy as a low strength ductile material, however, the failure modes and performances of these two materials might be quite different. The former has a large initiation energy and a small propagation energy, but the latter has a small initiation energy with a large propagation energy. These informations can not be obtained from a traditional Izod impact tester. In order to solve these problems and to further investigate the toughness of a material, instru-

mented impact testing machines have been developed (4-14).

In this research, the effects of carbon fiber content, carbon/glass hybrid system, and carbon/glass laminates types on the impact behavior of polystyrene-based thermpolastic composites have been studied. A comparison of the traditional Izod impact test data and an instrumented falling weight impact test results is also reported.

## 2.0  EXPERIMENTAL

### 2.1  Materials

#### 2.1.1  Thermoplastic Resin

Polystyrene (PS): Poly PG-79, Poly Petrochemical Inc., Taiwan, Republic of China.

#### 2.1.2  Fiber Reinforcements:

(1) Glass fiber mat: chopped strand glass fiber mat which is 0.5 OZ/$YD^2$ in weight.
(2) Carbon fiber mat: chopped strand carbon fiber mat, the weight of the mat is 1.0 OZ/$YD^2$.

Both mats are manufactured by the International Paper Company, Tuxedo, N.Y., U.S.A., which were treated with a high solubility thermoplastic binder and both are one inch in length.

#### 2.1.3  Processing Method:

258

Thermoplastic composites were laminated by simply compressing the alternating matrix sheet with chopped strand fiber mat. Processing conditions follow the suggested molding conditions from supplier. In this study, the following arrangements of hybrid lamination were used:

(1) Sandwich type laminate:
  (a) Carbon core and glass shell, (G/C/G).
  (b) Glass core and carbon shell, (C/G/C).
(2) Alternate type laminate.

2.1.4  Sample Preparation:

Notched Izod impact test specimens were prepared according to the standard of ASTM D256-81, and instrumented falling weight impact test specimens were 8 $cm^2$ in size. All specimens were cut from composite laminates by using a circular diamond tip saw.

2.2  Testing Instruments and
     Procedures:

2.2.1  Testing Instruments:

The instruments utilized in this study include an Izod impact tester and an instrumented falling weight impact tester. The Izod impact tester used is a traditional impact testing machine from TMI, Inc., U.S.A.. The instrumented falling weight impact testing system was designed by MRL, ITRI, Taiwan,

Republic of China. It has a piezo-electric type load cell and a power unit, a transient recoder stores the impact load signal which can feed back to provide a hard copy plot of data, thus the load and energy versus time curves can be obtained. The system has a striking tup which is 1.5 cm in diameter and with a hemispherical head as in Gardner test, or similar tests specified in ASTM D3029-82a procedure B. The tup and striker have a total weight of 6.7 Kg. A specimen support with a circular opening of 6 cm in diameter is used.

2.2.2  Testing Procedures:

Detail items and procedures of Izod impact test follow ASTM D256-81. Testing procedures of the falling weight impact testing are described as follows:
Test specimens were laid on the specimen support and unclamped. The height from the tip of the tup to the specimen is 0.86 m. Since

$$V_0 = \sqrt{2gh_0} \qquad (1)$$

$$E_0 = \frac{1}{2} mV_0^2 \qquad (2)$$

the velocity ($V_0$), and the total impact energy ($E_0$) before tup strikes the specimen is 4.1 m/s and 56.3J respectively. Symbols and notations are listed in Table 1. These values are large enough to fracture the specimens under test.

After striking the specimen, the transient recoder stores the impact load singal and then a hard copy plot was used to plot the detail curves of load and energy versus time.

From the curves of load and energy versus time and failure samples, one can investigate the impact energies, the maximum load, and the history of the impact process.

The testing conditions in this study are at room temperature and 50% R.H. humidity.

### 3.0   RESULTS AND DISCUSSION

3.1   Instrumented Falling Weight Impact Test versus Izod Impact Test:

Testing results from Izod impact tester and instrumented falling weight impact tester are summarized in Tables 2 and 3. Both testers can provide total impact energy of tested materials, however, the instrumented falling weight impact tester may provide more informations than the Izod impact tester does. Figure 1 shows typical curves of load and energy versus time obtained by the instrumented falling weight impact tester. From curves of the load and energy versus time, the initiation impact energy ($E_i$) is defined as the point where the load reaches a maximum, and beyond this point is the propagation of failure.

The total energy required to fracture the impact specimen is defined as the total impact energy ($E_t$), which is the area under the load-time curve up to the point " $\otimes$ " as shown in the curves. Both energies can be divided by the thickness of each specimen (d) to obtain their normalized values (5), thus:

$$U_t = E_t/d \qquad (3)$$

$$U_i = E_i/d \qquad (4)$$

From the total impact energy, the region between initiation energy and total impact energy is called propagation energy ($E_p$), which represents the energy required to propagate fracture. If the propagation energy is low, the specimen will break suddenly, while for a high propagation energy material, the fracture propagates in a progressive manner continuing to absorb energy at small load up to fail the specimen, hence, the propagation energy per unit thickness is given by:

$$U_p = U_t - U_i = E_t/d - E_i/d \qquad (5)$$

The ratio of $E_p$ to $E_i$ or $U_p$ to $U_i$ can be defined as "Ductility Index (DI)":

$$DI = E_p/E_i = U_p/U_i \qquad (6)$$

DI value ranges from zero to infinite, the smaller the DI value the more brittle the material. From these informations one can identify the ductile character of a material.

## 3.2 Effects of Hybrid Fiber Content and Type on Impact Strength:

Table 2 illustrates the comparative impact strength of polystyrene composites reinforced with various fibers. From figure 2, one can observe that the impact strength decreases with the increasing of carbon fiber content, and a positive deviation from the rule of mixture relationship is existed. After combining two types of fiber into the polystyrene matrix, the improvement of the impact strength is explicit. Since carbon fiber is a strong and brittle fiber with very low fracture strain, while, glass fiber is ductile and can tolerate very high fracture energy, composite reinforced by these two types of fiber will have a significant improved impact strength. Figure 3 shows that the impact behavior of glass fiber reinforced Polystyrene composite which is a ductile material, has a broad load-time curve and a higher DI value, i.e., it can absorb more impact energy and fails in a progressive manner. For a carbon/glass hybrid fiber reinforced composite, when the brittle component (e.g. carbon fiber was used in this study) content increases, the load-time curves become narrower and DI values are decreased as shown in Figures 4 to 8. Carbon fiber

reinforced composite has a very narrow load-time curve, i.e. its impact strength is very low and DI value is also low as illustrated in Table 2 and Figure 8. Furthermore, from the failure samples as shown in Figure 9, it was found that ductile material has a large failure zone, the more brittle the material the smaller the failure zone. The brittle material, such as carbon fiber composite, can be broken into pieces easily. In our previous research (15) showed that both tensile strength and flexural strength increase with the carbon fiber content. It is an important task for material engineer to tail the material in order to obtain optimum properties.

## 3.3 Effect of Hybrid Fiber Arrangement on the Impact Behaviors:

Table 3 summarizes the results of the comparative impact strength of carbon/glass (50:50,wt%) hybrid composites, which have the same composition but different fiber arrangements. Although the total impact strength of these materials are similar but their failure pattens are quite different. Figure 10 shows the impact damage record of a carbon core and glass shell sandwich type composite. Result exhibits that this composite has a ductile character. The possible explanation of these phenomena is that, when tup strikes the specimen, the failure of the carbon core is buckling and the material reaches its initial fracture immediately. The glass shell has high fracture toughness and fails in a progressing manner, hence, the composite material has a low initation energy and a high propagation energy, a high DI value and a ductile character formed.

Figure 11 illustrates the impact result of a glass core and carbon shell sandwich type composite. Result indicates that this hybrid fiber reinforced composite has a brittle character. After analyzing of this test, one can found three possible steps may occur. At first, when tup strikes the specimen, upper carbon shell was broken quickly. Since the glass core supports the load of the tup, the specimen did not reach its initiation fracture at this point. When tup breaks the glass core, the specimen reaches its initiation fracture at the same time. As the tup reaches the bottom carbon shell, since it is a brittle material, composite was broken suddenly. Hence, the material shows a very low DI value and brittle character.

After comparing Figures 5,10, and 11, one can find that, when the same polymer matrix was reinforced by hybrid fibers with various arrangements, composites show different ductility. Carbon core and glass shell sandwich type hybrid fiber reinforced composite show the most ductile character, while glass core and carbon shell sandwich type hybrid fiber reinforced composite shows more brittle than the composite reinforced with alternated laminate type hybrid fibers. Figure 12 also shows the impact damage of composites with various hybrid arrangements. The failure zone of carbon core and glass shell sandwich type hybrid fiber reinforced composite is larger than that of a glass core and carbon shell sandwich type hybrid fiber reinforced composite which has a small hole only. In other words, the impact damage of the material can be predicted by designing and tailoring the material in advance.

## 4.0 CONCLUSIONS

This research has shown that:
(1) The impact behaviors of carbon/ glass hybrid fiber reinforced polystyrene composites show positive hybrid effect.
(2) The effect of fiber reinforcement on the impact strength of hybrid fiber reinforced polystyrene composites were investigated. Both impact strength and ductility of the composites decrease with the increasing of carbon fiber content for carbon/ glass hybrid system.
(3) Carbon core and glass shell sandwich type of carbon/glass (50:50 wt%) hybrid composite

shows more ductile than that of alternating laminate type hybrid composite. However, glass core and carbon shell sandwich type hybrid composite shows brittler than alternating laminate type hybrid composite.

## 5.0 ACKNOWLEDGEMENTS

The authors gratefully acknowledge Mr. Chun-Hung Chen, MRL, ITRI, Taiwan, R.O.C., for his valuable assistance in instrumented falling weight impact testing, and Dr. Alex Y. Lou, the Phillips Petroleum Co., U.S.A., for helpful discussions.

## 6.0 REFERENCES

1. Myhre, J.B. and Baumann, J.A., "Glass Reinforced Thermoplastic Laminate/Sheets-Response to Evolving Technology", the 37th Annual Conference, Reinforced Plastics/Composites Institute, The Society of the Plastics Industry Inc., Sec. 25-D (1982).

2. Plastic Design Forum (Sept./Oct. ), pp. 41(1980).

3. Starita, J.M., "Impact Testing-what it can really tell you", Plastic World, April, 58(1977).

4. Davies, C.K.L., Turner, S., and Williamson, K.H., "Flexed Plate Impact Testing of Carbon Fiber Reinforced Polymer Composites", Composites, 16, 279(1985).

5. Lou, A.Y., "Instrumented Impact Testing of PPS-Based Stampable Composites", the 30th Nat'l

SAMPE Symposium, March, 786(1985 ).

6. Perry, J.L., and Adams, D.F., "Charpy Impact Experiments on Graphite/Epoxy Hybrid Composites ", Composites, 6, 166(1975).

7. Golovoy, A., "Instrumented Impact Testing of Glass Reinforced Polypropylene", Poly. Eng. Sci., 25, 903(1985).

8. Wnuk, A.J., Ward, T.C., and McGrath, J.E., "Design and Application of an Instrumented Falling Weight Impact Tester," Poly. Eng. Sci., 21, 313(1981).

9. Yeung, P., and Broutman, L.J., "The Effect of Glass-Resin Interface Strength on the Impact Strength of Fiber Reinforced Plastics", Poly. Eng. Sci. 18, 62(1978).

10. Zoller, P., "Instrumentation for Impact Testing of Plastic," Polymer Testing, 3, 197(1983).

11. Mallick, P.K., and Broutman, L.J., "Impact Behavior of Hybrid Composites", the 30th Anniversary Technical Conference, SPI RP/CI, Sec. 18-F(1975).

12. Gonzalez, H.Jr., and Stowell, W.J., "Development of an Autographic Falling Weight impact system," J. Appl. Poly. Sci., 20, 1389(1976).

13. Ma, C.C., et.al., 1984 Annual Technical Conference, SPE Inc., New Orleoms, Louisiana, U.S.A., 690(1984).

14. Chen, C.H., and Wu,R.Y., "The Impact Preperties of Carbon Fiber/Epoxy Composite Material

with Various Ply Arrangements by Instrumented Falling Weight Impact Test," the 31st International SAMPE Symposium, April 1549(1986).

15. Ma, C.C., Shieh, B.Y., and Shah, H.J., "Long Fiber Reinforced Stampable Thermoplastic Sheet Composites (II)," Proceeding of the 2nd China and Japan Seminar on Carbon Fiber, Nov., 1(1985).

## 7.0 BIOGRAPHIES

Dr. Chen-Chi Martin Ma is a Professor of Chemical Engineering at National Tsing Hua University, Hsin-Chu, Taiwan, Republic of China since 1984. He received his Ph.D. degree in Chemical Engineering from North Carolina State University at Raleigh, North Carolina, U.S.A., in 1978. He has worked in the past with Monsanto Company (1977-1979) as a senior research engineer and polymer rheologist, with Lord corporation (1979-1980) as a senior research associate, and with Phillips Petroleum Company (1980-1984) as a research engineer in the Advanced Materials Section at the Phillips Research Center. Professor Ma's research interests include the processing of high performance thermoplastic composites, pultrusion and polymer blends. Professor Ma is an author and coauthor of textbooks and more than 40 technical papers and patents.

Biing-Yng Shieh is an associate scientist of Material Research Laboratory (MRL) of Industrial Technology Research Institute (ITRI) in Hsin-Chu, Taiwan, Republic of China. He received his B.S.(1984) and M.S. (1986) in Chemical Engineering from National Tsing Hua University. His research interest is in the field of advanced composite material processing and applications.

TABLE 1. SYMBOLS AND NOTATIONS

$V_0$: Velocity before tup strikes the specimen (m/s)

$E_0$: Total impact energy before tup strikes the specimen(J)

g : Local acceleration of gravity (kg m/N $S^2$)

$h_0$: Height from the tip of the tup to the specimen(m)

m : Mass of tup (Kg)

$E_t$: Total impact energy(J)

$E_i$: Initiation impact energy(J)

$E_p$: Propagation impact energy(J)

d : Thickness of specimen(mm)

$U_t$: Normalized total impact energy (J/mm)

$U_i$: Normalized initiation impact energy(J/mm)

$U_p$: Normalized propagation impact energy(J/mm)

DI: Ductility index(dimensionless)

TABLE 2. THE COMPARATIVE IMPACT TEST RESULTS OF HYBRID FIBER REINFORCED PS COMPOSITES

| FIBER (carbon: Glass wt.ratio) | Total Fiber (wt%) | Carbon Fiber Fraction (vol%) | INSTRUMENTED FALLING WEIGHT IMPACT TEST | | | | | | IZOD IMPACT TEST |
|---|---|---|---|---|---|---|---|---|---|
| | | | MAX.LOAD (N) | $E_i$ (J) | $E_p$ (J) | $E_t$ (J) | NOR.$E_t$ (J/mm) | D.I. | NOR.$E_t$ (J/mm) |
| 0:1 | 29.1 | 0 | 977.4 | 1.46 | 5.80 | 7.26 | 4.40 | 3.97 | 0.30 |
| 1:3 | 29.4 | 32 | 695.0 | 1.33 | 4.60 | 5.93 | 3.78 | 3.46 | 0.26 |
| 1:1 | 27.5 | 59 | 749.3 | 1.49 | 3.18 | 4.67 | 3.07 | 2.13 | 0.24 |
| 3:1 | 29.3 | 81 | 727.6 | 1.02 | 2.45 | 3.47 | 2.09 | 2.40 | 0.21 |
| 1:0 | 27.7 | 100 | 510.4 | 0.71 | 0.87 | 1.58 | 0.92 | 1.23 | 0.12 |

TABLE 3. THE COMPARATIVE IMPACT TEST RESULTS WITH VARIOUS FIBER ARRANGEMENTS OF C/G (50:50 wt%) HYBRID FIBER REINFORCED PS COMPOSITES

| FIBER Arrange-ment | Total Fiber (wt.%) | Carbon Fiber Fraction (Vol.%) | INSTRUMENTED FALLING WEIGHT IMPACT TEST | | | | | | IZOD IMPACT TEST |
|---|---|---|---|---|---|---|---|---|---|
| | | | MAX.LOAD (N) | $E_i$ (J) | $E_p$ (J) | $E_t$ (J) | NOR.$E_t$ (J/mm) | D.I. | NOR.$E_t$ (J/mm) |
| ALT | 27.5 | 59 | 749.3 | 1.49 | 3.18 | 4.67 | 3.07 | 2.13 | 0.240 |
| G/C/G | 28.1 | 58 | 803.6 | 1.13 | 3.87 | 5.00 | 3.11 | 3.42 | 0.242 |
| C/G/C | 28.5 | 59 | 955.7 | 3.37 | 1.69 | 5.06 | 3.18 | 0.50 | 0.242 |

ALT : Alternating laminate type
G/C/G: Carbon core and glass shell type
C/G/C: Glass core and carbon shell type

265

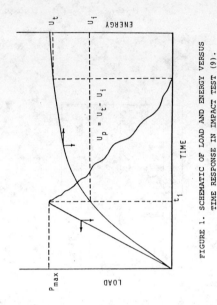

FIGURE 1. SCHEMATIC OF LOAD AND ENERGY VERSUS
TIME RESPONSE IN IMPACT TEST (9).

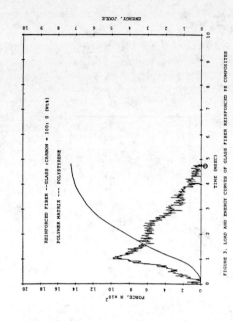

FIGURE 3. LOAD AND ENERGY CURVES OF GLASS FIBER REINFORCED PS COMPOSITES

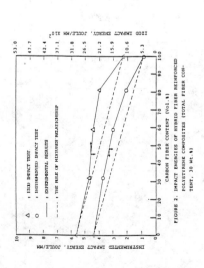

FIGURE 2. IMPACT ENERGIES OF HYBRID FIBER REINFORCED
POLYSTYRENE COMPOSITES (TOTAL FIBER CON-
TENT, 30 Wt.%)

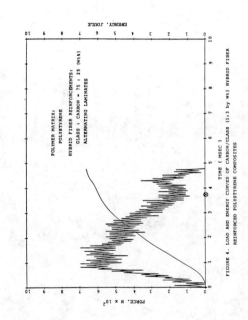

FIGURE 4. LOAD AND ENERGY CURVES OF CARBON/GLASS (11:3 by Wt) HYBRID FIBER
REINFORCED POLYSTYRENE COMPOSITES

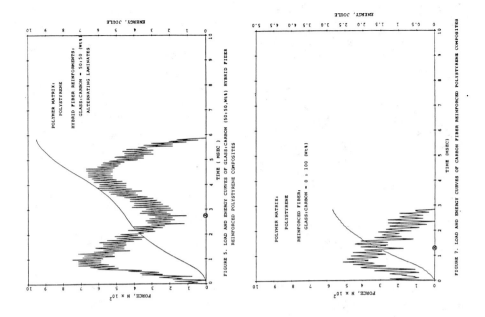

FIGURE 5. LOAD AND ENERGY CURVES OF GLASS/CARBON (50:50,Wt%) HYBRID FIBER
REINFORCED POLYSTYRENE COMPOSITES

POLYMER MATRIX:
POLYSTYRENE

HYBRID FIBER REINFORCMENTS:
GLASS:CARBON = 50:50 (Wt%)
ALTERNATING LAMINATES

FIGURE 7. LOAD AND ENERGY CURVES OF CARBON FIBER REINFORCED POLYSTYRENE COMPOSITES

POLYMER MATRIX:
POLYSTYRENE

REINFORCED FIBER:
GLASS:CARBON = 0 : 100 (Wt%)

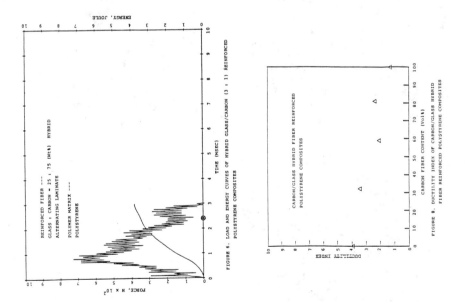

FIGURE 6. LOAD AND ENERGY CURVES OF HYBRID GLASS/CARBON (3 : 1) REINFORCED
POLYSTYRENE COMPOSITES

REINFORCED FIBER ---
GLASS : CARBON = 25 : 75 (Wt%) HYBRID
ALTERNATING LAMINATE

POLYMER MATRIX ---
POLYSTYRENE

FIGURE 8. DUCTILITY INDEX OF CARBON/GLASS HYBRID
FIBER REINFORCED POLYSTYRENE COMPOSITES

CARBON/GLASS HYBRID FIBER REINFORCED
POLYSTYRENE COMPOSITES

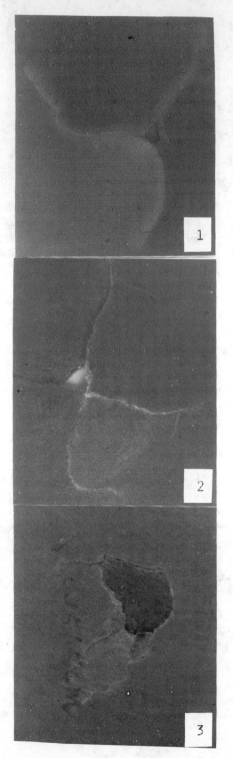

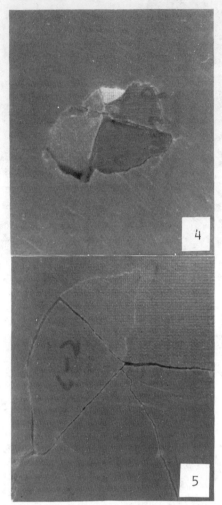

1. 100% GLASS FIBER
2. CARBON/GLASS (1:3)HYBRID
3. CARBON/GLASS (1:1) HYBRID
4. CARBON/GLASS (3:1) HYBRID
5. 100% CARBON FIBER

FIGURE 9. IMPACT DAMAGE OF
  HYBRID FIBER REINFORCED
  POLYSTYRENE COMPOSITES

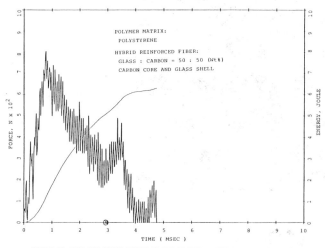

FIGURE 10. LOAD AND ENERGY CURVES OF CARBON CORE AND GLASS SHELL C:G (1:1)
HYBRID FIBER REINFORCED POLYSTYRENE COMPOSITES

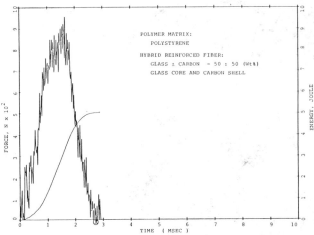

FIGURE 11. LOAD AND ENERGY CURVES OF GLASS CORE AND CARBON SHELL G:C (1:1)
HYBRID FIBER REINFORCED POLYSTYRENE COMPOSITES

ALTNATING TYPE LAMINATE

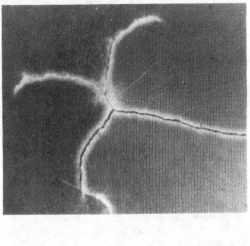

CARBON CORE AND GLASS SHELL
TYPE LAMINATE (G/C/G)

GLASS CORE AND CARBON CORE
TYPE LAMINATE (C/G/C)

FIGURE 12. IMPACT DAMAGE OF HYBRID FIBER REINFORCED COMPOSITES
WITH VARIOUS ARRANGEMENTS

FIELD REPAIR COMPOUNDS FOR
THERMOSET AND THERMOPLASTIC COMPOSITES

Edwin C. Clark and Kenneth D. Cressy
Furane Products Company
5121 San Fernando Road West
Los Angeles, CA 90039

## Abstract

There is currently a push by the aerospace industry to utilize more thermoset and thermoplastic materials in the construction of aircraft. This has created new demands for repair of these vehicles in the field. Restrictions imposed on repair materials in the field environment were used to evaluate current repair techniques. This evaluation uncovered the need for materials that perform at higher temperatures and have the ability to bond contaminated surfaces. These deficiencies were most pronounced in the repair of composite laminates and adhesive bonded structures. Syntactic materials currently being used to repair damaged honeycomb were found to meet repair requirements. There is also a lack of data on the use of field repair materials on the new generation thermoplastics. It is in the area of high temperature performance, surface insensitivity, and thermoplastic repair that future developmental work must be focused.

## 1. INTRODUCTION

Repair of composite structures in the aircraft industry can be found in three areas, manufacturing, depot, and field locations. Of these three, field repair is the most demanding in terms of time spent on, and simplicity of, repair. This is due to the short turn around time required for maximum equipment utilization and the limited facilities and personnel training found at these sites. Because of the increasing use of composites in aircraft, emphasis is being placed on the development of field repair materials by both the commercial and military sectors.

Materials and techniques for field repair have increased substantially in the last few years. Many papers have been given on the repair of specific types of damage. Others have focused on new materials that appear to meet the challenge out in the field. It is the intent of this paper to review three commonly encountered repair areas, namely damage to composite laminates, damaged honeycomb structures, and disbonds on thermoset and thermoplastic materials. This review will characterize the field level requirements for polymeric materials and the techniques used to make repairs and will profile the types of materials currently available for use.

## 2. FIELD LEVEL REQUIREMENTS

Of the three areas of composite pair, field repair has the lowest level of trained personnel and equipment. Because of these restrictions, field repair materials must perform functions that are not required of the original building materials. The following is a list of repair material requirements that are currently being sought by field personnel:[1]

1) Shelf life in excess of 1 year at room temperature.
2) Easy to use with repeatable results.
3) Curability with or without external heat.
4) Cure temperatures below 212°F.
5) Capable of achieving full properties within 2 hours after application.
6) Relatively non-hazardous.
7) Can be effective even with surface contamination.

This list of requirements will be used to evaluate currently available materials designed to repair damaged laminates, honeycomb, and disbonds between thermoplastic and thermoset substrates.

### 3. LAMINATE DAMAGE

Typical damage to composite laminates includes delamination between plies and destruction of material due to impact. Figure 1 shows a typical delamination between plies in a composite structure. The current field repair technique for repairing this type of damage is to remove the damaged material either through cutting or sanding as shown in Figure 2. The same type of fabric that was used in the original composite is then cut to shape and impregnated with a laminated resin. The plies are then laid in the same orientation as the original composite (Figure 3). The plies can then be cured either at room temperature or heated using heat blankets, hot air or I.R. lamps. Vacuum bagging can also be used to insure proper consolidation of the repaired material. Through the use of a scharf joint, ply alignment, and proper selection of laminating resin, over 60% of unflawed strength in the repaired laminate may be achieved. With design allowables in the range of 40-60% of unflawed strength, many times this type of repair will return the part to it's original capability.[2]

To profile the typical field repair laminating resin, a review of cur-

rently used materials was performed. Certain criteria were placed on the resin systems before they were considered. The restrictions included shelf life in excess of one year, materials capable of both room temperature and elevated temperature cure, and materials with low toxicity. The profile generated from this is shown on Figure 4. Due to the variety of tests used to characterize laminating resins, the mechanical part of the profile was limited to tensile and flexural strengths. Based on this profile, it appears that current resin systems can in many cases perform adequately. There are, however, areas where current materials cannot adequately be used to make repairs and where improvements need to be made.

Deficiencies of field repair laminating resins center on their relatively low temperature capability, their need to be simple to use, and their difficulty in bonding contaminated substrates. With requirements for long shelf life stability, short cure cycles, and low temperature cures, laminating systems used today can only operate efficiently up to 180°F. Due to the lack of equipment and the need for short repair times, repair resins must be more tolerant of surface contamination. Finally, in the field there is constantly changing personnel with various levels of training. With this situation, repair materials must be simple to use and have an inherently high level of reliability. This involves resin systems with convenient mix ratios, color coded components, and good wetting characteristics. These deficiencies need to be addressed if wet laminate repairs are to be used effectively in the active field environment.

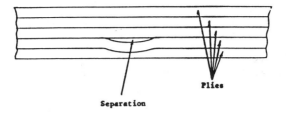

Figure 1.  Ply Delamination

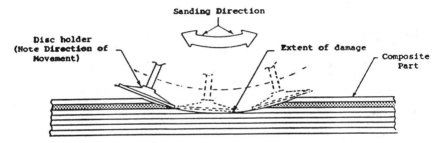

Figure 2.  Removal of Damaged Laminate

Figure 3.  Relamination of Damaged Area

273

Figure 4.    Typical Laminating
Repair Resin

Shelf Stability:  3-12 months, 77°F
Specific Gravity:  1.10 - 1.35 g/cc
Viscosity: 2,000 - 7,000 cps
Gel Time: 3-120 min, Ave. 30 min.
Cure:  2 Hrs. @ 200°F Max.

## Mechanical Properties:

12 Ply, Glass Reinforced Laminate

Resin Content: 26-36%
Tensile Strength @ 77°F: 30,000-50,000 psi
Flexural Strength @ 77°F: 45,000-80,000 psi
Hardness:  70-90  Shore D
Flexural Strength @ 180°F = 10,000-30,000 psi
Hot/Wet Flexural Strength (160°F, 95% R.H.):
  reductions in strength range from 20-70%

## 4.  HONEYCOMB STRUCTURE DAMAGE

Damage to honeycomb structures generally takes on the appearance of delamination between laminate and honeycomb, or deformation of the structure due to impact (Figure 5). Typically, there are three ways of repairing honeycomb structure in the field. The first method is to simply fill in the dented area with a body filler or syntactic type material (Figure 6). In this case the damage is minimal and repair brings the structure back to aerodynamic efficiency.   The second method is to remove the skin and honeycomb from the damaged area. After removal, the hole is filled with a structural syntactic foam then a new skin is built over it (Figure 7). In this case, the syntactic can either be cured before hand or co-cured with the skin. This type of repair can achieve original strengths in many cases, but may result in additional weight gains. The third type of repair involves removal of the damaged skin and honeycomb. A new piece of honeycomb is cut to replace the material that was removed.  The new honeycomb is then set into place with a splicing or high strength syntactic material (Figure 8).   Again the

skin is replaced with the syntactic either being cured before or with the skin.  The advantages of this method of repair is that it can frequently achieve original strengths  with minimal weight gains.

Currently, there are two major types of honeycomb used in aerospace manufacturing, a polyimide coated paper and aluminum. The densities of these materials range from 1.0 to 10.0 lbs./cu.ft. with ultimate compressive strengths up to 1600 psi. [3] The density of syntactics now being used in the field range from 20 to over 62 lbs./cu.ft. Based on their density, syntactics can have compressive strengths in excess of 15,000 psi.  Figure (9) reviews the typical properties of syntactics now in use in the field

An evaluation of the data in Figure (9) indicates that field requirements for honeycomb repair are being met with current syntactic materials. Even at elevated temperatures, the syntactic materials have higher compressive strengths than the original honeycomb.  They also meet the field requirements for a two hour cure at temperatures below 200°F.  The major disadvantage of using these materials is that in every repair case, additional weight is added to the aircraft.

Future development work should focus on obtaining higher compress-strengths with lower density materials.

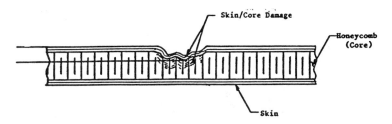

Figure 5.  Impact Damage to Laminate, Honeycomb Structure

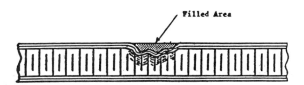

Figure 6.  Repair of Minor Impact Damage

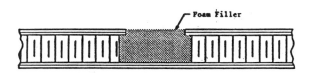

Figure 7.  Replacement of Honeycomb
with Syntactic Foam

Plug Patch

Figure 8. Splicing in Replacement Honeycomb

Figure 9.  Syntactic Foam Repair Materials

Work Life:  10-90 min.

Consistency: Non-Sag Paste

Cure Schedule: 48 hrs. @ 77°F or
2 hrs. @ 150°F

| Density (lbs/cu.ft.) | Compressive Strength(77°F) | Service Temp., (°F) |
|---|---|---|
| 31.2 | 4,000 | 250 |
| 34.3 | 4,500 | 250 |
| 37.5 | 5,000 | 250 |
| 43.7 | 6,500 | 250 |
| 49.9 | 8,000 | 350 |

## 5. REPAIR OF
## ADHESIVE DISBONDS

Adhesive disbonds generally occur due to environmental attack on the bondline, improper use of adhesive, poor surface preparation or due to assembly design strenths being exceeded. To repair disbonds, the area in question must be cleaned of all residual adhesive. The bonded surface is then pretreated by either solvent wiping, sanding, sand blasting, and/or the application of an adhesion promoter. The adhesive is then applied to both surfaces to be bonded, and then assembled. The bond must then remain immobile until the adhesive achieves sufficient strength to hold the part together.

A significant amount of work has recently gone into incorporating new thermoset and thermoplastic materials into commercial and military aircraft. The emphasis of this work has been on the development and utilization of these polymers to increase toughness, high temperature performance, and flame resistance. Figure (10) lists a number of the plastics that are currently being used or under development, and the bond strengths that can be achieved using current field repair adhesives.

After reviewing the information in Figure (10), it appears that current field repair adhesives perform adequately on many of todays more commonly used plastics. However, there are still two areas where deficiencies exist. The first involves the same restrictions that were discussed during the examination of laminating resin systems. This includes the need for higher temperature performance and the ability to bond to contaminated surfaces. The second area concerns the utilization of new thermoplastics. Very little data has been generated on these materials (polyetheretherketone, poly(amide-imide), polyester liquid crystals, and polyimide) using current field repair adhesives. It is anticipated that there will be problems adhesively bonding these materials due to their high temperature usage and restrictive solubility in existing solvents. It is in these two areas that work needs to be focused if field adhesive bonding is to advance to meet future performance requirements.

Figure 10. Bonding Capabilities of
Field Repair Adhesives

| Plastic Substrate | |
|---|---|
| Acrylic | Substrate Failure |
| Bismaleimide | Substrate Failure |
| Epoxy | Substrate Failure |
| Phenolic | Substrate Failure |
| Poly(amide-imide) | NA |
| Polycarbonate | Substrate Failure |
| Polyester Liquid Crystal(4) | 700 |
| Polyether ether Ketone | NA |
| Polyetherimide (5) | 1120 |
| Polyethersulfone | Substrate Failure |
| Polyphenylene Sulfide (6) | 1500 |

## 6. CONCLUSION

Field damage repair is currently in a state of change. With the utilization of new high performance materials and aircraft designs, new emphasis must be placed on repair. Within the restrictions placed on the materials by the field environment, repairs to damaged laminates, honeycomb, and adhesive bonds have been evaluated. Honeycomb repair materials have been found to perform very well with the exception of adding additional weight to the aircraft. In the areas of laminate and adhesive repair, materials are available that will selectively perform well. The limitations that currently exist are in high temperature performance and substrate sensitivity. These areas will have to be advanced in order to meet future field damage repair needs.

## 7. BIOGRAPHY

Edwin C. Clark: B.S. Chemistry(79), MBA-Marketing (85), Univ. of Akron. Work experience has been in the areas of adhesive, coatings and composite research and development, primarily with the Loctite Corp. He joined Furane Products Company in 1986 as a Technical Service Supervisor-Aerospace Products. Current responsibilities include research, development, and customer relations for Aerospace adhesives, syntactic, and laminating products.

Kenneth D. Cressy: B.S. Univ. of New Hampshire 1950. Worked 4 years at Dow Chemical Co., Freeport, TX. in Quality Control and Product Development. He joined Furane Plastics in 1955 and has continued in various roles of product development, Technical Service, and sales. He has delivered several papers at IEEE, SPE, SAMPE Conferences and is currently Market Mgr. of Aerospace Syntactics and Adhesives at Furane Products Company.

## References

1. Brown, H., Ed., Composite Repairs, SAMPE Monograph No. 1, SAMPE International Business Office, 1985, pp 14-24.

2. p. 29 of Ref. 2.
3. 'Dura-Core' Aluminum Honeycomb data sheets, American Cyamamid Company.
4. 'Epibond 88807-A/B' data sheet, Furane Products Company.
5. 'Epibond 1545-A/B' data sheet, Furane Products Company.
6. 'Adhesive Bonding Ryton PPS Composites' data sheet, Phillips Petroleum Company.

AN ADHESIVE INTERLEAF TO REDUCE STRESS CONCENTRATIONS
BETWEEN PLYS OF STRUCTURAL COMPOSITES

Raymond B. Krieger, Jr.
Mgr, New Developments, Adhesives
American Cyanamid Co., Engineered Materials Dept.
Old Post Road, Havre de Grace,MD 21078

## Abstract

This paper describes a special
high-strain, low-flow adhesive
for use between strategic plys of
structural composites. The purpose is to reduce inherent shear
stress peaks, or concentrations,
by providing a high-strain adhesive interleaf between plys. This
interleaf has high strain capability because of its formulation
and because it is much thicker
than the usual matrix layer between plys. Shear stress-strain
data is presented for the interleaf and for typical matrix resins.
A linear stress analysis is developed for the plane between plys.
This shows that stress peaks can
be reduced by as much as ten times,
in the linear range, and far more
in the ultimate strength range.
This interleaf technique can great
ly improve the strength and
fatigue durability of structural
composites.

## 1. INTRODUCTION

The advent of composite laminates
on the airframe scene has not always led to success comensurate
with the high promise of material
properties. In many secondary
structures, the designs have saved
weight and money. In primary
structure, glass and Kevlar have,
in some measure, given way to the
superior properties of carbon
fiber composites. The simpler
designs in carbon fiber have generally met their structural goals.
In more complex designs, premature
failures have been unexpected and
disconcerting. The reasons for
these failures are not to be found
in the fundamental properties of
the laminate, i.e., tension, compression, flexure, and interlaminar
shear. The author suggests that a
primary cause is the Achilles heel
of the transfer of shear due to
bending between the skins and
stringers of "box" beams in wings
and empenages. The same problem
waits in fuselage structure using
skin and stringers. These shears
have not been wholly overlooked by
stress analysts. Plies of adhesive
film are often introduced at strategic locations where high shear
flows are predicted. Nevertheless,
the author suggests there are
three questionable facets in the
contemporary approach, namely:
1. While shear flows are readily
calculated, the shear stress concentrations are not, because they
depend on resin stiffness in carrying the shear flow transfer.
2. The adhesive ply will mix with
the matrix resin. This increases
the adhesive stiffness, and so
raises the shear stress.
3. The hot-wet compression
strength is reduced because the
modulus of the matrix resin is lowered by mixing with the adhesive.

This paper addresses these problems
by A) introducing a new adhesive

film which will not mix with the matrix, and B) offering stress analysis techniques to more accurately calculate the shear stress concentrations in order to better predict the strength of the design.

## 2. CONCLUSIONS

1. It is possible to drastically reduce shear stress concentrations at critical locations in composite structure. This is done by introducing a ply, or interleaf, of adhesive film, with low stiffness, at the strategic location.
2. It is possible to preduct the shear stress concentrations because the requisite adhesive shear stiffness can be known and the stress formulae have been developed.
3. It is possible to obtain maximum reduction of shear concentration and minimum loss of hot-wet compression strength by use of an adhesive formulated so as not to mix with the matrix resin.

## 3. SHEAR TRANSFER IN LAMINATES

In beams made from laminated plies of composite there are often serious shear forces which must be transferred between plies. Since this plane is not directly crossed by fibers of the composite, the shear loads must be carried by the matrix resin, acting as a thin glue line. This resin plane, or glue line, can become overloaded becuase of shear stress concentrations. Shear failure on this plane can preclude the structural test or even trigger the collapse of the entire beam, well below design strength.

To understand the situation more clearly, we can begin with Fig. 1. This shows a square increment of laminate loaded with a shear flow around its edges. The shear flow is equal and opposite on parallel sides, and causes a shear deformation from a square to a diamond shape. Moving to Fig. 2, the thickness of the laminate has been abruptly increased over half of the increment. The shear flow

on the thinner half must now redistribute over the extra laminate thickness. This results in a sudden concentration of shear force at the edge of the thicker half. This can readily overload the matrix and cause splitting between the plies, as shown.

To visualize the shear concentration problem, it may help to add a layer of adhesive at the critical plane. Fig. 3 shows this glue line as far thicker than the matrix resin layer between plies. It is seen that the adhesive is deformed (loaded) at a maximum at the edge and fades to zero at some point when the laminate shear strain is equal through its entire thickness.

## 4. SHEAR FLOW IN A "BOX" BEAM

Fig. 5 is a schematic drawing of a possible box beam serving as the bending structure of an aircraft wing. The skins and spar webs carrying a shear flow which ultimately produces the tension and compression loads in the stringers and chords. A simplified picture of these shears is shown in Fig. 5. It is noted that the shear flow is variable around the box perimeter. The flow is a maximum in the spar webs and drops each time a chord or stringer is passed. The drop is the amount of shear needed to load the stringer in tension or compression. If we look closely at a typical stringer we may see the shear concentrations that can prematurely destroy the structure. Fig. 6 shows the deformations due to shear transfer at a stringer. The shear peaks (concentrations) are shown at several typical locations.

## 5. QUANTITATIVE CALCULATION OF SHEAR STRESS CONCENTRATIONS

In Fig. 3 it can be seen that the shear load transfer in the adhesive is stiffness driven in the same sense as in the classic metal to metal bonded skin doubler specimen, Ref. 1. Fig. 4 shows this specimen deformed when the skin

(or laminate) is under tension.
The adhesive shear stress distribution is shown.  The peak stress
is given by the equation

$$\frac{KP}{\sqrt{\dfrac{t\ t_a\ E}{G}}} \qquad (1)$$

where $K \cong .7$, $P$ = load per inch,
$t$ = laminate thickness, $t_a$ = glue
line thickness, $E$ = laminate tensile modulus and $G$ = adhesive
shear modulus.  (Ref. 1)

Fig. 3 shows the same principle
for adhesive shear stress distribution.  Equation (1) will apply
since all the stiffness parameters are the same, except of
course that E (laminate tensile
modulus) must now become $G_L$
(laminate shear modulus).

### 6.   REDUCTION OF SHEAR PEAK BY USE OF AN ADHESIVE INTERLEAF

The addition of the glue line,
shown in Fig. 3, is a great improvement in reducing shear stress
concentrations.  It is readily
seen by inspection of equation (1)
that the glue line stiffness is a
function of the glue line thickness ($t_a$) and the shear modulus
of the adhesive (G).  This makes
it possible to closely estimate
the improvement percentage, or
ratio, obtainable by adding the
adhesive interleaf.  For Fig. 2,
no adhesive, the matrix shear
modulus, G, is 200,000 psi and the
"glue line thickess" of the single
layer of matrix between plies is
.00004 inches.

Then $\dfrac{1}{\sqrt{\dfrac{t_a}{G}}}$ = 70,900 = stiffness factor   (2)

Next, for Fig. 3, adhesive interleaf, $t_a$ = .005 inches and G =
100,000

Then $\dfrac{1}{\sqrt{\dfrac{t_a}{G}}}$ = 4460 = stiffness factor   (3)

The shear stress reduction is
proportional to these stiffness

factors, and so the reduction is
70,900/4460  = 16 to 1.

It is recognized that equation (2)
may be too stringent.  The "stiffness factor" may be too high.  More
than one layer of matrix resin between plies is deforming in shear.
Let us assume a layer of composite
several plies thick, acting as
adhesive at the critical plane.
We may say $t_a$ = .005 inches and
G = 500,000 psi

so then $\dfrac{1}{\sqrt{\dfrac{t_a}{G}}}$ = 10,000   (4)

This factor may at least be doubled
since there will be a gradient of
shear strain peaking at the center
of the .005 inch "glue line" so,
we may say that the shear stress
reduction is 20,000/4460= 4.5 to 1.
It is noted that, in equations (2),
(3), and (4), the values for adhesive shear modulus, matrix shear
modulus, and composite shear
modulus were measured by KGR-1
extensometer (Ref. 2).

Equations (2), (3), and (4) are
for the linear range of shear modulus.  This means that the stress
reductions apply to the lower load
range of fatigue and creep.  For
ultimate load, the matrix modulus
will not change substantially, but
the adhesive exhibits very high
ultimate shear strain.  The
"effective modulus" could be 1/3
of the linear modulus and the
stress reductions as high as 27 to
1 and no lower than 8 to 1.

These calculations are seen to have
great scatter and to be based on
considerable assumptions.  Nevertheless, we offer that the minimum
improvement of 8 to 1 with the
adhesive interleaf is entirely
convincing.

### Ref. 1

"Evaluating Structural Adhesives
Under Sustained Load in Hostile
Environment", Oct. 1973, 5th
National SAMPE Conference.

Ref. 2

"Stiffness Characteristics of
Structural Adhesives for Stress
Analysis in Hostile Environment",
Oct. 1975, American Cyanamid Co.

Copies of each of these papers are
available from American Cyanamid
Company, Engineered Materials Dept.,
Havre de Grace, Maryland.


Raymond B. Krieger, Jr., graduated
M.I.T. with a B.S. in Aeronautical
Engineering in 1941, and from then
until 1953 was with the Glenn L.
Martin Company. Until 1956 he was
with Luria-Cournand as Chief
Engineer. From 1956 to the present
he has been with American Cyanamid
Company, Bloomingdale Aerospace
Products, first as Chief Engineer
then Sales Manager, now as Technical Manager. His duties encompass
criteria development and testing
of structural adhesives, technical
service to the airframe industry
including design, processing and
quality control aspects of structural bonding, metal to metal,
sandwich and structural plastics.
He is the author of numerous
papers on structural bonding.

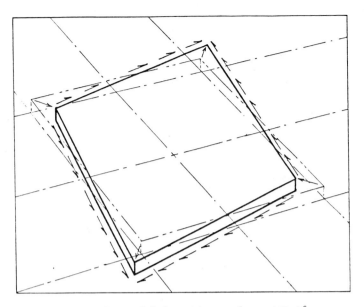

Fig. 1  Shear Flow and Deformation on Increment of
Composite Laminate

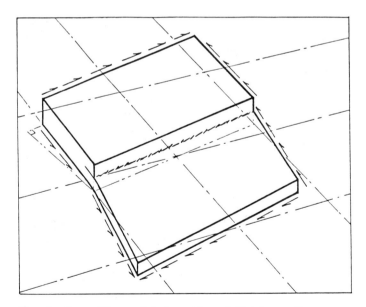

Fig. 2  Shear Flow and Deformation at Thickness Increase
in Laminate

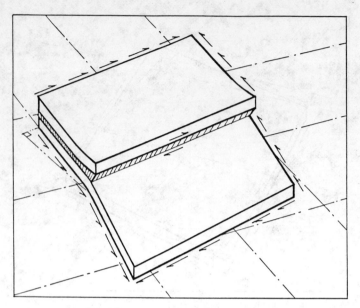

Fig. 3   Adhesive Shear Distribution When Laminate
Increment is in Shear

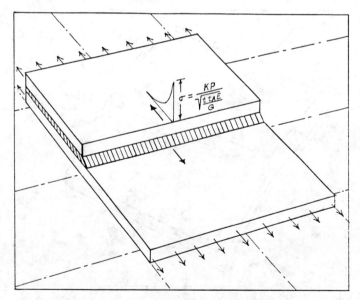

Fig. 4   Adhesive Shear Distribution when Laminate
Increment is in Tension

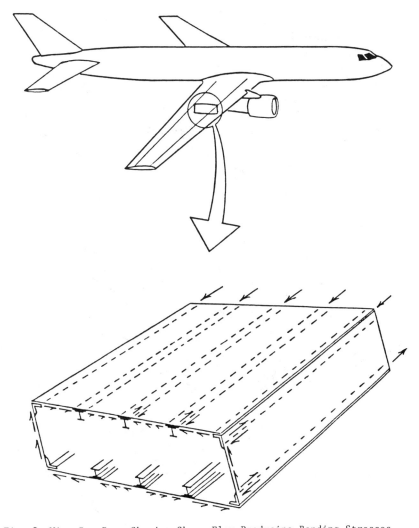

Fig. 5  Wing Box Beam Showing Shear Flow Producing Bending Stresses

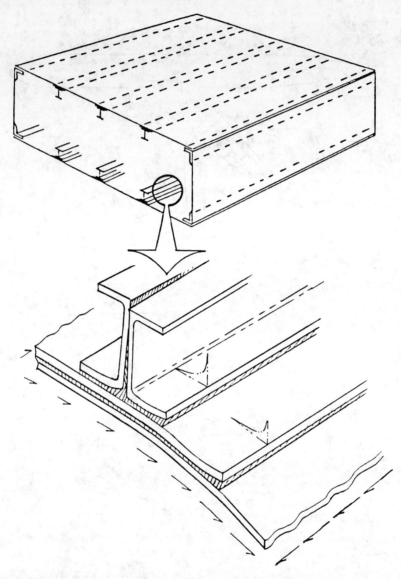

Fig. 6   Adhesive Shear Strains Showing Stress Concentration Points at
a Typical Stringer

EVALUATION OF 350°F CURING ADHESIVE SYSTEMS
ON ALUMINUM SUBSTRATES

D. Robert Askins
University of Dayton
Research Institute
Dayton, Ohio

## Abstract

Ten different 177°C (350°F) curing adhesive/primer systems were used to bond phosphoric acid anodized aluminum test specimens. The objective of the program was to compare the compatibility of these adhesive/primer systems with phosphoric acid anodized and sulfuric/dichromate acid etched aluminum surfaces. All the specimens were prepared by Douglas Aircraft Company during the PABST program and tested at the University of Dayton. Lap shear, peel, and stress-durability tests were carried out. The test results indicated several general types of behavior. Firstly, the predominant failure mode for most of the adhesive/primer systems in all three types of tests was an interfacial failure between the primer and adhesive. Secondly, little to no difference is observed between the results for the anodized and acid/dichromate etched surface treatments, although a few of the specific adhesive/primer/alloy/test type combinations exhibited better or worse properties for one surface type than the other.

## 1. INTRODUCTION

During the late seventies, a major research program entitled, "Primary Adhesive Bonding Structure Technology" (PABST) was carried out by the Douglas Aircraft Company under USAF funding. The thrust of the program was to advance the state of the art of adhesive bonding technology to the point where it could be confidently used in bonding primary aircraft structure without the use of mechanical fastening. A substantial portion of this work was concerned with establishing the compatibility of the phosphoric acid anodizing (PAA) surface preparation and 122°C (250°F) curing adhesives.

Test specimens for evaluating the compatibility of 177°C (350°F) curing adhesives with PAA were fabricated by Douglas Aircraft but

287

not tested because contractual funds were exhausted. These specimens were turned over to the Air Force Materials Laboratory (AFML) who sent them to the University of Dayton Research Institute (UDRI) for testing.

Three types of mechanical tests were performed (lap shear, peel, and stress-rupture) at various temperatures and after various exposure conditions. General observations are made regarding the performance of the various adhesive/primer systems.

## 2. MATERIALS AND TEST PROCEDURES

A total of nine different adhesives and ten different primers were used to prepare the specimens. Each adhesive was paired with one of the primers so that a total of ten adhesive/primer systems were in the program. Table 1 lists these adhesives and primers along with the respective suppliers.

All of the specimens tested in this investigation consisted of bare aluminum alloy (2024-T81 and 7075-T76) adherends whose surfaces were prepared for bonding with either the PAA or the Optimized Forest Product Laboratory (OFPL) surface treatments. These two surface treatments were in accordance with Boeing Aircraft Company specifications BAC 5555 (for the PAA) and BAC 5514 (for the OFPL).

All of the panels from which specimens were cut were nondestructively inspected by an ultrasonic "C-scan"

technique. Most of the panels exhibited high-quality void-free bond areas although a few exhibited bond area defects varying from minor to major.

### 2.1 Lap Shear Tests

Lap shear specimens of the "blister detection" type were fabricated and tested to the requirements of ASTM D3165. For some alloy/surface preparation/adhesive combinations the D3165 type lap shear specimens were not available. Extra specimens of the type used for stress-durability testing were available however and were used for generating lap shear data in these cases. The stress-durability type specimen, also referred to as a RAAB type specimen, has four test joints and is illustrated in Figure 1. When these specimens were used for lap shear data, only one of the four joints was broken.

Traditionally, bonded joint failures have been reported as adhesive, cohesive, or some combination of the two. However, it was felt that this method of reporting failure modes was inadequate. Accordingly, a different format for reporting failure mode was utilized during this investigation. This format is illustrated and explained in Figure 3.

The interpretation of bonded joint failure modes is very subjective. It is difficult, with the naked eye, to ascertain the exact failure mode unless it is totally cohesive (within the adhesive layer). While interfacial failure modes may appear obvious, one cannot be sure, short of resorting to expensive surface instrumental analysis, that a very thin layer of primer or adhesive has not remained adhered to an otherwise clean appearing surface. Since the primer layer is so thin, the only evidence of its presence, to the eye, is generally color. In this investigation, the only discriminations made regarding failure mode were those detectable by eye. Thus, it is to be recognized that regardless of the different presentation format, the failure modes reported in this document are still subjective.

Lap shear tests were conducted at -54°C (-65°F), 22°C (72°F), and 177°C (350°F) on dry unaged specimens. Each specimen was held at the test temperature for ten minutes prior to loading to insure thermal equilibrium. Generally, three replicate tests were conducted for each material and test condition.

## 2.2  Peel Tests

Metal-to-metal peel specimens were prepared in accordance with the requirements of ASTM method D3167 except that the specimens were one inch wide rather than one-half inch. All peel tests were carried out at -54°C (-65°F) after a ten minute soak at the test temperature to insure thermal equilibrium. Three replicate tests were carried

out for each material combination available for testing.

## 2.3  Stress-Durability Tests

Stress-durability specimens (Fig.1) were machined from full-area bonded panels measuring 4-inches wide by 9-1/2-inches long. Three samples were machined from each panel.

Stress-durability tests were conducted in accordance with ASTM method D2919. The specimens were loaded to 50% of their room temperature lap shear strength and placed in an elevated temperature, high humidity or salt spray environment until the specimen failed or the exposure period reached 2400 hours. In the event that the specimen survived the 2400 hour limit without failure, the fixture was removed from the environmental cabinet, the specimen unloaded and removed from the fixture, and tested for residual strength. Since the specimens used for these tests contained four test joints, up to four time-to-failure data points could be obtained for each specimen. In the event that one of the four test joints failed during aging, the fixture was removed from the environmental cabinet, the specimen removed from the fixture, a hole drilled through the center of the failed lap joint, the failed joint bolted together, the specimen remounted and reloaded in the test fixture, and the assembly returned to the environmental cabinet. Each time a test joint failed, this procedure was repeated with the recorded time-to-failure for each joint being the cumulative time in the aging environment while under load. Residual strength testing was conducted at room temperature and involved breaking of only one of the surviving test joints. Thus, all residual strength values presented here represent only one value. Stress-durability testing was carried out in two different environments: (a) a hot-humid environment of 60°C (140°F) and 95-100% relative humidity (RH), and (b) a salt-fog environment of 35°C (95°F) with a 5% salt spray solution.

### 3. DISCUSSION OF RESULTS

#### 3.1 Lap Shear Results

Figures 6-11 graphically illustrate the results of the lap-shear tests.

Inspection of the data listed presented in these figures and the failed specimens leads to the following observations.

- Most of the failures in the lap shear tests occurred within the primer layer or along the primer/adhesive interface. This was judged by visual inspection and was based on whether the primer could be visually observed (via color) on both or only one surface of the failed joint.

- Most of the failures judged to be within the adhesive layer (the traditional "cohesive" failure) occurred in the 177°C (350°F) tests.

- Three of the adhesive/primer systems (HT-424/HT-424F, FM-61/BR227, and FM-61/BR227A) exhibited 177°C (350°F) strength levels significantly lower than the other seven adhesive systems tested.

- The HT-424/HT-424F system exhibited significantly lower room temperature strength levels than the other nine systems on the OFPL prepared surfaces. The failure mode for this adhesive, however, was within the adhesive layer on both these low strength OFPL specimens as well as on the PAA specimens, which yielded reasonably high strength. The panels from which the HT-424/HT-424F specimens were obtained exhibited poor quality in the NDI C-scans for both the OFPL and PAA panels. Consequently, no plausible explanation is evident to explain the low OFPL results as opposed to the PAA results.

- The only test condition for which lap shear behavior could be compared on the PAA and OFPL surface preparations was for the room temperature test condition with no environmental aging. For this condition, no significant overall differences in behavior were apparent between the two surface preparations. Six of the ten adhesive/primer systems (RB398/RB500, EA9649/EA9205, AF130/EC3917, AF31/EC2174, FM61/BR227, and FM61/BR227A) exhibited no difference in either strength or failure mode between the two surface preparations. Two of the ten adhesive/primer systems (MB329/MB6725-1 and FM400/BR400) exhibited no difference in strength, but a lesser degree of primer failure on the OFPL surface than on the PAA surface. One of the ten adhesive/primer systems (PL729-3/PL728) exhibited a higher strength on OFPL etched specimens than on PAA specimens, but with no difference in failure mode. One of the ten adhesive/primer systems (HT424/HT424F) exhibited a higher strength on PAA specimens than on OFPL etched specimens, but with no difference in failure mode.

## 3.2 Peel Results

Figures 12 and 13 illustrate the peel data graphically. Inspection of these figures and the failed specimens leads to the following observations:

- Nearly all of the failures were interfacial between the primer and adhesive. Only the HT-424/HT-424F and FM-61/BR227A systems exhibited much deviation from this failure mode and in these cases, the failures were within the adhesive layer (the traditionally designated cohesive failure).

- Most all of the peel strengths were in the 2.5 to 4.0 lb/in range. The only exceptions were in those cases where the failures occurred in the adhesive layer, the HT-424/HT-424F and FM-61/BR227A systems cited above. In these cases, the peel strengths reached 6-7 lb/in for the FM-61/BR227A and 13-15 lb/in for the HT-424/HT424F.

- Comparisons between the peel behavior of specimens prepared with PAA and OFPL etched surfaces can be made on only eight of the ten adhesive/primer systems in the program, and then only for the -54°C (-65°F) unaged test condition. On six of these systems (RB398/RB500, MB329/MB6725-1, FM400/BR400, EA9649/EA9205, AF31/EC2174, and PL729-3/PL728) no difference in either strength or failure mode between the two surface preparations was evident. One of the systems (FM61/BR227) exhibited better strength with the OFPL etch surface preparation than with the PAA on the 2024 alloy, but with little difference in failure mode. Further, no difference between the two surface preparations was noted for the 7075 alloy with this adhesive/primer system. One of the systems (FM61/BR227A) exhibited both better strength and less primer related failure with the OFPL

etch than with the PAA treatment on both alloys.

## 3.3 Stress-Durability Results

Tables 2-5 present the results of the hot-humid stress-durability tests while Tables 6-9 present the results of the salt-spray stress-durability tests. Inspection of these data leads to the following observations:

- Nearly all systems survived the full 2400 hours without failure in both the hot humid and salt-spray environments except for the FM-61/BR-227 and FM61/BR227A systems. In the hot-humid environment, neither of these systems survived for more than 1250 hours. In the salt-spray environment the FM-61/BR227 system survived for the full 2400 hours while with the FM-61/BR227A system over half of the joints survived for the full 2400 hours.

- In most cases, the systems that did survive the full 2400 hours exhibited residual strengths that equaled or exceeded their original unaged room-temperature lap shear strength. In those cases where the residual strength was significantly less than the original room-temperature lap shear strength, the PL-729-3/PL728 and FM-61/BR-227A systems exhibited this strength degradation most frequently.

- The most common failure mode observed in the stress-durability tests was interfacial failure between the primer and adhesive layer or cohesive failure within the adhesive layer. Five systems (RB-398/RB-500, MB329/MB6725-1, AF-31/EC-2174, PL-729-3/PL-728, and FM-61/BR-227) exhibited a mixture of failure modes, including adhesive, primer-to-adhesive, and primer failures.

- In the hot-humid durability tests, no difference in durability, residual strength, or failure mode was observed for nine of the

291

ten adhesive/primer systems. The one exception (AF-130/EC-3917) exhibited higher residual strength and less primer failure on OFPL surfaces than on PAA surfaces. For the hot-humid durability tests, cohesive failure within the adhesive layer and interfacial failure between the primer and adhesive were the two predominant failure modes.

In the salt-spray durability tests, the difference (or similarity) between the behavior of the OFPL and PAA prepared surfaces is not clear cut. Three of the adhesive/primer systems (RB-398/RB-500, EA-9649/EA-9205, and FM-400/BR-400) exhibit no difference between the two surface preparations. One system (FM-400/BR-400) exhibits higher residual strength on both alloys with the PAA surface than with the OFPL surface. Six systems (MB-329/MB-6725-1, AF-130/EC3917, AF-31/EC-2174, PL-729-3/PL-728, HT-424/HT-424F, and FM-61/BR-227A) exhibited higher residual strength with one of the alloys (generally the 7075) for the PAA surface preparation than for the OFPL preparation. In these six cases the other alloy showed no differences between the two surface preparations. Only two adhesive/primer systems (PL729-3/ PL-728 and FM-61/BR-227A) exhibited higher residual strengths for OFPL etched surfaces than for PAA surfaces and in both cases this was for the 2024 alloy. In nearly all cases, the predominant failure mode for the salt-spray durability test specimens was interfacial between the primer and adhesive. This was true for both alloys, both surface preparations, nearly all of the adhesive/primer systems, and for both residual strength specimens as well as for those few specimens which failed during exposure.

## 3.4   Summary of Results

No consistent differences were observed between the performance of the various adhesive/primer, alloy, surface preparation combinations tested in this program. The goal of the testing was to evaluate the compatibility of 177°C (350°F) curing adhesive/primer systems with the PAA surface treatment. Most of the adhesive/primer systems exhibited equivalent performance with both the PAA and OFPL etch surface treatments for all types of tests and exposure conditions. Some of the adhesive/primer systems performed better with the OFPL etch surface treatment than with the PAA surface treatment, but in those instances where such a difference appeared, it was only for one of the three types of tests conducted.

In lap-shear, for example, the MB329/MB6725-1 and FM400/BR400 systems exhibited less primer failure on OFPL surfaces than on PAA surfaces but with no difference in strength. In peel and stress-durability these two systems exhibited equivalent behavior, both in strength and failure mode, for the two surface preparations.

In peel, the FM61/BR227A system exhibited better strength and less primer related failure with the OFPL etch than with the PAA but in lap-shear and stress-durability no difference in performance between the two surface preparations was noted for this system.

In stress-durability, the AF-130/EC-3917 system yielded higher residual strengths and less primer failure with the OFPL etch than with the PAA for the case of a humidity environment but no difference between the two surface preparations was observed for this system in lap-shear, peel, or in salt-spray stress-durability.

The results of the tests carried out in this program serve to indicate that, in general, there was no glaring incompatibility between any of the materials tested and the PAA surface preparation.

BIOGRAPHY

D. Robert Askins is Head of the Plastics, Adhesives and Composites Group at the University of Dayton Research Institute, where he has been employed since 1964. He earned BChE and MChE degrees from the University of Dayton in 1963 and 1974, respectively. Mr. Askins has been responsible for conducting many programs over the past 22 years involving adhesive bonding, advanced composite materials, and plastics. He is presently serving as principal investigator on four Air Force contracts related to these same technology areas as well as numerous industrially sponsored projects.

TABLE 1

ADHESIVE/PRIMER SYSTEMS

| Adhesive | Primer | Supplier |
|----------|--------|----------|
| RB 398 | RB 500 | Reliable (Ciba Geigy) |
| MB 329 | MB 6725-1 | Narmco |
| FM 400 | BR 400 | American Cyanamid |
| EA 9649 | EA 9205 | Hysol |
| AF 130 | EC 3917 | 3M |
| AF 31 | EC 2174 | 3M |
| PL 729-3 | PL 728 | B.F. Goodrich |
| HT 424 | HT 424F | American Cyanamid |
| FM 61 | BR 227 | American Cyanamid |
| FM 61 | BR 227A | American Cyanamid |

TABLE 2

HOT-HUMID STRESS-DURABILITY TEST RESULTS
2024-T81 BARE ALUMINUM

| Adhesive/Primer System | Stress During Exposure (psi) | Time to Failure (hrs) | No. of Joints Represented (4 joints/specimen) | Residual Shear Strength (psi) | Failure Mode[1] | | |
|---|---|---|---|---|---|---|---|
| | | | | | P | P/A | A |
| BAC 5555 (PAA) Surface Preparation | | | | | | | |
| RB-398/RB-500 | 1240 | 2400 | 8 | 2380 | 0 | 80 | 20 |
| MB-329/MB-6725-1 | 1160 | 2400 | 8 | 2820 | 0 | 60 | 40 |
| FM-400/BR-400 | 1320 | 2400 | 8 | 2680 | 0 | 0 | 100 |
| EA-9649/EA-9205 | 1270 | 2400 | 8 | 2900 | 60 | 0 | 40 |
| AF-130/EC-3917 | 770 | 2400 | 8 | 1670 | 100 | 0 | 0 |
| AF-31/EC-2174 | 2000 | 2400 | 8 | 3570 | 0 | 85 | 15 |
| PL-729-3/PL-728 | 1790 | 2400 | 8 | 3380 | 0 | 85 | 15 |
| HT-424/HT-424F | 1030 | 2400 | 8 | 1980 | 0 | 15 | 85 |
| FM-61/BR-227 | 1410 | 672-840 | 6 | --- | 0 | 0 | 100 |
| FM-61/BR-227A | 1370 | 624-1248 | 6 | --- | 0 | 100 | 0 |
| BAC 5514 (OFPL) Surface Preparation | | | | | | | |
| RB-329/RB-500 | 1090 | 2400 | 8 | 2510 | 40 | 10 | 50 |
| MB-329/MB-6725-1 | 1100 | 2400 | 8 | 2580 | 0 | 25 | 75 |
| FM-400/BR-400 | 1170 | 2400 | 8 | 2270 | 0 | 0 | 100 |
| EA-9649/EA-9205 | 1320 | 2400 | 8 | 2950 | 0 | 50 | 50 |
| AF-130/EC-3917 | 850 | 2400 | 8 | 2380 | 0 | 30 | 70 |
| AF-31/EC-2174 | 1910 | 2400 | 8 | 3230 | 0 | 100 | 0 |
| PL-729-3/PL-728 | 2020 | 2400 | 8 | 3400 | 0 | 90 | 10 |
| HT-424/HT-424F | 620 | 2400 | 4 | 700 | 0 | 0 | 100 |
| | | 48 | 3 | --- | 0 | 10 | 90 |
| FM-61/BR-227 | 1350 | 696-912 | 6 | --- | 0 | 5 | 95 |
| FM-61/BR-227A | 1390 | 696-1104 | 6 | --- | 0 | 60 | 40 |

[1]P = failure within primer layer.
P/A = failure along primer/adhesive interface.
A = failure within adhesive layer (the traditional "cohesive" failure).


TABLE 3

HOT-HUMID STRESS-DURABILITY TEST RESULTS
7075-T76 BARE ALUMINUM

| Adhesive/Primer System | Stress During Exposure (psi) | Time to Failure (hrs) | No. of Joints Represented (4 joints/specimen) | Residual Shear Strength (psi) | Failure Mode[1] | | |
|---|---|---|---|---|---|---|---|
| | | | | | P | P/A | A |
| BAC 5555 (PAA) Surface Preparation | | | | | | | |
| RB-398/RB-500 | 1200 | 2400 | 8 | 2650 | 0 | 70 | 30 |
| MB-329/MB-6725-1 | 1090 | 2400 | 8 | 2630 | 0 | 45 | 55 |
| FM-400/BR-400 | 1230 | 2400 | 8 | 2110 | 0 | 25 | 75 |
| EA-9649/EA-9205 | 1310 | 2400 | 8 | 2990 | 30 | 35 | 35 |
| AF-130/EC-3917 | 840 | 2400 | 8 | 1590 | 90 | 0 | 10 |
| AF-31/EC-2174 | 1860 | 2400 | 8 | 3570 | 0 | 80 | 20 |
| PL-729-3/PL-728 | 1720 | 2400 | 8 | 3240 | 0 | 90 | 10 |
| HT-424/HT-424F | 970 | 2400 | 8 | 2140 | 0 | 15 | 85 |
| FM-61/BR-227 | 1380 | 0-696 | 6 | --- | 0 | 5 | 95 |
| FM-61/BR-227A | 1330 | 216-888 | 6 | --- | 0 | 100 | 0 |
| BAC 5514 (OFPL) Surface Preparation | | | | | | | |
| RB-398/RB-500 | 1120 | 2400 | 8 | 2430 | 0 | 70 | 30 |
| MB-329/MB-6725-1 | 960 | 2400 | 8 | 2440 | 0 | 70 | 30 |
| FM-400/BR-400 | 1170 | 2400 | 8 | 2310 | 0 | 40 | 60 |
| EA-9649/EA-9205 | 1260 | 2400 | 8 | 2930 | 55 | 0 | 45 |
| AF-130/EC-3917 | 840 | 2400 | 8 | 2220 | 0 | 50 | 50 |
| AF-31/EC-2174 | 1860 | 2400 | 8 | 3320 | 0 | 100 | 0 |
| PL-729-3/PL-728 | 2040 | 2400 | 8 | 3320 | 0 | 90 | 10 |
| HT-424/HT-424F | 370 | 2400 | 8 | 1770 | 0 | 50 | 50 |
| FM-61/BR-227 | 1370 | 696-864 | 6 | --- | 0 | 15 | 85 |
| FM-61/BR-227A | 1400 | 720-960 | 6 | --- | 0 | 90 | 10 |

[1]P = failure within primer layer.
P/A = failure along primer/adhesive interface.
A = failure within adhesive layer (the traditional "cohesive" failure).

TABLE 4

**SALT-SPRAY STRESS-DURABILITY TEST RESULTS**
**2024-T81 BARE ALUMINUM**

| Adhesive/Primer System | Stress During Exposure (psi) | Time to Failure (hrs) | No. of Joints Represented (4 joints/specimen) | Residual Shear Strength (psi) | Failure Mode[1] | | |
|---|---|---|---|---|---|---|---|
| | | | | | P | P/A | A |
| BAC 5555 (PAA) Surface Preparation | | | | | | | |
| RB-398/RB-500 | 1240 | 2400 | 4 | 2570 | 10 | 80 | 10 |
| MB-329/MB-6725-1 | 1160 | 2400 | 4 | 2470 | 0 | 65 | 35 |
| FM-400/BR-400 | 1320 | 2400 | 4 | 2810 | 0 | 80 | 20 |
| EA-9649/EA-9205 | 1270 | 2400 | 4 | 2600 | 0 | 100 | 0 |
| AF-130/EC-3917 | 770 | 2400 | 4 | 2930 | 0 | 50[2] | 0 |
| AF-31/EC-2174 | 2000 | 2400 | 4 | 4120 | 0 | 75 | 25 |
| PL-729-3/PL-728 | 1790 | 2400 | 4 | 1920 | 0 | 80 | 20 |
| HT-424/HT-424F | 1030 | 2400 | 4 | 1330 | 0 | 0 | 100 |
| FM-61/BR-227 | 1410 | 2400 | 4 | 2190 | 0 | 90 | 10 |
| FM-61/BR-227A | 1370 | 720 | 1 | --- | 0 | 90 | 10 |
| | | 2400 | 3 | 1420 | 0 | 100 | 0 |
| BAC 5514 (OFPL) Surface Preparation | | | | | | | |
| RB-398/RB-500 | 1090 | 2400 | 4 | 2540 | 0 | 50 | 50 |
| MB-329/MB-6725-1 | 1100 | 2400 | 4 | 2490 | 0 | 75 | 25 |
| FM-400/BR-400 | 1170 | 2400 | 4 | 2420 | 0 | 45 | 55 |
| EA-9649/EA-9205 | 1320 | 2400 | 4 | 2770 | 0 | 85 | 15 |
| AF-130/EC-3917 | 850 | 2400 | 4 | 960 | 0 | 0 | 100 |
| AF-31/EC-2174 | 1910 | 2400 | 4 | 3790 | 0 | 95 | 5 |
| PL-729-3/PL-728 | 2020 | 2400 | 4 | 3140 | 0 | 80 | 20 |
| HT-424/HT-424F | 620 | ---[3] | 4 | --- | 0 | 20 | 40[2] |
| FM-61/BR-227 | 1350 | 2400 | 4 | 2210 | 0 | 60 | 40 |
| FM-61/BR-227A | 1390 | 648 | 1 | --- | 0 | 90 | 10 |
| | | 2400 | 3 | 2230 | 0 | 40 | 60 |

[1] P = failure within primer layer.
P/A = failure along primer/adhesive interface.
A = failure within adhesive layer (the traditional "cohesive" failure).

[2] These specimens had a very porous bondline. About 40-50% of failure surface appeared to be due to voids.

[3] Specimens failed in loading due to large void area in bonds.

TABLE 5

**SALT-SPRAY STRESS-DURABILITY TEST RESULTS**
**7075-T76 BARE ALUMINUM**

| Adhesive/Primer System | Stress During Exposure (psi) | Time to Failure (hrs) | No. of Joints Represented (4 joints/specimen) | Residual Shear Strength (psi) | Failure Mode[1] | | |
|---|---|---|---|---|---|---|---|
| | | | | | P | P/A | A |
| BAC 5555 (PAA) Surface Preparation | | | | | | | |
| RB-398/RB-500 | 1200 | 2400 | 4 | 2560 | 0 | 90 | 10 |
| MB-329/MB-6725-1 | 1090 | 2400 | 4 | 2520 | 0 | 80 | 20 |
| FM-400/BR-400 | 1230 | 2400 | 4 | 2320 | 0 | 50 | 50 |
| EA-9649/EA-9205 | 1310 | 2400 | 4 | 2580 | 0 | 100 | 0 |
| AF-130/EA-3917 | 840 | 2400 | 4 | 1520 | 10 | 90 | 0 |
| AF-31/EC-2174 | 1860 | 2400 | 4 | 3720 | 0 | 75 | 25 |
| PL-729-3/PL-728 | 1720 | 2400 | 4 | 3210 | 0 | 85 | 15 |
| HT-424/HT-424F | 970 | 2400 | 4 | 2030 | 0 | 30 | 70 |
| FM-61/BR-227 | 1380 | 2400 | 4 | 2400 | 0 | 100 | 0 |
| FM-61/BR-227A | 1330 | 2400 | 4 | 1610 | 0 | 100 | 0 |
| BAC 5514 (OFPL) Surface Preparation | | | | | | | |
| RB-398/RB-500 | 1120 | 2400 | 4 | 2630 | 0 | 60 | 40 |
| MB-329/MB-6725-1 | 960 | 1309 | 1 | --- | 0 | 50 | 50 |
| | | 1989 | 1 | --- | 0 | 85 | 15 |
| | | 2205[2] | 1 | --- | 0 | 85 | 15 |
| FM-400/BR-400 | 1170 | 2400 | 4 | 2140 | 0 | 40 | 60 |
| EA-9649/EA-9205 | 1260 | 2400 | 4 | 2530 | 0 | 90 | 10 |
| AF-130/EC-3917 | 840 | 2400 | 4 | 1460 | 0 | 100 | 0 |
| AF-31/EC-2174 | 1860 | 2400 | 4 | 1870 | 0 | 85 | 15 |
| PL-729-3/PL-728 | 2040 | 456 | 1 | --- | 0 | 85 | 15 |
| | | 912 | 1 | --- | 0 | 90 | 10 |
| | | 2400 | 2 | 3390 | 0 | 70 | 30 |
| HT-424/HT-424F | 370 | 2400 | 4 | 1350 | 0 | 0 | 100 |
| FM-61/BR-227 | 1370 | 2400 | 4 | 2450 | 0 | 100 | 0 |
| FM-61/BR-227A | 1410 | 648 | 3[3] | 260 | 0 | 85 | 15 |

[1] P = failure within primer layer.
P/A = failure along primer/adhesive interface.
A = failure within adhesive layer (the traditional "cohesive" failure).

[2] Test terminated after this joint failed because of excessive adherend corrosion.

[3] Three joints failed after 648 hours. Remaining joint tested for residual strength without further aging.

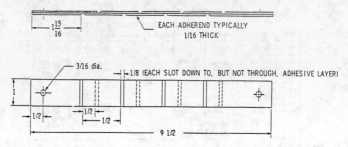

Figure 1.  Stress-Durability Test Specimen (RAAB Style).

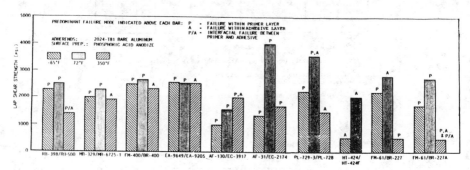

Figure 2.  Lap Shear Strength of Adhesives on PAA 2024-T81 Aluminum.

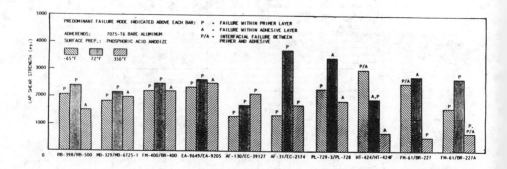

Figure 3.  Lap Shear Strength of Adhesives on PAA 7075-T6 Aluminum.

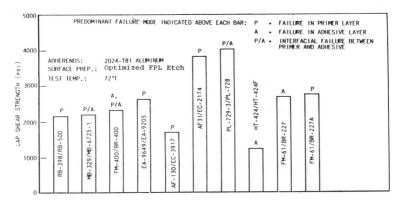

Figure 4. Lap Shear Strength of Adhesives on OFPL Etched 2024-T81 Aluminum.

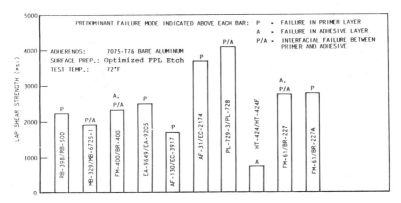

Figure 5. Lap Shear Strength of Adhesives on OFPL Etched 7075-T76 Aluminum.

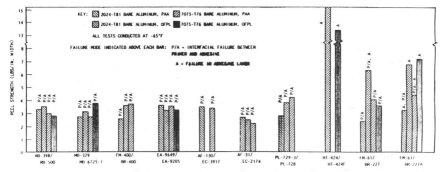

Figure 6. Peel Strength of Adhesives.

297

**AUTOMATED COMPOSITES CURE CONTROL IMPLEMENTATION**
**A CURE MODELING APPROACH TO AUTOMATION**
Richard W. Roberts
General Dynamics Convair Division
P. O. Box 85377, MZ 91-4710
San Diego, CA 92138-5377

## Abstract

The automation of composites curing
has been a proven factor in improving
the quality of composites parts.
However, the simple automation of
cure cycle profiles that are derived
by traditional methods does not take
the actual chemical considerations of
the resin into account for curing.
In order to account for the needs of
the resin system, industry has
developed a system based on
processing windows to attempt to
control the resin to ensure "proper"
cure. This method leads to
inconsistencies in the cure of the
composites and can result in poor
quality parts. This paper introduces
the cure modeling environment to
fully utilize the power of the
computer to accurately trace and
monitor the raw materials, from entry
into the processing facility through
cure, and to use this womb-to-tomb
resin condition data to predict the
cure requirements interactively
during the cure cycle. This will
ensure that the resin of the
composite is fully cured.

## 1. INTRODUCTION TO AUTOMATED CURING

The introduction of advanced
composites into the aerospace
industry forces a significant change
in the way that companies view and do
business. Undertaking the
manufacture of composite structures
causes the manufacturer to be not
only a material user, but also become
a material producer. As a user of
classical materials, such as
aluminum, a manufacturer influences
the actual properties of the material
very little, usually within the same
order of magnitude. For example, the
modulus in the as received state of
2024-T6 aluminum is 68,258 MPa and it
is 71,016 MPa after processing and
heat treating. With a composite, the
manufacturer now becomes responsible
for influencing the properties of the
finished product by several orders of
magnitude. For example the
transverse modulus of a composite is
approximately 1 Pa at receipt and
after processing it is 19,926 MPa, an
order of magnitude change of ten.

In order to account for the need for
control of the overall composites
process, industry has established a
set of processing windows to keep the
materials in 'acceptable' processing
state. These windows, out-time,
expiration dates, re-test, shelf
life, etc. are very restrictive and
often do not produce the desired
results. A batch of resin can be
procured that marginally passes the
test criteria of incoming inspection,
and is then given the normal shelf
life. This resin may not be
processible to the standard set of
cure conditions for this resin after
any significant shelf life. An
alternative to this parametric or
processing window environment will be
proposed in this paper. This

alternative called the establishment of the cure modeling environment, or cure modeling will account for the actual state of the resin and adjust the cure cycles appropriately. To aid in understanding the cure process, a brief history of automated curing is presented to show the value and importance of the cure process to successful composites.

The majority of the final properties of a composite are obtained during the cure of the matrix material of the composite. During the cure cycle, chemical reactions occur, which causes the matrix material (resin) to polymerize and obtain the desired properties of the material. The cure of the resin can be considered the most important phase of the manufacturing process. Before a resin cure phase, a considerable investment has been made into a composite component. If the part is incorrectly cured, then the outcome of the cure is irreversible and the part is usually scrapped. No amount of work before cure can ensure that an incorrectly cured part will be usable. It is apparent that the most critical step in obtaining good composites parts is the cure cycle.

Even though the cure cycle is the most important step in the composite process, until recently (1979-1980), no real work went into ensuring consistent, automated cure cycles. The use of cam-type mechanical or electronic controllers had been around for many years but lacked the feedback control required to adequately control the chemical reactions that occur during the cure process. Applied Polymer Technology Inc. developed a thermal feedback control system for LearFan in the 1980-1981 time frame. This was the first real-time feedback computer control based on part temperature delta, heat rate, and programmed temperature and pressure profiles that used chemical reaction related logic in its programming for control.

Since the development of the original automated controller there has also been considerable work done to monitor the rate of reaction and progression of the chemical reaction in autoclave curing conditions.

These are in the form of ultrasonic, dielectric, microdielectric and, very recently, Fourier Transform Infrared (FTIR) monitors which place sensors within the bag to monitor the chemical reaction of the resin. Considerable investment has been made trying to make dielectric monitoring feasible, with somewhat encouraging results up to the later stages of the cure process. In the late stages of the cure process, the dielectric readings are incapable of indicating meaningful physical characteristics of the resin system. Dielectric monitors measure the dielectric constant of the material which is only related to the state of the resin, not a physical characteristic. Ultrasonic monitoring has shown meaningful results all the way though the cure process by measuring the physical parameter, compression modulus.

These monitoring systems (dielectric and ultrasonic) have been coupled to automated computer control systems and used to trigger different steps within the cure cycle. The steps are generally pressure application, beginning of heat-up sequences and thermal soaks. The in-bag monitoring devices are somewhat cumbersome in the production environment. The sensors are sensitive to application method, relatively expensive, prone to damage, and they can fail for a number of reasons within the cure cycle, which will fail to trigger the desired action. Sensors require additional penetrations in the vacuum bag and could generate additional potential leak paths. Even when they are used, fail-safe thermal triggers are placed within the cure profile in case the sensor does not signal before a certain time-temperature relationship exists. Clearly, these sensors are not optimal for the production environment.

For the optimum cure, Two questions must be considered: 1) What benefit does the sensor data really provide? and 2) How can the sensor information be obtained without the problems associated with the sensor?

The sensor data is required to correctly and reliably cure the resin system. The cure of the system is dependant on many conditions including time at temperature, heat rate, and humidity exposure. Currently, the only way to determine how the cure cycle is affecting the resin is to place the above monitors on it during the cure cycle. With this type of data, the cure cycle can be adaptively adjusted to conform to the resin requirements instead of parametrically cured as it is today. The potential of increasing the success of the cure cycle process will occur two ways: 1) it will eliminate "marginal" cures, those cures where the parametric cure cycle does not adequately account for the condition of the resin system, e.g. excessive out-time, high humidity exposure, marginal resin batch chemistry, etc., and 2) it has the potential of increasing autoclave throughput by shortening and lengthening cure cycles as appropriate, depending on the resin requirements.

Similar data can be obtained without the use of sensors in production by the use of the cure modeling technique. To model a cure, mathematical models are used to predict the current state of the resin system. The resin state prediction is related to the resin degree of cure, alpha ($\alpha$) and can be modeled before, during, and after the cure process. Alpha indicates how far along the reaction path the resin system is at a given set of temperature and time conditions. Figure 1 shows a pictorial representation of how the degree of cure progresses as the resin is processed in the normal production environment. The implementation of the Automated Composites Cure Control (ACCC) system is dependant on the ability to accurately predict the degree of cure at any point on the line shown in Figure 1.

The physical characteristics that an acceptable composites laminate possesses are: 1) low to no void / porosity content, 2) correct degree of cure, 3) correct fiber volume / resin content, and 4) correct fiber orientation. Each element of the Automated Composite Cure Control System relates the processing to one or all of these goals to produce a good laminate. Consistently meeting the criteria for composite laminate is the goal for this paper and the process it describes.

## 2. AUTOMATED COMPOSITES CURE CONTROL (ACCC) SYSTEM INTRODUCTION

The elements that are involved in the implementation of the ACCC system are as follows:

1) Theoretical Model - Resin Degree of Cure
    a) Degree of Cure Completion (kinetics)
    b) Visco-Elastic Flow State
2) Meaningful Battery of Incoming Material Tests
3) Basic Cure Model for Each Resin System
4) Material Database
    a) Materials Properties
    b) Material History
    c) In-Process Material Tracking
5) Model Validation Requirements
6) Software
    a) Model
    b) Expert Shell
    c) Autoclave Control Unit
    d) Periodic Revalidation of Model
    e) Data Storage Methods for Historical Purposes

A schematic drawing of the proposed final ACCC system is shown in Figure 2. This schematic introduces the interplay of the elements to aid in the overall picture of how this system needs to function. Each of the elements and their interface to other elements will be described in detail in the following sections.

## 3. MODEL - RESIN DEGREE OF CURE

The major technology required for the implementation of ACCC is the fully developed model for the degree of cure of a subject resin system. Each resin system will require a specific model. The techniques of establishing each model is described in Section 5.

The model must be capable of performing two distinct functions: 1) degree of cure completeness based on incoming materials tests and resin advancement which can predict the

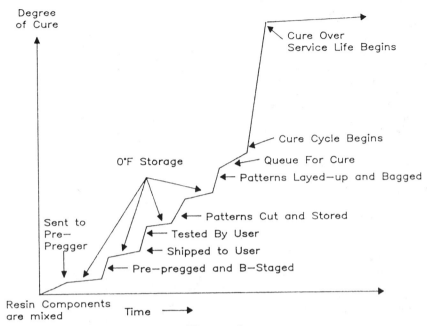

Figure 1
Resin Life Cycle

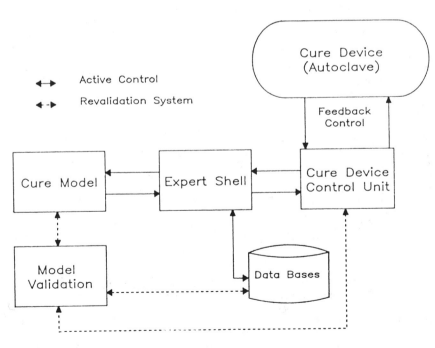

Figure 2
Cure Modeling Environment

kinetics of the reaction for temperature limits, the time-temperature pathway starting from resin advancement, and moisture / volatile content, and 2) visco-elastic flow state from time-temperature pathway and using information obtained from percent moisture/volatile component and resin starting advancement at the time of entering cure. The starting advancement refers to where the resin actually is on the line shown in Figure 1, at the time when the resin enters the manufacturing process.

The backbone of the system is the first function of the degree of cure model. The degree of cure model is generally in the form of an Arrhenius-type equation which can be formulated into control algorithms. The formula, which is pathway dependant, uses inputs of time, temperature, humidity exposure, and initial advancement conditions and outputs a predicted degree of cure for the subject resin. The manner in which a part reaches a temperature affects the predicted output. The modeled formula is used to gage progress through the cure process and for determining the completion point.

The second function of the model is the determination of the visco-elastic flow state of the subject resin. This model subset will use as inputs the moisture content of the material, the starting advancement of the resin system, and the time-temperature pathway. The visco-elastic flow state of the resin are important to the model in that they will allow for the correct heat rate determination and to determine the optimum time for application of pressure, reduction of vacuum, other viscosity oriented changes. It will also allow for prediction of final resin modulus and could be used to verify part cure integrity.

Another important consideration of the visco-elastic flow state model is the visco-elastic characteristics of the resin system can be monitored during cure. Monitoring these characteristics provides a method for checking the accuracy of the model prediction throughout the cure cycle and thereby can be used for model validation.

## 4. INCOMING MATERIAL TESTS TO SUPPORT MODELING

One of the most important steps in the modeling process will be the determination of the resin's state of advancement at the time it enters the manufacturing facility from the vendors. There are many variables that the resin could experience during transportation and storage prior to and after entering the facility where it will be used. These incoming tests will positively identify the material prior to it's acceptance for production.

Currently, most composite materials are mechanically tested during receiving inspection. Few chemical tests are performed on the materials. Mechanical tests fail to accurately evaluate the condition of the resin, the most variable entity within the composite. The cure modeling approach will require an accurate understanding of the resin system and its condition when it enters the in-house material control system. In order to accomplish this, the following tests need to be the basis for incoming material acceptance.

1) Material Identification
   a) HPLC (High Pressure
      Liquid Chromatography)'
   b) Infrared Spectroscopy
2) Reaction Dependant Behavior
   a) Rheometric Analysis of
      Visco-elastic behavior
   b) DSC (Differential
      Scanning Calorimetry)
3) Physical Form
   a) Resin Content (RC)
   b) Fiber Areal Weight (FAW)
   c) Volatile Content
4) Fiber Type
   a) Microscopic
   b) Tensile

Only one of these tests involves the making and breaking of any specimens and that test is on dry fibers. If the material passes these tests, it will, by definition, make good panels. Each type of test has a specific purpose, each major area will be covered in detail such that this purpose will be understood.

The material identification tests are to accurately identify the constituents of the matrix material. Each type of material possesses a unique fingerprint based on its constituent chemicals. Two methods of evaluating this fingerprint are by IR spectroscopy and HPLC. These methods examine the constituents of the material, show relative quantity by the position, magnitude of absorbtion of IR light or position and magnitude of migration in a medium. These techniques are quantitative and definitive. The results will give a very accurate breakdown of components, a ratio of reactants to products of reactions (an indicator of advancement), and will positively identify the material. These tests are easy to accomplish and require very small sample sizes.

The reaction dependant behavior tests checks the reaction behavior of the material, more specifically, how does the resin react during cure conditions? This is accomplished by rheometric measurements to accurately determine the resins visco-elastic response to temperature ramps and isothermal holds. The resultant data is compiled and is used to determine starting advancement. Differential Scanning Calorimetry (DSC) measures the heat evolved during the reaction of the resin. The resin is heated at several preset rates and left at isothermal conditions. The energy evolved or absorbed is measured. This data is vital in the determination of the advancement of the resin. It will also yield information on the greatest expected degree of cure for the resin batch.

The third test area is the physical form tests. These are the most simple tests to run and produce results which are important to the flow and reaction kinetic models. In order to produce laminates of the proper matrix / fiber ratio, resin content and fiber areal weight are important for the determination of bleeder requirements. Volatile content is necessary for moisture and volatile data for the cure model.

The final tests, fiber type, examine

the fiber identification and examination of condition. Since the fibers represent the majority of the load carrying capability of the laminate, it is important to ensure that the required fibers are present and properly prepared in the composite. Microscopic examination for surface finish and tensile testing for fiber strength will ensure that the fiber will meet the engineering design criteria for strength.

The acceptance window for each of these tests can be set at finite limits or specific numbers based on the tests used to create the models. Acceptance windows, as opposed to the processing windows currently used by industry, will directly relate to the chemical state of the resin. Acceptance criteria of this type will allow the acceptance of material that previously fell outside the older processing window environment. This can be accomplished by establishing criteria limits that either allow processing within the model or require processing changes by the model. This will make the model environment a very powerful tool in the reduction of material costs by allowing more latitude in the production of the composite materials, and by removing some of the previous constraints on material production.

All of these tests are relatively easy to perform, require small sample sizes, and produce meaningful results related to the physical and chemical characteristics of the resin. Mechanical tests run previously for incoming qualification produce only bulk data and yield insufficient information on the chemical condition and requirements of the resin. When this test battery is introduced, the elimination of the mechanical tests is possible with a resulting savings in processing, sample machining, and testing time. The equipment to run these new tests is expensive, but the reduction in test time and the increased reliability and quality of the data is an appropriate justification for its purchase.

## 5. NEW MODEL DEVELOPMENT

The introduction of any resin system to the cure modeling environment will require the creation of a model to reflect the processing requirements of that resin system. The parameters for each resin system can be derived by using data obtained in the same type of test battery used for incoming material and with several additional and expanded tests.

The new model development test battery will include: 1) Rheometric Analysis, 2) In-Process Fourier Transform Infrared Spectroscopy (FTIR) or functional group conversion analysis, 3) Moisture content versus time and temperature exposure, TGA (Thermogravimetric analysis), 4) HPLC, 5) DSC, and possibly 6) more advanced techniques such as Nuclear Magnetic Resonance (NMR) analysis and X-Ray diffraction for verification.

The primary source of model data is the use of the Rheometric measurement system to test the resin change in visco-elastic properties as a function of time and temperature conditions. Both ramp and isothermal conditions are modeled and the data generated produces the visco-elastic portion of the model.
DSC data is used to determine the expected degree of cure and the reaction of various pathways on resin kinetics. The use of a pressurized DSC system will allow the effect of pressure on the resin kinetics to be included in the resin model. This will allow the fluid hydrostatic pressure of the liquid resin to be considered in the model on resin degree of cure.

In-Process FTIR measurements during the cure of the resin will yield data on the conversion rate of functional groups into desired and undesired reactions. This data will generate minimum and maximum conditions such as heat rate and temperature soaks in order to avoid competing undesirable reactions.

Studies into the absorbtion rate of airborne moisture in the resin system will establish the model for determining percent moisture content after atmospheric exposure. This

information will be used to modify the starting condition of the resin at the beginning of cure. Then, the moisture effect on kinetics can be appropriately accounted.

The use of advanced analysis equipment such as NMR analysis is being studied in an effort to reduce the testing required to establish the model. NMR is a good possibility due to the detailed of quantitative data that is produced during a test. X-ray diffraction is capable of providing the data on short-range crystal structure of a polymer and will be useful in modeling the behavior of thermoplastics.

When all of the parameters are defined to complete the model, the new model will be placed in a standard format and algorithms and software will be developed to support that data. The estimated time from introduction of a new resin and its implementation into the ACCC system is thirty days. This is a significant change from the several month time frame to develop optimum cure cycle using the previous trial and error methods and testing of test coupons. After the model is developed, the optimum cure cycle can be run the first time in a production environment.

## 6. MATERIALS DATABASE

The basis for the cure modeling process is the availability and usage of material data. The system needs to track each individual batch and roll of material from the time it enters the facility and a starting advancement is determined to the time it is cured. This is required in order that the starting advancement can be updated to reflect the actual conditions the resin has seen during storage and processing. To this end, a material control system and data base will be established to accurately track any entity during the entire process. The elements of this system are 1) the materials properties database, 2) the material history and cure cycle historical data base, and most importantly 3) the work-in-process and raw stores inventory.

The materials properties database must be available for processing parameters and any troubleshooting that may be required. Currently, to obtain this type of data, much research is required with the existing literature and is very time-consuming. This database will contain much of the same physical properties data that is currently compiled for materials, such as tensile strength, the various moduli, flex strength, etc., but also the time-temperature response of the resin's degree of cure and other non-typical data. This will include the data used to develop the models and may also be expanded into semi-automated model development module. This will create a work history on the model development such that the data will be available for future analysis as required. This section of the material data base can be referred to as the cure model history file.

The cure history file actually stores the readings, predictions, and any validation data from each cure load. This will provide actual response data for the controlling expert system to evaluate choices based of previous runs. This file will store data on tooling response to heat rate, mass / surface ratio, problems with tooling, data relating to autoclave load arrangement, and tool compatibility. This file will be empirical, that is, actual readings will be stored, in order to reduce the need to model tooling thermal responses. All model validation data will be contained in this system in order to correlate known to predicted data. This will be the quality buy-off portion of the cure system. The predicted degree-of-cure will be compared to actual data points from the validation series to be described later.

The on-line material control and inventory system for all age dated materials is vital to the successful use of the materials database. This system is responsible for the tracking and accountability for all material in the cure modeling environment. The system will start tracking in the material receiving area and will continue tracking through the entire processing cycle. Tracking will conclude at the successful completion of the cure cycle.

Specific items that are tracked include but are not limited to: location, freezer / storage in-out times, humidity and temperature in storage and processing areas, kits, tools used, a unique serial number for each trackable entity, and any other applicable information. The unique serial number assigned to each entity will provide traceability to individual batch, roll, starting advancement, test data, and any other information for that material that is required for the cure model environment. Any retest or updated information for that batch will be added to all applicable serial numbers automatically.

After a part is cured, the information for every tracked entity will be stored in an archive file for that part number and serial number. That way, the part work history will be available at any time.

The humidity and temperature exposure data will allow for the automatic survey of all processing and storage areas. The data collected will be used to update the starting advancement and percent volatile / moisture content of the resin. This data will be converted by a formula to exposure units for ease of maintenance. The exposure units will be incremented to the exposed material every 6 hours so that the worst case computer failure would only cause 6 hours of uncertainty in a materials life. The development of a current loop based temperature / humidity sensor is required to accurately record this information.

The material control system is an important portion of the cure modeling environment because it will not only ensure traceability of the material but also it will provide a very accurate representation of the added ageing and advancement that a resin system has experienced since the determination of the starting advancement. This information will allow for the model to accurately predict the correct cure pathway for the resin's present condition.

## 7. MODEL VALIDATION REQUIREMENTS

Cure modeling represents a major shift in the way of curing composite structures. The final intent of having a fully cured laminate after cure remains the same, however, the method of approaching that goal is the difference. The cure model will be capable of predicting the degree of cure for a given resin system. That degree of cure prediction will be the basis for inspection buy-off of the finished parts. In order for the prediction of the degree of cure to be recognized as an accurate quality criteria, a validation of the cure model must occur.

In this validation phase, the cure model will be tested under varying conditions and its ability to accurately track the degree of cure will be compared to actual measurements. Ultrasonic sensors will be installed on several parts which will be subjected to differing cure conditions. The ultrasonic measurements can be correlated to compression modulus which is an indicator of degree of cure. This data will be compared to the output of the cure model to ensure that the predictions correctly reflect actual conditions.

When the model data correlates, then the model can be accepted as a accurate indicator of cure condition. At that point, several test cures are made with the cure model determining optimum processing profiles. Once again the predicted degree of cure and the actual monitored conditions are compared. If the correlation is acceptable, the model is ready for production implementation. While in production, the model will be subjected to periodic revalidation as described in the section on software.

## 8. SOFTWARE

The greatest challenge of the Automated Composites Cure Control environment is the development of the software to implement the system. The software is divided into five parts: Model, Expert Shell, Cure Device (autoclave, oven, or tool)

Control Unit, Periodic Revalidation of Model, and Data Storage Method for Historical Records. These are all separate entities to be run in a multi-tasking environment. Each entity will be described separately and then summarized to show how each they relate to each other.

The cure model module will contain the algorithms and information for the actual modeling of resin condition and desired cure pathway. It will use as inputs the starting advancement, the starting percent volatile and moisture, the age-temperature-humidity history from the material control system, the process equipment, capability, loading and tool data from the expert shell, and the specific resin system model parameters.

The output from the cure model module will be the optimum cure profile and expected degree of cure for the conditions presented in the cure model inputs. This data will be sent to the expert shell and the cure device control unit to initiate the cure cycle. During the run, if the actual cure cycle conforms to the optimum cycle, then the cure model does not have any more activity. If the actual deviates from the optimum cycle, due to part shadowing or loading constraints, then the model will rerun with the constraints that are occurring as inputs and a new cure profile will be generated to optimize the existing conditions and still obtain the desired degree of cure. If other constraints occur, the model is reiterated to simulate the changed conditions and assure that the highest degree of cure is obtained and the revised profiles are passed to the control unit through the expert shell.

The expert shell is the main control entity within the cure modeling environment. The expert system is responsible for all logic analysis and decision making. An example of a decisions that the system will be required to make includes: deciding if the parts in queue are compatible and can be run together in a single load. From this decision, the system must then decide what the worst case cure condition is that will allow all

306

parts to be run concurrently and insuring that the cure considerations of all parts are met. The system must use the database to access the correct information on which to base its decisions. The system will answer go/no go questions on marginal cases. The system will control the flow of information between the different portions of the cure control system, will require the capability to act on the conformance / non-conformance of the cure device control unit (described next) and cure device to the required cure profile. It must be capable of determining the point at which the cure cycle is jeopardizing the load and request an updated profile from the cure model entity.

The expert system is, in a sense, the cure operator and is tasked with many decisions throughout the cycle. The important difference is that the expert system will be allowed to adjust the cure cycle based on a set of rules based on the model. However, the system must also be able to determine when there is need for operator intervention and request such intervention. The expert system is designed to be the most dependable, most comprehensive, most attentive operator ever to run a piece of cure equipment. It will also have the ability to act appropriately to correct cure cycles based on the chemical models in its control.

The expert system will have a different criteria for curing parts and acceptance than the current standard of conformance to cure profile. The expert system will cure to degree of cure without constraint on profile or pathway. The attainment of the correct degree-of-cure will be the final acceptance criteria.

The Cure Device Control Unit is the software module which actually controls the equipment providing the cure conditions. This equipment can be autoclaves, ovens, self contained tools, or any other curing device. The software of the control unit will monitor and control these devices using the same type of thermal feedback control used by the Applied

Polymer Technology, Inc. CAPS system. The controller will adjust the cure conditions based on thermal feedback from the affected parts and will follow profiles generated by the cure model.

The control unit will constantly monitor part thermal conditions to ensure that they are conforming to modeled part temperature delta, required heat rates, temperature soaks, etc. as required by the model. The control unit will also track cure device parameters, vessel over temperature, and other independent cure device conditions that can affect the cure cycle.

The cure device control unit is responsible to identify any anomalies in the part temperature readings that deviate from the cure profile. These anomalies will be sent back to the expert system where an evaluation is done. In this evaluation, the expert system will either allow continued processing with the anomaly or it will have the cure model generate a revised profile. The rules section of the expert system will determine its course of action. If a new profile is required, it will be returned to the controller to affect a change.

The cure device interface unit is also responsible for generating the cure history file interactively while processing. This file will contain thermocouple readings, vessel temperature readings, a time related history for each part, the original profile and any changes to it, and comments on any anomaly identified during the run, such as a blown vacuum bag. It will also generate the data for the tooling file to be contained in the main data base. This section is the most input - output intensive section of the entire software package, and therefore, this section will have complete control of the data acquisition system. Any other program requiring data from the data acquisition system will interface through this program.

Since the quality of the parts produced by the cure model is dependant on the accuracy of the predicted degree of cure produced by the model, a periodic revalidation of the model will be required. This will be a separate software package designed to operate outside of the expert system between the control / data acquisition and the cure model portions of the system. The purpose of this package is to provide a real-time in-process comparison between the output of the cure model and the input from ultrasonic monitors installed on the parts. This will provide an in- process, no production impact method for revalidating the accuracy of the model. Similar comparisons to those in the initial validation can be made in actual cure conditions in the production environment, if desired. This revalidation system will assure acceptable quality control with minimal production impact. This section is analogous to periodic recalibration of precision equipment.

The data storage requirements for the ACCC system are extensive. An important part of the development task is to create a data storage method that will allow for the compressed storage of the generated data. This could be in the form of algorithm to describe the cure pathway taken and the degree of cure accomplished by that pathway or it could be a set of parameters that will be assigned by the expert system to describe the pathway. Either method would require less data be stored than simply archiving the raw data. Additionally, the analysis of the historical data by the expert system to support decisions will be much less cumbersome and time intensive.

## 9. SUMMARY

The overall picture of the Automated Composites Cure Control System involves the integration of several disciplines, Materials and Processes, Material Control, Manufacturing Technology, Engineering, Quality Assurance, Data Processing, Facilities, and Production. The System will require the inputs of each discipline to fully realize its potential. The required change in manufacturing philosophy is extensive. This is a completely new

way of doing business. It is however, a way of doing business that is consistent with the chemical processing needs of the product.

The implementation of the cure modeling environment and the Automated Composites Cure Control System is not a easy task. Much development work must be accomplished. The basic building blocks of the system are in place and only need to be expanded and interconnected. The concept of cure modeling has been proven. A simple model developed for Fiberite 934 resin system has been tested with ultrasonic monitoring and has predicted the correct pattern of degree of cure and visco-elastic properties. The model needs some expansion and refinement, but it does work.

The computer based autoclave control system based on intermediate processing power minicomputers are in place and only need software changes to provide the cure device control units. The minicomputers have the multi-tasking real-time processing capability that is required to implement the cure modeling environment. The establishment of the data bases and material control systems is underway and will be complete in time to support the implementation.

This method of approaching the curing of composites is the next logical step in the improvement of the reliability of composites and their processing methods.

## 10. BIOGRAPHY

Richard W. Roberts received his Bachelors Degree in Chemistry from University of California at San Diego in 1977. He has worked at General Dynamics, Convair Division for seven years and is currently a Manufacturing Engineer Specialist, Senior working on special projects in Composites Technology.

COMPUTER CONTROLLED RESIN IMPREGNATION FOR
FIBER COMPOSITE BRAIDING

A. H. Kruesi and G. H. Hasko
U.S. Composites Corp.
Rensselaer Technology Park, Troy, NY   12180

## Abstract

Textile processes such as braiding
may significantly reduce the cost of
advanced composites.  Braiding can
be readily automated and offers high
fiber deposition rates and geometric
versatility.  Previously, the use of
braiding was limited because of the
difficulty in wetting fibers moving
in a complex path.  A novel resin
impregnation system has been
developed to adapt braiding to
automated production of advance
composites.  A tubular braider has
been equipped with a computer
numerical control system.  Two axes
are used to drive pumps which feed
resin and catalyst to the resin
impregnation ring.  The mix ratio
and resin volume fraction can be
coordinated in software with braider
and mandrel speeds to fabricate
parts which have complex features.
Test specimens have been fabricated
with glass, Kevlar, and carbon
fibers.  Fiber wetting and void
content were evaluated from
microphotographs of parts cured on a
mandrel without compaction. Burn-off
tests indicated that fiber volume
fractions are in excess of 60
percent.  Applications for the
process include space station tubes,
rocket motor cases, launch tubes,
gun barrel reinforcement, rotor
blades, stiffeners, drill risers,
and similar structural components.
This work was supported by the U.S.
Army Materials Technology Lab under
the Small Business Innovation
Research (SBIR) program.

## 1.  INTRODUCTION

The fabrication of composite
structures using pre-impregnated
(pre-preg) fiber materials is now
highly developed.  However, the cost
of composites manufactured with
pre-pregs is prohibitively high for
many applications. Automated,
low-cost fabrication technologies
are needed to allow the use of
composites  in cost-sensitive
products such as ground vehicles,
bridging, and shelters.  Budgetary
constraints are also driving the use
of automated manufacturing for
flight vehicle components.
Textile processing techniques such
as braiding offer potential
advantages in composite fabrication.
For example, tubular braiding
equipment can weave a seamless fiber
ply around a mandrel at a precise
angle.  Braiders are readily
automated and achieve high
production rates.  Complex, non-
circular shapes can be braided, and
threaded plugs may be overbraided in
a single operation to form a high
reliability end fitting.  Braided
composites also have unique
properties which can be advantageous
in damage-tolerant and energy-
absorbing structures.  Examples of
braided composites are shown in
Figure 1.

McDonnell Douglas Astronautics Co.
initiated the use of braided

composites in high volume applications including fiberglass/epoxy launch tubes, ducting, and fuel lines (1,2). Recently other companies have used braiding to fabricate helicopter rotor spars, bridge truss tubes, and space station elements (3,4,5). U.S. Composites' engineers first employed braided composites in the development of a 7.5 m diameter wind turbine system (1980-1982); three braided rotor blades were built and successfully tested during 1,000 hours of operation over a two-year period. This experience with braiding led to the conclusion that braiding required a) automation and b) a reliable, on-line resin impregnation system in order to be used in production applications.

An innovative concept was developed by U.S. Composites to solve the resin impregnation problem: a ring shaped resin application device would be attached to the braider to wet the individual fiber tows prior to the braid convergence point. The ring surface is porous to allow uniform impregnation over 360° of the ring. Small beads of resin form over each of the pores. Because of the small pore diameter, the resin surface tension prevents uncontrolled dripping. As a fiber bundle passes over each pore, the surface tension is broken and resin wicks into the bundle. Wetting is assisted by contact pressure and by mechanical working due to the combined radial and tangential fiber motion. The porous surface is supplied with resin through a segmental plenum, so gravity effects on the ring are negated. U.S. Patent 4,494,436 has been awarded on the resin applicator system, and additional U.S. and foreign patents are pending. After completing some simple tests of the concept, U.S. Composites proposed a two-phase development program to the Army Materials Technology Laboratory, Watertown, Mass. under the Small Business Innovation Research (SBIR) program. This paper outlines the work performed under the resulting Phase I and II contracts.

## 2. PROGRAM OBJECTIVES

The objectives of the SBIR contracts were two-fold: a) to conduct a production-scale demonstration of the resin applicator ring and b) to generate needed data on the properties of wet-braided composites using high performance fibers and resins. It was first necessary to establish performance goals for the system. The primary criteria was the ability to consistently produce composites with a high fiber volume fraction and low void content when operating at normal braiding speeds. Technical literature was studied for comparative figures. Wet filament wound components can have void contents ranging from 3 to 8 percent(6). The lowest void contents in braided composites were 2.46 to 2.75 percent measured in a vacuum resin impregnation rotor spar(3). The goals subsequently established for the production braiding system are listed in Figure 2. These goals were for the lowest cost approach--wet braiding and oven curing on the mandrel without compaction.

### FIGURE 2

### PERFORMANCE GOALS FOR THE PRODUCTION COMPOSITE BRAIDING SYSTEM

- Apply liquid resin in a controlled manner during all modes of braiding: biaxial, triaxial, and bi-directional.

- Achieve the predicted fiber volume fraction within +4%. o Maintain consistency of fiber volume fraction with +2%.

- Achieve a void content of less than 3% without shrink tape, vacuum bags, or other means of compaction.

- Automate the system with computer numerical control.

- Allow for the optional use of resin heating or cooling and resin de-aeration.

## 3. SYSTEM DESCRIPTION

Standard tubular braiding machines manufactured by Mossberg Industries, Cumberland, R.I. have been adapted for use as composite braiders. A prototype production system was first tested at U.S. Composites on a 64 carrier machine (Figure 3). Later, a large scale (450 mm bore) resin applicator ring was installed on a 144 carrier braider installed in the Benet Weapons Laboratory at Watervliet Arsenal. A similar 144 braider at U.S. Composites is shown in Figure 4.

The resin applicator ring system consists of a laser-drilled porous ring with resin distribution plenum, adjustable mounting arms, and reversing ring assembly. The ring is supplied with resin by a pair of precision gear pumps driven by AC brushless servo motors (Figure 5). The resin and catalyst are mixed in a static mixing head on the back of the resin plenum, so the quantity of catalyzed resin is very small. The drive motors are AC induction motors controlled in a servo loop. This represents the latest technology in servo systems. Each motor has an optical shaft encoder for feedback control. The servos are commanded by an IBM PC equipped with a two-axis, motion- control board (two boards are used in the four-axis control system).

A program entitled "BRAID" was written to predict braiding parameters and thus minimize trial and error testing of new applications. Inputs include part diameter, fiber angle, braid speed, fiber type and denier, and bundle aspect ratio. The output includes predictions for ply thickness, fiber volume fraction, coverage, mandrel speed, resin flow rate, and weight of material deposited per ply. This information is easily entered into the braider control computer. The machine operator can start the automatic cycle, interrupt, and resume operation at will using a single pushbutton. Any number of control programs for specific parts can be stored on floppy disk for later use.

## 4. TEST PLAN AND RESULTS

A test plan was prepared to answer key questions about the resin applicator ring and the computer controlled pumps:

- How accurate is the pumping system over the range of component viscosities, mix ratios, and flow pressures?

- Is the segmented plenum effective in distributing resin evenly around the ring?

- How well are the performance goals (Figure 2) met by the system?

- What are the mechanical properties of wet-braided composites?

The pumping system was tested by measuring the volume flow rates using typical epoxy components: Epon 826 resin (6.5-9.5 Pa-sec) and MTHPA catalyst (.05-.08 Pa-sec). The results agreed with predictions within 1% for pressure differentials up to 275 kPa (40 psi). The resin mix ratio was verified by high pressure liquid chromatography (HPLC) tests of resin samples collected from the face of the ring. Four tests were conducted using two volume flow rates and two pumping pressures. Using a sample weighted with an analytical balance as a reference, the average error was 0.35% and the maximum error was 1%.

The flow distribution around the ring was tested by measuring the flow at the 12:00, 3:00 and 6:00 positions using epoxies with a mixed viscosity of 0.8, 1.9, and 2.9 Pa-sec. These tests showed that the plenum design enabled control of distribution to within 5% if desired.

The Army Materials Technology Laboratory had extensive data on the properties of $\pm45^\circ$ filament wound tubes made of S2 glass, Kevlar, and carbon fiber, and an anhydride cure epoxy, Epon 828/MTHPA/BDMA. To obtain comparative results, it was lecided to fabricate sets of wet-braided test specimens using the same materials, size, and fiber angle. Some additional specimens were fabricated using an 0/+45° triaxial construction and using other epoxy systems. Static tension, compression, and torsion tests are being conducted by the Army Materials Technology Laboratory and will be reported separately.

Operational tests have demonstrated that glass, Kevlar, and carbon yarns can be readily wet braided with any resin suitable for filament winding. Dow Tactix (TM) 138/H41 epoxy offered the best combination of physical properties and processing behavior of the various resins tested. Biaxial and triaxial braiding tests were conducted in both forward and reverse directions at normal braiding speeds. The only material that had proven difficult was untwisted carbon tow; excessive damage occurred as the fibers threaded through the carrier mechanism. Since then, modified carriers have been installed to allow braiding with carbon tows. The set-up and cleaning procedures for the resin applicator ring are not difficult. The use of computer controls has led to excellent consistency of results when tests are repeated.

A key question was the adequacy of wetting of individual filaments by the resin applicator ring. To answer this concern, a number of test specimens were sectioned and examined by microphotographs. A cross section of wet braided Kevlar/epoxy shows excellent wetting (Figure 6). Elliptical- shaped fiber bundles surrounded by resin rich zones are characteristic of braided composites (Figure 7). Voids that are found in the cross section are usually confined to the resin rich zones and are circular in shape. They appear to be caused by air bubbles present in the resin or by volatiles generated during cure. Resin de-aeration and vacuum bagging prior to cure are expected to reduce the void content to levels found in autoclave cured parts. Multiple ply laminates up to 12 mm thick have been wet braided without any sign of fiber wrinkling or other defect.

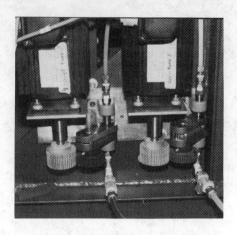

Figure 5    Servomotor driven resin and catalyst pumps.

A major use of braided composites is in rocket launch tubes. A typical acceptance test is that launch tubes must be leak free when pressurized to 552 kPa (80 psi). A two-ply, wet-braided, S2 glass/epoxy tube was tested at up to 690 kPa (100 psi) and found to be leak free. Compression loading is often a critical test for composite structures. To perform a simple test of compression strength, a triaxial S2 glass/Tactix (TM) epoxy cylinder 51 mm diameter with a 9.5 mm wall thickness was fabricated. This cylinder was subjected to eleven cycles at an external pressure of 69 MPa (10,000 psi), then subjected to one cycle at 138 MPa (20,000 psi) without collapsing.

5.   PRESENT APPLICATION AND FUTURE DEVELOPMENTS

Five composite braiders (from 16 to 144 carriers) are now installed at U.S. Composites; three are equipped with the resin applicator ring system and computer controls. A wide variety of applications have emerged during the course of this program. A high modulus graphite

312

complex geometry port is being wet-braided with a 320°C capable resin system. Wet braided, tri-axial, S2 glass/epoxy tubes have been successfully tested for launch tube and externally loaded pressure vessel applications. Long, column-type structures have been braided at ±17° using a 12 k carbon fiber. Prospective applications for the near future include a Kevlar/ceramic braided structure for electro-magnetic guns, carbon/epoxy energy absorbing composites, and glass/epoxy reverse osmosis tubes.

Other important developments have resulted from this program. Modified braiding approaches have extended the range of applications. A pultrusion/overbraided construction (patent pending) has been developed at U.S. Composites to produce straight and tapered tubes with up to 85 percent of the fibers oriented at 0°. Wet braided stiffeners can be layed up and co-cured with resin transfer molded panels (patent pending). This hybrid approach enables the fabrication of parts which could not be produced by either method alone. The computer controlled braider can be used to form complex dry fiber shapes for use as is or as a pre-form which is further processed.

The computer controlled pumping system has been found to be ideally suited for use in automated resin transfer molding (RTM) applications. The RTM process is emerging as another low cost manufacturing technology which can be adapted to the needs of aerospace users.

Additional development work is in progress. Experiments are being conducted with non-autoclave, low-cost compaction methods. The goal of this work will be to raise fiber volume fractions to the 65 to 70 percent range and reduce void contents to one percent or less. New resin systems will be tested, particularly those with special performance features such as high temperature capabilities or improved fracture toughness. The resin applicator ring concept is also applicable to other composite processes such as filament winding.

## 6. ACKNOWLEDGEMENT

The authors wish to express their gratitude to Mr. Noel J. Tessier of the Army Materials Technology Laboratory for his invaluable guidance and support.

## 7. BIOGRAPHIES

A. Hugo Kruesi founded U.S. Composites Corp. and serves as president. He has been awarded U.S. and international patents on the resin applicator ring system. He graduated from Rutgers University with a BSME and has taken composite courses at RPI and Yale.

Gregory H. Hasko is the chief engineer at U.S. Composites. He was formerly employed as a senior engineer at Combustion Engineering and the Hamilton Standard division of UTC. Mr. Hasko received his BS and MSME degrees from Rensselaer Polytechnic Institute, and is a registered professional engineer.

## 8. REFERENCES

(1) Post, R. J., "Braiding Composites--Adapting the Process," 22nd National SAMPE Symposium and Exhibition, Vol. 22, San Diego, April 26-28, 1977, pp 486-503.

(2) Sanders, L. R., "Braiding--A Mechanical Means of Composite Fabrication," SAMPE Quarterly, January 1977, pp 38-44.

(3) White, Mark L., "Development of Manufacturing Technology for Fabrication of a Composite Helicopter Main Rotor Spar by Tubular Braiding," AVRADCOM Report No. TR-81-F-9, April 1981.

(4) Tsuchiyama, Terry, "Braided Carbon/Epoxy Tubes for Large Space Structures," Materials for Space--The Gathering Momentum, SAMPE International Business Office, 1986, p. 24.

(5) Brookstein, David, "Design and Development of a High Stability

Truss Chord," Advancing
Technology in Materials and
Processes," SAMPE
International Business Office,
1985, pp 682-690.

(6)  Elegante, Thomas L., "Filament
Winding," <u>Mechanical</u>
<u>Engineering</u>, Vol. 108, No 12,
1986, pp 32-36.

FIGURE 8

INITIAL RESULTS OF WET BRAIDING TESTS

| Material | Fiber Volume Fraction % | | Void Content % | Test Method |
| | Predicted | Measured | | |
|---|---|---|---|---|
| Kevlar 49 (±45°) | 66% | 63.9% | 1.74% | D'Andreas |
| S2 Glass Yarn, 8 ends (±45°) | 64% | 62.9% | 2.11% | ASTM D2734 |
| Celion 12k Carbon tow (±45°) | 62% | 39.8% | 3.71% | D'Andreas |
| S2 Glass Roving (0\±45°) | 56% | 56.8% | 2.40% | ASTM D2734 |

Note:  D'Andrea's method is an acid digestion technique.

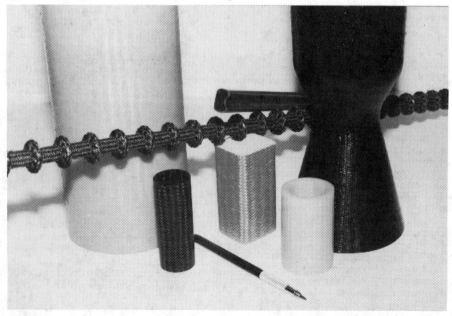

Figure 1    Examples of carbon, Kevlar, and glass fiber braided composites.

Figure 3    Prototype resin applicator ring being tested on a Mossberg 64
            carrier braider.

Figure 4    144 carrier composite braider with resin applicator ring and
            4-axis computer numerical control.

Figure 6    Microphotograph of wet-braided Kevlar/epoxy (404x).

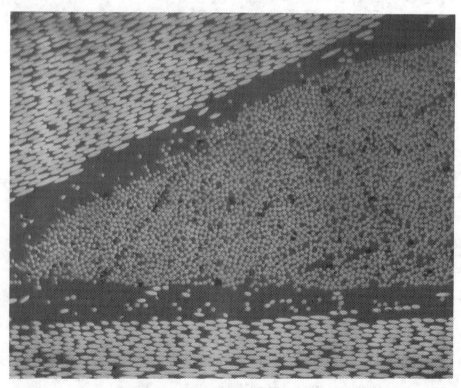

Figure 7    Wet-braided carbon/epoxy showing elliptical fiber bundles
surrounded by resin-rich areas (150x).

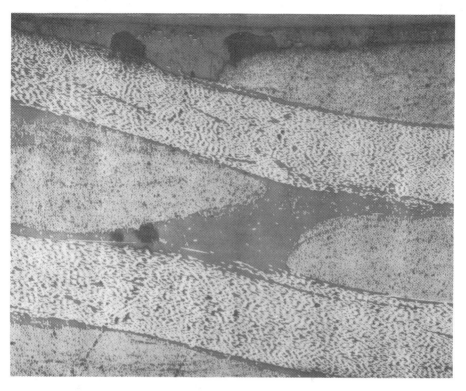

Figure 9    Circular shaped voids in resin rich areas of wet braided part
(50x).

UNIFIED LIFE CYCLE ENGINEERING IN COMPOSITES
C. Y-C Lee, H.M. Burte, and C.F. Browning
Air Force Materials Laboratory/MLBC
Wright-Patterson Air Force Base, Ohio

## ABSTRACT

Unified Life Cycle Engineering
(ULCE) is an engineering approach
whereby at the design stage,
issues associated with producibility and supportability can be
considered. Outlined in this
presentation are suggestions for
R & D needed and a program
strategy to achieve ULCE in
composites.

## INTRODUCTION

In most design activities of high
performance, advanced systems,
the primary objective is to
generate the details of an
article or system that can
satisfy specific performance
requirements. This is consistent
with the traditional view that
the benefits of an engineered
article are performance driven.
A more appropriate interpretation
of the benefit from an engineered
component should take into
consideration all of the burdens
associated with the part during
its life cycle. These burdens
are time loss in repair, cost of
manufacturing and design, logis-
tics difficulties etc. The life
cycle of an engineered com-
ponent/system is made up of four
main stages, phases or functions:
design, manufacturing, utility and
maintenance, and many subtasks
associated with each of these
functions. While gross benefit is
derived from the utility function
through the usage and performance
of the component, the net benefit
should be considered as the gross
benefit minus the burdens of
design, manufacturing, support-
maintenance. Unfortunately, the
latter two often receive too
little attention at the design
stage. There are numerous
activities on-going within the Air
Force that are addressing issues
on how to maximize the performance
benefit, or to minimize the
burdens of design, manufacturing
and support. With the advance of
electronics, computers are used
more and more in all these areas:
extensively in design for performance; growing in manufacturing;
and starting in maintenance and
other aspects of product support.
Even if not explicitly stated, the
net result of these activities is
to seek an improvement on the net
benefit of the component/system.
Mostly, however, they are activities that are concentrating on
individual functions. CAD is to
help with the design function and
CAM is to ease the manufacturing
process and so forth. Very few

activities are taking more than one function into consideration; and even if they are, the interaction between functions are limited to data transfer from design to manufacturing and product support with little feedback.

One can conceptualize a type of engineering activity that is aimed explicitly at improving the net benefit of an engineered component/system. The goal of this concept is not new as evidenced by the number of on-going activities that are aimed at reducing costs at different functional levels and improving performance capabilities of the component/system. The refreshing elements of this concept would be the explicit identification of the net benefit as the figure of merit, and to achieve this goal through activities that address the interrelationships of the functions.

The advocation of this engineering concept should be more than just heightening the awareness of the interfunctional relationship, whereby activities in one function can affect activities in other functions, which in turn will change the net benefit equation of the system. For example, a choice of a certain manufacturing process may minimize the manufacturing cost, buy may preclude the repair of the component, leaving the replacement of the component as the only alternative in the support function. This may increase or decrease the net benefit of the component depending on the situation. If the component is intended for short time usage, likely the net benefit will increase; but if the component will require repairs several times in its life cycle, the net benefit may decrease. The ways these selections affect the net benefit equation depend on the intended purposes, or

functional objectives of the component. Any trade-off studies in respect to net benefit considerations thus can only come after a careful and detailed definition of the functional objectives of the component.

A more aggressive stance of this advocation is to identify specific activities that one can do in one function that can lead to an improvement in the net benefit of the component through its effects on activities of the other functions. For example, one can identify utility practices that can reduce support/maintenance costs or increase the net benefit through an improvement in quality control in manufacturing that leads to higher performance capabilities.

One can think of many activities that can fit within the interpretation described above. It is our opinion that activities in one of the four functions can be uniquely fruitful in increasing the net benefit of a component. This function is the design function. Unified Life Cycle Engineering (ULCE) is aimed at performing the design function in such a fashion that the net benefit of the engineered system/component is maximized. This engineering approach is receiving serious consideration for future activities in the Air Force.

UNIFIED LIFE CYCLE ENGINEERING

During the design phase of advanced systems or their components, design specifics were often optimized to yield maximum performances, with the general idea that by doing so, the benefit of the designed articles is maximized. In the context of a net benefit consideration, the burdens of the other functions are often ignored. Yet the details specified in the design function can bound the scope of activities within the manufacturing and support/maintenance functions. Quite often, activities in the manufacturing and support/maintenance functions have to cope with this lack of consideration, and are unduly eroding the net benefits of the system.

The situation is increasingly recognized in the Air Force. More and more, considerations of "ilities" other than performance are demanded in new system development. Currently, such considerations are usually carried out in a serial fashion. The component is firstly designed for maximum performance. The manufacturing engineers then evaluate and define appropriate production approaches. Redesign may be requested at this stage to facilitate manufacturing. Subsequently, other support "ilities" will in turn receive consideration. At such stages, it may be difficult or too late to impact the design.

Instead of the serial approach, other "ilities" can be considered up front at the design stage. The time is ripe for such activities. Computer Aided Design (CAD) is being used not only for drafting but to rapidly and effectively analyze the performance of many design options as might be illustrated by a finite element analysis. Progress in this "performance" domain (area) has been rapid and is continuing. Similarly, significant progress, although at an earlier stage of development, is being made in the "manufacture" domain. "Computational plenty" is permitting developing adequate solutions for previously intractable mathematical models describing the unit operations or unit processes of manufacturing, such as the simulation of the heat flow in a thick laminate to prevent an exotherm. Similar models are being developed for the combination of unit operations into manufacturing centers and, for an entire factory.

Modeling in the "supportability" domain is in its infancy but its potential is indicated by early work which allows predicting the probability of a component requiring repairs as a function of the lay-up sequence of the composite laminate. This modeling which permits evaluation of the frequency of repair as a function of design details is an example of Computer Aided Support (CAS). Given tools such as these, the designer can consider performance, manufacture and supportability essentially concurrently rather than serially. A large number of design options can be rapidly and accurately evaluated from many points of view. For example, the lay-up sequence and the number of plies of a component might be changed, sacrificing some weight savings, to increase damage tolerance or to reduce the scrap rate of manufacturing. The goal of effectively considering the "ilities" during initial design can be approached.

The example given here is deliberately simplified to illustrate the concept. Analytical models may not be available or appropriate. The input information may be in the form of qualitative models such as a computer generated mockup which simulates repairability, or graphical or tabular data, or various forms of symbolic data such as lists of "do's and don'ts" or experience represented in expert systems.

Unified Life Cycle Engineering can be applied to many areas, e.g. electronics, metal and composite structural components. The status of models or effectively organized databases will differ, as well as the R & D necessary to develop and computerize them. In particular, the support/maintenance domain has received little attention and the problems of fully integrating the domains are unprecedented. Examples of the latter include the inherently heterogeneous nature of the data bases which must be included, providing real time response to a designer at his workstation, and providing optimization approaches to limit the search space of "what-if" possibilities or simulations to be considered.

A strategy of concurrent efforts ranging from quite fundamental research to attempts to apply a ULCE approach to realistic situations (windows) is recommended. The latter will provide specific problems and vectoring to the former, and increasingly demonstrate what can be put into use. Incremental usable progress is expected even if an approach to the full concept will require a decade or more of extensive work.

In the following we will suggest some windows for early exploration of what is possible, and areas where R&D will probably be desirable. Our purpose is not tutorial; it is to stimulate similar thinking in the composite community.

Advanced composites are being used increasingly in advanced Air Force systems. This material is "relatively" new, so Unified Life Cycle Engineering development activities in this area can substantially advance the technology and increase its utilization.

## "ILITY" MODEL DEVELOPMENT

A systematic pursuit will be necessary to identify the "ility" issues that can be affected by design changes. ULCE will not be able to solve all "ility" issues, but only those that can be influenced by the change of some design specifics. The measure of improvement or worsening of the "ility" issues, let's call it a merit indicator, needs to be quantified as a function of the

design changes if trade-off consideration and optimization are to be entertained. This may be accomplished by establishing a mathematical equation describing the merit-design relationship, or collecting sufficient engineering data so that the relationship can be tabulated. When these are not possible, at least "do and don't" rules should be established to avoid certain undesirable situations. While these rules can usually be formulated from past experience, and primarily for the purpose of avoiding past mistakes, optimization or trade off consideration cannot be achieved unless the relationship is quantified.

Even at the outset one can suggest important areas where models or databases are lacking. Research efforts should be invested in these areas to ensure all appropriate "ility" issues can be considered in a functional ULCE system. There is no comparable model in composites as in metals for specifying the probability of detecting damages as a function of design specifics. While the lay-up sequence of a composite laminate may not alter the probability of detecting delamination significantly, the locations and geometry of subassembly components can affect the effectiveness of NDE techniques. The effect of design specifics of composite components, in terms of geometry or lay-up sequence, on repair methods has thus far received little attention. For example, the role of ply drop-off areas in void formation during the manufacturing process has not been quantified. These examples by no means form a complete list of the missing elements that are needed for a fully functional ULCE system. Other lacking areas should be identified and appropriate models, being numerical or rule-based, or databases should be developed to remove these deficiencies. The anisotropic nature of composites has offered a unique tailoring capability for designing for performance. This unique characteristic should be explored for other "ilities" as well.

Another problem might be to investigate how to recast simulation models into forms that would enhance the capabilities of a ULCE system. Most simulation programs in all functions take design specifics as inputs for the simulation, and the simulated results are outputted. More appropriate for ULCE would be models using the same fundamental equations or programming logics to describe the simulated process, but taking acceptable conditions as input and produce design specifics that would satisfy these conditions. A more aggressive posture would be to demand these models to provide not just any design specifics that can satisfy the prescribed conditions, but an optimized one. These submodels should not be merely optimizing the relationship addressed in the model. Restraints from other "ility" issues should be appropriately addressed as a part of the input to the model, or interaction with other submodels can be an integral part of the optimization process and are built in within the program.

## ULCE MODEL ARCHITECTURE

The architecture of the design process using advanced structural composites is receiving increasing attention. It must be further explored and defined to provide a system architecture, or a blue print, whereby different computer based elements can be linked, and all "ility" functions can receive their appropriate levels of consideration. This would be a flow chart where the relationships between different submodels and databases can be defined, and the nature and scope of information transfer between elements specified. The sequence

of "ility" consideration in the design process should be optimized for efficiency and practicality. The subelements of each "ility" should be distributed appropriately at different design phases. Some subelements should be reconsidered repeatedly at different phases of the design process. Conditions and criteria for evoking specific simulations or rule-based "do's and don'ts" considerations should be carefully specified. All in all, an appropriate design process for ULCE needs to be developed.

It is also important to group relevant "ility" issues with the appropriate design specifics. In general, the design of a composite structure is not generated in one single step, but will go through a series of design stages, at each of which, certain design specifics will receive consideration. To facilitate ULCE, the "ility" submodels need to be distributed appropriately among the stages. Some "ility" issues may be relevant to more than one design specific and will be considered in more than one stage. The design architecture should be able to handle conflicts arising from different stages, and loop back mechanism should be provided so that design decisions at later stages do not have to be inflexibly restricted by decisions made in earlier stages.

Another example of ULCE architecture issues for designing composite components is the interface with other non-composite design activities. Composite components are usually just parts of a larger functional entity. So the design of the component cannot be considered in isolation. Similarly, many non-composite elements will likely be housed within the composite components in question, and the interface of their design activities with the composite component design needs to be addressed as well. Being a component of an overall system, the design of the component probably will have to accommodate a set of system constraints resulting from the overall design of the system. For example, these constraints may be stress and strain requirements, or physical dimensions and locations of assembly points. It seems that the net benefit can best be optimized by starting the ULCE process before the set of constraints is finalized. That may mean participating in the overall system design in a limited fashion by considering some relevant manufacturing and supportability issues relating to the components. This will ensure that the set of restraints will not be overly restrictive but is suitably optimized for both the overall system and the composite component.

## NET BENEFIT EQUATION

To truly achieve the objective of ULCE, complex trade-off studies and optimization will have to be implemented. Trade-off studies are not new, but to do so between different "ility" requirements and in the context of ULCE can be a whole area of research by its own right. One can list all the a-priori requirements for all the "ility" issues, but to translate this set of "ility" issues into a net benefit equation which can be useful for resolving conflicts is not a simple task. To do so may require a level of system definition of the article's life cycle benefit objectives that we are not accustomed to. Mathematical representation of a figure of merit may not be sufficient.

## OTHER RELATED TECHNOLOGIES

Advances will be required in the ability to integrate information from the many different models

and data bases which must feed into a ULCE system. Progress here can be generically useful and not restricted to composites. Integration of heterogeneous data bases can be a tremendous challenge. In some cases combinations of symbolic knowledge with quantitative models are necessary to allow proper consideration of some "ility" issues. Data transfer between incompatible computer systems can be cumbersome, and may require attention to allow a ULCE system to function within practical time restraints at the work station of the designer. A lot of these issues fall under the areas of computational science, both hardware and software, and artificial intelligence programming. Appropriate development in these areas is vital to ULCE.

## APPLICATION WINDOW

As soon as possible there should be efforts against realistic application possibilities (windows) to explore what is possible and stimulate its transition to use, and to provide focus for continuing R & D which will yield the next generation of possibilities.

One starting point for this task may be to carry out case studies of some chosen composite parts, eg. an F-16 tail skin assembly, or an F-15 speed brake. The interaction between the design for performance, manufacturing and support functions can be explored by computer simulation. This simulation would provide a base with which to study the nature and problems of functional interactions, to identify missing models or submodels, and to evolve a more sophisticated scheme of interaction that would achieve the desirable objectives of Unified Life Cycle Engineering of composites. In order not to be hampered by some missing technology ele-

ments, such a task would make use
of existing technology only.
Cumbersome data transfer between
functional models may have to be
utilized to study the flow within
the system and to study the
effects of the interaction between
submodels. "Make-believe" submodels may have to be constructed,
or look up tables may have to
substitute mathematical equations
in situations where these models
and equations are unavailable.

The identification of suitable
windows will be one of the more
challenging aspects in the further-
ance of ULCE technology. The
results of such demonstration and
validation tasks should be analyzed
to extract generic Unified Life
Cycle Engineering principles.
These principles would be stating
the fundamentals of Unified Life
Cycle Engineering but without the
specifics associated with a
particular application. These
principles could then be applied
to other application areas.
Similar principles should be
extracted from ULCE activities in
other application areas for the
advancement of ULCE technology in
general.

## SUMMARY

The advantages of being able to
consider all "ilities" at the
design stage are generally
accepted, and the states of
development of many related areas
are making such advantages a
realistic possibility. For
composites, the ability to tailor
the material for performance
should be exploited for other
"ilities" as well. The potential
of a ULCE approach is real. The
challenges are significant and
will require increasing attention
within the composite community.

AUTOCLAVE VS. NON-AUTOCLAVE COMPOSITE PROCESSING

R. Dave, J. L. Kardos, S. J. Choi, and M. P. Duduković
Department of Chemical Engineering and
Materials Research Laboratory
Washington University
St. Louis, MO  63130

## Abstract

The fact that autoclave processing of structural composites is clearly intensive in labor, energy, and capital investment has led to recent efforts to develop a non-autoclave process. The recent development and coupling models for kinetics, viscosity, flow, and void formation allows one to examine the relative merits of such a process in terms of producing a void-free laminate of the correct resin content. This paper utilizes a coupled model to identify processing cycles for both autoclave and non-autoclave processes, which lead to void-free laminates. In particular, it is shown that the non-autoclave process has a very narrow "window" in process parameter space, in which good quality laminates can be made. The strategy in autoclave processing should be to keep the pressure everywhere in the laminate greater than that which allows void growth via water diffusion. The non-autoclave process must either begin with nearly water-free prepreg, or utilize a cool-down cycle before final curing.

## 1.  INTRODUCTION

Historically, composite curing processes used autoclaves as a source of pressure and energy. The autoclave process is energy-, capital-, and labor-intensive. It is often the limiting factor in assembly production rates because curing parts one by one is very time consuming. An alternative process for composite fabrication is therefore a logical goal which has recently received considerable attention at Rockwell International Corp.[1]

In several earlier papers[2-4] we have developed models for void growth and resin transport during composite processing. Combining these models with the resin cure kinetics and viscosity models, we have developed a better understanding of the autoclave process, which in turn led to the development of guidelines for the optimum selection of an autoclave cure cycle[5]. In this paper we will utilize our models to examine the possibility of producing high quality laminates using the Rockwell non-autoclave process. These results will be compared and contrasted with the production of similar laminates using an autoclave process.

## 2. PROCESS DETAILS

### 2.1 The Autoclave Process

One of the often-used production processes for fabrication of high performance structural laminates is the Autoclave/Vacuum Degassing (AC/VD) laminating process. In this process, individual prepreg plies are laid up in a prescribed orientation to form a laminate. The laminate is laid against a smooth tool surface and covered with successive layers of release cloth, glass bleeder fabric, occasionally a caul plate, combination glass breather cloths, and finally a vacuum bag.

The curing process is accomplished by exposing the prepreg sandwich to temperature and pressure according to a given cure cycle. The processing parameters and cure cycle currently used in the fabrication of composite structures are usually derived from experience, primarily as a result of iterative experimental testing. Two major drawbacks

of the current process development approach are: (a) the need for extensive and expensive experimental testing to determine the proper cure cycle and (b) the fact that a cure cycle found to be suitable for a given material for one application is often inapplicable for a different material or for the same material in a differe-t part geometry.

The autoclave cure cycle analyzed in this paper is shown in Figure 1. The autoclave pressure and the bleeder bag pressures are maintained at 100 psig and zero psig, respectively, throughout the curing process.

In a commercial autoclave process, dams are usually placed around the curing laminate. This leads to the situation of varying and unknown boundary conditions during the cure cycle. To avoid this problem, we have assumed that "free-bleeding" (0 psig in the bleeder bag and negligible bleeder resistance to flow at the boundaries) exists. For the case of "free-bleeding," the pressure at the boundaries of the laminate is the same as the bleeder bag pressure.

The cure temperature during the various stages is maintained as follows:

Stage 1: The temperature is increased from 80°F (27°C) to 240°F (116°C) at 4°F (2.22°C)/min,

yielding a rise time of 40 min.
Stage 2: The temperature is held at
240°F (116°C) for 70 min.
Stage 3: The temperature is increased from 241°F (116°C) to
350°F (177°C) at 4°F (2.22°C)/min.;
the rise time taken is 27.5 min.
Stage 4: The temperature is held at
350°F (177°C) for 130 min., i.e.,
until the end of the cure cycle.

2.2  The Non-Autoclave Process

The non-autoclave process evolved
as a result of a series of experiments conducted at Rockwell[1] on
AS1-3501-5A graphite/epoxy materials placed on a heated plate within
a large bell jar. It was observed
on heating several unidirectional
and cross-ply laminates to approximately 250°F under a vacuum of
28-30 inches of mercury that (a)
violent outgassing began at about
150°F, (b) the majority of the bubble formation was interlaminar, and
(c) the most vigorous outgassing
occurred when the resin viscosity
was at a minimum. These degassed
laminates were then cooled to room
temperature under vacuum and subsequently cured by reheating to a
temperature of 350°F using only
vacuum pressure. Examination of
these cured laminates showed that
their mechanical properties, i.e.,
porosity, thickness, interlaminar
shear strength, etc., were similar
to those of autoclave-cured laminates.

Based on our prior modeling experience[2-5], we believe that the
violent outgassing in the Rockwell
bell jar experiments begins at
about 150°F because the emanating
gas was mainly water. This argument is justified on three counts.
(1) Grayson and Wolf[6-7] have used
the technique of precision abrasion
mass opectrometry to study the volatile compounds present within voids
located in a graphite/epoxy laminate; they detected water as the
main component in the voids.
(2) According to our void stability
equation[2] (details of which are
presented in section 3.1), pure
water voids will form in an epoxy
resin which has been initially
equilibrated under 25-40% relative
humidity, if the resin is heated to
150°F under a vacuum of 28 inches
of mercury (the same vacuum as that
applied under the bell jar).
(3) In work done at General Dynamics
[8] similar bell jar experiments on
layups of Fiberite T300/976 plies
showed that trapped air alone could
not cause interlaminar voids and
that water-stabilized air or water
alone was the source of the void
problem. If entrapped air does
exist, then in our previous paper
[2] we have shown that it can act
as a preferential site for the water
vapor void growth to occur. To
create an environment in a non-
autoclave process, which is similar
to the bell-jar process, simplified
tooling coupled with flat laminate
processing was used[1]. Tooling
consists of an inner bag over the
sandwiched laminate and an outer

metallic hard shell or hard tooling. The two chambers can be independently pressurized. Comparing this layout with the autoclave process, it is evident that the metallic hard shell over the bag really acts like an autoclave (transmitting the shell pressure to the resin) which can withstand low pressure or vacuum only and has no heating elements in it. Heating of the laminate is done by placing the entire tool with the hard shell and its contents into an oven. In the autoclave process, the autoclave acts as an oven and also transmits the pressure to the resin in the laminate.

Figure 2 is a typical non-autoclave cure cycle obtained from reference 1. After moving the entire tool (sandwiched bag and the hard shell) into an oven, vacuum is applied under the inner bag and under the outer hard shell, and the temperature is increased in the manner shown. As the temperature is increased, the resin viscosity decreases rapidly and chemical reaction begins. Also rapid degassing of water vapor takes place above a certain temperature. When the temperature reaches about 250°F, it is held constant for an initial dwell time under vacuum in both the inner bag and the outer hard shell. Subsequently, the outer hard shell is vented to atmosphere or pressurized up to 20 psig during the consolidation dwell. After the consolidation dwell, the laminate is heated to and held at 350°F for 60 minutes followed by cooling under vacuum. Alternatively, a cycle which more exactly matches the bell jar experiments involves a stage cycle cool-down after the consolidation dwell (see dashed curve of Figure 2). The laminate may subsequently be post-cured at 350°F for varying times.

It may be noted from FIgure 2 that the following processing parameters can be varied in the cure cycle up to stage cool-down.
1) Initial temperature ramp rate under vacuum
2) Initial vacuum in the bag and the hard shell
3) Dwell time under the initial vacuum
4) Consolidation dwell time
5) Consolidation pressure in the hard shell
6) Stage cool-down rate
7) Initial relative humidity exposure of the prepreg

In the Rockwell report[1], the values of the first four processing parameters under which high-quality laminates were made have been specified. However, the values of the latter three variables are missing. Therefore, in this paper, the effect of the latter three on void growth shall be analyzed with the other four variables held constant as follows:

328

1) Initial temperature ramp rate for temperature rise from 70°F to 250°F = 5°F/min.

2) Initial vacuum in the bag and the hard shell = 27 inches of Hg = 0.1 atm.

3) Dwell time under the initial vacuum = 30 min.

4) Consolidation dwell time = 15 min.

## 2.3 The Physical Situation

In our previous paper[5], we have shown that it is feasible to select an autoclave cure cycle such that conditions favorable for water vapor void growth are never allowed to exist at any point within the laminate during the cure. This is achieved by maintaining the resin pressure at every point in the laminate greater than the minimum resin pressure required to prevent void growth by diffusion of water.

In the case of the non-autoclave process, vacuum is initially applied within the bag as well as the hard shell. Therefore, there is no way to prevent void growth during the initial temperature ramp and the initial dwell period at 250°F. In fact, in the non-autoclave process, it is desirable to pull as large a vacuum as possible during the above-mentioned stages to drive out the dissolved moisture from the resin. However, the final laminate may still be made void-free by using the following procedure. During consolidation dwell, by releasing the vacuum

from the hard shell and applying pressure within the hard shell, the resin pressure increases from about 28-30 inches of Hg vacuum to about 20 psig. This increase in pressure combined with the subsequent cooldown can cause the voids to collapse under specific conditions. It is important that the cool-down should begin when the resin viscosity is fairly low and by the end of the cool-down period, the resin should be below the glass transition temperature without having gelled, or have gelled before reaching the glass transition temperature. Such chemo-rheological changes occur in an epoxy resin in accordance with Gillham's time-temperature-transformation theory[9]. Once the resin has solidified, neither void growth nor dissolution is possible. If gelation has occurred, no further void growth is possible upon reheating the laminate up to 350°F and post-curing it at that temperature.

## 3. MODELING

### 3.1 System Selection

In this study, void growth in Hercules 3501-5A resin is analyzed. The choice of this resin depended on the following considerations:

(1) Hercules ASI/3501-5A was one of the starting materials used by Rockwell[1], which produced high-quality laminates.

(2) The diffusivity and solubility data as a function of relative humidity and temperature are available in the literature[10].

The master model of any thermosetting matrix composite curing process consists of the heat transfer model, the resin kinetic model, the resin viscosity model, the void growth model, and the resin transport model. All of these models except that for heat transfer have been addressed in our previous papers[2-5]. Therefore, in this article, we shall only briefly review our models for void growth and resin transport insofar as they are pertinent to this study.

3.1 Void Growth Model

It has been argued that there are two possible sources of voids. One is the moisture dissolved in the resin during the prepreg layup procedure. The other is entrapped pockets of air. As the temperature of the resin is raised or if the pressure drops sufficiently, dissolved water will form a vapor void. Such vapor voids and the existing air voids can then grow by diffusion of moisture. This growth is possible if the concentration of the dissolved water in the resin (which is a function of the relative humidity to which the resin was exposed) exceeds the water concentration at the void resin interface (which is a function of water vapor pressure at that temperature and water mole fraction in the void). Since the driving force for diffusion rises with temperature, the only way to prevent void growth is by maintaining sufficient resin pressure. We have shown[2]

that for a prepreg equilibrated with moisture at a relative humidity, $(RH)_o$, at room temperature, the potential for void growth during the cure cycle can be eliminated if the resin pressure at any point satisfies the following inequality:

$$P_{min} \geq 4.962 \times 10^3 \exp\left(\frac{-4892}{T}\right) (RH)_o \quad (1)$$

where $P_{min}$ is in atm and T is in °K. To prevent voids from forming during the entire autoclave cure cycle, the autoclave and vacuum bag pressure should be adjusted to follow the temperature-time history in such a way that the resin pressure at every point in the laminate must be greater than $P_{min}$.

If void growth cannot be prevented, as in the case of the non-autoclave process, then we have shown[2] that void growth by diffusion of water vapor from the bulk resin into an isolated, spherical, stationary void can be described approximately by the differential form of the Scriven's equation[12].

Scriven's equation was used to calculate void growth during the cure cycle. Pressure and temperature were assumed to be uniform throughout the resin at any moment in time but to vary with time according to the prescribed cure cycle. The model input parameters such as the solubility of water in the resin, the diffusivity, the resin water concentration, the interface water concentration, and the gas density were obtained from the literature as functions of temperature and pressure during the cure cycle.

## 3.2 Resin Transport Model

Void growth requires a knowledge of the resin pressure within the laminate. This can only be obtained by solving the resin flow model which can predict the resin pressure at any point in the laminate at any time during cure.

Our model[3] is based on the following interpretation of the physical phenomena. When a porous fibrous bed, completely saturated with incompressible fluid (resin), is subjected to compression during the processing of graphite/epoxy composites, the phenomenon of flow through the fibrous bed due to consolidation is similar to the situation when a water-saturated soil is subjected to compressive stresses[13]. For both the fibrous bed as well as the water-saturated soil, the volume decrease is due almost entirely to a decrease in the liquid content, the volume decreases of the solid constituents and the liquid in the pores being negligible.

The generalized situation of resin flow through a porous fiber bed is now represented by one-dimensional consolidation with three-dimensional flow. The governing differential equation satisfying the consolidation of a porous bed

within a given time interval with three-dimensional flow and one-dimensional confined consolidation (no motion in x or y directions) involves the specific permeabilities in the x, y, and z directions, which depend on the stress level, the viscosity of the resin, the resin pressure, the time, and the coefficient of volume change, which describes the stress-strain behavior of the porous fiber bed.

The numerical solution is obtained by simultaneously solving this equation of continuity, with the initial and boundary conditions specified, and the following equations: 1) the equilibrium equation, 2) an experimentally obtained relationship between the void ratio and effective stress (due to the spring-like action of the fiber bed), 3) the equation for the coefficient of volume change, 4) an experimentally determined relationship for the specific permeabilities as a function of fiber diameter and length, void ratio and shape factor, and 5) a relationship for resin viscosity as a function of the extent of cure and temperature.

### 4. MODEL PREDICTIONS AND DISCUSSION

## 4.1 The Autoclave Process

As previously mentioned, by combining our void growth and resin transport models, it is now possible to select an autoclave cure cycle such that conditions favorable for void growth are never allowed to exist at any point within the laminate at any time during the curing cycle process. Figure 3 shows the profiles of time-at-temperature and the corresponding $P_{min}$, the minimum resin pressure required to prevent void growth by moisture diffusion at any point within the curing laminate, for the prepregs equilibrated at 60% relative humidity prior to autoclave processing. It should be noted that if the pressure at the laminate boundary is maintained at 0 psig, then the potential for void growth by moisture diffusion does not exist during the first temperature ramp or the first temperature hold when $P_{min}$ is below 0 psig. However, during the second temperature ramp, void growth by diffusion is possible if the resin pressure falls below $P_{min}$, which increases from about 0 psig to about 65 psig. Finally, during the second hold, neither void growth nor dissolution is possible because the resin has already gelled. One key conclusion that can be derived from the above discussion is that if the laminate boundary pressure is maintained at 0 psig throughout the cure cycle, and if the prepreg is equilibrated below 60% initial relative humidity exposure, then the potential for void growth by moisture diffusion would arise mainly during the second temperature ramp of the cure cycle when $P_{min}$ increases from below 0 psig to about 65 psig.

After analyzing several autoclave cure cycles, and obtaining a clearer insight into how flow-related phenomena are related to void growth, we suggest the following basic guideline for autoclave and boundary pressure selection such that void growth by moisture diffusion is prevented during the cure cycle. It is important to note that throughout this discussion we are referring to the laminate boundary pressure and not the bleeder bag pressure. The two would be equal only for the case of "free bleeding" (i.e., negligible bleeder cloth resistance to fluid flow). However, in a commercial autoclave process, this is usually not the case (later shown in Figure 6). For any given time-at-temperature cycle, a laminate boundary pressure cycle is chosen such that the boundary pressure at any time during cure is greater than $P_{min}$, the minimum resin pressure required to prevent void growth by moisture diffusion. Furthermore, the autoclave pressure is maintained above the boundary pressure by a positive amount throughout the cure cycle in order to prevent resin backflow. Figure 4 illustrates the use of this rule in the selection of the laminate boundary and autoclave pressures for the case of "free-bleeding." Figure 5 shows the resin pressures at point R and Q (see inset of Figure 5) for the pressure cycles shown in Figure 4 and the temperature-time history shown in Figure 1, as well as the minimum resin pressure required to prevent void growth by moisture diffusion in prepregs equilibrated at 60% initial relative humidity. It is clear from Figure 5 that the resin pressure at any point within the curing laminate is maintained greater than $P_{min}$ and therefore conditions favoring void growth by moisture diffusion do not exist throughout the cure cycle. It should be noted that while in Figure 5 the profiles of resin pressure have been plotted only at points R and Q, the computer code checks and verifies that conditions for void growth do not exist throughout the entire laminate during cure.

The above proposed guidelines have recently been verified experimentally[14]. Experiments show that laminates made from prepreg with 35% initial relative humidity exposure, using the cure cycle shown in Figure 6, were void-free whereas those made from prepreg with 85% initial relative humidity exposure, using the same cure cycle, were full of voids. Figure 6 shows the profiles of the resin pressure in the laminate and at the top boundary during cure. Also shown in Figure 6 are the corresponding profiles of $P_{min}$ for 35% and 85% initial relative humidity exposure of the prepreg. From Figure 6 it is seen that the resin pressure within the laminate fell below $P_{min}$ prior to

gelation when the prepreg was initially exposed to 85% relative humidity. This consequently led to void formation during cure. On the other hand, the resin pressure within the laminates with 35% initial relative humidity exposure was well below the corresponding $P_{min}$ throughout the cure cycle and, therefore, the final laminate was void-free.

4.2 The Non-Autoclave Process
Figure 7 illustrates void growth for a non-autoclave process having a cool-down cycle after the consolidation dwell (see Figure 2). The maximum hard-shell pressure of 20 psig was utilized and the initial prepreg relative humidity exposure was 60%. First of all it should be pointed out that void growth will occur for all situations until the hard-shell pressure is applied. Only in the case of near-zero water content would void growth via diffusion be forbidden. At the fastest cooling rate of 10°F/min., the voids cannot totally collapse before room temperature is attained (the calculation was stopped at room temperature). One must also remember that during the cool-down step the system may become glassy before the gel point is reached. At this point (glass transition temperature), any voids would be frozen in even though the system had not gelled. Thus, at the highest hard-shell pressure and the gastest cooling rate only a few minutes leaway is available between

333

complete dissolution of the voids and gelation which occurs at about 150 minutes. If the hard-shell pressure is lowered, or the relative humidity exposure is increased, or the cooling rate is decreased, the "window of non-porosity" will disappear.

Figure 8 depicts void growth for the cycle in Figure 2 in which, after the consolidation dwell, the temperature is ramped at 5°F/min up to 350°F and then held there for 45 minutes. Even at the relatively low prepreg RH exposure of 20%, the voids do not completely dissolve even when the applied hard-shell pressure is at its maximum of 20 psig (2.36 atm). Lower hard-shell pressures, higher initial prepreg RH exposures, and longer initial dwell times all would result in even worse situations.

Thus it would seem that in order for this particular non-autoclave process to yield void-free laminates, the initial prepreg has to be almost perfectly dry, or an intermediate cool-down cycle must be employed.

## 5. CONCLUSIONS

1. Although the non-autoclave process can produce void-free laminates under certain conditions, either a cool-down ramp must be employed after the consolidation dwell, or the prepreg must be essentially water-free at the beginning of the process. Since commercial prepregs are usually equilibrated at relative humidities at least as high as 35%, a cool-down ramp would have to be employed.

2. Even when a cool-down ramp is employed, the cooling rate-time "window" for void-free laminates is relatively small for the maximum hard-shell pressure. It is possible that different initial dwell temperatures and times might improve this situation somewhat.

3. The higher pressure capability of an autoclave makes it easier to prevent water-dominated void formation and growth than is the case for a non-autoclave process. Thus, a wider range of prepreg water contents can be accommodated. At the same time resin content can be optimized in an autoclave, which is not the case for a non-autoclave process in which very little resin flow is allowed to occur.

## 6. ACKNOWLEDGMENT

We are grateful to Mr. F. Campbell, Mr. D. Mallow, and Mr. F. Muncaster of the McDonnell Aircraft Company for many helpful discussions and for the experimental data in Figure 6.

## 7. REFERENCES

1. Burroughs, B. A., et al., "Manufacturing Technology for Non-Autoclave Composite Fabrication," Rockwell International Corp., Final Report, 10/1/80-4/1/84, Contract No. F33615-80-C-5080, AFWAL/MLBC, Wright-Patterson, AFB, OH 45433.

2. Kardos, J. L., Duduković, M. P., and Dave, R., *Advances in Polymer Science, Vol. 80-Epoxy Resins and Composites IV,* Springer Verlag, 1986, p. 101.

3. Dave, R., Kardos, J. L., and Duduković, M. P., "A Model for Resin Flow During Composite Processing: Part 1 - General Mathematical Development," J. Comp. Mat., in press.

4. Dave, R., Kardos, J. L., and Duduković, M. P., "A Model for Resin Flow During Composite Processing: Part 2 - Numerical Analysis for Unidirectional Gr/Ep Composites," J. Comp. Mat., in press.

5. Dave, R., Kardos, J. L., and Duduković, M. P., Proceedings of the American Society for Composites, 1st Technical Conference, Technomic Publishing Co., Lancaster, PA, 1986, p. 137.

6. Grayson, M. A. and Wolf, C. J., *Advances in Chemistry Series,* J. Koenig, Ed., American Chemical Society, Washington, DC, 174, 1979, p. 53.

7. Grayson, M. A., ACS Symposium Series, No. 132, American Chemical Society, Washington, DC, 1980, p. 449.

8. Brand, R. A., Brown, G. G. and McKague, E. L., "Processing Science of Epoxy Resin Composites," General Dynamics Convair Division, AFWAL-TR-83-4124, Final Report, Jan. 1984, AFWAL/MLBC, Wright-Patterson AFB, OH 45433, pp. 169-175.

9. Enns, J. B. and Gillham, J. K., *Advances in Chemistry Series,* M. J. Comstock, Ed., American Chemical Society, Washington, DC, 203, 1983, p. 27.

10. Loos, A. C. and Springer, G. S., J. Comp. Mat., 13, 131 (1979).

11. Lee, W. I., Loos, A. C. and Springer, G. S., J. Comp. Mat., 16, 510 (1982).

12. Scriven, L. E., Chem. Eng. Sci., 10, 1 (1959).

13. Lambe, T. W. and Whitman, R.V., *Soil Mechanics,* J. Wiley and Sons, NY, 1969.

14. Campbell, F. C., Mallow, A. R. and Ameudo, K. C., "Computer Aided Curing of Composites," McDonnell Aircraft Co., 7th Interim Report, 10/1/85- 12/31/85, Contract No. F33615- 83-C-5088, AFWAL/MLBC, Wright Patterson AFB, OH 45433.

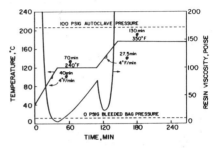

Fig. 1. Time-at-temperature, pressure and Hercules 3501-6 epoxy resin viscosity profiles for a typical standard commercial cure cycle for graphite/epoxy composites.

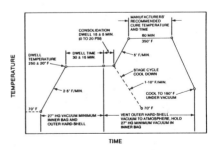

Fig. 2. Nonautoclave cure cycle. (obtained from reference 1)

335

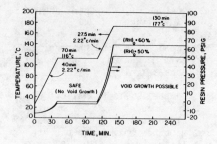

Fig. 3. Profiles of time-at-temperature and the corresponding $P_{min}$ curves showing the minimum resin pressure required to prevent void growth by moisture diffusion.

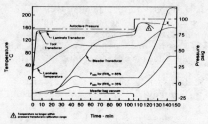

Fig. 6. Profiles of resin pressures at the bottom, middle, and top of the laminate for the time-at-temperature and pressure cycles shown in the figure. The profiles of the corresponding $P_{min}$ for 35% and 85% initial relative humidity exposure of the prepreg are also shown. (Experimental data from reference 13.)

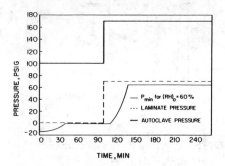

Fig. 4. Desirable profiles of autoclave and boundary pressures for void-free laminates, compared with that of $P_{min}$ for the time-at-temperature cycle shown in Figure 1.

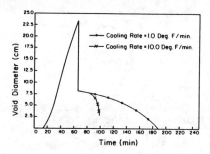

Fig. 7. Void growth and dissolution in neat resin as a function of cooldown rate for the curing cycle shown in Figure 2 but following the dashed line cool-down ramp. The initial RH exposure was 60% and the hard-shell pressure was 20 psig.

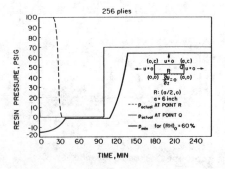

Fig. 5. Profiles of resin pressures at R and Q during autoclave processing using the time-at-temperature cycle shown in Figure 1 and the autoclave and boundary pressure cycles shown in Figure 4.

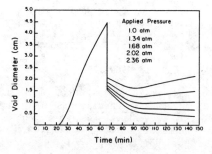

Fig. 8. Void growth and dissolution in neat resin as a function of hard-shell pressure for the curing cycle shown in Figure 2 without cool-down and with a 45 minute hold at 350°F. The initial RH exposure was 20%.

R. S. Dave is currently Research Associate at the Michigan Molecular Institute in Midland, MI where he is initiating research in the area of modeling of composite material processing. He recently obtained his D.Sc. from Washington University where he was awarded Second Place in the 1986 National Contest for Graduate Research sponsored by SAMPE. Dr. Dave has a M.S. degree from Washington State University in Materials Science and Engineering and a B.S. in Chemical Engineering from I.I.T., Kharagpur in India.

J. L. Kardos is Professor of Chemical Engineering and Chairman of the Graduate Program in Materials Science and Engineering at Washington University in St. Louis. He has over 20 years experience in the research and development of composite materials, including structure/property relations for short-fiber reinforced systems, interface characterization and modification, and, more recently, development of models for auto-clave processing of laminated composites. Dr. Kardos received a B.S.ChE. from Penn State University, a M.S.Ch.E. from the University of Illinois, and a Ph.D. from Case Western Reserve University in Polymer Science and Engineering.

S. J. Choi is a Graduate Research Assistant at Washington University in St. Louis, where he is doing research in the area of composite material processing. He has a B.S. degree in Chemical Engineering from Sung Kyun Kwan University in Korea and a M.S.Ch.E. from the University of Texas in Austin.

M. P. Duduković is Professor of Chemical Engineering and Director of the Chemical Reaction Engineering Laboratory at Washington University in St. Louis. His research interests include modeling and experimental investigations in the areas of fluid-solid non-catalytic reactions, 3-phase reactor systems, and transport phenomena in chemical engineering. He received his B.S.Ch.E. from the University of Belgrade in Yugoslavia and his M.S. and Ph.D. degrees in Chemical Engineering from Illinois Institute of Technology in Chicago.

MONITORING PROCESSING PROPERTIES OF HIGH PERFORMANCE THERMOPLASTICS
USING FREQUENCY DEPENDENT ELECTROMAGNETIC SENSING

D. E. Kranbuehl, S. E. Delos, M. S. Hoff, L. W. Weller,
P. D. Haverty, J. A. Seeley and B. A. Whitham
Department of Chemistry
College of William and Mary
Williamsburg, Virginia 23185

ABSTRACT

A nondestructive in-situ dielectric impedence measurement technique (DDA) has been developed for sensing (both in a research and production tool) the cure processing properties of high-temperature, high-performance thermoplastics and thermosets. The technique emphasizes continuous frequency dependent dielectric measurement and analysis throughout the cure cycle. The instrumentation can measure the complex permittivity, an intrinsic polymer property, over 9 orders of magnitude ($10^{-3}$ to $10^{6}$) and 6 decades of frequency (Hz to MHz) at temperatures to $400^{o}C$. The frequency dependence of the complex permittivity is used to monitor both ionic and Debye-like dipolar relaxation processes which in turn serve as molecular probes of the chemical and physical changes occuring during cure. The Hz to MHz frequency range is advantageous as it senses both ionic and dipolar diffusion processes and minimizes electrode polarization problems.

In this report the ability of DDA to monitor processing properties in several high performance thermoplastics is examined. Using as examples the aromatic polyimide-LaRC-TPI, PEEK, and poly (arylene ether) we have shown that the DDA can be used to monitor the viscosity, Tg, crystal melting, recrystallization, and residual solvent content and evolution. These processing properties can be used to help define an optimal processing window and to monitor the actual state of the polymer as a function of position, time and temperature in the processing tool.

1. INTRODUCTION

A nondestructive in-situ electromagnetic-dielectric impedence measurement technique has been developed for sensing (both in a research and production tool) the cure processing properties of high-temperature, high-performance thermoplastics and thermosets [1-5]. The technique emphasizes continuous frequency dependent dielectric measurement and analysis throughout the cure cycle. The instrumentation can measure the complex permittivity, an intrinsic polymer property, over 9 orders of magnitude ($10^{-3}$ to $10^6$) and 6 decades of frequency (Hz to MHz) at temperatures to $400^\circ$C. The frequency dependence of the complex permittivity is used in conjunction with dielectric theory to determine molecular parameters which monitor the ionic and dipolar diffusion processes. These parameters serve as molecular probes of the chemical and physical changes occurring during cure. The Hz to MHz electromagnetic frequency range is advantageous as it senses both ionic and dipolar diffusion processes and minimizes extraneous sensing problems.

In this report electromagnetic sensing is used to measure and monitor processing properties in several high performance thermoplastics. In other reports we have focused on the ability of our electromagnetic sensing techniques to measure thermoset properties such as viscosity and degree of cure in-situ in a production tool. Using as examples the aromatic polyimide-LaRC-TPI, PEEK, and poly(arylene ether) we show that electromagnetic sensing can be used to monitor the viscosity, Tg,

crystal melting, recrystallization, and residual solvent content and evolution. These processing properties can be used to help define an optimal processing window and to determine the actual state of the polymer as a function of position, time and temperature in a production tool.

The major objective of this work is the development of frequency dependent electromagnetic techniques as a smart sensor for quantitative nondestructive material evaluation and intelligent closed loop process control.

## 2. EXPERIMENTAL

Frequency dependent electromagnetic-dynamic dielectric measurements were made using a Hewlett-Packard 4192A LF Impedance Analyzer controlled by a 9836 Hewlett-Packard computer. Measurements at frequencies from 5 to $5 \times 10^6$ Hz were taken at regular intervals during the cure cycle and converted to the complex permittivity, $\epsilon^* = \epsilon' - i\epsilon''$. Measurements were made with a geometry independent Dek Dyne DDA Sensor.* A detailed description of the procedure has been given elsewhere [5].

Viscosity measurements were done using a Rheometrics System IV-dynamic mechanical spectrometer.

DSC measurements were made using a Perkin-Elmer DSC-7 differential scanning calorimeter.

The polyarylene ether, PEEK and LaRC-TPI thermoplastic samples

339

were furnished by NASA-Langley and used as received [6,7].

## 3. THEORY

Frequency dependent measurements of the complex impedance, characterized by C and G, were used to calculate the complex permittivity $\epsilon* = \epsilon' - i\epsilon''$

$$\epsilon' = \frac{C \text{ material}}{C_o}$$

(1)

$$\epsilon'' = \frac{G \text{ material}}{C_o 2\pi f}$$

This calculation is possible when using a probe whose geometry is invariant over all measurement conditions. Both the real and the imaginary parts of $\epsilon*$ can have a dipolar and an ionic component [8].

$$\epsilon' = \epsilon'_d + \epsilon'_i$$

(2)

$$\epsilon'' = \epsilon''_d + e''_i$$

The dipolar component arises from diffusion of bound charge or molecular dipole moments. The frequency dependence of the polar component may be represented by the Cole-Davidson function:

$$\epsilon* = \epsilon_\infty + \frac{\epsilon_o - \epsilon_\infty}{(1+i\omega\tau)^\beta}$$

(3)

where $\epsilon_o$ and $\epsilon_\infty$ are the limiting low and high frequency values of $\epsilon$, $\tau$ is a characteristic relaxation time and $\beta$ is a parameter which measures the distribution in relaxation times. The dipolar term is generally the major component of the dielectric signal at high frequencies and in highly viscous media.

The ionic component, $\epsilon_i*$, often dominates $\epsilon*$ at low frequencies, low viscosities and/or higher temperatures. The presence of mobile ions gives rise to localized layers of charge near the electrodes. Since these space charge layers are separated by very small molecular distances on the order of $A^o$, the corresponding space charge capacitance can become extremely large, with $\epsilon'$ on the order of $10^6$. Johnson and Cole, while studying formic acid, derived empirical equations for the ionic contribution to $\epsilon*$[9]. In their equations, these space charge ionic effects have the form

$$\epsilon'_i = C_o Z_o \sin\frac{(n\pi)}{2}\omega^{-(n+1)} (\frac{\sigma}{8.85 \times 10^{-14}})^2$$

(4)

where $Z* = Z_o(i\omega)^{-n}$ is the electrode impedance induced by the ions and n is between 0 and 1 [6-8]. The imaginary part of the ionic component has the form

$$\epsilon''_i = \frac{\sigma}{8.85 \times 10^{-14}\omega} - C_o Z_o \cos\frac{(n\pi)}{2}$$

$$\omega^{-(n+1)} (\frac{\sigma}{8.85 \times 10^{-14}})^2$$

(5)

where $\sigma$ is the conductivity (ohm$^{-1}$ cm$^{-1}$), an intensive variable, in contrast to conductance

340

G(ohm$^{-1}$) which is dependent upon cell and sample size. The first term in Eq. 5 is due to the conductance of ions translating through the medium. The second term is due to electrode polarization effects. The second term, electrode polarization, makes dielectric measurements increasingly difficult to interpret and use as the frequency of measurement becomes lower.

Analysis of the frequency dependence of $\epsilon^*(\omega)$ in the Hz to MHz range is, in general, optimum for determining both $\sigma$ and $\tau$. In turn, the ionic parameter, $\sigma$ and the dipolar parameter, $\tau$ are directly related on a molecular level to the rate of ionic translational diffusion and dipolar rotational mobility [8].

4. RESULTS AND DISCUSSION

The polyarylene ether thermoplastic film was ramped to 300$^{\circ}$C, held for 60 minutes and then cooled.

A plot of $\epsilon'$ versus time (Fig. 1) shows a sharp transition at the glass transition both on heating and cooling. The ability of the sensor to measure Tg is shown by the agreement of the DSC determined Tg of 150$^{\circ}$ and the rise of $\epsilon'$ at T = 150$^{\circ}$. An advantage of this sensing technique in addition to its in-situ production tool capability is that the magnitude of the inflection is not decreased for slow heating and cooling rates.

A plot of $\epsilon''$ x $2\pi$ frequency (Fig. 2) shows visually the dipolar and ionic contributions to the loss term. When the $\epsilon''$ x $2\pi$ frequency curves differ at each frequency, the molecular processes contributing to $\epsilon''$ are bound charge dipole-like diffusion processes. Overlapping curves indicate translational diffusion of charged species is the dominant molecular process determining the loss $\epsilon''$.

Below the glass transition, the rate of translational diffusion of ions is extremely slow and the value of the loss $\epsilon''$ is small. A frequency-dependent dipole-like relaxation can be observed just above T$_g$. The maximum in $\epsilon''$ at a given frequency corresponds to the relaxation rate, $\tau$, of the dipolar rotational diffusion process. At the temperature of the $\epsilon''$(f) maximum, $\tau(T) = 1/(2\pi f)$. As the temperature increases, the resin becomes more fluid so that the dipoles can orient faster. Thus $\epsilon''$ max for the higher frequencies occurs at the higher temperatures. As the temperature increases, ionic translation contributes to the loss at increasingly higher frequencies. The results are identical for the heat up and cool down curve, confirming that no reaction has taken place and the material is a true thermoplastic.

The magnitude of the specific conductivity $\sigma$(ohm$^{-1}$cm$^{-1}$) which reflects the rate of translational diffusion of charge is determined from equation 5 and the frequency dependence of $\epsilon''$. The ability of $\sigma$ to be used to measure the viscosity is shown by the results of simultaneous measurement of $\sigma$ and viscosity, Fig. 3.

The dipolar-like relaxation process observed just above Tg in Fig. 2 is the $\alpha$ relaxation process associated with the cooperative polymer chain motions which dominate diffusion and viscosity in the Tg to Tg + 50$^{o}$ region. The relaxation time $\tau_\alpha$ associated with the $\alpha$ relaxation process and its temperature dependence is determined as previously discussed.

The temperature dependence of $\sigma$ and $\tau$ (see Fig. 4) can be fit to both an Arrhenius and a WLF dependence. The results of this fit are shown below.

Arrhenius Fit

$E_{Ionic}$(kcal/mole)
31.8

$E_{Dipolar}$(kcal/mole)
95.4

WLF Free Volume Fit

| $c_1^I$ | $c_2^I$ | $c_1^D$ | $c_2^D$ |
|---|---|---|---|
| 16 | 26 | 11 | 50 |

The values of the activation energy are markedly higher for the dipolar diffusion processes most likely reflecting the larger size of the dipoles. The values of the WLF coefficients can be used to examine in greater detail the fractional free volume in the thermoplastic, which in turn can be related to toughness.

Next we examine the capability of frequency dependent electromagnetic sensing to detect the glass, amorphous, crystalline, and melt phase transitions. The poly (aryl-ether-ether-ketone) thermoplastic PEEK in its semi-crystalline state was heated to 360$^{o}$, held 30 minutes and then cooled. The $\epsilon$" x $2\pi$ frequency data taken in a press are shown in Fig. 5. The DSC data on a similarly treated sample are shown in Fig. 6. The electromagnetic sensor is able to detect Tg at 145$^{o}$, melting at 342$^{o}$, recrystallization of the supercooled melt at 295$^{o}$ and the cooling glass transition at 145$^{o}$.

Shown in Fig. 7 is the value of $\epsilon'$ and $\epsilon$" x $2\pi f$ at 5 kHz for an amorphous PEEK sample (less than 2% crystalline (10)) on heat up to 220$^{o}$, a hold at 220$^{o}$ and cool down.

Fig. 8 is the corresponding DSC plot for a piece of the amorphous sample on heating. The value of $\epsilon'$ senses at 166$^{o}$ a cool crystallization process and the accompanying polarization changes which occur over molecular dimensions as the polymer crystallizes. The onset of the cool crystallization process just above Tg at 160$^{o}$ is verified by the DSC plot Fig. 8. The capability of the frequency dependent electromagnetic-sensing to characterize the glass, amorphous, crystalline, states as well as phase changes of the polymer through quantitative measurement of $\tau$, $\sigma$ and $\epsilon$* is clearly possible. The advantage of the technique is not only that the measurement can be made in-situ in a production tool but that sensitivity is not affected by use of a slow annealing process.

Last, the ability of frequency dependent electromagnetic sensing to characterize the cure properties of a partially crystalline polyimide thermoplastic LaRC-TPI are shown in Fig. 9. The value of $\epsilon''$ at 5 kHz shows a large broad peak at $100^o$. The site of the peak, the fact that the peak does not occur on cool down, and the temperature indicates that moisture is being driven off. The rise in $\epsilon''$ at 188 $^o$ indicates the onset of the $\alpha$-relaxation and the glass transition. At $245^o$, $\epsilon''$ rises rapidly indicating melting. The slow drop in $\epsilon''$ at the 290 $^o$ hold indicates a decreasing rate of diffusion and recrystallization of the sample, results which are consistent with DSC data. The moisture, solvent, amorphous and crystalline content of LaRC-TPI and their effect on the changes which occur in LaRC-TPI are complex (11). The ability to sense the state and changes in state of LaRC-TPI both in the laboratory and during production is therefore important.

## 5. CONCLUSION

Frequency dependent electromagnetic sensing of $\epsilon^*$ in the Hz to MHz vision and determination of molecular parameters $\sigma$ and $\tau$ describing the diffusion rates of ions and dipoles is a sensing technique which can be used to characterize the solvent, glass, amorphous, and crystalline state of thermoplastics as well as phase changes therein. The advantage of the technique is that the measurement can be made in-situ in a production tool both in the laboratory and during production. Further on-line sensitivity is not affected by slow rates of change in temperature.

## 6. REFERENCES

1. Kranbuehl, D.E., Delos, S.E., Yi, E.C., Hoff, M.S. and Whitman, M.E., ACS Polym. Mater. Sci. and Eng. 52, 191 (1985).

2. Kranbuehl, D.E., Delos, S.E., Jue, P.K., Jarvie, T.P., and Williams, S.A., Nat'l SAMPE Symp. Ser. 29, 1261 (1984).

3. Kranbuehl, D.E., Delos, S.E., and Jue, P.K., National SAMPE Symp. Ser., 28, 608 (1983); SAMPE Journal, 19 (4), 18, (1983).

4. Kranbuehl, D.E., Delos, S.E., Jue, P.K., Polymer, 27, 11 (1986).

5. Kranbuehl, D., Delos, S., Hoff, M., and Weller, L. ACS Polym. Mater. Sci. and Eng. 53 535 (1986).

6. St. Clair, A.K. and St. Clair, T.L., SAMPE Quarterly, 13, (1), 20 (1981).

7. Hergenrother, P.M., Jensen, B.J., and Havens, S.J., Polym. Preprint 26 (2), 174 (1985).

8. Hill, N., Vaughan, W., Price, A., and Davis, M., Dielectric Properties and Molecular Behavior, Van Nostrand, London (1969).

9. Johnson, J. and Cole, R., J. Am. Chem. Soc. 73, 4536 (1985).

10. Kemmish, D. and Hay, J., Polymer 26, 905 (1985).

11. Burks, H.D., St. Clair, T.L., and Hou, Tan-Hung, SAMPE Quarterly <u>18</u>, 1, (1986).

*Inquiries regarding the Dek Dyne sensor probe and the DDA instrumentation should be directed to D. Kranbuehl.

The dielectric work was made possible in part through the support of the National Aeronautics and Space Administration-Langley Research Center research grant No. NAG 1-237 and support from the Lockheed Missles and Space Company.

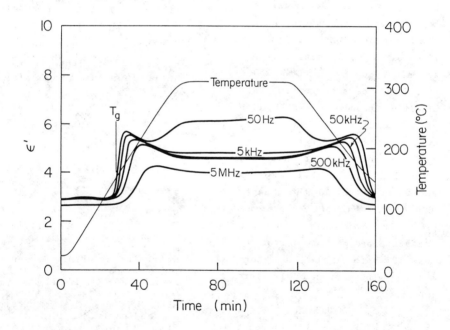

Figure 1.  Plot of $\varepsilon'$ vs time for polyarylene ether thermoplastic.

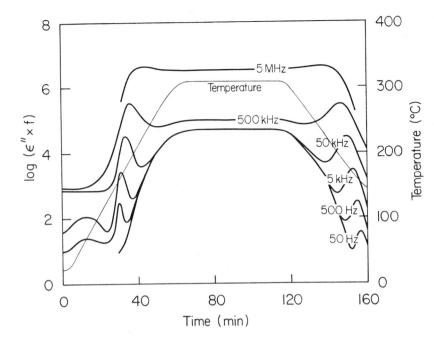

Figure 2.  Plot of log( ε" x 2πf) vs time for polyarylene ether
thermoplastic.

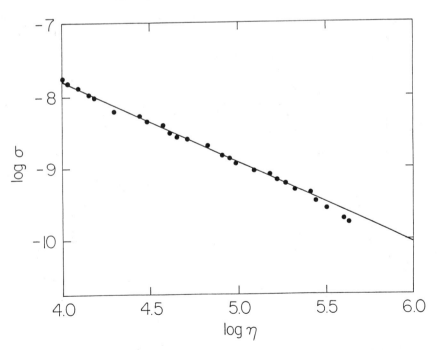

Figure 3.  Plot of log(specific conductivity) vs log(viscosity) for
polyarylene ether.

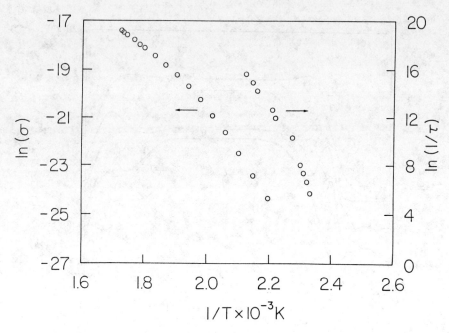

Figure 4.   Ln(σ) and ln(1/τ) vs 1/T for polyarylene ether.

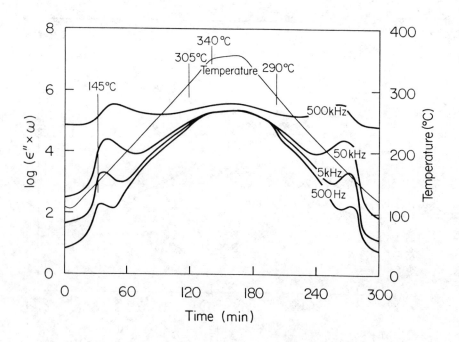

Figure 5.   Log(ε" x 2πf) vs time for crystalline PEEK.

346

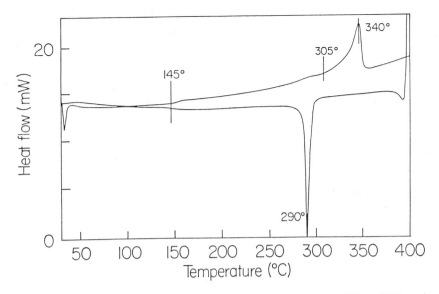

Figure 6.  DSC scan of heat up and cool down of crystalline PEEK.

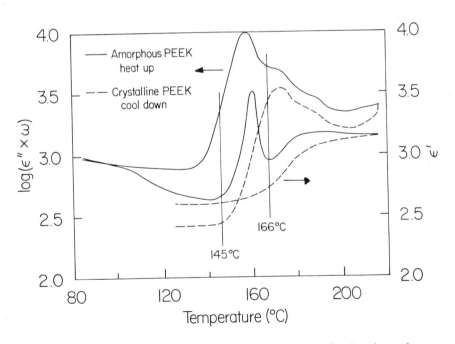

Figure 7.  Plots of ε' and ε" x 2πf vs temperature for heating and
cooling of an amorphous PEEK sample.

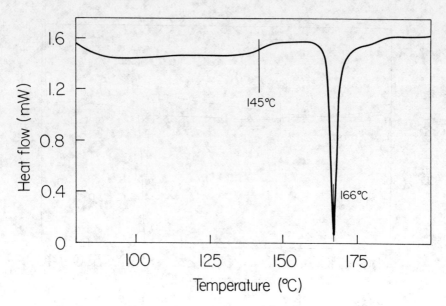

Figure 8.  DSC scan of an amorphous PEEK sample.

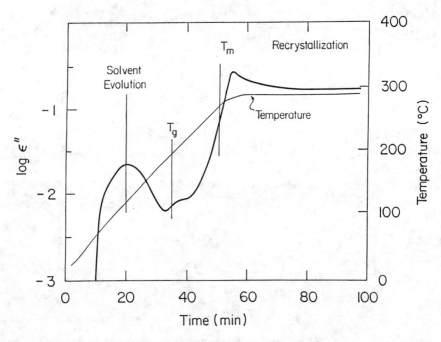

Figure 9.  Plot of log($\epsilon$") vs time for a LARC-TPI sample ramped
to 290C and then held for 60 minutes.

A RADICAL PROPOSAL FOR IN-PLANE SHEAR TESTING
OF FIBROUS COMPOSITE LAMINATES

by

**Dr. L. J. Hart-Smith**
**Douglas Aircraft Company**
**McDonnell Douglas Corporation**
**Long Beach, California**

## ABSTRACT

Because of inherent difficulties in testing composite laminates and laminae under in-plane shear loads, a tensile test of a $\pm45$-degree laminate is customarily substituted for a shear test of a 0/90-degree laminate. The tensile test strength is divided by two and called the in-plane shear strength, which is a resin-dominated property in this case. Such a correlation is based on the maximum-shear-stress failure criterion for ductile isotropic materials. This paper proposes that the fiber-dominated in-plane shear strength of a $\pm45$-degree laminate be likewise taken as half the tensile strength of a 0/90-degree laminate. If the tensile and compressive strengths differ appreciably, as with many of the new high-strain fibers, the shear strength would be evaluated as $F_{su} = F_c F_t / (F_c + F_t)$. This proposal is made because tensile and compressive tests are much easier to perform reliably than shear tests. Also, the answer so obtained would agree very well with the best of the composite in-plane shear tests and give higher allowables than most prior shear tests, which are known to be unreliable.

## INTRODUCTION

This paper is concerned with the in-plane shear strength of fiber-dominated composite laminates. This has been a problem in the past for two principal reasons. First, testing has usually been difficult and beset with premature failures because of substantial local stress concentrations. There is still no agreement on the best specimen to be used for in-plane shear testing. Second, most of the laminate failure theories, including the most popular maximum-strain theory, have grossly overestimated the in-plane shear strength – by as much as a factor of two for $\pm45$-degree laminates. The lack of universally accepted in-plane shear strengths has made it difficult to weed out inferior laminate theories. Likewise, the absence of agreement on laminate theories has resulted in what would, under other circumstances, be considered unnecessary testing.

It is proposed here that the off-axis tension test used to establish the resin-dominated shear strengths be generalized to fiber-dominated laminates as well. While it is recognized that such a proposal may seem radical at first, there is a great deal of justification for using this method (at least for carbon-epoxy laminates). Acceptance of this proposal would lead to industry-wide standardization based on a simple, reliable coupon yielding the same result that only the best of prior shear testing was able to attain. It would also focus attention on the discrepancies between the various published laminate strength theories and improve that situation.

## RESIN-DOMINATED SHEAR STRENGTHS

The shear testing of fibrous composite laminates has long been a difficult task (see Reference 1, for example). Even the short-beam-shear coupon (ASTM D-2344) for measuring interlaminar shear strength (a resin-dominated property) under three-point bending (Figure 1) is regarded with suspicion because the failure modes are often clearly influenced by failure of the fibers. Consequently, an alternative specimen has been developed (see Reference 2) in the form of the off-axis tensile test shown in Figure 2.

Stress created by an axial tensile load on a long ±45-degree coupon is equivalent to biaxial tension in the 0- and 90-degree directions, superimposed on equal and opposite tension and compression in the same directions, respectively. This latter component is equivalent to pure shear with

respect to the ±45-degree directions of this particular test coupon. That, in turn, is the same as in-plane shear with respect to 0- and 90-degree fibers, which is the resin-dominated property being sought. These equivalencies, explained in Figure 3, rely on the lack of an interaction with the uniform biaxial stress component, which is carried by the fibers, not the resin.

The in-plane shear strength is taken to be one half of the tensile stress at failure of the coupon described in Figure 2. That in-plane shear strength is, in turn, taken to be the interlaminar strength of fibrous composites as well as the in-plane shear strength of unidirectional monolayers. It is important to test a laminate with fibers in both the +45- and −45-degree directions to represent the shear strength of a 0/90-degree laminate, rather than to test a specimen with all of the fibers in one direction only. In the latter case, failure would be triggered by the onset of cracking in the resin, whereas, in a real structure, cross plies would arrest those initial interfiber cracks and delay the failure of the laminate until some fibers became overloaded or the crack density in the resin became excessive.

The failure criterion associated with the logic of off-axis testing is the maximum-shear-stress theory (see Figure 4) customarily used for ductile metal alloys. The theory is applied

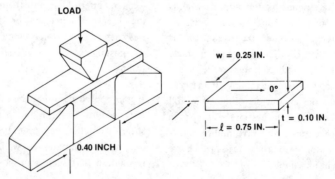

**FIGURE 1. SHORT-BEAM (INTERLAMINAR) COMPOSITE SHEAR TEST COUPON**

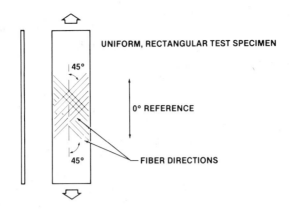

FIGURE 2. OFF-AXIS TENSILE TEST ON ± 45-DEGREE COMPOSITE COUPON TO SIMULATE SHEAR ON A 0/90-DEGREE COUPON

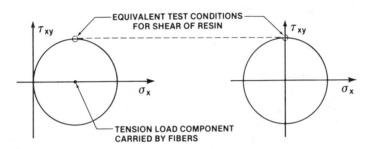

OFF-AXIS TENSION TEST                    PURE SHEAR TEST

FIGURE 3. MOHR CIRCLES FOR OFF-AXIS TEST AND PURE SHEAR TESTS

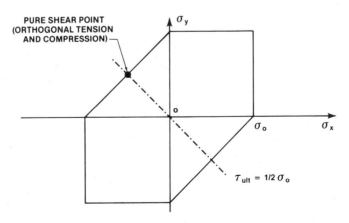

FIGURE 4. MAXIMUM-SHEAR-STRESS FAILURE THEORY FOR DUCTILE MATERIALS

351

in this case because the resin, which is failing, is isotropic.

The off-axis coupon in Figure 2 is becoming increasingly popular — it is now standardized by the ASTM as D-3518 — because it is simple to fabricate and test and it gives repeatable results. It is also preferred because it generates higher allowable strengths than other coupons employed for the same purpose. The Iosipescu (or double V-notched) coupon, illustrated in Figure 5, has also received considerable attention recently (see Reference 3). However, it is more difficult to fabricate, it requires special test equipment, and its strength is greatly affected by the geometry of the notch at each end of the test section (see Reference 4).

## FIBER-DOMINATED STRENGTHS

Testing for fiber-dominated shear strengths of composite laminates, particularly the in-plane shear strength of a ±45-degree laminate, has suffered from problems similar to those encountered in testing for resin-dominated strengths. There is no consensus on how best to perform such tests, although it is hoped that this paper will be instrumental in bringing that about.

The earliest tests for fiber-dominated shear strengths used either the untapered rail-shear coupon (Figure 6) or the picture-frame test, shown schematically in Figure 7.

Both specimens are unsatisfactory because the shear stresses are highly nonuniform throughout the test area. The rail-shear coupon inevitably has a stress concentration at each end, just like that associated with adhesively bonded joints. So also does the three-rail shear coupon (Figure 8) that has been used more recently. In addition, unless the coupon is bonded to steel rails, very premature failures are triggered by the stress concentrations associated with the bolt holes through which the load is introduced. The picture-frame test suffers from a severe nutcracker action in the corners of the specimen.

These deficiencies were well understood by the early 1970s when the author proposed the use of the bonded tapered rail-shear coupon shown in Figure 9. The superiority of this specimen was evident immediately; the first two test results (59.5 and 60.4 ksi) for a ±45-degree Narmco 5206 carbon-epoxy laminate were consistent and nearly twice as high as the typical result from the earlier coupons.

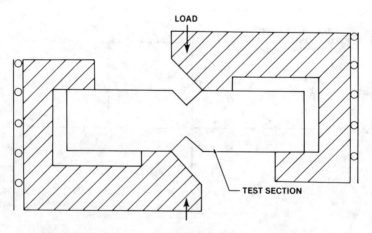

**FIGURE 5. IOSIPESCU IN-PLANE SHEAR TEST SPECIMEN**

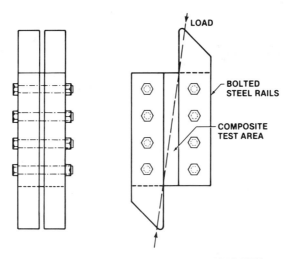

FIGURE 6. RAIL-SHEAR COMPOSITE TEST SPECIMEN

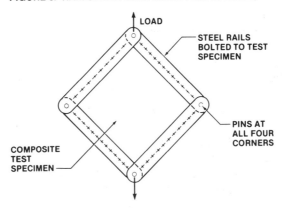

FIGURE 7. PICTURE-FRAME COMPOSITE SHEAR TEST SPECIMEN

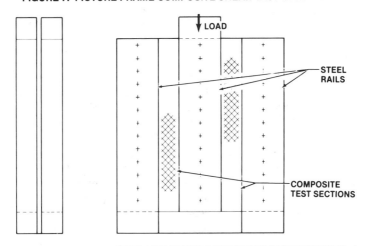

FIGURE 8. THREE-RAIL IN-PLANE SHEAR COMPOSITE TEST SPECIMEN

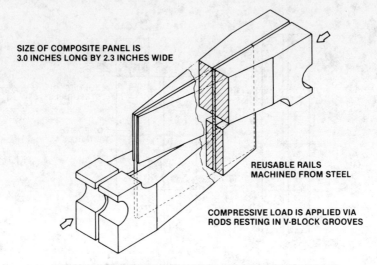

SIZE OF COMPOSITE PANEL IS
3.0 INCHES LONG BY 2.3 INCHES WIDE

REUSABLE RAILS
MACHINED FROM STEEL

COMPRESSIVE LOAD IS APPLIED VIA
RODS RESTING IN V-BLOCK GROOVES

**FIGURE 9. BONDED TAPERED RAIL-SHEAR TEST SPECIMEN**

(Ironically, they were still barely half as high as the contemporary laminate theories were predicting.) These bonded tapered rail-shear specimens were used for the NASA-funded DC-10 upper-aft composite rudder program (see Reference 5). Reference 6 contains a thorough analysis of this coupon, showing how a near-uniform stress distribution is obtained. This specimen originated from the author's earlier investigation into adhesively bonded composite joints (Reference 7). The basic idea is to copy the scarf joint and shear the load in uniformly from relatively rigid tapered rails, which are in any case at a constant stress.

The Douglas bonded tapered rail-shear specimen was purposely loaded in compression rather than tension because any local stress concentrations at the corners of the test area would be less of a problem than with a tensile load. However, that required that the reusable steel rails be kept in matched sets to avoid twisting the specimen during testing. The rails were recently redesigned to ensure that every specimen is automatically aligned properly during fabrication, as shown in Figure 10, eliminating the need to keep the rails in matched sets. The axis of the applied load remains along the diagonal of the test area, which is 3.0 inches long and 0.3 inch wide, to permit installation of a strain gage rosette.

Apart from the care needed to align the rails from side to side, the bonded tapered rail-shear coupon has given consistently good and believable results. Those results have, in turn, been instrumental in the belated development of a reliable theory for predicting the strength of cross-plied composite laminates (see References 8 and 9). It is now clear that the ductile shear failure theory for metals is just as applicable to fiber-dominated failures of composites (at least carbon-epoxy) as it is to the resin-dominated cases. In his latest composite workshops (Reference 10), Tsai uses the generalized von Mises failure criterion in conjunction with a more than three-fold reduction in transverse stiffness of the monolayer, because of matrix and interface cracking, to predict the strength of composite laminates. The maximum-shear-stress theory is only slightly conservative (by typically up to 14 percent) with respect to the generalized von Mises theory, whose use for composites was first proposed by Norris in Reference 11.

There is now agreement between test and theory for the in-plane shear strengths of

354

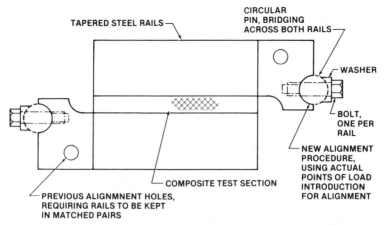

**FIGURE 10. IMPROVED BONDED TAPERED RAIL-SHEAR TEST SPECIMEN**

fibrous composites, even though the universities and much of the aerospace industry still use older, inappropriate laminate theories extensively. The reason for this continued use of inferior formulations seems to be that the grossly unconservative predictions of in-plane shear strength had been discounted because of the known unreliability of most of the early shear testing. Almost any theory could be shown to be satisfactory, with minor tweaking, provided that it had to be correlated against only one load condition – usually uniaxial tension or compression. A good in-plane shear test result for a ±45-degree laminate has long been needed as an additional check on the validity of laminate theories. Getting agreement on such a test specimen would then focus attention on discarding some of the old laminate theories and improving the accuracy of composite structures analysis.

## THE PROPOSED "SHEAR" TEST SPECIMEN

As explained in the preceding section, the best (and highest) measured in-plane shear strengths of ±45-degree carbon-epoxy laminates have been correlated with theoretical predictions based on measured tensile and compressive 0-degree monolayer strengths. The agreement is within half a percent.

What is significant about this is that the theoretical in-plane shear strength of the ±45-degree laminate is actually established as precisely half the predicted tensile strength of a 0/90-degree laminate. This obviously suggests that the tests could be interchanged also, at a tremendous decrease in cost and complexity. Further, doing so would be directly equivalent to the currently accepted use of a tension test on a ±45-degree laminate to simulate the in-plane shear strength of a 0/90-degree laminate. This proposal is explained in Figure 11. For some of the newer high-strain fibers, there is a significant difference between the tensile and compressive strengths. In such a case, the theory in Reference 8 would predict that the shear strength would be given by

$$F_{su} = F_c F_t / (F_c + F_t) \qquad (1)$$

which would require both tensile and compressive testing of the complementary fiber pattern. The tensile and compressive strengths $F_t$ and $F_c$ measured on a 0/90-degree laminate would be equivalent to shear ($F_{su}$) on a ±45-degree laminate, as explained in Figure 11. Only with a quasi-isotropic fiber pattern (with equal numbers of fibers in each of the 0-, +45-, 90-, and −45-degree directions) would there be no change in fiber pattern with the rotation of the axes.

355

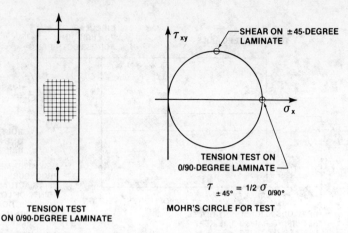

**TENSION TEST**
**ON 0/90-DEGREE LAMINATE**

SHEAR ON ±45-DEGREE
LAMINATE

$\tau_{xy}$

$\sigma_x$

TENSION TEST ON
0/90-DEGREE LAMINATE

$\tau_{\pm 45°} = 1/2\ \sigma_{0/90°}$

**MOHR'S CIRCLE FOR TEST**

**FIGURE 11. PROPOSED "SHEAR" TEST FOR ±45-DEGREE COMPOSITE LAMINATES**

Now, the simple conservative straight-line failure surfaces shown in References 8 and 9 would suggest that the combination of biaxial tension or compression with pure shear should decrease the shear strength of the laminate below that for shear alone. However, the appendix of Reference 8 illustrates that this need not be so when the failure is established more precisely by considering the stresses acting on the fibers and resin separately. Reference 8 also shows that the Norris failure theory for composites, which is essentially an orthotropic generalization of the von Mises criterion, is not very sensitive to this interaction between biaxial tension and shear either. Depending on the degree of approximation used in a given theory, there may be some minor discrepancy between the equivalence of pure uniaxial tension on the one hand and the combination of biaxial tension with shear rotated through 45 degrees on the other. However, that inconsistency can be resolved by physical reasoning and is not significant in the context of the issues raised here because any related error will be small and inherently conservative.

What cannot be resolved yet is whether the maximum-shear-stress theory should be preferred to the generalized von Mises criterion. The former would be inherently conservative, but not by much, if the latter were ever shown to be precise. The use of the former is advocated here, primarily because of simplicity and consistency with the author's laminate theory in References 8 and 9. However, the prime function of this paper is to encourage the adoption of axial testing of the 0/90-degree laminate to characterize the in-plane shear strength of a ±45-degree laminate. Nothing about this paper is intended to discourage the use of the generalized von Mises theory with the same test specimen shown in Figure 11. The difference would be that, for the maximum-shear-stress theory, the in-plane shear allowable would be precisely half the measured tensile strength while, for the generalized von Mises failure theory, the allowable would be 57 percent of the measured tensile strength [or its equivalent from Equation (1)].

Obviously, this issue would need to be resolved by pure shear testing, as with the bonded tapered rail-shear specimen (Figure 9). It could never be resolved by use of the tensile specimen advocated here (Figure 11). That would be begging the question.

The use of this proposal for other fiber patterns is straightforward, provided that the laminates are balanced. For example, the in-plane shear strength of a laminate pattern containing 20 percent 0-degree plies, 60 percent ±45-degree plies, and 20 percent 90-degree plies would be established by axial testing of a laminate containing 30 percent 0-degree plies, 40 percent ±45-degree plies, and 30 percent 90-degree plies. The shear strength of a (30,56,14) laminate would be simulated by axial tests on a (28,44,28) laminate. Those fibers in the ±45-degree directions in the reference laminate are shared equally between the 0- and 90-degree directions in the test laminate, and the sum of the 0- and 90-degree fibers in the reference laminate is shared equally between the +45- and −45-degree directions in the test laminate. The in-plane shear strength of unbalanced laminates would have to be established by the use of any theory that had previously been shown to be valid by correctly predicting the measured in-plane shear strengths of balanced laminates.

## CONCLUSION

This paper proposes that the in-plane shear testing of fibrous composite laminates, which has long been a difficult task, be replaced by the much simpler axial tensile (and compressive) testing of a laminate with the complementary fiber pattern. For example, the in-plane shear strength of a ±45-degree laminate would be established as nominally half the tensile strength of a 0/90-degree laminate.

Such a change in test specimen can be justified by two totally independent lines of reasoning. The first is that doing so would give the very same predictions of strength that have been obtained in the past with the best of the pure shear tests − the bonded tapered rail-shear coupon. Such a change would be justified by greatly improved strengths and reduced scatter of the tests with respect to inferior shear tests used in the past and still not fully discarded today. Such reasoning does not imply anything about accepting any specific laminate failure theory; empirical testing could lead to the same conclusion.

The second reason is that the maximum-shear-stress and generalized von Mises (or Norris) criteria, which are recognized as the best of the laminate failure criteria (at least for carbon-epoxy composites), would predict that such an interchange of test coupons would be permissible.

It is premature to propose the use of this specimen for shear testing of Kevlar fiber-reinforced laminates, because their more complex failure modes are not yet fully characterized. In the case of fiberglass-reinforced laminates, the proposed change would at least be conservative. However, it should first be justified by a comparison between test results with the best of the in-plane shear specimens and the axial test(s) proposed here. Obviously, the intent of using axial testing instead of in-plane shear testing is to generate higher allowable strengths more economically. However, this can be accomplished only after the change in coupons has already been justified by earlier pure shear testing, as has been reported here for HTS carbon-epoxy composites. Inevitably, such initial shear testing would always be needed for any new material before the change could be sanctioned.

Even for well-characterized composite materials, it is not yet acceptable to abandon testing for these shear strengths and to rely instead upon the predictions of laminate theories. Too many of those theories, including those most widely used, are known to be unacceptably unconservative, particularly for shear-dominated loads. However, they will not be discarded as long as there is only one test condition (uniaxial tension or compression) against which the theories are correlated. The characterization of any new composite material would necessarily require biaxial testing (such as pure shear) as well as uniaxial testing to establish appropriate failure criteria before any consideration could be given to the kind of revised testing proposed here.

The primary benefit of the advocated change in test procedures is that there could then be standardization on a simple, reliable, and reproducible test coupon for in-plane shear testing of all fiber-dominated composite laminates. In a way, this proposal can be looked upon as a generalization of the already accepted off-axis testing for resin-dominated shear strengths.

## ACKNOWLEDGEMENT

This paper was prepared as part of the author's duties with the MIL-17 Handbook committee, which was at the time under the control of the U.S. Army Materials Technology Laboratory, Watertown, Massachussetts. The author would like to express his appreciation for the committee's encouragement to present this new approach to in-plane shear testing for consideration by those involved in composite laminate testing. The preparation of this paper was performed as part of the Douglas Aircraft Company Independent Research and Development (IRAD) program.

## REFERENCES

1. J. M. Whitney, D. L. Stansbarger, and H. B. Howell, Analysis of the Rail Shear Test — Applications and Limitations, J. Composite Materials, Vol. 5 (January 1971), pp 24-34.

2. L. M. Lachman, D. O. Losee, J. A. Rohlen, and T. T. Matoi, Advanced Composites Data for Aircraft Structural Design, Volume IV: Material and Basic Allowable Development — Graphite/ Epoxy, North American Rockwell, U.S. Air Force Contract Report AFML-TR-70-58, September 1972.

3. J. M. Slepetz, T. F. Zagaeski, and R. F. Novello, In-Plane Shear Test for Composite Materials, AMMRC TR 78-30, July 1978.

4. V. Ramnath and S. N. Chatterjee, Composite Specimen Design Analysis, Material Sciences Corporation, Army Materials Technology Laboratory Contract Report MSC TFR 1701/1703 (Contract No. DAAG46-85-C-0058), March 1986.

5. G. M. Lehman, et al, Advanced Composite Rudders for DC-10 Aircraft — Design, Manufacturing, and Ground Tests, Douglas Aircraft Company, NASA Langley Contract Report NASA CR-145068, November 1976.

6. J. B. Black, Jr., and L. J. Hart-Smith, The Douglas Bonded Tapered Rail-Shear Test Test Specimen for Composite Laminates, Douglas Aircraft Company, Paper 7764, to be presented to 32nd International SAMPE Symposium and Exhibition, Anaheim, California, April 1987.

7. L. J. Hart-Smith, Adhesive-Bonded Scarf and Stepped-Lap Joints, Douglas Aircraft Company, NASA Langley Contract Report NASA CR-112237, January 1973.

8. L. J. Hart-Smith, Simplified Estimation of Stiffness and Biaxial Strengths for Design of Laminated Carbon-Epoxy Composite Structures, Douglas Aircraft Company, Paper 7548, presented to 7th DoD/NASA Conference on Fibrous Composites in Structural Design, Denver, Colorado, June 1985.

9. L. J. Hart-Smith, Simplified Estimation of Stiffness and Biaxial Strengths of Woven Carbon-Epoxy Composites, Douglas Aircraft Company, Paper 7632, presented to 31st International SAMPE Symposium and Exhibition, Las Vegas, Nevada, April 1986.

10. S. W. Tsai. Composites Design — 1968, Think Composites, Dayton, Ohio 1986.

11. C. B. Norris, Strength of Orthotropic Materials Subject to Combined Stress, U.S. Forest Products Laboratory, Report No. 1816, May 1962.

## BIOGRAPHY

Dr. Hart-Smith has 22 years of experience in the analysis, design, and manufacture of advanced composite structures. He has worked at the Douglas Aircraft Company since 1968, when he came to the United States from Australia. In addition to his work at Douglas, he has consulted extensively with other companies on composite and adhesively bonded metal structures. His specialty is the development of nonlinear joint analysis methods and computer programs, and he has presented many papers on these subjects. Dr. Hart-Smith has been program manager on several U.S. Government-funded contracts on joints in fibrous composites. He is currently employed as a senior staff engineer in the Airframe Technologies group of Engineering at McDonnell Douglas, Long Beach. Dr. Hart-Smith holds a bachelor's degree in Mechanical Engineering (with honors) from Melbourne University, Australia, and a doctorate in Engineering from Monash University, Australia.

# THE DOUGLAS BONDED TAPERED RAIL-SHEAR TEST SPECIMEN
# FOR FIBROUS COMPOSITE LAMINATES

by
J. Benson Black, Jr.
and
Dr. L. J. Hart-Smith

Douglas Aircraft Company
McDonnell Douglas Corporation
Long Beach, California

## ABSTRACT

The purpose of this paper is to present an analysis of the Douglas bonded tapered rail-shear specimen for measuring the in-plane shear strength of composite laminates. A detailed linear finite-element analysis for a ±45-degree laminate indicates that there are no local stress concentrations severe enough to prevent the specimen from developing the full shear strength of the laminate. Those minor stress concentrations that are predicted are diminished greatly in a reanalysis that simulates the nonlinear behavior of the flexible adhesive layer between the laminate and the steel rails. Actually, the adhesive is merely softened greatly in those areas in which it would have been strained beyond its elastic limit, and a new elastic analysis is performed. The stress concentrations are not totally eliminated because there is a physical reason for their existence, which is presented. Similar analyses for quasi-isotropic composite panels exhibited slightly different stress concentrations, for reasons which are explained.

However, again, these are shown not to be sufficient to cause premature failures. It is concluded that this specimen is quite suitable for its intended purpose and will give reliable test data with which to assess the accuracy of the various published composite laminate theories.

## INTRODUCTION

The in-plane shear testing of advanced composite structures has proved to be even more of a challenge than the same testing of ductile metal alloys. Unlike uniaxial tensile or compressive test specimens, which are simple and have no more serious problem than extraneous stresses caused by imperfect load-introduction tabs, all of the flat shear test specimens deviate significantly from a uniform pure shear state. Most have quite severe stress concentrations that prevent the attainment of full laminate strength. The alternative approach of testing tubular specimens in torsion has introduced the difficulty of making wraparound laminates as wrinkle-free as the flat laminates they are to represent. Reference 1 contains a more detailed discussion of test coupons for testing the in-plane shear strength of composite laminates.

There is, however, one in-plane shear specimen suitable for use with simple flat composite laminates. For more than a decade, this specimen has given test results nearly twice as high as those produced by the older untapered rail-shear specimens. (The actual difference between test results is not constant, but a function of the fiber pattern. This difference is most severe for the ± 45-degree laminate, less so for a quasi-isotropic laminate, and negligible for 0- or 90-degree laminates. This variation is also evident in the comparisons between test and theories in Reference 1.) This improved test specimen is the Douglas bonded tapered rail-shear coupon, first introduced during development of the DC-10 composite rudder under contract to NASA Langley. It was designed by Hart-Smith and A. V. Hawley in the early 1970s, based on the former's analyses of adhesively bonded joints, also under contract to NASA Langley. The specimen may be looked upon simplistically as a bonded scarf joint in which the "adhesive" has been replaced by the test section of the composite laminate. The stiff linearly tapered steel rails are intended to be uniformly stressed along their length in order to provide a uniform shear flow into the composite test section.

In addition, now that at least some fibrous composite laminate strength-prediction theories can accurately predict *biaxial* strengths (as well as the uniaxial strengths that can be predicted by many more theories), it is evident that reliable measurements of the in-plane shear strength will be needed. There is no other way to establish the limitations on composite laminate failure theories, some of which are quite unconservative for biaxial loadings, and to likewise establish the improved validity of other failure

theories. This distinction is impossible with uniaxial testing alone because those failures are completely dominated by the longitudinal stress in the fibers with only negligible transverse stress acting.

The objective of this paper, therefore, is to describe this novel specimen and the rationale behind its development, explain why and how it works, and to identify its limitations (which are few).

## HISTORICAL BACKGROUND

The Douglas bonded tapered rail-shear test coupon (Figure 1) was conceived at a time when other specimens being used to measure the in-plane shear strength of fibrous composite laminates were all failing prematurely because of severe stress concentrations of one form or another. A comparison between test results and various failure theories is given in Reference 1.

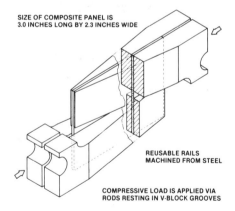

SIZE OF COMPOSITE PANEL IS
3.0 INCHES LONG BY 2.3 INCHES WIDE

REUSABLE RAILS
MACHINED FROM STEEL

COMPRESSIVE LOAD IS APPLIED VIA
RODS RESTING IN V-BLOCK GROOVES

**FIGURE 1. BONDED TAPERED RAIL-SHEAR TEST SPECIMEN**

Those specimens in which uniform rails, through which the load was sheared, were bolted directly to unreinforced composite laminates often failed because of the stress concentration at the bolt holes. When that problem was solved by first bonding uniformly thick metal doublers to the composite panel and bolting the resulting assembly to the load-introduction rails, it was found that the shear flow was not uniform along the length but peaked sharply at each end. These peaks are directly analogous to those at the ends of the overlap in adhesively bonded joints between uniformly thick adherends.

What was wanted was a new test coupon, devoid of stress concentrations and designed around a simple flat composite panel (because of the very high cost of advanced composites in those days). It seemed then that it would be too much to expect a perfectly uniform shear flow; indeed, it still seems that way today. The geometry of the then-new test specimen is described in Figure 2. The rails are tapered to introduce a more uniform shear flow in the test section than would have been the case with uniform rails. This distinction is analogous to the uniform shear stress in a bonded scarf joint in contrast to the highly nonuniform stress distribution (with sharp peaks at each end of the overlap) that occurs in bonded joints between adherends of constant thickness. The machined steel rails are intended to be reusable, of course, and are bonded on by an adhesive appropriate for the test environment.

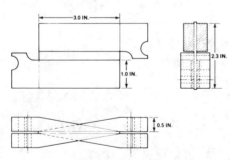

**FIGURE 2. GEOMETRY OF DOUGLAS BONDED TAPERED RAIL-SHEAR TEST SPECIMEN**

The precured laminate serves as its own alignment fixture for bonding provided that the

semicircular cutouts, through which the load is applied by circular rods resting in V-groove blocks, are properly aligned. The test setup is described in Figure 3.

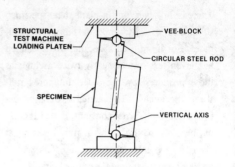

**FIGURE 3. TEST SETUP FOR BONDED TAPERED RAIL-SHEAR TEST SPECIMEN**

The omission of the rails across the short ends of the test section leaves no path for the complementary shear forces. Because of this, there can be no in-plane shear stresses on those free edges, and there is a small compressive stress across the specimen. It had been recognized at the outset that there would be a small zone at each end of the test section in which the shear stress would be much lower than the average. Although some local (uniaxial) stress concentrations in or near the corners were inevitable, it was thought that they would be alleviated by the flexibility of the adhesive layer. It was confidently felt that the failures would be dominated by the essentially uniform shear stress in the bulk of the interior of the test section.

The width of the test section (0.3 inch) was made only one-tenth of the length, barely sufficient to install a strain gage rosette. The line of action for the load was set to pass through the diagonal of the test area to approximate a pure shear condition as closely as the absence of the end rails would permit. The decision to load the specimen in compression rather than tension was deliberate and was based on two considerations. First, there was considerably more concern about adding any transverse tension stresses, which might peak locally at the ends, than there was about adding any equivalent compressive ones. Second, if such compressive stresses *did* develop, they might simply cause the laminate to buckle locally and thereby relieve any transverse loads. Since each steel rail was half an inch thick, the total footprint of the specimen was just over an inch wide, so that the test specimen should automatically be well stabilized and free from out-of-plane warping.

The magnitude of any nonuniformity in the shear flow was not characterized at the time, but it was clear that neglecting it would be conservative, by an amount estimated not to exceed 10 percent. The first two tests, for ±45-degree Narmco 5206 carbon-epoxy laminates, yielded 59.5 and 60.4 ksi as the average shear stresses. The scatter was small in comparison with prior tests on older specimen configurations, and the mean strength was almost double that previously measured. Similar testing of 90-degree laminates at that time was also very successful.

Obviously, the specimen would be too fragile to test the shear strength of a 0-degree laminate (0 degrees being the longitudinal reference axis). However, that strength was assumed to be identical to the measured shear strength of the 90-degree laminates. Intermediate laminate patterns, such as the quasi-isotropic one, were not tested then because those strengths could reasonably be estimated by interpolating between the results for the two patterns that were tested. (It should be noted also that contemporary laminate analyses had already identified the in-plane shear strength of the ±45-degree laminate as the most critical test case, even when allowance was made for premature failures of the earlier test specimens.)

At first, the new test specimen gave consistently good results. Later, however, when the matched sets of rails were mixed up, the unmatched alignment holes (not shown in Figure 1) caused the semicircular grooves to be misaligned from side to side. Consequently, the specimen then yielded some bad results and it fell into undeserved disfavor. That temporary problem has since been solved by a minor redesign that enables the load-introduction grooves to be used directly for alignment during bonding. Doing so thereby obviates the need to align the pairs of rails by means of a secondary hole (through which a quarter-inch bolt used to be installed). It also eliminates the need for keeping the rails in matched sets.

## ANALYSIS OF INTERNAL LOAD DISTRIBUTIONS

In a recent survey of various shear test specimens for composite laminates (Reference 2), Ramnath and Chatterjee attempted to quantify the stress concentrations associated with each of the designs. These included the Iosipescu specimen with and without radii in the notches, the uniaxial tension of $(\pm 45°)_s$ laminates, and the rail-shear test with both uniform and tapered rails. Based on their initial analyses, they redesigned the specimens and reanalyzed them to evaluate the effectiveness of the modifications. The survey concluded that the bonded tapered rail-shear specimen was among the best, if not the best, of those evaluated. What is really significant about that favorable evaluation, however, is that the survey did not, in fact, do justice to this specimen. Their two-dimensional analysis totally neglected the flexibility of the adhesive layer and thus introduced artificial singularities in two corners of the test section. The analyses presented here show that this specimen is even better than the analysis in Reference 2 suggested.

Figure 4 shows the finite-element model used in the present analysis. The idealization utilizes a

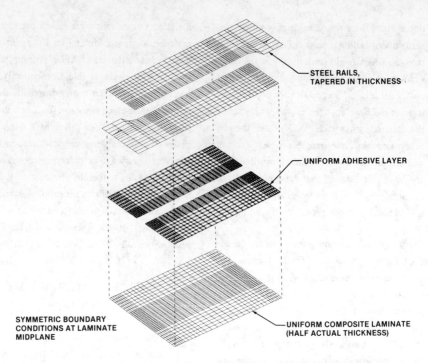

STEEL RAILS,
TAPERED IN THICKNESS

UNIFORM ADHESIVE LAYER

SYMMETRIC BOUNDARY
CONDITIONS AT LAMINATE
MIDPLANE

UNIFORM COMPOSITE LAMINATE
(HALF ACTUAL THICKNESS)

**FIGURE 4. FINITE-ELEMENT MODEL OF BONDED TAPERED RAIL-SHEAR COMPOSITE
TEST SPECIMEN**

combination of two-dimensional elements to approximate a three-dimensional analysis. The composite laminate and the tapered steel rails are represented by membrane elements (anisotropic and isotropic, respectively). The variation in location of the neutral axis of the tapered rails is ignored since the normal forces generated are simply reacted through the adhesive layers and composite laminate to the opposite rail and are thus of no interest here. The bond line is represented by an "egg crate" arrangement of rods and shear panels. The model represents half the specimen with the plane of symmetry at the middle of the composite laminate. The basic analysis was linear; however, the effects of the nonlinear characteristics of the adhesive were also investigated by selectively softening the shear panels comprising the bond line model. That refinement is similar to representing the adhesive by its secant modulus, where appropriate, instead of by its initial elastic modulus everywhere.

The results of the linear analysis of the test section for a $\pm 45$-degree carbon-epoxy laminate are shown in Figure 5, normalized with respect to the average shear stress acting on the test section. The key features of the shear-stress contours are the large interior area of essentially uniform shear stress and the local end zones in which the shear stress decays to zero (as was forecast when the specimen was originally designed).

There are, in addition, very localized regions of irregular shear flow near the areas of load introduction. The source of these irregularities is more easily identified if the laminate stresses are separated and plotted along the $+45$-degree and $-45$-degree directions, as has been done in Figure 5. Again, the stresses are normalized to the nominal average shear stress in the test section. The stresses in the $-45$-degree direction build steadily from near zero (actually slightly negative) at the ends of the specimen to just over

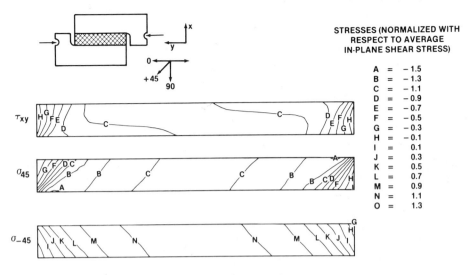

| | | |
|---|---|---|
| A | = | − 1.5 |
| B | = | − 1.3 |
| C | = | − 1.1 |
| D | = | − 0.9 |
| E | = | − 0.7 |
| F | = | − 0.5 |
| G | = | − 0.3 |
| H | = | − 0.1 |
| I | = | 0.1 |
| J | = | 0.3 |
| K | = | 0.5 |
| L | = | 0.7 |
| M | = | 0.9 |
| N | = | 1.1 |
| O | = | 1.3 |

**FIGURE 5. ANALYSIS OF LINEAR RAIL-SHEAR SPECIMEN FOR ±45-DEGREE COMPOSITE LAMINATE**

1.0 for the large middle zone (Figure 5). The stresses in the +45-degree direction are likewise just over 1.0 in the middle zone, but increase to a peak of just over 1.5 at a location a little inboard of the ends of the test section, before decaying toward zero at the very ends.

The question arises then as to whether failure will be triggered prematurely by this very local (and largely uniaxial) stress concentration, or if it will occur at the desired stress level as the result of the relatively uniform shear stresses in the interior of the test section. However, before that question need be asked, it is appropriate first to establish that these stress "concentrations" are real and not just quirks arising from the linear mathematical model of what is really a nonlinear adhesive layer. If these stress concentrations are real, there must be a physical cause for their existence.

In fact, they are real, even though their effect is exaggerated in Figure 5. And there is a sound reason why they do not extend to the ends of the test specimen. This is explained by the enlarged view of the corner of the test area, shown in Fig-

ure 6. The primary load in the ±45-degree fibers for a pure shear load is axial, either in tension or compression, even though the fibers are also subjected to an orthogonal stress of the opposite sign. (Although that transverse stress is necessarily much lower than the longitudinal stress, it cannot be neglected because the transverse strength of the fibers and resin is also very low.) The axial load in those fibers must then be transferred from the steel rails via the flexible adhesive layer.

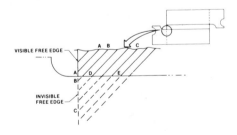

**FIGURE 6. DETAIL OF IRREGULARITIES IN STRESS FIELD AT CORNER OF TEST SECTION (+45-DEGREE FIBERS)**

Now, with reference to Figure 6, it is clear that the fiber AA must be unloaded by the time it has

365

reached the free edge of the laminate. The fiber CC must be unloaded at the same free edge, even though it is concealed between the steel rails. That unloading is effected through the substantial length of adhesive on the insides of the rails, and that fiber is fully loaded within the test area at E. However, the fiber BB does not extend far enough past the end of the test section to be fully loaded in area D. Consequently, there is a load deficiency at D, which is compensated for by an increase in the adjacent area E (and the appearance of the stress concentration at +45 degrees evident in Figure 5).

The phenomenon ignored in this linear analysis is that the adhesive layer adjacent to the edge of the rails will be strained far beyond the proportional limit before ultimate failure of the composite occurs. Therefore, the adhesive layer will be much softer there than the linear finite-element analysis would suggest, and the local stress concentrations in Figure 5 are exaggerated. This is confirmed by the results of a pseudo-nonlinear analysis (Figure 7), where the most highly strained adhesive elements were remodeled with much lower stiffnesses. Rather than softening the adhesive elements iteratively in an attempt to match the shear stress and stiff-

ness precisely, all of those adhesive elements that were overloaded in the linear analysis had their modulus reduced by a factor of 10. Also, to minimize artificial irregularities in the adhesive stresses, a transitional band of partially softened elements was inserted between the softened ones on the edge and the stiff elements in the interior. The reasoning behind this revolved around a desire to avoid an iterative solution process and the recognition that, otherwise, the second analysis would undoubtedly identify further adhesive elements that were overloaded and thus needed softening. To a limited extent this was still true but, nevertheless, the analysis is considered more than adequate and clearly shows the results that a precise nonlinear analysis would have established.

Figure 7, of course, does not show these shear flow irregularities (located at E in Figure 6) completely removed, but merely diminished since they really do exist.

At this point, it is appropriate to compare the likelihood of the failure occurring as desired in the bulk of the test area or prematurely at these irregularities in the shear flow. Making that assessment needs an estimate for the orthogonal

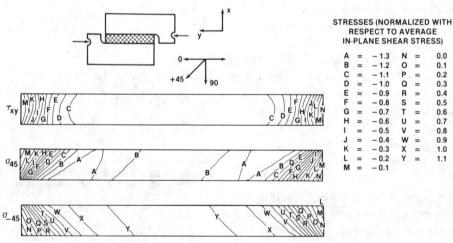

FIGURE 7. "NONLINEAR" REANALYSIS OF ±45-DEGREE IN-PLANE SHEAR TEST BY USING LOCALLY SOFTENED ADHESIVE LAYERS

stresses on the fibers as well as a reliable failure criterion for those fibers. Such a criterion is to be found in References 3 and 4. Reference 5 also provides a suitable failure criterion, provided that the transverse stiffnesses of the monolayer are reduced appropriately, as explained in Reference 3, and that the further modifications cited in the same reference are incorporated.

Now, for a ± 45-degree carbon-epoxy laminate loaded in shear, failure of the fibers in the basic test area will occur when the stress in the fibers is approximately half what it would take to fail those same fibers in a uniaxial tensile or compressive load situation, in the absence of transverse stress. (That factor of two varies with the fiber pattern of the laminate being tested, as explained in Reference 4.) So, to assess the severity of the local stress concentrations, we return again to Figure 5, which shows that the local stress concentrations occur essentially in the absence of transverse loads on those particular fibers.

The physical explanation for the absence of the transverse stresses can be seen in Figure 8. The fiber FF is obviously virtually unloaded because of the proximity of the free edge at the end of the test section. Likewise, the fiber GG is also very much less loaded than the interior fiber HH for the very same reason.

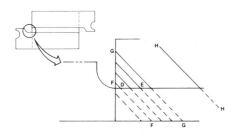

FIGURE 8. DETAIL OF IRREGULARITIES IN
STRESS FIELD AT CORNER OF
TEST SECTION
(–45-DEGREE FIBERS)

Consequently, since the local stress concentrations suggested in Figure 5 are in areas of essentially uniaxial loading, they would need to be nearly twice as intense as the average shear stress, and that is clearly not the case. The bonded tapered rail-shear specimen does not contain stress concentrations of a sufficient magnitude to cause premature failure. Indeed, the test result will be only slightly conservative, because of the ineffective end zones. And even they could be compensated for in terms of an analysis of the type presented here, if desired.

## COMPARISON WITH PRIOR ANALYSIS

The modest stress concentrations in Figure 5 have a physical origin, as explained in the preceding section, and are definitely not the result of any mathematical singularity of the type identified in the analysis performed in Reference 2. This was confirmed by modifying the analysis model used here to make the adhesive layers artificially rigid and thus eliminate their flexibility.

The results of this reanalysis are presented in Figure 9; they are very similar to those given in Figure 28 of Reference 2. While the previous analysis quoted a maximum shear stress concentration factor of 1.7984 in opposite corners of the test section, where the load is introduced, Figure 9 indicates that this artificial factor appears to be more like 1.2.

The actual value of the stress concentration is, in fact, established primarily by the number of elements used in the model — the finer the grid, the larger the erroneously predicted stress concentration in the corner. Since the two models did not use exactly the same number of finite elements, the two analyses for infinitely stiff adhesive layers cannot be compared directly. Only the presence of the spurious stress concen-

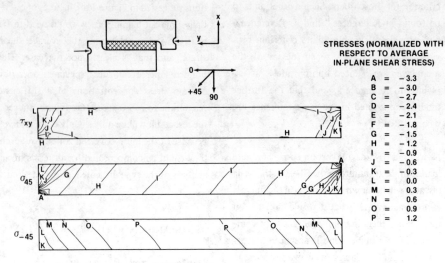

**FIGURE 9. REANALYSIS OF ±45-DEGREE IN-PLANE SHEAR TEST WITH "RIGID" ADHESIVE LAYERS**

trations in the corners of both test sections is significant. This stress concentration can reasonably be ignored based on the physical explanations of the real behavior of the test specimen given in Figures 6 and 8.

It is now possible to eliminate the caveats expressed in Reference 2 about this shear test specimen and to heed only the views expressed there about its merits.

Reference 2 also contains many other analyses about aspects of shear test specimens that are not discussed here. Their response to the problems created by the stress concentrations apparently revealed by their analyses was to seek design modifications to diminish their severity. Although it has been shown that those refinements were not really needed for this particular test specimen, the information associated with those modifications, such as changing the specimen to a parallelogram rather than a rectangle, might well be useful in relieving stress concentrations associated with cutouts and joints in real composite structures.

## OTHER CONSIDERATIONS

Taking the laminate shear strength to be the load divided by the length and thickness of the test section is not precise. Indeed, it is now known to be slightly conservative. However, any modification of the nominal test results to account for the ineffective end zones is not recommended here because the magnitude of such a correction would vary with the fiber pattern. (In any event, all such modifications would be relatively small.) This paper has dealt with the ±45-degree test since it is the most difficult to perform (by other means, anyway) and also the most important in terms of having reliable test results with which to accept or reject a cross-plied composite laminate theory.

The bonded tapered rail-shear specimen also has often been used to assess the in-plane shear strength of the quasi-isotropic laminate (typical of many structural composites), so that strength margins could be written during those times in which it was known that the contemporary laminate failure theories were inadequate, but it was not yet known what a satisfactory theory was. Those stress calculations were thus based on the laminate strength rather than computed on a ply-by-ply lamina basis.

The specimen is also quite suitable for testing 90-degree or 0/90-degree laminates in shear but, as noted earlier, is too difficult to handle for the 0-degree laminate. While these strengths are needed for ply-by-ply laminate strength analyses, such resin-dominated test results usually need to be substantially modified (typically by dividing the moduli by a factor of about three and multiplying the transverse tensile strength by a factor of about two). Such modifications permit the theory to account for the effective (or in-situ) transverse properties instead of projecting linearly from the initial very-small-strain behavior. Such projections usually result in erroneous solutions for all of the stress states that are not dominated by uniaxial loading with only small transverse stresses on the critical plies. There is no need to modify the test results for the shear strength of a ±45-degree laminate, because that strength is entirely fiber-dominated.

As noted earlier, the test section is 3 inches long and 0.3 inch wide, with a bond width of 1 inch per side. It is obviously necessary, therefore, that the thickness of the test laminate should be limited to ensure that the bond will not fail first. This situation is similar to, but not as severe as, the problem with testing honeycomb sandwich beams, in which the thickness of the face sheet must be restricted. For the bonded tapered rail-shear test specimen used to evaluate carbon-epoxy laminates, an effective thickness of no more than eight 0.005-inch unidirectional tape plies (or an equivalent thickness of woven cloth layers) should be used. Thus, a ±45-degree laminate would be 0.040 inch thick. However, a quasi-isotropic $(0°, +45°, 90°, -45°)_s$ laminate could be twice as thick (0.080 inch) because the additional 0-degree and 90-degree plies are so

relatively weak in shear. These limits on the thickness of the test laminate have been established for testing in the operational environments of subsonic transport aircraft. Other limits might well be necessary, because of different adhesive strengths, for appreciably different test environments.

The results from analyses of quasi-isotropic laminate shear tests show phenomena not exposed in the analyses of ±45-degree laminates in Figures 5 and 7. Figure 10 shows the in-plane shear stresses for a quasi-isotropic $(0°, ±45°, 90°, -45°)_{2s}$ laminate. Again, the large uniformly stressed interior region in the shear-stress contour diagram is clearly evident. However, the plot of the stresses in the transverse (90-degree) direction (Figure 10) shows stress concentrations at the ends of the test section that were not present for the ±45-degree laminate, because there were no 90-degree fibers in that pattern to develop such stresses. These 90-degree stresses occur because the 90-degree fibers at the ends of the test section try to act as the missing rails on the short edges. The irregularities in the corresponding plots of the uniaxial stresses in the +45- and −45-degree directions are less than those shown in Figure 5 because the 90-degree fibers here provide a stiffer load path to make up for the missing end rails.

The apparent peak 90-degree stresses, at 2.2 times the average shear stress, would appear to be sufficiently intense to cause premature failure of the laminate. However, the appropriate "nonlinear" analysis in Figure 11 shows that the peak 90-degree stresses are not really so intense. The uniaxial 90-degree stress concentration factor of only 1.45 shown in Figure 11, for which the adhesive had been softened in the manner of Figure 7, is not sufficient to cause a premature failure; shear failure of the quasi-isotropic laminate also will occur under the correct mode of failure.

369

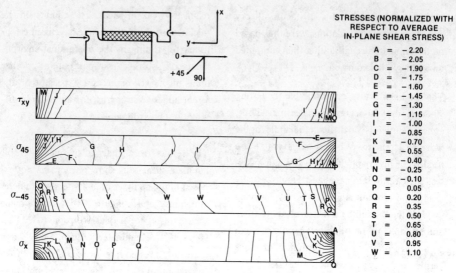

STRESSES (NORMALIZED WITH
RESPECT TO AVERAGE
IN-PLANE SHEAR STRESS)

A = −2.20
B = −2.05
C = −1.90
D = −1.75
E = −1.60
F = −1.45
G = −1.30
H = −1.15
I = −1.00
J = −0.85
K = −0.70
L = −0.55
M = −0.40
N = −0.25
O = −0.10
P = 0.05
Q = 0.20
R = 0.35
S = 0.50
T = 0.65
U = 0.80
V = 0.95
W = 1.10

**FIGURE 10. LINEAR ANALYSIS OF IN-PLANE SHEAR OF QUASI-ISOTROPIC COMPOSITE LAMINATE**

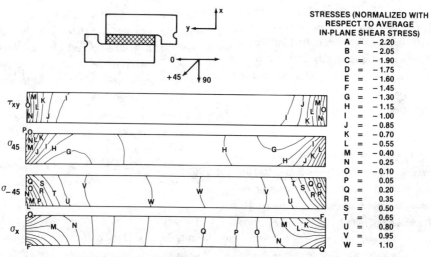

STRESSES (NORMALIZED WITH
RESPECT TO AVERAGE
IN-PLANE SHEAR STRESS)

A = −2.20
B = −2.05
C = −1.90
D = −1.75
E = −1.60
F = −1.45
G = −1.30
H = −1.15
I = −1.00
J = −0.85
K = −0.70
L = −0.55
M = −0.40
N = −0.25
O = −0.10
P = 0.05
Q = 0.20
R = 0.35
S = 0.50
T = 0.65
U = 0.80
V = 0.95
W = 1.10

**FIGURE 11. "NONLINEAR" ANALYSIS OF IN-PLANE SHEAR ON QUASI-ISOTROPIC COMPOSITE LAMINATE (USING SOFTENED ADHESIVE)**

## CONCLUSION

This paper has described the bonded tapered rail-shear test specimen for composite laminates developed at the Douglas Aircraft Company over a decade ago.

The logic behind its design has been documented for the first time.

A detailed analysis by finite elements has confirmed that the only parasitic stress concentrations are not sufficiently intense to cause spurious failures. Apart from lightly loaded (ineffective) end zones adjacent to the stress-free short edges of the test section, this specimen very closely approaches a uniform shear-stress state throughout the test area.

It is recommended that this specimen be considered very seriously if there is any attempt to standardize the in-plane shear testing of composite laminates.

It is also noted that the correct prediction of the in-plane shear strenth of $\pm 45$-degree composite laminates is a far more discriminating test for composite laminate theories than any simple tensile or compressive uniaxial test could ever be.

## ACKNOWLEDGMENT

This paper was prepared as part of the second author's activities with the MIL-17 Handbook committee, which was at that time under the control of the U.S. Army Materials Technology Laboratory, Watertown, Massachussetts. The authors would like to express their appreciation for that committee's encouragement to improve upon prior analyses and to publicize the specimen. The original design of the specimen and the preparation of this paper were both performed as part of the McDonnell Douglas Independent Research and Development (IRAD) program.

## REFERENCES

1. J. M. Whitney, D. L. Stansbarger, and H. B. Howell, Analysis of the Rail-Shear Test — Applications and Limitations, J. Composite Materials, Vol. 5 (January 1971), pp 24-34.

2. V. Ramnath and S. N. Chatterjee, Composite Specimen Design Analysis, Material Sciences Corporation, Army Materials Technology Laboratory Contract Report MSC TFR 1701/1703, (Contract No. DAAG46-85-C-0058), March 1986.

3. L. J. Hart-Smith, Simplified Estimation of Stiffness and Biaxial Strengths for Design of Laminated Carbon-Epoxy Composite Structures, Douglas Aircraft Company, Paper 7548, presented to 7th DoD/NASA Conference on Fibrous Composites in Structural Design, Denver, Colorado, June 1985.

4. L. J. Hart-Smith, Simplified Estimation of Stiffness and Biaxial Strengths of Woven Carbon-Epoxy Composites, Douglas Aircraft Company, Paper 7632, presented to 31st International SAMPE Technical Conference and Exhibition, Las Vegas, Nevada, April 1986.

5. C. B. Norris, Strength of Orthotropic Materials Subject to Combined Stress, U.S. Forest Products Laboratory, Report No. 1816, May 1962.

## BIOGRAPHIES

### J. B. BLACK, JR.

Mr. Black is a senior engineer in the Airframe Research and Development department at Douglas Aircraft Company. He joined the company in 1977 after graduating from Clemson University with a B.S. degree in Civil Engineering. Since that time he has contributed to the design and analysis efforts on a variety of R&D projects. This work has included design and analysis of a Kevlar wing/fuselage fillet for the MD-80 aircraft and a superplastic-formed/

diffusion-bonded (SPF/DB) titanium main landing gear door for the T-38 trainer. He has also provided analyses required for fitting winglets to the DC-10 wing, and has performed dynamic analyses of a concrete runway/pier structure interacting with aircraft taxi and landing loads. His more recent work has been concentrated in the area of advanced composites, specifically their application to fuselage structures for commercial transport aircraft.

## L. J. HART-SMITH

Dr. Hart-Smith has 22 years of experience in the analysis, design, and manufacture of advanced composite structures. He has worked at the Douglas Aircraft Company since 1968, when he came to the United States from Australia. In addition to his work at Douglas, he has consulted extensively with other companies on composite and adhesively bonded metal structures. His speciality is the development of nonlinear joint analysis methods and computer programs, and he has presented many papers on these subjects. Dr. Hart-Smith has been program manager on several U.S. Government-funded contracts on joints in fibrous composites. He is currently employed as a senior staff engineer in the Airframe Technologies group of Engineering at McDonnell Douglas, Long Beach. Dr. Hart-Smith holds a bachelor's degree in Mechanical Engineering (with honors) from Melbourne University, Australia, and a doctorate in Engineering from Monash University, Australia.

# ADVANCED COMPOSITE ENGINE DUCTS

Terry L. Stockham, Jerry A. Lunde, James W. Davidson

Rohr Industries, Inc.

## Abstract

A product application in jet engines where advanced composites are beginning to be used is for low pressure ducts and fairings. Current examples include the carbon/epoxy fan ducts for the Rolls-Royce Tay engine, the carbon/polyimide bypass ducts for the G.E. F404 engine and the carbon/polyimide interface fairing between the airframe and engine for the LAVI fighter with its PW1120 engine.

This paper discusses these and other applications where the use of advanced composite ducts continues to be investigated to take advantage of lower weight at competitive cost. As a structural material carbon fiber composites offer an ideal alternative to titanium and aluminum. The use of standardized material forms as well as the implementation of promising automated processing techniques to form engine duct shells with integral flanges, combined with the near net shapes produced will result in reduced manufacturing costs.

Both epoxy matrix systems for service temperatures in the 300°F range and polyimide systems for the 500°F range are either already production qualified or soon will be. Other matrix systems such as PPS and PEEK are now being investigated for application to ducts and offer the promise of additional reduction in labor and easier repair.

## 1. BACKGROUND

Although advanced composite materials have been used extensively in aerospace in many applications, their use in turbine engines has been slow in evolving. One of the reasons for this is that turbine engines operate in a demanding environment where performance is critical and material technology needs to be very mature prior to its use.

The main reason for investigating and applying advanced composite materials to engine components is the same as in other applications which is primarily lower weight.

Figure 1 - Carbon/Polyimide F100 Engine Augmentor Duct

With the advances in fabrication technology and falling material prices, the cost of using composite engine ducts has now become very competitive with metal.

Advanced composite engine ducts have another unique feature which can be important in many applications. That feature is the ability to tailor directional stiffness to modify dynamic natural frequencies. Advanced composite materials have generally higher damping coefficients than metal. These two factors can be made to work together to provide more durable ducts.

The serious efforts at applying advanced composite materials to engine ducts can be traced to the development of the carbon/polyimide F100 engine augmentor duct shown in Figure 1. This duct was developed by Rohr Industries under USAF sponsorship in 1978. It consists of a carbon fabric-polyimide resin matrix duct shell with titanium end flanges. The duct was satisfactorily tested on a ground test engine at Pratt and Whitney, but was damaged when overheated by an augmentor liner burn-through.

2. PRODUCTION READY APPLICATIONS

There are at least three major composite duct applications for current engines. These include the carbon/epoxy bypass fan ducts for the Rolls-Royce Tay engine, the carbon/polimide fan ducts for the

G.E. F404 engine and the carbon/polyimide engine interface fairing for the PW1120 engine installation on the LAVI fighter. Integral end flanges are featured in both of the fan duct applications. In the case of the F404 duct, there is also a longitudinal split flange creating upper and lower duct halves. All three of these structures are now qualified bill of material parts and are either in production or considered to be production ready. The F404 fan duct is shown in Figure 2, the Tay fan ducts in Figure 3, and the PW1120 fairing in Figure 4.

Figure 2 - Carbon/Polyimide Fan Duct for G.E. F404 Engine

Figure 3 - Carbon/Epoxy Fan Ducts on Rolls-Royce Tay Engine

Figure 4 - Carbon/Polyimide
Interface Fairing for PW1120

3. DEVELOPMENT ACTIVITIES

There are several development activities currently underway to improve the technology of advanced
composites in engine ducts. For
example, Rohr Industries produced
the filament wound duct shown in
Figure 5 to a configuration provided by Pratt and Whitney for e-
valuation and testing. The duct
features a filament wound carbon/
PMR-15 shell with integral flanges.
Tape and fabric prepreg materials
were layed up with the shell during winding to provide local reinforcement at the flanges and engine mount ring. The mount ring
and thrust fitting pads were layed
up and cured seperately and then
bonded to the precured duct shell
with FM-35 polyimide adhesive.

In another project a smaller duct
has been filament wound using car-
bon/epoxy material. This duct
shown in Figure 6 also features
integral flanges. In yet another
project, a vibration problem on an
engine duct will be solved without
an increase in weight through the
application of composites.

Figure 5 - Carbon/Polyimide Filament Wound Duct

Figure 6 - Carbon/Epoxy
Filament Wound Duct

## 4. DUCT CONFIGURATION

The first impulse of the engineer assigned to design a composite engine duct is to make it look like its metal predecessor except for the material. We all recognize that effective application of composites must go beyond this "black metal" syndrome in order to achieve the cost and weight benefits available. Features such as end flanges, ports for fuel, air and lubricants, fittings for hardware attachment, engine mounts and axial flanges for maintainance access on split ducts must all be reconsidered with a recognition of the strengths and weaknesses of composite construction.

End flanges present a particularly difficult problem for several reasons. First, fibers which run in the circumferential direction can't be stretched enough to make a typical flange upstanding leg. The simplest solution to this is to have all the fibers in the flange area oriented either radially or on the bias. Another is to use 0°/90° fabric in short enough overlapping segments to allow the weave to be deformed.

The second major problem with end flanges is the difficulty in fabricating and inspecting the 90 degree angle section of laminate. Voids, porosity, and high resin content, all symptomatic of poor compaction, are most likely to occur in this area. At the same time, the bending moments due to load eccentricity at the flange joint are at a maximum. Workable solutions include the use of throw-in blocks and other pressure enhancers during the cure to improve compaction, radius block washers to reduce the bending moments, or giving up some cost and weight to use separate flange details which can be joined to the duct body.

The third problem at the flange is to account for the composite laminate's poor peel, bearing and crushing strength. This requires a complete rethinking of the use of interference fit snap rings and shear pins as well as of the fasteners used to join the flanges of adjacent ducts. Molded snap rings in which the reinforcement is continuous provide one answer where alignment is critical or where adequate sealing cannot be provided by other means. When bolted flanges are used, special attention is required to assure that the bolt preload and working loads are distributed into the composite without causing damage to the area around the bolt hole.

Provisions for penetrations and attachments present another set of design problems. Local reinforcement of the duct shell is generally required either due to disruption of the duct load paths by holes or due to the application of local external loads. These requirements are usually met easily by the addition of local doubler plies of prepreg fabric.

A much more difficult problem is created by the need for threaded fasteners to attach external equipment to the ducts. Maintainability dictates that these fasteners be easily replaceable in the event that they are damaged. While workable solutions such as replaceable grommets installed in the duct are in use, they are very expensive. There is a need for a more cost effective solution to this problem.

Another challenging area is the configuration of "split" ducts, such as the F404 fan duct where the top and bottom halves must be independently removable and replaceable. Constraints include interchangeability, sealing, and high hoop loading. The use of a standard "L" type flange almost demands a fabric layup in order to close out the corners and be able to form the flanges. A filament wound duct with a shear joint of joggled, overlapping elements or a splice strap would be lighter, but interchangeability becomes a problem.

## 5. FINISHING OPERATIONS

Cure of the composite laminate is only the beginning of the engine duct fabrication process. The laminate must then be inspected to assure the structural integrity of the finished part. The most effective non-destructive technique is ultrasonic C-Scan. This provides a "map" of the part with its acoustic response characteristics recorded on a closely spaced geometric pattern. While both pulse echo and through transmission scans are useful, the through transmission technique is better for the detection and quantification of minor defects like porosity. With digital recording and processing of the C-Scan data, corelation with physical standards having known defects provides an accurate

assessment of part strength. Difficulties are still encountered in areas of abrupt contour changes such as flanges, where acoustic coupling between the transducer and the part is disrupted by such things as splashing and non-perpendicular alignment. Coupled with a good process control system and a limited amount of destructive evaluation of coupons and part trim material, current technology does provide for good methods of establishing part structural integrity.

The nature of engine ducts requires that many of their features be machined to assure assembly fit and interchangeability. Special provisions are required for the machining of carbon fiber reinforced composites. Dust is produced during cutting which is electrically conductive as well as environmentally undesirable. Therefore, dust collection and removal systems are required and electrical equipment must be enclosed. The types of cutters and the feeds and speeds used are very different from those used for metals, so machining equipment must be selected with this in mind. In addition, the tolerances and complexities of engine ducts demand a high level of automation in the machining process to avoid sizable scrap and rework costs.

## 6. FUTURE ADVANCES

The use of advanced composites in engine ducts will, over the next few years, result in many changes to the ways in which things are now made in composites. Among the material manufacturers there has been a strong move to explore the use of thermoplastics to replace the thermoset resin matrix materials. Carbon fiber reinforced materials with PPS and PEEK resins are suitable for service to 300°F. The driver in this move is cost. While the materials are still more expensive, the processing of thermoplastics is far simpler and repair techniques require only the local application of heat and pressure for a short time.

Matrix materials for high temperature service (above 500°F) will surely be improved. Processing complexities associated with PMR-15 demand that a better material be developed.

Molded parts with chopped fiber reinforcements will find much broader application in areas where strength is not a primary concern such as boss spacers, fairings, fillers, etc. Such parts provide satisfactory performance at a weight less than metals and cost less than those with continuous reinforcements.

## 7. CONCLUSION

We stand on the threshold of an era when many of the metal parts of jet engines will be replaced by composites. Fan ducts are an ideal place to start. Questions of durability and production repeatability can really only be answered by building a quantity of parts and putting them into service. With the Tay bypass ducts going into service at Rolls-Royce and the F404 fan duct going into production status at G.E., those answers are close at hand. The problems will be resolved and we will have lighter, less costly engines.

## 8. REFERENCES

Elkin, Robert A., "Advanced Composite Engine Static Structures Fabrication," AFML-TR-78-169.

Ziegler, E. Eugene, "F404-GE-400 Graphite/PMR-15 Composite Bypass Ducts," General Electric Co. Report Number R84AEB056.

## 9. BIOGRAPHIES

Terry L. Stockham is a Project Manager for Engine Component Development at Rohr Industries, Chula Vista, California. He received B.S. and M.S. degrees in Engineering from UCLA in 1956 and 1964, respectively. Mr. Stockham has managed a number of projects for the development and production implementation of advanced technology structures for aircraft engines and nacelles. These include titanium honeycomb ducts for the F100 engine, acoustically treated fan ducts and a titanium daisy mixer for the JT8D-209, brazed superalloy nozzle parts and advanced composite cowling.

Gerald A. (Jerry) Lunde is a Senior Project Engineer at Rohr Industries in the jet engines component group of Manufacturing Technology. During his 28 years in the industry he has developed many composites applications including F100 boron polyimide fan blades, graphite/epoxy fan exit guide vanes, and external nozzle flap, augmentor ducts, fan ducts and rotating cowling all of graphite polyimide for various engines.

James W. Davidson is Director of Engine Component Development at Rohr Industries, Chula Vista, California. He received B.S., M.S. and Ph.D. degrees in engineering from UCLA in 1966, 1969 and 1974, respectively. Dr. Davidson has been involved with the development of composite material products since authoring two chapters of the First Edition Air Force Design Guide for Advanced Composite Materials nearly 20 years ago at North American Rockwell. Prior to coming to Rohr Industries, Dr. Davidson has been responsibile for many other significant composite products, most recently having been Director of Research and Development for HITCO Fabricated Composites.

32nd International SAMPE Symposium
April 6-9, 1987

# SEMI-CRYSTALLINE THERMOPLASTIC MATRIX COMPOSITES FOR SERVICE AT 350°F

F N Cogswell        D C Leach          P T McGrail
H M Colquhoun[+]    P MacKenzie        R M Turner
                  ICI Materials Centre
PO Box 90, Wilton                  +PO Box 13, The Heath,
Middlesbrough,                      Runcorn,
Cleveland, TS6 8JE, UK              Cheshire, WA7 4QF, UK

## Abstract

A Thermoplastic Aromatic Polymer Composite with a High Temperature Crystalline matrix - APC(HTX) - has been developed as a structural composite with a service temperature of 350°F. Such temperatures are demanded by the requirements of the next generation of aircraft, in particular the USAF Advanced Tactical Fighter. Building on previous experience a linear chain, semi-crystalline, thermoplastic polymer, HTX, has been developed as the matrix for the composite. The glass transition temperature of HTX is 205°C (401°F) and the polymer has been tailored in terms of mechanical properties, and degree of crystallinity. These properties combined with a controlled fibre/matrix adhesion define the performance of the composite. Evaluations on the composite confirm that it has room temperature mechanical properties equivalent to those of APC-2, including toughness and damage tolerance. In addition the composite shows excellent resistance to aggressive environments. APC(HTX) retains >75% of its room temperature compressive and flexural properties at 180°C (356°F). Initial fabrication studies confirm that APC(HTX) can be formed into complex shapes using the processing techniques being developed for APC-2.

## 1.    INTRODUCTION

Thermoplastic matrix composites are gaining growing acceptance in the aerospace industry as structural materials (1). Foremost in this development has been the family of Aromatic Polymer Composites based on poly(ether-ether-ketone) ('Victrex' PEEK). The use of this high performance, semi-crystalline, thermoplastic polymer as the matrix, combined with controlled interfacial properties gives the composite many attractive features. Notable amongst these are toughness (2), environmental resistance (3) and the potential for a wide range of fabrication technologies (4) leading to reduced manufacturing costs (1).

The upper temperature performance of thermoplastic composites is controlled by the glass transition temperature (Tg) of the polymer matrix and whether it is semi-crystalline or amorphous (5). The Tg of 'Victrex' PEEK is 143°C (289°F) and the composites retain a high proportion of both short- and long-term properties at temperatures up to Tg. Above Tg the amorphous regions of the polymer can move cooperatively and hence there is a reduction in the polymer stiffness. However the crystalline units provide a constraint and hence a proportion of the composite properties are retained at temperatures above Tg. Practical experience suggests that the long-term service temperature of the composite is 25-30°C below the Tg of the matrix (5), and therefore the maximum long-term service temperature for APC-2 is probably in the region of 120°C (248°F).

A number of current applications require higher temperature performance. Notably materials for the proposed USAF Advanced Tactical Fighter (ATF) must retain structural performance at temperatures up to 176°C (350°F) in order to meet the service criteria defined for this aircraft. Initial work in developing thermoplastic composites with increased temperature performance has been presented previously (5). This paper discusses the development of a High Temperature Crystalline (HTX) matrix and the Aromatic Polymer Composites based on it - APC(HTX).

2.    DEVELOPMENT METHODOLOGY

The development of polymers for use in composites has been largely empirical until recently. Polymers were synthesised, their properties evaluated and then the properties of composites based on them evaluated. These properties would be compared with service requirements and the final polymer characteristics would be arrived at through an iterative process.

Improvements in our understanding of the influence of matrix and interface on composite properties, combined with advances in synthetic chemistry have enabled a new methodology to be followed in the development of APC(HTX). This methodology is illustrated in Figure 1. The first requirement was to define the end-use criteria of the composite through discussion with the aerospace industry. The composite service criteria that were established are summarised in Figure 2. The major aims were to retain the room temperature performance of APC-2, and to extend the service temperature to 350°F whilst still retaining the ability to fabricate parts rapidly using the techniques that are being developed for APC-2. These objectives represent a formidable scientific challenge and therefore required careful definition of polymer and interfacial requirements to achieve the correct balance of properties in the final composite.

The characteristics of the interface between the matrix and reinforcing fibre control the properties of the final composite. Therefore designing that interface is an integral part of the composite structure. This aspect has recently been discussed for thermoplastic composites based on 'Victrex' PEEK (6) and from this experience it is clear that for tough matrix systems the interface must be tailored to give the optimum fibre/matrix adhesion characteristics.

The influence of the matrix on composite mechanical properties has received considerable attention and basic guidelines are now established (7). In certain respects the room temperature properties of 'Victrex' PEEK exceed the requirements for a composite matrix, however achieving room temperature mechanical properties equivalent to those of 'Victrex' PEEK is a relevant target for the new matrix. To meet the service temperature of 176°C (350°F), the experience with APC-2 indicated that a polymer with a Tg of at least 200°C (392°F) was required. It is essential that composites used in aircraft primary structural applications have excellent resistance to service environments and hence a semi-crystalline thermoplastic was preferred. Meeting the demands of reduced manufacturing costs requires rapid fabrication and versatility in fabrication techniques. To achieve these targets we considered that it was essential to have true thermoplastic characteristics in which the processing cycle only involves the melting of the polymer, consolidation/ shaping, and then cooling under a variety of conditions to give the required microstructure.

From the polymer physical properties it was then possible to establish the fundamental polymer characteristics such as molecular sequences and molecular weight. The monomers and polymerisation conditions were identified and the polymer was made. The properties of the polymer and of composite made from it were then compared to requirements and the composite properties optimised through fine tuning of the polymer and interfacial characteristics.

The final aspect which should not be under-estimated, is the ability to scale-up the manufacture of polymer and composite to give a reproducible structural material. In this area ICI has considerable experience in aromatic polymers and composites based on them. This experience has been essential in developing a new material within the tight timescales dictated by the ATF project.

3.     POLYMER CHARACTERISTICS

In developing the HTX polymer it was necessary to control:

    Glass transition temp (Tg)
    Crystallinity
    Melt temperature (Tm)
    Viscosity
    Toughness
Crystallinity, Tg and Tm are controlled by the basic molecular structure of the polymer and hence it is necessary to understand the factors which influence this. That understanding underpins the whole range of 'Victrex' aromatic polymers (5). The

polymer properties can be controlled by the selection of suitable linking units between the aromatic rings and by deliberate control of the resulting sequences produced in the polymer chain.

In order to ensure that the polymer is capable of crystallising it is necessary to minimise the presence of bulky side groups and to have precise control over the sequences along the linear polymer chain. These sequences are achieved typically by reacting two or more monomers which contain the required units. To achieve the best possible molecular architecture a new monomer with the optimal configuration has been developed specifically for use in HTX.

The viscosity and toughness of the polymer are also influenced by the polymer structure but are also very strongly affected by the molecular weight of the polymer and the molecular weight distribution. A balance is usually required to give a polymer which has the required toughness but can still be readily processed.

4.     HTX POLYMER THERMAL
       PROPERTIES

HTX polymer is a semi-crystalline, aromatic thermoplastic. Thermal properties of polymers are commonly characterised by

Differential Scanning Calorimetry (DSC) and DSC scans of HTX are shown in Figure 3. The onset of the glass transition by DSC is at 205°C (401°F), that is 29°C (52°F) above the designated service temperature requirement. The specimen used for the heating scan had been annealed at 300°C to fully crystallise the polymer, and this shows as an annealing peak at 318°C (604°F). Two melting peaks are observed, the dominant one being at 365°C (689°F). The second scan shows a cooling curve on HTX polymer. The crystallisation exotherm is clearly seen, the minimum in the process (Tc) being at 285°C (545°F). The Tg process can be seen on both the heating and cooling scans. The polymer develops up to 25% crystallinity, the exact value depending on the cooling conditions used. The mechanical properties of the polymer are very similar to those of 'Victrex' PEEK and will be discussed in more detail in another paper.

## 5. APC(HTX) PROCESSING AND FABRICATION

### 5.1 Processing

APC(HTX) prepreg can readily be processed into panels using similar conditions to those developed for APC-2. The conditions for consolidation in a platten press are:-

Temperature : 400-410°C (752-770°F)
Pressure : 200-250 psi
Time : 5 minutes

Where a compliant layer is present, such as in autoclaving or diaphragm forming, then lower pressures can be used. During processing, the prepreg may be heated up at any rate and cooling rates of <20°C/minute are required to develop adequate crystallinity. If faster cooling rates are required then the full crystallinity can be achieved by a short, free-standing anneal at 300°C (572°F).

### 5.2 Fabrication

Initial fabrication experiments on APC(HTX) have been carried out using the diaphragm forming process (3). A moulding made from APC(HTX) using this process is shown in Figure 4. The conditions used to fabricate this component were:-

Forming Temperature: 400°C (752°F)
Pressure Differential: 120psi
                                    + Vacuum
Forming Time: 26 mins

These conditions are very similar to those recommended for APC-2. The component shown in Figure 4 is a complex, double-curvature shape and is much more demanding to fabricate than most practical components. That APC(HTX) can be fabricated into such a complex shape confirms its potential for fabrication using the techniques developed for APC-2.

## 6.    APC(HTX) PROPERTIES

### 6.1    Basic Mechanical Properties

The room temperature mechanical properties of APC(HTX) composite are almost identical to those of APC-2. Table 1 shows a comparison of properties for the two composites, both reinforced with Hercules 'Magnamite' AS4 carbon fibres. It can be seen from these results that the original targets of achieving equivalence between APC(HTX) and APC-2 in terms of utilisation of the fibre and in compressive strength have been achieved.

### 6.2    Toughness

The toughness has been assessed using a double cantilever beam (DCB) test to measure inter-laminar fracture toughness $(G_{1c})$ and using a damage tolerance test to measure compressive strength after low energy impact. These two tests give information on the basic fracture toughness characteristics and the practical resistance to damage of the material. Results from the two tests are shown in Table 2. The interlaminar fracture toughness of APC(HTX) is similar to that of APC-2. The high toughness of APC(HTX) is due to the ductility of the polymer and the strong fibre-matrix adhesion. This is illustrated in Figure 5 which shows a scanning electron micrograph of one of the DCB fracture surfaces. The polymer adheres strongly to the fibres and ductility of the matrix can be clearly seen. The damage tolerance test was conducted with an impact energy of 1500 in-lb/in. The high interlaminar toughness is reflected in the high residual compressive strength after the low energy impact, giving APC(HTX) excellent damage tolerance.

### 6.3    Environmental Resistance

The environmental resistance of APC(HTX) has been examined using a very aggressive environment (dichloromethane) for a short exposure time and a much less aggressive environment (boiling water) until saturation. For these experiments a transverse flexural test was used since, in our experience, this test is the most sensitive to any weakness of the resin or failure of the interface. These results are summarised in Table 3. The excellent solvent resistance is defined by the semi-crystalline nature of the polymer which characteristically gives resistance to environments, and through the controlled fibre-matrix interface.

## 6.4    Effect of Temperature on Mechanical Properties

One of the objectives for APC (HTX) is that the composite should retain adequate properties at the anticipated service temperature of 350°F (176°C). In discussion with the aerospace industry the most important property emerged as compressive strength at the service temperature, and 'adequate' retention is normally taken as in excess of 65% of the room temperature value. The compressive and flexural properties of APC (HTX) at 180°C (356°F) are summarised in Table 4. The anticipated effect of the 205°C Tg is indeed observed with 78% of the room temperature compressive strength and 75% of the flexural strength being retained at 180°C. This confirms that APC (HTX) retains the required level of the critical engineering properties at the anticipated service temperature.

## 7.    CONCLUSIONS

The development of APC (HTX) has been drawn by the current requirements for higher temperature composites which also combine good damage tolerance, environmental resistance and potential for reduction in manufacturing costs. This target represents a major technical objective and to meet it a new semi-crystalline thermoplastic polymer, HTX, has been developed to provide a matrix with the required balance of properties. The development of APC (HTX) has built on previous experience with aromatic thermoplastics and through a proprietary impregnation process. The properties of the polymer combined with the control of the interface give APC (HTX) the required property spectrum. Evaluations on the composite confirm that it combines the key aspects of good damage tolerance and resistance to aggressive environments with a high retention of properties at the 350°F service temperature. In addition APC (HTX) can be readily fabricated into complex shapes using the techniques developed for APC-2. The results presented here show that APC (HTX) meets the demanding service requirements of the next generation of military aircraft.

## 8.    ACKNOWLEDGEMENTS

We should like to thank our many colleagues in ICI's Materials Centre, APC Development Group, Chemicals and Polymers Group, Organics Division and Fiberite who have helped in the development of APC (HTX).

## 9.   REFERENCES

1  S Christensen, L P Clark, H Wu "Thermoplastic Composites for Structural Applications - An Emerging Technology", 31st SAMPE Symposium, p1747-1755, April 1986.

2  S L Donaldson "Fracture Toughness Testing of Graphite/ Epoxy and Graphite/PEEK Composites", Composites 16,2, p103, 1985.

3  A J Barnes, J B Cattanach "Advances in Thermoplastic Composite Fabrication Technology", 'Materials Engineering' London, UK November 1985.

4  W J Horn, F M Shaikh, A Soeganto "The Degradation of the Mechanical Properties of Advanced Composite Materials Exposed to Aircraft In-Service Environment", AIAA Conference, p353-361, 1986.

5  D C Leach, F N Cogswell, E Nield "High Temperature Performance of Thermoplastic Aromatic Polymer Composites", 31st SAMPE Symposium, p434-448, April 1986.

6  F N Cogswell, R M Turner "The Effect of Fibre Characteristics on the Morphology and Performance of Semi-Crystalline Thermo-plastic Composites", 18th SAMPE Technical Conference, October 1986.

7  N J Johnston "Synthesis and Toughness Properties of Resins and Composites" ACEE Composite Structures Technology Conference, Seattle, August 1984.

## 10.   BIOGRAPHIES

### F N Cogswell

After completing his education with the Royal Marines, Neil Cogswell joined ICI Plastics Division in 1959. He is at present a Senior Research Associate at the ICI Materials Centre, is an Honorary College Fellow in the Department of Applied Mathematics at the University College of Wales, and is the author of "Polymer Melt Rheology : a guide for industrial practice"

### H M Colquhoun

Howard Colquhoun graduated from the University of Cambridge in 1972. He was awarded a PhD from the University of London in 1975 for work on metasilanes and continued research on this topic as a Research Fellow at the

University of Warwick. He
joined ICI in 1977 to work
on co-ordination chemistry
applied to organic synthesis
and has been working on aromatic
monomer and polymer chemistry
since 1982.

## D C Leach

David Leach graduated from
Imperial College, University
of London 1979 and joined
ICI Plastics Division in the
same year. He has carried
out research on the physical
properties of engineering
polymers and their composites
and has been closely involved
in the development of continuous
reinforced thermoplastics.
He is currently leading the
ICI high temperature composite
development programme.

## P MacKenzie

Paul MacKenzie graduated from
Salford University in 1981
and went on to study for a
PhD in organic chemistry at
King's College, University
of London, which he obtained
in 1984. In that year he
joined ICI New Science Group
and is carrying out research
and development on polymer
composite matrices.

## P T McGrail

Dr P T McGrail is Work Group
Leader of the Composite Matrices
and Interfaces Group with ICI
New Science Group. He read
chemistry and completed his
PhD at Sheffield University.
He has extensive experience
in the areas of polymer chemistry
and film manufacture and coating.
Recently he has been specialising
in the development of high
performance polymers for both
thermoplastic and thermosetting
composite matrices, in conjunction
with studying the interfacial
science of such composite materials.

## R M Turner

Dr Roger Turner joined ICI
in 1968 having read Physics
at Birmingham University.
He has been involved in research
on short and continuous fibre
composites, in particular on
the processing and mechanical
properties of the materials.
He is currently a Senior Research
Scientist in the ICI Materials
Centre.

# Room temperature mechanical properties of APC (HTX)

| Property | Units | APC(HTX)/ D/AS4 | APC-2/ AS4 |
|---|---|---|---|
| Fibre volume | % | 61 | 61 |
| Longitudinal Tensile Modulus | GPa msi | 132 19.1 | 134 19.4 |
| Longitudinal Tensile Strength | MPa ksi | 1984 288 | 2130 309 |
| Longitudinal Flexural Strength | MPa ksi | 1775 257 | 1880 273 |
| Longitudinal Compressive Strength (IITRI Test) | MPa ksi | 1136 165 | 1100 160 |

Both composites reinforced with Hercules 'Magnamite' AS4 carbon fibre

Table 1


# Toughness properties of APC(HTX)

**Double Cantilever Beam Test**
Interlaminar fracture toughness $(G_{1c})$ 2.2 kJ/m$^2$
12.5 in-lb/in$^2$

**Damage Tolerance Test**
Impact energy 1500 in-lb/in
Residual compressive strength     301 MPa
43.6 ksi

Table 2

391

# Environmental Resistance of APC(HTX)

| Environment/ Exposure Conditions | Weight Change % | Transverse Flexural Strength | | Retention % |
|---|---|---|---|---|
| | | MPa | ksi | |
| Water/14 days/100°C | +0.29* | 93 | 13.5 | 102 |
| Dichloro-methane/ 24 hours/23°C | +0.23 | 95 | 13.8 | 105 |

*saturated

Table 3


# Effect of Temperature on Mechanical Properties of APC(HTX)

| Property | Units | Temperature | | Retention % |
|---|---|---|---|---|
| | | 23°C/73°F | 180°C/356°F | |
| Longitudinal Compressive Strength (ASTM D-695 Test) | MPa ksi | 1132 164 | 879 128 | 78 |
| Longitudinal Flexural Modulus | GPa msi | 131 19.0 | 131 19.0 | 100 |
| Longitudinal Flexural Strength | MPa ksi | 1775 257 | 1330 193 | 75 |

Table 4


# Performance Targets for APC(HTX)

**Room Temperature Properties Equivalent to APC−2**
- Full utilisation of Fibre Properties
- Good Damage Tolerance/Toughness
- Good Compressive Strength
- Excellent Solvent Resistance
- Strong Fibre/Matrix Interfacial Adhesion

**Service Temperature of 350F/176°C**
- Minimum Creep/Relaxation at 176°C
- Retention of Adequate Compressive Strength at 176°C
- Minimum changes in all other Properties in 23-176°C range

**Processing/Fabrication**
- To be Processed & Fabricated using techniques developed for APC-2 at 400-420°C

Figure 2

# APC(HTX) Development Methodology

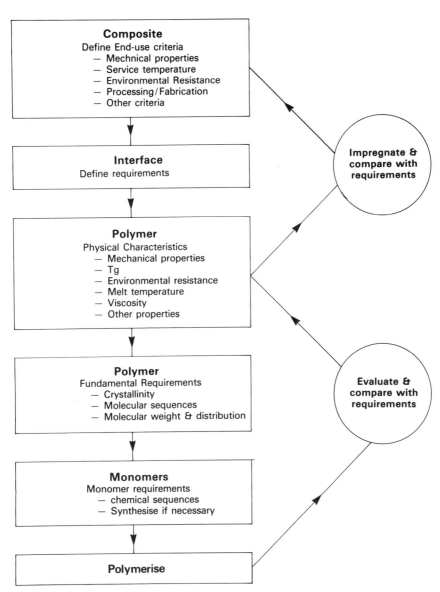

Figure 1

# DSC Scans on HTX Polymer

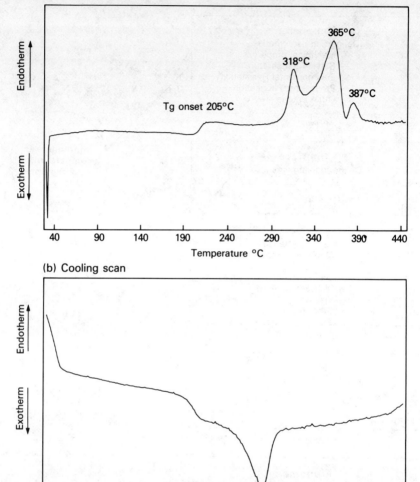

(a) Heating scan on polymer - annealed at 300°C

Endotherm

Exotherm

365°C

318°C

387°C

Tg onset 205°C

Temperature °C

(b) Cooling scan

Endotherm

Exotherm

285°C

Temperature °C

Figure 3

## APC(HTX) Moulding fabricated using the Diaphragm forming process

Figure 4


## APC(HTX) Interlaminar Fracture surface (Double cantilever beam test)

Figure 5

32nd International SAMPE Symposium
April 6-9, 1987

# EVALUATION OF PEEK MATRIX COMPOSITE

*T. Nagumo, *H. Nakamura, *Y. Yoshida and **K. Hiraoka

* Aircraft Engineering Division
Fuji Heavy Industries, Ltd.
Utsunomiya, Tochigi, Japan

**Technology Engineering Department
Japan Aircraft Development Corporation
Toranomon, Tokyo, Japan

## ABSTRACT

Recently thermoplastic matrix composite is one of the most attractive materials becaqse cost reduction will be expected and impact resistance is much more superior to conventional composite base on epoxy.
This paper reports the evaluation results of mechanical properties, and feasibility of PEEK Matrix composite for aircraft structual applications. In this evaluation, basic mechanical test of ICIS APC-2 panel and our panel laminated from APC-2 prepreg were carried out, and toughness of both panels were confirmed. Matched die forming process could be established, however we found more investigation and improvement would be necessary. New bonding test results gave a good bond strength about equivalent to adhesion, and suggested us to apply for PEEK matrix composite structures.
Finally experimental small size door structure was fabricated, which provided us the estimation of cost reduction ratio, and totally PEEK matrix composite has a sufficient potential to apply aircraft structures.

## 1. INTRODUCTION

Along with the progress and increment of composite Technologies, it is important that the weight reduction with composite materials and the analysis subjected to impact apply to the development of aircraft structures. While, in many material suppliers, conventional thermoset matrix composite, such as epoxy, bismaleimide and imide are improving to pocess the properties of high service temperature, high toughness and sufficient processability at the same time in order to keep up with some needs of manufacturing companies. Subsequent these conditions, in addition to general thermoplastics, the emerging of high temperature thermoplastics are considered to gradually become the turning point in the movement of composite applications. Because the thermoplastic matrix composite has many fascinated properties compared with the thermoset matrix composite. For example, the new thermoplastic matrix composite has not merely to reduce the manufacturing cost but also to achieve the high compression strength after impact or durability against exposed hot-wet enviroment considered to be the most severe evaluation. In this study, we selected the ICIS polyetheretherketone (PEEK) matrix resin as a superior material is toughness, as well as moisture and solvent resistance out of several commercial thermoplastics. Therefore, in oder to evaluate and grasp the PEEK matrix composite properties for actual application of aircraft structures, the

evaluations were carried out
and discussed the results.

2. EXPERIMENT

2.1 Mechanical evaluation.

Test specimens were prepared from
ICIs laminate panels and our own
manufacturing panels from APC-2
prepreg.
APC-2 prepreg was 145g/m² areal
weight, using AS-4 12K graphite
fibers. Test methods were mainly
according to ASTM and NASA 1092
Reference Publication procedure,
especially, toughness evaluation
method applied $G_{1C}$ and compression
after impact. And tests were carried
out in room temperature, 83 ℃ and
maximum 0.2 % moisture pick up
condition.

2.2 Press forming.

Preliminary forming evaluations were
carried out to estimate the optimum
condition during the forming of
access door illustrated in Fig.1.
Futher, several parameters of fiber
orientations and mold materials were
varied as shown in Table 1.
Plied fabric laminate were the press
formed with PEEK film and TOHO 8
harness HTA graphite fabric by Mitsui
Toatsu Chemicals Inc. Metal molds were
carbon steel SAE 1055, and rubber
molds were fluorine-contained rubber.

2.3 Radip bonding

Three rapid methodologies (Heating,
Vibration, Induction ) were selected
other than conventional adhesion
method.
Bonding concrete procedure is
illustrated in Fig.2.
In order to evaluate the bonding
quality, shear strength was measured
according to ASTM D1002.
Adhesion specimens were applied
surface treatment of sodium
dichromate-sulfuric acid etch and
were bonded by using adhesive film
3M AF163-2K.
Heating bond was carried out by
weld instrument made by ourselves.
Vibration bond was applied branson
vibration weld instrument
MODEL 2100 being able to vibrate
at 100HZ.
Induction bond was carried out by
applying stainless mesh 100
between test specimens with
Seidenshadenshi Kogyo UH-7K
being able to induced 400KHZ
electromagnetic field.

3. RESULTS AND DISCUSSION

3.1 Basic mechanical property
    evaluation of APC-2.

The results of basic mechanical
porperty test are shown in
Fig.3 (a), (b), (c) . Totally,
both panels properties were
equivalent at room temperature
and 83℃, WET condition, and further,
these properties were similar to the
conventional epoxy matrix composites
mechanical property used graphite
fibers of 360kg/mm² tensile strength.
However some deteriorations of
strength were observed in 83 ℃,
83 ℃ WET conditions, this extent
of strength reduction were not
noticeable compared with the
conventional epoxy matrix composite.
It is considered that PEEK resins
moisture pick up is only about
0.2% of composite in saturation,
and this low moisture pick up is
advantageous against the mechanical
degradations. While, in this result,
±45° tensile strength result was
different in ICIs panel and our
panel. Our panels strength was
about 10% lower. We analysed
this causes, our panel had the
area of graphite fiber waviness
on the surface after press forming,
because melting PEEK matrix resin
viscosity is never reduced less
than 10 poise and likely to move
with reinforced graphite fibers
during press forming at 360～400℃.
On the other hand, APC-2 panels
were about 10 times superior to
conventional epoxy matrix composite
and $G_{1C}$ fracture surface showed
the slow and fast crack areas in
the same specimen.
Therefore, we expected the high
compression strength after impact
and could recongize why APC-2 is
tough by SEM observations of
sticky fracture without brittle
fracture as shown in thermoset
matrix composites. Compression
after impact results also showed
the high toughness as being unable
to expected in thermoset matrix
composites. APC-2 strength
achieved about 30 kg/mm² after
1000LB IN/IN impact while the
conventional epoxy matrix composite
was about 14kg/mm² in our experiences.
According to the observation of
Fig.4, it is realized the crack
length is apparently shorter
after compression fracture and
the damage area is not so extented
during the impact loading.

## 3.2 Press-forming test of access door skin structure

In this study, we used 8 harness plied fabric laminate from our experiences of the epoxy prepreg laying, because it had been more effective to be able to attach to complex molds in laying up.
Table 2 summarizes the result of press formings. We settled 340 ~ 350℃ of preliminary heating temperature through press forming of 0° fabric laminate. The temperature of 340~350℃ seemed to be appropriate condition since the burned signs had been observed on the PEEK resin at more than 360 ℃ for about half hour. Mold temperature were carefully determined. We selected the extent of 310~ 320 ℃, so that a proper softness was obtained and mold (female and male) could attach uniformly including soften laminate without any bridges phenomena by occuring rocal rinkles. There existed a tendency that the appearance and quality of press forming get better as the mold temperature approaches to the extent 340~350℃. At more than melting point temperature, we found that it would be difficult to remove the formed parts and would not be economical. Because the molds have stuck together too well to be torn off when the majority of PEEK resin were flowed out of fabric laminate. High pressure was necessary to prevent from generating the inter layer voids with rinkle as reducing the settled temperature of the mold. It found the maximum pressure 54~ 56kg/cm² of our facility gave a good quality. Further, we tried to prevent from occuring the rinkle by three steps gradually press, so that the shear movement of inter layer not recognized but this three steps was useful to eliminate the fiber break of parts corners. In another method, we made the hole at the center of surrounding bead and applied the much of release agent in order to prevent from breaking the fibers. During these trial, we found the release agent could make the laminate move smoothly during pressing. On the other hand, the evaluation of 45° and 45°/0₂/45° fabric laminate had also same results except of the location of rinkle area (Table 3 ). But 45°/0₂/45° fabric laminates had about twice numbers of 45° and 0° fabric laminates rinkle respectively. In addition to these, unidirectional tape laminate suggested to establish a good quality without any rinkles. Unidirectional tape will be effective to move the inter layers at high temperature. Mold material evaluation results summarize in Table 4. We found rubber molds are required more stiffness and durability at the processing temperature. It will be necessary to develop the new mold rubber materials.

## 3.3 Development of new rapid bonding methodology.

The test results of four methods (Adhesion, Heating, Vibration, Induction) are shown in Fig.5 and Table 5. Fig.5 is the result of shear strength meaurments. This result provided following superiority, Adhesion >Heating > Induction >Vibration. In adhesion, the result of more than 3.0 kg/cm² depended essentially on the surface treatment of sodium dichromate surfuric acid. We confirmed the improved surface condition by SEM observation. Heating bond had been considered to be the most basic and reliable method. The result showed the highest strength as we expected in three methods other than adhesion. But it was observed the burned PEEK resin around the contacting area of heating press. So this rapid heating system will be necessary to improve and develop. Induction bond are unique method compared with the conventional adhesion. The stainless steel mesh is induced 400 KHZ electromagnetic wave and heated, so that PEEK resin is melted rapidly. However, the debonding area were observed between stainless steel wire and PEEK resin by SEM observation. So, the strength will increase more by improving of stainless mesh surface treatment. In vibration bond, the strength resulted the lowest in these method. By our investigation, this cause was the ramdom failure of the graphite fiber during the vibration friction between specimens. It will be difficult for these composites to be applied the structure bondings. Through above results, we summarize in Table 5 each bonding methodology. If these rapid bonding can replace with the convetional adhesion, we considered the following

398

superiority in points of cost,
utility, performance and prospect.
Induction>Heating >Vibration.

### 3.4 Study of manufacturing process and cost saving.

Finally, we challenged to fabricate
the small access door with some
knowledges of processability (Fig.6).
We could confirm the completion
of this door consists with our
designing specification without
any deflections.  In the fabrication,
we tried to apply the rapid bonding
such as induction, heating.
However, we had failed.
By our investigations, it was
necessary to develop the original
facility in oder to apply the
bonding of inner skin to outer skin
in fabrication.
Therefore, we fabricated the access
door using adhesion (Fig.7), we
could predict the bonding facility
will become a important subject
in fabrication.  We discussed the
reduction cost of this access door
based on the established process
in this study.  Fig.8 shows the
result of calculated cost reduction
rate in comparison with epoxy
matrix composite.
The only 3 % cost reduction was
established against our expecta-
tion. In our analysis, the price
of PEEK matrix composite became
about double of epoxy matrix
composite, and on the contrary,
the labor cost became about one
half.  So it is obviously desirable
to establish new low cost materials
along with the increment of a demand.

### 4. CONCLUSIONS

We conclude that PEEK matrix
composite provided us the promising
properties about mechanical,
rapid bonding, press forming and
cost reduction in this study, and
it requires more investigations of
forming and various bonding
technologies to apply the practical
aircraft composite structures.
It is desired thermoplastic
matrix composites including PEEK
should be the material being
more easy to process with high
toughness and high service
temperature at low cost.

### 5. ACKNOWLEDGEMENTS

This work reported here was
sponsored by Japan Aircraft
Development Corporation in
Tokyo, Japan.
The authors wish to thank
Mitsui Toatsu Chemical Inc,
Branson Ultrasonic Corporation
Japan and Seidensha Denshi Kogyo
Inc.

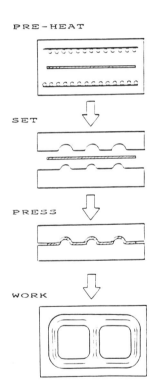

PRE-HEAT

SET

PRESS

WORK

Fig.1 forming Flow

# Table 1 Press-forming Test Condi-tions of PEEK Matrix Composite

| CON-DI-TION NO. | PRE-HEAT TEMP. [°C] | | MANDREL TEMP. [°C] | | | | PRESSURE [kg/cm²] | | | PRESS SPEED [cm/min] | | |
|---|---|---|---|---|---|---|---|---|---|---|---|---|
| | 340~350 | 360~370 | 200 | 250 | 300 | 350 | 30 | 40 | 55 | SLOW 1,STEP | SLOW 3,STEP | FAST |
| 1 | ◯ | | | ◯ | | | ◯ | | | ◯ | | |
| 2 | | ◯ | ◯ | | | | ◯ | | | | | ◯ |
| 3 | ◯ | | | ◯ | | | ◯ | | | | | ◯ |
| 4 | ◯ | | | | ◯ | | ◯ | | | | ◯ | |
| 5 | ◯ | | | | | ◯ | | ◯ | | | ◯ | |
| 6 | ◯ | | | | | ◯ | | | ◯ | | ◯ | |

ORIENTATION
 FABRIC
  (0°)₄
  (±45°)₄
  (45°/0°₂/45°)
 TAPE
  (±45°₂/90°/0°)s

MOLD
 METAL-METAL
 RUBBER-METAL
 METAL-RUBBER

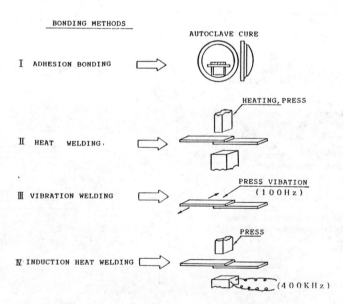

BONDING METHODS

Ⅰ ADHESION BONDING ⟹ AUTOCLAVE CURE

Ⅱ HEAT WELDING ⟹ HEATING, PRESS

Ⅲ VIBRATION WELDING ⟹ PRESS VIBATION (100Hz)

Ⅳ INDUCTION HEAT WELDING ⟹ PRESS (400KHz)

# Fig.2 Bonding Methodology of PEEK/Graphite-PEEK/Graphite

## TENTION-0°Direction

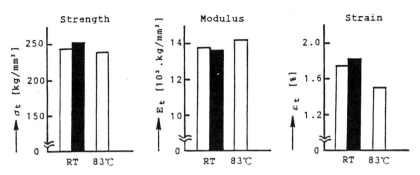

## TENTION-90°Direction

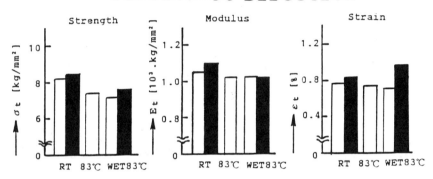

## TENTION-45°Direction

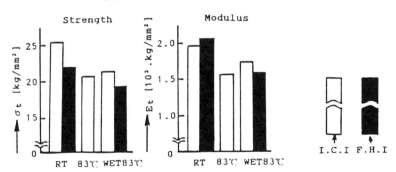

Fig.3(a) Mechanical Properties
of PEEK Matrix Composite

## COMPRESSION-0°Direction

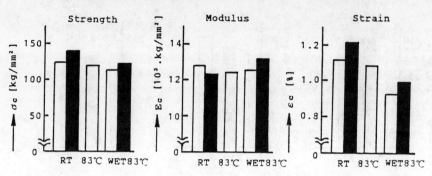

## COMPRESSION-90°Direction

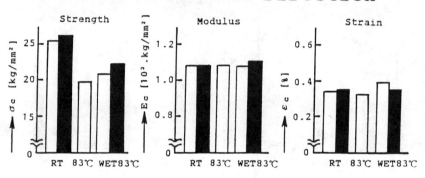

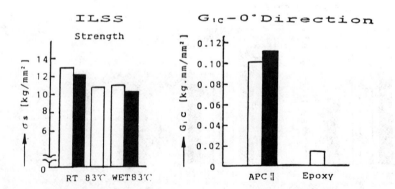

# Fig.3(b) Mechanical Properties of PEEK Matrix Composite

## CAI–[+45°/0°/–45°/90°] Direction

**Strength**

$\sigma_c$ [kg/mm²]

30

20

10

0

1000 1500
LB.IN/IN LB.IN/IN

**Modulus**

$E_c$ [10³·kg/mm²]

6.0

5.0

4.0

0

1000 1500
LB.IN/IN LB.IN/IN

**Strain**

$\varepsilon_c$ [%]

0.6

0.5

0.4

0

1000 1500
LB.IN/IN LB.IN/IN

Fig.3(c) Mechanical Properties
of PEEK Matrix Composite

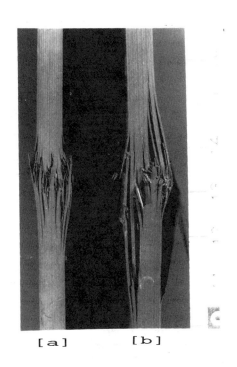

[a]          [b]

Fig.4 Failure Mode differences
between PEEK Matrix Composite[a]
and Epoxy Matrix Composite[b]

## Table 2 Press-forming Conditions

| ITEM | CONDITION |
|---|---|
| PREHEAT TEMP. | $340 \sim 350\,^{\circ}\mathrm{C}$ |
| MOLD TEMP. | $310 \sim 320\,^{\circ}\mathrm{C}$ |
| PRESSURE | $54 \sim 56\,\mathrm{kg/cm^2}$ |
| FORMING RATE | SLOW 3STEP |
| HOLE | NONE |
| RELEASE AGENT | APPLY |

## Table 3 Effect of Ply Orientation on Press-forming

| ORIENTATION | RESULT |
|---|---|
| FABRIC $(0^{\circ})_4$ | G |
| FABRIC $(\pm 45^{\circ})_4$ | G |
| FABRIC $(45^{\circ}/0^{\circ}_2/45^{\circ})$ | F |
| TAPE $(45^{\circ}_2/90^{\circ}/0^{\circ})s$ | G |

G:GOOD  F:FAIR

## Table 4 Effect of Mold on Press-forming

| MOLD (MALE-FEMALE) | RESULT |
|---|---|
| METAL-METAL | G |
| RUBBER-METAL | P |
| METAL-RUBBER | F |

G:GOOD  F:FAIR  P:POOR

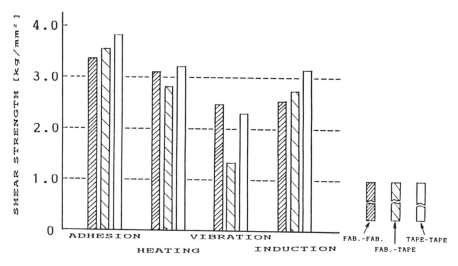

Fig.5 Shear Strength in the four
different Bonding Methods

Table 5 Comparison of the four
different Bonding Methods

| ITEM METHOD | COST | | UTIL-ITY | PERFORMANCE | | | PROS-PECT | TOTAL |
|---|---|---|---|---|---|---|---|---|
| | FABRI-CATION | JIGS | | APPEARANCE | STRENGTH | MORPHOLOGY | | |
| ADHESION | F | E | E | E | E | E | G | G |
| HEATING | G | E | G | P | G | E | G | F |
| VIBRATION | E | G | F | G | P | G | F | P |
| INDUCTION | E | G | G | G | F | E | E | E |

E : EXCELLENT
G : GOOD
F : FAIR
P : POOR

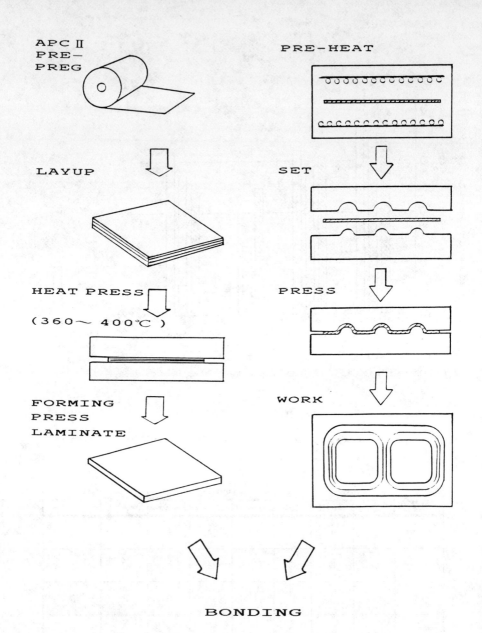

APC II
PRE-
PREG

PRE-HEAT

LAYUP

SET

HEAT PRESS

PRESS

(360～400℃)

FORMING
PRESS
LAMINATE

WORK

BONDING

Fig. 6 Manufacturing Flow of Door Structure

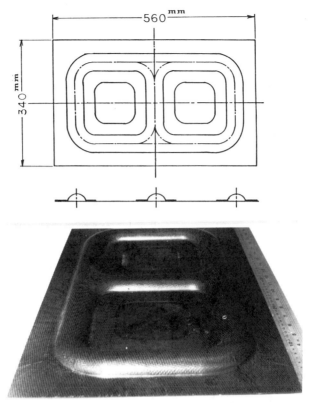

Fig.7　　　Door Structure

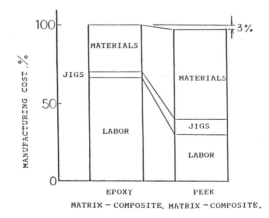

Fig.8 Comparison of Manufacturing
Cost

KETIMINE MODIFICATIONS AS A ROUTE TO NOVEL AMORPHOUS AND DERIVED
SEMICRYSTALLINE POLY(ARYLENE ETHER KETONE) HOMO- AND COPOLYMERS

D. K. Mohanty, R. C. Lowery, G. D. Lyle and J. E. McGrath*
Department of Chemistry and Polymer Materials and Interfaces Laboratory
Virginia Polytechnic Institute and State University
Blacksburg, VA 24061

*To whom correspondence should be addressed.

## Abstract

A series of amine terminal amorphous poly(arylene ether ketone) oligomers of controlled molecular weights (2-15 K) were synthesized. These oligomers have been found to undergo "self crosslinking" reactions upon heating above 220°C, via the reaction of the terminal amine groups with the in chain keto carbonyl functionalities. The resulting networks are ductile, chemically resistant and nonporous. The networks obtained via generated ketimine functionality were characterized by solid state NMR. They have also been found to be remarkably stable towards hydrolysis. Ketimine functional bishalide monomers have also been synthesized. Such monomers have been utilized to synthesize a wide variety of amorphous poly (arylene ether) ketimine polymers. A high molecular weight hydroquinone functional poly (arylene ether) ketimine has been acid treated to regenerate a poly(arylene ether ketone) backbone in solution. This novel procedure thus allows for the synthesis of important matrix resins under relatively mild conditions.

## 1. INTRODUCTION

Amine terminal polyarylene ether sulfones of controlled molecular weights have been synthesized in this laboratory and elsewhere[1,2]. Such oligomers have been successfully utilized as a crosslinking and toughening component of epoxy resins, bismaleimide matrix[3,4], and epoxy-carbon fiber composites[5]. Thus, one of the objectives of the present investigation was to synthesize a series of amine terminal poly (arylene ether ketone) oligomers of controlled molecular weight and to utilize these oligomers as a reactive component in the epoxy and BMI resin systems. This was of particular interest since fracture toughness values obtained for amorphous poly(arylene ether) ketones are significantly higher than that of the corresponding poly(arylene ether sulfone)[6]. However, this increase is accompanied by ~30°C reduction in the glass transition temperature of the ketone relative to the sulfone.

A series of oligomers of controlled molecular weight using bisphenol-A, 4,4'-difluorobenzophenone and p-amino bisphenol A have been synthesized using an NMP/Toluene/$K_2CO_3$ reaction system. Their general structure is listed in Scheme 1.

Surprisingly, we found that amine terminal oligomers of 5000 up to 15000 molecular weight can be transformed into insoluble, tough ductile non-porous plaques when heated above 220°C for at least two hours.

A part of the present discussion will focus on the chemistry and phenomena associated with "self crosslinking" behavior.

The other related aspect of our work reported herein is to describe a novel procedure for the synthesis of poly(arylene ether ketones). In an earlier publication, we had outlined a synthetic procedure for an amorphous copolymer with a pendant t-butyl group. The bulky t-butyl group could be partially cleaved via post reaction to yield a semicrystalline poly(arylene ether ketone) backbone structure[7]. Our synthetic strategy remained essentially the same, namely to synthesize an amorphous high molecular weight polymer in solution at a relatively low temperature and to convert it to a semicrystalline backbone via post reactions. We report herein the successful synthesis of an all aromatic ketimine functional bisfluoride monomer and subsequent polymerization with hydroquinone in an NMP/Toluene/$K_2CO_3$ reaction system. The very high molecular weight ductile polymer thus produced has been post treated with acid to yield a crystallizable poly(arylene ether ketone) backbone.

## 2.    EXPERIMENTAL

### 2.1  Materials

Bisphenol A, supplied kindly by Dow Chemical Co., was vacuum dried at 60°C for 24 hours prior to use. Polyester grade hydroquinone provided by Eastman Kodak Chemical division was crystallized from deoxygenated acetone. The bisfluoride, 4,4'-diflurobenzo-

phenone (Aldrich) was crystallized from boiling ether (m.p. = 105°C). The bis(4,4'-chlorobenzoyl) benzene (CBB) was synthesized from tere-phthaloyl chloride, chlorobenzene and aluminum chloride[8]. Aniline (Fisher) was distilled over calcium hydride ($CaH_2$) under reduced pressure. Toluene (Fisher) was used as received. N-Methyl pyrolidone (NMP) and N,N-dimethyl-acetamide (DMAC) purchased from Fisher were vacuum distilled over $CaH_2$. Anhydrous $K_2CO_3$ (Fisher) was dried overnight at 100°C was powdered in a blender. Molecular Sieves (3Å) (Fisher) were used as received without prior thermal treatment. Trifluoromethane sulfonic acid was purchased from Aldrich. The "end blocker" p-amino-bis A synthesized according to a previously described procedure[9] was crystallized from ethanol/ water. Decolorizing charcoal was used to remove all colored impurities.

### 2.2  Differential Scanning Calorimetry

Glass transition temperatures were determined with a Perkin Elmer Model-2 Differential Scanning Calorimeter, DSC. The baseline was checked for flatness at each heating rate and temperature cali-bration was achieved by using indium which has a melting point of 156.6°C and a known heat of fusion of 6.8 cal/gram. The heating rate was maintained at either 10 or 20° K/min.

### 2.3   Solid State [13]C NMR

Delrin[R] rotors packed tightly with the appropriate samples were used. All samples were spun at the magic angle and were cross polarized. Spectra were recorded in a JNM-FX-60Q instrument. Contact times of 2 millisecond were maintained and a total of 6000 scans/sample were collected.

### 2.4   Solution [13]C NMR Analysis

All [13]C solution NMR were run with a Bruker 53 MHz FT NMR spec-trometer. Samples were dissolved in CHCl$_3$ (10%). The peaks due to residual CHCl$_3$ were used as the internal standard. In general 300 scans per sample were recorded.

## 2.5 Measurement of Mechanical Properties

All stress-strain measurements were conducted with an Instron Model 1122 instrument at room temperature with dog bones cut out of compression molded films. A crosshead speed of 100% elongation/min. based on the original sample length was used.

## 2.6 Synthesis of Amine Terminal Polyarylene Ether Ketones Oligomers

The following is a procedure for the synthesis of a (Mn = 5000) oligomer. The same general procedure was followed for the synthesis of other oligomers. However, based on the target molecular weight of the oligomers, the ratio of bis-A to p-amino-bis-A was varied and the amount of 4,4'-difluorobenzophenone (0.1 moles) was always held constant. Alternatively, DMAC, p-amino-bis-A and bis(4,4'-chlorobenzoyl) benzene (CBB) were used as the solvent, "end blocker", and ketone monomer, respectively.

The reaction vessel consisted of a four-necked round bottom flask fitted with an overhead stirrer, a nitrogen gas inlet, a thermometer and a Dean-Stark apparatus fitted with a condenser with a drying tube. The following monomers were charged to the flask: Bisphenol-A 21.02 grams (0.0922 moles); 4,4'-difluorobenzophenone 21.8 grams (0.1 moles) and p-amino bis-A 3.76 grams (0.016 moles). All the monomers were weighed very carefully on Teflon coated aluminum pans. The pans were washed repeatedly with portions of NMP (N-methyl pyrrolidone) or DMAC to complete the transfer process. A total of 175 ml of NMP was used for this purpose. The entire operation was carried out under a constant purge of nitrogen. Anhydrous, K$_2$CO$_3$, 20.0 grams (40% molar excess) was transferred into the reaction vessel. The weighing pan and the funnel used for the entire transfer operation were washed with 75 ml toluene.

The entire reaction mixture was heated to the reflux temperature of 160°C and water (the by-product of the reaction) was removed via azeotropic distillation with toluene. Complete removal of water took approximately 3 to 4 hours. Usually more than the stoichiometric amount of water could be removed from the system. Toluene was then removed and the reaction temperature was held at 170°C for a period of 8 to 10 hours. Finally, the reaction mixture was cooled, diluted with 300 ml of THF and filtered through a fritted funnel to remove all residual salts. The polymer solution was then acidified with glacial acetic acid to pH = 7, precipitated in a 75/25 mixture of CH$_3$OH/water and dried in a vacuum oven at 80°C for 10 hours. For the complete removal of all salts and residual NMP from the polymer, the following procedure was followed. The dried polymer was redissolved in 300 ml THF, filtered through a Buchner funnel fitted with number two Whatman's filter paper. The filtrate, which is usually wine red in color, was acidified with methanol/water. The polymer was filtered and transferred into a beaker containing 1 liter of 50/50 mixture of methanol/water. The heterogeneous mixture was gently heated to 40°C with stirring for at least 5 hours. The polymer was finally filtered, washed with methanol and then vacuum dried at 80°C for at least 10 hours.

## 2.7 Synthesis of Fluorine and Hydroxyl Terminal Polyarylene Ether Ketone Oligomer

An identical procedure as described earlier for the amine terminal oligomer was used, except for the fact that the end-capper (p-amino-bis-A) was omitted. Also, depending on the nature of the end group of the oligomer (-F or -OH), and the number average molecular weight desired, either bisphenol-A or 4,4'-difluorobenzophenone was taken in excess. The exact molar excess was calculated based on the well known Carother's equation.

## 2.8. Synthesis of Hydroquinone Functional Poly(arylene Ether) Ketimine

To a 500 ml 4-neck flask fitted with an overhead stirrer, a nitrogen gas inlet, a thermometer and a Dean-Stark apparatus fitted with condenser with a drying tube, the following monomers were charged: Hydroquinone (0.1 moles, 11 gm.), 4,4'-difluorobenzophenone imine (0.1 moles 29.3 gm.), and 20 grams of anhydrous $K_2CO_3$. All starting materials were weighed on teflon coated aluminum pans and were carefully transferred with portions of NMP (N-methyl pyrrolidone). A total of 175 ml of NMP and 85 ml of toluene were used for this purpose. The synthetic procedure was similar to the one followed for the synthesis of amine terminal polyarylene ether ketones, and will not be elaborated upon.

## 2.9 Synthesis of N-Benzohydroxylidene aniline Model Compound

In a three-necked round bottom flask fitted with a thermometer, a stopper and a condenser was added benzophenone (18.2 grams, 0.1 moles), aniline (10.3 grams, 0.11 moles) and 150 ml of toluene. About 100 grams of 3Å molecular sieves was added to this homogeneous reaction mixture. The temperature of the reaction vessel was raised to a gentle reflux and the reaction was allowed to continue for 24 hours. Filtration, followed by evaporation of toluene under reduced pressure, yielded a bright yellow crystalline desired compound in 85% yield (m.p. = 111-113°; lit. m.p. = 113°C).

## 2.10 Synthesis of 4,4'-difluoro (N-benzohydroxylidene aniline)

The starting materials, 4,4'-difluorobenzophenone (21.8 gm., 0.1 mole) and aniline (10.3 gm., 0.11 mole) were dissolved in 150 ml toluene in a three-necked flask

with a condenser, a thermometer and a stopper. Approximately 150 grams of 3Å molecular sieves were added to the system. The reaction was allowed to continue for 24 hours under gentle reflux. It was filtered, evaporated under reduced pressure and crystallized from toluene (yield: 70%, m.p. = 114°C).

## 3. RESULTS AND DISCUSSION

### 3.1 Oligomer Synthesis

All oligomers were synthesized via nucleophilic aromatic substitution reaction in a dipolar aprotic solvent such as N-methyl pyrrolidone 4. The bis fluoro derivative of benzophenone (namely 4,4'-difluorobenzophenone) was chosen as the activated halide to conduct the substitution reaction because of the ease with which fluorine can be displaced under the reaction condition employed (see Experimental section).

However, prior to the actual oligomer synthesis, studies on model compounds were conducted. These studies were necessary in order to ascertain whether any side reactions occur involving highly electron poor carbonyl group of 4,4'-difluorobenzophenone and the relatively electron rich primary amine group of p-amino-bis-A during the polymer formation.

All model compound studies were conducted in NMP in the presence of toluene and anhydrous $K_2CO$ in order to simulate the actual reaction variables during the polymer syntnesis. Table I shows that there was no reaction of aniline with 4,4'-difluorobenzophenone, with benzophenone or with 4,4'-dimethoxybenzophenone. Subsequently, a series of amine terminal poly(arylene ether ketone) oligomers were successfully synthesized by using bisphenol-A, 4,4'-difluorobenzophenone and p-amino-bis-A (Scheme 1). The stoichiometric ratio between bis-A and p-amino-bis-A were calculated based upon the well known Carother's equation. Table II lists the target number average molecular weight and the titrated molecular weight of the amine terminal oligomers along with their intrinsic viscosity values. It is clear from the table that the titrated molecular weights are in good agreement with the target number average molecular weight. This also confirms the fact that no side reactions occur during the polymer formation.

## 3.2 "Self Crosslinking" Behavior of Amine Terminal Polyether Ketone

It was our initial intent to cure commercially available bismaleimide [(1,1'-methylene-4, 1-phenylene) bismaleimide)] (BMI) with the amine terminal PEEK to attempt to produce a toughened BMI matrix.

The curing reaction was conducted in a press at 220°C for various time periods, from 2 to 22 hours. Also, various molecular weight amine terminated oligomers in the absence of BMI were pressed under identical conditions. However, to our surprise, all the oligomers (in the absence of any other additive) were found to be insoluble after the thermal treatment. The formation of imine linkage via the reaction of the amine end groups with the carbonyl groups in the main behavior was postulated as shown in Scheme 2. In order to ascertain the role of the amine end group in conjunction with the carbonyl of the main chain, three oligomers were chosen as controls. They were, the fluorine terminated polyether ketone (Mn = 10,000), the hydroxyl terminated polyether ketone (Mn = 10,000) and the amine terminated polyether sulfone (Mn = 10,000)$^8$. The above mentioned oligomers were pressed at 220°C for 2 hours and were found to be soluble in all solvents in which they are soluble prior to thermal treatment. Thus, having identified the groups responsible for the "self-crosslinking" behavior of amine terminal polyether ketone, it was necessary to verify the presence of imine groups in the crosslinked polymer network. To achieve this, we utilized FT-IR and

$^{13}$C solid state magic angle NMR spectroscopy. The FT-IR of the amine terminated oligomer before and after crosslinking unfortunately exhibited very little difference. Our attempt at discerning the presence of a band due to C=N- group was not successful. This may partly be due to the presence of the strong carbonyl band centered around 1655 cm-1. This band usually commences at 1689 cm$^{-1}$ and ends at 1627 cm$^{-1}$. Solid state NMR was the only choice left to us to verify the presence of C=N-. At first, the solid state CP MAS NMR of the model compound synthesized from benzophenone and aniline (see experimental) was examined. The peak at 176.2 ppm was assigned to the C=N- group (Figure 4). The solution $^{13}$C NMR of the virgin polymer (Mn=5000) (prior to thermal treatment) is shown in Figure 2. Examination of the spectrum reveals certain interesting features. First, there is no peak between 160 and 196 ppm. The peak at 196 ppm is due to the keto carbonyl of the polymer backbone. Secondly, one cannot observe any peak due to the influence of the amine end groups on the benzene ring to which they are attached. This was expected in view of the low end group concentration. On the other hand the scenario is very different indeed when one examines the solid state CP MAS NMR (Figure 3) of the same polymer after the thermal treatment, namely 2 hours at 220°C. The peak at 176.5 ppm is clearly evident and can be assigned to - C=N- groups now present in the polymer after the thermal treatment.

## 3.3 Thermal and Mechanical Behavior of Amine Terminal Oligomers Prior to and After "Self-Crosslinking"

In Figure 4, the DSC thermograms of the amine terminal oligomer (Mn=10,000) before and after the thermal cure are shown. Only the second heat values are reported. Interestingly, in the first heat, in either case, no cure exotherm could be observed. This is possibly due to the fact that the ΔH of imine formation is very low indeed, due to concentration considerations. One does observe an increase of 7°C in the $T_g$ value after the sample has been cured for two hours at 220°C. Also the glass transition temperatures of samples cured for various time periods up to 22 hours at 220°C remain essentially unchanged.

The stress/strain behavior of the amine terminal polyether ketone (Mn=15,000) prior to and after curing area shown in Figure 5. Both polymers exhibit a yield point as expected. However, the cured sample exhibits higher stress at yield and also exceptionally high elongation at break. This can possibly be explained on the basis of pre-gelation chain extension due to inter-chain branching and crosslinking via the formation of imine, as well as the relatively low extent of (end-linked) crosslinking. Although generation of water during the imine "chain extension and crosslinking step" might be expected to generate voids, preliminary microscopy studies indicate that the specimen are solid, and not porous. We suspect that the extremely small amount of water generated is soluble in the resin under molding conditions. Interestingly, boiling the specimen for three days in water does not degrade the network. One concludes that a novel method of network generation has been uncovered.

The synthesis of semi-crystalline poly(arylene ether ketone) via an alternative methodology had been our other objective. In the past, polyarylene ethers have been synthesized by taking advantage of the activating influence of sulfone, keto and nitrile functionality ortho or para to the leaving group in a nucleophilic aromatic substitution reaction[10]. However, we were interested in utilizing an all aromatic ketimine functionality to activate the halogen group ortho or para to it. The appropriate monomer with ketimine functionality was synthesized according to Scheme 3, by reacting aniline with either 4,4'-dichlorobenzophenone or the corresponding fluoro analogue[11].

A wide variety of bisphenols were used with either the chloro or the fluoro functional ketimine monomer[12]. Our present discussion will center around hydroquinone as the bisphenol of choice. The chloroanalogue was not found to be reactive enough to yield a high molecular weight polymer. On the other hand, with the bisfluoro functional ketimine monomer very high molecular weight amorphous polymer could be obtained in NMP/$K_2CO_3$/Toluene reaction medium (Scheme 4).

The polymer synthesized afforded a transparent and ductile film either upon solvent casting or form compression molding. It was found to be soluble in THF, methylene chloride, chloroform, chlorobenzene and a host of dipolar aprotic solvents. It exhibited an intrinsic viscosity of 0.89 in THF at room temperature, indicating high molecular weight. The GPC chromatogram of the polymer is shown in Figure 6. It shows a unimodal distribution. The number and weight average molecular weight based on universal calibration curve are 69K and 134K, respectively. This indicates the high molecular weight nature of the polymer. Indeed it also establishes the activating influence of the all aromatic ketimine functionality towards step-growth nucleophilic aromatic substitution polymerization reactions[12].

The newly discovered thermoplastic was hydrolyzed in the presence of trifluoromethane sulfonic acid at room temperature to yield a poly(arylene ether ketone) backbone structure (see experimental). The hydrolysis is rapid and can be carried out in approximately 10 mins (Scheme 5). Our choice of trifluoromethane sulfonic acid was due to the following two reasons, firstly the starting ketimine functional polymer is soluble in the acid and secondly, the resulting semicrystalline material also remains in solution. Thus, complete conversion of the ketimine functionality to the desired ketone group could be achieved.

The DSC thermograms of the synthesized poly(arylene ether ketone) via the modified procedure are shown in Figure 7. An examination of Figure 7 reveals certain interesting facts. The synthesized PEEK shows a $T_g$ at 150°C, a crystallization exotherm with a maxima at 191°C and a melting point endotherm maxima at 322°C, whereas, for the commercial sample a $T_g$ at 150°C and a $T_m$ maxima at 342°C has been reported[7].

## 4. CONCLUSIONS

Novel ketimine linked networks were prepared by the "self-crosslinking" reaction of terminal amine groups with the in-chain ketone groups of poly(arylene ether ketone)s. The networks are ductile, solvent resistant and hydrolytically stable. Ketimine containing aromatic difluoride monomers were also synthesized and used to produce amorphous poly(arylene ether)s. Treatment with acids regenerates a crystalline, solvent resistant poly(arylene ether ketone).

## 5. ACKNOWLEDGEMENTS

We would like to thank NASA Langley Research Center for their support of this project, R. Subramamanian for the solid state NMR spectra and D. Tyagi for the stress-strain data.

## 6. REFERENCES

1. Kawakami, J. H., Kwiatkowski, G. T., Brode, G. L., Beduin, A. W. J. Polymer Sci. Polym. Chem. Ed. 12, 565 (1974).
2. Jurek, M. J., McGrath, J. E. Polym. Prep. 26(2), 283 (1985).
3. Hedrick, J. L., Jurek, M. J., Yilgor, I., Wilkes, G. L., McGrath, J. E. Polym. Prep. 26(2), 293 (1985).
4. Jurek, M. J., McGrath, J. E. 31st Natl. SAMPE Symp. 31, 913 (1986).
5. Cecere, J. A., Senger, J. S., McGrath, J. E., Steiner, P. A., Wong, R. S., Sachdeva, Y. 32nd Natl. SAMPE Symp. 32, 000 (1987).
6. Private communication, NASA Langley Research Center, Hampton, VA.
7. Mohanty, D. K., Lin,. T. S., Ward, T. C., McGrath, J. E. 31st Natl. SAMPE Symp. 31, 945 (1986).
8. Lyle, G. D., Jurek, M. J., Mohanty, D. K., Hedrick, J. C., McGrath, J. E. Polym. Prep. 28(1), 000, (1987).
9. Jurek, M. J., Ph.D. dissertation, VPI&SU (1987).
10. Johnson, R. N., Farnham, A. G., Clendinning, R. A. , Hale, W. F., Merriam, C. N. J. Polym. Sci., A-1 5, 2375 (1967).
11. The Chemistry of Carbon-Nitrogen Double Bond, S. Patai (ed.), Interscience Publisher, 64 (1970).
12. Mohanty, D. K., Lowery, R. C., McGrath, J. E. (in preparation).

## 7. BIOGRAPHIES

Dillip K. Mohanty was born on November 13, 1956 in Orissa, India. He received his M.Sc. in Organic Chemistry from Indian Institute of Technology, Kharagpur, India in 1976. In August of 1977, he received a diploma in High Polymer and Rubber Technology from the same school. A month later, he joined the State University of New York at Stony Brook. At Stony Brook, after qualifying for a Ph.D. degree, he chose to earn a Masters of Science instead, in order to change his area of research into Polymer Science. As a consequence, he continued his study towards a Ph.D. degree in Polymer Science at Virginia Polytechnic Institute and State University, one of the few schools in the U.S.A. offering a well defined polymer program. He earned his Ph.D. in June 1983. After a brief period in industry, he is presently serving as a research associate under Prof. J. E. McGrath.

Richard C. Lowery received his B.S. in Chemistry from VPI&SU in 1986 and is currently employed at the General Electric Co. Research and Development Center in Schenectady, NY.

Gregory D. Lyle received his B.S. in Chemistry from North Carolina State University and is currently completing an M.S. in Chemistry from VPI&SU.

James E. McGrath was born in Easton, New York and received his B.S. in Chemistry from Siena College in 1956. He was employed in cellulose fiber and film research by ITT Rayonier in Whippany, NJ until October 1959. At that time he joined the research division of the Goodyear Tire and Rubber Co. where he conducted research on synthetic rubbers. He obtained an M.S. degree in Chemistry from the University of Akron in 1964 and Ph.D. in Polymer Science from the same university in 1967. Professor McGrath joined the Union Carbide Corporation in August of that year. Professor McGrath joined VPI & SU in September 1975. He was named a tenured Associate Professor in 1977 and was promoted to Full Professor in 1979. In 1986, he was named as the first Ethyl Chair of Chemistry at VPI&SU. He is a consultant to industry and government. He is also a Co-Director of the Polymer Materials and Interfaces Laboratory. Professor McGrath has coauthored what is considered a definitive book on Block Copolymers (Academic Press, 1977), and has edited books on Anionic Polymerization (ACS Symposium Series No. 166, 1981), Ring Opening Polymerization (ACS Symposium Series No. 286, 1985) and Advances in Polymer Synthesis (with B. M. Culbertson) Plenum Press, 1986. He was the Chairman of the Polymer Division of the American Chemical Society in 1986. He has over 100 contributions in the literature, and is also an author or coauthor of 21 U.S. patents.

Scheme 1

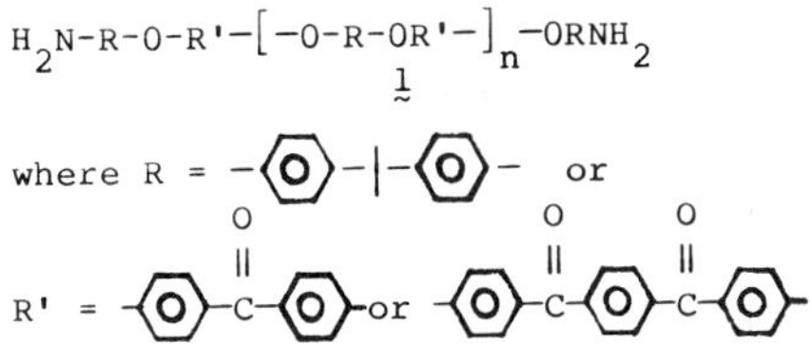

$$H_2N-R-O-R'-\left[-O-R-OR'-\right]_n-ORNH_2$$

1

where R = or

R' = or

SCHEME 2

CHEMISTRY

OF

"SELF CROSSLINKING"

BEHAVIOR

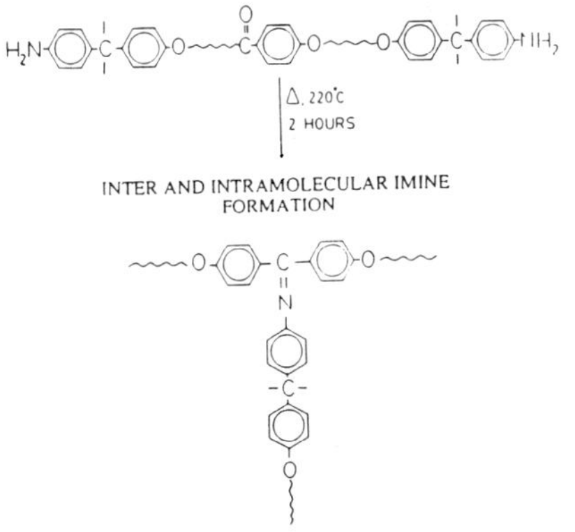

Δ, 220°C
2 HOURS

INTER AND INTRAMOLECULAR IMINE FORMATION

SCHEME 3

SYNTHESIS OF KETIMINE FUNCTIONAL MONOMERS

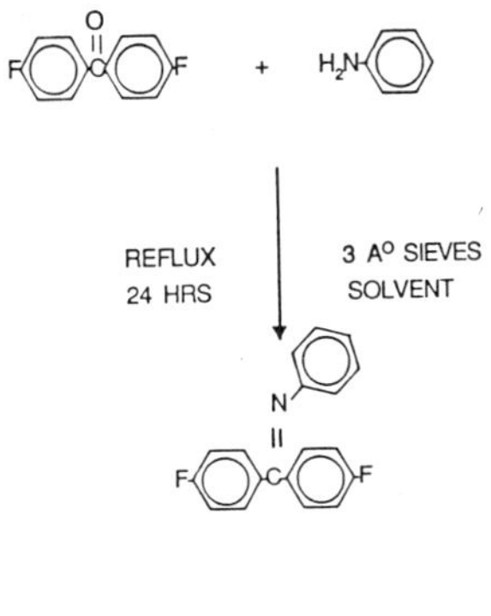

REFLUX
24 HRS

3 A° SIEVES
SOLVENT

415

SCHEME 4

SYNTHESIS OF HQ FUNCTIONAL
POLY(ARYLENE ETHER ) KETIMINE

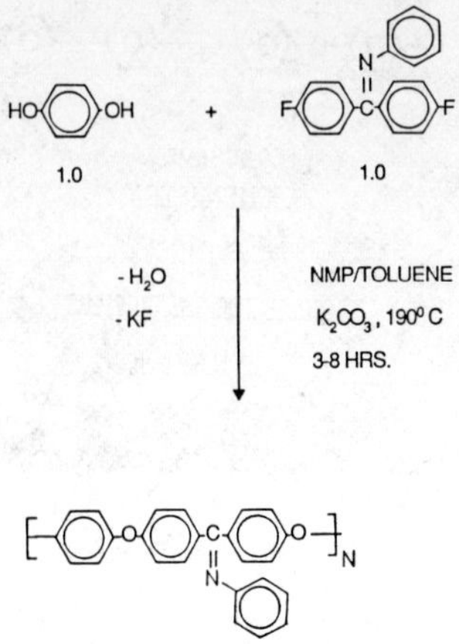

SCHEME 5

HYDROLYSIS OF
HYDROQUINONE - FUNCTIONAL
POLYARYLENE ETHER IMINE

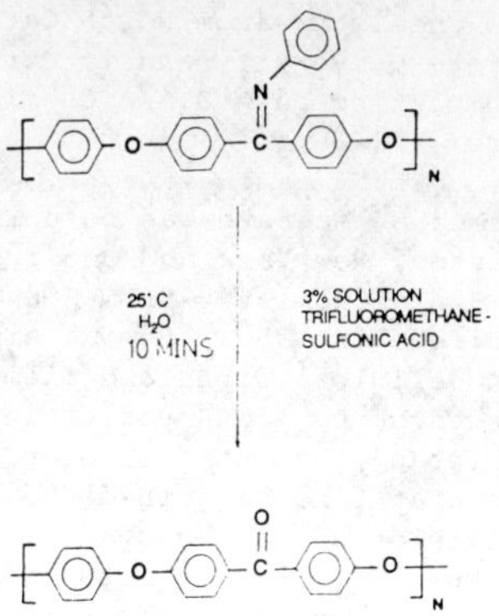

TABLE I

Observation Regarding Attempted Model Compound Synthesis

| Reactants | Reaction Temperature, °C | Time Hours | Observation |
|---|---|---|---|
| Aniline + Benzophenone | 170 | 7 | Starting Materials |
| Aniline + 4,4'-Difluorobenzophenone | 170 | 7 | Starting Materials |
| Aniline + 4,4'-Dimethoxybenzophenone | 170 | 7 | Starting Materials |

TABLE II

Intrinsic Viscosity and Number Average Molecular Weights

of Amine Terminal Polyarylene Ether Ketones

| $\bar{M}_n$ | | |
|---|---|---|
| Target | Titrated Value | $[\eta]^{THF}_{25°C}$ |
| 5000 | 6300 | 0.12 |
| 7500 | 8904 | |
| 10000 | 12606 | 0.21 |
| 13000 | 15040 | |
| 15000 | 17030 | 0.35 |
| 25000 | 31924 | 0.54 |

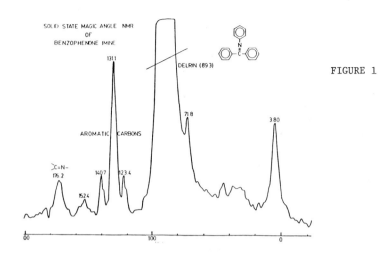

FIGURE 1

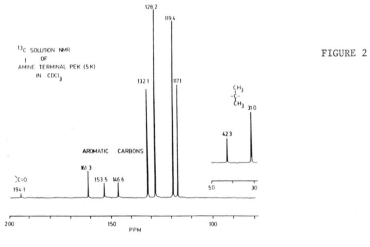

FIGURE 2

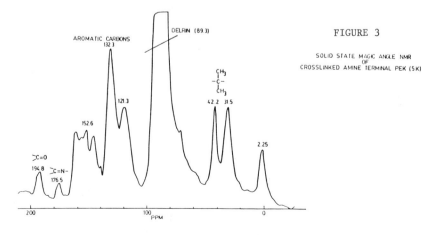

FIGURE 3

SOLID STATE MAGIC ANGLE NMR
OF
CROSSLINKED AMINE TERMINAL PEK (5K)

417

FIGURE 4

EFFECT OF CURING TIME
ON
THE GLASS TRANSITION TEMPERATURE.
[ AMINE TERMINAL PEK (10 K) CURED AT 220°C ]

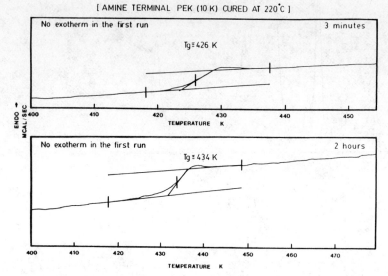

FIGURE 5

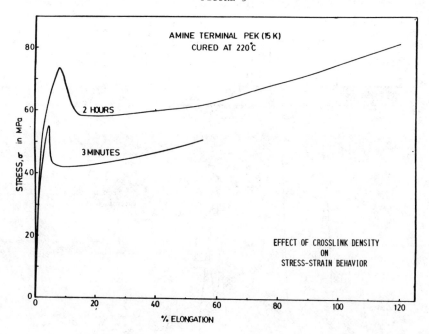

FIGURE 6

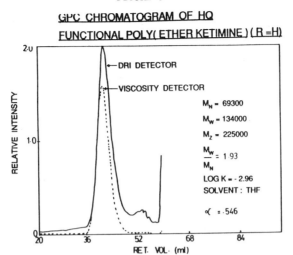

## GPC CHROMATOGRAM OF HQ
## FUNCTIONAL POLY( ETHER KETIMINE ) ( R = H )

FIGURE 7

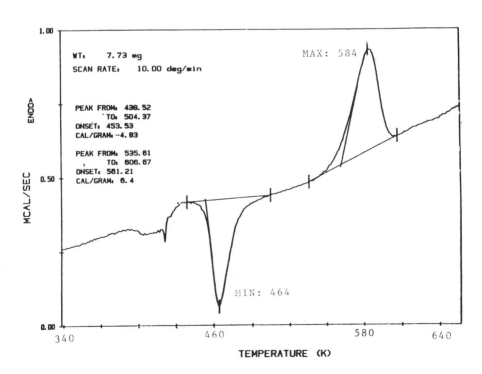

# THERMOPLASTIC MATRIX COMPOSITES BASED ON POLYETHERIMIDES

S.L. Peake, A. Maranci
Engineered Materials Technology,
American Cyanamid Company
P.O. Box 60
Stamford, CT  06904-0060


E. Sturm
Engineered Materials Technology
American Cyanamid Company
Old Post Road
Havre de Grace, MD  21078

## ABSTRACT

A new class of polyetherimides has been developed which shows a number of advantages as a matrix resin for advanced composites. These polyimides permit consolidation and forming under modest pressures, have acceptable hot/wet performance to 150°C (300°F), and are resistant to common aircraft fluids. They are true thermoplastics, and can be repeatedly reformed. Laminates have been formed using pressures as low as 60 psi and cycle times as low as 3 minutes. Their mechanical properties are comparable to conventional thermosets, but their performance extends to 150°C (300°F) wet. Toughness is outstanding, as evidenced by a residual compression strength, after 1500 in-lb/in impact, of 317 MPa (46 ksi) and a $G_{IC}$ of 23 kJ/m$^2$ (15 in-lb/in$^2$). This paper will describe the mechanical properties and processing of laminates based on this new polyetherimide.

## 1.  INTRODUCTION

Thermoplastic polymers are receiving considerable scrutiny by the aerospace community as matrix resins for advanced composites. The principle driver behind this interest is related to both cost and performance.

Thermoplastic resins lend themselves to rapid forming techniques which can greatly reduce the manufacturing costs of composite parts.

Increased performance is derived from the inherent ductility of many thermoplastic polymers. In addition, the greater durability of thermoplastic composites will lead to lower life cycle costs of thermoplastic components.

The thermoplastic resins studied to date can be divided into three categories: low Tg, semi crystalline polymers which behave as true thermoplastics; higher Tg, pseudo-thermoplastics whose processing more closely resembles thermosets; and amorphous engineering thermoplastics which have not found acceptance because they are aggressively attacked by common aircraft solvents.

A thermoplastic in the latter category is ULTEM® 1000 resin. Laminates made using ULTEM resin can be consolidated under autoclave pressures to high quality, void free laminates, which show good mechanical properties and toughness. In particular, the residual compression strength after impact is 44 ksi. However, ULTEM 1000 is not suitable for aircraft structure because it is attacked by a number of solvents. In cooperation with General Electric Plastics, American Cyanamid has developed prepreg based on a new polyetherimide resin which is resistant to common solvents and which retains the processing advantages and mechanical

properties of ULTEM 1000 resin.

The polymer shows no degradation of properties after exposure to methyl ethyl ketone, hydraulic fluid (Skydrol), or JP-4. The processing and properties of laminates made using this unique polyetherimide are described below.

## 2. RESULTS AND DISCUSSION

2.1 Resin Properties - The new polyetherimide has mechanical and physical properties which are very similar to the commercial polyetherimides, eg., ULTEM 1000 and ULTEM 6000, but solvent resistance is significantly increased. For example, whereas ULTEM 1000 dissolves readily in methylene chloride, the new polyetherimide swells slightly after 24 hours immersion. Glass transition temperature by DMA is 230°C (445°F). The neat resin properties are summarized in Table 1.

2.2 Laminate Fabrication - The solvent resistant polyetherimide is available in prepreg form under the name CYPAC™ X7005. The prepreg consists of neat imidized resin impregnated into continuous fiber in either tape or fabric form. No solvent is present in the prepreg and no reactions occur during laminate consolidation. Thus, the prepreg is stable indefinitely at room temperature and little sensitivity to thermal histroy during consolidation (i.e., heat-up/cool-down rates) is seen.

The imidized CYPAC X7005 prepreg is tackless and boardy. Because no chemistry is occurring during consolidation of imidized prepreg, laminates can be consolidated rapidly. Indeed, cycle times as low as 3 min. were demonstrated in our laboratories by placing plies of prepreg in a preheated press at 325°C (620°F), consolidating under 7 MPa (100 psi) then quickly transferring to a second press, applying 7 MPa and cooling with water under pressure. Compression molding conditions ranging from 60 to 150 psi at 304°C (580°F) to 343°C (650°F) have been successful. Processing above 375°C (700°F) leads to lower mechanical properties. Time at molding temperature is not critical since no cure is taking place. We have held the laminate at the molding temperature for 2 to 120 minutes and found no effect on mechanical properties.

Formed parts have been made at General Electric's Plastic Applications Center using preconsolidated laminates. The optimum conditions were: heat a blank to 285°C (550°F), transfer to a mold heated to between 120 to 180°C (250 to 350°F), close the mold under 4.6 MPa (670 psi) and hold for 10 sec. Examples of glass and graphite parts made using this procedure are shown in Figure 2.

One of the key advantages of a true thermoplastic is the ability to reform or reconsolidate. This has been demonstrated with CYPAC X7005 in two experiments. A 50 ply graphite fabric laminate was cut into four sections, stacked and remolded at 330°C (625°F) and 7 MPa (100 psi) to give a 200 ply, high quality laminate. C-scan showed excellent consolidation. A simple experiment was used to demonstrate repair: failed short beam shear coupons were remolded in a 330°C (625°F) press. The resulting specimens recovered all of their interlaminar shear strength.

An advantage of CYPAC thermoplastic prepregs is that they may be processed as either a thermoplastic or as a thermosetting polyimide. While thermoplastic processing is the driving force behind these materials, processing as a thermosetting polymer may offer some advantages in the short term, e.g., when volumes of a particular part are too low to justify the capital investment for thermoplastic processing equipment. Also, some parts may be of such complexity that hand lay-up and autoclave cure are more appropriate. Therefore, a tacky drapable form of the prepreg was developed. The handling of this prepreg, CYPAC X7000, is comparable to conventional epoxies, and it is "cured" in much the same manner as other thermosetting polyimides.

One must make allowances for the large volumes of solvent that are evolved during "cure." The maximum rate of volatile evolution occurs between 200 to 220°C (400 to 430°F). As can be seen from the GC-mass spectrum shown in Figure 1, water evolution from imidization occurs earlier than NMP evaporation. Satisfactory laminates have been produced using pressures as low as 4 MPa (60 psi), and the resulting material is a true thermoplastic. Mechanical properties of laminates derived from the tacky prepreg are indistinguishable from those made from the standard CYPAC X7005.

2.3 Mechanical Properties – Mechanical properties from CYPAC X7005 laminates are shown in Tables 2, 3 and 4. Particularly noteworthy is the high toughness, as shown by $G_{IC}$, $G_{IIC}$, and post impact compression. The $G_{IC}$ value is an order of magnitude greater than epoxy laminates, and the residual compression strength after 1500 in-lb/in impact is 46.4 ksi. Compressive, tensile, flexural and interlaminar shear properties are comparable to those seen for typical epoxy fabric laminates, but with performance up to 150°C (300°F)/wet.

The finish on graphite fibers seems to give adequate bonding between the fiber and the thermoplastic matrix. However, there is a drop in room temperature mechanical properties seen after exposure to water. This behavior can usually be ascribed to a poor resin-fiber interface. This is further supported by the observation that better retention of properties is seen on the Celion fiber than Hercules AS-4. On glass fibers a dramatic difference in wet performance was noticed between fiber finishes. After surveying commercially available sizings, a finish was located which showed negligible drop-off in mechanical properties at room temperature after exposure to water.

Even though the decomposition temperature of Kevlar fiber is close to the molding temperature of CYPAC X7005, excellent retention of tensile strength and tensile modulus is seen. In preliminary results, room temperature tensile strength and modulus were 435 MPa (63 psi) and 25 GPa (2.6 Msi), respectively. This compares to typical epoxy values of 480 MPa (70 ksi) and 28 GPa (4.0 Msi).

Solvent resistance is a key issue for thermoplastic polymers. Crystalline polymers in general show excellent resistance to aircraft solvents. Many amorphous polymers, are severely affected by common solvents. CYPAC X7000 and CYPAC X7005 show good resistance to JP-4, methyl ethyl ketone, hydraulic fluid, and methylene chloride. Solvent resistance was measured using two methods: creep in flexure (Figure 3) and retention of short beam shear strength after solvent exposure (Figure 4). In hydraulic fluid, methyl ethyl ketone, and JP-4, negligible effect is seen. In methylene chloride, some plasticization of the resin occurs. For example, 20% of the short beam shear strength is lost after exposure to methylene chloride for 24 hours at 23°C. After 7 days, over 70% of the short beam shear strength is lost; however, if the sample is allowed to dry overnight, the short beam shear strength recovers to nearly 80% of the unaged value. Soaking small coupons for a week in methylene chloride is a very severe test, and based on the increasing restrictions being placed on the use of methylene chloride, we believe that the solvent resistance of CYPAC X7005 to hydraulic fluid, JP-4 and methyl ethyl ketone will be adequate for aircraft primary structure.

One of the outstanding features of polyetherimides is their flammability characteristics. ULTEM 1000 resin itself has a Limiting Oxygen Index (LOI) of 47. In the NBS Smoke test, ULTEM 1000 gives a Ds @ 4 min of 0.7 and Dmax @ 20 min of 30. These values are almost an order of magnitude better than some phenolics currently used in aircraft interiors. CYPAC X7005 graphite and glass laminates gave LOI's of 71 and 54 respectively.

## 3.  SUMMARY

True thermoplastic characteristics are displayed by CYPAC X7005. Laminates can be consolidated rapidly at modest pressures. Reconsolidation and cold stamping have been demonstrated using preconsolidated laminates. The laminates show excellent toughness and mechanical properties with hot/wet performance up to 300°F. Because no cure chemistry is occurring during the consolidation process, the laminates show little sensitivity to thermal history and the prepreg is stable at room temperature, in principle, indefinitely.

Future work will be focused on higher $T_g$ polyetherimides which maintain the solvent resistance and processability of CYPAC X7005, but that have hot/wet performance in the 350 to 400°F range.

## 4.   ACKNOWLEDGEMENT

The authors would like to thank J. Male, D. Floryan and D. Olson of General Electric for their support. Mechanical testing and data analysis by Ann Cronin and Stan Kaminski is gratefully acknowledged.

## 5. BIOGRAPHIES

Steven Peake is the Group Leader for Matrix Resins in the Engineered Materials Technology Section of American Cyanamid. Prior to joining Cyanamid, he received a Ph.D. in Chemistry from the University of Wisconsin at Madison in 1979. His group's responsibilities include the development of new polymer and monomer chemistries for use in aerospace composites and adhesives.

Art Maranci is a Senior Research Engineer in the Matrix Resins Group at American Cyanamid, Stamford, CT. He received a Ph.D. in Material Science from the Massachusetts Institute of Technology in 1970. His research interests include impact resistance and damage tolerance of composites, processing of acrylic fibers and carbon fiber precursors, as well as, thermoplastic matrix composites.

Ed Sturm is a Composite Development Chemist in the Advanced Composites Group in the Engineered Materials Section at American Cyanamid Company, Havre de Grace, MD. Prior to joining Cyanamid in 1982, he received a B.S. in Chemistry from the State University of New York (Buffalo). He is currently involved in the qualification, database generation and technical support for CYCOM and CYPAC prepregs.

CYPAC is a trademark and CYCOM is a registered trademark of the American Cyanamid Company.

ULTEM is a registered trademark of the General Electric Company.

## TABLE 1
## NEAT RESIN PROPERTIES

|  | CYPAC X7005 | ULTEM 1000 |
|---|---|---|
| Tg (DMA) °C (°F) | 230 (446) | 230 (229) |
| Flex Creep (1000 psi) in methylene chloride | 3%/5 hrs | Fails 2 min. |
| Specific Gracity | 1.29 | 1.27 |
| Flex Modulus 23 °C, GPa (Msi) | 3.0 (0.43) | 3.3 (0.48) |
| Flex Strength MPa (ksi) | 128 (18.5) | 145 (21) |

**TABLE 2**
**CYPAC X7005 LAMINATE MECHANICAL PROPERTIES**
**Graphite Fabric**

| | 3K70P-AS-4 | | 3K135-8H-Celion | |
|---|---|---|---|---|
| Laminate Tg °C (°F) | | | | |
| DMA | 225 | (448) | 225 | (448) |
| Fiber Volume (%) | 54 | | 56 | |
| Compression Strength MPa (ksi) | | | | |
| 23 °C Dry | 503 | (73) | 572 | (83) |
| 23 °C Wet | 434 | (63) | 559 | (81) |
| 120 °C Dry | 400 | (58) | 476 | (69) |
| 120 °C Wet | 365 | (53) | 386 | (56) |
| 150 °C Dry | 414 | (60) | | |
| 150 °C Wet | 289 | (42) | 310 | (45) |
| 177 °C Dry | 322 | (46) | 427 | (62) |
| 177 °C Wet | 228 | (33) | 262 | (38) |
| Compression Modulus GPa (Msi) | | | | |
| 23 °C Dry | 56 | (8.0) | 58 | (8.3) |
| 150 °C Dry | 54 | (7.8) | 56 | (8.0) |
| 150 °C Wet | 54 | (7.7) | 53 | (7.6) |
| Tensile Strength MPa (ksi) | | | | |
| 23 °C Dry | 733 | (105) | 684 | (98) |
| 150 °C Dry | 768 | (100) | | |
| Tensile Modulus GPa (Msi) | | | | |
| 23 °C Dry | 57 | (8.1) | 53 | (7.6) |
| 150 °C Dry | 57 | (8.2) | | |
| Short Beam Shear Strength MPa (ksi) | | | | |
| 23 °C Dry | 66 | (9.5) | 46 | (6.6) |
| 23 °C Wet | 60 | (8.6) | 43 | (6.2) |
| 150 °C Dry | 41 | (5.8) | 31 | (4.5) |
| 150 °C Wet | 35 | (5.0) | 21 | (3.5) |
| Flexural Strength MPa (ksi) | | | | |
| 23 °C Dry | 838 | (120) | 628 | (90) |
| 150 °C Dry | 559 | (80) | 447 | (64) |
| 150 °C Wet | 405 | (58) | 377 | (54) |
| Flexural Modulus GPa (Msi) | | | | |
| 23 °C Dry | 56 | (8.0) | 52 | (7.5) |
| 150 °C Dry | 54 | (7.9) | 54 | (7.9) |
| 150 °C Wet | 53 | (7.7) | 53 | (7.2) |

TABLE 3

## CYPAC X7005 LAMINATE MECHANICAL PROPERTIES
### Glass Fabric

| | 120E | | 7781 | |
|---|---|---|---|---|
| Laminate Tg ºC (ºF) | | | | |
| DMA | 225 | (448) | 225 | (448) |
| Fiber Volume (%) | 53 | | 52 | |
| Compression Strength MPa (ksi) | | | | |
| 23 ºC Dry | 572 | (83) | 554 | (80.4) |
| 23 ºC Wet | 455 | (66) | 555 | (80.5) |
| 120 ºC Dry | — | — | — | — |
| 120 ºC Wet | — | — | — | — |
| 150 ºC Dry | 345 | (50) | 355 | (51.5) |
| 150 ºC Wet | 262 | (38) | 332 | (48.1) |
| 177 ºC Dry | — | — | — | — |
| 177 ºC Wet | — | — | — | — |
| Compression Modulus GPa (Msi) | | | | |
| 23 ºC Dry | 24 | (3.5) | 23.5 | (3.41) |
| 150 ºC Dry | 32 | (4.7) | 19.0 | (2.75) |
| 150 ºC Wet | 23 | (3.3) | 16.5 | (2.39) |
| Tensile Strength MPa (ksi) | | | | |
| 23 ºC Dry | 338 | (49.0) | 408 | (59.2) |
| 150 ºC Dry | 282 | (41.0) | 330 | (47.9) |
| Tensile Modulus GPa (Msi) | | | | |
| 23 ºC Dry | 21 | (3.0) | 22.5 | (3.26) |
| 150 ºC Dry | 21 | (3.0) | 21.8 | (3.16) |
| Short Beam Shear Strength MPa (ksi) | | | | |
| 23 ºC Dry | 70.3 | (10.2) | 70.3 | (10.2) |
| 23 ºC Wet | 66 | (9.6) | 70.3 | (10.2) |
| 150 ºC Dry | 39 | (5.6) | 38.7 | (5.6) |
| 150 ºC Wet | 31 | (4.5) | 28.8 | (4.2) |
| Flexural Strength MPa (ksi) | | | | |
| 23 ºC Dry | 413 | (60) | 595 | (86.3) |
| 150 ºC Dry | 269 | (39) | 418 | (60.6) |
| 150 ºC Wet | 269 | (39) | 352 | (51.1) |
| Flexural Modulus GPa (Msi) | | | | |
| 23 ºC Dry | 21 | (3.0) | 21.5 | (3.1) |
| 150 ºC Dry | 17 | (2.5) | 20.1 | (2.9) |
| 150 ºC Wet | 19 | (2.7) | 18.6 | (2.7) |

## TABLE 4
## CYPAC X7005 TOUGHNESS PARAMETERS
### 3K70P Graphite Fabric

| TEST | VALUE | |
|---|---|---|
| | (ksi) | (MPa) |
| Open Hole Compression | | |
| 23 °C Dry | 40.9 | 282 |
| 150 °C Dry | 26.1 | 180 |
| CSAI (MPa)* | | |
| 23 °C Dry | 46.4 | 320 |
| ± 45 ° Tensile: Modulus | 574 | 3960 |
| (from [0,90] fabric) | | |
| | in-lb/in$^2$ | (kJ/m) |
| Toughness | | |
| GIC (Double Cantilever Beam) | 23 | 4.0 |
| GIIC (End Notched Flexure) | 14 | 2.5 |

* Residual Compressive Strength after 1500 in-lb/in (6.67 J/mm impact)

FIGURE 1

# CYPAC X7000
## Thermogravimetric Mass Spectrum

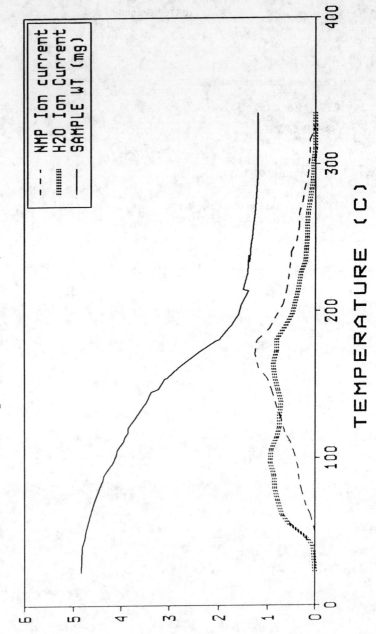

- - - NMP Ion current
▨▨▨▨ H2O Ion Current
—— SAMPLE WT (mg)

TEMPERATURE (C)

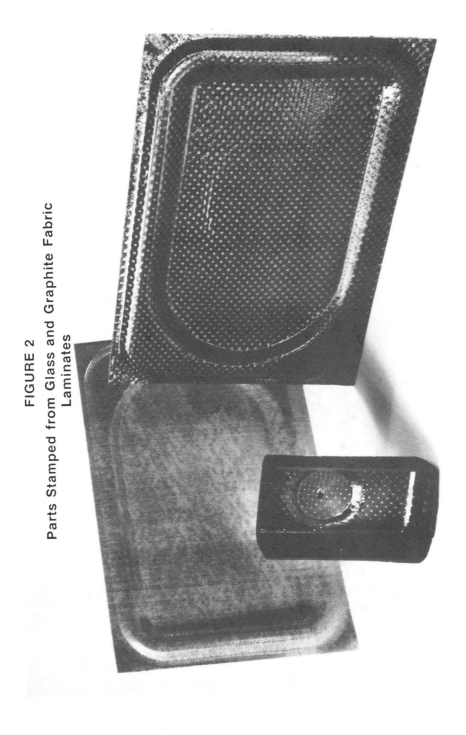

FIGURE 2
Parts Stamped from Glass and Graphite Fabric Laminates

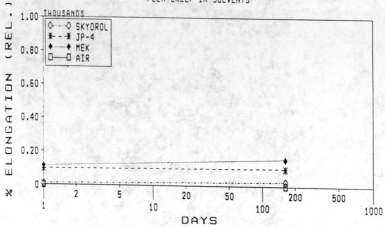

FIGURE 3
# CYPAC X7005
## FLEX CREEP IN SOLVENTS

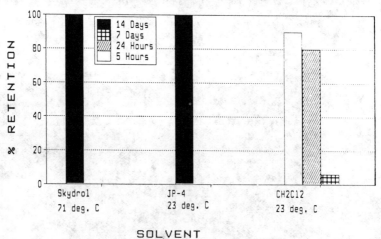

# Figure 4 - Cypac Solvent Resistance
## % Retention of Short Beam Shear

CERAMIC MATRIX AND RESIN MATRIX COMPOSITES:   A COMPARISON
Frances I. Hurwitz
NASA Lewis Research Center
Cleveland, OH 44135

## Abstract

The underlying theory of continuous fiber reinforcement of ceramic matrix and resin matrix composites, their fabrication, microstructure, physical and mechanical properties are contrasted. The growing use of organometallic polymers as precursors to ceramic matrices is discussed as a means of providing low temperature processing capability without the fiber degradation encountered with more conventional ceramic processing techniques. Examples of ceramic matrix composites derived from particulate-filled, high char yield polymers and silsesquioxane precursors are provided.

## 1.   INTRODUCTION

The increasing acceptance of resin matrix composites as structural materials for aerospace applications has led to a heightened interest in extending the use of composites to higher temperatures, particularly in engine environments. Ceramics, because of their stability at elevated temperatures, offer significant potential in engine applications; however, their use has been limited by their brittle fracture behavior and high degree of flaw sensitivity. The incorporation of particulates, whiskers or continuous fibers into a ceramic matrix could provide mechanisms for "toughening" the ceramic, increasing reliability by decreasing flaw sensitivity, and, in the case of continuous fibers, ameliorating the tendency toward catastropic failure.

The base of knowledge gained with resin matrix composites can in part, serve as a base for the development of structural ceramic composites. However, there are important differences between ceramic matrix and resin matrix composites in fabrication, microstructure, physical properties and mechanical behavior. This paper offers a comparison of the two, as well as a discussion of opportunities for the utilization of organometallic polymers as precursors to ceramic composites. The focus will be on continuous fiber reinforced materials.

## 2. THEORY

In resin matrix composites typically the fiber modulus and strength are much greater than those of the matrix. The fiber, therefore, provides stiffness and load bearing capability, while the matrix serves to distribute the load among the fibers. Often, especially in the case of epoxy matrix materials, a coupling agent is introduced to enhance the strength of the interface, increasing load transfer capability from matrix to fiber. The strain capability of the matrix typically exceeds that of the fiber; hence, initiation of fracture is from the failure of a fiber or group of fibers. The load is then redistributed to the remaining fibers until additional fiber breakage occurs.

In contrast, in a ceramic matrix composite the moduli of the fiber and matrix typically are very similar. The fiber therefore is not the source of stiffness or necessarily of strength for these materials (1). Monolithic ceramics offer sufficient strength for structural applications, so incorporation of fibers is not necessary to achieve load bearing capability. Rather, the fiber is relied upon to bridge cracks or flaws in the matrix. Because of differences in processing between resin and ceramic matrices, the ceramic matrix is anticipated to have a higher population of processing introduced flaws and microcracks than its resin counterpart. The strain capability of the fiber is expected to exceed that of the matrix. Hence, matrix failure precedes fiber fracture. Microcracking of the matrix may provide some toughening as well as some pseudoplasticity (2-4). The desired mode of failure is debonding at the interface. Because ultimate matrix strain, $\varepsilon_m$, is less than ultimate fiber strain, $\varepsilon_f$, fiber fracture would result in brittle failure of the material. By bridging microcracks or flaws, the fiber can increase the strain to failure of the composite over that of the monolithic (see Figure 1), imparting non-catastrophic failure. If the fiber/matrix interface is not strong, cracks can propogate around the fibers rather than through them. Therefore, the primary role of the fibers in a ceramic matrix composite is crack bridging and toughening by matrix crack blunting and debonding at the interface, as opposed to providing stiffness and strength in the resin matrix composite.

## 3. FABRICATION

Resin matrix composites typically are processed at relatively low temperature (<350°C). The polymer resin may be either a thermoplastic or thermoset, but typically will undergo viscoelastic flow, providing a means of fiber impregnation and matrix consolidation.

Glasses do become molten, and offer the advantage of forming glass matrix composites by hot pressing or transfer molding. A number of glass matrix composites have been studied (5-7). However, the melting points of many glasses are sufficiently high so that reaction with the fiber can become a problem. As with resin matrix composites, the use temperature of glass matrix composites will be well below their softening point. Carbon reinforced glass composites have been reported which offer zero coefficients of thermal expansion (7).

Ceramics, by contrast, do not melt flow at temperatures low enough for composite fabrication. Conventional sintering and hot pressing temperatures usually exceed those at which available fiber reinforcements degrade in strength.

Oxide ceramic matrices can be formed by sol-gel processing (8-9). Ceramic composites can also be fabricated by chemical vapor infiltration (CVI) or by chemical vapor deposition (CVD) (10-11). Silicon nitride matrix composites have been fabricated successfully by reaction bonding of silicon powder in a nitrogen atmosphere (12-13). Organometallic polymers also might serve as precursors which can be pyrolyzed to a ceramic char (14-24); these are discussed in greater

detail below. In addition, porous carbon chars can be formed by polymerization around a pore former, and the resulting porous carbon reacted with silicon to form silicon carbide (25). All of these techniques can be expected to produce matrices with rather high levels of porosity (up to 35 - 40 percent), as opposed to perhaps <3 - 5 percent porosity typically found in resin matrix composites. How well this porosity can be tolerated will depend on the uniformity of pore size and distribution and the diameter of bridging fibers.

4. FIBERS

Resin matrix composites can be fabricated from a wide variety of fibers, offering a broad choice of modulus and strength for tailoring to particular applications. The selection of fibers available for ceramic composites is much more limited, however, due to the higher temperatures required for composite processing and in use.

Carbon fibers have been used with glass matrices to temperatures of about 600°C (7). However, at higher temperatures carbon will react to form carbides and will be extremely susceptible to oxidation when used in a porous matrix.

SiC fibers would offer good modulus and high temperature properties. Large diameter (140 μm) SiC fibers produced by CVD are available with temperature capabilities to 1400°C (26), but their diameter precludes their use in woven structures

435

or in the formation of complex shapes. For strengthening a ceramic matrix, interfiber spacing less than the critical flaw size is needed; this requires fiber diameters to be smaller than the interfiber spacing. The ultimate matrix tensile strain also is influenced by fiber diameter and the strength of the fiber-matrix interface (2).

Available small diameter fibers include polymer-derived Nicalon SiC and oxide fibers such as Nextel 312, 440 and 480 (27), comprised of boria, alumina and silica, and FP-alumina (28). However, all of these become thermally unstable at temperatures from 1000 to 1200°C.

## 5. PHYSICAL AND MECHANICAL PROPERTIES

In any composite system where there is a difference in coefficient of thermal expansion (CTE) between fiber and matrix, residual thermal stresses will be expected to develop during both composite fabrication and on thermal cycling. For most fibers, expansion is anisotropic, differing along the fiber axis and the radial direction. A pure $\beta$-SiC fiber might prove to be the exception.

When graphite fiber is used as the reinforcement, the CTE along the fiber axis is slightly negative. Residual stresses developed during fabrication place the fiber in compression and the matrix in tension. In any ceramic composite fabrication approach in which matrix shrinkage takes place, and especially in polymer derived materials, the matrix will likely be in tension as well. However, since ceramic matrix composites can be expected to see high temperatures both in fabrication and in service, the residual stresses are expected to be much greater than in their resin matrix counterparts, and may give rise to microcracking on fabrication. In terms of designing with composites this leads to a significant difference between resin and ceramic matrix materials. Whereas resin, glass and probably RBSN matrix composites can be designed as a crossply layup of unidirectional lamina, ceramic matrix composites in which matrix shrinkage occurs in processing likely will have to be fabricated from 2D cloth or as 3D woven structures to minimize shrinkage cracking.

Some ceramic matrix composites might alleviate the problem of residual stresses by choice of fibers with high axial CTE's relative to the matrix, which would place the matrix in compression. However, if the CTE of the fiber is much greater than that of the matrix, debonding of the interface will result (3).

Decreased matrix strain capability relative to the fiber may decrease tensile strength, as the matrix cannot plasticly deform to accommodate stress concentrations. Behavior under flexural loading might become more complex. Prewo

436

(6) has studied Nicalon/epoxy and Nicalon/lithium aluminosilicate (LAS) composites in both tension and flexure, and notes that while the Nicalon/epoxy composites had the higher tensile strength, in flexure the LAS matrix material appeared stronger as it "yielded" due to microcracking, shifting the neutral axis of the beam toward the compressive side of the test specimen. Thus, ceramic composites may appear to have greater strength in flexure than their resin matrix counterparts as a result of both their higher matrix compressive strength and the pseudoductility imparted by microcracking.

Microcracking also can become a problem to the environmental stability of the fibers in air, leading to fiber degradation and fracture during use at elevated temperature.

6. POLYMERIC PRECURSORS TO
CERAMIC MATRICES

The use of organometallic precursors which can be pyrolyzed to a ceramic char provides a means for forming ceramic matrices utilizing the advantages of ease of fiber infiltration, control of rheology and low temperature processing typical of resin matrix composites. Choice of polymer is influenced by stoichiometry of the resulting char and shrinkage on pyrolysis. Ideally, the closer to stoichiometric SiC or $Si_3N_4$ the final product, the more thermally stable it may be expected to be. Also,

the smaller the weight loss and volumetric shrinkage on pyrolysis, the less the need for multiple re-impregnation cycles and the less the likelihood of large cracks.

A number of organometallic polymers have been studied; some of these are shown in Table I. The polycarbosilane work of Yajima (14-16) serves as the basis for Nicalon fibers. The char is rich in excess carbon. It also contains oxygen intentionally introduced as a cross-linking agent to stabilize the fiber structure prior to pyrolysis. The deviation from stoichiometry results in thermal instability above 1200°C.

Most of the polysilanes and polycarbosilanes listed undergo weight losses of 40-60 percent on pyrolysis. Higher char yields have been demonstrated for the polysila-zanes; 80-85 percent char yields have been reported by Seyferth (23). The polysilazanes are, however, moisture sensitive, and require composite processing to be carried out in inert atmospheres.

Recent work in our laboratory (29) has examined a group of silsesquioxanes (Figure 2) having the general structure $RSiO_{1.5}$, where R = methyl, phenyl, propyl, or vinyl, as SiC precursors. The silsesquioxanes melt flow, thermally crosslink, and then can be pyrolyzed at nominally 500°C. At higher temperatures they can be expected to undergo a carbothermal reduction to SiC with loss of CO (Figure 3). By controlling the starting ratio of

Si/C we hope to control the stoichiometry of the end product.

Microstructure of Nicalon/
silsesquioxane composites is shown
in Figures 4 and 5. Initial impregnation shows few voids. After
pyrolysis followed by heating to
1000°C in argon, matrix shrinkage
and pore formation are evident.
(Figure 4). The darkest phase seen
in the optical micrograph (Figure
4b) is epoxy which was vacuum infiltrated into the composite after
pyrolysis for polishing, and shows
the extent of matrix cracking. The
composite might be reimpregnated
with silsesquioxane and again pyrolyzed to increase matrix density.

Matrix surface cracks both
parallel and perpendicular to the
fibers also are observed (Figure 5),
with the more matrix rich surfaces
showing the higher extent of cracking. These cracks likely arise
from a combination of shrinkage on
pyrolysis, mismatch in CTE between
fiber and matrix and anisotropic
fiber expansion.

We also have studied a SiC
particulate filled, high char yield
carbon resin (30) which introduced
the particulate third phase as a
means of minimizing matrix shrinkage and cracking. We were able to
fabricate unidirectional composites
with very low pore volumes and few
cracks on a single impregnation and
pyrolysis cycle, provided that we
used small fiber tow sizes and
fibers which were not uniformly
cylindrical (compare Figures 6 and

7). However, strain capability of
the matrix was low ($\sim$.3%), and tensile specimens failed in shear in
the tab region. Stress-strain
behavior was linear to fracture.
Cross-ply layups showed extensive
matrix cracking and delamination,
the result of residual thermal
stresses.

Jamet et al. (31) have shown
that additives of a BN particulate
to a polyvinylsilane minimizes
linear shrinkage and weight loss on
pyrolysis. Thus, the concept of
particulate filled precursors would
seem to warrant further study.

New polymers which yield more
stoichiometric ceramic products, as
well as small diameter, thermally
stable fibers are needed, as is a
greater understanding of the pyrolysis process and resulting stresses
at the fiber-matrix interface, for
major advances in new ceramic composite materials to be achieved.

The author wishes to thank
Drs. Donald R. Behrendt and James A.
DiCarlo for helpful discussions.

## 8. REFERENCES

1. D. K. Hale and A. Kelly,
   "Strength of Fibrous Composite
   Materials", in Annual Review of
   Materials Science, Vol. 2,
   R. A. Higgins, R. H. Bube and
   R. W. Roberts, eds., 1972, pp.
   405-462.

2. J. Aveston, G. A. Cooper and
   A. Kelly, "Single and Multiple
   Fracture", in the Properties
   of Fiber Composites, Conference

438

Proceedings, National Physical Laboratory, pp 15-26, November 1971.

3. D. K. Shetty, "Ceramic Matrix Composites", in Metals and Ceramics Information Center Current Awareness Bulletin, No. 118, Battelle Columbus Laboratories, December, 1982.

4. J. Aveston and A. Kelly, "Theory of Multiple Fracture of Fibours Composites", J. Mater. Sci., 8, 352 (1973).

5. K. M. Prewo, "A Compliant, High Failure Strain, Fibre-Reinforced Glass Matrix Composite", J. Mater. Sci., 17, 3549-3563 (1982).

6. K. M. Prewo, "Tension and Flexural Strength of Silicon Carbide Fibre-Reinforced Glass Ceramics", J. Mater. Sci., 21, 3590-3600 (1986).

7. K. M. Prewo, J. J. Brennan and G. K. Layden, "Fiber Reinforced Glasses and Glass-Ceramics for High Performance Applications", Ceram. Bull. 65, 305-313, 322 (1986).

8. E. Fitzer and R. Gadow, "Fiber Reinforced Ceramic Composites Fabricated Via the Sol/Gel Route", Conference on Tailoring Multiphase and Composite Ceramics, Penn. State Univ., July 1985.

9. D. E. Clark, "Sol-Gel Derived Ceramic Matrix Composites", in Science of Ceramic Chemical Processing, L. L. Hench and D. R. Ulrich, eds. New York,

Wiley, 1986, pp 237-246.

10. E. Fitzer and R. Gadow, "Fiber-Reinforced Silicon Carbide", Am. Ceram. Soc. Bull., 65, 326-35 (1986).

11. P. L. Lamicq, G. A. Bernhart, M. M. Danchier and J. G. Mace, "SiC/SiC Composite Ceramics", Am. Ceram. Soc. Bull., 65, 336-38 (1986).

12. R. T. Bhatt, "Mechanical Properties of SiC Fiber-Reinforced Reaction-Bonded $Si_3N_4$ Composites", NASA TM 87085 (1985).

13. R. T. Bhatt, "Effects of Fabrication Conditions on the Properties of SiC Fiber Reinforced Reaction-Bonded Silicon Nitride Matrix Composites (SiC/RBSN), NASA TM 88814 (1986).

14. S. Yajima, U. S. Patent 4,052,430 (1977).

15. S. Yajima, "Development of Ceramics, Especially SiC Fibres, from Organosilicon Polymers by Heat Treatment", Phil. Trans. Soc. London - A294, 419-426 (1980).

16. S. Yajima, "Special Heat-Resisting Materials from Organometallic Polymers", Am. Ceram. Soc. Bull., 62, 893-898, 903, 915 (1983).

17. R. West, "Polysilastyrene: Phenylmethylsilane-Dimethylsilane Copolymers as Precursors to Silicon Carbide", Am. Ceram. Soc. Bull., 62, 899-903 (1983).

18. K. S. Mazdiyasni, R. West and L. D. David, "Characterization of Organosilicon-Infiltrated

Porous Reaction-Sintered $Si_3N_4$", *J. Am. Ceram. Soc.*, *61*, 504-508 (1978).

19. R. W. Rice, "Ceramics from Polymer Pyrolysis, Opportunities and Needs", *Am. Ceram. Soc. Bull.*, *62*, 889-892 (1983).

20. R. R. Wills, R. A. Markle and S. P. Mukherjee, "Siloxanes, Silanes and Silazanes in the Preparation of Ceramics and Glasses, *Am. Ceram. Soc. Bull.*, *62*, 904-11 (1983).

21. C. L. Schilling, J. P. Wesson and T. C. Williams, "Polycarbosilane Precursors for Silicon Carbide", *Am. Ceram. Soc. Bull.*, *62*, 912-915 (1983).

22. D. Seyferth, "A Liquid Silazane Precursor to Silicon Nitride", *Comm. Am. Ceram. Soc.*, C-13 (1983).

23. D. Seyferth and G. H. Wiseman, "A Novel Polymeric Organosilazane Precursor to $Si_3N_4$/SiC Ceramics", in Science of Ceramic Chemical Processing, L. L. Hench and D. R. Ulrich, eds., New York, Wiley, 1986.

24. B. E. Walker, "Preparation and Pyrolysis of Monolithic Composite Ceramics Produced by Polymer Pyrolysis", *Am. Ceram. Soc. Bull.*, *62*, 916-923 (1983).

25. D. R. Behrendt, "Porous Silicon Carbide as a Matrix for Ceramic Composites", NASA TM 88837 (1986).

26. Avco Specialty Materials, Lowell, MA.

27. 3M Corporation, Minneapolis, MN.

28. Dupont deNemours, Wilmington, DE.

29. F. I. Hurwitz, L. Hyatt, J. Gorecki and L. D'Amore, "Silsesquioxanes as Precursors to Ceramic Composites", to be published in *Ceram. Eng. Sci. Proc.* (1986).

30. F. I. Hurwitz, "Carbon-Rich Ceramic Composites from Ethynyl Aromatic Precursors", NASA TM 88812 (1986).

31. J. Jamet, J. R. Spann, R. W. Rice, D. Lewis and W. S. Coblenz, "Ceramic-Fiber Composite Processing via Polymer-Filler Matrices", *Ceram. Eng. Sci. Proc.*, *5*, 677-694 (1984).

## TABLE I

### Some organometallic precursors to ceramics

| Polymer | Structure | Reference |
|---|---|---|
| polycarbosilane | $(-\overset{\textstyle |}{\underset{\textstyle |}{Si}} - \overset{\textstyle |}{\underset{\textstyle |}{C}} -)_n$ | 14-16 |
| polyvinylsilane | $CH_3-\overset{\textstyle CH_3}{\underset{\textstyle CH=CH_2}{Si}}-(\,Si\,)_{\overline{n}}\,Si-CH_3$ | 21 |
| polyborosiloxane | $\Phi$ = phenyl | 16 |
| polysilastyrene | $(-\overset{\textstyle CH_3}{\underset{\textstyle \Phi}{Si}}-)_n$ | 17 |
| polysilazane | $R = CH_2$ | 22-23 |

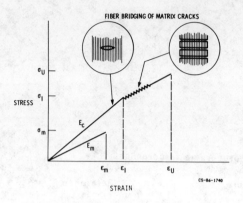

Figure 1 - Theoretical stress strain behavior for ceramic matrix composite fiber bridging of matrix cracks. (Figure courtesy of Dr. James A. DiCarlo.

$$2\,SiO_{1.5} + 5\,C \longrightarrow 2\,SiC + 3\,CO\uparrow$$

$$SiO_{1.5} + C\,(excess) \longrightarrow SiC + C + CO\uparrow$$

$$SiO_{1.5} + C\,(deficient) \longrightarrow SiC + SiO_2$$
$$+ SiO\uparrow + CO\uparrow$$

Figure 3 - Possible carbothermal reduction products of silsesquioxanes based on the Si/C ratio in the starting polymer.

SILSESQUIOXANES

$$RSiO_{1.5}$$

R = methyl, propyl, vinyl, phenyl

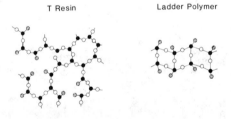

T Resin                    Ladder Polymer

Figure 2 - Silsesquioxanes may exhibit either an extended cross-linked (T resin) or double ladder structure, or a mixture of the two. The extended structure is more prevalent when R = phenyl.

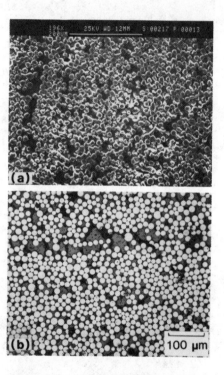

Figure 4 - Nicalon/silsesquioxane composite (a) as fabricated and (b) after pyrolysis followed by heating to 1000°C in argon.

442

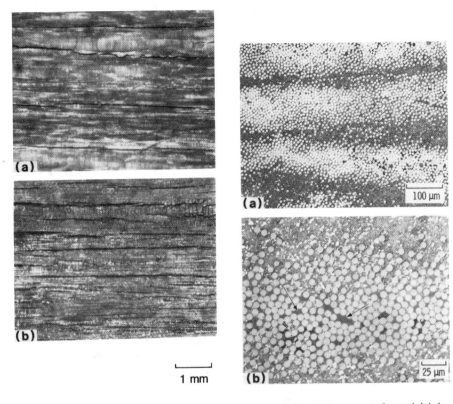

Figure 5 - Surface cracking of Nicalon/silsesquioxane composite (a) on pyrolysis at 525°C for 2 hours and (b) after heating to 1000°C in argon.

Figure 6 - Celion reinforced high char yield poly(arylacetylene) matrix filled with SiC particulate showing (a) layering of fiber rich and matrix rich regions and (b) matrix areas devoid of particulate (arrows).

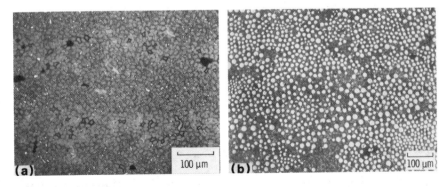

Figure 7 - (a) Nextel and (b) Nicalon reinforced SiC particulate filled poly(arylacetylene) matrix. Composite homogeneity is enhanced when fibers cannot close-pack (Compare Figure 6).

# A PROCESS OPTIMIZATION METHOD FOR LAP SHEAR BOND STRENGTH

Beth Anne Kaufman
Duane H. Stone
Texas Instruments Incorporated
Dallas, Texas 75266

## Abstract

The purpose of this paper is to review an experimental design method that was used to optimize the bonding process for a modified epoxy film adhesive. The optimum process for the adhesive is used to structurally bond a radar antenna. ASTM D 1002, "Strength Properties of Adhesives in Shear by Tension Loading," was selected as the test method for evaluating the lap shear bond strength at room and elevated temperatures. Statistical techniques were used to configure the samples and optimize the bonding process.

## 1. INTRODUCTION

The purpose of this paper is to demonstrate an experimental design technique that was used to optimize the bonding process of a modified epoxy adhesive. The process was required to yield high shear strength at room temperature and at 105°C. The resulting process was used for bonding a flat-plate antenna face sheet to a back plate.

Flat-plate antennas were being bonded using a modified epoxy film adhesive. The current bond process was not yielding sufficient shear strength data to meet the antenna's shear strength requirements at room temperature and 105°C. It was decided to optimize the bond process using Taguchi methods.

The Taguchi method is a problem-solving technique that is used in Japan. It was first introduced in the United States in 1984 and was used primarily in the automobile industry. The method is an improved version of classical experimental designs, and uses techniques such as analysis of the variance (ANOVA), orthogonal arrays, linear graphs, and fractional factorial analysis. It allows for efficient analysis of a large number of variables, making it practical for industry.

The initial step was to determine the process variables. These variables were divided into two sections: most critical variables and variables that could be controlled (Table 1). Two levels of performance were selected for the most critical variables. These levels are variable-dependent. Possible interactions between variables were determined and are listed in Table 1.

The variables and possible interactions were assigned positions on a fractional factorial analysis table (Table 2). This table uses orthogonal arrays to vary process variables. It allows for a minimum amount of process procedures to test with a maximum amount of variation.

The order for building the test specimens was then radomized, and the specimens were prepared. One of the etch processes used has a 72-hour window in which the bonding could be done. This constrained radomizing of the order for building the samples within each of the etch processes.

Lap shear specimens, per ASTM D 1002, were used to determine the shear stress of the specimens. These tests were performed at room temperature and 105°C. Thermocouples were bonded to the specimen to ensure proper temperature. A set of five samples was tested for each process procedure at each temperature.

The final data were then analyzed and an optimum process established. This final process was used to build a set of 25 samples to be tested at room temperature and 105°C. These samples were used to verify the final process with a degree of confidence of 95 percent.

## 2. SAMPLE PREPARATION

The lap shear samples used 6061–T6 aluminum alloy as the substrate and their dimensions were 4 by 7 inches. The vast majority of published data on lap shear strength of adhesives is from tests performed on samples prepared with 2024–T4 aluminum. However, 6061 was chosen for this series of tests since it is the alloy used in the antenna.

The bond surfaces of the test panels were prepared using two methods of finishing: chromate conversion coating and phosphoric acid anodizing (PAA). The chromate conversion coating was performed in-house, and the PAA process was subcontracted. The chromate conversion and PAA-processed panels were bonded within 72 hours of processing to reduce the risk of oxidation.

The thickness of the modified epoxy film adhesive was 0.001 inch. Because of the thickness of the adhesive system, it was necessary to carefully select the substrate. The upper and lower panels were a standard 0.063-inch thick, with the shim plate matching the lower plate to ±0.0002 inch.

The test specimens were assembled on a flat tooling plate. A strip of the adhesive was placed across the 7-inch direction. Alignment pins were used to ensure the standard ½-inch overlap of the bond area. A silicone rubber sheet was used as the bag.

A vacuum line was attached to the bag as needed.

The tooling plate was placed on a base plate, which was preheated per the fractional factorial table. The samples were cured in an autoclave using the temperatures, times, and pressures dictated by the table. The temperatures were monitored using thermocouples placed in the autoclave chamber, base plate, and tooling plate. The timing was started as soon as the bond plate reached 160°C. The tooling plate was allowed to cool below 65°C before the samples were debagged.

The panels were then cut into five 1-inch strips for testing. The outer edges of the panel were not used for samples because of wicking of the adhesive material. The samples were then marked per process and sent to testing.

## 3. TESTING

The specimens were tested on an Instron 1137 Universal Testing Machine with Microcon II microprocessor. The dimensions of the bond sample were measured and input to the microcon. The samples were clamped in wedge action grips and pulled at a speed of 0.05 inch per minute. The microcon recorded the peak load and calculated the peak stress of the sample from the cross-sectional area.

High-temperature testing was performed using a temperature chamber that surrounds the grips and the sample. Thermocouples were bonded over the lap joint with conductive cement. The sample was then placed into the grips. The sample was monitored until it reached the test temperature of

105°C. It was allowed to ambiate for 10 minutes before applying load.

## 4. ANALYSIS

The mean of the five samples tested was used to evaluate the effect each variable had on the shear stress. Table 3 shows the stress at room temperature and 105°C for each of the processes performed.

The sum of the levels for each variable was tabulated. Analysis of the variance was then used to evaluate the effect of each variable for each test temperature (Tables 4 and 5). Table 6 is an analysis of the variance for optimization between room temperature and 105°C. This is referred to as overall optimization.

After determining which variable had the most significant effect on the process, the optimum process for each test condition was determined. The level selected was based on which sum of levels was the highest number. Table 7 shows the process that was used for each optimization.

From the data, it was determined that a sample size of 25 was needed to have a 95-percent confidence in the results. The samples were then built for each optimization: room temperature, 105°C, and overall, to verify their performance (Table 8).

A student's T test at 99-percent confidence was used to compare the mean of the three verification results. The room temperature test results showed the mean to be equivalent for all three processes. The 105°C test results showed the room temperature and 105°C mean to be equivalent and greater than the overall optimization.

## 5. DISCUSSION

The critical variable for room temperature samples was determined to be etch process. This was determined by analysis of the variance as shown in Table 4. This affected the sample result by 53.51 percent. The second critical variable was pressure at 28.15 percent. Table 5 shows that the critical variable for 105°C was pressure, affecting the result by 56.76 percent. Iron time interacting with iron temperature, vacuum interacting with pressure, and cure time had approximately equal effects on the sample.

One of the assumptions made by classical experimental design is that the variance of a sample holds true no matter what the test condition. Taguchi methods assume that the variance of a sample is independent of the sample and can change for each test condition. This allows for optimization of a process by analysis of the variance that occurs between various test conditions. It can be applied to an infinite number of test conditions by applying the following equation to the test result for that process.

$$-10 \times \log\left(\frac{1}{n}\sum_i \frac{1}{y^2}\right) \qquad (1)$$

where

$n$ = number of tests
$y$ = value for the test.

Analysis of the variance was then applied to the resulting numbers from Equation (1) to yield an overall optimization process. The overall contributors to the process were cure temperature 52.79 percent and pressure 26.93 percent for optimization between test conditions.

It is interesting to note that the interaction of vacuum with pressure was not a major contributor to the process as was first thought. The interaction was evaluated by plotting the sums of Level One effects versus Level Two effects. Figure 1 demonstrates that using vacuum with pressure Level One increases the shear strength, while using vacuum with pressure Level Two decreases the shear strength.

When the initial levels were selected for cure temperature, cure time, and cure pressure, it was assumed a linear relationship existed within the levels for each of the variables. However, this is usually not true. Therefore, it was decided to perform another set of tests that would take into account a nonlinear relationship. This was performed on the three variables: temperature, time, and pressure. Three levels for each of these variables were evaluated (Table 9). The fractional factorial table with test results is shown in Table 10.

The second set of processes was performed using the best process recommended by the original evaluation, such as PAA etch, vacuum, and preheat temperature of 177°C.

Analysis of the variance was again applied to the results to determine the important contributor to the process. Room temperature showed pressure at 49.90 percent with cure time and cure temperature approximately equal. The 105°C critical variables were cure time 38.70 percent and pressure 28.18 percent. Overall optimization showed pressure at 39.65 percent and cure time at 39.25 percent contribution.

Twenty-five samples for verification were performed on the room temperature and 105°C optimum processes. The overall process selected as optimum was equivalent to the 105°C, since either the Level One or Level Two cure temperatures could be used. The final processes are shown in Table 11.

## 6. CONCLUSION

In conclusion, this experimental design technique was found to be an excellent method for optimizing a process. The analysis can be applied to multilevel process variables as well as simple two-level variables. The process also can be optimized to yield the optimum results needed to meet a variety of requirements.

## 7. ACKNOWLEDGMENTS

The authors would like to acknowledge the following people for their support in preparing and testing the samples: Gary Seifferman, Don Rice, Kirk Norris, Larry Henry, Eric Shields, and Brian Thompson.

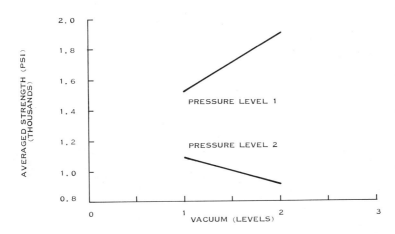

Figure 1
Interaction of Pressure and Vacuum

Table 1
Process Variables and Levels

| Code | Variable | Level 1 | Level 2 |
|------|----------|---------|---------|
| A | Iron time | <20 s | >30 s |
| B | Iron temperature | 158° to 171°C | 177° to 190°C |
| C | Bonding cure time | 1 hour | 2 hours |
| D | Bonding cure temperature | 166°C | 193°C |
| E | Cure pressure | 15 psi | 90 psi |
| F | Etch | PAA | Alodine |
| G | Vacuum | 0 | 28-mm Hg |
| H | Error | | |

Expected interactions

| | |
|---|---|
| Cure time and cure temperature | (C/D) |
| Iron time and iron temperature | (A/B) |
| Vacuum and pressure | (G/E) |

## Controllable Variables

Flatness of tooling plate

Thickness of lap shear panels

Repeatability of vacuum bagging

Handling of panels

Monitoring time and temperature

## Table 2
### Fractional Factorial Table

| Variable Code | G | E | D | C | A | F | C/D | B | H | e1 | H/E | e1 | A/B | e1 | e1 |
|---|---|---|---|---|---|---|---|---|---|---|---|---|---|---|---|
| **Process Number** | | | | | | | | | | | | | | | |
| 1 | 1 | 1 | 1 | 1 | 1 | 1 | 1 | 1 | 1 | 1 | 1 | 1 | 1 | 1 | 1 |
| 2 | 1 | 1 | 1 | 1 | 1 | 1 | 1 | 2 | 2 | 2 | 2 | 2 | 2 | 2 | 2 |
| 3 | 1 | 1 | 1 | 2 | 2 | 2 | 2 | 1 | 1 | 1 | 1 | 2 | 2 | 2 | 2 |
| 4 | 1 | 1 | 1 | 2 | 2 | 2 | 2 | 2 | 2 | 2 | 2 | 1 | 1 | 1 | 1 |
| 5 | 1 | 2 | 2 | 1 | 1 | 2 | 2 | 1 | 1 | 2 | 2 | 1 | 1 | 2 | 2 |
| 6 | 1 | 2 | 2 | 1 | 1 | 2 | 2 | 2 | 2 | 1 | 1 | 2 | 2 | 1 | 1 |
| 7 | 1 | 2 | 2 | 2 | 2 | 1 | 1 | 1 | 1 | 2 | 2 | 2 | 2 | 1 | 1 |
| 8 | 1 | 2 | 2 | 2 | 2 | 1 | 1 | 2 | 2 | 1 | 1 | 1 | 1 | 2 | 2 |
| 9 | 2 | 1 | 2 | 1 | 2 | 1 | 2 | 1 | 2 | 1 | 2 | 1 | 2 | 1 | 2 |
| 10 | 2 | 1 | 2 | 1 | 2 | 1 | 2 | 2 | 1 | 2 | 1 | 2 | 1 | 2 | 1 |
| 11 | 2 | 1 | 2 | 2 | 1 | 2 | 1 | 1 | 2 | 1 | 2 | 2 | 1 | 2 | 1 |
| 12 | 2 | 1 | 2 | 2 | 1 | 2 | 1 | 2 | 1 | 2 | 1 | 1 | 2 | 1 | 2 |
| 13 | 2 | 2 | 1 | 1 | 2 | 2 | 1 | 1 | 2 | 2 | 1 | 1 | 2 | 2 | 1 |
| 14 | 2 | 2 | 1 | 1 | 2 | 2 | 1 | 2 | 1 | 1 | 2 | 2 | 1 | 1 | 2 |
| 15 | 2 | 2 | 1 | 2 | 1 | 1 | 2 | 1 | 2 | 2 | 1 | 2 | 1 | 1 | 2 |
| 16 | 2 | 2 | 1 | 2 | 1 | 1 | 2 | 2 | 1 | 1 | 2 | 1 | 2 | 2 | 1 |

Table 3

Table 3
Resultant Stress Per Process

| Process Variable | G | E | D | C | A | F | C/D | B | H | e1 | H/E | e1 | A/B | e1 | e1 | Results Room Temp (psi) | 105°C (psi) |
|---|---|---|---|---|---|---|---|---|---|---|---|---|---|---|---|---|---|
| Process Number | | | | | | | | | | | | | | | | | |
| 1 | 1 | 1 | 1 | 1 | 1 | 1 | 1 | 1 | 1 | 1 | 1 | 1 | 1 | 1 | 1 | 4,001 | 1,623.0 |
| 2 | 1 | 1 | 1 | 1 | 1 | 1 | 1 | 2 | 2 | 2 | 2 | 2 | 2 | 2 | 2 | 4,285 | 2,132.0 |
| 3 | 1 | 1 | 1 | 2 | 2 | 2 | 2 | 1 | 1 | 1 | 1 | 2 | 2 | 2 | 2 | 4,353 | 1,720.0 |
| 4 | 1 | 1 | 1 | 2 | 2 | 2 | 2 | 2 | 2 | 2 | 2 | 1 | 1 | 1 | 1 | 4,390 | 2,245.0 |
| 5 | 1 | 2 | 2 | 1 | 1 | 2 | 2 | 1 | 1 | 2 | 2 | 1 | 1 | 2 | 2 | 3,002 | 691.6 |
| 6 | 1 | 2 | 2 | 1 | 1 | 2 | 2 | 2 | 2 | 1 | 1 | 2 | 2 | 1 | 1 | 3,232 | 663.3 |
| 7 | 1 | 2 | 2 | 2 | 2 | 1 | 1 | 1 | 1 | 2 | 2 | 2 | 2 | 1 | 1 | 3,801 | 1,605.0 |
| 8 | 1 | 2 | 2 | 2 | 2 | 1 | 1 | 2 | 2 | 1 | 1 | 1 | 1 | 2 | 2 | 3,257 | 675.4 |
| 9 | 2 | 1 | 2 | 1 | 2 | 1 | 2 | 1 | 2 | 1 | 2 | 1 | 2 | 1 | 2 | 3,142 | 1,693.0 |
| 10 | 2 | 1 | 2 | 1 | 2 | 1 | 2 | 2 | 1 | 2 | 1 | 2 | 1 | 2 | 1 | 2,740 | 1,169.0 |
| 11 | 2 | 1 | 2 | 2 | 1 | 2 | 1 | 1 | 2 | 1 | 2 | 2 | 1 | 2 | 1 | 3,364 | 1,499.0 |
| 12 | 2 | 1 | 2 | 2 | 1 | 2 | 1 | 2 | 1 | 2 | 1 | 1 | 2 | 1 | 2 | 3,016 | 1,550.0 |
| 13 | 2 | 2 | 1 | 1 | 2 | 2 | 1 | 1 | 2 | 2 | 1 | 1 | 2 | 2 | 1 | 2,607 | 1,209.0 |
| 14 | 2 | 2 | 1 | 1 | 2 | 2 | 1 | 2 | 1 | 1 | 2 | 2 | 1 | 1 | 2 | 2,164 | 571.9 |
| 15 | 2 | 2 | 1 | 2 | 1 | 1 | 2 | 1 | 2 | 2 | 1 | 2 | 1 | 1 | 2 | 2,810 | 1,059.0 |
| 16 | 2 | 2 | 1 | 2 | 1 | 1 | 2 | 2 | 1 | 1 | 2 | 1 | 2 | 2 | 1 | 2,952 | 1,294.0 |

Table 4

Analysis of the Variance for Room Temperature Tests

| | G 1 | E 2 | D 3 | C 4 | A 5 | F 6 | C/D 7 | B 8 | H 9 | e1 10 | H/E 11 | e1 12 | A/B 13 | e1 14 | e1 15 |
|---|---|---|---|---|---|---|---|---|---|---|---|---|---|---|---|
| Level 1 | 30321 | 29291 | 27319 | 25173 | 26662 | 26988 | 26495 | 27080 | 26029 | 26465 | 26016 | 26367 | 25728 | 26556 | 27087 |
| Level 2 | 22795 | 23825 | 25554 | 27943 | 26454 | 26128 | 26621 | 26036 | 27087 | 26651 | 27100 | 26749 | 27388 | 26560 | 26029 |
| Difference | 7526 | 5466 | 1765 | 2770 | 208 | 860 | 126 | 1044 | 1058 | 186 | 1084 | 382 | 1660 | 4 | 1058 |
| Sum of sqs | 3540042 | 1867322 | 194701 | 479556.2 | 2704 | 46225 | 992.25 | 68121 | 69960.2 | 2162.25 | 73441 | 9120.25 | 172225 | 1 | 69960.2 |
| Total | 6596534 | | | | | | | | | | | | | | |
| Pure variance | 3529841 | 1857121 | 184500 | 469355.4 | — | — | — | 57920.2 | 59759.4 | — | 63240.2 | — | 162024 | — | 59759.4 |
| Percent contribution | 53.51054 | 28.15298 | 2.79693 | 7.115182 | 0 | 0 | 0 | 0.87803 | 0.90592 | 0 | 0.95868 | 0 | 2.45620 | 0 | 0.90592 |
| Total error | 1.988392 | | | | | | | | | | | | | | |

Table 5

Analysis of the Variance for 150°C Tests

| | G | E | D | C | A | F | C/D | B | H | e1 | H/E | e1 | A/B | e1 | e1 |
|---|---|---|---|---|---|---|---|---|---|---|---|---|---|---|---|
| | 1 | 2 | 3 | 4 | 5 | 6 | 7 | 8 | 9 | 10 | 11 | 12 | 13 | 14 | 15 |
| Level 1 | 11355.3 | 13577 | 10663.7 | 9698.8 | 10511.9 | 11196.4 | 10865.3 | 11045.6 | 10224.5 | 9685.6 | 9668.7 | 10927 | 9533.9 | 10956.2 | 11307.3 |
| Level 2 | 9990.9 | 7769.2 | 9492.3 | 11647.4 | 10834.3 | 10149.8 | 10480.9 | 10300.6 | 11121.7 | 11660.6 | 11677.5 | 10419.2 | 11812.3 | 10390 | 10038.9 |
| Difference | 1364.4 | 5807.8 | 1171.4 | 1948.6 | 322.4 | 1046.6 | 384.4 | 745 | 897.2 | 1975 | 2008.8 | 507.8 | 2278.4 | 566.2 | 1268.4 |
| Sum of sqs | 116349.2 | 2108158 | 85761.1 | 237315.1 | 6496.36 | 68460.7 | 9235.21 | 34689.0 | 50310.4 | 243789 | 252204 | 16116.3 | 324444 | 20036.4 | 10055.2 |
| Total | 3673919 | | | | | | | | | | | | | | |
| Pure variance | 93535.23 | 2085344 | 62947.1 | 214501.1 | — | — | — | 1875.0 | 27496.5 | — | 229390 | — | 301630 | — | 77738.4 |
| Percent contribution | 2.545925 | 56.76076 | 1.71335 | 5.838482 | 0 | 0 | 0 | 0.32322 | 0.74842 | 0 | 6.24376 | 0 | 8.21003 | 0 | 2.11595 |
| Total error | 12.64824 | | | | | | | | | | | | | | |

452

Table 6

Analysis of the Variance for Overall Optimization

| | G 1 | E 2 | D 3 | C 4 | A 5 | F 6 | C/D 7 | B 8 | H 9 | e1 10 | H/E 11 | e1 12 | A/B 13 | e1 14 | e1 15 |
|---|---|---|---|---|---|---|---|---|---|---|---|---|---|---|---|
| Level 1 | 515.315 | 532.075 | 455.735 | 507.135 | 512.335 | 517.675 | 515.735 | 518.335 | 513.595 | 508.815 | 508.495 | 514.915 | 506.895 | 516.255 | 518.895 |
| Level 2 | 512.68 | 495.92 | 506.06 | 520.86 | 515.66 | 510.32 | 512.26 | 509.66 | 514.4 | 519.18 | 519.5 | 513.08 | 521.1 | 511.74 | 509.1 |
| Difference | 2.635 | 36.155 | 50.325 | 13.725 | 3.325 | 7.355 | 3.475 | 8.675 | 0.805 | 10.365 | 11.005 | 1.835 | 14.205 | 4.515 | 9.795 |
| Sum of sqs | 0.433951 | 81.69900 | 158.287 | 11.77347 | 0.69097 | 3.38100 | 0.75472 | 4.70347 | 0.04050 | 6.71457 | 7.56937 | 0.21045 | 12.6113 | 1.27407 | 5.99637 |
| Total | 296.1411 | | | | | | | | | | | | | | |
| Pure variance | — | 79.75616 | 156.345 | 9.830638 | — | — | — | — | — | — | 5.62653 | — | 10.6685 | — | — |
| Percent contribution | 0 | 26.93180 | 52.7940 | 3.319578 | 0 | 0 | 0 | 0 | 0 | 0 | 1.89995 | 0 | 3.60251 | 0 | 0 |
| Total | 5.904460 | | | | | | | | | | | | | | |

Table 7

Optimium Processes

| | Iron Time (s) | Iron Temp (°C) | Cure Time (hr) | Cure Temp (°C) | Cure Pressure (psi) | Preheat Temp (°C) | Etch | Vacuum |
|---|---|---|---|---|---|---|---|---|
| Room temperature | <20 | 158 to 171 | 2 | 166 | 15 | 177 | PAA | Full |
| 105°C | >30 | 158 to 171 | 2 | 166 | 15 | 177 | PAA | Full |
| Overall | >30 | 158 to 171 | 2 | 193 | 15 | 177 | PAA | Full |

Table 8

Verification Results

Test Temperature

| | Room Temperature (psi) | | 105°C (psi) | |
|---|---|---|---|---|
| | Mean | Standard Deviation | Mean | Standard Deviation |
| Room temperature optimization | 4,764 | 272.6 | 2,716 | 160.0 |
| 105°C optimization | 4,908 | 148.0 | 2,829 | 234.5 |
| Overall optimization | 4,920 | 154.1 | 2,493 | 148.0 |

Table 9

Three-Level Variables

| | Level 1 | Level 2 | Level 3 |
|---|---|---|---|
| Cure pressure (psi) | 15 | 30 | 45 |
| Cure temperature (°C) | 166 | 177 | 192 |
| Cure time (hours) | 1 | 2 | 3 |

## Table 10
### Results of Level-Three Processes

| Variable | Cure Pressure | Cure Temperature | Cure Time | e1 | Results |
|---|---|---|---|---|---|
| | | | | | Room Temp (psi) / 105°C (psi) |
| Process Number | | | | | |
| 1 | 1 | 1 | 1 | 1 | 4,509   2,131 |
| 2 | 1 | 2 | 2 | 2 | 4,852   2,626 |
| 3 | 1 | 3 | 3 | 3 | 4,838   2,843 |
| 4 | 2 | 1 | 2 | 3 | 5,159   2,214 |
| 5 | 2 | 2 | 3 | 1 | 4,743   2,514 |
| 6 | 2 | 3 | 1 | 2 | 4,368   1,907 |
| 7 | 3 | 1 | 3 | 2 | 4,372   2,519 |
| 8 | 3 | 2 | 1 | 3 | 4,080   1,750 |
| 9 | 3 | 3 | 2 | 1 | 3,735   1,460 |

## Table 11
### Final Process Optimization

| | Cure Pressure (psi) | Cure Temperature °C | Cure Time (hours) |
|---|---|---|---|
| Room temperature and overall | 30 | 166 | 3 |
| 105°C | 15 | 177 | 3 |

SURFACE PREPARATION OF POLYOLEFINS PRIOR TO
ADHESIVE BONDING

R. Rosty
D. Martinelli
A. Devine
M. J. Bodnar
J. Beetle
U.S. Army Armament Research, Development
and Engineering Center
Dover, NJ  07801-5001

## Abstract

The standard polyolefin surface
treatment for adhesive bonding
specifies using a potassium dichrom-
ate/sulfuric acid/water solution.
This solution is both carcinogenic
and polluting due to chromate inclusion.  A study was made of the treated surface of polyolefin versus a
control (untreated) surface in an
attempt to find an alternate chemical
polyolefin pretreatment for adhesive
bonding.  The new pretreatment
solution would have to be noncarcinogenic and nonpolluting.  Scanning
electron microscope photos of a
polyethylene surface treated in the
standard solution show pitting not
seen in the controls, similar to the
P2 etch effect on aluminum surfaces.
This pitting roughens the surface
and leaves a larger surface area for
adhesive bonding, suggesting one of
the reasons this pretreatment yielded
such good bond strengths.  Another
reason may be because it is an
oxidant.  Many nonchromate-bearing
alternative solutions were tested.
A good number of these were oxidizing
agents, patterning the standard solution.  Several test solutions resulted in bond strengths close in
magnitude to those resulting from
standard solution usage.

## 1.  INTRODUCTION

Untreated polyolefin surfaces are
difficult to adhesive bond because
of their nonpolarity and nonporosity
(1).  Polyolefins also have low surface and cohesive energies which are
poor adhesive bonding characteristics (2).  For these reasons, polyolefin surfaces must be treated
prior to adhesive bonding.

Various methods of treating polyolefins have been found to alter the
surface, making it more bondable.
These methods include the flame and
gas plasma treatments.  The chemical
method generally used is a potassium
dichromate/sulfuric acid/water solution (standard solution), although
chromates have been found to be
carcinogenic and polluting (1).

An alternate chemical solution that
is noncarcinogenic and nonpolluting
must be found to replace the old one.
Although other methods of treatment
are available, a chemical solution
is preferable in some cases where
the other treatments are not applicable or in the instances where a
polyolefin object has an irregular
shape.  The flame and gas plasma
treatments may yield a non-uniform
treatment on an irregularly shaped
object.  The secondary objective of
this work was to determine, through
various analytical procedures, the
reason why the standard solution was

such an effective polyolefin surface preparation for adhesive bonding.

## 2. DISCUSSION

### 2.1 Chemical Solution Testing

The search for an alternate chemical pretreatment for the adhesive bonding of a polyolefin material was initiated with a preliminary testing phase. This preliminary testing was used to screen numerous replacement candidates, some of which were strong oxidizing materials like the standard solution. The screening was accomplished with the use of the lap shear specimen shown in figure 1, which utilizes the experimentally treated polyolefin material as the substrate. This specimen was good as a quick screening tool for evaluating a potential replacement for the standard polyolefin pretreatment. It was not used for final analysis, however, because the polyolefin substrate material was not rigid as specified in ASTM specification D3163 (3). For this reason, the better pretreatment candidates were again analyzed with the adhesive joint shown in figure 2, as specified in ASTM specification D3164 (4). Although this specimen is more difficult to fabricate, the aluminum substrate is now rigid with the polyolefin material either treated or a control (unetched) serving as the sandwich material. The lap shear testing of this specimen gave shear strength data for the various treated polyolefin surfaces allowing comparison between the control, the standard pretreatment, and the new test polyolefin treatments. This was possible because the only variable was the polyolefin surface treatment. The aluminum surface treatment (P2 etch), the adhesive (Epon 828/V-140), and the adhesive curing conditions (4 hours at 70°C) remained constant throughout this testing.

Lap shear strengths resulting from treating the low density polyethylene material with various chemical solutions are diagramed in the bar chart in figure 3 (5). The polyethylene sandwich material was immersed in most solutions for 7 days at room temperature. These conditions were selected to allow the maximum amount of chemical exposure time in order to compare the solutions under the same treatment conditions. Higher solution temperatures were tested in some cases to see if better results could be obtained, as with the lead dioxide/sulfuric acid/water solution. Optimal polyolefin exposure conditions with the better pretreatments shown in this chart should be determined experimentally.

The standard solution was still the best pretreatment according to the lap shear results, but a solution of bleach/detergent and lead dioxide/sulfuric acid/water resulted in the same type of joint failure as the standard solution. This failure type was polyethylene cohesive failure within the polyethylene material itself. Both of these new solutions have safety problems associated with them and users should study them before using either solution.

The potassium iodate/sulfuric acid/water solution, which comes next on the bar chart, had good lap shear results, however, after long exposure times this pretreatment begins to dye the polyethylene pinkish in color.

### 2.2 Analytical Procedures

In an attempt to understand the reason why the standard solution, along with the new test solutions (bleach/detergent, lead dioxide/sulfuric acid/water, and potassium iodate/sulfuric acid/water) yield such high lap shear strengths compared to the other solutions tested, analytical work was performed. This analytical work included weight loss studies of the low density polyethylene versus time in the standard solution, as well as scanning electron microscope (SEM) and energy dispersive spectroscopy (EDS) analyses.

The graph resulting from the weight loss study of the low density polyethylene versus time in the standard solution is shown in figure 4. This data proves that the polyethylene is physically losing a portion of its total weight versus time in the standard solution. The weight loss rate is equal to approximately 0.1 percent of the total specimen weight every 40 hours for at least the first 160 hours of solution exposure.

Scanning electron microscope photographs were taken of both chemically treated and control (unetched) low density surfaces to discover if the chemical pretreatments resulting in high lap shear strengths were physically altering the surface in any way (6).

Figure 5 shows a control low density polyethylene surface at 3,000 times magnification. Scratches and irregularities are evident on the surface. Figure 6 shows the same surface type at 3,000 times magnification after treatment for 7 days at room temperature in the standard solution. A regular system of pitting is evident, which is not seen on the control specimen. This pitting is physical evidence that some of the surface material is being removed due to exposure in the standard solution and explains the results of our weight study. This solution appears to selectively etch the surface in a way similar to the P2 etch affect on aluminum surfaces when preparing them for adhesive bonding. This surface pitting helps the adhesion in that it roughens the surface, and in so doing, increases the surface bonding area and allows for a mechanical interlocking of the adhesive with the surface.

The following explanation may shed some light on what is actually occurring at the surface. According to reference 2, polyethylene is a melt crystallized polymer. Apparently, when the polymer is cooled, the higher molecular weight species crystallize at the liquid-solid interface. These higher molecular weight crystalline species stay in the bulk of the polymer, while the lower molecular weight species are rejected to the polyethylene surface where they remain there in an amorphous form. This low molecular weight amorphous species provides a low mechanical strength surface, resulting in poor adhesive bonds (2). The removal of this surface by the

standard solution, along with the resultant pitting appears to provide a good polyethylene surface for adhesive bonding.

Figure 7 is a 3,000 times magnified SEM photo taken of the low density polyethylene surface exposed to the bleach/detergent solution for 7 days at room temperature. The bleach/detergent solution showed this same type of pitting, but the pitting was not as extensive as that resulting from the standard solution.

Figure 8 is a 3,000 times magnified SEM photo of the polyethylene surface treated in the potassium iodate/sulfuric acid/water solution for 7 days at room temperature. The surface was pitted but in a different way. The pits were not as deep or large as resulted from the standard solution, but were much more extensive in number. Some dark spots were evident on some areas of the surface. An EDS analysis was run on one of these spots, but showed no evidence of iodine.

## 2.3  Sanding

Sanding the polyolefin surface prior to chemical treatment with the better solutions led to lap shear strengths with the ASTM 3164 specimen that were worse than if no abrasive treatment had been used prior to chemical treatment (7).

## 3.  CONCLUSIONS

A few strong oxidizing polyolefin pretreatment solutions were discovered that lead to adhesive bond lap shear strengths comparable with results from the standard solution.

Sanding prior to chemical pretreatment of the polyolefin did not prove effective.

Scanning electron microscope photos of the standard solution, as well as the bleach/detergent and potassium iodate/sulfuric acid/water solution, show evidence of polyethylene surface pitting.

## 4. REFERENCES

1. Landrock, A. H., Processing Handbook on Surface Preparations for Adhesive Bonding, Picatinny Arsenal Technical Report 4883, December 1975.

2. Baijal, M., Plastics Polymer Science and Technology, John Wiley and Sons, NY, 1982.

3. American Society for Testing and Materials, Standard Recommended Practice for Determining the Strength of Adhesively Bonded Rigid Plastic Lap Shear Joints in Shear by Tension Loading, ASTM D3163.

4. American Society for Testing and Materials, Standard Recommended Practice for Determining the Strength of Adhesively Bonded Plastic Lap Shear Sandwich Joints in Shear by Tension Loading, ASTM D3164.

5. Rosty, R., et al, Preparation of Polyolefin Surfaces for Adhesive Bonding, Volumes 1 and 2, Technical Report ARAED-TR-86005, ARDEC, Dover, NJ, May 1986.

6. Rosty, R., et al, Scanning Electron Microscopic and Energy Dispersive Spectroscopic Surface Analysis of Polyolefins Pretreated to Enhance Adhesive Bond Strengths, Technical Report ARAED-TR-86016, ARDEC, Dover, NJ, July 1986.

7. Rosty, R., et al, Preparation of Polyolefin Surfaces for Adhesive Bonding, Volume 3, Technical Report ARAED-TR-86005, ARDEC, Dover, NJ, June 1986.

## 5. BIOGRAPHIES

Roberta Rosty is a materials engineer with the Adhesives Section of the Organic Materials Branch at ARDEC. She is active in a number of projects involving adhesive bonding and durability of bonds. She received her BS and MS degrees in chemical engineering at New Jersey Insititute of Technology.

Dean Martinelli is a materials engineer with the Adhesives Section of the Organic Materials Branch at ARDEC. He is active in many adhesive bonding projects which include the bonding of polyolefins, rubber, composites, and various metals. He received his Bachelor of Engineering degree at Stevens Institute of Technology and is presently enrolled in a Master's program at Stevens.

Andrew Devine has been involved with adhesives bonding since joining Picatinny Arsenal's Feltman Research Laboratories in 1966. Over this time he has dealt with all aspects of adhesive bonding to include substrate surface preparations and design. Mr. Devine served as the Army liaison representative to ASTM committee D-14 on adhesives. He authored an Engineering Design Handbook on "Joining of Advanced Composites" published in 1978 and has worked on bonding problems associated with many Army items. Currently, he has a patent pending on noncarcinogenic chemical etches for polyethylene. Mr. Devine is a 1963 graduate of Manhattan College with a B.Ch.E. and was awarded an M.S.Ch.E. from Newark College of Engineering in 1969.

Michael J. Bodnar is Chief of the Adhesives Section, Organic Materials Branch at ARDEC. He is a graduate of Carnegie Tech and has been in materials engineering on adhesives since 1951 when he first joined Douglas Aircraft Company. He has published widely in the field and is editor of four books on structural adhesives bonding.

James Beetle is a metallurgist in the Materials Evaluation Section of the Metallic Materials Branch at ARDEC. He is a graduate of the University of Pennsylvania and has been an Army civilian employee for 20 years. He has been involved in research programs concerned with materials related deformation and failure mechanisms of armament materials for the past 17 years.

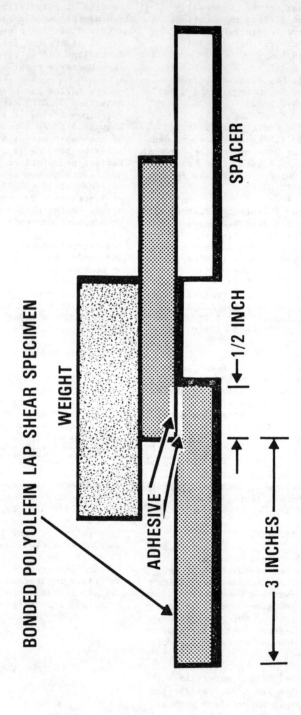

Figure 1. Lap shear specimen with a polyolefin substrate

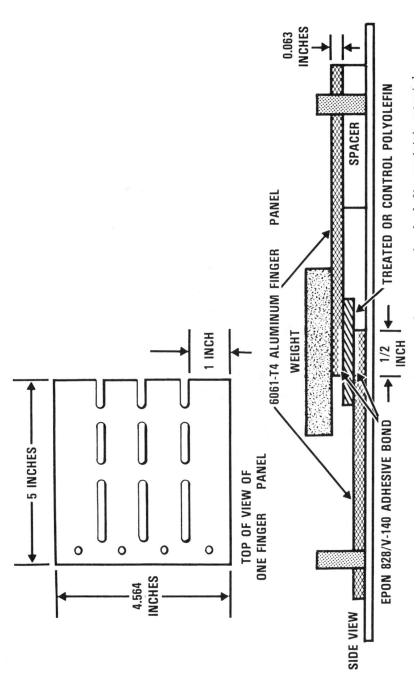

Figure 2. Lap shear specimen with aluminum substrates and polyolefin sandwich material

461

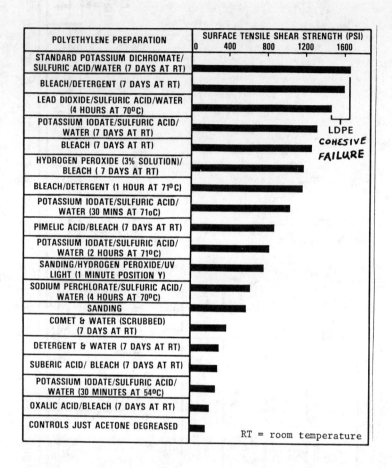

Figure 3.   Tensile shear strength versus polyethylene
surface treatments

462

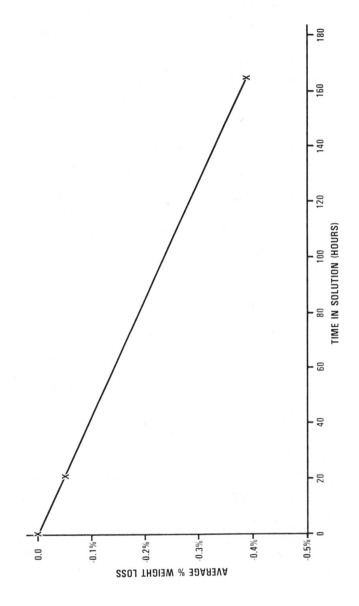

Figure 4.  Average weight loss of low density polyethylene versus time in the standard potassium dichromate/sulfuric acid/water solution pretreatment

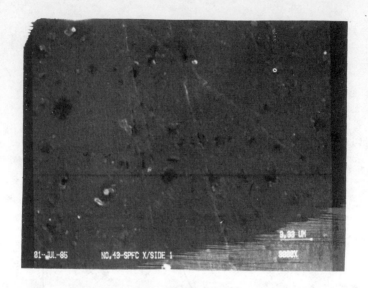

Figure 5.   SEM photograph of control (unetched) polyethylene surface,
3,000 times magnified

Figure 6.   SEM photograph of polyethylene treated in the
potassium dichromate/sulfuric acid/water
solution, 3,000 times magnified

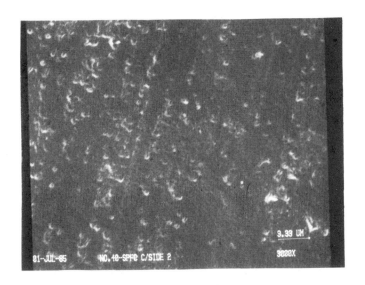

Figure 7.  SEM photograph of polyethylene treated in
bleach/detergent, 3,000 times magnified

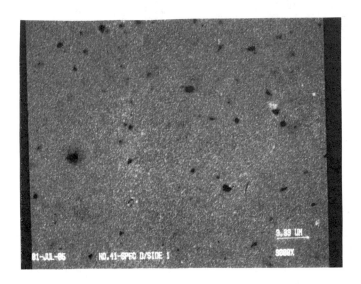

Figure 8.  SEM photograph of polyethylene treated in
potassium iodate/sulfuric acid/water,
3,000 times magnified

32nd International SAMPE Symposium
April 6-9, 1987

FROM SPACE PROCESSING TO MICROGRAVITY SCIENCE AND APPLICATIONS
A HISTORICAL PERSPECTIVE

John R. Williams
Center for Space and Advanced Technology
500 Boulevard South, Suite 104
Huntsville, AL 35802

## Abstract

This paper summarizes the history of the U.S. Microgravity Science and Applications (MSA) Program from its beginning as a strict applications program, through several changes in program philosophy, content and leadership, to its present status as a maturing, research program. In addressing each phase of the evolution of the MSA program, special attention is given to the effect of the interactive nature of program philosophy, program content and budget with acceptance by the scientific community, flight experiment/hardware development, and perceptions from the industrial community. This historical perspective provides the framework for understanding why we are, where we are, in the MSA program today with special emphasis on lessons learned that will enable the program to move forward in a more deliberate, and efficient manner.

## 1. INTRODUCTION

The early beginnings of today's MSA Program can be traced back to the initial considerations for space activities. In the late 1950's and early 1960's interest arose from several disciplines in the effects of the microgravity environment that man was preparing to enter. In most cases, this interest was created by a "need-to-know" in order to design systems that would operate in the "zero" gravity of space. The behavior of fluids in low-gravity became an important factor for designing propellant management systems for liquid upper stages. The design of fluid systems that would operate in low-gravity required attention to the control of bubbles and the development of phase separators. The first low-gravity solidification studies were prompted by the requirement to design phase change systems for spacecraft thermal control energy storage. Joining structural members by welding and brazing in space raised questions about the influence of gravity on these processes. Questions concerning fire safety in spacecraft motivated research in material flammability and flame propagation under low-gravity, oxygen-enriched environments. Facilities at both the Lewis Research Center (LeRC) and the Marshall Space Flight Center (MSFC) were used to investigate many of these phenomena. These facilities included drop towers and aircraft flying ballistic trajectories, which are still in use today.

As these operational problems were addressed by the various disciplines, engineers and scientists began to consider the application of the unique environment of low-gravity to the solutions of important practical and scientific problems. The characteristics of this environment include the drastic reduction of convective flows, virtual elimination of sedimentation or stokes setteling, absence of hydrostatic pressure, ability to suspend droplets or high temperature melts in a containerless environment, and the availability of an ultra-high vacuum with tremendous pumping capacity. These characteristics have always been the cornerstones of processing in low-gravity. It has been our approach to the understanding and utilization of these characteristics that has evolved over the years.

## 2. EARLY GROUND-BASED RESEARCH

During most of the decade of the 1960's, research concerned with the phenomena of microgravity was dominated by the "applications" environment of the time; which was heavily oriented toward the philosophy of manufacturing in space. The program budget began at a level of several tens of thousands of dollars and grew only to several millions of dollars by the end of the decade. Most of the research conducted during this period was within NASA at the field centers, predominately LeRC and MSFC. A number of contracts were let for outside research, but, for the most part, they were small and did not involve a significant portion of the scientific community. The program became known as Space Processing Applications which carried over into the middle of the 1970's.

## 3. EARLY FLIGHT RESEARCH

After almost a decade of small, exploratory ground based research efforts, the first opportunities to conduct experiments in low-gravity for periods longer than a few seconds came with the Apollo program. These space processing experiments were not originally part of the lunar program but were added to the flight manifests of Apollo 14, 16, and 17 as small carry-on or "suitcase" experiments. They were conducted by the crew during the trans-earth coast. The experiments consisted of a low temperature furnace to perform casting experiments on model composite systems, a heat flow and

convection experiment, a simple electrophoresis experiment to test the concept of performing electrophoretic separation of biological materials, such as cells and macromolecules, in the absence of buoyancy-driven flows, and a fluid pumping experiment to test different techniques for achieving and maintaining phase separation of stored liquids during the draining and filling of storage tanks. All the experiments operated successfully, but revealed the presence of more subtle, non-gravitational flows that needed further understanding before the microgravity environment could be used to advantage. A total of 5 space processing experiments were performed during the Apollo program.

With the Apollo program ending in the early 1970's, the Skylab program offered the next opportunity for conducting experiments in low gravity. It also provided the first opportunity to conduct these space processing experiments in a dedicated facility. The experiments consisted of several solidification experiments involving alloys and composites, welding and brazing investigations, a study of combustion in a diffusion controlled environment, crystal growth experiments, and an unsucessful attempt at containerless solidification. In addition to these planned experiments, a number of demonstration experiments were conducted late in the mission since the crew had more free time than expected. These demonstrations were conceived during the mission and were performed with equipment already on-board the spacecraft. For example, a liquid bridge was established using colored water suspended between two socket wrench extensions mounted in camera brackets. This simple experiment demonstrated that a liquid bridge, simulating a molten floating zone, could be extended to its theoretical Rayleigh limit (length equal to circumference) before it broke into two individual droplets, however, an unexpected "jump-rope" instability developed when the zone was rotated (1). A total of 23 experiments were performed during the Skylab program.

The Apollo-Soyuz Test Project (ASTP) was a joint U.S.-USSR mission which followed the Skylab program and provided an opportunity to fly several new experiments as well as to refly several of the Skylab experiments that had produced unexpected results. However, the resources available on ASTP were considerably less than on Skylab and constrained the types of experiments that could be performed. Several interesting scientific results were, nonetheless, obtained and can be found in the following references, (2,3,4,5,6,7, 8,9,10). A total of 13 space processing experiments were performed during the ASTP. This flight activity carried the Space Processing Applications Program up to the mid 1970's. The budget had grown to approximately eight (8) million dollars annually and the program philosophy continued to be dominated by the manufacturing in space concepts, i.e., the approach taken by many investigators tended to be somewhat Edisonian in nature. Also, flight schedules were short and did not permit a thorough ground based research program. For these, and other reasons, results from these early flight experiments can be used as indicators and not as conclusive results unto themselves.

A hiatus in manned space flight opportunities was anticipated between the conclusion of the ASTP and the initiation of space shuttle flights. As an interim or "stop-gap" measure, sounding rocket flights were initiated using Nike-boosted Black Brandt rockets. These were referred to as Space Processing Applications Rockets (SPAR). These rockets could carry several small experiment packages and provide up to 5 minutes of free fall time. From the mid 1970's to the early 1980's ten (10) SPAR flights provided opportunities to conduct additional experiments, test new concepts and develop new experiment apparatus. During this period acoustic levitators (both ambient and high temperature), an electromagnetic levitator furnace, and several general purpose rocket furnaces were developed and flown successfully on SPAR. It should be noted that the hardware developed

for SPAR was focused on conducting experiments in a relatively short time, i.e., 5 minutes of reduced gravity. Experiments conducted during SPAR flights served as additional indicators for experiment development for the space shuttle. A total of 24 experiments were performed in SPAR flights, and some of the results can be found in the following references (11,12,13,14, 15,16,17,18,19,20).

## 4. THE SPACE SHUTTLE ERA

In 1977 an Announcement of Opportunity (AO) was released by NASA for Materials Processing Investigations on Space Shuttle Missions (OA-77-03). This marked a change in several aspects of the program. The program name was changed to Materials Processing in Space (MPS), a new Program Director was named (Dr. John Carruthers), and the program philosophy was changed to incorporate more ground based research and a goal of developing commercial ventures in space. There were 117 proposals submitted under the AO. From this number, 14 were selected for flight (table 1) and 19 for further development (table 2). These selected investigations were under contract by 1979. During this same period, Dr. Fletcher, then Administrator of NASA, requested the National Academy of Engineering to organize a study to provide guidance to the Materials Processing in Space program. To perform this study, an ad hoc committee, the Committee on Scientific and Technological Aspects of Materials Processing in Space (STAMPS), was established. The overall objective of the study was to provide guidance for the future course of NASA's program of research and development on processing materials in the space environment. The principal objectives were:

An assessment and evaluation of the scientific and technological significance of what has been learned to date about processing materials in the space environment.

A judgement of the merit of a program on materials processing in space-- possible benefits, if any; values;

advantages and disadvantages.

Recommendations regarding the nature and scope of NASA's future program of experiments on materials processing in space, as well as on a program of complementary experiments in ground based facilities or theoretical studies designed to provide a sound scientific basis for the program.

The STAMPS report was released in 1978 and contained a number of both scientific and administrative conclusions (21). The results of this study were taken seriously by NASA and incorporated in the MPS program. The program philosophy moved from a manufacturing-in-space mentality toward establishing the program on a sound scientific base wherein flight experiments become a necessary means-to-an-end rather than an end unto themselves.

The Materials Processing in Space program made great strides in the late 1970's and early 1980's in establishing a stronger science base for the program. Science advisory groups were established to provide review and council for the evolving research activities, peer review groups were established to review and provide recommendations concerning proposed research for NASA sponsorship, and interaction and involvement with a broader segment of the scientific community was achieved. This brought a better understanding of the program and thus it became an accepted, creditable scientific research program. For instance, the number of papers published in refereed journals increased from a few dozen prior to 1977 to almost 200 per year by the early 1980's.

The MPS budget grew to better than 30 million dollars per year in 1979 and 1980, due largely to the evolving flight program and it's hardware development activities. The ground-based research base has slowly increased to approximately 11 million dollars per year; which continues in the present program today. While the research activities were steadily improving, problems were encountered in the hardware development. As the Principal Investigators, selected with the 1977 AO, were being placed under contract, NASA was selecting contractors to develop the flight hardware for conducting their experiments in the space shuttle. TRW was selected to develop three major facilities, two to be flown in the Spacelab module and one flown in the space shuttle bay. These facilities were the Fluid Experiment System (FES), the Vapor Crystal Growth System (VCGS) and the Solidification Experiment System (SES), respectively. The FES is a sophisticated holographic system designed to make real time holograms of a growing crystal and its surrounding media. Technical problems, reflected in significant cost overruns, forced NASA to take premature delivery of the subsystems from the contractor and complete the assembly, integration and test in-house at MSFC. The VCGS is a facility designed to establish a controlled thermal environment and provide for optical observation during the growth of $HgI_2$. This facility was also brought in-house NASA with the FES since they shared several subsystems, not due to technical problems. The VCGS is, for the most part, a space hardened version of the Principal Investigators ground-based hardware. The FES and VCGS flew on Spacelab 3.

The Solidification Experiment System (SES) was expected to be a furnace system which would meet the requirements of several flight investigators. To be more specific, it had to accommodate the requirements of Dr.'s Larson, Davidson, Crouch, Wiedemeier and Gelles as shown in Table 1. This became a technical and managerial nightmare since the collective requirements of these PI's were not completely compatible; they were, in fact, mutually exclusive in many areas. Therefore, to attempt to design one furnace that could meet all these requirements on one shuttle flight was a futile undertaking. As these problems arose and the cost overruns grew, NASA was forced to terminate this effort with very little to show for its investment. Efforts were initiated to accommodate the five PI's, planned for the SES, in other ways. Dr.'s Larson, Wiedemeier and Gelles were, and still are, being accommodated in modified furnaces from SPAR. These PI's have flown onboard the shuttle utilizing this dated hardware. A new furnace (Advanced Automated Directional Solidification Furnace, AADSF) is being developed which will accommodate Dr. Davidson (now Dr. Lehoczky's experiment) and Dr. Crouch. This means that it will be more than ten years from the time several of the PI's proposed their experiment to their first flight opportunity. An acceptable pattern for the future? I think not.

Another major element of the Materials Processing in Space program, initiated in 1977, was space commercialization. In concert with a sound scientific base, a long range goal of the MPS program was to provide the framework and basis for industry to get involved in space activities as a commercial venture. The premis was that industry would conduct space experiments, at its own expense, if the front end risks were reduced to a reasonable level by NASA and business arrangements, rather than contracts, could be developed wherein industry would be a partner with NASA and be provided with certain marketplace rights which would make the return on investment acceptable. In reducing the front end risk, it was recognized that NASA should conduct sufficient research to demonstrate feasibility, both scientific and ingineering, concerning various techniques, processes, etc. This is why the MPS program of yesterday, and the MPS program of today, is the basis upon which industry commits to a commercial space venture. It is crucial that the research program be scientifically sound, conduct research pertinent to industry, transfer research results to industry and support industry in the understanding, interpretation and utilization of the results. This was a key element in the MPS program and continues to be a key element of the MSA program today.

The business side of space commercialization required considerable attention in the late 1970's. NASA was very adept at administering contracts and grants, but in order to grant certain rights and exclusive to industry different forms of agreements were developed, i.e., Joint Endeavor Agreements (JEA), Technical Exchange Agreement (TEA), and Industrial Guest Investigator (IGI). The IGI was an industrial investigator assigned as part of a flight investigation team to work agreed to portions of the flight experiment. Industry supported the IGI and, in general, the flight results were shared and published. The TEA was an arrangement for industrial investigators to collaborate with NASA investigators which, in most cases, involved the use of unique NASA facilities, i.e., drop tubes, drop towers, aircraft, etc. No funds are exchanged and, in general, the results are published jointly. The JEA was the most sophisticated agreement in that it was a business agreement between NASA and industry, no funds exchanged, for flight activities where the quid quo pro was negotiated. For instance, this provided a mechanism for industry to get a number of flights on the shuttle and certain exclusivity in return for NASA utilization of their hardware. McDonnell Douglas signed the first JEA with NASA in 1980.

The MPS program carried into the early 1980's and saw a budget peak in 1979 of approximately 35 million dollars drop to approximately 21 million dollars in 1982. Most of the budget peak supported the cost overruns on the hardware development, therefore, the budget drop by 1982 came predominately out of the flight program. In addition to continuing budget constraints and hardware development problems, the program continued to be severly limited by the lack of access to long-duration low gravity. For instance, the first MPS experiment on the shuttle (the monodispersed latex reactor on STS-3) accumulated more time in low gravity during that one mission than all previous flight experiments, including SPAR, combined. It's obviously difficult to build a strong program with only one or two "data points" from low-g per year.

After the departure of Dr. Carruthers to private industry and an interim program director (Dr. Louis Testardi), Mr. Richard Halpern was named Director of MPS. In his congressional testimony in 1983, Mr. Halpern proposed a new name for the program: Microgravity Science and Applications (MSA). This new name signaled the expansion of the Divisions charter scientifically and its increased attention to applications. In addition to the materials processing in space activities, the MSA Division new includes the Physics and Chemistry Experiments (PACE). This broadened the science base to include not only materials science, but combustion science, cloud and aerosol physics, fluid dynamics and critical phase transitions. The goals of MSA not only continue to embrace a sound science base and a need to explore and determine potential applications for commercialization in space, but now include the utilization of a manned laboratory in the Space Station.

## 5. OBSERVATIONS

Although todays MSA program can be traced back over two decades, when viewed as a flight research program it is in a promising but embryonic state. The education acquired thus far has been costly, but not without significant value. In reviewing the evolution of the program philosophy and the manner in which the program has been implemented, if one word were used as a guide to the future it appears to be balance. The program has suffered from time-to-time in the past on an imbalance of emphasis on certain aspects. Over emphasis on applications can lead to inappropriate fly-and-try experiments at the expense of maintaining a sound science base. On the other hand an over emphasis on science can generate a research program which is out-of-touch with industry and without relevence. There must be a proper mix, or balance, between very basic research, which supports the continued buildup of the MSA knowledge base, and applied or directed research which focuses on specific problems relevent to industry.

Improper programmatic emphasis not only create problems such as indicated above, but contribute to many others. Too casual adherence to science requirements can allow one to develop hardware which can be only minimally utilized. Too rigid adherence to requirements, however, precludes proper engineering trades and charts a course frought with technical problems and cost overruns. Hardware development should, therefore, be directed toward meeting sets of specific, compatible requirements which have been iterated and are well understood by both the PI's and the hardware developer. One can develop multiuse hardware which can be easily reconfigured to accomodate different sets of requirements. One should not, in my opinion, attempt to build multipurpose hardware, such as the SES. An approach which embodies this development philosophy is one where considerable effort is spent in defining and understanding the requirements and their impact, verifying the necessary technology through modeling and breadboard testing and concept verification before commiting to the design, fabrication, assembly and testing of the flight hardware. This is, incidently, the process now used by the present MSA program.

To a large degree, much of the program evolution during the MPS era was based on proposals submitted to NASA and their subsequent peer review and acceptance. The potential danger in this approach is that the program can evolve into an aggregation of uncorrelated and uncoordinated research activities. This is especially true for this program since it is, by nature, a multidiscipline program. This speaks to the need for a program plan which projects the proper programmatic balance, and serves as a baseline and guide for growth in the program and a qualification document for future resource requirements.

## 6. REFERENCES

1. J.R. Carruthers, "Studies of Liquid Floating Zones," Proceedings of Third Space Processing Symposium, Skylab Results, Volume II, M-74-5, June 1974, p. 843.

2. A.F. Witt, et al., "Crystal Grown and Steady State Segregation under Zero Gravity: InSb," J. Electrochem. Soc. 122, 276 (1975).

3. H. Wiedemeier, F.C. Klaessig, E.A. Irene, and S.J. Wey, "Crystal Growth and Transport Rates of Ge Se and Ge Te in Microgravity Environment," J. Cryst. Growth 31, 36 (1975).

4. H. Wiedemeier, et al., "Morphology and Transport Rates of Mixed IV-VI Compounds in Microgravity," J. Electrochem. Soc. 124, 1095 (1977).

5. D. Chandra and H. Wiedemeier, "Chemical Vapor Transport and Thermal Behavior of the GeSe--GeI$_4$ System for Different Inclinations with Respect to the Gravity Vector, Comparison with Theorectical and Microgravity Data," J. Cryst. Growth 57, 159 (1982).

6. D.J. Larson, "Zero-G Processing of Magnets (MA-070)," Apollo-Soyuz Test Project Summary Science Report, Volume I, NASA SP-412, 1977, p. 449.

7. R.G. Pirich, "Characterization of Effects of Plane Front Solidification and Heat Treatment on Magnetic Properties of Bi/MnBi Composites," IEEE Trans. Mag. MAG-16, 1065 (1980).

8. D.J. Larson, "Zero-G Processing of Magnets, Exp. MA-070," Final Report, NAS8-30577, December 1976.

9. S. Takahashi, "Preparation of Silicon Carbide Whisker Reinforced Silver Composite Material in a Weightless Environment," 15th AIAA Aerospace Sciences Meeting, Los Angeles, CA, January 24-26, 1977 (AIAA Paper 77-195).

10. R.E. Allen, et al., "Column Electrophoresis on the Apollo-Soyuz Test Project," Sep. Purif. Meth. 6, 1-59 (1977).

11. C.F. Schafer and G.H. Fichtl, "SPAR I Liquid Mixing Experiment," AIAA Journal 16, 425 (1978).

12. T.G. Wang, M.M. Saffren, and D.D. Elleman, "Acoustic Chamber for Weightless Positioning," 12th AIAA Aerospace Sciences Meeting, Washington, D.C., 1974 (AIAA Paper 74-155).

13. R.R. Whymark, "Acoustic Field Positioning for Containerless Processing," Ultrasonics 13, 251 (1975).

14. R.T. Frost and C.W. Chang, "Theory and Applications of Electromagnetic Levitation," Materials Research Society Symposia Proceedings, Materials Processing in the Reduced Gravity Environment of Space, Volume 9 (G.E. Rindone, ed.), North Holland, 1982, p. 71.

15. M.H. Johnston and C.S. Griner, "The Direct Observation of Solidification as a Function of Gravity Level," Met. Trans. 8A, 77 (1977).

16. M.H. Johnston and R.A. Parr, "The Influence of Acceleration Forces on Dendritic Growth and Grain Structure," Met. Trans. 13B, 85 (1982).

17. S.H. Gelles and A.J. Markworth, "Microgravity Studies in the Liquid-Phase Immiscible System: Al-In," AIAA Journal 16, 432 (1978).

18. C. Potard, "Solidification of Hypermonotectic Al-In Alloys under Microgravity Conditions," Materials Research Society Symposia Proceedings, Materials Processing in the Reduced Gravity Environment of Space, Volume 9 (G.E. Rindone, ed.), North Holland, 1983, p. 543.

19. R.G. Pirich, "Studies of Directionally Solidified Eutectic Bi/MnBi at Low Growth Velocities," Metallurgical Transactions A, 1983.

20. D.J. Larson, R.G. Pirich, R. Silberstein, J. de Carlo, and C. Buscemi, "Effect of Applied Magnetic Field on Directional Solidification of Eutectic MnBi/Bi," Proceedings of ASM/AIME Fall Meeting, Philadelphia, PA, September 1983

21. "Materials Processing in Space,"
Report of the Committee on Scientific
and Technological Aspects of Materials
Processing in Space (STAMPS) of the
Space Applications Board, Assembly
of Engineering, National Research
Council, National Academy of
Sciences, 1978.

## 7. BIOGRAPHY

Mr. John R. Williams is Vice
President, Technical Operations,
for the Center for Space and
Advanced Technology (CSAT), Inc.
In this position, Mr. Williams is
Director of CSAT's Technical Center
in Huntsville, Alabama, and is responsible for overseeing the
technical aspects of CSAT's activities. Prior to joining CSAT, Mr.
Williams had over twenty years with
NASA. He was Manager of the Experiment Devices Office, responsible
for managing the scientific research
and flight hardware development for
the Microgravity Science and Applications program. He has been
involved in materials processing
and the space commercialization
activities since their inception,
and is still actively involved with
NASA and industry in these endeavors.

Investigations Selected for Flight

| Investigator/Organization | Experiment Title | Apparatus Required |
|---|---|---|
| Dr. Wagner/Arizona State Univ. | Solid Electrolytes Containing Dispersed Particles | Solidification Experiment Systems (SES) |
| Dr. Larson/Grumman Aerospace | Aligned Magnetic Composite | Solidification Experiment Systems (SES) |
| Dr. Gelles/S.H. Gelles Associates | Liquid Miscibility Gap Materials | Solidification Experiment Systems (SES) |
| Dr. Davidson/MSFC | Growth of Solid Solution Crystals | Solidification Experiment Systems (SES) |
| Dr. Crouch/LaRC | Semiconductor Materials Grown in Low-G Environment | Solidification Experiment Systems (SES) |
| Dr. Schnepple/EG&G | HgI2 Crystal Growth for Nuclear Detectors | Vapor Crystal Growth System (VCGS) |
| Dr. Weinberg/Owens-Illinois | Fining of Glasses in Space | Acoustic Levitator Furnace |
| Dr. Subramanian/Clarkson College | Phenomena in Containerless Processing | Acoustic Levitator Furnace |
| Mr. Happe/Rockwell International | Containerless Preparation of Advanced Optical Glasses | Acoustic Levitator Furnace |
| Dr. Dintenfass/Univ. of Sydney | Aggregation of Human Red Blood Cells | Experiment Unique Apparatus (PI Provided) |
| Dr. Shlichta/JPL | Crystal Growth in a Spacecraft Environment | Fluid Experiment System (FES) |
| Dr. Lal/Alabama A&M Univ. | Solution Growth of Crystals in Zero-Gravity | Fluid Experiment System (FES) |
| Dr. Wiedemeier/RPI | Vapor Growth of Alloy-Type Semiconductor Crystals | Solidification Experiment System (SES) |
| Dr. Vanderhoff/Lehigh Univ. | Large Particle Size Monidispersed Latexes | Experiment Unique Apparatus |

Table 1

Investigations Selected for Further Development

| Investigator/Organization | Investigation |
| --- | --- |
| Dr. Fowle/Arthur D. Little, Inc. | Marangoni Effect in Crystal Processing |
| Dr. Ostrach/Case Western Reserve Univ. | Surface Tension Driven Convection Phenomena |
| Dr. Leferer/USC | Convection in Floating Zone Process |
| Dr. Verhoeren/Ames-ERDA | Float Zone Experiments in Space |
| Dr. VanOss/SUNY-Buffalo | Electrophoresis of Human Pancreatic Cells |
| Dr. Bier/Univ. of Arizona | Harmone Purification by Isoelectric Focusing |
| Dr. Nyiri/Lehigh Univ. | Multipurpose Space Bioreactor System |
| Dr. Mieszkue/JSC | Tissue Culture Growth/Fermentation |
| Dr. Weinberg/Owens-Illinois | Investigation of Space Produced Glasses |
| Dr. Rembaum/JPL | Electrophoretic Cell Separation Based on Immunomicrospheres |
| Mr. Shafer/MSFC | Fluid Mixing Effects in Levitated Melts |
| Dr. Lacy/MSFC | Nucleation and Growth of Immiscible Phases |
| Dr. Johnston/MSFC | Directional Solidification of Immiscible Alloys |
| Dr. Chiovetti/UAB | Whole-Cell Electrophoresis Aboard Spacelab |
| *Dr. Neugebauer/GE | Ultravacuum Vapor Epitaxial Growth of Silicon |
| *Dr. Grunthaner/JPL | Ultrahigh Vacuum Semiconductor Thin Film Technology |
| *Dr. Schmidt/Ames-ERDA | Electrotransport of Solutes in Refractory Metals |
| *Dr. Schmidt/Ames-ERDA | Efficient Solar Cells by Space Processing |
| *Dr. Bunshah/UCLA | Ultrapure Metals Preparation in Space |

*Wake Shield Investigations

Table 2

475

EFFECT OF LASER RADIATION ON THE RAPID PROCESSING OF
GRAPHITE-REINFORCED COMPOSITES

C.M. Tung, D.S. Gnanamuthu, R.J. Moores and C.L. Leung
Rockwell International Science Center
Thousand Oaks, CA 91360

## ABSTRACT

The effect of laser radiation on the curing of graphite-reinforced composites is discussed. The study showed that by using a low-intensity $CO_2$ laser (10 to 30 $W/cm^2$), fiber wet-out for a four-ply composite was attained in 100-300 s. Laser curing was shown to be independent of fiber orientation, and therefore applicable to the processing of cross-ply laminates. Optical properties of neat resins and composites measured at 10.6 $\mu$m ($CO_2$ laser wavelength) indicated that laser energy was absorbed by the first layer of a multilayered composite. Hence, the effectiveness of laser curing depends on the use of graphite fibers as heat conductors in the laminate.

## 1. INTRODUCTION

The current methods for processing a composite from prepregs are based entirely on the application of external sources of heat and pressure, which is usually done in an autoclave. These techniques are time-consuming, energy inefficient, and hinder the development of new resins that, due to different polymerization characteristics, may require extensive modification of conventional autoclaves or presses.

Over the years, various forms of radiation have been used for effecting the polymerization of organic polymers. These comprise electromagnetic radiation such as gamma rays, UV rays, visible light rays, microwave, rf waves, particle radiation including electrons, alpha and beta particles, and neutrons. A majority of these methods are deemed inappropriate because of various deficiencies. For example, shielding and size of accelerators are problems with radioisotopes and high-energy electron methods. UV radiation lacks penetration ability for parts thicker than a few mils. Microwave radiations have shielding considerations as well as the inability to process cross-ply laminates because of polarization requirements.

The use of laser as the heating source can circumvent most of the objections posted by other radiation methodologies. Rapid heating and the ability to maintain a constant temperature with lasers offer the possibility of achieving uniform and complete cure in a very short period of time. Temperature is controlled by laser power density and processing time. Temperature does not depend on the size or shape of the part, since the process only affects the localized area where the laser beam is focused. In addition, the portability of the laser beam using either compact and portable lasers or optical fibers and mirror systems allows easy adaptation to applications such as field repair.

In this paper, the use of a low-intensity $CO_2$ laser beam in the curing of high-temperature thermosetting of thermoplastic composites is discussed. The effects of laser intensity and irradiation time on the cure process will be examined. Theoretical consideration based on the optical measurements of neat resins and composites will also be presented.

## 2. EXPERIMENTAL

### 2.1 Materials

A commercially available graphite-reinforced bismaleimide (BMI) prepreg, V388A (U.S. Polymeric), was used as the high-temperature thermosetting material. Polyetheretherketone graphite-reinforced composite (PEEK-APC-2), available from Imperial Chemical Industries (ICI), was used as the thermoplastic system.

### 2.2 Laser Processing

A Spectra-Physics Model 975 laser system was used. This system produces a cw $CO_2$ laser radiation at 10.6 μm wavelength. Depending on the beam size and the cross-sectional area needed to be processed, laser curing can be accomplished by either impinging the laser beam on a stationary workpiece or "sweeping" the workpiece under the laser beam. Dimensions of the cure zone are determined by absorbed laser beam power, beam size, exposure time (for the stationary workpiece), travel speed (for the moving workpiece), absorption coefficient, specific heat and diffusivity. Experiments were completed outside the pressurized chamber. Only small amounts of pressure were exerted on the sample as a means of holding the sample in place during laser heating.

## 3. RESULTS AND DISCUSSION

### 3.1 Theoretical Consideration

Consider a laser beam of uniform and constant intensity I ($W/cm^2$) incident normally on a plane boundary of an absorbing material with an absorption coefficient $\alpha$ for a time duration t. Depending on the thermal and optical properties of the material, two limiting cases may be distinguished, which are illustrated in Figs. 1a and 1b.[1]

Case 1. The optical absorption depth $\alpha^{-1}$ is small compared to the thermal diffusion length, $\alpha(2Dt)^{1/2} \gg 1$, where D is the thermal diffusivity. In this case, the energy absorbed is used to heat a layer of thickness $(2Dt)^{1/2}$. The average temperature rise in this layer is

$$\Delta T = (1-R)It/C_p \, \rho(2Dt)^{1/2} \quad , \qquad (1)$$

where R = reflectance, I = beam intensity, t = irradiation time, $C_p$ = specific heat, $\rho$ = density, and D = thermal diffusivity.

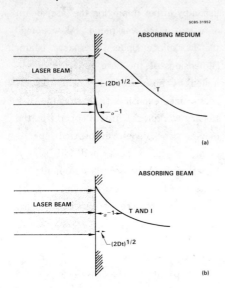

Fig. 1   Surface penetration by a laser beam.

After the laser treatment is terminated, the heat in this layer diffuses into the subsequent layers. Cooling time is again equal to t, the time needed for the heat to diffuse over a distance $(2Dt)^{1/2}$. Thus, the heating and cooling rates can be obtained by Eq. (2),

$$\Delta T/t = (1-R) It/C_p\rho (2Dt)^{1/2} \qquad (2)$$

Case 2. The optical absorption depth $\alpha^{-1}$ is large compared to the thermal diffusion length, $\alpha(2Dt)^{1/2} \ll 1$. The light absorption creates an exponential temperature profile, with characteristic length $\alpha^{-1}$ given by

$$\Delta T(z) = (1-R)\alpha \ I \ e^{-\alpha z} \ t/C_p\rho \qquad (3)$$

The heating rate is $\Delta T(z)/t$. Since the heat has to diffuse into the substrate over a length $\alpha^{-1}$, the cooling time is roughly $\alpha^{-2}/2D$. The cooling rate can be approximated by

$$\frac{dT}{dt} = (1-R) \ \alpha^3 \ (2DIt/C_p\rho) \qquad (4)$$

In graphite fiber-reinforced composites, assuming that the fibers are highly absorptive at the wavelength of the laser beam ($\approx 70\%$ by volume), their concentration will affect the amount of thermal energy propagating across the composite. The absorption can be estimated qualitatively by comparing the wavelength of the laser beam with the diameter of graphite fibers, as described by the following. The depth of penetration of an electromagnetic wave is proportional to its wavelength. In the case of graphite, the depth of penetration by a $CO_2$ laser beam is estimated to be $\approx 0.3 \ \mu m^2$ which is much shorter than the diameter of a graphite fiber ($\sim 8 \ \mu m$). The absorption of $CO_2$ laser energy by the graphite fibers hence should be very high. Thus, the interaction between the incident laser beam and a graphite-reinforced composite can be represented by case (1); i.e., the optical absorption depth is small compared to the thermal diffusion length. The absorbed laser energy therefore will be consumed primarily in heat generation in the graphite fibers and will then dissipate rapidly into the surrounding matrix and fibers. The direct interaction between the laser beam and the matrix thus becomes less important.

Therefore, the presence of graphite fibers will aid enormously in heat conduction during laser curing, as illustrated in Fig. 2. The processing variables that affect cure of composites are the thermal/optical properties of the fibers and the matrix, beam intensity, irradiation time, fiber diameter and fiber spacing.

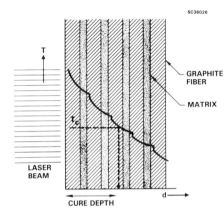

Fig. 2  Interaction of laser beams with graphite-reinforced composite.

## 3.2 Optical Property Measurements

Table 1 shows the optical properties of a thermosetting composite (Gr/BMI) and a thermoplastic composite (Gr/PEEK) measured at 10.6 μm. These results indicate that the optical properties of the composites are indeed dominated by the presence of graphite fibers. Although the BMI neat resin appears to be relatively transparent to the $CO_2$ laser beams, both composites show very low transmittance, indicating strong absorption by the graphite fibers. The results also indicate that most of the incident beam energy was absorbed by the first layer of a multilayered composite. Hence, the effectiveness of laser curing depends on the effectiveness of graphite fiber as the heat conductor.

## 3.3 Laser Curing

The effect of irradiation time on the cure of four-ply Gr/BMI composites with respective fiber orientations is shown in Figs. 3 and 4. Fiber wet-out on the backface of both composites is apparent after both were exposed to the $CO_2$ laser beam for approximately 64 to 128 s. The exact extent of the cure was not determined. The results are quite impressive

Table 1

Optical Properties of Graphite-Reinforced Composites (λ = 10.6 μm)

| Materials | Reflectance | Absorbance | Transmittance |
|---|---|---|---|
| Bismaleimide (neat resin) | ~ 4%[a] | ~ 20%[c] | ~ 76.0%[b] |
| PEEK (neat resin) | ~ 4%[a] | ~ 80%[c] | ~ 16.0%[b] |
| BMI/Gr (1 ply) | d | | 0.1%[e] |
| PEEK/Gr (1 ply) | d | | 0.3%[e] |

a.  Assumed value, based on the normal spectral reflectance value for polymethyl methacrylate.
b.  Measured by FTIR.
c.  Estimated from absorbance = 1 - (Transmittance + Reflectance).
d.  Too scattered to measure.
e.  Measured by a pyroelectric IR detector.

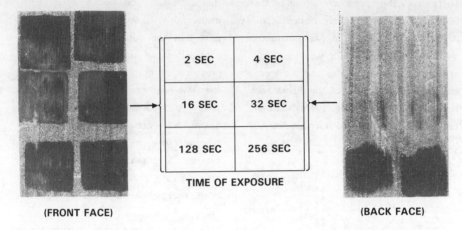

|  |  |
|---|---|
| 2 SEC | 4 SEC |
| 16 SEC | 32 SEC |
| 128 SEC | 256 SEC |

**TIME OF EXPOSURE**

(FRONT FACE)　　　　　　　　　　　　　　　　　　　(BACK FACE)

Fig. 3　Results of laser curing on a unidirectional four-ply graphite/bismaleimide laminate.

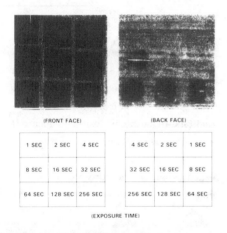

(FRONT FACE)　　　　　(BACK FACE)

| 1 SEC | 2 SEC | 4 SEC |
|---|---|---|
| 8 SEC | 16 SEC | 32 SEC |
| 64 SEC | 128 SEC | 256 SEC |

| 4 SEC | 2 SEC | 1 SEC |
|---|---|---|
| 32 SEC | 16 SEC | 8 SEC |
| 256 SEC | 128 SEC | 64 SEC |

(EXPOSURE TIME)

Fig. 4　Results of laser curing on a cross-ply four-ply graphite/bismaleimide laminate.

in that no preconsolidation was carried out prior to laser irradiation. Time for fiber wet-out also seems to be independent of fiber orientation. Similar experiments were also carried out on PEEK/Gr composites. The results indicate that with increased power density (20 or 30 W/cm$^2$) consolidation can be achieved. The time required for fiber wet-out, however, is not as clear as in the case for BMI/Gr.

Optical photomicrographs illustrating the graphite fiber/matrix interaction after laser curing are presented in Fig. 5. To obtain these photomicrographs, a cross section of the area within a laser-cured region was cut out and mounted in bakelite, exposing the region between the plies. The mounted specimen was ground and carefully polished to fully reveal the graphite fibers and the surrounding matrix.

Figure 5a shows the cross sectional view of a two-ply Gr/BMI composite along the interface between plies after laser curing. Figure 5b, taken at a higher magnification, clearly shows the regions along the interface between the plies have indeed come into intimate contact, and excellent fiber wet-out is observed. These photomicrographs clearly illustrate the excellent material consolidation obtained with low-intensity $CO_2$ laser beams.

480

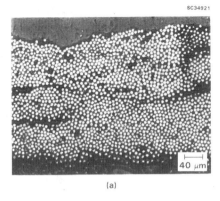 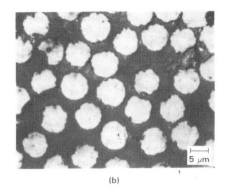

(a)  (b)

Fig. 5 Cross-sectional views of a laser-cured Gr/BMI composite in 4 s showing (a) the interface between the two plies, and (b) the wet-out of graphite fibers by the BMI matrix.

Laser energy has also been demonstrated to cure composites in sequential order. In one case, a stack of four plies of unidirectional BMI/Gr prepregs was first exposed to the laser at a power density of 10 W/cm$^2$ for 256 s. Three additional plies of BMI/Gr of the same fiber orientation were then placed on top of the previous laminate and exposed to the laser at the same intensity for an additional 256 s. In another case, a piece of PEEK neat resin film was sandwiched between two plies of PEEK/Gr prepreg. They were then exposed to the $CO_2$ laser with a power density of 10 W/cm$^2$ for 150 s, followed by an increased power density of 20 W/cm$^2$ for 150 s. In both cases, the BMI and PEEK laminates were cured to a certain extent, as evidenced by fiber wet-out on the backfaces and the hardening of the irradiated area.

## 4. CONCLUDING REMARKS

1. The feasibility of using laser radiation as an advanced processing concept for composites was demonstrated. It appears that laser processing has potential application for the field repair of damaged composite structures.

2. Graphite fibers absorb $CO_2$ laser beam and generate heat. In graphite-reinforced composite laminates, the heat generated along the graphite fibers would be conducted from the surface into subsurface layers. When sufficient heat flux is deposited from the fibers into the surrounding matrix, the process of curing or consolidation occurs.

3. The use of low-intensity $CO_2$ laser (10 W/cm$^2$) can effect the fiber

wet-out of a four-ply BME composite in 100-300 s. The same effect was seen in PEEK/Gr composites with increased power density (20 or 30 W/cm$^2$).

4.  Laser curing is shown to be independent of fiber orientation, and thus amenable to processing of cross-ply composites.

5.  Future investigation on laser processing of composites should deal with the effect of beam power, exposure time, travel speed on the quality of process composites, the differences in interaction between thermosets and thermoplastics with the laser beam, the effect of fiber orientation, mechanism of void generation, and residual thermal stresses.

## 5. REFERENCES

1.  N. Bloembergen, "Fundamentals of Laser-Solid Interactions," AIP Conf. Proc., ISSN:0094-243X/79/5000001, Amer. Inst. of Physics (1979).
2.  Handbook of American Institute of Physics, 3rd Ed., McGraw-Hill, 1972.

## 6. BIOGRAPHY

C.M. Tung: Member of the Technical Staff, Polymer Synthesis and Processing Department, Rockwell International Science Center. Ph.D., University of California, Los Angeles, 1973. Dr. Tung joined the Science Center in 1978. Dr. Tung has been extensively involved in the area of elucidating structure-process-property relationships of thermoset and thermoplastic materials. She was also responsible for developing high-temperature, fracture-tough BMI using the cure state/space concept. Recent research includes applying the cure state concepts to BMI composites and characterization of high-temperature thermoplastic materials.

D.S. Gnanamuthu: Member of Technical Staff, Metals Processing Department, Rockwell International Science Center. Mr. Gnanamuthu has 22 years industrial experience including 15 years in laser metalworking. He joined the Science Center in 1978 to lead research studies on laser processing of materials. Between 1963 and 1978, he was with Ford Motor Company Scientific Research Laboratories and AVCO Everett Research Laboratory, where as Chief Metallurgist he developed new methods for heat treating, surface alloying and cladding with laser beams. Mr. Gnanamuthu received his B.S. in Chemistry and M.S. in Physical Chemistry from the University of Madras, India, his M.S. in Metallurgical Engineering from the University of Missouri, and his MET.E. in Metallurgy from MIT.

R.J. Moores: Senior Research Specialist, Metals Processing Department, Rockwell International Science Center. Mr. Moores has 27 years of industrial experience, including 17 years in laser metalworking. He joined the Science Center in 1979 to conduct research and development on laser processing of materials. Between 1960 and 1979, Mr.

Moores was with AVCO Everett Research Laboratory and the IIT Research Institute's Laser Center. At AVCO, he primarily was responsible for operation, maintenance and research studies on several high-powered $CO_2$ lasers; as Applications Manager at IITRI, he carried out workpiece tooling design and procedure development, as well as laser beam diagnostics for welding, cutting, heat treating, alloying, and cladding with laser beams.

C.L. Leung: Manager, Polymer Synthesis and Processing Department, Rockwell International Science Center. Dr. Leung received his Ph.D. in Polymer Science from the University of Akron. His main research interest lies in the mechanisms of refinement, structure-properties relationship, fracture mechanics, crystallization and adhesion of polymers and composites. He has 20 publications on the fracture, moisture diffusion and cure mechanism of fibrous composites, and was an invited lecturer at the "Thermoplastics Elastomers: Properties, Processing and Uses" Short Course sponsored by the Plastics Institute of America and the University of Southern California.

MEDICAL ASPECTS OF ORBITAL SPACEFLIGHT
AND THEIR IMPLICATIONS FOR
MANUFACTURING IN SPACE
George T. Delli-Santi
Zimmer, Inc.
Warsaw, Indiana

## ABSTRACT

A general review of the biomedical consequences of prolonged orbital spaceflight is presented. Data from Apollo, Skylab and Soviet space programs is used to describe a number of the physiologic changes associated with adaptation to a microgravity environment. The author concludes by speculating on how these medical aspects may influence the design and development of orbital manufacturing facilities.

## 1.   INTRODUCTION

It has been nearly thirty years since the successful launch of Sputnik, the first artificial satellite to fly in space. Within such a brief expanse of time, great technological advances have been made in achieving Earth orbit on a somewhat routine basis. The feasibility of people living in space and working under microgravity conditions has been demonstrated over the last twenty-five years of manned spaceflight. Despite this experience, which includes data from the Skylab experiments during 1973 and 1974 and the recent Spacelab flights in 1983 and 1985, little data exists regarding how humans react to long-term weightlessness, such as would be experienced aboard a permanent space station or orbital manufacturing facility.(11) It is still unkown if human beings will be able to survive long-term exposure to microgravity conditions.(20)

Before the launch of Alan Shepard, our nation's first astronaut, on May 6, 1961, and despite the prior success of cosmonaut Yuri A. Gagarin's flight on April 12, 1961, doctors and scientists were making dire predictions about the physiological effects of spaceflight.(35) The list of prospective medical afflictions awaiting our astronauts was diverse, sometimes contradictory, and included: anorexia, disorientation, sleepiness, sleeplessness, euphoria, hallucinations, decreased gravity tolerance, gastrointestinal disturbance, urinary retention, diuresis, muscular incoordination, kidney stones, collapsed air sacs in the lungs, increased heart rate, hypertension, hypotension, cardiac arrhythmias, dehydration, infectious disease, nausea and motion sickness, muscle atrophy, demineralization of bone, reduced blood volume, weight loss, and decreased capacity for exercise.(35)

It is now known that some of these medical anomalies do, indeed, occur during spaceflight and especially under long-term microgravity exposure. Areas of particular concern in human adaptation to microgravity include Space Adaptation Syndrome ("space sickness"), deconditioning of the cardiovascular system, loss of muscle tissue, demineralization of bone, and metabolic abnormalities.(13;35;21)

## 1.1 Organization of This Paper

The medical aspects of spaceflight to be reviewed in this paper will be categorized according to the mechanisms of their origin. Although the human body is extremely complex and all physiologic systems are interrelated in some manner (e.g. hormonal control, direct nervous stimulation, chemical mediators, mechanical forces, etc.), the physiologic adaptation of the human body to spaceflight appears to be in response to two principle sequelae of microgravity: a shift in the distribution of bodily fluids and removal of the burden of one's weight from the musculoskeletal system. These influences are addressed in the second and third sections, respectively, of this paper.

Psychological factors are also of great concern in long-term spaceflight. For this reason, and because such factors inevitably affect the health of the crew, these factors are described separately in section four. Other health problems which are consequences of the body's adaptive changes are addressed in section five.

Section six will briefly summarize methods which have been used in order to diminish the physiologic effects of extended spaceflight. This paper concludes with a discussion, in section seven, of how future space manufacturing activities may be influenced by the medical limitations of spaceflight.

## 2. A SHIFT IN BODILY FLUIDS

During the Skylab experiments it was noted that the redistribution of fluids in the body was, "responsible for more trends and had greater impact physically of any other single effect of weightlessness."(4,p.33) The cardiovascular (CV) system contains much of the body's fluids and is attributed with having a large role in the process of adaptation to microgravity.(7)

The CV system is designed to resist the pull of gravity in order to maintain blood pressure in the brain. Vascular pressure sensors (baroreceptors) in the carotid arteries of the neck monitor the pressure of blood being pumped to the head. When there is a drop in this pressure, such as when we stand and blood rushes to our feet, these baroreceptors signal the body to compensate for the pull of gravity by increasing our heart rate and constricting the blood vessels of the lower body.(5)

These same compensatory, reflexive mechanisms initially overreact in the microgravity environment and cause one-half to two liters of blood to surge to the head. This results in symptoms characteristic of a head cold (i.e. stuffy sinuses, a nasal voice), facial swelling and a loss of circulation to the legs.(5;4,p.33)

Blood tends to pool in the chest, or thoracic cavity, during spaceflight. This reduces the breathing capacity of the lungs and increases upper body blood pressure to an extent which causes a false sensation that the level of total body fluids is too high.(5;20;4,p.33;45) This, in turn, increases the body's excretion of urine and depresses the sense of thirst. The former can create electrolyte imbalances and accelerate the loss of calcium from the bones while the latter can result in dehydration if astronauts do not force themselves to drink an adequate amount of water regularly.(5;20;4,p.45)

## 2.1 Electrolyte Imbalances

Electrolytes are substances dissolved in bodily fluids which exist as ions, e.g. sodium, potassium, calcium and phosphorus. They play an enormous role in important metabolic processes such as the conduction of electrical signals along nerves and the contraction of muscle tissue. Electrolytes are inevitably lost in the urine and in the process of perspiration. The increase in urine output from blood pooling in the chest can deplete the body of essential electrolytes and lead to electrolyte imbalances. Such imbalances were responsible for the heart arrhythmias experienced by Apollo 15 astronauts while on the moon.(4,p.45)

As shall be described, the bones of the skeleton begin to deteriorate and lose their calcium under microgravity conditions. The increased urine output stemming from bodily fluids shift compounds this problem by removing calcium from the bloodstream. This prevents the serum calcium concentration from achieving an inhibitory level, a process which would normally help to decrease the rate of bone demineralization.

Data from long-term missions to date have indicated that after many weeks and months in microgravity the human body does not achieve a new equilibrium in the balance of electrolytes.(15) It is questionable if such a new balance would even be desirable.

## 2.2 Neurovestibular Changes

The eyes, sensors in the extremities which tell us where our limbs are in relation to our bodies, and the gravity-sensing semicircular canals of the inner ear comprise the neurovestibular system of the body.(5) This complex system is responsible for our sense of balance and orientation and it is greatly affected by deprivation of normal Earth gravity.

The shift of fluids to the upper body affects the sense of balance.(4,p.23) Skylab astronauts used a rotating chair experiment to assess their sense of balance while in flight. On Earth, the experiment created feelings of nausea and dizziness as they moved their heads while rotating in the chair because such motion disturbs the fluid in the semicircular canals of the inner ear.(4,p.23) After the first few days of adaptation to microgravity, however, the crew became immune to the effects of the rotating chair.(4,p.23) It is believed that increased amounts of fluid in the inner ear caused a complete loss of balance in the Skylab astronauts and that this was responsible for their immunity to the dizzying effects of the test.(4,p.23,33)

Fish and spiders flown aboard Skylab demonstrated that the sense of sight can be used to replace a lost sense of balance.(4,p.71) These animals initially lost all ability to orient themselves but soon adapted to using visual clues for orientation. Such experiments have led to the theory that eyesight allows one to determine a sense of vertical in the absence of a sense of balance, the major determinant of vertical on Earth.(4,p.72)

There have been some indications that the loss of balance from exposure to microgravity may reach a point of irreparable damage. Some of the Skylab astronauts retained their spaceborne immunity to the rotating chair experiment for up to one month after their return to Earth.(4,p.181)

## 2.3 Space Adaptation Syndrome

"Space sickness", now known as Space Adaptation Syndrome (SAS), is the least life threatening, but also the most troublesome, side effect of relocation from Earth gravity to microgravity.(21) SAS consists of nausea, vomiting, anorexia, headache, malaise, drowsiness and lethargy, but such symptoms and their duration vary among spaceflight crew members for unknown reasons.(21)  For example, on Space Shuttle flights some crew members experienced episodic vomiting without any nausea while others had feelings of nausea only.(21)  Even more perplexing is that an Earthbound susceptibility to motion sickness does not correlate with susceptibility to SAS.(21)

According to Marvin R. Christensen, NASA Life Sciences Coordinator for the Space Station, problems with SAS tend to occur in the first few days of flight.(11)  Half of our astronauts have been victims of SAS so far and although the symptoms usuallly disappear in three to four days, it is unknown if SAS can recur during long-term flights.(20)

Cosmonauts were the first space travelers to experience SAS because of the relatively spacious Soviet spacecraft.(35)  SAS was never experienced during the Mercury and Gemini programs because the astronauts remained strapped to their seats and the capsules were tight and confining.(35;21)  Able to freely float in their larger Apollo cabins, however, our astronauts soon came to know the discomfort of SAS.(21)

Space Shuttle flights STS-4 through STS-8 determined that SAS does not produce illusions or disturbed occulomotor function and additional data has indicated that it does not originate in the central nervous system.(21)  Dr. William E. Thornton, a Discovery shuttle astronaut, states that, "The most compelling arguments are that an otolith organ-canal and visual conflict are responsible [for SAS]".(21)  The conflict theory of SAS was developed from ground-based investigations and data from STS-8, STS-11 and Spacelab-I shuttle flights.  In The conflict theory states that incoming sensory signals from the eyes and the balance organs are evaluated in comparison with previous information stored in the central nervous system.  When the two sets of information conflict with each other, a neural mismatch occurs which in turn affects the autonomic nervous system and produces the effects symptomatic of SAS.(21)

SAS symptoms seem to occur while the sense of balance is still active. After the balance mechanism is shutdown by the shift of body fluids to the head and chest, as previously described, SAS clears up.(4,p.36) During some Space Shuttle flights, however, SAS was responsible for notable lapses in productivity.(21)

2.4 Cardiovascular Deconditioning

After extended periods of weightlessness, the circulatory reflexes responsible for these fluid imbalances begin to shut down. During spaceflights lasting fifteen to sixty-three days among thirteen cosmonauts, three distinct stages were observed during CV adaptation to spaceflight and the changes associated with these stages coincided with the cosmonauts' well-being.(7)

First Stage:  During the first week, the increased central blood volume in the chest created discomfort associated with the fluids shift effects described previously in section one of this paper.(7)

Second Stage:  A relative stabilization of reactions occured during the second week. Neurohumoral regulatory mechanisms reduced the function of the circulatory system to preflight levels, resulting in less blood surge to the head and general improvement in the well-being and motor activities of the cosmonauts.(7)  This may be due in part to a desensitization of the carotid artery baroreceptors.(5)

Third Stage:  From the third week on, the function of the circulatory system is strained with concomitant symptoms, (e.g. an alteration of sleep patterns and increased fatigability), indicative of a deconditioning of the cosmosnauts' CV system.(7)

Other symptoms illustrative of CV deconditioning include changes to the heart muscle itself and alteration of red blood cells (RBCs). Measurements indicate that the heart shrinks during long spaceflights and decreases of eight to twenty percent in heart muscle volume of returning astronauts have been observed.(5) RBCs change shape, becoming thin, attenuated and flat.(4,p.51) RBCs are also lost at an accelerated rate, up to as much as one-half liter on long flights, and it is believed this effect is attributable to microgravity-induced changes of the spleen and premature destruction of RBCs.(21)

Deconditioning of the CV system increases the difficulty of re-adjusting to Earth's normal gravity. It is known that the CV system remains abnormal for months after returning to Earth.(21) What is still not clearly understood is how CV changes occur and whether, after an extended period of time in microgravity, the body will eventually lose the ability to regain normal CV function upon returning to Earth.(5)

## 3. MUSCULOSKELETAL UNBURDENING

The effects of microgravity on the skeleton and muscles of the human body are remarkably similar to the medical consequences of immobilization from conditions such as paraplegia and prolonged bed rest.(27;6;22;25) In spaceflight the body's musculoskeletal structure is no longer burdened by the inexorable pull of gravity. The available data suggest that reduced mechanical loading of the musculoskeletal (MS) system inevitably results in bone loss and muscle tissue breakdown.(27;21)

### 3.1 Normal Bone Physiology

Bone loss was one physiologic alteration which never stabilized during the Skylab missions.(4,p.182) It is understandable, therefore, that much attention has been paid to this consequence of life in space. The mechanisms of microgravity-induced bone loss are more easily comprehended in light of the normal physiology of bones functioning under Earth gravity.

Bone is a living composite material. It is composed of a complex calcium-phosphate salt, hydroxyapatite $[3Ca_3(PO_4)_2 \cdot Ca(OH)_2]$, and calcium carbonate, both of which are intimately deposited in an organic connective tissue matrix consisting of collagen, a fibrous protein.(23,p.758) Aside from its obvious functions of supporting the body and providing locomotion through its interaction with muscles, the skeleton stores more than ninety-eight percent of the calcium and eighty percent of the phosphorus in the body.(23,p.759)

Calcium is an essential mineral constituent of all tissues and participates in many important physiological processes including blood coagulation, maintenance of heart rhythms, control of cell membrane permeability, neuromuscular excitability, production of milk and formation of bone and teeth.(23,p.758) Phosphorous is also important physiologically and functions in the intermediary metabolism of food and in maintaining the pH of body fluids.(23,p.758)

Bone is maintained in a state of dynamic equilibrium: it is constantly being torn down and rebuilt. The processes of bone resorption and bone formation are controlled by three types of cells: 1) osteoblasts, which form collagen fibers into the templates which are calcified to form bone; 2) osteoclasts, which are involved in the dissolution of bone minerals and the destruction of bone collagen; and 3) osteocytes, or bone cells, which are capable of mediating the conflicting forces of bone formation and bone resorption.(5;23,p.759) Mechanical and hormonal factors can affect the balance of these forces and this results in a remodeling of bony architecture.

The mechanical theory of bone remodelling was suggested in 1982 by

488

Wolff (34), who showed that the internal structure of bones is defined by the extrinsic loads which they must support.(14) It is now thought that bones respond to the amplitude of the strains generated within them by such external loads.(14) For a given strain (the limit strain) bone resorption and deposition occur equally and a dynamic equilibrium is established. If this limit strain is exceeded, deposition prevails and, conversely, bone resorption is dominant if strains are below the limit value.(14) How bone tissue senses mechanical strain and "decodes" such information is still unknown, although many hypotheses have been suggested.(14) It would seem, however, that human experiences in microgravity are substantiating this theory's prediction that an overall decrease in the loads borne by the skeletal system will lead to demineralization and osteoporosis, a porous condition of bone tissue.(14)

Hormonal control of bone physiology is largely achieved through two secretions of the parathyroid glands: Parathyroid Hormone (PTH) and Calcitonin (CT).(23,p.757) These hormones have marked effects on the calcium levels of the body's tissue fluid and bone.

PTH is secreted in response to low serum calcium levels, or hypocalcemia, and results in calcium being released from the bones. Excessive secretion of PTH leads to the destruction of bone matrix, demineralization of bone tissue and a consequent softening of the skeletal system, and excessive amounts of calcium in the serum, or hypercalcemia.(23,p.760) CT acts on the same target tissues as PTH but in the role of an antogonist. CT is secreted in response to hypercalcemic conditions and results in inhibition of osteoclastic activity and the maintenance of bone.(23,p.761-2) In their respective roles, PTH and CT exert a precise, homeostatic control over serum calcium levels.

## 3.2  Bone Loss From Microgravity

As described in section 3.1 above, bone is maintained in a state of dynamic equilibrium and is affected by mechanical strains and hormones released in response to electrolyte imbalances. Although the exact mechanism of bone loss in the microgravity environment is not known, the consequences of the mechanism manifest themselves in osteoporosis and electrolyte imbalances: metabolic anomalies usually associated with disease states.

The effects of calcium loss are not equal in all bones of the body. Osteoporosis caused by unloading the skeleton seems to target weight-bearing bones, such as the calcaneus, or heel bone.(14) The spongy bone (trabecular bone) making up the internal structure of most flat bones and the ends of long bones usually exhibits disuse osteoporotic changes faster than the hard, compact bone (cortical bone) of the body.(27)

Demineralization of the body during weightlessness can severely undermine the structural integrity of the skeleton. A loss of twenty to twenty-five percent of normal bone mineral content can predispose bones to fracture upon return to Earth's normal gravity.(5;28,p.1) This amount of demineralization can occur in as little as three months of living in microgravity.(14) Studies using rats flown aboard the Soviet vehicle COSMOS 936 indicated that bone torsional strength and energy-to-failure decreased significantly in rats after just 18.5 days of spaceflight.(6)

The hypercalcemic condition which results from skeletal demineralization also has physiologic consequences. Excess serum calcium is normally excreted in the feces with only a small amount passed through the urine.(23,p.759) In a hypercalcemic state, however, the body begins to excrete greater amounts of calcium through the kidneys. The resultant elevated output of calcium-rich urine can lead to the formation of kidney stones (renal calculi) which are capable of causing kidney infection, ureter blockage or even kidney failure.(5;23,p.758;28,p.1) Abnormal conditions such as increased excretion and loss of phosphate, increased blood coagulability and decreased excitability of nerve tissue can also result from hypercalcemia.(23,p.760)

It has been theorized that the effects of bone demineralization may be permanent after long stays in space.(21) Medical data from bed-ridden patients have demonstrated that six months of total immobilization can lead to a new equilibrium value of bone density.(27) This suggests that permanent skeletal damage could be incurred from long spaceflights if, upon return to Earth, the body's calcium balance were restored to a new equilibrium value before the bones had replaced all of their lost minerals. Mineral deposition in the bones would then cease and the bones would be left permanently weakened.(21) Such permanent bone damage has been observed in rats which have spent three weeks, a significant portion of their normal life spans, in space.(21)

## 3.3 Muscle Tissue Changes

Increases in circulating nitrogen and creatinine kinase, indicators of muscle tissue destruction, have demonstrated that muscle loss occurs during spaceflight.(21) Furthermore, biopsies of rat muscle tissue pre- and postflight indicate that along with this loss, a transformation in muscle type is probably occurring.(21) Unburdening of the MS system appears to favor the "fast twitch", white muscle fiber type at the cost of the amount of "slow twitch", red muscle fiber.(21) The cause of this conversion of muscle fiber types is thought to be related to the fact that red muscle fibers are associated with exercise endurance and, therefore, see little use during spaceflight.(21)

## 4. PSYCHOLOGICAL FACTORS

Soviet physicians are studying the possible psychological consequences of extended spaceflight and much attention is being devoted to the problem of group compatibility among cosmonauts.(1;13) According to NASA's Christensen, the long-term duration limits for spaceflight are not going to be set by physiology, but rather psychology.(11) Most people can tolerate anything for about eight weeks, but thereafter behavioral changes seem to occur.(4,p.88)

## 4.1 Isolation

Experiences from Arctic and Antarctic expeditions, shipwrecks, and submarine voyages have indicated that, "the ability of small groups of persons to work together and live together during prolonged isolation is a difficult and complex process which does not always work out satisfactorily."(1) According to H. Clayton Foushee, Ph.D., a social psychologist with NASA's Ames Research Center, Human Factors Research Division, "...chronic stresses related to isolation and confinement are obvious problems for an environment such as the space station...In polar research stations there are numerous reports of group conflict, including violent physical aggression. In strategic nuclear submarines, there are consistent reports of higher than average incidences of hostility, depression, and anxiety that appear to manifest themselves in terms of cliques, physical conflict, and the development of pecking orders more reminiscent of prison environments."(12)

In extended tours of orbital flight this pattern may become modified. During the Skylab missions, conflict appeared not among the crew, but between the crew and Mission Control. The third Skylab crew went on strike at the end of the sixth week of their mission.(4,p.128) The second Skylab crew had taken a month to hit their stride, yet the third crew was never allowed time to fully

490

acclimate themselves by a Mission Control filled with unrealistic expectations for their performance.(4,p.87,128-9) Christensen states, "...hopefully the workload on the first few days in orbit can be minimized... in terms of getting lots of useful work done, the choice of a three month tour of duty seems a reasonable one."(11)

## 4.2 Spacecraft Ergonomics

Spacelab studies showed that spacecraft should be designed such that astronauts are not placed in a position where the location of the floor is unclear.(11) Astronauts tend to be reluctant to abandon their psychological need for a local vertical reference.(4,p.109) The docking adapter, for example, was deemed to be the least successful part of Skylab because it was without any sense of up or down, having equipment and consoles jutting radially from its cylindrical walls.(4,p.110) Skylab's Earth-like lower deck, conversely, was elected the most successful layout by most of the crewmembers because of its defined floor and ceiling.(4,p.110) Such a layout, however, uses only about one-sixth of the space made available by the unique locomotive situation presented by weightlessness.(4,p.109)

## 5. OTHER HEALTH ISSUES

Living in space will require new policies and facilities for basic health maintenance, according to Dr. Donald Trunkey of the Oregon Health Sciences University in Portland.(5) Impairment of the body's immune system is known to occur (20;21) but whether this ultimately results in a decreased resistance to infections is not known. The observed shift of bodily fluids affects how drugs are absorbed by the body and the manner in which pharmacologic agents become distributed, metabolized, and excreted in the body must be investigated before correct "weightless" dosages can be prescribed.(20)

Some common terrestrial health concerns are complicated by microgravity. Bone fractures incurred in space, for example, may not heal normally without gravity because load-bearing bones tend to heal vertically.(5) A theoretical study by the Los Alamos National Laboratories indicated that even common air pollutants such as dust, bacteria and smoke may become respiratory hazards in weightlessness.(12) Instead of settling out of inhaled air and depositing on lung surfaces (for removal by the mucosal system), such particles may instead fill up passages in the human respiratory tract.(12)

Quarantine from procreation may be required for male personnel returning from long-term flights if trends seen in biosatellite COSMOS 936 rat studies represent what will also occur in humans. Mature sperm exposed to microgravity fathered litters inferior to terrestrial controls with respect to neonate growth and development.(30) This effect is due to deconditioning, not genetic changes, of the sperm, however, and rats mated three months postflight fathered normal offspring.(30)

Outside of Earth's protective atmosphere, radiation levels also become of greater concern. Crew members in low Earth orbit can expect to recieve 100 millirems of radiation per day.(5) The Skylab astronauts even reported "seeing" blue-green flashes of light, the result of high-energy electrons impacting their eyes.(4,p.123)

## 6. PROPHYLACTIC MEASURES

A summmary of the many prophylactic regimes investigated for impeding or preventing the undesirable biomedical affects of microgravity is beyond the scope of this paper. Some notable failures worthy of brief review are the following.

Dietary supplements of phosphate (28,p.19), calcitonin administration (28,p.23), lower body negative pressure treatments (28,p.49), longitudinal compression of the long bones (4,p.101;28,p.17,18), and electrical stimulation (17) have all been demonstrated to be ineffective in arresting bone demineralization and osteoporosis. In-flight exercise does not deter bone loss or CV deconditioning. Skylab astronaut Charles Conrad estimated that one would have to work out five hours per day simply to equal the cardiovascular effort expended in merely existing on Earth.(4,p.104) Many other researchers are in agreement that formal exercise programs, which have been resisted by some crews, may not be effective measures for counteracting the biomedical effects of many months of microgravity.(13;3;12;14)

NASA flight surgeons and designers of space settlements concur that the surest way to control MS deconditioning on long flights would be to provide the spacecraft with gravity.(15;4,p.182) This has been substantiated by rat experiments aboard COSMOS 936 wherein a centrifuge was used to provide some of the rats with artificial gravity. The centrifuged rats had bones with mechanical properties no different from those of terrestrial controls after nearly nineteen days of spaceflight. (24;14;6) Their weightless comrades, however, returned with evidence of bony degeneration.(24)

Considered by both American and Soviet scientists, an on-board centrifuge for preventing bone atrophy was found to be both difficult and costly to build.(13) Artificial gravity could most simply be accomplished by rotating the spacecraft.(5;15) A rotating environment does have drawbacks because of its imposition of Coriolis forces which, at spin rates of several rpm, can cause disconcerting locomotion effects amd motion sickness.(15) If rotation rates are below three rpm, however, people may be able to adapt after prolonged exposure.(15)

## 7. DISCUSSION

The current duration record for manned spaceflight was set in 1984 aboard Salyut 7: cosmonauts Leonid Kizim, Vladimir Soloviov and Oleg Atkov spent 237 days, nearly eight months, in Earth orbit.(5) Despite this impressive record, it is still unknown if there is a "point-of-no-return" for weightless living, after which the human body is unable to re-adjust to Earth gravity.(20) Current trends seen in this area are disturbing. Salyut 6 cosmonauts Vladimir Lyakhov and Valeriy Ryumin spent 175 days aloft in 1979, but their bodies were deconditioned to such an extent that their return to Earth left them, "virtually paralyzed".(5)

Adapting to microgravity is often deceptively easy for crew members.(13) As reported in this paper, the adaptive mechanisms of the body decondition organ systems and render them less efficient upon returning to Earth.(20;21) Although only one in every ten humans who have flown in space has stayed longer than a month, postflight disturbances in the ability to stand and walk, gastrointestinal stability, and vestibular function have persisted for up to six weeks.(13) The Salyut 6 crew required six weeks of intensive rehabilitation for their CV and MS systems to recover fully from their six month stay in microgravity.(5;20) In light of these outcomes, a "safety period" of six to nine months has been suggested for spaceflights, after which authors Whedon, et al,(33) believe function of the MS system could be permanently impaired.

We are now planning to send scientists and other professionals, not career astronauts in peak physical condition, into space and at least one researcher believes the general public to be suffering some degree of osteoporosis under terrestrial conditions because of the lack of strenuous, daily exercise in modern lifestyles.(8) The available research data, therefore, may prove to be a very conservative estimate of the affects of long-term microgravity on human beings.

The medical aspects of orbital spaceflight described in this paper suggest that the need for human operators would greatly complicate an orbital manufacturing facility. Unmanned facilities, such as the Industrial Space Facility under development by Space Industries, Inc. and Westinghouse Electric Co. (10,p.28) are to be preferred. The proposed Space Station could serve as a base for maintaining and harvesting product from such autonomous orbital manufacturing facilities as well as provide laboratory facilities during the basic R&D stages of manufacturing process development.

The Space Station should also, "...have designated facilities for continued long-term study of humans and their environment," according to Robert H. Moser, M.D., former director of the American College of Physicians.(20) A rotating space station with artificial gravity would best meet this end. Such a configuration, although in conflict with the idea of manufacturing under microgravity conditions, would be ideal for studying the subject of long-term microgravity adaptation. Subjects could be weightless and then returned to 1g conditions without the expensive flights between orbit and Earth. Investigators would also have the tightest possible control of all experimental variables.

Such studies of orbital weightlessness could have profound benefits for orthopaedic medicine. Study of healthy subjects in microgravity might disclose the exact mechanism of osteoporosis and lead to a cure for this condition or prophylactic treatments. A cure for osteoporosis may be more easily discerned in such studies because there would be no superimposed disease state to complicate diagnoses and analysis.

Despite the inherent medical problems of long-term manned spaceflight, there can be little doubt of the advantages and benefits of human intervention in the exploitation of Earth orbit resources. McDonnell Douglas Corp. recently completed a twenty-six month study, called "The Human Role in Space" (THURIS), funded by NASA's Marshall Space Flight Center, in which guidelines for the best use of man and machine were developed.(16)

The THURIS study found that, "...man was the most cost effective way to deploy and assemble large structures, ...to repair and maintain machinery, [and perform] space experiments and everyday chores."(16) It was estimated in this study that the cost of a self-deploying orbiting space platform would be 2.4 million dollars whereas a similar astronaut-deployed structure would cost one-tenth that amount.(16) The THURIS study concluded that machines are better suited for jobs involving excessive information processing and numerical computations.(16)

A human's ingenuity, flexibility and versatility in managing unforeseen situations were stated as advantages over unmanned robotic systems in the THURIS study.(16) The first Space Shuttle crew, for example, repaired a faulty camera after receiving instructions from ground personnel.(16) Salyut 4 photographed Soviet landscapes automatically but cosmonauts often intervened to correct for dense cloud cover.(17)

Humans will be indispensable in setting-up orbital manufacturing facilities of commercial scale. NASA commercial space officials are currently investigating the theory that automated research in space is less efficient than human-tended research.(10,p.31) According to Isaac Gillam, NASA assistant administrator for commercial space, NASA can no longer afford to devote shuttle capacity to experiments which are only forty to fifty percent successful, such as the NASA

Microgravity Experiment Apparatus which was recently flown.(10,p.31) Space Shuttle astronauts may be required to spend more time operating industrial payloads in order to increase their efficiency.(10,p.31) The last Spacelab mission, for example, achieved a 100 percent success rate, performance which Gillam ascribed to that flight's emphasis on human-tended, rather than automated, experiments.(10,p.31)

## 8. SUMMARY

The biomedical aspects of orbital spaceflight described in this report indicate that, for the near-term, manned manufacturing facilities would be undesirable. Weightlessness alters appreciably the function of several important physiologic systems and such deconditioning could reach levels capable of greatly reducing work capacity and efficiency. The required adaptation period and limited duty period of crews would prevent realization of the maximum productivity of such a manufacturing facility. Autonomous, unmanned orbital manufacturing facilities, located away from the micro-atmospheric pollution and vibrations of manned spacecraft and free of the constraints which would be associated with the medical limitations of a human crew, are to be preferred. Manned spaceflight, howver, will play a large role in the pilot plant stage of orbital manufacturing facilities. Continued study and resolution of microgravity-induced medical problems is therefore essential for the continued progress of orbital manufacturing development activities.

## Bibliography

1. Alyakrinskiy, B.S., "Problems of Social Psychology in Space", in NTIS report Space Biology and Medicine, Vol 7, No 4, 116-120 (1973).

2. Belakovskiy, M.S., et al, "Effect of Excessive Phosphorus Intake on Some Aspects of Phosphorus-Calcium Metabolism...", USSR Report, Foreign Broadcast Information Service, Vol 16, No 5, Sept-Oct, 102-107 (1982).

3. Beregovkin, A.V., et al, "Cardiorespiratory System Reactions of Cosmonauts to Exercise Following Long-Term Missions...", USSR Report, Foreign Broadcast Information Service, Vol 14, No4, 9-13 (1980).

4. Cooper, H.S.F., Jr.: A House In Space, Bantam Books, NY (1976).

5. DeCampli, W.M., "The Limits of Manned Spaceflight", The Sciences, Sept-Oct, 47-52 (1986).

6. Donaldson, C.L., et al, "Effects of Prolonged Bed Rest on Bone Mineral", Metabolism, 19, 1071-1084 (1970).

7. Doroshev, V.G., et al, "Comparative Assessment of Circulatory Reaction During Work in Weightlessness and in Salyut Station Mockup", ref 2, 52-56 (1982).

8. Emiola, L., and O'Shea, J.P., "Effects of Physical Activity and Nutrition on Bone Density Measured by Radiographic Techniques", Nutrition Reports Int'l., June, Vol 17, No 6, 669-680 (1978).

9. Finch, E.R, Jr., and Moore, A.L.: Astrobusiness, Walden Book Co., Inc., Stanford, CT (1984).

10. Fink, D.E., and Quast, D., editors, Commercial Space, Vol 2, No 3, Fall, McGraw-Hill, NY (1986).

11. Giuffre, W.L., "Concerns Are Being Raised About Living in the Space Environment", Commercial Space, Winter, 69-71 (1986).

12. Goldsmith, M.F., "How Will Humans Act As Science Fiction Becomes Science Fact?", JAMA, Oct 17, Vol 256, No 15, 2048-2052 (1986).

13. Gunby, P., "Soviet Space Medical Data Grows; Other Nations Joining In", ref 12, 2026-2052 (1986).

14. Hinsenkamp, M., et al, "In Vivo Bone Strain Measurements: Clinical Results, Animal Experiments, and a Proposal for a Study of Bone Demineralization in Weightlessness", Aviation, Space and Environmental Medicine, Feb, 95-103 (1981).

15. Johnson, R.D., and Holbrow, C., editors, "Space Settlements - A Design Study", NASA, SP-413, 21-36 (1977).

16. Kandebo, S.W., "Technology Preview", Commercial Space, Spring,9 (1986).

17. Konovalov, B., "Weightless Orbit", Izvestiya, Feb 9, Nr. 34/17877, reproduced by NTIS, Springfield, VA, p.3 (1975).

18. Kreydich, Yu. V., et al, "Effect of Immersion Hypokinesia on Characteristics of Human Eye and Head Movements...", ref 2, 57-62 (1982).

19. Markelov V.V., and Chernykh, I.V., "Direct-Reading Dosimeters Used for Monitoring Radiation in Salyut Stations", ref 2, 120-124 (1982).

20. Marwick, C., "Physicians Called Upon to Help Chart Future Space Effort", ref 12, 2015-2025 (1986).

21. Merz, B., "The Body Pays a Penalty for Defying the Law of Gravity", ref 12, 2040-2052 (1986).

22. Minaire, P., et al, "Quantitative Histological Data on Disuse Osteoporosis", Calcif Tissue Res, 17, 57-73 (1974).

23. Moore, W.W., "Endocrine Control of Calcium Metabolism", from: Physiology, E.E. Selkurt, editor, 4th ed, Little, Brown and Co., Boston (1976).

24. Prokhonchukov, A.A., et al, "Comparative Study of Effects of Weightlessness and Artificial Gravity on... Calcified Tissues", ref 3, 30-34 (1980).

25. Rambaut, P.C., et al, "Skeletal Response", from: Biomedical Results of Apollo, NASA, SP-368, 303-322 (1975).

26. Rüegsegger, P., et al, "Bone Loss in Premenopausal and Postmenopausal Women...", JBJS [Am], 66, 1015-1023 (1984).

27. Rüegsegger, P., et al, "Disuse Osteoporosis in Patients with Total Hip Prostheses", Arch Orthop Trauma Surg, 105, 268-273 (1986).

28. Schneider, V.S., et al, "The Prevention of Bone Mineral Changes Induced by Bed Rest...", NASA Terminal Report, Contract #T-81070, NASA-CR-141453, Public Health Service Hospital, San Francisco.

29. Sergeyev, I.N., et al, "Role of 24,25-Dihydroxycholecalciferol in Bone Mineralization in Hypokinetic Rats", ref 2, 108-113 (1982).

30. Serova, L.V., et al, "Reproductive Function in Male Rats After Flight Aboard COSMOS-1129 Biosatellite", ref 2, 90-95 (1982).

31. Singh, M., et al, "Changes in Trabecular Pattern of the Upper End of the Femur as an Index of Osteoporosis." JBJS, April, Vol 52-A, No 3, 457-467 (1970).

32. Van Allen, J.A., "Myths and Realities of Space Flight", Science, Vol 232, May 30, 1075-1076 (1986).

33. Whedon, G.D., et al, "Mineral and Nitrogen Metabolic Studies, Experiment M071", from: Biomedical Results from Skylab, NASA, SP-377, 164-174 (1977).

34. Wolff, J.: Das Gesetz der Transformation der Knochen , Hirschold, Berlin (1892).

35. Ziporyn, T., "Aerospace Medicine; The First 200 Years", ref 12, 2010 (1986).

## Biography

George T. Delli-Santi is a 1981
graduate of Rensselaer Polytechnic
Institute, Troy, NY. He holds a
Master's of Engineering degree in
Biomedical Materials Engineering.
His undergraduate education consists
of a Bachelor of Science degree in
Biomedical Engineering, also from
RPI, and an Associate of Arts degree
in Biological Science from Union
College, Cranford, NJ.

Mr. Delli-Santi has been employed for
six years with Zimmer, Inc., Warsaw,
IN, a leading orthopaedic device
manufacturer. He has worked as a
Biomedical Research engineer, testing
total joint prostheses and
biomaterials, and as a senior
Advanced Technology research
engineer, testing and designing
advanced thermoplastic composite
implants, while associated with
Zimmer.

His work in the orthopaedic industry
involved addressing the problems of
disuse atrophy of bone and
osteoporosis. The combination of
this background and his interests in
aviation and manned spaceflight led
to the preparation of this paper. He
has been a member of SAMPE for over
three years and is also a member of
the Planetary Society and the Society
for Biomaterials.

SIMULATION OF FLUID FLOWS DURING GROWTH OF ORGANIC CRYSTALS
IN MICROGRAVITY

Gary D. Roberts, James K. Sutter, and R. Balasubramaniam*
National Aeronautics and Space Administration
Lewis Research Center
Cleveland, Ohio 44135
William Fowlis
National Aeronautics and Space Administration
Marshall Space Flight Center
Huntsville, Alabama 35812
M.D. Radcliffe and M.C. Drake
3M Corporation
St. Paul, Minnesota 55144

ABSTRACT

Several counterdiffusion type crystal growth experiments have been conducted in space. Improvements in crystal size and quality have been attributed to reduced natural convection in the micro-gravity environment. One series of experiments called DMOS (Diffusive Mixing of Organic Solutions) was designed and conducted by researchers at the 3M Corporation and flown by NASA on the space shuttle. Since only limited information about the mixing process is available from the space experiments, a series of ground based experiments were conducted to further investigate the fluid flow within the DMOS crystal growth cell. Solutions with density differences in the range of $10^{-7}$ to $10^{-4}$ $g/cm^3$ were used to simulate microgravity conditions. The small density differences were obtained by mixing $D_2O$ and $H_2O$. Methylene blue dye was used to provide flow visualization. The extent of mixing was measured photometrically using the 662 nm absorbance peak of the dye. Results indicate that extensive mixing by natural convection can occur even under microgravity conditions. This is qualitatively consistent with results of a simple scaling analysis. Quantitative results are in close agreement with ongoing computational modeling analysis.

1.    INTRODUCTION

Organic crystals have shown considerable potential for use in

_____

*NASA Resident Research Associate

nonlinear optical devices.[1] The usefulness of these materials in devices is often limited by the inability to grow large crystals with a sufficiently low density of defects. On Earth, temperature and composition gradients result in natural (bouyancy driven) convection, which can interfere with crystal growth.[2] One technique for growing crystals involves the counterdiffusion of two reactants (A and B) and precipitation of the salt (AB) in the interfacial region. This approach is particularly useful when the salt is not sufficiently soluble in common solvents to allow growth by precipitation from a homogeneous solution. In the counterdiffusion process natural convection can occur along with molecular diffusion as a result of density gradients within the container. The density gradients are a result of: (1) concentration gradients for components A and B, (2) the density difference between the solution and the solid salt crystal, and (3) thermal gradients resulting from heat released during salt formation and crystallization. These density gradients cannot be eliminated in practical crystal growth experiments. However, the natural convection caused by the density gradients can be considerably reduced if the crystal growth experiment is conducted in space.

Organic crystals were grown by researchers at the 3M Corporation in a series of space experiments called DMOS (Diffusive Mixing of Organic Solutions). Included in DMOS-2 were two experiments designed to measure the extent of solution mixing in microgravity. In order to examine the fluid dynamics more closely, a series of ground based experiments were designed to simulate the fluid flow which occurred during the DMOS experiments. These experiments were conducted at the NASA Lewis Research Center in collaboration with researchers at the 3M Corporation and the NASA Marshall Space Flight Center. This paper describes the details of the ground based experiments and results which apply both to the DMOS experiments and to other similar experiments with horizontal composition gradients.

2. EXPERIMENTAL PROCEDURES

The DMOS crystal growth apparatus consists of three cylindrical chambers placed end to end. Between the chambers a sliding gate valve assembly is used to control mixing. For the ground based experiments described in this paper, the simpler two chamber cell shown in Fig. 1 was used. The chambers of this cell are rectangular for better photographic flow visualization, but the gate valve assembly is identical to that used in the DMOS experiments.

In a typical experiment the left chamber is filled with a 0.01 g/l solution of methylene blue dye in water while the right chamber is filled with pure water. The density difference between fluids in the chambers is controlled by adding deuterated water ($D_2O$) to one chamber or the other. Qualitative information about the flow patterns during mixing was obtained using time lapse photography at 2 frames per second. Quantitative results were obtained by photometric analysis using the 662 nm absorbance peak of the methylene blue dye. This was done by closing the gate valve after the desired mixing time, shaking the apparatus to get a uniform dye concentration, and taking samples from each chamber for analysis. At sufficiently long times the mass fraction of dye in each chamber is 0.5. If f(t) is the mass fraction of dye in the right chamber at any time, t, the extent of mixing, Xm(t), is defined by Eq. 1.

$$\%Xm = 100X\left(\frac{f}{0.5}\right) \qquad (1)$$

Using this definition, the extent of mixing goes from 0 to 100 percent as the mixing time goes from zero to infinity. The temperature was maintained at 30±0.1 °C during the the mixing experiments by immersing the apparatus in a circulating water bath. A quartz thermometer was used to measure the temperature difference between various points in the bath. When

the density difference between the solutions was less than $10^{-5}$ $g/cm^3$, heating of the dye solution by the background lighting used for photography began to influence the fluid flow. Therefore, when quantitative data was taken, no background lighting was used, and the water bath was covered with a black cloth to exclude room light.

The diffusion coefficient of the methylene blue dye in $D_2O$ was measured by Prof. Ernst von Meerwall at the University of Akron using a pulsed gradient spin echo (PGSE), nuclear magnetic resonance (NMR) technique. Physical properties of $H_2O$ and $D_2O$ used in section 3 were obtained from standard tables using values at 30 °C.

3. RESULTS AND DISCUSSION

3.1 The Microgravity Environment

Experiments conducted on the shuttle experience both translational and rotational accelerations from many sources.[3] Atmospheric drag causes a steady acceleration of the center of mass antiparallel to the orbit velocity. A steady acceleration perpendicular to the orbit also exists for points not located on the orbital path of the center of mass. Oscillatory accelerations occur as a result of structural vibrations, and transient accelerations occur as a result of attitude maneuvers, satellite deployments, crew motions, etc. An experiment can be run during quiet periods, so that only steady accelerations are important.

In this paper only flows resulting from a steady acceleration perpendicular to the initial density gradient are considered.

## 3.2 Microgravity Simulation-Experiments

It will be assumed for now that the fluid flow which occurs during a space experiment can be simulated in a ground-based experiment if the product of the solution density difference and the effective gravitational acceleration is the same in each experiment. The validity of this assumption will be discussed in more detail in section 3.3. A detailed description of the accelerations which occurred during the DMOS-2 experiments will be published later. Preliminary analysis of accelerometer data indicates that the effective acceleration was on the order of $10^{-5}$ times that on Earth. Density differences in ground-based simulations must therefore be smaller than those in the space experiments by the same factor. The solutions used to obtain these small density differences must be different from the solutions used in the space experiments. In some cases solutions very similar to those used in the space experiment can be used in ground-based simulations if isotopic mixtures (such as $D_2O/H_2O$) are used to control solution densities. This allows the density difference between two solutions to be varied over many orders of magnitude with only a minor modification of chemical properties.

An experiment was conducted in the apparatus of Fig. 1 using a blue solution ($H_2O$ + dye) in the left chamber and pure water ($H_2O$) in the right chamber. A dye concentration of 0.01 g/l was needed for photographic flow visualization. When the gate valve was opened, the blue solution flowed from left to right through the three bottom channels (counterclockwise flow). This indicates that the dye causes an increase in the density of the water. However, the density change is too small to be measured directly by standard techniques. In order to determine the effect of the dye on the density of the water, a series of experiments were run in which the density of the solution in the right chamber was incrementally increased by adding various amounts of $D_2O$ to the water. Each experiment was stopped after 2 hr, and the extent of mixing was determined. Results of these experiments are shown in Fig. 2. Positive and negative values indicate flow in the counterclockwise and clockwise directions respectively. If ideal mixing is assumed, the density difference between the $H_2O/D_2O$ solution and pure $H_2O$ can be calculated using Eq. 2.

$$\Delta \rho = v_{D_2O}\left(\rho_{D_2O} - \rho_{H_2O}\right) \qquad (2)$$

$v_{D_2O}$  volume fraction of $D_2O$

$\rho_{H_2O}$    density of pure $H_2O$

$\rho_{D_2O}$    density of pure $D_2O$

According to Fig. 2, the effect of the dye is equivalent to a $D_2O$ volume fraction of $3.5 \times 10^{-5}$. Using this value for $v_{D_2O}$ and $(\rho_{D_2O} - \rho_{H_2O}) = 0.108$ g/cm$^3$ in Eq. 2, the density increase caused by adding 0.01 g/l of dye to $H_2O$ is calculated to be $3.8 \pm 0.1 \times 10^{-6}$ g/cm$^3$. Notice that it is possible to prepare solutions with density differences on the order of $10^{-6}$ g/cm$^3$ without knowing the absolute densities of the solutions to the same accuracy. This is possible only if the same source of water is used to prepare both solutions and no significant concentration of impurities is introduced during preparation of the solutions. The consistency and reproducibility of results obtained throughout this work suggest that these conditions were satisfied. Another potential source of error is thermally induced density gradients. The maximum temperature difference measured between any two points in the water bath just outside of the apparatus was 0.007 °C. Using $3.0 \times 10^{-4}$ K$^{-1}$ as the volumetric thermal expansion coefficient of water, the maximum possible thermally induced density difference is $2.1 \times 10^{-6}$ g/cm$^3$. Since the time averaged temperature difference between any two points outside of the apparatus is much less than 0.007 °C, thermally induced density differences inside the apparatus are probably on the order of $10^{-7}$ g/cm$^3$ or less.

In order to observe qualitatively how mixing occurs, a series of experiments with density differences ranging from $10^{-6}$ to $10^{-4}$ g/cm$^3$ were conducted. The flow was qualitatively the same throughout this range of density differences. The denser (blue) solution flowed from left to right through the bottom three channels of the gate valve assembly. When the blue solution entered the right chamber, it flowed like a waterfall in a layer about 0.2 cm thick along the vertical wall of the gate valve assembly to the bottom of the chamber. The flow continued until the levels of the blue solution were nearly equal in each chamber.

In some of the DMOS space experiments the density difference between solutions in different chambers was on the order of 1 g/cm$^3$. In order to simulate the flow which occurred in these space experiments, a series of ground based experiments with $\Delta\rho = 4.9 \times 10^{-6}$ g/cm$^3$ were run. The extent of mixing as a function of time is shown in Fig. 3. Nearly complete mixing is obtained in about 32 hr. At this time each chamber contains the blue solution in the bottom of the chamber and a clear solution at the top of the chamber with a diffuse horizontal interface separating the two solutions. No observable change in the dye distribution was observed between 32 and 56 hr.

### 3.3 Microgravity Simulation-Theory

In order to describe the fluid flow which occurs during experiments conducted in the apparatus of Fig. 1, it is convenient to consider flow through the valves and flow within the chambers separately. As a simplification, flow through the valves can be described as Poiseuille flow through a rectangular channel. The average velocity of fluid within the valves can then be calculated from Eq. 3.[4]

$$\frac{\Delta P}{L} = \left(\frac{\lambda}{d}\right)\left(\frac{\rho}{2}\right)u^2 \qquad (3)$$

u    average fluid velocity
ΔP   pressure difference
L    channel length
ρ    fluid density
d    hydraulic diameter
λ    friction factor

The hydraulic diameter is defined as 4 x (area)/(circumference) for the rectangular channels. Using the Reynolds ($Re = ud/v$) number based on the hydraulic diameter, the friction factor is approximately $\lambda = 64/Re$. If the hydrostatic pressure difference resulting from a density difference of $4.9 \times 10^{-6}$ g/cm$^3$ is used in Eq. 3, the fluid velocities through the bottom three channels are $3.02 \times 10^{-3}$, $2.14 \times 10^{-3}$, and $0.73 \times 10^{-3}$ cm/s respectively, beginning with the bottom channel. For these velocities the entrance length for fully developed flow $[(\ell e = (0.035)(d)(Re)]$ is much less than the channel length. This condition and an assumption of quasisteady flow are required in order to apply Eq. 3. If the total volume of fluid transferred from the left to the right chamber is calculated using these flow velocities, the extent of mixing after 3162 sec (log t = 3.5) is calculated to be 18.7 percent. This is close to the experimental value of 12.0 percent in Fig. 3. Numerical computations and computer models are currently being developed to provide a more accurate description of the flow.

Since the denser solution emerging from the channels flows like a waterfall to the bottom of the chamber, the vertical flow velocity must be much larger than the horizontal velocity. This can be explained using simple scaling arguments.[5] The relative importance of various transport processes is determined by calculating the magnitude of appropriate dimensionless parameters. The parameters of interest are the Grashof number, Gr, and the Schmidt number, Sc, which are defined below.

$$Gr = \frac{g(\Delta\rho)L^3}{\rho v^2} \qquad (4)$$

$$Sc = \frac{v}{D} \qquad (5)$$

g    gravitational acceleration
ρ    average density
Δρ   density difference

L     characteristic length

$v$     kinematic viscosity

D     diffusion coefficient of dye

Using $\Delta\rho = 4.9\times10^{-6}$ g/cm$^3$, $\rho = 1.00$ g/cm$^3$, g = 980 cm/s$^2$, L = 4.45 cm, $v$ = 0.00804 cm$^2$/s, and D = $3\times10^{-6}$ cm$^2$/s in Eqs. 4 and 5 gives, Gr = $6.6\times10^3$, Sc = $2.7\times10^3$, and ScGr = $1.8\times10^7$. An order of magnitude estimate of the vertical fluid velocity can be calculated from Eq. 6 or 7.[5]

$$v = Gr^{1/2}\left(\frac{v}{L}\right) \text{ (for } Gr \gg 1)\qquad(6)$$

$$v = Gr\left(\frac{v}{L}\right) \text{ (for } Gr \ll 1)\qquad(7)$$

Using Eq. 6 the vertical fluid velocity is 0.146 cm/s. As expected, this is much larger than the horizontal velocities calculated above. A boundary layer thickness ($\delta$) of 0.75 cm is estimated using Eq. 8.[5]

$$\frac{\delta}{L} = \frac{1}{Gr^{1/4}}\qquad(8)$$

This is, within order of magnitude agreement with the observed thickness of 0.2 cm.

So far it has been implicitly assumed that the motion of the dye molecules represents the motion of the fluid. The validity of this assumption depends on ScGr as described in equations 9 and 10.

$$ScGr \ll 1 \begin{pmatrix}\text{diffusion}\\\text{dominant}\end{pmatrix}\qquad(9)$$

$$ScGr \gg 1 \begin{pmatrix}\text{convection}\\\text{dominant}\end{pmatrix}\qquad(10)$$

Since ScGr $\gg$ 1, transport of the dye by molecular diffusion is much slower than by convection, and the motion of the dye does represent convection of the fluid. Diffusion is not important unless $(\Delta\rho)g \ll 2.7\times10^{-10}$ g/cm$^2$-s$^2$. This requires matching densities to within $10^{-7}$ g/cm$^3$ for a space experiment or $10^{-13}$ g/cm$^3$ for a ground based experiment. Use of such small density differences is not experimentally feasible. Furthermore, the time scale of a diffusion controlled experiment would be too long to be of practical interest. Therefore, changes in the geometry of the apparatus in Fig. 1 would be required in order to conduct a diffusion controlled experiment.

It remains to be determined whether or not matching the $(\Delta\rho)g$ product is sufficient for simulation of fluid flows. Within the chambers it has been shown that molecular diffusion is not important, so that the only important dimensionless group is Gr. For Poiseuille flow between the chambers the mass transfer rate is proportional to $(\Delta\rho)g/\rho v$. Since flow between chambers depends on viscosity as $1/v$ and flow within the chambers depends on viscosity as $1/v^2$, the overall process can not be simulated unless the solution viscosities are the same in both ground based and space experiments. However, both the viscosity and density of most organic and aqueous solutions containing moderate concentrations of low molecular weight solutes do not vary by more than one order of magnitude. As a result, matching the $(\Delta\rho)g$ product should provide reasonable (but not exact) simulation in most cases of interest.

## 4. CONCLUSIONS

The fluid flow which occurs during counterdiffusion type crystal growth experiments in space can be studied using ground based simulations. For isothermal mixing of solutions, approximate simulation can be accomplished by matching the $(\Delta\rho)$ product. Density differences as small as $10^{-6}$ g/cm$^3$ can be easily obtained using isotopic mixtures. For experiments with a geometry similar to that in Fig. 1, natural convection is expected to be the dominant mode of transport for both space and ground based experiments if the direction of the gravitational acceleration is perpendicular to that of the initial density gradient. The experimental approach to ground based simulation and the scaling analysis described in this paper can be generally applied to other systems with horizontal composition gradients in a vertical gravitational field.

## REFERENCES

1. Williams, D.J., ed.: Nonlinear Optical Properties of Organic and Polymeric Materials. American Chemical Society, 1983.

2. Hurle, D.T.J.: Crystal Growth. Materials Sciences in Space, B. Feuerbacher, H. Hamacher, and R.J. Naumann, eds., Springer-Verlag, 1986, pp. 390-393.

3. Hamacher, H.: Simulation of Weightlessness. Materials Sciences in Space, B. Feuerbacher, H. Hamacher, and R.J. Naumann, eds., Springer-Verlag, 1986, pp. 390-393

4. Schlichting, Hermann: Boundary-Layer Theory, 7th ed., McGraw-Hill, 1979, pp. 612-613.

5. Ostrach, S.: Low-Gravity Fluid Flows. Annual Review of Fluid Mechanics, Vol. 14, M. VanDyke, J.V. Wehausen, and J.L. Lumley, eds., Annual Reviews Inc., 1982, pp. 313-345.

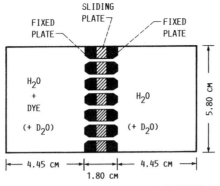

FIGURE 1.- CROSS SECTION OF 2 CHAMBER MIXING CELL
USED FOR GROUND-BASED EXPERIMENTS.

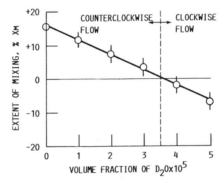

FIGURE 2.- EXTENT OF MIXING AFTER TWO
HOURS VERSUS VOLUME FRACTION OF $D_2O$
IN RIGHT CHAMBER. LEFT CHAMBER CON-
TAINS 0.01 G/l METHYLENE BLUE DYE.
$H_2O$ IS THE SOLVENT IN BOTH CHAMBERS.

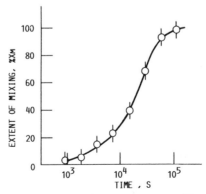

FIGURE 3.- EXTENT OF MIXING VERSUS
TIME FOR INITIAL DENSITY DIFFERENCE
BETWEEN CHAMBERS OF $4.9 \times 10^{-6}$ g/cm$^3$.

RESEARCH OPPORTUNITIES IN MICROGRAVITY SCIENCE
AND APPLICATIONS DURING SHUTTLE HIATUS
Bruce N. Rosenthal, Thomas Glasgow
National Aeronautics and Space Administration
Lewis Research Center
Cleveland, Ohio 44135
Richard E. Black
National Aeronautics and Space Administration
George C. Marshall Space Flight Center
Marshall Space Flight Center, Alabama 35812
Daniel E. Elleman
Jet Propulsion Laboratory
Pasadena, California 91109

## Abstract

The opportunity to conduct microgravity and related research still exists, even with the temporary delay in the U.S. Space Shuttle program. Several ground-based facilities are available and use of these facilities is highly recommended for the preparation of near and far term Shuttle or Space Station experiments. Drop tubes, drop towers aircraft, sounding rockets and a wide variety of other ground-based equipment can be used to simulate microgravity. This paper concentrates on the materials processing capabilities available at NASA Lewis Research Center (LeRC), NASA Marshall Space Flight Center (MSFC) and the California Institute of Technology Jet Propulsion Laboratory (JPL). Also included is information on gaining access to these facilities

## 1. INTRODUCTION

Gravity pervades all aspects of materials processing from the liquid or gaseous state. Gravity driven fluid flows redistribute dopant or alloy additions limiting crystalline perfection and device efficiency. The earthbound necessity of holding liquids in a container introduces both bulk contaminants and catalysts for

506

premature nucleation. Subtle phenomena such as thermocapillary flow or critical point behavior may be largely masked by bouyancy driven flows. Other processes are especially difficult in our usual one-g environment such as the growth of large diameter crystals by the float zone process or the production of stable metallic foams from the melt. For these reasons and many more, materials scientists, fluid physicists, combustion scientists, and biotechnologists view an orbital Shuttle or Space Station as an appropriate laboratory.

Given the high cost and limited accessibility of space it is axiomatic that ground-based research must proceed and accompany space processing. The early use of ground-based research facilities can sharpen the focus of microgravity research, enhancing the likelihood of profitable work in space. In some cases it is possible that ground-based research may even eliminate the need for on-orbit trials. In this paper ground-based facilities for microgravity materials science research are described. The use of these facilities is always recommended and recent difficulty with access to the Shuttle suggests greater attention be given to their potential.

The facilities described fall into one of three categories: those which provide an actual microgravity environment for a limited time such as rockets, planes, or drop towers and drop tubes; those which emulate one or more aspects of the microgravity environment such as electromagnetic or acoustic levitators; and those which may be used to further understand system behavior in or out of the microgravity environment such as model furnaces or computational facilities.

2. DROP TUBES AND TOWERS

Drop tubes are an expeditious way to study the effects of short term microgravity and/or containerless processing on melting and solidification of materials. Small material samples can be melted and then allowed to solidify during free fall. Most materials science drop tube experiments conducted thus far have studied the relationship between undercooling and material microstructure/properties. To accurately determine the direct amount of undercooling achieved, the temperature of the sample must be measured while it is in free fall. Optical pyrometry has been used because of its nonintrusive nature. A drawback is that the sensitivity of most detectors to thermal radiation below 1000 degrees kelvin is poor and limits experiments to higher melting point materials.

2.1 Drop Tubes

Currently there are five drop tubes available. These range in length from five to one hundred meters. The drop tubes are discussed below in order of descending magnitude of free fall time.

The 100 meter drop tube[1][5] located at the MSFC provides up to 4.6 seconds of low gravity. A stainless steel bell jar mounted atop the tube contains the melting apparatus used to process the material samples. The tube/jar atmosphere can consist of argon, helium, helium with 6% hydrogen, nitrogen or the tube can be evacuated to a pressure of $1 \times 10^{-7}$ kPa. These various gases are used to increase convection cooling required for lower melting point materials when radiation is not the predominant form of heat loss.

Acceleration of $10^{-6}$ g is obtainable during vacuum free fall. Larger accelerations ($10^{-3}$ g) are observed in gaseous environments. Various methods are available to receive the falling samples via a detachable catching device. Deceleration media include diffusion pump oil, a copper block and various foils[1].

Infrared detectors mounted along the length of the tube are used to determine the onset of nucleation by sensing the increase in sample temperature as the heat of fusion is released[5].

Three types of melting apparatus[1][5] are presently operational for use in conjunction with the drop tube. An electron beam bombardment furnace operates in the range of 1600 to 2500°C and is for use under vacuum type conditions. An electromagnetic levitation furnace operates over the range of 500 to 3000°C and can operate in either a vacuum or gaseous environment. Samples must be electrically conductive to electromagnetically levitate. A resistance heating capillary tube employing a quartz capillary crucible to contain the sample during melting may

508

be used for nonconductive materials.[1][5]

A 30 meter drop tube is also located at MSFC. Capable of up to 2.6 seconds of free fall time this tube possesses characteristics similar to the 100 meter drop tube previously discussed. An air tight bell jar located on top of the tube contains the sample processing apparatus which may either be an electron bombardment furnace, a resistance heating capillary tube or any furnace which can be adopted to the bell jar.[1][2][4]

The Jet Propulsion Laboratory maintains a forced free drop tube[1][4] that eliminates aerodynamic drag. By drawing air down the tube via a suction fan at the base, an acceleration is imposed on the air column due to the convergent nature of the tubes crossection. While in free fall the air and sample move at the same velocity. This helps eliminate aerodynamic drag and sample distortion although convective cooling is reduced. The facility provides up to 1.7 seconds of free fall time. Principle research activities include the study of fluid surface phenomena and the formation and spheroidization of metallic and glass microballoons.[3]

The cryogenic drop tube[1][3] located at JPL allows up to 1.7 seconds of free fall time under vacuum. Three separate temperature zones are consecutively located along the tube length for control of ambient gas temperature and density as a means of improving solidification. A balance between convection and radiation cooling is obtained by using cryogenic temperatures in one section of the tube. Internal gas pressure can be controlled between $10^{-5}$ and 200 kPa.

Located at LeRC is a 5 meter drop tube facility[6] capable of 1 second of vacuum free fall time. Temperatures from 700 to 4000$^0$C are measured optically by four high speed, 2 color optical pyrometers. An additional pyrometer mounted at the bottom of the tube will be used to observe samples while they fall during the entire drop. This is achieved by employing tracking optics and higher sensitivity pyrometers.

A 25 kW electromagnetic levitator (EML) is housed in a water cooled vacuum chamber attached to top of the drop tube. The drop tube/EML may be operated in a vacuum or an inert gas atmosphere. The chamber portion of this facility is approximately 75 centimeters in diameter and 45 centimeters deep. Different types of furnace devices, designed to fit in the vacuum chamber, are currently being developed for working with non-conductive samples. Up to 10 samples may be arranged on a computerized turntable assembly for serial processing during a single pump down cycle.

Currently under construction at JPL is a 14 meter drop tube. This stainless steel drop tube will be capable of operation at pressures in excess of 10 atmospheres or a vacuum of $10^{-7}$ kPa. External coolant tubes attached to the five lower drop tube sections provide control from room temperature to $-180^{\circ}$C. Each of these five 2.4 meter sections can be individually temperature-controlled.

## 2.2 Drop Towers

Drop towers, as opposed to drop tubes, can accommodate large experimental drop packages consisting of sample processing and data acquisition equipment. Construction of these facilities began in the late 1950's and early 1960's when research programs were geared towards the study of propellant sloshing, settling and draining[7]. At present there are three drop towers of varying capability available for low-gravity experimentation.

The two drop facilities at LeRC have been widely used for combustion and fluid physics studies. The 30 meter drop tower offers 2.2 seconds of free fall behind a drag shield in air; the 145 meter drop tower offers 5.2 seconds of free fall in vacuum, with residual gas drag approximately $10^{-5}$ g. Both facilities can accommodate large packages, up to 70 kg and 450 kg respectively. Typical packages include high speed cameras, actuating systems, onboard power (28 volts), and up to 18 channels of data transmission. Thus materials can be examined or affected (by radiation for example), during the drop. Technology has been developed for applying a very gentle controlled acceleration during drop to simulate the behavior of spacecraft. Evacuating the 145 meter long, 9 meter diameter tower limits the drops to one or two per day (the 30 meter tower can provide as many as ten drops per day).

A 100 meter drop tower, located at Marshall Space Flight Center, is currently used extensively for ground based research in microgravity science. The facility permits experimentation for a period of microgravity up to 4.27 seconds and is capable of accommodating experimental packages as large as .76 cubic meters and weighing up to 204 kg. The configuration includes a drag shield equipped with battery, electrical distribution and telemetry system to meet experimental requirements. An auxiliary thrust is provided to overcome aerodynamic drag during the free fall period.[8]

Current materials science interests in the drop facilities include the study of phase separation during polymerization, undercooling of metals, positioning of liquids for float zone solidification, containment of liquids for thermocapillary flow experiments, and weld pool behavior. Experiments at MSFC have shown that the containerless feature of drop tube processing is more important than the freedom from acceleration in attaining undercooling.[5]

### 3. AIRCRAFT AND SOUNDING ROCKETS

#### 3.1 Aircraft

NASA uses three aircraft for microgravity research: the KC 135 and F-104 both[1] stationed at MSFC, and a Model 25 Learjet[9] stationed at LeRC. The KC-135 cargo jet has the largest carrying capacity with a 3 meter by 16.4 meter by 1.8 meter bay size. The KC-135 is capable of up to 40 floats per flight with free float time ranging from 5 to 15 seconds. The Learjet can provide up to 6, twenty second, low gravity parabolic loops per day for small and medium size attached packages only. The investigator may accompany and operate experiments on the KC-135 and the Learjet. Only fully automated very small, e.g. 15 kg, packages can be carried in the F-104, but low acceleration time can last up to 60 seconds; one loop per flight is available.

A typical low-gravity flight profile follows a parabolic trajectory consisting of a slight dive requiring a 2 to 3 g pullup followed by the ascent. When the aircraft reaches about a 50 degree ascent angle, a pushover maneuver is performed at which time the low gravity conditions begin (typically $10^{-1}$ to $10^{-3}$ g are the best low-gravity conditions obtainable)[9].

## 3.2 Sounding Rockets

Originally used to develop and test experimental equipment, before the time of manned space flight, today the use of sounding rockets for microgravity materials science processing fills the gap between research performed on aircraft (maximum 60 seconds of low-gravity) and research performed on the Space Shuttle (days of low-gravity). A family of suborbital sounding rockets capable of up to 275 kilograms payload provide typically 4 to 7 minutes of accelerations remaining below $10^{-4}$ g.[11] These conditions exist during the unpowered coast phase after launch and before re-entry into the atmosphere.

## 4. ADDITIONAL FACILITIES

The facilities described above provide an actual microgravity environment for a limited amount of time. But, as will be explained, it is not necessary to be in a microgravity environment in order to study the effects microgravity has on material processing. For example, convection free solidification may be studied in orbit or in a 1 g environment. Emulating microgravity convection in conductive melts can be accomplished by the imposition of a steady magnetic field to dampen fluid flow. Bouyancy driven convection in binary liquid systems can be reduced by choosing liquids of nearly equal density.

The Microgravity Materials Science Laboratory[6] established at the NASA LeRC is available to scientists and engineers from industry, university and government agencies nationwide. The laboratory is equipped with pieces of experimental hardware which 1) are functional duplicates of those flown on the Space Shuttle or being developed for use on future Shuttle/Space Station missions; 2) provide a temporary low-gravity environment; 3) emulate conditions of the low-gravity environment; 4) or provide 1 g testing and research capabilities for developing and/or expanding microgravity experimental ideas. The laboratory also provides a place where the process in question can be studied in detail and mathematical modeling employed to remove or enhance the effect of gravity.

Currently under construction at the MMSL is a high temperature Directional Solidification Furnace designed to perform experiments on metal samples. The sample is sealed in an

512

ampoule (typically quartz) and the furnace coil/cooling assembly is moved along the length of the sample ampoule. The unique feature of this particular system is that the sample is exposed to a large magnetic field which is used to reduce convection as discussed above. Other MMSL equipment include electromagnetic and acoustic levitation systems, various resistance and induction melting systems and materials characterization support hardware.

The Jet Propulsion Laboratory hosts a variety of unique materials processing/handling equipment[1]. The vertical and horizontal quadrupole levitators permit fine particles or liquid droplets to be positioned and transported in the vertical and horizontal directions respectively. The levitators operate in a high AC voltage mode which produces the electrostatic forces necessary to center the sub-millimeter samples within the 4 pole fields. The electrostatic-acoustic hybrid levitator also located at JPL is a general purpose, room-temperature system by which drop oscillation, rotation and fission can be induced acoustically while a drop is being levitated in 1 g. Charged water or aqueous drops approximately 3 to 4 millimeters in diameter can be levitated with less than 3% loss in sphericity. Using this system experiments such as charged liquid drop dynamics or crystal growth from aqueous salt solutions can be performed. With minor modifications the present system can be adapted to a microgravity environment such as the KC-135 research aircraft. Presently under experimental study and development are a dual temperature acoustic levitator, a high temperature siren levitation facility and a high temperature, single-mode levitation facility.

Ground-based laboratory facilities located at MSFC support a multitude of material processing capabilities[1]. Containerless processing facilities include an airjet levitator which uses a compressed air stream to levitate material samples. During levitation the samples can be melted with a 700 watt $CO_2$ laser (up to $2700^0C$) and then allowed to resolidify without the physical contact of a containment vessel. An electromagnetic levitator with

electron bombardment heating
has been used successfully to
levitate and melt tungsten and
is capable of similar
experiments in containerless
processing of refractory
metals. A single axis
acoustic levitation system is
capable of levitating samples
as large as 3 millimeters
having a density of 3.8
grams/cm$^3$. Temperatures above
1000 c have been achieved.

In addition to these
ground based levitation
facilities, MSFC maintains
various microgravity apparatus
used aboard the Space Shuttle
for material processing
experimentation. Ranging from
crystal growth equipment to
devices for separation of
biological materials, many of
these systems may also be
operated in an earth-based
mode to develop and prepare
experiments for space flight.

## 5. CONCLUSIONS

The facilities available
for ground-based microgravity
research have been described.
Figure 1 contains a guide
intended to help those
interested individuals obtain
additional information
regarding the facilities and
equipment discussed in this
paper.

| FACILITY | LOCATION | CONTACT |
|---|---|---|
| **Additional Facilities:** | | |
| Microgravity Materials Science Laboratory | LeRC | Thomas Glasgow (216) 433-5013 |
| Marshall Space Flight Center | MSFC | Richard E. Black (205) 544-1983 |
| Jet Propulsion Laboratory | JPL | Daniel D. Elleman (818) 354-5182 |
| **Drop Tubes:** | | |
| 100 meter | MSFC | Michael B. Robinson (205) 544-7774 |
| 30 meter | MSFC | Kenneth R. Taylor (205) 544-0640 |
| 13.1 meter Force Free | JPL | Charles L. Youngberg (818) 354-3559 |
| 13.1 meter Cryogenic | JPL | Charles L. Youngberg (818) 354-3559 |
| 10 meter | LeRC | Thomas Glasgow (216) 433-5013 |
| **Drop Towers:** | | |
| 145 meter | LeRC | William Masica (216) 433-2864 |
| 100 meter | MSFC | William F. Kaukler (205) 544-7782 |
| 30 meter | LeRC | William Masica (216) 433-2864 |
| **Research Aircraft:** | | |
| KC-135 | MSFC | Peter Curreri (205) 544-7763 Robert E. Shurney (205) 544-2189 |
| Learjet | LeRC | William Masica (216) 433-2864 |
| F-104 | MSFC | Kenneth R. Taylor (205) 544-0640 |
| **Sounding Rockets:** | | |
| Contact Office of Space Commercialization | LeRC | Anthony F. Ratajczak (216) 433-2921 |

FIGURE 1.- TABLE OF MICROGRAVITY MATERIALS SCIENCE FACILITIES.

## REFERENCES

1. Microgravity Science and Applications: Experiment Apparatus and Facilities, NASA Brochure, Microgravity Science and Applications Division, NASA Headquarters, Code EN, Washington, DC 20546.

2. Robinson, M.B., Undercooling Measurements in a Low-gravity containerless environment, NASA TM-82423, May 1981.

3. Lee, M.C., et al, Sensational Spherical Shells, Aerospace America, Jan. 1986.

4. Lipsett, F.R., New Opportunities in Space: Research in Microgravity, Canadian Aeronautics and Space Journal, March 1983.

5. Bayuzik, R.J., et al, Microgravity Containerless Processing in Long Drop Tubes, NASA MSFC 2nd Symposium on Space Industrialization,pp 243-259.

6. Microgravity Materials Science Laboratory: Laboratory Description and Scientist and Engineer's Application Procedures For its Use, NASA Lewis Research Center Brochure, July 1986.

7. Aydelott, J.C., Symons, E.P., LeRC Reduced Gravity Fluid Management Technology Program, NASA Document N80-30380.

8. McAnelly, W.B., Covington, S.S., Low Gravity Fluid Mechanic Drop Tower Facility, NASA TM X-53817, Feb. 28, 1969.

9. Users Guide to Learjet for Low-Gravity Research, NASA Lewis Research Center Brochure.

10. Wallops Flight Center Sounding Rocket Program, NASA Brochure, June 1978.

11. Naumann, R.J., Early Space Experiments in Materials Processing, NASA TM-78234, July 1979.

516

STUDY OF DESIGN PARAMETERS OF
CYLINDRICAL COMPOSITE PRESSURE VESSELS

Ajit K. Roy
Air Force Materials Laboratory
AFWAL/MLBM
Wright-Patterson AFB, OHIO 45433-6533

## ABSTRACT

Design parameters for cylindrically
orthotropic thick pressure vessels are
studied. Solutions of thin wall and thick
wall are also compared. It is found that, thin
wall solution is limited to wall thickness
ratio, $b/a \leq 1.10$, in predicting burst pressure.
The radial displacement of orthotropic
vessels decreases with increasing winding
angle and becomes negative between 58° and
78°. Multiple layer construction, for thick
vessels, with optimum winding angle
combinations leads to efficient material use
and improves vessel performance. For
$b/a \leq 2.0$, multilayer vessels with 2 or 3
layers, for all practical purposes, yields an
efficient design.

## KEYWORDS:

Pressure Vessel, Cylindrical Orthotropy,
Burst Pressure, Strength Ratio, Vessel
Efficiency, Multilayer, Composites.

## 1. INTRODUCTION

Cylindrical Pressure Vessels are widely
used, both in commercial and aerospace
applications, ranging from small bottles of a
few inches in diameter to large storage tanks.
Its application is not only limited to the use
of thin shells; in many of its applications
thick shells with wall thickness ratio (ratio of
the outer radius to the inner one of the
cylindrical wall) of 2 or even higher are
being used. For example, the wall thickness
ratio of gun barrels can be of the order of 2
or so. Thick-walled vessels have also been
considered for submerssible vessels
subjected to high external pressure[1]

The emerging high modulus, high strength
composite materials are being used in
manufacturing pressure vessels for some
time. It is now known that use of composite
materials improves performance and offers a
significant amount of material saving over
that of the isotropic materials. A good
amount of literature is available on pressure
vessels. Tauchert[2] studied to obtain an
optimal fiber distribution through the
thickness to maximize failure pressure and to
minimize radial displacement. In his
analysis fibers are confined only in hoop
direction. Sherrer[3] reported a sensitivity
study of the resin properties to the failure of
the cylindres. He obtained optimum lay up
in two layers based on equal fiber stress.
Gerstle and Moss[4] showed that the
pressure carrying capacity of quasi-isotropic
spherical pressure vessel can be improved by
hybridizing the vessel with more than one
materials. However, all these studies lack in
taking full advantage of the orthotropic
properties of the composite materials,
particularly in thick wall construction. A
reliable design method taking full advantage
of the orthotropic properties of the composite
materials should be based on an accurate
stress analysis with the use of an appropriate
failure criterion. In this paper, the stress
analysis of thick cylindrical vessels is based
on linear elasticity solution. Analysis based
on elasticity solution is reliable and is not
limited to any structural geometry, e.g., one
such geometry can be wall thickness. For
thick pressure vessels, the state of stress or
strain is three dimensional. The use of
maximum stress or strain criterion in 3-D
stress or strain of orthotropic materials, in

particular, gives rise many vector equations. Moreover, none of these two criteria includes the interactions among the stress or strain components. Whereas a 3-D quadratic failure criterion yields only one scalar equation and it also includes the interactions among the stress or strain components. Thus, in this analysis, the quadratic failure criterion proposed by Tsai and Wu[5], analogous to maximum distortional energy, has been used.

In filament wound pressure vessel technology elastomeric (nonload bearing) or metalic (load bearing) liners are used to prevent leakage due to matix cracking. The load bearing liners are designed to operate to be fully plastic throughout the operational life of the vessel. Although the anlytical model discussed here does not include liner, the presence of load bearing liner will not alter the analysis because the liner always operate in plastic domain; only the amount of plastic load absorbed by the liner should be added to the predicted failure pressure to get the overall burst pressure. To keep the analysis simple, the effect of longitudinal bending deformation of the cyndrical wall due to the end closures is ignored. The vessel is assumed cylindrically orthotropic, i.e., the adjacent ($\pm\alpha$) angle wound plies are assumed to behave as an orthotropic unit. Even with these limitations, this model serves as an analytical tool to sort out important material parameters in the design of thick cylindrical vessels.

## 2. ANALYSIS

### 2.1 Thick Wall Vessel

The pressure vessel is assumed to be cylindrically orthotropic, i.e., one of the material symmetry axes is parallel to the longitudinal axis of the cylinder. The configaration of the vessel is given in Fig. 1. In this analysis, filament wound pressure vessels are assumed to have adjacent ($\pm\alpha$) angle lay ups and that adjacent ($\pm\alpha$) lay ups act as an orthotropic unit. A pressure vessel can be made up with several of such orthotropic units (Fig. 1B) wound one over another. An orthotropic unit of ($\pm\alpha$) angle lay ups will be referred, in this section, as an orthotropic layer of angle $\alpha$.

It is assumed that the length (L) of the vessel is such that the longitudinal bending deformation due to the end closures of the vessel is limited to only small end portions of the vessel compared to the overall length. The vessel is subject to axisymmetric internal ($q^{(i)}$) and external ($q^{(e)}$) pressures and axial force (F). Due to cylindrical orthotropy and axisymmetric loading, and ignoring the longitudinal bending deformation due to end closures, the problem can be treated as a generalized plane strain problem. The strain-stress relations for an axisymmetric multilayerd cylinder in cylindrical coordinates are given by

$$\varepsilon_r^{(k)} = S_{rr}^{(k)}\sigma_r^{(k)} + S_{r\theta}^{(k)}\sigma_\theta^{(k)} + S_{rz}^{(k)}\sigma_z^{(k)}$$

$$\varepsilon_\theta^{(k)} = S_{r\theta}^{(k)}\sigma_r^{(k)} + S_{\theta\theta}^{(k)}\sigma_\theta^{(k)} + S_{\theta z}^{(k)}\sigma_z^{(k)}$$

$$\varepsilon_z^{(k)} = S_{rz}^{(k)}\sigma_r^{(k)} + S_{\theta z}^{(k)}\sigma_\theta^{(k)} + S_{zz}^{(k)}\sigma_z^{(k)}$$

where $S_{ij}^{(k)}(i,j=r,\theta,z)$ are components of the compliance matrix. The superscript **k** refers to the k-th layer.
The strain in **z**-direction (axial direction) is assumed to be constant, i.e., $\varepsilon_z^{(k)} = \varepsilon_z^{\circ}$ (a constant). Using this generalized plane strain condition, the above strain-stress relations can be written as

$$\varepsilon_r^{(k)} = \beta_{rr}^{(k)}\sigma_r^{(k)} + \beta_{r\theta}^{(k)}\sigma_\theta^{(k)} + \upsilon_{rz}^{(k)}\varepsilon_z^{\circ}$$

$$\varepsilon_\theta^{(k)} = \beta_{r\theta}^{(k)}\sigma_r^{(k)} + \beta_{\theta\theta}^{(k)}\sigma_\theta^{(k)} + \upsilon_{\theta z}^{(k)}\varepsilon_z^{\circ}$$

where $\beta_{ij}^{(k)} = S_{ij}^{(k)} - S_{ij}^{(k)}S_{jz}^{(k)}/S_{zz}^{(k)}$, $(i,j=r,\theta)$; $\upsilon_{iz}^{(k)} = S_{iz}^{(k)}/S_{zz}^{(k)}, (i=r,\theta)$
The normal traction acting on the interface between k-th and (k+1)th layers is denoted by $q^{(k)}$ (Fig. 1B). Then the radial, $\sigma_r^{(k)}$, and hoop, $\sigma_\theta^{(k)}$, stresses are[6]

$$\sigma_r^{(k)} = A_k[(r/a_k)^{g(k)-1} - (a_k/r)^{g(k)+1}] + B_k[-(r/a_k)^{g(k)-1} + c_k^{2g(k)}(r/a_k)^{g(k)+1}] \quad \ldots (1)$$

$$\sigma_\theta^{(k)} = A_k g(k)[(r/a_k)^{g(k)-1} + (a_k/r)^{g(k)+1}] - B_k g(k)[(r/a_k)^{g(k)-1} + c_k^{2g(k)}(a_k/r)^{g(k)+1}] \quad \ldots (2)$$

518

and

$$\sigma_z^{(k)} = (\varepsilon_z^\circ - S_{rz}^{(k)}\sigma_r^{(k)}$$
$$-S_{\theta z}^{(k)}\sigma_\theta^{(k)})/S_{zz}^{(k)} \quad \ldots (3)$$

$$\tau_{r\theta}^{(k)} = 0$$

where, $A_k = (q^{(k-1)}c_k{}^{g(k)} + 1)/(1 - c_k{}^{2g(k)})$,
$B_k = q^{(k)}/(1 - c_k{}^{2g(k)})$, $c_k = a_k/a_{k-1}$ and

$g(k) = [\beta_{rr}^{(k)}/\beta_{\theta\theta}^{(k)}]^{1/2}$, $k$ within the parenthesis of $g(k)$ renders to $g$ being a function of $k$.

The non-zero displacement components are

$$u^{(k)}(r) = [\beta_{r\theta}^{(k)}\sigma_r^{(k)} + \beta_{\theta\theta}^{(k)}\sigma_\theta^{(k)}$$
$$+ \upsilon_{\theta z}^{(k)}\varepsilon_z^\circ]r$$

$$w^{(k)} = \varepsilon_z^\circ L$$

To determine $\varepsilon_z^\circ$, the axial stress, $\sigma_z^{(k)}$, is assumed to satisfy the axial traction on the average, i.e.,

$$\sum_{k=1}^{n} 2\pi \int_{a_{k-1}}^{a_k} \sigma_z^{(k)} r\, dr = \pi(q^{(i)} - q^{(e)})a^2 + F \quad \ldots (4)$$

where $F$ is the applied axial force and the first term on the right hand side of the above equation is the axial stress induced by the end closure due to internal, $q^{(i)}$, and external, $q^{(e)}$, pressures. Substituting $\sigma_z^{(k)}$ from equation (3) and the expressions for $\sigma_r^{(k)}$ and $\sigma_\theta^{(k)}$ from equations (1) and (2) into equation (4) and performing the integration, the expression for $\varepsilon_z^\circ$ is given by

$$\varepsilon_z^\circ = [(q^{(i)} - q^{(e)})a^2 + F/\pi$$
$$- \sum_{k=1}^{n} (q^{(k-1)}\delta_k + q^{(k)}\mu_k)]/\Delta$$

where
$$\delta_k = -2[a_k c_k{}^{g(k)+1}(S_{rz}^{(k)} + g(k)S_{\theta z}^{(k)})$$
$$(a_k - c_k{}^{g(k)}a_{k-1})/(1+g(k)) - a_{k-1}(S_{rz}^{(k)}$$
$$- g(k)S_{\theta z}^{(k)})(a_k c_k{}^{g(k)} - a_{k-1})/(1 - g(k))]$$
$$\{S_{zz}^{(k)}(1 - c_k{}^{2g(k)})\}$$
$$\ldots (5a)$$

$$\mu_k = -2[-a_k(S_{rz}^{(k)} + g(k)S_{\theta z}^{(k)})$$
$$(a_k - c_k{}^{g(k)}a_{k-1})/(1+g(k)) + a_k c_k{}^{g(k)}$$
$$(S_{rz}^{(k)} - g(k)S_{\theta z}^{(k)})(a_k c_k{}^{g(k)} - a_{k-1})$$
$$/(1-g(k))]/\{S_{zz}^{(k)}(1 - c_k{}^{2g(k)})\}$$
$$\ldots (5b)$$

and

$$\Delta = \sum_{k=1}^{n} (a_k{}^2 - a_{k-1}{}^2)/S_{zz}^{(k)}.$$

The interface normal tractions, $q^{(k)}$'s, which are unkown, are to be determined by satisfying the contact condition of the interfaces,

$$u^{(k)} = u^{(k+1)} \quad \text{at} \quad r = a_k$$

which gives a set of simultaneous equations to determine $q^{(k)}$'s:
$$q^{(k+1)}\phi^{(k+1)}a_{k+1} + q^{(k)}\gamma^{(k)}a_k$$
$$+ q^{(k-1)}\phi^{(k)}a_{k-1} + (1/\Delta)(\upsilon_{\theta z}^{(k+1)} - \upsilon_{\theta z}^{(k)})$$
$$\sum_{i=1}^{n} q^{(i-1)}\delta_i + q^{(i)}\mu_i)a_i = (1/\Delta)$$
$$(\upsilon_{\theta z}^{(k+1)} - \upsilon_{\theta z}^{(k)}) [(q^{(i)} - q^{(e)})a^2 + F/\pi]$$

where
$$\phi^{(k)} = 2g(k)\beta_{\theta\theta}^{(k)}c_k{}^{g(k)}/(1 - c_k{}^{2g(k)}) \quad ,$$

$$\gamma^{(k)} = -\beta_{r\theta}^{(k)} - g(k)\beta_{\theta\theta}^{(k)}(1 + c_k{}^{2g(k)}) +$$
$$\beta_{r\theta}^{(k+1)} g(k+1)\beta_{\theta\theta}^{(k+1)}$$
$$(1 + c_{k+1}{}^{2g(k+1)})/(1 - c_{k+1}{}^{2g(k+1)})$$

and $g(k+1) = [\beta_{rr}^{(k+1)}/\beta_{\theta\theta}^{(k+1)}]^{1/2}$.

The second term within the parenthesis of equations (5a) and (5b) contain $(1-g(k))$ in the demomenator. For winding angle $\alpha = 0$, $g(k) = [\beta_{rr}^{(k)}/\beta_{\theta\theta}^{(k)}]^{1/2} = 1$, then $\delta_k$ and $\mu_k$ become singular. In actual computation, this difficulty on singularity was averted by assigning a very small number to $\alpha(=0.01)$ when $\alpha = 0$.

519

## 2.2 Prediction of Burst Pressure

The Last Ply Failure (**LPF**) pressure is considered to be the burst pressure. The **LPF** pressure was calculated by degrading the matrix stiffness by a factor called Matrix Degradation Factor (**MDF**); **MDF**=E(m)/E(m)°<1.

Material of a filament wound pressure vessel is transversely isotropic; the isotropic plane of the material is perpendicular to the fiber direction. The 3-D quadratic failure criterion for the k-th layer in material symmetry coordinates is given by[5]

$$F_{zz}^{(k)}\sigma_z^{(k)2}+F_{rr}^{(k)}(\sigma_r^{(k)2}+\sigma_\theta^{(k)2}) +$$
$$F_{ss}^{(k)}\tau_{\theta z}^{(k)2}+2F_{rz}^{(k)}(\sigma_r^{(k)}+\sigma_\theta^{(k)})\sigma_z^{(k)}$$
$$+2F_{r\theta}^{(k)}\sigma_r^{(k)}\sigma_\theta^{(k)}+F_r^{(k)}(\sigma_r^{(k)}+\sigma_\theta^{(k)})$$
$$+F_z^{(k)}\sigma_z^{(k)}-1=0$$

where
$F_{rr}^{(k)}=1/(Y^{(k)}Y'^{(k)})$,
$F_{zz}^{(k)}=1/(X^{(k)}X'^{(k)})$,  $F_{ss}^{(k)}=1/S^{(k)2}$,
$F_r^{(k)}=1/Y^{(k)}-1/Y'^{(k)}$,  $F_z^{(k)}=1/X^{(k)}-1/X'^{(k)}$,  $F_{rz}^{(k)}=-0.5*[F_{rr}^{(k)}F_{zz}^{(k)}]^{1/2}$,
$F_{r\theta}^{(k)}=-0.5*F_{rr}^{(k)}$

and $X^{(k)}$, $X'^{(k)}$ are respectively longitudinal tensile and compressive strengths, $Y^{(k)}$, $Y'^{(k)}$ are those for transverse direction and $S^{(k)}$ is the shear strength of the k-th layer. Although $\tau_{\theta z}^{(k)}$ in vessel coordinates (Fig. 1) is zero, after taking stress transformation to material symmetry coordinates $\tau_{\theta z}^{(k)}$ will not be zero for winding angle, $\alpha \neq 0°, 90°$. Since the failure criterion is used in material symmetry coordinates, the term containing $\tau_{\theta z}^{(k)}$ is included in the failure criterion. The failure pressure can easily be calculated in terms of Strength Ratio, **R**, as suggested by Tsai[5] such that

$$\{\sigma\}_{max} = R\ \{\sigma\}_{applied}\ .$$

Substituting the above expression in the failure criterion, the expression for the positive root of **R** is given by

$$R = -(\xi/2\rho)+[(\xi/2\rho)^2+1/\rho]^{1/2}$$

where
$$\rho=F_{zz}^{(k)}\sigma_z^{(k)2}+F_{rr}^{(k)}(\sigma_r^{(k)2}+\sigma_\theta^{(k)2}) +$$
$$F_{ss}^{(k)}\tau_{\theta z}^{(k)2}+2F_{rz}^{(k)}(\sigma_r^{(k)}+\sigma_\theta^{(k)})\sigma_z^{(k)}$$
$$+2F_{r\theta}^{(k)}\sigma_r^{(k)}\sigma_\theta^{(k)}$$
$$\xi = F_r^{(k)}(\sigma_r^{(k)}+\sigma_\theta^{(k)})+F_z^{(k)}\sigma_z^{(k)}$$

The negative root for **R** does not have any physical meaning, and thus omitted. When **R**=1, failure occurs. In burst pressure calculation, first the stresses are calculated after degrading the stiffness of the matrix by **MDF**, then those stress values are used in the failure criterion to calculate burst pressure. In point stress analysis (as in the case of elasticity solution) the point of minimum value of **R** dictates the failure point. When a unit load is applied the the minimum value of **R** is the burst pressure.

## 3. RESULTS

Except for Table 1 all the material data used were obtained from reference [7]. The results presented here are for pressure vessels subjected to only internal pressure. If not mentioned otherwise, the composite used to study the design parameters was T300/N5208.

### 3.1 Displacement Distribution

Like the stress distributions, the distribution of radial displacement, **u(r)**, through the thickness with increasing b/a was nonlinear. However, the plot of outer displacement, **u(b)**, versus winding angle, $\alpha$, showed some interesting results. In Fig. 2, **u(b)** was plotted versus $\alpha$ for b/a=1.025 (a very thin cylinder), and for a fixed internal pressure. The displacement was normalized by its value at 0° winding angle. Both the thin and thick wall solutions (elasticity solutions) showed that the radial displacement decreases monotonically with increasing winding angle until it becomes negative between 58° and 78°. The negative displacement is due to poisson's effect at those winding angles. The **u(b)** for isotropic material, for instance, is not a function of winding angle and always positive, and does not show the above feature observed in composite materials. The distribution of **u(b)** in Fig. 2 is very interesting from the point of view of fastening the pressure vessel with other structural members. In other words, the radial displacement of the vessel can be made compatible with that of the fastening members by choosing a particular winding angle. Similar results were observed for thick vessels also.

## 3.2 Matrix Degradation Factor (MDF)

As mentioned in the analysis, to take the matrix cracking into account in predicting LPF, i.e., the burst pressure, macroscopically the stiffness of the matrix is degraded by **MDF**. Table 1 shows the comparison of the predicted burst pressure and failure strains with the available experimental results. The prediction was based on thick wall solution for two different MDF values (0.3 and 0.1). For thin cylinders, it could be seen from the table that the predicted failure pressure is not very sensitive to the change in the values of **MDF**. For a moderately thick cylinder, $b/a \approx 1.20$, [Massard[10], Table 1] the failure presure is somewhat sensative to the value of **MDF**. However, considering the two values used, **MDF**=0.3 gives overall a better prediction and thus is used for the rest of the results.

## 3.3 Thick and Thin Wall Solutions

In Fig. 3 the burst pressure, $P_b$, for both thick and thin walls was plotted versus increasing winding angle, $\alpha$. The burst pressure was normalized by its value at $0°$ winding angle. For a thin wall, $b/a$=1.025, both thick and thin wall solutions predict $54.75°$ as the optimum angle for maximum burst pressure. The agreement between these two solutions is very good, except for the value of the burst pressure at $\alpha$=54.75°, where the thick wall solution predicts the value a little higher than that of the thin wall solution. However, for a thick vessel, $b/a$=1.50 it was found that there is a significant difference between the two predictions in the neighborhood of optimum winding angle. The thick wall solution predicted the optimum angle as about $\alpha$=56° whereas the thin wall solution prediction is the same as in Fig. 3. Also thick wall prediction (e.g. for $b/a$=1.50) of maximum failure pressure is significantly lower than that of the thin wall. For thick vessels, it was found that there is not much of penalty in the loss of failure pressure due to some small error in optimum winding angle. Whereas in case of thin wall this penalty is significant, as can be seen from Fig. 3. To check the limitation of the thin wall solution on wall thickness, $P_b$ for both thin and thick wall was plotted versus increasing wall

thickness ratio in Fig. 4. Up to $b/a$=1.10 the thin wall solution agreed well with that of the thick wall. The agreement differed significantly for $b/a \geq 1.30$. It is interesting to point out that the thick wall solution did not predict any significant increase in failure pressure with increasing wall thickness for $b/a$>1.30. Physically this means that failure is not self arresting. Once failure starts at a particular pressure, it will propagate through the thickness and that can not be stopped by merely increasing the wall thickness.

## 3.4 Comparison of Vessel Performance of Some Materials

The burst pressures, $P_b$, of single layer ($\alpha$=55.5°) cylinders of T300, IM6, and KEV49 fibers in Epoxy matrix were compared with that of Aluminum (isotropic) in Fig. 5. All three composites considered in this figure show limiting values of their respective failure pressures; the limiting values were 225, 174, and 70MPa for T300, IM6 and KEV49 composites respectively with $a$=0.10m. Whereas the isotropic material (Aluminum in this case) does not show this limiting pattern. Although IM6/Epoxy composite is longitudinally more than twice as strong as the T300/Epoxy composite, T300/Epoxy carries more internal pressure than IM6/Epoxy. However, for thin vessel, $b/a \approx 1.10$, IM6/Epoxy offered better performance than T300/Epoxy because IM6/Epoxy carried more internal pressure than T300/Epoxy. To compare the overall performance of the pressure vessels made of these materials, the pressure vessel performance efficiency, $\eta$, was plotted in Fig. 6. The efficiency, $\eta$, is defined as the ratio of the product of burst pressure and internal volume to the vessel weight and $\eta$ has a dimension of length. As it is known, and Fig. 6 also revealed that the composite materials have much better efficiency than that of isotropic materials. As it was seen in Fig. 5, it could also be seen from Fig. 6 that at low pressure IM6/Epoxy has higher efficiency than that of other materials. However, near the limiting burst pressure the efficiency of all three composite materials droped very rapidly. Close to the limiting burst pressure, the increase of wall thickness (i.e., increase of material use) practically did not increase the burst pressure (Fig. 5), and thus the performance efficiency drops drastically near the respective burst pressures.

521

It has been observed so far that single layer (i.e., single winding angle) pressure vessels were not capable of carrying pressure beyond the limiting pressure irrespective of their wall thickness, and thus materials were not used efficiently at the limiting pressures. The vessel pressure carrying capability can further be increased by considering hybrid construction having more than one layer. The layers can be of different materials or of same material with different winding angles. To simplify the study of the design parameters multilayered vessels of same material were studied.

## 3.5 Study of Efficient Material Use

As discussed in section 2.2, the product of the Strength Ratio, $R$, and the applied load gives the strength (i.e., load carrying capacity) at any point. Thus the strength is proportional to the value of $R$. A material is said to be used very efficiently when the ratio, $R_{max}/R_{min}$, is close to unity. Ideally, if the ratio is unity, i.e., if $R$ has a constant value through the thickness, the material use efficiency is maximum. The deviation of the value of any of the above ratios from unity signifies non-efficient use of materials. It was found out that, like $R_{max}/R_{min}$, the ratio, $R(a)/R(b)$ can also be used to study the material use efficiency. Here, $R(a)$ and $R(b)$ are the values of $R$ at $r=a$ and $r=b$ respectively. Thus the ratio $R(a)/R(b)$ was used to study the material use efficiency because it was easy to calculate compared to that of $R_{max}/R_{min}$.

From Fig. 6 it was shown that the orthotropic materials have better performance efficiency than that of isotropic materials. However, to compare qualitatively the material use efficiency of orthotropic materials with that of isotropic materials, the ratio, $R(a)/R(b)$ is plotted vesus $b/a$ in Fig. 7. It was worth studying the material use efficiency of the composite materials at the optimum winding angle. Thus in Fig.7, for composites, the ratio, $R(a)/R(b)$, was plotted for $\alpha=55.5°$. This figure shows that with increasing wall thickness ratio, $b/a$, the material use efficiency in orthotropic materials drops drastically compared to that of the isotropic materials. Moreover, T300/Epoxy shows an interesting behavior. For $b/a \leq 1.15$ the ratio, $R(a)/R(b)>1$ and then for $b/a>1.15$, $R(a)/R(b)$ becomes less than unity. That means for $b/a \leq 1.15$ failure occured on the outer surface and the failure shifts to the inner surface for $b/a>1.15$. For $b/a$ close to 1.15, the ratio $R(a)/R(b)$ was very close to unity, and thus an efficient material use was already achieved. Any further significant improvement in material use was not possible, as was illustrated in Table 2 (for $b/a=1.15$).

For a multilayer T300/N5208 vessel, results in Table 2 were obtained for layers of equal thickness and the thickness ratio of each layer was $(b/a-1)/n$; n is the number of layers. The multilayer optimum winding angle combinations are obtained within a resolution of one degree. For low $b/a$, (Table 2) an increase in number of layers did not increase burst pressure by any significant amount. From column E of Table 2 was plotted in Fig. 8, and it could be seen from this figure that percentage increase in burst pressure is significant only for thick vessels and limited to only two layers. Fig. 8 also showed that there is some increase in % increase for 4 layers than 3 layers for $b/a=1.25$ and 2.00. In other words, it reveals that the optimum angle combination for 3 layers at these two $b/a$'s did not give the best burst pressure. In this conjecture, it is worth mentioning that the optimum angle combination in Table 2 is obtained by taking equal thickness of layers. Equal thickness consideration may not necessarily give the best burst pressure and this is evident here in 3 layer case. For a given number of layers, along with the angle optimization one should also optimize layer thicknesses, instead of taking equal thickness, to obtain the best burst pressure.

As is discussed earlier, in case of T300/N5208, failure point shifts from outside surface to inside one as the wall thickness increases. At about $b/a=1.15$, this transition occurs. Observation made on the multiple angle results in Table 2 revealed that the surface, where failure occurs, is to be softened by giving winding angle towards axial direction and the other surface is to be hardened by winding towards hoop direction. Thus, for $b/a \leq 1.15$ the optimum angle combinations in Table 2 showed angles close to hoop winding at inner layers and for higher $b/a$'s the hoop windings are on outer layers.

It was indicated before that the optimization based on equal layer thickness may not always predict the best burst pressure. To justify that, the optimum angle combinations, for 2 layer case, were taken to calculate burst pressure with increasing **b/a** and were plotted in Fig. 9 maintaining equal layer thickness for each layer. Although angle combinations (60/48) and (66/42) showed an increasing trend of burst pressure, however (40/66), (27/71), and (0/73) showed peaks in their respective curves. This means, the later three combinations were best for respective **b/a**'s where peaks occured and the peak for (0/73) was most interesting. For (0/73), peak occured at **b/a**=1.87 which was lower than the value for which this angle combination was optimized in Table 2. The optimization was done with equal thickness constraint; thus the above occurance of the peak at lower **b/a** justifies the fact that, to obtain the best burst pressure one should optimize both layer thickness and winding angle simultaneously.

A comparison was made in performance efficiency, $\eta$, of two layer vessels (in Fig. 10) with that of T300/Epoxy in Fig. 6 . Angle combination (60/48) and (66/42) showed better efficiency than that of the single layer at lower pressure, however, did not improve in pressure carrying capacity because the efficiency for two layer cases dropped close to zero at 230 MPa which was almost the same for single layer case. This was expected, because (60/48), (66/42) were optimized for lower pressures (45 and 95 MPa respectively). Although angles (40/66), (27/71), and (0/73) did not show great promise on efficiency, but had more pressure carrying capability than the two angles optimized for lower pressures. In particular, the limiting pressure for (0/73) improved from 225 MPa (single layer) to 280 MPa (2 layer). The efficiency of the later three angles can also be improved by optimizing more layers instead of only two layers.

In the first pargraph of this section 3.5, it was inferred that a constant value of Strength Ratio, **R**, through the thickness will result in best material use. To make the point claer, the distribution of **R** through the thickness was plotted in Fig. 11 for all four optimized angles obtained for **b/a**=1.50 in Table 2. It can be seen from the plot that, with the increasing number of optimized layers, the value of $R_{min}$ (i.e., the failure pressure) increased and the difference between the $R_{max}$ and $R_{min}$ decreased. With more number of optimized layers, the distribution of **R** can be brought more or less uniform through the thickness and a most efficient material use can be achieved. However, a

comparison of rate of convergence of the distribution of **R** toward the uniform distribution of **R** and the values of $P_b$ in Table 2, reveal that, for all practical purposes, pressure vessel with 2 or 3 optimized layers can offer very good performance.

## 4. CONCLUSIONS

Results of this study of design parameters, are presented for pressure vessels subjected to only internal pressure. Comparison of burst pressures calculated from thin wall and thick wall solutions reveals that the validity of the thin wall solution is limited to wall thickness ratio, **b/a**≤1.10 which in terms of radius to thickness ratio, r/t≥10.

The outer radial displacement of filament wound vessels decreases with increasing winding angle and becomes negative between 58° and 78°. This displacement variation with winding angle is very interesting from the point of view of fastening the vessel with other structural members. By choosing a proper winding angle the vessel displacement can be matched with that of the fastening members.

A single layer construction does not offer efficient material use, especially for thick vessels. Failure in single layer vessels is either in outer or in inner surface. It is observed that softening the surface where failure occurs, by giving winding towards axial direction, and hardening the other surface yields an efficient use of materials. Thus multilayer construction with optimized winding angles yields an efficient use of materials and inceases pressure carrying capability. The optimization based on both layer thickness and winding angle will lead to best use of materials and that also improves vessel performance efficiency. For vessel wall thickness ratio up to 2, it is shown that vessel with optimized 2 or 3 layers practically yields a very good design within a reasonable computer time.

## 5. ACKNOWLEDGEMENT

This work was performed under the Research Associateship Program of the National Research Council and Air Force System Command. The author is indebted to Dr. S.W. Tsai for introducing to this problem and for his constant encouragement. A time to time discussion with Dr. K.P. Rao has greatly benefited the author in preparing this paper.

## 6. REFERENCES

1.Bert, C.W., Analysis of Radial Filament-Reinforced Spherical Shells Under Deep Submergence Conditions, 2nd Intl. Conference on Pressure Vessel Technology, Part I, Design and Analysis, ASME, Oct 1973, pp 529-534.

2. Tauchert, T.R., Optimum Design of a Reinforced Cylindrical Pressure Vessel, Journal of Composite Materials, Vol 15, 1981, pp 390-402.

3. Sherrer, R.E., Filament Wound Cylinders with Axial-Symmetric Loads, Journal of Compisites Materials, Vol 1, 1967, pp 344-355.

4. Gerstle Jr., F.P. and M. Moss, Thick-Walled Spherical Composite Pressure Vessels, Composites in Pressure Vessels and Piping, Kulkarni, S.V. and C.H. Zweben eds., ASME, PVP-PB-021, 1977.

5. Tsai, S.W., A Survey of Macroscopic Failure Criteria for Composite Materials, Journal of Reinforced Plastics and Composites, Vol 3, Jan 1984, pp 40-62.

6. Lekhnitskii, S.G., Anisotropic Plates, translated by S.W. Tsai and T. Cheron, Gordon and Breach, New York, 1968.

7. Tsai, S.W., Composites Design 1986, Think Composites, Dayton, Ohio,1986.

8. Beckwith,S.W., et.al.,Filament Wound Case (FWC) Graphite/Epoxy Pressure Vessel Response to Environmental Conditioning, 31st Intl. SAMPE Symposium, April 7-10, 1986, Las Vegas.

9. Lark, R.F., Recent Advances in Lightweight, Filament-Wound Composite Pressure Vessel Technology, in Composites in Pressure Vessels and Piping, ASME, PVP-PB-021, 1977.

10. Massard, T., Commission on Atomic Energy of France (CEA) - private communication.

## 7. BIOGRAPHY

Ajit K. Roy is a visiting scientist at the U.S. Air Force Materials Laboratory under a Research Associate Fellowship from the National Research Council and the Air Force System Command.  He has been on this fellowship since October, 1985 which is awarded to scientists and engineers of unusual promise, based on an international competition.  Dr. Roy received his M.S. and Ph.D., both in Mechanics, from University of Minnesota in 1983 and 1985 respectively. His B.Tech.(Honors) was from Indian Institute of Technology, Kharagpur, India. His current research interests are in structural mechanics and material testing.

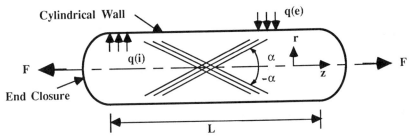

Fig. 1A.   Configuration of the Pressure Vessel

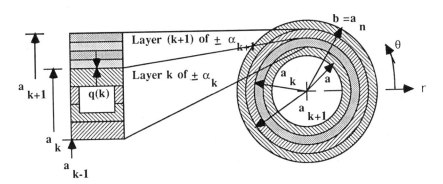

Fig. 1B.   Enlarged View of the Cross Section

| | A | B | C | D | E | F | G | H | I | J | K |
|---|---|---|---|---|---|---|---|---|---|---|---|
| 1 | Reference | Materials | | LAY UP# | Reported | Predicted | | Reported Failure | | Predicted Failure | |
| 2 | | Liner | b/a | (ANGLE)/ | Failure | Failure Pres. | | Strains (%) | | Strains(%)** | |
| 3 | | | | [% thickness | Pressure | MDF= | MDF= | eps(θ) | eps(z) | eps(θ) | eps(z) |
| 4 | | | | | (MPa) | 0.3 | 0.10 | | | | |
| 5 | Beckwith, | AS4/Epoxy | 1.03 | (24.5/89/24.5/ | 37.1 | 30.46 | 30.53 | 1.3 | 0.32 | 1.08 | 0.22 |
| 6 | et.al.[8]## | Elastomer | | 89/24.5) | | | | | | | |
| 7 | | | | [16/16/34/16/16 | | | | | | | |
| 8 | Lark [9] | Kev49/Epx | 1.01 | (0/90) | 14.22 | 11.55 | 12.69 | 2 | 1.6 | 1.90 | 1.40 |
| 9 | | Rubber | | [40/60] | | | | | | | |
| 10 | Massard [10] | AS/Epoxy | 1.20 | (5/11/18/24/ | 103.9 | 107.5 | 96.00 | - | - | 0.65 | 0.44 |
| 11 | | - | | 31/90) | | | | | | | |
| 12 | | | | [9/9/7/11/9/52] | | | | | | | |
| 13 | Note | | | | | | | | | | |
| 14 | # | Lay up anlges are (+/-) of each angle shown above. | | | | | | | | | |
| 15 | ** | Strain values for MDF=0.3. The failure strain is very insensative to the change of value of MDF | | | | | | | | | |
| 16 | ## | Numbers within [ ] represent reference numbers. | | | | | | | | | |

Table 1. Comparison of Experimental and Predicted Failure
Pressures and Strains.

| | A | B | C | D | E |
|---|---|---|---|---|---|
| 1 | | No of | Optimum Angle | Burst | % Change |
| 2 | b/a | Layers | Combination | Pressure | in Burst |
| 3 | | n | (inside to outside) | MPa | Pressure |
| 4 | | 1 | 54.5 | 42.93 | - |
| 5 | 1.05 | 2 | 60/48 | 45.38 | 5.71 |
| 6 | | 3 | 63/54/45 | 46.14 | 1.67 |
| 7 | | 1 | 54.5 | 89.92 | - |
| 8 | 1.10 | 2 | 66/42 | 95.56 | 6.27 |
| 9 | | 3 | 68/56/38 | 97.14 | 1.65 |
| 10 | | 1 | 54.75 | 141.80 | - |
| 11 | 1.15 | 2 | 84/34 | 142.27 | 3.32 |
| 12 | | 3 | 75/70/23 | 143.37 | 0.77 |
| 13 | | 1 | 55.5 | 184.26 | - |
| 14 | 1.25 | 2 | 40/66 | 211.19 | 14.61 |
| 15 | | 3 | 35/55/75 | 217.20 | 2.85 |
| 16 | | 4 | 36/48/67/65 | 224.80 | 3.49 |
| 17 | | 1 | 55.75 | 216.25 | - |
| 18 | 1.50 | 2 | 27/71 | 262.25 | 21.27 |
| 19 | | 3 | 23/41/81 | 284.48 | 8.48 |
| 20 | | 4 | 23/35/51/82 | 298.10 | 4.77 |
| 21 | | 1 | 57.75 | 224.55 | - |
| 22 | 2.00 | 2 | 0/73 | 272.50 | 21.34 |
| 23 | | 3 | 1/40/75 | 292.30 | 7.34 |
| 24 | | 4 | 2/32/66/52 | 316.90 | 8.42 |
| 25 | | | | | |
| 26 | | | | | |

Table 2. Prediction of Burst Pressure of several Multi-Layer
Vessels. Material: T300/N5208.

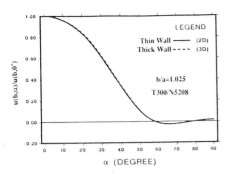

Fig.2. Outer Displacement, u(b,α),
for b/a=1.025. Material: T300/N5208 ;

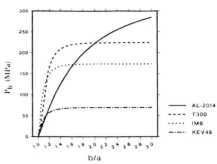

Fig.5. Burst Pressure, $P_b$, for several
materials. For composites,
α=55.5°.

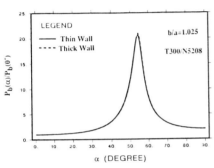

Fig.3. Burst Pressure, $P_b$, for
b/a=1.025. Material: T300/N5208.

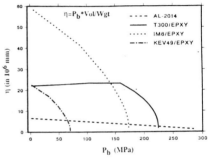

Fig.6. Vessel Performance Efficiency,
η, for several materials. For
composites, α=55.5°.

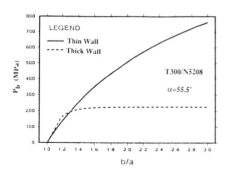

Fig.4. Burst Pressure, $P_b$, for
α=55.5°. Material: T300/N5208.

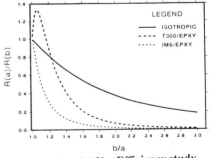

Fig.7. Material Use Efficiency study
for several materials. For
composites, α=55.5°.

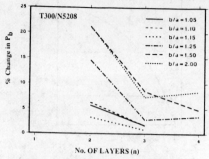

**Fig.8. Percentage Change of Burst Pressure versus No. of optimized Orthotropic Layers. Material: T300/N5208.**

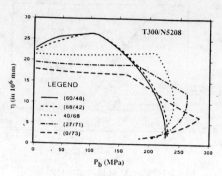

**Fig.10. Vessel Performance Efficiency, $\eta$, for Two Layer Vessel. Angle Combinations are from Table 2.**

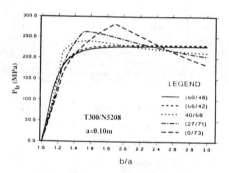

**Fig.9. Burst Pressure, $P_b$, for Two Layer Vessel. Angle Combinations are from Table 2.**

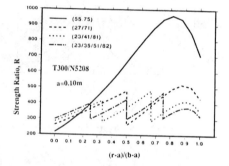

**Fig.11. Strength Ratio, R, for b/a=1.50 in Table 2.**

# SIMPLE DESIGN APPROACH AGAINST LOW-VELOCITY IMPACT DAMAGE

Peter Sjöblom
University of Dayton Research Institute
300 College Park
Dayton, Ohio 45469

## Abstract

Foreign object impact has been recognized as a severe threat to composite structures. From a structural viewpoint, the alarming possibility exists that a minor, low-velocity impact could reduce the compressive load-carrying capacity of a structure without showing any visible damage. This work has concentrated on the problem of defining threshold values for the damage initiation in graphite fiber composite plates impacted transversely at low-velocity. The goal was to be able to scale the results from laboratory-size specimens to larger and thicker real airframe structures. Variations in both the laminated plate size and thickness were studied.

Results are presented from tests on two different composite material systems, Hercules' AS4/3502 and ICI's AS4/APC-2. The test matrix consisted of Four different sizes of supports and Four different plate thicknesses. The findings confirmed that a threshold damage value for a specific material could be determined using a small test sample and that this value could be used to predict the corresponding value for plates of other thicknesses and sizes of the same material through simple straightforward formulas.

## 1. INTRODUCTION

During recent years there has been a growing concern about the susceptibility to impact damage in high-performance fiber composite structures. An impact is a very complex event. Figure 1 shows an attempt to visualize the possible interactions between material properties and the performance of an impacted structure. Normally the problem of designing against impact damage can be divided into two major fields, durability and damage tolerance. At first glance, one may not grasp the significant difference between the two. A closer examination will, however, show that having a damage tolerant structure does not necessarily guarantee a comparably high durability. The damage tolerance requirement is there to ensure the load-carrying capacity of the damaged structure. Durability, on the other hand, defines the amount of abuse the structure can take without any resulting damage. In other words, one might say that durability defines the threshold level for damage initiation and that the damage tolerance defines the performance of a damaged structure. The two different concepts are related, but there are no easy links between them. A high durability will not always imply a good damage tolerance. This investigation focused on the prediction of damage initiation during low-velocity impact of plates, a measure of the durability of a structure.

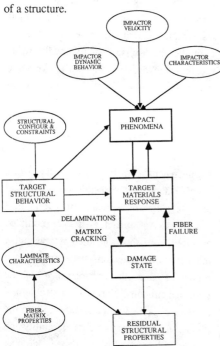

Fig. 1. Global view of impact.

Designers will face the difficulties of scaling impact test results from small-size laboratory specimens to larger-size structures. No standardized test method exists today. Even more serious is the fact that no established methods for evaluating the results from the methods used exist. A plate subjected to the impact of a rigid impactor will deform both locally under the impactor and globally as a plate under a concentrated load. The thickness, as well as the boundary conditions, will determine the resulting deformation and stress distribution in the plate. To characterize a material, the influence of these two geometrical variables must be understood. First then will it be possible to derive any material properties from an impact test.

## 2. THEORY

This paper deals only with the question of how to scale the impact results with respect to plate thickness and support size. The initial failure modes of impacted composites are often hard to identify. This problem arises when the damage is massive. The final fracture often disguises the initial. To find the origin of the failure one must examine specimens hit just above the level at which the initial damage occurs. Fractographic studies of impact damaged specimens suggested that the initial damage was either small matrix cracks or delaminations. One peculiarity of the matrix cracks was their orientation. Most often it was found that these cracks were slanted, i.e., the cracks were not perpendicular to the plane of the plies. Without further discussion, the conclusion is obvious that these types of cracks were formed by a stress field having a substantial amount of shear stress.

Delaminations can only be formed by normal tensile stresses or shear stresses. As it is hard to visualize any normal tensile stresses in the vicinity of the impact point, the conclusion must be that the delamination is initiated by shear stresses. This reasoning is supported by stress calculations [1]. It must be pointed out that spalling can not be responsible for the delaminations. To spall the plate the load must be applied during a very short time. The length of the impactor must be in the order of the plate thickness. For longer impactors the wave reflection pattern cancels out the dangerous transverse tensile stresses.

A very important parameter in all impacts is the velocity. In this paper the term "Low-Velocity" is used when the velocity is low enough to justify a static analysis, i.e., no effects of vibrations or wave propagation need to be considered, at least not to the first order. For most plates, such as wing panels, low-velocity ranges roughly up to about 10 m/s. This range of velocities will cover most of the tool drop situations and other handling impacts.

The simple model described below by no means claims to be complete. It is assumed to predict reasonably well the initiation force for certain combinations of plate thicknesses, support sizes, impactor geometries, and velocity ranges.

The Hertian contact law has been found to accurately describe the indentation of a hard sphere into a fiber composite plate [2]. It states that the contact force, P, is proportional to the depth of indentation, sometimes called the approach, $\alpha$, to the power of 1.5. $K_c$ is the contact stiffness.

$$P = K_c \, \alpha^{1.5} \tag{1}$$

For a rigid, perfectly plastic thick plate and a rigid spherical indenter of diameter D, Fig. 2, the relationship between the depth of indentation, $\alpha$, and radius of contact, r, assuming $\alpha \ll D$, is given by:

$$r = (D \, \alpha)^{0.5} \tag{2}$$

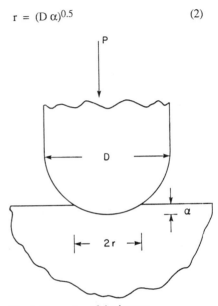

Fig. 2. Geometry of the impact.

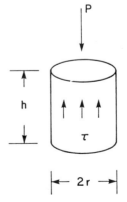

Fig. 3. Shear stress equilibrium at the edge of the contact area.

531

The maximum average shearing force occurs at the periphery of the contact area. A simple balance of forces, Fig. 3, gives

$$P = 2\pi rh\tau_{ave} \qquad (3)$$

Combining equations (1) and (2) yields

$$r = [D(P/K_c)^{2/3}]^{0.5} \qquad (4)$$

Finally, combining equations 3 and 4 give us

$$P = (2\pi h\tau_{ave})^{1.5} D^{0.75}/K_c^{0.5} \qquad (5)$$

Assuming that the average shear strength and the contact stiffness do not vary with the thickness of the plate and keeping the impactor diameter constant, a simple relation between the initiation force and the plate thickness can be established.

$$P_{init} = C^* h^{1.5} \qquad (6)$$

Equation (5) predicts that the initiation force should increase with both thickness and shear strength to the power of 1.5. The influence of impactor diameter and compressive stiffness are less pronounced. It is important to remember that Equation (5) only predicts the damage initiation. In the case of penetration resistance, the impactor size and shape play a completely different role. Another observation worth mentioning is that Equation (6) predicts the damage initiation force to be independent of the support size. This is a consequence of the assumption of the shear forces being the cause of the damage initiation.

## 3. EXPERIMENTAL APPROACH

Because of the relative low energy levels and velocities studied, a pendulum impactor was used. The advantages are the accurate velocity measurements and the ease of operation, i.e., avoiding multiple impacts. The setup can be seen in Fig. 4. The key ingredients making the system useful are the digital oscilloscope, the computer, and the velocity measurement gates.

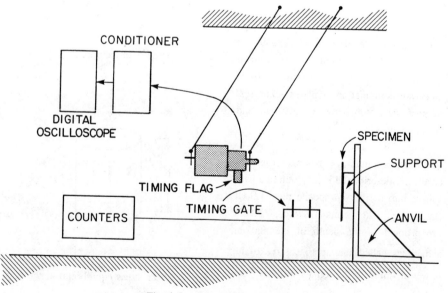

Fig. 4. Impact pendulum system

Instrumented impact has been used for many years. Most of the commercially available systems are designed to perform penetration tests in which the velocity during impact stays almost constant, or at least does not change sign during the impact. For this type of impact an electronic integration can be used. Hitting the plate at low enough energy to have a rebound makes it necessary to numerically integrate the energy as a function of time. The equations are:

$$V(t) = V(0) - 1/m \int_0^t F(\tau) \, d\tau \qquad (7)$$

$$E(t) = 0.5m \, (V(0)^2 - V(t)^2) \qquad (8)$$

where $V(t)$ is the velocity of the impactor and $E(t)$ is the work done on the impactor by the contact force F. Time zero is defined as the instant when the impactor hits the plate. E can of course also be interpreted as the energy absorbed by the target plate.

A small error in the impact velocity will have a large influence on the calculation of the absorbed energy. There is no way to get around this problem of induced error. One must simply measure the velocity very accurately. The pendulum has an additional advantages over drop towers in that the gravitational pull just before, during, and after the impact is virtually zero in the direction of movement. The end result is a low-velocity impact system with a precision in the velocity measurement better than 0.1 percent.

Force measurements during a rapid event can be affected by phenomena originating from wave propagation, strain rate sensitivity in material response, and differences in response of the electronic equipment used [3]. Statically calibrated load measuring instrumentation may give several percent error in the load reading. One way to correct for the potential errors is to make a dynamic calibration that matches the impulse with the momentum change, using the accurately known impact and rebound velocities.

Suppose that the true force, $F_a$, is proportional to the measured force, $F_m$.

$$F_a = C \, F_m \qquad (9)$$

The proportionality factor C is given by

$$C = m \, (V_i - V_r) / \int_0^{t_i} F_m \, dt \qquad (10)$$

where $t_i$ the total impact time and $V_i$ and $V_r$ are the impact and rebound velocities, respectively. We have found the indicated force, using the static calibration, to differ about 5 to 10 percent from the corrected one. Normally the load cell outputs a to high load, $C = 0.95$ typical. This phenomenon has also been observed by others using a piezoelectric load washer, [4]. A 10 percent error in the force measurement may render the calculation of the absorbed energy useless for impact situations where the damage is small, i.e., when the absorbed energy is small compared to the impact energy.

The identification of the damage load level is far from a closed issue. Fig. 5 defines different damage regions for the impacted laminates.

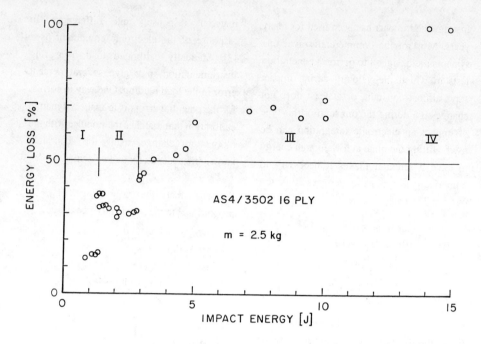

Fig. 5. Damage "Regions".

In Region I the damage is at most a few matrix cracks or very limited delamination in the vicinity of the impact point. In Region II we mainly find matrix cracks and larger delaminations. The beginning of Region III is determined by the appearance of back face damage and in Region IV the panel is penetrated. Of importance to this investigation is the fact that there is a very sharp boundary between Region I, the undamaged region, and Region II. In other words, the damage initiation is not a gradual process. A panel either received a measurable amount of damage, absorbed energy from the pendulum, or it was not affected to any measurable extent.

The sharp boundary can be explained by the fact that the damage nitiation is a very sudden event. In Fig. 6, a typical force vs time response is shown. Note the very rapid load drop associated with the initial damage formation. This effect is what makes it possible to identify the initiation point from the load trace alone. This approach to determining initiation is significantly faster then nondestructively examining the samples after the impact and then deciding what energy level to use for the next hit.

## 4. RESULTS AND DISCUSSION

Two parameters were studied, the plate thickness and the boundary conditions, i.e. the support size. For a check on the generality of the behavior, two different materials, AS4/3502 (epoxy) and AS4/APC-2 (PEEK), with different matrix properties were tested in parallel. Four different supports were used. The samples were simply supported on either a 25, 50, 75, or 100 mm diameter ring.

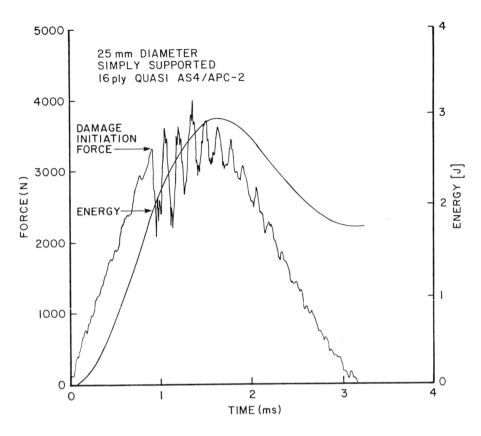

Fig. 6. Typical impact response

A total of at least four specimens of each material, thickness, and support size were tested. The impact velocity was kept at 1.5 ±0.1 m/s except for the last three 32 ply AS4/APC-2 panels on the 100 mm ring. The first impact showed that with the maximum pendulum weight the energy at 1.6 m/s was just at the borderline for damage initiation. A velocity of 2.0 m/s was used in the remaining three tests. All panels had a π/4 quasi isotropic layup.

The nominal thickness per ply was 0.127 mm for both materials. However, slight deviations were noticed and no real pattern could be established. Therefore the unit of thickness was chosen as the number of plies. In an exploratory investigation this should be accurate enough. The coefficients in Table 1 are the least-square best fit to the average experimental data.

TABLE 1. Coefficient $C^*$ in equation (6) in units of [Newton/(No plies)$^{1.5}$].

|  | Support diameter [mm] | | | |
|---|---|---|---|---|
|  | 25 | 50 | 75 | 100 |
| AS4/3502 | 30.2 | 29.1 | 28.4 | 28.1 |
| AS4/APC-2 | 35.8 | 37.2 | 43.1 | 42.0 |

Figure 7 shows the correlation between Equation (6), using the coefficients in Table A, and the experimental data.

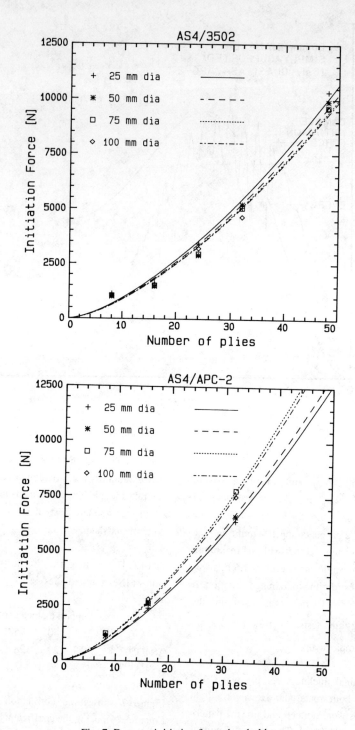

Fig. 7. Damage initiation force thresholds.

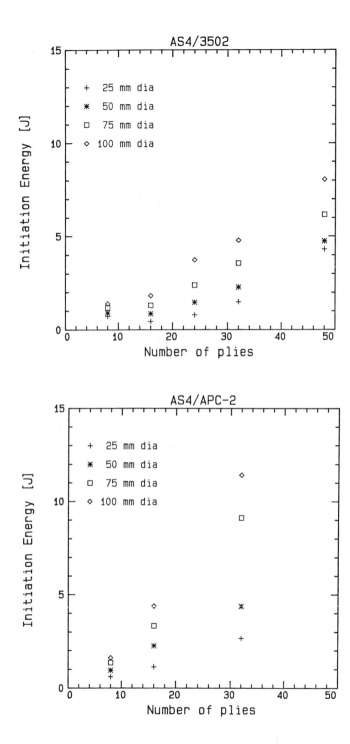

Fig. 8. Damage initiation energy thresholds.

The correlation is excellent for the epoxy material. The data for the PEEK material on the other hand shows a difference in C* between the two larger and the two smaller support sizes. Since the material for all support sizes was the same for each thickness, the discrepancy can not be explained by variations in material properties. It must be pointed out that even though the plates were hit at the same velocity the strain rate was not constant. As the support size is increased or the the thickness decreased, the strain rate will decrease. A lower strain rate for the larger plates may be a part of the explanation of the response of the PEEK material. Further investigations must be conducted to understand the behavior.

The experimental values of initiation energy as a function of thickness are shown in Fig. 8. Here the often claimed superior "impact resistance" of the PEEK matrix composite shows up very clearly. The use of a tough ductile matrix will improve the impact performance of the composite by two mechanisms. A stronger matrix material will raise the ultimate shear stress level and a more ductile matrix will increase the contact area and thereby reduce the maximum shear stress.

A common way to present impact performance is to normalize the initiation or penetration energy with respect to the laminate thickness. Linear scaling is the most commonly used but in [5] an empirical relation, claiming both the initiation and penetration energies to be proportional to the thickness to the power of 1.42, was used. According to Equation (5), one should not try to scale the initiation force linearly with either the thickness, the shear strength, or the diameter of the indenter. The initiation energy depends on the stiffness of the plate. There are at least two different deformation mechanisms involved, the nonlinear contact indentation and the linear plate bending. The picture can be even more complicated if shear and membrane deformations are involved. In this light, any good one-parameter or linear fit of the initiation energy to the plate thickness must be purely coincidental.

The search for analytical tools can be divided into two areas, the detailed stress calculation during an impact and the simple method for design purposes. A detailed stress calculation of a general laminate during the impact event has been shown to be quite involved. Finite element studies have, however, given some useful information on potential failure mechanisms [1]; but is not practical to run these types of calculations in the early design stage. A designer needs models simpler to use. A model that bounds the problem may be sufficient.

The results indicate that a scaling of low-velocity impact data is possible. This opens the possibilities for standardizing a test for durability qualification. One problem remaining is setting of the requirements. For a user of a product it has been much easier to specify a certain impact energy. It is not very easy to get a feel for impact energies alone. Any requirement must be specified so that it includes the shape and material of the impactor, the velocity, and the support. A possible solution is to establish a standard test and write the requirements as minimum performance of the material in that test.

In the wider perspective, a thorough understanding of the deformation and damage mechanisms is needed. The key to development of more impact resistant materials is the manipulation of the failure mechanisms.

## 5. CONCLUSIONS

Instrumented impact testing makes the identification of the damage initiation threshold possible with just ONE test.

The controlling parameter is not the impact energy. For the low-velocity impacts the impact force is controlling the failure initiation.

For the plates tested the initiation force did not depend on the support size used, thus supporting the assumption that the failure is initiated by shear stresses.

Durability threshold values for composite plates can accurately be predicted from simple tests on small scale laboratory specimens.

## 6. ACKNOWLEDGMENTS

This work has been performed under Air Force contract F33615-84-C-5070, with Dr James Whitney as the contract monitor. I would also like to thank Mr Tobey Cordell for the many constructive discussions and Miss Margaret Hudock for her devoted work on the experimental investigation.

## 7. REFERENCES

1. Joshi, S. P. and Sun, C. T., "Impact Induced Fracture in a Laminated Composite", J. of Composite Materials, Vol. 19, January 1985, pp 41-66.

2. Yang, S. H. and Sun, C. T., "Indentation Law for Composite Laminates", ASTM STP 787, 1982, pp. 425-429.

3. Ireland, D. R., "Procedures and Problems Associated with Reliable Control of the Instrumented Impact Test", ASTM STP 563, 1974, pp. 3-29.

4. Winkel, J. D. and Adams, D. F., "Instrumented Drop Weight Testing of Cross-Ply and Fabric Composites", Composites, Vol. 16, October 1985, pp. 268-278.

5. Carlile, D. R. and Leach, D. C., "Damage and Notch Sensitivity of Graphite/PEEK Composite", 15th National SAMPE Technical Conference, Cincinnati, Ohio, October 4-6, 1983, pp. 82-93.

## 8. BIOGRAPHY

Dr Sjöblom recieved a MS in Applied Physics 1978 and a PhD in Engineering 1983 from Linköping Institute of Technology, Linköping, Sweden. He worked in the field of strength predictions of composite materials as a NRC associate for two years before joining University of Dayton Research Institute in 1985. He is presently working with structural and material testing.

# DESIGN FOR ELASTIC STABILITY OF CORRUGATED/SANDWICH/STIFFENED COMPOSITE PANELS

K.P. Rao
USAF Materials Laboratory
AFWAL/MLBM
Wright-Patterson AFB, Dayton, Ohio
(on leave from Indian Institute of Science, Bangalore, India)

## Abstract

Laminated fibre reinforced sandwich/ stiffened composite panels are increasingly being used in several engineering industries. For example, in aerospace industry, typical components include wing skins, corrugated spar webs or inter-stages in rockets. Stability studies are needed to achieve good structural performance with minimum weight.

The work presented in this paper deals with the prediction of elastic buckling loads (through a computer program) for simply supported sandwich/stiffened composite rectangular panels under Kirchhoff-Love assumptions. The given panel is idealised as a homogeneous orthotropic plate whose equivalent constitutive law is determined based on various parameters of the given panel. The panels considered have sinusoidal core or hat type core or a regular grid core. Critical buckling loads are obtained using conventional orthotropic plate theory for various loading cases. A large class of 0/90/45/-45 lamination schemes leading to quadri-directional, tri-directional and bi-directional laminates is examined and merit listed from the buckling point of view. Results indicate that more the number of 45 degree layers, the higher is the buckling load for $(a/b) \geq 1$ and for $(a/b) < 1$, the corresponding angle is zero.

**Key Words:** composite structures; stability; laminated sandwich/corrugated/ stiffened plates; computer software for instability prediction.

# 1. INTRODUCTION

Fibre reinforced composites are increasingly being used in many engineering industries. In aerospace industry, to achieve minimum weight, sandwich panels with either corrugated core or honeycomb core or stiffened panel construction are employed. Stability studies of composite sandwich panels or stiffened panels are necessary to achieve good structural performance using minimum weight.

Considerable amount of work has been carried out in the past on statics, dynamics and stability of corrugated/sandwich/ stiffened isotropic panels. Timoshenko and Gere[1] and Bruhn[2] dealt with the problem of stability of isotropic stiffened plate and isotropic corrugated sheet/honeycomb core sandwich plates respectively. Stability of homogeneous orthotropic plates with or without stiffeners under different loadings and support conditions have been presented by Lekhnitskii[3]. A comprehensive survey on buckling of laminated composite plates and shell panels has been presented by Leissa[4]. Brunelle and Oyibo[5] use a double affine transformation in solving the classical buckling equation for specially orthotropic plates. The solutions are presented in graphical form as well as in the form of simple equations. However, these results are applicable to a restricted class $(D^* < 1)$ of orthotropic panels. For a general laminate with laminae having different orientations, it can be shown that $D^*$ is not necessarily less than unity and thus the simple formulae given in reference (5) are not applicable to such cases. Van Zelst[6] presented a hand calculation method by making simplifying assumptions in the buckling equations of composite cylindrical shell structures. As particular cases, by taking the radius as infinite, he discusses the stability of simply supported plates, corrugated plates and sandwich plates. Libove and Hubka[7] derived the elastic constants for corrugated core sandwich isotropic plates. Using these elastic constants Seide[8] solved the problem of stability under compression of corrugated core isotropic sandwich plates for some boundary conditions. In reference (9) static behaviour of sinusoidally corrugated sheets where the wave length and amplitude are small in comparison with the dimensions of the plate has been discussed. The procedure adopted is to treat the corrugated sheet as a homogeneous orthotropic panel with equivalent bending and twisting rigidities. Lau[10] showed by evaluating the complete elliptic integral of the first kind accurately, that the bending stiffness of a sinusoidally corrugated sheet is significantly under estimated if one uses the formulae given in reference (9).

The work presented in this paper deals with the prediction of elastic buckling loads (through a computer program) for simply supported sandwich/stiffened composite rectangular panels under Kirchhoff- Love assumptions. We resort to finding the nearly equivalent homogeneous orthotropic properties of the given panel and use the available analytical methods to obtain the buckling loads. A large class of 0/90/45/-45 lay-up schemes are examined and merit listed from the buckling point of view.

## 2. CONFIGURATIONS STUDIED

The following configurations are studied.

i) a sandwich plate with a sinusoidally corrugated core (Fig. 1).

ii) a sandwich plate with a hat type corrugated core (Fig. 2).

iii) a sandwich plate with regular grid as core (Fig. 4).

In all cases, core and face sheets are of laminated composite construction using mainly T300/N5208 plies.

## 3. PROCEDURE USED

The equation governing the buckling behaviour of an orthotropic plate[5] is

$$D_{11}(\partial^4 w/\partial x^4)$$

$$+ 2(D_{12} + 2D_{66})(\partial^4 w/\partial x^2 \partial y^2)$$

$$+ D_{22}(\partial^4 w/\partial y^4) - N_x(\partial^2 w/\partial x^2)$$

$$- 2N_{xy}(\partial^2 w/\partial x \partial y) - N_y(\partial^2 w/\partial y^2)$$

$$= 0 \qquad (1)$$

and the strain energy stored is[5]

$$U = 1/2 \int_0^a \int_0^b \{D_{11}(\partial^2 w/\partial x^2)^2$$

$$+ 2D_{12} (\partial^2 w/\partial x^2)(\partial^2 w/\partial y^2)$$

$$+ 4D_{66} (\partial^2 w/\partial x \partial y)^2$$

$$+ D_{22} (\partial^2 w/\partial y^2)^2$$

$$+ N_x (\partial w/\partial x)^2$$

$$+ 2N_{xy} (\partial w/\partial x)(\partial w/\partial y)$$

$$+ N_y(\partial w/\partial y)^2\}dxdy \qquad (2)$$

The object is to find first $D_{11}$, $D_{12}$, $D_{22}$, $D_{66}$ (1,2 are the laminate axes) of the homogeneous orthotropic plate which is nearly equivalent to the given sandwich/ stiffened panel and then use the conventional procedures using equations (1) and (2) to predict the buckling loads.

## 4. EQUIVALENT ORTHOTROPIC PROPERTIES

4.1. Sinusoidally Corrugated Sheet

In Fig. 1 is shown a typical sinusoidal corrugation $(z = f \sin \pi y/L)$ and the notation used. Using reference (9) ,we can write with reference to the middle plane of the homogeneous orthotropic plate equivalent to the corrugation,

$$[D_{11}]^{cor} = E_{equivalent}.I,$$

$$[D_{12}]^{cor} = 0,$$

$$[D_{22}]^{cor} = L/S[D_{22}]^{lam},$$

$$[D_{66}]^{cor} = S/L[D_{66}]^{lam},$$

$$[A_{11}]^{cor} = S/L[A_{11}]^{lam},$$

$$[A_{22}]^{cor} = 0,$$

$[A_{12}]^{cor} = 0$ ,

$[A_{66}]^{cor} = S/L[A_{66}]^{lam}$ (3)

where [ ]$^{cor}$ refers to the property of the equivalent plate and [ ]$^{lam}$ refers to the laminate property of the corrugation. Lau[10] showed by evaluating complete elliptic integrals of the first and second kind, that

$S = (2L/\pi)\sqrt{(1+\mu^2)}.E(k,\pi/2)$

$I = [\{2hf^2.\sqrt{(1+\mu^2)}\}/(3\pi k^2)].$

$\qquad [(2k^2-1).E(k,\pi/2)$

$\qquad\qquad + (1-k^2)F(k,\pi/2)]$ (4)

where

$\mu = f\pi/L,\ \beta = \pi y/L,\ k = \mu^2/(1+\mu^2)$

$E(K,\pi/2) = \int_0^{\pi/2}\sqrt{(1 - k^2\sin^2\beta)}.d\beta$

$\qquad = (\pi/2)[1 - (k^2/4) - (3k^4/64)$

$\qquad - (45k^6/2304)............]$

$F(k,\pi/2) = \int_0^{\pi/2}d\beta/\sqrt{(1 - k^2\sin^2\beta)}$

$\qquad = (\pi/2)[1 + (k^2/4) + (9k^4/64)$

$\qquad + (225k^6/2304)...........]$

### 4.2. Hat Type Corrugated Sheet

Fig. 2 shows a typical hat type corrugated sheet and the notation used. We can write equivalent properties as

$R = [b_1+b_2] / [b_1+b_2\sin\emptyset]$,

$[A_{11}]^{hat} = [A_{11}]^{lam} .R$,

$[A_{12}]^{hat} = 0$,

$[A_{66}]^{hat} = [A_{66}]^{lam}.R$,

$[A_{22}]^{hat} = 0$,

$[D_{12}]^{hat} = 0$ .

$[D_{22}]^{hat} = [D_{22}]^{lam}/R$

$[D_{11}]^{hat} = E_{equivalent}.$

$\qquad [I_1+I_2] / [b_1+b_2\sin\emptyset]$

$[D_{66}]^{hat} = [D_{66}]^{lam}.R$ (5)

where $I_1 = b_2^2.\cos^2\emptyset.b_1h/4$ and

$I_2 = b_2^3.h.\cos^2\emptyset/12$.

### 4.3 Laminated Composite Grid

Fig. 4 shows a typical grid with 'X' and 'Y' being uniform spacings between stiffeners in the x and y directions. Two possible layup schemes exist for the grid. (a). the rib laminae are parallel to the grid middle plane(Fig. 5a) and (b) the rib laminae are perpendicular to the grid middle plane (Fig. 5b). For the case (a),using reference (9) we can write with reference to the grid middle plane,

$[D_{11}]^{grid} = [D_{11}]^{lam}w/Y$,

$[D_{22}]^{grid} = [D_{11}]^{lam}w/x$,

$[D_{12}]^{grid} = 0$,

$[D_{66}]^{grid} = 0$,

$[A_{11}]^{grid} = [A_{11}]^{lam}w/Y$,

$[A_{12}]^{grid} = 0$,

$[A_{22}]^{grid} = [A_{11}]^{lam}.w/X$,

$[A_{66}]^{grid} = 0$ (6)

For the case (b) we have (see appendix for details),

$[D_{11}]^{grid} = [A_{11}]^{lam} D^3/(12Y)$,

$[D_{12}]^{grid} = 0$

$[D_{66}]^{grid} = 0$,

$[D_{22}]^{grid} = [A_{11}]^{lam}D^3/(12X)$,

$[A_{11}]^{grid} = [A_{11}]^{lam}D/Y$,

$[A_{22}]^{grid} = [A_{11}]^{lam}D/X$,

$[A_{12}]^{grid} = [A_{66}]^{grid} = 0$ (7)

## 4.4 Laminated Composite Facing Sheets

The properties of the top and bottom laminated facing sheets are represented with reference to their resepective middle planes as

$$
\begin{vmatrix} N^i \\ M^i \end{vmatrix} = \begin{vmatrix} A^i & B^i \\ B^i & D^i \end{vmatrix} \begin{vmatrix} \varepsilon^i \\ \kappa^i \end{vmatrix} \tag{8}
$$

where $i$ represents top or bottom face.

## 4.5 Equivalent Properties Of The Sandwich/Stiffened Panel.

Figure 3 shows the middle planes of the facing sheets and core or grid and also the location of the neutral plane. In order to find $[A]$, $[B]$, $[D]$ of the equivalent homogeneous plate with reference to the neutral plane, we use reference (11). Thus,

$$[A] = [A]^{top} + [A]^{grid/core} + [A]^{bot},$$

$$[B] = [B]^{top} + [B]^{grid/core} + [B]^{bot}$$

$$- [A]^{bot} Ecc - (Ecc\text{-}f) [A]^{grid/core}$$

$$- (Ecc\text{-}2f) [A]^{top},$$

$$[D] = [D]^{top} + [D]^{grid/core} + [D]^{bot}$$

$$- 2[B]^{bot} Ecc$$

$$- 2(Ecc\text{-}f)[B]^{grid/core}$$

$$- 2(Ecc\text{-}2f) [B]^{top}$$

$$+ (Ecc)^2 [A]^{bot}$$

$$+ (Ecc\text{-}f)^2 [A]^{grid/core}$$

$$+ (Ecc\text{-}2f)^2 [A]^{top} \tag{9}$$

where $Ecc$ is yet to be determined. In order to determine we make $B_{11} = 0$.

$$Ecc = \{[B_{11}]^{top} + [B_{11}]^{grid/core}$$

$$+ [B_{11}]^{bot} + 2f [A_{11}]^{top}$$

$$+ f [A_{11}]^{grid/core}\}/$$

$$\{[A_{11}]^{top} + [A_{11}]^{grid/core}$$

$$+ [A_{11}]^{bot}\} \tag{10}$$

Using the value of $Ecc$ given by Eqn. (10), $[D]$ is computed using Eqn. (9).

## 5. RESULTS AND DISCUSSION

A computer programme has been developed for the prediction of buckling loads for various layups, loading conditions and geometric parameters. The flow chart adopted is shown in Fig. 6. The cases studied are shown in Fig. 7. The relevant formulae for homogeneous orthotropic plates are given in reference (3). We consider the examples in the following sequence.

a) Isotropic sandwich/stiffened panels,
b) CFRP sandwich panels for which experimental results are available,
c) Application to general composite sandwich plates.

### 5.1 Isotropic Sandwich Plate

The following parameters are chosen for the problem of a square plate made of Aluminium subjected to uniaxial compression

a = 10", h = 0.03937" (1mm),
E = $10^7$psi, v = 0.3,

i) sinusoidal core: f = 0.25",
L = 1", h = 0.03937" (1mm)

ii) hat type core: Pitch = 2",
D = 0.5", ø = 30°,
h = 0.03937" (1mm)

iii) grid core: X = 2", Y = 2",
D = 0.5", w = 0.03937" (1mm)

The isotropic face sheets and the core are treated in the computer program as laminated plates consisting of 8 isotropic layers. The results are shown in Table 1. The results agree with those predicted by references (3) and (5).

### 5.2. Comparison With Available Experimental Results

Pearce and Webber[12] conducted experiments on simply supported square CFRP sandwich plates with the following parameters.

a = 228mm, core depth = 5mm,

ply properties:

Thickness: 0.125mm,
$E_L$ = 142000N/mm²,
$E_T$ = 9800N/mm²
$G_{LT}$ = 4300N/mm²,
$\nu_{LT}$ = 0.34

Table 2 shows the comparison of results by the present code with the experimental and theoretical results obtained by Pearce and Webber. Orthotropic solutions as also solutions, modified to take into account the presence of bending-twisting coupling, are presented. Empirical formulae due to Fogg as given in reference (4) to take into account bending-twisting coupling in symmetrical laminates (with $D_{16}$ > 0) are

$[Pcrit]^{anisotropic}/[Pcrit]^{orthotropic} =$

$[1-5.585(D_{16}/DBAR)^{k1}]$    for m=1
$[1-9.588(D_{16}/DBAR)^{k2}]$    for m=2
$[1-9.766(D_{16}/DBAR)^{k3}]$    for m=3

where
   $k1$ = 1.995,
   $k2$ = 2.135,
   $k3$ = 2.117,
$DBAR = (D_{11}D_{22}{}^3)^{k4} + D_{12} + 2D_{66}$,
   $k4$ = 0.25

These formulae have been arrived at by examining a large number of graphite/epoxy symmetric balanced laminates. Though these formulae are strictly applicable to symmetric balanced laminates, we apply them to sandwich panels considered in reference (12) which are symmetric but not balanced (ie. $A_{16}$ and $A_{26}$ of the equivalent homogeneous plate are not zero). It can be seen that there is a fairly good agreement between the modified solutions and the experimental results. Theoretical predictions of reference (12) wherein the shear deformation effects are taken into account, are also presented in the table.

### 5.3. Application to General Composite Sandwich Panels

In this section are considered typical examples of sandwich/stiffened composite panels. In all cases the material system chosen is T300/N5208 with the following properties,

$E_L$ = 0.263x10⁸psi ,

$E_T$ = 0.149x10⁷psi ,
$G_{LT}$ = 0.104x10⁷psi ,
$\nu_{LT}$ = 0.28 ,
ply thickness = 0.00492" (0.125mm)

The computer code is so programmed that it can take into account different layup schemes, number of layers and materials for the top face, bottom face and the core or ribs individually. However, in the present exercise we assume the facing sheets and core to have the same layup schemes and thicknesses. In addition the lay up angles chosen are **0°, 90°, 45°, -45°**.
We examine

i) **35**, 8-ply, quadri-directional laminates starting with the code
   **[5111],[4211]....[1115]**
ii) **40**, 6-ply, tri-directional laminates starting with the code
   **[4110],[3210]..[0114]**
iii) **18**, 4-ply, bi-directional laminates starting with the code
   **[3100],[2200],...[0013]**
iv) **6**, 2-ply, bi-directional laminates starting with the code
   **[1100],....[0011]**

where the laminate code designates the number of plies in the order of 0°, 90°, **45°, -45°**. For example **[5111]** designates **[0₅/90/45/-45]**. There are altogether 99 laminates under this class. First we consider the sandwich plate (case 1 of Fig. 7) with a grid core with the following parameters.

a = 10",  b = 10",
h = 0.03937" (1mm),
X = 2",  Y = 2",
D = 1",  w = 0.03937" (1mm)

We rank order the 99 laminates based on the buckling loads and the results are presented in Table. 3. It can be seen that putting more ±45° layers compared to 0° layers leads to an improved performance. For example [1115] gives a critical load which is 24% higher than that corresponding to [5111]. Also it is seen that the result is unaffected by by the actual number of layers chosen of +45 and -45 orientation for a given total number of 45° layers (ie. [1133], [1124], [1115], [1151]). Compared to [2222] laminate (quasi-isotropic) [1115] has 11.5% more buckling load. In many aerospace structures plies are dropped in zones of lower loads (ex. wing skin). Tables like the one presented here will help assess any ply drop scheme from

Table 4 shows the effect of a/b ratio for case 1 of Fig. 7. Here we consider only 8-ply laminates. It is found, for a given 'a' (=10") if a/b $\geq$ 1, that having more $\pm$45 layers is more beneficial whereas if a/b < 1, having more 0° layers is more advantageous. For example [5111] leads to 27% (for a/b=0.5) and 69% (for a/b=0.25) higher buckling loads compared to [1115]. The corresponding values by which [5111] is found to be superior to quasi-isotropic case [2222] are 32% ( for a/b = 0.5) and 58% (for a/b = 0.25). For uniaxial compression, biaxial loading or linearly varying load, computations done for the 10"x10" CFRP panel show that having -45° layers is more beneficial. Maintaining all the parameters given in Section 5.1 for the cases of sinusoidal core and hat type core, except the material and if T300/N5208 is used,the highest and lowest buckling loads obtained for 8-ply, 6-ply, 4-ply and 2-ply cases are given in Table 5.

## 6. CONCLUDING REMARKS

A computer program has been developed to predict the elastic buckling loads for simply supported sandwich/stiffened composite rectangular panels under Kirchhoff-Love assumptions. The panels considered have sinusoidal core or hat type core or regular grid core and the loadings considered are uniaxial compression, biaxial loading and linearly varying compression. A large class of lamination schemes examined show that for a/b $\geq$1, having more 45° layers leads to increased buckling loads whereas for a/b<1, having more 0° layers is more advantageous. Some parts of the work presented in this paper are also presented in the book "Composites Design-1987" by Stephen W. Tsai.

## 7. ACKNOWLEDGEMENTS

The author wishes to thank the US National Research Council for providing the Research Associateship tenable at Airforce Materials Laboratory, WPAFB, Dayton and expresses his sincere thanks to Stephen W. Tsai, Chief, Mechanics and Surface Interactions Branch, AFWAL/MLBM, WPAFB for many helpful discussions and suggestions. Thanks are also due to Cpt. Craig Stice, AFWAL/MLSC, WPAFB for the help given in computer programming.

## 8. REFERENCES

[1]  Timoshenko, S.P., and Gere, M., Theory of Elastic Stability, McGraw-Hill Book Company, 1961, pp 404-408.
[2]  Bruhn, E.F., Analysis and Design of Flight Vehicle Structures, Tri-State Offset Company, Cincinnati, 1965, C12.1- C12.38.
[3]  Lekhnitskii, S.G., Anisotropic Plates Translated by S.W. Tsai and T. Cheron, Gordon and Breach Science Publishers, 1968,  pp 445-526.
[4]  Leissa, Arthur W., Buckling of Laminated Composite Plates and Shell Panels, AFWAL - TR - 85 - 3069, June 1985.
[5]  Brunelle, E.J.,  and Oyibo, G.A., Generic Buckling Curves for Specially Orthotropic Rectangular Plates, AIAA Jl, Vol.21, No.8, 1983, pp1150-1156.
[6]  Van Zelst, R.F.P., Hand Calculation Method for Buckling of Composite Shell Structures, Paper Presented at ESA Workshop on Composite Design for Space Applications, Oct 15-18, 1985.
[7]  Libove, Charles,  and Hubka, Ralph, Elastic Constants for Corrugated Core Sandwich Plates, NACA - TN - 2289, Feb 1951
[8]  Seide, Paul, The Stability Under Longitudinal Compression of Flat Symmetric Corrugated Core Sandwich Plates with Simply Supported Loaded Edges and Simply Supported or Clamped Unloaded Edges, NACA-2679, April 1952.
[9]  Timoshenko, S., and Woinowski - Krieger, S., Theory of Plates and Shells, McGraw-Hill  Book Company, 1959, pp 367- 368
[10]   Lau, John, Stiffness of Corrugated Plate, Jl. of Engng. Mech. Div, ASCE, Vol.107, No EM1, 1981, pp 271-275.
[11]  Tsai, S.W., Composites Design-1986 Think Composites, 1986, pp 9.6-9.7.
[12]  Pearce, T.R.A., and  Webber, J.P.H., Experimental Buckling Loads of Sandwich Panels with Carbon Fibre Face Plates, Aero. Quarterly, Vol24, No.4, 1973, pp 295-312.

## 9. APPENDIX

Let us consider the behaviour of the rib shown in Fig. 5b. Object is to find the constitutive law for the rib corresponding to bending about the x axis and stretching along the y axis.

$$\varepsilon_y = \varepsilon^{\circ}_y + z\kappa_y$$

where $\varepsilon^{\circ}_y$ is the mid plane (xy plane) strain and $\kappa_y$ is the midplane change in curvature.

$$\sigma_y = Q_{11}[\epsilon^\circ_y + z\kappa_y]$$

$$wN_y = \sum_{i=1}^{N} \int_{-D/2}^{D/2} \int_{x_{i-1}}^{x_i} \sigma_y dz dx$$

$$= A_{11} D \epsilon^\circ_y$$

$$wM_y = \sum_{i=1}^{N} \int_{-D/2}^{D/2} \int_{x_{i-1}}^{x_i} \sigma_y z dz dx$$

$$= A_{11} D^3 \kappa_y / 12$$

where $N$ is number of laminae, and $x_i - x_{i-1}$ is the thickness of the $i$ th layer.

## 10. BIOGRAPHY

Dr. K. Prabhakara Rao is presently a National Research Council Senior Research Associate, working at the Materials Laboratory, Wright-Patterson AFB, Dayton, Ohio on leave from Indian Institute of Science, Bangalore, India. He received Master's degree in Mathematics from Sri Venkateswara University and Master's degree in Aeronautical Engineering from Indian Institute of Science. Subsequently, he received his Ph.D degree from Imperial College of Science and Technology, London, U.K. He has been working in the areas of composites, finite element method and shell structures. He is a professor in the Department of Aerospace Engineering of Indian Institute of Science, Bangalore, India.

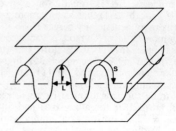

CORE DETAIL

FIG 1. A LAMINATED COMPOSITE SANDWICH PLATE WITH SINUSOIDALLY CORRUGATED CORE

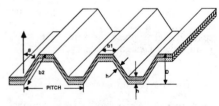

FIG 2 HAT TYPE CORRUGATION

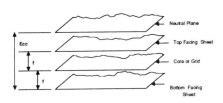

FIG 3. LOCATION OF THE NEUTRAL PLANE W.R.T. MID-PLANES
OF FACING SHEETS AND CORE

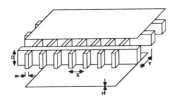

FIG 4    A LAMINATED COMPOSITE SANDWICH PLATE
WITH REGULAR GRID AS CORE

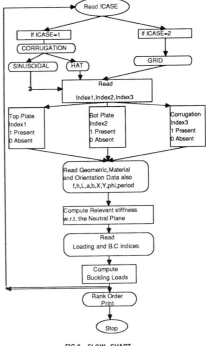

FIG 6  FLOW CHART

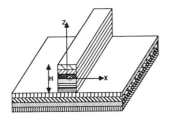

FIG 5(a). RIB WITH LAMINAE PARALLEL L TO THE GRID
MIDDLE PLANE

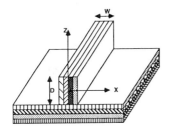

FIG 5(b). RIB WITH LAMINAE PERPEND ICULAR TO THE GRID
MIDDLE PLANE

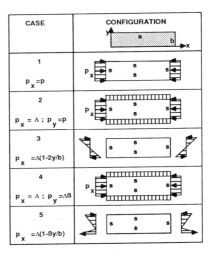

FIG 7. CASES STUDIED

547

| CONFIGURATION | $P_{CRITICAL}$ (lb/inch) | | |
|---|---|---|---|
| | SINUSOIDAL SHEET CORE | GRID CORE | HAT TYPE CORE |
| TOP FACE ONLY | $0.221*10^2$ | $0.221*10^2$ | $0.221*10^2$ |
| BOT FACE ONLY | $0.221*10^2$ | $0.221*10^2$ | $0.221*10^2$ |
| CORE ONLY | $0.132*10^4$ | $0.445*10^3$ | $0.221*10^4$ |
| TOP & BOT FACES | $0.214*10^5$ | $0.214*10^5$ | $0.214*10^5$ |
| SANDWICH PLATE | $0.227*10^5$ | $0.218*10^5$ | $0.236*10^5$ |

TABLE.1.BUCKLING LOADS FOR THE ISO TROPIC
SANDWICH PANEL UNDER UNIAXIAL COMP RESSION

| SANDWICH PANEL NUMBER | 2 B | 3 | 4 | 5 |
|---|---|---|---|---|
| LAYUP SCHEME | 3 PLY 0/90/0 | 3 PLY 45/0/45 | 4 PLY 45/0/0/45 | 2 PLY 0/90 |
| EXPERIMENTAL OVERALL BUCKL LOAD (N/mm) | 185 | 161 | 262 | |
| EXPERIMENTAL FAILURE LOAD (N/mm) | 213 | 188 | 283 | 117 |
| THEORETICAL PREDICTION (REF 12 ) (N/mm) | 152 | 199 | 238 | 105 |
| PRESENT SOLUTION * (ORTHOTROPIC) | 158 | 235 | 288 | 105 |
| MODIFIED SOLUTION (ANISOTROPIC) | 158 | 187 | 240 | 105 |

* SOLUTION OBTAINED ASSUMING ZERO STIFFNESS CORE

TABLE2. COMPARISON WITH EXPERIMENT AL RESULTS

| NO | CODE | P crit | | NO | CODE | P crit | | NO | CODE | P crit | |
|----|------|--------|-----|----|------|--------|-----|----|------|--------|-----|
| 1 | 1133 | 0.101 | D06 | 34 | 2411 | 0.787 | D05 | 67 | 4110 | 0.584 | D05 |
| 2 | 1124 | 0.101 | D06 | 35 | 1014 | 0.787 | D05 | 68 | 4101 | 0.584 | D05 |
| 3 | 1115 | 0.101 | D06 | 36 | 1032 | 0.787 | D05 | 69 | 3210 | 0.574 | D05 |
| 4 | 1151 | 0.101 | D06 | 37 | 1023 | 0.787 | D05 | 70 | 3201 | 0.574 | D05 |
| 5 | 1142 | 0.101 | D06 | 38 | 1041 | 0.787 | D05 | 71 | 2310 | 0.564 | D05 |
| 6 | 2132 | 0.964 | D05 | 39 | 1511 | 0.777 | D05 | 72 | 2301 | 0.564 | D05 |
| 7 | 2123 | 0.964 | D05 | 40 | 132 | 0.777 | D05 | 73 | 411 | 0.562 | D05 |
| 8 | 2114 | 0.964 | D05 | 41 | 123 | 0.777 | D05 | 74 | 22 | 0.555 | D05 |
| 9 | 2141 | 0.964 | D05 | 42 | 114 | 0.777 | D05 | 75 | 13 | 0.555 | D05 |
| 10 | 1232 | 0.954 | D05 | 43 | 141 | 0.777 | D05 | 76 | 31 | 0.555 | D05 |
| 11 | 1223 | 0.954 | D05 | 44 | 2022 | 0.741 | D05 | 77 | 1410 | 0.554 | D05 |
| 12 | 1214 | 0.954 | D05 | 45 | 2031 | 0.741 | D05 | 78 | 1401 | 0.554 | D05 |
| 13 | 1241 | 0.954 | D05 | 46 | 2013 | 0.741 | D05 | 79 | 1030 | 0.509 | D05 |
| 14 | 3131 | 0.916 | D05 | 47 | 1140 | 0.731 | D05 | 80 | 1003 | 0.509 | D05 |
| 15 | 3122 | 0.916 | D05 | 48 | 1104 | 0.731 | D05 | 81 | 103 | 0.498 | D05 |
| 16 | 3113 | 0.916 | D05 | 49 | 231 | 0.721 | D05 | 82 | 130 | 0.498 | D05 |
| 17 | 2231 | 0.906 | D05 | 50 | 213 | 0.721 | D05 | 83 | 2002 | 0.461 | D05 |
| 18 | 2222 | 0.906 | D05 | 51 | 222 | 0.721 | D05 | 84 | 2020 | 0.461 | D05 |
| 19 | 2213 | 0.906 | D05 | 52 | 3012 | 0.693 | D05 | 85 | 202 | 0.441 | D05 |
| 20 | 1322 | 0.896 | D05 | 53 | 3021 | 0.693 | D05 | 86 | 220 | 0.441 | D05 |
| 21 | 1313 | 0.896 | D05 | 54 | 2130 | 0.683 | D05 | 87 | 3010 | 0.412 | D05 |
| 22 | 1331 | 0.896 | D05 | 55 | 2103 | 0.683 | D05 | 88 | 3001 | 0.412 | D05 |
| 23 | 4112 | 0.868 | D05 | 56 | 1230 | 0.673 | D05 | 89 | 3100 | 0.352 | D05 |
| 24 | 4121 | 0.868 | D05 | 57 | 1203 | 0.673 | D05 | 90 | 2200 | 0.342 | D05 |
| 25 | 3212 | 0.857 | D05 | 58 | 321 | 0.663 | D05 | 91 | 1300 | 0.332 | D05 |
| 26 | 3221 | 0.857 | D05 | 59 | 312 | 0.663 | D05 | 92 | 301 | 0.323 | D05 |
| 27 | 2312 | 0.847 | D05 | 60 | 4011 | 0.644 | D05 | 93 | 310 | 0.323 | D05 |
| 28 | 2321 | 0.847 | D05 | 61 | 3120 | 0.634 | D05 | 94 | 11 | 0.278 | D05 |
| 29 | 1412 | 0.837 | D05 | 62 | 3102 | 0.634 | D05 | 95 | 1001 | 0.230 | D05 |
| 30 | 1421 | 0.837 | D05 | 63 | 2202 | 0.624 | D05 | 96 | 1010 | 0.230 | D05 |
| 31 | 5111 | 0.818 | D05 | 64 | 2220 | 0.624 | D05 | 97 | 101 | 0.220 | D05 |
| 32 | 4211 | 0.808 | D05 | 65 | 1302 | 0.614 | D05 | 98 | 110 | 0.220 | D05 |
| 33 | 3311 | 0.798 | D05 | 66 | 1320 | 0.614 | D05 | 99 | 1100 | 0.171 | D05 |

TABLE3. CRITICAL LOADS FOR 8,6,4,2   PLY SCHEMES

| SIZE | NO | CODE | P crit | NO | CODE | P crit |
|------|----|------|--------|----|------|--------|
| a=10 | 1 | 5111 | 0.405 D05 | 31 | 1313 | 0.195 D05 |
| | 2 | 4121 | 0.363 D05 | 32 | 1331 | 0.195 D05 |
| | 3 | 4112 | 0.363 D05 | 33 | 1412 | 0.172 D05 |
| b=40 | 4 | 4211 | 0.339 D05 | 34 | 1421 | 0.172 D05 |
| | 5 | 3131 | 0.321 D05 | 35 | 1511 | 0.149 D05 |
| a=10 | 1 | 5111 | 0.466 D05 | 31 | 1313 | 0.295 D05 |
| | 2 | 4121 | 0.440 D05 | 32 | 2411 | 0.281 D05 |
| | 3 | 4112 | 0.440 D05 | 33 | 1412 | 0.259 D05 |
| b=20 | 4 | 3122 | 0.415 D05 | 34 | 1421 | 0.259 D05 |
| | 5 | 3131 | 0.415 D05 | 35 | 1511 | 0.222 D05 |
| a=10 | 1 | 1133 | 0.404 D06 | 31 | 5111 | 0.327 D06 |
| | 2 | 1124 | 0.404 D06 | 32 | 4211 | 0.323 D06 |
| | 3 | 1115 | 0.404 D06 | 33 | 3311 | 0.319 D06 |
| b=5 | 4 | 1151 | 0.404 D06 | 34 | 2411 | 0.315 D06 |
| | 5 | 1142 | 0.404 D06 | 35 | 1511 | 0.293 D06 |
| | 1 | 1133 | 0.101 D06 | 31 | 5111 | 0.818 D05 |
| | 2 | 1124 | 0.101 D06 | 32 | 4211 | 0.808 D05 |
| a=10 | 3 | 1115 | 0.101 D06 | 33 | 3311 | 0.798 D05 |
| | 4 | 1151 | 0.101 D06 | 34 | 2411 | 0.787 D05 |
| b=10 | 5 | 1142 | 0.101 D06 | 35 | 1511 | 0.777 D05 |

**TABLE 4. BUCKLING LOADS FOR THE CF RP SANDWICH PLATES**

**SIZE EFFECT    ( CASE 1 - FIG.1 )**

| NUMBER OF PLIES | | SINUSOIDAL CORRUGATION | | HAT TYPE CORRUGATION | |
|-----------------|---------|------|------------|------|------------|
| | | CODE | $P_{crit}$ | CODE | $P_{crit}$ |
| 8 | Highest | 1142 | 0.101 D06 | 1142 | 0.101 D06 |
| | Lowest | 1511 | 0.785 D05 | 1511 | 0.794 D05 |
| 6 | Highest | 1041 | 0.780 D05 | 1041 | 0.787 D05 |
| | Lowest | 1401 | 0.551 D05 | 1401 | 0.555 D05 |
| 4 | Highest | 0022 | 0.551 D05 | 0022 | 0.554 D05 |
| | Lowest | 1300 | 0.327 D05 | 1300 | 0.329 D05 |
| 2 | Highest | 0011 | 0.275 D05 | 0011 | 0.277 D05 |
| | Lowest | 1100 | 0.168 D05 | 1100 | 0.171 D05 |

**TABLE 5.   CFRP SANDWICH PLATE WITH    SINUSOIDAL OR HAT**

**TYPE CORRUGATED SHEET AS CORE**

# A Computer Aided Aircraft Structural Composite Repair System

Forrest Sandow
Air Force Wright Aeronautical Laboratories
Wright-Patterson Air Force Base

## ABSTRACT

With the increased use of advanced composites for structural applications in new military aircraft, increased repair challenges are being faced by maintenance personnel. Portable personal computers can aid in solving many of the design problems that are being faced. The purpose of this development project is to develop a computer aided design system for analysis of damage, design of repairs and establishment of repair procedures.

The first phase of this project involves the development of computer programs which will analyze the damaged area of a composite structure. Using complex potentials, quadratic failure criterion, fracture based on point stress, and other theories, interactive computer programs are being developed to give stresses, strains and strengths around a damage area. Phase two of the project extends the theory to an analysis of potential repair methods for the damaged part. Analysis is being performed on plug repairs, bolted repairs, bonded repairs and any combinations of these concepts. This phase of the project also includes the fabrication, processing and curing of repairs. Consideration will be given to moisture content of damaged composite structures, available fabrication equipment and available repair materials. The final phase of the program involves establishing, packaging, and demonstrating the use of a small, portable, highly interactive personal computer for an information base for the repair personnel. This approach will result in a system which is easy to use, update and extend to other applications. The phase includes the integration of the damage assessment techniques, the repair analysis techniques and existing technical order information into an expert system to aid the repair personnel at the depot, field or Aircraft Battle Damage Repair level. With the simple input of the damage location, extent and available equipment, the computer will design a repair, recommend a fabrication procedure and define any limits on aircraft performance.

## Background on ABDR

In modern warfare, with missiles able to strike anywhere in the world in less than thirty minutes, and with the destructive power of these weapons, it becomes apparent that a war may easily be won within the first few days. Typical limited conflicts today are of very short duration, like the Arab-Israeli 6 day war. In this type of conflict, the war may be won or lost in the first few hours or days. For this reason, it is imperative to have as much usable equipment provided to the active troops in as short of a time as possible. In modern combat, aircraft are one of the most important weapons. The purpose of the Aircraft Battle Damage Repair (ABDR) concept is to determine how much damage had happened to the aircraft, whether it can be flown with only cosmetic repairs, whether it can be fixed for the next mission within 24 hours, whether it can be fixed at all, or whether it should be fixed just enough to send it to a rear area for complete repair. The goal is to have as many aircraft as possible available for the next mission. Any non-mission essential equipment may be left unrepaired or disabled. Also in the past, there has been the need to repair more aircraft than the manpower and resources could handle. This need has been met by the use of innovative repairs and substitutions. This concept is diametrically opposed to the normal maintenance operations of peacetime.

One of the problems faced in the assessment of damage and the repair of modern aircraft is the need to have repair personnel trained in the many new advanced materials used to obtain the high performance characteristics of the aircraft. Although much work has been done in the ABDR area and Technical Orders (T.O.) have been developed which address the

general techniques and information, there exists a need for simple and effective aids for the combat assessor who will be overloaded with work in the combat situation. All assessor candidates are trained first as technicians. Ideally, the assessors would be engineers, but with the current shortage of engineers in the workforce, there will likely be a shortage of trained people for this job. Also, with the many different types of damage, both structural, electrical, weapons and other systems, it is unlikely an engineer would have training in all the types of damage he would be have to assess. The TO guidelines are aids to the assessor but at present are large, difficult to use and lack flexibility for unique situations. As an example the F-4 -39 TO sections for structural repair consist of 278 pages covering both assessment and repair. The use of the bulky material is slow and confusing and can easily lead to errors in a combat situation. Some of this problem is presently being solved by making the information available to the assessor in a portable computer system.

In order to keep TO's to a reasonable size, areas of the aircraft are divided into smaller sections and a maximum stress is assumed throughout these sections. Sectioning of the F-16 horizontal is shown in Figure 1. This is a conservative approach. The repair must be designed to handle the maximum stress in the area. Damages larger than acceptable by TO guidelines may be repairable if the stress at an exact damage location are lower than the assumed for the larger section outlined. Also, the TO's do not have provisions to handle substitutions where the required repair materials or the

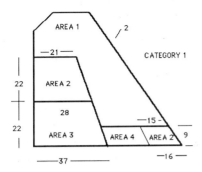

*Figure 1. Damage Sectioning of F-16 Horizontal Stabilizer.*

required repair fabrication equipment are not present. Ideally, the assessor would have the judgement and engineering experience to make many of these decisions, but as stated before, this may not be possible

The purpose of the work covered in the paper is to provide the assessor with a structural design consultant in the form of a computerized expert system to aid in his work. This system would not require a detailed knowledge of the analysis methods on the part of the user, but would work transparent to him. The code for this system could be integrated into the existing portable computer now being developed to supplement its operation.

**Composite Expert System**

The composite repair expert system would be broken down into four major blocks; a set of code to assess the damage, a method to analyze repair methods, an expert method to evaluate possible repairs and propose the best approach, and an output of the final repair design with documentation for future reference. The

overall approach to this problem and much of the code developed will have application to the repair of conventional metal aircraft components. Only the detailed analysis techniques used for the composite structures will be unique.

**A. Assessing the damage.**

The only requirement for the combat damage assessor would be to input the location and size of the damage to the structure. Evaluation techniques to analyze the effect of the damage and information on the parts construction and load requirements would be internal to the computer system.

The computers first step in analyzing the damage is to form a complete internal description of the part in the location of the damage and its loading requirements. Figure 2 shows the ply dropoffs of the F-16 horizontal for which the TO sectioning of the wing is shown in Figure 1. This shows a much more complex structure, with layups varying from 6 plies at the edge to 52 plies at the root. These variations can easily be stored and retrieved from a database in the computer system once the location of the damage is input to the computer. Using a similar method, the computer can store the strength and stiffness requirements of the laminate in various locations.

The next is for the the computer to do an analysis of the damage. The methods used would be internal and not require any knowledge of structural analysis or composite material design from the user of the system. Various approaches to this problem are possible. It is possible to do a finite element model of

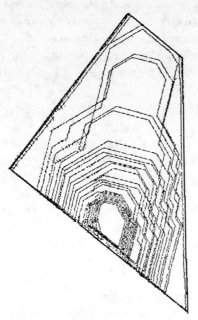

*Figure 2. Stepping of Plies on F16 horizontal Stabilizer*

shown in Figure 3, the elliptical hole may have any major (b) or minor (a) axis and be at any arbitrary angle ($\psi$) to the axis of the laminate. The code performs a ply-by-ply analysis. The code will give stress and displacements both at the hole and away from the hole. It also uses quadratic failure criteria to predict the remaining strength of the plies. This information can be used to see if any plies will fail due to the loading caused by the hole. If no failures are indicated the assessor will be informed that no structural repair beyond cosmetic or aerodynamic smoothing is required and no further action is required in a combat situation.

the part and the damage for a detailed analysis of the problem. But this approach requires a large amount of computer memory, which may not be available in a portable computer and a large amount of computer time. This approach to an analysis scheme would be useful in a peacetime environment but not where time and portable standalone equipment is required. For this reason, a point stress approach using complex potentials (Reference 1and 2) is being used. This work is being performed by Dr. Seng Tan in the Air Force Materials Laboratory. Using the complex potential approach for plates developed by Lekhnitskii (Reference 3), Dr. Tan has developed a computer program to derive the stress field around an arbitrary elliptical hole in a composite structure due to any given stress field. As

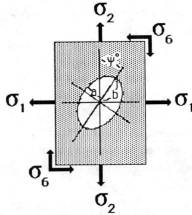

*Figure 3. Damage analysis*

**B. Repair Analysis**

Assuming that the damage assessment portion of the expert system decides that the structure requires repair, analysis will be turned over to a portion of the expert system which will be designed to design and analyze concepts. The analysis of the repair and decision of the type of repair is a complex problem. Consideration must be given to available equipment and personnel, time requirements, and mission requirements. As shown in Figure 4, there are many different approaches to the repair of composite damage. Repairs may also be any combination of these repair methods. At the present time, bolted repairs are in the most common use for battle damage situations because they are more reliable. Existing bonded repair methods require very high control of surface treatments, storage of temperature sensitive bonding materials and special equipment for processing of the bonded repair. But as materials technology advances, these may not be problems in the future, so the system which is being planned will have provisions for analysis of these types of repairs.

For its initial step, the expert system will be programed to design an optimum patch based on the expected materials on hand and with the purpose to restore full

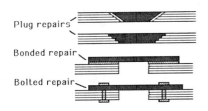

Plug repairs

Bonded repair

Bolted repair

*Figure 4. Repair Methods*

operational capability to the aircraft. The analysis methods presently being implemented for the system will include the use of the complex potentials method mentioned previously, the bolted repair analysis program (BREPAIR) which has

been developed by The McDonnell Aircraft Company under contract to the U. S. Government (Reference 4), and basic T.O. information which has been generated. If it is possible to design such a repair, an output will then be presented to the assessor, with a listing of required materials, personnel, time and processing equipment. This will then be implemented if possible. If it is not possible to design a fully operational repair, the system will inform the operator of a limited performance repair, with it limitations on the aircraft, or will inform the assessor that the damage cannot be fixed and the part should be scrapped. If any of the repair fabrication requirements cannot be met, the assessor will have the option of telling the computer to repeat its analysis, but to only use limited materials, time or personnel. The expert system will then be required to produce its analysis with the limits the user put on it. This will result in the design of a different repair which may, or may not restrict aircraft capability, but which will be able to be fabricated with existing equipment. The expert system would also be used to generate any required documentation of the repair required for future use, and relieve the assessor of this routine requirement, giving him time for critical needs.

C. Hardware Development.

Since the combat battle damage assessor will be working in forward battle damage areas, the environment will be harsh. The typical assessor may be required to work in areas subjected to both chemical and biological agents. It may be necessary for the assessor to preform his duties while wearing chemical defense ensemble (CDE). Due to these conditions, the type of computer equipment and the operator interface are extremely limited. To solve this problem, the Human Resources Laboratory at Wright Patterson Air Force Base is developing a Portable Computer Aided maintenance System (PCMAS). This is a small, graphics oriented, hardened computer system with a series of function keys as the user interface. The design of the composite repair expert will be menu driven through a series of function keys in order to be compatible with this system. This system is based on a 68010 microporcessor chip for its ability to handle high quality graphics

Present development of the composite repair expert system is being performed on various systems. The analysis code which is being developed by Dr. Tan is being written in Fortran 77 on a Prime computer. The BREPAIR analysis code has also been developed in Fortran 77. An initial demonstration of the artificial intelligence interface to the design tools has been performed by Blue Chip Computer under contract to The Materials Laboratory. Work is being done to integrate these various design tools into a demonstration system on the Apple Macintosh computer. This is a 68000 based microprocessor with similar graphics and coding to the PCMAS computer. If this can be shown to be feasible, the code will be ported over to the PCMAS computer system.

D. Demonstration System.

In order to demonstrate the usefulness of this approach it was decided to implement the complete approach for a simple, but representative aircraft component. The F-16 horizontal stabilizer was chosen as the part for this demonstration. As shown in Figure 5, the part includes a composite outer skin and an aluminum corrugated substructure. This part has an established ABDR T.O. for comparison to the analysis methods to evaluate the effectiveness of the detailed design method over the general established guidelines.

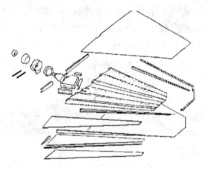

*Figure 5. F-16 Horizontial Stabilizer*

**Acknowledgements**

The author would like to thank Dale Nelson of the Aeronautical Systems Division who performed much of the initial work in conceptualizing an expert repair system while in the Flight Dynamics Laboratory and from whom many of the ideas in this paper were taken. The author would also like to thank Dr. Steve Tsai and Dr. Sung Tan who are providing the expert knowledge base in the analysis of composite structures.

## References

1. Tsai, Stephen, *Composite Design 1986*, Chapter 20, pgs. 20-1 through 20-12, 1986.

2. Tan, S. C., 'Notched Strength Prediction and Design of Laminated Composites Under In-Plane Loadings,' to appear in the *Journal of Reinforced Plastics and Composites*.

3. Likhnitskii, S. G., *Anisotropic Plates*, 2nd edition. Translated from Russian by S. W. Tsai and T. Cheron, Gordon and Breach, (1968).

4. Wilmarth, D. D., 'BREPAIR, Bolted Repair Analysis Program, Final Documentation,' Arthur D. Little Inc., April 1982.

32nd International SAMPE Symposium
April 6-9, 1987

SUPPRESSION OF INTERLAMINAR STRESSES OF THICK COMPOSITE LAMINATES
USING SUBLAMINATE APPROACH

C. S. Hong

Department of Mechanical Engineering

Korea Advanced Institute of Science and Technology

P. O. Box 150, Chongyang

Seoul, Korea

## ABSTRACT

Thick composite laminates are now of considerable interest as the application of composite is considered for primary structures. The design of thick laminated structure needs to determine stacking sequence as well as ply orientation and thickness. Thick laminate considered in this paper is not a clustered configuration but a repetition configuration of sublaminates. The sublaminate approach is introduced to investigate the reduction of interlaminar stresses with respect to the thickness of laminate, ply orientations, and stacking sequence. This approach is simple and economic to determine the stacking sequence for the thick laminate construction. The suppression of interlaminar stresses in thick laminate is presented for several cases of sublaminates to prevent the free-edge delamination.

## 1. INTRODUCTION

In recent years, considerable attention has been paid to the problem of thick composite laminates as the application of composite is considered for the primary structures. Experimental results on the interlaminar failure of thin and thick laminates are different according to the literature[1]. The question may be raised which laminate is more prone to the free-edge delamination than the other. It is well known that free-edge delamination is mainly attributed to the existence of interlaminar stresses which are dependent on

558

the stacking sequence of multi-layered laminates. The design of thick laminated structure needs to determine stacking sequenbe as well as ply orientation and thickness. It is important to use the proper stacking sequence of thick composite laminate with other design parameters. A great amount of analytic work [2-6] has been reported on the interlaminar stress distributions in thin laminates (e.g., 4-and 8-ply laminates). Most practical laminates, however, consists of considerably thicker than eight plies. It is impossible or difficult to calculate interbaminar stresses in thick laminates (e.g., 100-ply laminate) using numerical methods such as the finite element method for thin laminates. Some techniques have been reported to prevent the free-edge delamination by wrapping the laminate[7-8]. This method requires extra work and attachments like a reinforcing cap. The best way to prevent the free-edge delamination is to determine the optimum stacking sequence of thick laminates. In this study, it is not intended to accurately calculate interlaminar stresses but rank the thick laminates by selecting the stacking sequence. Since thick laminates can be constructed by repetition of sublaminate, thick laminate considered in this study is not a clustered configuration but a repetition configuration of sublaminates. This sublaminate approach is presented to determine the stacking sequence of sublaminate for thick laminate construction.

## 2. INTERLAMINAR NORMAL STRESS OF SUBLAMINATE

It is well known that interlaminar normal stress develops at the free edges and delamination fronts in composite laminates loaded in-plane due to the different mechanical and thermal properties of each anisotropic plies. Since the primary interest of this study is to rank the thick laminates based on the magnitude and sign of interlaminar stresses, the moment equilibrium approach is utilized with the coupling properties of unsymmetric laminates. Thick laminates can be constructed by repetition of sublaminate as shown in Fig. 1. The sublaminate is, therefore, considered to reduce free-edge interlaminar stresses by selecting the proper stacking sequence. A symmetric laminate loaded in-plane tension induces interlaminar stresses at the free edge boundary. By using classical laminate theory, moments in unsymmtric laminate can be written asw following;

$$\{M\}=[b]\;\{e\}+[D]\;\{k\}-\{M^t\} \quad (1)$$

where [B] is coupling stiffness, [D] is bending stiffness, $M^t$ is thermal moments in the laminate, e and k are strains and curvatures, respecively. One half of the symmetric laminate is taken with constrained boundary conditions, all components of curvature k=0, as shown in Fig. 2.

559

$$\begin{Bmatrix} M_1 \\ M_2 \\ M_{12} \end{Bmatrix} = \begin{bmatrix} B_{11} & B_{12} & B_{16} \\ B_{12} & B_{22} & B_{26} \\ B_{16} & B_{26} & B_{66} \end{bmatrix} \begin{Bmatrix} \varepsilon_1 \\ \varepsilon_2 \\ \varepsilon_{12} \end{Bmatrix} - \begin{Bmatrix} M_1^t \\ M_2^t \\ M_{12}^t \end{Bmatrix} \quad (2)$$

Equation (2) indicates that this unsymmetric laminate induces moments due to coupling components, $B_{ij}$. The induced moment can be ranked by changing stacking sequence of the sublaminate, since interlaminar normal stress has the following relationship with moment as following;

$$\sigma_z = g \, M_2(z)/h^2$$

where g is the scale factor of stress distribution, $M_2(z)$ is induced moment and h is ply thinkness. In Ref. [4], g was approximated as 3.2143 and we could use other values for other laminates. However, it is not intended to calculate accurate stresses but rank the stacking sequence of the laminate in this study. Moment $M_2$ in Equation (2) is related to the moment in Equation (3) at the midplane of the laminate. Therefore, the magnitude and the sign of moments in the laminate is taken into account to evaluate interlaminar normal stress.

It is very simple and easy to determine the best stacking sequence among the sublaminate groups to reduce interlaminar stress. As thick composite laminates are constructed by repitition of the sublaminate, coupling stiffness becomes small at the rate of 1/n, the number of repitition. Therefore, interlaminar stresses are substantially reduced to prevent the free-edge delamination of the thick composite laminate.

## 3. RESULTS AND DISCUSSION

Free edge interlaminar stresses should be suppresed to prevent free edge delamination of thick composite laminate without any mechanical attachments if possible. In this study, the economic and simple way is presented to select the proper stacking sequence of thick composite laminate to prevent free edge delamination.

As an example, quasi-isotropic sublaminate of graphite/epoxy was considered. There are twenty four possible combinations of plies in this laminate and onle twelve are distinct because of the interchangeability of the +45 and 45 plies. Among them only six cases are shown in Fig. 3. Various stacking sequences of these groups show distinct interlaminar normal stresses under uniaxial tension. Since this laminate is under in plane tensile loading, the worst laminate is [+ 45/0/90]s laminate. In case of compressive loading, the rank will be different and obviousely, [90/0/+ 45]s laminate is the worst one. For the concern of any possible loading conditions, [0/-45/90/+45]s and/or [+45/90/0/-45]s laminates are desirable. However,

560

thermal effect due to curing temperature should be taken into account for the analysis. Fig. 4 shows the effect of curing residual stress with mechanical loading (e= 0.17%). The discrepancy of these curves will be different from the ratio of mechanical loading and thermal loading. In Fig. 5, interlaminar normal stress through the thickness direction is given with the increased number of sublaminate group. It is clearly seen that the repitition of sublaminate decreases the interlaminar stresses.

## 4. CONCLUSION

The sublaminate approach is introduced to construct a thick laminate by repeating sublaminate groups which shows the optimum stacking sequence to prevent the free-edge delamination. The sublaminate is imposed the condition $k_i = 0$ along the upper and lower surfaces of a repeating sublaminate which induces moment resultants due to coupling stiffness. It is fast and economic to select the optimum stacking sequence in a sublaminate by evaluating the induced moment, since interlaminar stresses are influenced by the induced moment. Thick compsite laminate constructed by repeating the sublaminate groups with the appropriate stacking sequence show that interlaminar stresses decrease appreciably to prevent the free-edge delamination.

## 5. ACKNOWLEDGEMENT

This study was conducted at Air Force Wright Aeronautical Laboratories where the author was a visiting scientist in 1986. The author wishes to thank Dr. S. W. Tasi of AFML for his helpful discussion.

## 6. REFERENCES

1. Graber, D. P., "Tensile Stress-Strain Behavior of Graphite/Epoxy Laminates," NASA CR 3592, NASA, Washington, DC. Aug. 1982.

2. Pipes, R. B. and Pagano, N. J., "Interlaminar Stresses in Composite Laminates Under Uniform Axial Extension," J. Composite Materials, Vol.4, pp. 538-548.

3. Wang, A. S. D. and Crossman, F. W., "Some New Results on Edge Effect in Symmetric Laminates," J. Composite Materials, Vol.11, 1977, pp. 92-106.

4. Pagano, N. J. and Pipes, R. B., "Some Observations on the Intetrlaminar Strength of Composite Laminates," Int. J. Mechanical Science, Vol.15, 1973, pp. 679-688.

5. Whitcomb, J. D. and Raju, I. S., "Analysis of Interlaminar Stresses in Thick Composite Laminates With and Without Edge Delamination," Delamination and Debonding of Materials, ASTM STP 876, 1985, pp. 69-94.

6. Salamon, N. J., "An Assemssment of the Interlaminar Stress Problem in Laminated Composites," J. Compsite Materials, Vol. 14, 1980, pp. 177-194.

7. Kim, R. Y., "A Technique for Prevention of Delamination," Mechanics of Composite Review, 1981, Dayton, Ohio.

8. Heyliger, P. R. and Reddy, J. N., "Reduction of Free-Edge Stress Concentration," J. Applied Mechanics, Vol. 52, 1985, pp. 801-805.

## 7.  BIOGRAPHY

Dr. C.S. Hong is Associate Professor of Aeronautical Engineering at Korea Advanced Institute of Science and Technology, Seoul, Korea.  He received his B.S. (1967) M.S. (1971) degrees from Yonsei University and Ph. D. degree (1977) from Pennsylvania State University. He was an NRC-NASA research associate at Langley Research Center during 1977-1979.  He has held his position at KAIST since 1979 and has taught mechanics of composite meterials and conducted research on composite materials, especilly their mechanical behavior, fracture, and stress analysis of composite structures.  Recently, he was a visiting professor at Washington University in St. Louis and a visiting scientist at Air Force Wright Aeronautical Laboratories in Wright-Patterson Base.

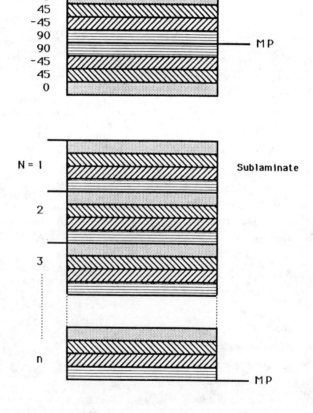

Fig.1   Thick laminate construction by sublaminates.

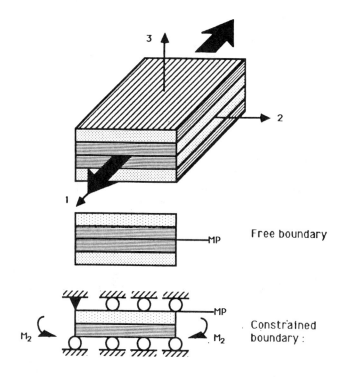

Fig. 2   Symmetric and unsymmetric laminate under in-plane tension
with free and constrained boundary conditions.

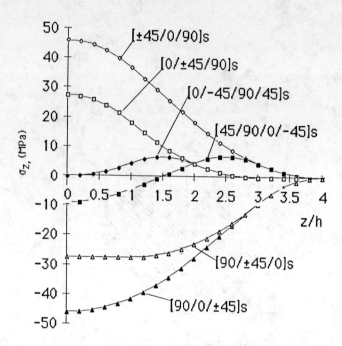

Fig.3 Interlaminar normal stress distributions of quasi-isotropic
laminates under mechanical loading (ε=0.17%).

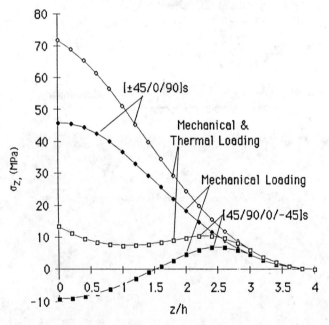

Fig.4 Interlaminar normal stress distributions under mechanical and
thermal loading (100 C).

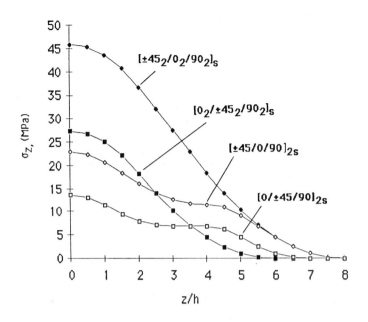

Fig. 5   Interlaminar normal stress distributions with number of
         sublaminates.

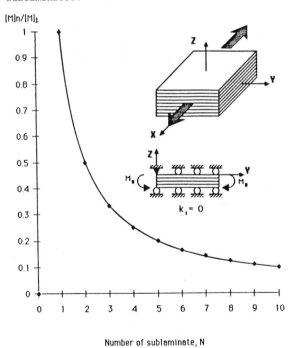

Fig. 6   Normalized moments with the increase of number of repititions in
         sublaminates.

32nd International SAMPE Symposium
April 6-9, 1987

# COMPARATIVE ANALYSIS OF STRUCTURAL AUTOMOTIVE COMPOSITES

Fred P. Isley
Knytex, Inc.
Seguin, Texas

## Abstract

A PC based software package that is not load driven has been developed for comparative analysis of multiple laminated structures. The package allows the user to compare laminate characteristics on a point by point basis without regard to input loads. Comparisons are made in five (5) primary areas:

- Cost
- Weight
- Thickness
- Strength
- Stiffness

This permits the substitution of materials, change of ratios, change of processing methods, etc. to be evaluated without having to recalculate input loads. The benefit of this program lies in its ability to rapidly sort through multiple laminate schedules in order to find the best fit at the lowest cost.

## I. INTRODUCTION

There are a myriad of programs available today to help engineers and designers specify and analyze composite structures. Generally these programs assume knowledge of or the ability to calculate the loads to be carried. This approach, while rational and conventional in engineering terms, does not answer the question of laminate optimization in terms of cost or alternative structures that may fulfill the design requirements.

Knytex has developed a laminate comparison program that allows the user to choose among several laminate constructions on cost, weight, thickness, strength, and stiffness criteria. This is a very useful tool when attempting to determine least cost for performance structures.

Several examples of various laminate constructions are compared in this paper to show the flexibility and utility of the program.

## II. DISCUSSION

The need for a quick/portable tool for analyzing composite structures was recognized in 1983. After reviewing the then available software, we realized that none of the existing programs matched our requirements.

The primary reason for this is that most programs were written for the aerospace industry. They are too large/expensive for every day use. They generally deal with pre-pregged unidirectional tapes. They assume the same laminating techniques. Cost was usually ignored. And most importantly, they assume that the loads are either known or can be calculated.

There are several problems with these assumptions if you attempt to apply these programs outside the aerospace environment, to the marine or automotive markets for example.

First, loads are not well known, well defined or easily calculated. The industry is very pragmatic. "If it works, use it." Until recently there has been no method of adequately analyzing or calculating the loads encountered in small marine structures.

Second, in both the marine and automotive markets there is a great deal of controversy over the "best" methods and materials to be used in any given structure. Further, since weight is not the primary concern, a much broader base of materials and methods is available for consideration. Each of these methods and materials has its own set of properties that are reflected in the final laminated structure.

Third, cost and cost control are primary motivational factors in any consumer product, and particularly in more competitive markets, such as marine and automotive.

Fourth, as suppliers, we do not wish to become involved in the primary structural design. With load driven programs the implication of fitness is inherent. This segment of the process is best left to those who have trained for it and whose job it is to develop the overall design.

Fifth, as suppliers, we are often attempting to replace existing laminate schedules. By using the comparative techniques in this package, it is easy to show the benefits of one schedule over another without

becoming involved in the design
loadings.

The Knytex series was developed to
allow more flexibility in the analysis process. With it you can examine an existing laminate with its
current resin/reinforcement ratio
and theoretically determine the
load carrying capability of the
structure. Additionally, the laminate weight, thickness and cost is
computed.

Reinforcement substitutions such as
ply sequence, reinforcement ratio,
material substitutions, etc., can
be made and new data calculated on
the proposed laminate. The new and
old laminates can then be compared
on a point basis. Since most changes involve a tradeoff of some type
-cost for weight, cost for strength,
weight for stiffness, etc., - an
informed decision of which schedule
to use then can be made.

This program, like any other has
its shortcomings. It does not analyze the structure for correctness.
It is based on beam analysis rather
than panel analysis. It will not
define the loads required. It does
not include impact or fatigue data.
The data base is somewhat limited
and assumes a linearity of properties over the entire range of
resin/reinforcement ratios.

III.  PROGRAM FEATURES

3.1 Item | Description/Comments

Weight — estimated laminate/ weight per square foot

Cost — estimated cost per square foot-materials only

Stiffness — EI of the laminate calculated in bending for first layer and "final failure" modes

Strength — bending moment at both first layer and "final failure" modes
— skin strength in tension and compression is also calculated

Thickness — laminate thickness in inches

Core Shear — shearing forces on the core for long slender laminates

Skin wrinkle — safety factor based on core compressive modulus

Section Modulus

Skin tensile and compressive modulus — calculated without the core

Fiber axis or orientations — allows the rotation of the fiber to various angles

Variable properties — can add or delete materials, change ratios reverse load direction etc.

IV.  EXAMPLES

The following examples compare .032"
double sided galvanized sheet steel
with various laminated structures.
In these examples the laminate modulus must be at least 1,500,000
pounds and the tensile and compressive strengths in the 20,000 psi
range.  From these criteria, we are
looking for the lowest cost and
lightest weight laminate.

The steel laminate has a modulus of
32,400,000 pounds and a strength of
40,659 psi.  The lower laminate
modulus was selected because the
overall steel structure will have a
lower stiffness due to welding.
The lower strength value was used
simply because the higher strength
is not needed.

Load values are used in these examples simply to make them more
realistic.  As you will see they
are not necessary to run the program.

Fred Isley is the Technical Director
of Knytex, Inc., Seguin, Texas.
Prior to joining Knytex he was the
QC/QA and later Operations Manager
of Certainteed Corp., Wichita Falls,
Texas.  At Knytex he is responsible
for product development and technical sales support.  Mr. Isley has
a B.S. in Textile Chemistry from
North Carolina State University and
an MBA from the University of North
Carolina.

AUTOMOTIVE THERMOPLASTICS: THEIR USE AND APPLICATIONS

Frederick S. Deans
Azdel, Inc.
Joint Venture between General Electric Company
and PPG Industries
Troy, Michigan

## Abstract

Passenger car and truck applications are being improved and expanded by the use of thermoplastic materials. Depending on material-type and process, increased value and performance is being offered to the transportation market. Key characteristics, such as performance, property retention, processability, and compatibility with government and OEM standards.

## MATERIALS AND PROCESSES

Thermoplastics, either in transparent or translucent form or reinforced "composite-types", are making steady inroads to most major automotive components. Exterior, body, interior, and under-the- hood parts are now being designed in various thermoplastics.

Two basic groups of thermoplastics are characterized by the following:

### Amorphous
- Wide softening range
- Limited chemical resistance
- Moderate heat resistance
- High impact resistance
- Creep resistance

### Crystalline
- Sharp melting point
- Good chemical resistance
- High heat resistance
- Lubricity
- Notch sensitivity
- Higher shrinkage

### Material Examples

| Amorphous | Crystalline |
| --- | --- |
| Polycarbonate (PC) | Polypropylene |
| Polyetheramides | PBT |
| ABS | PET |
| Polyphenylene oxide (PPO) | Nylons |

Copolymers, blends, or alloys, of amorphous and crystalline thermoplastics can produce materials with improved characteristics. For example, blending polycarbonate and PBT produces a material with high impact and good chemical resistance.

Additives, such as reinforcements, chopped, long, or continuous length fiberglass, fillers, pigments, fire retardants, and ultraviolet light stabilizers are incorporated to make composite thermoplastics for automotive applications.

Thermoplastic part processing has a direct bearing on part size, thicknesses, material selection, finished appearance, and economics. The following list highlights thermoplastic material/process classifications[1].

- Injection Molding
- Structural Foam Molding
- Hollow Injection Molding
- Sandwich Injection Molding (sequentially inject skin, then different core material)
- Injection/compression molding (inject into slightly opened tool)
- Thermoplastic Compression Molding (flow forming and sheet stamping - little or no flow)
- Thermoforming (non-matched tools)

- Blow Molding
- Twin-sheet forming

Today's thermoplastic materials offer engineers a number of performance/appearance characteristics that are required by automotive part applications. For example[2]:

- Thermal performance
- Impact/modulus
- Long-term property retention
- Processability
- Adaptability to secondary operations
- Performance to government and/or OEM imposed regulations

These key characteristics play a major influence on material selection. An examination of several automotive thermoplastic applications reveal these relationships.

Instrument Panels (I/P) - Early plastic I/P's (introduced in the mid-1960's) utilized relatively inexpensive materials, but performed with brittle impact behavior. Current I/P designs call for upgraded materials, such as polyphenylene oxide (PPO) and polycarbonate (PC) resins. These resins provide improved characteristics, especially in heat, impact, and processability.

Low installation angles and overall increase in automotive windshield areas have increased

the "greenhouse effect" on automotive I/P's. Materials must be
capable of withstanding 240°F heat
distortion temperatures (for
exposed surfaces) and 200°F for
lower I/P requirements. These
heat exposures also demand that
materials maintain long-term
property retention (heat aging).
Levels of impact resistance,
stiffness, and dimensional
stability must not be adversely
affected.

Impact performance and modulus
are related properties. Improved
impact performance is associated
with higher ductility (vs. brittleness) in I/P panels. However,
higher ductility generally meant
lower flexural modulus/stiffness.
Higher ductility and stiffness
have been developed by glass
filled composites. These materials can achieve modulii up to
300,000 to 900,000 psi.

Newer process techniques such as
structural foam, blow, and compression molding have the potential to bring weight reduction,
upgraded cross-section design,
and improved processability to
thermoplastic I/P panels.

## BUMPER SYSTEMS

Automotive bumpers represent a
rapidly growing application area
for thermoplastic composites.

Materials that offer attractive
alternatives to steel bumpers do
so by providing:[3]
- Weight reduction; up to 30% over
  steel systems
- Lower tooling costs - especially
  for complex geometries
- Faster timing for tooling
- Increased styling freedom
- Corrosion resistance
- Design integration and consolidation of parts, such as
  brackets, supports, energy
  absorbers, etc.

High impact/modulus performance of
thermoplastic composites provide
the basis for proper bumper
designs. Several types of thermoplastic bumper systems are:
- Injection molded systems -
  provide maximum styling and low
  weight; deflections in the range
  of 2.5 to 4.0 inches.
- Blow molding systems - low tool
  cost and weight; similar deflections as injection molded
  systems.
- Stamping or compression molding
  systems - maximum stiffness and
  load capabilities; deflections
  less than 2.0 inches.

Thermoplastic bumper systems must
maintain structural performance
over a wide temperature range.
Candidate materials are usually
evaluated at temperatures of -40°F
to 180°F. Thermal cycling must
not significantly degregate

impact, stiffness, and energy management performance.

Unlike thermosets, thermoplastic composites can be molded with fast cycle times, generally 1 minute or less. Tooling capacities for a single cavity mold are generally in excess of 200,000 units/year, for a 2 shift operation.

The adoption of Federal Motor Vehicle Safety Standards, FMYSS 205, has established the basis for bumper system design criteria. Low-speed impact criteria, i.e. 5 mph, have been adopted by the OEM industry.

Computer design programs have been established for plastic bumper systems.[4,5] These programs take form as bumper system models that provide quick materials evaluation and initial section development and finite element analysis (FEA) that fine tune design sections to establish tooling for prototype evaluation programs.

## LIGHTING SYSTEMS

Sealed beam, glass headlamp systems have long hindered engineers and stylists by their inflexible requirements for vertical installational angles. Highly transparent, heat stable, high impact polycarbonate thermoplastics are now leading the way

to replaceable-bulb lighting systems. These systems support highly styled, aerodynamic designs that are contoured with the front end "skins". These stylistic shaped and efficient lighting systems also offer weight reductions, better impact resistance (which reduces overall replacement costs) and accordingly, lower replacement costs vs. traditional sealed beam units.

High reflective base thermoplastic units, mounting structures, and lenses take advantage of high speed injection molding processes. New techniques and materials for compression molding thermoplastic materials will achieve further efficiencies for mounting base designs.

## NEW DEVELOPMENTS

Vehicles in the 1990's will be the beneficiaries of the dynamic expansion of thermoplastic applications, made possible by new heat resistant, smooth surface materials, improved processing, and an enlightened design community. Traditional automotive metal and thermoset parts are being targeted by thermoplastics. A few examples are:

Vertical & Horizontal Exterior Body Panels - injection molded thermoplastic fenders are currently specified on several fenders. Improved impact, reduced weight, and economics result in this application. Of course, the elimination of rust and corrosion is a given. Development efforts on higher modulus, Class A stampable thermoplastics will develop horizontal panels, such as hoods, roofs, and rear decks.

Large parts such as pick-up truck boxes are now possible in structural, high impact, and fast processing thermoplastic composites.

Under-the-Hood Components - high temperature, creep and chemical resistant thermoplastics have been developed for such applications as valve covers, oil pans, electrical and component covers. Process improvements in injection and compression molding will make thermoplastics attractive application candidates when compared to metal and thermosets.

Interior and Components - floor pans, load floors, seats, door panels, and firewalls are just a few application areas that thermoplastic composites provide increased values. These applications are also possible due to advances in material handling, adhesive and coating technology advances, and increased engineering capabilities.

Glazing - significant improvements in film hardening and thermal stability have stimulated the use of this thermoplastic coating of inner windshield surfaces. These "anti-lacerative" developments will lead to the goal of reducing abrasive impact injuries. Also, fixed quarter windows are currently being specified in thermoplastics. These windows offer reduced weight, improved styling, and, from an installation perspective, better fit and finish for OEM manufacturers.

ENVIRONMENTAL ISSUES

Studies indicate that plastic application volumes will almost double in 1990's vehicles from current levels. Increased thermoplastic useage will present society with end-product scrap and recycling opportunities.

Unlike thermosets, the use of thermoplastic materials for automotive applications provide scenarios where thermoplastic part recycling can provide reprocessing "blends" that could find life-cycle applications in other markets, such as packaging or housing. Recycling of automotive thermoplastic parts would be done by a new industry, consisting of plastic counterparts to scrap-metal dealers.[6]

## REFERENCES

1. "Plastic Exterior Body Panels: Materials/Process Requirements", by Murray, A.D. and Gentle, D.F., "Automotive Engineering," Vol. 94, No. 12, Dec. 1986, pp. 38-45.

2. "Advanced I/P Materials and Process," Dumouchelle, D.G. and Florence, R.A., SAE Truck and Bus Meeting and Exposition, Nov. 10-13, 1986, Paper No. 861968.

3. "Exterior Thermoplastics: New Needs, New Solutions," Risk, W.R., 1987 SAE International Congress and Exposition, Feb. 23-27, 1987, Paper No. 870282.

4. "Computer Aided Engineering Analysis of Automotive Bumpers," Glance, P.M., Society of Automotive Engineers, 1985, Paper No. 840222.

5. "Analysis Techniques for the Design of Thermoplastic Bumpers," Nimmer, Dr. R, Bailey, O., Paro, T., SAE International Congress and Exposition, Feb. 23-27, 1987, Paper No. 870107.

6. Wascher, U.S., Speech given at Modern Plastics International K'86 Pre-Show Conference, Dusseldorf, W. Germany, November, 1986.

# NOVEL SOLUBLE SILICONE-IMIDE COPOLYMERS

Chung J. Lee, Ph.D.
Manager, Polymer Technology
Technology Center
Occidental Chemical Corporation
Grand Island, New York 14072

## Abstract

Commercially available, high performance silicone-imide copolymers are found valuable in applications such as halogen-free cable jackets, high temperature sealant and dielectric insulators for microelectronics, etc. (See C.J. Lee, 39th National SAMPE Symposium, March 19-21, 1985, p. 64.) Some of these silicone-imide copolymers are diglyme soluble; however, their thermal stability is limited to 400°C/0.5 hours and their glass transition temperatures are far below 200°C (i.e. 125 ∽ 150°C). In comparison, the novel diglyme soluble silicone-imide copolymers, introduced herein, have a glass transition temperature of more than 200°C and thermal stability up to 400°C/2 hours with retention of excellent film strength. The compositions and properties of these novel diglyme soluble silicone-imide copolymers and other related photosensitive silicone-imide copolymers based on a similar chemistry will be discussed.

## 1. INTRODUCTION

In a recent review article, Dr. Reine, of the Naval Weapon Research Center stated that "..... The most promising area for further research in thermally stable elastomers is that of siloxane block copolymers". (1) From the performance/cost relationship, one of the most important classes of siloxane block copolymers is the silicone-imide block copolymer or the polyimidesiloxane. Polyimidesiloxane, as a new class of thermally stable polymers, can be prepared as a rigid thermoplastic, thermoplastic elastomer or elastomer, depending on: their imide block size, imide/ siloxane weight ratio and their siloxane block size. (2) These materials possess, in general, the best combinations of desirable properties derived from polyimides and polysiloxanes. For instance, many of these polyimidesiloxanes, in contrast to polyimides, can be processed by injection molding or extrusion at temperatures employed for processing common engineering plastics such as Nylon 6, Nylon 6,6 or polyethylene terephthalate. Compared to polysiloxanes, polyimidesiloxanes have superior impact resistance and their tensile strength ranges from 100 to 700 $Kg/cm^2$. In addition to their -80°C to -40°C glass transition temperature, they have an upper rigidity or glass transition temperature as high as 200°C. They also have better hydraulic fluid or jet fuel resistance as well as flame resistance compared to polyfluorosiloxanes. They have excellent thermal stability. For instance, an ⌐40% siloxane containing polyimidesiloxane can have a 100% tensile strength retention after a 200°C/200 hour or a 450°C/0.5 hour aging. A polyimidesiloxane containing more than 40% imide content also has good flame resistance. Therefore, polyimidesiloxane is currently the only class of halogen-free, non-flammable plenum jacket material. And, they are potentially the most desirable material for use as heat resistance gaskets or sealers based on both performance/cost criteria. Furthermore, it has been learned that polyimidesiloxanes have an excellent resistance to attack by or exposure to atomic oxygen. Therefore, they can be used as a protection coating for polyimides such as Kapton in low orbitor flight. They also have been used as a passivation layer in semiconductors, LCD alignment, and $\lambda$-particle shielding in VLSI applications.

Unfortunately, all the polyimidesiloxanes based on 3,3',4,4' benzophenone tetracarboxylic dianhydride (BDTA), pyromellitic dianhydride (PMDA) or 3,3',4,4'-biphenyl tetracarboxylic dianhydride (BPDA) are only soluble in high boiling and/or polar solvents such as N-Methyl pyrrolidinone, phenol or m-cresol, etc. In addition, the PMDA and BPDA based polyimidesiloxanes are only soluble before their imidization or in their polyamide acid siloxane stage. These polyamide acid siloxane solutions have to be kept at refrigeration temperatures ($< 4°C$) to prevent an irreversible depolymerization process. The complete removal of the solvents such as NMP from a microinch thin film requires temperatures as high as 250°C for 30 minutes. And a completion of imidization in a solid state may need temperatures as high as 300-350°C for at least 30 minutes. The removal of NMP in production of polyimidesiloxane requires extensive treatments such as precipitation and centrifugal filtration; or it uses very costly equipment such as a thin film evaporator. The need for high temperature drying or imidization process in coating applications is not only energy extensive and costly but also prohibits the usage of these polyimidesiloxanes or polyamide acid siloxane in some heat sensitive semiconductors or in thick film coating applications.

## 2. NOVEL DIGLYME SOLUBLE POLYIMIDESILOXANES

Polyimidesiloxanes which are soluble in low boiling and less polar solvents such as diglyme can be prepared using some organic diether anhydrides, such as that described in U.S. Patent 3,847,867 (3). However, all these polyimidesiloxanes have relatively low glass transition temperatures (i.e. 125°C to 150°C) and also limited thermal stability (400°C/ 0.5 hour with retention of film creasability or flexibility). More thermally stable polyimidesiloxanes which are soluble in diglyme or even THF or MEK are not currently available.

The Polymer Technology Group of Occidental Chemical Corporation has recently developed a new class of polyimidesiloxanes based on a proprietary organic dianhydride which is designated as Oxy-dianhydride in the present article. The Oxy-dianhydride route provides some polyimidesiloxanes with surprisingly good solubility and also high glass transition temperatures. The properties of these polyimidesiloxanes and their modi-fied products are presented here.

## 2.1 Mechanical Properties

The tensile strength and elongation at break of these polyimidesiloxane films in the range of 10-20 microinches are shown in the Figures 1 and 2 in comparison with polyimidesiloxanes derived from other organic dianhydrides. As can be seen, their tensile strength is a function of siloxane content, or (%) $G_m$ of the polyimidesiloxanes as previously reported.(2) The organic dianhydrides used in Figure 1 are B for 3,3',4,4'-benzophenone tetracarboxylic dianhydride; U for 2,2-bis[4-(3,4-dicarboxy phenoxy)phenyl] propane dianhydrides and H for 1,4-(3,4-dicarboxy phenoxy) benzene dianhydride and O for Oxydianhydride. Figure 1 shows that the tensile strength of these novel polyimidesiloxanes is as good as various polyimidesiloxanes derived from other organic dianhydrides. The elongation at break of the Oxy-dianhydride based polyimidesiloxanes are more similar to H and B based polyimidesiloxanes.

## 2.2 The Solubility Property and Other Modified Products

The comparative solubility of various polyimides and polyimidesiloxanes based on B, H, and O dianhydrides is summarized in Table 1. All these polyimides employed a common organic diamine, T; and also an $\alpha$,w-diaminosiloxane as a co-monomer in their polyimidesiloxanes.

Surprisingly, despite the insolubility of Oxy-dianhydride based polyimide, $(OT)_n$, versus that of H and B based polyimide $(HT)_n$ and $(BT)_n$ in NMP; the Oxy-dianhydride based polyimidesiloxane is very soluble in NMP, diglyme, and even THF.

2.2.a. Fully Imidized and Diglyme Soluble Photosensitive Polyimidesiloxane Products

The good solubility of Oxy-dianhydride based polyimidesiloxanes renders an opportunity for further chemical modifications of these copolymers in low boiling and inert solvents such as diglyme. Therefore, an acrylate-bearing polyimidesiloxane (AB-SIM) was prepared in diglyme. The AB-SIM can be cross-linked by thermal or photo methods. Contrary to commercially available NMP soluble acrylate-bearing polyamide acid or polyamide acid siloxanes based on BTDA, BPDA or PMDA, these novel photosensitive acrylate-bearing polyimidesiloxanes are fully imidized and soluble in diglyme. Structurally-wise commercially available photosensitive products have an acrylate group attached to the carboxylic group of the amide acid repeat units. In addition to the shelf life problem of all polyamide acids, these acrylate groups tend to degrade during the imidization process, which requires temperatures as high as 350°C/0.5 hours at solid state. A fully imidized and diglyme soluble

AB-SIM, on the other hand, can be cross-linked and the solvent can be removed at temperatures below 180°C; conditions under which the photo-patterning features will not be degraded. The cross-linked AB-SIM is not soluble in NMP or diglyme. The cured products also exhibit higher modulus at higher temperatures and have a higher glass transition temperature ( 50°C) than the uncured products.

2.3. Glass Transition Temperatures of Diglme Soluble Polyimidesiloxanes

Basically there are two distinctly different classes of diglyme soluble polyimidesiloxanes now available. One is based on the diether dianhydride, such as 2,2-bis[4-(3, 4-dicarboxy phenoxy)phenyl] propane or the U-dianhydride and the other class is based on Oxy-dianhydride or O-dianhydride. There are two distinct glass transition temperatures for a true imidesiloxane block copolymer which is also a new class of thermally stable, thermoplastic elastomer. These two distinct Tgs can be seen in a DMA trace of a imidesiloxane block copolymer as shown in Figure 3. The upper glass transition temperature of a polyimidesiloxane is dictated by the composition and the block size (see Figure 4) of the polyimide block. For a given polyimide composition and a defined imide block size, it has been found that the following general equation (4) can be used to esti-

mate the glass transition temperatures of the U and O-based polyimidesiloxane relative to that of a
BTDA or B based polyimidesiloxane:

$$Tg(O) = 0.96 \ Tg(B) \qquad (1)$$
$$Tg(U) = 0.73 \ Tg(B) \qquad (2)$$
or $Tg(O) = 1.32 \ Tg(U) \qquad (3)$

It is verifiable, therefore, that
for a given organic diamine and
$\lambda$,w-diaminosiloxane, the O-based
polyimidesiloxane will have an
upper glass transition temperature
of at least about 30% higher than
that of a U-based polyimidesiloxane. For instance, the $[(OT)_2G_9]$
polyimidesiloxane has an upper
glass transition temperature of
196°C which is 1.31 times that of
the $(UT)_2G_9$ polyimidesiloxane
(Table 2).

## 2.4. Thermal Stability and Flamma-
bility of Polyimidesiloxanes

The flammability of novel polyimidesiloxanes is predominately
controlled by (%) Gm, or the siloxane content as shown in Figure 5.
In addition, the preparation
method that affects the imide
block size and size distribution,
and the molecular weight of the
polyimidesiloxanes also influences
the flammability of the product.
In general, for a given organic
diamine and $\lambda$,w-diaminosiloxane
an O-based polyimidesiloxane is as
flame resistant as a B-based
polyimidesiloxane; but is far superior to that of an U-based polyimidesiloxane (Table 2).

The thermal stability of O-based
polyimidesiloxanes is also far
superior to that of the U-based
polyimidesiloxanes. For instance,
the $(UT)_2G_9$ polyimidesiloxane lost
its flexibility after 400°C/0.5
hour; whereas, the $(OT)_2G_9$ polyimidesiloxane still retained its
creasability after 400°C/2 hours
aging.

## 4. CONCLUSION

Some distinctly superior diglyme
soluble polyimidesiloxanes have
been developed based on a proprietary organic dianhydride, the Oxy-
dianhydride. These fully imidized polyimidesiloxanes have higher
upper glass transition temperature, better thermal stability and
flame resistance than the 2,2-bis-
[4(3,4-dicarboxy phenoxy)phenyl]
propane or the U-based polyimidesiloxanes. These novel polyimidesiloxanes have potential as halogen-free plenum cable jacket, substrate for flexible circuit board
and 3D molded wire board, permse-
lective membrane, passivation
layer and thick screen coating in
semiconductors, $\lambda$-particle shielding and dielectric insulators in
microelectronics and protective
coating and adhesive for polyimides such as Kapton, etc.

The diglyme soluble photosensitive
polyimidesiloxanes could prove
useful in mask manufacturing and
microchip patterning. They can
also be processed easier and more
economically than BTDA or PMDA
based materials.

## 3. REFERENCES

(1)  R. Rheine, "Thermally Stable Elastomers, A Review", AD A137, 914, N.T.I.S. (1984).

(2)  Chung J. Lee, "High Performance Silicone-Imide Copolymers", SAMPE Journal, Vol. 21, No. 4, pp. 34 (1985).

(3)  D.R. Heath and J.G. Wirth, U.S. Patent 3,847,867, Polyetherimides (1974).

(4)  Chung J. Lee, "On Assessement of Glass Transition Temperatures of Thermally Stable Polymers", to be published.

### BIOGRAPHY

Chung J. Lee has a B.S. Degree from National Taiwan University, a M.S. Degree from U.C. Berkeley and a Ph.D. Degree from Rensselaer Polytech Institute with a major in Polymer Physical Chemistry (1976). He is author of a review article (Rev. Macromol. Chem., 1978), 14 publications, and 10 patents regarding inventions in PET molding compounds, unsaturated polyester, epoxyacrylate, imide epoxy resins and silicone-imides. His prime interests are structure/property/processing relationships of high temperature polymers, hot melt adhesives, IPN and polyblends and theoretical assessments of Tg, $\Delta$Cp and Young's Modulus of Polymers. Presently, Dr. Lee is Manager of Polymer Technology for New Business Development at the Durez Division, Occidental Chemical Corporation, where he is in charge of New Polymeric Product Development and Polymer Physical and Testing Laboratory.

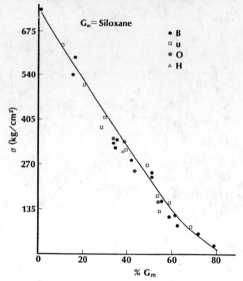

**FIGURE 1**

**Structure-Property Relationship
of Silicone-Imide Copolymers**

$G_m$ = Siloxane

● B
□ u
⊛ O
△ H

$\sigma$ (kg/cm²)

% $G_m$

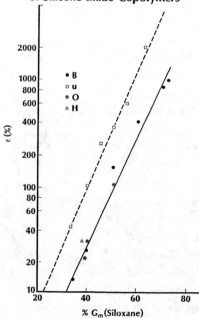

**FIGURE 2**

**Structure-Property Relationship
of Silicone-Imide Copolymers**

● B
□ u
⊛ O
△ H

$\epsilon$ (%)

% $G_m$ (Siloxane)

FIGURE 3

# Dynamic Mechanical Property of $\{(OT)_{2,5}Ga\}$

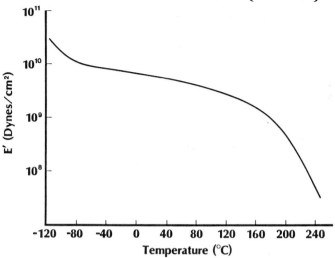

FIGURE 4

# THE GLASS TRANSITION TEMPERATURE
## OF $\{(OT)_nG_m\}$

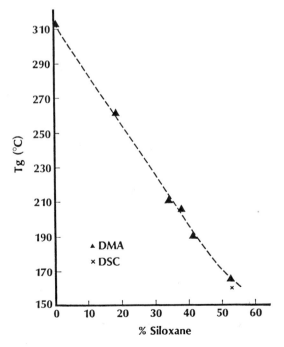

# FIGURE 5

## DEPENDENCE OF FLAMMABILITY ON SILOXANE CONTENT

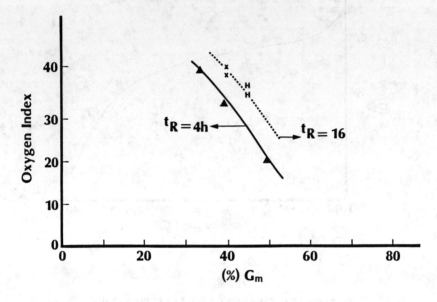

TABLE 1

# Comparative Solubility of Various Polyimides and Polyimide Siloxanes

| Designated Composition | Solubility (10 g/100 mil) | | | |
|:---:|:---:|:---:|:---:|:---:|
| | NMP | Diglyme | THF | MEK |
| $(BT)_n$ | + | − | − | − |
| $(OT)_n$ | − | − | − | − |
| $(HT)_n$ | + | − | − | − |
| $(UT)_n$ | + | − | − | − |
| $(BT)_2G_a$ | + | − | − | − |
| $(OT)_2G_a$ | + | + | + | ± |
| $(HT)_2G_a$ | + | − | ± | − |
| $(UT)_2G_a$ | + | + | − | − |

TABLE 2

# COMPARATIVE THERMAL PROPERTIES OF SIM

| | Oxygen Index | | | Glass Transition | |
|:---:|:---:|:---:|:---:|:---:|:---:|
| | 10 mil Film | 1/32″ | 1/16″ | $Tg_1$ | $Tg_2$ |
| $(BT)_2G_9$ | 40-41 | − | − | -80 | 205 |
| $(OT)_2G_9$ | 40-41 | 42 | 44 | -80 | 196 |
| $(HT)_2G_9$ | 36-37 | − | − | -80 | 170 |
| $(UT)_2G_9$ | 30-31 | − | − | -80 | 145 |

STRUCTURE PROPERTY BEHAVIOR OF
POLYIMIDE SILOXANE SEGMENTED COPOLYMERS

C. A. Arnold, J. D. Summers, R. H. Bott,
L. T. Taylor, T. C. Ward, J. E. McGrath*
Department of Chemistry and Polymer Materials and Interfaces Laboratory
Virginia Polytechnic Institute and State University
Blacksburg, VA  24061

*To whom correspondence should be addressed

ABSTRACT

Procedures were developed for preparing soluble fully imidized polyimide-polydimethyl siloxane segmented copolymers of wide ranging compositions. At low siloxane levels, the materials behave as modified polyimides. At higher concentrations, the materials are analogous to thermoplastic elastomers. Characterization by dynamic mechanical and thermal analysis methods will be reported along with an assessment of the bulk mechanical properties and the surface behavior. The surface behavior is particularly interesting since one can tailor the materials to have siloxane surfaces even at rather low siloxane contents. This influences a number of properties such as the coefficient of friction and, importantly, the degradation of these materials under aggressive oxygen environments (e.g. atomic oxygen, oxygen plasmas).

1.0 INTRODUCTION

Polyimides synthesized from aromatic momomers are of great interest for high performance applications due to their excellent thermal and mechanical properties. However, these polymers are often insoluble in their fully imidized form unless carefully designed. Many approaches to enhancing the solubility and processability of polyimide systems have been investigated which also attempt to maintain their high performance properties. Successful attempts to accomplish this goal have involved the incorporation of flexible bridging units into either the diamine or dianhydride monomers, which imparts mobility to the otherwise rigid polyimide backbone.[1-3] Incorporation of bulky side groups[4] and utilization of diamines containing meta linkages[5,6] also enhance processability.

Incorporation of flexible siloxane segments into the polyimide backbone can yield soluble, processable copolyimides while still maintaining fairly good thermal-mechanical properties.[7] In addition to enhanced solubility, incorporation of polysiloxane blocks imparts a number of other desirable characteristics, such as improved weatherability to aggressive environments, improved impact resistance, decreased sensitivity to water absorption and surface modification.[7,8] Such advantages render polysiloxane-modified polyimides attractive as protective coatings for aerospace and microelectronics applications.[9-13]

Much of the available literature regarding these copolymers is only available in the form of patents.

The thermal-mechanical properties of siloxane-modified polyimides are a function of the weight fraction of incorporated siloxane, molecular weight of the siloxane blocks and chemical architecture of both the siloxane and imide segments. Thus, copolymers with high concentrations of incorporated siloxane (>50%), where the siloxane is the continuous phase, behave as thermoplastic elastomers, whereas lower siloxane concentrations result in more rigid materials. At high polyimide compositions the upper glass transition value and most mechanical properties approach those of the unmodified controls. However, the surface structure can still be strongly dominated by the low surface energy polysiloxane microphase.

The great difference in solubility parameters of the siloxane and imide segments is a driving force for microphase separation, particularly when higher molecular weight siloxane oligomers are incorporated. Additionally, because the siloxane component possesses a relatively low surface energy, it will migrate to the air or vacuum interface, yielding a surface dominated by the siloxane component. This effect is observed even for low (~5 weight %) levels of siloxane incorporation. In aggressive oxygen environments, the surface siloxane segments are postulated to convert to a ceramic-like silicate ($SiO_2$) which provides a protective outer-coat to the bulk material. The conversion of polysiloxane to silicon dioxide in oxygen plasma has, in fact, been documented within the literature of the electronics industry.(14-17)

The thermal-mechanical and surface properties of a soluble, fully imidized poly(imide siloxane) segmented copolymer were investigated. The copolymer was based upon 3,3',4,4'-benzophenone tetracarboxylic dianhydride (BTDA), the meta-substituted diamine 3,3'-di-

aminodiphenyl sulfone (DDS) and aminopropyl-terminated polydimethylsiloxane oligomers of various molecular weights in the range of 950 to 10,000 grams per mole. The incorporated siloxane oligomer was varied from 5 to 70 weight percent. Two methods of imidization of the intermediate siloxane-modified poly(amic acid) were employed. The data presented herein are for the thermal imidization of the siloxane modified poly(amic acid). Conversion to the poly(imide siloxane) may also be accomplished at lower temperatures, vg, 150-170°C, by means of a solution imidization procedure. In all cases, tough, transparent, flexible, soluble films are obtained. Another paper presented at this meeting discusses the synthesis of these materials in depth(18).

2.0 EXPERIMENTAL

2.1 Materials

High molecular weight, soluble polyimide-polysiloxane segmented copolymers were prepared, as described in the accompanying paper by Summers, et.al.(18). A representative structure for the poly(imide siloxane) copolymers is depicted in Figure 1.

2.2 Characterization of the Poly(imide siloxane) Copolymer

2.2.1 Viscosity Measurements
Intrinsic viscosity measurements yielded relative molecular weights and were performed in NMP at 25°C using a Cannon-Ubbelohde viscometer.

2.2.2 Thermal Analysis
Differential scanning calorimetry (DSC) was used to determine the glass transition temperatures (Tg) of the copolymers with a Perkin-Elmer Model DSC-4. Scans were run at 10°C/min with a sensitivity of 10 mcal/sec in a nitrogen atmosphere. Reported values were obtained from a second scan after heating and rapid cooling. Some copolymer transitions were obtained by dynamic mechanical thermal analysis (DMTA) at a frequency of

1 Hz using a Polymer Laboratories instrument.

### 2.2.3 Thermal Gravimetric Analysis and Thermal Mechanical Analysis

TGA was performed on a Perkin-Elmer TMS-2 instrument on samples in film form. Scans were run at $10^\circ$C/min in an air atmosphere. TMA was performed on the same equipment with a $10^\circ$C per minute heating rate with a 50 gram quartz penetration probe.

### 2.2.4 Stress Strain Analysis

An Instron Model 1122 was used to determine the mechanical behavior of dogbone shaped specimens of the copolymers. Specimen gauge lengths were 10mm. Specimen widths were 2.76mm. The extension rate was 5mm per minute.

### 2.2.5 Contact Angle Measurements

Water contact angles were measured on samples prepared by spin-coating the amic acid intermediates in its reaction solution onto a ferrotype plate and then imidizing by thermal treatment. In some cases, recast fully imidized films were also examined.

### 2.2.6 X-ray Photoelectron Spectroscopy

Samples for XPS were prepared by spin-coating the amic acid intermediate in its reaction solution onto ferrotype plates which had been washed in hexane three times prior to coating. The samples were covered with a watchglass to prevent contamination during thermal imidization by the standard method in a forced air convection oven. After imidization, the samples were thrice washed in hexane and placed in clean glass containers with lids. Analysis was performed on a Kratos instrument at exit angles of $15^\circ$ and $90^\circ$.

### 2.2.7 Oxygen Plasma Stability

Samples of the siloxane-polyimide copolymers were analyzed in an oxygen plasma environment to determine qualitative physical degradation and weight loss. A Plasmod unit from the Tegal Corporation of Richmond, CA was used. Film samples were maintained in an oxygen-charged environment for 45 minutes, under a vacuum of 1.5 torr. The applied radio frequency was 50 KHz. Oxygen flow rate to the analysis chamber was approximately 30 cc per minute.

### 2.2.8 Initial Adhesion Results

Single lap shear specimens were prepared by sandwiching a scrim cloth (112 E glass) coated with the poly(imide siloxane) resin between two primed titanium adherends. The poly(amic acid siloxane) solution in diglyme/THF, which was molecular weight controlled by the addition of phthalic anhydride, was coated onto the scrim cloth and cured in a forced air convection oven using the following schedule:

RT - $100^\circ$C, for $\frac{1}{2}$ HR
100 - $150^\circ$C, for $\frac{1}{2}$ HR
150 - $200^\circ$C, for $\frac{1}{2}$ HR

This procedure was repeated until an overall thickness for the cloth and resin of 9-13 mils was achieved.

The titanium adherends were sandblasted, treated with Pasa Jell 107, ultrasonically cleaned, and immediately primed with a coating of amic acid solution to preserve the surface treatment. This primer coat was imidized using the same thermal schedule as the scrim cloth.

Single lap shear specimens ($\frac{1}{2}$ inch overlap) were prepared by pressing the coated scrim cloth between two primed adherends using the following bonding cycle:

- R.T. to $325^\circ$C, apply 200 PSI at $280^\circ$C
- Hold for 15 min. @ $325^\circ$C
- Cool under pressure

## 3. RESULTS AND DISCUSSION

### 3.1 Solubility and Molecular Weights

One of the major goals of this endeavor was to solubilize the intractable polyimides by incorporation of siloxane segments. Solubilities of the siloxane-modified polyimide copolymers were evaluated in a variety of solvents at a 1 or 2 (w/v) percent concentration. Copolymer solubility was primarily a function of the concentration of the siloxane oligomer. At concentrations equal to or greater than 10 weight percent, all copolymers were soluble in dipolar, aprotic solvents such as NMP and DMAC. Copolymers containing higher percentages of siloxane were soluble in a wider range of solvents, such as THF, diglyme and methylene chloride. Intrinsic viscosities of the copolymers are listed in Table 1 and range from 0.51 to 0.84 dl/g, indicating that high molecular weight copolymers have been obtained over the entire compos-ition and segment molecular weight range.

## 3.2 Thermal Analysis and Thermal Stability

Values of the upper glass trans-ition temperatures of the siloxane modified polyimides were found to be a function of both the level of incorporated siloxane as well as the siloxane molecular weight (Table 2). Generally, the upper transition temperature will increase with greater siloxane oligomer molecular weight and with decreasing siloxane incorporation. In many cases, the copolymers' upper transition temperature is depressed only slightly relative to that of the control, indirectly indicating that good microphase separation was achieved. The lower temperature siloxane transition is difficult to detect by DSC for low levels of siloxane incorporation, ie, <20 weight percent. At greater levels of incorporation, however, the transition is detected by both DSC and DMTA within the range -117 to -123°C, representing 20 to 50 weight percent siloxane, respec-tively.

Copolymer thermal stability also varied with both siloxane oligomer molecular weight (Figure 2) and level of siloxane incorporation (Figure 3), increasing with the former variable and decreasing with the latter. From this observation, one may conclude that thermal

degradation begins at the aliphatic n-propyl segments linking the sil-oxane oligomers to the polyimide matrix. As the siloxane oligomer molecular weight is increased, the concentration of n-propyl linkages in the copolymer backbone decreases, thus increasing the overall thermal stability. Copolymers containing higher amounts of siloxane show decreased resistance to probe penetration by TMA (Figure 4). The TGA traces indicate, however, that the copolymers possess fairly high stability values at elevated temperatures. Even at high sil-oxane levels of 60 weight percent, the copolymers maintained good thermal stability, although as the siloxane content is increased, degradation occurs at lower temper-atures. The char yield at high temperatures, proportional to the siloxane content, suggests that a silicate-type structure is the principal degradation product in an air atmosphere.

## 3.3 Mechanical Property Analysis

An initial investigation of the mechanical behavior of these copolymers indicated that this property is highly dependent upon the level of siloxane incorporation (Figure 5). While copolymers containing low to moderate amounts of siloxane maintained good rigidity and ductile mechanical properties, a significant decrease in modulus and an increase in elongation occurred for high amounts of siloxane, no doubt due to inversion of the imide matrix to a continuous siloxane phase in the latter case.

## 3.4 Surface Properties

Since the surface composition of these macromolecular materials is directly related to such properties as coefficient of friction and atomic oxygen stability, X-ray photoelectron spectroscopy (XPS or ESCA) was employed in order to characterize the surface compo-sition of the siloxane-modified polyimide copolymers. By varying the angle of the sample relative to

589

the analyzer, different depths of the polymer were sampled, such that a 15° grazing take-off angle characterizes molecules from the uppermost surface (~10 to 20 Å) more so than molecules from the bulk. The 90° take-off angle, on the other hand, yields compositional information more characteristic of the subsurface regions (~50 to 70Å).

The results of the XPS experiments are listed in Table 3. The results conclusively demonstrate that the siloxane component dominates the surface of the copolymer. Furthermore, the extent of domination is independent of the weight percent of the siloxane incorporated into the copolymer. Thus, one is able to achieve a surface characteristic of the siloxane component while tailoring the physical properties which are characteristic of the bulk.

These results are supported by the data obtained by contact angle measurements. Unlike XPS, however, contact angle measurements provide no direct quantitative information.

Water contact angle measurements are listed in Table 4. A significant increase in contact angle is observed between the unmodified control and copolymer samples containing only 5 to 10 weight percent siloxane. In fact, the copolymer contact angles approach those of pure polydimethylsiloxane. The water absorption is also significantly reduced upon siloxane incorporation. Both of these properties are attributed to the hydrophobicity of siloxanes.

3.5 Oxygen Plasma Stability

The copolymers examined were based upon BTDA-PSX-DDS as well as their oxydianiline (ODA) analogue, and had siloxane contents of 30 and 50 weight percent and siloxane segment molecular weights of 1000. Results obtained thus far are listed in Table 5. In every run, the DDS-based 50 weight percent siloxane samples lost no weight after exposure to the oxygen environment

and their ODA analogues performed similarly well. Interestingly, the ODA-based systems seemed to perform somewhat less successfully than the DDS-based copolymers. Additionally, the 30 weight percent siloxane copolymers lost more weight during exposure than the 50 weight percent siloxane analogues, in both the DDS and ODA systems. Consistently, Kapton lost more weight than either the DDS or ODA based siloxane-modified polyimides, and coating Kapton with the DDS-based copolymers enhanced the stability of the Kapton film under the aggressive environment for both the 30 and 50 weight percent siloxane levels. Under these latter conditions, the performance of the Kapton-coated film resembled the performance of the siloxane-modified polyimides which constituted the coated layer.

The scanning electron micrographs (1600 x magnification) in Figure 6 qualitatively depict the enhanced stability to oxygen plasma of the DDS-based poly(imide siloxane) copolymers. The control exhibits severe cracking and delamination whereas the addition of 10 weight percent siloxane appears to inhibit delamination, although cracking still persists. Even more significant improvement is obtained at the 20 weight percent level of siloxane incorporation where delamination has not occurred and cracking is minimal.

3.6 Initial Adhesion Results

Preliminary adhesive data on controlled molecular weight copolymers are shown as lap shear strengths in Table 6. The incorporation of low to moderate amounts of siloxane ($M_n$ = 950 g/m) does not significantly alter the adhesive characteristics of these materials. Only at high levels of siloxane do the lap shear strengths fall below 2000 psi, indicating the potential utility of these materials as atomic oxygen resistant structural adhesives. The effect of increasing the siloxane oligomer molecular weight on adhesion is being studied currently to determine if the release agent qualities of siloxanes are more pronounced at these compositions.

## 4.0 CONCLUSIONS

New, high molecular weight, randomly coupled poly(imide siloxane) soluble block copolymers have been synthesized from bis(amino propyl) polydimethylsiloxane equilibrates of various molecular weights, an aromatic meta-linked diamine, and BTDA. A thermal procedure was used to convert the poly(amic acid siloxane) intermediates to the corresponding polyimides, essentially quantitatively, although a novel solution imidization procedure has more recently been developed based upon the co-amide solvent NMP/CHP reacted at milder temperatures of 150-170°C (18). The poly(imide siloxane)s produced by the former method are flexible, tough, transparent, soluble and are possible candidates for environmentally stable structural matrix resins, coatings and structural adhesives. Upper glass transition temperatures of many copolymers approach those of the unmodified controls, (~272°C) indicating good microphase separation. Lap shear strengths achieved against surface treated titanium showed that relatively good adhesion could be achieved under practical bonding conditions. Moreover, the values were basically unchanged as a function of siloxane content at low (e.g., 10-15%) siloxane levels. Photoelectron spectroscopy (XPS or ESCA) demonstrated that the top surface (~10Å) is dominated by the siloxane. Additional "ashing" experiments in oxygen plasma reconfirm that weight loss is definitely reduced by this feature.

## 5.0 ACKNOWLEDGEMENTS

The authors would like to acknowledge the Environmental Research Branch of NASA Langley Research Center for funding of this project. We are also indebted to the late Mr. George Sykes for his enthusiasm and continued support throughout the course of this work.

## 6.0 REFERENCES

1. Burks, H. D. and St. Clair T. L. in "Polyimides: Synthesis, Characterization, and Applications", Vol. 1, K. L. Mittal Ed., Plenum, NY, 1984, pp. 117-135.
2. Young, P. R., Wakelyn, N. T. Proceedings from the 2nd International Conference on Polyimides. Ellenville, NY, 1985, pp. 414-425.
3. Critchley, J. P., White, M. A. J. Polym. Sci., Polym. Chem. Ed., 10, 1809 (1972).
4. Harris, F. W., Norris, S. O., Lanier, L. H., Reinhardt, B. A., Case, R. D., Varaprath, S., Padaki, S. M., Torres, M., Feld, W. A. in "Polyimides: Synthesis, Characterization, and Applications", Vol. 1, K. L. Mittal Ed., Plenum, NY, 1984, pp. 3-14.
5. St. Clair, A. K., St. Clair, T. L., Smith, E. N. Polym. Prepr. 17, 359 (1976).
6. St. Clair, T. L., Yamaki, D. A. in "Polyimides: Synthesis, Characterization, and Applications", Vol. 1, K. L. Mittal Ed., Plenum, NY, 1984, pp. 99-116.
7. Johnson, B. C., Yilgor I., McGrath, J. E. Polymer Preprints, 25(2), 54 (1984); Johnson, B. C. Ph.D. Thesis, VPI and SU 1984; Johnson, B. C., Summers, J. D. McGrath, J. E. J. Poly. Sci., in press, 1987; McGrath, J. E., Sormani, P. M. Elsbernd C. S., Kilic, S. Makromol. Chemie, accepted, 1986; Sormani, P. M., Minton, R. J., McGrath, J. E. in "Ring Opening Polymerization: Kinetics, Mechanisms and Synthesis", McGrath, J. E. Editor, ACS Symposium Series No. 286 (1985), Chapter 11, p. 147-161.
8. A Final Report for NASA Langley Research Center, Grant No. NAG-1-343, suppl. 8, Oct., 1986, McGrath, J. E., et.al.
9. Lee, C. L. "High Performance Silicone-Imide Copolymers", SAMPE Symposium No. 30, p. 52-63 (1985).
10. Berger, A. "Modified Polyimides by Silicone Incorporation", SAMPE Symp. No. 30, p. 64-73 1985.
11. Davis, G. Photosensitive Polyimide-Siloxanes, in "Polymers in Electronics", T. Davidson, Editor, ACS Symposium Series No. 242, 1984, p. 259-272.

12. Berger, A. U.S. Patent 4,395,527 (1983).

13. Summers, J. D., Arnold, C. A., Bott, R. H., Taylor, L. T., Ward, T. C., McGrath, J. E. Polym. Prep. 27(2), 403 (1986).

14. Chou, N. J., Tang, C. H., Paraczczak, J., Babich, E. Appl. Phys. Lett. 46(1), 31 (1985).

15. Thompson, L. F., Wilson, G. C., Bowden, M. L., Eds. "Introduction to Microlithography", ACS Symp. Series, No. 219, Washington, DC (1983).

16. Taylor, G. N., Wolf, T. M. Polym. Eng. Sci. 20(16), 1087 (1980).

17. Taylor, G. N., Wolf, T. M., Moran, J. M. J Vac. Sci., Technol. 19(4), 872 (1981).

18. Summers, J. D., Arnold, C. A., Bott, R. H., Taylor, L. T., Ward, T. C., McGrath, J. E. 32nd Natl. SAMPE Symp. 32, 000, (1987).

## 7.0 BIOGRAPHIES

Cynthia Arnold received a B.S. in Chemical Engineering from the University of California, Berkeley, in 1980. Through 1982, she worked at Raychem Corporation in Menlo Park, CA, doing process and product development in the Thermofit Division. She was later employed at Mercor, Inc. of Berkeley, CA, in pilot plant and manufacturing engineering operations. During her employment in industry, she simultaneously studied for a graduate degree in business, receiving her M.B.A. in May, 1985. Currently she is pursuing her Ph.D. in Materials Engineering Science at VPI&SU, where she is a Kodak Fellow. Her work has been in the area of structure-property behavior of engineering polymers.

John D. Summers, a native of Allentown, Pennsylvania, received his B.S. in Chemistry from Ursinus College in 1983. He entered the Ph.D. program in Chemistry at Virginia Polytechnic Institute & State University in the fall of 1983. His research has involved the synthesis and characterization of siloxane modified engineering thermoplastics and thermosets.

Richard H. Bott, a native of Freeland, Pennsylvania, was graduated from Ursinus College with a B.S. in Chemistry in 1983. He entered the Ph.D. program in Chemistry in the fall of 1983. His research has focused on the characterization of poly(imide siloxane) copolymers and addition curing polyimide oligomers in terms of adhesive, thermal, electrical and mechanical properties.

James E. McGrath was born in Easton, New York and received his B.S. in Chemistry from Siena College in 1956. After being employed at ITT Rayonier in Whippany, NJ and the research division of the Goodyear Tire and Rubber Co., he obtained an M.S. degree in Chemistry from the University of Akron in 1964 and Ph.D. in Polymer Science from the same university in 1967. Professor McGrath joined the Union Carbide Corporation in August of that year. Professor McGrath joined VPI & SU in September 1975. He was named a tenured Associate Professor in 1977 and was promoted to Full Professor in 1979. He is a consultant to industry and government. He is also a Co-Director of the Polymer Materials and Interfaces Laboratory. Professor McGrath has coauthored what is considered a definitive book on Block Copolymers (Academic Press, 1977), and has edited books on Anionic Polymerization, Ring Opening Polymerization and Advances in Polymer Synthesis (with B. M. Culbertson). He has over 100 contributions in the literature, and also 21 U.S. patents.

T. C. Ward, a native of North Carolina, received his B.S. in Chemical Engineering from North Carolina State University in 1963. His M.S. and Ph.D. in Physical Chemistry were granted from Princeton University in 1966. After receiving his Ph.D. he spent two years in England with Professor Manfred Gordon at the University of Strathclyde and the University of Essex. He then joined VPI & SU as an Assistant Professor in 1968. He was promoted to Associate Professor with tenure in 1972 and Professor in 1981. Professor Ward has received several teaching honors, and is a member of the Academy of Teaching Excellence at VPI & SU. He has also been a major contributor to the ACS polymer short course program since its inception in 1976. His publications have been principally in the areas of polymer characterization, testing, adhesion and long-term property evaluation of polymeric materials.

Larry T. Taylor joined the Virginia Tech faculty as an Assistant Professor in 1967 after receiving his B.S. and Ph.D. degrees from Clemson University in 1962 and 1965, respectively. In the interim, he spent 2 1/2 years as a National Institute of Health Postdoctoral Fellow at the Ohio State University working on the synthesis and properties of metal complexes containing macrocyclic ligands related to biological materials. A native of South Carolina, Dr. Taylor was promoted to Associate Professor in 1970 and Professor in 1978. During the summer 1976, he was a NASA-ASEE summer faculty fellow at the Langley Research Center, Hampton, Virginia. Dr. Taylor currently serves as a consultant for the Electric Power Research Institute, the Research Triangle Institute and as a referee for proposals submitted to the Department of Energy, the Research Corporation and the U.S.-Israel Binational Science Foundation. Approximately 90 coauthored manuscripts have been published. He recently completed a review entitled "Incorporation of Metal Related Materials into Electrically Neutral Polymers".

TABLE I

INTRINSIC VISCOSITIES OF
POLY (IMIDE SILOXANE) COPOLYMERS

| PSX WT% | PSX Mn | $[n]$ 25°C, NMP |
|---|---|---|
| 10 | 950 | 0.69 |
| 10 | 2100 | 0.66 |
| 10 | 5000 | 0.71 |
| 10 | 10,000 | 0.73 |
| 20 | 950 | 0.84 |
| 20 | 2100 | 0.79 |
| 20 | 5000 | 0.51 |
| 40 | 950 | 0.55 |
| 60 | 2100 | 0.57 |

TABLE II

UPPER GLASS TRANSITION TEMPERATURES
of POLY (IMIDE SILOXANE) COPOLYMERS

| PSX WT% | PSX Mn | Tg (°C) DSC | DMTA |
|---|---|---|---|
| CONTROL | --- | 272 | --- |
| 10 | 950 | 256 | 260 |
| 10 | 2100 | 267 | 267 |
| 10 | 5000 | 264 | --- |
| 10 | 10,000 | 264 | 266 |
| 20 | 950 | 246 | 248 |
| 20 | 2100 | 259 | --- |
| 20 | 5000 | 262 | --- |
| 40 | 950 | * | 225 |

* NO TRANSITION DETECTED IN DSC SCAN

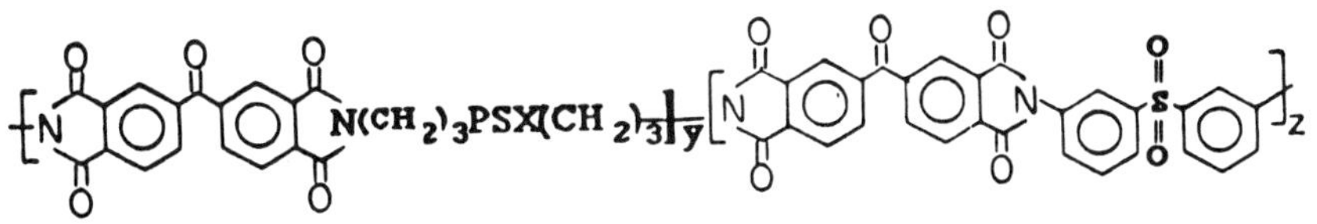

Figure 1: Representative Structure for Poly (Imide Siloxane) Segmented Copolymers

593

## TABLE III

### XPS ANALYSIS OF
### POLY (IMIDE SILOXANE) COPOLYMERS

| PSX WT% | PSX Mn | TAKE-OFF ANGLE | WT% PSX at SURFACE |
|---|---|---|---|
| 5 | 950 | 15° | 85 |
| 5 | 950 | 90° | 34 |
| 10 | 950 | 15° | 77 |
| 10 | 950 | 90° | 35 |
| 10 | 10,000 | 15° | 87 |
| 10 | 10,000 | 90° | 39 |
| 20 | 950 | 15° | 87 |
| 20 | 950 | 90° | 53 |
| 40 | 950 | 15° | 86 |
| 40 | 950 | 90° | 63 |

## TABLE IV

### WATER CONTACT ANGLE MEASUREMENTS
### of POLY (IMIDE SILOXANE) COPOLYMERS

| PSX WT% | PSX Mn | WATER CONTACT ANGLE (DEGREES) |
|---|---|---|
| 5 | 950 | 68 |
| 10 | 950 | 92 |
| 10 | 2100 | 100 |
| 10 | 5000 | 98 |
| 10 | 10,000 | 100 |
| 20 | 950 | 101 |
| 20 | 2100 | 102 |
| 20 | 5000 | 102 |
| 40 | 950 | 106 |
| 40 (RECAST) | 950 | 105 |
| 60 | 2100 | 106 |

## TABLE V

### OXYGEN PLASMA STABILIITY of
### POLY (IMIDE SILOXANE) COPOLYMERS

(PSX Mn of 1000; BASED UPON EITHER 3,3'-DDS or 3,3' ODA; AVERAGES of MULTIPLE RUNS)

| SAMPLE | WEIGHT LOSS (MG PER CM$^2$) |
|---|---|
| KAPTON | 0.86 |
| DDS-CONTROL | 0.86 |
| DDS- 10% PSX | 0.75 |
| DDS- 20% PSX | 0.67 |
| DDS- 30% PSX | 0.24 |
| DDS- 50% PSX | 0 |
| ODA-CONTROL | 1.13 |
| ODA- 30% PSX | 0.30 |
| ODA- 50% PSX | 0.14 |
| DDS- 30% PSX ON KAPTON* | 0.35 |
| DDS-50% PSX ON KAPTON* | 0.10 |

* SPRAY-COATING OF POLY (IMIDE SILOXANE) COPOLYMER
~ 0.03mm THICKNESS ON KAPTON

## TABLE VI

### PRELIMINARY ADHESIVE RESULTS

| PSX WT% | PSX Mn | [n] 25°C, NMP | LAP SHEAR STRENGTH (PSI) |
|---|---|---|---|
| CONTROL | --- | --- | 2510 |
| 5 | 950 | --- | 2480 |
| 10 | 950 | 0.40 | 2360 |
| 20 | 950 | 0.36 | 2110 |
| 40 | 950 | 0.34 | 1860 |

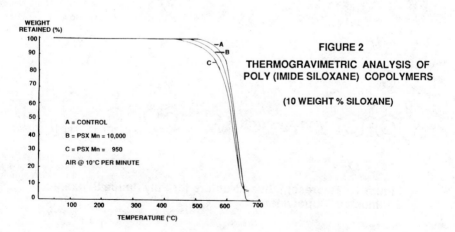

### FIGURE 2

### THERMOGRAVIMETRIC ANALYSIS OF POLY (IMIDE SILOXANE) COPOLYMERS

### (10 WEIGHT % SILOXANE)

A = CONTROL
B = PSX Mn = 10,000
C = PSX Mn = 950
AIR @ 10°C PER MINUTE

594

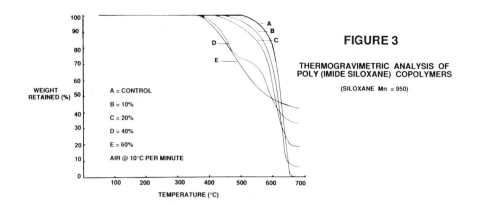

FIGURE 3

THERMOGRAVIMETRIC ANALYSIS OF
POLY (IMIDE SILOXANE) COPOLYMERS

(SILOXANE Mn = 950)

A = CONTROL
B = 10%
C = 20%
D = 40%
E = 60%
AIR @ 10°C PER MINUTE

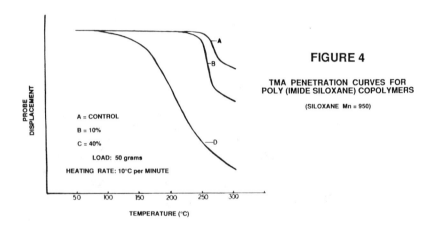

FIGURE 4

TMA PENETRATION CURVES FOR
POLY (IMIDE SILOXANE) COPOLYMERS

(SILOXANE Mn = 950)

A = CONTROL
B = 10%
C = 40%
LOAD: 50 grams
HEATING RATE: 10°C per MINUTE

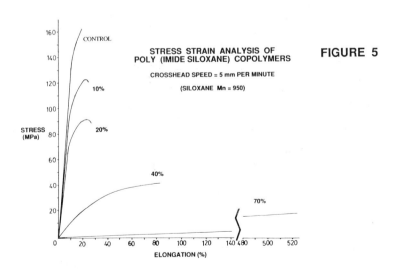

STRESS STRAIN ANALYSIS OF
POLY (IMIDE SILOXANE) COPOLYMERS

CROSSHEAD SPEED = 5 mm PER MINUTE

(SILOXANE Mn = 950)

FIGURE 5

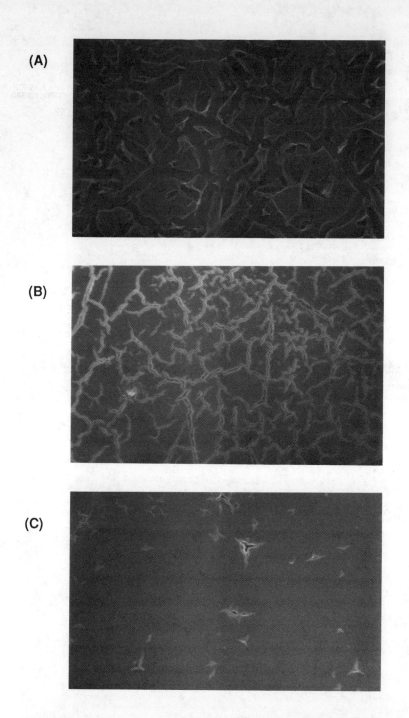

**FIGURE 6:** Scanning Electron Micrographs at 1600 X Magnification of DDS-based polyimides and siloxane-modified polyimides after exposure to oxygen plasma. (A) No Siloxane  (B)10 wt % Siloxane and  (C) 20 wt % Siloxane  (Siloxane Mn = 950).

32nd International SAMPE Symposium
April 6-9, 1987

ALLYLNADIC-IMIDES -
A NEW CLASS OF HEAT RESISTANT THERMOSETS

M.A. Chaudhari, Byung Lee, John King
CIBA-GEIGY Corporation
Ardsley, New York 10502
Dr. A. Renner
CIBA-GEIGY, Marly Switzerland

ABSTRACT

CIBA-GEIGY Corporation has been very instrumental in the advancement of high performance resin technology and has in the past introduced state of the art epoxy resin systems and more recently has also responded to the emerging needs of the composite industry by introducing bismaleimide and polyimide systems. As a result of our continuous interest and ongoing research in the high temperature resins, we have now developed a new class of unsaturated thermosetting polyimide resins, which we call "allylnadic-imides". This paper discusses the general synthetic procedure and the salient characteristics of these materials.

Essentially these resins are synthesized from dicyclopentadiene as a starting raw material. Materials ranging from low viscosity liquid to moderate melting point solids have been prepared on the laboratory scale and characterized. Upon thermal polymerization with or without the aid of a catalyst, the cross-linked polymers have glass transition temperatures in excess of 300°C with excellent thermal/mechanical properties.

1.   INTRODUCTION

Most widely used matrices in advanced composites are currently based on thermoset resins, particularly epoxy resins. The multifunctional epoxy (CIBA-GEIGY Araldite MY720 is the state-of-the-art resin) resins are the materials of choice for the current aerospace applications. However, the future aerospace and military applications require tough and high temperature resistant materials which can withstand more stringent environmental conditions.

In recent years considerable effort has been devoted to the development of high temperature resistant materials. The polyimide class of materials have been found to be the most desirable high temperature materials however, most of these materials require higher temprature and pressure conditions for processing. The processing difficulties encountered with most of the current polyimides have hindered their wider use. From the processing point of view, an ideal material would be a liquid or low melting solid, which can be processed like the state of the art epoxy systems but, upon final curing yield a high glass transition temprature.

CIBA-GEIGY's research team has developed a new class of thermosetting materials, "allylnadic-imides", which range from low viscosity liquids to low melting point solids. These materials are readily soluble in common organic solvents and can be processed into composites using conventional epoxy curing techniques and equipment and upon curing yield polymers with glass transition temperatures in excess of 300°C.

This paper gives the general synthetic procedure for this class of materials and lists some cured neat resin properties for the three experimental products. More detailed chemistry and composite data will be a subject of another paper at a later date.

## CHEMISTRY

### 2. GENERAL SYNTHETIC ROUTE

. Dicyclopentadiene is used as a starting material for these resins. Dicyclopentadiene is thermally cracked and the resulting monocyclopentadiene is transformed into the desired products via different routes. One general route is shown below:

## SYNTHESIS OF ALLYLNADIC-IMIDES

### 3. DEVELOPMENTAL PRODUCTS AND THEIR CHARACTERISTICS

As can be seen from the synthetic scheme, variety of allylnadic monomers with aliphatic, cycloaliphatic or aromatic substituents are possible. In fact, many dozens of such materials have been synthesized in the laboratory. These materials range from low viscosity liquids to low melting solids. Out of these large number of candidates, at this time, three experimental products have emerged and are at the process development stage.

The rest of this paper will concentrate on the processing and performance characteristics of the following three experimental materials:

. RD86-181
. RD86-182
. RD86-183

The neat resin properties presented in this paper were obtained by thermal homopolymerization of the allylnadic-imide material using the following cure cycle:

3 Hours @ 200°C +
3 Hours @ 225°C +
12 Hours @ 250°C +

Although the data shown here was obtained upon materials cured per this somewhat cumbersome cure schedule, this cure cycle is not optimized by any means, in fact we have now found that the allylnadic cure can be accelerated with cationic catalysts in general. In particular latent sulfonic acid catalysts are very efficient with these materials. Figures 1 and 2 are the DSC scans showing the effect of 1% sulfonic acid catalyst on the reactivity of the RD86-181. Neat properties obtained with 1% addition of the catalyst and curing for 2 Hr/ 250°C, are very comparable to the thermally cured materials using the extended cure cycle.

- All mechanical testing was done per ASTM procedures.

- Dupont's thermal analyzer unit 9900, equipped with DSC, TGA and TMA modules, was used for thermal properties.

## EXPERIMENTAL PRODUCT RD86-181

RD86-181 is a solid resin with 80 - 85°C softening point and is readily soluble in solvents such as Toluene, Xylene, MEK and THF etc. Typical chemical structure is:

CURED NEAT RESIN PROPERTIES:
(THERMAL HOMOPOLYMERIZATION)

Cure Cycle:  3 Hrs/200°C  +  3 Hrs./225°C + 12 Hrs/250°C

Flexural Strength    (ksi)  14.9
Deflection                  3.6%
HDT                         250°C
Glass Transition Temp. $T_g$  360°C
Water Pick up (2 Hr/100°C) 0.21%
TGA (Air) Decomposition Begins @ 396°C

10% Wt. Loss @ 436°C

THERMAL AGING RESULTS (IN AIR)
The specimens were aged in air atmosphere at 250°C and 275°C for 30 days prior to testing at room temperature.

|                    | 250°C | 275°C |
|--------------------|-------|-------|
| Flex Strength, ksi | 15.2  | 9.9   |
| Wt. Loss, %        | 2.4   | 3.8   |

EXPERIMENTAL PRODUCT RD86-182

RD86-182 is a very low viscosity liquid and is represented by the following chemical structure:

CURED NEAT RESIN PROPERTIES:

Cure Cycle:  1  Hr/200°C  +  1 Hr/220°C + 12 Hrs/250°C

Tg °C                        337

Flexural Strength, ksi  12.9

Deflection %             2.5

Water Pick Up
(1 Hr./100°C), %         0.68

Upon thermal or catalytic homopolymerization, the resultant polymer has glass transition temperatures in excess of 300°C. Because of its low viscosity and good thermal stability, RD86-182 can be used as a reactive diluent and/or a polymerizable solvent as a processing aid in other high temperature systems.

Experimental Product RD86-183

RD86-183 is a high viscosity resinous material at room temperature with 200-300 cps viscosity @ 80°C. Typical chemical structure is:

CURED NEAT RESIN PROPERTIES:

CURE CYCLE:  1 Hr/200°C + 1 Hr/220°C + 12 Hrs/250°C

R.T. Flexural Strength, ksi= 13.0
    R.T. Deflection, %    = 4.5
    HDT, °C                 247
    $T_g$, °C              = 302
    Thermal Decompositon in Air
              Begins @ 264°C
              10% Wt. Loss @ 419°C

Because of the aliphatic chain in the backbone, RD86-183 is comparatively less thermally stable.

## SUMMARY

Liquid or low melting unsaturated polyimide resins have been developed on the laboratory scale and scale up work is in progress. These materials can be processed comparatively easily and thermal or catalytic cure results in polymers which have extremely good thermal/mechanical performance characteristics. The exact curing mechanism is still under investigation and will be a subject of another paper at a later date, however, whatever the exact mechanism may be, cyclopolymerization, Diels-Alder rearrangement or most likely combinations of these, the resultant crosslinked polymers have extremely good heat resistance and very desirable mechanical properties. It should be noted that these materials are at research stages and the scale up work is in progress.

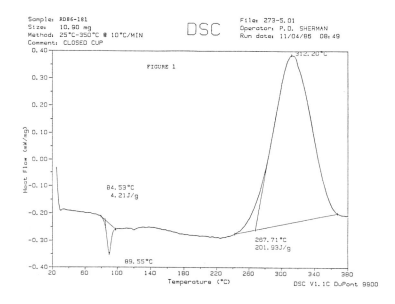

Sample: RD86-181
Size: 10.90 mg
Method: 25°C-350°C @ 10°C/MIN
Comment: CLOSED CUP

DSC

File: 273-5.01
Operator: P.D. SHERMAN
Run date: 11/04/86  08:49

FIGURE 1

312.20°C

84.53°C
4.21J/g

89.55°C

267.71°C
201.93J/g

DSC V1.1C DuPont 9900

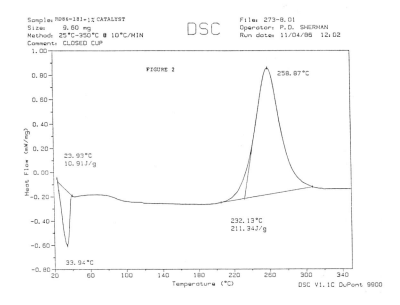

Sample: RD86-181+1% CATALYST
Size: 9.60 mg
Method: 25°C-350°C @ 10°C/MIN
Comment: CLOSED CUP

DSC

File: 273-8.01
Operator: P.D. SHERMAN
Run date: 11/04/86  12:02

FIGURE 2

258.87°C

23.93°C
10.91J/g

33.94°C

232.13°C
211.34J/g

DSC V1.1C DuPont 9900

PMR POLYIMIDE COMPOSITIONS
FOR IMPROVED PERFORMANCE AT 371°C
Raymond D. Vannucci
NASA Lewis Research Center
Cleveland, Ohio

## Abstract

Studies were conducted to identify
matrix resins which have potential
for use at 371°C (700°F). Utilizing PMR methodology, neat resin
moldings were prepared with various
monomer reactants and screened for
thermo-oxidative stability at 371°C
(700°F) under both ambient and four
atmospheres air pressure. The
results of the resin screening
studies indicate that high molecular weight (HMW) formulated resins
of first (PMR-15)[1] and second
(PMR-II)[2] generation PMR materials
exhibit lower levels of weight loss
at 371°C (700°F) than PMR-15 and
PMR II resins. The resin systems
which exhibited the best overall
balance of processability, Tg and
thermo-oxidative stability at 371°C
were used to prepare unidirectional
Celion 6000 and T-40R graphite
fiber laminates. Laminates were
evaluated for thermo-oxidative
stability and 371°C mechanical
properties. Results of the laminate evaluation studies indicate
that two of the resin compositions
have potential for use in 371°C
applications. The most promising
resin composition provided
laminates which exhibited no drop
in 371°C mechanical properties
and only 11 percent weight loss
after 200 hours exposure to 4
atmospheres of air at 371°C.

## 1. INTRODUCTION

The objective of the polymers
research being conducted at the
Lewis Research Center is to develop
technology for new generations of
organic polymers intended for
application in advanced aeropro-
pulsion systems. Studies to
improve aircraft engine performance
have indicated that advanced
designs will dictate higher thrust-
to-weight ratios than present day
levels. This can be achieved
through the use of advanced lightweight composite materials. To
meet the requirements for high
temperature advanced composites,
Lewis developed the PMR polyimides[1]. Today PMR polyimides are

the leading matrix resins for high temperature advanced polymer matrix composites and are commercially available from a number of suppliers. The technology for application of these advanced composite materials is developing rapidly. At the present time, PMR materials are being used in a number of composite engine components currently in production or soon to be introduced into production. While current engine applications of PMR materials have resulted in some significant cost and weight savings, their application has been limited to engine sections operating in the 200 to 300°C (392 to 572°F) temperature range. Further benefits can be realized by moving composites into the higher temperature regions on engines.

The purpose of this study was to investigate higher molecular weight (HMW) PMR resin formulations and to determine the effect of these formulations on resin and composite physical properties, thermo-oxidative stability, and mechanical properties after exposure in air at 371°C (700°F).

## 2. EXPERIMENTAL PROCEDURE

### 2.1 Resin Preparation

The monomer reactants used to prepare the resins investigated in this study are shown in Table I. All of the monomers except the dimethylesters were purchased from commercial suppliers. The dimethylesters (BTDE and HFDE) were prepared as 50 w/o methanol solutions by refluxing a suspension of the corresponding dianhydride in anhydrous methanol until all solids had dissolved and then for an additional two hours.

PMR reactant solutions were prepared at room temperature by dissolving the reactants in anhydrous methanol to form 30 - 50 w/o solids solutions. The stoichiometric ratio of the monomers used for the addition curing resins prepared was as follows: n moles of dimethylester, n + 1 moles of diamine, and 2 moles of NE.

Neat resin moldings were prepared by placing solutions containing 15 grams of solids into an air circulating oven set at 121°C (250°F) until all of the solvent had evaporated and the reactants were partially imidized. The dry material was then crushed and the oven temperature increased to 177 to 204°C (350 to 400°F), depending on resin formulated molecular weight (FMW), for 1 hour to complete the imidization to the end-capped prepolymer. The imidized material was then ground into a fine powder. Approximately 4 grams of powder was then placed into a 2 inch diameter metal die at room temperature. The die (with thermocouples attached) was placed into a press preheated to 316 to 343°C (600 to 650°F). When the temperature of the die reached 260°C (500°F), a pressure ranging between 6.9 to 13.8 MPa (1000 - 2000 psi) was applied. Final cure temperature and pressure was dependent on the FMW of the resin. Higher FMW's required higher cure temperatures and pressures. Final cure temperature and pressure was maintained for two hours. After curing, the molding were allowed to cool under pressure to 232°C. At 121°C (250°F) the moldings were removed from the die and postcured in an air circulating oven programmed to heat at 6°C/min. to 260°C and then 1°C to 371°C, followed by a 24 hour hold at 371°C.

## 2.2 Laminate Fabrication

All laminates were prepared from unidirectional graphite reinforced prepreg tape. The reinforcements used in this study included Celion 6000 and Amoco's T-40R graphite fiber. Prepreg tapes were prepared by brush application of resin solutions onto drum wound unidirectional fiber calculated to yield laminates having 58 v/o fiber after curing. The prepreg was allowed to dry on the drum under quartz lamps to a volatile content of 11-12 w/o. The prepreg was then removed from the drum and cut into 7.62 cm by 20.3 cm (3 in. by 8 in.) plies and a number of plies stacked unidirectionally into a preform staging tool to yield laminates having a cured thickness of .20 to .23 cm (080 to 090 in.). The prepreg stacks were then heated under 0.1 psi of pressure for 60 minutes at temperatures of 177 to 204°C (350 to 400°F). Laminates were then compression molded by placing the staged layup into a flat matched metal die at room temperature and then inserting the die into a press preheated to either 316°C (600°F) or 343°C (650°F). When the die reached 232°C (450°F) a pressure ranging from 3.45 to 17.3 MPa (500 to 2500 psi) was applied. When the die reached the final cure temperature, pressure and temperature were maintained for two hours. The laminates were than allowed to cool slowly to 232°C (450°F) (~45 minutes) under pressure and then fast cooled without pressure to room temperature. All laminates were than postcured according to the same postcure schedule used for neat resin moldings.

## 2.3 371°C Isothermal Aging

Isothermal weight loss measurements were performed on neat resin and laminates after exposure to air at 371°C under both 1 atm. and 4 atm. of pressure. A forced air oven was used for 1 atm. exposure and the air change rate employed was 100 cc/min. The air change rate for 4 atm. exposures was 5 air changes per hour in a 2.0 liter autoclave chamber.

## 3. LAMINATE EVALUATION

Prior to testing all laminates were inspected for porosity using an ultrasonic C-Scan technique.

## 3.1 Flexural and Interlaminar Shear Tests

Flexural strength and interlaminar shear strength (ILSS) tests were performed on specimens ranging in thickness from 0.20 to 0.23 cm (.080 to .090 in). Flexural strength tests were performed on 0.635 cm (.25 in) by 7.62 cm (3.0 in) specimens in accordance with ASTM D-790 at a constant span/depth of 28. ILSS tests were performed on 0.635 cm (.25 in.) wide specimens in accordance with ASTM D-2344 at a span/depth of 5. Elevated temperature tests were performed in an environmental heating chamber. Property values reported are averages of 3 to 6 tests.

4. RESULTS AND DISCUSSION

4.1 Resin Screening Studies

Table I lists the monomer reactants used to prepare the resins investigated. Table II identifies the composition of each of the resins. All of the resins contained NE. The stoichiometric ratio of the reactants used for each of the resins was in the ratio $n/(n + 1)/2$, where n = moles of dimethylester, n + 1 = moles of diamine and 2 = moles of NE. The value of n ranged between 1.67 and 14.5. The formulated molecular weights (FMW) ranged between 1270 and 7500 for the end-capped pre-polymers prior to final curing of the resin. The number included in the resin designation corresponds to the FMW of the resin, i.e., 15 = 1500, 30 = 3000. Two of the formulations contained a mixture of diamines and were designated MD. Also shown in Table II is the

weight percent of NE present in each of the resin formulations.

Table III lists the percent weight loss for each of the neat resins after exposure to air at 371°C under both one atmosphere and four atmospheres of pressure. Also shown are the resin glass transition temperatures (Tg's) after 24 hours and 50 hours exposure to air at 371°C and 1 atm. of pressure.

The 4 atm., 371°C exposure condition was selected to simulate the condition which composites would experience in an engine zone operating at 371°C.

The table shows that the 371°C oxidative stability of both PMR and PMR-II resin formulations improve as the FMW is increased. This is a result of the reduction in alphatic content due to the Nadic endcap with increasing FMW as shown in Table II. However, the data shows that PMR resin FMW's higher than 5000 offer no further improvements in 371°C oxidative stability. This is probably due to difficulty in processing resins having high FMW's. It can be seen that the PMR-II resins exhibit considerably better thermo-oxidative resistance (TOS) that the other formulations investigated. This is due to the more thermally stable reactants (HFDE, PPDA) used in the PMR-II resins. Note that PMR-II-13 having the highest NE content (25.2 w/o), exhibited comparable weight loss, under both exposure conditions, to that of the higher FMW PMR and MD resins which contained considerably lower NE content. The excellent TOS exhibited by the PMR-II-50 resin was achieved in spite of the presence of blisters which developed during the 24 hour 371°C postcure prior to the aging study. Blistering might be due to the presence of unreacted material and the low Tg exhibited by the HMW PMR-II-30 resin.

Comparing the Tg's of resins after 50 hours exposure under 1 atm. shows that the resins containing BTDE and a mixture of the diamines BDAF and PPDA (MD-60 and MD-64), exhibit significantly higher Tg's than the other formulations. While the NE content of these HMW formulations is ~5 w/o, there appears to be a high degree of oxidative cross-linking taking place as a result of the carbonyl and ether linkages present in BTDE and BDAF, respectively.

4.2 Laminate Studies

The resins selected for laminate studies included all the resins listed in Table II except MD-64 and PMR-II-50. The reinforcement used to prepare the laminates was unidirectional Celion 6000 graphite fiber.

Figure 1 compares the weight loss of the laminates as a function of exposure time in air at 371°C and 4 atm. of pressure. The figure shows that, as expected, the PMR-II-30 laminate exhibited the lowest weight loss (12.5 w/o) while the PMR-15 laminate exhibited the highest weight loss (24 w/o) after only 120 hours of exposure. Laminates prepared from all other resins exhibited comparable weight loss (23 - 24 w/o) after 200 hours exposure.

Figure 2 compares the retention of interlaminar shear strength (ILSS) of the laminates after exposure to 4 atm. of air at 371°C. The figure shows that the laminates prepared from PMR-15 and MD-60 resin, which exhibited the highest resin Tg's, also exhibited the highest initial ILSS when tested at 371°C. The MD-60 laminate retained 65 percent of its initial ILSS after 200 hours exposure, while the PMR-15 laminate retained only 50 percent of its initial 371°C ILSS after only 120 hours of exposure. The laminates prepared from PMR-II-13 and PMR-II-30 exhibited the lowest initial 371°C ILSS, but essentially retained that strength throughout the exposure time.

Figure 3 compares the retention of 371°C flexural strength for the same laminates exposed under the same conditions discussed above. The figure shows that the 371°C flexural strength retention of the laminates are in close agreement with the results shown for 371°C ILSS retention.

Based on the results of studies shown so far, it can be concluded that the PMR-II-30 resin offers the highest potential for use as a matrix resin for 371°C composite applications. It must be noted that while the laminates prepared from the MD-60 resin exhibited nearly twice the weight loss shown for the PMR-II-30 laminates during exposure to 371°C air, the laminates prepared from MD-60 resin exhibited the highest mechanical properties over very nearly the entire 371°C exposure time. However, MD-60 laminates as well as the laminates prepared from PMR-15, 30, 50 resins exhibited a significant amount of loose surface fiber after 100 hours exposure to 371°C air at 4 atm., while the PMR-II-30 laminate exhibited no loose surface after 200 hours of exposure. In order to determine any possible effect of the reinforcing fiber on the oxidative weight loss of the laminates, additional laminates were fabricated using Amoco's T-40R graphite fiber which exhibits significantly better oxidative stability during exposure to air at 371°C and 4 atm. pressure than Celion 6000 graphite fiber.

Figure 4 compares the weight loss of laminates prepared from MD-60 and PMR-II-30 resins with both Celion 6000 and T-40R fiber reinforcements during exposure to air at 371°C and 4 atm. of pressure. Also shown are the weight losses of Celion 6000 and T-40R bare fiber exposed to the same conditions for 120 hours. The figure shows that after 200 hours of exposure, the T-40R/MD-60 laminate exhibited 5 percent less weight loss than the Celion 6000/MD-60 laminate. The T-40R/PMR-II-30 laminate exhibited

3 percent less weight loss than the Celion 6000/PMR-II-30 laminate after 200 hours of exposure. The T-40R/MD-60 laminate still exhibited loose surface fiber after 100 hours exposure, but to a lesser degree.

Figure 5 compares the 371°C flexural strength retention of the laminates compared in Figure 4 as a function of exposure time in 371°C air at 4 atm. It can be seen that the flexural properties of the T-40R/MD-60 were slightly lower than those of the Celion 6000/MD-60 laminate over the entire exposure time. No appreciable differences are apparent for the flexural strength retentions shown for the two laminates prepared from the PMR-II-30 resin. Note that, again, at the end of 200 hours exposure, the MD-60 system seems to be losing strength rapidly, while the PMR-II-30 system curve remains flat.

Based on the results of this study, the following conclusions can be drawn:

1. A PMR-II resin composition has been identified (PMR-II-30) which has potential for use in engine zones operating at temperatures up to 371°C.

2. The use of higher FMW PMR compositions results in enhanced oxidative stability during exposure to air at 371°C.

3. Higher FMW PMR compositions require higher cure pressures to yield high quality laminates.

### 5. REFERENCES

1. Serafini, T. T., Delvigs, P.
   and Lightsey, G. R.: Thermally
   Stable Polyimides from Solutions of Monomeric Reactants,
   Journal of Applied Polymer
   Science, Vol. 16, No. 4, April
   1972.
2. Serafini, T. T., Vannucci, R. D.
   and Alston, W. B.:  Second
   Generation PMR Polyimides, NASA
   TM X-71894, 1976.

### 6. BIOGRAPHY

Raymond D. Vannucci has been
employed at NASA Lewis Research
Center since 1965.  He received a
B.S. in Engineering from Cleveland
State University.  His current
research involves the fabrication
and characterization of high temperature polymer matrix composite
materials.

Table 1.-MONOMERS USED FOR POLYIMIDE SYNTHESIS

| STRUCTURE | NAME | ABBREVIATION |
|---|---|---|
| | MONOMETHYL ESTER OF 5-NORBORNENE-2,3-DICARBOXYLIC ACID | NE |
| | DIMETHYL ESTER OF 3,3'4,4'-BENZOPHENONETETRACARBOXYLIC ACID | BTDE |
| | DIMETHYL ESTER OF 4,4'-(HEXAFLUOROISOPROPYLIDENE)-BIS(PHTHALIC ACID) | HFDE |
| | 4,4'-METHYLENEDIANILINE | MDA |
| | P-PHENYLENEDIAMINE | PPDA |
| | BIS(AMINOPHENOXY) PHENYLHEXAFLUOROPROPANE | BDAF |

Table II. COMPOSITIONS OF RESINS INVESTIGATED[a]

| Resin | Diester | Diamine | n | FMW | w/o NE |
|---|---|---|---|---|---|
| PMR-15(control) | BTDE | MDA | 2.09 | 1500 | 21.8 |
| PMR-30 | " | " | 5.2 | 3000 | 10.9 |
| PMR-50 | " | " | 9.3 | 5000 | 6.5 |
| PMR-75 | " | " | 14.5 | 7500 | 4.4 |
| MD-64 | " | BDAF/PPDA 3:2 | 9 | 6400 | 5.1 |
| MD-60 | " | BDAF/PPDA 1:1 | 9 | 6000 | 5.4 |
| PMR-II-13 | HERE | PPDA | 1.67 | 1270 | 25.2 |
| PMR-II-30 | " | " | 5 | 2980 | 10.9 |
| PMR-II-50 | " | " | 9 | 5050 | 6.5 |

(a) Resin stoichiometry:  NE/Diester/Diamine
             Moles:    2/   n   /n + 1

Table III.  RESIN WEIGHT LOSS DURING EXPOSURE TO ONE AND FOUR
ATMOSPHERES OF AIR AT 371°C

| Resin | Tg °C[a] | Tg °C[b] | Percent Weight Loss After: 300 hrs/1 atm. | 75 hrs/4 atm. |
|---|---|---|---|---|
| PMR-15(control) | 370 | 388 | 18.0 | 18.2 |
| PMR-30 | 365 | 375 | 12.0 | 14.0 |
| PMR-50 | 363 | 375 | 13.0 | 13.0 |
| PMR-75 | 358 | 370 | 16.5 | 13.8 |
| MD-64 | 370 | 398 | 16.0 | 17.0 |
| MD-60 | 370 | 403 | 12.2 | 14.0 |
| PMR-II-13 | 368 | 381 | 13.0 | 12.3 |
| PMR-II-30 | 345 | 368 | 8.0 | 6.4 |
| PMR-II-50 | 340 | 355 | 5.5 | 5.0 |

(a) After 24 hour postcure at 371°C
(b) After 50 hours of exposure, 371°C, 1 atm.

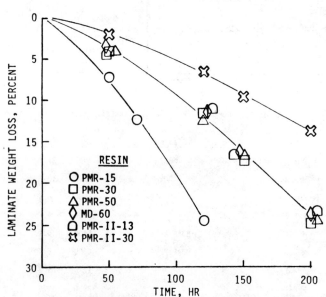

Figure 1.-Weight loss of Celion 6000 graphite/PMR
laminates exposed in 4 atm. of air at 371°C.

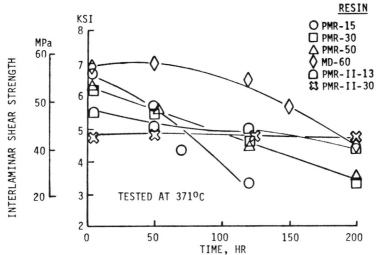

Figure 2.- Interlaminar shear strength of Celion 6000 graphite/PMR laminates after exposure to 4 atm. of air at 371°C

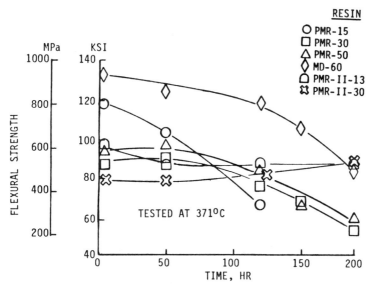

Figure 3.-Flexural strength of Celion 6000 graphite/PMR laminates after exposure to 4 atm. of air at 371°C

611

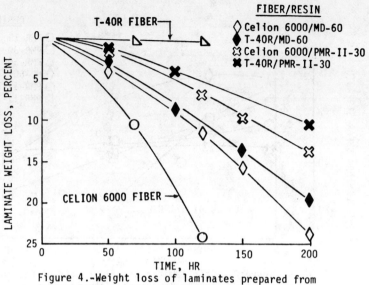

Figure 4.-Weight loss of laminates prepared from
Celion 6000 graphite and T-40R graphite fibers,
after exposure to 4 atm. of air at 371°C

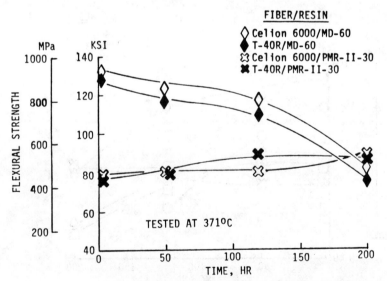

Figure 5.-Comparison of flexural strength retention of laminates prepared from Celion 6000 and
T-40R graphite fiber after exposure to 4 atm. of
air at 371°C

32nd International SAMPE Symposium
April 6-9, 1987

SYNTHESIS AND SOLUTION IMIDIZATION STUDIES OF
SOLUBLE POLY(IMIDE SILOXANE) SEGMENTED COPOLYMERS

J. D. Summers, C. A. Arnold, R. H. Bott,
L. T. Taylor, T. C. Ward, J. E. McGrath*
Department of Chemistry and Polymer Materials and Interfaces Laboratory
Virginia Polytechnic Institute and State University
Blacksburg, VA


*To whom correspondence should be addressed

ABSTRACT

Soluble meta-linked poly(imide
siloxane) segmented copolymers were
synthesized utilizing a THF/NMP
cosolvent system. The presence of
the dual solvent allows one to
reach high molecular weight in the
amic acid stage. Incorporation of
siloxanes at about 10 weight percent or higher enables the materials to be fully soluble in a
range of polar solvents even after
imidization. Imidization may be
achieved either by conventional
thermal methods on cast amic acid
films or in appropriate solvent/
azeotroping agent systems employing
moderate temperatures (150-170°C).
The imidization procedure has been
followed by FT-IR and NMR studies.
FT-IR studies show the solution
imidization follows first order
kinetics and proceeds to ~96%
completion. NMR studies of the
isolated products show residual
amic acid may be present after
solution imidization, but only at
very low levels. Properties of the
solution imidized materials compare
well with those obtained from
samples imidized as thin films.

1. INTRODUCTION

Many publications on siloxane
containing polyimides have shown
that the incorporation of siloxane
segments into a polyimide backbone
can permit the synthesis of copolyimides with fairly good thermal and
mechanical properties[1-4]. These
materials are of great interest in
a wide variety of applications,
including the electronics and aerospace industries[5-7]. Previous work
in our laboratories[8] has shown that
tough, soluble, poly(imide siloxane) segmented copolymers could
be synthesized from amine terminated siloxane equilibrates, meta-
linked diamines containing flexible
bridging units, and BTDA.

The most common synthetic route to
polyimides has been through the
cyclization of poly(amic acid)
intermediates with the loss of
water. The kinetics of this
process as it applies to the imidization of thin solution cast amic
acid films has been the subject of
many publications. The rate of
imidization has been reported to be
dependent on parameters such as the

casting solvent[9,12], film thickness[10], imidization temperature[9-14], and chemical structure[13]. In general, the imidization rate has been observed to slow as the reaction proceeds[8-14]. Solvent loss[14] and decreasing backbone flexibility[11,13] have been postulated as reasons for the decreasing imidization rate. Consequently, temperatures near or above the Tg of the fully cured polyimide must be employed to achieve complete cyclization.

Recently, Takekoshi and coworkers[15] have successfully employed solvent/azeotroping agent systems to solution imidize certain polyetherimides at temperatures of 160°C-180°C. This work suggests that the effects responsible for slowing the rate of imidization are absent if the imidizing species is solvated at all times. Consequently, quantitative imidization can be achieved in solution employing only moderate temperatures.

This publication describes the synthesis and characterization of soluble poly(imide siloxane) copolymers by the thermal cyclization of poly(amic acid siloxane)s in solution using a solvent/azeotroping agent system. The solution imidization technique is studied in a semi-quantitative manner using FT-IR. Imide formation and amic acid conversion are monitored as a function of time and temperature to determine various parameters of the process. Properties of the solution imidized copolymers are also compared with the properties of samples imidized by conventional thermal curing cycles to evaluate the effectiveness of the solution imidization process.

## 2. EXPERIMENTAL

The 3,3'-4,4'-benzophenone tetracarboxylic dianhydride (BTDA) was obtained in high purity from the Chriskev Company and subjected to a thermal treatment prior to use. 3,3'-diaminodiphenylsulfone (DDS) was obtained from FIC Corporation and recrystallized from a deoxygenated methanol/water solution

(90% yield, m.p. 172-173°C). The α,ω-aminopropyl polydimethylsiloxane oligomers were synthesized by methods previously disclosed[16] and their corresponding molecular weights were determined by potentiometric titration of the amine end groups. N-methylpyrrolidinone (NMP, Fisher), tetrahydrofuran (THF, Fisher), and N-cyclohexylpyrrolidinone (CHP, GAF) were distilled from calcium hydride under reduced pressure and stored in round bottom flasks sealed with rubber septa.

Synthesis of the poly(amic acid siloxane) intermediates was conducted in a 3-neck, 250mL round bottom flask fitted with a mechanical stirrer, nitrogen inlet, drying tube, and addition funnel. The actual cosolvent ratio used to synthesize the amic acids varied with the amount of siloxane oligomer to be incorporated. A representative synthetic procedure will be outlined to synthesize a copolymer containing 10 weight percent of a siloxane oligomer whose molecular weight is 950g/m.

The entire apparatus was assembled and flamed to remove residual moisture from the system. BTDA (11.3546g, 0.0352moles) was added to the flask and rinsed in with 40mL of NMP and 30mL of THF. Next, 2.0000g of the siloxane oligomer (0.0009moles) and 20mL of THF were placed in the addition funnel and added dropwise to the stirring dianhydride solution. After stirring for an additional 15 min, 6.7016g (0.0334moles) of 3,3'-DDS was added as a chain extender along with 20mL of NMP and 10mL of THF. The reaction was allowed to proceed for 8 hours. The resulting clear viscous solution was stored at ~10°C until needed.

"Curing" of the poly(amic acid siloxane) intermediate to the imide form was accomplished by two methods. In the first method, the amid acid solution was cast onto glass plates at thicknesses ranging from 5 to 50 mils and the plates were placed in a vacuum oven to remove the reaction solvents.

After this treatment, the films on the plates were imidized in a forced air convection oven at temperatures of 100°C, 200°C and 300°C, each for one hour.

The second imidization method was accomplished in solution using a solvent/azeotroping agent system at moderate temperatures. A multineck 500mL round bottom flask was fitted with a mechanical stirrer, nitrogen inlet, thermometer, Dean Stark trap, condenser, and drying tube. Next, 36mL of NMP and 24mL of CHP were added to the flask and heated to 150°C. The previously synthesized poly(amic acid siloxane) in its original NMP/THF solution was then added to the flask. The THF boiled rapidly and was efficiently removed from the system by the steady nitrogen flow and collected in the Dean Stark trap. The solution imidization was then allowed to proceed for 24 hours at 150°C, after which time the dark viscous solution was cooled to room temperature and diluted to 10% solids using NMP. The poly(imide siloxane) was then isolated by precipitation in methanol/water, followed by vacuum drying at 100°C. To further purify the final product, it was redissolved in DMAC and reisolated in the manner just described. Poly(imide siloxane) was obtained in 85% yield.

To monitor the extent of imidization, samples were removed from the reaction at certain time intervals during the course of the imidization and analyzed using a Nicolet MX-1 FT-IR. Samples were cast onto glass plates and solvent removal was accomplished in a vacuum oven at 70°C for one hour. The resulting thin films were removed from the glass and analyzed for imide and amic acid content.

Intrinsic viscosities of the imidized copolymers were determined in NMP at 24°C. Proton NMR spectra were obtained using an IBM 270MHz instrument. Samples were dissolved in deuterated DMSO for analysis. Glass transition temperatures of the poly(imide siloxane) copolymers were determined on a Perkin-Elmer

Model-2 DSC. Samples were run at 10°C per minute in a nitrogen atmosphere. All samples were run at least two times to accurately determine glass transitions. The thermal stabilities of the siloxane modified polyimides were investigated with a Perkin-Elmer System-2 Thermogravimetric/Thermomechanical analyzer. TGA scans were run at 5°C/min in an air atmosphere.

## 3. RESULTS AND DISCUSSION

The general synthetic scheme for the siloxane modified poly(amic acid) copolymers and their conversion to the imidized form is depicted in Figure 1. The removal of moisture from the reaction system was necessary to obtain high molecular weight amic acids. For this reason, monomers and solvents were dried before use and the distilled solvents were stored in sealed flasks. When needed, these solvents were handled using syringe techniques.

Maintaining solubility of the reactants and products throughout the polymerization was also a major factor in obtaining tough, transparent films of predictable composition. A single solvent polymerization media could not accomplish this goal due to the vastly different solubility characteristics of the siloxane oligomers and the aromatic monomers used. Therefore, a cosolvent system consisting of THF and NMP was developed to achieve complete solvation of all species during the polymerization. Capping the siloxane oligomer with excess dianhydride prior to chain extension with 3,3' DDS served to increase the oligomers solubility in NMP, thus increasing the efficiency of the THF/NMP cosolvent system.

### 3.1 Imidization
The poly(amic acid siloxane) copolymers were converted to the corresponding imidized materials by the conventional high temperature curing of poly(amic acid) films (bulk imidization) or by solution imidization techniques at moderate temperatures using solvent/azeo-

troping agent systems. To insure complete imidization when using the former technique, the amic acid films were exposed to an upper cure temperature of 300°C for one hour. This temperature was above the upper Tg of the copolymers and allowed complete imidization without decomposing the siloxane portion of the material. A proton NMR spectrum of a bulk imidized copolyimide containing 40 weight percent siloxane is shown in Figure 2. The spectrum shows a strong resonance at ~0ppm, indicative of siloxane methyls and a series of aromatic proton resonances. The absence of peaks above 10ppm due to residual carboxylic acid structures indicates quantitative conversion of the amic acid. Also, the amount of siloxane in the copolymer could be determined by ratioing the peak areas of the aromatic proton region and the siloxane methyls. The amount of siloxane actually incorporated into the copolymer agreed well with the amount charged during the copolymer synthesis.

The thermal solution imidization technique was studied in a semi-quantitative manner using FT-IR to determine various parameters of the process. Imide formation and amic acid conversion were monitored as a function of time to determine the amount of imidization and to see if the disappearance of amic acid closely paralleled the appearance of the imide moiety. The effect of temperature on the rate and extent of imidization was also studied. The data were then analyzed kinetically to obtain rate constants for the process at various temperatures.

A solvent/azeotroping agent system consisting of NMP and CHP was chosen as the imidization medium. In this system, NMP had the role of solvating the molecular species. CHP was used to effectively remove the water of imidization from the solution to prevent hydrolysis of any uncyclized amic acid. CHP is an interesting solvent that becomes immiscible with water at elevated temperatures, enabling it to act as an azeotroping agent. Early

attempts using more conventional azeotroping agents such as toluene were unsuccessful due to premature precipitation of partially imidized copolymers containing low amounts of siloxane. This problem was alleviated by the use of CHP.

FT-IR bands at $1778cm^{-1}$ (10) and $725cm^{-1}$ (11,14) were used to monitor the appearance of imide structures. The absorbance at $1546cm^{-1}$ (11,17) was studied to evaluate the disappearance of amic acid. Fortunately, these bands were well isolated and did not overlap with bands due to residual solvent. Small quantities of samples were removed during the course of the solution imidization and quickly cooled to room temperature to halt the imidization process. The solutions were then cast as thin films onto glass plates and the solvents were evaporated under vacuum at 70°C for one hour. This procedure removed much of the solvent from the polymer films without allowing significant amounts of further imidization to occur. The films were then removed from the glass and absorbtion FT-IR spectra were obtained for each sample.

Absorbance intensities were measured and ratioed to a reference band to normalize variations in film thickness. To utilize the IR absorbance ratio method, this reference band must depend only on film thickness and not on solvent content, degree of imidization, or side reactions[12]. Such a band seemed to occur in our system at $1480cm^{-1}$ and is probably assignable to an aromatic structure. A fully imidized sample was prepared by employing conventional thermal imidization methods using an upper cure temperature of 300°C. The FT-IR spectrum of this sample was used in conjunction with the spectrum of a poly(amic acid siloxane) film dried under vacuum at 70°C for one hour to set up standard reference states between which percent imidization was calculated.

The results of the FT-IR study on the solution imidization process are shown in Figure 3. In this figure, imide content is plotted against time at the three temperatures studied. Intensities of the imide band at 1778cm$^{-1}$ have been omitted for reasons of clarity, even though they are not inconsistent with the data that are presented. Several observations can be made based on this study. First, imidization proceeds almost quantitatively in solution (~96%) at temperatures that are only suitable to partially imidize the material as a bulk amic acid film. Also, imidization proceeds faster at higher solution temperatures. In many cases, the appearance of the imide linkage is closely paralleled by the disappearance of the amic acid species, suggesting the absence of significant side reactions and/or degradation mechanisms.

To further assess the extent of imidization, a proton NMR of the isolated solution imidized material was obtained, Figure 4. The NMR is very similar in appearance to the proton NMR of the bulk imidized material described earlier. The only difference in the the two spectra is the presence of a very small peak at ~11ppm in the case of the solution imidized material, indicating that a trace of amic acid may still be present, but only at very low concentrations.

The first order kinetic plots of the solution imidization data in terms of the appearance of imide band at 725cm$^{-1}$ are shown in Figure 5. The linear nature of these plots indicates the solution imidization process follows first order kinetics up to roughly 96% imidization. Kinetic interruption of the process does not seem to occur in solution as it does during bulk imidization, presumably because the imidizing species are solvated in the former case. First order rate constants were calculated for each temperature, and are listed in the figure. Rate constants at other temperatures must be obtained before the activation energy for the process can be accurately determined.

To further study the effectiveness of the solution imidization process, efforts were concentrated on characterizing the final products. Identical poly(amic acid siloxane) samples were imidized by both of the thermal methods described in this paper and were compared in terms of their intrinsic viscosities, upper glass transitions, and overall thermal stability.

Table 1 lists the intrinsic viscosities and the upper glass transition temperatures of the bulk and solution imidized materials for the limited amount of samples evaluated thus far. The numbers are observed to be fairly independent of the method of imidization, indicating again that solution imidization proceeds to near completion without degradation.

The overall thermal stability as judged by TGA was also fairly independent of the method of imidization, Figure 6. The solution and bulk imidized samples showed similar initial decomposition temperatures and little premature weight loss due to the evolution of residual solvent or incomplete cyclization.

A major consequence of the solution imidization technique is that the final products are much more soluble and moldable than the same materials cast as amic acids and imidized as thin films. The phenomena responsible for this observation are the subject of current studies.

4. CONCLUSIONS

High molecular weight, randomly coupled, soluble, segmented siloxane modified polyimides were synthesized using amine terminated poly(dimethyl siloxane) oligomers, an aromatic meta-linked diamine, and BTDA. A cosolvent system was employed to achieve complete solvation of all the components throughout the polymerization. Imidization was accomplished in solution using an NMP/CHP solvent/azeotroping agent system at temperatures ranging from 150-

170°C. The solution imidization proceeded to ~96% completion as judged by FT-IR and proton NMR studies. The FT-IR studies also suggested the solution imidization could be described entirely by first order kinetics. The intrinsic viscosities, glass transitions, and TGA traces of the solution imidized copolymers were similar to those obtained from samples imidized by conventional thermal curing cycles, indicating near quantitative solution imidization without copolymer degradation.

## 5. ACKNOWLEDGEMENTS

The authors would like to thank the researchers of NASA Langley Research Center, especially the late George Sykes, for their comments and financial support throughout the course of this work. We also thank J. M. Hoover for his assistance in FT-IR interpretation.

## 6. REFERENCES

1. Maudgal, S. and St. Clair, T. L. Int. J. Adhesion and Adhesives 4(2), 87 (1984).
2. Maudgal, S. and St. Clair, T. L. Prodeedings From The Second International Conference on Polyimides, Ellenville, NY, 1985, pp 47-73.
3. Babu, G. N. in "Polyimides: Synthesis, Characterization, and Applications", vol. 1, Mittal, K. L. ed. Plenum, NY, 1984, pp 51-66.
4. Johnson, B.C., Yilgor, I., and McGrath, J. E. Polym. Prepr. 25(2), 54 (1984).
5. Lee, C. L. "High Performance Silicone-Imide Copolymers", SAMPE Symposium No. 30, (1985), pp 52-63.
6. Berger, A. "Modified Polyimides by Silicone Incorporation", SAMPE Symposium No. 30, (1985), pp 64-73.
7. Davis, G. Photosensitive Polyimide-Siloxane in "Polymers in Electronics", Davidson, T., ed. ACS Symposium Series No. 242, (1984) pp 259-272.
8. Summers, J. D., Arnold, C. A., Bott, R. H., Taylor, L. T., Ward, T. C., and McGrath, J. E. Polym. Prepr. 27(2), 403 (1986).
9. Laius, L. A., Bessonov, M. I., Florinskii, F. S. Polymer Sci, USSR 13, 2257 (1971).
10. Ginsburg, R. and Susko, J. R. in "Polyimides: Synthesis, Characterization, and Applications", vol 1, Mittal, K. L. ed. Plenum, NY (1984) pp 237-247.
11. Numata, S., Fujisaki, K. and Ninjo, N. in "Polyimides: Synthesis, Characterization, and Applications", vol. 1, Mittal, K. L. ed. Plenum, NY (1984) pp 259-271.
12. Frayer, P. D. in "Polyimides: Synthesis, Characterization, and Applications", vol. 1, Mittal, K. L. ed. Plenum, NY (1984) pp 273-294.
13. Laius, L. A. and Isapovetsky, M. I. in "Polyimides: Synthesis, Characterization, and Applications", vol. 1, Mittal, K. L. ed. Plenum, NY (1984) pp 295-309.
14. Kreuz, J. A., Endry, A. L., Gay, F. P. and Sroog, C. E. J. Polym. Sci. A-1, 4, 2607 (1966).
15. Takekoshi, T., Kochanowski, J. E., Manello, J. S. and Webber, M. J. J. Polym. Sci.: Polymer Symposium 74, 93-108 (1986).
16. Sormani, P. M., Minton, R. J. and McGrath, J. E. in "Ring Opening Polymerization: Kinetics, Mechanisms, and Synthesis", McGrath, J. E. ed. ACS Symposium Series No. 286 (1985), chapt. 11, pp 147-161.
17. Krasovskii, A. N., Antonov, N. P., Koton, M. M., Kalinsh, K. K. and Kudryavtsev, V. V. Polym. Sci., USSR 21, 1038 (1980).

## 7. BIOGRAPHIES

John D. Summers, a native of Allentown, Pennsylvania, received his B.S. in Chemistry from Ursinus College in 1983. He entered the Ph.D. program in Chemistry at Virginia Polytechnic Institute & State University in the fall of 1983. His research has involved the synthesis and characterization of siloxane modified engineering thermoplastics and thermosets.

Cynthia Arnold received a B.S. in Chemical Engineering from the University of California, Berkeley in 1980. She has worked at Raychem Corp. in Menlo Park, CA, and Mercor, Inc. of Berkeley, CA, while working toward her M.B.A., which she received in 1985. Currently she is pursuing her Ph.D. in Materials Engineering Science at VPI&SU, where she is a Kodak Fellow. Her work has been in the area of structure-property behavior of engineering polymers.

Richard H. Bott, a native of Freeland, Pennsylvania, was graduated from Ursinus College with a B.S. in Chemistry in 1983. He entered the Ph.D. program in Chemistry in the fall of 1983. His research has focused on the characterization of poly(imide siloxane) copolymers and addition curing polyimide oligomers in terms of adhesive, thermal, electrical and mechanical properties.

James E. McGrath was born in Easton, New York and received his B.S. in Chemistry from Siena College in 1956. After being employed at ITT Rayonier in Whippany, NJ and the research division of the Goodyear Tire and Rubber Co., he obtained an M.S. degree in Chemistry from the University of Akron in 1964 and Ph.D. in Polymer Science from the same university in 1967. Professor McGrath joined the Union Carbide Corporation in August of that year. Professor McGrath joined VPI & SU in September 1975. He was named a tenured Associate Professor in 1977 and was promoted to Full Professor in 1979. He is a consultant to industry and government. He is also a Co-Director of the Polymer Materials and Interfaces Laboratory. Professor McGrath has coauthored what is considered a definitive book on Block Copolymers (Academic Press, 1977), and has edited books on Anionic Polymerization, Ring Opening Polymerization and Advances in Polymer Synthesis (with B. M. Culbertson). He has over 100 contributions in the literature, and also 21 U.S. patents.

T. C. Ward, a native of North Carolina, received his B.S. in Chemical Engineering from North Carolina State University in 1963. His M.S. and Ph.D. in Physical Chemistry were granted from Princeton University in 1966. After receiving his Ph.D. he spent two years in England with Professor Manfred Gordon at the University of Strathclyde and the University of Essex. He then joined VPI & SU as an Assistant Professor in 1968. He was promoted to Associate Professor with tenure in 1972 and Professor in 1981. Professor Ward has received several teaching honors, and is a member of the Academy of Teaching Excellence at VPI & SU. He has also been a major contributor to the ACS polymer short course program since its inception in 1976. His publications have been principally in the areas of polymer characterization, testing, adhesion and long-term property evaluation of polymeric materials.

Larry T. Taylor joined the Virginia Tech faculty as an Assistant Professor in 1967 after receiving his B.S. and Ph.D. degrees from Clemson University in 1962 and 1965, respectively. In the interim, he spent 2 1/2 years as a National Institute of Health Postdoctoral Fellow at the Ohio State University working on the synthesis and properties of metal complexes containing macrocyclic ligands related to biological materials. A native of South Carolina, Dr. Taylor was promoted to Associate Professor in 1970 and Professor in 1978. During the summer of 1976, he was a NASA-ASEE summer faculty fellow at the Langley Research Center, Hampton, Virginia. Dr. Taylor currently serves as a consultant for the Electric Power Research Institute, the Research Triangle Institute and as a referee for proposals submitted to the Department of Energy, the Research Corporation and the U.S.-Israel Binational Science Foundation. Approximately 90 co-authored manuscripts have been published. He recently completed a review entitled "Incorporation of Metal Related Materials into Electrically Neutral Polymers".

TABLE 1: INTRINSIC VISCOSITIES AND UPPER GLASS TRANSITIONS OF
BULK AND SOLUTION IMIDIZED POLY(IMIDE SILOXANE)
COPOLYMERS

| SILOXANE WT. % | SILOXANE Mn | IMIDIZATION METHOD | INTRINSIC VISCOSITY NMP, 24 $^{\circ}$C | Tg ($^{\circ}$C, DSC) |
|---|---|---|---|---|
| 10 | 2,100 | BULK | 0.78 | 261 |
| 10 | 2,100 | SOLUTION | 0.73 | 260 |
| 20 | 2,100 | BULK | 0.60 | 258 |
| 20 | 2,100 | SOLUTION | 0.57 | 252 |

FIGURE 1: SYNTHETIC SCHEME FOR POLY(IMIDE SILOXANE)
SEGMENTED COPOLYMERS

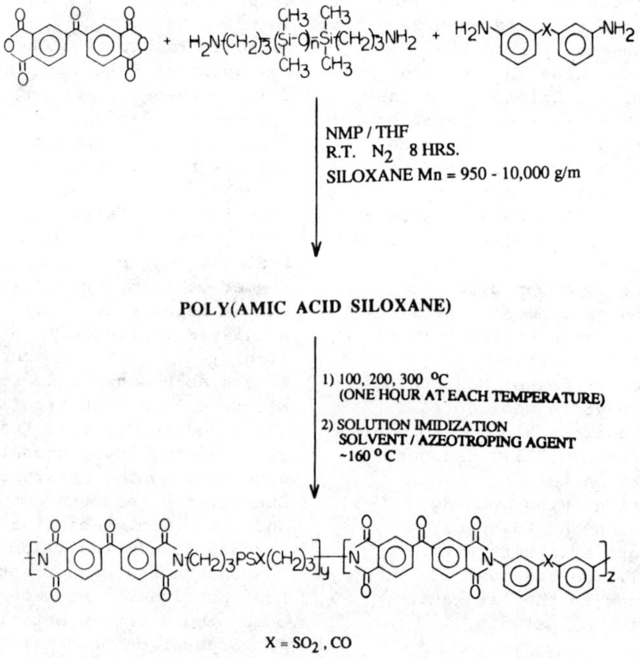

NMP / THF
R.T. $N_2$  8 HRS.
SILOXANE Mn = 950 - 10,000 g/m

POLY(AMIC ACID SILOXANE)

1) 100, 200, 300 $^{\circ}$C
(ONE HQUR AT EACH TEMPERATURE)

2) SOLUTION IMIDIZATION
SOLVENT / AZEOTROPING AGENT
~160 $^{\circ}$C

X = $SO_2$ , CO

ALL FILMS ARE TRANSPARENT AND CREASABLE

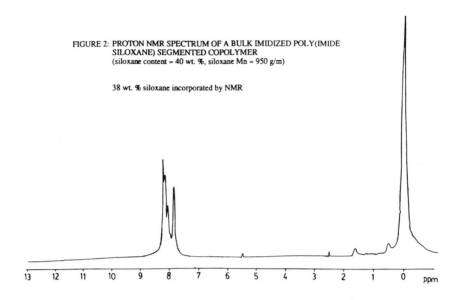

FIGURE 2: PROTON NMR SPECTRUM OF A BULK IMIDIZED POLY(IMIDE
SILOXANE) SEGMENTED COPOLYMER
(siloxane content = 40 wt. %, siloxane Mn = 950 g/m)

38 wt. % siloxane incorporated by NMR

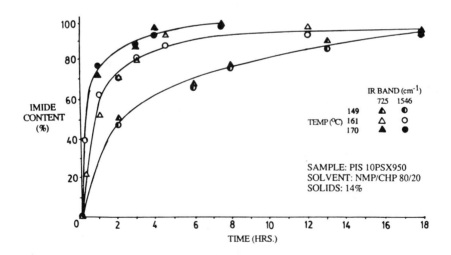

FIGURE 3: IMIDE CONTENT vs. TIME FOR POLY(IMIDE SILOXANE)
COPOLYMER SOLUTION IMIDIZATIONS
(siloxane content = 10 wt. %, siloxane Mn = 950 g/m)

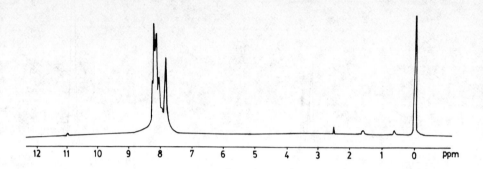

FIGURE 4: PROTON NMR SPECTRUM OF A SOLUTION IMIDIZED
POLY(IMIDE SILOXANE) SEGMENTED COPOLYMER
(siloxane content = 10 wt. %, siloxane Mn = 950 g/m)

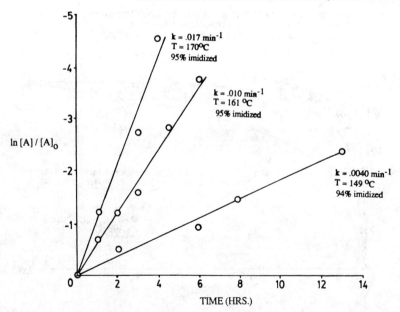

FIGURE 5: FIRST ORDER KINETIC PLOTS OF POLY(IMIDE SILOXANE)
COPOLYMER SOLUTION IMIDIZATIONS (725 cm$^{-1}$ imide band)

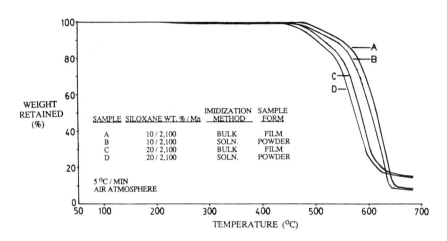

FIGURE 6 : THERMOGRAVIMETRIC ANALYSIS OF SOLUTION AND BULK
IMIDIZED POLY(IMIDE SILOXANE) BLOCK COPOLYMERS

| SAMPLE | SILOXANE WT. % / Mn | IMIDIZATION METHOD | SAMPLE FORM |
|--------|---------------------|--------------------|-------------|
| A | 10 / 2,100 | BULK | FILM |
| B | 10 / 2,100 | SOLN. | POWDER |
| C | 20 / 2,100 | BULK | FILM |
| D | 20 / 2,100 | SOLN. | POWDER |

5 °C / MIN
AIR ATMOSPHERE

# GLASS MATRIX COMPOSITES FOR HIGHER USE TEMPERATURE APPLICATIONS

Roger A. Allaire, Victor F. Janas,
Sharon Stuchly and Mark P. Taylor
Corning Glass Works
Corning, NY 14831

## Abstract

*Mechanical and thermal properties of reinforced glass and glass-ceramic composites for structural purposes are discussed in terms of their constituent fibers and matrices. High modulus fibers yielded high modulus composites; high strength fibers yielded high strength composites. Both glass and glass-ceramic composites are available which maintain their properties at elevated temperatures. Properties of graphite/ glass composites were maintained to $500°C$ $(930°F)$, while the properties of Nicalon/glass and Nicalon/glass-ceramic composites were maintained to $700°C$ $(1290°F)$ and $1250°C$ $(2300°F)$, respectively.*

## 1. Introduction

Major advances in civilization are usually precipitated by subtle advances in materials science. Examples range from the development of straw reinforced mud to the separation of atomic isotopes. Ceramics have long been prized for high temperature capabilities, oxidation resistance, chemical durability, and low density relative to metals. However, they have frustrated and charmed their fabricators with their one common trait: catastrophic failure.

In terms of materials science, catastrophic failure means that once a critical crack is formed, application of additional stress will cause the crack to propagate through the body. The resistance of a material to advancement of this crack is the fracture toughness of the material. It can be measured in a number of standard tests and ranges from about 0.7 MPa$\sqrt{m}$ for glasses to ~5 for ceramics to ~9-12 for toughened ceramics such as partially stabilized $ZrO_2$. By comparison, metals range from ~20 for aluminum to over 100 for superalloys.

The recent development of ceramic matrix composites, with apparent toughness between 10 and 30 MPa$\sqrt{m}$, has generated much excitement in the materials science community. One of the first ceramic matrix composites to be investigated was carbon fiber reinforced glass or glass-ceramic[1-6]. Due to carbon fiber's limited temperature capability in oxidizing environments, carbon fiber reinforced glasses have limited applications. For higher temperature applications, silicon carbide SiC fibers have been investigated[7-13].

The purpose of this paper is to describe the fabrication and properties of several glass and glass-ceramic matrix composites. An excellent update of glass and glass-ceramic composite technology is given in the review articles found in the *American Ceramic Society Bulletin*[14].

## 2. Experimental

### 2.1 Materials

Graphite fibers were acquired in the form of continuous tows from Hercules (A-4, and HM) and Union Carbide (P55, and P75). Nicalon (Ceramic Grade) fibers were acquired from Nippon Carbon.

Both glass and glass-ceramic matrices were examined. A glass-ceramic is a material which is formed as a glass with all the unique flow properties of a glass but can be given a subsequent heat treatment (called the *ceram*) which causes controlled crystallization. The cerammed body often has properties which are much different from the original glassy material.

A wide variety of glass compositions which are compatible with graphite and Nicalon fibers have been identified. These material systems are each intended to satisfy a unique range of product specifications. From the available composition menu, four composites have been chosen. These composites, representing the present range of material properties, have matrices which fit into three broad categories of composites; alkali borosilicate (ABS) glasses, such as Corning Code 7740 (PYREX), alkaline-earth aluminosilicates (AEAS), such as Corning Code 1723, and a calcium aluminosilicate (CAS) glass-ceramic.

### 2.2 Fabrication

The composite fabrication process used is shown schematically in Figure 1. Fiber yarn was unwound from the spool and dipped in a suspension of powdered glass in an aqueous or organic vehicle. This vehicle contained an organic binder to help hold the powder in place after drying. The yarn was then wrapped on a translating drum to form unidirectional prepreg plys.

After drying, the plys were removed from the drum and die cut to the correct size for the mold. The plys were stacked unidirectionally and the binder was burned out in a low temperature oven.

The powdery prepreg was then transferred to a carbon mold and heated in an inert atmosphere hot press until the glass softened. Pressure was applied through a hydraulic ram to squeeze the fibers together and consolidate the composite. Power was then shut off and the finished composite was removed from the mold. Glass-ceramic matrix composites may be heat treated to crystallize the matrix.

## 2.3 Composite Flexural Testing

All composite specimens were tested in four point bend flexure with span to depth ratios of at least 30:1. Specimen dimensions were 6.4 cm× 0.47 cm× 0.20 cm. All tests were performed in air environments, at a strain rate of $5 \times 10^{-5}$ sec$^{-1}$. Tensile test data is preferred for this type of composite, and studies have begun to gather design quality data on the more promising candidates.

## 3. Results and Discussion

### 3.1 Mechanical Properties

A typical flexural stress/strain curve for a carbon fiber reinforced glass is shown in Figure 2 and consists of several regions. The first is the elastic region where the curve is linear. This region ends at the *microcrack yield stress*, $\sigma_{MCY}$, where matrix microcracking begins. The strain at this point varies from as low as 0.2% to as high as 0.6%, depending on the matrix and fiber characteristics. The microcrack yield stress is regarded as the engineering limit of the material, although there have been almost no detailed studies of these materials to demonstrate this.

Between $\sigma_{MCY}$ and the *ultimate stress*, $\sigma_{ult}$, the curve is non-linear. The matrix is microcracking and the fiber is debonding.

The size of this region depends on the degree of bonding between fiber and matrix: strongly bonded systems have a relatively small debonding region whereas weakly bonded systems have a large region. The composite measured in Figure 2 failed in shear, suggesting that a stronger bond would increase the composite strength. Nevertheless, a bond which is too strong can cause the composite to fail catastrophically at $\sigma_{MCY}$ because matrix microcracks have no stopping mechanism. In this case, the stress intensity at the crack tip is higher than the strength of the carbon fiber and the crack propagates unimpeded through the composite.

The mechanical properties of some fiber reinforced glasses and glass-ceramics are shown in Table 1. The table shows that the composite properties are substantially governed by the properties of the fiber. High modulus fibers (HM, P55, and P75) yield high modulus composites, while lower modulus fibers (A-4 and Nicalon) yield lower modulus composites.

Composite strengths are also aligned with fiber strengths. High strength fiber (A-4) yields high strength composites, while lower strength fibers (HM, P55, and P75) yield lower strength composites.

It is important to note, however, that each fiber type presented in Table 2 required a unique and specific matrix composition to achieve these reported properties. Compatibility of fiber and matrix is a critical key to obtaining the composite properties desired.

Table 1 further shows the effect of matrix composition on the properties of Nicalon fiber composites. At room temperature, Nicalon fiber reinforced glass (AEAS) has a higher microcrack yield stress and a higher ultimate stress than the Nicalon fiber reinforced glass-ceramic (CAS). The advantages of glass-ceramics at higher temperatures will be discussed in the following section.

## 3.2 High Temperature Mechanical Properties

Several potential applications for glass and glass-ceramic matrix composites are in the area of high temperature structural materials. In an effort to assess the temperature limits of glass and glass-ceramic composites, the mechanical properties were measured in an air atmosphere high temperature flexural test apparatus. The results are shown in Figures 3 and 4.

Figure 3 is a plot of microcrack yield stress $\sigma_{MCY}$ versus temperature for A-4/ABS, Nicalon/AEAS, and Nicalon/CAS composites, while Figure 4 is a plot of ultimate stress $\sigma_{Ult}$ versus temperature for the same composites. The figures show that for each composite studied, high temperature properties are substantially governed by the anneal temperature of the glass. By definition, the anneal temperature is the temperature at which internal stress is reduced to 250 psi in approximately 15 minutes. At this temperature the viscosity of the glass is approximately $10^{13}$ poise.

In the following sections, the high temperature properties of graphite and Nicalon reinforced composites are discussed.

### 3.2.1 Graphite Composites

Figure 3 shows that A-4/ABS maintains a $\sigma_{MCY}$ in excess of 650 MPa (94 Kpsi) to at least 500°C (930°F), which is the anneal temperature of the ABS glass. Figure 4 shows that an ultimate stress in excess of 950 MPa (139 Kpsi) is also maintained to at least 500°C (930°F). So, the properties are maintained to temperatures far above the upper use temperatures of the best polymeric composites.

The thermal exposure time for the tests shown in Figures 3 and 4 were only 15 minutes, so they do not reflect long term stability. The long term stability of the graphite/glass composite was assessed by exposing mechanical test bars to a 400°C (750°F) air environment for various lengths of time, and then measuring their mechanical properties at room temperature. The results are shown in Table 2. No significant loss in properties was observed after 72 hours. Clearly, longer term and higher temperature tests are needed, but these results suggest that the glass matrix is protecting the carbon fibers from oxidation.

The thermal expansion of a Hercules A-4/ABS glass composite was also examined, and the results are shown in Figure 5. The fibers exert a large influence on the expansion because of their high modulus (207 GPa) relative to the glass (62 GPa). The resulting expansion is low, ranging from 0.0 to +0.135 ppm/°C over the range 25-150°C.

627

The figure also shows that when the thermal expansion was measured in heating and cooling, a hysteresis effect was observed along with an apparent permanent offset of ~20 ppm in the length of the material. Similar results were observed for HMS/borosilicate glass[15], GY-70/borosilicate glass[16], and Thornel Pitch type VS0054-0/borosilicate glass[16] composites. In addition, these graphite/glass composites had much lower transverse thermal expansions than corresponding graphite/epoxy and graphite/polyimide composites and that the expansion did not change as a result of thermal cycling.

Applications for graphite/glass composites will depend on the temperature capability of the material, its strength, and cost. Graphite fibers are almost an order of magnitude lower in cost than SiC fibers. Graphite/glass composites can be tailored to have either high moduli or high strength, while maintaining a low density ($\approx$ 2.0 gm/cc).

The low thermal expansion is well within the range of materials such as Invar and approaches that of ultra-low expansion optical materials such as titanium silicate glass (Corning Code 7971) or $\beta$-quartz glass-ceramic (Schott ZERODUR). This recommends the material for use in optical mountings structures and applications where precise dimensional control is required. In fact, it has been possible to bond a carbon/glass composite to an ultra-low expansion titanium silicate glass facesheet and polish the facesheet to the figure of a high perfor-

mance mirror. The observed hysteresis is recognized as a serious problem for ultra-precise optical systems such as telescopes.

Oxidation of the graphite fibers, particularly if the oxidation begins at the ends of the fibers and proceeds along the interface, will limit the long-term use of graphite/glass composites to low temperatures. Glass and glass-ceramic matrix composites using carbon fibers are expected to find applications at lower temperatures, especially in environments which are too hot or corrosive for organic matrices. In non-oxidizing environments, much higher use temperatures can be expected as the limit becomes the anneal point of the glass or glass-ceramic. End uses may include space structural applications and terrestrial corrosion resistant piping and structures for critical components.

### 3.2.2 Nicalon Composites

The properties of Nicalon/glass and Nicalon/glass-ceramic composites as a function of temperature are shown on Figures 3 and 4. Figure 3 shows that Nicalon/AEAS maintains a microcrack yield stress in excess of 500 MPa (74 Kpsi) to at least 700°C (1290°F). The ultimate stress (Figure 4) is maintained in excess of 1000 MPa (145 Kpsi) to at least 700°C (1290°F).

The figures show that at low temperatures, the mechanical properties of the A-4/ABS composite are in the range of the mechanical properties of the Nicalon/AEAS composite. However, above 500°C (930°F), the mechanical properties of the

A-4/ABS composite drop significantly. The more refractory glass (AEAS) found in the Nicalon/AEAS composite has a higher anneal temperature, and therefore the microcrack yield stress $\sigma_{MCY}$ and ultimate stress $\sigma_{Ult}$ hold to a higher temperature. So, more refractory glasses produce higher use temperature composites. The high ultimate strength of Nicalon/AEAS (Figure 4) makes this system one of the toughest glass matrix composites in inventory.

The figures also show that Nicalon/CAS has a lower $\sigma_{MCY}$ and a lower $\sigma_{Ult}$ than either A-4/ABS or Nicalon/AEAS at low temperatures. However, as shown in Figure 3, the microcrack yield stress remains fairly constant at a value in excess of 200 MPa (29 Kpsi) to at least 1250°C (2300°F). Likewise, the ultimate stress (Figure 4) remains fairly constant at a value in excess of 300 MPa (43 Kpsi) to the same temperature.

These findings further demonstrate the effect of matrix composition on composite properties, and the considerations which must be followed when choosing which composite to use in a specific application. The glass-ceramic matrix composite, with measurably lower properties at room temperature, manifests its superiority to the glass matrix composites when high temperature properties are examined. The room temperature property compromise is in part due to the limitation in the availability of materials which simultaneously: (i.) are chemically compatible with the fiber bonding constraints which result in tough glass matrix composites, and (ii.) maintain their properties at high temperatures.

Potential uses for Nicalon/glass and glass-ceramic matrix composites are those applications where long exposures to high temperatures are expected. A use temperature of 1250°C (2300°F) for Nicalon/CAS will allow this composite to find heat engine applications, and other high temperature structural applications. The oxidation resistance of the Nicalon fibers allows for longer high temperature applications than graphite fiber composites, with a density of approximately 30% higher ($\geq$ 2.6 gm/cc versus 2.0 gm/cc).

## 4. Summary

The fabrication and properties of several glass and glass-ceramic matrix composites were discussed.

Composite properties are greatly influenced by the properties of the fiber. High modulus fibers yield composites with high moduli; high strength fibers yield high strength composites.

Glass and glass-ceramic composites maintain their properties at high temperatures. Use temperatures for graphite fiber reinforced materials are limited by oxidation of the graphite fibers. The SiC reinforced composites will be useful to temperatures which approach the anneal point of the glass or glass ceramic.

Applications for glass and glass-ceramic composites ultimately will be governed by factors such as oxidation and economics.

## 5. Acknowledgments

We are pleased to acknowledge and thank Corning scientists Dr. Ron Stewart and Mr. Gene Nowlan for the strength measurements, Drs. George Hares and William Dumbaugh for suggestions of glass matrix compositions, and Messieurs Robert Fisk, Charles Smith, and William von Hagn for expert hot press technology.

## 6. References

1. Crivelli-Visconti, I., and Cooper, G. A., "Mechanical Properties of a New Carbon Fibre Material", *Nature*, **221** 754 (1969).

2. Sambell, R. A. J., Briggs, A., Phillips, D. C., and Bowen, D. H., "Carbon Fibre Composites with Ceramic and Glass Matrices: Part 2 Continuous Fibres ", *J. Mater. Sci.*, **7** 676 (1972).

3. Phillips, D. C., "The Fracture Energy of Carbon-Fibre Reinforced Glass ", *J. Mater. Sci.*, **7** 1175 (1972).

4. Phillips, D. C., Sambell, R. A. J., and Bowen, D. H., "The Mechanical Properties of Carbon Fibre Reinforced Pyrex Glass ", *J. Mater. Sci.*, **7** 1454 (1972).

5. Levitt, S. R., "High-Strength Graphite Fibre/Lithium Aluminosilicate Composites", *J. Mater. Sci.*, **8** 793 (1973).

6. Phillips, D. C., "Interfacial Bonding and the Toughness of Carbon Fibre Reinforced Glass and Glass-Ceramics" *J. Mater. Sci.*,**9** 1857 (1974).

7. Lindley, M. W., and Godfrey, D. J., "Silicon Nitride Ceramic Composites With High Toughness", *Nature*, **229** 193 (1971).

8. Prewo, K. M., and Brennan, J. J., "High-Strength Silicon Carbide Fibre Reinforced Glass-Matrix Composites" *J. Mater. Sci.*, **15** 463 (1980).

9. Prewo, K. M., and Brennan, J. J., "Silicon Carbide Yarn Reinforced Glass Matrix Composites", *J. Mater. Sci.*, **17** 1201 (1982).

10. Brennan, J. J., and Prewo, K. M., "Silicon Carbide Fibre Reinforced Glass-Ceramic Matrix Composites Exhibiting High Strength and Toughness", *J. Mater. Sci.*, **17** 2371 (1982).

11. Mah, T., Mendiratta, M. G., Katz, A. P., Ruh, R., and Mazdiyasni, K. S., "Room-Temperature Mechanical Behavior of Fibre-Reinforced Ceramic-Matrix Composites", *J. Am. Ceram. Soc.*, **68** C-27 (1985).

12. Mah, T., Mendiratta, M. G., Katz, A. P., Ruh, R., and Mazdiyasni, K. S., "High-Temperature Mechanical Behavior of Fibre-Reinforced Glass-Ceramic-Matrix Composites", *J. Am. Ceram. Soc.*, **68** C-248 (1985).

13. Prewo, K. M., "Tension and Flexural Strength of Silicon Carbide Fiber-Reinforced Glass Ceramics", *J. Mater. Sci.*, **21** 3590 (1986).

14. Schioler, L. J., and Stiglich, Jr., J. J., "Ceramic Matrix Composites: A Literature Review", *Am. Ceram. Soc. Bull.*, **65** 289 (1986).

15. Tompkins, S.A., "Thermal Expansion of Selected Graphite Reinforced Polyimide-, Epoxy-, and Glass-Matrix Composites", *NASA Technical Memorandum 87572* (1985).

16. Prewo, K. M., and Thompson, E. R., "Research on Graphite Reinforced Glass Matrix Composites", *NASA CR-159312* (1980).

## 7. Biography

R. A. Allaire, V. F. Janas, S. Stuchly, and M. P. Taylor work in the Research and Development Division of Corning Glass Works at CGW's Sullivan Park Research Center. They are involved in the development of composites for structural applications.

# TABLES

| Fiber | Matrix | $\sigma_{MCY}$ MPa (Kpsi) | $\sigma_{Ult}$ MPa (Kpsi) | Modulus GPa (Mpsi) |
|-------|--------|---------------------------|---------------------------|---------------------|
| Hercules A-4 | ABS | 670 (97) | 1104 (160) | 131.1 (19.0) |
| Hercules HM | ABS | 426 (67) | 649 (94) | 165.6 (24.0) |
| Union Carbide P55 | ABS | 518 (75) | 731 (106) | 140.1 (20.3) |
| Union Carbide P75 | ABS | 366 (53) | 649 (94) | 198.7 (28.8) |
| Nicalon | AEAS | 511 (74) | 1428 (207) | 135.9 (19.7) |
| Nicalon | CAS | 266 (39) | 752 (109) | 124.2 (18.0) |

**Table 1:** Room temperature flexural mechanical properties of uniaxial glass and glass-ceramic composites. Nominal 40 volume percent fiber.

| Exposure hours | $\sigma_{MCY}$ MPa (Kpsi) | $\epsilon_{MCY}$ % | $\sigma_{Ult}$ MPa (Kpsi) | $\epsilon_{Ult}$ % | Modulus GPa (Mpsi) |
|----------------|---------------------------|--------------------|---------------------------|--------------------|---------------------|
| 0 | 697 (101) | 0.56 | 1076 (156) | 0.96 | 125.6 (18.2) |
| 24 | 649 (94) | 0.50 | 1180 (171) | 1.00 | 127.7 (18.5) |
| 72 | 683 (99) | 0.53 | 987 (143) | 0.99 | 129.0 (18.7) |

**Table 2:** Room temperature flexural mechanical properties of Hercules A-4/ABS glass composite after 400°C (750°F) air exposures.

## SCHEMATIC DIAGRAM OF COMPOSITE FABRICATION

### PREPREG PROCESS

FIBER
SUPPLY

SIZING
BURN-OFF

SLURRY IMPREGNATION

TAKE-UP DRUM

PREPREG

### COMPOSITE CONSOLIDATION

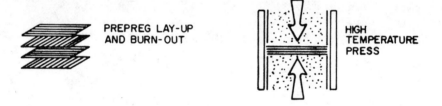

PREPREG LAY-UP
AND BURN-OUT

HIGH
TEMPERATURE
PRESS

**Figure 1:** Glass and glass-ceramic fabrication process.

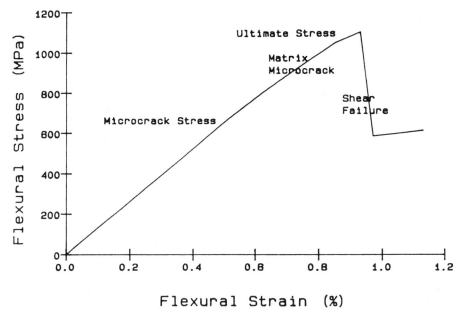

**Figure 2:** Flexural stress/strain curve for Hercules A-4/ABS glass composite. Room temperature, air atmosphere.

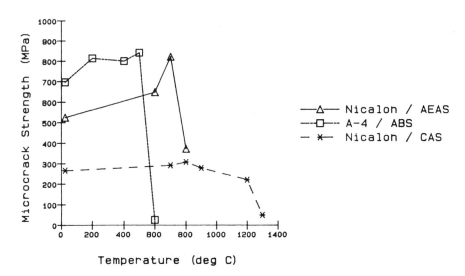

**Figure 3:** Microcrack yield stress versus temperature in A-4/ABS, Nicalon/AEAS, and Nicalon/CAS composites. Air atmosphere.

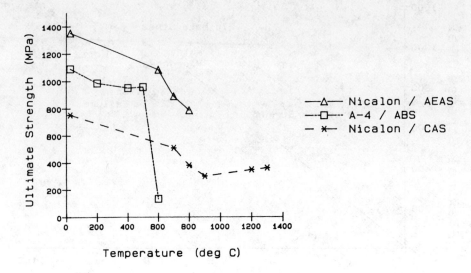

**Figure 4:** Ultimate stress versus temperature in A-4/ABS, Nicalon/AEAS, and Nicalon/CAS composites. Air atmosphere.

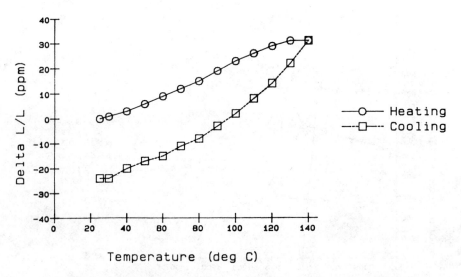

**Figure 5:** Thermal expansion ($\Delta$L/L, ppm) of a Hercules A-4/ABS glass composite. Air atmosphere.

# PRINCIPLES & PRACTICE OF ADHESION

David H. Kaelble
Arroyo Computer Center
Thousand Oaks, California 91320

## Abstract

The general principles of adhesion
include both the theory of adhesive
bonding and the fracture mechanics
of bonded adhesive joints. This
discussion outlines the current
theory and practice of adhesion.
The significant role of adhesion
tests in designing bonded structures is summarized. Recent developments in CAD/CAM(computer-aided
design & manufacture) of adhesives
will be presented, along with the
potentials for future trends in
this new frontier of adhesion
science.

## 1. INTRODUCTION

According to a report from the
Center for Adhesion Science at
Virginia Polytechnic Institute(VPI)
& State University[1], over ten
billion pounds of adhesives (five
billion dollars value) are currently
sold annually in the United States.
Structural adhesives provide a high
value fraction of this market with
two hundred million pounds annual
sales and a six to eight hundred
million dollar market value (three
to four dollars per pound). A
structural adhesive is designed to
sustain a permanent lap shear load
of greater than 1000psi. Other
extreme types of high technology
adhesives are the pressure sensitive
adhesives utilized in temporary
bonding products such as transparent
tapes, masking tapes, and labels
where ease of bonding and clean
interface removal are primary requirements.

Both structural adhesives and
pressure sensitive adhesives fall
into the area of high technology
polymers where -according to a
recent U.S. National Research
Council (NRC) report -Japan aims
for world leadership in the 1990's.[2]
The Japanese government and industry
have underway a well defined
national strategy for gaining world
leadership in this important field
of advanced materials technology.
The West German government - by
formation of the Max Plank Inst.

for Polymer Research in Mainz - is also focussing a national program of basic research which supplements already strong industrial efforts to develop a world leadership position. The United States experienced early international technology dominance in engineering polymers between 1955 to 1970, but since this period U.S. research investment has been curtailed and our leadership position has been diminished.

In an era of global competition for high technology markets, it appears that the U.S. industrial community marketing engineering plastics, adhesives and composite matrices needs to review both their internal research programs and support of national programs such as the Center for Adhesion Science at VPI. This problem of technical excellence and technological competition for advanced technology adhesives markets will be the special focus of this discussion.

## 2. PRINCIPLES OF ADHESION

The theory of adhesion has now evolved to a high level of detail and maturity. Important aspects of the theory of adhesion now fill book length reviews[3-5]. Three outwardly unrelated fields of study which are: 1) surface chemistry & wettability, 2) chemorheology and viscoelasticity, and 3) stress analysis & fracture mechanics are combined in any general theory of adhesion, such as developed by Kaelble[4]. It is

not surprising that this inherent complexity of adhesion science has confined adhesion studies to be a specialized subject within surface science and polymer science with only several English language journals specifically devoted to the science of adhesion[6,7].

The new methodology of dealing with the inherent complexity of adhesion science and technology through use of computer-aided design (CAD) of polymers, adhesives and composites has been summarized in a recent book by Kaelble[8]. A computer model for defining pressure sensitive "tack" in terms of 1) work of adhesion, 2) polymer structure and 3) fracture mechanics illustrates the implicit generality & efficiency of this CAD methodology[9]. Pressure sensitive tack is defined as the ability of an adhesive to display spontaneous bonding combined with dynamic bond strength under short term loading. In structural adhesives and composite matrices this property is termed good "green strength" response prior to curing. Analysis of the pressure sensitive tack or green strength properties thus provides an inquiry into the principles of adhesion for all polymeric adhesive and matrix materials.

The computer model for polymer tack, or green strength, presents the fundamental proposition that tack (TK) as measured by bond strength is the product of three functions which are: 1) the interfacial adhesion function A, the viscoelastic bonding function B, & the cohesive holding function H. Tack is thus defined by the following simple expression:

$$\text{Tack} = TK = A * B * H \qquad (1)$$

The parameters A and B of Eq. 1 are dimensionless with numeric ranges of zero to unity and H has the dimension of stress or strength. In this computer model the entire phenomena of surface chemistry & thermodynamics of adhesion are confined in the A function of Eq. 1. The rheology of bonding and wetting kinetics are confined in the B function. The remaining concepts of stress analysis and mechanical strength specific to the bond test method are entirely confined to the H function.

A moment's reflection by a person experienced in adhesive technology will confirm the power and intrinsic validity of Eq. 1. For example, all adhesive and matrix materials are bonded to a low surface energy release liner until final application. The function of the release liner as defined by Eq. 1 is to maintain the adhesion function A at controlled levels of $0.3 \leqslant A \leqslant 0.5$ to provide easy release and insure interfacial failure without roughening the soft adhesive. With pressure sensitive tapes and labels designed for temporary bonding, the

application of high polymer molecular weight and chain entanglements or very light crosslinking holds the bonding function B to a preferred range $0.6 \leqslant B \leqslant 0.8$. By this control of the bonding function B the permanently soft adhesive does not penetrate the cappillary structure of porous substrates like paper, and thereby fail to display the preferred interfacial failure upon subsequent removal. With structural adhesives it is an essential requirement during cure that both the adhesion function A and the bonding function B achieve unity values so that the structural state of tack defined by Eq. 1 can achieve the limiting value TK=H which defines completely cohesive failure within the adhesive phase and maximum bond strength and durability.

All of these technology constraints which specifically define structural adhesion or temporary pressure sensitive bonding are readily demonstrated by the computer model or by direct experiment. In summary, it is proposed that computer-aided design (CAD) models can provide easy and efficient implementation of detailed and complex adhesion theory and can effectively accelerate the development of new adhesive technologies.

## 3. PRACTICE OF ADHESION

The modern practice of adhesion

which utilizes the currently available theory requires a comprehensive plan of action to achieve optimum efficiency. One such logic flow diagram for adhesive research and development, by Kaelble[9], is shown in Fig. 1. Materials and process concepts are introduced at the top of Fig. 1 and a defined product performance analysis based upon chemical/mechanical requirements for both product reliability and durability emerge at the bottom of the flow chart. The upper portion of Fig. 1 defines a typical integrated program of process assurance testing. The lower portions of Fig. 1 describes a combined nondestructive and destructive test program for product assurance. The minimum critical path indicated by the heavy lined central arrow would define instrumented thermal analysis as a critical process assurance test and polymer mechanical property measurement as a critical product assurance test.

Along with polymer adhesive and matrix research and development the generous use of computer predictive modelling, indicated by the right hand arrow of Fig. 1, is now suggested as a new strategy to streamline and focus all other research efforts. The elimination of nonessential experiments and characterizations can accelerate progress and improve research productivity. The intrinsic productivity of this recommended computer-aided design (CAD) approach has been extensively demonstrated and documented[8-10].

## 4. SUMMARY

This brief review has touched on the central issues concerning the principles and practice of adhesion which may influence the U.S. position in the global market competition during the next decade. This discussion points out the need for a national appraisal of research funding for engineering plastics, advanced adhesives and polymer matrix materials. This discussion also points out the computer-aided design (CAD) of polymer adhesion properties is a new and feasible strategy to increase research creativity and productivity.

## 5. REFERENCES

1) H. F. Brinson, "Mechanics Applied to Adhesion Science", Applied Mechanics Review, 38, No. 9, (1985)

2) R.J. Seltzer, "Japan Aims for World Leadership in Advanced Polymers by the 1990's," Chem. & Eng. News, November 17, (1986), p.35-38

3) R.L. Patrick(Ed.), Treatise on Adhesion & Adhesives Part I:Theory, Marcel Dekker Pub., New York (1967)

4) D.H. Kaelble, PHYSICAL CHEMISTRY OF ADHESION, Wiley-Interscience, New York (1971)

5) S. Wu, Polymer Interface &
Adhesion, Marcel Dekker, Inc.,
New York (1982)

6) L.H. Sharpe (Ed.), The Journal
of Adhesion, Gordon & Breach Pub.,
New York (1970-)

7) K.L. Mittal & W.J. van OOij(Eds.)
Journal of Adhesion Science &
Tech., VNU Science Press, Utrech,
Netherlands (1986-)

8) D.H. Kaelble, "Computer-Aided
Design of Polymers & Composites",
Marcel Dekker, Inc. New York (1985)

9) D.H. Kaelble, "Viscoelastic
Theory of Pressure Sensitive Tack",
Proc. 30th Nat. SAMPE Symposium,
SAMPE, Covina, CA (April 1985)

10) D.H. Kaelble, "Computer-Aided
Design of Polymers & Composites",
Proc. 31st Int. SAMPE Symposium,
SAMPE, Covina, CA (April, 1986)

## 6. BIBLIOGRAPHY

David H. Kaelble is Director of
Arroyo Computer Center, a small
business devoted to consulting and
development of chemically based
models for the computer-aided
design and manufacture (CAD/CAM)
of polymers and composites. Mr.
Kaelble is author of the standard
text, PHYSICAL CHEMISTRY OF ADHESION,
Wiley-Interscience, New York (1971)
and a new book entitled COMPUTER_
AIDED DESIGN OF POLYMERS &
COMPOSITES, Marcel Dekker, Inc.,
New York (1985).

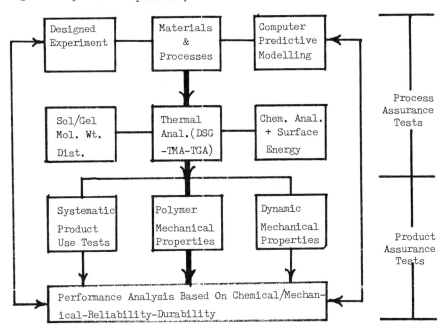

Figure 1    Logic Flow Diagram For Quantitative Product
& Process Characterization

A COMPUTER MODEL FOR RADIATION (UV or EB)
CURING OF POLYMERS

B. K. Bordoloi
Avery International Research Center
Pasadena, California
D. H. Kaelble
Arroyo Computer Center
Thousand Oaks, California

## ABSTRACT

A combined model and computer analysis of ultraviolet (UV) and electron beam (EB) curing of polymers is demonstrated. The sol-gel properties of radiation-cured silicones, polyvinyl chloride, styrene-ethylene-butylene block copolymers and polyether acrylic isocyanate oligomers are analyzed. This model incorporates the Charlesby-Pinner relations[1] in a twofold computer analysis applicable to both UV and EB cure processes to quantify the crosslink (Gc) and scission (Gs) yields. The role of the multifunctional radiation crosslinkers is analyzed by use of the computer model.

## INTRODUCTION

A general characterization scheme for computer-assisted analysis of radiation curing of polymers is introduced and discussed. The significant symbols and meanings of terms utilized in this discussion are summarized in Table I.

Table I: Nomenclature for UV-EB Curing

| Computer Symbol | Physical Symbol | Meaning |
|---|---|---|
| R2 | $R_0^*$ | Irradiation dosage for incipient gelation without chain scission |
| R1 | $R_0$ | Irradiation dosage for incipient gelation with chain scission |
| GC | $G_c$ | Crosslinks/100 EV dosage |
| GS | $G_s$ | Scissions/100 EV dosage |
| S | S | Sol fraction = $1-g$ |
| G | g | Gel fraction |
| NØ | $N_0$ | Avagadro's number= $6.023 \cdot 10^{23} \frac{\text{molecules}}{\text{mole}}$ |
| MW | $M_{wo}$ | Original wt. avge. molecular weight before irradiation |
| T2 | $t_0^*$ | Irradiation time for incipient gelation without chain scission |

T1    $t_0$    Irradiation time for incipient gelation with chain scission

T    t    Irradiation time

Dosage Equivalents

1.0 (Mrad) = 10.0 kGy = 10.0 $\frac{watt\ sec}{gm}$

        = 6.24 · $10^{19} \frac{EV}{gm}$

1.0 (rad) = 100 $\frac{erg}{gm}$ (absorbed)

1.0 (Mrad) = 1.0 · $10^6$ rad

Computer Model

The critical equations which enter the computer model are summarized as follows:

EB Relations

$$R_0^* = R_0 [1 - (\frac{G_S}{4\,G_C})] \qquad (1)$$

$$S + \sqrt{S} = \frac{G_S}{2\,G_C} + \frac{2\,R_0^*}{R} \qquad (2)$$

$$S + \sqrt{S} = \frac{G_S}{2\,G_C} + \frac{2\,R_0}{R} [1 - (\frac{G_S}{4\,G_C})] \qquad (3)$$

$$G_C\,(\frac{crosslinks}{100\ EV}) = \frac{(100\ N_0)}{2\,M_{wo}\,R_0^*} \qquad (4)$$

UV Relations

$$t_0^* = t_0 [1 - (\frac{G_S}{4\,G_C})] \qquad (5)$$

$$S + \sqrt{S} = \frac{G_S}{2\,G_C} + \frac{2\,t_0^*}{t} \qquad (6)$$

$$S + \sqrt{S} = \frac{G_S}{2\,G_S} + \frac{2\,t_0}{t} [1 - (\frac{G_S}{4\,G_C})] \qquad (7)$$

The unique features of the above relations is that all expressions are uniformly developed in terms of the radiation yields of inter-chain crosslinks ($G_C$) and scission ($G_S$) per unit energy absorbed (see Table I for definitions). The computer model utilizes a listing of measured sol fraction S versus EB dosage R.

The UV computations utilizes an input of the measured values of sol fraction, S, of the irradiated polymer versus the UV exposure time, t. Within the computation, the effective UV dosage in units of watt sec/gm (see Table I) can be utilized to directly correlate the nominal UV and EB exposures in equivalent physical units.

The computer model utilizes Eq. 2 to convert the measured EB irradiation variable S and R into a linear least-squares analysis of the dimensionless ratio of chain scission to crosslinking in terms of $G_S/2\,G_C$ as the extrapolated intercept of Eq. 2 at 1/R = zero. The slope analysis of Eq. 2 provides the value of $2R_0^*$ which defines the critical radiation dosage for incipient gelation for pure crosslinking without scission. By the linear least-squares plot of S + $\sqrt{S}$ versus 1/R to S + $\sqrt{S}$ = 2.0, the value of $1/R_0$ is also obtained as shown by application of Eq. 3. If the initial weight average molecular weight of the unirradiated polymer $M_{wo}$ is known, then the absolute values of $G_C$ and $G_S$ can be defined by use of Eq. 4. The relation between the critical dosages $R_0$ and $R_0^*$ wherein the effects of scission are isolated is defined in Eq. 1.

The UV relations shown in Eqs. 5-7 parallel the EB relations but define nominal exposure times $t_0^*$ and $t_0$ for incipient gelation for non-scission reaction ($t_0^*$) and combined crosslink-scission ($t_0$) curing processes. Here the experimental data input to the computer analysis are measured values of sol fraction S versus UV exposure time t.

The computer model presents tabular outputs of the above process variables and analyzed results. In addition, the computer output presents two graphical displayed outputs which provide an immediate overview of the radiation cure process. These two graphical displays present two frames of information which correlate theory and experiment in semi-independent fashion. The lower frame of the graphic output displays the results of the linear least-squares analysis defined by either Eq. 2 of EB curing or Eq. 6 for UV curing. While useful, the coordinates of this lower frame unnaturally distort the experimental values of the input data.

To present a more physically meaningful correlation of theory and experiment, a second upper frame of graphic output is generated by the computer model. The graphical coordinates of this upper frame are presented in familiar bilogarithmic plots of sol fraction S versus either EB dosage R or UV exposure time t. This data display is consistent with the classical Charlesby plots.[2];[3] In the upper frame, one may wish to obtain the best fit between the experimental S versus R (or t) data and one of the computer

generated curves of varying $G_s/G_c$ values.

Table 2: Experimental Data for Computer Analysis of EB and UV Curing

| Polymer Type | Silicone | Polyvinyl Chloride | |
|---|---|---|---|
| $M_{wo}$ | 650,000 | 83,200 | |
| Cure Process | EB | UV | |
| Cure Data: | S | R | S | t |
| | (fract.) | (Mrad) | (fract.) | (hr) |
| | 0.767 | 0.55 | 0.973 | 13.3 |
| | 0.364 | 0.99 | 0.823 | 16.1 |
| | 0.254 | 1.67 | 0.665 | 18.2 |
| | 0.189 | 3.41 | 0.484 | 20.4 |
| | 0.181 | 5.03 | 0.255 | 22.2 |
| | 0.165 | 10.1 | 0.169 | 47.6 |
| | 0.160 | 20.2 | 0.156 | 34.5 |

Computed Cure Parameters

$R_0 = 0.391$ Mrad   $t_0 = 13.11$ hr

$R_0{}^* = 0.274$ Mrad   $t_0{}^* = 14.54$ hr

$G_s/G_c = 1.195$   $G_s/G_c = -0.4363$

$G_c = 2.71 \dfrac{\text{crosslinks}}{100 \text{ EV}}$

$G_c = 0.29 \dfrac{\text{crosslinks}}{100 \text{ EV}}$

DISCUSSION

Representative examples of EB and UV curing data published respectively by Corfield, Astill, and Clegg[4] and by Kwei[5] are summarized in Table 2. The computer analysis of the EB curing of the silicone polymer is summarized in the lower portion of Table 2 and indicates a relatively low EB dosage of $R_0 = 0.391$ Mrad to achieve incipient gelation. The computer analysis of the UV curing of polyvinylchloride shown

642

in lower Table 2 indicates a non-physical result of a negative ratio of scission to crosslink yield ($G_s/G_c = -0.4363$) where theory requires a zero or positive result. The graphical displays of the experimental data and computer analyzed curves provides a rapid and convenient clarification of this apparently anomalous result for PVC.

In Fig. 1 is displayed the two graphic frames for the EB and silicone polymer where agreement between the computer model and experimental data is evident in the Charles-Pinner plot (lower view). The computer generated data displays of Fig. 2 (lower frame) for UV-cured polyvinyl chloride shows the difficulties encountered by the computer model in analyzing the data with low $S < 0.48$ and high $t > 20$ hr. The linear least squares fit program of the computer model forces an extrapolation to negative values of $S + \sqrt{S}$ at $1/R = 0.0$ as shown in the lower frame of Fig. 2. Inspection of this lower frame of Fig. 2 shows that a small portion of the experimental UV data of Table 2 is responsible for this peculiar result. The computer model permits convenient and rapid recalculation using edited data to permit an improved correlation between theory and experiment which excludes potentially anomalous data points. The lower frames of Figs. 1 and 2 thus expose the specific interactions of the theory and experiment.

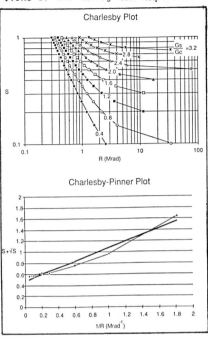

Fig. 1: EB curing analysis of Polydimethyl siloxane

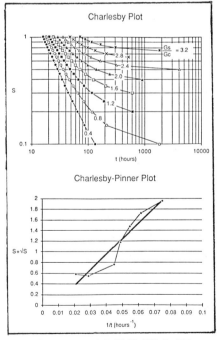

Fig. 2: UV curing analysis of PolyVinyl Chloride (PVC)

The upper frames of Figs. 1 and 2 present the generalized Charlesby plots of computed log S versus log R or log t for varied scission/ crosslink yield represented by ratios of $G_s/G_c$. These upper Charlesby maps may include the experimental S vs R for EB cure or S vs t for UV cure data. These curves are generated from the computer-based curve fitting shown in lower Fig. 1 and Fig. 2 wherein the values of $R_o^*$ or $t_o^*$ for idealized incipient gelation without scission are obtained. The objective of the upper frames of Fig. 1 and 2 is to more closely define processing options as defined by varied ratios of $G_s/G_c$ = scission/crosslink yields.

The oral presentation of this paper will present more specific illustrations of the value of the computer model when the ratio of $G_s/G_c$ is shown as a variable of EB dosage as evidenced in the study of Zue-Teh and Liang-Chang[6] on polyether acrylic isocyanate oligomers. The interesting study of Zurawski and Sperling[7], when analyzed by the computer model of EB curing clearly shows the influence of low concentrations of crosslinking monomer on the $G_s/G_c$ ratios in styrene block-ethylene/butylene block co-polymers. The addition of chemical additives to polyvinyl chloride is readily shown to modify UV cure responses in the study of Kwei[5] and examples of computer analysis of this data will be discussed as an extension of the illustrations of Fig. 2 for pure polyvinyl chloride polymer.

## SUMMARY AND CONCLUSIONS

This discussion has introduced and demonstrated a new graphics enhanced computer model for radiation curing of polymers. Specific examples for EB-cured polydimethyl siloxane and UV-cured polyvinyl chloride illustrate the versatility of the model. The ease by which the computer displays of theory and experiment are generated permits testing of varied assumptions concerning the cure path defined by $G_s/G_c$ = scission/crosslink ratios as a function of degree of cure. This detailed cure-path information provided by the computer model can be utilized in both reformulation of materials and adjustment of curing conditions to achieve optimized radiation curing.

## ACKNOWLEDGEMENTS

The authors gratefully acknowledge the support and encouragement of this research by Avery International Research Center and the kind assistance of Mr. R. Baggarley in the development of the computer-aided models.

## REFERENCES

[1]A. Charlesby and S. Pinner, Proc. Roy. Soc. (London) A249, 367 (1959).

[2]A. Charlesby, Proc. Roy. Soc. (London), 215A, 187 (1952); Ibid., 222A, 60 (1954); Ibid., 224A, 120 (1954).

[3]A. Charlesby, J. Polymer Sci., II, 513, (1954).

[4]G. C. Corfield, D. T. Astill and D. W. Clegg, "Radiation Stability of Silicone Elastomers" in Materials for Microlithography, American Chemical Society, Washington, D.C., (1984), pp 473-480.

[5]K-P. S. Kwei, J. Applied Polymer Sci., 12, Pp. 1543-1550 (1963).

[6]M. Zue-Teh and D. Liang-Chang, Radiation Phys. Chem, 22, 1001, (1983).

[7]D. E. Zurawski and L. H. Sperling, Polymer Eng. and Sci., 23, 510, (1983).

# INTRODUCTION TO PRESSURE-SENSITIVE ADHESIVES

Robert J. Rona

Swedlow, Incorporated

Garden Grove, California

## Abstract

This paper defines what a pressure-sensitive adhesive is, and gives general requirements in terms of glass transition temperature and molecular weight. The two major chemical types of PSA's are reviewed, as well as the three primary coating forms. Physical testing of PSA's is also reviewed.

## 1. INTRODUCTION

A pressure-sensitive adhesive (PSA) may be thought of as a permanently "tacky" (sticky) polymeric material carried by a backing (called the "facestock") which is more rigid than the adhesive. The facestock may be any of a wide variety of materials such as paper, cloth, polymeric films, or foils. PSA's are unique because of their ability to form an immediate bond upon contact with another rigid surface (called the "substrate").

An important requisite for this immediate contact is a glass transition temperature (Tg) some $20^\circ$ to $60\,^\circ C$ below use temperature. Besides having a Tg low enough to wet out and develop a bond to the substrate, the polymer used must be of sufficiently high molecular weight to impart cohesive strength. A broad molecular weight distribution helps improve adhesion. It is well known to those in the industry that as adhesive strength increases, cohesive strength generally decreases, and vice-versa.

Successful PSA formulations rely on a proper balance of adhesive and cohesive properties, between "hardness" and "softness". This will depend on the desired application. For example, an office mending tape requires an immediate aggressive build up of adhesion, while a vinyl marking film requires repositionability at first with a more gradual build-up. An adhesive note pad requires low initial adhesion with minimal increase in adhesion for removability.

## 2. CHEMICAL CLASSES OF PSA'S

The two classes of polymers which comprise most PSA's are rubbers and acrylics.

### Rubber

Natural rubber consists of polyisoprene (all cis), while synthetic polyisoprene is a mixture of cis and trans. Other synthetic rubbers used in PSA applications are block copolymers of styrene with either isoprene or butadiene, homopolymers of methyl vinyl ether or isobutylene, or copolymers such as ethylene/vinyl acetate.[1] Other combinations of these are also possible. Carboxylated rubbers incorporate an unsaturated acid monomer, such as acrylic acid, into the formulation.[2] Compounding with tackifiers (low-molecular-weight resins) to increase adhesion is a common practice with rubber-based adhesives.[3]

### Acrylics

Esters of acrylic acid whose alkyl chain length is from four to twelve carbons have glass transition temperatures in the desirable range. Acrylic PSA's are most frequently copolymers of two or more of these esters in a wide variety of combinations, and are almost never homopolymers. This makes acrylics especially versatile as PSA's. Common acrylic monomers are butyl acrylate, 2-ethylhexyl acrylate, and iso-octyl acrylate. A small amount (usually 5% or less) of a polar comonomer such as acrylic acid or hydroxyethyl acrylate is often incorporated into the polymer to aid in adhesion and crosslinking.

## 3. TESTING OF ADHESIVES

Two primary ways are used to describe failure modes in PSA's: adhesive and cohesive. Adhesive failure occurs because of debonding between the adhesive layer and either the substrate ("panel" failure) or the backing ("facestock" failure). Cohesive failure results from insufficient internal strength in the adhesive layer and is characterized by splitting of the adhesive between the bonding surfaces.

With the advent of sophisticated measuring devices in recent years, information about viscoelastic properties of PSA polymers has become more available. However, adhesive testing of PSA's involves three primary properties: peel, shear, and tack. Below is a brief description of each.

### Peel Testing

A 2.5 x 20 cm test sample of the adhesive, bonded to the face material, often paper or polyester film, is applied to a standard substrate, typically stainless steel. A standard pressure is then applied to the sample using a roller. After a certain dwell time, the sample is placed in a machine which measures force as the sample is peeled off the substrate at a constant rate, usually at either a 90-or 180-degree angle. Peel strength is measured as force per unit width of the sample.

### Shear Testing

A 1 x 5 cm sample of the adhesive, is

applied to the substrate as above, except that only a 1 x 1 cm area at the end of the sample is applied to the bottom edge of the substrate. A standard weight is hung from the sample and the time to drop is measured.

## Tack Testing

Tack measures how quickly adhesion builds up and involves contact with the substrate for a relatively short time. This test may take various forms. For example, in a probe tack test machine, a small test sample is applied to a weight with a round hole in the center. A metal probe then comes in contact with the sample for a short time, typically a few seconds or less. The machine then measures the force required to debond the sample from the probe.

## 4. COATING OF PSA'S

Pressure-sensitive adhesives are typically coated from solution, emulsion, or bulk ("hot melt"). Either rubber or acrylic polymers may be coated in any of these forms.

## Solutions

The first PSA's were solvent-based. A solution acrylic polymer is usually prepared by addition of appropriate monomers and thermal initiators (such as azo or peroxide compounds) to an organic solvent at or near the boiling point. Although superior adhesive properties are obtained from solution PSA's, flammability, toxicity, and pollution are major drawbacks. As a result, their use has declined steadily during the past ten years.

## Emulsions

These water-based adhesives consist of microscopic particles of polymer stabilized by surfactants and/or colloids. Emulsion polymers are often made via water-soluble redox type initiators, and thus require relatively little heating during polymerization. Emulsions have some obvious advantages over solvent-based polymers. However, they are more difficult to coat, and require more energy in drying. With emulsions, it is also more difficult to obtain adhesive performance as high as with solutions. Nevertheless, the use of emulsion PSA's has grown tremendously in recent years.

Crosslinking compounds such as metal complexes or difunctional organics can be added to solutions and emulsions prior to coating to improve the cohesive strength of the dried polymers

## Hot Melts

These adhesives are polymerized in bulk (i.e., without solvent). Since there is no carrier, they must be applied at temperatures high enough that their viscosity will be suitably low. Unfortunately, if the polymer is of coatable viscosity (even at elevated temperatures), the molecular weight is frequently too low to give good adhesive performance. Therefore, an extra step involving radiation curing (ultraviolet or electron beam) is often included in the coating operation to cause crosslinking of the polymer.

Pressure-sensitive adhesives are not usually coated directly onto the facestock. Instead, they are applied onto a paper liner precoated with a release material such as silicone or polyethylene. In the case of solution and emulsion polymers, a thin layer of wet polymer of controlled thickness is applied to the liner, and the solvent (or water) is removed by oven drying. This leaves a film of adhesive which is then laminated to the desired facestock.

## 5. CONCLUSION

Pressure-sensitive adhesives have a broad variety of end use applications ranging from removable low-tack to aggressive and permanent. With the equally broad range of materials available in the formulator's "arsenal", tailoring an adhesive and a coating process to meet requirements should be limited only by the chemist's skill and imagination.

## REFERENCES

1. Krenceski, M.A., Johnson, J.F., Temin, S.C., J. Macromol. Sci. -- Rev. Macromol. Chem. Phys., C26(1), 143 (1986).

2. Midgley, A., Adhesives Age, 17 (Sept. 1986).

3. Foley, K.F., Chu, S.G., Adhesives Age, 24 (Sept 1986).

## BIOGRAPHY

Robert J. Rona was born in Ulm, Germany. He received his B.S. degree from the University of California, Los Angeles, and his M.S. from California State University, Long Beach. Following receipt of his Ph.D. in chemistry from Rutgers University, he was employed for five years as a Research Chemist at Avery International, where his primary research involved development of new pressure-sensitive adhesives. Since 1983, he has been employed at Swedlow, Incorporated, where his research involves development and manufacture of transparent acrylic polymers used as windows in commercial and military aircraft.

COMPOSITE DRIVE SHAFTING APPLICATIONS
Brian E. Spencer
Brunswick Corporation
Lincoln, Nebraska

## Abstract

This paper discusses the recent increase in the use of composite drive shafts due to advancements in design and processing techniques. The use of composite drive shafts has been discussed for many years for commercial (automobile) and aerospace (helicopter) applications. The reasons for considering composites are obvious: weight, corrosion resistance and unsupported length. User industries have been reluctant to embrace composite shafts for several valid concerns: cost, impact resistance, thick-wall laminate quality and attachment techniques. Several applications to be discussed have brought about significant improvements which have the potential to eliminate some of the previous concerns. Advances in resin technology have improved composite impact resistance. Automation and lower material prices have helped to reduce shaft costs.

Brunswick's work with Boeing-Vertol in developing high torque, thick-wall shafts resulted in manufacturing processes that can efficiently produce low void (1 percent), thick-wall, filament-wound laminates.

The NCF (no cut fiber) coupling concept was developed because of the inefficiencies of available attachment techniques. The NCF coupling is a proven, efficient method for transmitting torque to and from composite shafts. A second attachment method has also been developed which uses a trapped fitting concept.

The most recent advancement has been the development of an analytical model to predict the residual stresses in filament-wound cylinders. This model allows processing optimization to ensure production of an acceptable part. It predicts final cylinder dimensions which allows tooling and attaching components to be properly designed.

## 1. INTRODUCTION

Composites are a combination of two

or more materials. Advanced composite usually means a combination of oriented continuous fibers and a resin binder. Table I shows the properties of common fibers and resins. Composites are orthotropic, which allows tailoring the laminate design to meet requirements that may be conflicting for metals. (An orthotropic body has material properties that are different in three mutually perpendicular directions and three mutually perpendicular planes of material symmetry).

Fiber-reinforced plastics are finally being considered the material of choice for many applications. Power transmission shafts exploit the many advantages of fiber-reinforced plastics. The ability to design a lightweight shaft with a high axial modulus means no intermediate support is necessary, as may be needed with a metal shaft. Composite shafts can be easily developed to have higher lateral-critical frequencies than steel shafts of comparable dimensions. This advantage is seen in terms of the specific modulus (modulus divided by density) in Figure 1. This figure compares several unidirectional fiber composites and metals. Figure 1 also shows the specific strength advantages of composites over metals.

These advantages, along with those listed below, are the reasons why the use of composites is increasing:

1. Decreased weight
2. Corrosion resistance
3. Fatigue resistance
4. Nonconductivity
5. Part count reduction
6. Reduction in torsional stiffness
7. Increased strength

2. DRIVE SHAFT APPLICATIONS

Table II lists some of the power transmission shafts Brunswick has produced or is now developing. A new application in a related area is torsion tubes for suspension systems. Composite torsion tubes offer reduced weight and improved fatigue resistance while providing the equivalent torsional stiffness of steel torsion bars.

Brunswick's first composite power shafts were manufactured for David Taylor Naval Ships Research and Development Center, Annapolis, Maryland. These shafts were test articles for a small research ship. The shafts were hybrid graphite-fiberglass with an epoxy matrix. Subscale drive shafts for a naval destroyer were also built and tested. These shafts demonstrated the feasibility of using composite shafts for naval propulsion systems. In addition to saving weight, the composite shafts reduce noise transmission.

This work was followed by the development of the center section

651

for the aft rotor shaft on the Boeing-Vertol Model 234 helicopter [1]. This thick-wall shaft shown in Figure 2, the first used as a primary structure on an aircraft, demonstrated the potential of composites by increasing service life from 1,600 to 5,000 hours and reducing weight by 30 percent over the aluminum component it replaced. Ballistic tests on these shafts show the benign failure mode of composites. Metal shafts fail catastrophically. Composites, especially graphite-fiberglass hybrids, retard crack growth from ballistic damage. Total component failure occurs over a period of time.

This shaft was also tested for its ability to transmit lightning strikes. This shaft and fitting attachment pins survived a 270,000 amp maximum intensity test without damage.

From the Model 234 shaft, the forward and aft shafts for Boeing-Vertol's Model 360 helicopter were developed. Where the Model 234 shaft has only a composite center section (1 meter in length, 38 inches), the Model 360 shafts are composite from gearbox to rotor hub.
These shafts, like the Model 234 shafts, are graphite-fiberglass hybrids with an epoxy matrix. The aft shaft uses high modulus graphite 345 GPa (50 msi) and the forward shaft standard graphite 234 GPa (34 msi).

Brunswick has also developed the pitch actuation shaft for the X-Wing aircraft. This shaft requires a layer of conductive material on the outer surface for lightning protection. The basic load carrying portion of the shaft is made of high modulus graphite; and the outer layer is made of aluminized fiberglass to provide the required conductivity. This high speed, small diameter shaft demonstrates the specific modulus advantages of composites.

One of the newest applications for composite drive shafts is the interconnect shafts for the V-22 Osprey aircraft. Bell will be using thin-wall composite shafts to connect the two wing-tip-mounted engines. These high speed shafts not only have to pass fatigue requirements but also demonstrate performance with impact damage.

In addition to the aerospace applications mentioned, automobile manufacturers have been using composite drive shafts for selected models. Along with the advantages cited for using composites, the composite shafts attenuate power train vibrations. This improves ride quality and reduces bearing wear.

A new related application for composite shafts is torsion tubes for vehicle suspension systems. Composite shafts can provide the stiffness and strength of steel in the same volume while saving substantial weight. One torsion bar being developed by Brunswick is designed to react 58,750 N•m (520,000 in.-lb.) of torque with a 7.62 cm (3.0 inch) external diameter. This tube uses intermediate modulus graphite (modulus of 276 GPa, 42 msi, tensile strength of 5.52 GPa, 800 ksi). The weight savings for this application is over 50 percent.

### 3. DRIVE SHAFT PROCESSING TECHNIQUES

The manufacturing techniques used in fabricating Brunswick's first drive shafts used conventional processing methods. The shafts were wet wound on a steel mandrel, oven cured, then ground to diameter on the exterior surface. This straightforward approach provided shafts of adequate quality and strength.

As performance requirements increased, the processing required changes to reduce voids and eliminate delaminations in thick-wall applications. For the Boeing-Vertol Model 234 shaft, the part was wet wound and cured in stages. This increased production time but was successful in reducing voids and eliminating delaminations. The completed shaft has a wall thickness of approximately 2.5 cm (1 inch).

For the Boeing-Vertol Model 360 shafts it was decided that process improvements were needed to reduce manufacturing time. Processing was developed to allow filament winding the entire 5 cm (2 inch) thick shaft without intermediate cures [2]. The processing was not developed without overcoming major obstacles, as shown by the first part built in Figure 3. Process optimization resulted in low void parts. The finished parts are shown in Figure 4.

With the emphasis in thick-wall shafts it has become important to be able to accurately predict residual stresses. Test parts have been built to demonstrate that transverse and radial residual stresses can exceed laminate strength. Several investigators have presented ways of modeling the filament winding process [3, 4]. These models are now becoming available and will aid in designing composite cylinders of all types. These models account for the majority of variables in the filament winding process. These variables are listed below:

1. Constituent material properties, including temperature effects for laminate and tool.
2. Filament winding parameters: wind tension, bandwidth, wind angle, fiber volume, etc.

3. Resin cure kinetics.
4. Viscous resin flow.
5. Heat transfer during curing.

The optimized process for filament winding thick-wall parts includes using preimpregnated fiber to control resin content and compaction steps to remove excess resin and trapped air. The design and winding parameters are important to final part quality and need to be considered as discussed in references 3 and 4. Thick-wall parts can either be oven cured or cured with an internally heated source. The manufacturing process used for most thin-wall shafts consists of wet winding the composite on a steel mandrel. The part is cured with steam heat inside the mandrel; internal steam heat allows quicker curing than an oven cure. Multiple shafts are wound on a single mandrel where feasible. Problems are rarely encountered when winding thin-wall tubes.

## 4. ATTACHMENT METHODS

The joining of tubular composite structures to adjacent components has been the "weak link" in many designs. The two most widely used approaches are bonding and mechanical fastening. For all the drive shafts described above, the end fittings were attached to the composite tube with rivets or pins except for the torsion tube which uses the NCF (no cut fiber) coupling. Adhesively bonded joints are adequate for many applications but can degrade with time and exposure to some environments. Mechanical fasteners, rivets and bolts, load the composite in bearing. Bearing allowables are low for composites as compared to metals, as illustrated in Table III. This can result in the need for additional composite material to increase the bearing area.

An innovative approach for attaching end fittings to composite tubes without drilling holes or relying on a bond joint has been developed. This concept is very efficient, as demonstrated by tests [5]. Figure 5 shows the NCF coupling concept. This design traps the composite and transfers the loads through interface pressure and friction.

The NCF joint capitalizes on mandrel-wound fabrication with no machining required after part removal from the winding mandrel. A number of unique features make the NCF coupling advantageous to the designer. They are listed below:

1. Maximum utilization of fiber elements.
2. Inspectability.
3. Replaceability.
4. Low stress concentration.
5. Minimum exposure to manufacturing error during assembly.
6. Minimum inspection requirements.

7. Ease of assembly and disassembly.

8. Broad flexibility in design applications and options.

By contrast, a design which uses through-the-wall pins is complicated. The procedures required when installing through-the-wall pins are listed below and are pictorially presented in Figure 6. The example shown is for a helicopter rotor shaft.

1. Precise machining of the composite tube (a stepped bore is used in the particular example shown).

2. Insertion of the metal fitting. (This requires the soaking of the metal fitting in an alcohol and dry ice mixture and the use of a press.)

3. Machining of the pin holes. (This is a multistep process requiring a precise setup and the use of special cutting tools and reamers.)

4. Insertion of the through-the-wall pins. (This procedure requires the use of a press which can induce delaminations.)

5. The applying of a sealant around the pins.

6. Multiple inspections to assure proper fit of the fittings and pins.

## 5. SUMMARY

The cited examples illustrate the advantages of composite over metal drive shafts. These examples have pointed out improvements both in processing and design concepts. These improvements have included laminate quality in thick-wall components and attachment techniques. Composite designs usually result in components that weigh less and last longer.

As process techniques become well characterized and automated equipment reduces labor content, the cost of composite drive shafts will decrease. Graphite fiber costs are decreasing. It is now possible to purchase 234 GPa (34 msi) graphite fiber for less than $33/kg ($15/pound).

Epoxy resins are being modified to improve impact resistance. Thermoplastic tubes are now being filament-wound. Thermoplastics do offer advantages in impact tolerances over epoxies.

With the continued improvements in fabrication equipment, materials, and design concepts and education of the engineering community in composite design, applications of the filament winding process will increase.

## 6. REFERENCES

1. Faust, H., et. al., "A Composite Rotor Shaft for the Chinook", presented at the

National Specialist's Meeting on Composite Structures, Sponsored by the Mideast Region of the American Helicopter Society, Philadelphia, PA, March 1983.

2. Spencer, B.E. and Steel, L.F., "Thickwall Structures - Design and Manufacturing Techniques", presented at the 29th National SAMPE Symposium, April 1984.

3. Spencer, B.E., "Prediction of Manufacturing Stresses in Thickwalled Orthotropic Cylinders", presented at the 31st International SAMPE Symposium and Exhibition, April 1986.

4. Calius, E., and Springer, G.S., "Optimization of the Cure Window for a Large Filament Wound Case", Report to Chemical Research Projects Office, NASA-Ames Research Center, November 1984.

5. Rumberger, W., and Spencer, B.E., "The NCF (No Cut Fiber) Coupling", presented at the Seventh DOD/NASA Conference on Fibrous Composites in Structural Design, Denver, Colorado, June 1985

## 7. BIBLIOGRAPHY

Brian E. Spencer is Program Manager for funded development at Brunswick Corporation, Lincoln, Nebraska. Mr. Spencer received a B.S. in Agricultural Engineering from the University of Nebraska-Lincoln in 1970; an M.E. in Industrial Engineering from the University of California-Berkeley in 1971; and an M.E. in Mechanical Engineering from the University of California-Davis in 1980.

Table I - Fiber and Resin Properties

--------------------------FIBER PROPERTIES--------------------------

| | S-2 Glass[1] | 234 GPa (34 MSI) Graphite[2] | 276 GPa (40 MSI) Graphite[3] | 345 GPa (50 MSI) Graphite[4] | Kevlar[5] |
|---|---|---|---|---|---|
| Density, g/cm$^3$ (lb./in$^3$) | 1.5 (.090) | 1.8 (.065) | 1.8 (.065) | 1.8 (.065) | 1.5 (.054) |
| Strand Tensile Strength GPa (KSI) | 4.59 (665) | 4.14 (600) | 5.65 (820) | 2.41 (350) | 3.62 (525) |
| Modulus of Elasticity, GPa (MSI) | 86.8 (12.6) | 241 (35) | 275 (40) | 344 (50) | 132 (19.2) |
| Coefficient of Thermal Expansion per -°F $\times 10^6$ | 3.1 | -.3 | -.3 | -.4 | -1.1 |
| Tensile Elongation, percent | 5.4 | 1.7 | 2.0 | .6 | 2.7 |

Source: Manufacturers Product Literature

1. Owens Corning S-2 Fiberglass
2. Union Carbide T-600
3. Union Carbide T-40
4. Union Carbide T-50
5. DuPont

MSI = MODULUS $\times 10^6$ PSI

RESIN PROPERTIES

| | Epon 828 Based Epoxy | |
|---|---|---|
| Density, g/cm$^3$, (lb/in$^3$ ) | 1.203 | (.0433) |
| Tensile Strength, GPa (KSI) | .083 | (12) |
| Tensile Modulus, GPa (KSI) | 3.03 | (440) |
| Percent Elongation | | 7.0 |
| Compressive Strength, GPa (KSI) | .14 | (20) |
| Compressive Modulus, GPa (KSI) | 4.55 | (660) |
| Shear Strength, GPa (KSI) | .055 | (8) |
| Shear Modulus, GPa (KSI) | 3.24 | (470) |
| Coefficient of Thermal Expansion, per °F $\times 10^6$ | | 30 |

ALL PROPERTIES AT ROOM TEMPERATURE

Table II - BRUNSWICK COMPOSITE DRIVE SHAFT APPLICATIONS

| Application | Size | Remarks |
|---|---|---|
| Aft Rotor Shaft Boeing Model 234 Helicopter | Inside Diameter: 26 cm (10.2") Thickness: Center: 2 cm (0.8") Ends: 3 cm (1.2") | Maximum Torque: 135,600 N·m (1.2 x 10⁶ in.-lb.) Combination of S-2 fiberglass/ epoxy and graphite/epoxy |
| Navy Destroyers; Propeller Drive Shaft, Subscale | Inside Diameter: 7 cm (2.78") Thickness: Center: 1.6 cm (0.62") Ends: 2.2 cm (0.88") | Combination of S-2 fiberglass/ epoxy and graphite/epoxy |
| Rotor Shaft Boeing Model 360 Helicopter Aft and Forward | Inside Diameter: 16.5 cm / 26.7 cm (6.5") / (10.5") Wall Thickness: 5.1 cm (2") Length: 241 cm (95") 83.8 cm (33") | Maximum Torque: 282,400 N·m (2.5 x 10⁶ in.-lb.) Combination of S-2 fiberglass/ epoxy and graphite/epoxy |
| Drive Shaft Amph. Vehicle | Outside Diameter: 7.6 cm (3.0") Wall: .34 cm (.135") | Maximum Torque: 4,600 N·m (41,000 in.-lb.) Combination of S-2 fiberglass/ epoxy and graphite/epoxy |
| X-Wing Aircraft Pitch Actuation Shaft | Outside Diameter: 5.6 cm (2.2") Wall: .19 cm (.075") | Maximum Torque: 203 N·m (1,800 in.-lb.) Speed = 3,300 RPM. Combination of S-2 fiberglass/aluminized fiberglass/graphite and epoxy resin |

657

Table III - Bearing and Shear Strengths of Engineering Materials

| Material | Ultimate Bearing Strength, MPa (KSI) | | Shear Strength, MPa (KSI) | |
|---|---|---|---|---|
| Aluminum 2024-T852* | 690 | (100) | 296 | (43) |
| Aluminum 6061-T651* | 690 | (100) | 207 | (30) |
| Titanium Ti-6AL-4V* | 2070 | (300) | 690 | (100) |
| Steel 4340 (200 KSI)* | 2415 | (350) | 896 | (130) |
| Steel A286* | 2070 | (300) | 620 | (90) |
| Graphite/Epoxy Cross-Ply Laminate** | 380 | (55) | 62 | (9) |
| Kevlar/Epoxy Cross-Ply Laminate** | 345 | (50) | Very Low | |
| Glass/Epoxy Cross-Ply Laminate** | 470 | (68) | 62 | (9) |

NOTE: All values are at room temperature

* Reference: Aerospace Structural Metals Handbood, 1984
** Reference: Brunswick test data

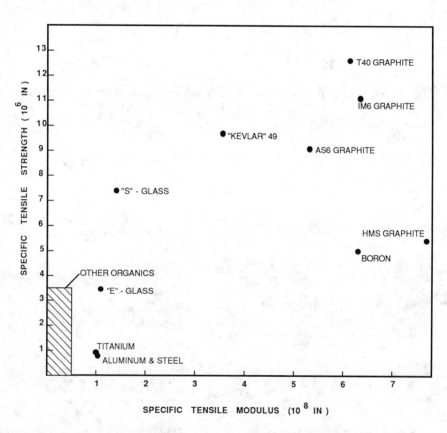

Figure 1.   Specific Tensile Strength and Specific Tensile
Modulus of Reinforcing Fibers

Figure 2.  Model 234 Rotor Shaft

Figure 3.  Delaminated Shaft Model 360

Figure 4.  Forward and Aft Model 360 Shafts
           Final Design

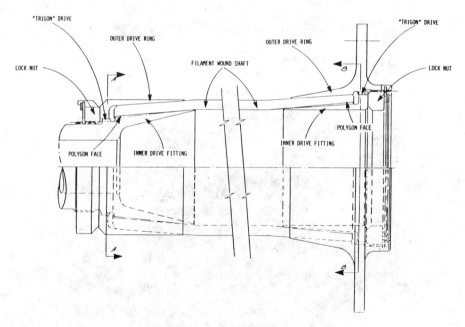

Figure 5.  NCF (no cut fiber) Coupling Concept

Figure 6.   Manufacturing Steps Required to Install
            Pinned Joints

# FILAMENT WOUND THERMOPLASTIC MATRIX PRESSURE VESSELS
## Ritch D. Hollingsworth and Donald R. Osment
### Aerojet Strategic Propulsion Company, Sacramento, California

## ABSTRACT

The availability of high temperature thermoplastic resins has created interest in the application of these systems to filament wound pressure vessels (chambers) for solid propellant rocket motors. Material property data obtained from hydroburst tests are compared for a standard carbon fiber/epoxy system against a thermoplastic matrix/carbon fiber system.

The filament winding of high temperature thermoplastic matrix/carbon fiber pressure vessels is discussed. The equipment used to fabricate the ASTM D-2585 pressure vessels as well as the design features of this test vessel are described. Data obtained from the hydroburst tests is presented. This data includes maximum pressure, composite and fiber tensile strength and performance (PV/W). Consolidation of the filament wound composite was investigated and the results are discussed. Photomicrographs of the pressure vessels are included.

## INTRODUCTION

The increased availability of carbon fiber impregnated with high temperature thermoplastic resin has stimulated great interest in manufacturing engineering circles on how best to process these materials. Aerojet Strategic Propulsion Company, as a manufacturer of filament wound chambers for solid propellant rocket motors, has a vested interest in prepreg material systems for filament winding. This is one of the most cost effective (cost per pound to manufacture) of all the continuous fiber reinforced composite manufacturing methods known. Aerojet's goal is to provide our customers with higher quality, and lower life cycle cost systems and thermoplastic matrix composites may prove to be a major factor in obtaining this goal.

Aerojet's capabilities in thermoplastic matrix composites are two fold. First as a fabricator of filament wound structures and secondly as a prepreger of fiber tow with thermoplastic resins. This report looks only at the fabrication of pressure vessels with commercially available thermoplastic matrix carbon fiber prepreg. Phillips Ryton AC40-60, 12K, single tow was chosen as a baseline due to history, availability, and relatively moderate processing temperatures. This material is Hercules AS-4 12K carbon fiber impregnated with Phillips polyphenylene sulfide (PPS) resin. To provide an accurate data base to compare the thermoplastic materials, parallel tests were conducted using a basic epoxy system and AS-4 12K fiber.

Mechanical properties, void content, and processing parameters were studied using ASTM D-2585 5.75-in. test bottles for the thermoplastic matrix composites noted. NOL rings were fabricated from thermoplastic and epoxy prepreg for short beam shear testing.

## EXPERIMENTAL

The primary vehicles used for these tests were the ASTM D-2585 5.75-in. bottle and NOL rings. The bottles were used for hydroburst tests and compaction analysis of the wound composite. This analysis included short beam shear and photomicrograph specimens in both hoop and axial directions. The NOL rings were used for shear specimens to compare the 5.75-in. bottle shear properties and to provide an even comparison to the thermoplastic "as wound" test.

The 5.75-in. bottles were fabricated using two different mandrel and bladder systems. The thermoplastic bottles were wound on sodium silicate mandrels with silicone rubber bladders. This system is capable of withstanding the relatively high temperatures (550- 600°F) required to process the thermoplastic materials. PVA/sand mandrels were used with chloro-butyl bladders for the epoxy bottles due to the lower processing temperatures required for this system.

The thermoplastic bottles were filament wound on a tumble winder in ASPC's Engineering Development Laboratory. The wind pattern consisted of 4 polar plies and 6 hoop plies using a 0.285-in. band width which yielded a stress ratio of 0.81. Phillips Ryton AC40-60, 12K, single tow prepreg was used. This is Hercules AS-4 12K carbon fiber impregnated with Phillips polyphenylene sulfide resin at a ratio of 37 percent resin and 63 percent fiber by weight. A heat source was used during winding to tack the fiber in place as it contacted the mandrel. The bottles were

vacuum bagged to prevent oxidation of the polyphenylene sulfide resin during the post-wind consolidation process. The consolidation cycle was 15 minutes at 575°F. Following heat treatment, the mandrel system was washed out and the bottles were prepared for testing.

The epoxy bottles were wound on the same equipment as the thermoplastic bottles. However, the design was changed to accommodate the different materials. Again, Hercules AS-4 12K carbon fiber was used, only this time it was used with an epoxy/amine resin. The fiber was impregnated with the resin at an "A" stage condition using in-house equipment. The wind pattern consisted of 4 polar plies and 6 hoop plies. Two tows were used for the polars at a 0.285-in. band width and one tow was used on the hoops with a 0.142-in. band width which yielded a stress ratio of 0.81. After winding, the bottles were "B" staged for 3 to 4 hours at 120°F, and final cure was accomplished at 250°F for 4 hours. Following cure, the PVA/silicate mandrels were washed out and the bottles were prepared for testing.

Hydroburst tests were conducted on four of the filament wound 5.75-in. bottles (see Figure 1). Two of the bottles were thermoplastic (Ryton PPS/AS4) and two were epoxy (SRF-210/AS4). As can be seen in Table I, there is a large difference in the burst pressures obtained between the epoxy and thermoplastic bottles due merely to a change in winding patterns. The epoxy bottles, while having the same stress ratio, contained more fiber than the thermoplastic ones due to a thinner band width. The thermoplastic bottles exhibited a lower fiber strength, thus, a lower PV/W than their epoxy counter parts. This was somewhat of a surprise. Further investigation through visual inspection of the hydroburst bottles and photomicrographs of sectioned thermoplastic bottles revealed thin areas in the composite wall (see Figure 2). The thin areas caused higher fiber stresses and, thus, lower burst pressures which were not taken into account when the calculations were made. The thin areas were caused by gaps in the polar plies due, in part, to variations in the tow width as supplied from the vendor. This can be corrected on future bottles by adjusting the wind pattern to accommodate for the variable tow width, or by tighter controls at the vendor's facility.

Figure 1.   5.75-in. Bottle PPS/AS4

TABLE I.   5.75-IN.   BOTTLE HYDROBURST DATA

|  | Burst Pressure, psi | Hoop Strength, ksi | Polar Strength, ksi | $\frac{PV}{W}$ | Stress Ratio |
|---|---|---|---|---|---|
| TP 1 | 1950 | 356.9 | 288.1 | 1.07 | 0.81 |
| TP 2 | 1660 | 303.8 | 245.2 | 0.83 | 0.81 |
| EX 1 | 5300 | 488.1 | 394.1 | 1.33 | 0.81 |
| EX 2 | 4775 | 439.7 | 355.0 | 1.22 | 0.81 |

Figure 2.   Cylinder Section 5.75-in. Bottle

Figure 3 shows an end view of Phillips Ryton AC40-60, 12K, tow magnified 100 times. Notice the even distribution of the resin around the fibers. This is not always the case with other prepregs. Photomicrographs were also taken of sectioned bottles to visually evaluate the void content and consolidation of the thermoplastic composite and to compare them with their epoxy counterparts. Figure 4 shows the relatively low void content of the thermoplastic bottle when compared to the standard epoxy bottle. While the void content is not as low as the bismaleimide system, it is acceptable for this early in the development program.

Short beam shear tests per ASTM D-2344-76 were conducted on sectioned epoxy and thermoplastic 5.75-in. bottles and thermoplastic NOL rings. While it can be argued whether or not short beam shear tests are a viable method for testing thermoplastic composites, the data obtained is valuable when comparing different thermoplastic processing parameters. First looking at the epoxy data (see Tables II and III) these specimens were obtained from 5.75-in. bottles. The specimens were prepared in two different orientations, hoop and axial. The hoop short beam shear data from the epoxy/carbon system, Table II, is acceptable for an epoxy matrix system indicating a mean shear stress of 6,240 psi and a standard deviation of 256. Table III shows the data obtained from the axial short beam shear specimens. Again, the data is good with a mean of 5,990 psi and a standard deviation of 312. Thermoplastic composite specimens taken from the subscale pressure vessels did not fail in shear, but underwent plastic deformation under load.

Thus, thermoplastic short beam shear specimens were prepared from NOL rings. The rings were prepared in two different versions, the "as wound" and "consolidated." The "as wound" specimens were filament wound using a heat source and our best compaction techniques to melt the resin and consolidate the plies while winding. The consolidated specimens were subjected to an oven post wind heat treatment under vacuum pressure. As can be seen in Table IV, it was difficult to obtain a shear failure in the as wound specimens. The best shear stress obtained was 4,540 psi. The consolidated specimens fared much better. Of the 10 specimens only 4 failed to shear (Table V). The mean shear stress for the 6 specimens was 5,300 psi. While this figure is somewhat lower than the epoxy comparison at 8,721 psi (Table VI), it is an acceptable value. The performance of the as wound shear specimens indicate an opportunity to improve the thermoplastic filament winding process to achieve full compaction while fabricating.

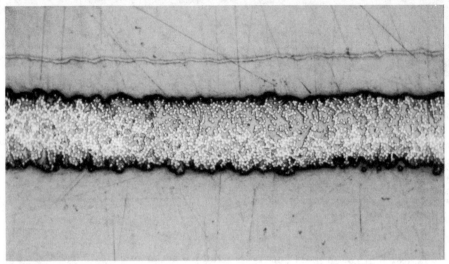

Figure 3.   AC 40-60 12K Tow 100X

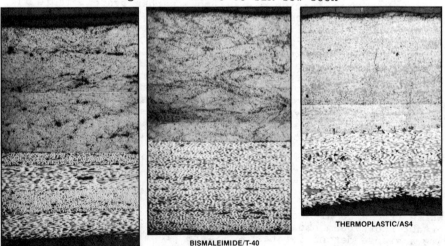

EPOXY/IM6

BISMALEIMIDE/T-40

THERMOPLASTIC/AS4

Figure 4.   Photomicrographs 5.75-in. Bottle Sections

## TABLE II.  5.75-IN. BOTTLE HOOP SHEAR, EPOXY/AS4

|        | Width, in. | Thick, in. | Ultimate Load, lb | Shear Stress, psi |
|--------|-----------|-----------|-----------|-----------|
| EXH 1  | 0.246     | 0.090     | 182       | 6170      |
| EXH 2  | 0.248     | 0.090     | 184       | 6180      |
| EXH 3  | 0.249     | 0.089     | No Shear  |           |
| EXH 4  | 0.247     | 0.091     | 183       | 6110      |
| EXH 5  | 0.247     | 0.090     | 183       | 6170      |
| EXH 6  | 0.246     | 0.091     | 175       | 5860      |
| EXH 7  | 0.249     | 0.091     | 187       | 6190      |
| EXH 8  | 0.248     | 0.091     | 202       | 6710      |
| EXH 9  | 0.248     | 0.089     | 180       | 6120      |
| EXH 10 | 0.248     | 0.089     | 196       | 6660      |

## TABLE III.  5.75 IN. BOTTLE AXIAL SHEAR, EPOXY/AS4

|        | Width, in. | Thick, in. | Ultimate Load, lb | Shear Stress, psi |
|--------|-----------|-----------|-----------|-----------|
| EXA 1  | 0.262     | 0.085     | 178       | 5990      |
| EXA 2  | 0.253     | 0.086     | 180       | 6200      |
| EXA 3  | 0.252     | 0.092     | 183       | 5920      |
| EXA 4  | 0.248     | 0.090     | 162       | 5440      |
| EXA 5  | 0.248     | 0.089     | 165       | 5610      |
| EXA 6  | 0.250     | 0.088     | 180       | 6140      |
| EXA 7  | 0.260     | 0.087     | 180       | 5970      |
| EXA 8  | 0.253     | 0.091     | 178       | 5800      |
| EXA 9  | 0.252     | 0.092     | 204       | 6600      |
| EXA 10 | 0.259     | 0.091     | 195       | 6210      |

## TABLE IV.  NOL RING SHEAR TESTS PPS/AS4 (AS WOUND)

|        | Width, in. | Thick, in. | Ultimate Load, lb | Shear Stress, psi |
|--------|-----------|-----------|-----------|-----------|
| TP 1   | 0.235     | 0.116     | 165       | 4540      |
| TP 2   | 0.240     | 0.114     | No Shear  |           |
| TP 3   | 0.241     | 0.113     | No Shear  |           |
| TP 4   | 0.238     | 0.112     | No Shear  |           |
| TP 5   | 0.238     | 0.115     | No Shear  |           |
| TP 6   | 0.239     | 0.114     | No Shear  |           |
| TP 7   | 0.239     | 0.114     | 146       | 4020      |
| TP 8   | 0.242     | 0.112     | No Shear  |           |
| TP 9   | 0.243     | 0.113     | No Shear  |           |
| TP 10  | 0.236     | 0.114     | 119       | 3320      |

TABLE V.   NOL RING SHEAR TESTS PPS/AS4 (CONSOLIDATED)

|        | Width, in. | Thick, in. | Ultimate Load, lb | Shear Stress, psi |
|--------|------------|------------|-------------------|-------------------|
| TPC 1  | 0.244      | 0.129      | No Shear          |                   |
| TPC 2  | 0.244      | 0.127      | 210               | 5080              |
| TPC 3  | 0.244      | 0.127      | 222               | 5370              |
| TPC 4  | 0.245      | 0.125      | 215               | 5270              |
| TPC 5  | 0.245      | 0.125      | 228               | 5580              |
| TPC 6  | 0.245      | 0.126      | No Shear          |                   |
| TPC 7  | 0.245      | 0.128      | 223               | 5330              |
| TPC 8  | 0.245      | 0.126      | No Shear          |                   |
| TPC 9  | 0.243      | 0.126      | 210               | 5140              |
| TPC 10 | 0.243      | 0.128      | No Shear          |                   |

TABLE VI.  NOL RING SHEAR TESTS, EPOXY/AS4

|       | Width in. | Thickness in. | Breaking Load lbs | Shear Strength psi |
|-------|-----------|---------------|-------------------|--------------------|
| EP 1  | .249      | .105          | 332               | 9523               |
| EP 2  | .250      | .104          | 303               | 8740               |
| EP 3  | .251      | .107          | 306               | 8545               |
| EP 4  | .250      | .107          | 326               | 9140               |
| EP 5  | .249      | .106          | 305               | 8666               |
| EP 6  | .250      | .106          | 301               | 8518               |
| EP 7  | .250      | .105          | 301               | 8600               |
| EP 8  | .250      | .104          | 290               | 8365               |
| EP 9  | .250      | .105          | 299               | 8542               |
| EP 10 | .250      | .106          | 303               | 8575               |

## CONCLUSIONS

The filament winding of pressure vessels with high temperature thermoplastic matrix carbon fiber composites has been investigated. ASTM D-2585 5.75-in. bottles were filament wound using Phillips Ryton AC40-60/AS-4 12K tow. These were compared with data obtained from epoxy/carbon bottles. Data obtained from hydroburst, short beam shear, and photomicrographs has been examined. This data shows the fiber strength of the PPS/AS-4 to be 71 percent of the baseline epoxy/AS-4 system. Short beam shear strength of the PPS/AS-4 consolidated system obtained from NOL ring specimens was 61 percent of a typical epoxy/AS-4 composite material. The lower value obtained from the as wound PPS/AS-4 system indicates a need to develop improved thermoplastic matrix filament winding methods to produce filament wound articles that require no further oven or autoclave post treatment, thereby reducing both time and labor hours. Photomicrographs reveal a low void volume and good resin distribution of the filament wound PPS/AS-4 composite. This was obtained with only vacuum-bag compaction, and is nearly equivalent to a state-of-the-art bismaleimide prepreg system.

The data obtained from this study indicates a promising future for thermoplastic matrix composites in filament wound pressure vessels. While development work needs to continue on manufacturing methods, the initial results show promise for a thermoplastic, filament wound composite niche. As the material suppliers develop improved resins and consistent processing methods the future will continue getting brighter for thermoplastic matrix composites in the filament winding field.

## ACKNOWLEDGMENTS

The authors wish to thank Bob Bertolucci of the Aerojet Strategic Propulsion Company Engineering Development Laboratory for his hard work and dedication in fabricating the subscale pressure vessels. Also appreciated was the generous and expedient assistance from Robert Shue, Phillips Petroleum, Composites Division in supplying material and technical data.

## BIOGRAPHY

Ritch Hollingsworth received his Bachelor of Science degree in Industrial Technology from California State University, Chico, in 1976. Prior to joining Aerojet Strategic Propulsion Company in 1979, he had his own company which designed and manufactured crash worthy fuel systems for race cars, and later working as a member of American Motors Corporation's Service and Reliability group.

Ritch has over ten years' experience in the development and manufacturing of products from fiber reinforced thermoplastic and thermosetting materials. His current responsibilities as a senior manufacturing engineer with Aerojet's Advanced Manufacturing Department are the development of manufacturing processes for thermoplastic matrix composites and internal insulation.

Donald Osment has over six years of research experience in the organic composite fields. Before joining Aerojet one and a half years ago, Donald worked at Morton-Thiokol's Wasatch Division in the filament winding R&D area. He attended the university of Arizona majoring in the chemistry and engineering fields. Mr. Osment is currently investigating cost saving measures for composite rocket motor manufacturing. This includes high temperature composite reliability, new materials investigations, and advanced cure coupled with improved processing techniques.

32nd International SAMPE Symposium
April 6-9, 1987

VALIDATION OF THE FILAMENT WINDING PROCESS MODEL

Emilo P. Calius and George S. Springer
Department of Aeronautics and Astronautics
Stanford University, Stanford, California 94305
and
Brian A. Wilson and R. Scott Hansen
Aerojet Strategic Propulsion Corporation
Sacramento, California 95813

## Abstract

Tests were performed towards validating the WIND model developed previously for simulating the filament winding of composite cylinders. In these tests two 24 in long, 8 in i.d. and 0.285 in thick cylinders, made of Hercules IM-6G fibers and Hercules HBRF-55 resin, were wound at $\pm 45°$ angle on steel mandrels. The temperatures on the inner and outer surfaces and inside the composite cylinders were recorded during oven cure. The temperatures inside the cylinders were also calculated by the WIND model. The measured and calculated temperatures were then compared. In addition, the degree of cure and resin viscosity distributions inside the cylinders were calculated for the conditions which existed in the tests.

## 1. INTRODUCTION

It is well recognized that the procedures employed during filament winding of composite cylinders have a significant effect on the quality and the cost of the finished product. Therefore, the procedures used during manufacture must carefully be selected for each application. This means that the appropriate process variables must be employed during fabrication. In case of filament winding these process variables include the crosshead and mandrel speeds, fiber tension, and heat applied during winding and subsequent curing (Figure 1). As has been demonstrated in recent years, these process variables can best be established by the use of analytical models. For this reason, a model applicable to filament winding has been developed Calius and Springer.[1-3] In this model, a process is

considered in which the following parameters are given

a) mandrel and composite cylinder dimensions

b) mandrel and composite material properties

c) mandrel angular velocity

d) cross-head speed

e) initial fiber tension

f) heat (or temperature) applied during and after winding.

For specified values of these parameters the model provides the following information as functions of position and time

a) temperature inside the mandrel and composite cylinder

b) degree of cure inside the composite cylinder

c) viscosity inside the composite cylinder

d) fiber tension

e) fiber position

f) residual (curing) stresses and strains

g) porosity

The model, referred to as "WIND", is composed of five submodels. The thermochemical submodel provides the first three of the aforementioned parameters, namely the temperature, degree of cure, and the viscosity. The fiber motion submodel yields the fiber position and the fiber tension. The stress submodel gives the residual stresses and strains, and the porosity submodel computes the changes in void diameter.

Table 1
Principal Characteristics of the Composite Cylinders
and the Manufacturing Process

| Cylinder | I | II |
|---|---|---|
| internal diameter (in) | 7.974 | 7.950 |
| length (in) | $24 \pm 0.01$ | $24 \pm 0.01$ |
| length including domes (in) | $32.8 \pm 0.01$ | $32.8 \pm 0.01$ |
| bandwidth (in) | $0.315 \pm 0.005$ | $0.315 \pm 0.005$ |
| final thickness (in) | $0.289 \pm 0.005$ | $0.279 \pm 0.005$ |
| average time to wind a helical layer pair (min) | 19.3 | 19.3 |
| thickness build-up rate $\times 10^3$ (in/min) | 0.935 | 0.903 |
| total winding time (hour: min) | 9 : 43 | 6 : 28 |
| winding tension (lb) | $24 \pm 0.5$ | $18 \pm 0.5$ |
| cure temperature (°F) | 250 | 250 |
| oven ramp rate (°F/min) | 3 | 1.5 |
| oven cooling rate (°F/min) | 1.5 | 0.5 |
| total oven time (hr) | 11 | 11 |

The "WIND" model has been shown to provide information useful in selecting the process variables. However, the validity of the model has not yet been demonstrated. Therefore, tests were performed to validate the thermochemical submodel which is the root, and hence most important, component of the entire filament winding model.

## 2. TESTS

In order to validate the thermochemical submodel, two cylinders were wound on a 24 in long 8 in o.d., 0.5 in thick steel mandrel. There was a 4 in radius hemispherical aluminium dome on each end of the mandrel, The composite was made of Hercules IM-6G fibers and Hercules HBRF-55 resin. The wind angle for each cylinder was ±45°. The principal differences between the two cylinders were their winding tension and cure temperature cycle. The dimensions of the cylinders, as well as the winding rates, initial fiber tension, and oven heating and cooling rates are summarized in Table 1.

The winding was performed at room temperature on an unheated mandrel. Once the winding was completed the mandrel-cylinder assembly was placed in a temperature controlled oven, without removing the domes. The actual temperatures on the inside surface of the mandrel and the outer surface of the cylinder were measured by J type thermocouples. Note that there were significant differences between the oven air temperature, programmed as shown in Table 1, and the boundary surface temperatures recorded by the aforementioned thermocouples. In addition, one thermocouple was mounted at the mandrel-composite interface. Thermocouples were also imbedded in the composite at two radial locations. These thermocouples were all located halfway between the two ends of the cylinder. The data recorded by the thermocouples are shown in the next section.

## 3. RESULTS

The temperatures of the inside surface of the mandrel $T_m$ and the outer surface of the cylinder $T_s$ are shown in Figure 2. Using these temperatures, the temperatures inside the composite were calculated by the thermochemical submodel. Since this submodel was described previously,[1, 4] it is not given here in detail;

it is outlined briefly in the Appendix. The calculations were performed with the "WIND" computer code.[1-3]

The temperatures calculated inside each cylinder are compared to the temperatures measured inside each cylinder in Figures 3 and 4. As can be seen, the calculated and measured temperatures are in excellent agreement, lending support to the validity of the thermochemical model.

In Figures 3 and 4 only the variations of temperature with time during the oven cure are shown. During winding the thermocouples on the rotating cylinders were not connected to the data-acquisition system. Room temperature averaged 70°F and calculations indicated the absence of significant temperature gradients through the thicknesses of the cylinder and the mandrel during winding.

For the recorded surface temperatures (Figure 2), the degrees of cure $\alpha$ and viscosities $\mu$ inside the cylinder were also calculated by the WIND code. Results of these calculations are presented in Figures 5 and 6. Note that for the resin, winding thickness and temperatures employed in this study, both the degree of cure $\alpha$ and the viscosity $\mu$ vary slowly at first and then increase rapidly as the cure temperature is approached. Complete cure is reached after about $1\frac{1}{2}$ to $2\frac{1}{2}$ hours in the oven. For the present cylinders any additional curing time appears to be unnecessary.

The foregoing figures illustrate the type of information provided by the thermochemical submodel. The variations in the degree of cure $\alpha$ and the viscosity $\mu$ with position and time would of course be altered by the use of different oven cure temperature cycles. With the WIND code, it is possible to examine the effects of the process input parameters and to select the winding and cure parameters which result in a manufacturing process which the user judges to be optimal.

## 4. ACKNOWLEDGMENTS

This work was supported in part by NASA Marshall Space Flight Center.

## 5. REFERENCES

1. Calius, E. P. and Springer, G. S. "Modeling the Filament Winding Process," *Proceedings of the Fifth International Conference on Composite Materials*, San Diego, CA, July 29–Aug. 1, 1071–1088 (1985).

2. Calius, E. P. and Springer, G. S. "Selecting the Process Variables for Filament Winding," *Material Sciences for the Future: Proceedings of the 31st SAMPE Symposium*, 891–899 (1986).

3. Springer, G. S. "Modeling the Cure Process of Composites," *Material Sciences for the Future: Proceedings of the 31st SAMPE Symposium*, 22–27 (1986).

4. Loos, A. C. and Springer, G. S. "Curing of Epoxy Matrix Composites," *Journal of Composite Materials*, Vol. 17, 135–169 (1983).

5. Bhi, S. T., Hansen, R. S., Wilson, B. A., Calius, E. P., and Springer, G. S. "Degree of Cure and Viscosity of Hercules HBRF-55 Resin," *Advanced Materials Technology '87, 32nd SAMPE Symposium* (this volume).

Appendix

## THERMOCHEMICAL SUBMODEL

The model is based on the energy equation

$$\frac{\partial(\rho C T)}{\partial t} = \frac{1}{r}\frac{\partial}{\partial r}\left(r\kappa\frac{\partial T}{\partial r}\right) + \rho\dot{H} \quad (1)$$

where $t$ is time, $r$ is the radial coordinate, $T$ is the temperature, $\rho$ is the density, $\kappa$ is the thermal conductivity, and $C$ is the specific heat of the composite. $\dot{H}$ is the rate at which heat is generated or absorbed by chemical reactions[4]

$$\dot{H} = \left(\frac{d\alpha}{dt}\right)H_r \quad (2)$$

where $\alpha$ is the degree of cure, $\frac{d\alpha}{dt}$ is the rate at which the cure is proceeding, and $H_r$ is the total heat of reaction.

The applicable initial and boundary conditions are as follows. The initial temperature and degree of cure of the material as it is added to the cylinder being wound is

$$T = T_o; \quad \alpha = \alpha_o \quad \text{at} \quad t = t_o \qquad (3)$$

where $T_o$ is the temperature of the material (which is generally the same as the temperature of the room), while $\alpha_o$ is the degree of cure the resin has acquired before it is deposited on the cylinder

The boundary conditions are

$$T = T_s \quad \text{at} \quad r = R + h_s \qquad (4)$$
$$T = T_m \quad \text{at} \quad r = R - h_m \qquad (5)$$

where $T_s$ is the temperature of the cylinder's outer surface, $T_m$ is the temperature at the mandrel's inner surface, $h_m$ is the effective thermal thickness of the mandrel, and $h_s$ is the cylinder's instantaneous thickness of the cylinder at time $t$. Note that $T_s$, $T_m$, and $h_s$ vary with time.

To complete the definition of the problem, an expression is needed which provides a rela- tionship between the rate of cure $\frac{d\alpha}{dt}$, the degree of cure $\alpha$, and the temperature $T$. A suitable relationship for the material used in this study is given in Ref. (5).

Expressions are also needed relating the thermal conductivity $\kappa$ and the thermal heat capacity per unit volume $C' = \rho C$ to the degree of cure $\alpha$ and the temperature $T$. The chosen approximate expressions are[1]

$$\kappa = \alpha \kappa^c + (1 - \alpha) \kappa^{uc} \qquad (6)$$
$$C' = \alpha C'^c + (1 - \alpha) C'^{uc} \qquad (7)$$

where the superscripts $c$ and $uc$ refer to the properties of the cured ($\alpha = 1$) and uncured ($\alpha = 0$) states.

Solutions to Eqs. (1)–(7) provide the temperature and the degree of cure as functions of position and time. From this information the viscosity $\mu$ can also be determined, provided the viscosity as a function of temperature and degree of cure is known. The viscosity for the resin used in this study is given in Ref. (5).

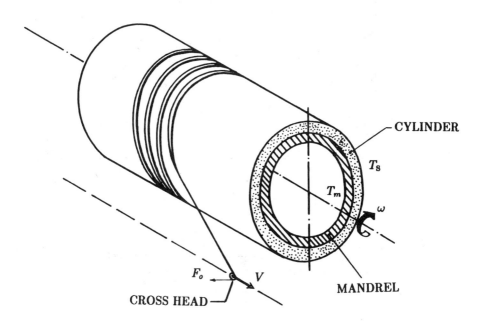

**Figure 1.** Description of the filament winding process.

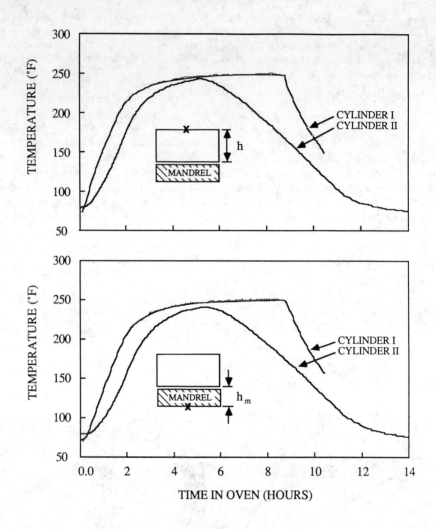

**Figure 2.** Measured temperatures at the inner (Tm) and outer (Ts) boundary surfaces of composite cylinders I and II during oven cure. Cross marks the thermocouple location.

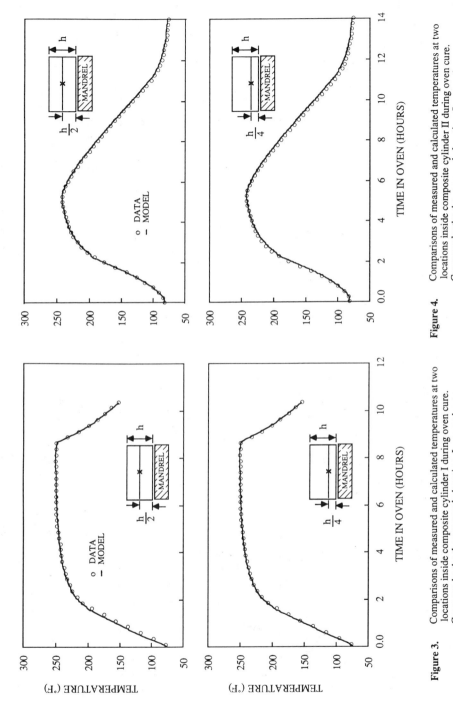

**Figure 3.** Comparisons of measured and calculated temperatures at two locations inside composite cylinder I during oven cure. Cross marks the thermocouple location. Inner and outer surface temperatures are shown in Fig. 2

**Figure 4.** Comparisons of measured and calculated temperatures at two locations inside composite cylinder II during oven cure. Cross marks the thermocouple location. Inner and outer surface temperatures are shown in Fig. 2

675

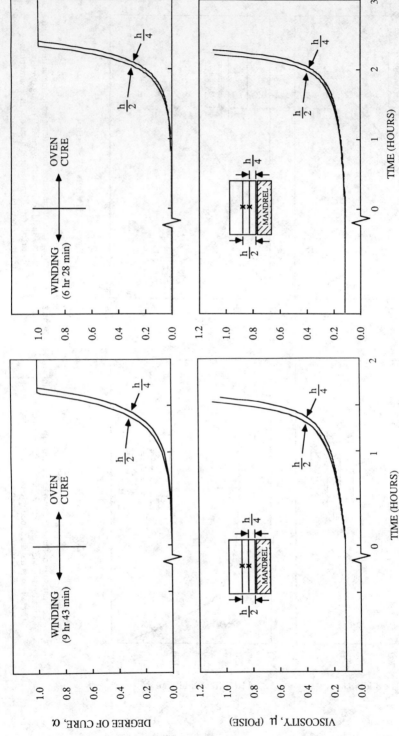

**Figure 5.** The degree of cure and viscosity calculated by the model at two locations inside cylinder I. Inner and outer surface temperatures are shown in Fig. 2

**Figure 6.** The degree of cure and viscosity calculated by the model at two locations inside cylinder II. Inner and outer surface temperatures are shown in Fig. 2

## GRAPHITE EPOXY PRESSURE VESSEL DOME REINFORCEMENT STUDY

James D. Erickson and James A. Yorgason
Morton Thiokol, Inc.
Wasatch Operations
P.O. Box 524
Brigham City, Utah

### Abstract

One of the most highly-stressed areas of a filament-wound composite rocket motor case is the dome/polar boss interface. During motor operation, all the pressure acting across the polar opening must be transferred to the composite in this area. A proven method of preventing failures at the polar boss interface is by local composite hoop-wound wafer reinforcement. This method allows a high helical to hoop stress ratio in the cylinder section of the motor, which minimizes helical material weight and maximizes pressure vessel efficiency $(PV/W)$. To attain a high efficiency motor case design made of stiff graphite fibers, this study was conducted to identify the optimum dome wafer reinforcement technique.

Using test vessels and finite element models, a number of dome reinforcement concepts were compared to determine which configuration produced the most improvement in dome strength. The selected configuration was then used in verification pressure vessels to demonstrate performance repeatability. As a result, wafer design and processing parameters were identified which produce consistently high dome strengths. Improvements in dome strengths as high as 17 percent over the unreinforced baseline were demonstrated with a minimal weight increase.

### 1. INTRODUCTION

The continuing trend toward use of high strength, high modulus graphite fibers has accented the need to characterize any peculiarities involved with their effective use in motor cases. The purpose of this study was to develop an improved reinforcement technique for the polar boss buildup area in filament-wound pressure vessels wound with these fibers as opposed to glass or Kevlar® materials.

In the past, filament-wound cases made of glass/epoxy materials proved insensitive to these stress concentrations and required no dome reinforcement. Later, the stiffer and more discontinuity-sensitive Kevlar®/epoxy material showed the need for dome reinforcement and the hoop-wound wafer reinforcement concept was developed. Today's graphite/epoxy materials are much stiffer than either glass or Kevlar® and have proven to be very sensitive to discontinuities at the polar boss flange interface.

In a rocket motor all of the internal pressure acting across the polar opening must be transferred to the motor case within the width of the boss flange. This contributes to making the dome/polar boss interface a very highly-stressed area. To prevent premature structural failure without reducing the case efficiency, some local reinforcement must be used. A dome reinforcement which increases the dome strength by 20 percent by reducing local stress peaks could reduce the total case weight for a first stage high performance ICBM by about 10 percent.

This study used 45.7 cm (18 in.) diameter test bottles wound with high polar/hoop stress ratios to assure burst failures in the dome/polar boss area. The study was initiated to determine which dome reinforcement techniques or parameters produce the most improvement in motor cases made of graphite fibers. Union Carbide T-700/6K with a tensile strength of 4.47 GPa (649 ksi) and a modulus of 257 GPa (37.3 msi) was used for the bulk of this study. Another Union Carbide fiber, T-40XS/12K with a tensile strength of 5.12 GPa (741 ksi) and modulus of 285 GPa (41.4 msi), was later substituted because of its higher strength. The selected dome reinforcement concept was verified with both fiber materials.

Since nearly all of the tests were conducted using a single test specimen, finite element analyses were used to augment the results. Finite element models were constructed and analyzed to verify the indi-

cated trends for each of the different design modifications.

## 2. DISCUSSION

For this study several dome reinforcement alternatives were tested to determine their structural effects on the pressure carrying capability of a graphite/epoxy vessel. The test vessel design had a 45.7 cm (18 in.) inside diameter with a 25.4 cm (10 in.) cylinder section and a polar opening of 10.2 cm (4.0 in.). The basic winding sequence used is shown in Figure 1. The basic reinforcement choice was the flat hoop-wound wafer; however, other dome reinforcements were studied to determine the vessel dome response produced. Results from each test were used to support and modify the subsequent testing. This method helped to produce an acceptable dome reinforcement and to identify critical processing parameters.

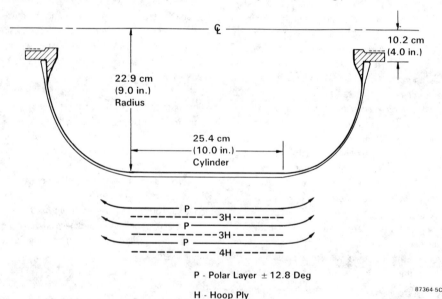

P - Polar Layer ± 12.8 Deg

H - Hoop Ply

87364-5C

Figure 1. Test Vessel Design

### 2.1 Stepped-Back Layer

In a filament-wound pressure vessel all of the helical or polar wraps pass next to the polar boss. This creates a local buildup of roving material which causes the outer windings to bridge from the buildup area to the dome, leaving a high void area near the stress critical boss flange tip. We attempted to reduce this local buildup by winding the second polar layer stepped back from the polar

boss one band width (Figure 2). The effect of this stepped-back design with no other reinforcement was a 25 percent reduction in dome strength at the polar boss. This strength reduction was attributed to the decrease in total polar fibers crossing over the polar boss flange. When one hoop wafer was used with this stepped-back design, we observed a 14 percent reduction in strength compared to a baseline design also with one wafer.

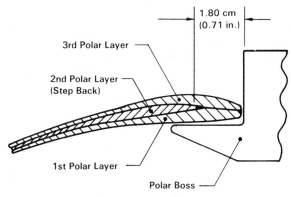

**Figure 2. Step Back Layer Configuration**

This observation indicated that there was a minimum amount of such crossover material required to retain the polar boss in the vessel dome. While a step back can be used to control dome material buildup, it is important to consider total boss retention material (polar fibers plus hoop reinforcements).

## 2.2 Polar Wafers

A more direct method of reinforcing the polar fibers over the polar boss flange was felt to be a polar wound wafer or partial dome cap. This wafer was wound on a regular bottle mandrel and cut out at a 30.5 cm (12 in.) outer diameter with an epoxy adhesive film as a carrier. The resulting wafer shown in Figure 3 was 1.5 mm (0.06 in.) thick at the outer diameter.

A finite element analysis was conducted to evaluate the reinforcement efficiency of this design. The analysis results, shown in Figure 4, predicted that the wafer would reduce fiber stress levels up to the polar boss flange. However, it also showed a bending stress concentration at the wafer tip. The test vessel with this wafer failed at the wafer tip at a pressure which was 12 percent below that of the baseline vessel.

An attempt was made to reduce the stress concentration at the wafer tip by reducing the wafer outer diameter thickness (Figure 5). The desired cross sectional thickness was achieved by winding with the machine advance increased enough to produce edge-to-edge band laydown at the wafer tip 25.4 cm (10 in.). For this wafer, the edge thickness was reduced to about 0.089 mm (0.035 in.) compared to 0.15 mm (0.06 in.) for the first design. This test vessel also failed at the wafer tip at the same pressure level as the first polar wafer vessel.

These polar wafers, regardless of edge thickness, had a fiber orientation nearly perpendicular to their edge. This orientation caused a much higher meridional stiffness discontinuity at the outer edge than did the hoop-oriented wafer with an even thicker edge profile. Furthermore, the local dome thickness was less at the tips of the polar wafers than at the tips of the hoop wafers and thus more discontinuity-sensitive, as shown in the stress analysis.

## 2.3 One Hoop Wafer

One hoop-wound wafer placed between the first and second polar layers increased the dome strength by 6 percent over the baseline vessel. It was during the fabrication of this vessel that several of the critical wafer processing parameters were developed. The first attempts at fabricating a vessel with the one hoop wafer demonstrated the need to have minimal wafer wrinkles and a good bond between the wafer and dome.

Since our wafers were wound flat and subsequently formed to the contour of the vessel dome, wrinkles could develop in the wafer fibers. Also, as polar layers are wound over a wafer, the advancing wind pattern sometimes pushes a wave into the wafer and creates distortions in the fibers. Several times these distortions were covered by the advancing band and were not discovered until the post-burst examination of the vessel. We were able to reduce the amount of wafer wrinkling by preforming the wafers to the shape of the vessel dome prior to placement.

Another critical parameter for wafer dome reinforcing is the need to have a good bond between the wafer and dome material. The wafers for this study were fabricated with a high viscosity prepreg resin and the case was fabricated with a typical wet winding resin. This difference in materials may have contributed to a low interlaminar shear strength which would reduce the load sharing capability between the dome and wafer. Examination of several burst tests revealed debonded wafers which were subsequently correlated with lower dome strengths than had been seen in vessels with well bonded wafers. This wafer/dome bond was improved by using an epoxy adhesive film between the wafer and dome. In addition to increasing the bond strength, the adhesive film also helped to make wafer placement more accurate and reduce the amount of wrinkling.

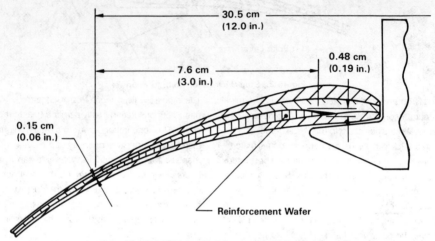

**Figure 3. Polar Wound Reinforcement Wafer Configuration**

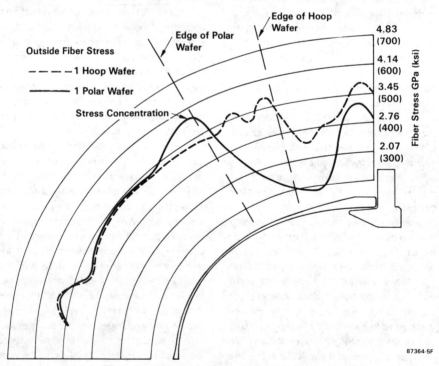

**Figure 4. Analytical Stress Prediction of Hoop-vs-Polar Wafer**

680

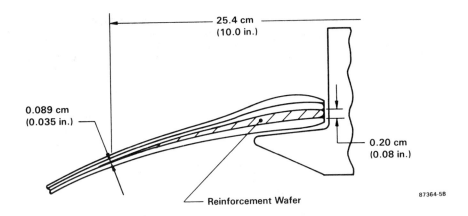

Figure 5. "Thin" Polar Wound Reinforcement Wafer

## 2.4 Two Hoop Wafers

For our test vessel the optimum or full wafer configuration consisted of 2 thin wafers (Figure 6) which extended to the edge of the composite opening and produced an increase in dome strength of 9 percent over the baseline. For a test vessel with a modified (flightweight) polar boss, the increase was 17 percent over a baseline test. This two-wafer design was one which had a wafer/polar material ratio which represented an optimum design since it reinforced the polar boss enough to force failures into the vessel cylinder and knuckle at a stress ratio of 0.92.

## 2.5 Short Flange Boss

The short flange or "flightweight" polar boss was modified to have a pressure distribution ratio (PDR) which was more in line with those found in a normal flightweight design. The PDR is the ratio of total opening area to the boss flange interface area. A typical flightweight polar boss has a PDR in the range of 2.5 to 3.5 in order to provide a sufficient boss to composite interface area so as to prevent local overstressing of the dome near the polar opening. Before modification, our test vessel had a PDR of 1.8, indicating an excessively wide flange for the boss size. After modification, the PDR for this polar boss was 3.27. This indicates that a favorable weight trade can be made between polar boss flange size and wafer material. Figure 7 compares the dome stresses for vessels with and without the modified boss, both with wafers and without.

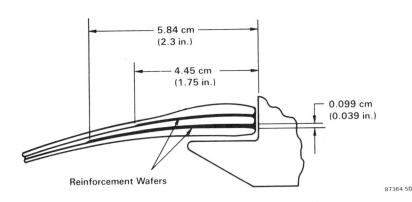

Figure 6. Final Selected Hoop Wafer Reinforcements

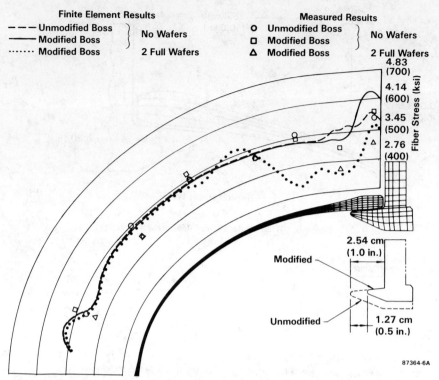

**Finite Element Results**
- – – Unmodified Boss
- —— Modified Boss  } No Wafers
- ······ Modified Boss    2 Full Wafers

**Measured Results**
- O  Unmodified Boss
- □  Modified Boss  } No Wafers
- △  Modified Boss    2 Full Wafers

4.83 (700)
4.14 (600)
3.45 (500)
2.76 (400)

Fiber Stress (ksi)

2.54 cm (1.0 in.)

Modified

Unmodified

1.27 cm (0.5 in.)

87364-6A

**Figure 7. Effect of Wafer Reinforcement and Boss Flange Modifications on Fiber Stress**

## 2.6 Wafer Placement

Just as important as the amount of wafer reinforcement is the positioning of the wafers. If wafers are placed so that the outboard edges are too far away from the buildup area, two things will occur: (1) since the buildup thickness decreases with increased radial distance from the boss, the geometric discontinuity caused in the polar fibers as they bridge over the wafer tip is magnified and (2) since the wind angle decreases rapidly with increased radial distance from the boss, the wafer's outer edge fibers approach perpendicularity to the polar fibers resulting in a severe stiffness discontinuity. This high angular orientation between wafer and polar fibers also causes strain incompatibilities which may lead to delaminations and premature vessel failure.

The most critical area of the dome which requires reinforcement is at the polar boss flange. All of the pressure retained by the polar boss must be transferred to the case in this boss flange interface area, causing a stress concentration in the inner fiber layers at the flange tip.

The purpose of the wafer is to reduce the overall stress level in the adjacent fibers so that the additional flange tip concentration does not cause this area to exceed other cylinder or dome stress levels. The most effective positioning of the wafer is directly over the boss flange with the outer edge terminating at a radial distance coinciding with a helical wind angle which is not so low that hoop loads become minor and edge discontinuities become prominent. Figure 8 shows the stress level reduction at the boss due to this full (two-wafer) wafer design compared to the baseline design.

This study used hoop wafers which were only wound flat with the same thickness throughout. However, the capability exists to fabricate wafers with a contour which more closely matches the shape of the dome on which they will be used. This capability helps to reduce the tendency of a wafer to wrinkle and buckle when being placed in a dome. Along with the contour, a wafer can also be made with a continuously varying thickness. By varying the wafer thickness discontinuities can be reduced and stresses more closely balanced.

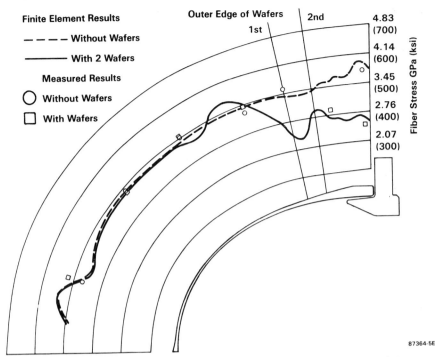

**Finite Element Results**

− − − − **Without Wafers**

———— **With 2 Wafers**

**Measured Results**

◯ **Without Wafers**

▢ **With Wafers**

Outer Edge of Wafers

1st | 2nd

4.83 (700)
4.14 (600)
3.45 (500)
2.76 (400)
2.07 (300)

Fiber Stress GPa (ksi)

87364-5E

**Figure 8. FE Predictions and Results of Wafer Reinforced-vs-Unreinforced**

## 2.7 Finite Element Models

Figure 8 shows a sample of the type of correlation which was achieved between the finite element analysis results and the measured strain data. This correlation, with few exceptions, was within ± 15 percent of the measured fiber stresses of the outside surface shown. The nominal stress levels predicted were very valuable in indicating trends which each successive design change would cause. These predicted trends were verified by the experimental burst pressures and measured strains. The excellent correlation between analytic and experimental data not only shows the value of using a finite element analysis for trade studies, but also adds credibility to the test result trends.

## 3. SUMMARY

This dome reinforcement study met the objectives of developing a reinforcement scheme for graphite/epoxy motor cases that would maximize delivered fiber strength as evidenced by mixed polar and hoop failure modes attained at a relatively high polar/hoop stress ratio of 0.92. The study shows that proper placement of wafers is essential to performance and that there is an optimum amount of wafer material above which no

additional vessel performance will be recognized. However, properly formed and placed wafers can substantially improve the efficiency of graphite/epoxy motor cases. This gives the test vessel a near minimum weight which translates into high PV/W values (pressure x volume/weight). The study also shows significantly greater improvements in dome strength due to wafer reinforcements in 18-in. vessels with a "modified" or "flightweight" short flange polar boss as compared to the "standard" 18-in. polar boss flange design.

Hoop-wound wafers are very effective in increasing the efficiency of graphite pressure vessels. However, the following four conditions must exist to achieve maximum benefits: (1) wafers must be placed next to the polar opening and extend over the polar boss flange tip; (2) wafers must be free of any slack in the hoop direction due to fibers wrinkling during placement or forming to the dome; (3) wafers must be adequately bonded to adjacent polar plies in order to transfer load from the hoop component of the polar fibers; and (4) the material modulus of the wafers must match that of the dome material system for them to be most effective. Figure 9 shows a comparison of the dome strengths for each of the major reinforcement options evaluated.

Polar-wound wafers (partial dome cap) are not as efficient as smaller hoop-wound wafers. However, the failure with polar wafers initiates at the wafer tip instead of the boss flange tip. The tip of a polar wafer creates an abrupt bending load discontinuity in the polar fibers.

A pressure vessel using a 50 percent shorter polar boss flange, which is more comparable to a common flightweight boss profile, without wafer reinforcement burst at a lower pressure than the baseline. However, wafer reinforcement successfully restored the dome to the equivalent strength of the standard boss design with wafers. This constitutes an improvement in dome strength of over 17 percent due to wafers.

Using one stepped-back polar layer with two regular layers and no wafers causes polar boss failures at lower pressures than the basic three-layer design.

Some additional factors which must be considered when designing a dome wafer reinforcement scheme are: polar boss torsional stiffness, dome contour, fiber wind angle, and ratio of boss opening diameters to the case diameter. The results shown in this study can be applied to other case designs with modifications made as required by changes in the above factors.

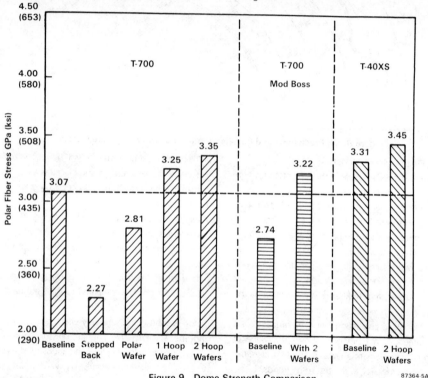

Figure 9. Dome Strength Comparison

87364-5A

### 4. BIOGRAPHIES

The authors are associated with the Composite Products Department of Morton Thiokol, Inc., Wasatch Operations.

James D. Erickson is supervisor of the Strategic Composite Structures Section and has nine years of experience in structural design and analysis. He received a B.S. degree in Applied Physics in 1976 from Weber State College and a Master of Engineering degree from the University of Utah in 1978.

James A. Yorgason is an associate engineer in the Composite Products Materials Department. He received a B.S. degree in Chemical Engineering from Brigham Young University in 1983. He has over four years of experience in composite materials and process development.

# THE SEA LANCE CAPSULE

C. T. Golden, Sea Lance Missile/Capsule Design Manager
E. E. Spear, Sea Lance Capsule Design Engineer
Boeing Aerospace Company

## Abstract

The design and development of the Sea Lance antisubmarine warfare standoff weapon capsule from its inception in 1980 through 1986 is presented herein. The Sea Lance weapon is an encapsulated missile system that can be launched from submerged U.S. attack submarines, float to the surface uncontrolled, broach, initiate missile/capsule separation and flight, fly to a targeted point downrange, separate payload and booster sections and release a warhead to enter the water. The capsule is designed as a graphite filament wet wound cylinder of sandwich construction to provide protection of the missile to natural and induced environments. Significant design drivers include environments during assembly, transportation and handling; hydrostatic and launch pressures; shock loads during stowage and launch in the submarine; floating to the surface; broach; and missile flyout. The design requirements, combined with the features of the capsule graphite composite construction, lead to several unique design decisions and fabrication processes that will be discussed herein.

## 1.0 Introduction

The Sea Lance was conceived as a two payloaded weapon system to provide a defensive attack capability against enemy submarines. The Navy identified the two potential payloads and Boeing won the opportunity to develop "common" propulsion, guidance and capsule components of the All-Up Round (AUR) for the Naval Sea Systems Command.

In conjunction with the overall concept, the following was specified:

1. Targeting base upon the submarine fire control system.

2. Interface totally with the submarine's torpedo handling, storage and launch capability.

3. After launch from the submarine,

float to the surface within a maximum time under various sea states through broach, capsule opening and missile flyout without propulsion or control features.

4. Fly down-range to the preselected separation point, separate the payload to enter the water and destroy the target.

This paper presents the design and development of the capsule through the Demonstration and Validation (D&V) and initial Full Scale Development (FSD) phase of the Sea Lance Program.

## 2.0 Requirements

The first consideration was that in order to float to the surface, the weapon must be buoyant. Considering the torpedo tube volume and a desired bouyancy of ten percent, a maximum weapon weight of 2800 pounds was established. An initial split of about 600 pounds for the capsule and 2200 pounds for the missile was baselined. Now, what kind of a capsule can be built for 600 pounds? A conventional metal tube approach was not going to meet the target weight. A solution was proposed for a sandwich construction of wet wound composite laminate skins with a honeycomb core. A second sandwich with a 0.30 inch core under an exterior kevlar laminate (outer skin) would be added to protect the basic structure. The basic capsule structural concept was expanded, detailed by analysis and presented as the Boeing proposal baseline. See Figure 1.

The Navy provided the weapon system

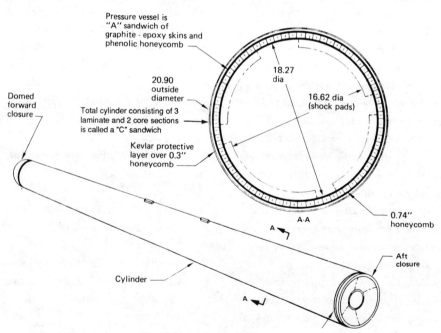

*Figure 1. Capsule Concept*

specification describing performance and associated requirements. The requirements were based upon the weapon interfacing with existing Navy equipment (shore based, tenders and submarines), transportation and handling criteria, the stockpile to target sequence, and the desired missile capability. A flow-down of requirements to the missile and capsule developed the two prime item specifications that control the design of the these major system elements.

### 3.0 Development Program

A development capsule was fabricated in parallel with Boeing's proposal effort but failed in test. Figure 2 presents the initial series of development tests for resin/fiber selection, laminate winding and fabrication process development (including cure cycle definition). These

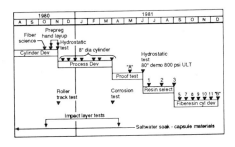

*Figure 2. Early Development Schedule*

development tests expanded our data base, however, pointed out more "don'ts" than "do's". The first success occurred during an attempt to validate the design concept with a hand layup of a pre-preg cylinder. The hydrostatic test had satisfied the pressure require-

ments, however, the fabrication process as expected, proved to be too labor intensive and expensive. The next step was to wind a series of seven skins on an eight-inch mandrel to develop techniques and processes applicable to a full size wet wound cylinder.

An update of the manufacturing process specification was prepared and fabrication of a hydrostatic test cylinder initiated. Cylinder A fabrication proceeded without difficulty until the finished cylinder was removed from the mandrel. It was evident that the "wax" parting agent between the aluminium mandrel and the inner laminate had diffused into the laminate and destroyed the structural integrity of the inner skin. Coupon testing substantiated the problem, however, pointed to quality laminates, bonding and strength elsewhere.

The process document was updated and an 80-inch hydrostatic test cylinder was wound and tested to its hydrostatic ultimate. This test established the feasibility of the approach but left several questions unanswered. Analytically, the test specimen should have carried more pressure and the failure seemed concentrated adjacent to the ends of the capsule. No end closure to cylinder "transitions" were built into the capsule, however, the cylinder was machined to interlock with an extension on the end closures. See Figure 3.

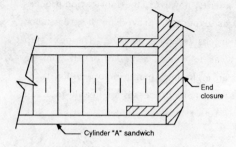

*Figure 3. End Closure Attachment*

After this success with the Applied Plastics Co. (APCO)T-300 capsule, three tests were conducted to assess resin and closure transitions (means to attach end closures to cylinder). Fiberesin, APCO and Dowden resin systems were compared. Delaminations occured in the APCO specimen. Fiberesin was selected based on wetability, cleanliness and minimum void content.

A series of seven full diameter, 48-inch Fiberesin cylinder sections were fabricated. Within the series, compaction, toe tension, cure cycle temperature, time and transition design were varied. The results were inconsistent, including delaminations during cool-down, skin buckling, low adhesive properties and voids in transitions. The test results lead to the development of a manufacturing plan combining the best features and then test Capsule B was fabricated.

Capsule B was a full- length (20-foot) C sandwich capsule with built-in solid transitions (intermediate transition sections were built in so the final

capsule could be cut into three test specimens). The through-the-thickness ultrasonic inspection (TTU) of the "A" sandwich showed a structurally "good" capsule. The "C" sandwich TTU showed extensive voids and delaminations. The cylinder was judged unacceptable for hydrostatic test and an investigation of the failure was performed. The investigation determined that the resin/adhesive interface had a time dependent phenomena in which the resin migrated into the core adhesive, leaving the laminate dry and of inadequate structural integrity.

Based on the data developed during the investigation of the Capsule B failure and sample tests, it was decided to return to the APCO resin system. Full diameter 48-inch long test cylinders, 12 through 16, were fabricated and the winding process and transitions were investigated. See Figure 4.

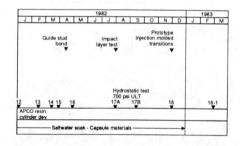

*Figure 4. Second Phase Development Schedule*

Toe tension, compaction methods, cure cycles (time and temperature) and transitions were varied. Each cylinder resulted in an improvement in cylinder quality. "Wait" periods were incorporated in the winding process to mimic the time phasing in the 48-inch test parts representative of winding a full length cylinder. Toe tension was varied and a compacting roller, riding on the wet fiber was used. Upon completion of Cylinder 16, a major revision of the manufacturing process was made and Cylinder 17 was fabricated.

Cylinder 17 was made full length with end and intermediate fiberglass transitions to produce two, ten-foot test parts. The 20-foot "A" sandwich was cut into two, ten-foot sections and the "C" sandwich layup completed using two approaches resulting in a three-cure cycle on 17A and a five-cure cycle on 17B. Cylinder A, with inserts potted in the transitions and closures installed, was hydrostatically tested to ultimate load, exceeding the required strength capability. Cylinder 17B was tested to verify that the mechanical strength requirements of the capsule could be met.

With the comparative success of Cylinder 17, two 48-inch test cylinders were built to improve cylinder processing. Cylinder 18 was processed to eliminate one cure cycle and reduce the residual stress in the cylinder buildup. The results were positive in that internal stresses were reduced and high quality skins were produced.

In this same time period, the Boeing Manufacturing, Research and Development Organization was investigating injection molded parts in various plastic materials. We pursued molded parts for use as reinforcements and transitions because of the significant cost reduction potential. The proposed reinforcement inserts were to be placed in the "A" sandwich core (0.74-inch thick). A materials investigation was initiated to select the best material for a part this size. Nylon 6-10 was selected and parts with no cracks and minimum flow lines were built. Test Cylinder 18-1 was built with injection molded transitions (Nylon 6-10), three cure cycles and various compaction techniques.

As a result of the success of Cylinder 18-1, the floating launch cylinder was designed and the process document was upgraded. Nylon 6-10 transitions and stowage rack reinforcement inserts were incorporated in the fabrication of a prototype cylinder to support the floating launch test. A surprising series of events occured when it became impossible to reproduce quality reinforcement inserts in nylon 6-10. Cracks, prominent flow lines and warped parts resulted, however, a culling operation selected what were felt to be adequate parts after rework. The cylinder fabrication appeared to be totally successful, however, the TTU of the cylinder indicated a total disbond between the nylon reinforcement and the cylinder inner skin. Several casual factors were hypothesized: (1) incompatible through the thickness Coefficient of Thermal Expansion (CTE), and (2) failure of the adhesive to wet the Nylon 6-10 to graphite laminate interface. Due to the difficulty and cost of producing the injection-molded parts (machining inner surface required) it was decided to discontinue the injection-molded reinforcements, repair the flawed cylinder to support the floating launch test and continue to develop a better approach to cylinder reinforcement.

Westinghouse Electric Corporation, Marine Division, was under contract to provide the Shock Isolation System (SIS) for the capsule. A Teflon-coated polyurethane pad approximately 0.80-inch thick was developed and Westinghouse analysis indicated their design might provide the necessary lateral shock protection. In addition, a system of seals was developed to limit the pressure build-up along the light weight interstage due to rocket motor ignition and burn as the missile "flies" out of the capsule.

Using this approach, the floating launch capsule was completed. A dummy missile with a one second short burn rocket motor was built and the prototype AUR was instrumented. The AUR was assembled on-site at San Clemente Island, California and successfully launched in April 1984, demonstrating capsule performance and missile flyout.

In parallel with the completion of the floating launch cylinder development, work continued in the investigation of transitions and reinforcements. Because of the Cylinder 17A hydrostatic test and the location of the failure, the approach had been to strengthen and increase the stiffness of the joint from cylinder to end closure. At this point though, the design approach for the "transition" was changed to look at the possible benefit of softening and making the joint more flexible. A softer end connection was developed using machined titanium rings, with a hard contact point and a rubber seal sandwiched in between to allow relative angular motion between the closure and the cylinder. A test part was fabricated from a previously built section of a cylinder and staked inserts were used (see Figure 5). The part was tested hydrostatically without leaking and the recorded test data indicated a significant loads reduction at the joint.

In parallel with this decoupler development, a study was conducted to select a better reinforcement insert in the 0.74-inch core at the stowage rack locations (load calculated at about 30,000 lbs). Several types of machined and formed honeycomb cores were investigated with final selection being to use a 1/8-inch cell 3AL-2.5V spot welded titanium honeycomb developed by Rohr (10 inches wide by 168 degrees) located on the cylinder lower centerline. In addition, the loads from the stowage rack support points (6 inches x 6½ inches) were projected to exceed the crushing strength of the 0.30-inch thick outer sandwich core material so rubber pads, designed to soften the shock loads, were built-in at these locations. The tactical configuration cylinder design drawings and process specifi-

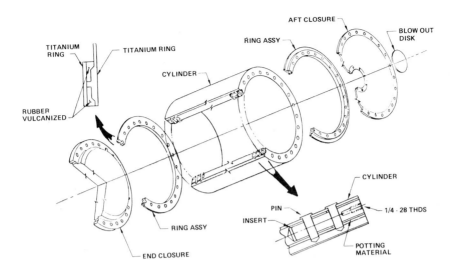

*Figure 5. Closure Attachment Design*

cation were updated to reflect these new features.

Cylinder 19 was fabricated for hydrostatic test. A complete TTU of the assembled cylinder showed there were essentially no defects, delaminations or disbonds. The capsule, with decoupler rings installed and test unique forward and aft closures, was tested at the Boeing Tulalip site to a hydrostatic ultimate approximately 40% above the required level. Test coupons were cut from excess cylinder trim material to determine flatwise tension, inner-laminer shear, core shear, and lash-down reinforcement ring strength values. All values were in excess of the requirements. After the hydrostatic test, the decision was made to upgrade the tactical closures to an ultimate strength comparable to the demonstrated cylinder strength. No other design changes were proposed.

The broach sensor is a forward closure equipment item that required its own development program. The broach sensor's function is to sense the capsule passing through the water to air interface and initiate closure removal (ordnance firing), rocket motor ignition and missile flyout. Various pressure sensor's were proposed and analyzed but combining sea state, float-up orientation and other parameters, it became evident that a zero pressure pulse (activating the sequence) could occur 50 feet under the surface — totally unacceptable. After an analyses an electrical system that "talks to itself" under the water and then is interrupted when the link has to travel through air was developed. The broach sensor probe is extended at submarine launch (lanyard pull) approximately four inches ahead of the capsule nose. This precludes carrying a bubble of water over the sensor as it broaches, thus delaying the signal. When the probe, travels into the air a no signal (much increased resistance condition) occurs and the programmed sequence of events is initiated to provide missile flyout within the required timeline. Testing was performed by the Gould Company of various probe lengths, orientations, capsule broach angle etc. and a workable configuration was established.

Initially in the D&V program, developmental test work was not required on the capsule end closures. An extensive trade study selected cast titanium closures, machined at equipment interfaces as the cost effective solution. The hemispherical forward closure would be separated circumferentially by ordnance near its attachment to the cylinder and thrust clear of the cylinder bore by an ordnance powered push rod. Development testing by the ordnance vendor showed a possible alternative. The tactical system design now cuts the closure into two pieces (circumferentially and over the dome) blowing them out of the way with the force of the ordnance cutting charge. The aft closure is designed as a concave diaphragm with the necessary equipment attachment features. A development program was initiated to size and perfect removal of a blow out port in the center of the closure required to control initial missile flyout. Size (approximately 6.6 in. diameter) and separation features (tabs which shear off under initial engine ignition pressure build-up) were tested and the design confirmed by the Capsule No. 1 floating launch test.

In mid 1984 it was becoming more evident that the dynamic conditions associated with the stowage rack and in tube shock loads were potentially the most significant design drivers. Westinghouse continued to analytically define the dynamics and resulting loads based on their shock isolation system design. Both Boeing and the Navy were aware that shock input data was needed. Previous Navy test programs were assessed and the data was found inadequate. A chance to perform a preliminary MIL-S-901C test was proposed and immediately accepted. The Preliminary Stowage Rack Shock Test (PSRST) was initiated based on the design of Capsule 19 and using the second generation of the SIS pads developed by Westinghouse. Cylinder 2 was fabricated and a comparatively sophisticated dynamic instrumentation system was installed on both the prototype missile and capsule. The tests were conducted during April and May of 1985 at Hunters Point, California. A series of 9 test shots were run varying the distance from charge to test barge from 60 to 20 feet. Each run was made with a mass simulated Mk 48 round on the adjacent position on both the high and low frequency rack locations. The series was totally successful in that no structural damage occurred to the capsule. The environments measured defined the input shock loads at the various locations on both the missile and capsule to be used as requirements for the hardware components.

The Shock Isolation System (SIS) was designed to limit the maximum shock loads to values equivalent to or less than what would be seen during flight. Maximum pad pressures were based upon the inherent capability of the rocket motor and payload. Maximum shear and bending capability of the missile and capsule were used to see if the shock load could be accommodated within the flight design load conditions. Pressure profiles of the pads were established under the expected shock load and maximum structural loads were calculated. Pad stiffness and distribution was established limiting the loads to the desired level. (See Figures 6 and 7).

The one "free" variable was the expected "g" level vibration at various equipment locations. Rather than limit these to some predetermined level, the approach was to define the level and then modify the equipment design to survive the load. This results in a penalty (weight/size) on some equipment items, however, all electrical, structural and ordnance elements can be reinforced to carry the design loads. The load levels will be verified by test during shock qualification of the AUR.

As a continuation of the long-range development plan, the next capsule (Capsule 3) was to explore the AUR launch performance as affected by the envelope of submarine launch motion (speed, depth, pitch-yaw-roll, etc.). An instrumented AUR was fabricated with external features and mass properties matching the current tactical configuration weapon. The fabrication and assembly of Capsule 3 progressed without incident. The TTU of the completed cylinder showed essentially flaw free fabrication. SIS pads were installed and a dummy missile duplicating the tactical missile mass properties was built, instrumented and installed in

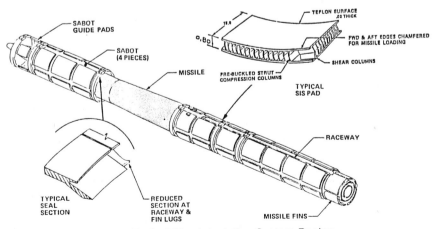

Figure 6. Typical Shock Isolation System Design

693

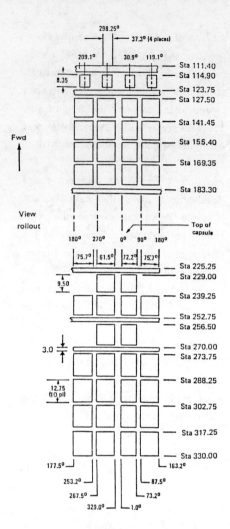

*Figure 7. Tactical SIS Pad Attachment*

the capsule. The test series consisted of launches at various speeds, submarine motion and depths to assess the AUR performance from launch through clearing the shutterway. The tests were conducted on both 637 and 688 class submarines. During the first few launches on the AUTEC Range, Bahama Islands, some scrapes and bumps were identified indicating impact of the AUR with the submarine launch tube or shutterway structure.

On the maximum speed/maximum depth launch, the AUR impacted the forward guide stud (bolted to a plate bonded to the cylinder's structural laminate) on the upper plate of the shutterway. The subsequent buckling of the capsule caused an implosion sinking the round in about 5000 feet of water. Subsequent recovery of the AUR was successful and the failure mode was defined. A program to reinforce the cylinder structure and monitor the impact loads was initiated. Four aluminium launch shapes (mass simulated) are being built as well as four increased strength cylinder test specimens. The program to define the launch environment and the capsule strength requirements and capability will be conducted in 1987.

Within the currently planned FSD program, three additional "development" type test series are planned. They consist of the following:

1. Sub Clearance Testing — Following the completion of the launch shape and increased strength cylinder tests, two prototype AUR's will be built and tested to confirm the weapon system's performance over the realm of submarine motions during launch.

2. Float-Up/Dynamic Launch — A prototype AUR consisting of a capsule and a prototype missile with a live short-burn rocket

motor will be built. Three missile floatups will be run to describe time, orientation, broach, etc., of the AUR launch from the submarine. The fourth float-up will include ordnance removal of the forward closure at broach, rocket motor ignition, missile release and a short rocket motor burn to separate the missile from the capsule.

3. Preliminary In-Tube Shock Test — The preliminary stowage rack shock test was so successful in defining the design conditions associated with this weapon system that a comparable test for the In-Tube Shock condition (wet and dry) will be conducted in early 1988.

drawings and the capsule fabrication process specification. The baseline capsule description in these two documents is being continually updated as a result of test results and analysis programs. A summary of the design is shown in Figures 8 through 10. As of now, the capsule has grown from a concept in 1982 to a workable design meeting the major requirements of the Sea Lance Weapon System and Capsule Prime Item Specification with a relative low risk, cost competitive design and manufacturing process.

At this time, the baseline capsule is described by tactical configuration

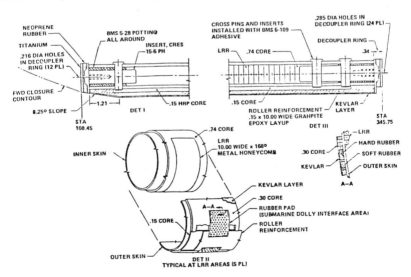

Figure 8. Tactical Cylinder Details

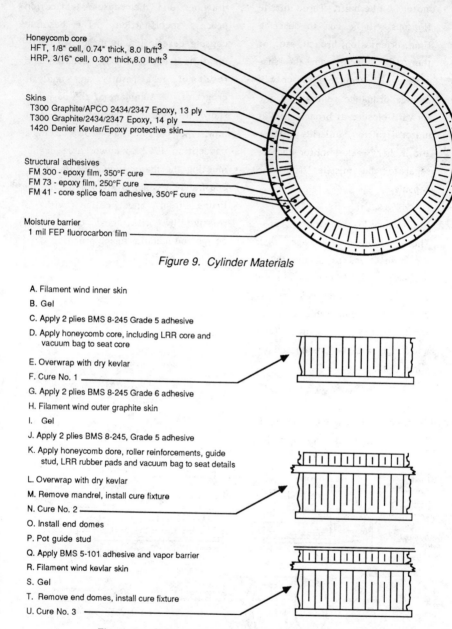

Honeycomb core
  HFT, 1/8" cell, 0.74" thick, 8.0 lb/ft$^3$
  HRP, 3/16" cell, 0.30" thick, 8.0 lb/ft$^3$

Skins
  T300 Graphite/APCO 2434/2347 Epoxy, 13 ply
  T300 Graphite/2434/2347 Epoxy, 14 ply
  1420 Denier Kevlar/Epoxy protective skin

Structural adhesives
  FM 300 - epoxy film, 350°F cure
  FM 73 - epoxy film, 250°F cure
  FM 41 - core splice foam adhesive, 350°F cure

Moisture barrier
  1 mil FEP fluorocarbon film

*Figure 9.  Cylinder Materials*

A. Filament wind inner skin
B. Gel
C. Apply 2 plies BMS 8-245 Grade 5 adhesive
D. Apply honeycomb core, including LRR core and
   vacuum bag to seat core
E. Overwrap with dry kevlar
F. Cure No. 1
G. Apply 2 plies BMS 8-245 Grade 6 adhesive
H. Filament wind outer graphite skin
I.  Gel
J. Apply 2 plies BMS 8-245, Grade 5 adhesive
K. Apply honeycomb dore, roller reinforcements, guide
   stud, LRR rubber pads and vacuum bag to seat details
L. Overwrap with dry kevlar
M. Remove mandrel, install cure fixture
N. Cure No. 2
O. Install end domes
P. Pot guide stud
Q. Apply BMS 5-101 adhesive and vapor barrier
R. Filament wind kevlar skin
S. Gel
T. Remove end domes, install cure fixture
U. Cure No. 3

*Figure 10.  Cylinder Manufacturing Process Summary*

# BIOGRAPHY

**CHARLES T. GOLDEN,** received a BS in Aeronautical Engineering and a BS in Mathematics from California State Polytechnic in 1957. Received a MS degree in Aeronuatical Engineering from the University of Washington in 1960. He has 29 years experience with The Boeing Company, primarily in structures and mechanics. Currently assigned to the Navy Sea Lance Program (an antisubmarine standoff weapon weapon system) as Missile/Capsule Design Manager since 1982. Primarily concerned with the requirements, design, development, manufacture and testing of the missile/capsule. Prior to this was the Program Manager for the Space Telescope/Optical Telescopic Assembly Structures Program for the Hubble Space Telescope. Boeing designed, fabricated and tested the metering truss, the focal plane structure and fine guidance sensor optical benches for the Perkin Elmer Company, Optical Division, and the wide field and planetary camera optical bench for JPL in controlled expansion graphite composite structure.

**EDWIN E. SPEAR,** graduated Washington State University 1949, BS in Civil Engineering. Professional Engineer, State of Washington, 32 years with The Boeing Company, primarily in the structures and system engineering areas. Worked since 1982 in the Navy's Sea Lance Program (a submarine launched missile system) on the development, design, manufacture and test of the capsule. Previously worked on the Optical Telescope Assembly Structure Program. Boeing built, under contract to the Perkin-Elmer Company, the metering truss (secondary mirror support assembly), the focal plane structure (supported the seven scientific instruments) and fine guidance sensor optical benches for the Hubble Space Telescope of controlled graphite composite structure (coefficient of thermal expansion controlled to or less than $\pm$ 0.025 x $10^{-6}$ in/in/$^{\circ}$F). Thermal expansion of metering truss (2.4m diameter x 5.2m long) held to 0.1 the diameter of a human hair.

# MECHANICAL BEHAVIOR OF CYLINDRICAL COMPOSITE TUBES
## UNDER TRANSVERSE COMPRESSIVE LOADS

Fu-Kuo Chang[*] and Zafer Kutlu[†]

Department of Aeronautics and Astronautics
Stanford University, Stanford, California 94305

### Abstract

An investigation was performed to study the mechanical response of cylindrical composite tubes subjected to transverse attacks. During the investigation, an analytical model was developed, which is capable of assessing damage in the laminates and predicting residual stiffness and strength of the composite cylinders subjected to two types of transverse compressive loads. In the model, deformations, stresses and strains in the laminates were analyzed based on the large deformation theory in elasticity. Damage accumulation, and the residual stiffness and strength of the composite tubes were evaluated by a combinaton of appropriate failure criteria with a proposed property degradation model. Based on the model, a nonlinear finite element code was developed. Experiments were conducted to verify the proposed model and the computer simulations. It is expected that the model can be extended to more complicated loading conditions and geometries.

## 1. INTRODUCTION

Recent advances in composite technology have led the application of composite materials to more and more sophisticated structural designs. One notable application is the construction of composite cylinders or curved panels for such applications as transport fuselage structures, missle launch tubes and pressure vessels, etc.[1-6]

Curved composite structures, on a large scale, are vulnerable to transverse attacks. It is

known that curved composites, unlike flat laminates, are much more sensitive to out-of-plane loading.[7] Therefore, predicting the mechanical behavior and residual stiffness and strength of curved composite structures during and after transverse attacks is extremely important. Knowledge of this information is critical to the application of composites for large scale structures.

Unfortunately, most of the related research on curved composites has been experimental. Few analytical models have been developed and the ones that have, could only be applicable to the areas where the structures acted within the undamaged range.[8-10] In this investigation, we focused our attention on modeling mechanical response of composite cylindrical tubes subjected to transverse crushing loads. The objective of this investigation was to develop an analytical model to assess damage development in laminates, to predict the residual stiffness and strength of the cylinders subjected to transverse compressive loads and to evaluate the maximum transverse load (referred to as crushing load) that the tube can sustain.

## 2. PROBLEM STATEMENT

Consider a long cylindrical tube made of fiber-reinforced laminated composites subjected to transverse compressive loadings as shown in Figure 1. The ply orientation of the laminates can be arbitrary, but must be balanced. Two types of loading conditions were considered in this investigation: first, the compressive loads were applied through two rigid blocks; second, two lines of loads were introduced by two thin rigid plates (see Fig. 1). To simplify

---

[*]Assistant professor, member of SAMPE.
[†]Graduate student.

the problem, the loads were applied uniformly throughout the longitudinal axis of the tube. It was desired to obtain the following information:

1. the initial failure load $P_i$ that caused initial damage,

2. the extent and types of damage (matrix cracking, matrix compression failure and fiber breakage, etc.) as the applied load became greater than the initial load $P_i$,

3. the mechanical response of the cylinders during loading and

4. the maximum load that the cylinder could sustain before catastrophic failure.

## 3. METHOD OF APPROACH

In order to achieve the proposed objectives, the progressive damage model previously developed by the author and K. Y. Chang was considerably extended for the present analysis.[11-13] The previous model was developed for laminates containing open or pin-loaded holes and subjected to in-plane tensile loading; i.e., only flat plates under in-plane loading conditions were considered. However, the present damage model was developed for curved composite panels subjected to out-of-plane loadings.

Since the material could undergo substantial out-of-plane deformation before catastrophic failure occurred, the large deformation theory was considered in the analysis. The proposed progressive damage model consists of stress analysis and failure analysis. In the following, only a brief description of the model will be given, since the detailed analysis of the model was rather lengthy and, hence, has been presented elsewhere.[14]

### 3.1 Stress Analysis

In this analysis, it was assumed that the length of the cylinder was considerably larger than its thickness. Accordingly, a two-dimensional plane strain condition was adopted, since the applied loads were also uniformly distributed along the longitudinal axis of the cylinder. The adoption of this condition could significantly simplify the analysis; otherwise, a three-dimensional analysis would be required. The

free edge effect on the mechanical behavior of the cylinder was neglected.

With the consideration of large deformations, the analysis was developed based on the updated Lagrange formulation. Thus, the equilibrium equations in weak form (incremental virtual work) at the current configuration is[11-18]

$$
\sum_{m=1}^{M} \left\{ \int_{v_{n-1}} \Delta e_{kl} \, {}^{m}C_{ijkl}^{n-1} \, \Delta e_{ij} \, dv \right.
$$
$$
\left. + \int_{v_{n-1}} {}^{m}\sigma_{ij}^{n-1} \, \Delta \eta_{ij} \, dv \right\}
$$
$$
= \sum_{m=1}^{M} \left\{ \int_{s_{\sigma}} \bar{T}_i \, \Delta u_i \, da \right.
$$
$$
\left. - \int_{v_{n-1}} {}^{m}\sigma_{ij}^{n-1} \, \Delta e_{ij} \, dv \right\}
$$

$$(1)$$

Where $e_{ij}$ are the infinitesimal deformations, $\eta_{ij}$ are the rotations and $\sigma_{ij}$ are the Cauchy stresses. $S_\sigma$ is the portion of the surface where surface tractions $\bar{T}_i$ are prescribed. ${}^{m}C_{ijkl}^{n-1}$, the reduced moduli, were the material properties of the $m$-th layer based on the plane strain condition. The expression of ${}^{m}C_{ijkl}^{n-1}$ based on the plane strain condition can be found in Ref. 7.

Note that depending on the ply orientation and the current configuration, the material properties ${}^{m}C_{ijkl}^{n-1}$ vary from position to position and from layer to layer. In order to solve Eq. 1, a nonlinear finite element code was developed during this investigation. The solutions of Eq. 1 strongly depend on both the ply orientation and the boundary conditions (force and displacements). In this investigation, only two types of loading conditions were considered as described in Figure 1. However, it is worth mentioning that the analysis developed in this investigation is not restricted only to these conditions, but can be applied to other loading conditions as well.

The finite element analysis was used to calculate the stresses, strains and deformations of composite cylinders under the given loading conditions.

## 3.2 Failure Analysis

As the applied loads continue to increase, damage could occur, resulting in degradation of mechanical properties. In order to evaluate the residual stiffness and strength of the cylinder due to the damage, a failure analysis was developed which contained a set of failure criteria and a property degradation model.[14] Damage and the corresponding failure mode were predicted by the failure criteria and the reduction of mechanical properties was evaluated based on the proposed property degradation model.

In this investigation, four different failure modes were considered; matrix cracking, matrix compression failure, fiber breakage and fiber-matrix shearout. The reason for considering only these four modes was because they were the ones observed most from the experiments on thin-walled composites under the given loading conditions. It is expected that other failure modes, such as delamination and fiber buckling, could occur in thicker laminates. Inclusion of the other modes will be considered in the further development of the model.

Once damage occurs, the material may degrade. The degree of property degradation strongly depends on the failure mechanism resulting from the damage. Hence, based on the mode of failure, the property reduction model was proposed. In principle, the proposed failure analysis is similar to the one previously developed by the author and K. Y. Chang.[11-13]

## 4. NUMERICAL ANALYSIS

Once damage occurs, the stresses and strains in the laminate will redistribute due to the reduction of mechanical properties in the damaged area. The redistributed stresses outside the damaged area could cause additional damage resulting in damage propagation. The procedures of recalculating stresses and strains and evaluating the extent of the damage were developed during this investigation and have also been given in Ref. 14, hence, they will not be repeated here.

The numerical procedures described in the foregoing have been implemented into a computer code, designated PDTUBE-1. This code may be obtained from F. K. Chang.

## 5. EXPERIMENTS

Fiberite T300/934 Graphite/Epoxy prepreg tapes were used to fabricate the cylindrical tubes. The tapes were appropriately sized and wrapped on an aluminum mandral with a two inch outer radius and a thickness of 0.25 inches. A flexible thin aluminum curved panel was then wrapped and tightened over the prepregs along with bleeders to provide a uniform pressure on the prepregs while they were being cured according to the manufacturer's recommended procedures.

Four different ply orientations were selected for the experiment. The ply orientation and the dimensions of each test configuration are listed in Table 1 and the material properties of T300/934 are given in Table 2. Each cylinder was then placed in between two thick steel plates which were mounted on an MTS testing machine. Each test was performed by pressing the steel blocks together until the cylinder totally collapsed. The load and the distance that each block traveled were recorded. Two specimens were tested for each selected configuration. Due to limited space here, only the test results of $[-30/30/0]_s$ composite tubes were presented in Figure 2. Addtional test results can be found in Ref. 14.

## 6. RESULTS AND COMPARISONS

Numerical calculations were performed by using the computer code to simulate the mechanical response of the cylinders under the test conditions. The results of the calculations based on the proposed model were compared with the data. Details of discussions of the results and comparisons are given in Ref. 14. An excellent agreement has been found between the data and the numerical simulations.

For example, the results of the calculations for $[-30/30/0]_s$ composite tubes are presented by a solid line in Figure 2, along with the experimental data indicated by solid and open

Table 1

Material properties used in the calculations :

Fiberite T 300/934 graphite epoxy prepreg tape

| | | |
|---|---|---|
| Ply longitudinal modulus | $E_{xx} = 2130000$ | psi |
| Ply transverse modulus | $E_{yy} = 1650000$ | psi |
| Out of plane modulus | $E_{zz} = 1650000$ | psi |
| In plane shear modulus | $G_{xy} = 897000$ | psi |
| Out of plane shear modulus | $G_{xz} = 634615$ | psi |
| Out of plane shear modulus | $G_{yz} = 634615$ | psi |
| Poisson's ratio | $v_{xy} = v_{xz} = v_{yz} = 0.3$ | |
| Ply longitudinal tensile strength | $X_t = \quad 25100$ | psi |
| Ply longitudinal compressive strength | $X_c = \quad 200000$ | psi |
| Ply transverse tensile strength | $Y_t = \quad 5000$ | psi |
| Ply transverse compressive strength | $Y_c = \quad 38900$ | psi |
| Ply shear strength | $S = \quad 19400$ | psi |

Table 2

Configuration and geometry of the test specimens

| Lay up | Inner Radius (in) | Thickness (in) | Length (in) | |
|---|---|---|---|---|
| $[-30/30/0]_s$ | 2.0 | 0.030 | 8.1 | 5.3 |
| $[30/-30/0]_s$ | 2.0 | 0.033 | 6.0 | 6.8 |
| $[0_2/45_2/-45_2]_s$ | 2.0 | 0.055 | 6.5 | 6.6 |
| $[30/-30/-30/30/0/0]$ | 2.0 | 0.032 | 7.2 | 6.0 |

701

circles. As a comparison, numerical solutions generated from the model without the inclusion of the proposed failure analysis are also shown in the figure by a dashed line. The ordinate in the figure represents the amount of the applied load per unit length and the horizontal axis indicates the distance (deflection) that each rigid block traveled during loading. Only the load-deflection curve was compared in the figure.

As indicated in Figure 2, the numerical calculation based on the proposed model agrees well with the data, from the initial loading through the damage stage to the final catastrophic failure. The predicted maximum compressive load in the figure was conservative and agreed with the data within 10%. On the contrary, the curve calculated without the consideration of damage deviated from the data once damage occurred in the laminates. The difference between the data and the calculated curve increased, as the applied load continued to increase. It is noted that the latter solutions gave no indication on the maximum failure load.

To illustrate the capability of the computer code, the deformations of the cylinder corresponding to the test case in Fig. 2 at various loading levels were plotted and are presented in Fig. 3. It is interesting to note that the numerical simulations were very consistent with the changes in the specimen configuration observed during the experiment. As noted, the information on the extent and types of damage at a given loading stage can also be provided by the code. Detailed discussions of the damage development and propagation predicted by the code have been given in Ref. 14.

Numerical simulations of the mechanical response of the $[-30/30/0]_s$ composite cylinders under the second type of loading condition were also performed and are presented in Figures 4–5.

## 7. CONCLUDING REMARKS

In this investigation, mechanical behavior of cylindrical composite tubes under two types of transverse compressive loads was studied. An analytical model and a corresponding com-

puter code were developed for this study. The computer code can be used to predict the behavior of the composite tubes subjected to transverse compressive loads, from the initial loading stage to the final catastrophic failure. The following information can be provided by the code.

1. the stresses, strains and deformations inside the laminate at any given loads,

2. the extent and types of damage in laminates at a given load,

3. the residual stiffness and strength of the cylinders after a preload, and

4. the ultimate transverse crushing load.

Although only two types of loading conditions were considered in this investigation, the analysis was fundamental and can be easily extended to other loading conditions. The information provided by the code could be very important in designing composite cylinders or curved panels which may be subjected to transverse attacks.

## 8. ACKNOWLEDGEMENTS

The support of the National Science Foundation through Contract MSM-8515610 is gratefully acknowledged.

## 9. REFERENCES

1. Carden, H.D., "Impact Dynamics Research on Composite Transport Structures," ACEE Composite Structures Technology Conference, Aug. 13-16, 1984, Seattle, WA.

2. Davis, G.W., Sakata, I.F., "Design Considerations for Composite Fuselage Structure of Commercial Transport Aircraft," Lockheed-California Company, Burbank, CA, NAS1-15949, NASA CR-159296, March 1981.

3. Chang, S.G., Mar, J.W., "The Catastrophic Failure of Pressurized Graphite/Epoxy Cylinders Initiated by Slits at Various Angles," 25th Structures,' Structural Dynamics and Materials Conference, May 14-16, 1984, Palm Springs, CA, pp. 123-129.

4. Vizzini, A.J., Lagace, P.A., "The Role of Ply Buckling in the Compressive Failure of Graphite/Epoxy Tube," 25th Structures, Structural Dynamics and Materials Conference, May 14-16, 1984, Palm Springs, CA, pp. 342-350.

5. Thornton, P.H., Edwards, P.J., "Energy Absorption in Composite Tubes," J. Composite Materials, Vol. 16, No. 6, Nov., 1982, pp. 521-545.

6. Graves, M.J., Lagace, P.A., "Damage Tolerance of Composite Cylinders," SAE Technical Paper Series 830766, April, 1983.

7. Chang, F.K., Springer, G.S., "Strength of Fiber-Reinforced Composite Bends," J. Composite Materials, Vol. 20, (1986), pp. 30-45.

8. Bauld, N.R., "Experimental and Numerical Analysis of Axially Compressed Circular Cylindrical Fiber-Reinforced Panels with Various Boundary Conditions," AFWAL-TR-81-3158, Air Wright Aeronautical Laboratory, Wright-Patterson AFB, OH, Oct., 1981.

9. Johnson, E.R., Hyer, M.W., and Carper, D.M., "Response of Long Hollow Cylindrical Panels to Radial Line Loads," presented at 25th Structural Dynamics and Materials Conference, Palm Springs, CA, May 14-16, 1984.

10. Almroth, B.O., Brogan, F.A., "The STAGS Computer Code," NASA CR-2950, (1978).

11. Chang, F.K., Chang, K.Y., "A Progressive Damage Model for Laminated Composites Containing Stress Concentrations," J. Composite Materials, (will appear in July, 1987 ed.).

12. Chang, F.K., Chang, K.Y., "Post-Failure Analysis of Bolted Composite Joints in Tension and Shear-out Mode Failure," J. Composite Materials, (will appear in July, 1987 ed.).

13. Chang, F.K., "A Progressive Damage Model for Laminated Composites Containing an Open or Pin-loaded Hole," Proceedings of 18th International SAMPE Technical Conference, Vol. 18, (1986), Seattle, WA, pp. 360-371.

14. Chang, F.K., Kutlu, Z., "Mechanical Response and Strength of Cylindrical Composite Tubes Subjected to Transverse Crushing Loads," J. Composite Materials, (submitted).

15. Fung, Y.C., Foundations of Solid Mechanics, Prentice-Hall, Inc., New Jersey, (1965).

16. Zienkiewicz, O.C., The Finite Element Method, McGraw-Hill, New York, (1977).

17. Horrigmoe, G., Bergan, P.G., "Incremental Variational Principles and Finite Element Models for Nonlinear Problems," Computer Methods in Applied Mechanics and Engineering, Vol. 7, pp. 201-217.

18. Bathe, K.J., Ramm, E. and Wilson, E.L., "Finite Element Formulations for Large Deformation Dynamic Analysis," International Journal for Numerical Methods in Engineering, Vol. 9, (1975), pp. 383-386.

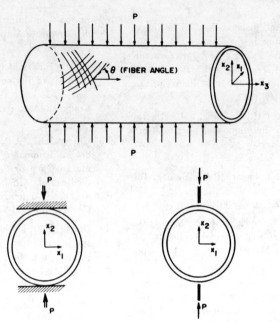

Figure 1. Description of the problem. (top) a cylindrical composite tube subjected to transverse compressive loads. (below) Loading conditions: (a) the cylinder pressed by two rigid blocks (area contact), (b) the cylinder pressed by two thin rigid plates ( point contact).

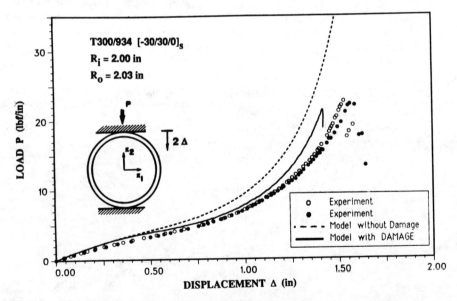

Figure 2. Load-deflection curve of [-30/30/0]$_s$ cylindrical composite tubes pressed by two rigid blocks. Comparisons between the data and the results of the present model both with and without the consideration of damage inside the laminates.

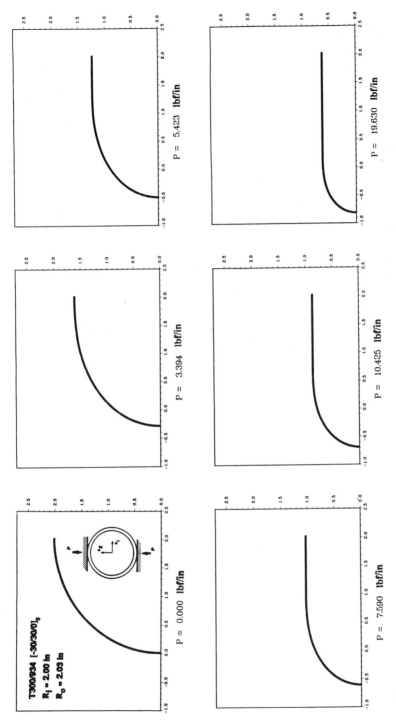

Figure 3. Illustration of the deformed configurations of the cylindrical composite tubes in Figure 2 at different loading levels.

705

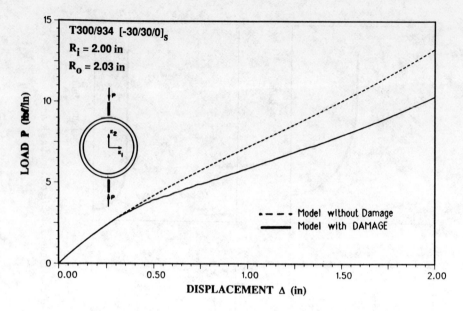

Figure 4. Load-deflection curve of $[-30/30/0]_s$ cylindrical composite tubes pressed by two lines of loads. Comparison of the results calculated from the present model with and without the consideration of damage inside the laminates.

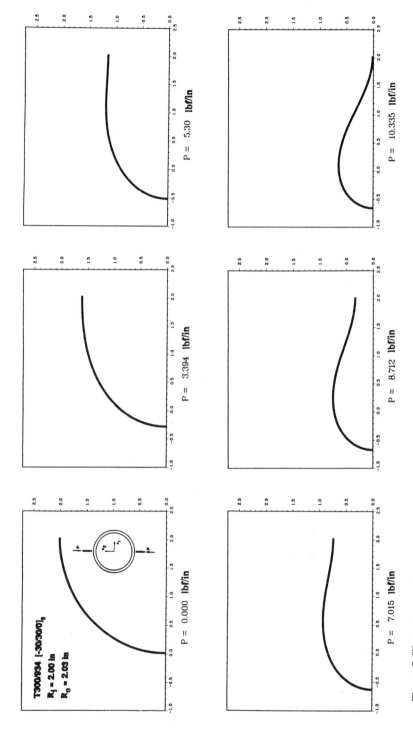

Figure 5. Illustration of the deformed configurations of the cylindrical composite tubes in Figure 4 at different loading levels.

# MODELING THE TENSILE STRENGTH OF RANDOM FIBER COMPOSITES

Selim Yalvaç
Dow Chemical U.S.A.
Central Research, 1712 Bldg.
Midland, MI 48674

## ABSTRACT

An analytical model based on the Kelly-Tyson equation modified with Chen's strength efficiency factor, Halpin-Tsai equations, Halpin-Pagano laminate analogy, Tsai-Hill failure criterion, Lavengood-Chen idealized fiber arrangement, and gradual failure model of lamina was modified and tested for estimating tensile strength of short, random fiber reinforced composite systems. Chen's strength efficiency factors were determined using Halpin-Kardos equations. The laminate analogy was applied twice, once to consider the effects of fiber length distribution and a second time to account · for the fiber orientation distribution. The distribution of fiber lengths was modeled as log-normal or the average fiber lengths were used. A Kelly-Tyson type equation, modified by Chen's strength efficiency factor was used to determine the longitudinal strength of each of the sub-laminae. The strength of the lamina was found using a rule of mixtures approach, summing the strengths of the sublamina, weighted by the fiber length distribution. The laminate analogy and the gradual failure model of lamina failure were used to determine the strength of the composite. For lamina failure, maximum stress and Tsai-Hill failure criteria were tested. The experimental data were obtained on aqueous wet laid polypropylene and polyethylene based quasi-isotropic composites, however, the model is valid for any thermoset, thermoplastic, or inorganic matrix random fiber composite system. Good agreement was obtained between the predicted and experimental values.

## INTRODUCTION

Analytical equations based on physical models are very useful when reliable predictions of properties of composite materials can be obtained using the corresponding constituent materials. These models not only play an important role as a design tool, but they also give a better understanding of the behavior of the material under various engineering conditions.

Randomly distributed fiber reinforced materials constitute many naturally occuring and man made composites. In many cases, the spatial orientation of short fibers is intermediate between a truly three dimensional distribution and a perfectly planar random array, depending on the thickness of the lamina, fiber length and the process used to prepare the prepregs. Such materials can be mathematically modeled as a laminate which consists of layers of unidirectional fiber composites with the fiber volume fraction in each layer oriented at an angle θ in the actual material.

Historically, tensile strength has been one of the most difficult to model mechanical properties of the random fiber composites.The objective of this work is to test the existing strength theory to cover randomly distributed, constant length fiber reinforced composites, however, this model can be used to calculate ultimate tensile strength of composites with both fiber length and/or fiber orientation distribution. This extends the range of use of the model to processes such as injection molding and extrusion, where excessive shear forces result in fiber aspect ratio degradation. When a fiber length distribution is present, either an average fiber length or two parameter log-normal distribution data is/are asked to be input from the user. Viscoelastic effects are not included, but they can be taken into consideration with some major modifications to the model. The user is given a choice of two models from which the ultimate random-in-plane tensile strength can be calculated. These are: (i) maximum stress failure criterion and (ii) Tsai-Hill failure criterion.

The computer program used in this study was originally written by Loughlin and Tucker (1) of the University of Illinois. The program was recently modified to run random-in-plane composites by inputting characteristic angles.

**THEORY**

At the present time, the only exact solutions available for the modulus properties of composites with aligned short fibers are numerical solutions of elasticity equations (2). The Halpin-Tsai equations (3) are generalized approximations which show very good agreement with the numerical solutions up to fiber volume fractions of 0.70. The modulus in the fiber direction, $E_{11}$, is given by:

$$E_{11}/E_m = (1 - \zeta \eta v_f)/(1 - \eta v_f) \qquad (1)$$

where

$$\eta = [(E_f/E_m) - 1]/[(E_f/E_m) + \zeta] \qquad (2)$$

and

$$\zeta = 2(l/d) \qquad (3)$$

where, $E_f$ and $E_m$ are the elastic moduli of the fiber and matrix, respectively, $v_f$ is the volume fraction of fibers in the layers, and $l$ and $d$ are the length and diameter of the fibers, respectively. The in-plane shear modulus, $G_{12}$, is estimated from

$$G_{12}/G_m = (1 - \eta v_f)/(1 - \eta v_f) \qquad (4)$$

where

$$\eta = [(G_f/G_m) - 1]/[(G_f/G_m) + 1] \qquad (5)$$

where $G_f$ and $G_m$ are the shear moduli of the fiber and the matrix, respectively. The transverse modulus is denoted by $E_{22}$ and can also be estimated using equation 1, with $l/d$ set equal to unity:

$$E_{22}/E_m = (1 + 2\eta v_f)/(1 - \eta v_f) \qquad (6)$$

where

$$\eta = (E_f/E_m - 1)/(E_f/E_m + 2) \qquad (7)$$

The Poisson's ratio of the composite is estimated by noting that it must be between the Poisson's ratio of a particulate composite and a continuous fiber composite. For large values of $l/d$, $\upsilon_{12}$ may be approximated using the rule of mixtures:

$$\upsilon_{12} \cong \upsilon_f + \upsilon_m(1 - v_f) \qquad (8)$$

where $\upsilon_f$ and $\upsilon_m$ are the Poisson's ratios of the fiber and matrix, respectively.

Halpin and Pagano (4) postulated that a random fiber composite can be mathematically treated as a material composed of layers with aligned fibers. The relative thickness of each of these layers will be equal to the fraction of fibers having the same orientation in the original material. In the special case of random-in-plane composite, all the layers have the same thickness, while in a composite with a particular fiber orientation distribution, the layers will have different thicknesses. This analogy is best known as the Halpin-Pagano Laminate Analogy.

The stress distribution used by Kelly and Tyson (5) is based on a shear lag type of analysis and neglects interactions between neighboring fibers. Chen (6) and Lavengood and Chen (7) have studied idealized fiber arrangements and interactions between neighboring fibers. Chen defined a strength efficiency factor, $\xi$, as the ratio of short fiber composite strength to the strength of a corresponding continuous fiber composite. The strength efficiency factor depends on a fiber aspect ratio and achieves a plateau at high aspect ratio values. The strength efficiency factor is plotted as a function of the ratio between fiber and matrix moduli in Figure 1. The strength efficiency factor can also be calculated using the Halpin-Kardos equation (8):

$$\xi = 0.5 + (E_f/E_m)^{-0.87} \quad \text{for } E_f/E_m > 5 \qquad (9)$$

Tucker (9) applied the plateau value of Chen's strength efficiency factor to Kelly-Tyson equation to estimate the tensile strength of aligned short fiber composites in the fiber direction:

$$\sigma = \xi[\sigma_f v_f + \sigma_m(1 - v_f)] \qquad (10)$$

where $\sigma$, $\sigma_f$, $\sigma_m$ are the strengths of the composite, fiber and the matrix, respectively. The transverse strength, Y, and the shear strength, T, of a layer is assumed to be controlled by the matrix material:

$$Y = \sigma_m(1 - v_f) \qquad (11)$$

$$T = 0.58\sigma_m \qquad (12)$$

Here, shear strength is calculated using the von Mises yield criterion.

Application of the laminate analogy for prediction of composite strengths requires models that can predict layer failure under complex loading conditions. Two biaxial strength theories used in this study are maximum stress theory and Tsai-Hill theory.

In the maximum stress theory, the fracture is said to have occured if any of the inequalities below is not satisfied. For tensile loading:

$$\sigma_1 < \sigma_c \qquad (13)$$

$$\sigma_2 < Y \qquad (14)$$

$$|\tau_{12}| < T \qquad (15)$$

In this criterion, interaction between different modes of failure are *not* taken into consideration.

Tsai-Hill theory is an extension of von Mises' isotropic yield criterion to anisotropic materials. According to this criterion, the layer fails when the following equation is satisfied:

$$\sigma_1^2/\sigma_c^2 - \sigma_1\sigma_2/\sigma_c^2 + \sigma_2^2/Y^2 + \tau_{12}^2/T^2 = 1 \qquad (16)$$

where $\sigma_1$, $\sigma_2$, and $\tau_{12}$ are the layer stresses in the principal material directions.

A pictorial representation of composite/lamina/sublamina structure of the Halpin-Pagano Laminate Analogy is presented in Figure 2. It is also assumed that all the fibers lie in the plane of the lamina (xy-plane), fiber length distribution, when present, is independent of fiber orientation distribution, and matrix shear strength is weaker than the fiber/matrix interfacial shear strength.

The sublamina elastic constants are used to calculate the stiffness matrix, $Q^{SL}$, of the sublamina:

$$Q^{SL} = \begin{bmatrix} Q_{11}^{SL} & Q_{12}^{SL} & 0 \\ Q_{12}^{SL} & Q_{22}^{SL} & 0 \\ 0 & 0 & Q_{66}^{SL} \end{bmatrix} \qquad (17)$$

where

$$Q_{11}^{SL} = E_{11}^{SL}/(1 - v_{12}^{SL}v_{21}^{SL}) \qquad (18)$$

$$Q_{22}^{SL} = E_{22}^{SL}/(1 - v_{12}^{SL}v_{21}^{SL}) \qquad (19)$$

$$Q_{12}^{SL} = Q_{11}^{SL}v_{21}^{SL} \qquad (20)$$

$$Q_{66}^{SL} = G_{12}^{SL} \qquad (21)$$

Classical laminate theory gives the lamina stiffness matrix as the summation of the sublamina stiffnesses, weighted by the fiber length distribution function

$$Q^L = \sum_{i=1}^{N} Q_i^{SL} f_i \qquad (22)$$

$$\sum_{i=1}^{N} f_i = 1.0 \qquad (23)$$

where $f_i$ is the fraction of fibers contained in the $i^{th}$ sublamina.

If present, fiber length distribution can be either input by calculating average fiber length or using a log-normal distribution model (10)

$$f(l_f) = (1/\sqrt{2\pi}\ \delta l_f)exp\left[-1/2\left\{\left(ln(l_f) - \lambda\right)/\delta\right\}^2\right]$$

$$\text{for } 0 \le l_f \le \infty \qquad (24)$$

The laminate theory is applied once again to determine the composite stiffness matrix, which is the summation of the lamina reduced stiffnesses, weighted by the fiber orientation distribution function:

$$Q^c = \sum_{i=1}^{M} Q'_i g_i \qquad (25)$$

$$\sum_{i=1}^{M} g_i = 1.0 \qquad (26)$$

where M is the number of laminae and $g_i$ is the fraction of fibers oriented in the direction of the $i^{th}$ lamina. The engineering constants are computed from the inverse of the stiffness matrix, $s^c$:

$$s^c = \left[ Q^c \right]^{-1} \qquad (27)$$

$$s^c = \begin{bmatrix} s_{11} & s_{12} & s_{16} \\ s_{12} & s_{22} & s_{26} \\ s_{16} & s_{26} & s_{66} \end{bmatrix} \qquad (28)$$

The engineering constants are given by

$$E_{11} = 1/s_{11} \qquad (29)$$
$$E_{22} = 1/s_{22} \qquad (30)$$
$$G_{12} = 1/s_{66} \qquad (31)$$
$$\upsilon_{12} = -s_{12}E_{11} \qquad (32)$$
$$\upsilon_{21} = -s_{12}E_{22} \qquad (33)$$

A complete laminate strength calculation could be done for each lamina, however, this would be very inefficient. Since all the fibers in the lamina are oriented, there is no shear coupling between sublaminae. This makes it possible to use the rule of mixtures approach for the lamina strength in the longitudinal direction. To calculate the strength of each sublamina in the fiber direction, the Kelly-Tyson equation

modified with Chen's strength efficiency factor is used:

$$S = \xi\left[\sigma_m(1 - v_f) + v_f \int_{l_f=0}^{l_f=\infty} \sigma_{f'} f(l_f) \, d(l_f)\right] \qquad (34)$$

where $f(l_f)$ is the fiber length distribution function and

$$\sigma_{f'} = \sigma_f(l_f/2l_c) \qquad 0 < l_f < l_c \qquad (35)$$
$$\sigma_{f'} = \sigma_f(1 - l_c/2l_f) \qquad l_c \leq l_f \leq \infty \qquad (36)$$

The strain in the lamina at failure is calculated from:

$$\varepsilon_x = X/E_{11}^L \qquad (37)$$

where X is the lamina strength in the fiber direction. The stress in the matrix at failure is approximated by

$$\sigma_m \approx \varepsilon_x E_m \qquad (38)$$

The contribution of the matrix to the lamina strength will then be:

$$\sigma_m \approx X E_m/E_{11}^L \qquad (39)$$

which is substituted in Equation 34 to determine the lamina strength in the fiber direction.

(30)    The laminate analogy with shear coupling is used to determine the strength of the composite. The gradual failure model is employed to lamina failure. It is assumed that the failed laminae continue to support the load that they were supporting at failure, until all laminae fail. The stress in the composite at any point is then given by:

$$\sigma_{xc} = \sum_i g_i \sigma_{fi} + \sum_i g_i (Q'_{11i}{}^L \varepsilon_x + Q'_{12i}{}^L \varepsilon_y + Q'_{16i}{}^L \gamma_{xy}) \qquad (40)$$

where $\sigma_{xc}$ is the composite stress in the X-direction, $Q'_i{}^L$ is the reduced stiffness matrix in the $i^{th}$ lamina, and $\varepsilon_x$, $\varepsilon_y$ and

$\gamma_{xy}$ are the composite strains.

For both the maximum stress criterion and the Tsai-Hill failure criterion, the computer program calculates the lamina stresses perpendicular, parallel and in shear relative to the fiber orientation. These are calculated for each layer by calculating the overall strains and using the layer stiffness matrix to find the layer stresses. Finally, these are used with either of the above failure criteria to see if the layer fails under the applied load conditions.

## EXPERIMENTAL

Random fiber reinforced composites were prepared according to the process described in U. S. Patent 4,426,470. This process involves making a random fiber mat from a slurry of finely divided polymer particles and short fibers on a commercial paper machine. The nature of the process prevents fiber breakage, therefore, the fiber aspect ratio stays constant throughout the process. Furthermore, the reinforcing fibers form a perfect random array, resulting in an ideal composite material where the fiber length is accurately known. Finally, the average fiber length is much greater than the thickness of the composite mat, a requirement of the laminate theory that needs to be satisfied if the material is to be considered as quasi-isotropic.

A number of these mats are then stacked on top of each other and compression molded to desired thickness.

Thickness of the laminate can be controlled by varying the number of mats stacked.

The variables studied were volume fraction and length of the glass fiber and the type of matrix resin. The fiber volume fraction was varied from about 0.04 upto 0.445 and the fiber length was varied from 4.75 mm (3/16 in) to 25 mm (1 in). The matrix resin was either polyethylene or polypropylene. Glass fiber content in each sample was determined by burning-off the resin at 600°C and weighing the residue.

Five sample bars were tested according to ASTM D-638 using an Instron Model 1127 test machine and the average was reported as the tensile strength. All tests were conducted at room temperature .

The interactive computer program (1), written in FORTRAN, was run under the NOS subsystem of a CDC CYBER 180 computer. Only the characteristic angles (0°/90°/±45° or 0°/±60°) were used to define the random-in-plane composite.

## RESULTS AND DISCUSSION

Table 1 summarizes the properties of the constituents of the composite materials and important parameters used in the computer program. Measured and predicted tensile strengths are presented in Table 2 and plotted in Figures 3 through 8. The data presented were computed using the Tsai-

Hill failure criteria. Predicted strengths were slightly higher than the measured values for 4.75 mm and 12.7 mm glass fiber reinforced PP systems. This can be explained by the higher degree of z-direction orientation of the shorter fibers which is not accounted for in the model. Comparison of Figures 4, 6, 7, and 8 shows an increasingly better prediction of the tensile strength with increasing fiber lengths. Agreement of the measured strength values with experimental data were very good for all other formulations. The measured tensile strength, however, starts to deviate from the predicted values once the glass fiber volume fraction reaches the range of 0.25 to 0.30.

The model, although very elaborate, still has many shortcomings. First of all, non-planar orientation of the fiber (z-direction orientation) is not taken into account in the model. Therefore, as discussed above, the model tends to overpredict the strength if there is significant amount of fibers in non-planar orientation. Furthermore, the presence of any residual porosity in the material is also not accounted for in the model. In other words, as one approaches a resin starved system, the measured tensile strength will level off or drop, but the predicted tensile strength will keep increasing with increasing fiber volume fraction. Figure 9 shows the density defect ratio, the ratio of the apparent density to the theoretical density, as a function of the volume fraction of the

fiber for the random fiber composite system studied. Residual porosity in the composite material starts to increase only after the fiber volume fraction exceeds 0.20. Resin starvation is clearly observed at about 0.35 fiber volume fraction. This explains why, after the fiber volume fraction range of 0.25 to 0.35, the predicted strengths were greater than the measured strengths. Finally, Lavengood-Chen idealized fiber arrangement is only an approximation of the stress field interactions of the neighboring fibers. In a real random fiber laminate, stress field interactions are much more complex, therefore, the use of an idealized arrangement over-predicts the tensile strength.

Besides the dicussed shortcomings of the model, there were also many simplifications which might have influenced the predictions. It was assumed that the matrix shear strength was weaker than the fiber/matrix interfacial shear strength. The shear strength of the composite, T, is calculated using equation 12, which is based on von Mises yield criterion. This provides a single value of shear strength which is independent of the fiber length in the composite. This does not appear to be a good approximation, since the experimental results indicate an increase in shear strength with an increase in fiber length.

Clumping of fibers, when present, leads to a change in the effective aspect ratio, requiring a modification of the geometry factor in the Halpin-Tsai equation. This was of minimal importance for the composites of this study, for one of the advantages of the process is to debundle the fiber bundles into individual filaments.

The gradual failure model assumes each of the failed laminae to carry the load that they were carrying at failure. The load carrying characteristics were examined in the direction of applied stress only and shear and transverse load carrying characteristics of the failed laminae were ignored. When a lamina fails, its stiffness contribution is removed from the composite stiffness matrix. In a real laminate, failure of a lamina in one direction does not immediately remove its stiffness contributions in all directions, therefore, simultaneous failure cannot occur.

## CONCLUSIONS

1. Predicted tensile strengths were in good agreement with the experimental values especially for fiber volume fractions less than 0.25 and fiber lengths greater than or equal to 12.7 mm.

2. The discrepancy between the experimental data and the predicted tensile strength values for fiber volume fractions greater than 0.25 is explained by residual porosity in the composite due to resin starvation.

3. Higher degree of accuracy in the predictions of tensile strength of composites with longer fibers is due to reduced z-direction orientation in the prepreg.

4. Despite the presented shortcomings of the model and the room for further improvement, it is a very useful and efficient design tool.

## ACKNOWLEDGEMENTS

The author would like to acknowledge Professor C. L. Tucker, III, of the University of Illinois at Urbana-Champaign, colleagues M. Edens, J. Huber, L. D. Yats, and J. D. Zawisza of the Dow Chemical Company for their help and discussions at various stages of the research and the Dow Chemical Company for permitting publication of this work.

## REFERENCES

1. Loughlin, P. T., M.Sc. Thesis, "Stiffness and Strength Predictions for Short Fiber Composites with Planar Fiber Orientation Distribution", University of Illinois, 1979.

2. Adams, D. F. and Doner, D. R., J. Comp. Mater., 1:4, 152(1967).

3. Halpin, J. C. and Kardos, J. L., Polym. Eng. Sci., 16:5, 344(1976).

4. Halpin, J. C. and Pagano, N. J., J. Comp. Mater., 3, 720(1969).

5. Kelly, A. and Tyson, W. R., J. Mech. Phys. Solids, 13, 329(1965).

6. Chen, P. E., Polym. Eng. Sci.,11:1, 51(1971).

7. Chen, P. E. and Lavengood, R. E., "Stress Fields Around Multiple Inclusions", in Structure, Solid Mechanics and Engineering Design, Part 1, M. Te'eni (ed.), Wiley-Interscience, New York, 1971.

8. Kardos, J. L., Halpin, J. C. and Chang, S. L., "Rheology, Vol.3: Applications", edited by Astarita, G., Maruvvi, G. and Nicolais, L., Plenum, New York, N.Y., 1980.

9. Tucker, C. L., "Reaction Injection Molding of Polymeric Parts", Ph.D. Thesis, MIT, 1978.

## BIOGRAPHY

*Selim Yalvaç* holds B.S. and M.S. degrees in Chemical Engineering from Middle East Technical University, Ankara, Turkey, and a Ph.D. degree in Chemical Engineering from the University of Michigan, Ann Arbor, Michigan. His area of interest includes process development, characterization and evaluation of properties and fabrication of polymeric and inorganic matrix composites. Currently, he is a project leader in the Polymeric Materials Laboratory of Central Research Department of Dow Chemical U.S.A. He was the recipient of a four year Fullbright-Hayes award in 1976.

**Table 1.** Properties of Polypropylene, Polyethylene, Glass Fiber and Program Parameters Used for Modeling the Random Fiber Composites

| Property | Polypropylene | Polyethylene | Glass Fiber |
|---|---|---|---|
| Tensile Strength, MPa (ksi) | 31.03 (4.5) | 24.13 (3.5) | 3,447.5 (500) |
| Tensile Modulus, GPa (Msi) | 1.517 (0.22) | 0.758 (0.11) | 72.40 (10.5) |
| Poisson's Ratio | 0.33 | 0.39 | 0.22 |
| Shear Strength, MPa (ksi) | 21.72 (3.15) | 20.2 (2.93) | - |
| Strength Efficiency Factor, $\xi$ | 0.58[†] | 0.52[†] | - |
| Layer Shear Strength Ratio, (Shear Strength/Tensile Strength) | 0.70 | 0.84 | - |

[†] From Figure 1

**Table 2.** Experimental and Predicted Tensile Strengths for Various Random Fiber Composites

| Description* | Fiber Weight Fraction | Fiber Volume Fraction | Tensile Strength, MPa (ksi) Experimental | Predicted |
|---|---|---|---|---|
| 4.75/PE | 0.108 | 0.044 | 46.47 (6.74) | 37.14 (5.39) |
| 4.75/PE | 0.164 | 0.069 | 59.85 (8.68) | 45.18 (6.56) |
| 4.75/PE | 0.219 | 0.096 | 70.33 (10.20) | 55.01 (7.98) |
| 4.75/PE | 0.284 | 0.130 | 78.95 (11.45) | 67.70 (9.82) |
| 4.75/PE | 0.291 | 0.134 | 92.53 (13.42) | 69.15 (10.03) |
| 4.75/PE | 0.310 | 0.145 | 87.01 (12.62) | 73.16 (10.61) |
| 4.75/PE | 0.318 | 0.149 | 94.19 (13.66) | 74.90 (10.86) |
| 4.75/PE | 0.319 | 0.150 | 78.19 (11.34) | 75.13 (10.90) |
| 4.75/PE | 0.320 | 0.151 | 92.39 (13.40) | 75.36 (10.93) |
| 4.75/PE | 0.324 | 0.153 | 77.50 (11.24) | 76.21 (11.05) |
| 4.75/PE | 0.327 | 0.155 | 83.29 (12.08) | 76.88 (11.15) |
| 4.75/PE | 0.327 | 0.155 | 88.67 (12.86) | 76.88 (11.15) |
| 4.75/PE | 0.328 | 0.156 | 85.77 (12.44) | 77.11 (11.18) |
| 4.75/PE | 0.330 | 0.157 | 83.43 (12.10) | 77.55 (11.25) |
| 4.75/PE | 0.330 | 0.157 | 93.50 (13.56) | 77.55 (11.25) |
| 4.75/PE | 0.331 | 0.157 | 94.39 (13.69) | 77.78 (11.28) |
| 4.75/PE | 0.332 | 0.158 | 93.15 (13.51) | 78.00 (11.31) |
| 4.75/PE | 0.337 | 0.161 | 85.36 (12.38) | 79.12 (11.48) |
| 4.75/PE | 0.337 | 0.161 | 91.70 (13.30)) | 79.12 (11.48) |
| 4.75/PE | 0.344 | 0.165 | 86.05 (12.48) | 80.72 (11.71) |
| 4.75/PE | 0.344 | 0.165 | 94.39 (13.69) | 80.72 (11.71) |
| 4.75/PE | 0.349 | 0.168 | 85.15 (12.35) | 81.84 (11.87) |
| 4.75/PE | 0.350 | 0.169 | 105.36 (15.28) | 82.10 (11.91) |
| 4.75/PE | 0.350 | 0.169 | 81.12 (11.75) | 82.10 (11.91) |
| 4.75/PE | 0.356 | 0.173 | 106.60 (15.46) | 83.48 (12.11) |
| 4.75/PE | 0.360 | 0.175 | 82.33 (11.94) | 84.42 (12.24) |
| 4.75/PE | 0.361 | 0.176 | 85.70 (12.43) | 84.64 (12.28) |
| 4.75/PE | 0.361 | 0.176 | 94.53 (13.71) | 84.64 (12.28) |
| 4.75/PE | 0.365 | 0.178 | 93.22 (13.52) | 85.57 (12.41) |
| 4.75/PE | 0.368 | 0.180 | 70.47 (10.22) | 86.29 (12.52) |
| 4.75/PE | 0.373 | 0.183 | 106.94 (15.51) | 87.49 (12.69) |
| 4.75/PE | 0.384 | 0.190 | 96.53 (14.00) | 90.15 (13.07) |
| 4.75/PE | 0.396 | 0.198 | 100.94 (14.64) | 93.11 (13.50) |
| 4.75/PE | 0.398 | 0.200 | 106.87 (15.50) | 93.59 (13.57) |
| 4.75/PE | 0.403 | 0.203 | 109.70 (15.91) | 94.87 (13.76) |
| 4.75/PE | 0.410 | 0.208 | 90.19 (13.08) | 96.63 (14.02) |
| 4.75/PE | 0.421 | 0.215 | 107.22 (15.55) | 99.49 (14.43) |

| | | | | |
|---|---|---|---|---|
| 4.75/PE | 0.448 | 0.234 | 85.77 (12.44) | 106.67 (15.47) |
| 4.75/PE | 0.474 | 0.254 | 107.15 (15.54) | 113.92 (16.52) |
| 4.75/PE | 0.524 | 0.294 | 80.33 (11.65) | 128.92 (18.70) |
| 4.75/PE | 0.529 | 0.298 | 73.78 (10.70) | 130.47 (18.92) |
| 4.75/PE | 0.582 | 0.344 | 98.60 (14.30) | 148.13 (21.48) |
| 4.75/PE | 0.631 | 0.392 | 76.53 (11.10) | 166.19 (24.10) |
| 4.75/PE | 0.680 | 0.445 | 76.53 (11.10) | 186.15 (27.00) |
| 4.75/PP | 0.114 | 0.044 | 38.54 (5.59) | 47.14 (6.84) |
| 4.75/PP | 0.224 | 0.093 | 48.68 (7.06) | 67.02 (9.72) |
| 4.75/PP | 0.312 | 0.139 | 56.40 (8.18) | 85.64 (12.42) |
| 4.75/PP | 0.367 | 0.171 | 79.78 (11.57) | 98.88 (14.34) |
| 4.75/PP | 0.410 | 0.198 | 59.02 (8.56) | 110.15 (15.98) |
| 4.75/PP | 0.483 | 0.249 | 63.57 (9.22) | 131.43 (19.06) |
| 12.7/PE | 0.120 | 0.049 | 51.85 (7.52) | 39.99 (5.80) |
| 12.7/PE | 0.244 | 0.109 | 77.22 (11.20) | 63.20 (9.17) |
| 12.7/PE | 0.337 | 0.161 | 83.84 (12.16) | 84.33 (12.23) |
| 12.7/PE | 0.347 | 0.167 | 112.39 (16.3) | 86.81 (12.59) |
| 12.7/PE | 0.375 | 0.185 | 91.77 (13.31) | 93.97 (13.63) |
| 12.7/PE | 0.376 | 0.185 | 102.32 (14.84) | 94.21 (13.66) |
| 12.7/PE | 0.381 | 0.189 | 95.50 (13.85) | 95.55 (13.86) |
| 12.7/PE | 0.406 | 0.205 | 110.46 (16.02) | 102.28 (14.83) |
| 12.7/PE | 0.409 | 0.207 | 120.11 (17.42) | 103.10 (14.95) |
| 12.7/PE | 0.421 | 0.215 | 115.84 (16.80) | 106.49 (15.44) |
| 12.7/PE | 0.502 | 0.276 | 125.49 (18.20) | 131.14 (19.02) |
| 12.7/PE | 0.619 | 0.380 | 126.87 (18.40) | 174.00 (25.24) |
| 12.7/PP | 0.219 | 0.091 | 53.85 (7.81) | 69.23 (10.04) |
| 12.7/PP | 0.298 | 0.131 | 66.40 (9.63) | 87.06 (12.63) |
| 12.7/PP | 0.301 | 0.133 | 78.81 (11.43) | 87.82 (12.74) |
| 12.7/PP | 0.316 | 0.141 | 79.02 (11.46) | 91.51 (13.27) |
| 12.7/PP | 0.330 | 0.149 | 79.98 (11.60) | 95.03 (13.78) |
| 12.7/PP | 0.364 | 0.169 | 86.12 (12.49) | 104.00 (15.08) |
| 12.7/PP | 0.366 | 0.170 | 79.50 (11.53) | 104.49 (15.16) |
| 12.7/PP | 0.391 | 0.186 | 88.39 (12.82) | 111.47 (16.17) |
| 12.7/PP | 0.397 | 0.190 | 87.08 (12.63) | 113.22 (16.42) |
| 12.7/PP | 0.400 | 0.192 | 91.36 (13.25) | 114.07 (16.54) |
| 12.7/PP | 0.402 | 0.193 | 88.53 (12.84) | 114.66 (16.62) |
| 12.7/PP | 0.402 | 0.193 | 86.95 (12.61) | 114.66 (16.63) |
| 12.7/PP | 0.403 | 0.194 | 106.60 (15.46) | 114.93 (16.67) |
| 12.7/PP | 0.408 | 0.197 | 97.77 (14.18) | 116.41 (16.88) |
| 12.7/PP | 0.414 | 0.201 | 98.46 (14.28) | 118.15 (17.14) |
| 12.7/PP | 0.415 | 0.201 | 100.80 (14.62) | 118.47 (17.18) |
| 12.7/PP | 0.418 | 0.203 | 94.94 (13.77) | 119.37 (17.31) |
| 12.7/PP | 0.418 | 0.203 | 106.25 (15.41) | 119.37 (17.31) |
| 12.7/PP | 0.419 | 0.204 | 103.63 (15.03) | 119.68 (17.36) |
| 12.7/PP | 0.427 | 0.208 | 105.77 (15.34) | 121.35 (17.60) |
| 12.7/PP | 0.428 | 0.210 | 120.18 (17.43) | 122.38 (17.75) |
| 12.7/PP | 0.429 | 0.211 | 81.02 (11.75) | 122.70 (17.80) |
| 12.7/PP | 0.429 | 0.211 | 85.84 (12.45) | 122.70 (17.80) |
| 12.7/PP | 0.431 | 0.212 | 103.01 (14.94) | 123.28 (17.88) |
| 12.7/PP | 0.434 | 0.214 | 92.74 (13.45) | 124.22 (18.02) |
| 12.7/PP | 0.440 | 0.218 | 89.08 (12.92) | 126.06 (18.28) |
| 12.7/PP | 0.440 | 0.218 | 109.98 (15.95) | 126.06 (18.28) |
| 12.7/PP | 0.441 | 0.219 | 113.22 (16.42) | 126.38 (18.33) |
| 12.7/PP | 0.451 | 0.226 | 89.57 (12.99) | 129.52 (18.79) |
| 12.7/PP | 0.452 | 0.227 | 99.77 (14.47) | 129.84 (18.83) |
| 12.7/PP | 0.452 | 0.227 | 107.08 (15.53) | 129.84 (18.83) |
| 12.7/PP | 0.456 | 0.230 | 91.15 (13.22) | 131.14 (19.02) |
| 12.7/PP | 0.463 | 0.235 | 83.02 (12.04) | 133.40 (19.35) |
| 12.7/PP | 0.492 | 0.256 | 101.77 (14.76) | 143.08 (20.75) |
| 12.7/PP | 0.495 | 0.258 | 87.98 (12.76) | 144.12 (20.90) |
| 12.7/PP | 0.509 | 0.269 | 90.39 (13.11) | 149.04 (21.62) |
| 12.7/PP | 0.614 | 0.361 | 121.63 (17.64) | 190.56 (27.64) |
| 12.7/PP | 0.635 | 0.382 | 91.43 (13.26) | 200.02 (29.01) |
| 19.0/PP | 0.249 | 0.105 | 76.47 (11.09) | 76.42 (11.08) |
| 19.0/PP | 0.268 | 0.115 | 72.10 (10.50) | 80.82 (11.72) |
| 19.0/PP | 0.325 | 0.146 | 93.22 (13.52) | 94.79 (13.75) |
| 19.0/PP | 0.370 | 0.173 | 97.43 (14.13) | 106.80 (15.49) |
| 19.0/PP | 0.372 | 0.174 | 111.15 (16.12) | 107.39 (15.58) |
| 19.0/PP | 0.384 | 0.181 | 98.18 (14.24) | 110.75 (16.06) |
| 19.0/PP | 0.426 | 0.209 | 109.29 (15.85) | 123.23 (17.87) |
| 19.0/PP | 0.435 | 0.215 | 100.60 (14.59) | 126.01 (18.28) |
| 10.0/PP | 0.439 | 0.218 | 115.01 (16.68) | 127.24 (18.45) |
| 19.0/PP | 0.455 | 0.229 | 125.35 (18.18) | 132.40 (19.20) |
| 19.0/PP | 0.479 | 0.246 | 119.14 (17.28) | 140.35 (20.36) |
| 19.0/PP | 0.505 | 0.266 | 124.04 (17.99) | 149.46 (21.68) |
| 19.0/PP | 0.516 | 0.275 | 117.35 (17.02) | 153.40 (22.25) |
| 19.0/PP | 0.526 | 0.283 | 123.35 (18.89) | 157.11 (22.79) |
| 19.0/PP | 0.541 | 0.295 | 130.94 (18.99) | 162.79 (23.61) |
| 19.0/PP | 0.555 | 0.307 | 124.18 (18.01) | 168.25 (24.40) |
| 25.4/PP | 0.394 | 0.188 | 105.15 (15.25) | 114.26 (16.57) |
| 25.4/PP | 0.407 | 0.196 | 120.25 (17.44) | 118.12 (17.13) |
| 25.4/PP | 0.463 | 0.235 | 124.52 (18.06) | 135.82 (19.70) |

* Fiber Length in mm/Matrix Type

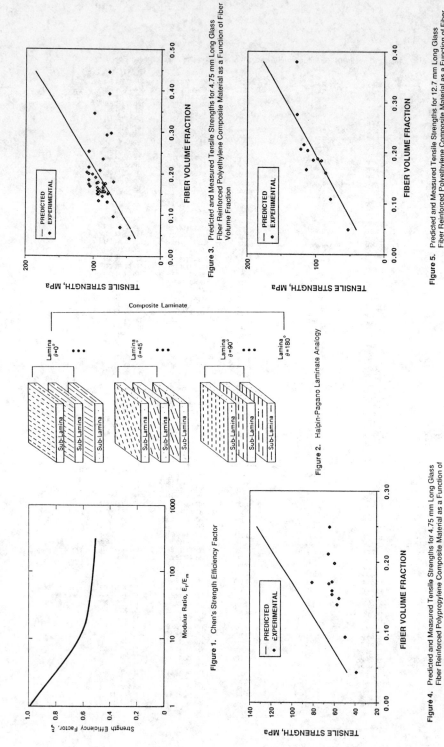

**Figure 1.** Chen's Strength Efficiency Factor

**Figure 2.** Halpin-Pagano Laminate Analogy

**Figure 3.** Predicted and Measured Tensile Strengths for 4.75 mm Long Glass Fiber Reinforced Polyethylene Composite Material as a Function of Fiber Volume Fraction

**Figure 4.** Predicted and Measured Tensile Strengths for 4.75 mm Long Glass Fiber Reinforced Polypropylene Composite Material as a Function of Fiber Volume Fraction

**Figure 5.** Predicted and Measured Tensile Strengths for 12.7 mm Long Glass Fiber Reinforced Polyethylene Composite Material as a Function of Fiber Volume Fraction

718

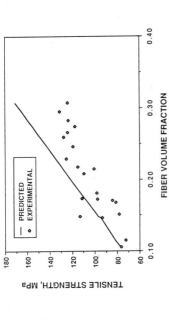

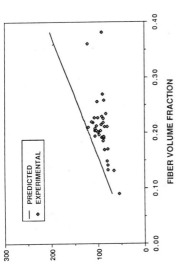

**Figure 6.** Predicted and Measured Tensile Strengths for 12.7 mm Long Glass Fiber Reinforced Polypropylene Composite Material as a Function of Fiber Volume Fraction

**Figure 7.** Predicted and Measured Tensile Strengths for 19.0 mm Long Glass Fiber Reinforced Polypropylene Composite Material as a Function of Fiber Volume Fraction

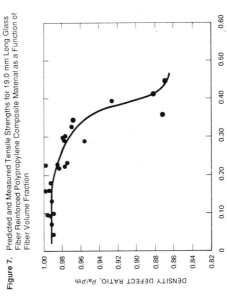

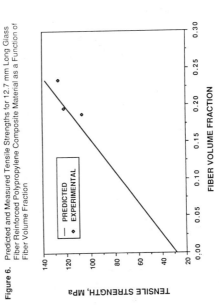

**Figure 8.** Predicted and Measured Tensile Strengths for 25 mm Long Glass Fiber Reinforced Polypropylene Composite Material as a Function of Fiber Volume Fraction

**Figure 9.** Density Defect Ratio as a Function of Glass Fiber Volume Fraction

719

# TENSILE RESPONSE OF LAMINATES TO IMPLANTED DELAMINATIONS

Paul A. Lagace and Douglas S. Cairns
TECHNOLOGY LABORATORY FOR ADVANCED COMPOSITES
Department of Aeronautics and Astronautics
Massachusetts Institute of Technology
Cambridge, MA 02139

## Abstract

A series of investigations was performed to determine the tensile response of graphite/epoxy and kevlar/epoxy laminates to implanted delaminations. Laminates with and without notches were tested with various configurations of implanted delaminations. The great majority of work was conducted on laminates manufactured from unidirectional tape in $[\pm45/0]_s$, $[0/\pm45]_s$, and $[+45/0/-45]_s$ configurations with circular delaminations located at the specimen center. The results show that delaminations located at only one ply interface have virtually no effect on either the strain field or the fracture stress. However, when multiple ply interfaces are delaminated at the same in-plane location, a reduction in strength of the unnotched specimens does occur. This is attributed to uncoupling of the plies which allows easier formation of splits in the individual plies without the capabililty of sharing load with the neighboring unbonded plies. In the case of specimens with notches (drilled holes), this uncoupling and subsequent splitting relieves local stress concentrations and results in an increase in notched strength over specimens without implanted delaminations. However, the implanted delaminations do not alter the strain field significantly prior to damage. A final set of experiments was conducted on notched (holes and slits) graphite/epoxy fabric laminates in a four-ply quasi-isotropic configuration. Both coupons in tension and pressurized cylinders were tested. Neither case showed any change in response due to the presence of implanted delaminations despite the fact that localized bending is important in the failure of the pressurized cylinders. The lack of effect is explained, in part, by the fact that the fabric configuration inherently restricts the formation of splits and thus the stress-relieving mechanisms exhibited in the tape laminates is not present in these cases.

## 1. INTRODUCTION

Modern advanced composite materials can offer significant increases in performance especially when considered on a weight basis[1]. This tremendous potential has accelerated application of these materials from secondary to primary structural components as these materials are being used in virtually all new aerospace programs. However, due to the complicated failure mechanisms present in laminated composite materials, a large lag in the understanding of basic failure mechanisms in advanced composites has occurred.

Unlike metals, limited fracture data exists for composite materials. This is compounded by the fact that many modes of failure can occur in composite materials. Consequently, a practical engineering tool equivalent to fracture mechanics for metals does not presently exist. Before such a tool can be developed, a thorough understanding (or at least description) of the mechanisms contributing to failure needs to be developed.

The first step in understanding the fracture behavior of composites is to understand failure on the ply level and how a multi-directional laminate is affected by such. The damage which occurs in (continuous filament) laminated composite materials can be loosely grouped into three catagories. These are fiber failure, matrix cracking and out-of-plane delamination. All of these phenomena are addressed in this study with particular emphasis on delamination. To develop models to describe the effects of damage, a good understanding of each of these phenomena alone and in combination needs to be addressed.

Delaminations are an important and complicated type of damage. This common yet poorly understood type of damage is an important design consideration[2] whether the delamination results during manufacturing or in service. Laminated composites are especially susceptible to delamination owing to their relatively poor out-of-plane properties. Impact can often cause delamination and may cause such in combination with other types of damage (e.g. in-plane damage such as fiber failure and matrix cracking).

For preliminary design assessment, many studies have addressed the behavior of delaminations under compression[3]. These studies are based on the premise that delaminations affect the failure of composites by causing sublaminates to first buckle out-of-plane. This buckling and subsequent growth of the delamination is highly dependent on the geometry of the delamination and the constitutive properties of the material (laminate and sublaminates).

In this study, the delamination is treated as an independent mode of damage. The laminates were tested in tension to prelude buckling and to isolate the delamination to study its effects on initial and final fracture. This study was conducted over a range of laminate configurations and materials.

## 2. SCOPE

The philosopy used in this study was to first develop an understanding of the influence delamination has as a type of damage. Then, once a preliminary understanding was developed, additional damage (i.e. notches) was introduced to examine the influence of combined damage.

For examination of isolated delamination, nonporous teflon inserts were placed between plies to create delaminations (the specific procedure is discussed later). For the first series of tests, these delaminations were placed at the midplane of an AS4/3501-6 graphite/epoxy $[+45/0/-45]_s$ laminate. This test isolates the delamination as an independent type of damage. The ply configuration prevents formation of transverse cracks prior to final fracture. Thus, the delamination is isolated in its effects on fracture. While it may be argued that no stress intensity can occur around the delamination on a medium mechanics level (defined as mechanics on a ply level treating the ply as a homogeneous orthotropic material) due to compatibility[4], no such claim can be made on the micromechanics level (where the individual fibers and matrix material are recognized) or after any additional damage develops.

Subsequent to these tests, a series of tests were conducted on Hercules AS4/3501-6 unidirectional tape in a $[\pm 45/0]_s$ configuration. In these laminates, transverse cracks can form prior to final fracture at high strain levels due to the presence of the delamination as there is no inhibitor to the growth of splits (defined as matrix cracks which form parallel to the fibers). Consequently, delaminations with transverse cracks were studied in a laminate where transverse cracks do not form at low strain levels as in a laminate configuration with 90° plies. Two configurations of delamination were used: delaminations were placed at the midplane; and delaminations were placed at each ply interface.

In the next series of tests, a delamination with a notch was studied in graphite/epoxy in a fabric configuration. The addition of this stress concentrator allowed the study of the interaction of the implanted delaminations and the notch on the final fracture. The woven nature of the fabric, illustrated in Figure 1, prevents the formation of transverse cracks. A four ply quasi-isotropic configuration was used, $(45,0)_s$, where the angle refers to the direction of the warp fibers. The results are compared to notched coupons without delaminations.

To further allow for the effects of matrix cracks and thus study the effects of all three types of damage previously mentioned, SP-328 kevlar/epoxy unidirectional prepreg in a $[+45/0/-45]_{2s}$ laminated configuration was studied. While a significant data base does not exist for this material, it was chosen for its unique sensitivity to transverse cracks. In kevlar laminates, unlike graphite laminates, transverse cracking can actually damage fibers[5]. The results of these tests are compared with notched laminates with no implanted delaminations.

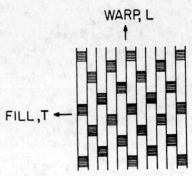

Figure 1 Construction of fabric prepreg.

Finally, in order to study the effects of delaminations in a structural configuration and allow for possible growth, pressurized cylinders were tested with implanted delaminations. The justification for the use of this specimen is twofold. First, a significant data base exists on this specimen for many notch configurations without implanted delaminations[6]. Second, this specimen allows out-of-plane (due to bending induced by pressurization) stresses to develop around the delamination. In this manner, the ability of the delamination to grow under high in-plane loads (without having to first buckle as in compression loading) is allowed.

A summary of the tests conducted is presented in Table 1. The type of specimen along with the parameters isolated are shown. While some gaps do exist, as a preliminary approach to resolving the phenomena, the matrix is reasonably extensive and isolates the effects of delaminations under tensile loading as a single initial damage mode and in conjunction with notches.

## 3. EXPERIMENTAL TECHNIQUES

The flat coupons were manfactured according to standard procedures. The preimpreganted unidirectional tape is supplied in a roll 305 mm in width while the preimpregnated fabric is supplied as 1 meter wide broadgoods. The material is cut into plies of the

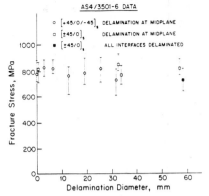

Figure 5 Fracture stress versus delamination diameter for tape specimens without notches.

delaminated, there was a measurable effect on the fracture stress as these points fall as much as 15 percent below the nominal unflawed data. This may be attributed to the lack of a mechanism for a ply to carry any transverse load once a split has formed. In the case of an undelaminated ply, the interply matrix layer can transfer load via shear into the adjoining ply around such a matrix crack. This capability is removed when the ply is delaminated. Consequently, a small reduction in strength may be found in this most severe case. However, more data need to be taken to better substantiate this effect as well as to quantify the specific underlying mechanisms.

### 4.2 Notched Case

Tests were subsequently conducted on A370-5H/3501-6 $(45,0)_s$ laminates to determine the effect of notches with delaminations implanted at the midplane interface. The results in terms of fracture stresses are summarized in Figure 6. For a notch configuration of a 6.35 mm diameter hole with a wide range of delamination sizes, the presence of the delamination had no influence on the ultimate fracture strength.

For the tests of this laminate conducted with various notch sizes and a midplane circular delamination with a 25.4 mm diameter, the data was correlated using the Mar-Lin equation[12]. This fit is based on the singularity of a crack in the matrix material terminating at a fiber/matrix interface and is of the form:

$$\sigma_f = H_c \, (2r)^{-m} \qquad (1)$$

where $\sigma_f$ is the fracture stress, $H_c$ is the composite fracture parameter, $2r$ is the notch length, and m is the power of the singularity. This equation has been shown[13] to yield an excellent correlation of the fracture stresses of notched composites which fail due to in-plane mechanisms. This value of m, determined analytically, is 0.28 for AS4/3501-6 graphite/epoxy and 0.35 for the SP-328 kevlar/epoxy system.

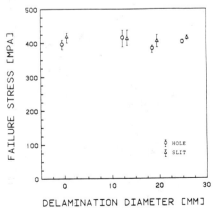

Figure 6 Failure stress versus delamination diameter for fabric specimens with a 6.35 mm notch.

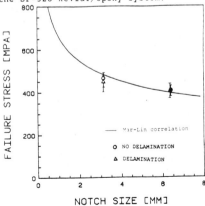

Figure 7 Predicted and measured failure stresses versus notch size for fabric specimens with a 25.4 mm circular delamination at the midplane.

The data for the $(45,0)_s$ graphite/epoxy fabric laminate, along with the correlation, is shown in Figure 7. The low scatter and consistency in the data indicates that the influence of the delamination on the fracture of a fabric laminate with a notch is negligible. It should again be noted that the fabric configuration prevents the formation of transverse cracks and thus the effect of the notch coupled solely with a delamination was isolated.

In the third series of coupon tests, conducted using kevlar/epoxy, transverse cracks were expected to influence the overall laminate strength. Interestingly, the presence of the delamination at each interface actually seemed to "help" the laminate with respect to tensile strength as can be seen in Figure 8. While the two data sets can each be correlated using the Mar-Lin equation with the same value of the exponent m, the composite fracture parameter, $H_c$, increases by 50% for the delaminated case over the undelaminated case. This is attributed to formation of axial splits in the 0° plies which relieve the stress concentrations in these major load-carrying plies. Also plotted in Figure 8 is the line of net stress

defined as the load divided by the net area (width minus hole diameter). This curve represents a line of "notch insensitivity". This indicates that the presence of the implanted delaminations causes the laminates to be "notch-insensitive". Again, the primary effect is attributed to the formation of the axial splits.

## 4.3 Pressurized Cylinders

The results of the tests conducted on the cylinders are plotted in Figure 9 in terms of the pressure at failure versus the diameter of the delamination implanted at the midplane. Two notch configurations were utilized: holes and slits. In each case the notch dimension was 50.8 mm. Virtually no influence of the delamination on fracture stress is seen. This is similar to the findings from the fabric coupons with notches despite the fact that additional bending loads at the notch are introduced in the cylinder case. This is discussed further in the following section.

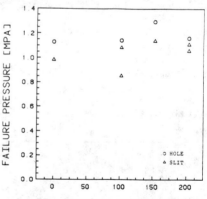

Figure 9 Failure pressures versus delamination diameter for fabric cylinders with a 50.8 mm notch.

## 5. BASIC MECHANISMS

The single midplane delamination configuration does not create a strength reduction under tensile loading in the laminates tested. Transverse cracks which formed in the laminates prior to fracture did not influence the laminate strength.

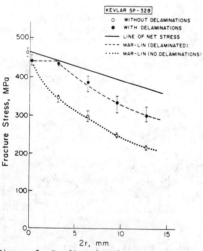

Figure 8 Predicted and measured failure stresses versus notch size for kevlar/epoxy tape specimens with and without delaminations.

However, in laminates where delaminations were introduced at each ply interface, there is no mechanism to transfer load around a transverse crack which occurs in a ply except at the edge of the delamination. This creates a large ineffective zone where the ply cannot carry transverse or shear load thus creating a reduction in overall laminate strength as seen in Figure 5. Although this mechanism is qualitatively explained and documented, additional experiments and analysis need to be performed to quantify the effect in this "worst case" scenario of a delamination at each ply interface.

In laminates where transverse cracking is suppressed (i.e. the fabric laminates), the combination of delaminations and notches has no influence on the ultimate strength of the laminate. In these laminates, ultimate fracture appears to be governed by the complicated damage region which develops adjacent to the main load-carrying fibers. In all cases, the presence of a midplane delamination was negligible over a wide range of delamination and notch sizes as was illustrated in Figures 6 and 7. This lends further credence to the explanation of the reduction in laminate strength due to the formation of transverse cracks. Furthermore, the fact that fracture is governed by the local behavior of the fiber-matrix interaction in these cases is further illustrated by the fact that holes behave similarly to slits. This has also been noted previously for undelaminated configurations[14].

The case where transverse cracks were allowed to form in the presence of notches shows that the early formation of an axial split can actually be beneficial. The presence of the delamination allows longer longitudinal splits to form at the points of maximum stress concentration since the shear lag mechanism present in undelaminated notched specimens is no longer present. As a consequence, the fracture stresses for the SP-328 kevlar/epoxy specimens fall nearly on the line of net stress. The curves plotted for the data still, however, follow the Mar-Lin correlation based on the predicted singularity, but the presence of the delamination and resulting axial splits serves to shift the curve upwards. This may be interpreted as increasing the composite fracture parameter, $H_c$. This also illustrates that the minimum notch size which affects laminate strength is increased (drawn as a horizontal line in Figure 8).

The delaminated cylinder case represents an interesting combination of mechanisms contributing to the fact that no influence of the delamination on fracture stress was found. The delamination creates a significant reduction in bending stiffness around the perimeter of the notch. Since load introduction is provided by internal pressure, this reduction in bending stiffness is expected to increase the stress peaks around the hole. Saeger[15] showed that even though the peak stresses are much higher at the edge of the notch for the delaminated tube, membrane stresses more quickly dominate in this region, with the combined effect resulting in the predicted laminate failure pressure being the same in the delaminated and undelaminated cases for the thin-walled pressure vessel tested. This may not be true, however, for thick walled cylinders, where geometry and load introduction are more complicated.

## 6. SUMMARY

A wide range of tests were conducted to examine the influence of delaminations under tensile loading. The results show that the presence of a single delamination has no influence on the ultimate tensile strength. However, delaminations implanted at all ply interfaces of a tape laminate cause a slight decrease in strength due to the total decoupling of all plies and the subsequent formation of large transverse cracks. This phenomenon warrants further investigation.

727

In notched fabric laminates where the formation of transverse cracks is suppressed due to the fabric configuration, the presence of the delamination has no influence on the notched tensile fracture. In the case of the unidirectional tape laminates where transverse cracks were allowed to form, the presence of the delamination suppressed fiber damage development at the notch edge by allowing matrix damage which relieved the stress concentration. In fact, in these laminates with implanted delaminations the results closely follow the line of notch insensitivity. This interesting phenomenon needs to be more carefully examined, especially with the current strong emphasis on "tough" (e.g. high strain-to-failure) matrix systems where matrix damage is inhibited.

The cylinders tested showed no influence from the presence of the delamination when transverse cracks are suppressed. Tests need to be conducted on unidirectional tape layups to obtain further understanding of the influences of geometry and load introduction.

The program conducted indicates that the presence of delamination does not have a significant effect in and of itself on the tensile fracture strength of composite laminates. However, the existence of these delamination allows large transverse cracks to occur in the delaminated unidirectional plies which causes a stress redistribution and results in a change in the fracture strength of the laminate. This change can be either a decrease or an increase depending upon the particular structural configuration as well as on the properties of the material. These phenomena warrant further investigation as their effect can be important as illustrated herein.

## 7. ACKNOWLEDGEMENTS

The authors wish to acknowledge the students who helped conduct this work over the past two years: Mary Bayalis, Thomas Grapes, Masami Kageyama, Karl Kowalski, and Leonard Robichaux. This work was supported by a joint Federal Aviation Administration/Navy program under contract N00019-85-C-0090.

## 8. REFERENCES

1. D.F. Adams, "Wind Turbine Blade Materials/Design Technology", Final Report, Faculty Research Participation Program, National Science Foundation, Southwest Research Institute, August, 1984, pp. 40-62.

2. J.E. McCarty et al., "Damage Tolerance of Composites", Interim Report No. 3 for Period 1 October 1983 – 29 February 1984, Air Force Wright Aeronautical Laboratory, Dayton, Ohio, March, 1984.

3. H. Chai and C.D. Babcock, "Two Dimensional Modelling of Compressive Failure in Delaminated Laminates", Journal of Composite Materials, Vol. 19, January, 1985, pp. 67-98.

4. A.E. Green and I.N. Sneddon, "The Distribution of Stress in the Neighborhood of a Flat Elliptical Crack in an Elastic Solid", Proceedings, Cambridge Philosophical Society, Vol. 46, 1950, p. 159.

5. G. Dorey, G.R. Sidey, and J. Hutchings, "Impact Properties of Carbon Fibre/Kevlar 49 Fibre Hybrid Composites", Composites, January, 1978, pp. 25-32.

6. P.A. Lagace and K.J. Saeger, "Damage Tolerance Characteristics of Pressurized Graphite/Epoxy Cylinders", presented at the Sixth ASME International Symposium on Offshore Mechanics and Arctic Engineering, March, 1987.

7. M.J. Graves, "The Catastrophic Failure of Pressurized Graphite/Epoxy Cylinders", TELAC Report 82-10, Massachusetts Institute of Technology, September, 1982.

8. L.M. Robichaux, "The Effects of Implanted Delaminations on the Tensile Behavior of Graphite/Epoxy Laminates", TELAC Report 86-4, Massachusetts Institute of Technology, January, 1986.

9. T.F. Grapes, "Sensitivity of Graphite/Epoxy to Implanted Delaminations under Tensile Load: Initial Study", TELAC Report 85-21, Massachusetts Institute of Technology, July, 1985.

10. M.K. Bayalis and K.G. Kowalski, "Damage Sensitivity of Kevlar/Epoxy Laminates in Tension", TELAC Report 85-14, Massachusetts Institute of Technology, December, 1985.

11. M. Kageyama, "The Effects of Delaminations on the Failure of Pressurized Graphite/Epoxy Cylinders", TELAC Report 86-18, Massachusetts Institute of Technology, June, 1986.

12. J.W. Mar and K.Y. Lin, "Fracture Mechanics Correlation for Tensile Failure of Filamentary Composites with Holes", Journal of Aircraft, Vol. 14, 1977, pp. 703-704.

13. P.A. Lagace, "Notch Sensitivity and Stacking Sequence of Laminated Composites", Composite Materials: Testing and Design (Seventh Conference), ASTM STP 893, American Society for Testing and Materials, 1986, pp. 161-176.

14. J.C. Brewer, "Tensile Fracture of Graphite/Epoxy with Angled Slits", TELAC Report 82-16, Massachusetts Institute of Technology, December, 1982.

15. K.J. Saeger, "Damage Tolerance of Composite Cylinders with a High Strain-to-Failure Matrix System", TELAC Report 86-11, Massachusetts Institute of Technology, May, 1986.

## 9. BIOGRAPHIES

Dr. Paul A. Lagace is an Associate Professor of Aeronautics and Astronautics at the Massachusetts Institute of Technology. He received his S.B. (1978), S.M. (1979), and Ph.D. (1982) degrees from M.I.T. and then joined the faculty. He is currently Co-Director of the Technology Laboratory for Advanced Composites where he conducts research on composite materials, especially their fracture, damage tolerance, and longevity characteristics. In addition to SAMPE, he is a member of the American Society for Composites, AIAA, and ASTM, where he is an active member of Committee D30 on High Modulus Fibers and their Composites.

Douglas S. Cairns is currently a doctoral candidate in the Department of Aeronautics and Astronautics at M.I.T. He received his BSME (1978) and MSME (1981) from the University of Wyoming where he also worked as a full-time staff engineer from 1979 to 1980. From 1981 to 1984, he was a senior engineer at Hercules, Inc. (Aerospace Division) involved with the design, testing, and analysis of composite materials. He is a member of ASME.

# TRANSIENT THERMAL STRESS ANALYSIS
## OF FIBER COMPOSITE MATERIALS
Hongsheng Wang
Research and Development Division
Chengdu Aircraft Company
P. O. Box 609, Chengdu, CHINA
Tsu-Wei Chou
Center for Composite Materials
University of Delaware
Newark, DE 19716 USA

## Abstract

This paper reports the transient temperature and stress distributions that develop within a thermally and elastically orthotropic medium with a rectangular boundary. The material initially held at a constant temperature is suddenly subjected to an arbitrary temperature variation along one of its edges. And then, the material is rapidly cooled from the steady-state thermal environment under the prescribed surface temperature down to room temperature. Based upon the solution of transient temperature field for heating and cooling situations, the thermal stress analysis is performed by using an elastic displacement-potential approach. Numerical results for the case simulating a varieties of unidirectional fiber composite have demonstrated high concentration of transient thermal stresses far exceeding those generated in a steady-state thermal environment. The authors have assessed the relative performance of more material system under thermal transient conditions. The techniques developed in Refs.(3) and (4) are adopted for the analysis. Numerical results show that the effects of such a orthotropy on thermal stresses are very sensitive.

## 1. INRTRODUCTION

In recent years the thermal behavior of anisotropic elastic materials has attracted considerable interest due to the increasing engineering usage in severe thermal environments. Fiber-reinforced composites are certainly well known examples of anisotropic materials. Fiber composite materials are anisotropic not only in their elastic properties but also in their thermal behavior. The thermal conductivities of a fiber-reinforced composite are usually quite different in its principal material axes. For composite GY70 Gr/Ep, the ratio of the conductivities in the transverse direction and the fiber direction is 0.0155, for example. Furthermore, the thermal expansion coefficients of the materials have the similar situation. To investigate the two-dimensional transient thermo-mechanical behavior of the materials is of considerable technological importance not only to dimensional stability but also to safety of structures. However, owing to the mathematical complexity in dealing with both the heat transfer and thermoelastic problems of anisotropic media, exact theoretical solutions have been obtained for only very limited classes of problems. Experimental work in this aera is very scarce.

More recently, Wang, Chou and Kulacki[1J], Wang and Chou[2J], Wang, Pipes and Chou[3J], Wang and Chou[4J], investigated the transient two-dimensional thermal behavior of a finit orthotropic slab with prescribed surface temperature change. The authors reported the general solutions for the heating and cooling problem,respectively.

The purpose of this paper is to study further the thermal orthotropic properties influence on transient temperature and stress distributions for more material system. The techniques developed in Refs. [3] and [4] are adopted for the analysis.

## 2. THEORY

The two-dimensional diffusion process taking place in a solid rectangular material(Figure 1) with prescribed thermal conditions can be represented by the Fick's law,as

$$\partial^2 T/\partial y^2 + K^2 \partial^2 T/\partial z^2 = \partial T/D\partial t$$

(1)

$$K^2 = K_z/K_y$$

in which $T(y,z;t)$ is the temperature; $K_z$ and

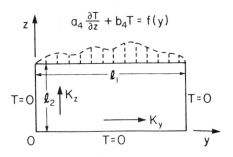

Fig. 1 An orthotropic elastic slab.

$K_y$ are the conductivities in the $z$ and $y$ directions, respectively; D is the thermal diffusivity in the y direction;$D,K_y$ and $K_z$ are assumed to be independent of temperature; t denotes time. In Figure 1,the thermal boundary conditions on edges $y=0,y=l_1$ and $z=0$ are for zero surface temperature or heat flow,whereas the nonhomogeneous boundary condition is for an arbitrary variation $f(y)$ of surface condition along the edge $z=l_2$. Temperature equation (1) can be solved for heating and cooling boundary conditions, respectively. Then,based upon the tempe rature solution, the induced stress can be solved by a displacement-potential approah. In the case of plane strain, the stress components in the yz plane are related to the in-plane displacements, $v(y,z;t)$ and $w(y,z;t)$, and temperature, $T(y,z;t)$ by substituting the strain-displacement relation into the stress-strain equation. Finally, the results are

$$\sigma_{yy} = A_{22}\partial v/\partial y + A_{23}\partial w/\partial z - \beta_2 T(y,z;t)$$

$$\sigma_{zz} = A_{23}\partial v/\partial y + A_{33}\partial w/\partial z - \beta_3 T(y,z;t)$$

$$\sigma_{yz} = A_{44}(\partial v/\partial z + \partial w/\partial y)$$

$$(2)$$

where $A_{ij}$ and $\beta_i$ are, respectively, the elastic stiffness coefficients and stress-temperature coefficients of the elastic medium.

The transient temperature and thermal stress distributions for the certain thermal and mechanical boundary conditions in heating and cooling cases can be found in Refs (3) and (4), respectively. Both the temperature and stress solutions are presented in series form.

## 3. NUMERICAL EXAMPLES AND DISCUSSION

To illustrate the foregoing analysis, the transient temperature and stress distributions of the slab in Figure 1 due to heating and cooling is considered as an example. The slab is made of a unidirectional fiber-reinforced composite material with fibers oriented parallel to the y-axis. The initial temperature of the slab is $T = 0$. Then the temperature raise at the upper edge($z = l_2$) is adopted as following

$$a_4 = b_4 = 0 \qquad (3)$$
$$T = f(y) = T_0 \sin(\pi/l_1)y$$

while the temperature over the remainder of the boundary is maintained at the initial value. All edges of the rectangle are taken to be traction-free. For the purpose of stress analysis, it is often reasonable to represent such a composite as a macroscopically homogeneous, orthotropic material. And the cooling problem can also be solved by using the similar techni-

ques.

In the following, we focus our attation on assessing the relative performance of a varieties of combinations of fiber and matrix materials under thermal transient conditions. The thermal properties of the materials adopted in numerical examples are in Table 1.

Table 1

| | K | $\beta_2/\beta_3$ |
|---|---|---|
| Gy70 Gr/Ep | 0.125 | 0.70 |
| Boron/Ep | 0.1 | 1.73 |
| Gr/Al | 0.995 | 2.5 |
| $Al_2O_3$/Mg | 1.56 | 2.86 |
| $Al_2O_3$/Al | 0.876 | 2.89 |
| Sic/Ti | 0.543 | 3.96 |

For showing the sensitivity of the stress to changes in the ratio K, and $\beta_2/\beta_3$, the elastic properties are adopted as the same as following

$$A_{22} = 208.92 Gpa$$
$$A_{33} = 27.86\ Gpa$$
$$A_{44} = 7.79\ Gpa$$
$$A_{23} = 26.06\ Gpa.$$

For the convenience of presenting the numerical results, the dimensionless quantities are introduced for temperature, stress, time and slab geometry.

Figures 2 and 3 show the z-direction variation of the longitudinal stress at cross-section, $y = l_1/2$ for a Boron/Ep material, and a Sic/Ti

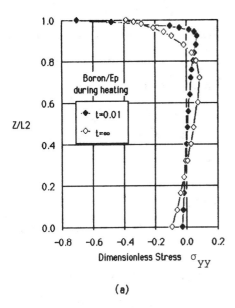

(a)

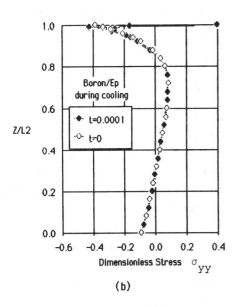

(b)

Fig. 2 Transient thermal stress distributions
for K=0.1, ß=1.73.

slab, respectively. In the heating case, the
maximum stresses occure at the time interval,
t=0.01. In the cooling case, t=0.0001. It can
be seen that large longitudinal stresses occur
in the vicinity of the heated boundary, where
the relative large temperature gradients,
$\partial T / \partial z$, exist. In Figure 2(a), the maximum
transient compressive stresses near the
upper edge is 82% higher than that generated
in steady-state(t=∞). In Figure 3(a), it is
1600% . The raising range is in 30%-1600%
for more material system in Table 1. It also
can be seen that large longitudinal stresses
occure in the vicinity of the cooled boundary.
In Figure 2(b), the maximum transient
tensile stress near the upper edge is 328%
higher than that generated in steady-state
(t=0). In Figure 3(b), it is 1800%. The
raising stress range is in 50%-1800% for
the materials in Table 1. The raising range
only shows the sensitivity of the stress to
changes in the relative thermal performance.
It seems clear that the transient compressive
stresses exhaust a significant portion of the
compressive strength of fiber composite
materials in situations involving severe
heating environments. On the orther hand,
the transient tensile stresses exhaust a
significant portion of the tensile strength
of fiber composite materials in situations

733

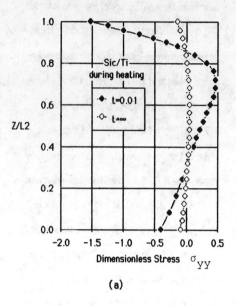

(a)

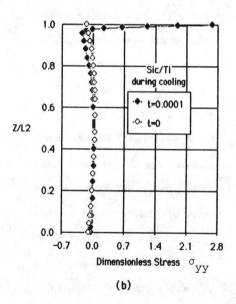

(b)

Fig. 3 Transient thermal stress distributions
for K=0.543, ß=3.96.

involving severe cooling environments.

## 4. CONCLUDING REMARKS

1) Many epoxy matrix or metal matrix fiber
composites in situations involving severe
thermal environments might have transient
thermal stress concentration far exceeding
those generated in a steady-state thermal
environment.

2) The parameters, K and $\beta_2/\beta_3$, characterizing the relative thermal and elastic anisotropic
properties, have significant effect on the temperature and induced thermal stress distributions.

3) Because of important implication of
transient thermal effects, an array of problem
involving, for example, boundary layer stress
concentrations, interfacial damage of laminates
under severe thermal conditions all deserve
in-depth assessment and examination in the
future.

## REFERENCES

1) H. Wang , T. W Chou, and F. A. Kulacki,
Proceedings of the 21st Annual Meeting of
SES, Blacksburg, Virginia, U. S. A.( 1984).
2) H. Wang and T. W Chou, J. of Comp. Mat.,
Vol.19 ,No. 5, pp.424-441 (Sept.1985).
3) H. Wang and T. W. Chou , AIAA Journal,
Vol. No. 4, pp.664-672 (April 1986).
4) H. Wang , R. B. Pipes, and T. W. Chou,
Metallurgical Transactions A, Vol.17A,
pp.1051-1055 (June 1986).

## BIOGRAPHY

Mr. H. wang received both undergraduate and

graduate education at the Northwestern Poly-
technical University, China in 1962 and 1966,
respectively. His speciality is in solid mechnics,
design and analysis, composite materials. Wang,
formerly Visiting Scholar at the Center for
Composite Materials, University of Delaware,
USA,during 1983-1985, is a Research Engineer,
Research and Development Division, Chengdu
Aircraft Company,P.O.Box 609,Chengdu, China.

Tsu-Wei Chou , Professor,Department of Mecha-
nical & Aerospace Engineering and Center for
Composite Materials, Materials Science and
Metallurgy Faculty, Ph.D. Stanford University,
is with The University of Delaware, Newark,
DE 19716. His speciality : Continuum mechanics,
Mechnical behavior of Polymeric and Metal metrix
Composites, crytal defect theory, fracture of
mechacs.

32nd International SAMPE Symposium
Apris 6-9, 1987

## TENSILE FRACTURE BEHAVIOR OF GRAPHITE COMPOSITE
## LAMINATES CONTAINING LARGE CENTER CRACKS

**S. J. Bradley,** Graduate Student
The Wichita State University

**J. G. Avery,** Chief of Advanced Programs
Boeing Military Airplane Company

**J. Chaudhuri,** Assistant Professor, Mech. Engineering
The Wichita State University

### Abstract

This paper presents an overview of a
program to investigate the static
tensile fracture of graphite
reinforced polymer laminates containing large through-the-thickness
cracks. The objective of the study
is to assess the applicability of
several available fracture models
for predicting residual strength of
laminates containing large cracks.

Much of the data used in developing
current fracture models were
obtained from testing panels containing cracks of one inch or
smaller. However, there are important applications requiring models
that are valid for larger cracks.
For example, in prior research Avery
and coworkers showed that the
fracture of composite panels containing large damages induced by
ballistic penetrators could be
correlated with the fracture of
panels containing mechanically induced cracks. An effective flaw
correlation was developed that
equates ballistic penetrator damages
to cracks for fracture analysis.

The current investigation will
provide new parametric test data
needed to help clarify the range of
applicability of residual strength
analysis methods for laminates with
large cracks. The test program
includes panels ranging in size from
4×20 inches to 16×32 inches. The
panels contain mechanically applied
through center cracks ranging from
one to four inches in length.
Smaller coupons were used to obtain
ultimate strength data. The investigation includes two polymer systems,
PEEK (polyetheretherketone from ICI)
and 3501 (epoxy from Hercules).
Hercules AS4 fibers were employed
using several layup configurations
in combinations of 0, +45, -45 and
90-deg plies. All panels were loaded
to failure in tension.

### Introduction

A question often raised in technical
circles over the past 15 years is:
Can linear elastic fracture mechanics (LEFM) be usefully applied to
laminated fiber composite materials?
Although there are many issues yet
to be resolved, it has been shown
that the fracture characteristics of
certain laminates containing high-
energy impact damage (HEID) can be
correlated using LEFM techniques
within an important range of
applications.

The objective of this paper is to
describe an in-progress investigation to assess the range of applicability of LEFM analysis methods for
fiber composites containing through-
the-thickness flaws. First, an
important application for fracture
analysis in composites will be discussed: the prediction of residual

strength following perforation by kinetic energy projectiles causing high energy impact damage (HEID). Summaries of several prediction approaches will be presented. Finally, a description will be given of a test program undertaken by the Boeing Company and The Wichita State University to validate LEFM models for composite laminates containing large damages.

## High-Energy Impact Damage (HEID) in Graphite Fiber-Reinforced Composites

High-energy impact damage (HEID) in composite laminates results from impacts by projectiles having sufficient mass and impact velocity to penetrate the laminate. Rocks, hailstones and projectiles generated by combat weapons can induce HEID. Figure 1 shows a schematic of the characteristics of HEID in graphite/ epoxy laminates. The impacted surface contains a relatively clean perforation in the shape of the penetrator surrounded by a small delaminated region. The exit surface exhibits more extensive delamination and, in the case of tape laminates, peeling of the surface plies. Extensive ultrasonic investigations (Ref. 3) have shown that virtually all of the induced damage can be detected by visual inspection of both surfaces of the laminate. There is only a small amount of internal delamination which is not indicated on the exit surface. This is in contrast to low energy impact damage (LEID) which can produce extensive internal delaminations extending well beyond visible damage indications.

In application, HEID can range in size from about one inch (a small rock or bullet) up to ten inches (high-explosive weapon damage). Therefore, the analysis methods used to predict residual strength of structure with HEID must address flaws in that size range. Considerable residual strength testing of ballistic damaged laminates (Refs. 1,2,3,4) indicated that the inverse square root relationship between damage size and fracture strength associated with LEFM provides good engineering correlations over an important range of laminate configurations. Figure 2 shows representative results for 0/+-45/90 and +-45 graphite/ epoxy containing high energy impact damage.

The straight line prediction shown in the figure reflects the well-known Waddoups, Eisenmann and Kaminski (Ref. 5) LEFM approach employing a characteristic dimension for an intense energy region assumed to exist at the crack tips, $a_0$. This dimension can be considered as a measure of an "inherent flaw". However, in the range of flaw sizes of interest the inherent flaw only has a significant effect for +-45 laminates, for which $a_0$ is about one-half inch. For laminate configurations with more than about 20 percent 0-degree plies, the inherent flaw is sufficiently small relative to HEID damage size that it is often incidental to the prediction.

Because of the experimentally indicated similarity between the fracture response of composites with HEID and with sharp-edged cracks, it has been useful to develop quantitative equivalencies between ballistic damage size and crack length as suggested above. This led to the development of effective flaw size models that compute flaw size as a function of impact conditions, as discussed in References 3,4 and shown schematically in Figure 3. The effective flaw analysis requires validated fracture prediction models for application to design. The following section discusses some of the candidate residual strength models which have appeared in the literature.

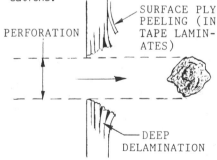

PERFORATION

SURFACE PLY PEELING (IN TAPE LAMINATES)

DEEP DELAMINATION

FIGURE 1. TYPICAL FEATURES OF HIGH-ENERGY IMPACT DAMAGE IN LAMINATED FIBER COMPOSITES

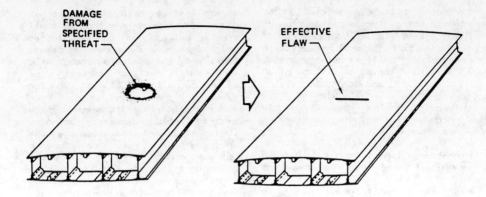

FIGURE 3. EFFECTIVE FLAW CONCEPT PROVIDES
A METHOD FOR ESTIMATING FRACTURE WITH HEID

## Residual Strength Prediction Methods for Laminated Composites

Testing by many researchers has shown that the notch sensitivity of graphite fiber reinforced epoxy laminates depends on the ply orientations. The capability to predict notch sensitivity in terms of such laminate parameters is desirable, since it would provide a means of tailoring laminate design to achieve toughness goals. Two approaches have been taken in developing residual strength predictions for laminates containing through-the-thickness flaws: extensions of linear elastic fracture mechanics (LEFM), and stress or "strength of materials" approaches stemming from the work of Whitney and Nuismer (Ref. 13). This paper will concentrate on LEFM approaches.

Before presenting selected residual strength prediction models, it is important to present some background on the failure modes seen in graphite/epoxy laminates containing through-the-thickness flaws. Figure 4 shows four primary failure modes identified by Bradley and Avery (Ref. 3). The first two are simple crack propagation failure modes. Crack propagation is transverse in laminates having balanced percentages of 0, +-45 and 90-deg plies or having a high percent of 90-deg plies. Propagation is diagonal for predominantly +-45-deg laminates.

The third and fourth failure modes occur with increasing percentages of 0-degree plies. The third failure mode is a combination of transverse crack growth and longitudinal tearing. The fourth failure mode represents a net-area (or nearly net area) response in which the matrix spits longitudinally on either side of the damage, then the panel fails at the ultimate strain of the uncut fibers. Laminates which display splitting failures do not lend themselves to currently available LEFM-type residual strength prediction models, since these models do not predict splitting nor post-splitting residual strength. The models which follow are therefore assumed to apply only for laminates with the first two failure modes.

As mentioned in the previous section, one of the earliest extensions of LEFM methods to fracture of fiber composites was presented by Waddoups, Eisenmann and Kaminski in 1971 (Ref. 5). WEK proposed that fracture test data could be correlated using an equation of the form:

$$K_Q = \sigma_c \sqrt{\Pi (a + a_o)}$$

$$K_Q = \sigma_u \sqrt{\Pi a_o}$$

738

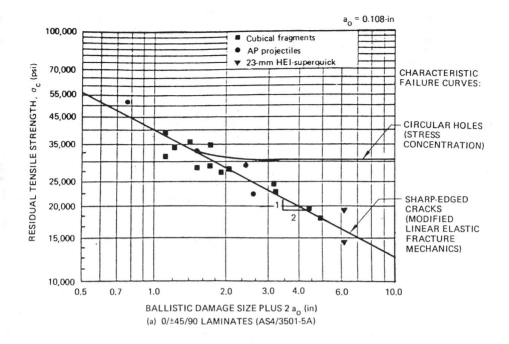

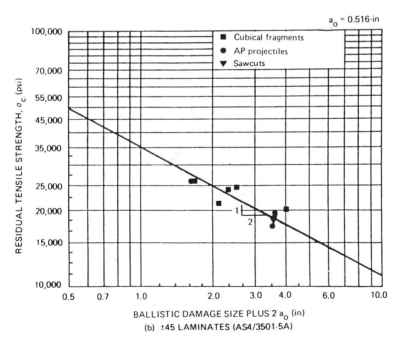

FIGURE 2. VARIATION OF RESIDUAL TENSILE STRENGTH
WITH BALLISTIC DAMAGE SIZE IN GRAPHITE/EPOXY[2]

739

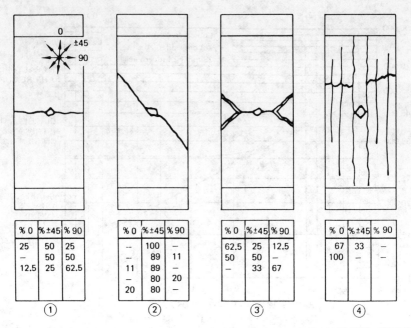

FIGURE 4. EFFECT OF PLY ORIENTATION ON TENSILE FAILURE MODES

**①**

| % 0 | %±45 | % 90 |
|-----|------|------|
| 25 | 50 | 25 |
| — | 50 | 50 |
| 12.5 | 25 | 62.5 |

**②**

| % 0 | %±45 | % 90 |
|-----|------|------|
| -- | 100 | — |
| -- | 89 | 11 |
| 11 | 89 | — |
| — | 80 | 20 |
| 20 | 80 | — |

**③**

| % 0 | %±45 | % 90 |
|-----|------|------|
| 62.5 | 25 | 12.5 |
| 50 | 50 | — |
| — | 33 | 67 |

**④**

| % 0 | %±45 | % 90 |
|-----|------|------|
| 67 | 33 | — |
| 100 | — | — |

where:

$K_Q$ = Mode I critical stress intensity factor;

$a$ = one-half the crack length;

$a_o$ = one-half the inherent flaw length;

$\sigma_c$ = fracture stress of laminate with crack;

$\sigma_u$ = ultimate undamaged laminate strength

In principal, this model requires only two tests for each laminate considered, an ultimate strength test and a fracture test, to determine $K_Q$ and $a_o$. However, it does not have the capability to define fracture toughness in terms of laminate properties.

Avery, Porter and Bradley (Refs. 6,7) proposed a correlation between fracture toughness and laminate ultimate tensile stress, and provided a "rule-of-thumb" discrete correlation between fracture toughness and laminate elastic properties. They proposed that:

$$K_Q = C_N \sigma_u$$

where:

$C_N$ = an empirical constant related to the ply orientations in the laminate.

The corresponding inherent flaw size can be found by relating $C_N$ to the WEK model:

$$\frac{K_Q}{\sigma_u} = C_N = \sqrt{\Pi a_o}$$

The model can be summarized by:

$$\sigma_c = \frac{C_N \sigma_u}{\sqrt{\Pi(a+a_o)}}$$

Based on considerable data using ballistic flaws, reasonable correlation for AS4/3501 graphite/epoxy was achieved using the following average values:

| | 0-Dominated Laminates | +-45 Dominated Laminates |
|-----|------|------|
| $C_N$ | 0.6 | 1.2 |
| $a_o$ | 0.115 | 0.46 |

0-deg dominated laminates are those which contain more than 20- and less than 50-percent 0-deg plies.

+-45-deg dominated laminates contain 20-percent or less 0-deg plies and a high percentage of +-45-deg plies.

Poe (Ref. 8) significantly extended this approach by introducing a general fracture toughness parameter which depends only on the fiber ultimate tensile strain. This parameter can be used along with laminate elastic properties to predict $K_Q$ of a laminate. His model incorporates the laminate parameters needed for laminate tailoring in design applications. The Poe model is given by:

$$K_Q = Q_c \frac{Ey}{\xi}$$

$$Q_c = 1.5\varepsilon_{tu}\sqrt{mm} = 0.3\varepsilon_{tu}\sqrt{in}$$

$$\xi = \left[1-\mu_{yx}\sqrt{\frac{Ex}{Ey}}\right]\left[\sqrt{\frac{Ey}{Ex}}\sin^2\alpha + \cos^2\alpha\right]$$

where:

$Q_c$ = a general fracture toughness parameter, independent
of ply orientations;

$Ey, Ex$ = laminate moduli in the
direction of and transverse
to the loading direction;

$\mu_{yx}$ = major Poisson's ratio;

$\alpha$ = direction of principal load
carrying fiber relative
to loading direction.

The final model that will be discussed is that presented by Mar and Lin (References 9 and 10). They proposed a new approach whereby fracture test data for fiber-reinforced laminates could be correlated using:

$$\sigma_c = H_c(2a)^{-m}$$

where:

$H_c$ = composite fracture toughness;

$m$ = the order of the singularity
of the crack tip.

The composite fracture toughness $H_c$ is a function of material and laminate layup. In Reference 9 Mar and Lin proposed that the value of $m$ is 0.26 for graphite/epoxy and 0.33 for boron/aluminum. The value of $m$ is related to the shear moduli and Poisson's ratios of the fiber and resin, and is the order of the singularity of a crack with its tip at the interface of the two materials.

Experimental Test Program

The models discussed above, and other important models available in the literature, were developed semi-empirically from fracture test data. However, with a few exceptions (Refs. 1,3,6,8), most of the testing to develop and validate the models used small specimens containing flaws of one inch or less in length. Since high energy impact damage can range up to ten inches, it is important that the models be validated for large cracks. Therefore Boeing initiated a program to perform fracture testing on composite panels with large mechanically applied central cracks. The results will be compared with fracture prediction models. This program is jointly funded by Boeing Military Airplane Company and the Kansas Advanced Technology Commission and is underway at The Wichita State University.

Table 1 summarizes the 56 tests included in the program. Panels of AS4/PEEK and AS4/3501 were fabricated by Boeing in layup configurations employing combinations of 0, +45, -45 and 90-degree plies. PEEK (polyetheretherketone) is a thermoplastic polymer matrix material which Imperial Chemical Industries (ICI) produces in a tape prepreg form with Hercules AS4 fiber under the trade name APC-2. PEEK is of interest for aircraft applications because of its improved impact damage tolerance relative to conventional epoxy. Hercules AS4/ 3501 graphite/epoxy provides a baseline for comparison of the graphite/thermoplastic with the same fiber. The 0, 45, -45 and 90 ply orientations are those most commonly found in aircraft structure designs, and the selected layup configurations simulate those found in actual designs.

The cracked panels have a crack size to panel width ratio of 1:4. The 1:4 ratio produces only about a four percent increase in the crack tip stress intensity compared to an infinitely wide panel (width correction method from Reference 11). Panel free-length to width ratios varied from 1.3:1 to 2.1:1.

741

## TABLE 1. TEST MATRIX FOR COMPOSITE FRACTURE STUDY

| PANEL GROUP | MATERIAL | PLY ORIENTATIONS | PANEL SIZE Width × Length* mm (in) | | CRACK LENGTH mm (in) | REPLI-CATIONS |
|---|---|---|---|---|---|---|
| A | AS4/PEEK | $(\pm 45)_{4s}$ | 406×560 | (16×22) | 102 (4) | 1 |
| | | | 305×508 | (12×20) | 76 (3) | 2 |
| | | | 203×432 | (8×17) | 51 (2) | 1 |
| | | | 102×178 | (4×7) | 25 (1) | 2 |
| | | | 25×203 | (1×6) | None | 2 |
| B | AS4/PEEK | $(0/45/0/-45/0_2/90/0)_s$ | 406X560 | (16×22) | 102 (4) | 1 |
| | | | 305×508 | (12×20) | 76 (3) | 2 |
| | | | 203×432 | (8×17) | 51 (2) | 1 |
| | | | 102×178 | (4×7) | 25 (1) | 2 |
| | | | 25×203 | (1×6) | None | 2 |
| C | AS4/PEEK | $(+45/-45/90/0)_{2s}$ | 406X560 | (16×22) | 102 (4) | 1 |
| | | | 305×508 | (12×20) | 76 (3) | 2 |
| | | | 203×432 | (8×17) | 51 (2) | 1 |
| | | | 102×178 | (4×7) | 25 (1) | 2 |
| | | | 25×203 | (1×6) | None | 2 |
| D | AS4/PEEK | $(\pm 45/90/\pm 45/0/\pm 45)_s$ | 406×560 | (16×22) | 102 (4) | 1 |
| | | | 305×508 | (12×20) | 76 (3) | 2 |
| | | | 203×432 | (8×17) | 51 (2) | 1 |
| | | | 102×178 | (4×7) | 25 (1) | 2 |
| | | | 25×203 | (1×6) | None | 2 |
| E | AS4/PEEK | $(\pm 45_2/90/\pm 45_2)_s$ | 406×560 | (16×22) | 102 (4) | 1 |
| | | | 305×508 | (12×20) | 76 (3) | 2 |
| | | | 203×432 | (8×17) | 51 (2) | 1 |
| | | | 102×178 | (4×7) | 25 (1) | 2 |
| | | | 25×203 | (1×6) | None | 2 |
| F | AS4/3501 | $(\pm 45/90/\pm 45/0/\pm 45)_s$ | 406×560 | (16×22) | 102 (4) | 1 |
| | | | 305×508 | (12×20) | 76 (3) | 2 |
| | | | 203×432 | (8×17) | 51 (2) | 1 |
| | | | 102×178 | (4×7) | 25 (1) | 2 |
| | | | 25×203 | (1×6) | None | 2 |
| G | AS4/3501 | $(0/45/0/-45/0_2/90/0)_s$ | 406×560 | (12×20) | 102 (4) | 1 |
| | | | 305×508 | (12×20) | 76 (3) | 2 |
| | | | 203×432 | (8×17) | 51 (2) | 1 |
| | | | 102×178 | (4×7) | 25 (1) | 2 |
| | | | 25×203 | (1×6) | None | 2 |

Total Tests: 56

* Panel length is the free, unsupported portion of the panel
not including the length within the friction grips.

For the largest panels it was not possible to increase the length to width ratio due to specimen size limitations in the 60-kip Baldwin Test Machine used for the tests. A strain survey was done using strain gage rosettes placed across the centerlines of undamaged large panels to calibrate the influence of grip proximity on the panel stress distribution.

It is appropriate here to make a special note about the sizing of the panels to obtain a valid fracture test. Although width correction factors are widely used in published test results, the influence of the absolute value of crack size is sometimes neglected. This can lead to erroneous conclusions about the fracture toughness of a particular laminate, especially if the laminate is compliant, as in the case of the +-45 layup. As shown in Figure 5, adapted from Fedderson (Ref. 12), if a fracture test results in a failure stress greater than 2/3 of the undamaged laminate strength, the crack size was too small to yield a true indication of $K_Q$.

Figure 6 shows a schematic of the mechanically applied cracks. The cracks were created using a three-stage process. An initial cut was made to a length 10-mm shorter than the final crack length with a 1-mm diameter graphite cutting wire in a numerically controlled graphite cutting machine. Then the crack was extended 5-mm at each end with a 0.254-mm thick saw blade. Finally, the end of the crack was "knicked" with a specially prepared 0.125-mm blade.

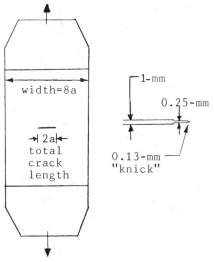

FIGURE 6. SCHEMATIC OF MECHANICALLY APPLIED THROUGH CENTER CRACK

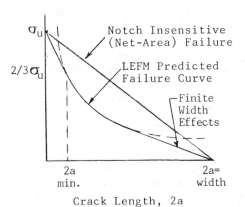

Crack Length, 2a

FIGURE 5. CRACK SIZING CRITERIA FOR A VALID FRACTURE TOUGHNESS TEST (per Fedderson[12])

Figure 7 shows a panel installed in the Baldwin test machine. Special features of the test setup include the friction gripping system, where high tensile strength bolts provide the clamping force for holding the panel in friction between two steel grip plates, plexiglass anti-buckling plates to prevent out-of-plane buckling at the crack, a crack-opening-displacement (C.O.D.) gage mounted on steel holders glued to the panel, and an extensometer to

743

monitor overall panel deformation. The friction grips are pin-connected through a clevice to a loading rod which is seated in the Baldwin head by a spherical nut for self-alignment. A load washer placed between the upper load train provided load signal input to X-Y plotters monitoring panel extension and crack opening displacement. The entire load train and gripping system was designed and fabricated especially for these tests, and the friction grips have been validated to the 60-kip capacity of the test machine without slipping or bearing failure at the bolts.

FIGURE 7. TEST ARRANGEMENT FOR COMPOSITE
FRACTURE STUDY

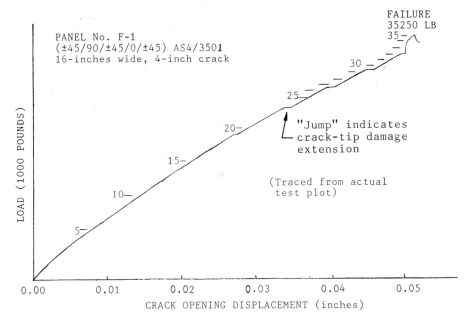

FIGURE 8. SAMPLE LOAD-C.O.D. PLOT FROM COMPOSITE FRACTURE STUDY

Figure 8 shows an example of a load-C.O.D. plot. The jumps in the C.O.D. axis correspond to intermediate failures and crack tip damage extension. Poe et al (References 4,8) have examined these intermediate failures using radiographic techniques and analyzed the effects. Figure 9 shows some of these failure modes which serve to indicate that crack tip damage extension is more complex than implied by the WEK inherent flaw model. Although radiographic techniques are not being used in the current program, the C.O.D. plots provide substantial information on crack-tip damage extension prior to final catastrophic panel failure.

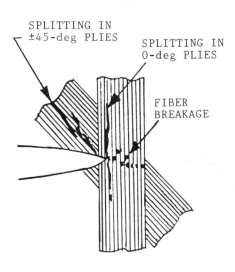

FIGURE 9. CRACK TIP DAMAGE MECHANISMS

Figures 10 through 13 show overall panel failure modes for four types of the panels. Figure 10 shows a +-45 graphite/thermoplastic panel which failed along 45-degree planes. In Figure 11, a quasi-isotropic graphite/thermoplastic panel shows a very irregular fracture surface. Figure 12 shows a graphite/epoxy panel with 75 percent +-45 plies and 12.5 percent 0-degree plies. These panels also fail along 45-degree failure planes. Finally, a highly directional (62.5 percent 0-deg plies) panel is shown in Figure 13. These panels split vertically at the crack, then failed in net section in the grips.

FIGURE 11. TENSILE FRACTURE MODE, $(\pm 45/90/0)_{2s}$ AS4/PEEK PANEL

FIGURE 10. TENSILE FRACTURE MODE FOR $(\pm 45)_{4s}$ AS4/PEEK PANEL

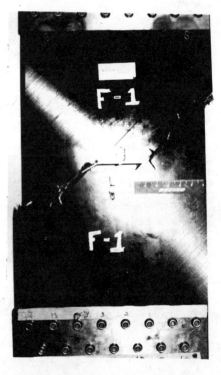

FIGURE 12. TENSILE FRACTURE MODE, $(\pm 45/90/\pm 45/0/\pm 45)_s$ AS4/3501

FIGURE 13. TENSILE FRACTURE MODE
$(0/45/0/-45/0_2/90/0)_s$ AS4/3501

## Summary and Conclusions

Testing was still in progress at the time this paper was written, and it was felt that inclusion of quantitative test data and analysis results would be premature. However, this paper has shown that there is an important application for LEFM techniques in composite structure which provides the rationale for testing of composite panels with large through cracks. An overview of some of the relevant analysis methods for fracture of composite laminates was presented, as well as a description of the test program. Quantitative results and conclusions will be presented at the SAMPE symposium in April, 1987 in Anaheim California, and in future published papers.

## References

1. J.G. Avery, S.J. Bradley, K.M. King, "Battle Damage Tolerant Wing Structural Development Program", Final Report, NASC Contract N00019-75-C-0178, 1979.

2. J.G. Avery, S.J. Bradley, R.J. Bristow, "Survivable Composite Structure in Combat Aircraft", AFFDL-TR-79-3132, 1979.

3. S.J.Bradley, J.G.Avery, "Survivable Composite Structure Test and Analysis Report", AFWAL-TR-84-3014, 1984.

4. S.J.Bradley, J.G.Avery, "Design Guide for Survivable Structure in Combat Aircraft", AFWAL-TR-84-3015, 1984.

5. M. Waddoups, J. Eisenmann, B. Kaminski, "Macroscopic Fracture Mechanics of Advanced Composite Materials", J.Composite Materials, Vol.5, 1971.

6. J.G.Avery, T.R.Porter, "Comparisons of the Ballistic Impact Response of Metals and Composites", ASTM STP 568, 1975.

7. J.G.Avery, S.J.Bradley, "Applications of Fracture Mechanics Analysis to Fiber Composite Structure", S.A.E. AeroTech '86, Long Beach CA, 1986.

8. C.C.Poe,Jr.,"A Unifying Strain Criterion for Fracture of Fibrous Composite Laminates", Engineering Fracture Mechanics, Vol.17, No.2, pp153-171, 1983.

9. J.W.Mar, K.Y.Lin, "Fracture Mechanics Correlation for Tensile Failure of Filamentary Composites with Holes", J.Aircraft, Vol.14, No.7, pp703-704,1977.

10. J.W.Mar, K.Y.Lin, "Fracture of Boron/Aluminum Com- posites with Discontinuities", J.Composite Materials, October 1977.

11. O.L.Brown, J.E.Srawley, "Plane Strain Crack Toughness Testing of High Strength Metallic Materials", ASTM STP 410, 1966.

12. C.E.Fedderson, "Evaluation and Prediction of Residual Strength of Center Cracked Tension Panels",ASTM STP 486, 1971.

13. J. Whitney, R. Nuismer, "Stress Fracture Criteria for Laminated Composites Containing Stress Concentrations", J. Composite Materials, Vol.8, 1974.

## Author's Biographies

**Susan J. Bradley** is currently doing graduate work in structures and materials at Wichita State University. Prior to her educational leave, she was a research analyst in structures technology for the Boeing Military Airplane Company. She served as principal investigator on several key Air Force and Navy contractual programs that have provided industry-wide guidelines for enhancing the damage resistance and tolerance of advanced fiber composite structure. These results include predictive models for high-energy impact damage size in composite laminates and the associated degradation of tensile and compressive strength. Much of this work is documented in AFWAL-TR-84-3015 "Design Guide for Survivable Structures in Combat Aircraft".

**John G. Avery** directs advanced structural development within the structures technology staff of Boeing Military Airplane Company, Wichita, Kansas. Development responsibilities include advanced fiber composite structural applications, advanced concepts, analysis methods and design allowables. Mr. Avery has over thirty professional publications in the fields of damage tolerance and survivability of fiber composite aircraft structures. He served as coordinator of the AGARD Structures and Materials working group on Impact Damage Tolerance of Structure.

**Jharna Chaudhuri** is an assistant professor of mechanical engineering at Wichita State University. She has been involved in research for the Navy, Air Force and National Science Foundation (NSF) in the areas of structure-property relationships in metal alloys, and advanced composite and electronic materials. Dr. Chaudhuri has held summer faculty research appointments at the Naval Research Center, Washington D.C. and the AFOSR Aeronautical Branch, Ohio. She is currently doing research in composite materials under funding from the Institute of Aviation Research and Development and in electronic materials under NSF in addition to the Boeing supported program described in this paper.

SPACECRAFT SOLAR ARRAY SUBSTRATE DEVELOPMENT

Thu P. Stankunas, W. I. Greenway, G. R. Holmquist
Lockheed Missiles & Space Company
Sunnyvale, California
(408) 743-1561

## ABSTRACT

A materials and processes development effort was performed to support the design, fabrication and
evaluation of solar array substrate
composite structures for a spacecraft application.  For stiffness,
weight and other design requirements the basic sandwich structure
incorporated a high modulus graphite fiber/epoxy laminate (P-75/
934) facesheet (+60, -60, 0, 0,
-60, +60 orientation) with Nomex
honeycomb core (at 1.5 and 4.0
lbs/cu. ft.).  A film adhesive FM-
73M was utilized for facesheet/core
bonding and as a dielectric protective layer on the surface of the
substrate structure.  Process
studies determined cure cycles,
fabrication techniques, bonding/
assembly procedures and other
related processing to achieve the
substrate properties needed to
satisfy design requirements.  Test
coupons were fabricated and evaluated for mechanical properties,
bonding/structural integrity,
dielectric characteristics, thermal
expansion concerns, influence of
long term thermal cycling in simulated space environments, and other
property characterization.  The
effort provided relevant data with
respect to solar array substrate
configuration, processing parameters and material behavior in
long term cyclic thermal environment.

## 1.0  INTRODUCTION

Solar arrays consisting of photo-
voltaic solar cells converting
sunlight to electrical energy are
essential for the successful operational performance of many spacecraft missions.  The solar array
configuration includes a substrate
on which solar cells are supported
or bonded.  Depending on each
specific design application, the
substrate can either be flexible or
rigid, vary in configuration and
utilize materials selected for the
specific design intent.  Materials
and processing effort was recently
performed to support the design,
fabrication and evaluation of a developmental solar array substrate
for a specific spacecraft mission.
The design required a rigid substrate with high stiffness, stability in a long term space environment, lightweight, dielectric
protection, off-the-shelf materials
and other related characteristics.
The selection of the substrate
configuration and materials, the
fabrication process study and the
evaluation of the substrate in
simulated space thermal environments are presented.

## 2.0  SUBSTRATE CONFIGURATION &
##       MATERIAL SELECTION

For the specific intended application, the design effort resulted in

a 1.5m x 1.8m x 3.84 cm (5 ft. x 6 ft. x 1.5 inch) sandwich honeycomb structure configuration. A cross-section of the structure selected as the baseline for the development solar array substrate panel is presented in Figure 1. The panel design is unique in that solar cells are to be mounted on both sides of the substrate and both 0.024 and 0.064 g/cc (1.5 and 4.0 lbs/cu. ft.) honeycomb are utilized for minimum weight considerations.

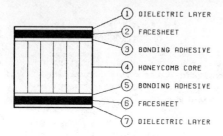

① DIELECTRIC LAYER
② FACESHEET
③ BONDING ADHESIVE
④ HONEYCOMB CORE
⑤ BONDING ADHESIVE
⑥ FACESHEET
⑦ DIELECTRIC LAYER

FIG. 1  SUBSTRATE CONFIGURATION

① Dielectric Layer, (American Cyanamid FM 73M, 147 gm/sq. m or 0.030 lbs/sq. ft. epoxy film). Required to electrically isolate the solar cell system from the facesheets and to provide a bonding surface for the cells.

② Facesheet (very high modulus graphite fiber P-75 with epoxy resin 934 - Fiberite HyE 2034D). Required to provide a high stiffness to weight ratio to withstand bending loads during deployment; fiber orientation of (+60, -60, 0)$_s$ selected to provide a "pseudo-isotropic" configuration.

③ Bonding Adhesive (American Cyanamid FM 73M, 147 gm/sq. m or 0.030 lbs/sq. ft. epoxy film adhesive). Required to bond core to facesheets with adequate strength to withstand panel bending stresses without failure over the operating temperature environments.

④ Honeycomb Core (Hexcel Nomex HRH-10-1/4-1.5 and HRH-10-3/16-4.0); with lightweight, 0.024 and 0.064 g/cc (1.5 and 4.0 lbs/cu. ft) and reduced thermal and electrical conductivity; the heavier core (approx. 30% of the area) used in areas supporting loaded hinges, and other inserts.

3.0  MATERIAL PROPERTIES

During the developmental effort, mechanical properties, thermal expansion, low temperature characteristics and other design information were generated for the graphite/epoxy facesheet and the adhesive bonding materials. The low temperature properties were of design concern since under certain operational modes, loads were to be imposed upon the structure at the lower temperatures.

a.  Facesheet Material:

The unidirectional basic properties of the P75/934 unidirectional and (+60, -60, 0)$_s$ laminates are presented in Table I.

b.  Adhesive Bonding Material:

The FM 73M adhesive material was evaluated for tensile, compressive and shear properties at various temperatures in the range of -90°C to +121°C (-130°F to +250°F). The results indicated that at -90°C (-130°F) the FM-73M has increased strength and modulus but decreased in ultimate tensile elongation from 5.7% to 3.8%. The results also showed approximately 50% reduction of the coefficient of thermal expansion (CTE) from 84.1 to 43.6 $\mu$m/m °C. The reduction in elongation and CTE at the low temperature end (-90°C) was suggested to contribute to the surface cracking of the substrate presented later on. In light of the concern for the thermal mismatch between the composite facesheet and the FM 73M on top of it, a test was conducted to determine the thermal expansion ($\Delta \ell/\ell$) of FM 73M and the facesheet matrix

TABLE I   P75/934 LAMINATE PROPERTIES

| PROPERTIES | UNIDIRECTIONAL LAMINATES 182°C (360°F) CURED | | $(+60.,-60.,0)_s$ LAMINATE | |
|---|---|---|---|---|
| | FIBER VOL.(%) | AVERAGE VALUE | 143°C (290°F) CURED | 183°C (360°F) CURED |
| LONGITUDINAL TENSION<br>STRENGTH. MPA (KSI)<br>MODULUS. GPA (MSI)<br>ULT. STRAIN. $\mu$M/M ($\mu$IN/IN)<br>POISSON RATIO | 61.7<br>↓ | 964   (139.8)<br>279   (40.4)<br>3170   (3170)<br>.032 | 355 (51.5) | 362 (52.5) |
| LONGITUDINAL COMPRESSION<br>STRENGTH. MPA (KSI)<br>MODULUS. GPA (MSI)<br>ULT. STRAIN. $\mu$M/M ($\mu$IN/IN) | 58.7<br>↓ | 417   (60.5)<br>232   (33.6)<br>2420   (2420) | | |
| LONGITUDINAL CTE. $\mu$M/M °C<br>($\mu$IN/IN °F) | | −0.97 (−0.54)<br>(+10→+93°C (50→+ 200F)) | | |

resin Fiberite 934. The results obtained by thermo-mechanical analysis (TMA) are presented below in Table II.

TABLE II   THERMAL EXPANSION. $\Delta l /l$, M/M (IN/IN)

| | TEMPERATURE RANGE | |
|---|---|---|
| | −62°C → −90°C<br>(−80°F → −130°F) | −51°C → +54°C<br>(−60°F → +130°F) |
| FM-73M ADHESIVE | $1.200 \times 10^{-3}$ | $8.93 \times 10^{-3}$ |
| FIBERITE 934 RESIN | $0.500 → 1.000 \times 10^{-3}$ | $5.10 \times 10^{-3}$ |

## 4.0   PROCESS INVESTIGATION

Process investigations were performed for the fabrication of the facesheet and the sandwich assembly. The selected processes were described as follows.

### a.   Facesheet Fabrication:

The graphite/epoxy P-75/934 tape laminate consisting of 6 layers at +60, -60, 0, 0, -60, +60 were laid up in the sequence described in Figure 2. The layup was cured under 100 psi pressure at 182°C (360°F) for 2 hours. The porous release fabric (TX 1040) was removed from the facesheet just prior to bonding to the core. The fabric also resulted in a proper impression on the surface to accommodate bonding of the solar cells. The seven (7) facesheet laminates fabricated for the development substrate coupons resulted in:   thickness of 0.76 − 0.81 mm (0.030 − 0.032"); resin content of 30.2 − 33.6 wt. %; and void volume of 0.0 − 1.4%.

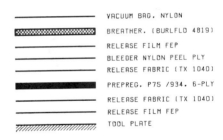

VACUUM BAG. NYLON
BREATHER. (BURLFLD 4819)
RELEASE FILM FEP
BLEEDER NYLON PEEL PLY
RELEASE FABRIC (TX 1040)
PREPREG. P75 /934. 6-PLY
RELEASE FABRIC (TX 1040)
RELEASE FILM FEP
TOOL PLATE

FIG. 2   FACESHEET PROCESS LAY-UP

Processing improvements were learned as the fabrication progressed. For example, when the layup was begun with the +60° ply, gaps occurred between the adjacent 12 inch tape pieces because they tended to shift. The problem was resolved by starting the layup with the two 0° plies and then adding the 60° plies to each side. This provided a surface for tacking the 60° ply to prevent movement. A second problem is that a partial layup tended to wrinkle when left overnight. This was resolved by vacuum bagging the uncompleted layup overnight.

b. Sandwich Assembly:

The sandwich substrate incorporated the graphite/epoxy facesheets with the Nomex honeycomb core. The core was slotted to provide a venting path for the panel (nodal notch in one end and non-nodal notch in the opposite end). The assembly layup for the cure is shown in Figure 3. The assembly was cured under vacuum in an oven at 121°C (250°F) for 1 hour. The vacuum pressure cure was selected to prevent damage to the light-weight honeycomb structure.

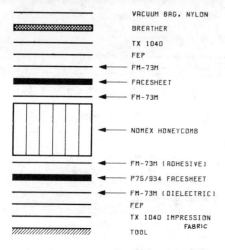

| | VACUUM BAG, NYLON |
| | BREATHER |
| | TX 1040 |
| | FEP |
| | FM-73M |
| | FACESHEET |
| | FM-73M |
| | NOMEX HONEYCOMB |
| | FM-73M (ADHESIVE) |
| | P75/934 FACESHEET |
| | FM-73M (DIELECTRIC) |
| | FEP |
| | TX 1040 IMPRESSION FABRIC |
| | TOOL |

FIG. 3  SANDWICH LAY-UP SEQUENCE

c. Mechanical Properties:

Mechanical properties were determined on specimens prepared from the cured sandwich development substrate panels. For design in-formation and comparative purposes, test panels were processed from both slotted and unslotted honeycomb core. Results of the tests are presented in Table III.

Test results showed expected values. There is no significant difference between the slotted and unslotted honeycomb. For flatwise tensile test the lower density honeycomb specimens failed in the core whereas the higher density core specimens failed in the facesheets, at the interface between +60 and -60 layers.

d. Dielectric Testing:

In order to prevent electrical shorting of the solar cells, the dielectric substrate surface must meet a specific requirement of breakdown strength. Dielectric testing was performed on the dielectric layers of the substrate panels. Breakdown strength of the dielectric surface was determined by applying a test voltage across electrodes separated by the epoxy layer. During the test the negative electrode was connected to the graphite layer, the positive was attached to a movable probe with a wet sponge sweeping across the substrate surface. Areas of low dielectric strength would result in a spark penetrating the dielectric and by the decrease in resistance. The fabricated substrate panels readily passed 500 volt dielectric test with only few sparks at the locations where the FM-73M was slightly thinner or punctured by the defects on the tool surface.

TABLE III  MECHANICAL PROPERTIES OF SUBSTRATE

| PROPERTIES (SPECIMEN)/METHOD | .024 G/CC (1.5 PCF) CORE | | .064 G/CC (4.0 PCF) CORE | |
|---|---|---|---|---|
| | UNSLOTTED | SLOTTED | UNSLOTTED | SLOTTED |
| FACING COMPRESSION STRESS, MPA (PSI) (5.1 CM X 55.9 CM (2.0" X 22.0") ASTM-C393) | 50.1 (7261) | 49.2 (7130) | 136.5 (19790) | 135.3 (19620) |
| CORE SHEAR STRESS, MPA (PSI) (7.6 CM X 20.3 CM (3.0" X 8.0") ASTM-393) | .216 (31.3) | .201 (29.2) | .892 (129.4) | .845 (122.5) |
| FLATWISE COMPRESSION STRENGTH, MPA (PSI) (5.1 CM X 5.1 CM (2.0" X 2.0") ASTM-C365) | .575 (83.4) | .475 (68.9) | 3.27 (474) | 3.46 (502) |
| FLATWISE TENSILE STRENGTH, MPA (PSI) (5.1 CM X 5.1 CM (2.0" X 2.0") ASTM-C397) | .924 (134) | .827 (120) | 1.400 (203) | 1.82 (264) |

752

## 5.0 THERMAL CYCLING TEST

The solar array panel substrate is subjected to various thermal environments both on-earth during processing and on-orbit in space shadow/sun resulting in low/high temperature cyclic conditions. The differences in the coefficient of thermal expansion and the fatigue capabilities of the constituent material systems used need to be evaluated when exposed to the cyclic stresses imposed by the thermal environment. For evaluation of the developmental substrate concepts, various configurations were exposed to particular thermal cyclic environments that may be encountered by the substrate both during ground preparation and in space. During the developmental phase, three (3) sets of specimens were subjected to three (3) thermal cyclic exposures. These were:

- Long Term Cycling: A 17,000 cycle exposure of baseline sandwich substrate panels with attached solar cells.

- Short Term Cycling: A 138 cycle exposure of baseline

sandwich panels and facesheet laminates with and without the dielectric layer.

- Alternate Facesheet Cycling: Alternate facesheet material systems exposed to modified Short Term Cycling.


a. Long Term Cycling:

The sandwich substrate with solar cells bonded to both surfaces were evaluated under the thermal cyclic environment. The thermal exposure consisted of an initial bake-out at 250°F for 24 hours to remove outgassing materials (moisture, etc) that may be in the substrate, 15 cycles of -80°C to +110°C (-112°F to +230°F), 10000 cycles of -57°C to +38°C (-70°F to +100°F) and 7000 cycles of 0°C to 35°C (+32°F to +95°F). A profile of the cycling is shown in Figure 4. This exposure represented a particular thermal environment for anticipated ground and space conditions.

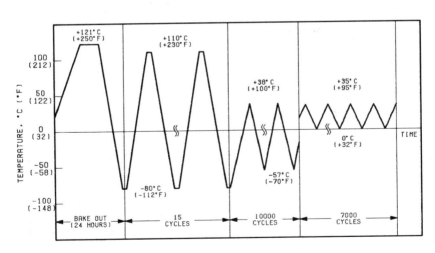

FIG. 4 LONG TERM THERMAL EXPOSURE
(HEAT RATE ≤ 10°C/MIN (18°F/MIN), DWELT = 5 MINUTES)

Visual and photographic examinations were performed at a predetermined numbers of cycles. The first observation at 2000 cycles of -57°C to +38°C (-70°F to +100°F) revealed micro-cracking of the substrate surface around the bonded solar cells. The cracking ran parallel to the fiber direction. Specimens were removed for additional photographic documentation at 4000, 6000, 10000 cycles. The frequency and dimensional size of the cracking appeared to increase with the increasing cycles. However, no further change was observed for the additional 7000 cycles of 0°C to +35°C (+32 to +95°F).

At the completion of the 17,000 cycles, the panel was sectioned for photographic documentation. Scanning electron micrograph (SEM) showed surface cracking along the fiber (Figure 5). No graphite fiber protruding from the substrate was observed.

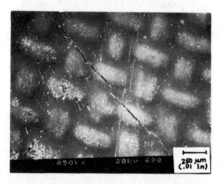

Fig. 5    Substrate Surface Crack After 10000 Cycles -57°C→+38°C (-70°F to +100°F) & 7000 Cycles 0°C→+35°C (+32°F to +95°F)

The microstructure of the substrate surface cross-section is presented in Figure 6. The dielectric layer appeared to be uniform in thickness with some waviness resulting from the impression of the release cloth TX 1040 during the sandwich bonding process. Moder-

ate porosity was noted throughout the thickness of the dielectric layer. The porosity is attributed to the foaming tendency of the FM-73M material under the vacuum processing conditions. The porosity did not appear to initiate the dielectric cracking since no through-thickness cracks were observed at the higher porosity sites.

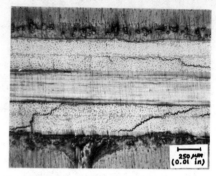

Fig. 6    Substrate Facesheet Cross Section After 10000 Cycles -57°C→+38°C (-70°F to +100°F) & 7000 Cycles 0°C→+35°C (+32°F→+95°F)

The observed cracking of the graphite/epoxy facesheet is suggested to be attributable to resin/fiber thermal coefficient mismatch and the thermal fatigue capability of the resin matrix and the FM 73M. It should be noted that specimen grinding/polishing may have contributed partially to the cracking. To determine the thermal cycling influence on the structural integrity of the substrate, flatwise tensile and compression specimens were prepared from the long term exposed panel and tested. The results compared with the as fabricated properties are presented in Table IV. No significant changes were observed. Although cracking was observed in the substrate, it appeared that adequate structural integrity was maintained for the intended application.

TABLE IV  MECHANICAL PROPERTIES BEFORE AND AFTER LONGTERM THERMAL EXPOSURE

| PROPERTIES | PRE EXPOSURE | | POST EXPOSURE | |
|---|---|---|---|---|
| | STRENGTH MPA (PSI) | FAILURE MODE | STRENGTH MPA (PSI) | FAILURE MODE |
| (b) SUBSTRATE FLATWISE TENSILE STRENGTH, MPA (PSI) (ASTM-C397, 5.1CM X 5.1CM) | | | | |
| o .024 G/CC (1.6 PCF) CORE | .827 (120) | CORE TENSION | .717 (104) | CORE TENSION + FACESHEET DELAMINATION |
| o .064 G/CC (4.0 PCF) CORE | 1.820 (246) | FACESHEET DELAMINATION | 1.572 (228) | FACESHEET DELAMINATION |
| (c) FACESHEET COMPRESSION STRENGTH KGM/CM (LBS/IN) (ASTM-03410, 0.64CM X 14.0CM) | 99.8 (558) | LAMINATE COMPRESSION & BUCKLING | 108.6 (610) | LAMINATE COMPRESSION & BUCKLING |

NOTE: (a) 10000 CYCLES (-57°C ➛ + 38°C) + 7000 CYCLES (0°C ➛ + 35°C)
      (b) AVERAGE DATA OF 3-5 SPECIMENS
      (c) FACESHEET FM-73M/GR/E/FM-73M

b.  Short Term Cycling:

Short term exposure was performed to determine the influence of the thermal exposure
on the constituent parts of
the baseline substrate.  The
specimens, 20.3 cm x 4.0 cm
included:  Facesheet laminates
with and without the dielectric layer and the sandwich
panels with and without the
dielectric layer.  Visual and
microscopic examinations were
performed to determine crack
initiation.  The testing was
conducted concurrently with
the Long Term Cycling.  The
cycling profile for the Short
Term Cycling was as shown in
Figure 7.  The cycle represents anticipated more severe
thermal ground exposures (bake-
outs) than the previous Long
Term Cycling.

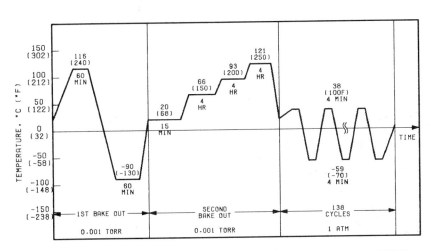

FIG. 7  SHORT TERM THERMAL CYCLING (HEAT RATE ≤ 10°C/MIN (18°F/MIN))

SEM analysis indicated the initiation of microcracking in the bare (without dielectric layer) facesheet laminates prior to the thermal exposure. These microcracks (barely visible under 10X magnification) appeared as a group of discontinuous short, thin lines parallel to the fiber length. The cracking increased in number and intensity after the thermal cycling. The surface of a bare facesheet laminate is shown in Figure 8. Microstructural analysis performed on the laminate with the dielectric layer on both sides showed resin matrix cracking through the thickness of the 0° plies after thermal exposure (Figure 9). The analysis indicated that the dielectric layer appeared to act as a covering layer for the cracking occurring in the facesheet laminate. However, due to thermal expansion and fatigue considerations the cracking in the facesheet appeared to propagate into and induce cracking in the outer dielectric layer. Examination of the specimens also indicated that the honeycomb core did not influence the cracking behavior of the facesheet.

Fig. 9 Facesheet Cracking After Exposure to 138 Cycles -57°C→+38°C (-70°F→+100°F)

c. Alternate Facesheet Cycling:

Since the evaluations suggested that the facesheet appeared to be the major contributing factor to the panel cracking, an accelerated investigation of alternate facesheet concepts was performed. The selection of alternate facesheets were within the constraints of design limitations and schedule. The alternate facesheet laminates were tested in comparison with the baseline. The thermal test profile was the same as the Short Term Cycling shown in Figure 7 except the number of -57°C to +38°C (-70°F to +100°F) cycles was varied.

o Alternate Facesheet Orientation:

The baseline (+60, -60, 0, 0, -60, +60) with 182°C (360°F) cure and two alternate orientation candidate laminates were fabricated to attempt to provide potential stress relief. The laminate orientation and cure temperature were as shown in Figure 10. Alternate Laminate B (+60, 0, -60, -60, 0, +60) separated the two 0° layers in the primary loading direction of the solar array. Laminate C (90, +45, 0, -45, -45, 0, +45, 90) added two additional layers to achieve smaller ply angle between each ply.

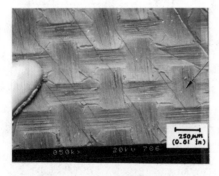

Fig. 8 Continuous Surface Cracks on Gr/E Laminate 138 Cycles -57°C→+38°C (-70°F→+100°F)

LAMINATE A ▬▬▬▬▬ (+60, -60, 0, 0, -60, +60), 182°C (360°F) CURE

LAMINATE B ▬▬▬▬▬ (+60, 0, -60, -60, 0, +60), 182°C (360°F) CURE

LAMINATE C ▬▬▬▬▬ (90, +45, 0, -45, -45, 0, +45, 90)143°C (290°F) CURE

FIG. 10  BASELINE LAMINATE & CANDIDATE LAMINATES

Microscopic examination of both laminates in the as-fabricated condition showed microcracks in the baseline (+60, -60, 0)$_s$. No microcracks were observed in the (90, +45, 0, -45)$_s$ laminate, in the as-fabricated and after 2 "bake-outs" and 55 thermal cycles of -57°C → + 38°C (-70°F to +100°F) conditions. The smaller angle between the two outer consecutive layers in this laminate and the lower cure temperature appeared to reduce residual stress and in turn reduce cracking.

o  Kevlar Fabric Addition:

Two facesheets incorporating the baseline laminate and the (90, +45, 0, -45)$_s$ laminate with Kevlar/epoxy fabric (Fiberite 120 weave/ 934 resin) and the dielectric layer were fabricated for thermal exposure. The configuration and process details were as shown in Figure 11. The fabric was added as as buffer between the facesheet and the dielectric layer in an attempt to prevent crack propagation to the exterior dielectric surface.

BASELINE

| DIELECTRIC | — FM-73M ADDED, 121°C (250F) CURE |
| KEV/E | — 1 PLY, COBONDED @ 143°C (290F) |
| GR/E | — (+60, -60, 0, 0, -60, +60) 182°C (360F) CURE |
| KEV/E | — 1 PLY, COBONDED @ 143°C (290F) |
| DIELECTRIC | — FM-73 ADDED, 121°C (250F) CURE |

ALTERNATE

FM-73M ADDED, 121°C (250F) CURE
1 PLY

(90, +45, -45, 0)$_s$  > COCURED
                          143°C (290F)

1 PLY
FM-73 ADDED, 121°C (250F) CURE

FIG. 11  KEVLAR ADDED FACESHEETS

757

Visual examination after 211 cycles of the Kevlar fabric added facesheets showed no cracks on he surface. Apparently, cracking had not propagated outward to the dielectric from the inner laminate.

To accelerate the exposure, the specimens were subjected further to reduced cycle times (30 minutes vs. 1 hr/ cycles). Visual inspection at 6055 cycles −57°C to +38°C (−70°F to +100°F) showed FM 73M dielectric crazing with very fine micro-cracks under 36X magnification. These micro-cracks which were not visible to the naked eye, ran along the Kevlar fiber direction and formed a fabric-like craze pattern as seen in Figure 12. No evidence of fiber protrusion from the facesheet surface was observed.

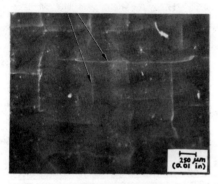

Fig. 12   Cracks Along Kevlar Fiber Direction, 6055 cycles −57°C→+38°C (−70°F +100°F)

Microstructural analysis of the two facesheet specimens after the cyclic exposure showed both graphite/epoxy and Kevlar/epoxy layers exhibited some cracking. However, the dielectric layer to be uniform on top of the Kevlar layer with no obvious through the thick-ness cracking. The Kevlar fabric addition functioned satisfactorily as a buffer to prevent any major crack-ing to be propagated to the surface. However, to as-sure the dielectric insula-tion layer between the solar cells and the composite substrate, a layer of Kap-ton film (0.05 mm) was ad-ded on top of the FM 73M dielectric/Kevlar.

6.0   SUMMARY

The process for fabrication of solar array substrates with graphite/epoxy (P-75/934 tape) facesheets (60, -60, 0)$_s$, Nomex core and FM 73M as a dielectric layer and a bonding adhesive was established. The substrate ex-hibited good structural integrity, excellent dielectric properties and solar cell bonding capability in the as-fabricated condition.

Upon exposure to cyclic thermal environment all laminates showed some degree of cracking along the fiber direction. The cracking propagated outward to the dielec-tric surface as cycle time in-creased. However, no significant changes attributable to the thermal exposure in the examined strength properties were evi-denced. Adequate mechanical properties appeared to have been retained for the particular sub-strate application.

Although improved performance was indicated by the (90, +45, 0, -45)$_s$ 8-ply facesheet laminate, design considerations (mechanical properties, weight, verification and other areas of concern) pre-cluded the immediate change to this orientation. Because of ap-plicability of the previously gen-erated design data and the in-dicated adequate test performance, the baseline laminate (60, -60, 0)$_s$ with the 290°F cocured Kevlar fabric was selected for the full-scale solar array substrate evalu-ations.

## BIOGRAPHY

Mrs. Thu Stankunas is a Materials and Process Engineer at Lockheed Missiles and Space Company, Space Systems Division of Sunnyvale, California. She is currently responsible for development/evaluation and problem resolution associated with composite application in spacecraft structures. Before joining LMSC in 1984, Mrs. Stankunas spent 5 years at Pratt & Whitney Aircraft where she was responsible for process development of composite materials for gas turbine engine/nacelle system. She holds a B.S. degree in Chemical Engineering from the University of California, Berkeley and M.S. degree in Polymer Science and Engineering from the University of Massachusetts, Amherst.

Mr. Warren I. Greenway is a Materials & Process Engineer with Lockheed Missile & Space Company, Space Systems Division of Sunnyvale, California. He has been with Lockheed, both CALAC and LMSC since 1967 where he as been associated with several aircraft and spacecraft design and development projects. Prior to joining Lockheed, he was involved with the development and application of non-metallic materials for aircraft application at Boeing. He has a B.S. in Chemical Engineering from South Dakota School of Mines & Technology.

Mr. Gary R. Holmquist is a Materials and Process Engineer with the Lockheed Missiles and Space Company, Space Systems Division of Sunnyvale, California. He has been with LMSC since 1974 where he has been associated with several spacecraft design and development projects. Prior to joining LMSC he worked for Pratt & Whitney Aircraft where he was involved with the development of military jet engines. He has a M.S. in Metallurgical Engineering from the University of Illinois and a B.S. in Metallurgy from Michigan State University.

## ACKNOWLEDGEMENT

The authors express their thanks to Mr. Doug Izu for program direction and helpful comments/suggestions, and to Mr. Pat J. McCormick for specimen fabrication and testing.

NOVEL COMPOSITE MATERIALS
FOR SPACE STRUCTURES AND SYSTEMS
Edward J. A. Pope and John D. Mackenzie
Dept. of Materials Science & Engineering,
University of California, Los Angeles.

## Abstract

Some novel composite materials that could be useful for space structure applications are reviewed. Criteria utilized in the selection of such materials are considered. Novel materials, such as hollow fiber/ resin composites, transparent sol- -gel derived glass/polymer and new "triphasic" composites made of ceramic, glass, and polymer phases are presented. Specific strength, specific modulus, vibrational dam- ping, and other properties are examined. Comparisons between the novel materials under development at UCLA and established materials, such as graphite/epoxy, are made on the basis of selection criteria.

### 1. INTRODUCTION

Although the many components of a large precision space structure perform different functions, most of them must be designed to with- stand the "hostile" environment of space. Some obvious conditions are high vacuum, radiation, and cyclic temperature changes. Desirable materials properties for such applications include; high specific modulus and strength, good radiation resistance, high vibrational damping characteristics, low thermal expan- sion, and low density[1,2]. Often, a single-phase material is unable to meet the stringent requirements of space applications and composite materials must be used[1]. Recent- ly, many new engineering materials (including composites) have been studied and a number of new prepa- ration processes have also been de- veloped, such as sol-gel processing. A good example of a new composite material are the high modulus fiber-inorganic glass matrix and glass-ceramic matrix composites developed by K. M. Prewo and co- -workers[3-6]. For graphite fiber- -borosilicate glass composites, for example, the average thermal expan- sion coefficients are very low from 25°C to 150°C, the density is only approximately 2 gm/cc. and the elastic modulus is over 200 GPa (over $30 \times 10^6$ psi). A number of other new monolithic materials and composites have been reported which appear to be promising. The objective of this paper is to examine the feasibility of these relatively new materials on a preliminary basis for use in space structures.

## 2. SOME NEW MATERIALS, PROCESSES, AND CONCEPTS.

The word "new" is naturally relative. In this paper, it is meant to imply that applications to precision space structures have not been carefully evaluated.

### 2.1 Some New Glasses and Glass-Ceramics

#### 2.1.1 Glasses based upon copper aluminosilicates [7-10]

Glasses from the system $Cu_2O-Al_2O_3-SiO_2$ have low thermal expansion coefficients from 0°C to 300°C. Expansion coeeficients similar to that of fused silica have been reported. The densities are in the range of 2.7 to 2.9 gm/cc. Melting can be made at 1550°C, considerably less than that for silica or $TiO_2-SiO_2$ glasses. With minor additions of such fluxes as $B_2O_3$, melting can be done at 1450°C without serious effects on the expansion coefficients. Although no elastic modulus values are available, the hardness of these glasses are some 40 to 50% higher than silica glass and 100% higher than that of Corning 7740(a low expansion glass with an expansion coefficient of $3.2 \times 10^{-6}/°C$). It is anticipated that the copper aluminosilicate glasses will have relatively higher elastic modulus. The softening temperatures can be as low as the 7740 glass. It is thus anticipated that they can be used as matrix materials for high modulus fibers with improved stiffness and less thermal distortion. Glass fibers have also been made from these compositions for potential use in epoxy matrix composites.

#### 2.1.2 Glass-ceramics based on copper aluminosilicate glasses [10-12]

The copper aluminosilicate glasses can be easily nucleated and crystallized to produce glass-ceramics of low, zero, or negative thermal expansion coefficients. Their potential roles will be similar to those for the parent glasses. It is conceivable that glass-ceramic fibers can also be made.

#### 2.1.3 Lithium aluminosilicate Glass-Ceramic Fibers

Glasses based on $Li_2O-Al_2O_3-SiO_2$ can be crystallized to give glass-ceramics of extremely low thermal expansion coefficients. Further, the bulk glass-ceramics can be strengthened through ion-exchange [13]. Theoretically, it is possible to increase the elastic modulus through ion-exchange due to surface compression[14]. Recently, fibers based on lithium aluminosilicate were converted into glass-ceramic fibers[15]. The fibers have low or zero expansion coeeficients, high strengths, and are transparent in the visible. Preliminary experiments at UCLA have confirmed that they can be ion-exchanged in a $KNO_3$

761

melt to increase both strength and elastic modulus.

## 2.2 Sol-Gel Derived Glasses, Ceramics, and Composites

The sol-gel method has been widely studied for the preparation of bulk glasses, thin films, and porous solids[16]. Unlike traditional glass and ceramic processing techniques, in which powders are melted or reacted at high temperatures, the sol-gel process relies upon hydrolysis and polymerization reactions in liquid solution at temperatures near ambient. Typically, a precursor, solvent, water, and a catalyst are mixed into solution. As the polymerization reaction proceeds, the solution viscosity increases until a "wet solid" called a gel is formed. The gel is composed of a continuous matrix phase and an alcohol and water liquor. Upon drying, the porous gel shrinks as the liquor in the capillaries is evaporated away. Upon heat treatment, the porous gel can be converted into a dense glass or ceramic or maintained as a porous solid. Dense glasses from sol-gel appear to have identical physical properties to melt-formed glasses [16]. Glass-ceramics can also be formed this way. Thus, the sol-gel approach can be utilized to form glass, glass-ceramic, or composite matrix composites. For example, porous composites of SiC in silica glass have been fabricated[17,18]. Moreover, a wide variety of second phases, such as SiC whiskers, TiC,

$Si_3N_4$, $Al_2O_3$, metal dust, and micro balloons can be added to the initial gel solutions[17,18]. Metallic particles can also be reduced in the pores of porous gels[19]. Lightweight, non-porous composites can be fabricated from the in situ polymerization of organic monomers yielding transparent composites of relatively high strength and low density(see figure 1)[17,18]. Also exciting is the preparation of "triphasic" composites based upon the polymer impregnation of porous composites[17,18]. The volume fraction porosity in the dried gel and, hence, the volume fraction of polymer phase in the final composite can be controlled by adjusting the catalyst in the initial solution chemistry[20].

### 2.2.1 Aerogels

Aerogels are fabricated by the hypercritical drying of wet gels in an autoclave. This class of porous solids can have up to 99% porosity, constituting one of the lowest density solids in existance. Bulk densities of 0.05 to 0.20 gm/cc are not uncommon. Moreover,

Figure 1: Transparent sol-gel derived PMMA-Silica composite.

aerogels are often quite transparent due to their small pore
diameters(approximately 200A).
Transparent silica aerogels, such
as the one shown in figure 2, also
possess the same low thermal expansion of silica glass. Such ultra-
light materials, albiet very fragile, may be ideal for insulation
of space structures due to their
low thermal conductivity and high
dampening properties[21]. Aerogel
tiles could be applied to the
external surfaces of space station
assemblies to reduce heat loss. A
mirrored metallic surface coating
could also be applied to reduce
heat absorption on surfaces facing
the sun(figure 3).

Figure 2: 98% porous, transparent
silica aerogel prepared at UCLA
(special thanks to Mr. M. Borden).

## 2.3 Exploitation of Fiber Geometry

### 2.3.1 Hollow Fibers

A hollow glass fiber can be easily
made by the drawing of thick-walled
glass tube. In a recent research
program supported by AFOSR at UCLA,
hollow glass fibers having an i.d.
of 10 microns and an o.d. of 30
microns have been fabricated. The
i.d. to o.d. ratio is easily controlled. Hollow fibers can also

be drawn directly from bushings
by melt drawing. Hollow fibers
offer weight reduction of over
30% compared to solid glass fibers
typically used in glass-epoxy
composites, while still retaining
most of their stiffness and specific
strength. Hollow fibers can also
be filled with polymers or metals.
Fine conducting metallic core fibers
could eliminate problems of static
charge build-up with only minimal
increases in fiber density.
Aluminum core glass fibers have
been successfully drawn in a con-

SPACE
STRUCTURE
SUBSTRATE

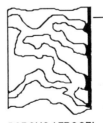

REFLECTIVE
SURFACE
COATING
(METAL OR CONDUCTING
OXIDE)

POROUS AEROGEL
ULTRAFINE PORES AND
OVER 90% POROSITY.

Figure 3: Schematic Diagram of aerogel tile employed as external insulation
and protection of space structures and stations.

tinuous drawing process.

### 2.3.2 Oval Fiber

Many glasses are easily fabricated into a variety of shapes because of their advantageous viscosity-temperature relationships. Round glass fibers and thin glass tapes are commercially available. There is no reason why an oval-shaped glass fiber cannot be drawn continuously through a specially shaped bushing. For two glass fibers with the same cross-section area, the stiffness of the oval fiber in the long direction can be significantly higher than that of the round fiber. For example, a round fiber with a 10 micron diameter has almost the same cross--sectional area as an oval fiber with long and short dimensions of 20 and 5 microns, respectively. In such a case, the stiffness along the 20 micron direction is FOUR times greater than the 10 micron round fiber. The alignment of the oval fibers in a matrix should not be difficult. In a sense, the "effective" elastic modulus of a glass fiber in the long direction can be $40 \times 10^6$ psi, even though the true modulus is only $10 \times 10^6$ psi.

### 2.4 Glass Microballoons

Glass microballoons, known as "eccospheres" and "cenospheres", have been commercially available for a long time. They have been used as fillers for organic resins[22]. External diameters can be varied from 20 to 200 microns. Both borosilicate and silica microballoons

are available.[23]. The density of individual balloons can be as low as 0.25 gm/cc. Microballoons have been self-bonded or bonded with ceramic frit to give light-weight heat-insulating, and high-temperature stable bodies[24]. Low density composites could be fabricated by incorporating microballoons in a glass or glass-ceramic matrix. The sol-gel method is particularly well suited because the composite can be fabricated at relatively low temperatures.

### 3.0 PROPERTIES OF SOME NOVEL MATERIALS

### 3.1 Thermal Expansion

The thermal expansion coefficients of a number of materials were determined from -200°C to 100°C. An apparatus was designed and constructed for this purpose. The average expansion coefficients for selected materials are shown in table 1. Some significant points to note are: a) for the resin-microballoon composites, because the resin is the matrix phase, it has a larger effect on the expansion despite the fact that the glass spheres may be the major constituent by volume; b) the copper aluminosilicate glass has an expansion coefficient similar to that of silica glass, and; c) $SiO_2$ gels, although only fired to 400°C, already exhibit the low expansion of silica glass.

### 3.2 Elastic Moduli

The elastic moduli(Young's modulus, shear modulus, and bulk modulus)

Table 1: Average Thermal Expansion Coefficient for Selected Novel Materials except for Results from the Literature.

| Material | Thermal Expansion Coefficient $(\times 10^{-7}/°C)$ |
|---|---|
| Resin + microspheres | |
| 0% glass spheres | 624 |
| 20%      "          " | 561 |
| 50%      "          " | 430 |
| 60%      "          " | 412 |
| 70%      "          " | 368 |
| Melt Cast Silica Glass | 5 |
| Melt Cast Copper Aluminosilicate | 5 |
| Alumina(polycrystalline) | 88* |
| Zirconia(stabilized) | 100* |
| SiC | 47* |
| $B_4C$ | 45* |
| TiC | 74* |
| Porous Silica Gel | |
| (fired at 400°C) | 5 |
| (fired at 600°C) | 5 |
| (fired at 800°C) | 5 |
| Silica-Silicon Carbide Composite | |
| (fired at 600°C) | 37 |
| (fired at 1000°C) | 12 |
| Porous Glass + 33%PMMA | 138 |

*from literature

and the Poisson's ratio were determined from transverse and longitudinal sounde velocity measurement and density as based upon standard methods using small rod samples[25]. In addition to the composite samples whose thermal expansion was determined, some polymer-SiC-silica gel "triphasic" composites were studied. Measurements were made at room temperature. Results for Young's modulus and Poisson's ratio are shown in table 2. Because of the very thin-walled glass microballoons used, the glass content in these composites is very small. This resulted in very low elastic moduli. Similar results were obtained by Lee and Westmann[26] who also developed equations for the estimation of elastic properties

of such composites. The silica gels studied were porous. Samples fired at 200, 400,600, and 800°C had porosities of 58%, 46%, 44%, and 33%, respectively. With the exception of the 200°C sample, which did not have an apparent density of 2.20 gm/cc, the results seem to obey the relationship

$$E = E_o(1-1.9P + 0.9P^2) \qquad (1)$$

where $E_o$ is the elastic modulus of silica glass[27].

3.3 Vibrational Damping

Vibrational damping was studied for some of our composites with a low frequency resonance method[28]. An apparatus was designed and constructed and is shown in figure 4. Results are shown in table 3. The basic equation for calculating the relative damping constant b/c is

Table 2: Elastic Moduli, Poisson's Ratio, and density for Selected Materials at Room Temperature

| Material | Density (gm/cc) | Young's Modulus (x $10^5$ psi) | Shear Modulus (x $10^5$ psi) | Bulk Modulus (x $10^5$ psi) | Poisson's Ratio |
|---|---|---|---|---|---|
| Glass Microballoons + Resin | | | | | |
| 0 % glass balloons | 1.134 | 6.10 | 2.26 | 6.87 | 0.35 |
| 20 " | 1.076 | 5.61 | 2.08 | 6.15 | 0.35 |
| 50 " | 0.787 | 4.37 | 1.63 | 4.45 | 0.35 |
| 60 " | 0.771 | 4.02 | 1.47 | 4.27 | 0.35 |
| 70 " | 0.624 | 3.03 | 1.10 | 3.36 | 0.35 |
| Silica Glass | 2.21 | 106.0 | 45.0 | 54.0 | 0.17 |
| Pyrex Glass | 2.23 | 95.0 | 39.0 | 58.0 | 0.23 |
| Lithium Aluminosilicate Glass-Ceramic(pyroceram 9608) | 2.60 | 124.0 | 50.0 | 83.0 | 0.25 |
| Copper Aluminosilicate Glass | 2.73 | 98.0 | 43.0 | 60.0 | 0.23 |
| High Purity Alumina(polycryst.) | 3.60 | 392.0 | 148.0 | 363.0 | 0.32 |
| PMMA + porous glass | 1.83 | 25.6 | 10.3 | 16.4 | 0.24 |
| PMMA + SiC/$SiO_2$ Gel Composite | | | | | |
| (600°C) | 1.56 | 15.7 | 6.16 | 11.8 | 0.28 |
| (1100°C) | 1.98 | 52.2 | 21.4 | 30.7 | 0.22 |
| Resin + Hollow Glass Fibers(50%) | 1.33 | (I) 0.3 - 0.5 | | | |
| | | (II) impossible to measure sound velocity due to damping. | | | |
| Resin + Aluminum/Glass Fibers | 1.74 | 56.0 | 20.2 | | |
| Porous Silica Gel(800°C) | 1.49 | 48.1 | | 26.1 | 0.19 |

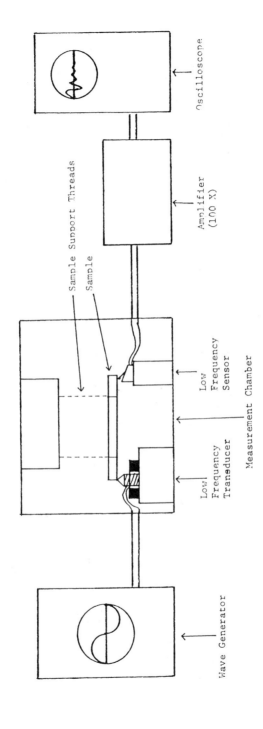

Figure 4: Schematic Diagram of Low Intensity Vibrational Damping Measurement System.

$$b/c = F_m/Af_o \qquad (2)$$

where $F_m$ is the driving force at resonance, $f_o$ is the resonant frequency of the sample and A is the amplitude of signal response at resonance. The damping coefficient is b and c is a system coupling constant. This technique was selected because of its relative simplicity and demonstrated successful application[28]. The last column of table 3 shows the specific damping constant, which is b/c over the density of the material. The composite composed of 50% glass balloons in resin seems to have the highest damping coefficient. From published literature, graphite fiber/epoxy composites appear to be the most promising candidate material for space structures(table 2 of ref. 1). This is because of high specific modulus and good damping behavior. The damping behavior in table 1 of the report by Trudell, et al, is expressed as a "loss factor" whereas our results are obtained in the form of damping constants. An approximate conversion, however, has been made to permit the comparison of the current material and the graphite/epoxy composites. It would appear that of those materials on which experimental measurements have been made, graphite/epoxy composites have the highest specific modulus and relatively good damping behavior.

4.0 RECOMMENDATION

Although the present study is preliminary in nature, it does reveal some promising future generations of composites which could be superior, especially in some properties, to the graphite/epoxy materials currently known. For example, an "ideal" candidate could be a composite made up of hollow ceramic fibers embedded in a sol--gel derived matrix of similar expansion coefficient. The sol--gel matrix, because of its interconnecting pores, could be further filled with an organic resin or polymer to obtain improved damping. Such a proposed structure is shown in figure 5. If the hollow ceramic fiber has a similar expansion coefficient to that of the matrix material, then the expansion of the composite should be nearly isotropic. The amount of resin necessary could be significantly less than that used in graphite/epoxy composites. The resin is also protected by the ultrafine pore structure of the sol-gel derived matrix.

5.0 ACKNOWLEDGEMENTS

This work was supported by the Air Force Office of Scientific Research(AFOSR), Directorate of Chemical and Atmospheric Sciences. We also wish to thank Dr. H, Nasu and Ms. A. Nakata for their assistance in obtaining much of the data herein contained.

Table 3: Comparison of Specific Modulus and Damping Properties of Selected Materials

| Material | Specific Modulus (x $10^6$ psi) | Relative Damping Constant (b/c) | Calculated Loss Factor | Relative Specific Damping Constant |
|---|---|---|---|---|
| Inconel X-750 | 136 | 1.64 | 0.0004 | 0.2 |
| Alumina | 301 | 1.50 | 0.0004 | 0.38 |
| Silica Glass | 125 | 1.13 | 0.0003 | 0.50 |
| PMMA(poly methyl methacrylate) | 10 | 4.10 | 0.0032 | 3.40 |
| Resin | 9 | 12.10 | 0.0260 | 10.70 |
| Resin/50% glass microballoons | 16 | 12.10 | 0.0260 | 15.40 |
| Resin + 50% hollow glass fibers | | 6.50 | 0.0080 | 4.90 |
| Resin + 50% Aluminum/Glass Fibers | 80 | 3.80 | 0.0030 | 2.10 |

| Epoxy/Graphite Composite Unidirectional | | Calculated Relative Damping Constant | Measured Loss Factor | Calculated Relative Specific Damping Constant* |
|---|---|---|---|---|
| Type I (// direction) | 407 | 5.5 | 0.0060 | 3.90 |
| Type II (// direction) | 466 | 5.8 | 0.0070 | 4.14 |
| Type II (⊥ direction) | 19 | 8.8 | 0.0140 | 6.30 |

*Density for Graphite/Epoxy Composite Assumed to be 1.40 gm/cc.

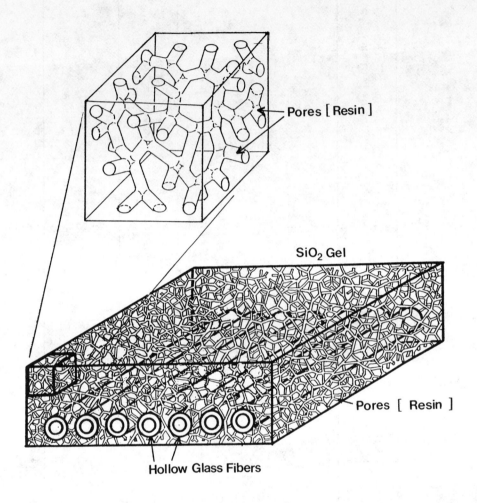

Figure 5: Schematic Representation of Potential New Composite Material Composed of Unidirectional Hollow Glass-Ceramic Fibers Embedded in a Sol-Gel Derived Porous Glass-Ceramic or Glass Matrix that has been Impregnated with Polymer.

## 6. ABOUT THE AUTHORS

Edward J. A. Pope received his Bachelor of Science in Engineering in 1983 and his Master of Science in Engineering in 1985 from the University of California, Los Angeles. He is currently pursuing his Ph.D. under the direction of Professor Mackenzie. His current research is primarily focused on sol-gel derived composite and optical materials.

John D. Mackenzie is Professor of Engineering in the Dept. of Materials Science & Engineering at UCLA. He has published over 250 papers in the areas of glass structure, infrared transmitting glasses and fibers, high modulus glass-ceramics and fibers, chalcogenide and chalcohalide glasses, and sol-gel processing.

## 7. REFERENCES

1. R. W. Trudell, et al, Proc. AIAA-80-0677 (1980).

2. M. F. Card, et al.,Astronautics and Aeronautics,16,48(1978).

3. K. M. Prewo and J. J. Brenana, J. Matls. Sci.,15, 463(1980).

4. K. M. Prewo, et al.,Proc. AIME/ASM Composites Conf.,New Orleans, Louisiana, Feb. 20-21, 1979.

5. K. M. Prewo and J. J. Brennan, J. Mat. Sci, 17, 1201(1982).

6. J. J. Brennan and K. M. Prewo, J. Mat. Sci.,17(2371(1982).

7. N.T.E.A. Baak and C.F. Rapp, U. S. Patent No. 3,779,781, Dec. 19, 1973.

8. K. Matusita and J. D. Mackenzie, J. Non-Cryst. Sol.,30,285(1979).

9. K. Matusita, et al.,J. Am. Cer. Soc., 66,33(1983).

10. J. Ko, M. S. Thesis, University of California, Los Angeles(1979).

11. K. Matusita, et al.,pp.277-86 in Advances in Ceramics, V. 4, ed by Simmons, Uhlmann and Beal, Am. Ceram. Soc.(1982).

12. J. D. Mackenzie, NASA Tech. Report, NCA-2-OR-390-803 (1982).

13. B. R. Karstetter and R.O. Voss, J. Am. Ceram. Soc,50,133(1967).

14. J. D. Mackenzie and J. Wakaki, J. Non-Cryst. Sol.,38, 385(1980).

15. E. T. Wu, et al.,pp. 237-43 in Advances in Ceramics, V.4, Am. Ceram. Soc.(1982).

16. J. D. Mackenzie,J. Non-Cryst. Sol.,48,1(1982).

17. E. J. A. Pope & J. D. Mackenzie, Proc. 1986 Spring Meeting Mat. Res. Soc. no. 73, pp.809-814.(1986).

18. E. J. A. Pope & J. D. Mackenzie, Proc. 21st Univ. Conf. on Ceram. Sci.,Penn. State Univ. Press (1985)

19. E. J. A. Pope & J. D. Mackenzie, "Ultrafine Metallic Particles in Silica Glass", in press.

20. E. J. A. Pope & J. D. Mackenzie, J. Non-Cryst. Sol.,87,185-198(1986).

21. J. Fricke, Universität Würzburg Report no. E12-0585-1(1985).

22. R. H. Whrenberg, Mat. Eng. 2 Oct.(1978).

23. Emerson & Cuming, Inc. Fact Sheet, Canton, Mass.,3M Co.,St. Paul, Minnesota.

24. A. Tobin, et al.,NASA report under contract NASI-10713 by Grumman Aerospace Co.(1972).

25. E. Schreiber, et al.,"Elastic Constants and Their Measurement", McGraw-Hill,NY(1973).

26. K. J. Lee and R. A. Westmann, J. Comp. Mat.,4,242(1970).

27. J. K. MacKenzie,Proc. Phys. Soc.,(London)B63,2(1950).

28. Y. Yamaguchi. Bull. Cer. Soc. Japan,18,1008(1983).

29. R. Resnik and D. Holliday, Physics, 3rd ed.,J. Wiley & Sons, NY(1977).

# FABRICATION AND ASSEMBLY OF AN
## ADVANCED COMPOSITE SPACE STATION TETRATRUSS CELL

Michael J. Robinson
McDonnell Douglas Astronautics Company
Huntington Beach, California

### Abstract

In a Space Station Phase B development effort, McDonnell Douglas Astronautics Company has fabricated a deployable tetratruss
cell made almost entirely of graphite/epoxy
composite materials. The cell is a prototype
which establishes a precedent for manufacturing large arrays of composite truss structure for the Space Station which will take
advantage of the superior stiffness and
strength-to-weight characteristics of these
materials. The cell consists of 24 struts, each
ten feet long, 15 of which bend in the middle
to allow the cell to collapse. The strut tubes
were fabricated with unidirectional preimpregnated graphite tape. Hinge fittings, tube
end fittings, and truss nodes were compression molded using tape and chopped fiber
molding compounds. The fully deployed cell
measures approximately 20 feet by 17.5 feet
by 8.5 feet and collapses into a bundle 10 feet
long and about 20 inches in diameter. The
manufacturing processes employed in this
project and discussed in this presentation include composite material layup and molding,
assembly tooling design and fabrication, subassembly drilling and adhesive bonding operations, and final assembly.

## 1. INTRODUCTION

Large truss structures will serve as the backbone of the Space Station which NASA will
deploy some time in the next decade. The
overall truss configuration, most likely a combination of one-dimensional beams and large
planar arrays, will form a network on which
the habitation modules of the station and all

auxiliary structures and equipment will be
mounted. Among the issues unresolved at the
onset of this program were the truss configuration type (erectable vs. deployable) and
materials (metal vs. composites). These issues have been the subject of considerable
research and development at NASA and in
the industry.

As a part of this research and development,
NASA-JSC selected the McDonnell Douglas
Astronautics Company-Huntington Beach
(MDAC-HB) to build a full-scale deployable
tetratruss cell made almost entirely of advanced composites. The purpose of the program was to verify the capability of these
high-strength, high-stiffness, lightweight materials for use in Space Station truss
applications.

The cell, illustrated in Figure 1, is based on
a tetrahedral geometry that can be extended

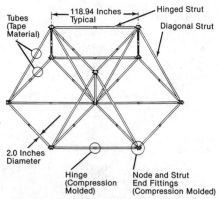

**Figure 1. Tetratruss Cell**

in two dimensions and in which all the struts are the same length.

As indicated in the figure, all the planar or face struts have hinges in the middle; the diagonal struts are rigid. The hinged struts allow the entire structure to collapse into a bundle, the intent being that large sections of the truss could be assembled and checked on the ground, collapsed for storage and transportation to space in the shuttle cargo bay or some other vehicle, and then deployed with a minimum of astronaut extra-vehicular activity (EVA) time.

The struts consist of 2-inch diameter graphite epoxy tubes with adhesive bonded end fittings and hinges. The end fitting consists of two laminated clevises bonded to a cruciform-shaped molded centerpiece called a lock body. The tubes and clevises were made from Fiberite AS4/976 unidirectional prepreg tape. The end fittings have a locking mechanism that locks the strut in the fully deployed position. The hinges are spring loaded and consist of two molded pivot fittings, a graphite epoxy latch bar which is depressed to unlock the hinge, and several small metal parts. The lock bodies, pivot fittings, latch bars, and nodes were made from Fiberite E-21718/X49566 graphite/ epoxy chopped fiber molding compound. The end fitting and hinge are illustrated in Figures 2 and 3, respectively.

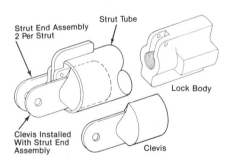

**Figure 2. Strut End Fitting**

The composite tetratruss design was adapted from an existing aluminum structure design provided by the Structures and Thermal division of NASA-JSC. The changes from the aluminum design were the result of special manufacturing and structural considerations of the advanced composite materials.

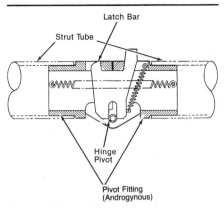

**Figure 3. Hinge Fitting**

## 2. PRODUCIBILITY

Producibility was a major issue in the design and fabrication of the tetratruss components. Considerations from the standpoint of manufacturing both this single cell and large arrays of composite truss structure influenced the design and choice of fabrication methods.

### 2.1 Tubes

Four tube fabrication methods were evaluated: braiding, pultrusion, filament winding, and roll wrapping. The evaluation was based on an Independent Research and Development (IR&D) project conducted at MDAC-HB in 1985. The key technical and cost issues identified for fabrication method with respect to materials and producibility were established. Demonstration tubes were fabricated by each process, and compression moduli were determined. The test results are given in Table 1. Tables 2 and 3 outline the merits of the four fabrication methods with regard to dimensional stability and damage tolerance.

The braiding process produced generally good quality tubes with highly repeatable properties at low cost. Braiding, however, requires specially formulated "no-tack" resins and high-strength (typically lower modulus) fibers due to mechanical loads imposed during processing. This presents a drawback for the current braiding technology since the Space Station trusses will be stiffness critical and will require high-modulus tubes. The no-tack resins have low toughness and are therefore subject to impact damage and microcracking.

**Table 1. TUBE TEST RESULTS**

| Process | Orientation | Fiber/resin | Compression* modulus (msi) | Theoretical modulus (msi) | Fiber volume (%) | Void volume (%) |
|---|---|---|---|---|---|---|
| Pultrusion | 0° | P-55 Hetron 9929200 | 17.0 | 33.0 | 46.2 | 5.7 |
| Braided | $(0_2/ \pm 20_6)$ | Hysom XAS/Newport Composites NC-76 | 12.2 | 17.6 | 56.1 | 0 |
| Filament wound | $(\pm 10/90/ \pm 10/90/ \pm 10)$ $(\pm 10)_4$ | AS-4/Fiberite 976 AS-4/Fiberite 976 | 13.5 15.2 | 15.6 19.8 | 62.8 61.3 | 0.84 2.5 |
| Roll wrapped | $(+30/0/-30/0/+30/0/- 30/\bar{0})_s$ | P-75/Union Carbide ERLX1962 | 27.7 | 31.2 | 51.9 | 3.4 |
| *Adjusted to 60% FV | | | | | | |

JOB1 H1611$$$T5

**Table 2. DIMENSIONAL STABILITY ISSUES**

| | As fabricated straightness | Microcracking | Fiber distribution |
|---|---|---|---|
| Roll wrapping | Good – mandrel dependent | Low microcracking possible; tough resins available; can vary ply thickness | Good |
| Pultrusion | Poor | Low in all 0-deg laminate; limited choice of resin | Thickness variation |
| Filament winding | Good – mandrel dependent | Low microcracking possible; tough resins available; can vary ply thickness | Good |
| Braiding | Good – mandrel dependent | Limited quantity of prepreg resins | Good |

JOB1 H1611$$$T3

**Table 3. DAMAGE TOLERANCE ISSUES**

| | Matrix "toughness" | Fiber orientation | Void content |
|---|---|---|---|
| Roll wrapping | Wide range of tough resins | Unlimited | Low |
| Pultrusion | Limited to very low viscosities, quick curing matrices – not usually tough | Primarily 0-deg; very low transverse mechanical properties | High |
| Filament winding | Wide range of tough resins | 10-deg–90-deg; crossovers increase toughness | Low |
| Braiding | "No tack" resin; limited selection of matrices; not usually tough | 15-deg–90-deg; crossovers increase toughness | Low |

JOB1 H1611$$$T4

The pultruded tubes were the first of their type to be made with ultra-high modulus (UHM) fibers. The information acquired during this effort significantly enhanced the probability of fabricating high quality pultruded tubes in future efforts. The tubes produced in the IR&D study, however, had an intolerably low fiber volume, resulting in a structural efficiency far below the target values. They also had a 1-inch bow over their 10-foot length and had slight eccentricities. These drawbacks excluded pultrusion as a method for use on the tetratruss tubes. As with braiding, however, the technology for this process continues to improve, and both methods are already cost effective due to automation and may prove useful in future applications.

Filament winding and roll wrapping of pre-impregnated composite tow and tape are well-established methods of fabricating high quality tubes with a variety of materials and winding angles. The IR&D study further confirmed this, and these two methods were deemed technically suitable for production of the tetratruss tubes. The tube subcontractor, DuPont, opted for roll wrapping because of their greater experience with that technique. Roll wrapping, however, is labor intensive. For larger quantities, an automated process would be far more cost efficient as illustrated by the cost comparison of the four fabrication methods given in Figure 4.

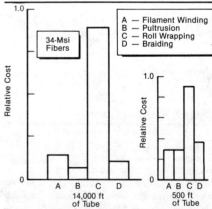

A — Filament Winding
B — Pultrusion
C — Roll Wrapping
D — Braiding

**Figure 4. Tube Fabrication Cost Comparison**

## 2.2 Nodes and Fittings

The non-planar, discontinuous geometries of the truss nodes and fittings shown in Figure 5 for the most part precluded fabricating with continuous fibers. With the exception of the clevises, graphite/epoxy chopped fiber molding compound was used to mold these parts to near net shape.

In another 1985 IR&D project conducted at MDAC-HB, molded end fitting clevises were designed, fabricated, and tested. A matched metal compression molding die was designed and built; test panels and end fitting clevises were compression molded using chopped fiber carbon/epoxy molding compound, carbon/epoxy fabric, and unidirectional tape prepregs. Coupon specimens and bonded truss tube fitting assemblies were evaluated for physical and mechanical properties. The results are given in Table 4. The coupon test results verified the much higher strengths and moduli (both tensile and compression) that

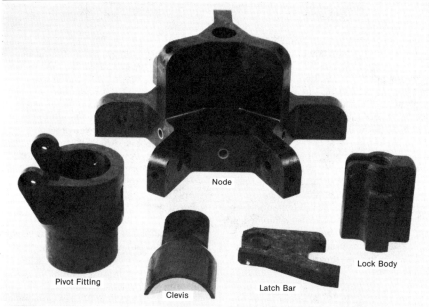

**Figure 5. Molded Parts**

**Table 4. MECHANICAL PROPERTIES OF CANDIDATE MATERIALS FOR END FITTINGS**

| Description | Material type | | | |
| --- | --- | --- | --- | --- |
| | Unidirectional tape[1] | Woven fabric[2] | Chopped molding compound[1] | |
| | 5-mil thick, T300/CE 321 (Ferro) | 8 harness satin AS4/3501-6 (Fiberite) | 1/2-in. chopped roving (Fiberite) | |
| | | | (A) 50 x 10⁶ psi fibers | (B) 33 x 10⁶ psi fibers |
| Orientation in flat panels | $[\pm 10_2 / \pm 45_3 / \pm 30]_s$ | $[0 \text{ to } 90]_c$ | Random | |
| Mechanical properties | | | | |
| Tensile ultimate strength (ksi) | 72.5 | 70.0 | 7.6 | 11.2 |
| Tensile Modulus (msi) | 8.7 | 8.5 | 3.8 | 4.3 |
| Compression ultimate strength (ksi) | 64.9 | 67.5 | 18.7 | 26.0 |
| Compression modulus (msi) | 9.5 | 7.2 | 6.0 | 4.5 |
| Bearing, 0.3-in. dia pin (ksi) | 79.2 | — | 28.9 | 37.3 |
| Physical properties (typical): | | | | |
| Specific gravity | 1.59 | 1.56 | 1.54 | 1.52 |
| Fiber (% vol) | 60 | 60 | 54 | 48 |

Notes:
1. Flat compression molded panels, tape tested in 0-deg direction (tape and fabric).
2. Fabric data are minimum average allowables, of flat panels tested in 0-deg direction MDAC material specification.

are expected with the continuous fibers in the tape and fabric. The processing time for the parts made with chopped fiber and fabric was significantly lower than for the tape (Table 5), making the fabric the most attractive combination of mechanical integrity and cost. For the tetratruss, however, in order to achieve coefficient of thermal expansion (CTE) compatibility with the truss tubes and take full advantage of the reliable prepreg tapes, the clevises were laminated using the same material and ply orientations as the tubes.

## 3. TOOLING

A variety of fabrication and assembly tooling was required for production of the tetratruss.

A steel mandrel was required for the tubes due to tight tolerances on diameter and straightness. Only one mandrel was used, since the short tubes for the hinged struts were made by cutting long tubes in half.

Steel matched die molds were employed for the clevises, lock bodies, pivot fittings, and nodes. They were designed to provide ease of part fabrication and minimal secondary operations. Bolt or pin holes, required in each of these parts, could have been molded in place, but because of the small quantities required for the tetratruss, it was cheaper to drill them in the cured part. Several drill jigs were designed and built for precise location drilling.

Four assembly tools were required: a steel combination bonding fixture/drill jig that was used to bond and match drill the three-piece end fitting subassembly; another drill jig, also steel, to match drill the pivot fittings; and two bonding fixtures for bonding tubes and fittings to make the two types of struts. The tools, as used in their associated subassembly operations, are illustrated in Section 5.

## 4. COMPONENT FABRICATION

The subcontractors and the parts manufactured by them are listed below:

DuPont                   Tubes
Wilmington, DE

Reynolds & Taylor        Clevises
Santa Ana, CA            Lock Bodies
                         Latch Bars

American Automated       Nodes
Huntington Beach,        Pivot Fittings
CA

As mentioned above, DuPont fabricated the graphite/epoxy tubes by roll wrapping which is a hand lay up process in which the preimpregnated graphite tape is laid out at specified angles and then rolled onto the mandrel. The ply configuration was [±10°/±30°/±10°]. At regular intervals in the layup process, the prepreg was debulked by hand adjustment and stretch tape wrapping. Small samples were cut from the ends of each tube from which void volumes were determined by acid digestion. Tubes with void volumes greater than 2 percent were rejected.

In a development study, DuPont cured some of the tubes in an autoclave and some in an oven to compare the void volumes achievable with the two methods. As expected, the autoclave-cured tubes were of high quality with an average of less than 0.5 percent voids and no rejections. The oven-cured tubes averaged 1.6 percent voids and had a rejection rate of 26 percent.

The end fitting clevises were fabricated by cutting and laying up the graphite prepreg on the matched die mold and then curing with the addition of heat and pressure. A significant development period was required for

Table 5. END FITTING FABRICATION SUMMARY

| Material | | | | Premolding processing time (min) | Physical dimensions | | |
|---|---|---|---|---|---|---|---|
| | | | | | Weight (gm) | Thickness (in.) | |
| Fiber | Resin | Fiber form | Orientation | | | Tang | Tubular section |
| Union Carbide T-300 | Hercules 3501-6 | 8-harness satin-weave fabric | 0/90 plus 2 sets of ±45 in tang | 15 | 19.8 | 0.121 | 0.107 |
| Union Carbide T-300 | Ferro CE321R | Unidirectional tape | [±10/±30/±10 /±30/±10] plus 3 sets of ±45 in tang and sloped section | 120 | 18.4 | 0.119 | 0.103 |
| Union Carbide T-300 | Fiberite 49560 | 0.5-in. chopped fibers | Random | 5 | 20.2 | 0.130 | 0.120 |

clevis fabrication due to porosity in the flat area of the first parts, caused partly by a lower-than-expected cured ply thickness of the prepreg. Additional plies and experimentation with processing parameters led to good, low-porosity clevises.

The hinge latch bar and pivot fitting were made from 1/4-inch fiber material and the node and lock body from 1/2-inch fiber material. Due to the small size, simple geometry, and small quantity of the latch bars on the tetratruss, it was cost effective to machine them from a plate of material molded to the specified latch bar thickness. It was decided that since the other components were being molded, there would be no additional information to be gained by making a latch bar die; therefore, the added cost was avoided.

The other parts were made using steel compression molds and established molding techniques. Some development was required for the molding of the lock bodies because of difficulty in getting the composite material to flow into the shape of the relatively fine internal threads (.5625- 12UNC-2B) of the part. As with the clevises, experimentation with processing parameters led to good quality threads and an acceptably low part rejection rate.

## 5. ELEMENTS AND SUBASSEMBLIES

### 5.1 Hinge Assembly

The first step in the hinge assembly was drilling the pivot pin hole. Pairs of the androgynous pivot fittings were match drilled to ensure precise location of the two fittings with respect to each other in the completed strut and to allow the hinge to open properly and completely. This was done with the drill jig in which the two fittings were placed and mated in the precise position (Figure 6). A specified force was applied to the fittings to hold them in position and to ensure complete mating. A stepped drilling technique was used whereby a hole slightly larger than the existing pilot hole was drilled, followed by two progressively larger ones and then reamed to the final .3125-inch diameter. This technique assured quality holes free of chips and delaminations and allowed the use of standard carbide spade drills. Holes were first drilled through one side of the matched pair. The tool was then inverted to drill the other side.

Figure 6. Match Drilling Pivot Fittings

After the pivot fittings were matched drilled, steel pins were bonded into pre-drilled holes at the back of each fitting and by the latch bar hole of one fitting. These pins were used to anchor the two springs that pull the hinge together and the single spring that pulls the latch bar into the closed position. After the adhesive cured, the hinge was easily assembled with no special tools. An assembled hinge is shown in Figure 7.

Figure 7. Hinge

### 5.2 End Fitting Assembly

The end fitting, consisting of two clevises and a lock body, was formed by adhesive bonding with the aid of a bonding fixture (Figure 8). The fixture held all three parts securely in the correct position utilizing the threaded hole in the lock body and the shape of the clevises. Particular attention was paid to achieving the correct fitting diameter so that the desired bond line would be achieved when the fitting was installed in the tube. This was accomplished with special clevis locators built into the tool. After the adhesive cured, the tool was used as a drill jig for match drilling the

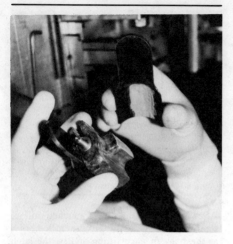

Figure 8. Clevis Bonding

bolt holes in the clevises through which the strut would be bolted to the node. The drilling was accomplished using the same stepped technique used to match drill the pivot fittings.

Two end fitting bonding fixtures were built to expedite the assembly of the large number of end fittings required. Accelerated curing of the epoxy adhesive saved additional time. The specified cure for this adhesive (Hysol EA 934 NA) is 5 to 7 days at 77°F or one hour at 200°F. By placing the tool and fitting, after adhesive application and part location, into an oven at low temperature (120°F) for one hour, the adhesive set up sufficiently for the end fitting to be safely removed from the tool. In this way, one tool was used to bond several end fittings per day, thus increasing throughput of assemblies.

## 5.3 Strut Bonding

Before the end fittings and hinges were bonded into the tubes, a method of assuring a uniform bond line with complete adhesive coverage was sought. Bonding concentric surfaces such as these is difficult because the parts must be slid into place, possibly squeezing out the adhesive and/or producing an uneven bond line. The selected method was to apply a few drops of adhesive to an end fitting, allow them to cure and then sand them down to a thickness equal to the required bond line. After excess adhesive was applied to the fittings and tubes, the cured adhesive "spots" served as shims which assured an adequate, uniform gap throughout the circumference as the fitting slid into place and the adhesive cured. This procedure is illustrated in Figure 9.

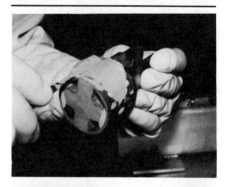

Figure 9. End Fitting Bonding

After the fittings were slid into the tubes, the assembly was installed in the bonding fixture (Figure 10) to precisely locate the end fittings and hinge utilizing the end fitting bolt holes and the hinge pivot pin. The strut bonding

Figure 10. Assembled Struts in Bonding Fixtures

was done in a controlled environment at a constant temperature (65°F) to assure no variation in strut length due to thermal expansion of the long aluminum box beam bonding fixture. As with the bonding of the end fittings, a method was used to accelerate the cure of the adhesive. The tubes were wrapped with heating tapes at the end fitting and hinge locations to apply heat directly at those places only and to prevent any significant expansion of the tool.

The last subassembly operation was installing the threaded insert, tapered pin, spring, and release handle that make up the locking mechanism in the end fitting (Figure 11). This was a simple operation requiring only hand tools.

## 6. FINAL ASSEMBLY

Final assembly of the cell was straightforward. It involved bolting the struts to the nodes in the correct configuration. The cell

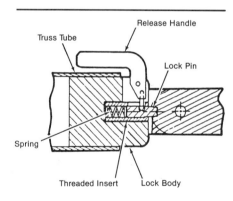

Figure 11. Strut Locking Mechanism

is shown in various stages of deployment in Figures 12, 13, and 14. Fully deployed, it measures approximately 20 feet by 17.5 feet by 8.5 feet and it collapses into a bundle less than 20 inches in diameter.

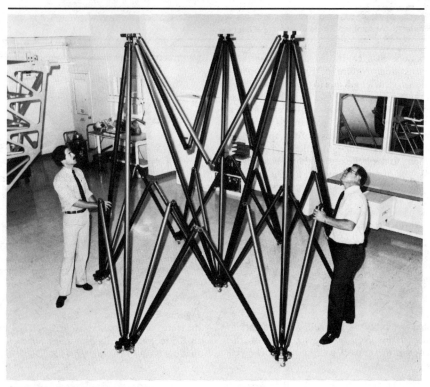

Figure 12. Fully Collapsed Cell

## 7. SUMMARY AND CONCLUSIONS

The completed truss was delivered on 21 July 1986, 11 months after ATP. In that time a basic design was provided, materials selected, purchase orders released, fabrication and assembly tools designed and built, components manufactured and assembled, and the unit assembled and demonstrated successfully.

The fabrication of this composite tetratruss cell established a precedent for a multi-cell structure. Although the issue of configuration has been resolved by NASA in favor of an erectable type, the issue of materials, specifically the decision between composites and aluminum, still has not yet been fully resolved. This program contributed to answers of questions regarding the feasibility of using composite materials in the truss structure. It was found that graphite/epoxy tubes and fittings can be made to the dimensions required, with the precision necessary, and with the desired properties for the Space Station truss structure.

Figure 13. Partially Deployed Cell

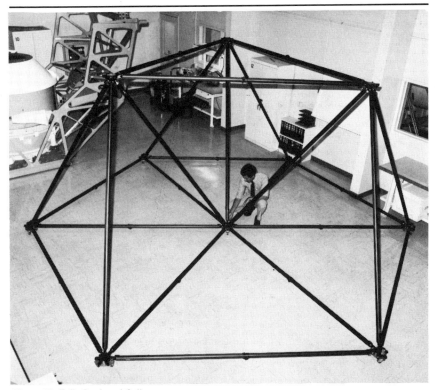

**Figure 14. Fully Deployed Cell**

## 8. BIOGRAPHY

Mr. Robinson is an Advanced Manufacturing and Process Engineer for the McDonnell Douglas Astronautics Company in Huntington Beach, CA. He has worked on the development of several advanced composite structures. He holds a BS degree in chemistry from Colorado State University and a MS degree in chemical engineering from the Colorado School of Mines.

ELECTRICALLY CONDUCTIVE, BLACK THERMAL
CONTROL COATINGS FOR SPACECRAFT APPLICATIONS
3.  PLASMA-DEPOSITED CERAMIC MATRIX

V. F. Hribar, J. L. Bauer, T. P. O'Donnell
Mechanical and Chemical Systems Division
Jet Propulsion Laboratory
California Institute of Technology

ABSTRACT

Five black, electrically conductive thermal control coatings have
been formulated and tested for
application on the Galileo spacecraft.  The coatings consist of
both organic and inorganic systems
applied on titanium, aluminum, and
epoxy fiberglass surfaces.  The
coatings were tested under simulated space environment conditions.  Coated specimens were
subjected to thermal radiation,
convective, and combustive heating
and cryogenic conditions over a
temperature range between -320 °F
(-196 °C) and 1000 °F (538 °C).

Mechanical, physical, thermal,
electrical, and thermooptical
properties are presented for one
of these coatings.  Selection,
preparation of plasma powder,
application, and surface preparation of substrate and methods of
determining electrical resistance
are also given.

This paper describes the preparation, characteristics, and spraying of iron titanate on titanium
and aluminum and presents performance results.

1.0  INTRODUCTION

A black, electrically conductive,
heat resistant thermal control
coating was needed for the Galileo
(GLL) spacecraft.[1,2]  Surface
electrical conductivity is critical for the electrostatic dis-
charge (ESD) requirement for this
deep space mission spacecraft.  A
development program to meet
Galileo spacecraft requirements
was therefore implemented.

Concurrent with the development of
a black graphite/silicate coating,
the concept of plasma-spraying
iron titanate $FeTiO_3$ (Ilmenite) and silicon carbide (SiC) on
titanium foil was conceived.  These
materials were found to be electrically conductive.  Of these
materials, iron titanate was
chosen because silicon carbide did
not meet required thermooptical
properties.

This paper describes the properties and results of tests performed on plasma-sprayed iron titanate on pure titanium foil 2 mils and 5 mils thick.

Ceramic powders other than iron titanate were also evaluated. These included boron carbide ($B_4C$), zirconium carbide (ZrC), and silicon carbide (SiC). Results of these material tests are excluded here. A comparison of property data between iron titanate and silicon carbide is given in Table 1.

The areas of potential application of plasma-sprayed iron titanate on the Galileo spacecraft were the titanium 400-Newton engine plume shield, titanium 10-Newton engine thruster shields, and aluminum shunt radiators (Figure 1).

## 2.0 REQUIREMENTS

Coating requirements for various parts of the Galileo spacecraft have been described previously and include performance to 1000 °F.[1,2] Requirements include electrical conductivity, defined thermooptical properties, adhesion, vibration resistance, resistance to flammability, outgassing, and space radiation effects.

Table 1. Properties of Plasma-Coated Titanium Foil with Iron Titanate and Silicon Carbide

| Property | Iron Titanate | Silicon Carbide |
|---|---|---|
| Electrical resistivity | | |
| Over bare Ti, ohms | 5.0 | 0.5 |
| Over Tiodized Ti, $\Omega/\square$ | $1.5 \times 10^4$ | $1.0 \times 10^3$ |
| Thermooptical | | |
| Emittance ($\varepsilon_n$) | 0.82 | 0.65 (not acceptable) |
| Absorptance ($\alpha_s$) | 0.89 | 0.62 (not acceptable) |
| Adhesion | | |
| Before thermocycling | Passes | Passes |
| After thermocycling[1] | Passes | Passes |
| Bend test, 180° flexing over 1.0" mandrel before and after thermocycling | No effect | No effect |

[1]Heat to 1000 °F (±100 °F) for one hour.
Cool to room temperature.
Immerse in $LN_2$ for 10 minutes.

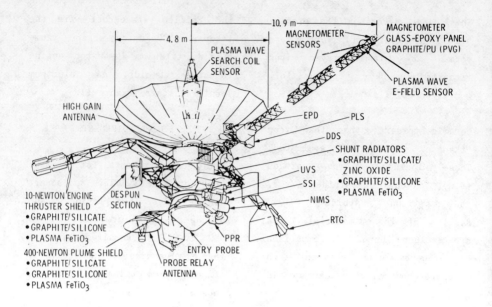

Fig. 1.    Areas of application of some thermal control coatings on the Galileo spacecraft.

### 3.0  FORMULATION

#### 3.1  Chemical Composition

Iron titanate (Ilmenite), $FeOTiO_2$ in its natural form, is procured as coarse iron-black granules. The chemical composition is given below.[3,4]

Percent by Weight

| | |
|---|---|
| Titanic oxide ($TiO_2$) | 63 |
| Ferrous oxide (FeO) | 21 |
| Silicon oxide ($SiO_2$) | 9 |
| Other elements | 7 |

#### 3.2  Melt Temperature

The melt temperature for iron titanate is 1400 °C. A phase dia-gram showing its characteristics is presented in Figure 2.[5]

#### 3.3  Electrical Resistivity

The electrical resistivity of natural iron titanate varies from $10^5$ ohm cm at 100 °K to 10 ohm cm at 500 °K.

#### 3.4  Particle Size and Configuration

The size of iron titanate aggregate in the "as-received" condition is between 50 and 60 mesh. For plasma spraying, the aggregate was ball milled to between 230 and 325 mesh. Scanning electron microscope images illustrated in Figure 3 show the physical characteristics of the aggregate before and after ball milling.

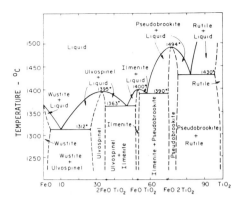

Fig. 2.  Melt temperature for iron titanate (Ilmenite).[5]

100X

AS RECEIVED: 50-60 MESH

100X

AFTER BALL MILLING: 230-325 MESH

Fig. 3.  Iron titanate particle size and characteristics.

## 3.5  Application

For this evaluation, the plasma coat was applied over the same substrate materials used for the plume shields (pure titanium per AMS 4901) and external shunt radiators (aluminum per QQ-250/11 6061 T6). All specimen sample dimensions were standardized as follows: for titanium, 2 inches by 6 inches by 0.002 inches and 0.005 inches in thickness; for aluminum, 2 inches by 6 inches by 0.032 inches in thickness.

## 3.6  Surface Treatment

Several surface treatments on titanium and aluminum were tested for their effectiveness on plasma coat adhesion. A surface treatment requirement was that it be compatible with the configuration of the fabricated part. A simple surface cleaning procedure for titanium and aluminum was devised, as described below.

● For rough surface abrasion: Degrease with isopropyl-alcohol then scour the bare metal surface using a slurry of Alconox/Ajax/ SiC powder (240-320 grit) with a 3M "Scotchbrite" pad. The ratio of Alconox to Ajax is 50-50 by weight. Sufficient SiC is added to promote abrasion. The surface is rinsed successively with tap and dis-

785

tilled/DI water until a water-free break surface is obtained.

- For mild surface abrasion: Follow the same procedure as above, but use an Alconox/ Ajax mixture only, and scour with "Rymple" cloth.

- For surfaces that cannot tolerate abrasion: Rub the bare metal surface with Alconox only, using an oil-free clean cloth, such as "Rymple" or "Clean Care."

Texture characteristics of aluminum and titanium surfaces produced by various chemical and mechanical methods were examined by scanning electron microscope (SEM) techniques. Abrasive treated titanium (Group 4) is presented in Figure 4. Other treatments, not shown here, were also reviewed.[1,2]

The effect of surface treatments on adhesion of $FeTiO_3$ plasma-coated titanium and aluminum is given in Table 2. Paint adhesion failure occurred when the bare metal surface was degreased only.

## 4.0 TEST METHODS AND RESULTS

Tests were devised and performed on the coated titanium panels to

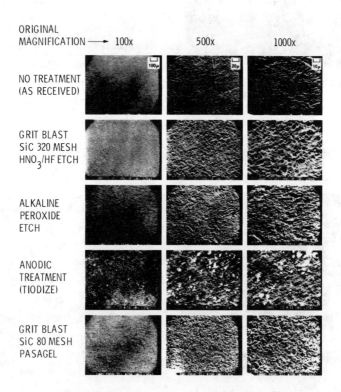

Fig. 4.    Surface texture produced by chemical and grit blast on titanium (Group 4) - SEM photo.

Table 2. Surface Treatment vs. Adhesion for Plasma-Coated Iron Titanate on Titanium and Aluminum[1,2]

| Surface Treatment No. | Description | Coating Thickness[3] Inch (Avg.) | Adhesion[4] |
|---|---|---|---|
| 1 | Degrease "as-received" surface with et alc | .0018 | fail |
| 2 | Degrease Tiodized surface with et alc | .0018 | pass |
| 3 | Silicon carbide grit blast | .0020 | pass |
| 4 | Glass bead grit blast | .0018 | pass |
| 5 | HF/HNO$_3$ etch | .0020 | pass |
| 6 | Alkaline peroxide etch | .0018 | pass |
| 7 | Alconox/Ajax/SiC/3M Scotchbrite rub over bare surface | .0018 | pass |
| 8 | Alconox/Ajax/Scotchbrite rub over Tiodized surface | .0022 | pass |
| 9 | Alconox/Ajax/SiC/3M Scotchbrite rub over bare surface | .0022 | pass |
| 10 | Alconox/Ajax/3M Scotchbrite rub over bare surface | .0020 | pass |

Notes:

(1) Titanium No's 1-8; Aluminum No's 9 and 10

(2) 2.0 inches by 6.0 inches by 0.005 inches thick titanium and 0.032 inches thick for aluminum

(3) Average of 3 specimens

(4) X-cut peel test per ASTM D-3359 Method A tested at 77° (±5°F)

assure performance to the defined requirements. These tests have been described and include convective heating in an inert atmosphere to 1000 °F, radiant heating in a vacuum, direct flame impingement, solvent resistance, electrical resistivity, optical properties, and coating thickness.[1,2]

4.1 Convective Heating in Argon Atmosphere

Two sizes of plasma-sprayed titanium foil (2 inches by 6 inches and 6 inches by 12 inches) were subjected to convective heating in a furnace at a temperature of 1000 °F (±50 °F) in an argon atmosphere for one hour to determine the effect of sample size on paint adhesion. No scaling, peeling, or apparent signs of coating degradation occurred. The X-cut tape peel adhesion tests performed on these samples were acceptable.

A prototype aluminum external shunt heater panel was plasma-sprayed with $FeTiO_3$ and then thermally cycled from 500 °F to -320 °F followed by vibration and humidity tests. Tests results showed no apparent signs of peeling, blistering, or coating degradation.

## 4.2 Conductive Heating – "Hot Plate" Blister Tests

Hot plate blister tests were performed on plasma-sprayed $FeTiO_3$

on aluminuim and titanium to determine the effect of instantaneous heating by conduction at a hot plate temperature of 1000 °F (±50 °F) for titanium and 800 °F (±50 °F) for aluminum. The plasma-sprayed coating showed no degradation when heated by conduction at these temperatures with minimal changes in thermooptical and electrical resistance properties.

## 4.3 Radiant Heating In Vaccum

Titanium samples were subjected to rapid radiant heating in a vacuum furnace to 1000 °F at $10^{-3}$ torr absolute pressure.

The sample was positioned for thermal cycling, heated to 1000 °F, cooled to 75°F, and then submerged into $LN_2$ at -320 °F.

Results of radiant heat tests conducted on plasma-sprayed titanium

Table 3. Property Changes of Plasma-Coated $FeTiO_3$ on Titanium as a Result of Radiant Heating in a Vacuum (1000 °F ±100 °F at $10^{-3}$ Torr)[1]

| Coating Thick. Inch | Thermooptical Properties | | | | Electrical Resistivity | | Adhesion[3] | |
| | $\alpha_s$ | | $\varepsilon_n$ | | $\Omega/\square$ | | | |
| | B | A | B | A | B | A | B | A |
| 0.0015[2] | 0.88 | | 0.84 | | $1.5 \times 10^4$ | | pass | |
| 0.0012 | 0.86 | 0.85 | 0.82 | 0.82 | $1.8 \times 10^5$ | $1.8 \times 10^5$ | pass | pass |
| 0.0018 | 0.85 | 0.85 | 0.83 | 0.83 | $1.4 \times 10^3$ | $1.8 \times 10^3$ | pass | pass |
| 0.0020 | 0.87 | 0.85 | 0.81 | 0.82 | $1.9 \times 10^5$ | $1.9 \times 10^5$ | pass | pass |

B = Before heating, A = after heating
1 = Over 2 mil titanium foil plasma-sprayed on both sides
2 = Control
3 = X-cut peel test per ASTM D-3359

at 1000 °F in a vacuum are given in Table 3. Radiant heating did not appreciably change the plasma thermooptical or electrical resistance properties. The plasma coating showed no visual signs of blistering (Figure 5).

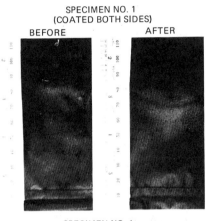

SPECIMEN NO. 1
(COATED BOTH SIDES)
BEFORE          AFTER

SPECIMEN NO. 1
MAGNIFIED VIEW OF ABOVE
BEFORE              AFTER

5X                5X
SPECIMEN NO. 2
BEFORE            AFTER

5X                5X

Fig. 5.  Effect of thermal-vacuum (radiant heat) tests performed on plasma-coated iron titanate on 2 mil titanium foil 1000 °F ±100 °F for 1 hour at $10^{-3}$ torr.

## 4.4  Flame Impingement Test – Atmospheric Conditions

Plasma-sprayed $FeTiO_3$ over titanium was subjected to direct flame impingement to determine its effect on coating degradation. After flame impingement, the specimen was visually examined. No apparent surface change and no flaking or spalling occurred after a 180° bend. The sprayed specimen passed the X-cut tape peel test.

Electrical surface resistance after the test was 3.5 x $10^5 \Omega/\square$ compared to 5.0 x $10^5 \Omega/\square$ before the test.

Thermooptical property changes were:

|            | Before | After |
|------------|--------|-------|
| $\alpha_s$ | 0.80   | 0.85  |
| $\varepsilon_n$ | 0.85 | 0.85 |

In addition to heating the specimen at 1000 °F, the titanium plasma-sprayed specimen was inadvertantly heated to approximately 1800 °F (the specimen glowed cherry red) and held at this temperature for 15 seconds. No visual signs of coating degradation occurred after cooling to room temperature.

## 4.5  Solvent Resistance

Solvent wiping tests performed on plasma-sprayed $FeTiO_3$ over titanium showed good resistance for the solvents tested.[1]

789

## 4.6  Thermooptical Measurements

The normal emittance $(\varepsilon_n)$ of sample coatings was determined by infrared reflective measurement. The solar absorptance $(\alpha_s)$ of the sample coating was determined by measuring reflectivity with a solar reflectometer. Results are reported in Table 3.

## 4.7  Coating Thickness

Coating thickness was determined by a hand micrometer or digital micrometer, as applicable.

The recommended coating thickness for plasma-sprayed $FeTiO_3$ to achieve the proper thermooptical properties was from 0.0015 inches to 0.0030 inches. A cross-section photomicrograph of plasma-sprayed $FeTiO_3$ on each side of 2 mil titanium foil is shown in Figure 6.

## 4.8  Cryogenic Tests

Sprayed specimens were immersed in a liquid bath of $LN_2$ for periods varying from 5 to 10 minutes. Electrical measurements were also performed at cryogenic temperatures with the specimen submerged in $LN_2$. No peeling or spalling was observed on plasma-sprayed specimens.

## 4.10  Humidity Tests

Humidity tests were performed on plasma-sprayed titanium and alumi-num panels. The specimens were subjected to the following conditions:

Relative humidity: 58 ($\pm 5\%$)
Temperature: 77 °F ($\pm 5$ °F)
Duration: 1, 2, and 3 weeks

At the end of each succeeding week, the specimens were air dried and the surfaces visually examined at 20x magnification. Results showed no apparent corrosion on the surface of the plasma-sprayed $FeTiO_3$ on titanium or aluminum. No appreciable change occurred in the thermooptical or electrical properties.

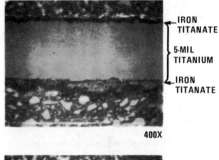

IRON TITANATE
5-MIL TITANIUM
IRON TITANATE

400X

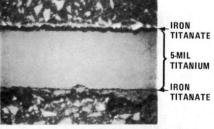

IRON TITANATE
5-MIL TITANIUM
IRON TITANATE

400X

Fig. 6.  Cross-sections of plasma-sprayed $FeTiO_3$ on titanium showing thickness of $FeTiO_3$ and metallurgical bonds.

## 4.11 Radiation

Because the Galileo spacecraft will orbit the planet Jupilter for 20 months, the surface materials must withstand wide temperature excursions as well as radiation fields.

Space-exposed thermal control coatings on the Galileo spacecraft also will be subjected to high-energy proton radiation. Although radiation tests were not performed, the coating is required to sustain twice the Jupiter Orbit Insertion (JOI) plus 12-orbit dose, i.e., $2.8 \times 10^{10}$ rads (Si). Changes in physical, thermooptical, and electrical conductivity should be <25% after exposure.

Ceramic material, such as iron titanate $FeTiO_3$, is an inert material and should not be subject to degradation by radiation. Therefore, iron titanate is considered acceptable for Galileo as well as other spacecraft applications.

## 4.12 Outgassing

Iron titanate $FeTiO_3$ is inert and passes the volatile condensible material (VCM) and total weight loss (TWL) tests consistent with ASTM-E-595.

## 4.13 Adhesion Tests

All adhesion tests were performed at room temperature per ASTM D-3359 Method A, using 3M No. 250 tape. Acceptance criteria were based on no evidence of surface coating removal or deposits on the tape, except for small amounts that remained on the edges of the crosscut (see Table 2).

## 4.14 Bend Test (Flexibility)

Titanium foil (0.002 inches and 0.005 inches in thickness), plasma-sprayed with $FeTiO_3$ (0.0015 inches to 0.002 inches in thickness), was bent 180° over a 1-inch mandrel at ambient room temperature without failure.

## 5.0 SUMMARY AND CONCLUSIONS

The concept of plasma spraying titanium foil with an inert ceramic material has been developed to produce an electrically conductive surface for spacecraft application. The coating has extremely high temperature resistance in addition to acceptable thermooptical properties. Plasma deposition on titanium foil was accomplished while maintaining backface temperatures of 300° F (±50 °F) to overcome part/surface deformation due to thermal stress. Plasma-sprayed iron titanate on titanium can be used where convective, conductive, or radiative type heating and mild abrasive forces are encountered.

The plasma coating met the electrical, thermooptical, and therm-

791

al requirements for the Galileo 400-Newton and 10-Newton engine plume and thruster shields, respectively.

The advantages of plasma-sprayed inorganic ceramic material as a thermal control surface for spacecraft far outweigh organic coatings (see Table 4). This technology is likewise applicable for other types of ceramic materials to produce stable and reliable thermal control surfaces for future spacecraft.

ACKNOWLEDGEMENTS

The research described in this paper was performed by the Jet Propulsion Laboratory, California Institute of Technology under contract with the National Aeronautics and Space Administration.

The development of this electrically conductive black thermo-optical coating would not have been successful had it not been for the assistance of the following individuals for environmental, electrical, thermooptical, and physical tests: Marge Bickler, Phillip Stevens, and Don Bacigalup.

REFERENCES

1. Hribar, V.F., Bauer, J.L., and O'Donnell, T.P., "Electrically

Table 4. Plasma-Sprayed Ceramics vs. Organic Paints

| Parameter | Plasma Coated Ceramic | Graphite/Silicate and Organic Matrix |
|---|---|---|
| Application | Backface temperature must be controlled[1] | Air spray - ease of application |
| Adhesion | Metallurgical bond | Surface bond |
| Moisture uptake | None | Absorbs moisture |
| Chemical | Inert | Water-based or organic resin |
| Abrasion | Resistant | Susceptible |
| Formulation | None required | Requires careful compounding and application |
| Outgassing | None | Must meet space TWL-VCM requirements |
| Flammability | None | Some organic coatings are flammable |
| Thermo-optical | Stable | Subject to change |
| Radiation | Stable | Subject to change |

[1]Use dry ice ($CO_2$) or $LN_2$ on backface during plasma deposition.

Conductive, Black Thermal Control Coatings for Spacecraft Applications, 1. Silicate Matrix Formulations." Society for the Advancement of Materials and Process Engineering, Vol. 31, April 1986.

2. Hribar, V.F., Bauer, J.L., and O'Donnell, T.P., "Electrically Conductive, Black Thermal Control Coatings for Spacecraft Application, 2. Silicone Matrix Formulation." Society for the Advancement of Materials and Processing Engineering, Vol. 18, October 1986.

3. Dana's System of Minerology, 7th Edition, John Wiley and Sons, Inc., 1958.

4. Spectrographic Analysis on Ilmenite Lot No. 1T6016, JPL, April 7, 1982 (internal document).

5. Lever, et al., Phase Diagrams for Ceramists (1969 Supplement), compiled by NBS, The American Ceramics Society, 1969.

## Biographies

Frank Hribar is a graduate of the University of Minnesota with a B.S. in Chemical Engineering. He attended graduate school at UCLA and USC. From 1950 to 1952, he was assigned to the Office of Naval Research with the rank of Commander, USNR. He was a member of the technical staff for Hughes Aircraft Co. from 1952-1960 and a member of the technical staff for the Aerospace Corporation from 1960-1970, assigned to Reentry and Launch Systems. From 1978 to 1984, he was a member of the technical staff, Jet Propulsion Laboratory, Materials and Applications Group, assigned to the Galileo Project, IRAS (Infrared Astronomical Satellite), and Deep Space Network (large antennas). He is presently a consultant in the aerospace industry.

Jerry Bauer is a member of the technical staff at JPL where he has been employed since 1977. He earned a B.S. and M.S. in Chemical Engineering at the University of Dayton and Pennsylvania State University, respectively. He also completed academic requirements for a Ph.D. at Ohio State University. An active SAMPE member, Mr. Bauer has been chairman of the San Diego and Los Angeles chapters, a member of the board of directors for 10 years, and national treasurer for four years. Mr. Bauer has also been employed by the University of Dayton, Ferro Corporation, General Dynamics, Lockheed, and Furane.

Tim O'Donnell is presently Technical Group Supervisor of the Materials Technology Group at JPL. In his 10 years at JPL he has been involved in a variety of metallic and nonmetallic material applications on spacecraft and energy projects. Recently he has been the materials cognizant engineer for the Galileo Project. Mr. O'Donnell holds a B.S. in Metallurgical Engineering from California Polytechnic University, San Luis Obispo, and an MBA. His professional society memberships include SAMPE, American Society for Metals (ASM), and American Ceramic Society (ACS).

# CURE CYCLE OPTIMIZATION OF COMPRESSION MOLDED HIGH MOLECULAR WEIGHT CELAZOLE® PBI RESIN

Mary E. Harb, Jon W. Treat
McDonnell Douglas Astronautics Company - St. Louis Division
P.O. Box 516, St. Louis, MO  63166

Bennett C. Ward
Celanese Advanced Technology Company, PBI Business Unit
P.O. Box 32414, Charlotte, NC  28232

## Abstract

A designed experimental program was conducted evaluating the effects of various processing parameters on high molecular weight Celazole® polybenzimidazole (PBI) molding resin. One objective of this effort was cure cycle development for this relatively new molding resin. A two level partial factorial design was used to screen the relative effects of the eight processing variables investigated. Our approach to this problem also provides identification of two factor interactions as well as significant linear terms.

Independent parameters such as temperature, time and pressure were evaluated with respect to the room temperature mechanical properties tensile, elongation, modulus, flexural strength and hardness. Although outstanding ambient temperature properties were found for this resin, tensile tests on the resin at 316°C (600°F) were characterized by pronounced swelling and blistering that may be attributable to a flaw in the manufacturing tooling. Mechanical test results obtained for this effort have provided sufficient information to satisfy design goals of the experiment and supply indications for the next experimental and testing phases. Conclusions regarding optimum molding parameters are provided.

## 1.  INTRODUCTION

Polybenzimidazoles (PBI) are a class of well-known, heterocyclic polymers possessing outstanding thermal stability (1).  They are the only polymeric matrix materials capable of maintaining load bearing properties for short periods of time at temperatures up to 648°C (1200°F) (2,3). In spite of this advantage, however, large scale structural applications for PBI have yet to be developed, in part because, until recently, PBI resins have either been unavailable or posed problems in handling, or because adequate processing techniques had not been found.

applications. Because of the relatively low degree of polymerization of the prepolymers, and consequently, very high levels of volatiles during cure, applications for low molecular weight neat resin in compression molding have been limited. The high molecular weight resin, however, exhibits much less offgassing during cure, which makes it more suitable for large scale compression molding systems (4), Figure 2.

M53-00312

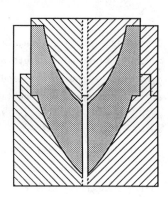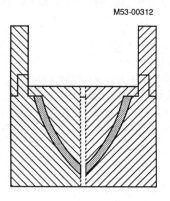

$$\text{COMPACTION RATIO} = \frac{\text{PRECOMPACTED VOLUME}}{\text{FINAL VOLUME}} = \frac{\text{FINAL DENSITY}}{\text{BULK RESIN DENSITY}}$$

COMPACTION RATIO ~ 2 TO 3

**FIGURE 1. COMPRESSION MOLDING TO A KNOWN VOLUME**

Recently, Celanese Corporation has commercialized Celazole® molding resin, which is high molecular weight poly (2,2' - (m-phenylene) -5,5' - bibenzimidazole). This material does not have the melt and flow behavior of low molecular weight PBI prepolymers, which have found specialized use as high temperature matrices in composite

Recent reports have described initial compression molding trials for high molecular weight Celazole® resin. At molding temperatures of 315–427°C (600–800°F), pressures of 13.79 MPa (2000 psi), and final hold times of over 1 hour, specimens were molded with ambient temperature tensile strengths of

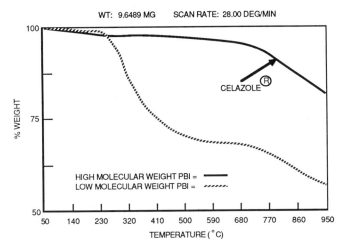

WT: 9.6489 MG    SCAN RATE: 28.00 DEG/MIN

% WEIGHT

CELAZOLE ®

HIGH MOLECULAR WEIGHT PBI = ———
LOW MOLECULAR WEIGHT PBI = ·······

TEMPERATURE (°C)

**FIGURE 2. THERMOGRAVIMETRIC ANALYSIS OF
CELAZOLE ® AND LAMINATING PBI RESINS**

48.28 MPa (7 ksi) (5,6).
Enhancement of tensile strengths
to 64.83 MPa (9.4 ksi) was found
upon postcuring the molded article
at 850°F for 2 hours (5).  A
subsequent paper (4) reported
excellent mechanical properties,
including tensile strengths of up
to 110.3 MPa (16 ksi), for several
developmental Celazole® PBI
resins molded at 418°C (785°F) and
34.48 MPa (5000 psi).  This paper
details optimization of the
molding and cure cycle for
commercially available Celazole®
PBI resin, obtained via a partial
factorial trial evaluating peak
molding temperature, hold time and
pressure, plus intermediate
temperature holds and pressures
during the cycle.

2.  CHOICE OF MOLDING PARAMETERS

As with many experiments, this
processing study used the results
of those earlier experiments to
develop a starting point for the
design.  McDonnell Douglas
Astronautics Co. (MDAC-STL) has
also done considerable processing
work with low molecular weight
(Inherent Viscosity ≤.05) PBI
resin for laminating and some
conclusions were inferred from
that work.

Rheologic and thermal properties
of the Celazole® powder most
greatly influenced initial molding
parameters.  The resin has no
apparent melting point at any
temperature; furthermore, it
retains its granular character
except under very high normal

pressure. Earlier experiments established flow conditions; resin flow does occur at around 418°C (785°F). Work with lower molecular weight laminating resin indicated that 468°C (875°F) and 13.79 MPa (2000 psi) operating conditions were required to achieve resin integrity, that is, low void, with a dark amber, glass-like appearance. It was found that for Celazole® pressures below 20.69 MPa produced pieces that were somewhat cake-like and more void-prone.

Molded panels were made using the determined flow conditions to produce mechanical test specimens. From those experiments a satisfactory compression molding cycle was developed. The experiments also provided a qualitative understanding of the important process variables, the basis of this factorial design. Table I shows some early mechanical test results.

In addition to the flow temperature and pressure, there were other factors to consider. The fineness of the powder, the need for precompaction, the effect of releasing pressure, the need for an intermediate hold, and other variables also seemed significant.

A fractional factorial design was chosen for this process study because it included the most factors for study in the fewest number of experiments. Table 2A is a 16 run fractional factorial that is effective for screening up to eight factors (7). An estimate of the composite effects of groups of two factor interactions may be obtained from the unassigned factors, E1-E7. These columns will indicate which factors have combined or interaction effects. A full factorial to study 8 factors at two levels each would require a $2^8$ or 256 experiments. Once the fractional factorial is

## TABLE 1
## COMPRESSIVE STRENGTHSOF EARLY, UNFILLED,
## UNOPTIMIZED CELAZOLE RESIN. SAMPLES MADE IN 1984 (MDC)

M53-00318

| TEMPERATURE | | YIELD STRENGTH | | ULTIMATE STRENGTH | |
|---|---|---|---|---|---|
| (°C) | (°F) | (MPa) | (Ksi) | (MPa) | (Ksi) |
| 22 | 72 | 372 | 54 | - | - |
| 94 | 201 | 317 | 46 | - | - |
| 202 | 395 | 200 | 29 | 407 | 59.0 |
| 316 | 600 | 96.6 | 14 | 400 | 58.0 |
| 371 | 700 | 55.2 | 8 | 372 | 54.0 |
| 427 | 800 | 13.8 | 2 | 279 | 40.4 |
| 538 | 1000 | 13.8 | 0.2 | 4.8 | 0.7 |

NOTE: SPECIMENS EXPOSED 10 MINUTES TO TEMPERATURE PRIOR TO TEST

# TABLE 2
## A TWO LEVEL PARTIAL FACTORIAL SCHEME WAS USED TO GENERATE SIXTEEN EXPERIMENTS TO STUDY CELAZOLE PBI RESIN PROCESSING

M53-00319

TABLE 2A

| TRIAL | $X_1$ | $X_2$ | $X_3$ | $X_4$ | $X_5$ | $X_6$ | $X_7$ | $X_8$ | $E_1$ | $E_2$ | $E_3$ | $E_4$ | $E_5$ | $E_6$ | $E_7$ |
|---|---|---|---|---|---|---|---|---|---|---|---|---|---|---|---|
| | DESIGN COLUMNS | | | | | | | | UNASSIGNED FACTORS | | | | | | |
| 1 | - | - | - | + | + | + | - | + | + | + | - | - | - | + | - |
| 2 | + | - | - | - | - | + | + | + | - | - | - | - | + | + | + |
| 3 | - | + | - | - | + | - | + | + | - | + | + | - | + | - | - |
| 4 | + | + | - | + | - | - | - | + | + | - | + | - | - | - | + |
| 5 | - | - | + | + | - | - | + | + | + | - | - | + | + | - | - |
| 6 | + | - | + | - | + | - | - | + | - | + | - | + | - | - | + |
| 7 | - | + | + | - | - | + | - | + | - | - | + | + | - | + | - |
| 8 | + | + | + | + | + | + | + | + | + | + | + | + | + | + | + |
| 9 | + | + | + | - | - | - | + | - | + | + | - | - | - | + | - |
| 10 | - | + | + | + | + | - | - | - | - | - | - | - | + | + | + |
| 11 | + | - | + | + | - | + | - | - | - | + | + | - | + | - | - |
| 12 | - | - | + | - | + | + | + | - | + | - | + | - | - | - | + |
| 13 | + | + | - | - | + | + | - | - | + | - | - | + | + | - | - |
| 14 | - | + | - | + | - | + | + | - | - | + | - | + | - | - | + |
| 15 | + | - | - | + | + | - | + | - | - | - | + | + | - | + | - |
| 16 | - | - | - | - | - | - | - | - | + | + | + | + | + | + | + |

TABLE 2B

| TRIAL | $P_0$ MPa | $P_1$ MPa | $T_1$ ($O_C$) | $t_1$ (HRS) | $P_2$ MPa | $T_2$ ($O_C$) | $T_3$ ($O_C$) | $t_3$ (HRS) |
|---|---|---|---|---|---|---|---|---|
| 1 | .345 | .345 | 204 | 2 | 55.2 | 410 | 441 | 3.0 |
| 2 | 6.89 | .345 | 204 | 0 | 27.6 | 410 | 468 | 3.0 |
| 3 | .345 | 6.89 | 204 | 0 | 55.2 | 371 | 468 | 3.0 |
| 4 | 6.89 | 6.89 | 204 | 2 | 27.6 | 371 | 441 | 3.0 |
| 5 | .345 | .345 | 316 | 2 | 27.6 | 371 | 468 | 3.0 |
| 6 | 6.89 | .345 | 316 | 0 | 55.2 | 371 | 441 | 3.0 |
| 7 | .345 | 6.89 | 316 | 0 | 27.6 | 410 | 441 | 3.0 |
| 8 | 6.89 | 6.89 | 316 | 2 | 55.2 | 410 | 468 | 3.0 |
| 9 | 6.89 | 6.89 | 316 | 0 | 27.6 | 371 | 468 | 1.5 |
| 10 | .345 | 6.89 | 316 | 2 | 55.2 | 371 | 441 | 1.5 |
| 11 | 6.89 | .345 | 316 | 2 | 27.6 | 410 | 441 | 1.5 |
| 12 | .345 | .345 | 316 | 0 | 55.2 | 410 | 468 | 1.5 |
| 13 | 6.89 | 6.89 | 204 | 0 | 55.2 | 410 | 441 | 1.5 |
| 14 | .345 | 6.89 | 204 | 2 | 27.6 | 410 | 468 | 1.5 |
| 15 | 6.89 | .345 | 204 | 2 | 55.2 | 371 | 468 | 1.5 |
| 16 | .345 | .345 | 204 | 0 | 27.6 | 371 | 441 | 1.5 |

complete and the three or four most important variables are determined, a full factorial may be performed.

Eight factors were chosen for study and their levels were set based on previous experience. Table 2B shows the conditions for each experiment. Po is the precompaction pressure. This factor was chosen to determine the effect of consolidating the resin powder before starting the cure. Pl is an intermediate pressure chosen to see if releasing the pressure will have any effect. A pressure drop might "burp" the resin, allowing volatiles and entrapped air to escape before final compaction.

T1 is the first hold temperature and t1 is the length of the hold. This hold was also intended to allow volatiles to escape. P2 is the compaction pressure and T2 is the compaction temperature. These are the conditions where flow occurs. T3 is the ultimate cure temperature and t3 the final hold time. This hold is necessary to develop good mechanical properties and also to allow remaining volatiles to diffuse. The experimental cure cycles range from a short cycle with a 1.5 hour hold at 441°C (825°F) to a long cycle with a 2.0 hour intermediate hold and a 3.0 hour hold at 468°C (875°F). Experiments with PBI laminating resin have shown that ablation occurs with cycles that are longer than three hours at that temperature. Figure 3 shows two typical cure cycles and graphically illustrates the "burping" step.

## 3.  EXPERIMENTAL SETUP

A five piece stainless steel mold was designed and machined for this processing study, Figure 4. The tool consisted of a base with a circular cavity. A split ring was inserted in the cavity and bolted to the base. A steel plate fit in the ring and prevented resin from flowing between the ring and the base. The cavity was filled with resin. A steel piston was used to compact the resin into a circular disk 23.5 cm in diameter. This type of mold is easy to assemble and disassemble and makes part removal easy. A thermocouple port was machined into the tool for monitoring the part temperature. The whole set up was placed in a hydraulic press with heated platens. The operator set the temperatures and pressure controls

M53-00316

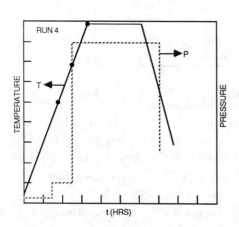 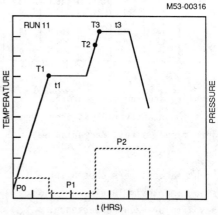

**FIGURE 3.  CURE CYCLE SCHEMATIC  TYPICAL OF THE EXPERIMENTS**

according to the conditions of the experimental design. Heat up rate was uniform at 2.8°C/min. The experiments were randomized so they were not run in the order listed in Table 2. Each panel was marked according to its cure cycle and set aside for machining of the test specimens. Each plate was approximately 0.613 cm thick.

Tooling for fabrication of these disks was machined to such tight tolerance that no flash from the resin was found even under 55.17 MPa (8000 psi) press cycles, although such is not normally the case. Sixteen parts were fabricated. The snugness of the tooling probably did not allow adequate ventilation of the phenol and water produced during resin cure, and thus the purpose of including an intermediate hold cycle for volatilization may have been partially thwarted.

## 4. TESTING

All room temperature tests and machining was performed at Celanese Research Facility in Summit, New Jersey under Mr. Frank Haimbach's direction. Tensile tests on 8.9 cm resin dogbones were used to measure yield, load to break, modulus and per cent elongation. Three point bending was performed on flexure bars. Table 3 reports the results of tensile and flexural testing for each trial corresponding to Table 3. The values reported are the simple average of four test specimens. Density determinations for these experiments show that the specimens were uniformly compacted and this measurement was not used to analyze the experiments.

## TABLE 3
## ROOM TEMPERATURE AND
## 316°C TEST RESULTS

M52-00317

| TRIAL | SPECIFIC GRAVITY | TENSILE (MPa) | ELONGATION (%) | T. MODULUS ($10^2$ MPa) | FLEXURAL BREAK (MPa) | FLEXURAL MODULUS (MPa) | 311°C TENSILE (MPa) |
|---|---|---|---|---|---|---|---|
| 1 | 1.31 | 75.9 | 1.2 | 64.8 | 158 | 68.3 | 14.4 |
| 2 | 1.30 | 124 | 2.2 | 58.6 | 145 | 64.8 | 30.0 |
| 3 | 1.31 | 117 | 2.0 | 59.3 | 158 | 66.2 | 21.0 |
| 4 | 1.30 | 75.9 | 1.2 | 60.7 | 117 | 66.5 | 12.5 |
| 5 | 1.30 | 124 | 2.2 | 60.7 | 221 | 65.5 | 33.9 |
| 6 | 1.31 | 69.0 | 1.2 | 60.7 | 110 | 66.2 | 15.2 |
| 7 | 1.30 | 75.9 | 1.3 | 60.0 | 152 | 66.2 | 15.2 |
| 8 | 1.30 | 110 | 2.0 | 59.3 | 152 | 64.8 | 28.7 |
| 9 | 1.30 | 96.6 | 1.6 | 60.7 | 145 | 64.8 | 23.4 |
| 10 | 1.30 | 75.9 | 1.1 | 66.2 | 145 | 68.3 | 17.6 |
| 11 | 1.30 | 75.9 | 1.2 | 62.8 | 103 | 64.8 | 16.4 |
| 12 | 1.30 | 96.6 | 1.6 | 58.6 | 131 | 66.2 | 20.0 |
| 13 | 1.30 | 75.9 | 1.2 | 60.7 | 96.6 | 65.5 | 13.9 |
| 14 | 1.30 | 89.7 | 1.6 | 59.3 | 138 | 65.5 | 22.0 |
| 15 | 1.30 | 96.6 | 1.5 | 63.4 | 145 | 62.1 | 18.6 |
| 16 | 1.31 | 62.1 | 1.0 | 62.8 | 103 | 66.9 | 13.7 |

## EVALUATION OF RESULTS

Tensile and tensile elongations values demonstrate a linearity in material performance typical of a brittle system. Experiments numbers produced resin disks with extremely good values for an unfilled resin system. Results of the high temperature testing (316°C tensile tests) are also documented in Table 3.

Data averages for the sixteen trials were evaluated according to the method outlined in Reference 7, and with SAS regression analysis. Table 4 summarizes the results of each analysis' identification of important factor effects and possible interactions.

temperature, $T_3$. Pressure effects appear as second order effects for tensile properties. Initial compaction pressure, $P_0$, is a prominent factor effect for room temperature flexure properties. It is not clear that $P_0$, $P_1$, $P_2$ effects are separable and distinct. $P_2$ effects are known to greatly affect compaction and void content. Intermediate hold conditions $t_1$, $T_1$ have mainly second order effects; the temperature ($T_2$) at which ultimate compaction pressure is applied is also significant.

The data distribution shows that values tend to be clustered, but this can be considered typical of the data produced by a two level

**TABLE 4**
**FACTORS, INTERACTIONS, WHICH MOST AFFECT MOLDED CELAZOLE QUALITY**

M53-00313

| DEPENDENT VARIABLE | SAS II [*] | FACTORIAL ANALYSIS [**] |
|---|---|---|
| R. T. TENSILE | $t_3$, $T_3$ | $T_3$, $t_3$, $T_3 t_3$, $T_1 t_1$, $P_1 P_2$ |
| 316°C TENSILE | —— | $T_3 t_3$, $T_1 t_1$, $P_1 P_2$ |
| R. T. ELONGATION | $t_3$, $T_3$ | $T_3$ |
| R. T. TENSILE MODULUS | $t_1$, $T_2$, $t_3$, $T_3$ | $t_1$, $T_2$, $P_1 T_1$, $T_2 T_3$ |
| R. T. FLEXURE | $P_0$, $t_1$, $T_3$, $t_3$ | $P_0$, $T_2$, $T_3$ |
| R. T. FLEX MODULUS | $P_0$, $P_2$, $T_3$ | $P_0$, $P_2$ |

[*] 85% CONFIDENCE
[**] 90% CONFIDENCE

Analysis shows that the most prominent linear effects are from final curing time, $t_3$, and

screening design. However, since the spread is not evidently very wide, considering the variety of experiments that were performed, this clustering of the data can be indicative of at least one or a combination of several other considerations. The following list is a thoughtful construct of some possibilities that can help to explain the data derived from the study and supply an indication for the progress of further work.

1. This 'screening design' was a take-off from various early experiments--the lack of scatter could indicate that the 'best' cycle is already at hand.

2. High and low levels could have been spaced more broadly. However the variable levels used were based upon practicality and earlier work. It may be valuable to examine longer and shorter cure cycle times and at least one higher cure temperature.

3. On the flip side, it could be that for a given parameter there is no effect on resin properties as long as the variable remains above a certain threshold, and thus linear regression techniques would not necessarily measure the true effect of that variable. For example, compaction pressure and intermediate hold cycles in these experiments were not identified as having prominent first order effects, yet it is known from earlier experimentation that compaction pressure has

significant bearing on density and resin flow. An intermediate hold cycle is generally required where thermal lag in the tooling or part is a problem. This experiment was not structured to measure such behavior. This on-off behavior can characterize the behavior of many kinds of manufacturing conditions that can otherwise go unaccounted for in a formal processing evaluation, Figure 5, although, data and its distribution could definitely be affected.

5.  HIGH TEMPERATURE RESIN TESTING

This topic deserves special attention in this article because the results of 316°C tensile testing of Celazole® resin dogbones may teach an important manufacturing lesson. A significant mechanical property test at or near the apparent Tg (410°C) is obviously a desirable screening tool for high use temperature materials. However during the performance of 480°C tensile tests, disturbing phenomena occurred. Gross swelling and blistering of the cured resin occurred at temperatures as low as 315°C in less than three minutes. These phenomena are not limited to samples from this experiment because the same phenomena have occurred with PBI resin nose cones manufactured in 1985 at MDC.

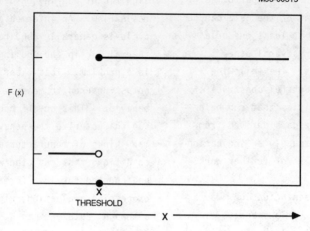

**FIGURE 5. COMPACTION PRESSURE ($P_0$, $P_1$, $P_2$) APPEARS TO BEHAVE AS A VARIABLE WHICH, AT A CRITICAL THRESHOLD, GREATLY AFFECTS DENSITY AND STRENGTH. ABOVE THE THRESHOLD FURTHER GAINS IN RESIN QUALITY ARE SMALL.**

Studies are underway to precisely pinpoint the cause. Celanese has been able to compression mold Celazole® resin with a similar cure cycle and a different pressing, tooling and curing arrangement. Mass spectroscopic analysis on the resin has been performed. Mass spec shows the presence of significant levels of phenol and water at 400°C, but no other solvent materials were found. The tight tooling may be preventing adequate offgassing ventilation. Excessive dissolved volatiles will cause plasticization and blistering of the resin.

In addition, because of the hygroscopic nature of this resin, a drying technique will be evaluated. Parts made by Celanese after drying the resin in ventilated tooling do not exhibit thermoplastic characteristics at elevated temperature to the same degree. Another possibility is the requirement for a slow postcure to allow the diffusion of volatiles to occur.

6. SUMMARY

This screening design has been effectively used to increase room temperature mechanical properties of neat molded Celazole® PBI resin. Molding cycles have been discovered which can yield 124 MPa (18 ksi) in tension, 221 MPa (32 ksi) in flexure and 372 MPa (54 ksi) in compression. High temperature testing of tensile dogbones may have helped identify significant process and tooling characteristics that are required for excellent resin performance properties, i.e., resin drying and tooling ventilation.

Ultimate cure time, $t_3$ and temperature, $T_3$, were identified as the most significant cycle conditions for producing good unfilled Celazole® resin parts. Compaction pressure and some second order parameter interactive effects have also been identified for study. A logical course of experimentation could be performed that would be used to mathematically model this cure cycle and produce an equation which quantitatively relates each cure cycle parameter to mechanical properties.

e.g.,

$$FLEXURE = aP_0 + bT_2 + cT_3 + CONSTANT$$

Unfilled molded Celazole® resin has very high mechanical properties--these could be improved with the addition of an appropriate fiber system to add dimensional stability and even higher modulus. Although resin swelling at high temperatures could prove to be a limiting factor in the applications of the resin system, there have been insufficient investigations into the causes of that phenomenon to pass final judgment.

ACKNOWLEDGMENTS

The authors wish to thank Dr. Frank Haimbach, Celanese Research Company, for invaluable assistance in physical testing and data interpretation.

## Biographies

Mary E. Harb is an engineer with McDonnell Douglas Astronautics Company in St. Louis. She is developing various high temperature materials and processes for advanced composite missile airframe and space applications. Her expertise is characterization of polybenzimidazole, polyimide processing, and novel resin matrix systems. Mary has co-authored two other publications on PBI and a fourth one on modelling of an extractive metallurgy process.

Jon W. Treat has a B.S. in Chemical Engineering. He spent two years working on process development and characterization of high temperature composites at MDAC-STL. Currently he is supporting graphite/epoxy production for the AFT Propulsion Subsystem (APS) program.

Bennett C. Ward was born in Laurel, Delware, and received his B.S. in Chemistry from Duke University in 1976. He received a Ph.D. in Organic Chemistry from

the University of North Carolina –
Chapel Hill in 1980, after which
he joined Celanese Corporation,
where he is currently a research
chemist with the PBI Business
Unit. He is working on process
and product development of
Celazole PBI resin for compression
molding and matrix applications.
He is author of three patents and
10 publications.

## REFERENCES

1. Levine, H. H., Encycl. Polymer
   Sci. Technol., 11 188 (1969).

2. Powers, E. J. Serad, G. A.,
   "History and Development of
   Polybenzimidazoles," ACS
   Symposium on the History of
   High Performance Polymers, Apr
   1986.

3. Cassady, P. E., "Thermally
   Stable Polymers, Synthesis and
   Properties," (Dekker, New
   York, 1980), pp. 168-173.

4. Ward, B. C., Fabricating
   Composites Group, Baltimore,
   MD (Sep 1986) EM86-704.

5. Treat, J. W., Jones, J. F.,
   Fountain, R., "Compression
   Molding of Polybenzimidazole,"
   High Temple Workshop V,
   Monterey, CA, 1985.

6. Jones, J. F., Fountain, R.,
   International Conference on
   Composite Materials IV, AIME,
   Warrendale, PA (1985) p. 1591.

7. Strategy of Experimentation,
   Revised Edition, Oct 1975, E.
   I. DuPont DeNemours & Co.,
   Wilmington, Delaware, 1969

DEVELOPMENT OF DAMAGE TOLERANT ACETYLENE TERMINATED RESIN/CARBON FIBER
COMPOSITES: II

Dr. Paul A. Steiner
Jim Browne
Michele T. Blair
John M. McKillen
Hysol Aerospace & Industrial Products Division
A Division of The Dexter Specialty Chemicals Group
2850 Willow Pass Road
Pittsburg, California 94565

ABSTRACT

In research sponsored by the U.S.
Air Force (Proposal No. F33615-84-
R-5050, AT Technology Transition)
formulated acetylene terminated
bisphenol A (ATB) and acetylene
terminated sulfone (ATS) resins
have been developed which exhibit
dramatic improvements in fracture
toughness while maintaining room
temperature mechanical properties
equivalent to present epoxy/graphite
composites and 450°F dry mechanical
properties equivalent to bismaleimides. The hydrophobic nature of
the m-ATB and m-ATS resins results
in a relatively moisture insensitive
cured resin with good properties
retention after exposure to humidity.
Neat resin and composite mechanical
test results are presented on formulated m-ATB and m-ATS resins. The
characteristic low moisture absorption of the AT resins has been
preserved so that these formulations
display 400°F to 425°F wet service
capability.

1. INTRODUCTION

The development of acetylene terminated (AT) resin systems began at
the Air Force Wright Aeronautical

Laboratories in the early 1970's [1].
After several materials had been
synthesized and initially characterized, two materials 4,4'-bis
(3-ethynlphenoxy) isopropylidenedi-
phenol (m-ATB) and 4,4'-bis
(3-ethynlphenoxy) diphenylsulfone
(m-ATS) were identified as the
members of that family of resins
which were most likely to possess
the desired uncured resin properties
(low Tg or melt temperature),
processing characteristics comparable to epoxy resins, high
temperature composite mechanical
properties, and the potential for
low cost synthesis [3,4]. At a
resin composition which is approximately 75% monomer and 25% oligomer
the bisphenol A backbone gives a
resin which is a viscous liquid at
ambient temperature while the sulfone backbone gives an amorphous low
melting solid. The acetylene terminated resins display excellent pot
life and can be thermally homopoly-
merized via free-radical addition
reactions leading to a conjugated
polyene without the evolution of
volatile by-products. The polymer
network present during the early
stages of cure is pictured in
Figure 1. The cured resins have
high thermal and thermal oxidative

stability making them suitable for long term use at 400-450°F and short term use at 550-600°F [5]. Their hydrophobic nature results in a relatively moisture insensitive cured resin with good properties retention after exposure to high humidity conditions. State-of-the-art structural epoxies and bismaleimides have had drawbacks in the area of moisture aging and hot/wet use temperatures. This has limited the use of epoxies and bismaleimides to 180°F and 350°F respectively in humid environments. AT resins are not subject to these limitations and therefore display acceptable hot/wet mechanical properties at 400-450°F [6,9]. Unfortunately, like epoxies and bismaleimides the AT resins exhibit a high degree of brittleness making the composite susceptible to damage by impact. The intent of this Air Force sponsored research was to toughen the m-ATB and m-ATS resins so that they may be considered for use in primary aircraft structural applications. What follows is a discussion of the physical and mechanical properties of formulated resins based on the m-ATB and m-ATS resins.

## 2. EXPERIMENTAL

### 2.1 Cure/Postcure Cycle Development

Since these AT resins have been extensively studied by the Air Force, a cycle already exists [2,8]. This cure cycle consists of one hour at 80°C followed by 15 hours at 140°C, 5 hours at 170°C, and then a 4 hour postcure at 250°C for m-ATB and 300°C for m-ATS. In an attempt to optimize the cure cycle and eliminate the 15 hour hold at 140°C the following cure cycle was developed:

Pressurize to 100 psi
Heat-up to 121°C at 1°C/minute
Hold at 121°C for 120 minutes
Heat-up to 179°C at 1°C/minute
Hold at 179°C for 300 minutes
Cool down at 1°C/minute
240 minute post cure in an oven
  unrestrained at 250°C for
  m-ATB and 300°C for m-ATS

The above cure cycle gave equivalent m-ATB neat resin properties and produced high quality void free laminates as judged by SEM photomicrographic inspection.

### 2.2 Characterization and Evaluation of m-ATB and m-ATS

Ideally a composite matrix resin should have a minimum viscosity which is low enough to allow for good fiber wet out during the production of the prepreg, but high enough to allow for only controlled flow during the autoclave cure of a prepreg laminate. Both m-ATB and m-ATS have minimum viscosities of less than 500 centipoise (Table 1) which results in good fiber wet out. However, the combination of this extremely low viscosity with the long gel times (Table 2) gives prepreg that is difficult to process due to excessive flow. For the prepreg to exhibit good tack, drape and handleability, it is necessary for the resin to have an uncured Tg less than ambient temperature. M-ATB with a Tg below ambient (Table 1) produces a very tacky prepreg while m-ATS, a low melting solid, produces a boardy prepreg with no tack.

Of particular interest are the thermal properties of these resins. Due to their largely aromatic structure and the high crosslink density of the cured resins, m-ATB and m-ATS would be expected to have very good elevated temperature properties.

From the neat resin modulus (Figure 6) obtained from the DMTA, m-ATB retains 64.9% and 42.9% of its ambient temperature modulus at 400°F and 450°F respectively. Replacement of the isopropylidene group in m-ATB with the sulfone group results in a significant improvement in the thermal oxidative stability of the cured resin as indicated by the thermal gravimetric analysis in Table 1. This translates to a retention of 72.4% and 62.9% of the ambient temperature modulus at 400°F and 450°F respectively. M-ATS still maintains 50% of its ambient temperature modulus at 500°F.

The structure of the AT monomer (Figure 1) and its cure mechanism is indicative of a very non-polar cured resin. The hydrophobic nature of the resin results in an extremely low equilibrium moisture of 0.6% for m-ATB and 1.09% for m-ATS when exposed to immersion in 71°C (160°F) water. The wet modulus versus temperature is given in Figure 7 and correlates well with the low moisture absorptions. The wet Tg's were difficult to obtain since the specimen tends to dry as the temperature is ramping upward. The wet Tg for m-ATB was 8°C below the dry Tg while the wet Tg for m-ATS actually came out greater than the dry (due to specimen drying). More indicative of the elevated temperature wet performance were the wet modulus values. M-ATS had the best hot/wet performance maintaining 69% of its ambient temperature dry modulus at 350°F versus 59% for m-ATB. At 425°F these values decrease to 61% for m-ATS and 36% for m-ATB. The wet properties of these AT resins may be considered excellent as compared to BMI's and epoxies which typically have wet Tg's 50 to 100°C lower than the dry Tg.

The neat resin fracture toughness was measured by use of the compact tension specimen of ASTM E399. The specimen utilized is pictured in Figure 5. Prior to testing, a sharp fatigue crack was tapped into the specimen using the notch as a guide for a heavy duty razor blade. The $G_{IC}$'s calculated from these resins appear in Table 1. At values of approximately 35 J/m$^2$ (0.2 in-lbs/in$^2$), m-ATB and m-ATS may be considered extremely brittle resins. As a comparison, typical unmodified epoxies or BMI's have $G_{IC}$'s in the range of 50-70 J/m$^2$ (.3-.4 in-lbs/in$^2$).

## 2.3 M-ATS and m-ATB Composite Mechanical Testing

Prepreg was fabricated from the m-ATB and m-ATS resins applied to Hysol Grafil XAS high strain carbon fiber. The fiber had a modulus of approximately 231 GPa (33.5 Msi) and an average tensile strength of 4.5 GPa (650 Ksi) and 4.3 GPa (618 Ksi) for the m-ATS and m-ATB based batches respectively. In addition these fibers received standard oxidative surface treatments followed by the application of an epoxy ("A" size) which was approximately 1.3% by weight. The prepreg physical properties are given in Table 2.

As expected the m-ATB produced a prepreg with excessive tack but good drape and handleability and the m-ATS a boardy prepreg with no tack. The AT resin prepregs were processed in a manner similar to carbon/epoxy prepregs with an identical bagging procedure. The cure/postcure cycle used was the one described previously in Section 2.1.

All mechanical property tests were performed using test machines complying with ASTM E4. Specimens tested at room temperature were tested at 75+ 10°F. Prior to initiating the test load dry specimens were held at temperature for 10+ 3 minutes and wet specimens, with the exception of shear and flexure, for 2+ 1 minutes. Wet interlaminar shear and flexure were held at temperature for 5+ 1 minutes. The complete composite test results appear in Tables 3 and 4.

When a specimen is tested in flexure one face is under compression while the other is in tension so that the flexure strength and modulus are a good measure of overall composite properties. The elevated temperature dry and wet flexure properties should parallel the neat resin DMTA modulus results. Flexure strength versus temperature is graphically presented in Figure 11. At room temperature the m-ATS flexure strength is approximately 48% greater than m-ATB. This was most likely due to the stronger fiber/resin interface present in the m-ATS composite (see Section 3.0). With increasing temperature both resins exhibit excellent retention of their room temperature properties maintaining approximately 78% and 66% of their room temperature properties at 350°F and 450°F respectively. From our neat resin results, we would have expected m-ATS to have superior elevated temperature flexure properties in comparison to the m-ATB.

Both m-ATS and m-ATB have low equilibrium moisture absorptions so that one might expect their wet composite mechanical properties to be only slightly lower than the dry values. The wet flexure strength (Figure 12) versus temperature gives a good representation of the elevated temperature wet composite mechanical properties for these two resins. While m-ATB resin absorbs only .6% water at equilibrium the ambient temperature wet flexure strength is only 68% of the dry whereas it is almost 91% for the m-ATS which absorbs 1.09% water at equilibrium. At 350°F and 425°F wet m-ATB retains a greater percentage of its ambient temperature dry properties than m-ATS, but m-ATS retains greater absolute flexure strengths. From the wet flexure results, it is obvious that the neat resin moisture absorption values alone are insufficient to predict the wet composite mechanical properties. This could be due to fiber/resin interface differences between the two resins. Interface differences are especially critical in hot/wet tests where water could potentially wick along the interface and thus complicate predictions of wet composite mechanical properties based upon neat resin moisture absorption values.

During compression, a matrix dominated property, the matrix must stabilize the fibers to prevent microbuckling. This becomes most critical under elevated temperature wet conditions. At 350°F the 0 degree wet compression strength of m-ATS is more than 50% greater than that of m-ATB, but the m-ATB does retain a greater percentage of its ambient temperature dry value. The m-ATS 350°F/wet compression strength

is superior to that of state-of-the-art BMI's and it also maintains a value at 425°F/wet which would still be superior to most BMI's at 350°F/wet. These wet results are paralleled by the +45° wet shear modulus ($G_{I2}$) results.

As the neat resin results reported, m-ATS and m-ATB are brittle. This was confirmed by their double cantilever beam composite $G_{IC}$ values and also their edge delamination strengths. The high $G_{IC}$ of 490 J/m$^2$ for m-ATS was due to a large amount of fiber bridging during the test. The actual value is more likely down at the 193 J/m$^2$ level obtained for the m-ATB. The edge delamination strength was also used to measure the interlaminar fracture toughness. The lay up utilized gives a fracture toughness, Gc, which is 57% $G_{IC}$ or crack opening and 43% $G_{2C}$, or shear opening. Values of 71 MPa and 55.8 MPa for m-ATB and m-ATS respectively are extremely low especially when compared to unmodified epoxies which would typically get values of 170-200 MPa. The low values may in part also be due to the weak AT resin/carbon fiber interface present in these composites.

## 2.4 Formulation

For the purposes of toughening the AT resins, two basic approaches were established [7]. The first was to attempt to decrease the crosslink density by the addition of monofunctional unsaturated modifiers which react with the AT resins and/or the addition of high glass transition temperature (Tg) ductile polymers which remain in a continuous phase with the cured resin. The second approach was to increase the toughness by the addition of a more ductile second phase as is commonly accomplished by the addition of elastomers or tough glassy thermoplastics to epoxies. Additional reactive diluents were added when necessary to increase the tack, drape and handleability of the prepreg. Two formulated m-ATB resins, AF-4 and AF-8, and two formulated m-ATS resins, AF-11 and

AF-20, were developed.

## 2.5 Characterization and Evaluation of m-ATB and m-ATS Formulations

The goal of the formulations was to improve the deficiencies present in m-ATB and m-ATS. As stated previously, the minimum viscosities of m-ATS and m-ATB are too low for prepreg resins. All four formulations have minimum viscosities in the range of 7.9-14.6 poise (Table 1) which would be considered optimum for a prepreg resin. Unlike the unmodified m-ATB and m-ATS all four formulated prepreg resins had good tack, drape and handleability. Differential Scanning Calorimetry (DSC) run on AF-4, AF-8, AF-11, and AF-20 at a heat-up rate of 10°C/ minute exhibit slightly higher peak exotherms than the unmodified resins, but the formulated resins may still be cured with a 350°F autoclave cure cycle followed by the higher temperature post cure.

A Cahn TGA was used to measure the thermal and thermal oxidative properties of the cured resins. The finely ground powdered resins were heated at a rate of 10°C/minute and the polymer decomposition temperature was taken as the point of 10% weight loss. There is some variation in the values for the formulated resins, but they are all comparable to the m-ATB or m-ATS controls.

The upper service temperature for a thermoset resin is typically limited by the glass transition temperature. Figure 5 illustrates the neat resin modulus temperature profile for the formulated resins and controls m-ATB and m-ATS. The Tg's for the formulated m-ATB resins, AF-4 and AF-8, were 262°C and 273°C respectively which was greater than the control. With the m-ATS formulations, AF-11 and AF-20, the Tg's were slightly lower than the control at 305°C and 302°C respectively. From Figure 5 it would appear that AF-8 had better retention of its elevated temperature properties than AF-4 retaining 56.3% of its ambient temperature

modulus at 400°F and 32.1% at 450°F. The two m-ATS formulations had nearly identical elevated temperature properties maintaining 76.5% at 400°F and 69.8% at 450°F of their ambient temperature modulus.

The equilibrium moisture absorption values in Table 1 indicates that the formulations have moisture absorptions that are more than double the values obtained on the controls. These values, however, are still far below the 4-5% typically obtained for epoxies or bismaleimides. The high wet Tg's (relative to dry Tg's) are indicative of the low moisture absorptions. Of the m-ATB formulations, AF-8 retains 59% of the ambient temperature dry modulus at 350°F versus 49% for AF-4 and these values drop to 27% and 25% respectively for AF-8 and AF-4 at 425°F. As with the m-ATS control, the formulations have superior hot/wet properties relative to m-ATB. AF-20 and AF-11 retain 71% and 65% respectively of their ambient temperature dry modulus at 350°F and 56% and 44% respectively at 450°F.

Of critical importance is whether there was any increase in the neat resin $G_{IC}$ of these formulations. Figures 2-4 show the SEM photomicrographs of the resins. The only resin to exhibit some evidence of a second non-continuous phase is AF-4 in which the particles are typically less than 5 microns in diameter. All of the formulations, however, do exhibit highly textured fracture surfaces indicative of toughened resins. The $G_{IC}$'s generated (Table 1) are in agreement with the SEM observations. The $G_{IC}$'s range from a low of 112 J/m$^2$ (.64 in-lbs/in$^2$) for AF-11 to a high of 165 J/m$^2$ (.94 in-lbs/in$^2$) for AF-4. This represents an increase in toughness from 300% to almost 500% over the control AT resins.

## 2.6 m-ATB and m-ATS Formulations Composite Mechanical Testing

As mentioned in Section 2.5 all of the formulated AT resin prepregs have good tack, drape and handleability. The prepreg physical data presented in Table 2 is consistent with 100% solids prepreg resin systems. The relatively high volatiles content reported for AF-11 and AF-20 is due to the presence of reactive diluents. These diluents polymerize during the autoclave cure and therefore do not cause any problems with laminate quality. Producing a tacky prepreg from m-ATS took more effort since the resin itself is a solid at room temperature.

Figure 11 illustrates the flexure strength versus temperature for the formulations. The room temperature flexure strengths for the formulations have shown dramatic increases over the controls. This was especially apparent in AF-4 and AF-8 which had nearly a 60% increase in strength over m-ATB. These improvements are most likely due to the strengthened AT resin/carbon fiber interface present in the formulations. The elevated temperature flexure strengths show that m-ATS had the greatest values followed by AF-8, AF-4, AF-20 and AF-11. In this instance the more extensive modifications that had to be performed on the m-ATS formulations for improved tack and toughness resulted in decreased high temperature flexure strengths. A similar trend was seen in the elevated temperature interlaminar shear strengths. However, here AF-20 falls just below AF-8 in high temperature performance.

The 0° compression test was conducted under hot/wet conditions to measure the performance of the formulations under these conditions. While AF-4 and AF-8 had moisture absorptions more than double the control m-ATB, they had wet compressive strengths which were respectively 9.3% and 17.5% greater than m-ATB at 350°F. At 425°F the improvements in compressive strength for AF-4 and AF-8 were more dramatic being 100% and 112% greater respectively than the control under the same conditions. The formulated

m-ATS resins had the greatest compressive strengths of any of the formulations at 350°F/wet. The values were only slightly lower than m-ATS itself under these conditions. AT 425°F, however, the compressive strengths of AF-11 and AF-20 fell below those values obtained with AF-4 and AF-8.

The +45° in-plane shear modulus, $G_{12}$, was used as another measure of the hot/wet properties of the formulations. AF-11 had the greatest shear modulus of any of the formulations at 350°F/wet and this value was only 5% lower than that of m-ATS. The results obtained with the +45° in-plane shear modulus as would be expected tended to parallel the wet compression results.

Of particular interest were the composite $G_{IC}$'s for these formulations which were generated from the double cantilever beam specimen. A review of the data in Tables 3 and 4 indicates that the m-ATS $G_{IC}$'s were significantly greater than the m-ATB results. This difference, however, was likely due to the high degree of fiber bridging present in the m-ATS, AF-11 and AF-20 test specimens. As a result of this fiber bridging it would be difficult to make a valid comparison of the $G_{IC}$'s for the m-ATS and m-ATB formulations. AF-4 and AF-8 gave increases in toughness of 89% and 63% respectively. These results correlate well with the neat resin $G_{IC}$'s which demonstrated that AF-4 is tougher than AF-8. AF-11 gave the highest $G_{IC}$ of the m-ATS formulations. Assuming an equivalent degree of fiber bridging AF-11 and AF-20 were 42.9% and 32.1%, respectively, tougher than the control m-ATS. From our neat resin results AF-20 was expected to give the greater composite $G_{IC}$. These double cantilever beam results, however, were in agreement with the edge delamination test results.

2.7 Interphase

Qualitatively, the strength of the AT resin/carbon fiber interphase can be judged by the amount of resin adhering to fibers and the degree of fiber pullout present in SEM photomicrographs of Figures 8-10. These SEMs represent the failed tensile side of a 0° composite flexure specimen. M-ATB with the greatest amount of fiber pullout and the least amount of resin adhering to the fibers has the weakest interface. This was confirmed by the low interlaminar shear strength of 75.8 MPa at room temperature. With slightly more resin adhering to the carbon fibers in the SEM photomicrograph of the m-ATS composite, m-ATS has a stronger interface than m-ATB. This is supported by the m-ATS interlaminar shear strength which is 13.6% greater than that for m-ATB.

The phenomenon of the weak AT resin/carbon fiber interface is due to the interaction of the relatively non-polar resin with the highly polar fiber surface. The formulated resins had greater polarity than the base resins and thus a stronger interface. This is clearly illustrated in Figures 9 and 10 by comparing the SEM photomicrographs of the formulated resins with the appropriate AT control resin. The most dramatic improvements were seen for AF-4 and AF-8 where the interlaminar shear strength improved from 75.8 MPa (for m-ATB) to 108 MPa and 97.9 MPa respectively. While the SEM photomicrographs of AF-11 and AF-20 appear to have slightly more resin adhering to the fibers than in the m-ATS control, this difference was not significant. The resulting AF-11 and AF-20 interlaminar shear strengths were nearly identical to the m-ATS control. This lack of improvement in the m-ATS interface could explain why the edge delamination strengths for AF-11 and AF-20 were not significantly greater than the control and also why the AF-11 and AF-20 composite results were generally lower than expected.

3.0  CONCLUSION

From the neat resin and composite results of the AF-4 and AF-8 formulations it may be concluded that dramatic improvements have been made

in the fracture toughness of m-ATB with little sacrifice in its high temperature properties. Under dry conditions AF-4 and AF-8 would be suitable for long term service at 450°F. One of the attractive features of m-ATB is its low moisture absorption and this quality was largely preserved in AF-4 and AF-8. As a result these resins would be considered capable of 400-425°F wet performance. All of the composite properties for the m-ATB control were low and this could be due to the poor carbon fiber/resin interface. This is especially true under wet conditions where the water could potentially wick along the interface. The improved interfaces in the AF-4 and AF-8 composites were partially responsible for the improved performance of the composite. While improved, the AF-4 and AF-8 interlaminar shear strengths were still not as high as those observed with epoxies or BMI's. Further strengthening of the interface would likely translate into even greater composite properties.

The interface strength was more of a concern with the m-ATS formulations which showed no significant improvement over the m-ATS control. When comparing the neat resin results, m-ATS and its formulations AF-11 and AF-20 gave significantly better high temperature properties than m-ATB and its formulations. In addition, the neat resin $G_{IC}$'s of AF-11 and AF-20 were comparable to AF-4 and AF-8. However, these improvements did not translate into the composite mechanical properties. This phenomenon could have been due, in part at least, to the weak AF-11 and AF-20 carbon fiber interface. With no further improvements in the interface AF-11 and AF-20 should be capable of long term service at 450°F under dry conditions and 375° to 400°F service under wet conditions

The goal of this research has been accomplished through the development of two m-ATB and two m-ATS formulations. While these formulations were not meant to be optimum formulations, they do represent the first attempts to formulate and toughen m-ATB and m-ATS. As such, the formulations have demonstrated composite mechanical properties superior to present state-of-the-art toughened bismaleimides achieving useful mechanical properties up to 425°F under wet conditions. The weak fiber/resin interface present in the unmodified resins has been strengthened through formulation, but more research needs to be focused in this area to further enhance the composite mechanical properties of these acetylene terminated resins. As more experience is gained working with these resins, their composite mechanical properties should improve and appear even more favorable, especially when compared with the alternative resins presently available.

## 4.0 ACKNOWLEDGEMENTS

This program is supported by the Air Force Wright Aeronautical Laboratories, Materials Laboratories, under contract F 33615-84-C-5115, monitored by Dr. Frederick L. Hedberg.

Special thanks goes to Cindy Megerdigian of Hysol for her work in providing the SEM photomicrographs.

## 5.0 REFERENCES

1. C. Y-C. Lee, I.J. Goldfarb, T.E. Helminiak and F.E. Arnold, "Air Force Review of Acetylene Terminated Resin Technology", National SAMPE Symposium, Vol. 28, 1983, pp. 699-710.
2. C.L.Leung, "Acetylene Terminated Resin Mechanical Characterization III. Effect of Cure States on Tensile Properties", National SAMPE Symposium, Vol. 28, 1983, pp. 590-595.

3. C. Y-C. Lee, L.R. Denny, and I.J. Goldfarb, "Characterization of a Bis-Phenol-A Based Resin with Terminal Acetylene Groups", Am. Chem. Soc. Polym. Prepr., Vol. 24, 1983, pp. 139-140.

4. F.L. Abrams and C.E. Browning, "Influence of Molecular Structure on Mechanical Properties of Acetylene Terminated Resins", Am. Chem. Soc. Org. Plast. Chem. Prepr., Vol. 48, 1983, pp. 909-913.

5. G.A. Loughran, B.A. Reinhardt and F.E. Arnold, "Thermal and Thermal Mechanical Properties of Acetylene Terminated Phenylene Systems", Am. Chem. Soc. Org. Coat. & Plast. Prepr., Vol. 43, 1980, pp. 777-782.

6. P.A. Steiner, "AT Technology Transition, Interim Tech Report", Contract F33615-84-C-5115, Hysol Hysol Grafil, Dec. 1,1985 - Feb. 18, 1986.

7. P.A. Steiner, M.T. Blair, and J. Browne, "Toughened Matrix Resins For Damage Tolerant Composite Structures", ASTM Symposium on Toughened Composites, 1985.

8. P.J.Dynes, "Evaluation of New Resin Systems, Final Report", Contract No. F33615-82-C-5064, Rockwell International Science Center, 1985.

9. P.A. Steiner, J. Browne, M.T. Blair, J.M. McKillen, "Development of Damage Tolerant Acetylene Terminated Resin/Carbon Fiber Composites", International SAMPE Technical Conference, Vol. 18, 1986, pp. 193-208.

## 6. BIOGRAPHIES

Dr. Paul A. Steiner is a Project Leader for Composites R&D for the Hysol Aerospace and Industrial Products Division. He has a Ph. D. in organic chemistry from the University of Washington. Dr. Steiner has worked for the Boeing Commercial Airplane Co. as a Materials Engineer and for the past five years has worked for the Hysol Aerospace and Industrial Products Division of the Dexter Corporation. His responsibilities have included the development and toughening of epoxy, bismaleimide and acetylene terminated matrix resins. He has a

total of 10 years experience working
in the fields of composites and
polymer chemistry.

Jim Browne is the Group Manager of
Composites R&D for the Hysol
Aerospace and Industrial Products
Division.  He has 16 years
experience in composites and
structural plastics.

Michele Blair is a Chemist for Hysol
Aerospace and Industrial Products
Division and holds a Bachelor of
Science degree in chemistry from the
University of California at Davis.
She has 6 years experience in hot
melt adhesives and composite matrix
resins.

John McKillen is a Senior Laboratory
Technician for Composites R&D for
Hysol Aerospace and Industrial
Products Division and holds a
Bachelor of Science degree in
chemical engineering from the
University of California at
Berkeley.  He has 4 years experience
in composite matrix resins.

early stages of cure

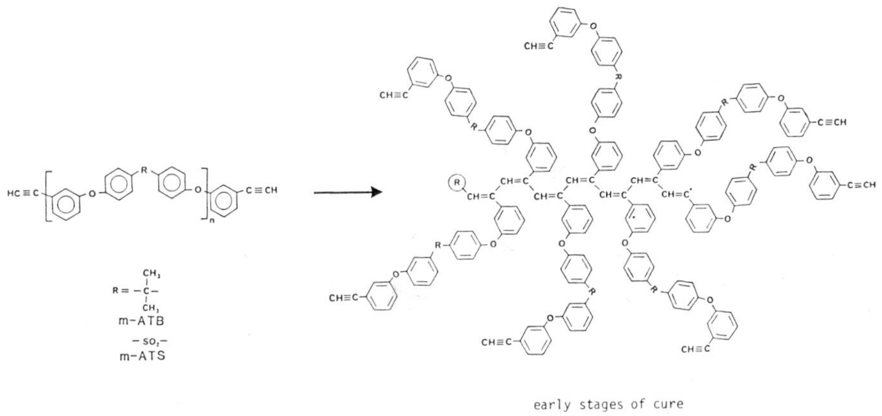

Figure 1.  AT Resin Chemical Structure and Cure Mechanism

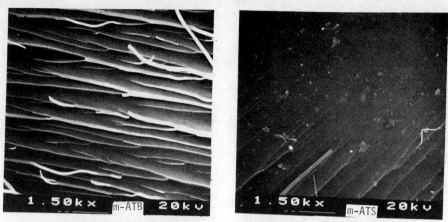

Figure 2. SEM Photomicrographs of Cured m-ATB and m-ATS

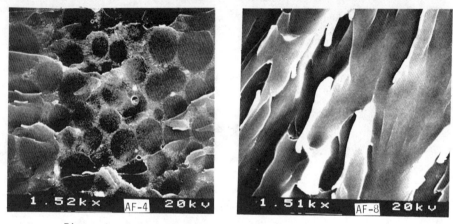

Figure 3. SEM Photomicrographs of Cured AF-4 and AF-8

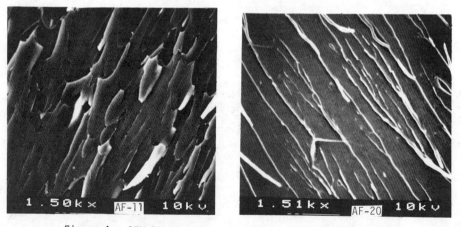

Figure 4. SEM Photomicrographs of Cured AF-11 and AF-20

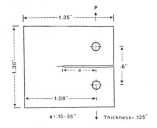

Figure 5.  Compact Tension Neat Resin $G_{1c}$ Specimen

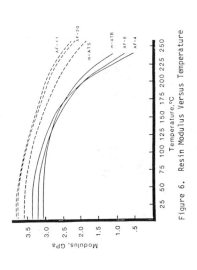

Figure 6.  Resin Modulus Versus Temperature

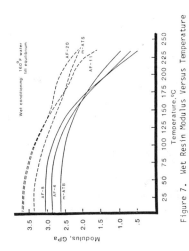

Figure 7.  Wet Resin Modulus Versus Temperature

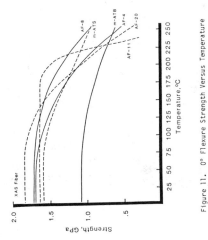

Figure 11.  0° Flexure Strength Versus Temperature

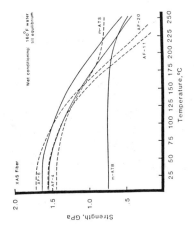

Figure 12.  0° Wet Flexure Strength Versus Temperature

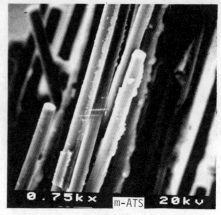

Figure 8. SEM Photomicrographs of m-ATB and m-ATS on XAS Carbon Fiber, 1.3% "A" Size

Figure 9. SEM Photomicrographs of AF-4 and AF-8 on XAS Carbon Fiber, 1.3% "A" Size

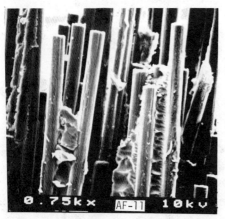

Figure 10. SEM Photomicrographs of AF-11 and AF-20 on XAS Carbon Fiber, 1.3% "A" Size

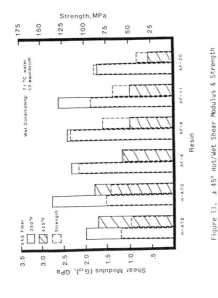

Figure 13. ± 45° Hot/Wet Shear Modulus & Strength

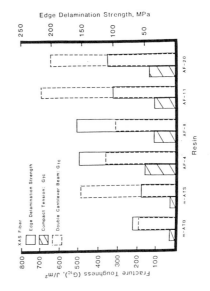

Figure 14. Relationship Between Composite Fracture Toughness and Resin $G_{1c}$

Table 1. NEAT RESIN PROPERTIES

| NEAT RESIN PROPERTY | FORMULATION | | | | | |
|---|---|---|---|---|---|---|
| | m-ATB | m-ATS | AF-4 | AF-8 | AF-11 | AF-20 |
| DSC, °C | | | | | | |
| Onset | 171 | 204.3 | 208.1 | 207.9 | 213.6 | 195.0 |
| Peak Exotherm | 239 | 227.5 | 254.9 | 254.7 | 240.7 | 230.8 |
| Uncured Tg °C | 14° | 25° | 3° | -6° | 5° | 11° |
| Minimum Viscosity, (Poise ) °C | .25 at 160°C | .2 at 147° | 8.4 at 179° | 14.6 at 171° | 7.9 at 186° | 14.5 at 150° |
| Tg, Dry °C | 260° | 320° | 262° | 273° | 305° | 302° |
| Young's Modulus, [a] 23°C,GPa(Ksi) | 3.07 (445) | 3.55 (515) | 3.37 (489) | 3.22 (467) | 3.17 (460) | 3.75 (544) |
| Eq. Moisture Absorption, % | .60 | 1.09 | 1.40 | 1.43 | 2.00 | 1.82 |
| Tg, Wet °C [b] | 252° | 328° | 258° | 251° | 288° | 300° |
| Young's Modulus, [a],[b] 23°C,GPa(Ksi) | 2.59 (376) | 3.36 (487) | 2.80 (406) | 2.99 (434) | 3.98 (577) | 3.90 (566) |
| TGA, °C at 10% Wt. Loss | | | | | | |
| Air | 365° | 426° | 397° | 350° | 407° | 399° |
| N2 | 446° | 453° | 458° | 421° | 481° | 469° |
| $G_{IC}$, J/m² (in-lbs/in²) | 35 (.2) | 35 (.2) | 165 (.94) | 114 (.65) | 112 (.64) | 140 (.80) |

a.  Calculated using Young's Modulus (RT) from DMTA
b.  Specimens immersed in 160°F water bath till equilibrium
    (about 14 days)

Table 2. PREPREG PHYSICAL DATA

| PREPREG PHYSICAL PROPERTIES | FORMULATION | | | | | |
|---|---|---|---|---|---|---|
| | m-ATB | m-ATS | AF-4 | AF-8 | AF-11 | AF-20 |
| Fiber Areal Weight, g/m² | 145 ± 4 | 145 ± 4 | 145 ± 4 | 145 ± 4 | 145 ± 4 | 145 ± 4 |
| Resin Content, percent weight | 30 | 32 ± 2 | 31.9 | 34.9 | 36 ± 3 | 36 ± 3 |
| Volatile Content, percent weight | .68 | 1.7 | .79 | .65 | 6.8 | 5.6 |
| Flow, percent weight | 14.8 | 11 | 6.8 | 8.7 | 11 | 15 |
| Gel Time at 170°C, minute | 19 | 18 | 7 | 7 | 50 | 47 |
| Out-time at 75°F | >2 months | None | 3 weeks | 3 weeks | 2 weeks | 2 weeks |
| Tack (self adhesion) | Pass/Very Good | None | Pass/Good | Pass/Good | Pass/Good | Pass/Good |

TABLE 3. COMPOSITE MECHANICAL PROPERTIES:M-ATB

| TEST | M-ATB Strength,MPa(Ksi) | M-ATB Modulus,GPA(Msi) | AF-4 Strength,MPa(Ksi) | AF-4 Modulus,GPa(Msi) | AF-8 Strength,MPa(Ksi) | AF-8 Modulus,MPa(Msi) |
|---|---|---|---|---|---|---|
| 0-deg. Tension | | | | | | |
| -67°F | 1613 (234) | 144 (20.9) | 2020 (293) | 134 (19.5) | 2289 (332) | 143 (20.8) |
| R.T. | 1696 (246) | 142 (20.6) | 1993 (289) | 134 (19.5) | 2220 (322) | 143 (20.8) |
| 0-deg. Flexure | | | | | | |
| R.T. | 1103 (160) | 126 (18.3) | 1744 (253) | 120 (17.4) | 1779 (258) | 121 (17.5) |
| R.T./Wet | 752 (109) | 125 (18.1) | 1606 (233) | 127 (18.4) | 1689 (245) | 121 (17.6) |
| 350°F | 855 (124) | 123 (17.8) | 1282 (186) | 114 (16.6) | 1510 (219) | 122 (17.7) |
| 350°F/Wet | 779 (113) | 128 (18.6) | 993 (144) | 100 (14.5) | 1007 (146) | 117 (16.9) |
| 425°F/Wet | 614 (89) | 115 (16.7) | 558 (81) | 91 (13.2) | 786 (114) | 108 (15.6) |
| 450°F | 731 (106) | 112 (16.3) | 800 (116) | 101 (14.7) | 1110 (161) | 113 (16.4) |
| 0-deg. Interlaminar Shear | | | | | | |
| R.T. | 75.8 (11.0) | | 108 (15.7) | | 97.9 (14.2) | |
| R.T./Wet | 31.7 (4.6) | | 103 (15.0) | | 106 (15.4) | |
| 350°F | 33.1 (4.8) | | 47.6 (6.9) | | 55.2 (8.0) | |
| 350°F/Wet | 18.6 (2.7) | | 33.1 (4.8) | | 44.8 (6.5) | |
| 425°F/Wet | 13.8 (2.0) | | 21.4 (3.1) | | 33.1 (4.8) | |
| 450°F | 26.9 (3.9) | | 26.9 (3.9) | | 38.6 (5.6) | |
| Edge Delamination Strength, (±30°2, 90°2)s | 71.0 (10.3) | | 156 (22.6) | | 160 (23.2) | |
| 90-deg. Flexure | 26.4 (3.83) | strain .25% | 68.9 (10.0) | strain 1.0% | 80.0 (11.6) | strain 1.0% |
| In-plane Shear,+45° Tension | | | | | | |
| 350°F/Wet | 60.1 (8.71) | 2.00 (.29) | 108 (15.7) | 2.34 (.34) | 117 (17.0) | 2.41 (.35) |
| 425°F/Wet | 48.3 (7.01) | 1.72 (.25) | 57.2 (8.3) | 1.17 (.17) | 80.0 (11.6) | .97 (.14) |
| 0-deg. Compression | | | | | | |
| R.T. | 855 (124) | | 1538 (223) | | 1241 (180) | 152 (22.1) |
| 350°F | 758 (110) | | 986 (143) | | 855 (124) | |
| 350°F/Wet | 669 (97) | | 731 (106) | | 786 (114) | |
| 425°F/Wet | 345 (50) | | 689 (100) | | 731 (106) | |
| 450°F | 572 (83) | | 841 (122) | | 724 (105) | |
| Double Cantilever Beam GIc, J/m2 (in-lbs/in2) | 193 (1.1) | | 368 (2.1) | | 315 (1.8) | |
| Thermal Aging | severe microcracking | | no microcracking | | no microcracking | |
| Thermal Spikes, % wt. gain | 0.06 | | 0.08 | | 0.12 | |

TABLE 4.    COMPOSITE MECHANICAL PROPERTIES: M-ATS

| TEST | M-ATS Strength, MPa(Ksi) | M-ATS Modulus, GPA(Msi) | AF-11 Strength, MPa(Ksi) | AF-11 Modulus, GPa(Msi) | AF-20 Strength, MPa(Ksi) | AF-20 Modulus, MPa(Msi) |
|---|---|---|---|---|---|---|
| 0-deg. Tension | | | | | | |
| -67°F | 2016 (292.4) | 141 (20.4) | 1737 (252.0) | 132 (19.1) | 1975 (286.4) | 143 (20.7) |
| R.T. | 2126 (308.4) | 137 (19.8) | 1855 (269.0) | 138 (20.4) | 2071 (300.4) | 141 (20.4) |
| 0-deg. Flexure | | | | | | |
| R.T. | 1628 (236.1) | 121 (17.5) | 1711 (248.1) | 122 (17.7) | 1912 (277.3) | 123 (17.8) |
| R.T./Wet | 1478 (214.3) | 117 (16.9) | 1595 (231.3) | 117 (16.9) | 1731 (251.1) | 122 (17.7) |
| 350°F | 1273 (184.6) | 125 (18.2) | 1356 (196.7) | 123 (17.8) | 1283 (186.1) | 122 (17.7) |
| 350°F/Wet | 856 (124.2) | 114 (16.5) | 652 (94.6) | 104 (15.1) | 693 (100.5) | 106 (15.4) |
| 425°F/Wet | 799 (115.9) | 103 (14.9) | 233 (33.8) | 35.2 (5.1) | 314 (45.5) | 52.4 (7.6) |
| 450°F | 1064 (154.3) | 126 (18.3) | 487 (70.6) | 72.4 (10.5) | 757 (109.8) | 114 (16.5) |
| 0-deg. Interlaminar Shear | | | | | | |
| R.T. | 86.2 (12.5) | | 86.9 (12.6) | | 84.1 (12.2) | |
| R.T./Wet | 66.1 (9.6) | | 60.7 (8.8) | | 47.6 (6.9) | |
| 350°F | 55.2 (8.0) | | 56.5 (8.2) | | 55.2 (8.0) | |
| 350°F/Wet | 42.1 (6.1) | | 45.5 (6.6) | | 33.1 (4.8) | |
| 425°F/Wet | 35.9 (5.2) | | 27.6 (4.0) | | 24.1 (3.5) | |
| 450 °F | 40.0 (5.8) | | 22.1 (3.2) | | 32.4 (4.7) | |
| Edge Delamination Strength, (±30°2, 90°2)s | 55.8 (8.1) | | 101 (14.7) | | 69.6 (10.1) | |
| 90-deg. Flexure | 59.3 (8.6) | strain = .67% | 53.0 (7.7) | strain = 1.06% | 57.6 (8.4) | strain = .83% |
| In-plane Shear ±45° Tension | | | | | | |
| 350°F/Wet | 76.5 (11.1) | 2.76 (.40) | 93.1 (13.5) | 2.62 (.38) | 89.6 (13.0) | 1.72 (.25) |
| 425°F/Wet | 70.3 (10.2) | 1.79 (.26) | 68.3 (9.9) | .97 (.14) | 41.4 (6.0) | .55 (.08) |
| 0-deg. Compression | | | | | | |
| R.T. | 1423 (206.4) | 130 (18.9) | 1335 (193.6) | 134 (19.4) | 1269 (184.1) | 132 (19.1) |
| 350°F | 1047 (151.9) | 140 (20.3) | 1091 (158.2) | 125 (18.1) | 861 (124.9) | 127 (18.4) |
| 350°F/Wet | 1013 (146.9) | 124 (18.0) | 854 (123.9) | 137 (19.8) | 896 (130.0) | 118 (17.1) |
| 425°F/Wet | 821 (119.1) | 126 (18.3) | 206 (29.8) | | 232 (33.7) | 123 (17.9) |
| 450°F | 928 (134.6) | 137 (19.8) | 243 (35.2) | 137 (19.9) | 486 (70.5) | 123 (17.9) |
| Double Cantilever Beam, GIC, J/m² (in-lbs/in²) | 490 (2.8) | | 700 (4.0) | | 648 (3.7) | |
| Thermal Aging | severe microcracking | | minor microcracking | | minor microcracking | |
| Thermal Spikes, % wt. gain | .06 | | .06 | | .06 | |

821

EFFECTS OF PMR-15 POLYIMIDE VARIABILITY
Sidney W. Street
Northrop Corporation
Advanced Systems Division

## Abstract

Effects of varying the PMR-15 polyimide resin reaction time, prepreg staging temperature and triester level on the rheology, autoclave processing and prepreg aging were studied on fiberglass and carbon fabric and carbon unitape composites.

Low triester levels result in lower viscosity of the PMR-15 resin through the entire temperature range, shift viscosity peaks to lower temperatures and reduces the tendency for blister formation in thick composites. Very high triester levels tend to reduce viscosity in the low and middle temperature regions but significantly increase viscosity in the high temperature region during gelation. The higher viscosity appears to greatly increase the tendency to form blisters in the composites.

The use of "low" triester prepregs with cure cycle modifications appears to enhance the ability to more consistently manufacture thick composites.

Dynamic viscosity can be determined on PMR-15 prepregs by Rheometric Dynamic Spectroscopy to verify prepreg reproducibility and predict successful composite manufacture.

## 1. INTRODUCTION

PMR-15 is a combination condensation and addition polyimide used for the manufacture of low void, composites with excellent high temperature strength retention. PMR is an acronym for Polymerizable Monomer Reactants. The 15 refers to a theoretical molecular weight of 1500. The three major reactants are benzophenone tetracarboxylic diester, (BTDE) norbornene monoester (NE) and methylene dianiline (MDA). Also present in the as received prepreg are other impurities and reaction products of the ingredients. All esters are methyl esters. The triester of BTDA is a major and important impurity which will be discussed. Fabric or fiber reinforcement is impregnated with the PMR-15 resin containing approximately 10 percent excess methyl alcohol and supplied in rolls up to 60 inch width at $0°F$ or below.

The first reaction step consists of the reaction of the alicyclic NE with the aromatic MDA diamine. This reaction proceeds at ambient temperature. Excessive aging of the prepreg at ambient temperature is believed to result in the formation of short molecular chains rather than the desired higher molecular weight chains which are more heat stable. The second major step is the reaction of the aromatic BTDE with MDA which is essentially nonreactive at ambient temperature.

Water and methyl alcohol are "condensed" out during these reaction steps as the temperature is increased to form the respective imides. The maximum evolution of water occurs about 300°F and methyl alcohol at about 375°F.

As the temperature is increased, the "addition" or endcap reaction begins at about 500°F. This reaction proceeds through the norbornene group and is generally accepted to be a reverse Diels-Alder mechanism. Small amounts of volatiles continue to be given off during the endcap reaction. The curing of PMR-15 composites is usually completed at approximately 600°F for about four hours under about 200 psi. Postcures to develop minimum high temperature properties usually consist of four to 10 hours at 625°F. Additional cross-linking continues to take place during aging at 450 to 600°F for period of two to four weeks. This additional curing results in a more brittle polymer with reduced mechanical properties at ambient temperature and increased mechanical properties at elevated temperature.

Variables Studied

Since PMR-15 is supplied as a mixture of monomeric reactants, the gel and flow tests normally run on epoxy prepregs to verify consistency are not applicable. High Pressure Liquid Chromatography (HPLC) and Rheometric Dynamic Spectroscopy (RDS) are relatively new techniques which were used to characterize the chemistry and viscosity of PMR-15 prepreg.

PMR-15 prepregs were prepared by U. S. Polymeric on 3K 8HS Celion carbon fabric, 7781 E-glass fabric, and 12K Celion unitape. PMR-15 resin was manufactured under three conditions as follows with corresponding prepreg treater designations.

a. Reduced reaction time    A
b. Standard reaction time   B-2
c. Extended reaction time   C

In addition, prepregs from resin "B" were staged 10°F lower and 10°F higher than standard. These prepregs were designated B-1 and B-3 respectively.

The effect of higher and lower benzophenone tetracarboxylic triester (BTTE) was also studied. Triester content was adjusted during the manufacture of the BTDE monomer subassembly, which was then reacted under standard PMR-15 manufacturing conditions. Prepregs at a high and low BTTE level were then prepared under standard prepreg conditions and designated D and E respectively. All variations were run on the 3K 8HS carbon fabric prepreg. Resin and prepreg staging variations were run on the 7781 E-glass. Resin variations only were run on the unitape prepreg.

Thick and thin composites of each prepreg were laid up and cured to determine processability using a standard cure cycle. The 3K 8HS carbon fiber prepregs, in addition to initial zero day testing, were aged for 10, 15 and 20 day intervals in sealed foil bags. The aged prepregs were tested for RDS, HPLC and processability.

In addition, composites prepared from the aged prepregs were tested for compressive interlaminar shear at ambient and 600°F initially and after aging 125 hours at 600°F.

2. DISCUSSION OF RESULTS

Rheology Effects

Rheology testing was conducted by Dr. Fred Meyers at Armco Research and Technology using a Rheometrics Dynamic Spectrophotometer (RDS). Prepreg samples were cooled to -75°C then heated at 2°C/minute to 400°C. Operating conditions were 40 radians/second, 0.4 percent nominal strain amplitude and initial tension of 40 grams. Data was obtained for $G'$, storage modulus, $G''$, loss modulus and $G^*$ complex shear using a specimen size of approximately 13mm$^2$ L/W ratio of 1/1. 3K 8HS carbon fabric prepreg was run on 1 ply, 7781 E-glass fabric on 2 ply and unitape fabric on 4 ply specimens. Multiple specimens were compacted at ambient conditions.

G\*, the complex shear stiffness combines elements of G' and G" and is related to complex viscosity through the vibrational frequency.

$$G* = \frac{(G')^2 + (G'')^2}{2}$$

$$= \frac{\text{Nominal Shear Stress Amplitude}}{\text{Nominal Shear Strain Amplitude}}$$

Figure 1 illustrates an RDS semilog plot of G', G" and G\* versus temperature for the standard B-2 3K 8HS prepreg. A monomer melt peak on the G" plot is evident, peaking at $19^{\circ}C$. Melting continues to a viscosity minima or valley at $100^{\circ}C$. Prepreg weight loss has also been superimposed on the plot. Weight loss at $100^{\circ}C$ is approximately four percent. As heating is continued, two maxima or peaks for G" are noted. The first occuring at approximately $140^{\circ}C$ may indicate the maximum imidization of the alicyclic NE/MDA component. The second maxima occuring at approximately $208^{\circ}C$ may be due to the maximum imidization of the aromatic BTDE/MDA component. Volatile loss continues at a high rate through $150^{\circ}C$ and then reduces in rate as the temperature continues to increase. The maximum evolution rate of water occurs at approximately $150^{\circ}C$ while the maximum evolution rate of methyl alcohol occurs at about $190^{\circ}C$. The imides formed in situ continue melting and reach a minimum at about $280^{\circ}C$. Autoclave pressure of approximately 200 psi is normally applied at about $250^{\circ}C$ after a 30 minute isothermal hold. Since a pressure cell for the RDS instrument is not available, the effect of pressure application on Rheology has not been determined. The effect of isothermal holds on rheology are readily observable with RDS. Viscosity reduction due to higher temperature melting is essentially stopped and viscosity increase due to crosslinking is noted. (These effects will be illustrated later.) The rate of volatile loss continues to decrease and the rate of addition polymerization continues to increase as the temperature rises to a G" peak

at about $338^{\circ}C$. The volatile losses listed here were reported by B. Hunter, Boeing Aerospace Corp.(1) Northrop cure cycles are usually conducted at a maximum of $316^{\circ}C$. RDS plots were conducted on all prepregs studies. A summary of Valleys (V) and Peaks (P) illustrating the effects of resin reaction times for the 3K8HS/PMR-15 prepreg is listed in Table 1 for G", loss modulus. Numerical "viscosity" values are listed at the temperature at which they occurred.

Several observations may be noted:

1. Increasing the resin reaction time, A, B-2, C tends to reduce viscosity in almost all regions.

2. The peak height for the monomer melt area is significantly reduced as reaction time is increased.

3. High triester (D) versus the standard (B-2) exhibits a lower monomer melt peak and lower viscosity in the lower and middle temperature regions, eliminates the valley at approximately $175^{\circ}C$, and increases viscosity in the high temperature region of maximum crosslinking or gelation.

4. Increasing prepreg staging temperature B-1, B-2 and B-3, shifts monomer melt peak temperature higher and reduces viscosity in almost all regions.

5. The low triester (E) exhibits significantly lower viscosity in the entire range and shift peaks and valleys to lower temperatures. Table 2 lists valleys and peaks for G\*, complex shear stiffness.

Several trends are also noted:

1. As with G", viscosity is significantly reduced as resin reaction times (A, B-2 and C) are increased, particularly in the imidization region at $165^{\circ}C$

2. High triester (D), Figure 2, is much lower in viscosity in the low and middle regions and again exhibits a significant increase in the high temperature crosslink region at $365^{\circ}C$.

824

3. Low triester (E) is very much lower in viscosity than any of the other prepregs and remains low even in the high temperature region. It should also be noted that it reaches its maximum at about 14°C lower than the standard B-2 resin. The lower viscosity peaking at a lower temperature of the low triester PMR-15 in the high temperature crosslinking region correlates with the increased ability to cure thick, blister free, PMR-15 composites. The lower viscosity in this region probably aids in allowing the remaining volatiles to escape more readily than higher viscosity, high triester systems. Figure 3 illustrates the low viscosity exhibited by (E). The very low viscosity in the low temperature region suggests a hold at about 66°C in order to provide a higher viscosity than would be obtained with a uniform 2°C/minute heat rise. The effect of this hold and a hold at 220°C are illustrated in Figure 3.

## Prepreg Aging Effects on Rheology

Tables 3 and 4 provide the effects of prepreg aging at 70°F on rheology for prepregs A, B-2 and C. G" loss modulus, in Table 3 exhibits a slight increase in peak and valley values and temperatures through the temperature ramp after the 10 day aging period. No chemical evidence is presently available to indicate why these occur. Values then generally decline through the 15 and 20 day aging periods. The monomer melt peak, P-1, decreases in intensity and shifts upward in peak temperature which suggests a reduction in the high crystalline nadic ester and methylene drainline components which probably react to form an ester amide at ambient temperature. P-1 for lot C exhibits the smallest peak at the highest temperature as would be expected since it was exposed to the most heat history.

Table 4 which contains G* complex modulus effects on prepreg aging lists similar trends as noted in Table 3.

## Prepreg Aging Effects on Compressive Interlaminar Shear (ILS)

Table 5 lists compressive ILS values for carbon fabric prepregs A, B and C variants which were aged at 70°F for 0, 10, 15 and 20 days. Prepregs were cut, laid up and aged in sealed foil bags prior to cure. Notched compressive ILS, which is a very matrix dominant test were run at ambient and 600°F after the initial cure and postcure then rerun at 600°F after aging 125 hours at 600°F. The 125 hour, 600°F aging resulted in a nearly 2X increase in 600°F strength retention in all cases. Since no deterioration is evident in 600°F compressive ILS, it would appear that aging the prepreg at 70°F max. for up to 20 days prior to cure does not cause a reduction in hot strength retention.

## Effect of Prepreg Variables on Thick Laminate Processing

Table 6 lists C-scan rankings of 8, 14 and 20 ply carbon fabric laminates cured after aging 0, 10 and 20 day aging at 70°F. As may be noted satisfactory laminates were prepared for all prepreg variants for 8 and 14 ply laminates except the high triester (D). No satisfactory laminates could be prepared from the very high triester prepreg (D). Satisfactory 20 ply thick laminates could not be made from any of the preg variants except the low triester (E). Insufficient low triester prepreg was available to conduct aging studies.

7781 E-glass fabric prepreg produced satisfactory composites in all cases except the high triester (D), which blistered during cure. Low triester 7781 and unitape prepreg were not included in the aging study.

## HPLC Effects Of Prepreg Aging

High Pressure Liquid Chromatography (HPLC) tests were conducted by Fred Bancroft at Hitco Materials using an isocratic, size exclusion method, DuPont SE 60 columns and a 2:1, water/tetrahydro furan mobile phase. The method was developed by B. Hunter. The prepregs were aged at 70°F in foil sealed bags for 0, 5, 10, 15 and 20 day periods and frozen at 0°F prior to the HPLC test. This method does not provide for the analysis of methylene dianiline or its reaction products.

The three diesters of BTDA (BTDE) the triester (BTTrE), the nadic ester (NE) and post reaction products are analyzed by this method. Tetraester can also be determined, however none was present in the prepregs studied. Results of the analysis are listed in Table 7 as a percent of total ingredient peak area. As may be noted, there is a minimum change in the concentration and ratio of the three aromatic diesters and triesters during the 70°F, 20 day aging of the prepregs tested. There is a significant reduction in the concentration of the aliphatic, nadic ester and increase in the post reaction products during the first five day aging period.

The rate of the decrease in the nadic ester peak continues to slowly decline through the 20 day aging. A substantial increase in the amount of post reaction peaks is noted after the fifteen day aging. These peaks probably consist principally of the reaction products of the NE and MDA components. Some of these ingredients tend to collect within the column which may help to explain the anomalies listed in the post reaction products.

The data suggests very little reaction of the aromatic esters aged at 70°F. The alicyclic NE however, appears to react readily at 70°F probably to form ester amides with the MDA which are probably exhibited in the post reaction product peaks.

The effect of resin and prepreg variations on ingredients analyzed by this HPLC test appears to be minimal. The "A" resin (with the lower temperature processing) however, does tend to exhibit slightly higher NE and lower post reaction peaks initially as would be expected. The "D" and "E", high and low triester prepregs, respectively, were not included in the HPLC test.

## 3. CONCLUSIONS

Small variations of time and temperature in the manufacture of PMR-15 resin and prepreg are observable with rheometric dynamic spectroscopic (RDS) analysis.

Effects of triester concentration on RDS are also very apparent and significant. Very low triester levels permit the more facile fabrication of thick PMR-15 composites, while high levels cause severe blistering even in thin laminates. Very low levels of triester result in much lower viscosity during the entire heat-up and cure of PmR-15 composites and require cure cycle modifications to obtain good processing characteristics. Formation of ester amides and other low temperature reaction products may also contribute to viscosity and processing variability and should be studied.

RDS is a useful test to determine and verify the viscosity and processing characteristics of PMR-15 prepregs and should be incorporated in PMR-15 material specifications after appropriate limits are defined.

HPLC is also an important test since major ingredients, particularly nadic ester and triester components can be identified and quantified. Newer HPLC tests which also identify and quantify MDA and its reaction products may prove even more valuable in obtaining consistent PMR-15 prepregs and process consistent composites.

PMR-15 prepreg aging of the manufacturing variables studied for 20 days at 70°F did not appear to exhibit any adverse effect on 125 hour aging of PMR-15 composites studied.

### 4. REFERENCES

1. NASA CR-168217, "Characterization of PMR-15 Polyimide Resin and Prepreg." 1983.

### 5. BIOGRAPHY

Mr. Sidney Street has a 5 year degree, Bachelor of Chemical Engineering from Ohio State University and Certificate of Business Management from UCLA Graduate School of Business. He joined Northrop ASD in August of 1984 and is group leader and coordinator of research activities for the M&P Advanced Composites Group. Previously worked for U. S Polymeric, Division of Armco Steel for 9 years as Manager of New Product Development, where he formulated a number of new prepreg matrix materials including V378A, a new bismaleimide resin currently used in the manufacture of high temperature composite parts for AV8B Harrier and composite wing structure for the F-16 XL. He was awarded several patents and the SAMPE award for excellence for this effort. He also developed several high temperature toughened epoxy composite systems and helped commercialize PMR-15 polyimide during his stay at U. S. Polymeric.

Mr. Street spent the previous five years at Ferro Composites as Manager of New Products and Advanced Composites, was responsible for the development of CE9000, a 350 F moisture resistant epoxy composite used for fiberglass high temperature composites on the B-1, and several other epoxy, phenolic and polyimide composite prepregs. He also worked for Narmco Materials as Assistant Manager, Research and Engineering, on development of epoxy film adhesives and prepregs, and at Reichold Chemicals as West Coast Manager, Epoxy Resins and as Manager of Technical Service. He began his career with Shell Chemical Corporation where he worked in research and technical sales.

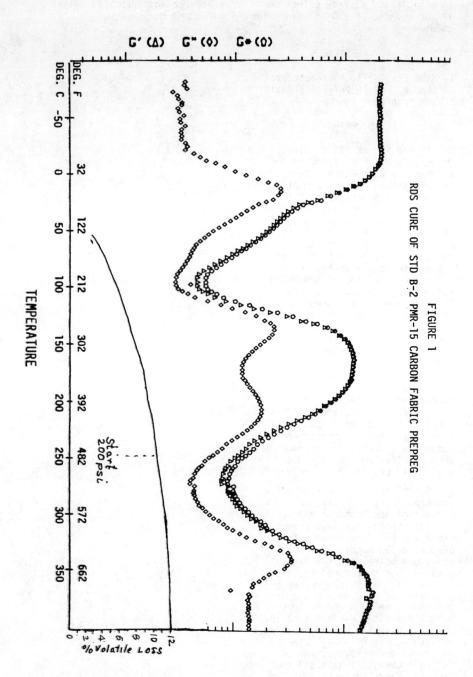

FIGURE 1

RDS CURE OF STD B-2 PMR-15 CARBON FABRIC PREPREG

G' (Δ)     G" (◇)     G* (O)

DEG. F

DEG. C

TEMPERATURE

Start
200 psi.

% Volatile Loss

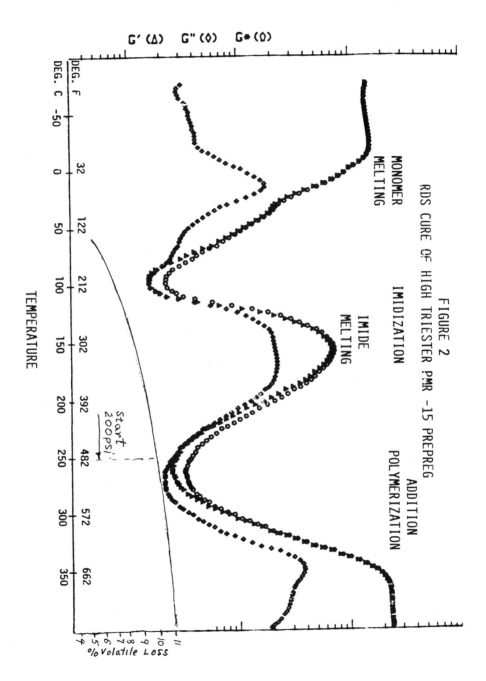

FIGURE 2
RDS CURE OF HIGH TRIESTER PMR -15 PREPREG

G' (Δ)   G" (◊)   G* (O)

MONOMER MELTING

IMIDIZATION

IMIDE MELTING

ADDITION POLYMERIZATION

TEMPERATURE

DEG. F
DEG. C    -50    0    50    100    150    200    250    300    350
          32   122   212   302   392   482   572   662

Start 200 psi

% Volatile Loss
4  5  6  7  8  9  10  11

829

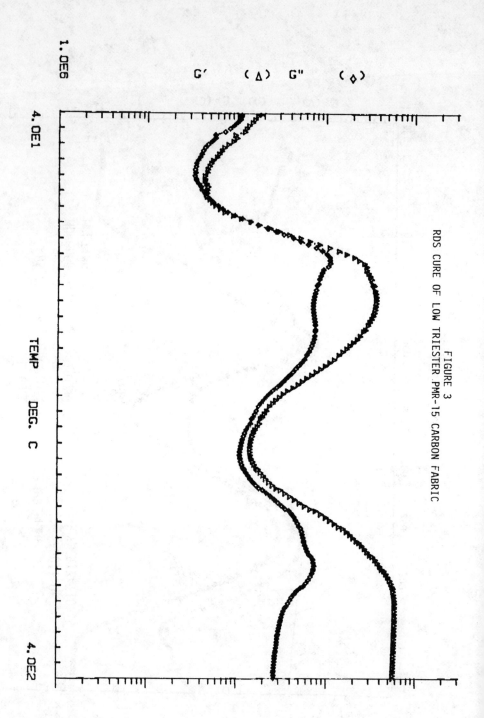

FIGURE 3
RDS CURE OF LOW TRIESTER PMR-15 CARBON FABRIC

G' (Δ)  G" (◇)

1.0E6

4.0E1    TEMP    DEG. C    4.0E2

## TABLE 1

### RDS (G") PMR-15/3K 8HS CARBON FABRIC PREPREG (AS RECEIVED)
### EFFECT OF RESIN REACTION TIME AND PREPREG STAGING TEMPERATURE

| RESIN | P-1/°C | V-1/°C | P-2/°C | V-2/°C | P-3/°C | V-3/°C | P-4/°C |
|---|---|---|---|---|---|---|---|
| A | 31.2/11 | 2.49/93 | 33.2/136 | 16.8/178 | 22.7/193 | 3.22/276 | 38.9/341 |
| B-2 | 24.4/19 | 2.83/99 | 22.0/139 | 11.4/174 | 17.2/208 | 4.31/281 | 33.1/338 |
| C | 9.96/14 | 2.06/84 | 15.9/137 | 5.47/174 | 9.64/215 | 2.62/279 | 29.4/336 |
| D | 17.6/14 | 1.57/97 | 18.9/138 | NO VALLEY | 22.2/169 | 2.11/272 | 37.8/348 |
| E | N/A | 0.33/80 | 11.0/135 | 7.12/162 | 0.59/162 | 1.10/259 | 7.85/327 |

| PREPREG STAGE | P-1/°C | V-1/°C | P-2/°C | V-2/°C | P-3/°C | V-3/°C | P-4/°C |
|---|---|---|---|---|---|---|---|
| B-1 | 21.2/11 | 2.47/74 | 27.5/136 | 13.4/180 | 17.2/208 | 4.58/272 | 53.9/341 |
| B-2 | 24.4/19 | 2.83/99 | 22.0/139 | 11.4/174 | 17.2/208 | 4.31/281 | 33.1/338 |
| B-3 | 7.17/20 | 1.99/90 | 12.0/140 | 6.11/165 | 10.3/205 | 1.98/273 | 18.5/336 |

## TABLE 2

### RDS (G*) PMR-15/3K 8HS CARBON FABRIC PREPREG (AS RECEIVED)
### EFFECT OF RESIN REACTION TIME AND PREPREG STAGING TEMPERATURE

| RESIN | V-1/°C | P-1/°C | V-2/°C | P-2/°C | RESIN REACTION TIME |
|---|---|---|---|---|---|
| A | 3.95/93 | 166/162 | 4.86/273 | 295/360 | 10' LESS |
| B-2 | 5.33/102 | 118/164 | 8.66/270 | 184/365 | STD |
| C | 3.30/93 | 66.9/168 | 5.24/273 | 162/363 | 10' MORE |
| D | 2.23/97 | 72.7/157 | 3.26/261 | 231/364 | STD HI TRIESTER |
| E | 0.538/80 | 36.7/158 | 1.77/255 | 58.9/351 | STD LO TRIESTER |

| PREPREG STAGE | V-1/°C | P-1/°C | V-2/°C | P-2/°C | PREPREG STAGE TEMPERATURE |
|---|---|---|---|---|---|
| B-1 | 3.41/92 | 119/164 | 8.08/268 | 298/362 | 10°F LOWER |
| B-2 | 5.33/102 | 118/164 | 8.66/270 | 184/365 | STD |
| B-3 | 3.25/93 | 67.8/171 | 3.74/270 | 123/358 | 10°F HIGHER |

## TABLE 3
## RDS (G") EFFECTS OF OUTTIME ON PMR-15 3K 8HS PREPREG

| PREPREG | P-1/°C | V-1/°C | P-2/°C | V-2/°C | P-3/°C | V-3/°C | P-4/°C |
|---|---|---|---|---|---|---|---|
| LOT A | | | | | | | |
| 0 DAY | 17/18 | 1.9/91 | 18/137 | 12/170 | 16/197 | 1.7/271 | 25/337 |
| 10 DAYS | 16/25 | 2.6/96 | 19/132 | 10/166 | 16/197 | 2.3/266 | 39/333 |
| 15 DAYS | 13/30 | 2.4/80 | 16/136 | 11/165 | 16/196 | 2.1/274 | 25/335 |
| 20 DAYS | 8/32 | 2.0/97 | 10/141 | 6/170 | 9/195 | 1.2/273 | 19/341 |
| LOT B-2 | | | | | | | |
| 0 DAY | 23/18 | 1.1/98 | 20/139 | 12/170 | 17/198 | 1.7/276 | 28/339 |
| 10 DAYS | 15/30 | 4.1/97 | 19/135 | 11/173 | 20/211 | 5.3/273 | 40/333 |
| 15 DAYS | 10/30 | 2.6/85 | 13/135 | 7/166 | 13/204 | 2.2/273 | 21/336 |
| 20 DAYS | 12/32 | 1.7/98 | 13/139 | 8/167 | 11/192 | 1.3/273 | 20/337 |
| LOT C | | | | | | | |
| 0 DAY | 17/28 | 1.9/95 | 17/139 | 8/175 | 10/191 | 1.71/275 | 25/336 |
| 10 DAYS | 14/22 | 2.8/88 | 21/136 | 12/170 | 21/202 | 3.2/274 | 37/338 |
| 15 DAYS | 13/32 | 1.6/98 | 15/139 | 8/170 | 12/196 | 1.8/278 | 26/338 |
| 20 DAYS | 5/35 | 2.6/89 | 12/140 | 7/170 | 11/202 | 1.4/276 | 19/339 |

## TABLE 4
## RDS (G*) EFFECTS OF OUTTIME ON PMR-15 3K 8HS PREPREG

| PREPREG | V-1/°C | P-1/°C | V-2/°C | P-2/°C |
|---|---|---|---|---|
| **LOT A** | | | | |
| 0 DAY | 3.4/101 | 97/163 | 3.6/271 | 109/352 |
| 10 DAYS | 4.2/48 | 100/163 | 3.8/263 | 227/358 |
| 15 DAYS | 5.5/92 | 84/161 | 4.5/263 | 161/360 |
| 20 DAYS | 3.9/104 | 51/167 | 2.7/265 | 106/363 |
| **LOT B-2** | | | | |
| 0 DAY | 1.7/104 | 105/167 | 3.6/273 | 124/355 |
| 10 DAYS | 6.4/101 | 107/170 | 11.2/262 | 275/369 |
| 15 DAYS | 5.0/91 | 78/170 | 4.4/264 | 158/379 |
| 20 DAYS | 1.7/98 | 63/167 | 2.6/268 | 103/359 |
| **LOT C** | | | | |
| 0 DAY | 3.2/95 | 82/159 | 3.0/275 | 124/357 |
| 10 DAYS | 4.58/91 | 117/170 | 6.0/268 | 247/366 |
| 15 DAYS | 2.9/101 | 75/167 | 3.1/274 | 147/362 |
| 20 DAYS | 5.2/101 | 68/166 | 3.6/272 | 127/366 |

## TABLE 5

## EFFECT OF PREPREG AGING ON COMPRESSIVE INTERLAMINAR SHEAR

| PREPREG VARIANT | PREPREG AGING DAYS @ 70°F MAX. | COMPRESSIVE INTERLAMINAR SHEAR AS RECEIVED R.T. | COMPRESSIVE INTERLAMINAR SHEAR AS RECEIVED 600 F | COMPRESSIVE INTERLAMINAR SHEAR, PSI AFTER AGING 125 HOURS @ 600 F — 600 F |
|---|---|---|---|---|
| A-1 | 0 | 9820 | 4890 | 7030 |
| B-1 | 0 | 7870 | 4180 | 6480 |
| B-2 | 0 | 9800 | 3990 | 5930 |
| B-3 | 0 | 8690 | 4520 | 5810 |
| C | 0 | 8350 | 4470 | 6020 |
| A-1 | 10 | 8520 | 3970 | 5410 |
| B-1 | 10 | 8680 | 3790 | 6440 |
| B-2 | 10 | 8810 | 3720 | 6300 |
| B-3 | 10 | 8680 | 3890 | 6310 |
| C | 10 | 9120 | 3730 | 6510 |
| A-1 | 15 | 9850 | 3920 | 6840 |
| B-1 | 15 | 8630 | 3970 | 5900 |
| B-2 | 15 | 9700 | 4310 | 7240 |
| B-3 | 15 | 8970 | 4330 | 6470 |
| C | 15 | 8490 | 4420 | 6410 |
| A-1 | 20 | 11,010 | 4110 | 7090 |
| B-1 | 20 | 9020 | 3540 | 6790 |
| B-2 | 20 | 9490 | 3710 | 7010 |
| B-3 | 20 | 9770 | 3390 | 6700 |
| C | 20 | 8630 | 3960 | 6790 |

## TABLE 6

### PMR-15/3K 8HS C FABRIC % C-SCAN AFTER 0, 10 AND 20 DAY AGE

| | A | B-1 | B-2 | B-3 | C | D | E |
|---|---|---|---|---|---|---|---|
| **0 DAY** | | | | | | | |
| 8 PLY | 100% | 100% | 100% | 100% | 100% | 0% | 100% |
| 14 PLY | 100% | 100% | 100% | 95% | 95% | 0% | 100% |
| 20 PLY | 5% | 5% | 10% | 15% | 12% | 0% | 100% |
| **10 DAY** | | | | | | | |
| 8 PLY | 100% | 100% | 100% | 100% | 100% | 0% | - |
| 14 PLY | 100% | 100% | 100% | 100% | 100% | 0% | - |
| 20 PLY | 2% | 1% | 6% | 2% | 3% | 0% | - |
| **20 DAY** | | | | | | | |
| 8 DAY | 100% | 100% | 100% | 100% | 100% | 0% | - |
| 14 DAY | 100% | 100% | 100% | 100% | 100% | 0% | - |
| 20 PLY | 3% | 1% | 35% | 1% | 1% | 0% | - |

## TABLE 7
## PMR-15, HPLC PER NASD METHOD T-202
## PEAK AREA % OF INGREDIENTS

| AGE DAYS | RESIN LOT | BTDE-1 | BTDE-2 | BTDE-3 | NE | BTTrE | POST |
|----------|-----------|--------|--------|--------|------|-------|------|
| 0 | A | 17 | 38 | 32 | 8.1 | 2 | 3 |
|   | B-1 | 17 | 38 | 31 | 7.5 | 1 | 5 |
|   | B-2 | 17 | 38 | 31 | 7.1 | 2 | 4 |
|   | B-3 | 17 | 38 | 30 | 8.2 | 2 | 4 |
|   | C | 16 | 37 | 30 | 8.0 | 2 | 5 |
| 5 | A | 17 | 36 | 30 | 6.9 | 1 | 9 |
|   | B-1 | 17 | 36 | 30 | 6.3 | 1 | 9 |
|   | B-2 | 17 | 37 | 30 | 6.3 | 1 | 8 |
|   | B-3 | 16 | 36 | 30 | 6.7 | 2 | 9 |
|   | C | 16 | 36 | 29 | 6.7 | 2 | 10 |
| 10 | A | 16 | 34 | 28 | 6.7 | 2 | 12.5 |
|   | B-1 | 16 | 35 | 29 | 6.3 | 1 | 11.9 |
|   | B-2 | 16 | 35 | 28 | 6.1 | 2 | 12.0 |
|   | B-3 | 16 | 36 | 29 | 5.7 | 1 | 11.0 |
|   | C | 16 | 35 | 29 | 6.1 | 1 | 12.0 |
| 15 | A | 15 | 33 | 26 | 6.2 | 2 | 17 |
|   | B-1 | 15 | 33 | 26 | 6.4 | 1 | 25 |
|   | B-2 | 16 | 35 | 26 | 5.6 | 1 | 16 |
|   | B-3 | 16 | 34 | 28 | 5.8 | 1 | 15 |
|   | C | 15 | 33 | 27 | 6.0 | 1 | 17 |
| 20 | A | 15 | 33 | 25 | 5.4 | 1 | 19 |
|   | B-1 | 15 | 32 | 25 | 5.5 | 1 | 21 |
|   | B-2 | 15 | 33 | 26 | 5.5 | 1 | 19 |
|   | B-3 | 15 | 33 | 26 | 5.6 | 1 | 18 |
|   | C | 15 | 32 | 25 | 6.9 | 1 | 19 |
| 0 | D | 15 | 38 | 33 | 3.0 | 11 | 0 |
| 0 | E | 19 | 44 | 31 | 4.1 | 1 | 0 |

## TABLE 1
### RDS (G") PMR-15/3K 8HS CARBON FABRIC PREPREG (AS RECEIVED)
### EFFECT OF RESIN REACTION TIME AND PREPREG STAGING TEMPERATURE

| RESIN | P-1/°C | V-1/°C | P-2/°C | V-2/°C | P-3/°C | V-3/°C | P-4/°C |
|---|---|---|---|---|---|---|---|
| A | 31.2/11 | 2.49/93 | 33.2/136 | 16.8/178 | 22.7/193 | 3.22/276 | 38.9/341 |
| B-2 | 24.4/19 | 2.83/99 | 22.0/139 | 11.4/174 | 17.2/208 | 4.31/281 | 33.1/338 |
| C | 9.96/14 | 2.06/84 | 15.9/137 | 5.47/174 | 9.64/215 | 2.62/279 | 29.4/336 |
| D | 17.6/14 | 1.57/97 | 18.9/138 | NO VALLEY | 22.2/169 | 2.11/272 | 37.8/348 |
| E | N/A | 0.33/80 | 11.0/135 | 7.12/162 | 0.59/162 | 1.10/259 | 7.85/327 |
| **PREPREG STAGE** | | | | | | | |
| B-1 | 21.2/11 | 2.47/74 | 27.5/136 | 13.4/180 | 17.2/208 | 4.58/272 | 53.9/341 |
| B-2 | 24.4/19 | 2.83/99 | 22.0/139 | 11.4/174 | 17.2/208 | 4.31/281 | 33.1/338 |
| B-3 | 7.17/20 | 1.99/90 | 12.0/140 | 6.11/165 | 10.3/205 | 1.98/273 | 18.5/336 |

## TABLE 2

### RDS (G*) PMR-15/3K 8HS CARBON FABRIC PREPREG (AS RECEIVED)
### EFFECT OF RESIN REACTION TIME AND PREPREG STAGING TEMPERATURE

| RESIN | V-1/°C | P-1/°C | V-2/°C | P-2/°C | RESIN REACTION TIME |
|---|---|---|---|---|---|
| A | 3.95/93 | 166/162 | 4.86/273 | 295/360 | 10' LESS |
| B-2 | 5.33/102 | 118/164 | 8.66/270 | 184/365 | STD |
| C | 3.30/93 | 66.9/168 | 5.24/273 | 162/363 | 10' MORE |
| D | 2.23/97 | 72.7/157 | 3.26/261 | 231/364 | STD HI TRIESTER |
| E | 0.538/80 | 36.7/158 | 1.77/255 | 58.9/351 | STD LO TRIESTER |

| PREPREG STAGE | V-1/°C | P-1/°C | V-2/°C | P-2/°C | PREPREG STAGE TEMPERATURE |
|---|---|---|---|---|---|
| B-1 | 3.41/92 | 119/164 | 8.08/268 | 298/362 | 10°F LOWER |
| B-2 | 5.33/102 | 118/164 | 8.66/270 | 184/365 | STD |
| B-3 | 3.25/93 | 67.8/171 | 3.74/270 | 123/358 | 10°F HIGHER |

AN OUT TIME STUDY ON PMR-15 POLYIMIDE RESIN
Tuyet Vuong
ROHR INDUSTRIES, Inc.
Riverside, California

Abstract:

PMR-15 polyimide is currently proposed for use in aircraft for high temperature applications. Advancement of the resin matrix during out time before cure has been a great concern since the resin is stored in the monomer stage. A study was performed to determine the effects of prepreg out time on chemical, physical, and mechanical properties of the cured laminates. Several analytical techniques were used to monitor changes in chemical composition of uncured resin during out time.

Advancements of the resin matrix were clearly detected by high performance liquid chromatography and dynamic mechanical analysis. The advancements of the resin resulted in increased viscosity and reduced resin flow. Interlaminar shear and compression tests indicated that twenty days out time of prepreg did not adversely affect mechanical properties of thin laminates.

1. INTRODUCTION

PMR-15 polyimide is currently evaluated at Rohr for 450 to 600 Deg.F applications, a region in which epoxies cannot be used since their strength rapidly deteriorates due to oxidation and polymer breakdown.

The thermosetting polyimide PMR-15, developed at the NASA Lewis Research Center (1), is available from several commercial sources. The PMR approach is unique in that an alcohol solution of three monomeric reactants is impregnated into a reinforcement, dried to a desirable volatile content, and stored without B staging. The three monomers comprised in PMR-15 are 4-4' methylene dianiline (MDA), dimethylester of 3,3', 4,4' benzophenone tetracarboxylic acid (BTDE) and a reactive end-cap, monomethyl ester of 5 norbornene 2,3 dicarboxylic acid (NE). The suggested reaction sequences for the system are shown in Figure 1.

Because the resin systems are formulated and stored in the monomer stage (stage A), the prepolymer and crosslinking reactions are thermally induced in situ during processing. It is reasonable to expect that the monomer solution and preimpregnated fiber are susceptible to advancement by thermal conditions during preparation of prepreg and long storage times.

It has been shown in previous studies (2,3) that reactions between the monomers did indeed occur during storage. Such advancement of the resin systems, varied with storage conditions, could potentially cause a shift in the competing condensation reactions which occur during imidization, thus resulting in changes in properties of the cured resin. A study was performed to determine the effects of prepreg out time on the processability and mechanical properties of PMR-15 cured laminates. Advancements of the uncured resin matrix were characterized by several analytical techniques.

## 2. EXPERIMENTAL PROCEDURES

### 2.1 Conditioning of Samples

Celion fiber C3K24X23, 8 harness satin weave fabric preimpregnated with PMR-15 resin was used for this study. The prepregs were cut into 8 pieces, 10 inch by 10 inch, for each panel. They were stacked with a separator film on each side, and individually stored in sealed bags. All the samples were exposed to an ambient temperature of approximately 24 Deg.C (75 Deg.F) for periods of 2, 5, 10, 15, 20, and 30 days consecutively. The samples were re-stored in a -18 Deg.C (0 Deg.F) freezer after reaching their required out time.

When the last sample reached 30 days, all the previously aged samples were removed from the freezer. They all were laid up and cured at the same time to minimize variations due to processing. It was suspected that drying of the free solvent (methanol) could occur after a long time exposure at room temperature. Therefore, a second set of samples was prepared simultaneously with those in the first set of samples starting with ten days out time. The second set of samples was sprayed with methanol a day prior to lay up. The samples from the first set were laid up "as is" and were designated with the suffix A in Table 1. The second set was designated with the suffix B. Six plies from each condition were used for fabrication of test panels with a zero degree orientation. The two remaining plies were used for chemical characterization. All the samples were cured in the autoclave using the following cure cycle: dwell at 81 Deg.C (180 Deg.F) for 60 minutes, followed by 205 Deg.C (400 Deg.F) for 45 minutes, then 302 Deg.C (575 Deg.F) for 180 minutes. Vacuum was applied from room temperature up to 249 Deg.C (480 Deg.F), then vented. A pressure of 1.20 MPa (175 psi) was applied, starting at 232 Deg.C (450 Deg.F), slowly over a 15 minute period. Post cure was performed in an oven at 315 Deg.C (600 Deg.F) for 10 hours.

## 2.2 Prepreg Characterization

Advancement of the resin matrix was monitored by infrared spectrophotometry (IR), high performance liquid chromatography (HPLC), reverse phase, and gel permeation chromatography. The behavior of the resin, during processing, was monitored by dynamic mechanical analysis (DMA). Resin and volatile contents were determined by solvent extraction.

## 2.3 Laminate Evaluation

Prior to testing, all laminates were examined for gross defects using an ultrasonic C-scan technique. Standard metallographic procedures were used to evaluate each panel for microporosity and uniformity of composite structure.

Resin and void contents were determined by the standard acid digestion method. Glass transition temperature was obtained by DMA technique.

The laminates were tested in compression and interlaminar shear at room temperature and 260 Deg.C (500 Deg.F). Interlaminar shear tests were performed using a three point fixture with a cross head speed of 1.28 mm (0.05 inch)/min. The span was adjusted to give a span to depth ratio of 3.5:1. The thickness of the test specimens

ranged from 2.0 to 2.2 mm (0.078 to 0.085 inches).

Elevated temperature tests were performed in an environmental heating chamber. The load was applied when the test specimen reached the specified temperature, with the aid of a thermocouple placed adjacent to the test specimen.

Compression tests were performed with a 25 mm (1 inch) gage length and a crosshead speed of 0.64 mm (0.025 inch)/minute. A Rohr designed compression test fixture was used with a mechanical gage linear variable differential transducer (LVDT) for recording strain. For elevated temperature, the load was applied when the specimens reached the required temperature.

## 3. RESULTS AND DISCUSSION

## 3.1 Prepreg Characterization

3.1.1 Resin and Volatile Contents – The resin and volatile contents of the prepregs are shown in Table 1. Resin content was determined by extraction of resin with tetrahydrofuran (THF) and was reported as the percent of extracted resin weight divided by the original sample weight (wet resin content). No correlation between the resin content vs. the out time was observed, probably due to the variation of resin content at different locations in the prepreg roll. The resin content varied from 42.9 to 46.9 percent.

The quantity of free solvent, methanol, was determined by drying the samples under vacuum at room temperature for 24 hours. The free solvent varied from 1.03 to 1.28 percent and did not reduce with out-time as expected, probably because the samples were aged inside a sealed bag and with the separator films in place. Therefore, the loss of solvent was minimized.

The total volatile contents, a combination of free solvent and reaction volatiles from the condensation reaction, were determined after exposure to 315 Deg.C (600 Deg.F) for 15 minutes. A slight decrease in total volatile contents was noted, 11.5 percent vs. 9.2 percent which indicated that some condensation reaction had taken place.

3.1.2 Infrared Spectrophotometry - Figure 2 illustrates the infrared spectra of samples aged for 0 and 30 days at 25 Deg.C. The ester band at 1725 cm-' decreased with duration of out time. The cyclic imide band at 1765 cm-' was observed as a shoulder on the ester band after 10 days. This band was more visible after 30 days. This indicates that imidization occurs slowly at room temperature. The analysis showed that IR is not very sensitive to the advancement of the resin matrix.

3.1.3 Partition Chromatography (Ion Pair Reverse Phase) - Advancement of the resin matrix is clearly illustrated in the chromatograms, obtained by reverse phase, shown in Figure 3. More comprehensive discussion and technique development

can be found in Reference 2. The area percents of monomers and reaction products vs. out time are summarized in Table 2 and Figure 4. The area percent is the percentage ratio of individual peak area over total areas under the curve with no attempt to account for the difference in response factors.

The reaction product of NE and MDA, peak number 3, eluted at 11.9 minutes, and designated as NE-MDA I, decreases with out time. This adduct is believed to be the mono-nadamide of MDA, which converts to the more stable cyclic imide, which is peak number 7, eluted at 17.1 minutes. This stable mono-adduct of NE and MDA, designated as NE-MDA II, was formed very fast at the first 15 days and at a slower rate thereafter. As noted in the previous studies (2), the rate of NE-MDA II adduct formation depends strongly on the unreacted NE concentration. The concentration of NE depleted quickly with out time, resulting in a slower formation rate of NE-MDA adducts toward the end of the out time.

The adducts of BTDE and MDA were formed slowly and continued to grow. The trio of adducts, peaks number 8, 9, and 10, at elution times from 18 to 23 minutes, were formed at the same rate throughout the course of the out time exposure, and probably are the three isomeric MDA-BTDE monoamides. The rate of formation of the pair of adducts, peak numbers 16 and 17, appeared to increase with out time, and probably are the cyclic imides of BTDE-MDA. Since the cyclic imide is symmetrical, two isomers should be produced.

3.1.4 Gel Permeation Chromatography (GPC) -- Gel permeation chromatography was used to follow the change in the molecular weight of these samples. The separation technique is based on the molecular size of each component, and species with high molecular weight will have shorter retention time.

The molecular weight distribution of samples aged for 0 and 30 days are shown in Figure 5. The shift to higher molecular weight with increased out time was observed. BTDE and MDA, with molecular weights of 386 and 198, are represented by peaks D and G, respectively. NE, molecular weight of 196, would elute at the same time with MDA, however, due to its very low absorptive characteristic at this wavelength (235 nm), its presence may not be detectable. Peaks E and F are mono-adducts of NE and MDA, while peak C is speculated to be mono-adducts of BTDE and MDA and probably bis-adducts of NE and MDA. The higher molecular weight components A and B are believed to be a combination of adducts of the three monomers.

The increase in their molecular weights is clearly shown. MDA and NE, peak G, reduce with out time and consequently, their reaction products, (NE-MDA) peaks E and F, increase in intensity. BTDE, peak D, decreases at a slower rate than MDA and the reaction products of

BTDE and MDA are revealed in peak C. An increase of the higher molecular weight adducts, peaks A and B, is also noted.

3.1.5 Dynamic Mechanical Analysis (DMA) -DMA is essentially a characterization of molecular responses of a material subjected to an oscillatory deformation at a constant low strain amplitude. The measured resonant frequency of oscillation is directly related to the elastic modulus of the material while the energy needed to maintain constant amplitude oscillation is a measure of damping within the material. Due to the uncertainty associated with the sample dimensions and the non-homogeneous nature of the material, the data in this report are presented in terms of frequency in Hertz (Hz), rather than tensile modulus, and damping in millivolt (mV).

Figure 6 shows the DMA of the prepregs, scanned through a temperature region of 25 Deg.C (77 Deg.F) to 480 Deg.C (864 Deg.F) at a constant heating rate of 3 Deg.C (5.4 Deg.F)/minute. There are several regions of interest. The prepreg begins to soften at the temperature below 100 Deg.C as a result of melting of the monomers. Subsequently, the frequency reaches a minimum at approximately 135 Deg.C (243 Deg.F) and then increases over a relatively narrow temperature interval. The increase in stiffness that occurs from 135 Deg.C (243 Deg.F) to 180 Deg.C (324 Deg.F) is associated with the imidization reactions in the PMR-15 resin.

Approaching 180 Deg.C (324 Deg.F), the frequency begins to decrease and is accompanied by a maximum in damping. This transition is associated with the softening of the imidized resin. The glass transition temperature in this region is believed to depend on the degree of imidization.

The second increase in stiffness which begins near 273 Deg.C (491 Deg.F) is attributable to the cross-linking reaction in PMR-15. At approximately 363 Deg.C (653 Deg.F), the stiffness begins to decrease. This transition is believed to be the softening region of the inter-mediate crosslinking polyimide. This glass transition temperature region (Tg) is also coincident to the Tg region observed on the cured laminate. The final increase in stiffness is a result of continuance of the crosslinking process. More comprehesive cure rheology of PMR-15 can be found in ref. 4.

The DMA curves of samples with different aging conditions: 0, 5, 10, 20, and 30 days are overlaid for ease of comparison. In the tempera-ture region of monomer melt, below 120 Deg.C (216 Deg.F), the viscos-ities increase with prolonged out time. This observation, consistent with weight loss of panels during cure, implies that the longer the out time, the lesser the flow at the early stage of the cure cycle. Imidization and crosslinking shift

to lower temperatures with increased prepreg out time. The stiffness maxima in the imidization region appeared to decrease with out time.

Figure 7 illustrates the behavior of PMR-15 preimpregnated fabric subjected to a simulated cure cycle. The cure profile shows the three isothermal steps associated with the removal of free methanol, imidization and crosslinking.

Similar interesting regions were observed. The monomers melt during the first heat up step. The viscos-ity slightly increases during the first dwell, partially as a result of the removal of the free methanol and mostly from early stage of con-densation and imidization. In the second heat up step, (81-105 Deg.C), the material goes through imidiza-tion, then softening of the imidized components. During the second dwell, the stiffness slightly increases as imidization proceeds. As the temperature increases (a third heat up step), the material completes its softening transition, then passes through a minimum viscosity before continuing through the crosslinking region.

Similar behaviors were noted when comparing samples with different out time. Curves 1, 3, 6, and 7 represent the stiffness vs. cure cycles of the prepreg aged for 0, 5, 20, and 30 days, respectively. The viscosities decreased with out time and a higher condensation reaction rate occurred during the first isothermal step (83 Deg.C).

The glass transition temperature of the imidized components (180-240 Deg.C) appeared to be lower for the aged samples, which implied that the imidized oligomers probably have lower average molecular weight, due to monomer stoichiometry changes as a result of aging. The maxima at imidization step appeared to decrease as the material aged. The aged samples crosslinked at lower temperature than the fresh samples.

## 3.2 Test Panel Fabrication

The weights of the panels were obtained before and after cure and post cure. The weight loss during cure (after deflashing) decreased with increasing out time, 21.7 percent vs 17.6 percent for 0 and 30 days out time, respectively, as shown in Figure 8 and Table 1. This indicated that the samples with lesser degree of advancement had a better flow; therefore the resin tends to bleed out more. Another factor which contributed to lower weight loss of the highly staged samples was, that a portion of the reaction volatile was probably released due to condensation reaction of the monomers during aging. Spraying of methanol appeared to improve resin flow as shown by the higher weight loss. Minimal difference was observed in the weight loss during post cure, which ranged from 0.23 percent to 0.66 percent.

## 3.3 Cured Laminate Characterization

### 3.3.2 Microstructures -

Microstructures of the polished cross sections of panels 1 and 6 (control and 20 days out time) are shown in Figure 9. The panels fabricated from the highly aged preimpregnated fabric appeared to have slightly more compact fiber bundles, i.e. less resin between individual fibers in the same bundle, and larger resin rich areas between the adjacent fiber bundles. This observation, consistent with weight loss of panels during cure and DMA scans of uncured prepregs, implies that the longer the out time, the lesser the flow in the early stage of the cure cycle.

3.3.3 Resin and Void Contents - Results of the resin and void contents of the cured laminates, determined by acid digestion, are given in Table 1. The resin content has a slight trend to increase with aging. The trend, however, is not very significant, probably due to the variation of the resin content of the prepreg and the location where the samples were taken in the panels. The void contents in all the panels are low and varied from 0.6 to 1.40 percent.

3.3.4 Dynamic Mechanical Analysis (DMA) - A representative DMA thermogram for the PMR-15 cured sample is shown in Figure 10. The samples were scanned from 25 Deg.C (77 Deg.F) to 420 Deg.C (788 Deg.F) at 3 Deg.C (5.4 Deg.F)/min. The two quantities actually measured were the resonant frequency which is proportional to the modulus of the sample and the damping which measures the energy dissipation due primarily to internal molecular motion.

The elastic modulus decreased slightly with increasing temperature over the course of the experiment. The only dramatic feature was the precipitous decrease in modulus at temperatures above 300 Deg.C (572 Deg.F), accompanied by an intense peak in the damping spectrum. This transition is the well known glass transition. The glass transition temperature (Tg) is defined as the temperature at the maxima in the damping curve. This point was selected based on its precision and reproducibility. Figure 10 shows an interesting comparison of the two samples, one before and the other after post cure. As illustrated, the glass transition temperature shifts to higher temperature after post cure (334 Deg.C vs. 353 Deg.C), indicating that post curing increases the crosslink density of the resin matrix. Aging of uncured preimpregnated fabrics did not appear to affect the glass transition temperatures in the final cured state. The Tg's after post curing varied from 351 Deg.C (663 Deg.F) to 358 Deg.C (677 Deg.F) and are tabulated in Table 3.

3.3.5 Interlaminar Shear Tests - Interlaminar shear test results for the postcured laminates are tabulated in Table 3. A non-postcured laminate, fabricated from unaged preimpregnated fabric was also tested. The data presented are averages of seven tests at each condition. The results indicate that out time of PMR-15 preimpregnated fabrics do not significantly affect the interlaminar shear strength.

Non-postcured panels had a slightly higher shear strength at room temperature, but lower shear strength when tested at elevated temperature in comparison to the postcured samples. Methanol addition did not appear to affect interlaminar shear strength.

3.3.6 Compression Tests - The results of the compression tests for the postcured laminates are summarized in Table 3. The compression strength of a non-postcured laminate is also included. The data presented are averages of seven tests for each condition. A slight reduction in compression strength was observed after 20 days, or longer, exposure time at 25 Deg.C. Methanol addition appeared to slightly increase the compression strength of the aged samples. No difference was observed between the postcured and non-postcured samples.

4. CONCLUSIONS

Advancements of the resin matrix are clearly detected by HPLC and DMA. IR is less sensitive to the aging condition than the other techniques.

The resin flow, at the early stage of the cure cycle, reduced with increasing out time. More resin bleed and more compact panels were obtained for the less aged samples.

Methanol addition appeared to improve resin flow of the aged preimpregnated fabric.

The out time appeared to have little or no adverse effect on the quality of the thin panels in this study. Effects on a thick panel are not known.

Mechanical tests showed only a slight reduction in compressive strength for the six ply panels fabricated from prepreg aged over 20 days. The results of the mechanical tests, together with the materials analysis of the uncured samples imply that the advancement of the resin matrix is not great enough to significantly affect laminate quality and therefore their mechanical properties.

## 5. REFERENCES

(1) T. T. Serafini, P. Delvigs, and G. R. Lightsey, U.S. Pat. 3,745,149, July 10, 1973.

(2) T. Vuong, "Characterization of PMR-15 by Reverse Phase Liquid Chromatography", RHR 86-012, February 1986.

(3) G. D. Roberts and R. D. Vannucci, SAMPE Journal March/April 1986, pp24-28.

(4) F. A. Meyers, "Cure Rheology of PMR-15 Prepreg" SPE Proc. 2nd International Tech. Conf. on Polyimides, Ellenville, NY, Oct 1985.

## 6. ACKNOWLEDGEMENTS

The author wishes to thank C. A. Alagar, J. M. Lambert, J. E. McGreavy, C. T. Torres and the Mechanical Laboratory for their significant contributions to this work.

BIOGRAPHY

Tuyet Vuong is a Senior Research Engineer at Rohr Industries, Inc. with ten years experience in polymeric analytical techniques, polymeric properties and applications.

She received a B.S. in Chemical Engineering from the National Institute of Technology, Saigon - VN and an M.S. in Materials Engineering from the University of Wisconsin, Milwaukee.

TABLE 1: Physical Properties of Prepreg and Cured Laminates vs. Out Time

| Panel No. | Aging Time) (days) | PREPREG | | | CURED PANEL | | PROCESSING | |
|---|---|---|---|---|---|---|---|---|
| | | Resin Content (%) | Free Volatile (%) | Total Volatile (%) | Resin Content (%) | Void Content (%) | Wt. loss during cure (%) | Wt. loss during PC (%) |
| 1A | 0 | 44.8 | 1.09 | 11.46 | 31.92 | 0.81 | 21.7 | 0.42 |
| 2A | 2 | 46.9 | 1.28 | 11.73 | 31.88 | 1.13 | 21.1 | 0.42 |
| 3A | 5 | 45.2 | 1.04 | 10.24 | 35.61 | 0.82 | 19.98 | 0.48 |
| 4A | 10 | 46.1 | 1.08 | 10.63 | 35.28 | 0.83 | 19.4 | 0.50 |
| 5A | 15 | 45.0 | 1.24 | 9.73 | 34.79 | 1.04 | 18.64 | 0.42 |
| 6A | 20 | 42.9 | 1.03 | 9.74 | 32.93 | 1.37 | 18.32 | 0.43 |
| 7A | 30 | 44.0 | 1.09 | 9.13 | 33.53 | 0.63 | 17.58 | 0.60 |
| 4B | 10 | | 1.16 | 9.70 | 33.56 | 1.00 | 20.5 | 0.42 |
| 5B | 15 | | 1.18 | 9.98 | 33.40 | 1.40 | 19.74 | 0.38 |
| 6B | 20 | | 1.38 | 9.91 | 32.10 | 1.21 | 19.75 | 0.42 |
| 7B | 30 | | 0.96 | 9.03 | 33.06 | 0.81 | 18.06 | 0.23 |

TABLE 2

HPLC Area Percent of PMR-15 Resin Products vs. Prepreg Out Time at 75 Deg.F

### HPLC Area Percent

| Peak Identification | | Control | 2 days | 5 days | 10 days | 15 days | 20 days | 30 days |
|---|---|---|---|---|---|---|---|---|
| 1, 2 | MDA | 27.12 | 26.12 | 23.47 | 20.30 | 18.93 | 18.20 | 16.40 |
| 3 | NE-MDA amide | 8.48 | 4.42 | 4.17 | 2.33 | 2.69 | 3.16 | 1.38 |
| 6 | NE | 1.08 | 1.21 | 0.99, | 0.33 | 0.38 | 0.59 | 0.64 |
| 7 | NE-MDA imide | 8.31 | 12.56 | 15.74 | 18.98 | 18.90 | 19.56 | 22.40 |
| 8,9,10 | BTDE-MDA amide | 2.03 | 2.64 | 3.19 | 4.25 | 3.91 | 4.86 | 4.69 |
| 12,13,14 | BTDE | 46.90 | 47.16 | 45.23 | 45.52 | 43.39 | 44.26 | 42.92 |
| 16, 17 | BTDE-MDA imide | 0.51 | 0.62 | 1.11 | 1.46 | 2.54 | 2.18 | 6.13 |
| 11,15,21 | NE-MDA-NE | 1.95 | 1.71 | 2.91 | 2.15 | 3.92 | 2.60 | 2.62 |

TABLE 3

Properties of Cured Laminates ($0_6$) vs. Prepreg Out Time at 75 Deg.F

| Sample | Prepreg Aging Time (days) | Tg Deg.C (Deg.F) | Interlaminar Shear Strength at | | Compressive Strength at | |
|---|---|---|---|---|---|---|
| | | | 25 Deg.C MPA (KSI) | 260 Deg.C MPA (KSI) | 25 Deg.C MPA (KSI) | 260 Deg.C MPA (KSI) |
| 1A | 0 | 354 (669) | 79.0 (1.15) | 43.7 (6.34) | 732 (106.2) | 49.4 (71.6) |
| 2A | 2 | 353 (667) | 79.7 (1.16) | 42.3 (6.14) | 723 (104.9) | 55.2 (80.0) |
| 3A | 5 | 357 (675) | 74.1 (1.08) | 39.9 (5.79) | 724 (105.0) | 50.8 (73.7) |
| 4A | 10 | 356 (672) | 77.2 (1.12) | 42.9 (6.22) | 731 (105.6) | 51.4 (74.5) |
| 4A | 15 | 353 (667) | 80.1 (1.16) | 42.5 (6.16) | 767 (111.3) | 55.0 (79.8) |
| 6A | 20 | 355 (671) | 75.4 (1.09) | 43.2 (6.27) | 686 ( 99.5) | 51.3 (74.4) |
| 7A | 30 | 351 (663) | 75.5 (1.10) | 42.9 (6.23) | 664 ( 96.3) | 46.9 (68.0) |
| 4B | 10 | 358 (677) | 77.2 (1.12) | 42.9 (6.23) | 745 (108.1) | |
| 5B | 15 | 353 (677) | 78.5 (1.14) | 42.2 (6.12) | 767 (111.2) | |
| 6B | 20 | 358 (676) | 75.2 (1.09) | 43.2 (6.26) | 728 (105.6) | |
| 7B | 30 | 356 (672) | 74.2 (1.08) | 43.1 (6.25) | 715 (103.8) | |
| NPC | 0 | 334 (634) | 85.8 (1.25) | 39.7 (5.76) | 746 (108.2) | |

A:   "as is" samples
B:   methanol added samples
NPC: non-postcured samples

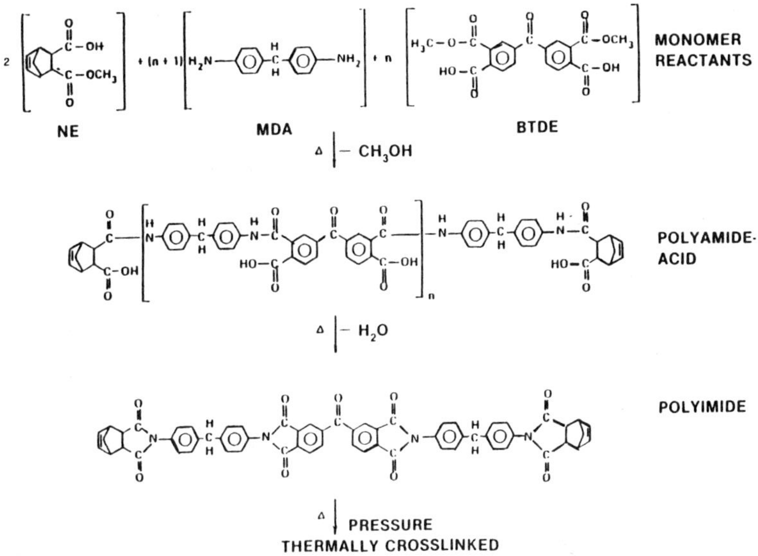

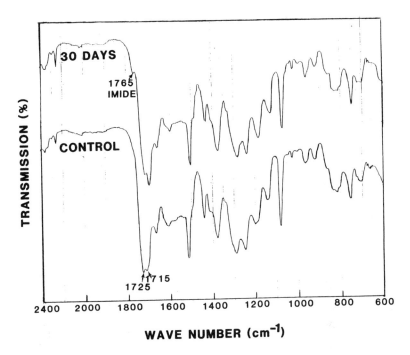

Fig. 1   Reaction Sequence for PMR Polyimide Resin

Fig. 2   IR Spectra of PMR-15 Prepreg Exposed to
0 and 30 Days at Room Temperature

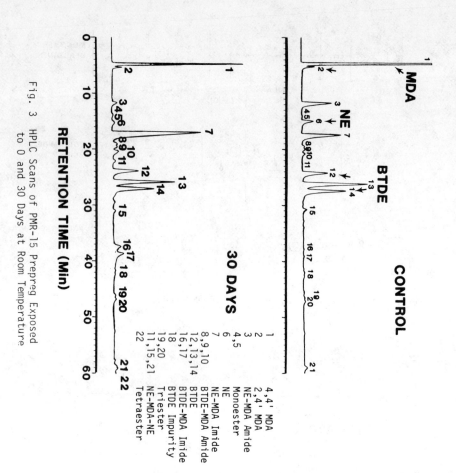

Fig. 3 HPLC Scans of PMR-15 Prepreg Exposed
to 0 and 30 Days at Room Temperature

| | |
|---|---|
| 1 | 4,4' MDA |
| 2 | 2,4' MDA |
| 3 | NE-MDA Amide |
| 4,5 | Monoester |
| 6 | NE |
| 7 | NE-MDA Imide |
| 8,9,10 | BTDE-MDA Amide |
| 12,13,14 | BTDE |
| 16,17 | BTDE-MDA Imide |
| 18 | BTDE Impurity |
| 19,20 | Triester |
| 11,15,21 | NE-MDA-NE |
| 22 | Tetraester |

Fig. 4 HPLC Area Percent of PMR-15
Monomers and Products vs.
Out Time

850

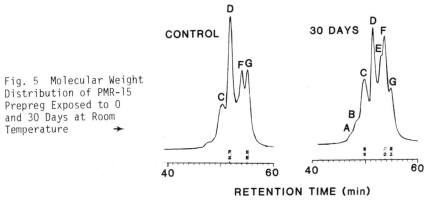

Fig. 5 Molecular Weight Distribution of PMR-15 Prepreg Exposed to 0 and 30 Days at Room Temperature ➔

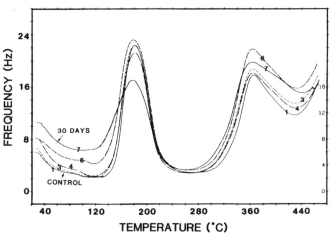

Fig. 6 Relative Stiffness of PMR-15 Prepreg Exposed to 0, 5, 10, 20, and 30 Days at RT vs. Temperature (Constant Heating Rate of 3°C/Min.)

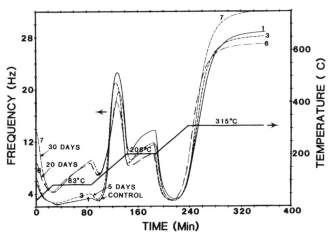

Fig. 7 Relative Stiffness of PMR-15 Prepreg Exposed to 0, 5, 20, and 30 Days at RT vs. Temperature (Typical PMR-15 Cure Cycle as Shown)

851

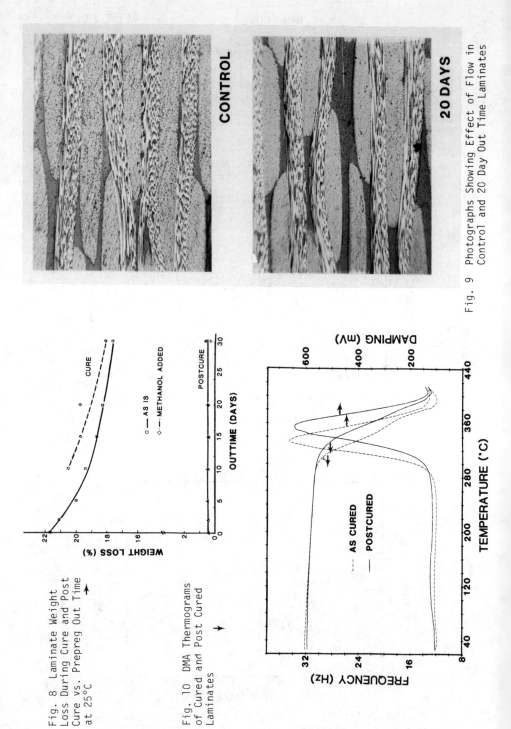

CONTROL

20 DAYS

Fig. 9  Photographs Showing Effect of Flow in Control and 20 Day Out Time Laminates

Fig. 8  Laminate Weight Loss During Cure and Post Cure vs. Prepreg Out Time at 25°C

Fig. 10  DMA Thermograms of Cured and Post Cured Laminates

32nd International SAMPE Symposium
April 6-9, 1987

EFFECTS OF MOLECULAR WEIGHT AND CURE CYCLE ON THE PROPERTIES OF
COMPRESSION MOLDED CELAZOLE PBI RESIN

Bennett C. Ward
Celanese Advanced Technology Company, PBI Business Unit
P. O. Box 32414, Charlotte, NC  28232

Abstract

A series of partial factorial
compression molding trials was
conducted evaluating the effects
of resin molecular weight and the
cure cycle parameters temperature,
pressure and hold time on Celazole
polybenzimidazole (PBI) molding
resin.  Optimization was based on
evaluation of the dependent
parameters tensile strength,
elongation and modulus at room
temperature.  Tensile strength and
elongation were directly related
to molecular weight and peak
molding temperature, with tensile,
elongation and modulus as high as
23 ksi, 3% and 0.9 msi observed.
Tensile strength was directly
related to elongation, with
constant modulus, for every sample
tested.  A wide range of
mechanical, physical, thermal and
electrical properties are also
reported for the first time for
parts molded from high molecular
weight PBI resin. Properties such
as 57 ksi compressive strength, 32
ksi flexural strength, Shore D
hardness of 99, and volume
resistivity of greater than $10^{16}$
ohm-cm were observed.

1.  INTRODUCTION

Polybenzimidazoles (PBI) are a
class of well-known, heterocyclic
polymers possessing outstanding
thermal stability[1] which contain
the benzimidazole functionality in
the polymer backbone.

The synthesis of wholly aromatic
polybenzimidazoles was reported in
1961 by Vogel and Marvel[2], who also
provided the first reports of the
polymer's outstanding thermal
stability.  Since that time,
several comprehensive reviews on
polybenzimidazole have been

published[3-6]. The polymers are conventionally synthesized via a high temperature reaction of aromatic bis-o-diamines and dicarboxylates (acids, esters or amides). A wide variety of aromatic variants have been investigated, although the particular polymer cited most often and used in all commercial and most developmental applications is poly[2,2'-(m-phenylene)-5,5'-bi-benzimidazole], which is referred to throughout this article as PBI.

PBI has been studied primarily in two molecular weight ranges: prepolymeric forms with an inherent viscosity (IV, 0.4% solution in 97% sulfuric acid) of 0.05 to 0.10 dl/g, and high molecular weight polymer, with IV above 0.5 dl/g. The relationship between IV and absolute molecular weight (determined by laser light scattering) is shown in Figure 1.

PBI has been found to be the only polymeric matrix material capable of maintaining load bearing properties for short periods of time at temperatures up to 1200°F[3,4]. In spite of this advantage, however, large scale structural applications for PBI have yet to be developed, in part because, until recently, PBI resins have either been unavailable or posed problems in handling, or because adequate processing techniques had not been found. PBI prepolymers, available from Aerotherm Division of Acurex Corporation as PBI 2801, 2803, PBI 1850 Laminating Material, and PBI 850 Adhesive[7], have been used extensively as matrix materials for composites because of their good melt and flow behavior[8], but are normally unsuitable for compression molding of large, neat specimens due to the relatively large amounts of offgassing exhibited during cure. The high molecular weight resin, however, exhibits much less offgassing during cure, which makes it much more suitable for large scale, compression molding systems[9].

High molecular weight Celazole PBI is produced as a resin form by Celanese. Variants being studied include 4 mesh, free flowing powder with an IV of 0.8 dl/g, and 100 mesh powder with an IV of 0.6 dl/g. These materials are fully polymerized and are essentially free of residual monomers. Polymer with an IV from 0.6 to 1.2 dl/g is currently being developed by Celanese, while PBI with IV between 0.1 and 0.6 dl/g, and above 1.2 dl/g, is regarded as research material only.

Recent reports described initial compression molding trials for high molecular weight Celazole resin. Workers at McDonnell Douglas indicated[10,11] that at molding temperatures of 600-800°F, pressures of 2000 psi, and final hold times of over 1 hour, specimens were molded with ambient temperature tensile strengths of 7 ksi. Enhancement of tensile strengths to 9.4 ksi was found upon postcuring the molded article at 850°F for 2 hours[11]. Celanese subsequently reported[9] excellent mechanical properties, including tensile strengths of up to 16 ksi, for several developmental, Celazole PBI resins molded at 785°F and 5000 psi. A concurrent paper at this symposium[12] details optimization of the molding and cure cycle for Celazole PBI resin, and reports tensile strengths of up to 18 ksi at mold temperatures of 875°F, hold times of 3 hours and pressures of from 4000 to 8000 psi.

## 2. COMPRESSION MOLDING OPTIMIZATION

### 2.1 Experimental

Celazole PBI resin samples were measured for particle size and inherent viscosity before each series of experiments. Unless otherwise indicated, the resin was dried in a vacuum oven at 110-140°C at 15 in Hg for at least 4 hours before use. All samples were molded on a 30 ton Dake press, equipped with top and bottom electrically heated 6x6 inch platens. Samples were formed in a 2-1/4 inch diameter matched metal mold, equipped with a thermocouple to monitor temperature. Temperatures reported are those measured at the mold itself. The pistons were coated with Freekote 440 mold release agent before each cycle.

Mold cycle specifics are described below. Samples were always cooled under pressure in the mold. Densities were determined by weight and physical measurement. Specimens for mechanical testing were obtained by cutting appropriately-sized test parts from the disk with a carbide band saw and finishing with a Tensilkut router. Photomicrograph analysis of mechanical testing specimens confirmed that, in all cases reported below, fracture occurred within the specimen, and was not a surface, or machining, artifact. A minimum of 3 specimens was molded and tested for each experimental condition, with the results reported being the averaged value for the test measurements.

### 2.2 Effect of IV (Molecular Weight) and Molding Temperature on Tensile Properties in Developmental Celazole Resins

Earlier studies[9] indicated a direct relationship between PBI resin molecular weight and subsequent tensile properties, when molded at a peak temperature of 785°F. In addition, these trials showed that increasing molding temperatures up to 785°F led to an increase in tensile properties. Additional experiments were run to determine if this effect was still present at higher molding temperatures, and if there was an upper IV limit, or cap, to the mechanical property increases observed.

Molding experiments were run evaluating the effect of a range of resin IV's (0.53, 0.69, 0.70 and 0.98 dl/g) at two different peak molding temperatures (825 and 875°F (440 and 470°C)). Resin particle size (smaller than 35 mesh or 500 microns), resin dryness (volatile content 1-3%), mold pressure (5000 psi) and plateau temperature hold time (90 minutes) were kept constant. A temperature and pressure profile for the molding cycle is shown in Figure 2.

The results are graphically presented in Figure 3. The highest tensile properties obtained were tensile strength/strain/modulus of 21 ksi/2.5%/0.91 msi respectively, which resulted on specimens molded from the highest IV resin (0.98) at the highest peak temperature (875°F). Both high polymer IV and high molding temperature correlated directly with higher

tensile and elongation. In addition, tensile modulus was relatively high (0.85 msi) and was not related to any independent variable evaluated.

Resin with ultra high IV (1.41 dl/g) was then evaluated at the high temperature molding conditions (875°F peak) and resulted in a tensile / strain / modulus of 18 ksi / 2.3% / 0.84 msi, which is lower mechanical strength than observed for lower IV resin molded under the same conditions. A plot of resin IV vs. tensile strength and elongation is found in Figure 4. The decline in strength indicates that there is a ceiling in the positive effect IV has on mechanical strength. This may be due to a higher processing temperature needed for extremely high molecular weight resins, or the previously observed higher degree of crosslinking in these materials.

When a plot was made of elongation (strain) vs. tensile strength for every sample tested up to this point (both this work and references 9 and 12), shown in Figure 5, it was found that there was a direct and linear relationship between these two parameters. In addition, tensile modulus for every sample tested was nearly identical, averaging 0.85 msi ± 0.1. This evidence, plus evaluation of the break topography by scanning electron microscopy, plus a linear stress-strain curve to break, is indicative of a brittle failure mode for molded PBI at ambient temperature.

Experiments were also run evaluating a combination of molding high IV (0.98 dl/g) resin under the optimum molding conditions determined by McDonnell Douglas[9]. Two sets of conditions, found in Table 1, were run. (The latter molding regime as found in Table 1 is graphically illustrated in Figure 6). These results indicated that implementation of temperature hold steps in the molding cycle improves tensile strength by 2 ksi over that observed for the simpler cycle described above.

## 2.3 Large Part Sintering Program

Celanese and Alpha Precision Plastics, Inc., Houston, Texas, have developed a proprietary process to make large formed articles and stock shapes via sintering of Celazole 100 mesh, 0.55 IV polymer. Solid rods and cylinders of PBI have been sintered as large as 7 inches o.d. and 12 inches long. These parts are fully fused, have densities of 1.28, and may then be machined to form a wide variety of parts, such as gaskets, chevron rings, valve seats, and disks. Applications are now being developed for these materials in industries such as oil drilling, chemical process equipment, and high temperature / high performance aerospace end uses.

## 2.4 Performance Data for Molded PBI

A variety of physical data have been generated for molded PBI, and are found in Table 2. Data were determined on as-molded parts made from 100 mesh Celazole resin molded either at Alpha Plastics or McDonnell Douglas.

Initial mechanical data, recently acquired at ambient temperature conditions for unfilled PBI resin, indicate that the material has relatively low density compared to other heterocyclic resins, has extremely high strength, especially for an amorphous polymer, and is somewhat brittle. Note that the value reported for tensile strength, 18 ksi, is for 100 mesh Celazole resin, molded via the optimized compression molding process described above at 875°F, and also sintered via the proprietary Alpha Plastics process. As reported above, tensile strengths of as high as 23 ksi have been obtained for developmental PBI resin variants with higher IV.

The compressive strength of 57 ksi is extremely high for an unfilled, thermoplastic material, and compares well with unfilled compressive strengths of 66 ksi reported by Levine for cured PBI prepolymer[1]. Unlike tensile and flexural tests, where brittle failure was evident, plastic deformation, or "barreling" was observed for molded Celazole resin under extreme compressive load. This result was unexpected given the brittle failure mode observed under tensile and flexural stress, and is so far unexplained. Both tensile and flexural modulus (0.85 mpsi) are relatively high compared to most unfilled molded thermoplastics, and are indicative of a stiffness level achieved with other materials only by addition of fillers. Tensile fatigue results indicate that a molded PBI part can be expected to last for one million cycles at 1 Hz at a stress level of 6.3 kpsi (35% of the 18 kpsi ultimate). Hardness and coefficient of friction values for molded Celazole were also quite interesting, indicative of an extremely hard and slick material, possibly suited for bearing applications.

Heat deflection temperature at 264 psi was measured to an instrument limit of greater than 300$^{\circ}$C (570$^{\circ}$F). Previous results gathered on unreinforced, cured PBI prepolymer indicated the heat deflection temperature to be 435$^{\circ}$C (815$^{\circ}$F)[1]. This value is higher, by 100$^{\circ}$F, than any commercially available, formable thermoplastic polymer known. This extremely

high value also correlates well to the glass transition temperature (Tg), which was found to be 450$^{\circ}$C (840$^{\circ}$F) by dynamic mechanical analysis (DMA). It is interesting to note here that DMA Tg values for bone dry and water saturated (5%) molded PBI samples were identical.

Regarding water absorption, immersion of a PBI ASTM flex bar in boiling water for 4 days results in a weight gain, due to water uptake, of 5% (Figure 7). This moisture can be removed by subsequent drying, under vacuum, at 125$^{\circ}$C (260$^{\circ}$F) for an equivalent period of time. Work is currently in progress to define, in full, the moisture adsorption properties of molded PBI, in both aqueous and hot, humid environments, and to determine the full effect of moisture on mechanical properties, and dimensional stability upon rapid thermal exposure.

Electrical properties, determined on samples conditioned according to ASTM D618, and containing approximately 1% volatiles, were quite varied, and were indicative of additional work that needs to be done to define the effect of voids and moisture on these values. For example, volume resistivity was greater than $10^{16}$, which classifies PBI as an insulator. The dielectric strength value, 300 V/mil, was actually an average of five determinations ranging from

150 to 500 V/mil, caused most likely by small differences in void content and location, not detectable by density determinations. The relatively high dissipation factor at high frequencies could be a result of high residual volatiles levels in the molded specimen. Thermal performance data is currently limited to semi-quantitative measurements of blistering and dimensional distortion after short term (5 minutes) exposure to the target temperature in air. Tests were run on specimens made from 100 mesh Celazole resin via the Alpha Plastics sintering process and the McDonnell Douglas compression molding process, and on specimens compression molded at Celanese from developmental PBI resin of larger particle size (40 mesh) and higher molecular weight (0.98 IV). Results are presented in Table 3, and indicate that PBI specimens molded by the sintering process exhibit much lower residual volatiles content, and, consequently, better blistering resistance than PBI molded by standard compression molding techniques. In addition, post drying of the specimens to remove residual volatiles markedly improves the resistance of the material to blistering in every case. McDonnell Douglas conducted GC/Mass Spec analysis of compression molded PBI specimens, and found that the primary volatile

component is phenol, with smaller amounts of water. This would account for the blistering propensity of poorly cured specimens, because phenol is known to be a good PBI plasticizer[9].

## 3. CONTINUING RESEARCH AND APPLICATIONS DEVELOPMENT

Additional data generation is underway to determine the impact of elevated temperatures on molded Celazole PBI mechanical properties, and to establish the effect of various hot and wet environments on moisture adsorption, and its effect on mechanical, thermal and electrical properties. In addition, end-use specific data are being gathered, such as coefficient of thermal expansion and thermal conductivity values. These data will be applied to end uses under development for aerospace applications, particularly leading edges requiring short term, extreme high temperature stability, and also to industrial applications, such as chemical process valves and seals, and oil well apparatus, where the high temperature and chemical resistance properties of PBI are required.

## 4. TRADEMARK

Celazole is a registered trademark of Celanese Corporation.

## 5. ACKNOWLEDGMENTS

The author wishes to thank
Dr. Frank Haimbach, Celanese
Research Company, Summit, NJ, for
invaluable assistance in physical
testing and data interpretation.

## 6. BIOGRAPHY

Bennett C. Ward was born in
Laurel, Delaware, and received his
B.S. in Chemistry from Duke
University in 1976. He received a
Ph.D. in Organic Chemistry from
the University of North Carolina -
Chapel Hill in 1980, after which
he joined Celanese Corporation,
where he is currently a research
chemist with the PBI Business Unit
of Celanese Advanced Technology
Company in Charlotte, NC. He is
working on product and
applications development of
Celazole PBI resin for compression
molding and as a matrix material.
He is author of three patents and
10 publications.

## 7. REFERENCES

1. Levine, H. H., Encycl. Polymer
   Sci. Technol., 11 188 (1969).
2. Vogel, H. Marvel, C. S., J.
   Polymer Sci., 50, 511 (1961).
3. Powers, E. J., Serad, G. A.,
   "History and Development of
   Polybenzimidazoles", ACS
   Symposium on the History of
   High Performance Polymers,
   April, 1986.
4. Cassady, P. E., "Thermally
   Stable Polymers, Synthesis and
   Properties", (Dekker, New
   York, 1980), pp. 168-173.
5. Jones, J. F., Macromol. Sci,
   C2, 303 (1968).
6. Korshak, V. V., "Heat
   Resistant Polymers"
   Izdatel'stvo "Nauka," Moscow,
   1969; English Translation:
   Keter Press, Jerusalem 1971,
   pp. 244-248.
7. Technical Bulletins, Acurex,
   Aerotherm Division, 485 Clyde
   Avenue, Mt. View, CA: PBI
   prepolymers 2801 and 2803, PBI
   1850 Laminating Material, and
   PBK 850 Adhesive.
8. Levine, H. H., "Fusible
   Polybenzimidazoles," US Patent
   3,386,969 (Whittaker Corp.),
   June 4, 1968.
9. Ward, B. C., Fabricating
   Composites '86, SME Composites
   Group, Baltimore, MD
   (September, 1986) EM86-704.
10. Treat, J. W., Jones, J. F.,
    Fountain, R., "Compression
    Molding of Polybenzimidazole",
    High Temple Workshop V,
    Monterey, CA 1985.
11. Jones, J. F., Fountain, R.,
    International Conference on
    Composite Materials IV, AIME,
    Warrendale, PA (1985) p. 1591.
12. Harb, M. E., Treat, J. W.,
    Ward, B. C., "Cure Cycle
    Optimization of Compression
    Molded High Molecular Weight
    Celazole PBI Resin", 32nd
    SAMPE Symposium, Anaheim, CA,

April, 1987.

13. Harb, M. E., McDonnell Douglas
Astronautics Company, Private
Communication.

Table 1:

Optimal Molding and Molecular Weight Conditions for Celazole PBI Resin

| Trial | 1 | 2 |
|---|---|---|
| Initial plateau temp, $^oF$ ($^oC$) | 400 (204) | 600 (315) |
| Initial pressure, psi (MPa) | 1000 (6.8) | 50 (0.3) |
| Initial plateau hold time, hr | 2 | 2 |
| Temp at pressure increase, $^oF$ ($^oC$) | 770 (410) | 600 (315) |
| Final pressure, psi (MPa) | 4000 (28) | 4000 (28) |
| Final hold temperature, $^oF$ ($^oC$) | 875 (470) | 875 (470) |
| Tensile strength, psi (MPa) | 23,000 (160) | 23,000 (160) |
| Strain, % | 3.0 | 3.0 |
| Modulus, $10^6$ x psi (1000 x MPa) | 0.85 (5.8) | 0.85 (5.8) |

Table 2:

Molded PBI Properties (100 Mesh Celazole Resin, Unfilled)

| Property | ASTM Method | Value | |
|----------|-------------|-------|---|
| Density, g/cc | | 1.3 | |
| Strength, psi (MPa) | D638 | 18,000 (124) | |
| Elongation, % | D638 | 2.4 | |
| Tensile Modulus, $10^6$ x psi (1000 x MPa) | D638 | 0.85 (5.8) | |
| Tensile Fatigue, % of stress to failure at 1 million cycles | | 35%, or 6.3 ksi | |
| Flexural Strength, psi (MPa) | D790 | 32,000 (220) | |
| Flexural Modulus, $10^6$ x psi (1000 x MPa) | D790 | 0.85 (5.8) | |
| Compressive Strength, psi (MPa) | D695 | 57,000 (390) (yield) 50,000 (340) (10% strain) | |
| Hardness | | | |
|    Rockwell M | D785 | > 125 | |
|    Rockwell E | D785 | 104 | |
|    Shore D | D2240 | 95 | |
| Izod Impact Strength, notched ft lb/in (J/m) | D256 | 0.4 (21) | |
| Coefficient of Friction | | Static | Dynamic |
|    Aluminum | | 0.14 | 0.16 |
|    Steel | | 0.18 | 0.17 |
|    Brass | | 0.16 | 0.18 |
| Heat Deflection Temperature, °F, 264 psi (°C, 1.8 MPa) | D648 | > 570 815[1] | (> 300) (435)[1] |
| Dielectric Strength, V/mil (kv/mm) | D149 | 300 | (11.5) |
| Volume Resistivity, ohm-cm | D257 | > $10^{16}$ | |
| Dissipation Factor | D150 | | |
|    1 kHz | | 0.006 | |
|    10 kHz | | 0.016 | |
|    0.1 MHz | | 0.11 | |
| Dielectric Constant | D150 | | |
|    1 kHz | | 3.4 | |
|    10 kHz | | 3.4 | |
|    0.1 MHz | | 3.3 | |

Table 3:

Short Term Resistance to Blistering Exhibited by Celazole Resin
Molded By Various Processes

| Molding Process | Residual Volatiles (%) | Blistering Temperature ($^{o}$F)[1] | |
| --- | --- | --- | --- |
| | | As-Molded | Dried[2] |
| Alpha Sintered | 0.7 | 900 | >900 |
| Celanese High IV Compression Molded | 1.5 | 700 | 900 |
| McDonnell Douglas Compression Molded | 1.4 | 700 | 900 |

Notes:

1.  Defined as appearance of surface irregularities, or density loss over 5%, after exposure of ASTM flex bars to air at the given temperature for 5 minutes. Test run at temperatures of 600 to 900$^{o}$F, by 100$^{o}$F increments.

2.  Specimens dried for 48 hours at 260$^{o}$F (120$^{o}$C) in a vacuum oven (15 inches Hg).

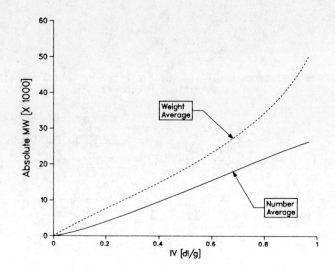

**Figure 1.** The relationship of PBI Resin Inherent Viscosity (IV, 0.4% solution in 97% sulfuric acid) to number and weight average molecular weight, determined by laser light scattering.

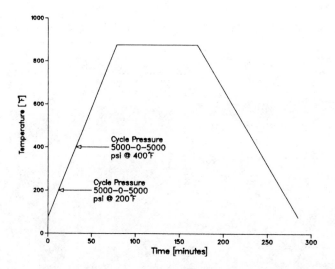

**Figure 2.** Typical temperature and pressure profile for unoptimized compression molding cycle. Peak temperature illustrated is 875°F (470°C). Pressure maintained at 5000 psi except where noted.

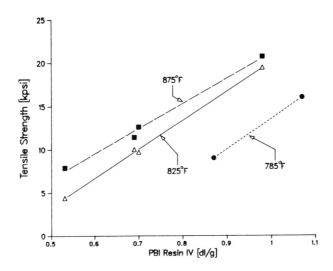

**Figure 3.**   Relationship between the independent variables peak molding temperature and resin IV, and the dependent variable tensile strength.  Molded under conditions outlined in Figure 2.

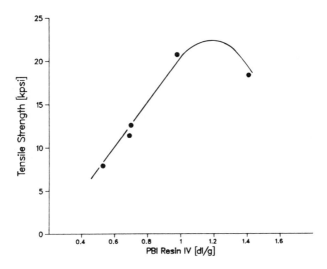

**Figure 4.**   Relationship between resin IV and tensile strength.  Peak molding temperature is 875°F (470°C), as illustrated in Figure 2.  Note decline in tensile strength at resin IV above 1.1 dl/g.

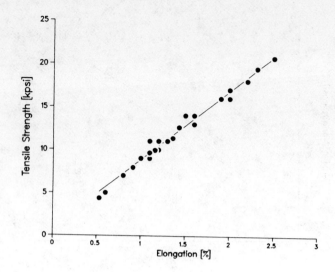

**Figure 5.**    Relationship of molded Celazole PBI tensile strength to elongation. Data points are taken from every observation tested, and indicate the brittle nature of the tensile failure.

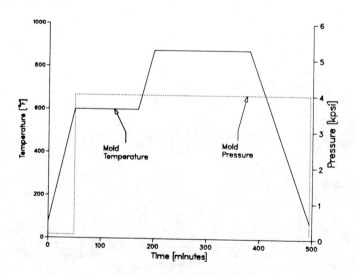

**Figure 6.**    Optimized PBI molding cycle as determined by McDonnell Douglas[12], and corresponding to trial number 2 in Table 1.

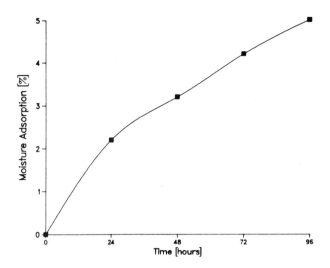

**Figure 7.**    Weight gain vs. time due to moisture uptake of a molded PBI specimen (ASTM flex bar) upon immersion in boiling water at atmospheric pressure.

# NEW INSIGHTS INTO CHEMISTRY AND PROCESSING
## OF CONDENSATION CURING GRAPHITE/POLYIMIDE COMPOSITES

Sean A. Johnson
Nancy K. Roberts
Advanced Manufacturing and Materials Technology
General Dynamics/Pomona Division

## Abstract

Development of consolidation techniques for condensation cure polyimide prepreg systems yielding high quality laminates has traditionally relied upon empirical methods. Chemical, rheological, and kinetic characterization of cure reactions occurring in commercial prepreg (T300/Skybond 703®) enabled development of a methodology to optimize consolidation procedures. Chemical characteristics of the resin as fabricated from the supplier and of resin extracted from prepreg was determined by high performance liquid chromatography (HPLC) analysis. Neat resin displayed very good batch to batch consistency whereas resin extracted from prepreg exihibited variability. Rheological characteristics of prepared panels were monitored using an Audrey ADR-380 Dielectric Rheometer. Volatiles were condensed and collected periodically during cure and postcure for gas chromatography (GC) analysis of composition. Identification of the volatile products allowed determination of reactions occurring, reaction mechanisms, and correlation to points in the laminate thermal profile. Viscosity, volatile evolution rate, and amount of volatiles remaining in the laminate were related to the thermal profile via GC and dielectric response. Taking all these factors into consideration, it was determined that during cure the imidization step occurs almost concurrently with formation of the polyamic acid and that the chemical mechanism of resin cure is affected by the degree to which the prepreg has been B-staged and by the processing mode.

## 1. INTRODUCTION

The material system studied was HMF 133/944 prepreg manufactured by Fiberite West Coast Corporation. This system is composed of Thornel® 300 graphite fiber in an eight-harness satin weave, impregnated with Skybond® 703 condensation curing aromatic polyimide resin manufactured by Monsanto. The system produces laminates possessing high strength and a high degree of thermo-oxidative stability. However, the high boiling solvent, N-methyl pyrrolidone (NMP), and the reactive volatiles generated during cure make successful processing of complex laminates with good structural integrity difficult.

Processing difficulties include fluctuation of yields, varying amounts of voids and porosity, delaminations, microcracking, and "brickdust." Aromatic polyimdides are notable for the formation of a powdery, opaque, orange to brown phase which resembles the dust from ground-up bricks. Normal appearance for a well processed aromatic polyimide is transparent, vitreous, and moderate to dark amber in color. Brickdust is a serious problem if it occurs extensively within the matrix.

## 1.1 Characterization Methodology

Frequent occurrence of the process difficulties defined above seriously hamper laminate production. In a factory setting, this situation is cause for serious concern. These concerns were responsible for institution of the material and process characterization program, the results of which are herein reported.

The segmented approaches utilized in the past to resolve process problems proved ineffective. An integrated methodology was adopted which considered the entire progression from prepreg precursors to commercial prepreg and through to finished laminates. Utilizing materials from each processing step a spectrum of analytical techniques were performed. The results were combined to give an overall view of material and processing characteristics.

## 1.2 Primary Concerns

In order to fully describe material behavior it was necessary to resolve the following issues:

•The fundamental relationships between prepreg chemistry, fabrication, and processing.

•Whether classical prepreg Quality Assurance procedures such as resin solids content, resin flow, gel time, and volatile content identify and control all properties essential for successful processing; if not, those procedures that fully identify the essential properties.

•The effect of solvent content in processing and on finished laminate quality.

## 1.3 Parameters Necessary to Resolve Primary Concerns

Specific resin and prepreg parameters required to resolve the full property and processing profile included:

•The amounts and composition of the volatiles generated during cure.

•Factors which are responsible for volatiles generation within the laminate.

•The mechanics of volatiles transport from the laminate and ways to control timing and amounts of volatiles generation.

•How viscosity varies during the cure cycle and how viscosity is influenced by heating rate, chemical reactions, and volatiles transport.

•The molecular weight distribution of the neat resin and resin as impregnated and how the molecular weight distribution affects processing and physical state of the laminate.

## 1.4 Nominal Resin Chemistry

The Skybond® 703 resin system is composed of a mixture of esters of benzophenone tetra-carboxylic acid (BTCE's), isomers of methylene dianaline (MDA's), solvent (NMP), and low molecular weight oligomers. Figure 1 shows the desired reactants and reaction sequence. The BTCE diesters and the 4,4'-MDA con-dense to form polyamic acid and ethanol as a volatile by-product. Subsequently, the adjacent amide and carboxylic acid groups of the polyamic acid undergo ring closure. This forms the polyimide structure with water as the volatile by-product. Finally, the cross-linking of polymer chains occurs, although there has been disagreement as to the specific sites and conditions at which this occurs[1].

## 2. EXPERIMENTAL TECHNIQUES

Resolution of the previously noted primary concerns is difficult, requiring the use of several analytical techniques and integration of the data provided by the individual techniques into a coherent, unified whole. In consideration of this, the material characterization program was comprised of three main sequences,

each centered about a particular set of analytical techniques.

## 2.1 Sequence 1

This activity was concerned with the characterization of cure processes and products and consisted of three basic efforts:

### 2.1.1 Dielectric Analysis

Multi-ply panels were constructed and cured in a Tetrahedron MTP-8 platen press. Dielectric properties were monitored using an ADR-380 Dielectric Rheometer. This provided data on laminate viscosity and volatiles generation and transport throughout the cure cycles.

Three laminate configurations were used: 4"x 4.5" twenty plies thick, 7"x 7.5" twenty plies thick, and tiered panels with base plies 7"x 7.5" (ten plies) and tiers 5"x 5" (ten plies). Electrodes were always placed in the laminate center. In most of the large panels, a second electrode was also utilized (via a multiplex arrangement) at the laminate edge.

Two types of initial cure cycles were utilized, one with a fast heating rate (greater than 5° C/min) and another with a slow heating rate (less than 3° C/min) which incorporated an intermediate dwell. Both cure cycles terminated with a one hour dwell at 180° C (356° F). Laminates which underwent postcure were step cured to 316° C (600° F).

### 2.1.2 Volatiles Analysis

Concurrent with the dielectric analysis during laminate cure, volatiles generated during the cure processes were condensed and collected at predetermined time intervals. Figure 5 illustrates the cure apparatus as modified for volatiles collection.

The volatiles collected were analyzed via gas chromatography to determine the types and amounts of each molecular species present at various intervals during the cure. Viscosity and chemical composition data were reduced and normalized. The resultant data were plotted on the same graphs with the thermal profiles used in order to easily correlate all three sets of parameters. Figures 2 to 4 are samples of this data with the thermal profiles omitted.

## 2.2 Sequence 2

Chemical composition and molecular weight distribution of neat resin and resin extracted from prepreg was determined.

### 2.2.1 High Performance Liquid Chromatography (HPLC)

Monomers of BTCE's, MDA's, and oligomers were resolved and quantified using a gradient elution technique. Oligomer/monomer ratios were calculated and used as an indication of degree of resin advancement.

### 2.2.2 Size Exclusion Chromatography (SEC)

Molecular weight distribution and weight average molecular weights (polystyrene equivalent) were determined for neat resin and resin extracted from prepreg. The data generated from SEC also provided an indication of degree of resin advancement and suggested chemical configurations for oligomers observed.

## 2.3 Sequence 3

This sequence consisted of analytical activities aimed at characterizing the physical states of the prepreg and finished laminates.

### 2.3.1 X-ray Diffraction

Neat resin was cured in the laboratory in such a manner as to retain NMP solvent. Heavy brick-dusting was produced by this method. The resultant resin was dried and analyzed in a powder X-ray diffraction unit. The objective was to determine presence of crystallinity in the cured resin.

### 2.3.2 Scanning Electron Microscopy (SEM)

SEM analysis was performed on uncured prepreg and cured laminates to determine resin morphology and wetting of the fibers by resin. This was important to consideration of brickdusting, microcracking, and volatile transport mechanism within the laminate.

## 3. RESULTS

The following results presented are based upon correlation of the data generated from each analytical sequence. Due to the large amount of data obtained during the course of this program, a large number of individual results were generated. Due to space constraints, only a limited amount of the supporting data is reported here. The authors felt that illustration of the methodology used and general usefulness of the results obtained warranted an initial general treatment emphasizing results and conclusions at the expense of raw data. Future papers will provide a more in-depth treatment of the individual activities.

## 3.1 Sequence 1

• Variations in B-staging of material between prepreg lots, rolls, and different areas within the same roll were observed with respect to dielectric response.

• Moisture peaks in the dielectric data coincided with times of major volatiles generation within the laminate. Figure 6 illustrates.

• Tests performed with two sets of electrodes (laminate edge and center) multiplexed within the same laminate confirmed volatiles accumulate in the center of laminates, as shown in Figure 7.

• There is no distinct imidization reaction temperature regime. The imidization reaction occurs almost concurrent with polyamic acid formation. This is shown in Figure 3 by the very close tracking of the water and ethanol condensate curves.

• The mechanism for mass transport from the laminate is due to thermally activated diffusion along a concentration gradient. Volatile transport was observed to depend heavily upon the amount of thermal energy supplied to the laminate. Transport was observed, via the dielectric data (Figure 6), to increase during the temperature ramps and to decrease during the dwells. The GC data confirm this effect; however, a lag behind the dielectric response is evident due to time required for diffusion from the matrix (Figure 4). There is no evidence either physical or chemical that supports a "gas pipe"

mechanism of volatiles evolution (diffusion along carbon fibers within the laminate) for this material. In addition to the data above, SEM analysis of prepreg and finished laminates showed the fibers to be well wet by the resin, leaving no path for diffusion other than through the matrix.

• Statistically significant differences (99% confidence level) in the magnitudes of the water/ethanol ratio for the initial cure cycle were observed. Substantial variations in the water/ethanol ratio during initial cure are not due to variations in ethanol generation but are due instead to variations in the water generation during cure. The variations in water generation during cure are the result of a shift in the selectivity of the basic reaction sequence within the prepreg. Evidence indicates that an undesired reaction mechanism is primarily responsible for increased water/ethanol ratios seen in panels fabricated using high heating rates and/or constructed so as to trap volatiles. This undesired reaction mechanism is a result of a shift from the amine-ester (MDA's and BTCE's) reaction mechansim for formation of the amide (polyamic acid) to an amine-acid reaction mechanism (MDA's and BTCA or the carboxylic acid functionalities on the BTCE's). Figure 8 illustrates operation of the shifted mechanism. The increase in the water/ethanol ratio represents an increase in the amount of volatiles which would otherwise be present. HPLC data confirm that the proper components are present in the correct amounts to allow operation of the reaction mechanism proposed.

## 3.2 Sequence 2

•HPLC provided satisfactory definition of the amount of monomers and oligomers present. The values were consistent in similarly aged resin lots with oligomer/monomer ratios of approximately 0.26. No impurities were detected.

•SEC of similarly aged resin lots gave number average molecular weights of approximately 420.

•HPLC of resin extracted from different lots of prepreg (at ambient temperature) showed an increase in oligomer content with oligomer/monomer ratios ranging from 0.99 to 3.37.

•SEC of resin extracted from various lots of prepreg (the same lots upon which the HPLC analyses were performed) exhibited number average molecular weight values of between 1059 and 1579.

## 3.3 Sequence 3

•X-ray diffraction showed some degree of crystallinity in cured resin that exhibited extreme brickdusting.

•As shown in Figure 9, the brickdust phase is composed of a network of resin nodules with interconnected (open) porosity. This network structure present in brickdust yields a weaker resin phase than the desired vitreous, homogeneous phase; providing numerous sites for initiation of cracks. Based on the morphology of the cracks (Figure 10) and also that brickdusting always

accompanied microcracking, matrix microcracking was determined to be induced by stresses caused by thermal contraction mismatch between the matrix resin and the graphite fibers[2]. The cracks extend only between individual lamina and are not associated with breakage of any graphite fibers. The cracks are not large and isolated; they are small, numerous, and fairly evenly distributed. The decrease in strength and numerous crack initiation sites resulting from pervasive brickdusting in the matrix were observed to be requisites for occurrence of extensive microcracking.

## 4.CONCLUSIONS

Based on the results obtained from the characterization study, the following conclusions were derived:

•Degree of B-staging and chemical composition cannot be determined from the chemical properties tests generally performed on prepreg as part of the normal quality conformance inspections. HPLC was demonstrated to be the only procedure capable of providing positive quantitative indications of degree of staging and chemical composition. The competing effects of amount of NMP present and degree of B-staging interact to determine the rheometric profile (i.e.; flow, tack, drape, viscosity) of the prepreg. The lack of an accurate method to determine the amount of B-staging or amount of NMP in prepreg has in the past permitted large uncertainties in the amount of NMP present, while chemical properties such as percent flow may display relatively little fluctuation. Taken together, the above facts indicate that the amount of NMP retained in the finished prepreg may be quite substantial and quite variable.

•Volatiles transport within the laminate is due to thermally activated diffusion against a concentration gradient through the resin matrix.

•Contrary to previously published information on this system[3], imidization occurs almost concurrently with formation of polyamic acid.

•The chemical mechanism of resin cure is affected by the processing mode. This is evidenced by water/ethanol ratios significantly greater than unity. This was demonstrated to be an indication that the resin reaction mechanism has been shifted from one dominated by ester-amine reactions to one dominated by acid-amine reactions. The shift of reaction mechanism is caused by accumulation of ethanol within the laminate during the cure. This results when the rate of diffusion of volatiles from the laminate is slow in comparison to the rate of volatiles generation. The increased reaction rates associated with high heating rates early in such cure cycles aggravate the situation by increasing the rate of volatile build up. Experimental evidence indicates that an increase in water/ethanol ratio represents a real increase in the number of moles of volatiles generated during the cure with respect to the situation in which the water/ethanol ratio is approximately one.

•Temperature conditions in the prepreg treater oven are sufficiently severe to stage the resin; however, the large free surface and short diffusion path of the prepreg permit the rapid escape of water and ethanol generated and the ideal reaction sequence (one mole of water generated for each mole of ethanol generated) prevails. When the prepreg is highly staged, the desired monomer reactants are depleted, leaving relatively more of those functionalities which react via the acid-amine mechanism (e.g., the carboxylic acid groups on BTCA and BTCE monoester). During high heating rate cure cycles with highly staged material the greater concentration of water generating functionalities results in significantly higher water/ethanol ratios. This may further contribute towards the tendency of the laminate to brickdust. The combination of large amounts of NMP present in the highly staged prepreg and the detracting influence of brickdust formation on the properties of the laminate plus the amount of water and ethanol generated via the altered reaction mechanism within the laminate may act to damage the forming laminate or prevent adequate consolidation at crucial points in the cure cycle. This may lead to delamination and low product yields. The higher water/ethanol ratios associated with more highly staged prepreg were confirmed for the panels

874

prepared as part of the study. To a confidence level of better than 98 percent (derived from statistical decision theory and small sampling theory[4]), the prepreg lot determined to be the most highly staged lot used during the study had a higher mean water/ethanol ratio than the mean of panels from all other lots prepared using a high heat rate cure cycle.

•Preliminary calculations based upon the HPLC and SEC results for neat resin and prepreg extracts indicate that the oligomers present in prepreg lots with the lowest amounts of staging to prepregs displaying the highest amounts of staging, ranged from trimers of amic acid to pentamers of amic acid; respectively. Amic acid monomer is represented as the first reaction product in Figure 1.

•The viscosity profile obtained is dependent upon the temperature profile used. High heating rates during intial cure tend to produce deep, narrow viscosity profiles that dwell for only short times near minimum viscosity, exhibit rapid build-up of viscosity after the minimum, and experience the minimum viscosity early in the cure cycle. Slow heating rates during the initial cure produce a more shallow, wide viscosity profile with the minimum viscosity occurring much later in the cure cycle and dwelling near the minimum value for a much longer time. Volatiles which are generated and retained within the laminate will also either reduce the effective viscosity or the rate at which it is increasing.

•Greater retention or generation of volatiles produces a greater amount of brickdust in the finished laminate. This was confirmed during the study, with large laminates displaying much more extensive brickdust than smaller panels and the laminate centers showing more severe brickdusting than the laminate edges.

•Matrix microcracking is a result of thermally induced stresses caused by the difference in relative rates of contraction between the graphite fibers and the matrix resin. Severe microcracking occurs when the matrix resin exhibits extensive brickdusting which reduces the strength of the matrix and provides a multitude of crack initiation sites due to the nodular, porous microstructure.

5. SUMMATION

The following observations are offered as considerations which may be beneficial to processors of condensation curing polyimide prepreg materials:

•In view of the detrimental effects of NMP with respect to finished laminate properties; i.e., brickdusting, delamination, microcracking, increased volatiles content, etc.; NMP content of the impregnating resin should be restricted to the absolute minimum amount possible. Impregnation with neat resin or dilution of impregnating resin with a lower boiling solvent should be explored.

•The amount of B-staging of prepreg should be controlled. The best method available at the present time to monitor the amount of B-staging is HPLC. Incorporation of HPLC as a quality conformance inspection procedure not only defines the amount of B-staging, but also provides data on chemical purity of the resin system.

•The material and the processing are not independent. To insure good yields and optimize the process: the consistency of the prepreg must be monitored and controlled, heating rates and cycles must be considered, the amounts and times of pressure application must be set to allow maximum release of volatiles (dielectric monitoring techniques will be helpful in this respect).

•A methodology similar to that developed to execute this program should be applied to all new organic structural composite materials development programs. Characterization of chemistry, physical properties, processing, and their interrelationships before committment to production can save huge amounts of money over the lifetime of an application.

6. ACKNOWLEDGEMENTS

The authors gratefully acknowledge the contributions of Dr. Walter Baumgartner of Tetrahedron Assoc. for dielectric work performed and Dr. Richard Wilson of Monsanto for HPLC analysis.

7. REFERENCES

1. "G. Lubin, ed.; Handbook of Composites; Van Nostrand Reinhold Co.; 1982."
2. "D. Wilson, et al; Preliminary Investigations into the Microcracking of PMR-15/Graphite Composites-Part 1 Effect of Cure Temperature; 18th International SAMPE Technical Conference; 1986."
3. "D.A. Scola; The Chemistry, Processing Characteristics and Properties of Commercial Polyimide Resins-A Review; 31st International SAMPE Symposium; 1986."
4. "G. Hahn and S. Shapiro; Statistical Models in Engineering; John Wiley and Sons, Inc.; 1967."

8. BIOGRAPHIES

Sean Johnson is a Senior Manufacturing Development Engineer in Advanced Manufacturing Technology at General Dynamics Pomona Division. His background includes experience in materials and processing of elastomers and organic, metal, and ceramic composites. He holds a B.S. degree in Chemistry from UCLA and a B.S. degree in Chemical Engineering from California Polytechnic University, Pomona.

Nancy Roberts is a Research Engineer in Materials Technology at General Dynamics Pomona Division. Her background includes experience in materials and processing of adhesives, elastomers, and composites. She recived a B.S. degree in Chemistry from UC Irvine and has attended California Polytechnic University, Pomona, School of Engineering.

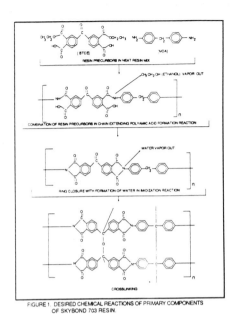

FIGURE 1. DESIRED CHEMICAL REACTIONS OF PRIMARY COMPONENTS OF SKYBOND 703 RESIN.

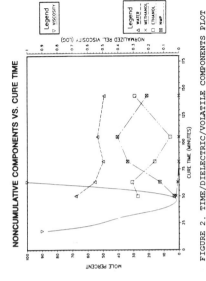

FIGURE 2. TIME/DIELECTRIC/VOLATILE COMPONENTS PLOT OF HIGH HEATING RATE PANEL INITIAL CURE.

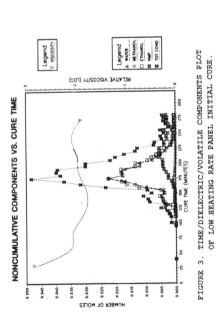

FIGURE 3. TIME/DIELECTRIC/VOLATILE COMPONENTS PLOT OF LOW HEATING RATE PANEL INITIAL CURE.

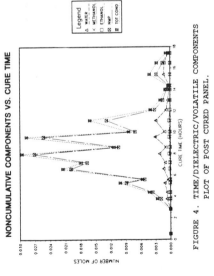

FIGURE 4. TIME/DIELECTRIC/VOLATILE COMPONENTS PLOT OF POST CURED PANEL.

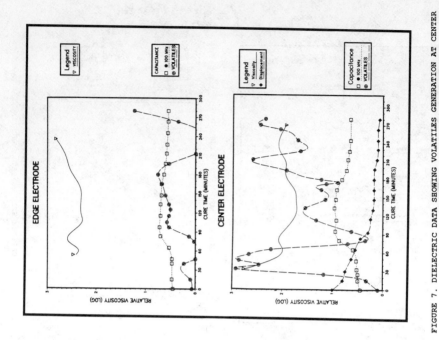

FIGURE 7. DIELECTRIC DATA SHOWING VOLATILES GENERATION AT CENTER
AND EDGE ELECTRODES OF PANEL DURING INITIAL CURE.

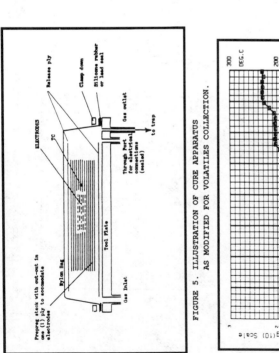

FIGURE 5. ILLUSTRATION OF CURE APPARATUS
AS MODIFIED FOR VOLATILES COLLECTION.

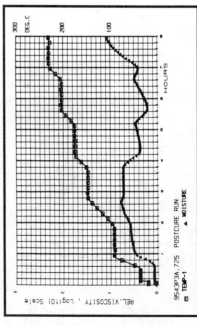

FIGURE 6. DIELECTRIC DATA SHOWING MOISTURE PEAKS CORRESPONDING
TO VOLATILE COMPONENTS SHOWN IN FIGURE 4.

878

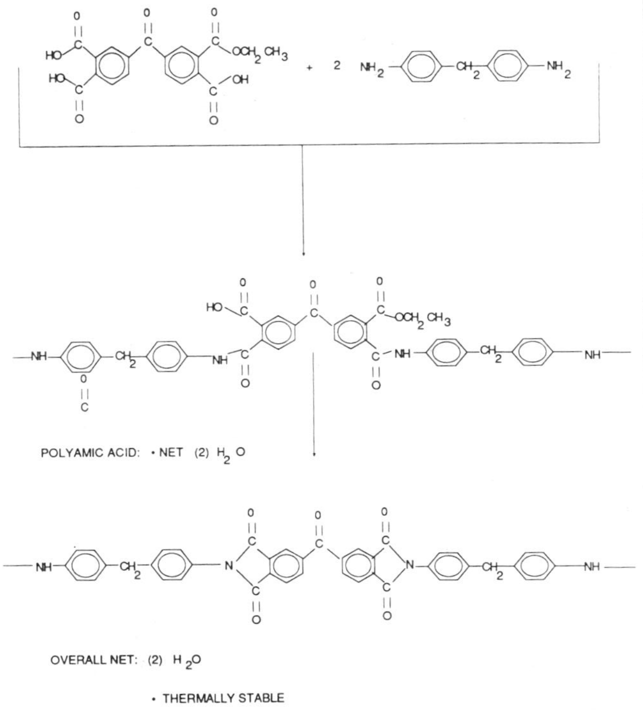

FIGURE 9.  SEM PHOTOGRAPH SHOWING NODULAR BRICKDUST.

FIGURE 10.  LAMINATE EXHIBITING MICROCRACKING.

POLYAMIC ACID:  · NET (2) $H_2O$

OVERALL NET: (2) $H_2O$

· THERMALLY STABLE

FIGURE 8. SHIFTED REACTION MECHANISM FOR RESIN CURE

# ALUMINIUM ALLOY METAL MATRIX COMPOSITES FOR ENGINEERING APPLICATIONS

Dr. K.G. Satyanarayana, Dr. B.C. Pai,
Dr. M.R. Krishnadev [**] and Dr. C.G. Krishnadas Nair [*]
Regional Research Laboratory, Trivandrum 19
(India)

## Abstract

A number of methods have been employed by various investigators in developing metal matrix composites. Regional Research Laboratory, Trivandrum have specialised in the development of Aluminium based metal matrix composites using Foundry route. Special equipment has been designed and developed to prepare particulate and fibre reinforced aluminium alloys of consistent properties for automobile and other engineering applications. A variety of particulates such as graphite, zircon and chopped fibres of carbon have been used in the preparation of different aluminium alloy matrix composites and their properties have been evaluated. Relations between property, type of reinforcement, and mechanical properties are discussed. Specific engineering applications of each composite system are highlighted.

## 1. INTRODUCTION

Metal matrix composites (MMC) have immense potentials as structural materials in aerospace industry for elevated temperature applications in view of their high temperature stability compared to conventional alloys systems and fiber reinforced plastics. Earlier R&D activity in MMC was mainly confined to continuous fiber reinforced systems (Al-carbon/graphite/SiC/$Al_2O_3$) with very little work on the use of expensive SiC Whiskers. This was so since the composites were mainly looked from an angle for improved mechanical properties which is possible by the reinforcement of the matrix by the continuous fibers/Whiskers. In recent years the focus has been shifted to use whiskers/short fibers and even particulates which are available at lower costs, since by choosing proper processing techniques with the dispersoids these composites can exhibit desired mechanical properties in addition to posessing special properties such as wear/abrasion resistance/self lubricating etc. with potential for applications in general engineering particularly the automotive and electromechanical systems.

Two major routes available for the synthesis of MMC are powder metallurgy (PM) and the casting route. The former though has advantages of incorporating higher amount with uniform distribution of dispersoids has the limitations such as high initial capital investments, and size and shape of the component to be fabricated from economical point of view. On the other hand, the casting route has advantages of making any size and shape of the component and on the whole the method is cheaper than P/M. Although the properties of MMC in the as-cast condition are inferior to that of P/M MMC, it is possible to use these in the as-cast condition itself for certain applications. Further, the properties can be enhanced by hot working, such as extrusion. A number of methods have been attempted by various investigators to synthesise the MMC which are mentioned in recent reviews[1-3]. Regional Research Laboratory, Trivandrum (RRL-T) has developed a simple and an inexpensive method of preparing cast composite for incorporating particulates or short fibres.

This paper highlights the R&D carried out at RRL-T and briefly outlines the procedure and composites developed in this laboratory. These include composites containing dispersoids such as graphite/coconut shell char/zircon/glass/fly ash and short carbon fibers synthesised by liquid metallurgy (Vortex method) technique, discusses various factors identified which affect the distribution of dispersoids and lists the observed properties and possible applications of these composites.

## 2. PREPARATION OF CAST METAL MATRIX COMPOSITES

The method developed in RRL-T as shown in Fig.1 is briefly described here. An aluminium alloy upto

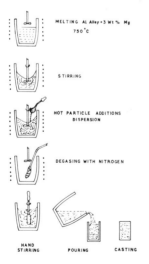

MELTING Al Alloy + 3 Wt % Mg
750°C

STIRRING

HOT PARTICLE ADDITIONS
DISPERSION

DEGASING WITH NITROGEN

HAND STIRRING   POURING   CASTING

Fig. 1: Schematics of steps involved in the preparation of MMC by liquid metallurgy

10 kg is melted in a crucible. Dispersoids such as graphite/zircon/fly ash/glass and short carbon fibers were incorporated into the melt at a suitable rate ( 0.001 kg/S ) while stirring was continued using a stirrer at a suitable rate (200-450 rpm). Then the melt was degassed using $N_2$ gas to reduce gas porosity and stirring continued before the composite melt is cast in the permanent moulds.

The consistency and reproducibility of properties of the composites depends on the extent of uniformity of dispersion of dispersoids which in turn depends on various factors such as size of the dispersoids, stirring method etc.

A systematic study[4] carried out with the aid of transparent medium model experiments has identified these factors including design of blades, use of baffles and two step heating of dispersoids etc. and the results of this are employed in RRL-T in the actual preparation of MMC using both particulate(p) and short fibers(f) dispersoids which are presented in the following sections:

### 2.1 Wetting of the dispersoid with the matrix

This is achieved by surface modification of the dispersoids by the pre-treatment of dispersoids at suitable temperatures (200-600°C), addition of reactive elements such as Mg to the melt prior or during the incorporation of dispersoids. Use of Mg helps in promoting wetting by decreasing the inter-facial energy of the melt or inducing wettability by a chemical reaction. This also helps partly in the dispersion of finer particles (< 20 µm)/short fibers which otherwise is very difficult to disperse.

### 2.2 Density difference

This factor between the matrix and the dispersoid determines the gravity segregation during solidification of the composite melt particularly in gravitydiecasting. Thus lighter particles such as graphite/carbon fibers float to the top while heavier particles such as zircon would settle down to the bottom. However, desired distribution of suspended particles in the matrix can be achieved by suitably controlling the flow of fluid and solidification parameters.

## 2.3 Stirring method and duration of contact between dispersoids and the melt

Recent study[5] carried out in RRL-T has established that type, shape, location of the blade in the melt and speed of rotation of the stirrer have significant effect on the distribution of the dispersoids in the melt. In addition time of contact between the dispersoids and the melt is found to be critical for getting good dispersion. The increase in mixing time would help in promoting wetting and bonding. Thus Fig.2(a) shows a macrophotograph of Al-graphite composite ingot having non-uniform distribution of graphite obtained by vortex method without resorting to the above mentioned parameters while Fig.2(b,c) show longitudinal and cross section of the same composites having uniform distribution of the particles obtained by proper stirring mechanism, optimum time of mixing, use of baffle etc. An

inert gas shielding during melting and dispersion was found to reduce excess drossing.

X 0.2

(b)

X 0.2

(c)

Fig. 2: Macro photograph of Al-graphite composites
(a) Nonuniform distribution
(b&c) Uniform distribution

(a)         X 0.2

## 2.4 Remelting of the composite

It has been observed[5] that more uniform distribution of the dispersoids takes place by simple stirring during remelting of the composites even having non-uniform distribution of the particles. This could be due to improved wettability of the dispersoid during remelting. This process would facilitate the large manufacturers to prepare composite ingots and supply them to small scale industries for use as in the case of alloy ingots. Secondly, master alloy concept can also be used by preparing composites having higher amount of dispersoids which can be further diluted by adding the matrix alloy to the required level.

## 2.5 Choice of casting method

Permanent mould casting resulted in good distribution of particles (Fig. 2 b-c) while sand castings gave gravity segregation. This can be overcome by resorting to pressure die casting/squeeze casting. It is found that by pressure die casting higher amount of dispersoids (upto 60 wt% in the case of zircon, 15-18% in the case of glass/flyash) can be incorporated compared to gravity die casting of the composites (upto 30 wt% of zircon and 8 wt% of glass and fly ash). This also helped in the reduction of casting defects such as porosity.

Centrifugal casting of the composite resulted in the dispersion of dispersoids upto 60-70 wt% as in the case of carbon micro-ballons[3] at specific surfaces (inner/outer periphery) in the castings which can be wear resistant or abrasion resistant.

Compocasting technique is now attempted in the laboratory to get more uniform distribution of particles in large castings (upto 30 kg level). In this, dispersoids are introduced in the partially solid slurry of the melts held between the liquidus and solidus by vigorous stirring. Rheocasting/ compocasting followed by squeeze castings for processing both particulate and shortfibers incorporated aluminium alloy matrix composites has shown great promise.

## 3. SOME IMPORTANT AL-ALLOY BASE CAST COMPOSITE SYSTEMS DEVELOPED AT RRL -T

Extensive work has been carried out in Al-graphite system. Other systems which have been studied are Al-coconut shell char, Al-zircon, $Al-TiO_2/ZrO_2$, Al-glass, Al-flyash, Al-carbon fibers. Important findings of a few of the above systems synthesised and studied are discussed below.

Some typical micro structures of the composites are given in Fig. 3 (a-e) while typical mechanical properties and possible applications of each system are given in Table 1. It can be seen that though generally the mechanical properties of composites are lower than the matrix, but adequate enough for general engineering applications.

## 3.1 Al-graphite system

The work on Al-base particulate composites in RRL-T was initiated with this system. Graphite particles of size (100-150 μm) and upto 5 wt% have been dispersed in LM6/ LM25 alloy using optimum conditions of preparation mentioned earlier. Castings of size 200 mm dia and 200 mm height as well as actual components such as bearings, electrical contacts and automobile parts (Fig. 4) have been prepared in the laboratory. Performance evaluation of these composites is going on at the users' site. Graphitic aluminium composites have exhibited good wear resistance by graphite acting as solid lubricant. The coefficient of friction was ranged between 0.21 and 0.31 and wear rate was about $3.3 \times 10^{-12}$ $m^3/m$ with Al-12.5 Si as matrix with 3% graphite content. Graphitic aluminium composites were able to run under boundary lubrication without any oil[6]. Preliminary studies with LM-13 piston containing about 3 wt% graphite have

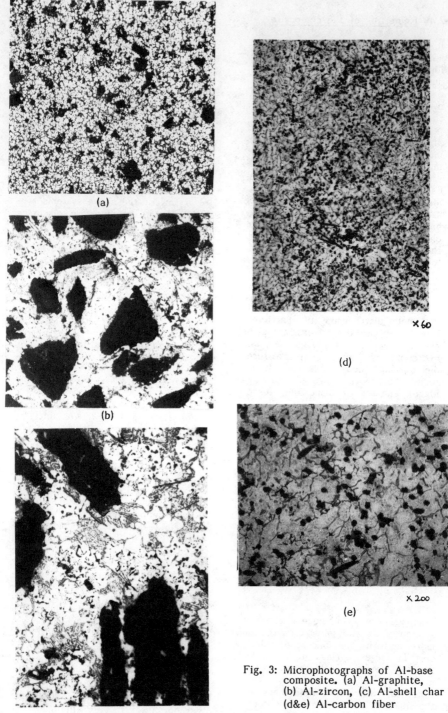

(a)

(b)

(c)    X 200

X 60

(d)

(e)    X 200

Fig. 3: Microphotographs of Al-base composite. (a) Al-graphite, (b) Al-zircon, (c) Al-shell char (d&e) Al-carbon fiber

884

## COMPARISON OF MECHANICAL PROPERTIES OF VARIOUS COMPOSITES IN AS-CAST CONDITION

| Properties | Al-graphite | Al-shell char | Al-zircon | Al-glass | Al-flyash |
|---|---|---|---|---|---|
| 1. % Dispersoid (wt %) | 1 - 6 | Upto 8% | 1 - 10 | 2 - 10 | 2.5 - 10 |
| 2. Density $(Kg/m^3)x10^2$ | 27 | 26 - 24 | 27 - 29 | - | - |
| 3. Yield strength$(MN/m^2)$ | 76 - 71 | - | 70 - 88 | 53 - 41 | 47 - 46 (0.2% P.S.) |
| 4. U.T.S. $(MN/m^2)$ | 279-121 | 130-90 | | 83 - 69 | 79 - 72 |
| 5. % Elongation | 1.0 - 0.3 | 3 - 2 | 7 - 1.6 | 14 - 5 | 12 - 4 |
| 6. Hardness (BHN) | 60 - 62 | 70 - 52 | 52 - 120 | 31 - 65 | 33 - 71 |
| 6. Compressive strength (Mpa) | 450 | 530-200 | - | 124-143 | 119-167 (0.5 Compr. P.S.) |
| 8. Weare rate x $10^8$ c.c./cm | 3.5 - 4.5 | 32 - 66 | - | - | - |
| 9. Fracture toughness (MPa m $^{-1/2}$) | - | - | - | 15 - 17 | 15 - 17 |
| 10. Applications | General bearings pistons, . piston liners, & any type of friction parts | Bearings, seals and electrical contacts | Abrasion resistance surfaces and parts | Extruded composites for structural applications | Extruded composites for structural applications |

P.S. = Proof Stress

Fig. 4: Graphitic aluminium bearing
x 0.2

indicated a saving of 5% in specific fuel consumption with good oil spreadability[7,8]. Also initial field trials of these composites in IC Engines as Journal bearings and in other automobile parts where friction is involved have indicated possibility of the systems to replace conventional babbits, Al-Sn bearings and even bronze and gun metal due to its superior performance as bearing material[9].

## 3.2   Al-coconut shell char system[10]

Shell char of size ~ 125 μm was dispersed upto 60 vol% in Al-11.8% Si alloy. These composites showed decrease in density and mechanical properties but superior wear resistance under adhesive wear conditions compared to the cast alloy. Friction coefficient values of the composite decreased with increasing volume fraction of char particles in the composite. Journal bearings of this composte were found to run under boundary lubrication conditions. These light weight composites were successfully tried as bush material in electrical contacts.

## 3.3   Al-zircon system[11]

Hard zircon particles of (size 40-200 μm) upto 60 wt% were dispersed in Al-Si alloy. To avoid gravity segregation, melts containing 30 wt% zircon were pressure die cast. It was found that mechanical properties of these composites including hardness, yield strength and abrasive wear resistance increased with increasing volume fraction of zircon. Al-11.8% Si-4% Mg alloy containing about 50 wt% zircon has been reported to have hardness equal to that of carbon steel and abrasion resistance comparable to that of brass. By centrifugal casting of the composite resulted in zircon rich surface (upto 70% by wt) at the outer periphery of the cylinder giving a hard surface. This can be used for a shaft requiring hard surface with good abrasion resistance. However, these composites having high hardness are difficult to cut and special tools are required.

## 3.4   Al-glass and Al-fly ash systems[12]

Upto 15 wt% of glass (size 20 - 140 μm) and about 18 wt% fly ash (size 10 - 100 μm) have been dispersed in Al-alloy matrix containing 3 wt% Mg by pressure diecasting. Adhesive wear rate of Al-glass composite was found to be less than the corresponding matrix while abrasive wear rate of Al-fly ash composite was found to be higher that the base alloy at any given test conditions. Gravity diecast composites were subjected to hot rolling, hot forging and extrusion. It was observed that while machinability of the composite was good, hot working of these composites improved the strength properties without change in ductility by nearly 300% over the as cast condition. This enhancement is attributed to various factors including fiberization/formation of stringers of the dispersoids (Fig. 5). Al-glass composite could be extruded at lower pressure since glass acted as lubricate during extrusion. Fracture toughness ($K_{Ic}$) of these hot forged composites showed a value pf 15-17 MPa m$^{-1/2}$ comparable to the values obtained in similar composites and Al-Si or Al-Cu alloys. Further 5 wt% of flyash dispersed 7075 alloy under wrought conditions exhibited higher UTS compared to the base alloy at the expense of ductility. These studies thus suggest possibility of preparing low density high strength composites for special applications such as lubricating bearings/structural materials.

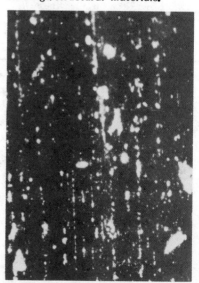

Fig. 5:   Microphotograph of extruded Al-glass composites showing fiberization        X 200

## 3.5 Al-carbon fiber system

About 3 wt% short carbon fibers obtained from M/s. Hysol-Grafil were dispersed in an indigenous Al-alloy equivalent to 7010 alloy by surface modification to the fibers. Fig. 4(d&e) show the microstructure of these composites having uniform distribution of fibers in the matrix. Further work is in progress to optimize conditions for incorporating higher volume fraction of the fibers ( > 30 wt%), and also to evaluate formability of the composites including extrusion and to study their properties.

## 4. CONCLUSIONS

Aluminium alloy matrix particulate composites having a wide variety of properties are prepared by liquid metallurgy technique by choosing a suitable dispersoid. Uniform distribution of the dispersoids is very essential for consistent properties and this is achieved by improving the wettability of the reinforcements with the matrix alloy and by proper mixing during dispersion. Graphite-Al composites developed have potential for self lubricated bearing systems specially in automotive industries while Al-zircon composite has potential for abrasion resistant applications. Fly ash and glass particle dispersed composites exhibited superior mechanical properties after hot extrusion and they have good potential for structural applications.

## Acknowledgement

The authors wish to acknowledge the help of Mr. Ramesh Upadhyaya, Mr. S.G.K. Pillai and Mr. P. Vijayakumar of RRL-T for their help during the preparation of this paper.

## 6. REFERENCES

1. SiC-reinforced aluminium metal matrix composites, Nair, S.V., Tien, J.K., and Bates, R.C., International Metals Reviews, 30, 275 (1985).

2. Fibre-reinforced metal matrix composites, Chow, T.W., Kelly, A., and Okura, A., Composites, 16, 187 (1985).

3. Solidification, structures, and properties of cast metal-ceramic particle composites, Rohatgi, P.K., Asthana, R. and Das, S., International Metals Reviews, 31, 115 (1986).

4. Modelling studies on dispersion of particles in Transponent liquids, Shaji, M.C., et al, (to be published).

5. Solidification of graphite-aluminium composite ingots, Roschen, et al., (to be published).

6. Seizure resistance of cast aluminium alloys containing dispersed particles of different sizes, Rohatgi, P.K. and Pai, B.C., Trans. ASME-J. Lubrication-Tech., 101, 376 (1979), also Wear, 59, 323 (1980).

7. Aluminium-graphite particulate composites as an internal combustion engine bearing material, Krishnan, B.P., et al., Wear, 60. 205 (1980).

8. Mechanism of improvement in oil spreadability of aluminium alloy graphite particle composites, Krishnan, B.P., et al., Tribology International, 301 (1981)

9. Bearing performance of graphite-aluminium particulate composite materials, Biswas, S., et al., Tribology International, 171 (1980)

10. Preparation and properties of cast Al-coconut shell char composite, Murali, T.P., et al., Met. Trans., 13B, 485 (1982).

11. Cast Al-alloys containing dispersions of zircon particles, Banerjee, et al., Met. Trans., 14B, 273 (1983).

12. Preparation and properties of Al-glass and Al-flyash particulate composites, Keshavaram, B.N., Ph.D. Thesis submitted to Faculty of Engineering, University of Kerala, June 1986.

Dr. K.G. Satyanarayana

Head, Materials Division of Regional Research Laboratory, Trivandrum, got his M.Sc. (Physics) from Bangalore University and Ph.D. (Physics-Met. Engg.) from Banaras Hindu University, Varanasi, and Fellow of the Institution of Engineers (India). Has about 16 years of R & D experience in Material Science and worked in the area of alloy development/natural fibers composites and material characterization. Taught Physics (1965-68) for undergraduates and guiding students for M.E. and Ph.D. Degrees. Visiting scientist at the Tohoku University, Japan (1975-76). Has a patent, about 100 publications, and edited two books on Material Science and Technology. Joint recipient of WIPO Gold Medal and NRDC Award. Member of Technical Committee of Non-Conventional Energy Sources, R & D Committee of Coir Board (Govt. of India), Task Group on Minerals and Metallurgy KSCSTE, Board of Tecnical Education, Board of Studies in job oriented courses, University of Kerala and also member of a number of professional bodies.

Dr. B.C. Pai

Scientist at Regional Research Laboratory, Trivandrum. Got his B.Sc. from Bangalore University, B.E. and M.E. in Metallurgy from Indian Institute of Science, Bangalore and Ph.D. from the University of Sheffield, UK, Worked as Scientist at National Aeronautical Laboratory, Bangalore (1974-83). Has about 14 years of R & D experience in the area of foundry/solidification. Member of a number of professional bodies. Has 40 publications. Areas of interest are metal matrix composites, premium quality aluminium alloy castings and newer casting techniques.

* Dr. C.G. Krishnadas Nair

General Manager, Engines Division of Hindustan Aeronautics Ltd., Bangalore. Got his M.Sc. and Ph.D. in Metallurgical Engineering from the University of Saskatchewan, Canada, and Fellow of Institution of Engineers (India) and NDT Society of India. Has over 22 years of R & D experience in Materials, Processes, and Product Development. Was Asst. Professor at Karnataka Regional Engineering College (1968-69), visiting Faculty at the University of Sheffield, UK (1969-71) and University of Laval, Quebec, Canada (1982). Guided students leading to M.E./Ph.D. and has over 110 publications and edited books. Recipient of National Metallurgists Award. Member of ARDB, Scientific Advisory Committee of NML (India), Board of Studies in Metallurgical Engineering, Mangalore University.

** Dr. M.R. Krishna Dev

Professor at the Laval University, Canada.

# DEVELOPMENT OF PARTICULATE REINFORCED HIGH STRENGTH ALUMINIUM ALLOY FOR AEROSPACE APPLICATIONS

Krishnadas Nair C.G*, Krishnadev M.R**, Dutta D*

## Abstract

A SiC particulate reinforced high strength Zr. refined Aluminium-Zn-Mg-Cu alloy has been developed through Powder Metallurgy route. Powder prepared by gas atomisation is mixed with SiC particulates ( $\approx 10\,\mu$ ) canned in vacuum followed by hot pressing and extrusion. Microstructure and properties have been of the composite are discussed. The resulting alloy is forgeable and heat-treatable to develop high strength. The matrix can be made superplastic by thermomechanical treatment and this makes the composite more versatile. The alloy is economical to produce and is considerably cheaper than the Al-Li. alloys. The paper discusses the development, processing, properties and potential applications of the alloy. Both optical and Scanning Electron Microscope have been used to study the microstructure and fracture.

## 1. INTRODUCTION

Airframe designers and manufacturers are constantly on the look out for higher performance materials to meet the demanding requirements of modern high performance aircrafts. Conventional high strength Aluminium IM alloys are therefore facing stiff competition from superior and costlier Titanium alloys and epoxy composites. Such a competition paved the way for development of newer Aluminium alloys & Aluminium alloy based metal matrix composites by Powder Metallurgy. Since high specific strength and stiffness are the prime requirements for an aircraft structural material, development in the recent past have taken place in introducing new Al.alloys and composites possessing higher strength and elastic modulii compared to the conventional high strength Aluminium alloys while retaining the density of conventional aluminium or even lower-

* Hindustan Aeronautics Limited, Bangalore, India.
** Laval University, Quebec, Canada.

ing it.

The new Al-Li. alloys which have been developed recently possess higher strength and modulus and also lower density compared to the high strength 7XXX alloys. A density reduction of 15% and modulus enhancement by 10% over the 7XXX alloys have been achieved by Al-Li. which resulted in a combined specific modulus enhancement of 30%.

Modulus of aluminium can also be dramatically increased, while retaining its density by artificially dispersing non-continuous SiC fibres or particulates in a high strength aluminium alloy matrix by powder metallurgy methods. Typical modulus values upto 130 GPa have been reported by reinforcing aluminium upto 30 volume percent of SiC whiskers[1,2]. In addition to the above 'E' enhancements achieved, fibre or particulate SiC bearing composites also exhibit higher resistance to wear and improved elevated temperature strength properties. Because of the above encouraging achievements among the various metal matrix composites, SiC fibre and particulate reinforced Al.alloy composites have raised considerable interest in recent times. A number of alloys, including 6061, 2024 and 7075 types have been investigated upon as possible matrix materials[3,4].

This paper describes the fabrication of a SiC particulate reinforced Zr.refined Al-Zn-Mg alloy composites and its mechanical properties, deformation and fracture characteristics.

## 2. EXPERIMENTAL PROCEDURE

The matrix Al.alloy was produced by conventional melting and casting into ingots of nominal composition 6%Zn-2.3%Mg-1.8%Cu and remainder commercially pure Aluminium. During melting, the alloy was refined by addition of Al-Zr. Master alloy. Alloy ingots were then re-melted and gas atomised into alloy powders. The composition of the resultant powder and its sieve analysis are presented in Table-1. Apparent density measurements of the various sieve fractions were carried out using a Hall Flowmeter and Al.alloy powder of minus 250 BS mesh fraction was chosen for the experiment to match the apparent density of as received green $\alpha$ -SiC grains of 1000 grit (average particle size 8 $\mu$m). Composite powder mixes with 5 and 15 volume percent of SiC were prepared by ball milling. The mixes were then canned in Aluminium canister and vacuum degassed at 550°C at a pressure of $2 \times 10^{-5}$ torr for 2 hours.

Vacuum degassing of the mixes was found necessary as earlier experiments carried out by conventional pressing and inert gas sintering produced composites of poor strength, ductility and particle-matrix bonding due to presence of gas in the as sintered and as extruded composites. The aluminium canisters were then sealed by hot crimping.

The aluminium cans were then hot pressed in an available extrusion con-

890

tainer, fitted with a blind die, at a temperature close to the solidus of the matrix alloy. Hot pressing was done at pressure between 18 and 20 tsi. to obtain nearly 100% density in the billets produced. The hot pressed billets were then decanned by machining and extruded at billet temperatures between 450 and 480°C at an extrusion ratio of 14.

The extrusions were then given T6 heat-treatment (solutionised for 2 hrs. at 475°C followed by ageing at 120°C). The ageing sequences for both 5% SiC and 15% SiC bearing composites were followed by hardness measurements at specific intervals. The ageing curves are presented in Fig.1 and compared with that of the unreinforced matrix alloy.

Room temperature mechanical properties of both the composite compositions were determined using a standard tension test specimen of 25 mm gauge length and 5 mm gauge diameter. Electrical strain gauge extensometers mounted axially along the specimen with suitable spring type clamps were used for recording extension. 0.2% Offset Proof Stress, UTS, Percentage Elongation and Elastic modulus values were determined. Brinnel hardness measurements $(P/d^2 = 10)$ were also carried out on the composites in the as extruded and heat-treated condition. Elevated temperature tensile tests were carried out on the composites at 200°C after 1 hour of hold at temperature using a flat tensile spe-

cimen of 30 mm gauge length, 2 mm thickness and 8 mm width. 0.2% Proof Stress, UTS and Percentage Elongation were determined and compared with the corresponding values of the unreinforced alloy tested at 200°C.

Optical and Scanning Electron Microscopic examination of longitudinal sections from the extrusions were done and fractography of failed tensile specimens were also carried out using SEM.

## 3. RESULTS AND DISCUSSION

### 3.1 Structure & Properties

Fig.2 shows an optical micrograph revealing the as polished microstructure at 100X of hot pressed billet. A random distribution of SiC particles in the Al.alloy matrix is observed. Fig. 3(a) and (b) show the microstructures at 100X of longitudinal polished sections from extrusions of 5 V/o SiC and 15 V/o SiC bearing composites extruded at a extrusion ratio of 14. Considerable orientation of SiC particles is observed in both the composites as a result of extrusion.

Fig.4(a) and (b) show the SEM Micrographs at 100X of longitudinal polished sections from 5 V/o and 15 V/o SiC bearing extrusions. Orientation of SiC particles along the extrusion direction is noticed. Also noticed is a pronounced surface relief of the SiC particles. One more feature which is observed is that the particles of high aspect ratio, in particular, have an orientation with their long axis aligned along the extrusion direction which shows that the particles of this shape are contributing to non-continuous fibre reinforcement.

Room temperature mechanical properties such as Hardness, 0.2% Proof Stress, UTS, Percentage Elongation and Modulus of Elasticity of 5 V/o and 15 V/o SiC bearing composites are presented in Table-2. The corresponding mechanical property values of the unreinforced PM matrix alloy made by the same PM processing method, are also presented.

It may be observed from a comparison of the mechanical properties of the unreinforced and the SiC reinforced alloy that the strength and modulus values increase with increase in SiC reinforcement volume fraction, however, it is also observed that the ductility of the composites, represented by percentage elongation values, falls steeply with increase in reinforcement volume fraction. Improvements in billet synthesis and usage of superior raw materials such as inert gas atomised aluminium alloy powder and finer SiC particles are expected to address the problem of low ductility and also lead to further improvements in strength and stiffness properties. Further experiments are being carried out to achieve these goals.

Table-3 presents a comparison of mechanical properties, such as UTS, Modulus of Elasticity, Specific strength and Specific modulus of the 5 V/o and 15 V/o SiC reinforced alloys and some

of the advanced aluminium alloys i.e. Aluminium-Lithium alloys. It can be observed that the specific strength and modulus values of the SiC reinforced Al.alloy composites are comparable to the Al-Li.alloys.

Elevated temperature mechanical properties of the 5 V/o and 15 V/o SiC bearing composites are presented in Table-4 and compared with the corresponding values of the unreinforced alloy. UTS and 0.2% Proof Stress of the reinforced alloys are observed to be higher than those of the unreinforced alloy, particularly in case of 15 V/o SiC bearing alloy.

3.2  Superplasticity

Some studies were also carried out with the objective of imparting superplasticity to the matrix alloy after giving a suitable thermo-mechanical treatment. A schematic diagram showing the time-temperature sequence of the thermo-mechanical treatment is shown in Fig.5. The fine grained microstructure which resulted from the above TMT is shown in Fig.6(a) which is contrasted with the microstructure of the original alloy before TMT shown in Fig.6(b). The thermo-mechanically treated alloy when tested for superplastic behaviour at 500°C and at a strain rate of $10^{-4}$ Sec$^{-1}$ gave elongation values to fracture of more than 200 percent as against values of 60 percent elongation obtained in the conventionally forged alloy under the same testing conditions. The superplasticity was retained both in the 5 V/o and 15 V/o percent Aluminium-Silicon composites. SEM studies of the fracture also corroborate this as discussed in the next section. However, the effect of higher percentage of SiC is to be still assessed.

## 3.3 Fracture

SEM fractograph of tensile fracture surfaces of 5 V/o and 15 V/o SiC reinforced composites are shown in Fig. 7(a) and (b) respectively. It is observed that although the elongation values of the composites were very low, localised matrix ductility characterises the fractures of 5 V/o and 15 V/o SiC composites which is shown by fine microvoid dimples. A few SiC particles are also noticed on the fracture surface (Fig.7(b) but their quantity does not correspond to the volume fraction present in the composite. One more feature which is observed is that the microvoid dimples are quite fine sized in both the fractographs which points out that the microvoid growth stage in the composites is limited and that void nucleation and growth takes place towards the final stages, just before failure. Fig.7(c) shows a fractograph of a failed sample of 5 V/o SiC bearing composites which was tested for superplastic behaviour at 500°C after being given TMT as per Fig.5. The fracture surface is seem to be characterised by a profuse dimple morphology showing a high ductility of this material after the TMT treatment and under superplastic conditions, as

compared to Fig.7(a) and(b). SEM micrographs of longitudinal sections perpendicular to the fracture surface of 5 V/o and 15 V/o SiC reinforced composites, presented in Fig.8(a) and (b) respectively. It is observed from both the micrographs that voids are nucleated at the particle-matrix interfaces in the direction of testing around particles located very close to the fracture surface. Some voids are also created due to interfacial decohesion at these particle-matrix interfaces in other directions as well. The fracture path probably propagates from particle to particle by linking up of these voids. Fig.8(c) shows such a fracture path in the 15 V/o SiC composite.

## 4. APPLICATIONS

The Al.alloy-SiC composites in view of their high attainable specific strength and stiffness have potential applications in aerospace industry. Formability, forgeability and weldability of the composites are quite good. Fig.9 illustrate typical forgings where the composites have potential for use.

In view of the superplasticity of the matrix, it is also considered that sheets made from the composites could find potential applications in the fabrication of parts of intricate shapes requiring high strength & stiffness, such as aelirons, tail plane rudder and other control surfaces.

Presence of hard SiC particles also imparts high wear resistance to the composites which therefore can be used for making wear resistant components such as amphibious vehicle track shoes, automobile pistons, cylinder liners, gears and gear racks, etc.

Further work is required to be carried out to explore the potential of the Al-SiC composites in the above areas of application.

## 5. REFERENCES

1. Divecha A.P, Fishman S.G, Karmakar S.D., J.of Metals, 9 (1981),pp.12.

2. Arsenault R.J, Material Science and Engineering, 64 (1984), pp.171.

3. McDanels D.L, Met.Trans. 16(a), (1985), pp.1105.

4. D.W.A.Composite Specialities Inc., Superior Street, Chatsworth, Ca 91311-4393, Personal Communication (1986).

## BIOGRAPHY

Dr.C.G.Krishnadas Nair is General Manager, Foundry & Forge Division, Hindustan Aeronautics Limited, Bangalore and is responsible for managing R&D in Materials and Processes and also for the manufacture of special alloys, castings, forgings & extrusions. He is a graduate in Metallurgy from IIT, Madras andPost-graduate leading to M.Sc. & Ph.D. from University of Saskatchewan, Canada. He has worked as Assistant Professor in Karnataka Regional Engineering College, India and as Visiting Faculty at the University of Sheffield, U.K. and University of Laval, Canada. He has over 20 years of experience in R&D and production management and has published over 110 technical/research papers in alloy development, import substitution, testing and evaluation of materials and in management. He is currently one of the Vice-Presidents of The Aluminium Association of India and also its Chairman, Executive Committee.

Mr.D. Dutta is Deputy Manager, Powder Metallurgy Shop, Foundry & Forge Division, HAL, Bangalore. He obtained his B.Tech degree in Metallurgical Engg. from Institute of Technology, Banaras Hindu University in 1979 and at present doing M.Sc. (Engg.) from Indian Institute of Science, Bangalore. He joined HAL as a Management Trainee in 1979 and attended a Post-Graduate course in Aeronautical Engineering and Production Technology at IIT Madras and Basic Management Programme at HAL Staff College as a part of his training. At HAL worked as an Inspection Engineer and Development Engineer before taking up the current assignment. He has completed a Proficiency course on NDT at Indian Institute of Science, Bangalore in 1982 and is an Associate Member of Indian Institute of Metals. His major interests are in development of materials and processes and Powder Metallurgy.

Dr. M.R. Krishnadev is Professor in the Department of Mining and Metallurgy of Universite Laval, Quebec, Canada. He obtained B.E. Degree in Mechanical Engineering in 1962 from University of Mysore, India and M.Sc. in Physical Metallurgy in 1967 and Ph.D. in Mechanical Metallurgy in 1969 both from University of Saskatchewan. He has

over 22 years of experience in Research & Laboratory work in the field of
Alloy Development, Structure & Properties of Aluminium alloys and HSLA
Steels. He is a Specialist in the area
of Instrumented Impact, Slow-bend and
drop-weight tear testing and well
versed in SEM, TEM, Auger and EDAX
techniques. He is a Consultant in the
area of Application of Fracture
Mechanics techniques to brittle materials, HSLA Steels and Composites.
He has travelled widely and has worked
in close collaboration with many
National and International Companies
on problems relating to the development and characterization of materials
for conventional and new energy technologies. Has published over 50 papers
in the area of Processing - Property -
Microstructure - Fracture relations. He
is a Member of QEO, AIME, IMS, ASTM,
ISIJ (Japan) and ASNT.

## TABLE-1

### COMPOSITION AND SIEVE ANALYSIS OF ALUMINIUM ALLOY POWDER

| \multicolumn{3}{CHEMICAL COMPOSITION} | | | \multicolumn{3}{SIEVE ANALYSIS} | | |
|---|---|---|---|---|---|
| Sl. No. | Element | Weight (%) | Sl. No. | B.S. Mesh of Sieves | % Retention by Weight on Sieves |
| 1. | Copper | 1.8 | 1. | 80 | Nil |
| 2. | Zinc | 6.7 | 2. | 100 | 7 |
| 3. | Magnesium | 2.06 | 3. | 150 | 10 |
| 4. | Zirconium | 0.18 | 4. | 200 | 20 |
| 5. | Iron | 0.12 | 5. | 250 | 20 |
| 6. | Aluminium | Remainder | 6 | 300 | 35 |
| | | | 7. | 325 | 2 |
| | | | 8. | -325 | Balance |

## TABLE-2

### MECHANICAL PROPERTIES OF ALUMINIUM ALLOY-SiC COMPOSITES

| Composition and Extrusion ratio | Density (% T.D) | | Hardness (HB,10) | | 0.2% P.S. (MPa) | UTS (MPa) | E (%) | Youngs modulus (GPa) |
|---|---|---|---|---|---|---|---|---|
| | Hot Pressed | Extru-ded | Extru-ded | Hot treated | | | | |
| Al.Alloy - 5 V/o SiC - T6 E.R. 14:1 | 99.2 | 100 | 95.0 | 179 | 510 | 565 | 6 | 77.2 |
| Al.Alloy -15 V/o SiC - T6 E.R. 14:1 | 99.0 | 100 | 99.5 | 183 | 560 | 582 | 2 | 85.7 |
| Al.Alloy (Al-Zn-Mg-Zr) E.R. 14:1 Unreinforced | - | 100 | 90.07 | 169 | 470 | 530 | 8 | 70.0 |

## TABLE-3

### COMPARISON OF PROPERTIES OF SiCp REINFORCED AL.ALLOY AND ALUMINIUM-LITHIUM ALLOYS

| Property | SiC$_p$ reinforced Al-Mg-Zn-Zr Alloy | | Al-Li PM Alloy | Al-Li IM Alloys Developed by ALCAN (UK) | |
|---|---|---|---|---|---|
| | 5 V/o SiC$_p$ | 15 V/o SiC$_p$ | (Al-Mg-Li) Developed by INCO Alloy Products | 8090 ( DTD XXXA) (Al-Li-Cu-Mg | 8091 (DTD XXXB) Al-Li-Cu-Mg |
| Density (g/cm$^3$) | 2.81 | 2.84 | 2.57 | 2.53 | 2.55 |
| Tensile Strength (MPa) | 565 | 582 | 500 | 475 | 525 |
| Elastic Modulus (GPa) | 77.2 | 85.7 | 80.6 | 79.5 | 79.9 |
| Specific Strength (MPa.cm$^3$/gm) | 200.5 | 204.93 | 194.5 | 187.7 | 205.9 |
| Specific Modulus (GPa.cm$^3$/gm) | 27.4 | 30.18 | 31.36 | 31.4 | 31.33 |

## TABLE-4

### HOT TENSILE PROPERTIES OF Al-SiCp COMPOSITES

| TEST TEMPERATURE - 200ºC (1 HOUR HOLD) | | | |
|---|---|---|---|
| Alloy | 0.2% Proof Stress (MPa) | Ultimate Tensile Strength (MPa) | Elongation (%) |
| Al. Alloy (Al-Zn-Mg-Zr unreinforced | 290 | 330 | 18 |
| Al.Alloy - 5 V/o SiC - T6 | 335 | 366 | 15 |
| Al.Alloy - 15 V/o SiC - T6 | 370 | 410 | 10 |

897

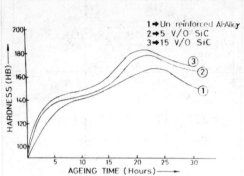

Fig.1
Ageing curves of Al.Alloy - SiC$_p$
composites

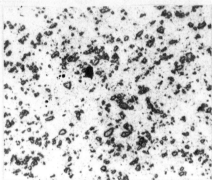

Fig. 2        X 100
Microstructure of Hot Pressed Billet
(As Polished)

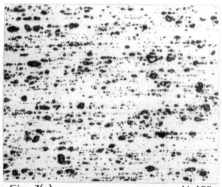

Fig. 3(a)        X 100
Microstructure of Al.Alloy 5 V/o SiC
Composite after Extrusion
(As Polished)

Fig. 3(b)        X 100
Microstructure of Al.Alloy 15 V/o SiC
Composite after Extrusion
(As Polished)

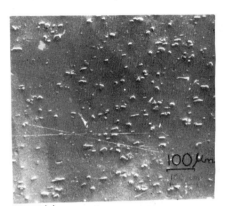

Fig. 4(a)
SEM Microstructure of Al.Alloy
5 V/o SiC

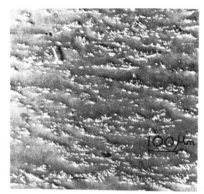

Fig. 4 (b)
SEM Microstructure of Al.Alloy
15 V/o SiC

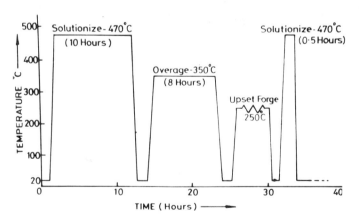

Fig.5
Four step Thermomechanical treatment fr Al.Allcy
(Al-Zn-Mg-Cu-Zr)

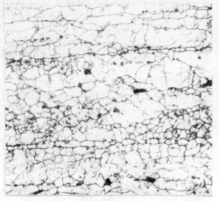

Fig. 6(a)                          X 400
Microstructure of Conventionally forged
      Al. Matrix Alloy (Etched)

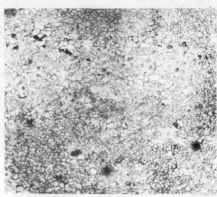

Fig. 6(b)                          X 400
Microstructure of Thermo-mechanically
Processed Al. Matrix Alloy (Etched)

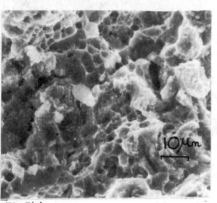

Fig.7(a)
SEM Fractograph of Tensile  fracture
surface of Al. Alloy - 5 V/o SiC

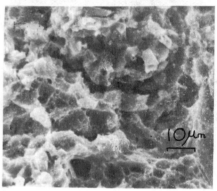

Fig. 7(b)
SEM Fractograph of Tensile fracture
  surface of Al. Alloy - 15 V/o SiC

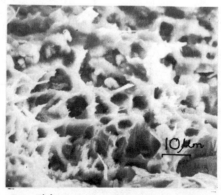

Fig. 7(c)
SEM Fractograph of Fracture Surface of Super-
plastically tested Al.alloy - 5 V/o SiC Composite

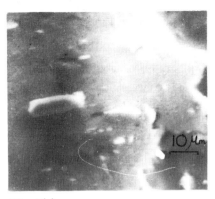

Fig. 8(a)
SEM Micrograph of longitudinal section
perpendicular to Tensile fracture surface of Al. Alloy - 5 V/o SiC

Fig. 8(b)
SEM Micrograph of longitudinal section
perpendicular to Tensile fracture surface of Al.Alloy - 15 V/o SiC

Fig. 8(c)
SEM Micrograph of Longitudinal section perpendicular to Tensile fracture surface of Al. alloy -
15 V/o SiC showing fracture path propagation
from particle to particle.

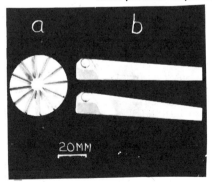

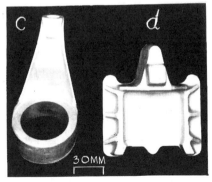

Fig. 9
Typical Die Forgings - Applications of Al-SiC Composite
(a) Compressor Wheel ;  (b) Fin ;  (c) Blade Horn Forging ; (d) Fitting

ARALL LAMINATES - RESULTS FROM A COOPERATIVE TEST PROGRAM
R. J. Bucci, L. N. Mueller, R. W. Schultz and J. L. Prohaska
Alcoa Laboratories, Alcoa Center, PA   15069

## Abstract

ARALL, Aramid Aluminum Laminate is a new hybrid material system consisting of thin aluminum sheets bonded by adhesive impregnated with strong aramid fibers. Uniting metal and fibers in this manner produces a lightweight material with the high strength and the excellent fatigue resistance of epoxy matrix composites, while retaining the traditional advantages of metals; namely, plasticity, impact strength, formability, easy machining and supportability. The material can be cut, sawed, drilled and joined by conventional mechanical fastening or adhesive bonding procedures. During 1985, Alcoa supplied laboratory samples to firms agreeing to share characterization data for the purpose of guiding R&D on this promising new material. This paper summarizes results from the first group of companies that examined ARALL Laminates under this cooperative program.

## 1. INTRODUCTION

ARALL Laminates are a family of new structural composite materials that show promise for use in weight sensitive, fatigue and fracture critical structures (1-7). The material consists of alterna-ting layers of thin aluminum sheet bonded by adhesive impregnated with high strength aramid fibers. A schematic representation is shown in Figure 1.

ARALL Laminates have outstanding fatigue properties when compared to monolithic aluminum. Under cyclic loading, cracks which initiate first in the metal layers

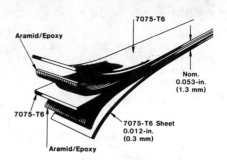

**Schematic of Alcoa 3/2 ARALL®-1 Layup**
**Figure 1**

tend to grow very slowly or arrest. This is attributed to load transfer from the metal to the stronger unbroken fibers which bridge the crack and restrain its opening. Additional fatigue enhancement can be achieved by prestraining the composite layup after cure. This imparts a favorable compressive residual stress that retards crack initiation in the metal.

ARALL Laminate mechanical properties, like those of fiber-resin composites, are directional as dictated by the fiber orientation. Advantages of ARALL Laminates over carbon/epoxy composites include plasticity, controlled residual stresses, lower material cost and applicability of standard metal fabricating and repair procedures (e.g., mechanical fastening, machining and forming operations). Additionally, the outer metal layers serve as barriers to moisture, offer lightning protection, resist impact and provide sensitivity for damage inspection by conventional techniques. ARALL Laminates have advantages over monolithic aluminum as well. These include lower density, higher directional strength and superior fatigue life.

Potential applications envisioned for first generation ARALL Laminates include aircraft lower wing skin, fuselage skin and tail skins. Attendant structural weight reductions of 15 to 40 percent over traditional airframe construction have been demonstrated (3-6). ARALL Laminates have also shown two to four times better vibrational damping ability than monolithic aluminum sheet. Consequently, airframe manufacturers are considering using ARALL Laminates where acoustic fatigue is a problem.

The ARALL Laminate concept originated at the Delft Technical University (1-6), and is now being commercialized by Alcoa (7-11). To broaden experience and obtain guidance for R&D, Alcoa initiated an ARALL Laminate cooperative test program during 1985. Under this program, Alcoa distributed samples of laboratory fabricated material to numerous aerospace companies for their evaluation. This paper summarizes test results from the first group of firms completing their evaluation, Table 1.

Table 1
ARALL-1 Laminates
Cooperative Test Program Participants

| Lab | Organization |
| --- | --- |
| 1. | Alcoa Laboratories, Alcoa Center, PA |
| 2. | Boeing, Seattle, WA |
| 3. | British Aerospace, Woodford, England |
| 4. | Lockheed-Georgia, Marietta, GA |
| 5. | McDonnell-Douglas, Long Beach, CA |
| 6. | McDonnell-Douglas, St. Louis, MO |
| 7. | NASA Langley, Hampton, VA |
| 8. | Naval Research Labs, Washington, DC |
| 9. | Northrop Aircraft Co., Hawthorne, CA |
| 10. | Rockwell - NAAO, Los Angeles, CA |

2.    MATERIAL

Many ARALL Laminate variants can be made by combining different fiber-resin systems and sheet alloys, by using different stacking sequences and fiber orientations, (such as uniaxial and cross-ply), and by using different surface preparation techniques. Also, the number and thicknesses of plies and level of prestraining may be varied. Although several laminate systems have been successfully fabricated, this investigation considers only a single configuration of the ARALL family. This configuration, designated as 3/2 ARALL-1 Laminate, has 0.053-in. (1.3 mm) nominal thickness and consists of three 0.012-in. (0.3 mm) thick 7075-T6 aluminum sheet layers and two layers of Enka unidirectional aramid-fibers impregnated with 3M Company AF-163-2, 250°F (121°C) cure adhesive. The prepreg layer fiber-adhesive ratio is 50/50 by weight, and the fiber axis and aluminum sheet rolling directions coincide. The standard 3/2 ARALL-1 configuration has about 18 percent lower density 0.083 lb/cu.in. (2.29 g/cu.cm) than monolithic 7075 aluminum 0.101 lb/cu.in. (2.79 g/cu.cm).

ARALL panel manufacture was performed at the Alcoa Technical Center. Prior to lay-up and cure, the aluminum surfaces were chromic acid anodized and primed. Curing was performed in a heated platen press. The curing process leaves a residual stress state with the metal layers in tension and fibers in compression. Consequently, the bonded laminate was given a 0.5 percent permanent stretch to reverse the residual stresses state to compression in the aluminum layers and tension in the higher strength aramid fibers. Final panels supplied to the program participants were nominal 1 x 2 ft. (30 x 60 cm) with the longer dimension being parallel to the fiber (0 degree) direction.

3.    RESULTS AND DISCUSSION

Because of the limited material availability and diverse interests of participating firms, each participant performed tests suited to their needs. These tests included tensile tests, compression tests, bearing tests, residual strength or fracture toughness tests, fatigue and fatigue crack growth tests and several other nonstandardized tests.

## 3.1 Tensile and Compressive Properties

Table 2 summarizes room temperature tensile and compressive property data supplied by the participants. The tensile properties show significant directionality, which is to be expected due to the longitudinal (0 degree) fiber reinforcement. The compressive properties in both longitudinal and transverse (90 degree) specimen orientations are similar. Again, this is expected since the fibers provide little compressive reinforcement.

The test reproducibility between laboratories is good, considering that not only is material variability a factor, but also the interlaboratory variability in test procedures. With the exception of Lab 6, all labs used dog-bone type specimens for tension testing. Lab 6 used a straight-edged specimen with bonded end tabs. All labs experienced some grip end tensile fractures, which in part explains why interlaboratory variability is greater for longitudinal tensile ultimate strength than for longitudinal tensile yield strength and other compressive and transverse properties. The lowest ultimate tensile property values were recorded by Labs 9 and 10, both of which consistently encountered failures in the specimen radius. The highest tensile ultimate strengths were reported by Lab 4, which elected to stack the ARALL Laminate between two aluminum plates during specimen machining. Conventional aluminum machining

Table 2

3/2 ARALL-1 Laminate(a) Room Temperature Tensile and Compressive Properties(d)

| Lab No. | Test Dir. | No. Tests | UTS ksi | TYS ksi | % Elong. (b) | % Strain to Fail. | Modulus msi | Poisson Ratio | No. Tests | CYS ksi | Modulus msi |
|---|---|---|---|---|---|---|---|---|---|---|---|
| | | | | | | | **Tensile** Avg. (Spread) | | | **Compressive** Avg. (Spread) | |
| 1 | L | 31 | 114 (111-119) | 92 (87-101) | 0.6 (0.4-0.8) | 1.9 (1.5-2.1) | 9.9 (9.8-10.1) | 0.34 | 2 | 54 (54-54) | 9.9 (9.8-9.9) |
| | LT | 31 | 55 (54-58) | 48 (46-49) | 6.9 (6.5-8.0) | 7.8 (6.1-8.5) | 6.9 (6.9-7.0) | 0.24 | 3 | 56 (55-57) | 7.4 (7.4-7.4) |
| 3 | L | 7 | 112 (95-119) | 95 (92-98) | ---- | ---- | 10.2 (9.4-11.3) | - | - | ---- | ---- |
| 4 | L | 3 | 122 (121-123) | 96 (94-99) | ---- | ---- | 11.0 (10.9-11.1) | - | 3 | 54 (54-55) | 10.3 (9.9-10.9) |
| 6 | L | 2 | 115 (109-121) | 94 (93-95) | ---- | ---- | 9.5 (9.3-9.6) | - | - | ---- | ---- |
| | LT | 2 | 56 (56-56) | 47 (47-47) | ---- | ---- | 7.0 (6.8-7.2) | - | - | ---- | ---- |
| 9 | L | 6 | > 102(c) (93-111) | 94 (91-97) | ---- | > 1.4(c) (1.2-1.6) | 10.0 (9.8-10.4) | 0.32 | 4 | 49 (47-51) | 10.0 (9.6-10.2) |
| | LT | 2 | 55 (55-55) | 46 (45-46) | ---- | ---- | 7.2 (6.9-7.5) | 0.22 | 1 | 54 | 7.5 |
| 10 | L | 3 | > 108(c) (107-109) | 96 (95-96) | ---- | ---- | ---- | - | 3 | 54 (53-56) | ---- |
| | LT | - | 56 (55-56) | 46 (45-47) | ---- | ---- | ---- | - | 3 | 54 (53-54) | ---- |

Grand Avg. & (Mean Spread):

| | | | | | | | | | | | |
|---|---|---|---|---|---|---|---|---|---|---|---|
| L | | | 112 (102-122) | 95 (92-96) | 0.6 ---- | 1.7 (1.4-1.9) | 10.1 (9.5-11.0) | 0.33 | | 53 (49-54) | 10.2 (9.9-10.3) |
| LT | | | 55 (55-56) | 47 (46-48) | 6.9 ---- | 7.8 ---- | 7.0 (6.9-7.2) | 0.23 | | 55 (54-56) | 7.5 (7.4-7.5) |

(a) Standard ARALL-1 Laminate: Nom. 0.053 in. thick; 3 layers 7075-T6 (0.012 in. thick) and 2 layers prepreg.

(b) Plastic strain determined by extrapolation of the elastic slope from the point of fracture on the test record.

(c) All specimens tested failed in the specimen radius. Consequently, actual properties are equal to or greater than property value reported.

(d) SI conversion: 1 ksi = 6.895 MPa.

operations were otherwise employed by the participants. Several laboratories indicated that elongation measured by the usual practice of fitting failed specimen halves produced inaccuracies because of fiber end splitting. Consequently, the only elongation values reported are those determined by extrapolation of the elastic slope from the point of fracture on the test record. The compressive failure mode was buckling in all cases, except for one test where Lab 9 reported a matrix malfunction.

Table 3 compares ARALL-1 Laminate room temperature tensile and compressive property averages to typical properties of 7075-T6 aluminum sheet and two common graphite-epoxy systems.

ARALL-1 Laminate tensile and compressive properties obtained at -30°F (-34°C), 180°F (82°C) and at room temperature after exposure to various aggressive environments are shown in Table 4. Tensile

properties remained stable under all conditions examined, except for the 26 percent strength loss at 180°F (82°C) reported by Lab 9, and the 15 percent loss after one week exposure at 212°F (100°C) reported by Lab 3. In contrast, the ARALL-1 Laminate showed a small, but consistent, compressive yield increase after presoaking. Based on residual strain data obtained before and after the presoak, Lab 9 concluded that the compressive yield increase is caused by partial relaxation of the metal residual compression introduced by the prestretch. For example, after the six-hour, 250°F (121°C) presoak about a 50 percent compressive stress relaxation was observed. However, work by Verbruggen (12) demonstrated that ARALL bondline shear property degradation, and hence residual stress relaxation, is not significant after a variety of wet and dry exposures at temperatures up to 160°F (71°C) and for durations up

Table 3

Mechanical Properties of 3/2 ARALL-1 Laminate Compared to
Monolithic 7075-T6 Aluminum and Graphite-Epoxy Composite

| Average Property | Test Direct. | ARALL-1(a) | 7075-T6(b) | Test Direct. | Graphite/Epoxy - 60% fiber cont. (c) | | |
|---|---|---|---|---|---|---|---|
| | | | | | Unidirectional (d) | Typical Structure (e) | |
| Tens. Ultimate | L | 112 (772) | 83 (572) | 0 degree | 180 (1241) | 95 (655) | |
| Strength, ksi (MPa) | LT | 55 (379) | 83 (572) | 90 degree | 8 (55) | 40 (276) | |
| Tens. Yield | L | 95 (655) | 74 (510) | 0 degree | ----- | ----- | |
| Strength, ksi (MPa) | LT | 47 (324) | 72 (496) | 90 degree | ----- | ----- | |
| Tens. Elong. | L | 0.6(f) | 11 | 0 degree | ----- | ----- | |
| % | LT | 6.3(f) | 11 | 90 degree | ----- | ----- | |
| Strain to | L | 1.7 | 12 | 0 degree | 0.9 | 0.5 | |
| Failure, % | LT | 7.8 | 12 | 90 degree | 0.5 | 0.5 | |
| Tens. Elastic | L | 10.1 (69.6) | 10.3 (71.0) | 0 degree | 21.0 (145) | 11.0 (75.8) | |
| Modulus, msi (GPa) | LT | 7.0 (48.3) | 10.3 (71.0) | 90 degree | 1.7 (11.7) | 5.0 (34.5) | |
| Poisson Ratio | L | 0.33 | 0.33 | 0 degree | ----- | ----- | |
| (Tension) | LT | 0.23 | 0.33 | 90 degree | ----- | ----- | |
| Compr. Yield | L | 53 (365) | 73 (503) | 0 degree | 180 (1241) (g) | 95 (655) (g) | |
| Strength, ksi (MPa) | LT | 55 (379) | 76 (524) | 90 degree | 30 (207) (g) | 42 (290) (g) | |
| Compr. Elastic | L | 10.2 (70.3) | 10.5 (72.4) | 0 degree | ---- | ----- | |
| Modulus, msi (GPa) | LT | 7.5 (51.7) | 10.5 (72.4) | 90 degree | ---- | ----- | |
| Density | | 0.083 (2.29) | 0.101 (2.79) | | 0.056 (1.55) | 0.056 (1.55) | |
| lb./cu.in. (g/cu.cm) | | | | | | | |

---

(a) Standard 3/2 ARALL-1 Laminate, 0.053 in. (1.3 mm) thick; Properties from Table 2.

(b) Typical values 0.06 in. (1.5 mm) thick sheet.

(c) Ref. "DoD/NASA Advanced Composites Design Guide Vol. 1-A" - high strength graphite/epoxy F180 (generic description for Hercules AS/3501-6 and Celanese T300/5208).

(d) 100% fibers in 0 degree direction.

(e) Fiber orientations: 42% 0 degree, 50% 45 degree and 8% 90 degree.

(f) Plastic strain determined by extrapolation of the elastic slope from the point of fracture on the test record.

(g) Compressive ultimate strength.

Table 4
Effect of Temperature and Presoak Condition on
Tensile and Compressive Properties of 3/2 ARALL-1 Laminate

| Lab No. | Presoak Conditions | Test Temp. | Property of Interest (a)/Room Temp. Property (b) | | | | | | | |
|---|---|---|---|---|---|---|---|---|---|---|
| | | | Tensile | | | | Compressive | | | |
| | | | UTS | | E, (c) | | CYS | | Ec, (c) | |
| | | | L | LT | L | LT | L | LT | L | LT |
| 1 | 100% RH @ 140°F (60°C) | | | | | | | | | |
| | for: 1 Mo. | RT | 1.01 | 0.99 | 1.01 | 0.99 | 1.07 | 1.07 | 0.99 | 0.98 |
| | 3 Mo. | RT | 1.01 | 0.99 | 0.99 | 0.99 | 1.03 | 1.04 | 0.95 | 0.97 |
| | 6 Mo. | RT | 0.99 | 1.01 | 0.99 | 1.05 | 1.04 | 1.00 | 0.98 | 1.01 |
| 1 | Hot-Wet/Cold/Hot-Dry, Cycled Daily (d) | | | | | | | | | |
| | 1 Mo. | RT | 1.01 | 0.99 | 0.95 | 1.00 | 1.04 | 1.00 | 0.97 | 0.91 |
| | 3 Mo. | RT | 0.98 | 0.97 | 1.00 | 1.01 | 1.05 | 1.01 | 1.04 | -- |
| 1 | Cont. 5% NaCl Spray | | | | | | | | | |
| | for: 500 hrs. | RT | 1.01 | 0.97 | 0.99 | 1.00 | 1.00 | 0.98 | 1.05 | 0.99 |
| | 1000 hrs. | RT | 0.99 | 0.89 | 0.99 | 0.99 | 1.04 | 1.05 | 0.99 | 0.99 |
| 3 | 212°F (100°C) Lab Air | | | | | | | | | |
| | for: 1 hr. | RT | 1.04 | -- | -- | -- | -- | -- | -- | -- |
| | 1 day | RT | 1.06 | -- | -- | -- | -- | -- | -- | -- |
| | 1 week | RT | 0.85 | -- | -- | -- | -- | -- | -- | -- |
| 9 | 250°F (121°C) Lab Air for: 6 hrs. | RT | 0.99 | -- | -- | -- | 1.20 | -- | 1.10 | -- |
| 9 | None | -30°F (-34°C) | -- | -- | -- | -- | 1.12 | -- | 1.04 | -- |
| 9 | None | 180°F (82°C) | 0.74 | -- | 0.74 | -- | -- | -- | -- | -- |
| 9 | 95% RH @ 170°F (77°C) | 180°F (82°C) | -- | -- | -- | -- | 1.02 | -- | 0.81 | -- |

(a) Average of triplicate tests for Lab 1, duplicate tests for Lab 9 and single test for Lab 3.
(b) Average room temperature property as measured by same laboratory; Ref. Table 2.
(c) Modulus values estimated from test record (Class B extensometer used).
(d) Daily cycle of 18 hrs. to 100% RH @ 140°F (60°C)/5 hrs. @ -10°F (-23°C)/1 hr. @ 180°F (82°C).

to 12 weeks. Both Labs 1 and 9 detected weight gains in specimens exposed to wet environments (moisture absorption). In all cases, however, the percentage weight increase was negligibly small, even for transverse oriented specimens with a greater number of fiber ends exposed. Since moisture can only gain entry at the free edges, it is reasonable to expect that sealants could be relied upon for protection in actual structures. Moreover, real structure will have a much lower free edge/volume fraction than the small test coupons considered herein.

## 3.2  Shear and Bearing Properties

There is presently no standard test practice for shear characterization of monolithic sheet. Historically, Alcoa has used the blanking (punch) shear test, and most of the aluminum sheet data appearing in Mil-Hdbk-5 (13) was developed by this method. The test procedure consists of punching out a nominal 2.75 in. (7.0 cm) diameter hole in a 4 x 4 in.(10.2 x 10.2 cm) sheet sample. Shear strength is calculated by dividing the maximum load by the sheared area of the hole. When tested by the same practice, 3/2 ARALL-1 Laminate has a blanking shear strength of 39 ksi (269 MPa), which is about 20 percent less than the 47 ksi (324 MPa) minimum (13) for comparable gauge 7075-T6 sheet. This decrease is expected since the aramid/epoxy layers constitute about 30 percent of the sheared area.

In-plane shear properties of the ARALL-1 Laminate and monolithic 7075-T6 were also evaluated by Laboratory 1 using a three rail shear test, Method B of ASTM D4255-83 (14). Shear yield strength (0.2 percent offset) values determined by this approach were 24 ksi (165 MPa) for ARALL-1 Laminate and 32 ksi (221 MPa) for

7075-T6 sheet, 0.050 in. (1.3 mm) thick. All specimens tested failed by out-of-plane buckling, thus, ultimate shear strengths were not obtainable. The shear modulus (G12) of the ARALL-1 Laminate was estimated to be about ten percent less than that of 7075-T6.

Short beam interlaminar shear (ILS) tests were conducted per ASTM method D2344 (15) by Lab 6. Initiation of resin failure was assumed at the point of slope change on the load versus head-deflection test record, and the corresponding load was used for calculation of ILS strength. The ILS strength determined for the ARALL-1 Laminate was 4.7 ksi (32.4 MPa), which is similar to that obtained for state-of-the-art epoxy based composite materials. Fiber orientation had no effect on ILS strength of the ARALL-1 Laminate, and the specimens exhibited yielding rather than catastrophic failure.

Pin-load bearing tests per ASTM method E238 (16) were performed on the ARALL-1 Laminate by Labs 1 and 6. The results of these tests are given in Table 5. The cause of the differences in bearing ultimate strengths reported by the two laboratories is unclear, though differences in hole preparation technique is one plausible explanation. Lab 1 reported two distinct failure types: a single mode in which the bearing area was sheared in one solid piece, and a mixed failure mode in which aluminum tearing and crushing were accompanied by delamination and fiber breakage. The former failure mode is common to monolithic aluminum, while the latter is unique

to laminates and can be aggravated by edge delamination introduced during either specimen manufacture or solvent cleaning of the machined hole. Instead of the ultrasonic cleaning practice recommended in ASTM E238, Lab 1 used a cotton swab dipped in acetone to remove contamination around the specimen hole.

### 3.3. Fracture Toughness and Static Notch Strength

ARALL-1 Laminate fracture toughness depends on the type of flaw present. ARALL panels containing cracks, developed by fatigue, show substantially higher fracture toughness (residual strength) than monolithic panels containing comparable size fatigue cracks (4-6); see also Figure 2. In the ARALL case, load transfer to unbroken fibers in the wake of fatigue cracking relaxes the crack tip stress in the metal layers (stress intensity decrease). If instead the crack is introduced as a saw-cut (severing the fibers), then Figure 2 shows that ARALL-1 Laminate fracture strength approaches that of monolithic 7075-T6 sheet. Table 6 gives raw data from sawed center-slot panel tests of ARALL-1 Laminate and 7075-T6 sheet. The Kc determinations from these tests show ARALL-1 and 7075-T6 fracture toughnesses to be comparable in their respective L-T (0 degree) and T-L (90 degree) test orientations.

Figure 3 contrasts residual strength data of ARALL-1 panels containing centrally located open holes and slots (fibers cut). The results show that when the ratio of hole diameter or slot length to

Table 5
3/2 ARALL-1 Laminate Pin-Load Bearing Test Results

| Laboratory No. | Test Direction | e/D (a) | Bearing Ultimate Strength, ksi (MPa) | Bearing Yield Strength, ksi (MPa)(b) |
|---|---|---|---|---|
| 1 | L | 1.5 | 95 (655) | 85 (586) |
| | | 2.0 | 106 (731) | 102 (703) |
| | LT | 1.5 | 102 (703) | 88 (607) |
| | | 2.0 | 105 (724) | 97 (669) |
| 6 | L | 2.0 | 84 (579) | 82 (565) |
| | LT | 2.0 | 85 (586) | 83 (572) |
| 7075-T6 | L | 1.5 | 124 (855) | 105 (724) |
| Minimums (c) | LT | 2.0 | 160 (1103) | 122 (841) |

(a)  e/D = (distance from edge of specimen to center of pin)/(pin dia.).

(b)  Calculated at hole deformation equal to 2 percent pin dia.

(c)  Ref. (13), Nom. 0.050 in. (1.27 mm) sheet gauge. The L and LT bearing properties of 7075 are typically taken as equal.

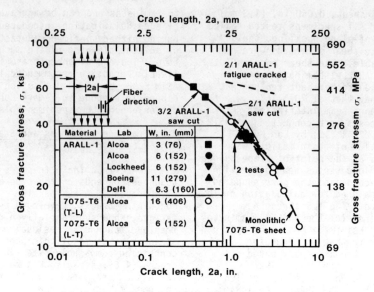

**Residual Strength Data from ARALL-1 Laminate and
0.063 in. (1.6 mm) Gage 7075-T6 Sheet Center Slotted Panels**

**Figure 2**

Table 6
Test Results on 3/2 ARALL-1 Laminate and
7075-T6 Sheet Center Slotted Panels

| Material | Lab No. | Test Dir. | Panel Dimensions Width in. (mm) | Thick. in. (mm) | Orig. Crk. Length, 2a in. (mm) | Gross Failure Strength ksi (MPa) | Fracture Tough., Kc ksi √in. (MPa √m ) |
|---|---|---|---|---|---|---|---|
| ARALL-1 | 2 | L-T | 11 (279) | 0.053 (1.3) | 3.57 (91) | 24.5 (169) | 64.1 (71) |
| ARALL-1 | 1 | L-T | 6 (152) | 0.053 (1.3) | 1.46 (10) | 34.5 (238) | 64.9 (71) |
|  |  |  |  |  | 1.46 (10) | 34.1 (235) | 62.1 (68) |
| ARALL-1 | 4 | L-T | 6 (152) | 0.053 (1.3) | 2.06 (52) | 30.1 (208) | 68.4 (75) |
|  |  |  |  |  |  | Avg. | 64.9 (71) |
| ARALL-1 | 1 | L-T | 3 (76) | 0.053 (1.3) | 0.125 (3.2) | 77.2 (532) | (b) |
|  |  |  |  |  | 0.250 (6.4) | 69.5 (479) | (b) |
|  |  |  |  |  | 0.375 (9.5) | 60.9 (420) | (b) |
|  |  |  |  |  | 0.500 (13) | 55.2 (381) | (b) |
| ARALL-1(a) | 1 | L-T | 6 (152) | 0.063 (1.6) | 1.42 (36) | 33.9 (234) | 66.5 (73) |
|  |  |  |  |  | 1.47 (37) | 35.8 (247) | 68.8 (76) |
|  |  |  |  |  |  | Avg. | 67.7 (74) |
| ARALL-1(a) | 1 | T-L | 6 (152) | 0.063 (1.6) | 1.47 (37) | 28.0 (193) | 63.9 (70) |
| 7075-T6 | 1 | L-T | 6 (152) | 0.063 (1.6) | 1.44 (37) | 35.4 (244) | 64.2 (71) |
|  |  |  |  |  | 1.43 (32) | 36.0 (248) | 69.9 (77) |
|  |  |  |  |  |  | Avg. | 67.1 (74) |
| 7075-T6 | 1 | L-T | 16 (406) | 0.063 (1.6) | ----- | ----- | 68 (75) (c) |
|  |  | T-L |  |  | ----- | ----- | 62 (68) (c) |

(a)  Substituted 0.016 in. (0.4 mm) thick 7075-T6 sheet for 0.012 in. (0.3 mm) thick 7075-T6 sheet in standard 3/2 ARALL-1 laminate configuration.

(b)  Not applicable, specimen width too small.

(c)  Typical value; Many tests Alcoa standard 16 in. (406 mm) wide panel.

specimen width is small, the specimen residual strength becomes insensitive to the discontinuity shape. In this case, failure is predicated upon the area of uncut cross-section. As discontinuity size increases relative to the panel width, then the static notch strength becomes increasingly sensitive to the stress concentration acuity; the more severe the stress concentration, the greater the strength loss. Continuous fiber-resin composites exhibit similar behavior, and fracture criteria for these occurrences are well established, e.g. (17-20).

Table 7 shows notched (open hole) specimen data obtained for the ARALL-1 Laminate. The 77 ksi (531 MPa) ambient notch tensile strength measured by Lab 9, is about 24 percent greater than that estimated for equivalently tested monolithic 7075-T6, 62 ksi (427 MPa). The ARALL-1 Laminate compressive yield strength of 47 ksi (324 MPa) is about 15 percent less than that of 7075-T6, 55 ksi (379 MPa), estimated by equating net section stress to the compressive yield strength from Table 3. The Lab 9 results also show that ARALL-1 sustains good tensile and compressive notch strength at 180°F (82°C) and at 180°F (82°C) after a 250°F (121°C) presoak respectively. The compressive failure mode was plastic buckling for the two temperatures considered.

Related work (4-6) has shown that fracture toughness and blunt

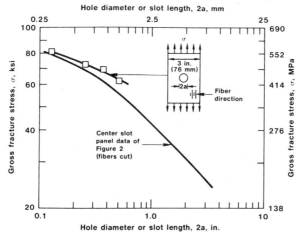

**Residual Strength Comparison of ARALL-1 Laminate Panels, 3/2 Layup, 0.053 in. (1.3 mm) thk., Containing Centrally Located Open Holes and Slots (Fibers Cut)**

Figure 3

Table 7

ARALL-1 Laminate (3/2 Layup, 0.053 in. (1.3 mm) thick)
Static Notch Strength Data - Longitudinal

| Lab No. | Specimen Width in. (mm) | Hole Dia. in. (mm) | Presoak Condition | Test Temp. | Tens. Fail. Strength (ksi) (MPa) | Comp. Yld. Strength (ksi) (MPa) |
|---------|-------------------------|--------------------|--------------------|------------|----------------------------------|---------------------------------|
| 1 | 3.0 (76) | 0.125 (3.2) | None | RT | 82 (565) | -- |
|   |          | 0.250 (6.4) |      |    | 73 (503) | -- |
|   |          | 0.375 (9.5) |      |    | 69 (476) | -- |
|   |          | 0.500 (13)  |      |    | 62 (427) | -- |
| 9 | 0.25 (6.4) | 0.06 (1.5) | None | RT | 77 (531)(a) | -- |
|   |            |            | None | 180°F (82°C) | 75 (517)(a) | -- |
| 9 | 0.75 (19) | 0.06 (1.5) | None | RT | -- | 47 (324)(a) |
|   |           |            | 6 hr. @ 250°F (121°C) | 180°F (82°C) | -- | 41 (283)(a) |

(a) Average of duplicate tests.

909

notch residual strength of ARALL-1 Laminate can be improved significantly by substituting tougher aluminum alloys 2024-T3 (ARALL-2 Laminate) or 7475-T6 (ARALL-3 Laminate) for the 7075-T6 sheet plies used in ARALL-1.

## 3.4 S-N Fatigue

ARALL-1 Laminate smooth and notched specimen fatigue behaviors are shown in Figures 4 and 5, respectively. Lab to lab variability in this data can in part be explained by the variety of specimen configurations employed by the participants. Contrasted to baseline 7075-T6 sheet, the ARALL-1 Laminate offers superior fatigue resistance when tested parallel to the fiber direction, and the magnitude of improvement is greater for tension- tension loading (R = 0.1) than for fully reversed tension-compression loading (R = -1.0). When loading is transverse to the fiber direction, the ARALL-1 cyclic lifetimes are shorter than those of the 7075-T6 baseline tested at comparable stress. Again, the latter two observations are expected since aramid/epoxy layers occupy about 30 percent of the specimen cross-section, and limited reinforcement is provided when loading is either compressive or oriented 90 degrees to the fibers.

Crack development in outer metal layers of the ARALL laminate often was observed well before termination of testing. Life to aluminum crack initiation [approx. 0.10 in. (2.5 mm)] in the longitudinal ARALL-1 specimens generally exceeded the failure life-time of baseline 7075-T6. Cracks emerging from notches typically arrested or grew very slowly, depending on the test conditions. Longitudinal unnotched specimens surviving approx. 10 million cycles at 60 and 63 ksi (414 and 434 MPa) were tension tested by Lab 10. In both cases, the post fatigue specimens retained the ultimate tensile strength of as-received material. Additional damaged specimens when tension tested possessed considerable residual strength, despite fatigue cracks being present in the outer metal layers.

Limited testing conducted by two of the participants demonstrated that good ARALL-1 Laminate fatigue strength is maintained after presoaking at elevated temperature and exposure to salt environment. Lab 3 showed that notch fatigue performance (Kt = 3) after a 168 hour presoak at 100°C (212°F) is equal to that of as-received material (Figure 5). Flexure fatigue tests (Kt = 1, R = -1) conducted in a humid salt environment by Lab 8 (21) showed ARALL-1 performance to be better than that of 7075 in a stress corrosion resistant T73 temper. In these tests, the ARALL specimen edges were masked with adhesive prior to salt exposure.

Relating coupon fatigue tests to actual structure for any composite material is difficult. However, good ARALL Laminate fatigue performance has been verified in tests simulating structural details (4-6). For example, Lab 5 fastener joint test results in Figure 6 show a large fatigue improvement for ARALL-1 over baseline 7075-T6.

## 3.5 Fatigue Crack Propagation

The fatigue crack growth characteristics of ARALL Laminates are their most important property improvement. Lab 1 conducted center slotted panel tests on baseline 7075-T6 sheet and on ARALL-1 material in the 0.5 percent stretch and no stretch conditions. In each of the specimens, a starting slot was introduced by jeweler's saw, and ensuing crack growth was monitored visually on both the front and back metal surfaces. The test procedure followed ASTM Method E647 (22) used for metals. The results of these tests plotted as crack length versus cycles are shown in Figure 7. Relative to the 7075-T6 baseline, the unstretched ARALL-1 Laminate shows about a ten-fold life improvement, and the stretched ARALL-1 Laminate shows more than a thousand-fold life improvement. The unstretched specimen data, when contrasted with that of the 7075 baseline, shows that the ARALL-1 Laminate is able to support a much larger crack than monolithic 7075.

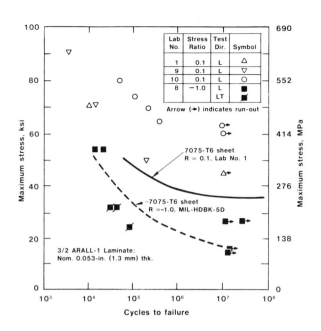

**ARALL-1 Laminate Smooth Specimen**
**Fatigue Behavior (K$_t$ = 1)**
Figure 4

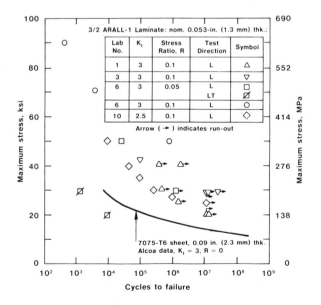

**ARALL-1 Laminate Notch Fatigue Behavior**
Figure 5

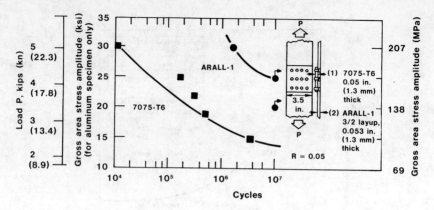

**Joint Fatigue Strength of ARALL-1 Laminate and 7075-T6 Aluminum Sheet**
(Courtesy McDonnell Douglas, Long Beach, CA)
Figure 6

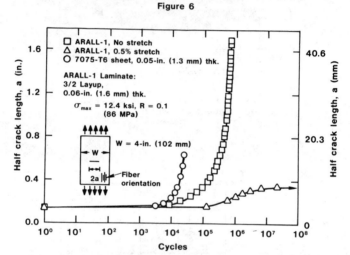

**Fatigue Crack Growth of Stretched
and Unstretched ARALL-1 Laminate and 7075-T6 Sheet
(Laboratory No. 1)**
Figure 7

Crack growth rate curves for ARALL-1 Laminate and baseline 7075-T6 are shown in Figure 8. Note that crack growth rate behavior of the 90 degree (transverse) specimen overlaps monolithic 7075-T6 data. To generate the ARALL curves stress intensity factor was calculated as is done for ordinary metals. Unlike metals, however, the ARALL crack growth rate curves are geometry and load dependent since these factors influence stress transfer to the un-broken fibers. This is illustrated in Figure 8 by differences in the stretched ARALL-1 growth rates obtained by Lab 1 and Lab 6. To normalize ARALL fatigue crack growth rate descriptions, fiber loading and local delamination effects must be taken into account as suggested by Marissen (23).

Lab 1 crack growth data shown in Figures 7 and 8 were obtained from a slight variation of standard 3/2 ARALL-1 Laminate. The non-standard version had nominal thickness of 0.062 in. (1.6 mm), and was made with 0.016 in. (0.4 mm) thick 7075-T6, as opposed to

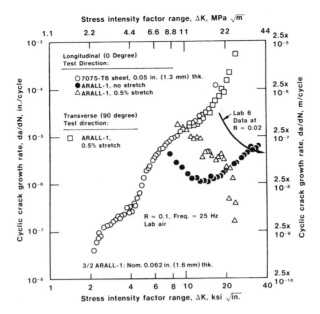

Stress intensity factor range, ΔK, MPa √m

**Comparison of ARALL-1 Laminate and 7075-T6
Fatigue Crack Growth Rate Behaviors
(Laboratory No. 1)**
Figure 8

0.012 in. (0.3 mm) thickness sheet used in the standard configuration. Laboratories 2, 4, 5, 6, and 9 tested standard ARALL-1 Laminate, and each verified the outstanding fatigue crack growth resistance evidenced in Figures 7 and 8.

### 3.6 Compressive Buckling/Crippling

Prior work on complex stiffened panels (3, 5) demonstrated that compressively loaded ARALL panels possess structural efficiencies equivalent to panels constructed from monolithic aluminum. Crippling tests performed by Lab 6 and compressively loaded stiffened panel tests performed by Labs 4 and 6 showed that ARALL-1 Laminate exhibits failure modes similar to monolithic aluminum. Lab 4 buckled ARALL stiffened panels to the point of first delamination. Subsequent tension tests conducted from material in the delaminated areas showed no apparent strength loss from that of the as-received material.

### 3.7 Low Energy Impact

Although damage by impact is not a major concern with monolithic aluminum, fiber resin-matrix composites are sensitive to delamination even by low energy impacts. The work of Lab 7 (24) and others (5, 6, 25, 26) has shown that percent strength loss after low energy impact is slightly greater for ARALL Laminate than for monolithic aluminum, but ARALL retains its superior crack propagation resistance. Relative to monolithic aluminum, ARALL Laminate ability to absorb impact energy is reduced by aluminum splitting parallel to the uniaxial fibers (24). Therefore, to maximize impact resistance a cross-ply arrangement would be preferred.

ARALL Laminate post impact performance has been contrasted to that of state-of-the-art graphite/ epoxy composite (6). When subjected to the same impact, ARALL consistently outperformed graphite/epoxy on a percent strength loss basis. Moreover, since the outer metal layers deform plastically, impact detection is better for ARALL Laminates than for graphite/epoxy. Static indentation tests performed by Lab 7 showed that impressions left by a 1 in. (2.5 cm) diameter steel ball

are visually detectable at applied
loads much lower (by about 4X) than
load levels required to initiate
loss of compliance (onset of actual
damage) in constrained ARALL-1
panels. When the same test was performed on state-of-the-art graphite/
epoxy composite, visual damage could
only be detected at loads exceeding
the onset of damage. The load associated with damage initiation in
both materials was about the same.
In the same investigation, impact
damaged ARALL Laminate and graphite/
epoxy panels were subjected to equivalent fatigue loadings. Tension
tests then followed to determine
strength degradation resulting from
the combined damage processes. In
general, the graphite/epoxy residual
strength surpassed that of the
ARALL-1 Laminate; however, the post
impact fatigue crack rates remained
low in the ARALL-1 specimens. Additional testing conducted by Lab 7
confirmed prior Delft University
results (6, 24, 25) showing that
prestraining enhances ARALL Laminate
post impact properties, and that
static and fatigue strengths after
impact are improved by substituting
tougher 2024-T3 for 7075-T6 in the
ARALL Laminate.

## 3.8 Damping Characteristics

Adhesively bonded metal laminates typically show vibrational
damping properties superior to those
of their monolithic counterparts
(27-30). The same good damping
qualities should apply to ARALL
Laminates, but with the advantage
that reinforcing fibers add stiffness and strength so that damping
augmentation is not at the expense
of other properties. Consequently,
ARALL Laminates may have possible
structure advantages where sound or
vibration transmission is a problem.
Lab 8 studied ARALL-1 Laminate
damping properties over a frequency
range of 1-hz to 1-khz (30). Their
results, given in Figure 9, show
that ARALL-1 has about two to three
times better damping than monolithic
aluminum and that damping parallel
to the fibers is about 30 percent
greater than cross fiber damping.
Lab 5 established the aeroacoustic
fatigue response of stiffened panel

| Material | Loss Factor $\eta$ (X $10^{-4}$) |
|---|---|
| 3/2 ARALL-1 Laminate (Longitudinal) | 21.0 |
| 3/2 ARALL-1 Laminate (Transverse) | 16.0 |
| 2024 Aluminum | 7.2 |
| 6061 Aluminum | 6.7 |

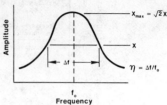

**Damping Characteristics
ARALL-1 Laminate vs. Monolithic Aluminum
(Laboratory No. 8)
Figure 9**

assemblies made from ARALL-1 Laminate and monolithic aluminum. The
ARALL-1 panels had about twice the
damping ability of aluminum panels,
which translated to significant life
improvement under the acoustic excitation.

## 4. CONCLUSIONS

ARALL Laminates are a new
system of composite materials that
could offer significant structural
and weight saving advantages, particularly in tension dominated,
fatigue and fracture critical applications. This ten laboratory
cooperative test program has shown
that ARALL Laminates can be tested
and characterized using existing
procedures. Reported mechanical
properties on first generation
product, 3/2 ARALL-1 Laminate, are
in good agreement with each other
and with the original University of
Delft work. The property values
obtained can be rationalized from
present understanding of metal and
fibrous composite material behaviors. The reinforcing fibers
provide significant directional tensile and fatigue strength increase
over monolithic aluminum, even after
exposure to various temperatures,
environments and damage agents.
Transverse tensile, compressive and
bearing properties follow expectations dictated by the volume of
metal present, and buckling/
crippling failure modes are similar
to those of metals. Various types

of fatigue and fatigue crack growth
tests all demonstrated significant
life improvement over monolithic
aluminum, even after damage by impact. Center slotted panel tests
with fibers severed showed equivalent damage tolerance to
monolithic aluminum. Good
structural damping characteristics
were demonstrated, and application
is envisioned where sonic fatigue is
a problem.

In summary, ARALL Laminates
provide an attractive blend of metal
and composite material properties.
Additionally, because much of the
technology is in place, manufacture
of structural hardware should not
require great new investments. More
detailed design and cost tradeoff
studies will now have to be done to
verify that ARALL Laminates are competitive with other advanced
aerospace materials.

### Acknowledgments

This paper compiles the work of ten labora-
tories, and contributions of the following
are acknowledged: G. H. Narayanan - Boeing;
A. Simpson and J. Fray - British Aerospace;
M. D. Goodyear and H. S. Pearson - Lockheed,
Georgia; W. Leodolter and A. Velicki -
McDonnell Douglas, Long Beach; E. A. Heckman
and V. M. Vasey - McDonnell Douglas, St.
Louis; C. R. Crowe, D. F. Hasson,
N. R. Nutter and R. G. Kasper - U.S. Navy;
W. S. Johnson - NASA Langley; S. W. Averill,
G. Turk and D. Kane - Northrop; G. R. Martin
- Rockwell. Also acknowledged are Alcoa
ARALL Laminates project team members;
A. A. Abercrombie, A. Govada, A. G. Holsing,
T. R. Kipp, S. Lee, K. E. Luyk, R. B. Sirkoch
and G. Sowinski, and the efforts of
J. W. Davis, 3M Company.

### 5. REFERENCES

1. Marissen, R. and Vogelesang, L. B.,
   "Development of a New Hybrid Material:
   ARALL," at Intercontinental SAMPE
   Meeting, Cannes, Jan. 1981.
2. Vogelesang, L. B., Marissen, R. and
   Schijve, J., "A New Fatigue Resistant
   Material: Aramid Reinforced Aluminum
   Laminate (ARALL)," at 11th ICAF Sym-
   posium, Noordwijerhout, May 1981.
3. Gunnink, J. W., Vogelesang, L. B. and
   Schijve, J., "Application of a New
   Hybrid Material (ARALL) in Aircraft
   Structures," ICAS-82-2.6.1, 13th Cong.
   Internat. Council of the Aeronautical
   Sciences, Seattle, Aug. 1982, p. 990.
4. Vogelesang, L. B. and Gunnink, J. W.,
   "ARALL, A Material for the Next Genera-
   tion of Aircraft; A State of the Art,"
   Delft Univ. Rpt. LR-400, Aug. 1983.
5. Gunnink, J. W., Verbruggen, M. and
   Vogelesang, L. B., "ARALL, A Light
   Weight Structural Material for Impact
   and Fatigue Sensitive Structures,"
   Vertica, Vol. 10, No. 2, 1986, p. 241.
6. Vogelesang, L. B. and Gunnink, J. W.
   "ARALL: A Challenge for the Next Gen-
   eration of Aircraft," at International
   Federation of Airworthiness, Amsterdam,
   Nov. 1985.
7. Mueller, L. N., Prohaska, J. L., and
   Davis, J. W., "ARALL (Aramid Aluminum
   Laminates): Introduction of a New Composite Material," at AIAA Aerospace Eng.
   Conf., and Show, Los Angeles, Feb. 1985.
8. Davis, J. W., "ARALL - From a Development to a Commercial Material," Progress
   in Advanced Materials and Processes:
   Reliability and Quality Control,
   Barthelds, G. and Schliekelmann, eds.,
   Elsevier Pub., Amsterdam, 1985, p. 41.
9. "ARALL: an Aluminum and Aramid Laminate," Aero. Eng., May 1985, p. 35.
10. "Laminated Metals Provide Tailor-Made
    Properties," Matls. Eng., Nov. 1985,
    p. 23.
11. Schultz, R. W., Gregory, M. A.,
    Teply, J. L., "Lightweight Structural
    Hybrid Laminates," Proc. of the First
    OMAE Specialty Symp. on Offshore and
    Arctic Frontiers, ASME, 1986.
12. Verbruggen, M., "Durability of ARALL,"
    Delft Univ. Rpt. LR-423, Mar. 1984.
13. MilHdbk 5D, "Metallic Materials and
    Elements for Aerospace Vehicle
    Structures," U.S. Dept. of Defense, Rev.
    May 1, 1985.
14. ASTM D4255-83, "Std. Guide for Testing
    Inplane Shear Properties of Composite
    Laminates," 1985 Ann. Bk. of Stds., Sec.
    15, Vol. 15.03, Am. Soc. for Test. &
    Matls., p. 262.
15. ASTM D2344-84, "Std. Test Meth. for
    Apparent Interlaminar Shear Strength of
    Parallel Fiber Composites by Short Beam
    Method," 1985 Ann. Bk. of Stds., Sec.
    15, Vol. 15.03, Am. Soc. Test. & Matls.,
    p. 55.
16. ASTM E238-84 "Std. Meth. for Pin-Type
    Bearing Test of Metallic Material," 1986
    Ann. Bk. of Stds., Sec. 3, Vol. 03.01,
    Am. Soc. Test. & Matls., p. 406.
17. Whitney, J. M. and Nuismer, R. J.,
    "Stress Fracture Criteria for Laminated
    Composites Containing Stress Concentra-
    tions," J. of Composite Materials, Vol.
    8, July 1984, p. 253.

18. Nuismer, R. J. and Whitney, J. M., "Uniaxial Failure of Composite Laminates Containing Stress Concentrations," Fract. Mech. of Comp., ASTM STP 593, Am. Soc. Test. & Matls., 1975, p. 117.

19. Mar, J. W. and Lin, K. Y., "Fracture of Boron/Aluminum Composites with Discontinuities," J. Composite Materials, Vol. 11, Oct. 1977, p. 405.

20. Caprino, G., "On the Prediction of Residual Strength for Notched Laminates," J. Materials Sci., Vol. 18, 1983, p. 2269.

21. Hasson, D. F. and Crowe, C. R., "Flexure Fatigue of Aramid Reinforced Aluminum 7075-T6 Composites in Lab and Salt Contaminated Moist Air," at ASM Matls. Week 86, Lake Buena Vista, FL, Oct. 1986.

22. ASTM E647-83, "Std. Test Meth. for Constant-Load- Amplitude Fatigue Crack Growth Rates Above $10^{-8}$ m/Cycle," 1986 Ann. Bk. of Stds., Sec. 3, Vol. 03.01, Am. Soc. Test. & Matls., p. 714.

23. Marissen, R., "Flight Simulation Behavior of Aramid Reinforced Aluminum Laminates," Eng. Fracture Mech., Vol. 19, No. 2, 1984, p. 261.

24. Johnson, W. S., "Impact and Residual Fatigue Behavior of ARALL and AS6/5245 Composite Material," NASA Langley TM 89013, Sept. 1986.

25. Roebroeks, G., Fatigue Behavior of ARALL After Impact," Document B2-85-01, Dept. Aero. Eng., Delft Univ., Feb. 1985.

26. Chen, D., "Drop Weight Impact Test on Some Sheet Materials," Doc. B2-86-03, Dept. Aero. Eng., Delft Univ., Jan. 1986.

27. Hyer, M. W., Anderson, W. J., and Scott, R. A., "Nonlinear Vibrations of Three-Layer Beams with Viscoelastic Cores, I: Theory," J. Sound and Vibration, Vol. 46, No. 1, 1976, p. 121.

28. Hyer, M. W., Anderson, W. J., and Scott, R. A., "Nonlinear Vibration of Three-Layer Beams with Viscoelastic Cores, II: Experiment," J. Sound and Vibration, Vol. 61, No. 1, 1978, p. 25.

29. Ely, R. A., "Laminated Damped Fuselage Structures," Vibration Damping 1984 Workshop Proc., AFWAL TR-84-3064, Nov. 1984, p. 00-1.

30. Nutter, N. R., Gray, R. A., Crowe, C. R. and Kasper, R. G., "Damping Characteristics of SiC/Al Metal Matrix Composites and Aramid Reinforced Aluminum Laminates (ARALL-1)," at ASM Matls. Week 86, Lake Buena Vista, FL Oct. 1986.

BIOGRAPHIES

R. J. Bucci received a B.S. from Northeastern, M.S. from Brown and Ph.D from Lehigh. He is a Technical Specialist in the Alloy Technology Div. of Alcoa Laboaratories. Since joining Alcoa in 1973, he has been involved in structural materials characterization and the associated methodologies. He is a member of ASTM and AIME and has over 50 publications.

L. N. Mueller joined Alcoa Laboratories in 1978 and is presently a Staff Application Engineer involved with the implementation of Alcoa's aerospace materials. He received his B.S. and M.S. degrees from Univ. of Missouri and is a registered professional engineer.

R. W. Schultz is a Sr. Engineer in the Alloy Technology Div. of Alcoa Laboratories. He joined Alcoa in 1984 after receiving his M.S. degree from Virginia Polytechnic Inst. His main efforts involve mechanical characterization of fibrous composite materials, and he is a member of ASTM.

J. L. Prohaska began his Alcoa career in 1963 and has had assignments at Alcoa's Davenport, New Kensington, and Cressona production plants. He holds a B.S. degree in electrical engineering, and is currently manager of the ARALL Laminate R&D project at Alcoa Laboratories.

PRD-149:  A NEW ULTRA-HIGH MODULUS FORM OF
KEVLAR® ARAMID FOR LOW STRAIN REINFORCEMENT
AND LOW MOISTURE REGAIN APPLICATIONS

James E. Van Trump and Jacob Lahijani
Textile Fibers Department
E. I. du Pont de Nemours & Co., Inc.
Wilmington, Delaware  19898

## ABSTRACT

The morphological and physical
properties of a new ultra-high
modulus form of Kevlar®* aramid,
designated PRD-149, are discussed.
The product differs significantly
from previous Kevlar® products in
its higher crystallinity and unique
super-crystalline structure.  The
yarn has tensile tenacity/modulus
of 2.3-2.6/162-175 GPa, compared to
commercial Kevlar® 49 with a ten-
acity/modulus of around 2.8/121 GPa
(depending on type and denier).
The high crystallinity also imparts
substantially reduced creep and
moisture regain relative to currently commercial grades of Kevlar®
aramid.  Yarn properties translate
well into the composite form with
epoxy strand moduli approaching
those of high strength carbon
fiber.  It is anticipated that this
material will find utility in

applications ranging from optical
fiber cable reinforcement to hybrid
Kevlar® aramid/ carbon composites.
* Du Pont registered trademark.

## 1.  INTRODUCTION

Para-aramid fiber was first commercialized by the Du Pont Company
under the Kevlar® trademark in 1971,
and a short time later a heat-
treated high modulus version,
Kevlar® 49 (K-49) was introduced[1].
K-49 aramid has since been used in
those applications, mainly advanced
composite, where specific modulus
and strength are paramount and where
damage tolerance, cost or other consideration give it an advantage over
ceramic, carbon or other high modulus fiber. The fiber now enjoys a
fairly diverse market base extending
from composites to ballistics,
tailored to its unique combination
of properties.  In common with most
organic fibers, it forms highly

damage tolerant composite structures because it generally fails in bending by a compressive rather than a tensile mode. However, this early compressive failure also limits the fiber's application in structural composites. Hybrid structures containing both aramid and carbon fibers have been shown to combine moderate levels of both compressive strength and damage tolerance[2,3] but at reduced tensile strength. This tensile strength loss is caused by the tensile modulus mismatch between Kevlar® 49 and carbon fiber, which reaches its break point (or, more realistically, its design limit) before the aramid is strained enough to sustain a substantial share of the load. A higher modulus version of Kevlar® would be very desirable in this application.

Additionally, there are tensile applications, fiber optic cable reinforcement for example, where the absolute fiber strength is less important than the sustainable tension at a strain level imposed by some other element, in this case the break elongation of the glass optic fiber. In such cases, traditional measures of fiber performance, for example "initial" modulus, become of limited value in predicting fiber utility, and use parameters must be carefully defined. This parameter will be discussed in more detail later.

Finally there are applications where improved stability is important, in hydrolysis resistance, or in creep, or in moisture regain and lateral expansion. All these applications should in principle be met by a more highly crystalline Kevlar® aramid product. This paper describes such a new experimental Kevlar® product, designated PRD-149. This work has led to the development of a commercial fiber Kevlar® 149, which was introduced to the trade in June and is now available. We will describe the more important properties of the experimental fiber PRD-149 and make some observations on its structure and how it differs from K-49. While the fiber can be made in a range of deniers, the bulk of the work reported here was done with 400 denier yarns, and conditions and properties reported refer to that denier unless otherwise specified. The property database on the production fiber Kevlar® 149, which is not identical to PRD-149, is still being developed. A current database of commercial Kevlar® 149 properties may be obtained from the author.

2. PROPERTIES

Table 1 summarizes the yarn properties of PRD-149 relative to Kevlar® 29 and 49. The yarn can be produced with varying modulus levels up to about 35% higher than K-49. There is, in general, a trade-off between tenacity and modulus with about a 5% tenacity drop for every 10% modulus gain vs. K-49. Figure 1 shows the stress-strain curves for a fully developed PRD-149 and for commercial aramids. Organic fiber stress-strain curves are normally characterized by a relatively high initial tensile modulus, a region at the beginning of the stress-strain curve where the modulus is at a local maximum. PRD-149 is unusual in that there is no local maximum initial modulus as there is for most organic fibers, nor a flat modulus response as is common for inorganic fibers. Instead, we see a steady increase in modulus with strain level as shown in figure 2. "Modulus" therefore requires careful definition especially since it is common practice to define it at a set fraction of full scale-to-break; a carry-over from the days of early computerization. For a yarn of this type, with a constantly varying slope to the stress-strain curve, the point of slope measurement must be clearly defined. Since the applied stress is more easily measured by conventional techniques than absolute strain, and to simplify comparison with our existing data base, we have defined the modulus to be the slope of the stress-strain curve between 2000 and 2450 grams stress for 44 tex (400 denier) yarns. The moduli for other deniers were taken at approximately equivalent specific stress levels. Under these conditions, we have made PRD-149 samples with tested modulus of over 178 GPa (1400 g/den) in yarn. For composite applications,

tensile measurements are commonly made via strand test, where the yarn is impregnated with epoxy before breaking and the yarn properties back calculated from the strand properties. PRD-149 performs relatively better in strand test (adapted from ASTM D2343-67) than K-49 with calculated fiber tenacity around 3.4 GPa (500 kpsi) and modulus up to 207 GPa (30 mpsi) as compared to K-49 at 3.6 GPa (525 kpsi) tenacity and 125-130 GPa (18-19 mpsi) modulus[4].

Secondary properties are no less unique. The fiber is around 2% more dense for example and its moisture regain is approximately one third that of K-49, presumably because of a relatively more compact structure. While not an important parameter in some applications, lower moisture regain is particularly important in composites in that the lateral expansion of the fiber with moisture pickup and the local stress this can impose on the composite matrix structure is expected to be correspondingly reduced. Thus, this fiber would be preferred for those uses where extremes, especially cyclic extremes, in environmental temperature and humidity might reasonably be encountered.

Further benefit from the more compact structure and overall molecular interaction comes in the form of fiber creep resistance. This has been measured at .012% per decade compared with K-49 aramid yarns which range from .02 to .04% per decade (50% of break load, 22°C). This property is of critical importance in any permanently tensioned system, especially in any application where reset is difficult or perhaps not possible at all-wound pressure vessels for example or flywheel systems. Creep resistance should be expected to become better relative to most competitive organic fibers (for example high modulus polyethylene or Teijin's HM-50) as use temperature is elevated, even on a transient basis.

Strength conversion in composite form is excellent, generally slightly better than K-49, presumably because of the shorter effective gauge length in the composite form compensates somewhat for that part of lower yarn strength generally ascribed to more frequent defects[7,8]. As noted earlier, strand testing in low temperature cure epoxy resin shows a modulus of up to 207 GPa (30 mpsi), sufficient to provide near perfect modulus matching with most high strength carbon fibers. We therefor expect substantial interest in hybrid composite end uses.

3. STRUCTURE

The equatorial wide angle X-ray spectrum shows two major peaks at 20 and 23 degrees, presumably corresponding to the Northolt type structure[5] characteristic of PPD-T and a number of PPD-T copolymer fibers. However, the peak definition indicates a substantially larger apparent crystallite size than for current fibers, as well as a higher crystallinity index (an approximate measure of overall fiber crystallinity). Also, as shown in table II, the unit cell dimensions for K-49 and PRD-149 differ significantly[9], the PRD-149 being somewhat larger in overall volume giving a theoretical density very close to that observed. Examining the meridional x-ray trace of a number of Kevlar® yarns we have found the paracrystalline distortion factor gII and the crystalline orientation angle correlate with modulus up to about 130 GPa modulus in general agreement with Barton[10]. Above this modulus these parameters remain essentially flat. Also in general agreement with Barton, we see no correlation between apparent axial crystallite length and modulus.

Longitudinal supercrystalline features of the fiber are revealed by optical and scanning electron microscopy (figure 3). None of the highly regular "pleat structure"[11] characteristic of the incumbent fibers is shown although larger, irregularly spaced density and/or orientational differences appear.

A wealth of transverse fiber structural features are accessible by caustic hydrolysis in the method of Horio et al.[12], which permits

920

brittle fracture of the fibers and shows the difference in supercrystalline fiber structure of PRD-149 vs. K-29 and 49. We see in figure 4 that degraded K-29 fiber fractures cleanly across the pleat face, presumably due to faster hydrolysis of the relatively voided, low crystallinity region at the pleat. The fracture plane steps up or down in integral pleat steps as resistance is encountered. The face appears to be composed of tightly packed discrete particles, presumably the ends of the microfibrillar fiber elements which comprise the fundamental units of the supercrystalline network. These microfibrils appear to be organized in a tangential pattern in spite of the well-established radial order of the crystallites themselves[13]. The K-49 fiber (fig. 5) has an intermediate structure with this pleated structure confined to the fiber core while the skin becomes a more resistant structure without clean fracture faces. PRD-149 itself (fig. 6) appears to have a substantially different, fine grained radial structure without definable microfibrils or pleats.

In addition to the substantial other differences between PRD-149 and current Kevlar® fibers, the inherent viscosity (IV) of PRD-149 is significantly higher than that of normal aramids which are generally in the 4.5 to 5.5 range. IV values (0.5%,

$30°C$, $H_2SO_4$) usually lie above 6.5 for PRD-149, and can, depending on the exact process, be as high as 14, where the polymer becomes insoluble in 100% sulfuric acid.

## 4. APPLICATIONS

Since this is an experimental fiber we have only limited applications data. However, we expect this material to enjoy fairly wide utility in applications where its unique properties can be used to advantage. We feel that its high modulus, coupled with the damage tolerance inherent to organic fibers, will give it usefulness in hybrid composites with carbon fibers, in sporting goods and perhaps in fiber optic cable reinforcement. Its low creep should be useful in belt reinforcement, as tension members in large fabric structures, metal lined pressure vessels, and, perhaps, rotating space structures. The fiber's low moisture regain should reduce stress-cracking in high altitude aircraft composite surfaces and it should be useful as a surface scrim for high specific strength composites in general. In short, we anticipate a diverse market base.

## 5. REFERENCES

1. Blades, H., U.S. Patent 3,869,430, "High Modulus, High Tenacity Poly(p-Phenyleneterephthalamide) Fiber" (1975).

2. Riewald, P. G. and Zweben, C. H., "Hybrid Composites for Commercial and Aerospace Applications", 30th Annual Conference,

Reinforced Plastics/Composites
Institute, SPI, Washington, DC,
Feb. 6, 1975 Sect. 14B, p. 1.

3. Norman, J. C., "Damage Resis-
tance of High Modulus Aramid
Fiber Composites in Aircraft
Applications, "SAE Preprints
#750532, 1389 (1975).

4. Du Pont Kevlar® 49 Data Manual,
page II-B.

5. Northolt, M. G., European
Polym. J., 10, 779 (1974).

6. Kwolek, S. L., in preparation.

7. Zweben, C., Smith, W. S. and
Wardle, M. W., ASTM Special
Tech. Pub. 674 (S. W. Tsai
Ed.), 236, (1978).

8. Bacon, R. and Schalomon, W. A.,
Appl. Polym. Symp. 9, (J.
Preston Ed.) ACS, 285 (1969).

9. Barton, R., personal communication.

10. Barton, R., J. Macromol. Sci.-
Phys., B24(1-4), 125 (1986).

11. Panar, M., Avakian, P., Blume,
R. C., Gardner, K. H., Gierke,
T. D. and Yang, H. H., J.
Polym. Sci, Polym. Phys. Ed.
21, 1955 (1983).

12. Horio, M., Kaneda, T.,
Ishikawa, S. and Shimamura, K.,

Sen_I Gakkaishi, 40, T-286
(1984).

13. Dobb, M. G., Johnson, D. J. and
Saville, P. B., Phil. Trans. R.
Soc. Lond. A294, 483 (1979).

## 6. BIOGRAPHY

Dr. James E. Van Trump is a Research
Associate with the Textile Fibers
Department of Du Pont where he has
been employed since 1975. He received his BA (1964) from the University of Wyoming and PhD (1975) in
Physical Organic Chemistry from the
University of California at San
Diego. During the past 11 years he
has worked in the general area of
fibers, both melt-spun and lyotropic
liquid crystalline, with short
detours into silica-based pigment
and composites.

Dr. Jacob Lahijani is a Senior
Research Chemist with the Textile
Fibers Department of E. I. du Pont
where he has been employed since
1980. He received his BS (1975)
from the University of Portland in
Science, MS (1977) from Pennsylvania
State University in Chemical
Engineering and PhD (1980) from
Pennsylvania State University in
Polymer Physics. During the past
three years he has been working in
the area of PAN and pitch-based
carbon fibers.

## TABLE I

### YARN PROPERTIES

| | PRD-149 | | K-49* | | K-29* | |
|---|---|---|---|---|---|---|
| | GPa | g/denier | GPa | g/denier | GPa | g/denier |
| Tenacity | 2.4 | 19 | 2.8 | 22 | 2.8 | 22 |
| Elongation (%) | | 1.3 | | 2.5 | | 3.5 |
| Modulus | 165 | 1300 | 125 | 970 | 83 | 650 |
| Density (g/cc) | | 1.48 | | 1.45 | | 1.45 |
| Moisture Regain[+] (%) | | 1 | | 4 | | 7 |
| Inherent Viscosity | | 6.6-Insol. | | 5.6 | | 5.6 |

[+] 70°, 65% RH
* Kevlar® 49 and Kevlar® 29 are Du Pont Registered Trademarks

## TABLE II

### CRYSTALLINE PARAMETERS

| | Kevlar® 49 | PRD-149 |
|---|---|---|
| O.A. | 12° | 12° |
| C.I. | 60 | 75 |
| 20° A.C.S. | 65Å | 120Å |
| 22° A.C.S. | 55Å | 80Å |
| a Å | 7.73 | 7.88 |
| b Å | 5.18 | 5.18 |
| c Å | 12.82 | 12.94 |
| Calc. D g/cm$^3$ | 1.54 | 1.50 |

FIGURE 1
KEVLAR® STRESS-STRAIN CURVES

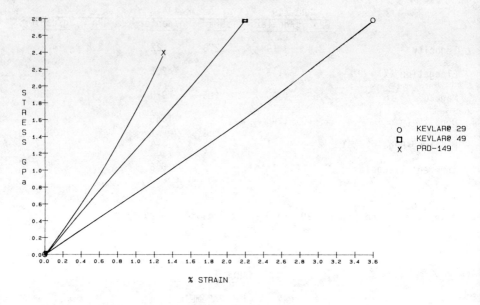

FIGURE 2
KEVLAR® MODULUS-STRAIN CURVES

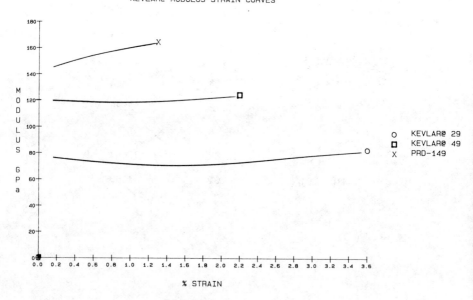

# FIGURE 3

## LONGITUDINAL PLEAT STRUCTURES

SEM                      OPTICAL, CROSSED POLARS

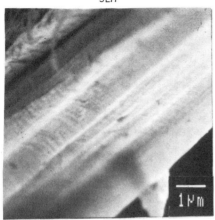

PRD-149

PRD-149

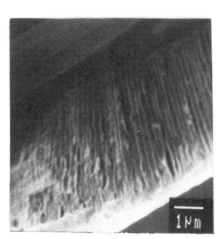

(Heat-Treated 400°C, 10 sec.)

K-49

FIGURE 4                    FIGURE 5

KEVLAR® 29                  KEVLAR® 49

Low Magnification           End View

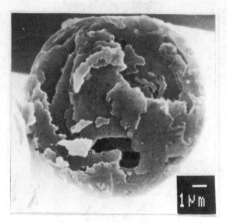

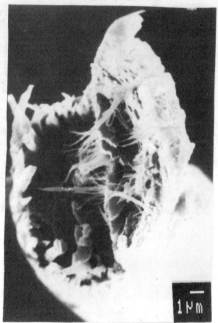

High Magnification

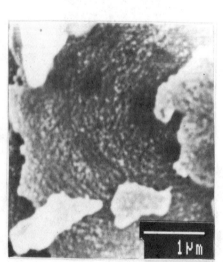

FIGURE 6

PRD-149

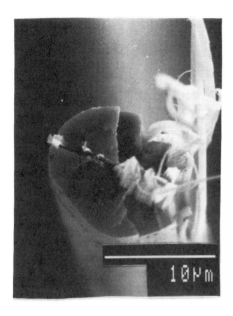

Low Magnification

High Magnification

32nd International SAMPE Symposium
April 6-9, 1987

"TORAYCA" T1000 ULTRA HIGH STRENGTH CARBON FIBER
AND ITS COMPOSITE PROPERTIES

Shoji Yamane, Tohru Hiramatsu and Tomitake Higuchi
Ehime Laboratory, Fibers and Textiles Research Laboratories,
Toray Industries, Inc.
1515, Tsutsui, Masaki-cho, Iyo-gun, Ehime-ken 791-31, Japan

## Abstract

High strength and intermediate
modulus carbon fiber "Torayca"
T1000 was developed. Tensile
strength of fiber measured with
resin impregnated strand
specimens was higher than 7GPa (one
million psi), tensile modulus of
elasticity was 294GPa (42msi) and
ultimate strain at failure was
2.4%.
High tensile strength of the fiber
was well reflected to strength of
T1000 composite. When combined
with adequate resin systems, 3.5 to
3.8GPa (500ksi or higher) of
composite tensile strength and more
than 2.0% of strain at failure were
achieved.

## 1. INTRODUCTION

High specific strength and modulus
of elasticity are unique
characteristics of CFRP, and
therefore, it has been replacing
metals to some extent for the
applications which are required to
reduce their weight. Since tensile
properties of a reinforcing fiber
dominate those of the composite
materials, increase of strength and
modulus of carbon fiber is
fundamental to improve the
performances of composites.
Since 1971, Toray Industries has
developed the carbon fibers of
versatile tensile properties and
those are shown in Figure 1. In
1984, intermediate modulus-high
strength fiber "Torayca" T800H was
added to the family of its products
and research work to obtain higher
strength fiber has been continued.
A challenging target of the
strength of fiber was set up to
higher than 7GPa, and modulus was
focused on so-called intermediate
modulus of 300GPa. The fiber which

928

met this target was successfully developed and it was named T1000, "Million Fiber" from its strength level of one million psi.
Fiber properties were introduced herein. Since the fiber should perform well as reinforcement in the composites, mechanical properties of the composites with various epoxy resin matrices were evaluated.

## 2. EXPERIMENTAL

### 2.1 Carbon Fiber

For this study, T800H and T300 were used as baseline materials, which are products of Toray Industries. Single filament strength was measured with gage length of 5 to 50mm and strain rate of 4%/min in accordance with ASTM D3379-82. Resin impregnated strand properties were measured as indicated in ASTM D4018-81 Method II. Impregnating resin was "Bakelite" ERL4221-BF$_3$MEA, which was recommended in the ASTM, and curing conditions were 130°C for 35min.

### 2.2 Composite

Five 180°C (350° F) curable epoxy resin systems were used. Their properties are shown in Table 1. #3601, #3630 and RX850 systems were limited to investigate the effect of resin on 0° composite tensile strength. Test fibers were impregnated with these resins to form unidirectional prepreg tapes. The prepregs were cut and laid up to meet required size and thickness for each testing item.
Preparation of specimens and composite testings were carried out in accordance with the test method specified by Boeing C.A.C. (1).

### 2.3 Modification of 0° Tensile Testing Method

Initially tensile testing of T1000 composite was performed by using a conventional wedge-action type grip in accordance with above method (1). The resulting tensile strength, however, was not so good as expected from strand tensile strength of T1000. In this testing, delamination in between layers of glass fabric composite tabs occurred, and failure of CFRP laminate followed within tab section, as shown in Figure 2. These phenomena were not encountered in case of T300 and rarely observed in case of T800H. When a specimen was strained near 2%, delamination at taper area of the tab occurred probably due to lower toughness of tab material. The failure within tab section might come from excessive clamping force through the wedge caused by high axial load of T1000 composite. To solve these problems, following modifications were made;
(1) use of longer tabs with

tougher resin system in tab
material.

(2) decreasing laminate thickness,
or use of hydrostatic grips without
wedge.

Table 2 shows the results of
improvements.  Tensile tests were
conducted using No. 2 method in
Table 2 for T1000 composite.

### 3.  RESULTS AND DISCUSSION

#### 3.1  Fiber Properties

Table 3 summarizes fiber properties
of T1000.  7GPa of tensile
strength, 294GPa of modulus and
2.4% of strain were achieved for
strand specimens.

To verify validity of strand
strength, single filament strength
was tested.  The results were
summarized in Figure 2 as a
function of gage length.  Filament
strength of T1000 is superior to
those of T800H and T300.

According to Rosen (2), strand
strength of filaments in composites
is given by the tensile strength of
bundles at an ineffective length.
Assuming ineffective length of
0.6mm after Noguchi and coworkers (3), tensile strength was
calculated.  The result is shown in
Figure 3.  Strand strength of 7GPa
for T1000 is proved to be
reasonable from data of single
filaments.

#### 3.2  Composite Properties

Composite properties of T1000 are
shown in Table 4, compared with
those of T800H and T300.

0° tensile strength of T1000
composites was 3.5GPa, almost twice
of that of T300, and strain at
failure was as high as 2.0%.

It is noted that strength depends
on matrix resin combined with the
fiber.  Strength measured with
various resin systems are plotted
in Figure 4 as a function of
ultimate strain of neat resin.  The
composite with resin of 12% strain
showed tensile strength of 3.85GPa
(at fiber volume of 60%).

Use of adequate resin system is,
therefore, a necessary condition
for T1000 to reflect its high
strength on composites.  It is also
worthy to note that the strength
values for T1000 composite were
obtained by the testing with
modified load-introducing sections.

Filament wound high pressure
vessels are one of promising
applications for T1000 fiber, and
evaluation programs are underway.

Other properties, such as 0°
compressive strength, 90° tensile
strength, interlaminar shear
strength are also shown in Table 4.
They were almost same as those of
existing fiber composites.

## 4. SUMMARY

1. High strength and intermediate modulus carbon fiber "Torayca" T1000 was developed.
Resin impregnated strand tensile strength of the fiber was 7GPa, tensile modulus was 294GPa and strain at failure was 2.4%.
2. 0° tensile strength of T1000 composite was 3.8GPa, and more than 2% strain at failure was achieved.

## 5. REFERENCES

1) Boeing Commercial Airplane Co. Boeing Material Specification BMS 8-276.

2) B.W. Rosen, AIAA J., 2, 1985 (1964).

3) K. Noguchi, K. Murayama, I. Matsubara, 5th FRP Symposium p.1. (1976 Osaka Japan); K. Morita et al., Pure and Applied Chem., 58, (3) 455 (1986).

## 6. BIOGRAPHIES

Shoji Yamane is a research associate in Ehime Laboratory of Fibers and Textiles Research Laboratories of Toray Industries. He is engaging in developing high quality carbon fibers. He holds a M.S. degree in chemistry from Kyoto University.

Tohru Hiramatsu is a research associate in the same laboratory of Toray Industries. He is also engaging in developing high quality carbon fibers. He holds a M.S. degree in chemistry from Kyoto University.

Tomitake Higuch is manager of Ehime Laboratory of Fibers and Textiles Research Laboratories of Toray Industries. He is responsible for high quality carbon fiber development.
He holds a B.S. degree in physics from Tokyo Kyoiku University.

Table 1 Properties of Cast Neat Resin

| Resin System | Strain at Failure % | Modulus of Elasticity GPa | Tg °C |
|---|---|---|---|
| #3601 | 2.1 | 3.7 | 230 |
| #3620 | 4.0 | 3.4 | 220 |
| #3632 | 4.9 | 3.7 | 215 |
| #3630 | 6.0 | 3.5 | 180 |
| RX850 | 12.0 | 3.0 | 161 |

Table 2  Tab Shape of Specimen and 0° Tensile Properties

| Case No. | Type of Loading Grip | Tab Dimension* L mm | a mm | Laminate Thickness mm | Tensile Strength GPa | Fracture Mode*** At Gage Section | At Tab Section | Delamination in Tabs |
|---|---|---|---|---|---|---|---|---|
| control | Wedge Action | 51 | 10 | 1.1 | 3.31 | 2/9 | 7/9 | severe |
| 1 | (lg*=55mm) | | | 1.1 | 3.49 | 2/5 | 3/5 | |
| 2 | | | | 0.83 | 3.49 | 6/6 | 0/6 | slight |
| 3 | | 85 | 20 | 0.55 | 3.41 | 9/9 | 0/9 | |
| 4 | Non Wedge | | | 1.1 | 3.49 | 5/5 | 0/5 | |
| 5 | (lg=100mm) | | | 0.83 | 3.48 | 3/3 | 0/3 | |

** Specimen T1000/#3620

*** Ratio to number of whole specimens tested

**** Testing materials

Tab ;     GFRP fabric 0/90 lay-up

Tab adhesives ;  AF126 (3M)

*

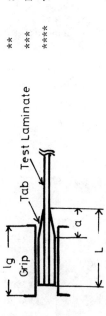

lg

Grip   Tab   Test Laminate

a

L

Table 3  Properties of Carbon Fibers

| Property | unit | T1000 | T800H | T300 |
|---|---|---|---|---|
| Filament Diameter | μm | 5.3 | 5.2 | 7.0 |
| Density | g/cm$^3$ | 1.82 | 1.81 | 1.77 |
| Filament Count | - | 12000 | 12000 | 12000 |
| Mass per Unit Length | tex | 480 | 445 | 800 |
| Tensile Strength* | GPa | 7.06 | 5.59 | 3.53 |
| Tensile Modulus of Elasticity* | GPa | 294 | 294 | 230 |
| Strain at Failure* | % | 2.4 | 1.9 | 1.5 |

* Resin Impregnated Strand Method

Table 4  Mechanical Properties of Unidirectional Composites

| Carbon Fiber Property | unit | T1000 | | T800H | | T300 |
|---|---|---|---|---|---|---|
| | | #3620 | #3632 | #3620 | #3632 | #3620 |
| 0° Tensile Strength | GPa | 3.49 | 3.55 | 2.95 | 2.94 | 1.70 |
| Modulus of Elasticity | GPa | 158 | 158 | 160 | 165 | 135 |
| Strain at Failure | % | 2.0 | 2.0 | 1.7 | 1.7 | 1.2 |
| 0° Compressive Strength | GPa | 1.70 | 1.65 | 1.60 | 1.70 | 1.60 |
| 0° Interlaminar Shear Strength | MPa | 110 | 95 | 120 | 115 | 125 |
| 90° Tensile Strength | MPa | 60 | 50 | 60 | 60 | 60 |
| Modulus of Elasticity | GPa | 9 | 9 | 9 | 9 | 9 |
| Strain at Failure | % | 0.7 | 0.6 | 0.7 | 0.7 | 0.7 |

1) Fiber volume fraction : 60%

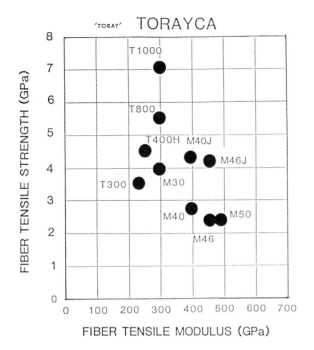

Figure 1 "Toray" carbon fibers

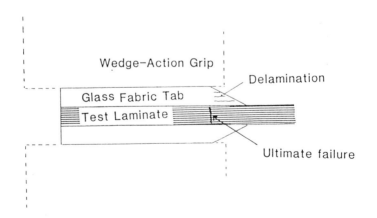

Figure 2 Test specimen and location of failure

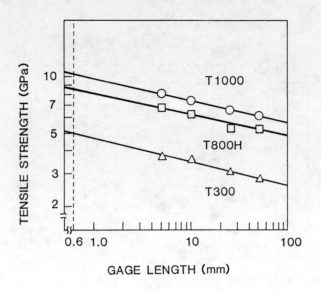

Figure 3 Strength of single filament

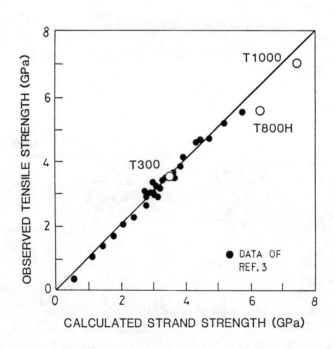

Figure 4 Relation between observed tensile strength
in strand testing and calculated value

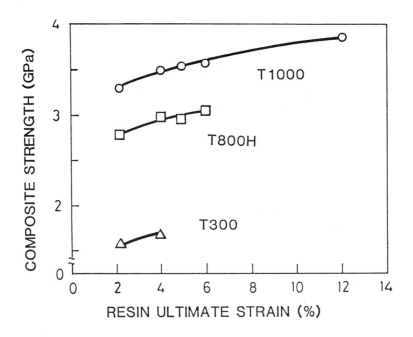

Figure 5  Relation of composite tensile strength
and  resin ultimate strain

CARBOFLEX®:  A NEW GENERAL PURPOSE PITCH-BASED CARBON FIBER
John W. Newman
Ashland Carbon Fibers Division
Ashland Petroleum Company
Ashland, Kentucky 41114

Abstract

A general purpose pitch-based carbon fiber, CARBOFLEX®, manufactured
in the U.S.A. is commercially
available.  A unique petroleum
pitch feedstock has been utilized
in a patented process to produce
a general purpose pitch-based
carbon fiber suitable for asbestos
replacement in selected applications.  A brief description of
the unique petroleum pitch feedstock,
the process by which the feedstock
is converted into a general purpose
carbon fiber, and some applications
of this fiber as an asbestos
replacement is presented.  Potential
applications for this multipurpose
carbon fiber possessing the properties of electrical conductivity,
chemical inertness, and the strength
of steel at one-fifth the weight
in the asbestos replacement area
include:  brakes (aircraft, auto,
truck, off-road); cement; pipe;
sheet; gaskets; seals; pump packing;
flooring (home, hospital, industrial); rust preventative coatings;
and furnace insulation.  Some
potential high technology applications for CARBOFLEX® carbon
fiber include x-ray tables,
space platforms, surgical implants,
storage batteries, health applications, thermoblast protection,
insulation, chemical warfare
filters, biological warfare filters,
disc antennas, space antennas,
machine parts, miscellaneous
sporting goods, computer housings,
anti-static brushes, lightening
dispersal, and surface activated
applications.

1.  INTRODUCTION

Due to its unique consistency,
A-240 petroleum pitch has been
given  considerable attention
in academic, government and industrial laboratories.[1, 2, 3, 4, 5,
6]
Perhaps the most important
development in carbon chemistry
took place in 1965 when two Austra-
lian coal researchers, Brooks and
Taylor, noticed, while examining

various coal samples under the polarized light microscope, that when the polarization of light was changed, areas of certain coals appeared optically active.[7] Subsequent research by Dr. J. L. White at the Aerospace Corporation and Dr. Harry Marsh at the University of Newcastle upon Tyne in England, and others at industrial research, government, and academic laboratories has shown that this observation explains the source of the structure seen in cokes and graphites. [8, 9] Prior to coking and graphitizing, the mesophase carbon sphere, or nematic liquid crystal, is formed by the coming together of large aromatic molecules. In the early stage of its formation, the process is reversible. Manufactured petroleum pitch is an excellent mesophase precursor.

The pioneering mesophase research of Brooks and Taylor has provided insight into the transformation of pitch and coke raw materials into high quality carbon and graphite electrodes and has led to advances such as thermally shock resistant graphite and carbon/graphite fibers made from manufactured petroleum pitch. [10, 11, 12] Manufactured petroleum pitch was introduced into nuclear reactor technology through the NERVA rocket engine development program sponsored by the U.S. government in the early 1970's.[13, 14, 15] A superior graphite based on A-240 petroleum pitch was developed and investigations carried out under this program showed that graphites ranging from anisotropic to isotropic could be made by combining various amounts of only two starting materials; manufactured petroleum pitch and a furan resin. Although the NERVA project was discontinued prior to flight testing, the value of manufactured petroleum pitch in nuclear applications was established. Shortly after the NERVA work was discontinued, a successful project was initiated at Oak Ridge National Laboratory to develop thermally shock resistant graphite for defense-related applications. Also at Oak Ridge in the late 60's, a research project was undertaken to develop an efficient method of manufacturing fuel sticks for the graphite-based, high temperature, gas cooled nuclear power reactor.[15, 16, 17] This work culminated in the successful development of a process in which manufactured petroleum pitch is utilized as the matrix for the nuclear fuel particles. Current research programs are focused on graphite utilization in the Tokamak fusion reactors.[18]

Considerable attention is being given in both academic and industrial laboratories to pitch-based carbon graphite fibers.[19]

Thomas Edison was an early pioneer in carbon fiber research. Early carbon and graphite fibers were made from rayon and polyacrylonitrile and, when first developed in the mid-60's, sold for more than $500/lb. Technological advances have dropped these prices in the range of $20/lb., and lower for some grades; still far too expensive for large scale usage in industries such as transportation and construction. The literature documents considerable progress both in the U.S. and Japan on developing a more cost effective carbon and graphite fiber starting with a relatively inexpensive pitch rather than organic fibers.[20] Resins such as polyesters, polysulfones, epoxies and a host of others (reinforced with carbon and/or graphite fibers) are being evaluated for applications such as noncorrosive tanks and doors, hoods, deck lids, fenders, bracket springs and engine components for automobiles and trucks.

## 2. DISCUSSION

A recent U.S. patent describes the use of A-240 petroleum pitch as a feedstock for the manufacture of general-purpose carbon fibers, trade named CARBOFLEX[TM].[21]

CARBOFLEX is the result of more than 11 years of research begun to explore new uses for A-240 petroleum pitch. The goal of this work was to produce a low cost carbon fiber for industrial applications including asbestos replacement. This multipurpose carbon fiber, while relatively inexpensive, has many of the unique properties of carbon, electrical conductivity, chemical inertness, the strength of steel at one-fifth the weight, and finds many potential applications where the high strength and modulus fibers have been priced out of the market. A multitude of potential applications are evident and include, in the asbestos replacement area: brakes (aircraft, auto, truck, off-road); cement (pipe, sheet); gaskets, seals (pump, packing); flooring (home, hospital, industrial); rust preventative coatings (Tectyl[TM] type, asphalt, asbestos cement); furnace insulation. In the electrically conductive plastics area: shielding for electronic equipment (computer/calculator cases); applications requiring electrostatic painting (auto, etc.). In the carbon/carbon composites area: electrodes, aerospace, high performance brakes. In the textiles area: fire resistant applications, and active carbon fibers for military protective clothing. Unique physical and chemical properties of CARBOFLEX which make them attractive for these applications include: light weight, electrical conductivity, chemical resistance, heat resistance, dimensional stability and abrasion resistance.

Potential applications for general-purpose carbon fibers which could be classed "high tech" include: X-ray tables, space platforms, surgical implants, storage batteries, health applications, thermal blast protection, chemical warfare filters, biological warfare filters, dish antennas, space antennas, machine parts, miscellaneous sporting goods, computer housings, antistatic brushes, lightening dispersal, and surface activated applications.

CARBOFLEX® carbon fibers are available in various forms including mat, chopped and milled; and many other forms (yarns, felts, wovens, papers) are being developed. Individual fiber properties include tensile strength in the 100,000 psi range, tensile modulus in the 7MM psi range, and fiber diameters of about 12 microns.

Composite materials are becoming of great interest in industrial as well as aerospace applications, and combining carbon fibers with engineering plastics are of special interest.

As a result of considerable research, a new commercial product, AEROCARB®, has also been introduced. AEROCARB® is a highly advanced pitch product exhibiting a high carbonizing content. AEROCARB® is suitable for specialty electrode, specialty graphite, and carbon/carbon fiber composite applications. AEROCARB® has an initial boiling point above 1000°F, melts at 400°F, and has a char yield (ASTM D-2416) of about 75 wt.%. By utilizing the high carbon yield of AEROCARB®, the carbon/carbon composite manufacturer can realize considerable savings versus conducting multiple impregnations with lower carbon yielding material. AEROCARB® has a sharp viscosity curve resembling a pure compound. It can be readily handled at proper temperatures and, if desired, can be readily converted into high quality mesophase. It is this unusual ability to convert to mesophase, and from there on into a high performance graphite, which has caused AEROCARB® to be of commercial interest to many companies.

Advanced composite materials include carbon/carbon nosetips used on missiles, rocket nozzles, high performance brakes, and oxidation resistant turbine wheels. The availability of a low cost, high yield carbonizing material in combination with a low cost carbon fiber, provides industry with materials that might be utilized in new ways to make high volume products such as inexpensive carbon/carbon fiber composite brakes for off-road vehicles, trucks, and passenger car use.[22] It may be possible to produce an inexpensive brake utilizing these new

941

high technology materials which
will last the life of the vehicle.

CARBOFLEX® is currently being
used as a high performance, cost
effective replacement for asbestos
in industrial brakes, and its
performance is outstanding.  In
addition, AEROCARB® promises to
provide the brake industry with
an essential ingredient of cost
effective, carbon/carbon fiber
brakes for both industrial and
automotive applications.  Racing
cars are already using carbon/carbon
composite brakes.  Although of
high cost, they represent the
leading edge of aerospace technology applied to civilian use.

## 3. CONCLUSION

Carbon technology has evolved
in a relatively short span of
years from the black art and
laboratory curiosity to an exact
science that will make a major
impact on many different industries in the future.  Petroleum
pitches are new routes to advanced
solid materials which will be
the cornerstone of many industries
of the future.

## 4. REFERENCES

(1) Newman, J. W., "What is
    Petroleum Pitch?," ACS
    Symposium Series Number 21,
    "Petroleum Derived Carbons,"
    ACS 1975, p. 52.  Deviney,
    M. L., O'Grady, T. M.

(2) Ball, G. L., et al., "Petro-
    leum Pitch--A Major New
    Carbon Source," Proceedings
    of Sessions - 101st AIME
    Annual Meeting, San Francisco,
    California, February, 1972,
    Published by AIME, 345 East
    47th Street, New York,
    NY 10017.

(3) Smith, W. E., et al., "Charac-
    terization and Reproducibility
    of Petroleum Pitches," Oak
    Ridge Y-12 Plant, Published
    by U.S. Department of Commerce,
    Report No. Y-1921.

(4) Newman, J. W., "Petroleum
    Pitch - A Consistent Carbon
    Precursor," 11th Biennial
    Conference on Carbon, Extended
    Abstracts, 1973, p. 104,
    Published by NTIS, U.S.
    Department of Commerce,
    Springfield, Virginia 22151.

(5) Grint, A., Marsh, H., and
    Sweitlick, U., "The Modifi-
    cation of the Carbonization
    of Coals by the Addition
    of Ashland A-200 Petroleum
    Pitch," 14th Biennial Con-
    ference on Carbon
    Abstracts, 1979.

(6) Alexander, C. D., Bullough,
    V. L., and Pendley, J. W.,
    "Laboratory and Plant
    Performance of Petroleum
    Pitch," paper presented at
    TMS-AIME Fall Meeting, New
    York City, 1971, Paper
    No. A71-29.

(7) Brooks, J. D. and Taylor,
    G. H., Carbon, 3, p. 185, 1965.

(8) White, J. L., et al., "Mecha-
    nisms of Formation of Needle
    Coke," 12th Biennial Confer-

ence on Carbon Extended Abstracts, 1975, p. 221.

(9) Marsh, et al., 14th Biennial Conference on Carbon Abstracts, 381, 1979.

(10) Kennedy, C. R. and Eatherly, W. P., "The Development of Thermal Shock Resistant Graphite," 11th Biennial Conference on Carbon Abstracts, p. 131.

(11) Horne, O. J. and Kennedy, C. R., "Graphites Fabricated From Green Petroleum Pitch Cokes," 12th Biennial Conference on Carbon Extended Abstracts, 1975, p. 221.

(12) Didchenko, R., et al., "High Modulus Carbon Fibers from Mesophase Pitches, Parts One and Two," 12th Biennial Conference on Carbon Extended Abstracts, 1975, pp. 329-332.

(13) Overholser, L. G., "Nerva Fuel Element Development Program Summary Report - July, 1966 through June, 1972," Oak Ridge Y-12 Plant, Published by U.S. Department of Commerce, Report No. Y-1857.

(14) Bradley, R. A., and Sease, J. D., "The Slug Injection Process for Fabricating HTGR Fuel Rods," 11th Biennial Conference on Carbon Abstracts, p. 239.

(15) Newman, J. W., and Sawran, W. R., "Manufactured Petroleum Pitch for Nuclear Applications," Carbon '76

Preprints, Deutsche Keramische Gesellschaft, Baden-Baden, 1976.

(16) Hamner, R. L., Robbins, J. M., and Coobs, J. H., "Development of Continuous-Matrix Fuel Rods for Advanced High-Temperature Gas-Cooled Reactors," 11th Biennial Conference on Carbon Abstracts, NTIS, U.S. Department of Commerce, 1973.

(17) Hamner, R. I., Robbins, J. M., and Eatherly, W. P., "Properties of Extruded Fuel Rods for HTGR Applications," 14th Biennial Conference on Carbon Abstracts, The American Carbon Society and the Pennsylvania State University, University Park, PA., 1979.

(18) Langley, R. H., and Eatherly, W. P., "Graphite Application in Tokamak Fusion Reactors," 14th Biennial Conference on Carbon Abstracts, The American Carbon Society and the Pennslyvania State University, University Park, PA., 1979.

(19) Netherlands Patent Application 73-04398, "Process for Preparation of Carbon Fibers from Pitch in the Mesophase State with a High Modulus of Elasticity and a Great Strength," Date open to inspection October 2, 1973.

943

(20) Hettinger, W. P. Jr., "New Low Cost Pitch-Based Carbon Fibers and Carbonizing Materials," Presented at New Directions in Carbon and Graphite and Other High-Performance Fibers and Fabrics Conference at Clemson University, Clemson, South Carolina, February 4-5, 1986.

(21) Sawran, W. R., et al., "Process for the Manufacture of Carbon Fibers," U.S. Patent 4,497,789, February 5, 1985.

(22) Hettinger, W. P., Jr., Newman, J. W., Krock, R. P., and Boyer, D. C., "CARBOFLEX® and AEROCARB® - Ashland's New Low Cost Carbon Fiber and Carbonizing Products for Future Brake Applications," presented at the SAE 1986 Earthmoving Industry Conference, Peoria, Illinois, April 7-9, 1986.

## 5. BIOGRAPHY

John W. Newman, a graduate of the University of Cincinnati with a degree in Chemical Engineering, is Vice President of the Carbon Fibers Division of Ashland Petroleum Company, a Division of Ashland Oil, Inc. He joined Ashland Petroleum in 1956 and has served in a number of capacities including Analytical Laboratory Supervisor, Pilot Plant Engineer, Project Evaluator, Senior Research Engineer, Group Leader and Research Manager.

Mr. Newman has many years of experience in the carbon technology area, active in the development and commercialization of Ashland 240 petroleum pitch and CARBOFLEX® pitch-based carbon fibers. He is a coinventor on nine U.S. patents in the areas of petrochemicals and carbon technology, and has presented papers before various technical organizations including The American Carbon Society, The American Chemical Society, AIChE, and AIME. He is a coeditor of "Petroleum-Derived Carbons," a book published in 1986 by The American Chemical Society.

FROM PAN-BASED PRECURSOR POLYMERS TO CARBON FIBERS:
EVOLUTION OF STRUCTURE AND PROPERTIES

A. S. Abhiraman
School of Chemical Engineering
Georgia Institute of Technology
Atlanta, GA 30332

## Abstract

The structure and properties of carbon fibers are determined by the nature of the precursor polymers and the evolution of morphology through the three major stages of processing, viz., precursor fiber formation, oxidative stabilization and carbonization. Critical interactions exist between the processing conditions and the physico-chemical changes that occur at each stage. This presentation will attempt to provide a comprehensive view of the PAN-based carbon fiber process which includes

1. Morphology and relevant morphological parameters in PAN-based precursor fibers.

2. Stress-Environment-Material interactions during oxidative stabilization.

3. A mathematical model of the kinetics of oxidative stabilization.

4. Evolution of structure and properties in carbonization.

Methods by which high lateral and orientational order can be obtained in the precursor and the processing conditions which can minimize orientational relaxation during stabilization will be presented. Prospects for new PAN-based precursor materials and fiber formation processes will also be discussed.

---

Carbon/graphite fiber-reinforced composites have emerged as the most important among advanced structural materials, with physical and mechanical properties to meet a variety of highly specialized needs. Among the major commercial precursors for these fibers, cellulose-, pitch- and polyacrylonitrile (PAN)-based, the PAN-based precursor fibers have assumed a dominant position. Recent fundamental investigations pertaining to the technology of precursor fiber formation have yielded acrylic fibers of significantly higher order which are being converted to carbon fibers of superior mechanical properties in commercial processes.

Many isolated aspects of the conversion of acrylonitrile based precursors to carbon fibers have been studied and reported in the literature (1). However, a high degree of empiricism exists still in relating various material and process contributions to the structure and properties of carbon fibers. The research effort in our laboratories is aimed at minimizing empiricism and improving fundamental knowledge of the evolution of structure and properties of these critical high performance fibers.

The primary objective in our research is to provide rational directions for advance in precursor structures and process configurations to extend the range of properties which can be obtained in polyacrylonitrile-based carbon fibers. The emphasis is on the chemical and morphological evolution from precursor polymers, through fiber formation, drawing and oxidative stabilization, to carbonization. Recent research in our laboratories has yielded a number of significant results pertaining to the morphology of acrylic precursor fibers, enhancement of precursor order and chemical and morphological changes through solid state oxidative stabilization and low temperature carbonization (2-10). These results are summarized in the following.*

## 1. Model for Precursor Morphology

A comprehensive set of evidence (Table 1) based on x-ray scattering (WAXD and SAXS), thermal analysis, thermal deformation and stress responses, and sonic pulse propagation points clearly to the presence of a connected sequence of oriented laterally ordered and oriented but laterally disordered domains, which is the structure proposed by Warner, et al. (11) for fibrils in oriented acrylic fibers.

## 2. High Order Through High Temperature Deformation

When precursor acrylic fibers are exposed rapidly to temperatures in the stabilization range, significant morphological rearrangements and changes in mechanical properties are observed well before the onset of reactions, in less than 10 seconds. The rearrangements reveal a spontaneous tendency toward increase in lateral order. In the absence of a significant level of tensile stress, it is also accompanied by large scale disorientation of the laterally disordered fraction. Based on the segmental mobility and the tendency toward increase in lateral order, a high temperature deformation process has been proposed to generate highly ordered precursor fibers. For example, the data in

---

* Experimental details and extensive data will be presented at the conference.

Table 2 show clearly the significantly higher orientational and lateral order which can be obtained by deformation through a high temperature oven of a fiber which has been only partially drawn through a hot water bath. The high temperature deformation step proposed by us has been utilized in commercial production to yield carbon fibers of significantly superior mechanical properties.

## 3. Multi-Zone Oxidation

Based on chemical as well as morphological considerations, multi-zone oxidation with independent control of stress and environment in each zone has been proposed by us for maximizing the translation of orientational order from precursor fibers to carbon fibers through the stabilization step. Initial batch experiments indicated clearly that a higher level of orientational order and mechanical properties could be retained with higher levels of tensile stress in stabilization (see, for example, figure 1). Also, the maximum stress level which can be applied advantageously changes during the course of stabilization, suggesting the need for a multi-zone process with independent control of stress in each zone. It may also be advantageous to provide a sequence of inert and oxidizing atmospheres during the course of stabilization. With the combination of internally initiated reactions and those initiated by the species arising from

diffusion-controlled incorporation of oxygen, an inhomogeneous shrinkage stress distribution will result in the fiber cross-section. Also the maximum extent to which ladder sequences can be formed through intramolecular nitrile polymerization can be shown to decrease with incorporation of oxygen before nitrile polymerization is completed. Such inhomogeneities can be reduced through nitrile polymerization under inert atmosphere in the initial stages, followed by stabilization in air.

A four-zone stabilization line, with computer control of stress/deformation in each zone, has been constructed to facilitate experimental investigation of multi-stress, multi-environment stabilization (Figure 2).

## 4. Mathematical Model of Oxidative Stabilization

Developing a mathematical model of solid-state oxidative stabilization of PAN-based precursors is extremely complex because of the multitude of events that occur in this process. Among the factors to be considered in this process are (i) initiation reactions by different species in the precursor polymer, such as comonomers and defect structures, (ii) reactions initiated by species from reaction of oxygen with the backbone, (iii) multiple options for reaction paths, (iv) intra- and inter-molecular reactions between similar species, (v) transport of species such as $O_2$, *OH, etc., and (vi) morphological changes and constraints on molecular mobility accompanying the reactions, which should alter the rate constants for diffusion reactions. If all of the possibilities are considered, it leads to a large number (>30) of coupled partial differential equations, with the associated boundary conditions and material constants (9). We have reduced this general set of equations to five equations by lumping similar reactions together and developed a numerical procedure for solving them. Trial solutions of this simplified set have been obtained with estimates of rate constants from published information. The predicted responses have been of the same order of magnitude as those observed in stabilization processes. Experiments are being conducted to compare theoretical predictions with global and, if possible, local concentrations of major reaction species. It is hoped that this procedure would lead ultimately to elimination of the essentially trial and error methods used currently for establishing stabilization conditions.

## 5. Continuous Low-temperature (1200°C) Carbonization

Several aspects pertaining to the evolution of structure and properties in low temperature (1200°C) carbonization have been studied. These are

(i) Two different mechanisms have been recognized for the hollow core which can occur in carbonization. One is through burning-off an incompletely stabilized core due to inadequate combination of time and temperature in diffusion-controlled stabilization. The other can be due to consolidation of structure inwards from the skin when a well stabilized fiber is raised rapidly to carbonization temperatures. Whether such dual mechanisms are operational at carbonization temperatures higher than 1200°C remains to be explored (8,10).

(ii) Among the parameters which may be examined as criteria for adequate stabilization, density has been observed to be a consistent parameter which can also be measured easily. Attaining a composition-dependent critical density, independent of precursor filament size and morphological parameters, appears to be necessary in order to avoid core blow-out in carbonization (8,10).

(iii) By analyzing filaments which are withdrawn rapidly from the feed end of a continuous carbonization process, the paths for evolution of the chemical and morphological structures in carbon fibers have been studied (8,10). For example, results from low temperature carbonization indicate "aromatization (inferred from H/C ratio) - basal plane formation (from N/C ratio) - development of strength and stiffness" to be the sequence in the evolution of structure and properties.

(iv) Apparent bulk density, measured by immersion techniques, increases rapidly to a maximum through the initial stages of carbonization, but then decreases rapidly to a lower asymptotic value (see, for example, figure 3). This drop is not accompanied by corresponding changes in linear density and diameter through the course of carbonization. The mechanism for the apparent drop in density appears to be related to conversion of some of the open or accessible pores to closed ones through consolidation of the carbonized structure around them.

## REFERENCES

1. Jain, M.K., and Abhiraman, A.S. J. Mater. Sci. (in press).
2. Jain, M.K., and Abhiraman, A.S. J. Mater. Sci. $\underline{18}$, 179 (1983).
3. Abhiraman, A.S., Balasubramanian, M., and Tincher, W.C., Proceedings of the XVI Biennial Conference on Carbon, 497 (1983).
4. Jain, M.K., Desai, P., and Abhiraman, A.S., p 517 of ref. 3.
5. Jain, M.K., et al., Proceedings of the XVII Biennial Conference on Carbon, 310 (1985).
6. Balasubramanian, M., Jain, M.K., and Abhiraman, A.S., p. 312 of ref. 5.
7. Jain, M.K., et al., J. Mater. Sci., (in press).
8. Balasubramanian, M., et al., (to be published).
9. Grove, D., Mathematical model of solid state thermo-oxidative stabilization of acrylonitrile-based precursors to carbon fibers, M.S. Thesis, Georgia Institute of Technology (1986).
10. Jain, M.K., Physical and morphological changes during the conversion of acrylonitrile-based precursors to carbon fibers, PhD Thesis, Georgia Institute of Technology (1985).
11. Warner, S.B., Uhlmann, D.R., and Peebles, L.H., Jr., J. Mater. Sci. $\underline{14}$, 1983 (1979).

## Biography

Professor A. S. Abhiraman was granted his PhD in Fiber and Polymer Science from N.C. State University in 1975. He worked in American Enka Co. till 1979 when he joined Georgia Institute of Technology. His current fields of interest are formation of high performance fibers, especially carbon, graphite and ceramic fibers, thermodynamics and kinetics of crystallization in polymers, fiber formation, polymer rheology, fiber-matrix interaction in composites, etc.

## Acknowledgements
The study was supported by the Office of Naval Research.

TABLE 1. MORPHOLOGICAL MODEL AND DRAWN PRECURSOR FIBER

| OBSERVATIONS | INFERENCES |
|---|---|
| 1. WAXD Pattern | Presence of an Oriented Laterally Ordered Phase |
| 2. Enthalpy Changes in Plasticized Heating | The Laterally Ordered Phase Consists of "True" Crystals i.e., Products of First Order Transition |
| 3. Spontaneous Shrinkage upon Free Annealing Without Loss of Orientation in the Laterally Ordered Phase<br><br>4. Development of Thermal Stress Upon Constrained Annealing<br><br>5. Large Spontaneous drop in Sonic Modulus ONLY when Shrinkage is Allowed During Annealing | Presence of an Oriented but Less Ordered Phase with Chain Segments Bridging the Laterally Ordered Crystals Along the Fiber Direction |
| 6. Constant Density of Fibers with Different Extents of Lateral Order | Ordered and Disordered Phases of the Same Density (?) |
| 7. SAXS after Diffusion of Electron-dense Molecules | Repeating Sequence of Oriented Laterally Ordered (LO) and Laterally Disordered (LD) Phases |

TABLE 2. HIGH TEMPERATURE DRAWING AND PROPERTIES
OF PRECURSOR FIBERS

|  | HWD | HTD |
|---|---|---|
| Jet Stretch | 0.7 | 0.9 |
| Draw Ratio (Hot Water) | 7.1 | 3.0 |
| Draw Ratio (High Temperature) | --- | 2.3 |
| Oven Temperature ($^{\circ}$C) | --- | 252 |
| Denier/filament | 1.6 | 1.4 |
| Sonic Modulus (g/denier) | 120 | 180 |
| Crystal Orientation function | 0.7 | 0.92 |
| Average Crystal Size (nm) | 5.4 | 13.0 |

HWD : Drawing in Boiling Water
HTD : High Temperature Drawing

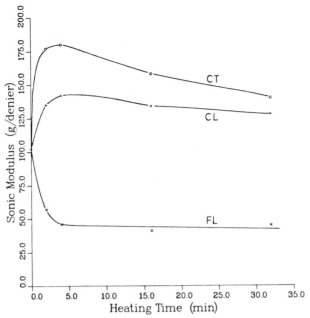

Figure 1.  Progression of Sonic Modulus during
           stabilization at 265°C (FL-free length;
           CL - Constant length; CT - Constant
           tension of 0.1 g/denier)

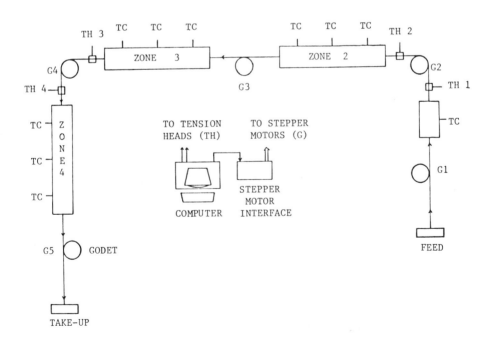

Figure 2.   Schematic of Multi-Stage Stabilization Line.

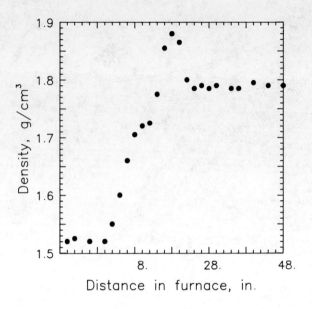

Figure 3.  Progression of Density during Carbonization.

NEW HIGH PERFORMANCE DOMESTICALLY PRODUCED CARBON FIBERS
Ian M. Kowalski
Amoco Performance Products Inc.
Parma Technical Center, 12900 Snow Road, Parma, Ohio  44130

## Abstract

Recent technical advances have led to improved properties and product forms available in THORNEL® pitch-based and PAN-based carbon fibers. Since the mid-70's, Amoco Performance Products, Inc.* has been producing pitch-based carbon fibers in a fully integrated facility in Greenville, South Carolina. In 1981, the first totally domestic PAN-based carbon fiber facility started production of T-300. The advances in pitch-based carbon fibers include ultra-high modulus (120 Msi), lower denier (500 filament tow) for thin space structure, high thermal conductivity (640 watt/m K), and low CTE. The numerous advances in PAN-based carbon fibers include T-650/42 intermediate modulus fiber, T-500 (560 ksi tensile strength), and fibers with superior oxidation resistance (T-300R, T-40R). Composite properties for both the PAN-based and pitch-based carbon fibers are reported in state-of-the-art epoxy resins and damage tolerant epoxies.

* Amoco acquired the Specialty Polymers and Composites Division from Union Carbide Corporation on June 20, 1986, and formed a new division, Amoco Performance Products, Inc.

THORNEL® is a registered trademark of Amoco Performance Products, Inc.

## 1.  INTRODUCTION

New and improved PAN-based and pitch-based carbon fibers are being developed by Amoco Performance Products, Inc.  These improvements encompass many areas including strength, modulus, thermal conductivity and oxidation resistance.  Through an increased understanding of and an ability to control the entire carbon fiber manufacturing process, carbon fibers with superior properties are being produced.

Carbon fiber properties are determined by their structure. The structure of the carbon fiber is developed throughout the process as controlled by the manufacturing parameters. This paper highlights the key areas of the process which determine the structure of the carbon fiber. Properties of new high performance carbon fibers and their composites are given.

2. THE CARBON FIBER PROCESS

Carbon fiber is typically made from one of three precursor materials: rayon, polyacrylonitrile (PAN) or mesophase pitch. PAN is the precursor of choice for very high strength carbon fiber while pitch precursor is used for very high modulus carbon fiber. Rayon-based carbon fibers, the earliest structural carbon fibers, are used chiefly in carbon/carbon applications and will not be discussed here.

The first step in the process is to polymerize the precursor material. This involves thermally or chemically treating petroleum pitch to form mesophase (liquid-crystalline) pitch or, in the case of the PAN process, the free radical polymerization of acrylonitrile (AN) and suitable comonomers to form PAN. The next step is to spin the precursor polymer into a multifilament strand, typically containing between 1,000 and 12,000 individual filaments. Mesophase pitch is easily melt spun, while PAN is typically dry or wet spun in solution. Single or multiple drawing stages are used to attenuate the fiber and improve the orientation of the polymer chains along the fiber axis. The drawing step is straightforward for the pitch-based fibers which remain soft from the residual heat of melt spinning. The drawing and consolidation of PAN-based fibers requires multiple stages and precise control of operating parameters. The resultant white fiber precursor visually resembles fiberglass roving.

The fibers are next thermally stabilized via an oxidation process which employs temperatures up to about 400°C. This involves crosslinking of the polymer chains that infusibilizes the fiber so that it will not melt when subjected to increasingly higher temperatures. The PAN-based fiber changes from white to black during stabilization. The next step of the process is carbonization which takes place in an inert atmosphere at temperatures between 1000°C and 2000°C. To achieve very high modulus, a graphitization step at 2000°C to 3000°C is used. The carbon or graphite fibers have achieved their final strength, modulus, and density at this stage of the process. To aid bonding of the fibers to composite matrix materials, a surface treatment is performed that removes a weak surface layer from the fiber and functionalizes the fiber surface to enhance chemical bonding with the matrix. The final steps in the process are the application of a sizing to aid handleability and winding of the fiber on spools.

## 3. CARBON FIBER STRUCTURE

There are three major structural parameters of a carbon fiber: orientation, crystallinity, and defect content (Figure 1). Orientation is an average measurement of how well the graphite layer planes are aligned with the axis of the fiber. Crystallinity is a lumped measurement of both crystalline perfection and crystallite size. The degree of perfection is related to how close the parallel spacing of the graphite layer planes approaches the theoretical value of 3.35 angstroms. The crystallite size is the average stack height of parallel graphite layer planes. Both orientation and crystallinity are measured by X-ray diffraction. Defect content of carbon fibers is easily seen by examining the fracture surfaces of carefully

broken single filaments. Defects can be internal (voids, inclusions) or external (pits, gouges, extraneous material).

### 3.1 Effect of Fiber Structure on Fiber Properties

Orientation in carbon fibers has a major effect on the final carbon fiber properties. In general, increasing the degree of orientation improves the longitudinal properties and decreases the transverse properties. Specifically, longitudinal tensile strength, tensile modulus, thermal conductivity, electrical conductivity and the negative CTE are increased. In the transverse direction, both tensile strength and tensile modulus decrease.

As the degree of crystallinity is increased many carbon fiber properties are affected. The following properties increase: thermal conductivity, electrical conductivity, the longitudinal negative CTE, and oxidation resistance. The following properties decrease: longitudinal tensile strength, compressive strength and shear modulus, as well as the transverse strength and modulus.

The removal of defects from carbon fibers has only positive effects. As defects are removed, tensile strength, thermal conductivity, electrical conductivity, and oxidation resistance are all increased.

## 3.2 Controlling Carbon Fiber Structure

The carbon fiber structural parameters are controlled by a number of processing steps. The fiber orientation is controlled by the fiber spinning (pitch-based), fiber drawing (PAN-based), and the heat treatment. The crystallinity is governed by the precursor chemistry and the heat treatment. The defect content is controlled by the precursor polymer purity, and throughout the process, by careful handling and protection from contaminants.

In order to produce the desired carbon fiber structure and, thereby, control the final properties, it is important to have control over all of the process elements from incoming polymer through spinning, drawing, and final heat treatments.

## 4. PAN-BASED CARBON FIBERS

Amoco has been producing PAN-based carbon fiber in its Greenville, SC facility since 1981. It is a fully integrated plant (AN monomer to carbon fiber) having a nominal carbon fiber capacity of approximately 1,000,000 pounds a year.

Products for sale include T-300, T-500, and intermediate modulus T-650/42. Table I summarizes the average fiber certification properties of recent PAN-based fiber production. T-40 intermediate modulus carbon fiber and T-50 high modulus carbon fiber produced by Toray Industries, Inc. in Japan and marketed by Amoco in America are included. All of the PAN-based fibers have been surface treated. Typical composite properties of these fibers are given in Table II. The tensile strength is measured in a brittle MY-720/DDS aerospace type 350°F cure resin as well as a ductile Amoco ERL-1962 epoxy which gives superior tensile strength translation. The difference between the tensile modulus and compressive modulus is due to the nonlinearity of the carbon fiber stress-strain curve. Carbon fiber modulus reversibly increases as the fiber is tensioned[1] and reversibly decreases as the fiber is compressed.

In December, 1986, Amoco introduced T-650/42, the first totally domestic high strain intermediate modulus carbon fiber. The typical fiber and composite properties are given in Table III. This fiber is certified to have a minimum lot average tensile strength of

4480 MPa (650 ksi) with typical strengths being much higher.

## 5. PITCH-BASED CARBON FIBER

Amoco has been producing pitch-based carbon fiber in its Greenville, SC facility since the mid-70's. The plant has a nominal carbon fiber capacity of 500,000 pounds a year. Products for sale include P-25, P-55S, P-75S, and ultra-high modulus P-100 and P-120. The ultra-high modulus fibers are also available with surface treatment. Table IV gives the average fiber certification properties of recent pitch-based fiber production. Typical composite properties of these fibers are given in Table V.

## 6. CARBON FIBERS FOR UNIQUE APPLICATIONS

Several carbon fibers have been developed to fill unique needs in new and existing applications.

### 6.1 Oxidative Stability

It is generally true that the oxidation resistance of carbon fibers increases as carbon assay is increased. Two new domestic fibers, T-300R and T-40R, have been produced with the goal of keeping their modulus the same while significantly increasing their oxidation resistance. Table VI gives the oxidation resistance of these and other carbon fibers in air at 316°C

(600°F) and 375°C (707°F).[2-3] Standard T-300 is recommended for service temperatures up to 232°C (450°F). T-300R and T-40 are suited for polymide use temperatures up to 316°C (600°F). T-40R is suitable for use temperatures up to 427°C (800°F). T-300R has the same mechanical properties as T-300. The current strength of T-40R is 3450 MPa (500 ksi) with a future strength goal of 4140 MPa (600 ksi) at the current 290 GPa (42 Msi) modulus.

### 6.2 Low Denier Fiber for Thin Structures

There are several space applications requiring ultra thin prepreg for making thin wall multi-axial composite structures. Low denier fibers have been developed to enhance spreading during thin prepreg processing. Both P-100 and P-75 with 1000 filaments/tow and 500 filaments/tow are available. The typical strength and modulus levels of the standard 2k tow count are maintained. In the case of P-75, prepreg with a thickness of only 0.035 mm (0.0013 inch) has been successfully made.

### 6.3 High Thermal and Electrical Conductivity

In general, as the modulus of carbon fibers increases both the thermal and electrical conductivity increase. This is primarily due to the increased orientation and crystallinity of the fibers. Figure 2 shows the relationship between these two properties and compares carbon fiber with metals.[4-5] The P-120 pitch-based carbon fiber has a thermal conductivity of 640 watt/mK, about 1.6 times that of copper. When normalized to account for the density differences, the specific thermal conductivity of P-120 is over six times that of copper. An experimental 965 GPa (140 Msi) modulus fiber has a thermal conductivity twice that of copper. Efforts are underway to further increase the thermal conductivity of pitch-based graphite fiber to approach that of highly oriented pyrolytic graphite (HOPG).

6.4 Zero CTE Structure

Space structure has a requirement of withstanding large thermal gradients while maintaining dimensional stability. Through the use of carbon fibers with negative coefficients of thermal expansion (CTE), composites with near zero CTE in critical directions can be designed. The various PAN-based and pitch-based carbon fibers offer a range of CTE's that can be used to achieve zero CTE structure (Table VII),[6] when they are combined with a matrix having a positive CTE.

7. ACKNOWLEDGMENTS

The author gratefully acknowledges the valuable contributions of several colleagues who supplied data for this paper: R. Bacon, B. H. Eckstein, C. D. Levan and D. A. Schulz.

8. REFERENCES

1. Kowalski, I. M., "Characterizing the Tensile Stress-Strain Nonlinearity of PAN-Based Carbon Fibers," ASTM 8th Symposium on Comp. Mat. Testing and Design, April 29, 1986, In Press.

2. Eckstein, B. H., Fiber Sci. and Tech. 14, 139 (1981).

3. Eckstein, B. H., Proc. 18th Int'l. SAMPE Tech. Conf., Seattle, WA, 1986, pp. 149-160.

4. Schulz, D. A., "Advances in Ultra-High Modulus Carbon Fibers," 16th Nat'l. SAMPE Tech. Conf. (1984).

5. Schulz, D. A., "Advances in UHM Carbon Fibers: Production, Properties, and Applications," SAMPE Journal, In Press.

6.  Bacon, R., et al., "Thermal
    Expansion Behavior of
    Unidirectional Composites of
    Ultra-High Modulus Carbon
    Fibers in Epoxy Resin," First
    European Conf. on Comp. Mat.,
    European Assoc. for Comp.
    Mat., Bordeaux, France,
    Sept. 25, 1985.

## 9. BIOGRAPHY

Ian M. Kowalski is a Development
Scientist in the Carbon Fiber
Applications Group with Amoco
Performance Products at their
Parma, Ohio, R & D Center.  He
received a BS in Mechanical
Engineering from the
University of Rhode Island and an
MS from Rensselaer Polytechnic
Institute.  Ian is Manager of the
Fiber Composites Laboratory at
Parma.

Table I:  PAN-Based Carbon Fiber Property Uniformity

|        | Strength MPa (ksi) | CV   | Modulus GPa (Msi) | CV   | Density (gm/cc) | CV   | Yield (gm/m) | CV   |
|--------|--------------------|------|-------------------|------|-----------------|------|--------------|------|
| T-300  |                    |      |                   |      |                 |      |              |      |
| 3k     | 3650 (529)         | 1.0% | 230 (33.3)        | 1.0% | 1.78            | 0.2% | 0.200        | 0.5% |
| 6k     | 3690 (535)         | 2.5% | 227 (33.0)        | 0.8% | 1.76            | 0.4% | 0.399        | 0.4% |
| 12k    | 3440 (499)         | 2.7% | 231 (33.5)        | 1.0% | 1.77            | 0.4% | 0.801        | 0.7% |
|        |                    |      |                   |      |                 |      |              |      |
| T-500  |                    |      |                   |      |                 |      |              |      |
| 3k     | 3860 (560)         | 0.4% | 245 (35.5)        | 0.6% | 1.78            | 0.1% | 0.195        | 0.2% |
| 6k     | 3930 (570)         | 0.5% | 253 (36.7)        | 0.4% | 1.78            | 0.1% | 0.401        | 0.1% |
| 12k    | 3790 (550)*        | 3.2% | 243 (35.2)        | 1.3% | 1.79            | 0.3% | 0.781        | 1.4% |
|        |                    |      |                   |      |                 |      |              |      |
| T-40*  |                    |      |                   |      |                 |      |              |      |
| 12k    | 5670 (822)         | 1.7% | 290 (42.0)        | 1.1% | 1.81            | 0.2% | 0.445        | 1.5% |
|        |                    |      |                   |      |                 |      |              |      |
| T-50*  |                    |      |                   |      |                 |      |              |      |
| 3k     | 2900 (420)         | 4.8% | 394 (57.2)        | 0.8% | 1.80            | 0.2% | 0.182        | 0.6% |
| 6k     | 2850 (413)         | 5.2% | 392 (56.9)        | 0.7% | 1.82            | 0.2% | 0.365        | 0.3% |

* Toray produced carbon fiber.

Table II:  Typical PAN-Based Carbon Fiber Composite Properties

| | T-300 | T-500 | T-40* | T-50* |
|---|---|---|---|---|
| **0° Tensile (Brittle Resin)** | | | | |
| Strength MPa (ksi) | 1860  (270) | 2030  (295) | 2860  (415) | 1310  (190) |
| Strain % | 1.30 | 1.30 | 1.53 | 0.53 |
| Modulus GPa (Msi) | 138 (20.0) | 150 (2.17) | 175 (25.2) | 240 (35.0) |
| **0° Tensile (Ductile Resin)** | | | | |
| Strength MPa (ksi) | 2140  (310) | 2410  (350) | 3240  (470) | 1410  (205) |
| Strain % | 1.50 | 1.50 | 1.70 | 0.56 |
| Modulus GPa (Msi) | 138 (20.0) | 150 (2.17) | 175 (25.2) | 240 (35.0) |
| Major Poisson's Ratio | 0.31 | 0.33 | 0.32 | 0.28 |
| **0° Compression** | | | | |
| Strength MPa (ksi) | 1720  (250) | 1720  (250) | 1720  (250) | 965  (140) |
| Modulus GPa (Msi) | 125 (18.1) | 135 (19.6) | 155 (22.6) | 230 (33.5) |
| **SBS MPa (ksi)** | 120   (17) | 120   (17) | 120   (17) | 75   (11) |

- Properties are normalized to 62% fiber volume.
- Tensile modulus is a secant from 0.1% to 0.6% strain except for T-50 which is 0.1% to 0.2% strain.
- Compression modulus is a secant from -0.1% to -0.3% strain except for T-50 which is initial tangent.

* Toray produced carbon fiber.

Table III:  Typical Properties of T-650/42 12k
Intermediate Modulus Carbon Fiber

| Tensile Strength | = | 4930 MPa  (730 ksi) | CV | = | 1.2% |
|---|---|---|---|---|---|
| Tensile Modulus | = | 286 GPa (41.5 Msi) | CV | = | 0.6% |
| Density | = | 1.784 gm/cc | CV | = | 0.1% |
| Yield | = | 0.4403 gm/m | CV | = | 0.6% |

| | Brittle Resin | Ductile Resin |
|---|---|---|
| **0° Tensile Resin** | | |
| Strength MPa (ksi) | 2430  (352) | 3080  (446) |
| Strain % | 1.37 | 1.66 |
| Modulus GPa (Msi) | 169 (24.5) | 174 (25.2) |
| **0° Compression** | | |
| Strength MPa (ksi) | 1670  (242) | 1630  (237) |
| Modulus GPa (Msi) | 152 (22.1) | |
| **90° Tensile** | | |
| Strength MPa (ksi) | 70 (10.2) | |
| Strain % | 0.74 | |
| Initial Modulus GPa (Msi) | 10.1 (1.47) | |
| **SBS MPa (ksi)** | 125 (18.0) | |

- Properties are normalized to 62% fiber volume.
- Tensile modulus is a secant from 0.1% to 0.6% strain.
- Compressive modulus is a secant from -0.1% to -0.3% strain.

Table IV:  Pitch-Based Carbon Fiber Property Uniformity

|  | Strength MPa (ksi) | CV | Modulus GPa (Msi) | CV | Density (gm/cc) | CV | Yield (gm/m) | CV |
|---|---|---|---|---|---|---|---|---|
| P-55S |  |  |  |  |  |  |  |  |
| 2k | 1790 (260) | 2.8% | 384 (55.7) | 2.2% | 2.01 | 0.9% | 0.319 | 1.2% |
| 4k | 1765 (256) | 1.9% | 388 (56.3) | 2.8% | 2.01 | 0.8% | 0.694 | 1.3% |
| P-75S |  |  |  |  |  |  |  |  |
| 2k | 1905 (276) | 4.1% | 518 (75.1) | 2.9% | 2.03 | 1.5% | 0.311 | 0.3% |
| P-100 |  |  |  |  |  |  |  |  |
| 2k | 2455 (356) | 12.8% | 766 (111) | 3.2% | 2.16 | 0.4% | 0.313 | 2.0% |
| P-120 |  |  |  |  |  |  |  |  |
| 2k | 2240 (325) |  | 830 (120) |  | 2.18 |  | 0.318 |  |

The "S" suffix indicates that the fiber has been surface treated.

Table V:  Typical Pitch-Based Fiber Composite Properties

|  | P-55S | | P-75S | | P-100S | | P-120 | |
|---|---|---|---|---|---|---|---|---|
| 0° Tensile (Brittle Resin) |  |  |  |  |  |  |  |  |
| Strength MPa (ksi) | 730 | (106) | 860 | (125) | 1140 | (165) |  |  |
| Strain % | 0.31 | | 0.27 | | 0.24 | |  |  |
| Modulus GPa (Msi) | 230 | (34) | 325 | (47) | 480 | (70) |  |  |
| 0° Tensile (Ductile Resin) |  |  |  |  |  |  |  |  |
| Strength MPa (ksi) | 900 | (130) | 930 | (135) | 1210 | (175) | 1280 | (185) |
| Strain % | 0.38 | | 0.28 | | 0.24 | | 0.25 | |
| Modulus GPa (Msi) | 230 | (34) | 325 | (47) | 480 | (70) | 505 | (73) |
| Major Poisson's Ratio | 0.34 | | 0.30 | | 0.31 | | 0.34 | |
| 0° Compression |  |  |  |  |  |  |  |  |
| Strength MPa (ksi) | 495 | (72) | 430 | (62) | 280 | (41) | 260 | (38) |
| Modulus GPa (Msi) | 190 | (28) | 310 | (45) | 500 | (73) | 540 | (79) |
| SBS MPa (ksi) | 62 | (9) | 55 | (8) | 35 | (5) | 28 | (4) |

- Properties are normalized to 60% fiber volume.

- Tensile modulus is a secant from 0.1% to 0.2% strain.

- Compression modulus is initial tangent.

Table VI: Representative Values for the Oxidative
Weight Loss of Carbon Yarns in Air at One Atmosphere

| | Percent Weight Loss in 1000 Hours | |
| | 316°C (600°F) | 375°C (707°F) |
| --- | --- | --- |
| T-300 | 7 | --- |
| T-300R | 3 | --- |
| T-500 | 6 | --- |
| T-40* | 1-3 | 80 |
| T-40R | 0.2 | 3 |
| T-50* | 0.6 | 0.8 |
| P-25 | <0.6 | 6 |
| P-55 | 0.3 | 0.5 |
| P-75 | <0.2 | 1.5 |
| P-100 | <0.1 | 0.3-1 |

* Toray produced carbon fiber.

Table VII: Coefficient of the Thermal
Expansion for Carbon Fibers

| | CTE $10^{-6} \cdot C^{-1}$ |
| --- | --- |
| T-40 | -0.56 |
| T-50 | -0.85 |
| P-55S | -1.15 |
| P-75S | -1.24 |
| P-100S | -1.29 |
| P-120S | -1.33 |

Measured at 65°C

Figure 1

# CARBON FIBER PROPERTIES ARE DETERMINED BY THEIR STRUCTURE:

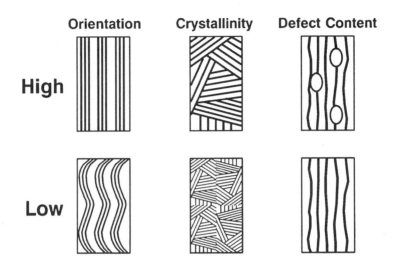

Figure 2

# ELECTRICAL AND THERMAL CONDUCTIVITY COMPARISON OF CARBON FIBERS AND METALS

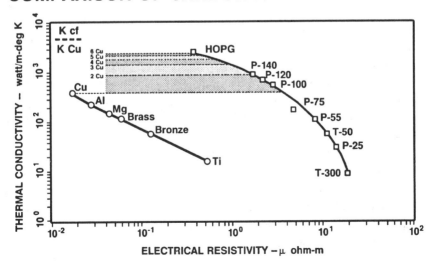

# IMPACT DAMAGE IN CARBON/EPOXY COMPOSITE CYLINDERS

A. P. Christoforou and S. R. Swanson
Department of Mechanical and Industrial Engineering
University of Utah
Salt Lake City, Utah 84112

S. C. Ventrello and S. W. Beckwith
Hercules Aerospace Division
Materials Technology (CTR)
Magna, Utah 84044

## Abstract

An experimental investigation was conducted to evaluate the effects of lateral, low velocity impact on Hercules IM7-12K/HBRF-55A carbon/epoxy cylinders. The cylinders were impacted using an instrumented falling weight to simulate low velocity impact which might occur in manufacturing and maintenance operations. The cylinders were impacted while empty and also while being supported internally with an RTV rubber or aluminum core to simulate various service conditions. After impact the specimens were inspected with C-scan and subjected to internal pressure to determine the strength loss. The results showed the degree of strength loss as a function of impact energy and type of internal support during impact. It was found that the RTV rubber support produced the largest strength loss at a given impact energy level. The damage to the cylinders depended directly on the type of internal support and

could be considered to be due to a combination of bending and direct contact stresses for the RTV support condition, bending for the empty cylinder, and local contact stresses for the aluminum support condition. An analytical model, based on a single degree of freedom spring-mass system, was used to predict the force and deflection histories of the cylinders during impact.

## 1. INTRODUCTION

Advanced fiber-reinforced materials such as carbon/epoxy have been succesfully employed as structural materials in the aerospace and aircraft industries. The performance of these composites has shown their superiority over metals in applications requiring high strength, high stiffness, excellent fatigue and corrosion resistance, as well as low weight. However, their relatively poor impact resistance has become a major concern in recent years. It is well known (1-5) that composites, unlike metallic materials, can suffer a significant reduction in compressive and tensile strength after impact.

Damage to composite structures can occur in a variety of forms, including surface cuts, delaminations, and broken fibers. The problem of foreign body impact is well known to have the potential of inflicting subsurface damage; thus, it is relatively important to be able to develop a predictive capability relating the type and probable extent of damage to the nature and parameters of the impact. A number of investigations have considered the impact problem from the viewpoint of predicting the resulting damage. In general, the nature of the damage depends on the velocity and energy level of the impact, the shape of the impacting body, and the stiffness of the impacted structure. Low velocity impacts have been shown (6,7) to predominately excite only the low frequency response waves of the structure so that the structural response of the impacting object and structure may be treated as a spring-mass system. The resulting impact force can then be calculated from structural dynamics considerations. It has also been suggested that the impact problem can be treated as an equivalent static concentrated force problem (8).

Although a significant amount of work is being done to characterize damage for the flat laminate type components, very little effort has been reported on critical damage parameters for pressure vessels subjected to low velocity impact. Most of the published work addresses the response of plates and beams to low velocity transverse impact by a hard object. In early studies (9,10) impact of laminated plates was studied based on small deflection thin plate theory. Recent studies (6,11), however, have shown that during low velocity impact, the plate undergoes large deflections and transverse shear deformation. The deformation characteristics of the plate can be represented by springs and the impactor and plate represented by rigid masses in a spring-mass model to simulate the impact response of the plate. The response of pressure vessels will have a number of features in common to that of flat plates, but the curvature of the shell will also change the situation somewhat, principally by adding a natural boundary condition.

The objective of the present investigation was to study the impact response and evaluate the effect of low velocity impact on the load carrying ability of pressurized composite cylinders. The impact response of the cylinders was calculated using a single degree of freedom spring-mass model, with inputs consisting of the nonlinear bending stiffness of the cylinder obtained from experimental data and the mass of the heavy impactor (assuming that the mass of the cylinder is negligible). Finally, the degradation of the ultimate strength of the composite cylinder was evaluated as a function of the kinetic energy of the impactor and type of internal support during impact.

## 2. EXPERIMENTAL

The 9.652 cm (3.8 in.) inner diameter cylinders employed in this study were filament wound by Hercules Aerospace Division using the Hercules IM7-12K carbon fiber and HBRF-55A epoxy in a [90,±22]$_2$ layup. The HBRF-55A epoxy is an amine-cure resin which has room temperature mechanical properties of 86 kPa (12.5 ksi) tensile strength, 3.2 GPa (460 ksi) tensile modulus, and 7.8% elongation. The cylinders had a wall thickness of 1.524 mm (0.06 in.). The cylinders were inspected by C-scan and then subjected to impacts at various energy levels using a 5.03 kg (11.1 lb) drop weight tester. The cylinders were held in a wooden cradle that supported the entire length of the cylinder during the impact, as shown in Fig. 1. The cradle extended over the bottom half of the circumference of the cylinder, but had only a moderately tight fit. An estimate of this fit is that a radial gap of approximately 1.588 mm (1/16 in.) existed on both sides of the cylinder in the cradle. Three different internal support conditions were employed: the empty cylinder, a cylinder with an RTV rubber core, and a hollow aluminum core, to simulate various service conditions. Both of the cores were inserted by hand, but had a reasonably snug fit.

After impact, the cylinders were again inspected by C-scan and delivered to the University of Utah Mechanics of Composites Laboratory. The cylinders were modified by the addition of stress relief reinforcements and end grips. The design of the tubular test specimen is descibed in (12, 13) and is illustrated in Fig. 2. The specimens were then subjected to a combination of internal pressure and axial load until the specimen failed. While in general the ratio of axial load to internal pressure load can be varied arbitrarily in the cylinder test procedure, in the present tests the apparatus was modified so that the load ratio produced a membrane stress ratio of exactly 2 to 1, as in the usual closed end pressure vessel. This was accomplished by designing a new test fixture that produced the axial stress as well as the hoop stress with internal pressure. Previous test fixtures used internal pressure for the hoop stress, with axial stress produced by an axial force test machine.

## 3. RESULTS

A summary of the test results is given in Table 1. The results consist of burst pressures for the three control cylinders and the cylinders impacted with various drop-weight energy levels and internal support conditions. The strength reduction due to impact can be seen in Figs. 3-5. Although some scatter in the results is evident, particularly for the impacted empty tubes, it is clear that the strength of the impacted tubes is degraded by increasing the impact energy level.

Lines indicating a fit of the burst pressure response are shown in Figs. 3-5 and compared in Fig. 6. On this basis it is seen that the cylinders with the RTV core show the most reduction in burst strength with impact energy, while the empty cylinders and the cylinders with the internal aluminum core show less degradation.

The force and energy during impacts were measured and recorded as a function of time. A typical trace during impact for the empty internal support condition is shown in Fig. 7.

Records of the delamination size as measured by the post-impact C-scans were also obtained. The axial and hoop delamination lengths for the empty and RTV supported cylinders are plotted in Figs. 8 and 9 as a function of the impact peak force. No delaminations were observed in the aluminum supported cylinders.

## 4. MODELING OF IMPACT

In the case of the empty cylinder, it was considered likely that the impact could be modeled by means of a single degree of freedom spring mass system, with the 5.03 kg (11.1 lb) impactor constituting the mass and the cylinder acting as a spring of negligible mass. The impact force and deflection can be calculated by means of the elementary equations of dynamics for a single degree of freedom system. The governing equation for the system is then

$$m\ddot{u}+k(u)u=0 \qquad (1)$$

where m is the mass of the impactor, u is the deflection and k is the nonlinear shell bending stiffness. Using initial conditions of u=0 and $\dot{u}$=Vo at t=0 the deflection time response of the impacted tubes was calculated with a numerical solution (Runge-Kutta) of Eqn. 1. The force time response can then be calculated by

$$F(t)=ku(t) \qquad (2)$$

The nonlinear bending stiffness of the cylinder was obtained by cross-plotting the force vs. time and deflection vs. time measured during impact as shown in Fig. 10.

The force deflection response was also obtained by statically loading the cylinder (mounted in the cradle) in an MTS testing machine. A comparison of the statically measured stiffness and the one obtained by cross-plotting the impact data is shown in Fig. 11. Comparisons of the calculated deflection time and force time response are shown in Figs. 12 and 13.

## 5. DISCUSSION

It is clear that major damage is done to the cylinders at the higher impact energy levels. This appears to be true for all three support conditions. The type of damage produced is likely to be dependent on the internal support, as evidenced by the extensive delamination for the empty cylinder and cylinder with the RTV rubber core, and no delamination for the cylinder with the aluminum core. Visual inspection of the tubes after impact did not show any permanent local deformation for the empty case which suggests that the tubes were damaged due to structural bending. In the RTV supported cylinders a combination of delamination and local permanent deformation was observed. The damage mechanism in this case was a combination of bending and direct contact stress. In the aluminum supported cylinders only local permanent deformation was observed and thus damage in this case was due to local contact stress.

In spite of this difference in type of damage, the effect of the damage appears to be quite similar for the three support conditions. In view of the scatter of the results, it is probably not justified to consider the relative degree of damage produced as firmly established. However, it does appear from the present tests that the RTV supported cylinder shows the largest reduction in burst strength at a given impact energy level. The scatter in the burst pressure results is largest for the empty cylinders. In particular, two tests appear to be out of line with the others and are noticeably low in burst pressure. No explanation for these low burst strengths is available. It was noted for both of these cylinders that visible cracks could be seen on the inside surface after impact. Cracks of this type were not observed in other cylinders impacted at similar energy levels.

The success of the single degree of freedom spring-mass model applied to the impact of the empty cylinder is quite encouraging. In the present tests it appears to be justified to neglect the mass of the cylinder in comparison to the mass of the impactor. However, the nonlinear spring stiffness used was obtained from the experimental data. Hence, it becomes evident that an analytical method is needed to predict the nonlinear force deflection response of the cylinder.

## 6. SUMMARY AND CONCLUSIONS

Pressure tests have been carried out on composite cylinders furnished by Hercules Aerospace Division, which had previously been subjected to impact. During the impact the cylinders were supported in a cradle, and were either empty, filled with an RTV rubber core, or had an aluminum tube core. Three additional cylinders were pressure tested as control specimens.

The test results showed definite reductions in burst strength for all of the impact support conditions. The burst strength results for the cylinders that were empty during impact were somewhat scattered. However, the present results appear to show that the most strength reduction occurred in the cylinders impacted with the RTV rubber internal support.

An analytical model was developed to predict the force-deflection response during impact of the empty supported cylinders. The model was based on a single degree of freedom spring-mass model, and gave excellent agreement with the experiment if a nonlinear spring stiffness was employed.

## 7. ACKNOWLEDGEMENT

The first two authors were supported in part by a contract with Hercules Aerospace Division.

## 8. REFERENCES

1. Labor, J.D., "Impact Damage Effects on the Strength of Advanced Composites," ASTM STP 696, Nondestructive Evaluation and Flaw Criticality for Composite Materials, 172-184 (1979).

2. Starnes, J.H., Jr., Rhodes, M.D., and Williams, J.G., "Effects of Impact Damage and Holes on the Compressive Strength of a Graphite/Epoxy Laminate," ASTM STP 696, Nondestructive Evaluation and Flaw Criticality for Composite Materials, 145-171 (1979).

3. Sharma, A.V., "Low Velocity Impact Tests on Fibrous Composite Sandwich Structures," ASTM STP 734, Test Methods and Design Allowables for Fibrous Composites, 54-70 (1981).

4. Williams, J.G., and Rhodes, M.D., "Effect of Resin on Impact Damage Tolerance of Graphite/Epoxy Laminates," ASTM STP 787, Composite Materials: Testing and Design (Sixth Conference), 450-480, (1982).

5. Cabrino, G., "Residual Strength Prediction of Impacted CFRP Laminates," J. Comp. Mtls., 18, 508-518 (1984).

6. Elber, W., "Failure Mechanics in Low-Velocity Impacts on Thin Composite Plates," NASA Technical Paper 2152 (1983).

7. McQuillen, E.J., and Gause, L.W., "Low Velocity Transverse Normal Impact of Graphite Epoxy Laminates," J. Comp. Mtls., 10, 79-91 (1976).

8. Bostaph, G.M., and Elber, W., "Static Indentation Tests on Composite Plates for Impact Susceptibility Evaluation," AMMRC MS 82-4, U.S. Army, 288-317 (1982).

9. Sun, C.T., and Chattopadhyay, S., "Dynamic Response of Anisotropic Plates Under Initial Stress Due to Impact of Mass," ASME J. of Applied Mechanics, 42, 693-698 (1975).

10. Dobyns, A.L., "Analysis of Simply-Supported Plates Subject to Static and Dynamic Loads," AIAA Journal, 19, 642-650 (1981).

11. Shirakumar, K.N., Elber, W., and Illg, W., "Prediction of Impact Force and Duration Due to Low-Velocity Impact on Circular Composite Laminates," ASME J. of Applied Mechanics, 19, 674-680 (1985).

12. Swanson, S.R., and Christoforou, A.P., "Response of Quasi-Isotropic Carbon/Epoxy Laminates to Biaxial Stress," J. Comp. Mtls., 20, 457-471 (1986).

13. Swanson, S.R., and Christoforou, A.P., "Progressive Failure in Carbon/Epoxy Laminates under Biaxial Stress," ASME Paper No. 85-WA/MATS-12 (1985).

## 9. BIOGRAPHIES

Andreas P. Christoforou is a PhD candidate in Mech. Eng. at the Univ. of Utah. He received his BS and MS degrees from the Univ. of Utah. He has been involved in programs to determine failure properties of advanced composites.

Stephen R. Swanson is Associate Professor in the Dept. of Mech. and Ind. Eng. at the Univ. of Utah. He received a BS degree from the Univ. of Minnesota, and MS and PhD degrees from the Univ. of Utah. He has been involved in programs to determine failure properties of advanced composite materials.

Scott W. Beckwith is Manager of the Materials Characterization group at Hercules Aerospace Division, Magna, Utah. He has been engaged in research and development programs pertaining to solid rocket propellants and composite materials. He received a BS degree from Texas A&M Univ., a MS degree from the California Inst. of Technology and a PhD degree from Texas A&M University.

Sandra C. Ventrello is an Engineer in the Materials Characterization group at Hercules Aerospace Division, Magna, Utah. She received a BS degree in Math and Physics from the Univ. of Honduras, and a BS degree in Mechanical Engineering from the University of Utah. She has been involved in programs to determine composite damage assessment and failure properties.

Table 1.
Test Results for Impact Study Cylinders

| Specimen No. | Internal support | Impact Energy Joules (ft-lbs) | Burst Pressure MPa (psi) |
|---|---|---|---|
| DA-4-SC-1A | Control | 0.0 | 31.0 (4496) |
| DA-4-SC-1B | Control | 0.0 | 26.5 (3840) |
| DA-4-SC-2A | Control | 0.0 | 31.4 (4560) |
| DA-4-SC-7A | Empty | 16.5 (12.17) | 29.2 (4228) |
| DA-4-SC-4D | Empty | 16.8 (12.39) | 23.1 (3350) |
| DA-4-SC-14D | Empty | 19.3 (14.25) | 24.4 (3541) |
| DA-4-SC-14A | Empty | 19.4 (14.30) | 27.6 (3998) |
| DA-4-SC-3A | Empty | 20.8 (15.31) | 7.70 (1117) |
| DA-4-SC-4C | Empty | 20.9 (15.38) | 7.19 (1043) |
| DA-4-SC-7D | Empty | 26.2 (19.34) | 19.7 (2862) |
| DA-4-SC-4A | Empty | 26.2 (19.33) | 28.1 (4072) |
| DA-4-SC-8A | Empty | 34.4 (25.37) | 18.1 (2631) |
| DA-4-SC-8B | Empty | 34.4 (25.38) | 26.2 (3804) |
| DA-4-SC-14B | Empty | 47.5 (34.99) | 21.0 (3051) |
| DA-4-SC-2D | RTV | 20.7 (15.30) | 26.2 (3800) |
| DA-4-SC-3C | RTV | 20.6 (15.22) | 24.4 (3536) |
| DA-4-SC-4B | RTV | 20.9 (15.38) | 25.5 (3693) |
| DA-4-SC-7B | RTV | 29.9 (22.03) | 17.1 (2474) |
| DA-4-SC-6A | RTV | 29.7 (21.90) | 10.8 (1573) |
| DA-4-SC-6B | RTV | 29.7 (21.90) | 11.7 (1690) |
| DA-4-SC-1D | Aluminum | 21.2 (15.60) | 22.6 (3278) |
| DA-4-SC-2C | Aluminum | 20.9 (15.38) | 30.0 (4358) |
| DA-4-SC-3D | Aluminum | 20.6 (15.17) | 27.3 (3960) |
| DA-4-SC-5D | Aluminum | 26.0 (19.21) | 24.1 (3490) |
| DA-4-SC-6C | Aluminum | 47.8 (35.23) | 17.0 (2465) |
| DA-4-SC-6D | Aluminum | 47.8 (35.23) | 17.9 (2600) |

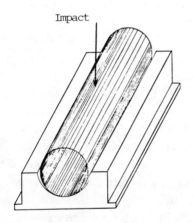

Figure 1.  Schematic of impact cradle.

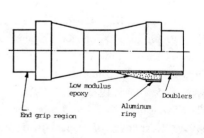

Figure 2.  Schematic of biaxial stress specimen.

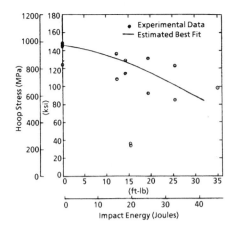

Figure 3. Effect of Impact Energy on the Hoop Stress at Failure for Empty Carbon/Epoxy Tubes.

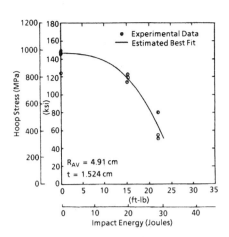

Figure 4. Effect of Impact Energy on the Hoop Stress at Failure for RTV-Supported Carbon/Epoxy Tubes.

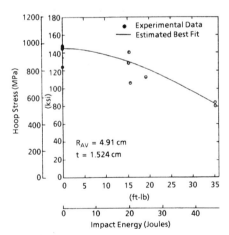

Figure 5. Effect of Impact Energy on the Hoop Stress at Failure for Aluminum-Supported Carbon/Epoxy Tubes.

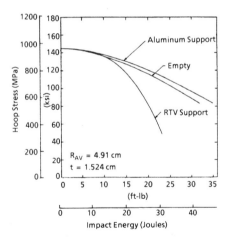

Figure 6. Effect of the Internal Support Conditions and Impact Energy on the Hoop Stress at Failure for Carbon/Epoxy Tubes.

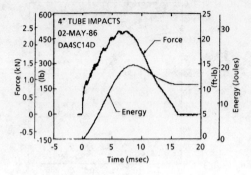

Figure 7. Measured Load Energy History for an Impact Event on an Empty Carbon/Epoxy Tube.

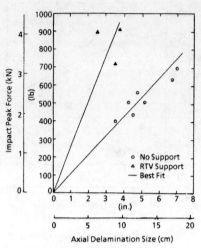

Figure 8. Effect of the Support Condition and Impact Peak Force on the Delamination Size for Carbon/Epoxy Tubes.

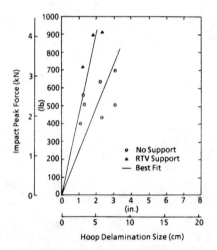

Figure 9. Effect of the Support Condition and Impact Peak Force on the Delamination Size for Carbon/Epoxy Tubes.

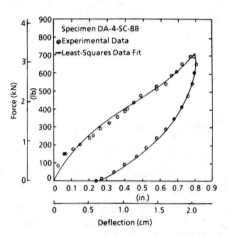

Figure 10. Measured Force-Deflection Response for an Empty Carbon/Epoxy Tube.

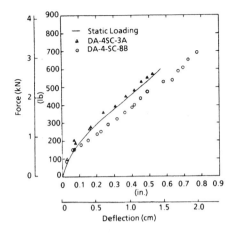

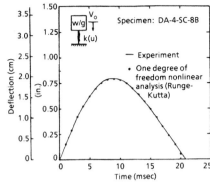

Figure 11. Comparison of Measured Static and Dynamic Force Deflection Response for an Empty Carbon/Epoxy Tube.

Figure 12. Comparison of Predicted and Measured Deflection for an Empty Carbon/ Epoxy Tube.

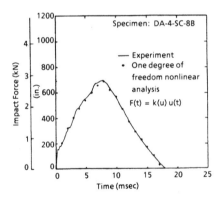

Figure 13. Comparison of Predicted and Measured Force for an Empty Carbon/ Epoxy Tube.

973

A METHOD FOR COMPUTER  GENERATED PATHS FOR OFF-LINE
PROGRAMMING OF FILAMENT WINDING PATTERNS
James V. Kelly
Computer Control Systems
Dan Higginbotham
Automation Dynamics, Inc.

ABSTRACT

This paper presents a method
for generating a precise
filament winding pattern by
utilizing a mathematical
model of the desired shape
loaded into a computer
programmed to calculate and
control all axes required to
produce the desired path.
Along with the computer
generated geodesic path,
other paths may be selected.
Using this off-line method to
calculate the geodesic path
for winding mandrels gives
the designers a valuable tool
that can be used
interactively to instantly
evaluate a new design.
Although the system is
intended to operate in real
time, it is possible to
generate a "pre-programmed
drive file" or a mechanically
read template that can be
used to run existing machines
that operate along pre-
programmed paths.

2. EVOLUTION OF CONTROLS

Filament winding of
structural fibers with binder
resins to produce a three
dimensional shape has evolved
through many types of
machines.  The basic machine
concept is for a system that
will synchronize and ratio
two or more axes of motion to
generate the path of a fiber
delivery device so that it
will lay the filament
precisely on the surface of a
three dimensional shape with
no tendency to slip from the
path.  By controlled advance
of the pattern,  subsequent
wraps will lay alongside
preceding wraps until the

shape is uniformly covered with a "cocoon" of geometrically placed, continuous fiber.

The simplest wind, one done in 1800 by Napoleon's arsenals to increase the burst strength of a cannon breach, is to wrap a cylinder at nearly 90 degrees to its longitudinal axis. The ratio of the longitudinal delivery eye axis to the rotating mandrel axis is simply one band width per revolution. This mechanical relationship is easy to produce.

Cylinders with domed ends are wrapped with a pattern, not only to increase burst strength along a longitudinal line but to prevent failure along a radial line. This necessitates a helical pattern at an angle much lower than the 90 degrees of the simple hoop wrap. The ratio between axes remains simple but the need to change fiber direction at the end of the wrap requires a slightly more complex mechanism.

As part shapes became more complex and designers strove to add localized fiber buildups to strengthen or stiffen unique portions of the part surface, machines were built with many more axes than were, at first glance, deemed necessary. Each axis contributed to the success or failure of the filament winding operation. As these parts were designed, the simple mechanical ratioing of two axes was inadequate to meet the product designer's needs. With additional axes, the relationship of the axes became even more complex. This lead to the development of the servo controlled machines. Servo controlled machines give users the ability to utilize variable ratioing of two or more axes. As diameters change along the axis of the part the wind angle must change. This is accommodated by changes in ratio between two or more axes. Servo machines vary axis ratio by utilizing an input from a template that is repeatedly "read" either mechanically or optically through each cycle of the winding process. While allowing great flexibility the servo machine is usually slow and subject to setup difficulties. The template is extremely critical to machine performance and was frequently created by a setup man having the machine "learn" the desired path.

"Self teaching" is an expedient of filament winding machine setup initiated with the advent of the servo controlled machine. The simplest way to develop the winding pattern for other than a cylindrical part is to draw the desired path on the winding mandrel and then to "fly" the delivery eye over the path causing a strand of fiber to lay along that path. As the machine axes are slowly moved to cause the fiber to lay along the desired path, a template blank is marked with a series of points which are then connected with a fair line to form the template outline. This becomes the "memory" for the servo machine to follow to be able to repeat the winding pattern. Mechanical ratioing of the template controlled axis to a master axis produces a repeating pattern and incorporates a band width advance.

With the development of low cost, powerful computers and high resolution, powerful electric servo motors, the template controlled servo machine has become obsolete. Machines have been developed that continue many of the characteristics of the template controlled servo machines. The template "memory" has been replaced with an electronic memory. The complete path of each winding axis is stored as a series of coordinates that form a "line" for the respective axes to follow. For simple cylindrical shapes the computer can be programmed to produce a uniform helix but the turnaround at the end of the shape remains somewhat of a compromise and experimentation with the winding setup is frequently called for before a satisfactory pattern is produced.

The "self teach" system used with the earlier servo machines is still employed to develop a setup for intricate shapes. This method has been somewhat refined in that, as the operator jogs the various axes to lay a fiber on the desired path, the coordinates of various axes are committed to memory and their relationship to one another is maintained. Because the operator cannot coordinate several axes simultaneously in a perfectly fair path, computer programs have been incorporated to average the operator generated path and produce a smoother, more nearly fair path for the actual winding operation.

## 3. TODAY'S WINDING MACHINE

The computer controlled servo machine of today is still a close relative to its predecessor, the mechanically synchronized servo machine. The machines exhibit a great deal of flexibility and are limited only by the imagination and knowledge of the user. There are some drawbacks however. Programming inputs are heavily weighted with empirical information that can, if not accurately derived, compromise the performance of the filament winding operation. Slow and sometimes erratic operation is experienced with filament winding machines that rely on following a path that has been developed empirically. Following a line of pre-programmed points that are not intimately "known" by the computer due to lack of computer "intelligence" inevitably requires much time spent in checking and correcting the paths of the various axes.

## 4. MATHEMATICAL MODELING

The winding machine computer software must be able to clearly identify the shape of the part being wound. At a very minimum, the computer must be able to recognize cylinders and cones. From these shapes an approximation of any shape can be made and described by diameter and length. An optimal program would also recognize a toroid and an ellipsoid. By joining the various shapes by their mathematical formulas and dimensions, the desired shape can be recognized by the machine computer. It is also noteworthy that diameters of the product become critical in the "helical" wrap whereas in the "circumferential" wrap they are inconsequential. The ARC-SINE of the ratio of the smallest diameter at the tank ends (or pole diameter) to any diameter along the path is the winding angle measured from the longitudinal axis at that point. This is the angle required to produce a geodesic path for the part.

## 5. INTELLIGENT WINDING MACHINE

The modern computer controlled filament winding

machine must be "intelligent" enough to accept a mathematical model of the shape that is to be wound, compute the optimum winding path, provide information of value to the product designer, and control all machine axes so as, at first try, to accurately and rapidly wind a part that is what the product designer had anticipated. If all of this criteria is met, the machine becomes more than a simple production or research tool; it becomes a knowledgeable contributor to the designer by creating geodesic paths where possible and identifying changes in part shapes that will allow production of a part that consists of totally geodesic paths. The "intelligent" computer operating from a mathematical model can be programmed to automatically generate the path of all axes with little additional input from a setup man. Beyond simple inputs the computer program recognizes that the fiber delivery eye path is not the same as the fiber path but is some distance ahead of it and on the line of tangency where the fiber makes contact with the mandrel surface. The computer program will then take over and calculate the motions of all axes to assure that the fiber will fall on the calculated winding path line.

## 6. DELIVERY EYE POSITIONING

One of the major problems in controlling a filament winder is knowing where to put the delivery eye. Knowing what wind angle to put on a mandrel surface is only part of the solution. The difficulty is in determining where to locate the delivery eye to achieve the required wind angle.
Using a machine with a cylinder for the mandrel and a pen that marks the mandrel will illustrate the difficulty of knowing where to place the delivery eye. When the pen is moved from right to left one inch and the mandrel is turned so the pen moved one inch on the surface of the mandrel in the direction of rotation, a line is drawn that has a forty-five degree angle with respect to the longitudinal axis of the mandrel. This is the wind angle. The wind angle is controlled by the ratio for the pen movement from left to right and the mandrel rotation. When the pen is replaced by a delivery

eye with filament going through it, the control of the delivery eye becomes more complex. The delivery eye path is not the same as the filament path because it is some distance ahead of it on the line of tangency where the filament makes contact with the mandrel surface. The problem is not too complex for a cylinder because the ratio of the delivery eye movement and mandrel is constant. The problem becomes extremely complex when a varying ratio is used. To solve the problem and to correctly control the point at which the filament is laid on the mandrel, the position of the filament as it leaves the mandrel must be mathematically extended from the point of tangency along the tangent line to the final delivery eye position.

## 7. DESIGN CRITIQUE FEEDBACK

The properly programmed computer accepting a mathematical input can be asked to create a geodesic path, in which case it will choose all winding angles that will assure the condition is met. The designer may have, as is usually the case, a desired wind angle which he can input to the computer. If the angle or angles produce a non geodesic path, the computer can be programmed to suggest changes in part dimensions or wind angles that will return the pattern to a geodesic condition. In this way the computer becomes a powerful design tool in that the designer can rapidly determine the optimum winding pattern for any proposed shape.

## 8. FACTORS IN THE DESIGN

### 8.1 Use of Position Equations

The input required for most of the position equations is very simple. Most of the position equations require only the distance, diameter, wind angle and the type of shape. The geodesic position equations do not need the wind angle because the wind angle is defined by another equation based on the diameter.

The mandrel is divided into sections at definition points. When all the sections are added together this will form one circuit. Each section will be defined as a cylinder, cone, toroid

or an ellipsoid. The path
along the section is then
defined at every point. This
path can be given an
equation. Thus far the
position equations that have
been developed are for
cylinders, cones, toroids and
ellipsoids. It is possible
to define position equations
for other shapes. The wind
angle can be held constant or
allowed to vary at a uniform
rate from one end of the
section to the other. The
wind angle may also be
generated to give the
geodesic path from one end of
a section to the other.
These position equations give
the mandrel position in
degrees everywhere along the
path. Also the position
equations give the total
number of degrees the mandrel
will rotate if the section
end point is input to the
equation. For example a
conical section with a given
wind angle and a given length
will rotate a given number of
degrees from one end of the
section to the other.

Any number of axes can be
added to the system by
defining the extra axis
position at the start and end
of the section. The position
of the extra axis can be made
to follow a predescribed

equation or change uniformly
from the start to the end of
the section.

8.2 Developing the Pattern

The computer generated path
has two major requirements.
The first requirement is that
the path along a surface,
such as a path on a cone or
cylinder, must be defined at
all points. The second
requirement is that the
pattern repeat at some time
with a reasonable bandwidth.
These two requirements can be
made to work with each other,
even along a geodesic path.

For every mandrel shape and
wind angle combination there
is a mathematical
relationship for bandwidth
advance, circuits-per-
pattern, rotations-per-
pattern, circuits-per-layer
and pattern-per-layer. All
of these values interact on
each other. If one of these
values is changed one or more
of the other values will
change. An example would be
to enter the circuits-per-
layer and calculate the
bandwidth advance. The other
values can be entered to
achieve the best possible
wind very quickly with an
"intelligent" computer.

## 9. HOW A COMPUTER WILL HELP

In deciding whether a computer generated filament winding program can be of value in solving a specific manufacturing problem, there are several questions to ask. If a computer system is used to control the filament winder, does the computer know what the mandrel looks like? Is the wind angle considered as one of the inputs? Is the bandwidth advance easy to control? If needed can the system produce the geodesic path as a starting path? Does the system automatically generate a smooth path? To what extent would improved speed and accuracy benefit the manufacturing program? Would greater speed and accuracy give desired cost efficiency?

In many cases an existing filament winder can be adapted to use a computer generated program, resulting in substantial savings of cost and time. In other cases, a new machine specifically designed to solve a unique problem would be a more advantageous solution.

A computer can be used to generate a winding path for filament winders by defining the shape needed by the machine, not each point. The number of points that are needed to define a wind pattern is reduced from hundreds of thousands to just a few. The use of computers to calculate winding paths can minimize the time required to generate these paths and produce the parts while, at the same time, producing a far better product.

# GLOSSARY

## 1. REAL TIME WINDING

A system that will do the required mathemathics while the part is being wound is described as real time.

## 2. OFF LINE

A system that will do the required mathematics at some time other than when the part is being wound is described as off line. The generated data is then stored for later use.

## 3. SHAPE

The profile of the part being wound is shown as a cross sectional cut along the longitudinal axis made up of cones, cylinders etc.

## 4. DEFINITION POINT

Primarily used to define the limits of the individual geometric shapes within the mandrel shape, the definition points are the stations along the longitudinal axis where the various geometric shapes making up the mandrel shape meet. Any location along the longitudinal axis of the mandrel where a change in command is defined is also a definition point.

## 5. TURN-AROUND

The place where the delivery eye stops going one direction along the longitudinal axis and then goes the other direction is called the turn-around point.

## 6. CIRCUIT

The path of the delivery eye from the one end of the winding mandrel to the other end and back is called a circuit.

## 7. PATTERN

A pattern is defined as a winding path that repeats itself after one or more circuits. It is usual to modify the pattern with the addition of an increment of advance equal to a bandwidth so succeeding patterns will lie adjacent to one another.

## 8. CIRCUITS-PER-PATTERN

The number of circuits required to complete a pattern is defined as the circuits-per-pattern. In a three circuit-per-pattern wind the delivery eye will travel from the first end to the second end and back three times before the pattern repeats itself.

## 9. BANDWIDTH ADVANCE

The bandwidth advance is defined as an increment of mandrel rotation added to the pattern so succeding patterns will be adjacent to one another.

## 10. LAYER

A layer is when the path closes to cover the complete surface.

## 11. CIRCUITS-PER-LAYER

The number of circuits needed to cover the surface of the mandrel or to make one layer is said to be the circuits-per-layer. If the circuits-per-layer is not an integer, the last circuit put on will have some overlap on the first circuit. With a computer the circuits-per-layer can be input as an integer. This will cause the pattern to close by changing the bandwidth slightly. An example of this would be a mandrel with 345.3 circuits-per-layer. The last circuit would cover the first circuit by .7 of a bandwidth. If the

circuits-per-layer were changed to 345.0 the layer would close with only a small change in the bandwidth advance.

## 12. PATTERNS-PER-LAYER

This is the number of patterns needed to cover the surface.

## 13. ROTATIONS-PER-PATTERN

The number of mandrel rotations needed for the pattern is rotations-per-pattern.

## 14. AXIS - MOVEMENT

The movement of the two major axes (rotation and translation) are defined by the shape and wind angle.

## 15. EXTRA AXIS - MOVEMENT

The movement of additional axes (delivery eye infeed, eye-rotation, fiber-tension, etc) is defined with respect to one of the two major axes. The position or value of the extra axis is entered at each definition point. The position or value of the extra axis will follow a predescribed equation or change uniformly from the one definition point to the next definition point.

## 16. THE GEODESIC PATH

The geodesic line is defined
as the shortest line lying on
a given surface and
connecting two given points.
A geodesic path has no
slippage and will allow the
greatest tension and highest
rate of winding speed.

## 17. HOOP WIND

A hoop wind is defined as a
nearly 90 degree wrap where
the delivery eye will advance
one bandwidth for each
revolution of the mandrel.
This bandwidth can be held
constant or it can vary
uniformly from one definition
point to the next to create a
varying wall thickness.

James Kelly received his BSEE
degree from California State
Polytechnic University,
Pomona, in 1975. He has been
active in the design of
numerous control systems. He
developed his math-model for
helical filament winders
eight years ago on a control
system for Hughes
Helicopters, Inc.. Based on
the multi-axis synchronized
motion hardware and software
he developed for McDonnell
Douglas and his experience in
filament winding, he has
developed an advanced control
system that will do
mathematical modelling in
real time for filament
winders. A filament winding
machine using this system was
introduced at the 32nd
International SAMPE
Exhibition.


Dan Higginbotham graduated
from California Polytechnic
College in 1958 with a degree
in mechanical engineering.
He has been designing and
building machinery since that
time. In 1971 he founded
Automation Dynamics, a
company which has been
continuously dedicated to
designing and building
filament winding and prepreg
production machines.

THE EFFECT OF SMALL ANGULAR FIBER MISALIGNMENTS
AND TABBING TECHNIQUES ON THE TENSILE STRENGTH
OF CARBON FIBER COMPOSITES

Peter W. Manders*
Ian M. Kowalski**
* Amoco Performance Products Inc.
P. O. Box 409
Bound Brook, New Jersey 08805
** Amoco Performance Products Inc.
12900 Snow Road
Parma, Ohio 44130

### Abstract

Small angular misalignments of the fibers in typical tensile coupons by as little as 1° are sufficient to dramatically reduce the measured strengths by over 30%. Alternating misalignment of successive plies is less detrimental than when all fibers are parallel, but mis-oriented to the tensile axis of the coupon. Fibers with lower levels of surface functionality are more sensitive to misalignment because their lower adhesion to the matrix promotes a splintering mode of failure. Compliant ± 45° glass fiber/epoxy end tabs with a shallow 5° taper are effective in preventing premature debonding and splintering in high strength composites.

### 1. INTRODUCTION

Manufacturers of carbon fiber have made significant progress in raising the strength of their products so that they now have some of the highest specific tensile strengths attainable in practical engineeering materials. For example, Thornel®carbon fibers are commercially available with strengths exceeding 5.5 GPa (800 kpsi) (over 2% strain to failure), and over 7 GPa (1,000 kpsi) in development quantities. These fibers have been developed for applications, such as filament-wound rocket motor cases, where specific tensile strength is at a premium. The quoted strengths, which have been obtained in tensile tests on single resin-impregnated strands of 6,000 or 12,000 filaments, are close to what can be obtained in a fabricated part under optimum conditions. The term "strength translation" is used to describe the ratio (usually a percentage) of strength realized in a test coupon

Thornel is a registered trademark of Amoco Corporation

or structure, to the strand strength, and is typically in the range of 85% to 105%. (Translation is based on the loads carried by the fibers).

Strength translation efficiency depends on many factors, among them:

- Size of the test coupon or structure. Generally, larger structures have lower strengths because they have a greater probability of containing a strength-limiting defect of low strength

- Type of structure and loading. For example, filament-wound pressure vessels may be designed to fail in the hoop windings, but transverse tensile cracking in the polar windings, and the biaxial stress state, create very different loading from a strand test.

- Resin system. Strand tests are typically performed by the fiber manufacturer in a standardized quality control procedure. The resin system used may not result in optimum strengths with all fibers, and this explains how translations over 100% may be obtained.

The strengths of tensile coupons prepared from laminated prepreg have been found to depend on the type of end tab used to transfer load from the test machine grips, and particularly upon the alignment of fibers within the sample. This paper presents a study of the effects of various types of misalignment, and shows that some fibers are more sensitive to off-axis loading than others.

## 2. EXPERIMENTAL TECHNIQUE

### 2.1 Failure Mode

This study was prompted by the observation that high strength composite coupons invariably failed by progressive splitting, often initiated by failure of the adhesive bond between tab and carbon fiber laminate. Figure 1 contrasts the brittle failure mode associated with lower strength fibers in brittle resins, with a typical splitting failure. Even after tab failures had been eliminated by the techniques described in the next section, high strength composites with actual strengths exceeding 400 kpsi still failed by a splitting mechanism. A stress strain curve for T40 fiber in ERL-1962 resin is reproduced in Figure 2. The jagged portion of the curve close to the point of failure indicates that splitting was occurring. The strain increases, but the load remains approximately constant as more and more of the specimen splinters away. The propensity to fail by splitting is related, at least in part, to the degree to which the surface of the fiber has been chemically treated to promote adhesion to the matrix, and is discussed further below. In order to obtain consistently high translation of the strand strength in coupons extreme care must be taken to maintain fiber alignment. Some of the results in this paper were obtained on experimental fibers which are identified as T40 EXP; they are not commercially available.

## 2.2 Tabbing Technique

The purpose of tabs is to provide a gripping surface, and to achieve smooth transfer of load into the gauge section of the coupon. Through experience, Amoco Performance Products Inc. (APPI) has arrived at the design of coupon shown in Figure 3. It features tabs with a shallow 5° taper to a knife-edge which reduces the stress concentration sufficiently to eliminate premature debonding. The length of the untapered portion should correspond to the length of the test machine grip faces to ensure smooth and uniform gripping action. However, it is important that none of the untapered portion extends beyond the grip faces because this promotes peeling of the tapered portion of tabs.

The tab material used by APPI is a 9-ply glass fiber/epoxy laminate about 2mm (0.085 inches) thick. A ± 45° fiber orientation minimizes stiffness in the tensile direction, and facilitates grinding of the taper.

Both room-temperature curing epoxy adhesive (3M type 2216 gray) and autoclave-cured film adhesives (FM 123 and FM 300) have been used to bond the tabs. Curing at room temperature has the advantage of eliminating residual thermal stress, but film adhesives are more convenient to handle. Whichever is used, the objective is to form a bond which remains intact until ultimate failure of the specimen. Pressure is conveniently applied during cure by a vacuum bag as shown in Figure 4, and this results in a smooth glue fillet in the critical region at the end of the taper. Figure 5 presents the results of a comparison of "standard" 0°/90° glass fiber/epoxy tabs, which have a blunt end approximately 0.13mm (0.005) inches thick, with the same layup ground to a point, and with ± 45° orientation tabs. About 350 MPa (50 kpsi) improvement in strength can be obtained with the ± 45° knife-edge tapered tabs when the failure mode is by splitting, as was the case for the T700 fiber in this study. A similar comparison of tabs on the more brittle and lower strength T300/3501-6 system showed no effect.

## 3. RESULTS AND DISCUSSION

### 3.1 Type of Misalignment

Two studies were performed. The first used an experimental T40 fiber to compare the effects of different types of fiber misalignment that can occur in the fabrication and testing of coupons or structures.

a)  All fibers misoriented by the same amount from the tensile axis. This occurs when prepreg is correctly laid up into a laminate, but tensile coupons are machined at a small angle to the fiber direction.

b) Alternating misalignment of plies by ± some small angle. Hoop windings in a filament wound structure have this sequence of orientations, and laminates may have small random misorientations of the plies.

c) Misalignment of correctly fabricated and machined coupons in a test machine. A similar type of misalignment exists if the grips on a test machine are not co-linear.

The results in Table 1 show that misalignment of the fibers within the coupon (a) is by far the most detrimental type of misalignment, resulting in a 38% loss of strength compared with the same laminate machined accurately in the fiber direction. The misalignment of 1° was chosen because it is the maximum allowable in a majority of aerospace industry material specifications. In practice it is quite easy to see such a misalignment if the fiber direction is clearly marked on a laminate, and when the specimens fail there is no longer any doubt that misorientation of the fibers is responsible for the splitting and low strengths. We have found it convenient to mark a direction at 90° to the fibers on test laminates before they are cured, and subsequently use this line to establish fiber direction when

specimens are machined. A strip of adhesive tape, or a line scored with a knife blade, are both effective, and can still be seen through the surface imprint left by woven peel-plies and release cloths. Some test laboratories attempt to establish fiber direction by splitting a narrow strip off the edge of a cured laminate. In our experience this is not as accurate because the crack does not always follow the fiber direction, particularly if the strip is twisted as it is pulled away, and because the fibers close to the edge are often misaligned with respect to the majority of the panel due to resin flow.

Once the fiber direction has been established, it is obviously essential that all the test coupons should be cut parallel. This is most easily accomplished on a surface grinder using a fine grit (80 mesh or smaller) diamond-coated blade with water cooling. Some widely used alternative procedures, such as band sawing individual coupons followed by surface grinding their edges, do not guarantee correct alignment of the fiber within the coupon, even though the sides are parallel to each other. Alternating misalignment, type (b), resulted in only 26% strength loss compared to the 38% when all the fibers were misaligned in the same direction. Splitting occurred at higher loads, and on a finer scale, suggesting that positive angle plies constrain crack propagation in the negative angle plies, and vice-versa. In a fracture mechanics model, one set of plies would inhibit the relaxation of tensile strain in the other set of plies, thereby reducing the principal driving force for crack growth. However, modelling along these lines was not pursued because the crack initiating defect is not at all well defined. In some instances it appears that splintering originates where locally highly misoriented fibers, such as those in wrinkles or around inclusions, intersect the edge of the specimen. In other cases it is probably tensile failure of small groups of fibers that initiate splitting, and this would account for the fact that the onset of splintering is approximately related to fiber strand strength. Although most splinters come off the edges of the coupons, careful polishing after machining did not increase strength.

Misalignment of coupons by 1° to the tensile axis of an Instron Model 1331 servohydraulic test machine with manual wedge-lock grips did not have a major effect on strength (6% decrease). An angular offset of this magnitude is quite apparent to the eye, at least in an unobscured gripping configuration, and is unlikely to occur in practice. It corresponds to an offset of more than 4mm (0.15 inches) over the length of a 25cm (10 inch) specimen. That it has so little effect on the strengths measured is due to the rotation which is possible as the serrations in the grip faces bite into the tabs, and the lateral flexibility of the load train. The test machine pulls itself into alignment during the test, and springs back noticeably when the specimen fails.

3.2 Effect of Fiber Type

The degree of adhesion between fiber and matrix is the major factor controlling the amount of splintering in typical composite tensile failures. Undoubtedly the ductility of the matrix, and the inherent strength and modulus of the fiber, which govern the energy release at failure, are also factors, but fibers with low adhesion invariably lead to splintering in the composite. A proprietary surface treatment (shear treatment), is applied to Thornel brand carbon fibers to promote adhesion to typical polymeric matrix materials. It chemically alters the surface, and results in the incorporation of elements other than carbon, primarily oxygen and nitrogen. Fibers with greater surface functionality display better adhesion, and are less susceptible to misorientation in composites. This was explored with three types of fiber:

   i) T40EXP - An experimental
       intermediate modulus fiber

having the lowest level of surface functionality.

ii) T40 - A commercial intermediate modulus fiber with a higher level of surface functionality than T40EXP.

iii) T500 - A commercial fiber with standard modulus. It has the highest level of surface functionality.

Table 2 summarizes the mechanical and surface properties of these fibers. The level of surface functionality is expressed as a percentage of that on the T500 fiber. The measurements were made by X-ray photoelectron spectroscopy (XPS or ESCA), which gives the chemical composition of the outermost 100 A of fiber. The atomic percentages of all detectable elements other than carbon were summed, and then ratioed to the sum for T500, to arrive at the relative levels listed in Table 2.

These fibers were prepregged in ERL-1962, a proprietary damage-tolerant epoxy resin formulation, and were laminated into 0° panels taking particular care to maintain fiber alignment. Sets of 5 tensile coupons were cut at 0°, 1°,.... etc. and their strengths, normalized to 100% fiber content, are plotted as a function of the angular misalignment in Figure 6. This figure also includes the strand strengths. The T40EXP, with the lowest level of surface functionality, shows the most rapid loss of strength as misalignment is increased. The two commercial fibers, T40, and T500, show much less sensitivity to misalignment, but the loss of strength, for what are usually considered small misorientations, is obviously severe.

The translation of tensile strength from strand to correctly aligned composite is close to 100% for the T500 fiber, but only 85-90% for the T40 fibers. There was a difference in failure modes between strands, which all failed in a brittle manner, and composites, which failed by splitting in the case of the T40 fibers, and in an intermediate mode with T500. It is tempting to speculate that this difference in failure mode between strand and composite accounts for the lower translation of the T40 fibers. This may be the case, but it does not follow that T40 type fiber strength will be improved if the composite failure mode is made to be more "brittle" in nature.

4. CONCLUSIONS

1) Misalignment of the fibers in tensile test coupons by as little as 1° causes severe loss of strength. Alternating misalignment between plies of ± 1° is less detrimental, and misalignment of coupons in a typical servo-hydraulic test machine has minimal effect.

2) Good adhesion between tabs and the test laminate is essential to prevent premature debonding and splintering of the gauge

section.

3) Commercial T40 fiber, which has a higher level of surface functionality and adhesion to the matrix, has less sensitivity to misalignment than an experimental grade with lower surface treatment

## 5. ACKNOWLEDGEMENTS

We gratefully acknowledge the following contributions:

D. H. Hecht - Prepreg Production
W. C. Harris - Surface Spectroscopy

## 6. BIOGRAPHIES

Peter W. Manders is a Project Scientist with Amoco Performance Products at their Bound Brook, New Jersey, R&D Center. He received a BA and MA in Materials Science from the University of Cambridge, and a Ph.D. from the University of Surrey, U.K. His dissertation was on the strength of mixed fiber composites, after which he pursued postdoctoral research on statistical aspects of tensile failure in composites at the University of Delaware. Dr. Manders manages the Composite Materials Science Laboratory at Bound Brook.

Ian M. Kowalski is a Development Scientist in the Carbon Fiber Applications Group with Amoco Performance Products at their Parma, Ohio, R&D Center. He received a BS in Mechanical Engineering from the University of Rhode Island and an MS from Rensselaer Polytechnic Institute. At Parma, Ian is Manager of the Fiber Composites Laboratory.

TABLE 1

Effect of Fiber Misalignment on Tensile
Strength of T40EXP/ERL-1962

| Test Case | Strength GPa(ksi) | COV (%) | V$_f$ (%) | Relative Strengths (%) |
|---|---|---|---|---|
| Cut 1° off axis, Tested on axis | 1.97(286) | 2.1 | 68.6 | 62 |
| Alternating ±1°, Tested on axis | 2.36(342) | 3.0 | 69.3 | 74 |
| Cut on-axis, Tested 1° off axis | 2.99(433) | 1.9 | 68.6 | 94 |
| Cut on axis, Tested on axis | 3.17(460) | 3.1 | 68.6 | 100 |

TABLE 2

Fiber Mechanical and Surface Properties

| Fiber | Modulus GPa(mpsi) | Strand Strength GPa(kpsi) | Elongation (%) | Relative Surface Functionality (%) |
|---|---|---|---|---|
| T500 | 241(35) | 3.79(550) | 1.5 | 100 |
| T40 | 290(42) | 5.59(811) | 1.9 | 63 |
| T40EXP | 290(42) | 5.17(750) | 1.8 | 43 |

Fig. 1 Comparison of "Brittle", "Intermediate",
and "Splitting" Failure Modes

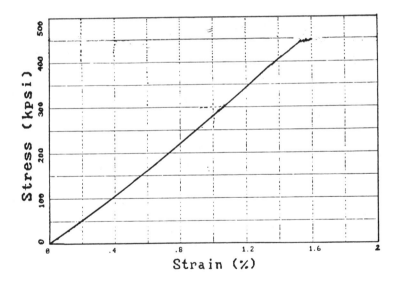

Fig. 2  Stress/Strain Curve for T40/ERL-1962 Illustrating
Splitting Before Ultimate Failure

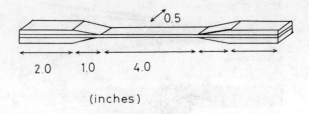

0.5

2.0    1.0    4.0

(inches)

Fig. 3  Tabbed Tensile Test Coupon

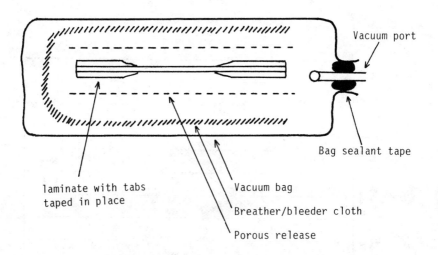

Vacuum port

Bag sealant tape

laminate with tabs
taped in place

Vacuum bag

Breather/bleeder cloth

Porous release

Fig. 4  Envelope Bagging Technique for Bonding Tabs

T-700/ MY 720/ DDS

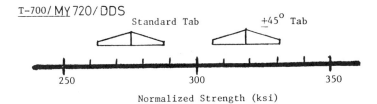

Standard Tab          +45° Tab

Normalized Strength (ksi)

250                300                350

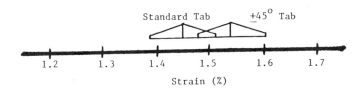

Standard Tab      +45° Tab

1.2      1.3      1.4      1.5      1.6      1.7

Strain (%)

T-700/ MY 720/ DDS          0/90 Tab Ground          +45° Tab
Standard Tab          to a Point

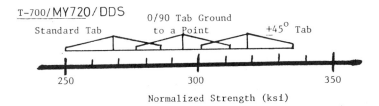

Normalized Strength (ksi)

250                300                350

0/90 Tab Ground   Standard Tab      +45°
to a Point                          Tab

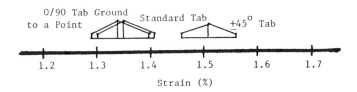

1.2      1.3      1.4      1.5      1.6      1.7

Strain (%)

95% confidence intervals shown about the treatment averages.

The stress and strain axes are scaled to be equivalent for a modulus of 20.8 Msi.

Fig. 5  Comparison of Blunt-ended Standard 0°/90° Tabs, With the Same Lay-up, and ± 45° Orientation Tabs, Both Ground to a Knife-edge

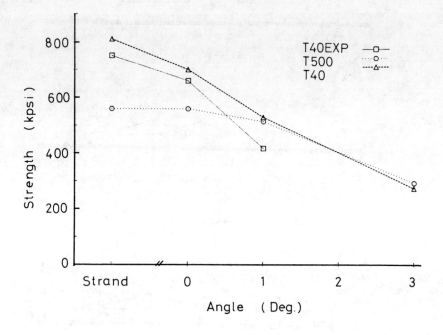

Fig. 6  Effect of Fiber Misalignment on Tensile Strength

A MACROSCOPIC APPROACH TO MATERIAL CHARACTERIZATION AND
NUMERICAL ANALYSIS OF METAL MATRIX COMPOSITES

A.R. Leewood, J.F. Doyle[1], and C.T. Sun[2]
Advanced Composite Engineering, Inc.
350 Sagamore Parkway, Suite 6
West Lafayette, IN  47906

## ABSTRACT

The basic premise of macroscopic characterization of metal matrix composites are that they can be modeled by classical anisotropic plasticity theory. Uniaxial tensile tests are conducted for a Boron/Aluminum composite at various load-fiber angle orientations and the anisotropic plasticity coefficients derived. The validity of the proposed anisotropic material model is verified for the uniaxial case.

The elastoplastic constitutive relations are rewritten and implemented into a 2-D plane stress finite element program, ANPLAST. The basic features of the program are highlighted. It is shown that ANPLAST solves the nonlinear problem by a straightforward incremental solution technique.

An experiment is conducted which evaluates the predictive capabilities of ANPLAST under a complex state of yielding. A Boron/Aluminum composite specimen with a centered circular hole is uniaxially loaded. The strains are recorded during the entire elastoplastic loading path and compared to the ANPLAST predictions.

The ANPLAST program is used to predict the growth of the plastic zone in the vicinity of an edge-cracked Boron/Aluminum composite. The effect of fiber orientation and load level on local yielding and subsequent plastic flow is highlighted. The strongly anisotropic behavior of the Boron/Aluminum composite is contrasted to the more familiar isotropic elastoplastic response.

---

1. Purdue University, W. Laf, IN.
2. Purdue University, W. Laf., IN

## 1. INTRODUCTION

The increasing use of metal matrix composites (MMC) requires that a design methodology be developed that captures the unique nonlinear behavior of this class of composites. Following the approach taken for elastic characterization of classical polymer composites[1], a macroscopic view of metal matrix composite behavior can be used. Bypassing the complex matrix-fiber interactions and passing directly to gross elastoplastic lamina behavior will provide the design engineer with a practical approach to analyzing structures made of these materials.

## 2. ELASTOPLASTIC MATERIAL CHARACTERIZATION

In order to make a decision as how to proceed with the characterization of a Boron/Aluminum (B/Al) metal matrix composite, a working knowledge of its basic behavior had to be established. A number of tensile tests were performed in order to determine principal direction material response.

The $0^o$ and $90^o$ (load-fiber orientation) uniaxial stress-strain response was measured and are illustrated in Fig. 1. The unloading sequences shown for the $90^o$ specimen confirms that Boron/Aluminum exhibits plasticity in the classical sense. That is, the unloading is elastic with a slope nearly that of the initial Young's modulus. Further, the reloading curve is elastic up to the previous maximum stress and then after yielding, work-hardening occurs at the previous rate. It is clear from the uniaxial response that the degree of plasticity is a function of load-fiber orientation. Therefore, a logical approach to characterization of the nonlinear behavior of Boron/Aluminum metal matrix composite is to assume it can be modeled in terms of anisotropic plasticity with work-hardening.

### 2.1 Elastic Orthotropic Formulation

It is assumed that the elastic response of MMC can be represented by the generalized Hooke's low for anisotropic materials. For the case of thin laminates, a plane stress (orthotropic) approximation is adequate and can be represented by a 3x3 matrix equation given below.

$$\begin{Bmatrix} e_{11} \\ e_{22} \\ 2e_{12} \end{Bmatrix} = \begin{bmatrix} c_{11} & c_{12} & \\ c_{12} & c_{22} & 0 \\ 0 & 0 & c_{66} \end{bmatrix} \begin{Bmatrix} \sigma_{11} \\ \sigma_{22} \\ \sigma_{12} \end{Bmatrix} \quad (1)$$

The subscripts 1,2 denote principal material directions. The matrix coefficients can be written in terms of engineering constants as

$$C_{11} = \frac{1}{E_1} \qquad C_{22} = \frac{1}{E_2}$$

$$\qquad (2)$$

$$C_{66} = \frac{1}{G_{12}} \qquad C_{12} = \frac{-\nu_{12}}{E_1} = \frac{-\nu_{21}}{E_2}$$

### 2.2 Plastic Orthotropic Formulation

The derivation of the orthotropic plasticity relations begin with the

998

statement of the flow rule for plasticity[2].

$$de_{ij}^p = \frac{\partial g}{\partial \sigma_{ij}} d\lambda \qquad (3)$$

Note that the superscript "p" indicates plasticity. The plastic flow function g is assumed quadratic in $\sigma_{ij}$. Taking advantage of material symmetry of an orthotropic body and using contracted notation, the plastic flow rule can be written in 3x3 matrix equation form as

$$\begin{pmatrix} de_{11}^p \\ de_{22}^p \\ de_{12}^p \end{pmatrix} = [r] \begin{pmatrix} \sigma_{11} \\ \sigma_{22} \\ \sigma_{12} \end{pmatrix} \frac{3}{2} \left(\frac{d\bar{e}^p}{d\bar{\sigma}}\right) \left(\frac{d\bar{\sigma}}{\bar{\sigma}}\right) \qquad (4)$$

where

$$[r] = \begin{bmatrix} r_{11} & r_{12} & 0 \\ r_{12} & 1 & 0 \\ 0 & 0 & r_{66} \end{bmatrix}$$

$$\bar{\sigma} = \sqrt{\frac{3}{2}(r_{11}\sigma_{11}^2 + \sigma_{22}^2 + 2r_{12}\sigma_{11}\sigma_{22} + 2r_{66}\sigma_{12}^2)}$$

$\left(\dfrac{d\bar{e}^p}{d\bar{\sigma}}\right)$ = instantaneous slope of work-hardening relationship

$r_{11}, r_{12}, r_{66}$ = plasticity coefficients

Note that the incremental plastic strains are orthotropic with respect to the current stress state.

A possible interpretation of effective stress $\bar{\sigma}$ is a single surface in space for a single point on the work-hardening curve. Therefore, one can say that yield will occur at a value of effective stress, say $\bar{\sigma}_y$, and is a function of plastic strain. In general, $\bar{\sigma}$ defines the yield

surface for an orthotropic material under a multiaxial state of stress.

## 2.3 Plastic Material Characterization

The plasticity coefficients ($r_{11}$, $r_{12}$, $r_{66}$) describe the strength of the plastic anisotropy and can be derived from uniaxial tests. The plastic Poisson ratio for an arbitrary load-fiber orientation $\theta$ is defined by KENAGA[2], and for three special orientations the expression simplifies to

$$\theta = 0° : \quad \nu_0^p = r_{12}/r_{11}$$

$$\theta = 90°: \quad \nu_{90}^p = r_{12} \qquad (5)$$

$$\theta = 45°: \quad \nu_{45}^p = \frac{r_{11} - 2r_{66} + 2r_{12} + 1}{r_{11} + 2r_{66} + 2r_{12} + 1}$$

Therefore, it is possible to derive the plasticity coefficients $r_{11}$, $r_{12}$, and $r_{66}$ from these uniaxial tests where both the transverse and longitudinal plastic strains have been recorded. Alternatively, the plastic Poisson ratios could be measured for many different orientations, and then the plasticity coefficients derived in a least squares sense.

Plastic work-hardening is defined by the relation between effective plastic strains and the effective stress

$$\bar{e}^p = h(\bar{\sigma}) \qquad (6)$$

A general form for nonlinear work-hardening can be written as

$$\bar{e}^p = \begin{cases} 0 & \bar{\sigma} \leq \bar{\sigma}_y \\ (\bar{\sigma}/\alpha)^\beta - (\bar{\sigma}_y/\alpha)^\beta & \bar{\sigma} \geq \bar{\sigma}_y \end{cases} \qquad (7)$$

Where $\alpha$ and $\beta$ are curve fitting parameters and $\bar{\sigma}_y$ is the effective yield stress. Other work-hardening relations are available and can be used to provide the best fit of the experimental data.

A complete elastoplastic characterization of a B/Al metal matrix composite conducted by KENAGA[2] yielded

$$E_1 = 29.4 \times 10^6 \text{psi.} \quad \nu_{12} = 0.169$$
$$E_2 = 19.1 \times 10^6 \text{psi.} \quad G_{12} = 7.49 \times 10^6 \text{psi.}$$
$$r_{11} = 0.001 \quad r_{12} = -0.01 \quad r_{66} = 1.9$$
$$\alpha = 60000.0 \quad \beta = 5.8$$
$$\bar{\sigma}_y = 13,500 \text{ psi.}$$

For comparison, the plasticity coefficients for an isotropic material would be

$$r_{11} = 1.0 \quad r_{12} = -0.5 \quad r_{66} = 1.5$$

It is interesting to note that the orthotropic effective stress expression reduces to the Von Mises yield criterion when these isotropic values are substituted into Eq. 4.

The uniaxial test results for B/Al at various load-fiber orientations and the orthotropic elastoplastic material model predictions are plotted together in Fig. 2. Clearly, the material model effectively captures the orthotropic elastoplastic behavior of the B/Al metal matrix composite.

## 3. ANPLAST FINITE ELEMENT PROGRAM
The design of MMC structures requires the use of a practical tool that can handle irregular geometries and multiaxial loading. To address this important requirement, the finite element program ANPLAST[3] was developed. This program implements the anisotropic plasticity theory previously described and incorporates a first order lamination theory.

### 3.1 Program Overview
The ANPLAST finite element program is designed for plane stress anisotropic elastic-plastic analysis. It solves the non-linear problem by dividing the total applied load into small load increments, updating all pertinent quantities (such as stress, strain, displacement, and nodal loads), as needed. The global stiffness matrix is updated at the beginning of every loading increment to account for the inherent material nonlinearity associated with plasticity.

### 3.2 Restart Capabilities
A typical nonlinear analyses conducted with ANPLAST require 10 to 15 increments. This can result in considerable computation time. The size of each loading increment must be estimated prior to the analysis and may turn out to be inappropriate. Also, the amount of data for each increment can be sizeable.

Because of these complexities, a "restart" option was built into ANPLAST. The "restart" capability permits the analysis to be stopped and started again at any previous increment and all pertinent results recovered.

## 3.3 Cyclic Loading

The ANPLAST program is designed for cyclic loading and unloading. Therefore, each element is permitted to yield, work-harden, unload back into the elastic region, and then re-load back to the the new yield surface. This transition from elastic to elastic-plastic is calculated on the element level and is not constrained by the direction of loading at the remote location. There is no limit to the number of times an element may pass through the elastic-plastic transition. This capability makes possible a cycle by cycle analysis when considering fatigue crack propagation.

## 3.4 Finite element Formulation

The linear displacement constant strain triangle (CST) element was chosen as the primary finite element in ANPLAST due to its simplistic formulation. Though higher order isoparametric elements were tested, they offered no apparent advantage in efficiency.

The incremental element stiffness matrix [K], as derived by Castigliano's Theorem[3] is written as

$$[K] = \iint [A]^T [C][A] \, da \qquad (8)$$

where

$$\{e\} = [A]\{q\} \qquad \{de\} = [C]\{d\sigma\}$$

q = nodal displacements

$\iint da$ = element area

Note that when an element yields, [C] contains the material non-linearity and consequently [K] becomes stress/strain dependent. Therefore, the global stiffness matrix is updated at the beginning of every increment.

## 3.5 Incremental Plasticity

The plane stress orthotropic plastic flow rule, Eq. 4, can be rewritten so that it is compatible with a numerical format. After considerable manipulation, the incremental plastic stress-strain relation becomes[3]

$$\begin{pmatrix} de_{11} \\ de_{22} \\ de_{12} \end{pmatrix}^p = [a] \begin{pmatrix} d\sigma_{11} \\ d\sigma_{22} \\ d\sigma_{12} \end{pmatrix} \frac{9}{4} \bar{\sigma}^{-2} \left( \frac{d\bar{e}^p}{d\bar{\sigma}} \right) \qquad (9)$$

where

$$[a] = \begin{bmatrix} a_{11}a_{11} & a_{11}a_{22} & a_{11}a_{33} \\ a_{22}a_{11} & a_{22}a_{22} & a_{22}a_{33} \\ a_{33}a_{11} & a_{33}a_{22} & a_{33}a_{33} \end{bmatrix}$$

$$a_{11} = r_{11}\sigma_{11} + r_{12}\sigma_{22}$$
$$a_{22} = r_{12}\sigma_{11} + \sigma_{22}$$
$$a_{33} = 2r_{66}\sigma_{12}$$

In the program, $r_{11}$, $r_{12}$, and $r_{66}$ are assumed to be constants. In symbolic matrix notation, Eq. 9 can be written as

$$\{de^p\} = [C^p]\{d\sigma\} \qquad (10)$$

The total material matrix [C] defined in Eq. 8 is obtained by superposition of elastic and plastic material response.

$$[C] = [C^e] + [C^p] \qquad (11)$$

All of the material nonlinearities enter into the stiffness matrix through $[C^p]$ for each element that is currently yielded.

## 3.6 Work-Hardening Options

There are currently three functional forms of work hardening available in ANPLAST. The values of $\alpha$ and $\beta$ must be specified.

1) Linear Work Hardening

$$\frac{d\bar{\sigma}}{d\bar{e}^p} = \frac{\alpha}{\beta} \qquad (12)$$

2) Ramberg-Osgood (Isotropic)

$$\frac{d\bar{\sigma}}{d\bar{e}^p} = \frac{E}{\beta\alpha\left(\dfrac{\bar{\sigma}}{\sigma_y}\right)^{\alpha-1}} \qquad (13)$$

3) Exponential Power Law (B/Al)

$$\frac{d\bar{\sigma}}{d\bar{e}^p} = \frac{\alpha}{\beta}\left(\bar{e}^p + \bar{e}^p_o\right)^{\frac{1}{\beta}-1} \qquad (14)$$

where

$$\bar{e}^p_o = \left(\frac{\sigma_y}{\alpha}\right)^{\beta}$$

## 3.7 Lamination Theory

Many applications of metal matrix composites require the lamination of unidirectional layers at different fiber orientations. Therefore, a first order lamination theory was included in ANPLAST.

Considering only the extensional stiffness of the laminate, the incremental laminate load deformation equations[1] can be written as

$$\begin{Bmatrix} dN_x \\ dN_y \\ dN_{xy} \end{Bmatrix} = \begin{bmatrix} A_{ij} \end{bmatrix} \begin{Bmatrix} de^o_x \\ de^o_y \\ de^o_{xy} \end{Bmatrix} \qquad (15)$$

where

$$A_{ij} = \sum_{k=1}^{N} [\bar{Q}_{ij}]_k t_k$$

$$[\bar{Q}]_k = [\bar{Q}]^e_k + [\bar{Q}]^p_k$$

$de^o_x$, $de^o_y$, $d\gamma^o_{xy}$ = laminate strains

$dN_x$, $dN_y$, $dN_{xy}$ = laminate forces per unit width

$[\bar{Q}]^e$ and $[\bar{Q}]^p$ are the transformed inverse of matrices of Eq. (1) and (9) for the $k^{th}$ layer of a multi-layered laminate.

## 4. EXPERIMENTAL VERIFICATION OF ANPLAST

An experimental study of the elasto-plastic response of a unidirectional B/Al MMC strip was conducted by RIZZI[4]. This particular experiment involved loading a flat B/Al specimen as shown in Fig. 3. Multiple strain gage recordings were taken in the vicinity of the hole. The test was run by increasing the load until certain preset breakpoints were reached, at which time the specimen was unloaded to zero.

The elastoplastic material properties used in the analysis are the same as those presented earlier. A preliminary elastic analysis revealed that initial yielding would occur at the horizontal hole perimeter at a remote stress of 4000 psi. Therefore the first ANPLAST loading step went straight to 4000 psi., and then the remainder of the elastic-plastic loading steps varied in size to coincide with experimental unloading points, but were never larger than 500 psi. The desired residual stresses and strains were obtained by restarting the analyses at the unload points and conducting an elastic unloading.

The longitudinal and lateral strain distributions for three elastic-plastic remote stresses (6000 psi, 7700 psi, and 10700 psi) are plotted against the experimental results and presented in Fig. 4. The residual strains produced after unloading are included. The agreement between the numerical solution and the experimental results are quite good.

Due to the very stiff nature of the reinforcing fibers, the lateral strains ($e_{xx}$) are small in magnitude and both the analysis and experiment demonstrate this. The longitudinal residual strains ($e_{yy}$) demonstrate that ANPLAST does a good job of predicting the plastic component of deformation.

Figure 5 highlights the stress-strain response at the innermost strain gage during the complete load-time history of the test. It is clear that ANPLAST predicts the overall elastoplastic load-deformation response of a B/Al metal matrix composite.

## 5. ANALYSIS OF EDGE CRACKED PANEL.
It is well known that plasticity plays an important role in crack growth in metals. The effect of plastic anisotropy is considered in the analysis of an edge-cracked B/Al panel to gain a better understanding of the differences from the more well known isotropic response.

An isotropic elastic-plastic analysis of an edge-cracked panel was conducted with ANPLAST. Work-hardening was expressed by the Ramberg-Osgood relationship. The progression of the plastic zone is illustrated in Fig. 6. The plastic zone representation is based on centroidal locations of yielded elements. A distinctive lobe at about $60^0$ measured counterclockwise from the horizontal is observed and coincides with the maximum Von Mises stress predicted in the near field theoretical solution. At a remote stress of 24500 psi, the plastic zone has progressed more than halfway across the width of the specimen, indicating large scale yielding.

The progression of the plastic zone for a B/Al MMC transverse edge cracked panel was conducted next. The plastic zone visualization is presented in Figs 6 and 7 for the case where the fibers are parallel and perpendicular to the crack. As was the case for the isotropic edge crack, large scale yielding is considered.

The plastic zone shape for the B/Al panel with the fibers parallel to the crack is quite similar to the isotropic example. When the fibers are orientated $90^0$ to the crack, a significant departure in behavior is observed. The plastic deformation occurs in a vertical band directly above the crack tip. A small zone of plastically deformed material resides directly in front of the crack tip, but for the most part the reinforcing action of the fibers retards progression parallel to the crack. It has been demonstrated in the laboratory[5] that the fatigue behavior of B/Al MMC was indeed a function of fiber orientation and parallels in behavior could be drawn to the resulting extent of plasticity.

## 6. CONCLUSIONS

An orthotropic elastoplastic material model has been presented that successfully captures the macroscopic lamina behavior of metal matrix composites. The elastoplastic material properties for a unidirectional Boron/Aluminum metal matrix composite were derived. A finite element program, ANPLAST, was shown to solve orthotropic elastoplastic problems involving complex geometries and stress states. An experimental analysis of a B/Al metal matrix composite sheet with a circular hole verified the ANPLAST predictions. The plastic zone in a B/Al edge cracked panel was illustrated with ANPLAST and compared to the isotropic case. It was found that fiber orientation has a strong influence on the resulting plastic deformation.

## 7. REFERENCES

1. Jones, R.M., Mechanics of Composite Materials, Scripta Book Company, Washington, D.C., 1975.
2. D, Kenaga, J.F. Doyle, and C.T. Sun, "The Characterization of Boron/Aluminum Composite in the Nonlinear Range as an Orthotropic Elastic-Plastic Material", to be published, J. of Composite Materials.
3. A.R. Leewood, J.F. Doyle, and C.T. Sun, "Finite Element Program for Analysis of Laminated Anisotropic Elastoplastic Materials", to be published, J. of Computers and Structures.
4. S.A. Rizzi, A.R. Leewood, J.F. Doyle, and C.T. Sun, "Elastic-plastic Analysis of Boron/Aluminum Composite Under Constrained Plasticity Conditions ", to be published, J. of Composite Materials.
5. C.T. Sun, J.F. Doyle, and D. Kenaga, "Fatigue Crack Growth and Overload Effects in Metal-Matrix Composites", AFWAL, Contract No. F33615-82-K-3218.

## 8. ACKNOWLEDGEMENTS

The authors would like to thank AFWAL for their support of this research. Contract No. F33615-82-3218.

## 9. BIOGRAPHY

**Dr. Alan R. Leewood**
Dr. Leewood is Vice President of Advanced Composite Engineering, Inc, West Lafayette, IN. He received his Ph.D. from Purdue University in 1985.

**Prof. J.F. Doyle**
Professor Doyle is an Associate Professor in the School of Aeronautics and Astronautics at Purdue University. He received his Ph.D. from University of Illinois in 1977.

**Prof. C.T. Sun**
Professor Sun is a Professor in the School of Aeronautics and Astronautics at Purdue University. He received his Ph.D. from Northwestern University in 1967.

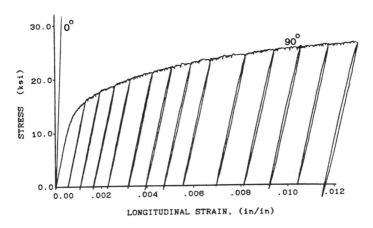

Figure 1. Uniaxial Experimental Response of a B/Al
Metal Matrix Composite.

axial stress (ksi)

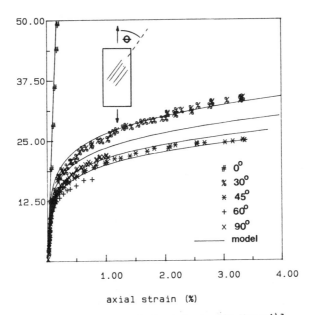

axial strain (%)

Figure 2. Comparison of the Anisotropic Material
Model with Uniaxial Experimental Results
for B/Al.

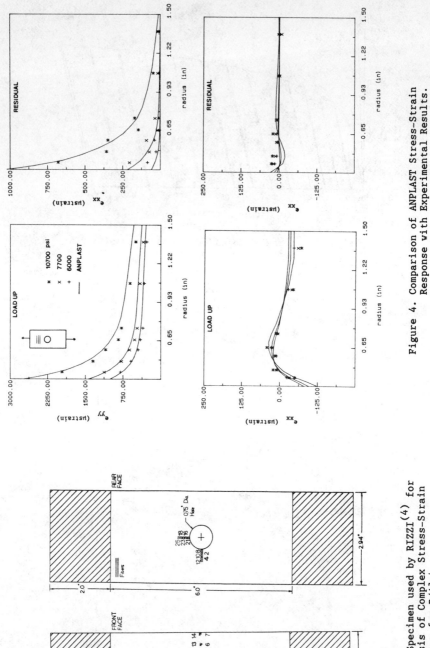

Figure 4. Comparison of ANPLAST Stress-Strain Response with Experimental Results.

Figure 3. Test Specimen used by RIZZI[4] for Analysis of Complex Stress-Strain Behavior for B/Al.

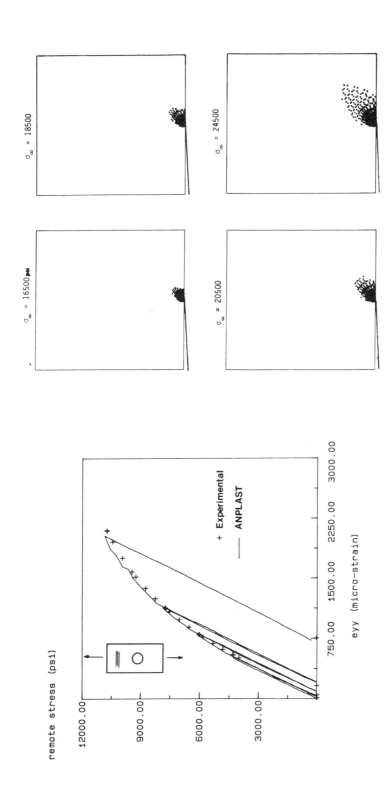

σ_∞ = 18500

σ_∞ = 16500 psi

σ_∞ = 24500

σ_∞ = 20500

Figure 6.  Plastic Zone Visualization of a
Transverse Edge-Cracked Panel
(isotropic) Subjected to Uniaxial Remote
Loading.

remote stress (psi)

12000.00

9000.00

6000.00

3000.00

+ Experimental

——— ANPLAST

750.00      1500.00      2250.00      3000.00

eyy (micro-strain)

Figure 5.  Time-History Comparison of Stress-Strain
Response with Experimental Results.

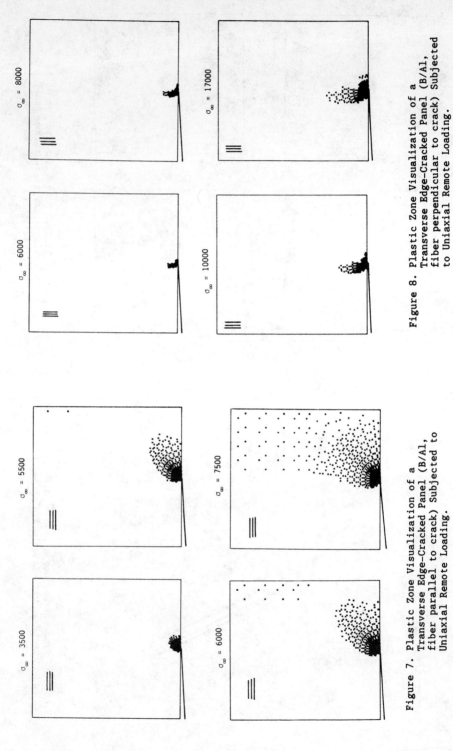

$\sigma_\infty = 8000$

$\sigma_\infty = 6000$

$\sigma_\infty = 17000$

$\sigma_\infty = 10000$

Figure 8. Plastic Zone Visualization of a Transverse Edge-Cracked Panel (B/Al, fiber perpendicular to crack) Subjected to Uniaxial Remote Loading.

$\sigma_\infty = 5500$

$\sigma_\infty = 3500$

$\sigma_\infty = 7500$

$\sigma_\infty = 6000$

Figure 7. Plastic Zone Visualization of a Transverse Edge-Cracked Panel (B/Al, fiber parallel to crack) Subjected to Uniaxial Remote Loading.

## CONTOURED/TAPERED COMPOSITE REINFORCEMENTS FOR
## HIGH-PERFORMANCE COMPOSITE MOTOR CASES

K. M. Tezak
Morton Thiokol, Inc. Wasatch Operations
P.O. Box 524
Brigham City, Utah
(801) 863-3739

### Abstract

A major design/manufacturing challenge with filament-wound solid rocket motors is the retention of the polar bosses (port openings). The typical solution has been to incorporate flat, constant-thickness, spiral hoop-wound reinforcements (wafers) in the area adjacent to the port openings. This solution in turn causes its own problems, such as stress discontinuities at the wafer tip and wrinkles due to forming a flat surface to the dome. These stress discontinuities can lead to early failure in the dome area of the case and lower pressure performance.[1]

One solution to the problem is to lower the helical dome stresses by adding additional material to the domes (i.e., design to a low helical-to-hoop stress ratio). However, this also results in a less than optimum performance.

The ideal solution is to design a high-stress ratio using dome reinforcement contoured to the dome's shape and tapered towards the edge of the reinforcement so that stress discontinuities are minimized.

This paper discusses the process development, equipment design, and software generated to produce a low-cost, four-axis winding machine to manufacture this ideal reinforcement. The reinforcements have been used with excellent results on ten different graphite epoxy case designs ranging from 10 to 75 in. in diameter.

## 1. INTRODUCTION

The use of flat, spiral-wound reinforcements to reinforce the polar boss region of a high-performance pressure vessel (case) has always caused significant problems, which ultimately result in a lower burst strength than the analysis predicted, along with higher variability. This in turn requires adding more helical layers and/or mats, dome caps, and fillers which increases weight and cost.[1] The main reasons for using flat, constant-thickness wafers are the ease and speed of fabricating the wafers and the low cost of tooling (two flat disks and spacer).

The three major problems associated with the use of flat wafers are:

1. Fiber Wrinkling. Forming flat wafers to a dome contour requires that the fibers be moved outward or wrinkles will form, which is usually the case to some degree (Fig. 1). A wrinkled wafer reduces the load-carrying capabilities of the fiber. This operation is entirely operator dependent and can lead to high variability of the case's ultimate burst strength. The time required to form these wafers is usually considerable and directly proportional to the quality.

2. Stress Discontinuities. A constant-thickness wafer causes stress discontinuities at adjacent helical layers because the helical layers bridge over the abrupt, outside diameter edge of the wafer (Fig. 2). Stress discontinuities at the tips of nontapered

87282-1

wafers are self-evident by the localized failure mode of dome-failed cases, in that they almost always fail at the wafer tip (Fig. 3).

3. Case Winding Reversal Buildup. Optimum reinforcement thickness, with respect to the stresses and composite buildup near the polar boss, cannot

Figure 1.    Wrinkling of Flat Wafer From Forming to a Dome

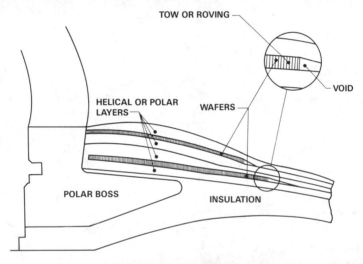

Figure 2.    High Stresses Caused by Abrupt Ending of Radial Full-Thickness Fibers

Figure 3.    Test Bottle Showing Localized Failure at Wafer Tip

be achieved. Composite buildups around the port opening where the winding band reverses cause a condition for the filament winding band to bridge on the outer layers. Wafers may also be used to flatten out this area to prevent winding band bridging (Fig. 4).

Obviously, these problems would not exist if the reinforcements were fabricated to the contour of the case and were wound to an optimum cross section thickness and tapered at the outside edge (Fig. 5). Fabrication of improved reinforcements requires a specialized numerically-controlled winding machine since there is nothing presently available (Fig. 6).

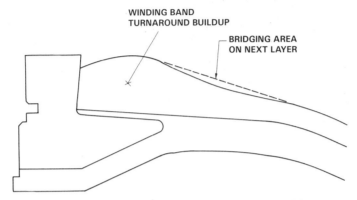

Figure 4.    Bridging Caused by Winding  Reversal Buildup                A004406

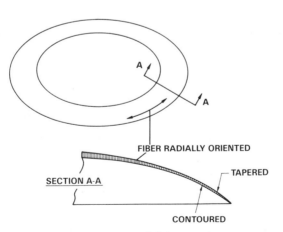

Figure 5.    Ideal Port Reinforcement                A004407

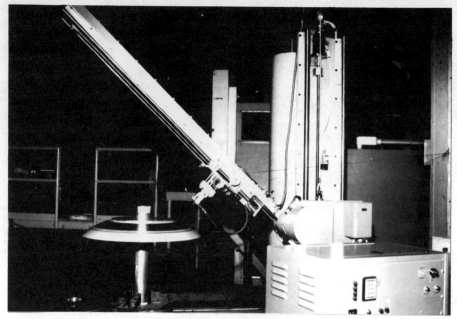

**Figure 6.   Contoured Tapered Winding Machine**

## 2. DEVELOPMENT OF THE PROOF-OF-CONCEPT WINDING MACHINE

Winding on a contoured mandrel similar to a tape lay-down machine is an obvious solution. A crude 2-axis mechanical gear-driven mechanism was designed and built for proof of concept and development of the following process parameters.

• Fiber payoff (delivery) onto the mandrel.

• Application of adhesive film to the mandrel.

• Heating and cooling of the prepreg fiber tow.

Several proof-of-concept, conical (5- to 40-deg), tapered reinforcements were wound on this machine, verifying this concept and defining the aforementioned process parameters.

## 3. DEVELOPMENT OF A FOUR-AXIS WINDING MACHINE

### 3.1 Machine/Material Considerations

Material development tests have clearly demonstrated the performance improvements that a graphite epoxy (Gr/E) case has over cases made from other filament materials. Because of Gr/E's stiffness, it is even more sensitive to discontinuities than Kevlar® or glass.

To achieve the highest potential in a pressure vessel with the new high-performance Gr/E materials, contoured tapered reinforcements will be required. The performance can be expressed by PV/W (pressure x volume/weight).[1]

To produce a spiral-wound reinforcement on a contoured mandrel and at different payoff rates, a computer-controlled winding machine would be required with the following four axes:

• Two coordinated axes that are slaved to the mandrel rotation through the computer program. They control the payoff rate (wafer thickness) and maintain the payoff wheel tangent to the mandrel surface.

• One axis for the arm pivot that maintains fiber delivery payoff pressure.

• One axis for the rotation.

### 3.2 Development Effort

The development effort included the following:

• Machine conceptualization and design.

• Computer drive design, wiring, and assembly.

- Software development for contour path generation and payoff rate.
- Software development for machine input/output and electrical drives.
- Proving and testing equipment.
- Actual winding of contoured/tapered wafers into a burst test case.
- Burst of test cases and evaluation of data.

### 3.3 Machine Requirements

A basic machine requirement was that it be capable of winding reinforcements up to 72 in. in diameter for large (160 in. dia) booster motors. This requirement was the main design driver behind the arm concept. This design also lends itself to easy fabrication and programming (Fig. 6).

It was decided early on that a microprocessor-controlled, stepper motor drive system was ideal for this application because of the low cost, accuracy, and ease of programming, along with the low torque requirements of the winding machine. The control microcomputer is a rugged Motorola 6809*-based board, which interfaces to a standard stepper motor indexer drive module.

### 3.4 Software Development

The software to determine the machine's coordinates with respect to the contour, and the payoff rate with respect to the mandrel rotation is done with FORTRAN on the mainframe computer. The equations for defining the path are simple trigonometry functions. The reason for performing the calculations on the mainframe is that the winding control tables that define the machine path can easily be calculated directly from the CAD/CAM drawing of the reinforcement, in effect creating a CAD/CAM loop. The output is then downloaded directly into a PROM memory device from the mainframe. Because of the critical function that these reinforcements have in a solid rocket motor, a nonvolatile, extremely dependable PROM memory board is used. (A PROM is a programmable read-only memory device that premanently stores the binary information.)

### 3.5 Machine Control Software

Software to program the machine microcomputer control board was developed with a Motorola*

EXORSET-110 development system in 6809* Motorola assembler. The operation uses the following flow logic:

1. Read switch number to load the address of the winding table into memory.
2. Drive the horizontal and vertical carriages to their home reference location.
3. Drive carriages to reinforcement start location.
4. Wind reinforcement.

The major difficulty of the machine development was the microprocessor control software, which ended up using 700 lines of assembler code. This included error traps, interrupt routines, error indicators, and a manual mode operated from a key pad, which has proved to be very useful during initial setup.

### 3.6 Winding Evaluations

The winding machine was installed onto an existing spin station and the rotation encoder was interfaced to the computer.

Initial winding evaluations demonstrated the sensitivity of the winding setup. To effectively produce high quality wafers, the critical parameters are:

1. Payoff wheel design.
2. Prepreg tack.
3. Advance of the resin at winding.
4. Temperature of the prepreg prior to payoff.
5. Cooling of the prepreg immediately after payoff.
6. Winding speed.
7. Payoff pressure.
8. Tack of the adhesive film.

When a new resin fiber is used, the foregoing factors would need to be reevaluated.

Due to the successful development effort, winding evaluations produced usable reinforcements for a hydroburst test case. Three distinctly different designs for each end were fabricated. Hydrotesting of this case was a success in that the case failed in the cylinder and the strain gauge readings located in the reinforcement area were very low. A PV/W of $2.0 \times 10^6$ was obtained from this 37-in.-

diameter pressure vessel, which is one of the highest ever produced in the solid rocket motor industry.

## 4. CONCLUSIONS

Because of this machine's capabilities, designers now have the freedom to design the optimum reinforcements. Our typical high-strength case uses three distinctly different reinforcement designs on each end, tailored to each particular layer's strength, stiffness, and contour requirements.

To date, these reinforcements have been successfully used on ten different high-strength graphite case designs, ranging from 18 to 75 in. in diameter. In all instances these reinforcements have:

- Reduced variability, increasing the reliability. This is very important when considering satellite booster motors and ballistic missiles.

- Increased PV/W by allowing a higher stress ratio, decreasing weight, and increasing performance.

- Decreased the stress in the critical dome/polar boss region.

- Decreased cost and both material usage and winding time of the case, because fewer helical layers are required due to the higher stress ratio.

## 8. REFERENCES

1. Michael H. Young and Ben A. Lloyd, "Rocket Motor Performance Optimization," 17th National SAMPE Technical Conference. October 1984.

## 9.BIOGRAPHIES

Kenneth M. Tezak is a process development engineer in the Advanced Composite Process Development Group. He graduated with a B.S. degree in manufacturing engineering from Weber State College in 1978. He has seven years experience in process development and manufacture of composites. He was the responsible manufacturing engineer for Stage-III MX and the STAR 75 satellite transfer motor that was successfully static fired in 1986.

COMPARISON OF ASTM STANDARD COMPRESSION TEST METHODS
OF GRAPHITE/EPOXY COMPOSITE SPECIMENS

Christina C. Gedney, Clyde R. Pascual,
Faysal A. Kolkailah, Ph.D.
Aeronautical Engineering Department
California Polytechnic State University
San Luis Obispo, CA 93407
and
Brian A. Wilson
Aerojet General Corporation

## Abstract

The focus of this study is to provide results which would support the selection of a reliable method of compressive testing of graphite/ epoxy composite materials. Three methods were examined for testing Toolrite MXG-7620, Graphite Prepreg Fabric, Style 2534 and Style 2577; ASTM D695; Union Carbide's Modified ASTM D695; and ASTM D3410, also known as the Celenese Method.

Aerojet Strategic Propulsion Company, funded this study by providing the composite materials along with the fixtures and jigs to test the specimens to failure in compression.

In order to determine the most reliable method of compression testing, the Ultimate Compression Strength of the specimens and the simplicity of each test method were measured and observed. CAEDS (a program developed by IBM), was employed to determine the Stress Concentrations for each specimen. Through CAEDS, compressive necking was apparent on all the specimens. The Celenese method provided the best results, but required the most effort and skill to perform. The ASTM D695 method required the least amount of effort but produced scattered results. The modified ASTM D695 method yielded results comparable to the Celenese method and was a simple test to conduct.

## 1. INTRODUCTION

The use of composites have become widespread in the Aerospace industry. This is a result of the conflicting requirements of high strength and low weight. There has arisen a need to establish a set of testing procedures by which the mechanical properties of a com- posite material can be determined, simultaneous to the application of composites.

"Before these high performance composites can be widely accepted in long term, critical high stress applications and under adverse environmental extremes, reliable methods must be established to characterize the fundamental properties. . . . Methods of analysis have been developed which reflect the homogeneity of the metals. These methods cannot anticipate the numerous variables in the composite constituents or in the structure of the composite-variables which have significant influence on the behavior of the composite structure."[1]

In order to develop a design, extensive compression testing must be conducted. Compressive properties of composite materials are among the most difficult to determine. Slight variations in specimen geometry result in the eccentricity of the applied load. Thus, complex loading fixtures and specimen geometries have been developed in order to achieve accurate measurement of the compressive strength of a composite material (Figures 1, 2, 3 and 4).

Aerojet Strategic Propulsion Company has contracted the Aeronautical Engineering Department at California Polytechnic State University to find the method of compression testing which yields the best results with the least effort. The tests compared were the ASTM D695, Union Carbide's modified ASTM D695, and the ASTM D3410, also known as the Celenese method. The composite specimens were fabricated from Toolrite MXG-7620 graphite prepreg fabric style 2534 and 2577. The specimens were supplied by Aerojet (Figure 1). The test specimens for the ASTM D695 were nominally 0.5 in. x 0.5 in. x 0.045 in. The modified ASTM D695 were 3.125 in. x 0.5 in. x 0.045 in. Half of the modified specimens had four fiberglass tabs bonded to the ends. The Celenese specimens were 5.5 in. x 0.25 in. x 0.125 in. The Celenese specimens had four 2.5 in. x 0.25 in. steel tabs bonded to the ends.

## 2. PROCEDURE

### 2.1 ASTM D695

For the ASTM D695 method, the procedure was quite simple. A Tinius-Olsen Test Machine was used along with the compression fixture shown in Figure 2. The compression fixture was placed between the Tinius-Olsen platens, and the platens were lowered to within an inch of the top of the compression fixture.

The specimens to be tested were measured and the dimensions were recorded and checked with the required tolerances. In order to produce consistent data, the specimens must be within the required tolerances. Next, the test specimen was placed between the surfaces of the compression fixture. Time was taken to align the centerline of the specimen to the centerline of the compression fixture plunger, and to ensure that the ends of the specimen were parallel with the surface of the compression fixture.

The Tinius-Olsen load indicator was set to zero and the loading process was begun. Loading continued until the specimen failed. The ultimate load for the specimen was shown by the load indicator of the Tinius-Olsen, and was recorded.

2.2 Modified ASTM D695

Two different types of specimens (tabbed and untabbed) were tested using the modified ASTM D695 method. The test procedures for the two types of specimens were similar. A Satec Test Machine was used along with a support jig shown in Figure 3.

Each specimen was measured, dimensions recorded and checked with the tolerances. The measurement of the tabbed section as well as the untabbed section was necessary for the tabbed specimens. The specimen was centered in the support jig, and the screws were finger tightened. The jig was placed on the bottom platen and the crosshead of the Satec Test Machine was lowered until it came into contact with the jig/specimen. Adjustment must be made to the jig/specimen so that it is centered and perpendicular to the platen surface.

The load indicator was set to zero. For the tabbed specimen, the load was applied until failure, while for the untabbed specimen the load was applied to one-third of the ultimate load of the tabbed specimen. The ultimate load was recorded.

2.3 ASTM D3410 - Celenese

The ASTM D3410 (the Celenese method) was conducted on an Instron Testing Machine. The machine was zeroed and calibrated according to standard Instron procedures. The test specimens were mounted into the grip cavities of the split collet-type grips shown in Figure 4. The grips were manually closed and fit into the sleeves which had an inside taper matching the taper of the grips. A one-half inch spacer was placed between the top and bottom grips. All surfaces were well lubricated, and the entire assembly was inserted into the cylindrical shell (Figure 4). The fixture was then placed between the platens of the Instron. The crosshead was lowered to within one-quarter inch of the fixture. The chart drive was turned on, the crosshead speed was set at 0.05 in./min., the spacer was removed, and loading was begun. The freeness of the cylindrical shell was assessed by moving it up and down vertically to ensure that it did not carry any load. The specimens were loaded to failure, and the ultimate load was read from the load chart. The specimen was unloaded, and the fixture was broken apart and the next specimen was inserted.

2.4 CAEDS

CAEDS is an IBM product developed by Structural Design

Research Corporation (SDRC). CAEDS is a design system that combines capabilities to model an object, create a finite element mesh on the object, translate the information to a finite element analyzer and display results. There are six major modules within CAEDS and a user can lease those which will be needed. CAEDS can also interact with other engineering tools such as CADAM, CATIA, and NASTRAN. The ease of using one major tool to combine the capabilities of several engineering tools greatly improved efficiency. CAEDS can drastically reduce the cost of designing parts and design cycle time.

For the scope of this investigation, four of the six modules of CAEDS were used. The graphics module offers interactive menu-driven finite element modeling using the pre-processor programs Model Creation and Enhanced Mesh Generation. GRAPHICS interfaces with CADAM for transferring a model drawing. The post-processing was done with Output Display which allows the user to display the Finite Element Analysis (FEA) results, and gives the capability to interactively define the output to be displayed. Output Display shows the user mode shapes, stresses, strains, strain energy, temperatures, reaction forces and kinetic energy. These results can be displayed using several display forms: contour, XY graphs, report, and arrow plots.

The Finite Element Solver (FES) was interfaced with CAEDS GRAPHICS. FES is a general purpose linear, 3-dimensional, finite element analysis program which addresses static, dynamic, and heat transfer problems. FES has pre-processing and post-processing capabilities and can easily interface with CAEDS GRAPHICS. Although FES does not have a large spectrum of applications, it has the advantage of being easy to use.

SYSTAN is an interactive system modeling and analysis program. This system was used to combine analytical data from FEA codes given by FES with experimental data from modal test results to determine the entire system response. The results were eigenvalues and eigenvectors, frequency response, and transient response which were stored in a SYSTAN database.

CAEDS offers many convenient utilities to increase design speed and efficiency. The File Translator was used to write the model information such as node and element data in a format readable by the FES. This utility greatly reduces the time spent preparing a FES job. The Data Loader allowed the output from FEA to be read into a universal file which CAEDS GRAPHICS can understand. Data Loader has the ability to allow the user to write their own interface with any finite element code. The system allows the user to view the results on the CADAM scope.

### 3. DISCUSSION OF RESULTS

In order to select a reliable method of compression testing, one test machine should be utilized. However, due to interdepartmental conflicts between the Aeronautical Engineering and Civil Engineering Departments, three different test machines were used. All of the Instron Test Machines on campus were unavailable for use due to other student projects or mechanical problems. For the ASTM D695 (style 2577), a Tinius-Olsen Test Machine was used, for it was compatible to the Tinius-Olsen compression fixture. A Satec Test Machine (similar to the Tinius-Olsen Test Machine) was used for the modified ASTM D695 (both styles), because the Tinius-Olsen machine was unavailable due to student projects. However, for the ASTM D3410 (Celenese) and the ASTM D695 (style 2534), the Civil Engineering Instron Test Machine was available. Originally the Aeronautical Engineering Instron Test Machine was to be used for all of the tests, but as a result of a faulty low speed clutch it was inoperable for the duration of the project.

The ASTM D695 method was conducted on two different test machines. Style 2577 was tested using the Tinius-Olsen Test Machine, producing an ultimate strength of 20 ksi with a standard deviation of 8 ksi for a coefficient of variation of 0.39 with 11 specimens. An Instron Test Machine yielded an ultimate strength of 45 ksi with a standard deviation of 5 ksi for a coefficient of variation of 0.11 with 9 specimens of style 2534. The results generated by the use of the Instron Test Machine were superior to the results found by use of the Tinius-Olsen Test Machine. A 72% reduction in the coefficient of variation clearly demonstrates the supremacy of the Instron Test Machine.

The Satec Test Machine was used for the modified D695 method. Style 2534, in which 8 specimens were tested, yielded an ultimate strength of 68 ksi with a standard deviation of 18 ksi for a coefficient of variation of 0.27. Nine specimens were tested of style 2577. An ultimate strength of 46 ksi with a standard deviation of 10 ksi for a coefficient of variation of 0.22 was obtained. The untabbed specimens failed at loads less than one-third of the tabbed specimens' ultimate load in end-crushing (brooming). Strain gauges were not applied to the untabbed specimens for the surface of the jig was flush to the surface of the specimen. Thus, strain readings would not be obtainable. Style 2534 yielded an ultimate strength of 19 ksi with a standard deviation of 3 ksi for a coefficient of variation of 0.16, with 8 specimens. Style 2577 yielded an ultimate strength of 16 ksi with a standard deviation of 3.6 ksi for a coefficient of variation of 0.22 for 9 specimens. If the modified ASTM D695 test, with the tabbed specimens, had been conducted using the Instron Test Machine, more accurate results would have been obtained, as shown by the percent difference in the coefficient of variation in the ASTM D695 test.

The ASTM D3410 (Celenese method) was conducted on the Civil Engineering Instron Test Machine. Strain readings from the Celenese specimens for the modulus of elasticity were unable to be obtained for the strain gauges came off when the leads were connected to a strain indicator apparatus. This could have been due to improper application materials or procedures, or faulty lead attachment. Style 2534 gave an ultimate strength of 62 ksi for a standard deviation of 7 ksi with a coefficient of variation of 0.11, with seven specimens being tested. For style 2577, seven specimens were also tested, an ultimate strength of 43 ksi for a standard deviation of 3 ksi with a coefficient of variation of 0.07 was obtained. The Celenese method yielded good results; however, due to the complexity of the fixture, it was difficult to prevent the sleeve of the fixture from carrying part of the load.

Given the material properties for the MXG-7620, graphite prepreg fabric - style 2534 (Table 1), the Finite Element Analysis (CAEDS) for each type of specimen was performed.

Using a CADAM scope, the model was created. Model creation produced a mesh, and attached nodes and elements to each model (Figures 5, 8).

After the model was solved, CAEDS gave the graphical representation of the stresses and strains (displacements). Viewing of the displacement (Figures 7, 10) and the stress concentration (Figures 6, 9) yields an accurate understanding of what the specimen experiences. From Figure 6, part of the specimen is under tension (Top). This tensile force is a result of the specimen deformation during testing. It can be deduced that as a result of the deformation of the specimen (Figure 7), the upper part of the specimen is in tension. Thus, the tension created by the compression process contributes significantly to the failure of the specimen.

4. CONCLUSIONS

As a result of this investigation, it has been found that the Union Carbide modified ASTM D695 with tabbed specimens yields the most accurate results with the least amount of effort. This can be accomplished when the test is conducted on a machine which is easily calibrated, such as an Instron Test Machine. The support jig could be improved by the drilling of holes to accommodate strain gauge instrumentation of the specimen. All methods, when conducted on an Instron Test

Machine, would yield satisfactory results. However, the modified ASTM D695 method combines precision and simplicity.

5. REFERENCES

1. Delmonte, J., Technology of Carbon and Graphite Fiber Composites, New York, Van Nostrand Reinhold Co., 1981.

2. Kincis, T., Static Test Methods for Composites, New York, Van Nostrand Reinhold Co., 1985.

3. Whitney, J. M., Daniel, I. M., and Pipes, R. B., Experimental Mechanics of Fiber Reinforced Composites, Englewood cliffs, New Jersey, Prentice-Hall Inc., 1982.

Table 1

Material Properties

| Specimen Style | Modulus of Elasticity E (psi) | Ultimate Stress (psi) | Poisson's Ratio |
|---|---|---|---|
| 2534 | $6.6 \times 10^6$ | 70,400 | .0887 |
| 2577 | $8.88 \times 10^6$ | 41,200 | - - - |

Table 2

Results

| Test | Specimen Style | Mean Ultimate Stress (lb./in$^2$) | Standard Deviation (lb./in$^2$) | Coefficient of Variation |
|---|---|---|---|---|
| ASTM D695 | 2534 | 44,517.2 | 4,819.4 | 0.1083 |
| | 2577 | 20,326.4 | 7,943.8 | 0.3908 |
| Modified ASTM D695 Tabbed | 2534 | 67,867.8 | 18,477.2 | 0.2723 |
| | 2577 | 45,947.6 | 10,369.4 | 0.2257 |
| Modified ASTM D695 Untabbed | 2534 | 18,588.9 | 3,082.8 | 0.1658 |
| | 2577 | 16,097.1 | 3,600.7 | 0.2237 |
| ASTM D3410 | 2534 | 61,870.8 | 7,078.5 | 0.1144 |
| | 2577 | 43,445.0 | 3,066.5 | 0.0706 |

1021

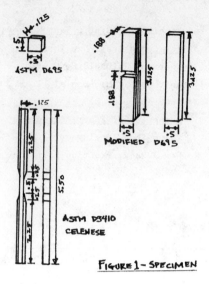

.5 ← .125
.5
ASTM D695

.188
3.155
3.125
.188
.5    .5
MODIFIED D695

.125
2.15
.5
1.5
.25
3.125
5.50

ASTM D3410
CELENESE

FIGURE 1 - SPECIMEN

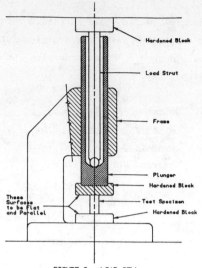

Hardened Block
Load Strut
Frame
Plunger
Hardened Block
These Surfaces to be Flat and Parallel
Test Specimen
Hardened Block

FIGURE 2 - LOAD CELL

.14 ← → ← .25
← .94 →

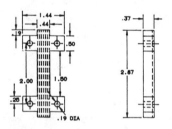

1.44
.44
.9
.50
2.00    1.50
.25
.19 DIA
.37
2.67

FIGURE 3 - SUPPORT JIG

.12    2.50    .50    2.50

Specimen
Collet Grip
Tapered Sleeve
Gage Section
Cylindrical Shell
Steel Tab

2.25    .25    .25    .14

FIGURE 4 - CELENESE FIXTURE

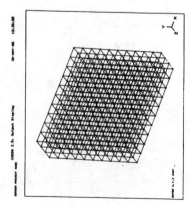

Figure 5 - ASTM D695 MODEL

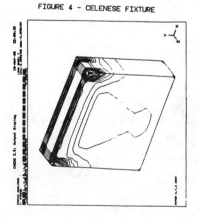

Figure 6 - ASTM D695 - STRESS AT ULTIMATE LOAD

1022

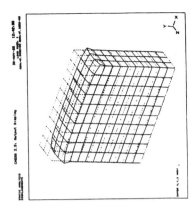

Figure 7 - ASTM D695 - DISPLACEMENTS

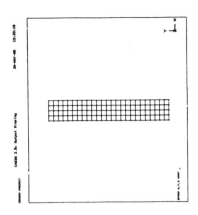

Figure 8 - Modified ASTM D695 - Model

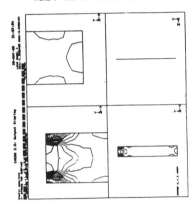

Figure 9 - Modified ASTM D695 - Stress
AT ULTIMATE LOAD

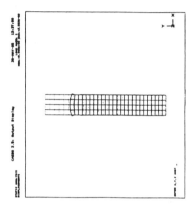

Figure 10 - Modified ASTM D695 - Displacements

## 6. BIBLIOGRAPHY

### Christina Gedney

Christina Gedney will be awarded her Bachelor of Science degree in Aeronautical Engineering on December 1986, from the California Polytechnic State University, San Luis Obispo. While at Cal Poly, she was the co-founder and president of the Cal Poly SAMPE Student Chapter. At this printing, Ms. Gedney's future employer has yet to be determined. However, she will probably be employed with an aerospace company in the composite material area.

### Clyde R. Pascual

Clyde R. Pascual will be awarded his Bachelor of Science degree in Aeronautical Engineering on December 1986, from California Polytechnic State University. He has extensive experience in Finite Element Analysis, particularly CAEDS and CADAM.

### Brian A. Wilson

Mr. Brian A. Wilson has over 25 years of technical responsibility in the aerospace industry for Bendix, Rockwell, Brunswick, and Aerojet in a variety of mechanical and material disciplines. During the past eight years, Mr. Wilson has been responsible for design, material research, and advanced development in strategic and tactical solid rocket motors.

He is a member of AIAA, SAMPE, SME, and SEM. He is the International Secretary of SAMPE, and a past chairman of the Sacramento chapter.

Mr. Wilson received his B.Sc. in Civil and Mechanical Engineering at St. Andrews University, Scotland, and his M.S. in Mechanics and Materials at California State University, Northridge. Currently, he is writing the dissertation for his Ph.D. at the University of Nebraska.

### Faysal A. Kolkailah

Dr. Faysal A. Kolkailah received his B.Sc. in Aerospace Engineering at Cairo University, his M.Sc. in Aerospace Engineering at University of Cincinnati, and his Ph.D. in Mechanical Engineering at Louisiana State University.

He taught at the University of Cincinnati, Louisiana State University, and University of New Orleans. Currently, he is a Professor of Aeronautical Engineering at Cal Poly.

Dr. Kolkailah has strong industrial experience as an aeronautical engineer and as a consultant to several private industries.

He has considerable research experience in composite structure, fracture mechanics, optimum design, aircraft design, and numerical methods in solid mechanics. He has published considerably in these areas.

Dr. Kolkailah has been a principal investigator or co-investigator on a number of research projects funded by NASA/ Ames Research Center, Rockwell International, and Aerojet Techsystems.

Dr. Kolkailah is a member of SAMPE and other professional societies. He is the founder and advisor of the SAMPE student chapter at Cal Poly.

HYDROTEST OF ADVANCED COMPOSITE PRESSURE VESSELS TO ASSESS
EFFECTS OF STRESS RATIO ON PERFORMANCE

S. Stagliano and F. A. Kolkailah, Ph.D.
Aeronautical Engineering Department
California Polytechnic State University
San Luis Obispo, CA 93407

## Abstract

The main objective of this project was to begin manufacturing bottles, hydroburst them, and analyze the results, setting up standard procedures for the Air Force Rocket Propulsion Laboratory, and solving hidden problems as they arise during the processing. The task chosen was to characterize the performance of T40/5245c graphite filament wound ASTM standard test and evaluation bottles for various longitudinal and hoop stress ratios and compare the results to theoretical calculations. The bottles were wound with stress ratios varying from 0.3 to 0.81. The netting equation was employed to calculate the stress ratios.

Two ways to change stress ratio are to change the wind angle or the ratio between the number of the longitudinal and the hoop piles. The second method was employed in this study. Bottle performance varied from 0.5 for a stress ratio of 0.3 to 1.2 for a stress ratio of 0.81.

Three sets of bottles were manufactured and tested. The first two sets fell short of the predicted theoretical values. The third set of bottles tested correlated well with theory as winding procedures improved, verifying that experimentally changing ply thickness effects bottle performance in the same manner as varying wind angle.

Within this study other parameters which could dramatically effect bottle performance, such as the effects of the tying of the tows upon completion of winding and the effects of interspersment were recorded. These will be looked at individually in further studies.

## 1. INTRODUCTION

The Air Force Rocket Propulsion Laboratory (AFRPL) is establishing a facility for research and development of composite materials used in weapon delivery systems. This is one in a series of projects pertaining to the construction and performance of filament wound pressure vessels to be undertaken at AFRPL.

Filament winding is a process by which continuous reinforcements, or rovings, are wrapped around a mandrel in such a way as to completely cover it to the desired thickness and orientation. The fibers are oriented in such a way so as to achieve a high strength. The rovings are impregnated with a thermoset resin which acts as a binding material. The filament wound structure is then heated until the resin is cured. The matrix serves to bind the fibers together and to transfer tensile loads from fiber to fiber.

The main objective of this project was to begin manufacturing bottles, hydroburst them, and analyze the results, setting up standard procedures, solving hidden problems as they arise during the processing, and building up the AFRPL competence in winding. Two spools of T40 fibers, preimpregnated with 5245c resin were employed.

1.1 Mandrel Fabrication

Figure 1 shows the ASTM mandrel mold used in this experiment. A rubber bladder was inserted into a mold half and sand, with Plastilease 512 B as a binder was packed in. The mold half was cured. Two mold halves were assembled to form the whole mandrel and a "belly band" was placed around the center holding the two halves together. This was cured again. A polar boss was glued to each end and the mandrel was sanded down to rid it of rough edges that could cause stress concentrations. Many mandrel halves were discarded because of improper curing, wrinkles, holes, and poor "belly bands." Thirteen mandrels were used for this experiment. Two were wrapped with Kevlar 49 and 11 with T40 graphite (1 failed in the polar region and was discarded.)

1.2 Filament Winding

An ENTEC MODEL T36-42 Tumble Winder was used for winding. The bottles were put on rods and inserted on the spindle of the winder and tilted to an angle of 11.5 degrees. The prepreg fibers were fed through the machine and tied to the bottles. The polar plies were applied first as shown in figure 2. The winding arm tumbled the bottle while at the same time rotating it. The angle that the fibers were laid was a constant based on geometry. The bottle rotation was very critical. For the graphite tows, a speed of 1.7 degrees rotation per tumble was selected as optimum.

There is an interesting phenomenon when winding the polar plies, that is, in order to wrap fully around the polar boss region, the center of the bottle is wrapped twice! For that reason, it is common to see polar plies in multiples of 2 (i.e., 2, 4, 6, 8). when the full number of polars were placed (two plies or 360 degree rotation) the hoop plies were placed. The hoop plies take most of the hoop stress in the bottle. They were wrapped circumferentially on the cylinder portion of the bottle. The rate chosen was 0.07 inches per 360 degrees rotation for the graphite fiber. A loop was made from a scratch piece of yarn and wound over about a half an inch. The yarn was cut and inserted into the loop and tightened. Further study on the effects of tie off on bottle performance is warranted.

1.3 Sand Removal

To soften the sand, the bottles were soaked in water for 15 to 20 minutes. The sand was drained out from the hole in the polar boss. A scraper was necessary to get out excess sand.

1.4 Bottle Testing

The bottles were filled with water and pressurized. Great care was taken to bleed out all the air in the bottles. The pressure was then increased until the bottle broke. The pressure at the time of burst was recorded. The rate of pressurization was not considered critical for this experiment because of the low rate so it was not recorded.

1.5 Evaluation of Data

The bottles were designed to break the fibers in the hoop direction without breaking the fibers in the polar direction. Netting analysis was used to calculate ultimate stress of the hoop fibers. The equation is:

$$P_{burst} = \sigma_H t_H / \{R(1+\epsilon)(1-\tan^2\theta/2)\} \quad (1)$$

(See appendix for derivations from the Netting equations.)

2. RESULTS

This experiment was separated into three separate tasks: mandrel fabrication, filament winding, and hydrobursting. Each task will be looked at separately.

2.1 Mandrel Fabrication

The first step in the fabrication of filament wound pressure vessels was the fabrication of the mandrels. Two separate pieces of 0.06 in. thick rubber had to be fit into the mandrel halves. A mixture of 30 grit sand and poly vinyl alcohol (PVA) was then packed into the molds and compressed. A pressure gage was not available on the press so the pressure was estimated. The actual pressure was not critical.

The mandrels were put into the oven at 325 F for 3 hrs. The mandrels were extracted from the molds while still hot to keep them from sticking to the mold halves and tearing. Mandrel halves that were not fully cured or were not fully compacted or had major wrinkles or defects were discarded. Two halves were joined together by a strip of 0.03 inch rubber "belly band" applied to the center of the mold. Acetone was used to make it sticky until the mandrels could be cured in an oven at 250 F for 3 hrs. This cured the "belly bands" to the bladder. The mandrels were then sanded smooth, any gaps or groves were filled with a silicone

1027

rubber and smoothed over. The mandrels at this time were ready to wind.

Thirty-seven halves were made. Thirty-two were used in bottles, five were discarded. Sixteen bottles were made (11 were graphite, two were kevlar, one was used for practice, two were discarded.)

2.2 Filament Winding

"Shear plys" were applied to the boss-mandrel junctions (see figure 15). These were pieces of 0.03 inch rubber that helped smooth out discontinuities at the junction and eliminate stress risers. This rubber was not cured before winding. The bottles were put on threaded rods and attached to the tumble winder. The longitudinal plies were laid first followed by the hoop plies. There was no interspersment used.

Because of the limited amount of material available, it was decided to limit the experiment to three stress ratios. The perform-ance curve vs. stress ratio is close to a straight line so three points is sufficient for accurately deter-mining the line. The number of polar plies was limited to two because of the small quantity of material available. Due to the fact that the minimum number of hoop plies that can be laid is one, a minimum stress ratio of 0.3 was used. Two hoop plies correlated to a stress ratio of 0.61, and three hoop plies to a stress ratio of 0.81. A maximum stress ratio of 0.85 was necessary to insure hoop fiber failure only. Above 0.85 there is a tendency to get mixed failures (hoop and polar) which would be difficult to analyze at this time.

The first practice bottle was used to determine the proper wind angle, longitudinal speed, hoop speed, and tension for future bottle's. The values used were: wind angle = 11.5 degrees, longo = 1.6 degrees, hoop = 0.07 inches, tension = 8-12 lbs. The bottles were wound in four distinct groups to separate them in case processing problems were encountered. The first group consisted of two wet would kevlar bottles. These were used to qualify the hydrotest cell. The second group consisted of two graphite bottles. These were wrapped early to foresee any problems that might be encountered later in the graphite bottles. Three problems were encountered. First, the prepreg tended to stick to the rollers that guided the prepreg through the machine. The rollers has to be adjusted to minimize that. Second, the fibers tended to rub in the guides causing fraying of the fibers. This could not be completely eliminated. Third, the fibers tended to twist in the winding machine and onto the bottle. That was minimized with practice. This set of bottles was wrapped with a stress ratio of 0.61.

The third group consisted of

four bottles. Two bottles had a stress ratio of 0.3 and the other two bottles had a stress ratio of 0.81. This gave a total, so far, of two at 0.3, three at 0.61, and two at 0.81.

The fourth set of bottles was saved until all the other bottles were burst and analyzed. These three bottles were to be used to fill in data at critical locations. As it ended, all three stress ratios had wide scatter. Therefore, one bottle was wrapped at each of the three stress ratios, giving a final total of three bottles at 0.3, four bottles at 0.61, and 3 bottles at 0.81.

2.3 Hydroburst

Hydrobursting bottles is the most widely used method in characterizing fiber/resin systems. The system used was developed specifically for RPL. The two kevlar bottles were used to qualify the system before any graphite bottles were tested. Pressure - time traces were recorded as well as peak pressure during the test. A 5000 lb. pressure transducer was attached to the bleed off line and measured the true pressure in the bottle. The results are summarized in table 1.

Figures 3 through 6 show that although the experimental data followed the same basic trend as the theoretical curve, the data consistently fell below the predicted value. The interesting feature of these figures is the fact that the

last series of tests performed exceeding above the average, seemingly to suggest that as we built up our confidence level, our ability to wrap "good" bottles increased. The bottle performance on the last series of tests increased 14% above the average values.

Another observation of the figures shows that the percent deviation of the experimental curve from the theoretical curve increased as stress ratio increased. Since the number of polar plies was kept constant throughout all the bottles, the only change in bottle construction was in the number of hoop plies.

From table 1 it is seen that a polar boss loosened in 60% of the tests. Closer observation shows that the tests where the bosses blew out showed the highest burst pressures and that the tests where they did not blow out had consistently lower pressures. This seems to suggest that the upper limit on burst strength of these bottles was limited by stresses induced by the polar bosses. Further study should be conducted to study the effects of polar boss tear out, and how to eliminate it all together. The ultimate stress decreased on all tests with increasing stress ratio. Bottle with a stress ratio of 0.3 approached the theoretical stress.

3. CONCLUSIONS

This was the first opportunity for the components laboratory at

AFRPL to conduct research on filament winding of ASTM standard test and evaluation bottles used by the aerospace industry to characterize fiber/resin systems. It enabled us to establish standard procedures for the processes involved in mandrel making and curing, bottle winding and curing, and hydrobursting of bottles. It familiarized us with some of the problems involved in this process and allowed us to gain expertise in solving them.

Bottle performance fell below the theoretical values as stress ratio increased, but as competence increased, so did bottles performance.

Certain trends were realized in this study that will lead to greater performance in the future. The first trend was that the deviation from the theoretical curve of the experimental curve increased as the stress ratio increased leading to the conclusion that the stress was not distributed properly between the plies as the number of plies increased. In discussions with the manufactures we found that our bottles consistently weighed more than those made by industry by 35%. This was primarily due to our selection of longitudinal and hoop distances in laying fibers. The increase in weight overshadows any positive effect of increased pressure on bottle performance. Reducing the weight can be accomplished in three major ways:

adjusting the resin, adjusting the fiber, and adjusting the fiber density laid on the bottle. The resin plays an important part in transferring stress between fibers. Our cure cycle did not match the manufactures cure cycle exactly and is a probable cause of low performance.

The second trend was that the bosses set the upper limit for bottle performance. It was observed that in all failures which set upper limits for bottle performance, a polar boss was jarred loose.

As further studies, some of the factors effecting the bottle performance including tow tying, interspersment, cure cycle, resin type, tow tension, weight, and polar boss/bottle junction are under investigation at AFRPL.

## 4. REFERENCES

Tsai, S. W. and Hahn, H. T. Introduction to Composite Materials, Technomic Publishing Company, Westport, Connecticut, 1980.

Garg, S. K. et. al., Analysis of Structual Composite Materials, Marcel Dekker Inc., New York, 1973.

Lubin, E., Ed., Handbook of Composites, Van Nostrand Reinhold Co., New York, 1982.

Wolcott, E. et. al., Composite Case Design and Analysis Materials Processing, Morton Thiokol, Inc., April, 1985.

Mumford, N. A. and Kirschner, P. T., ACCME Status Review: Advanced

Composite Case Materials Evaluation, F04611-83-C-0036, July, 1985. ASTM Standard Method D2585, Preparation and Tension Testing of Filament Wound Pressure Vessels, 1974.

Clayton, K. I., Hydrotest of Pressure Vessels to Assess Effects of Processing on Performance of GR/PI Wound Structures, paper presented at 1984 JANNAF CMCS/S & MBS JOINT SUBCOMMITTEE MEETING, Pasadena, CA, November, 1984.

## 5. APPENDIX

### Netting Equations

$$t_\alpha = \frac{PR(1 + \varepsilon\theta)K_\alpha}{2F_\alpha f\cos^2\alpha} + \frac{M(1 + \varepsilon\theta)K_\alpha}{\pi R^2 F_\alpha f\cos^2\alpha} - \frac{TK_\alpha}{2\pi RF_\alpha f\cos^2\alpha(1 + \varepsilon\theta)} \tag{2}$$

$$N = \frac{t_\alpha}{A_E \rho_\alpha K_\alpha} \tag{3}$$

$$t = PR(1 + \varepsilon\theta)(1 - \frac{TAN^2\alpha}{2})K\theta/Fe_f \tag{4}$$

$$N = \frac{t\theta}{A_E \rho \, \theta K\theta} \tag{5}$$

WHERE

$t_\alpha$ = HELICAL/POLAR COMPOSITE THICKNESS (IN)

$t\theta$ = HOOP COMPOSITE THICKNESS (IN)

$K_\alpha$ = HELICAL/POLAR BULK FACTOR

$K\theta$ = HOOP BULK FACTOR

$\alpha$ = HELICAL/POLAR WRAP ANGLE (DEG)

$\varepsilon\theta$ = MAXIMUM HOOP STRAIN AT BURST (IN/IN)

$P$ = CASE MEOP X FACTOR OF SAFETY REQUIREMENT (PSIG)

$R$ = CASE RADIUS (IN)

$M$ = CASE EXTERNAL BENDING MOMENT (IN-LBS)

$T$ = CASE THRUST AT MEOP X FS REQUIREMENT (LBS)

$F_{\alpha f}$ = HELICAL/POLAR FIBER STRESS ULTIMATE ALLOWABLE (PSI)

$Fe_f$ = HOOP FIBER STRESS ULTIMATE ALLOWABLE (PSI)

$A_E$ = FIBER AREA (IN$^2$/END)

$\rho_\theta$ = HOOP BAND DENSITY (ENDS/IN/PLY)

$\rho_\alpha$ = HELICAL BAND DENSITY (ENDS/IN/PLY)

$N_\alpha$ = REQUIRED NUMBER OF HELICAL/POLAR PLIES

$N_\theta$ = REQUIRED NUMBER OF HOOP PLIES

$$M \doteq 0 \tag{6}$$
$$T \doteq 0$$

$$N = \frac{PR(1 + \varepsilon\theta) K_\alpha}{2F_{\alpha f}\cos^2 A_E \rho_\alpha K_\alpha} \tag{7}$$

$$F_{\alpha f} = \frac{PR(1 + \varepsilon\theta)}{2N_\alpha \cos^2\alpha A_E \rho_\alpha} \tag{8}$$

$$t\theta = PR(1 + \varepsilon\theta)(1 - \frac{\tan^2}{2}) \frac{K\theta}{Fe_f} \tag{9}$$

$$N\theta = \frac{t\theta}{A_E \sigma\theta K\theta} \tag{10}$$

$$N\theta = \frac{PR(1 + \varepsilon\theta)(1 - \tan^2\alpha) K\theta}{2Fe_f A_E \rho\theta K\theta} \tag{11}$$

$$Fe_f = \frac{PR(1 + \varepsilon\theta)(1 - \frac{\tan^2\alpha}{2})}{2N\theta A_E \rho\theta} \tag{12}$$

$$\frac{F_{\alpha f}}{Fe_f} = \frac{PR(1 + \varepsilon\theta)}{2N_\alpha \cos^2\alpha A_E \rho_\alpha} \frac{2N\theta A_E \rho\theta}{PR(1 + \varepsilon\theta)(1 - \frac{\tan^2\alpha}{2})} \tag{13}$$

$$\frac{F_{\alpha f}}{Fe_f} = \frac{N\theta}{N_\alpha} \frac{\rho\theta}{\rho_\alpha} \frac{1}{\cos^2\alpha (1 - \frac{\tan^2\alpha}{2})} \tag{14}$$

$$\sigma_L = F_{\alpha f} \qquad \sigma_H = Fe_f$$

1032

$$\text{Stress Ratio} = \frac{\sigma_L}{\sigma_H} = \frac{N_H}{N_L} \quad \frac{1}{\cos^2\alpha \ (1 - \frac{\tan^2\alpha}{2})} \quad \frac{P\theta}{P_\alpha} \tag{15}$$

Figure 1: ASTM mold

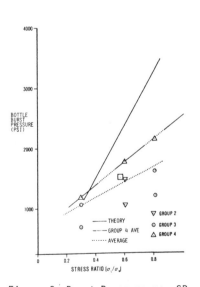

Figure 2: Burst Pressure vs. SR

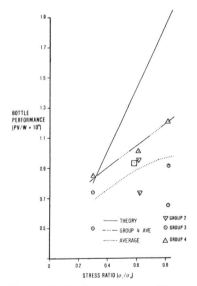

Figure 3: Performance vs. SR

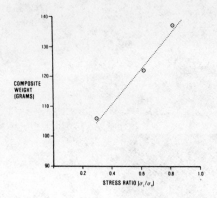

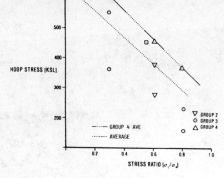

Figure 4:  Weight vs. Sress Ratio        Figure 5:  Hoop Stress vs. SR

Table 1

Summary of Experimental Data

| Bottle | Plies Hoop | Stress Ratio | Burst Pressure (PSI) | PV/W x10$^6$ | HOOP x10$^3$ (PSI) | Material Type | Comments |
|--------|-----------|--------------|----------------------|--------------|---------------------|---------------|----------|
| M2  | 2 | 0.45 | 1000 | 0.6  | ---  | T-40   | 1   |
| M5  | 2 | 0.57 | 1500 | 0.93 | 451  | T-40   | 1,3 |
| M1  | 2 | 0.64 | ---  | ---  | ---  | Kevlar | 4   |
| M6  | 2 | 0.64 | ---  | ---  | ---  | Kevlar | 4   |
| M10 | 2 | 0.61 | 1455 | 0.95 | 376  | T-40   | 2,3 |
| M13 | 2 | 0.61 | 1040 | 0.69 | 287  | T-40   | 2   |
| M11 | 1 | 0.30 | 1050 | 0.74 | 556  | T-40   | 2,3 |
| M8  | 1 | 0.30 | 700  | 0.50 | 364  | T-40   | 2   |
| M9  | 3 | 0.81 | 1625 | 0.92 | 284  | T-40   | 2,3 |
| M12 | 3 | 0.81 | 1200 | 0.65 | 210  | T-40   | 2   |
| M14 | 1 | 0.30 | 1170 | 0.85 | 608  | T-40   | 2   |
| M15 | 2 | 0.61 | 1755 | 1.08 | 456  | T-40   | 2,3 |
| M16 | 3 | 0.81 | 2140 | 1.20 | 371  | T-40   | 2,3 |

1.  Dome failure due to incomplete polar wind
2.  Hoop failure
3.  Boss tearout
4.  Wet wind used to qualify system

## 6. BIBLIOGRAPHY

### Scott Stagliano

Scott Stagliano received his Bachelors degree in Aeronautical Engineering at California Polytechnic State University at San Luis Obispo, California. He is currently working towards his Masters degree at the same institution. He has been employed by the Air Force Rocket Propulsion Laboratory since his sophomore year as a cooperative education student and has performed research on flat carbon phenolic panels, filament wound pressure vessels, and acoustic emissions on 2-D carbon-carbon composites.

### Dr. Faysal A. Kolkailah

Dr. Faysal A. Kolkailah received his B.Sc. in Aerospace Engineering at Cairo University, his M.Sc. in Aerospace Engineering at University of Cincinnati, and his Ph.D. in Mechanical Engineering at Louisiana State University.

He taught at the University of Cincinnati, Louisiana State University, and University of New Orleans. Currently, he is a Professor of Aeronautical Engineering at Cal Poly.

Dr. Kolkailah has strong industrial experience as an aeronautical engineer and as a consultant to several private industries.

He has considerable research experience in composite structure, fracture mechanics, optimum design, aircraft design, and numerical methods in solid mechanics. He has published considerably in these areas.

Dr. Kolkailah has been a principal investigator or co-investigator on a number of research projects funded by NASA/Ames Research Center, Rockwell International, and Aerojet Techsystems.

Dr. Kolkailah is a member of SAMPE and other professional societies. He is the founder and advisor of the SAMPE student chapter at Cal Poly.

POLYIMIDE PROCESSING ADDITIVES

J. R. Pratt,* T. L. St. Clair, H. D. Burks and D. M. Stoakley
NASA Langley Research Center
Hampton, Virginia  23665-5225

## Abstract

A method has been found for enhancing the melt flow of thermoplastic polyimides during processing. A high molecular weight 422 copoly(amic acid) or copolyimide was fused with approximately 0.05 to 5% by weight of a low molecular weight amic acid or imide additive, and this melt was studied by capillary rheometry. LARC-TPI and 6F-BDAF polyimides were also studied. Excellent flow and improved composite properties on graphite resulted from the addition of a PMDA-aniline additive to LARC-TPI. Solution viscosity studies imply that amic acid additives temporarily lower molecular weight and, hence, enlarge the processing window. Thus, compositions containing the additive have a lower melt viscosity for a longer time than those unmodified.

## 1. INTRODUCTION

Aromatic thermoplastic polyimides are increasingly being developed as matrix resins for fiber-reinforced composites and hot-melt adhesives. These engineering thermoplastics are in competition with metals for many aerospace applications because of their combination of high strength, stiffness, heat resistance and toughness [1]. Unfortunately, most of these thermoplastic polyimides are difficult to process into useful shapes, requiring high temperature and pressure and extended time to achieve a limited degree of flow.

This research was conceived to elucidate the effect that small amounts of certain amic acid or imide additives might have on the

*PRC Kentron, Inc., at Langley Research Center

processing characteristics of high molecular weight polyimides. The initial assumption was that the additive would plasticize the polyimide and enhance its melt flow. π-Complex formation in polyimides is known [2-5]. This electron donor-acceptor interaction between the additive and certain segments of the polyimide backbone could lessen the interchain attractive forces in the polyimide. The resulting composition should remain inert because of its all-imide structure, yet process more readily.

## 2. EXPERIMENTAL

### 2.1 Polymers

The 422 copoly(amic acid) was prepared by modifying the procedure of Burks et al. [6] (Scheme I). Sublimed 4,4'-oxydianiline (ODA) (49.08 g, 0.2451 mol) was dissolved under nitrogen in 2-methoxyethyl ether (diglyme) in a flamed 2 L reaction kettle equipped with a mechanical stirrer and lid. 4,4'-Bis(3,4-dicarboxyphenoxy)diphenylsulfide dianhydride (BDSDA) (250.0 g, 0.4902 mol) was added in one charge, washing traces of the remaining dianhydride into the kettle with diglyme. A total of 1305 g of this solvent was required. Stirring was maintained at 300 rpm. After 25 min. m-phenylenediamine (26.47 g, 0.2448 mol) was added, resulting in the formation of a stringy, solid slurry.

Stirring was continued for 18 hours, during which time these suspended particles dissolved and the solution became dark and viscous. Phthalic anhydride (0.7260 g, 0.0049 mol) dissolved in 1.70 g of diglyme was then added to endcap the poly(amic acid) and stirred for 1 hour. The resulting poly(amic acid) solution at 20.0% solids had an inherent viscosity of 0.67 dl/g in N,N-dimethylacetamide (DMAc). The solution was stored at 0°C.

Precipitation of the resin was conducted in a large household blender by pouring a stream of the DMAc solution into the stirred water. The precipitated, off-white solid was filtered through cheese cloth, reslurried in distilled water in the blender, refiltered, and dried at room temperature for 2 days. This solid was dried further at room temperature in vacuo to constant weight. The yield of poly(amic acid) powder was 364.5 g (theoretical yield, 326.3 g), indicating a residual amount of solvent and/or water.

A 170 g sample of the dried copoly(amic acid) resin was imidized in a forced air oven for 1 hour at 100°C and 1 hour at 220°C in a large baking dish. The resulting solid was homogenized in a blender to yield 136 g of imidized powder, which is referred to as the 422 copolyimide or baseline polyimide.

LARC-TPI (3,3'4,4'-benzophenonetetracarboxylic dianhydride-3,3'-diaminobenzophenone) was obtained commercially as a 29-30% poly(amic acid) solution ($n_{inh}$, 0.52 dl/g) in diglyme [7]. The solution was stored at 0°C until used. The 6F-BDAF poly(amic acid) ($n_{inh}$, 1.1 dl/g) was prepared as reported [8] and also stored at 0°C. A sample was endcapped with 2.00% by weight of sublimed phthalic anhydride. After stirring overnight at room temperature, the inherent viscosity had fallen to 0.76 dl/g. Scheme I shows the structures of all polymers.

2.2  Di(amic acid) and Diimide Additives

Scheme II gives examples of the synthesis of the three types of additives used in this study: the di(amic acid) of pyromellitic dianhydride (PMDA) and two moles of aniline (An), the di(amic acid) of p-phenylenediamine (p-PDA) and two moles of phthalic anhydride (PA), and the diimide from 2,2-bis(3,4-dicarboxyphenyl)hexafluoropropane dianhydride (6F) and two moles of m-trifluoromethylaniline (3-TFMAn).

The PMDA-An·2NMP, di(amic acid) additive, for example, was prepared as follows. A solution of resublimed PMDA (21.8 g, 0.100 mol) was prepared in a mixture of 125 ml of N,N-dimethylformamide (DMF), 125 ml of 1-methyl-2-pyrrolidinone (NMP), and 125 ml of diglyme with slight heating to effect solution. After cooling, redistilled aniline (18.6 g, 0.200 mol) was added portionwise over a several minute period. After stirring over the weekend, a colorless precipitate was filtered and rinsed with toluene. The filtrate was stirred with 750 ml toluene for several days to afford additional product. The total yield of colorless product was 44.1 g (73.3%) after vacuum drying overnight at room temperature. The DTA (Figure 1) displayed a sharp endotherm at 176°C (melting); a moderately broad endotherm at 205°C (loss of NMP); a broad exotherm extending past 265°C (imidization), and a sharp endotherm at 444°C (melting of the diimide). The NMR ($d_6$-DMSO) gave the following spectrum: 12 $\delta$ (2H, s, carboxylic acid H), 8.2 $\delta$ (2H, s, pyromellitic aromatic H), 6.9 - 8.0 $\delta$ (10H, m, aromatic H), 3.4 $\delta$ (4H, t, $CH_2$ of 2-pyrrolidinone ring), 2.8 $\delta$ (6H, s, N-$CH_3$), 1.4 - 2.6 $\delta$ (8H, m, -$CH_2$-$CH_2$- of the 2-pyrrolidinone ring). The carboxylic acid protons did not integrate fully due to partial exchange in the solvent, and their signal disappeared on addition of $D_2O$. The two amide protons were not observed, due to rapid exchange in the solvent. Although the 60 MHz NMR experiment did not establish the isomeric purity of this additive, the compound appears to be essentially all the 2,5-di(amic acid) isomer. The melting points, elemental analyses, and FTIR spectra are reported in Table I.

The synthesis of the PMDA-An, diimide additive is given as an example below. PMDA (6.0 g, 0.028 mol) and redistilled aniline (5.1 g, 0.056 mol) were refluxed for 3 h with 75 ml of NMP. The resulting solution was cooled to ambient temperature, and the product crystallized. After drying in vacuo a 7.6 g (86%) yield of product, m.p. 449-53°C, was isolated. Lit. reports m.p. 444-46°C [2]. Melting points, elemental analyses, and FTIR spectra of all four imide additives are reported in Table II. Other diimide additives were similarly prepared in refluxing DMF or glacial acetic acid [2].

2.3 Preparation of the 422 Copolyimide Compositions

Scheme III outlines two procedures used to mix the polymer and additive. The 422 copoly(amic acid) powder and the powdered additive were mixed on a weight percent basis by dissolving or slurrying them in DMAc. The composition was then precipitated from water, air dried overnight, and heated in a forced air oven for 1 h each at 100°C and 220°C. Alternately, the two imidized powders were mixed in a CRC Homogenizer® or a blender. Fourcombinations of additive and polymer were studied: the amic acid additive in the 422 copoly(amic acid) heated 1 hour each at 100° and 220°C; the amic acid additive in the 422 copolyimide heated 1 hour each at 100° and 220°C; the imide additive in the 422 copoly(amic acid) heated 1 hour each at 100° and 220°C; and the imide additive in the 422 copolyimide with no further heating. In addition all imidized compositions were held in the barrel of the capillary rheometer at 350°C for 0.5 h before measurements were made.

2.4 Capillary Rheometry of the 422 Copolyimide-Additive Melts

The apparent viscosities of the melted polyimide compositions were determined on an Instron Model 3211 capillary rheometer at 350°C and strain rates between 0.40 and 135 $sec^{-1}$ [9]. The capillary had a length-to-diameter ratio of 33 (ID, 0.0603 in. and length, 2.0058 in.), hence no correction for wall drag was needed. A plot of the apparent viscosity (stress/strain, Pa-sec) vs. strain rate ($sec^{-1}$) data on log-log graph paper was made for each composition. In addition the best fit for each line was calculated and graphed using a linear regression analysis (Figures 2-3) on log-log paper.

## 2.5  Characterization

Infrared spectra were recorded on a Nicolet 60SX Fourier Transform Infrared spectrometer from KBr pellets. The inherent viscosities (in dl/g) were determined with a Cannon-Ubbelohde viscometer at a concentration of 0.5% (wt./vol.) in DMAc at 35°C. The NMR spectrum was recorded on a Varian EM-360A 60 MHz spectrometer. Imidization and cure studies were conducted in a forced air oven; the temperature was accurately monitored with a thermocouple placed directly above the sample. Time-temperature profiles for curing the films were strictly controlled.

## 2.6  Solution Viscosity Study of 6F-BDAF/PMDA-An Additive Compositions

Four compositions were prepared for study. One contained 5.0% by weight of the PMDA-An·2 NMP, di(amic acid) additive in the unendcapped 6F-BDAF poly(amic acid); another identical composition was endcapped with 2.0 mol % of phthalic anhydride. Endcapped and unendcapped poly(amic acid) solutions in DMAc without the additive were also prepared. Films from all four solutions were cast from approximately 15% by weight solutions and dried overnight under low humidity. They were cured in a forced air oven for 1 h each at 150°C and 250°C, then further sequentially cured at 300°C for up to 110 h, periodically taking small samples for inherent viscosity measurements. The percent of polymer solids in each partially cured film was determined by thermogravimetric analysis (TGA) so that solution concentrations could be calculated. Inherent viscosity vs. time of imidization curves were then plotted (Figure 5) for each composition until approximately 1% of the polymer film had become insoluble in DMAc at room temperature.

## 3.  RESULTS AND DISCUSSION

The incorporation of two molecules of NMP in the PMDA di(amic acid) was not anticipated, although such solvent complexes are well known. This complex, no doubt, exists through hydrogen bonding of one NMP molecule to one carboxylic acid group. The NMP molecules do not appear to detrimentally affect the imidization; a programmed TGA (2.5°C/min in air) showed that 84% of the theoretical volatiles had been lost by 250°C (programmed increase from 25°C to 250°C for 1.5 h). The formation of similar solvate complexes with the other amic acids was not observed.

The theoretical structures of the three polyimides evaluated in this study are given in Scheme I. The 422 copolyimide, a random polyetherimide of BDSDA (4 mol), 4,4'-ODA (2 mol), and m-PDA (2 mol), was initially chosen for study based on earlier results in this laboratory [6] which showed that it exhibited the lowest melt viscosity by capillary rheometry of three copolyimides at all strain rates studied. After observing the effect of low levels of amic acid and imide additives on the melt viscosity of this 422 copolyimide, our studies were extended to include LARC-TPI, a commercial thermoplastic polyimide. Preliminary results on this project have appeared [10,11]. Both of these polyimides were difficult to process without additives; our objective was to improve the melt processability without degrading other polymer properties.

In order to explain the significant enhancement in flow of these polyimide-additive compositions, our studies were extended to the 6F-BDAF polyimide, a polymer that is fully soluble in DMAc after a 300°C cure. The question was whether changes in molecular weight, as evidenced by changes in solution viscosity, could be used to explain the enhanced processability of the polyimide-additive compositions.

3.1 Capillary Rheometry Study of the 422 Copolyimide and Additive Mixtures

Figures 2 and 3 show the apparent viscosity (Pa-sec) - strain rate (sec$^{-1}$) graphs of the 422 copolyimide-additive melts.

As shown in Figure 2 the addition of a small amount of di-(amic acid) or diimide additive to the 422 copoly(amic acid), followed by imidization of the mixture for 1 hour each at 100° and 220°C, gave compositions having decreased melt viscosities at all strain rates studied. In one experiment [5.0% by weight of p-PDA-PA, di-(amic acid), imidized] the melt viscosity decreased over three orders of magnitude (approximately 1 X 10$^6$ Pa-sec) at the lowest strain rate. Other additives were effective to a lesser degree. In general the amic acid additives were more effective in reducing melt viscosity than the corresponding imide additives. This tentative conclusion must be confirmed by further testing with other additives. When the original 422 copoly(amic acid) powder was redissolved in DMAc, precipitated from water and imidized to give a new base-line polyimide powder, its apparent viscosity graph was shifted to slightly lower values. This probably resulted from some hydrolysis of the poly(amic acid) during work up and imidization.

Similar lowering of melt viscosities of additives blended with 422 copolyimides is shown in Figure 3. Diimide additives (and one di(amic acid)) were added to the base-line resin by mixing the powders in a blender, followed by heating these mixtures for 1 hour each at 100° and 220°C. An inert component, naphthalene, had no effect on the melt viscosity. Furthermore, when one compares the effect of the additive as amic acid with that as imide (compare the 5.0% PMDA-An of Figures 2 and 3), little difference is noted. Unfortunately, apparent viscosity-strain rate graphs based on the mole % of additive were not obtained, hence a molecular comparison of the effects of the different additives is difficult to make. This preliminary data, however, appears to show a strong plasticization effect by amic acid and imide additives in the 422 copolyimide base-line resin.

3.2 Composite Properties

Table III gives the short beam shear, flexure strength, and flexure modulus data of graphite composites made from LARC-TPI and LARC-TPI containing 2.5 wt. % of the PMDA-An·2NMP, di(amic acid) additive after heating in a press to 350°C and 300 psi for 1 h [11]. The LARC-TPI unidirectional composites containing no additive had relatively poor melt flow properties and poor ultrasonic C-scans, even though the panels appeared well consolidated. Short beam shear strengths and flexural properties were poor.

Composites made from the additive modified LARC-TPI showed improved melt flow. Unidirectional composites were well consolidated and had good C-scans, short beam shear strengths, and flexural properties at all temperatures tested.

Photomicrographs of two LARC-TPI-graphite composites are shown in Figure 4 [12]. They show cross sections at 100X of 12 ply, unidirectional laminates made from (a) LARC-TPI reinforced with Celion 6000® carbon fibers and (b) the same material with 2% by weight of the PMDA-An·2NMP, di-(amic acid) additive. The cross section view of the unmodified composite shows resin-rich channels which give the composite inhomogeneity. The doped composite, however, appears quite homogeneous. Sufficient flow has occurred such that the individual layers in the prepreg are indistinguishable. Hence each composite property from this material will likely be more uniform.

3.3 Solution Viscosity-Imidization Study of the 6F-BDAF Poly(amic acid)/Polyimide Containing 5.0% by Weight of the PMDA-An·2NMP Di(amic acid) Additive

Young and Chang [13-15] have shown by inherent viscosity measurements that the staged imidization of several high molecular weight unendcapped poly(amic acids) proceeds via an apparent intramolecular break of amide groups in the poly(amic acid) backbone to generate anhydride and (probably) amine terminated chains. Further elevated temperature treatment allows chain-extension to occur to yield high molecular weight polyimide. Endcapped poly(amic acids) similarly treated also show a minimum in viscosity; however, there is little increase in viscosity on further elevated temperature treatment.

In an attempt to understand the behavior of the additives, a related solution viscosity study of polymer films using isothermal cure conditions was conducted (Figure 5). The 6F-BDAF polyimide was chosen for study because it remains soluble in DMAC after 300°C imidization, whereas both LARC-TPI and the 422 copolyimide are insoluble. The poly(amic acid) films were cured as described in the Experimental. Changes in the inherent viscosity of the aged films were measured for four systems: the unendcapped and the 2.0 mole % phthalic anhydride endcapped polyimides and the same materials which had been precipitated with the additive. A direct relationship between solution and melt viscosities involving the molecular weight of the polyimide was assumed; however, we know of no study showing this relationship.

The isothermal results in Figure 5 appear to confirm those of Young and Chang. The inherent viscosity of both the endcapped and unendcapped 6F-BDAF films fell moderately during early imidization. After 1 hour at 300°C the viscosity had begun to increase. The imidization was continued at 300°C, and samples were periodically taken for analysis until approximately 1% of the polyimide became insoluble in DMAc at room temperature. At this point the final viscosity was recorded. Those films containing the additive displayed a larger initial decrease in inherent viscosity, followed by a much slower and incomplete recovery. The unendcapped polyimide with additive remained fully soluble in DMAc at room temperature after 110 hours at 300°C.

Based on these results, it appears that the improved melt processability of these compositions may be due to a drop in the molecular weight of the polyimide, as well as the longer recovery time needed for the molecular weight to approach that of the fully cured material. This results in a larger "processing window" during which time the polyimide can be formed into useful shapes and molded into uniform and well-consolidated composites. Our original explanation involving charge transfer complex formation between segments of the polyimide and the additive still seems plausible; however, the results of this study imply that changes in molecular weight may primarily control processability.

## 4. CONCLUSIONS

The melt viscosity of a 422 copolyimide as determined by capillary rheometry can be significantly lowered with small amounts of amic acid or imide additives. Graphite composites of LARC-TPI with one of these additives show greatly improved processability. They process with much better flow-out of the resin into the fibers to give well-consolidated test panels. Ultrasonic C-scans were markedly improved while composite properties such as short beam shear strength, flexure strength, and flexure modulus were increased. A solution viscosity-time study during imidization of a soluble 6F-BDAF polyimide revealed that the additive increases the size of the processing window of the molten polymer, probably through a transient lowering of polymer molecular weight. If confirmed, these types of additives may provide a major and novel way to improve the processability of polyimides without destroying their high performance characteristics.

## 5. ACKNOWLEDGEMENTS

The authors wish to thank Maura Hamrick and Roslyn Smith for their technical assistance on this project.

## 6. REFERENCES

1. V. Wigotsky, Plastics Engineering, Jan., 1985, p. 37.

2. R. A. Dine-Hart and W. W. Wright, Makromol. Chem. 143, 189 (1971).

3. T. A. Gordina, B. V. Kotov, O. V. Kolninov, and A. N. Pravednikov, Vysokomol. soyed. B15: 378, 1973 ("Charge Transfer Complexes of Certain Aromatic Polyimides," NASA TM-77974, Dec., 1985).

4. B. V. Kotov, T. A. Gordina, V. S. Voishchev, O. V. Kolninov, and A. N. Pravednikov, Polymer Science USSR, 19(3), 711 (1977) (Engl. Transl.); Vysokomol. soyed., A19(3), 614 (1977).

5. M. Fryd in Polyimides: Synthesis, Characterization, and Applications (K. L. Mittal, ed.), Vol. 1, p. 377 (1984), Plenum Press, NY.

6. H. D. Burks, T. L. St. Clair and D. J. Progar, "Synthesis and Characterization of Copolyimides with Varying Flexibilizing Groups," Proceedings of the Second International Conference on Polyimides, Ellenville, NY, Oct. 1985, p. 529; T. L. St.

Clair and H. D. Burks, NASA
Invention Disclosure Case
LAR-13354-1.

7. LARC-TPI Product Literature,
Mitsui Toatsu Chemicals,
Inc., NY.

8. A. K. St. Clair, T. L. St.
Clair, and K. I. Shevket, ACS
Polymeric Materials: Science
and Engineering, Proceedings,
51, 62(1984).

9. L. E. Nielsen, "Polymer
Rheology," Marcel Dekker,
Inc., NY, 1977, Ch. 2, p. 12.

10. D. M. Stoakley, T. L. St.
Clair, A. K. St. Clair, J.
R. Pratt and H. D. Burks,
Polymer Preprints, 27(2),
406 (1986).

11. N. J. Johnston and T. L. St.
Clair, International SAMPE
Conference, Preprints, 18,
53(1986).

12. A. Falcone, "Advanced
Thermoplastic Resins,"
Bimonthly Report under NASA
Contract NAS1-17432 to Boeing
Aerospace Co., Oct. 1986.

13. P. R. Young and A. C. Chang,
SAMPE Journal, 22(2),
70(1986).

14. P. R. Young, N. T. Wakelyn
and A. C. Chang,
"Characterization of a
Soluble Polyimide,"
Proceedings of the Second
International Conference on
Polyimides, Ellenville, NY,
414(1985).

15. P. R. Young and A. C. Chang,
"Cure Study of Soluble
Aromatic Polyimide Films,"
SAMPE Preprints, 32nd Inter-
national SAMPE Symposium,
Anaheim, CA, April 1987.

## SCHEME I. PREPARATION OF POLYMERS

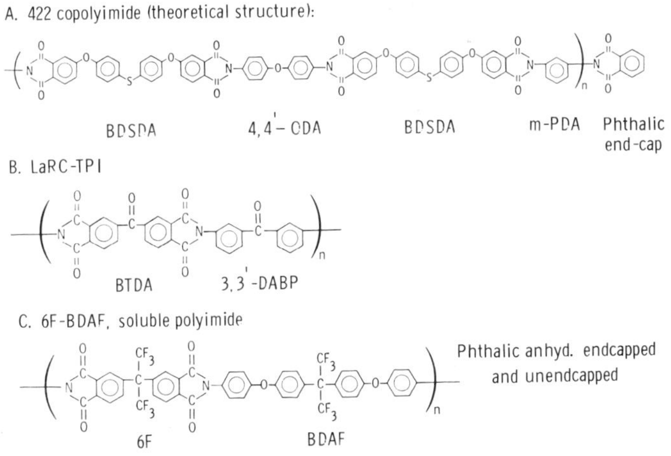

A. 422 copolyimide (theoretical structure):

BDSDA    4,4'-ODA    BDSDA    m-PDA    Phthalic end-cap

B. LaRC-TPI

BTDA    3,3'-DABP

C. 6F-BDAF, soluble polyimide

6F    BDAF    Phthalic anhyd. endcapped and unendcapped

## SCHEME II. PREPARATION OF ADDITIVES (EXAMPLES)

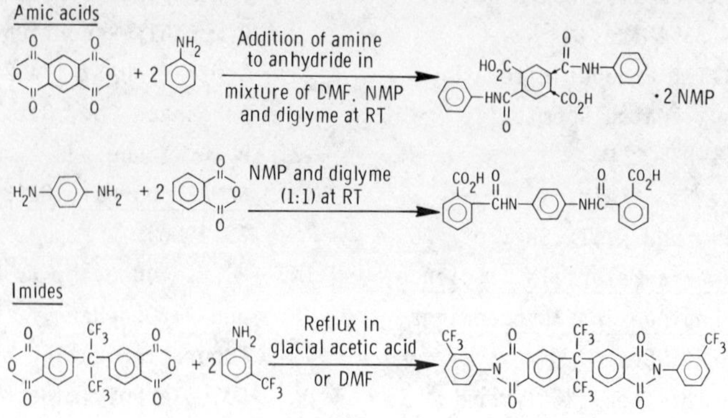

Amic acids

Addition of amine to anhydride in mixture of DMF, NMP and diglyme at RT · 2 NMP

NMP and diglyme (1:1) at RT

Imides

Reflux in glacial acetic acid or DMF

## SCHEME III

## PREPARATION OF 422 COPOLYIMIDE COMPOSITIONS

422 poly(amic acid) solution → $H_2O$ ppt. Air dry then vac. dry at RT → Poly(amic acid) powder **A** → Δ → Polyimide powder **B**

A or B + additive → DMAc → Mixture (solution or slurry) → $H_2O$ ppt. Air dry then Δ in air oven → Copolyimide composition

## TABLE I. DIAMIC ACID ADDITIVES

| Structure/symbol | m.p. ($^0$C) | Analysis Calc | Found | FTIR (cm$^{-1}$) |
|---|---|---|---|---|
| 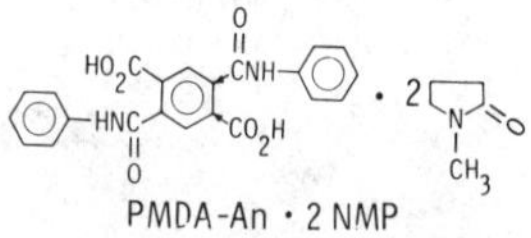 PMDA-An · 2 NMP | 176 | C 63.78<br>H 5.69<br>N 9.30 | 63.90<br>5.87<br>9.26 | 2300-3700 carboxylic acid OH dimer<br>1719 carboxylic acid carbonyl<br>1628 amide I of amic acid<br>1681 amide I of NMP<br>3327 amide NH |
| 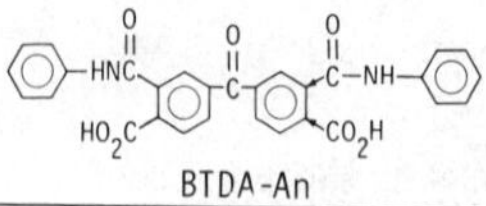 BTDA-An | 364.5-67 | C 68.50<br>H 3.94 | 68.65<br>4.22 | 3040 carboxylic acid dimer max<br>1718 carboxylic acid carbonyl<br>1700 amide I<br>1650 benzophenone carbonyl |
| 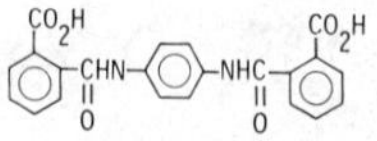 p-PDA-PA | 194 | C 65.34<br>H 3.99<br>N 6.93 | 64.63<br>4.35<br>7.05 | 2300-3700 carboxylic acid OH dimer<br>1714 carboxylic acid carbonyl<br>1661 amide I<br>3325 amide NH |

## TABLE II. DIIMIDE ADDITIVES

| Structure/symbol | m. p. ($^0$C) | Analysis Calcd. | Found | FTIR(cm$^{-1}$) | |
|---|---|---|---|---|---|
| | | | | | Ref( 2 ) |
| PMDA-An | 449-53 | C 71.74<br>H 3.28 | 71.73<br>3.45 | 1745<br>1723<br>1395 | Imide I<br>Imide I<br>Imide II |
| 6F-3,3'-TFMAn, R = -CF$_3$ | 6F-An,<br>R = -H<br><br>182.5-<br>184.5 | C 62.63<br>H 2.71<br>N 4.71 | 62.24<br>2.63<br>4.65 | &1782<br>1715<br>1375<br>1101<br>711 | Imide I<br><br>Imide II<br>Imide III<br>Imide IV |
| | 145-48 | C 54.26<br>H 1.93<br>N 3.84 | 54.05<br>2.07<br>3.73 | &1785<br>1730<br>1375<br>1118<br>720 | Imide I<br><br>Imide II<br>Imide III<br>Imide IV |
| 4,4'-ODA-PA | 279 | C 73.04<br>H 3.50<br>N 6.08 | 72.36<br>3.83<br>6.16 | &1780<br>1710<br>1381<br>1115<br>718<br>1231 | Imide I<br><br>Imide II<br>Imide III<br>Imide IV<br>Aryl ether |

## TABLE III. COMPOSITE PROPERTIES[°]

| Short beam shear | | | | Flexure strength & modulus | | | | | |
|---|---|---|---|---|---|---|---|---|---|
| LaRC-TPI | | LaRC-TPI with 2.5 wt. % PMDA - An ·2 NMP, Di(amic acid) | | LaRC-TPI | | | LaRC-TPI with 2.5 wt % PMDA-An ·2 NMP, Di(amic acid) | | |
| Temp($^0$C) | KSI | Temp($^0$C) | KSI | Temp($^0$C) | KSI | MSI | Temp($^0$C) | KSI | MSI |
| RT | 9.2 | RT | 13.5 | RT | 137 | 12.8 | RT | 242 | 14.7 |
| 93 | 8.8 | 93 | 12.0 | 93 | 121 | 12.6 | 93 | 213 | 14.4 |
| 149 | 8.4 | 205 | 8.4 | 149 | 111 | 11.1 | 205 | 160 | 14.1 |
| 177 | 7.8 | | | 177 | 99 | 12.1 | | | |

[°]Unsized Hercules AS-4 12K carbon fiber.
Results from ref ( 11 ).

## FIGURE 1

### DTA OF THE PMDA-An · 2 NMP Di(AMIC ACID) ADDITIVE

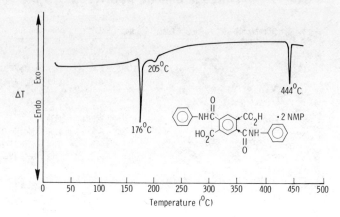

## FIGURE 2

### CAPILLARY RHEOMETRY STUDY
### OF ADDITIVES IN THE 422 POLY(AMIC ACID), IMIDIZED

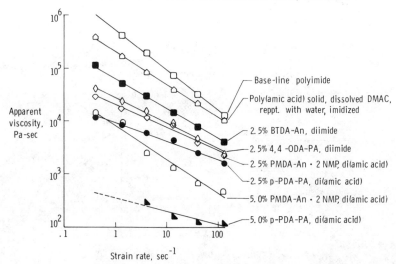

## FIGURE 3

## CAPILLARY RHEOMETRY STUDY
## OF ADDITIVES IN THE 422 POLYIMIDE

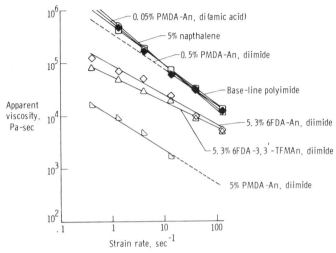

FIGURE 4. PHOTOMICROGRAPHS OF TWO LARC-TPI-GRAPHITE COMPOSITES.

Unmodified LaRC-TPI (100X)

LaRC-TPI modified with
2% by weight PMDA-An • 2 NMP,
di(amic acid) (100X)

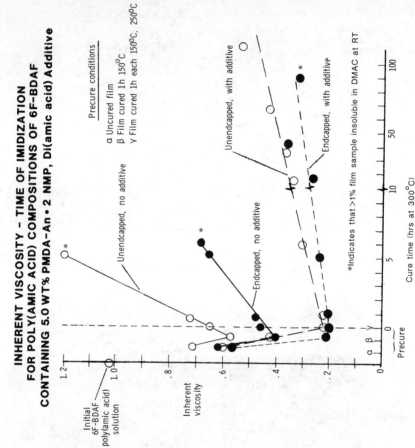

FIGURE 5

INHERENT VISCOSITY – TIME OF IMIDIZATION
FOR POLY(AMIC ACID) COMPOSITIONS OF 6F-BDAF
CONTAINING 5.0 WT% PMDA–An • 2 NMP, Di(amic acid) Additive

CURE STUDY OF SOLUBLE AROMATIC POLYIMIDE FILMS

Philip R. Young and A. C. Chang*
NASA Langley Research Center
Hampton, Virginia  23665-5225

## Abstract

Several soluble aromatic poly(amic acid) films were staged at intervals to 325°C and characterized by infrared spectroscopy and various solution property techniques. A series of films in which the polymer had been endcapped in an effort to control chain extension was also examined. Much of the behavior observed is consistent with an interpretation that a reduction in molecular weight occurred during cure before the ultimate molecular weight was achieved as a polyimide. The results of this study serve to increase our fundamental understanding of how polyimides are formed from precursor polymers. Implications of how this understanding may impact polyimide processing are discussed.

## 1. INTRODUCTION

In a previous study, an aromatic poly(amic acid) was found to undergo an apparent reduction in molecular weight during cure before achieving its ultimate molecular weight as a polyimide.[1,2] That soluble polymer, based on the reaction of 4,4'-bis(3,4-dicarboxyphenoxy)- diphenyl sulfide dianhydride (BDSDA) with 2,2-bis[4-(4-aminophenoxy)phenyl]hexafluoropropane (BDAF), was staged at 25°C intervals to 325°C. It was characterized by infrared spectroscopy, gel permeation chromatography, inherent viscosity, and number average molecular weight. The present study is an attempt to confirm that original observation by examining additional polymer systems. Such a confirmation

*PRC Kentron, Inc., Hampton, Virginia  23666

would not only increase our understanding of how polyimides are formed from precursors, but could also impact how these materials are processed into useful articles.

In the present study, the poly(amic acid) obtained from 2,2'-bis(3,4-dicarboxyphenyl)-hexafluoropropane dianhydride (6F) and 3,3'-diaminodiphenylsulfone (3,3'-DDS) was characterized.

## SCHEME I

This polymer is soluble in dimethylacetamide at all stages of cure from poly(amic acid) to polyimide.[3] Films were cast, staged at 25°C intervals, and characterized by various techniques.

One conclusion of the earlier study was that molecular weight behavior could be followed by monitoring a band in the infrared spectrum at 1850 $cm^{-1}$, attributed to anhydride, which appeared during cure. However, the origin of this band was not clear. Thus, 6F-3,3'-DDS poly(amic acid) was synthesized, endcapped with aniline, and characterized. Since endcapping in this manner should remove residual anhydride endgroups from the polymer chain, an effective infrared study of the

anhydride region could be conducted.

To complement analyses on the 6F-3,3'-DDS series of films, additional data is included on a third soluble polyimide obtained from the reaction of 6F with

## SCHEME II

BDAF. Results of the characterization of these soluble aromatic polyimides are reported in this paper and compared with similar results for the original BDSDA-BDAF resin.

2. EXPERIMENTAL

2.1 Materials. Starting materials and solvents were obtained from commercial sources. The 2,2'-bis(3,4-dicarboxyphenyl)-hexafluoropropane dianhydride (6F) was purified by refluxing for 2.5 hr. in 50/50 toluene/acetic anhydride, m.p. 243°C. Diamine monomers 3,3'-diphenylsulfone (3,3'-DDS) and 2,2-bis[4-(4-aminophenoxy)phenyl]hexafluoropropane (BDAF) were recrystallized from water, m.p. 172°C, and ethanol with charcoal, m.p. 156°C, respectively. Redistilled aniline was used.

2.2 Preparation of Poly(amic acid). Solid 6F dianhydride (0.02014 mol) was added over 15 min. at room temperature to a stoichiometric amount of 3,3'-DDS dissolved in N,N-dimethylacetamide (DMAc) in a dry serum bottle equipped with a stopper and a magnetic stirrer. The 15% (w/w) resin solution was stirred for 20 hr. ($n_{inh}$ = 0.44) and refrigerated until used. The 6F-BDAF resin was obtained under similar conditions ($n_{inh}$ = 1.06).

Endcapped 6F-3,3'-DDS was prepared identically to the unendcapped resin except that, after stirring for 20 hr., 2% molar excess of solid anhydride was added. After stirring an additional 19 hr. at RT, 4% molar excess of aniline was added dropwise over a 1 hr. period. The endcapped resin ($n_{inh}$ = 0.39) was also refrigerated until used. A 0.5% molar excess of aniline was added directly to an aliquot of 6F-BDAF resin and stirred for 25 hr. at RT to yield the endcapped polymer ($n_{inh}$ = 0.96). A 2% molar excess of aniline was similarly added to yield a second endcapped 6F-BDAF polymer ($n_{inh}$ = 0.89).

2.3  Preparation of Film Specimens. Films were cast by pouring the poly(amic acid) onto soda-lime glass plates in a dust free chamber. A doctor blade set at 15 mils was used to spread the resin. After drying for 24 hr. at RT under a slight air flow, the plates were placed in a vacuum

desiccator at 40°C for 72 hr. Approximately 1 mil tack-free unendcapped 6F-3,3'-DDS and 6F-BDAF and endcapped 6F-BDAF films were easily removed from the plates. Endcapped 6F-3,3'-DDS film was removed in pieces.

Films were laid in baking dishes and placed in a preheated forced air oven for 15 min. periods. Upon removal from the oven, sufficient sample was retrieved for future analyses. The oven was then taken to the next temperature and the procedure repeated. Samples staged at 25°C intervals from 75° to 325°C as well as the 40°C sample were prepared. Films of unendcapped 6F-BDAF staged for 15 min. at 150°, 250°, and 325°C were also prepared. Finally, films of unendcapped 6F-3,3'-DDS and endcapped 6F-BDAF (2%) were prepared using the standard polyimide cure of 1 hr. each at 100°, 200°, and 300°C.

2.4  Characterization.  Infrared spectra were obtained on a Nicolet 60SX Fourier Transform Infrared system using a diffuse reflectance technique.[4]  Gel permeation chromatography (GPC) analyses were conducted on a Waters Associate high pressure liquid chromatograph using a $10^6/10^5/10^4/10^3$Å µStyragel column bank and a refractive index detector. Inherent viscosity ($n_{inh}$) was measured at 0.5%

(w/v) concentration at 25°C with a Cannon-Ubbelohde viscometer. Intrinsic viscosity $[\eta]$ was also determined on 6F-BDAF films staged at 40°, 150°, 250°, and 325°C. DMAc (Fluka AG) and 0.1M anhydrous LiBr in DMAc were used as solvents for solution property measurements.

Number average molecular weight measurements in DMAc were made by membrane osmometry by a commercial laboratory. Weight average molecular weight determinations in 0.1M LiBr/DMAc were conducted on an LDC/Milton Roy CMX-100 Low Angle Laser Light Scattering (LALLS) photometer.

3. RESULTS AND DISCUSSION

3.1 6F-3,3'-DDS Characterization. The molecular weights of the unendcapped and endcapped 6F-3,3'-DDS resins, as evidenced by their respective 0.44 and 0.39 inherent viscosities, were not as high as had been desired. However, several polymerization attempts did not improve these values. As a result, poly(amic acid) films were brittle, especially the end-capped film. A qualitative increase in brittleness was observed for films staged to 200°C. Above that temperature, the films were more flexible. This behavior with polyimides has been previously reported.[5] Low viscosity was not expected to influence the overall results or conclusions of this investigation. However, this low viscosity did prompt the additional studies on higher viscosity 6F-BDAF resins.

As found in earlier studies,[1,2,6] a band attributed to anhydride developed at approximately 1850 cm$^{-1}$ during thermal imidization and disappeared at elevated temperature. Figure 1 shows the region of the infrared spectrum from 2000-1600 cm$^{-1}$ for the 6F-3,3'-DDS series of polymers. The band at 1856 cm$^{-1}$ was present in 175°, 200°, and 225°C spectra of unendcapped polymer. It was also present in 150°, 175°, and 200°C spectra of endcapped polymer. The fact that this aniline-endcapped polyimide, which should contain no unreacted anhydride groups, exhibited this band during thermal imidization, suggests that the polymer chain experienced an intramolecular break to generate anhydride and, presumedly, amine. We found no direct spectroscopic evidence for the presence of amine for this polyimide. However, indirect evidence has been reported for a different polyimide where an aromatic aldehyde was added to react with terminal amino groups to produce a unique UV chromaphore so that molecular weight could be calculated.[7]

Figure 2 shows results of inherent viscosity measurements made at room temperature on films staged to 325°C. Both series exhibit a lowering in viscosity for films heated to approximately 150°C. While this effect is not large, considering previous work[1,2] and low initial resin viscosity, the effect is believed to be real. The most striking feature of this figure is the rapid viscosity increase above 225°C for the unendcapped polymer. This series is obviously chain-extending at elevated temperature while the endcapped series, as anticipated, is not.

## 3.2  6F-BDAF Characterization.

Both unendcapped and endcapped 6F-BDAF resins were higher viscosity and, presumedly, higher molecular weight than the two 6F-3,3'-DDS resins. Thus, films from this series were characterized to a greater extent. 6F-BDAF films also developed a band ~1850 $cm^{-1}$ during cure. However, this band did not always completely disappear after staging to 325°C.

Extensive GPC analysis of unendcapped 6F-BDAF revealed behavior attributed to polyelectrolyte effects. Figure 3 shows curves for the 40°, 150°, 250°, and 325°C films using both neat DMAc and 0.1M LiBr/DMAc as mobile phases. The four upper curves, obtained in DMAc, appear as higher molecular weight distributions than the four lower curves of the same specimens but using 0.1M LiBr/DMAc. Like charges along the polymer chain repel in dilute solutions. This increases the hydrodynamic volume of the polymer molecule and affects the GPC distribution. Suppression of these polyelectrolyte effects using dilute salt solutions is accepted practice with polyimide precursors.[8] Highly purified solvents have also been used.[9] The fact that polyelectrolyte effects are observed on thermally staged films in DMAc led to an earlier erroneous interpretation on the applicability of GPC for monitoring the cure of soluble polyimides.[1,2] The use of GPC in the present study to differentiate between various stages of cure was difficult when polyelectrolyte effects were suppressed.

Figure 4 shows inherent viscosity data for 6F-BDAF films staged for 15 min. at the indicated temperature. This figure shows the same general trends with temperature as Figure 2. However, the method used to endcap this polymer did not appear to be as effective in controlling molecular weight as the method used to endcap 6F-3,3'-DDS.

Intrinsic viscosity was determined for unendcapped 6F-BDAF films staged at 40°, 150°, 250°, and 325°C. Although the same resin batch was used to prepare all films, approximately 60 days elapsed between the time films were cast for inherent viscosity studies and when they were cast for intrinsic viscosity studies. Figure 5 gives experimental data for the 325°C sample. Polyelectrolyte effects reported by Cotts[9] were observed at concentrations below 0.25 dL/g for all four samples. These data points are not included in the figure. A 0.1M LiBr/DMAc solvent suppressed these effects. It should be noted that polyelectrolyte effects were also observed in the GPC portion of this work, where concentrations were very low, and that dilute salt solution suppressed them. The minimal slope of the ln $\eta_{rel}/c$ line in Figure 5 was observed for all four samples, suggesting that inherent viscosity at 0.5 dL/g is a good approximation of intrinsic viscosity for this polymer.

Intrinsic viscosity for the 40°, 150°, 250°, and 325°C 6F-BDAF films was 0.63, 0.41, 0.51, and 0.57 dL/g, respectively. These values show the trend of relatively high viscosity for the 40° poly(amic acid) sample, significantly lower viscosity for the

followed by viscosity increases for the 250° and 325° samples.

3.3 Correlation with BDSDA-BDAF. Experimental results in this study compare favorably with previously reported results for BDSDA-BDAF.[1,2] The 6F-3,3'-DDS, 6F-BDAF, and BDSDA-BDAF poly(amic acid) films exhibit behavior indicative of an initial decrease in molecular weight during cure followed by a molecular weight increase as imidization is achieved. However, temperatures at which various events occur are different for the three polymers. In two of three cases, an inflection occurred in the viscosity curve in the vicinity of the glass transition temperature of the material. Figure 6 gives data for BDSDA-BDAF.[2] Note the inflection between 200°-225°C. A similar inflection is noted between 250-275°C in Figure 2 for unendcapped 6F-3,3'-DDS. $T_g$ for these two polyimides is 210° and 279°C.[3] No apparent inflection is noted in Figure 4 for 6F-BDAF. Tg of that polymer is 263°C.[3] Further work may establish an empirical relationship between polyimide Tg and the most rapid viscosity increase of dissolved cured films.

3.4 Molecular Weight Determination. Table I gives number average molecular weight ($\overline{M}_n$)

data as a function of cure temperature on three different polyimides. This data was obtained in DMAc by membrane osmometry. Polyelectrolyte effects were not expected to be a factor at concentrations used for these measurements. These results clearly establish the increase in molecular weight at elevated cure temperatures. However, except for BDSDA-BDAF resin 1, they did not support the contention of an initial decrease in molecular weight at low cure temperatures. This fact caused considerable concern during the course of this research.

A possible explanation for this behavior was derived from inherent viscosity measurements made over a period of time on 6F-BDAF unendcapped film specimens. A poly(amic acid) film that received only the 40°C treatment decreased in viscosity from 0.63 to 0.38 dL/g over an 8-month period in a desiccator at RT. A second 6F-BDAF poly(amic acid) film from a different batch of resin decreased from 0.47 to 0.28 dL/g over a similar period. The aged films were less flexible. This decrease in solution viscosity with time for poly(amic acid) films is probably not surprising considering the known instability of poly(amic acid) resins at RT. Refrigerated resin yielded poly-

(amic acid) film viscosities of 0.63 and 0.64 dL/g over the 8-month period. Further, the viscosity of films heated to 150° and above was essentially constant with time at RT.

Table I also lists the date that the poly(amic acid) film was cast and dried, and the date on which the commercial laboratory actually made $\overline{M}_n$ measurements. Considering the viscosity discovery, there likely was a degradation in molecular weight between the time the film was cast and the time it was analyzed. There probably was no such degradation for the thermally staged films. Thus, due to the time delay, 40°C values in Table I may not be true indications of the molecular weight of the original film. With this reasoning in mind, all amic-acid films for weight average molecular weight measurements were refrigerated until analyzed.

Table II gives weight average molecular weight ($\overline{M}_w$) values obtained in 0.1M LiBr/DMAc by low angle laser light scattering (LALLS) photometry. Several constant multipliers used to calculate these values from LALLS measurements were not known precisely for these polymer/solvent combinations and had to be estimated. Further work will include determining these factors so that more accurate values can be calculated. In most cases, film specimens submitted for $\overline{M}_w$ analysis were not the same ones submitted for $\overline{M}_n$. Thus, the data in Table II are reported only for the purpose of showing trends.

As previously discussed, samples for $\overline{M}_w$ were refrigerated between the time they were cast and analyzed. Thus, 40°C $\overline{M}_w$ values should be better estimates of that parameter than 40°C $M_n$ values, where some polymer degradation was suspected. A decrease in $\overline{M}_w$ is indicated during the initial stages of imidization for each of the three polymers. The minimum in molecular weight appears to occur after staging to approximately 150°C. This observation is in general agreement with viscosity data given in Figures 2, 4, and 6.

3.5 Implications to Polyimides in General. Polymers discussed in this paper were soluble in DMAc and, thus, could be characterized by solution property measurements. Most conventional polyimides of more immediate interest to the aerospace community are insoluble, except for the poly(amic acid) and perhaps the fully imidized stage, and cannot be adequately characterized during cure. However, infrared spectroscopy may provide a link between soluble and insol-

uble polyimides. The behavior of the 1850 cm$^{-1}$ band likely provides an estimate of polymer molecular weight behavior during cure.

All evidence in this work indicates that polyimides go through a molecular weight minimum at approximately 150°C. This evidence should be considered when planning processing and cure cycles for manufactured articles. Traditionally, pressure is applied during processing above $T_g$ of the polymer. This work shows that molecular weight, and presumedly, resin viscosity, have rapidly increased by this temperature. This suggests that application of pressure at lower temperature might result in greater flow and consolidation. An earlier bonding study in this laboratory on a polyimide sulfone adhesive indicated that superior properties were achieved when pressure was applied at 160°C during processing rather than at 280°C, which is close to $T_g$ of the adhesive.[10] We have also found that polyimide molding powder is more easily prepared when the amic-acid resin is staged at 150°C and ground, as opposed to attempting to grind unstaged amic-acid or fully imidized material. Preliminary results indicate that moldings of the 150°C-staged material are of superior quality.

## 4. CONCLUDING REMARKS

Several soluble aromatic polyimides have been characterized
during cure by infrared and solution property measurements. The
behavior observed was indicative
of an initial reduction in molecular weight during cure followed by
a molecular weight increase as
imidization proceeded. Both
unendcapped and endcapped poly-
(amic acid) films developed a band
~1850 cm$^{-1}$ during cure, attributed to anhydride, suggesting
that the molecular weight reduction was due to a break in the
polymer backbone. An instability
of cast poly(amic acid) film was
also noted.

A relationship may exist between
the glass transition temperature
and the temperature region in
which the most rapid molecular
weight buildup is achieved. The
greatest change may occur just
prior to $T_g$. Results of this
study serve to increase our fundamental understanding of how polyimides are formed from precursor
polymers. How they may impact
processing should be investigated
in a systematic manner.

## 5. ACKNOWLEDGEMENT

The authors gratefully acknowledge
the effort of Judith R. J. Davis,
PRC Kentron, Inc., in conducting
the LALLS portion of this
research.

## 6. REFERENCES

1. P. R. Young and A. C. Chang;
   SAMPE Symposium, 30,
   889(1985).
2. P. R. Young, N. T. Wakelyn,
   and A. C. Chang; Proc. Second
   Intl. Conf. on Polyimides,
   Ellenville, NY, Oct. 30 -
   Nov. 1, 1985, p. 414.
3. A. K. St. Clair, T. L. St.
   Clair, and K. I. Shevket;
   Polym. Matls. Sci. Engin.
   Preprints, 51, 62(1984).
4. P. R. Young and A. C. Chang;
   SAMPE Symposium, 28,
   824(1983).
5. R. A. Rine-Hart and W. W.
   Wright, J. Appl. Polym. Sci.,
   11, 609(1967).
6. P. J. Dynes, R. M. Panos, and
   C. L. Hamermesh; J. Appl.
   Polym. Sci., 25, 1059(1980).
7. E. V. Kamzolkina, P. P.
   Nechaev, V. S. Markin, Ya.S.
   Vygodskii, T. V. Grigor'eva,
   and G. E. Zaikov; Dokl.Akad.
   Nauk. USSR, 219, 650(1974).
8. M. L. Wallach; J. Polym.
   Sci., A-2, 5, 653(1967).
9. P. M. Cotts; in "Polyimides,"
   Vol. 1. K. L. Mittal, Ed.,
   Plenum Press 1984, p. 223.
10. J. F. Dezern and P. R. Young;
    Intl. J. Adhesion and
    Adhesives, 5, 183(1985).

# TABLE I. NUMBER AVERAGE MOLECULAR WEIGHT [1]

| DESCRIPTION | DATE CAST | DATE ANALYZED | CURE TEMPERATURE, °C MOLECULAR WEIGHT | | | |
|---|---|---|---|---|---|---|
| 6F-3,3'-DDS (UNENDCAPPED) | 5/1/85 | 11/19/85 | 40° 9150 | 150° 12,800 | 200° 9,900 | 300° 28,700 |
| 6F-BDAF (UNENDCAPPED) | 11/14/85 | 3/21/86 | 40° 11,100 | 150° 20,700 | 250° 27,800 | 325° 31,000 |
| BDSDA-BDAF (RESIN 1) | 4/1/85 | 6/18/85 | 40° 22,200 | 125° 19,800 | 175° 21,200 | 300° 41,600 |
| BDSDA-BDAF (RESIN 2) | 4/22/85 | 11/19/85 | 40° 11,400 | 125° 11,400 | 175° 21,750 | 300° 43,700 |

[1] SOLVENT: DMAC

# TABLE II. WEIGHT AVERAGE MOLECULAR WEIGHT TRENDS [1-3]

| DESCRIPTION | DATE CAST | DATE ANALYZED | CURE TEMPERATURE, °C MOLECULAR WEIGHT | | | | |
|---|---|---|---|---|---|---|---|
| 6F-3,3'-DDS (UNENDCAPPED) | 10/15/86 | 10/23/86 | 40° 38,400 | 100° 27,100 | 150° 22,100 | 200° 78,000 | 300° 114,700 |
| 6F-BDAF (UNENDCAPPED) | 11/14/85 | 11/4/86 | 40° [4] 53,600 | | 150° 32,000 | 250° 42,800 | 325° 65,800 |
| BDSDA-BDAF (2% ENDCAPPED) | 9/18/86 | 10/17/86 | 40° 34,800 | 100° 26,200 | | 200° 36,800 | 300° 62,800 |

[1] SOLVENT: 0.1M LiBr/DMAC
[2] 40° FILM SPECIMENS STORED AT 10°C
[3] $dn/dc$ ASSUMED TO BE 0.15
[4] RECAST FROM ORIGINAL RESIN ON 8/4/86

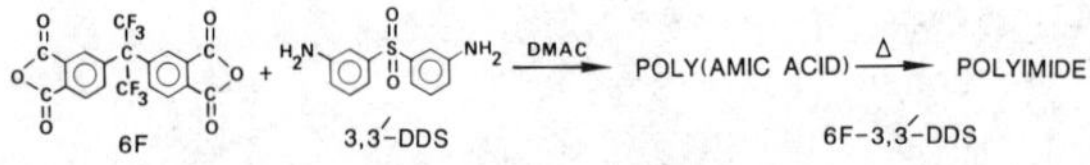

## SCHEME I

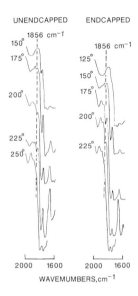

SCHEME II

Figure 1.   Infrared spectra of thermally staged unendcapped
and endcapped 6F-3,3'-DDS film.

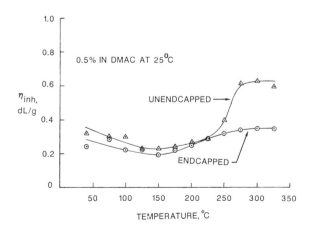

Figure 2.   Inherent viscosity of thermally staged 6F-3,3'-DDS film.

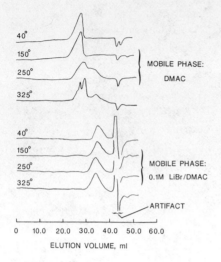

Figure 3. Gel permeation chromatography of thermally staged unendcapped 6F-BDAF film. Conditions: Column - $10^6/10^5/10^4/10^3$Å μ Styragel; Mobile Phase - as shown; Rate - 1.5 ml/min; Size - 40 μl (0.25%); Detector-Refractive Index.

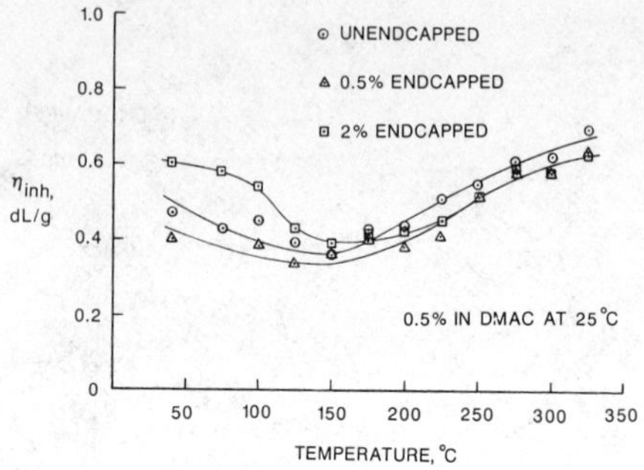

Figure 4. Inherent viscosity of thermally staged 6F-BDAF film.

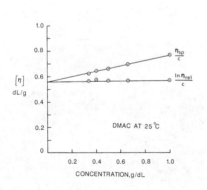

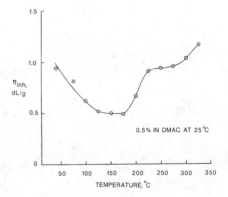

Figure 5. Intrinsic viscosity of 325° unendcapped 6F-BDAF film.

Figure 6. Inherent viscosity of thermally staged BDSDA-BDAF film.

THERMAL PERFORMANCE OF CYANATE FUNCTIONAL
THERMOSETTING RESINS
D. A. Shimp
9800 E. Bluegrass Parkway
Louisville, KY 40299

## Abstract

Polycyanurate thermoset plastics prepared by the cyclotrimerization of aryl dicyanates are compared to toughened bismaleimides and a tetra-functional epoxy for both short and long term aspects of thermal performance. Oligomers of three aryl dicyanates, including a flame retardant sulfur-linked bisphenol derivative, developed superior cured-state properties when catalyzed with specific metal acetylacetonates and phenolic catalysts. Performance exceeded that of the BMI and epoxy reference materials with respect to the onset of rapid degradation in air (~400°C), moisture absorption in boiling water (1-3%), peel strength to 250°C, and impact resistance. Performance was intermediate with respect to modulus retention (wet and dry), and inferior to BMI's for long term stability in air at temperatures exceeding 200°C.

## 1. INTRODUCTION

The successful application of high temperature plastics in the electronics and structural composite fields creates demand for new materials with either escalated thermal performance requirements, or improved moisture resistance, toughness, adhesion, dimensional stability, purity, processibility, dielectric constant (lower), as well as combinations of the above. Some needs are for short term performance at 200-350°C processing temperatures. Multilayer printed wiring boards (1), for example, require dimensional stability during solder application, particularly in field repair where momentary laminate temperatures of ~350°C can blister flame retardant resins employing halogenated components (1). Improved toughness (drillability) and adhesion to conductors are additional needs not fully met by today's polyimides (2).

Composites used in supersonic air-craft must maintain mechanical strength and stiffness for a poorly defined period at temperatures as high as 232°C (dry) and 177°C (wet). Engine applications require long term (thousands of hours) perform-ance at generally higher tempera-tures. Associated matrix resin needs are for improved damage tolerance and faster processibility. The development of structural adhesives for high temperature composites has lagged innovations in the matrix resin.

Several publications (3-5) describe the chemistry and properties of polycyanurates formed by the cyclotrimerization of aryl di-cyanates. Key features are:

- Addition polymerization.
- Hot-melt processibility.
- Cureable at 177°C.
- Low shrinkage.
- Tg values 240-290°C.
- Tensile strain-at-break 2-6% range.
- Saturated $H_2O$ abs. 1-3%.
- Low $D_K$ (3.0-3.5).
- Excellent adhesion.

Thermal performance has not been well characterized, particularly with respect to chemical structure and catalyst effects.

## 2. EXPERIMENTAL

### 2.1 Cyanate Resins

Three aryl dicyanates of >99.5% purity are chemically described in Figure 1. The following acronyms will be used to represent these

structural variants:

BADCy - Bisphenol A precursor
METHYLCy - Methyl shielded functionality (hydrophobic)
THIOCy - Sulfur linkage (flame retardant)

Amorphous prepolymers are prepared by partially trimerizing the crystalline dicyanates (Table 1). These resins were catalyzed and de-aired neat at 80-100°C for the preparation of 1/8 inch sheet cast-ings.

Metal acetylacetonate catalysts (6) are sparingly soluble, but easily incorporated into solvent-free formulations by dissolving first into the nonylphenol co-catalyst at temperatures up to 150°C. Refer-ences (3,5,7) describe improved trimerization catalysts comprising solutions of transition metal carboxylates or chelates in hydroxyl functional, non-volatile liquids.

### 2.2 Bismaleimide Resins

Commercially available, toughened BMI resins were prepared by adduct-ing (chain extending) 4,4' Bis-maleimidodiphenylmethane (BMI) with:

a] Methylene dianiline (MDA) Kerimid 601 type (8)
b] Diallylbisphenol (DAB) XU-292 type (9)

Because MDA adducts of BMI have insufficient pot life in the molten state for casting, it was necessary to pour a molten solution of the components into the mold to advance at 150°C prior to cure. A mole ratio of 2.5 BMI to 1.0 MDA was

employed in this study.
BMI was dissolved into the liquid diallylbisphenol A component at 150°C (mole ratio 1.00:0.87), deaired, and gelled at that temperature. Neither of the BMI resins were catalyzed, in conformance with recommendations.

## 2.3 Epoxy Reference

An MY720 (TGMDA)/Epi-Rez SU8/DDS system in 69.2/7.6/23.2 weight ratio was cast to represent the class of tetrafunctional epoxy matrix widely used in commercial composites.

## 2.4 Curing Conditions

The following gelation and cure schedules were employed:

|  | Cyanate | BMI | Epoxy |
|---|---|---|---|
| Gel Temp.,°C | 105 | 150 | 150 |
| Cure, Hours |  |  |  |
| at 177°C | 1 | 1 | 2 |
| at 232°C | - | 2 | 2 |
| at 250°C | 2 | 6 | - |

## 2.5 Thermal Analysis

A DuPont 1090 analyzer was employed with the following specimen heat-up rates:

| Test Mode | °C/Min. |
|---|---|
| DMA | 5 |
| TGA | 10 |
| TMA | 10 |

A flat probe weighted with 2 grams was used for the TMA determinations.

## 2.6 Degree of Cure

Cured cyanate castings were evaluated for extent of cyclotrimerization via FTIR (see Fig. 2). A small sample was pulverized with KBr and pressed into pellets. Ratios of cyanate absorbance (2270 cm$^{-1}$) to methyl C-H stretch absorbance (2875 cm$^{-1}$) were proportioned to that of an unheated dicyanate standard for calculation of % conversion.

## 2.7 Other Test Conditions

Heat deflection temperatures were determined at 264 psi "fiber" loading. Bars were conditioned for wet HDT testing by sealing in a chamber containing water at 92°C for 64 hours (equivalent to ~80% of moisture saturation).

Wet flexural modulus was measured on bars immersed 48 hours in boiling water. Dummy bars with embedded thermocouples were found to require 2-3 minutes heat soak in a chamber set 15°C above the target temperature, prior to initiating crosshead travel.

Isothermal ageing in air was conducted in a Young Brothers forced-draft oven.

## 3.1 Cure Catalysts

Reference (3) describes the ring-forming mechanism together with the roles of a transition metal ion (gathers cyanate functionality via coordination complexes) and an active hydrogen component (ring closure via proton transfer). Screening studies on transition metal carboxylates taught the avoidance of metals which promote transesterification and hydrolysis reactions, i.e. Sn, Pb, Sb, and Ti.

Table 2 compares the effect of six metal acetylacetonate catalysts with two levels of nonylphenol (hydroxyl source) on reactivity, extent of trimerization, moisture absorption and thermal stability. In general, the metal ion has a significant effect on stability in 235°C air and moisture plasticization (wet HDT). Preference is in the order $Cu^{+2}=Co^{+2}>Zn^{+2}>Mn^{+2}>Fe^{+3}>Al^{+3}$. The level of hydroxyl source is a more important contributor to conversion than the metal. In general 2-4 phr nonylphenol (13 to 26 meq./cyanate) is preferred when 250°C postcure is feasible, increasing to 4-6 phr for 177°C max. cure temperature. The acetylacetonate form of most metal catalysts is preferred over the naphthenate or octoate for maximum thermal stability. The following catalysts were employed for curing the cyanate resins evaluated in this paper:

| Resin | Metal/ppm | NP,phr |
|---|---|---|
| BADCy | $Cu^{+2}/360$ | 2 |
| METHYLCy | $Zn^{+2}/150$ | 2 |
| THIOCy | $Cu^{+2}/50$ | 2 |

3.2 Basic Property Comparisons

Mechanical, thermal and flammability properties of three cyanates, two bismaleimides and the tetrafunctional epoxy system are compared in Table 3. Toughness indicators (strain-at-break, Izod impact, and crack propagation energy) are significantly higher for the polycyanurates while thermo-mechanical limits (HDT and Tg) are intermediate. The TGA onset of thermal degradation in air (Fig. 4) is significantly higher for the polycyanurates (400-

411°C) than the toughened BMI's (369-371°C) and the epoxy (306°C). Thus short-term performance at temperatures well over Tg values, e.g. thermal spike situations like solder application, is indicated for the cyanate homopolymers. Tensile storage moduli (DMA, Fig.6) are generally lower at RT for polycyanurates, however BADCy and THIOCy homopolymers exhibit a comparatively flat thermal response.

Char values are significantly higher for THIOCy (46%) and BMI-MDA (48%), accounting for the V-0 flammability ratings achieved at 1/8" thickness.

3.3 Moisture Absorption

Fig. 3 plots weight gain over a 500 hour immersion period in boiling water. Moisture absorption at saturation is considerably lower for the polycyanurates (1.5-2.5%) than for bismaleimides (4-5%) and TGMDA-DDS (6%). Previous testing of the cyanate ester homopolymers in boiling water (850 hours) and 15 psig steam (100 hours) produced no indication of hydrolysis. Flexure strength retention after these conditioning periods was >90%. Cyanate homopolymers are remarkably resistant to hot water and steam when the conversion of cyanate functionality to triazine rings exceeds 90%, and the recommended metals, $Cu^{+2}$, $Co^{+2}$ and $Zn^{+2}$, are employed as cure catalysts.

3.4 Thermal Stability

While short term exposure to high temperatures (see TGA plots in Fig.

4). indicated remarkable stability, long term service in air at $\geq 200°C$ can cause polycyanurates to fail prematurely by outgassing. Weight loss plots in air at 235°C (Fig. 5) illustrate degradation rates comparable to the toughened bismaleimides over a 500 hour period, with THIOCy homopolymer notably superior. Swelling is noted, however, at relatively low weight loss values, indicating a degradation mechanism other than chain scission is a limiting factor in long term service at temperatures within about 50°C of HDT values. Shown below are hours at air temperatures resulting in (a) swelling and (b) 50% flexure strength retention for the three cyanate homopolymers:

|  | BA | METH | THIO |
|---|---|---|---|
| 200°C |  |  |  |
| Swelling | 2400 | 3500 |  |
| 50% ret. | 3000 | 4200 |  |
| 235°C |  |  |  |
| Swelling | 300 | 400 | 400 |
| 50% ret. | 400 | 450 | 500 |

The cause of swelling is believed to be evolution of $CO_2$ from carbamates formed when residual (sterically isolated) cyanate functionality encounters moisture in air at temperatures producing segmental mobility. Outgassing has not been observed in ageing tests conducted below 200°C.

3.4 Hot-Wet Properties

Modulus values of flexure bars conditioned 48 hours in boiling water (about 80% moisture satura-tion) and tested at temperatures up to 177°C are compared in Fig. 7. The exceptional hydrophobicity of the methylated polycyanurate permits little if any moisture plasticiza-tion. (Note that wet and dry HDT values are essentially identical - Table 3). Wet modulus values for the METHYLCy homopolymer are well within the range of the toughened bismaleimides at temperatures of 100-177°C.

3.5 Adhesive Properties

Oligomers of BADCy have been used as laminating resins in printed wiring boards since 1978, with reports (10) of excellent copper peel strength maintained to 250°C. Flame retardant laminates prepared with THIOCy oligomers in our labora-tory produced the following peel strengths (1 ounce copper cladding):

| TEMP.,°C | Lbs./Inch |
|---|---|
| 25 | 11.0 |
| 100 | 10.7 |
| 200 | 10.2 |
| 230 | 9.0 |

By comparison, the BMI-MDA polyimide laminates evaluated in ref. (10) report 8.0 lbs./in. at 25°C, falling to 6.1 lbs./in. at 230°C.
Lap shear strengths in the range of 4,000-7,000 psi are reported for aluminum, stainless steel and titanium adherends, with $\geq 80\%$ retention to 232°C, in ref. (11).

4. CONCLUSIONS

Homopolymers of three aryl dicyanates exceed the thermal capabilities of tetrafunctional epoxy resins and compare favorably with toughened bismaleimides. While short term (TGA) stabilities of the polycyanurates were highest in the series evaluated, long-term ageing in air at ≥200°C resulted in premature failure via outgassing.

Moisture absorption at saturation in boiling water occurs at values approximately one half those of bismaleimides. The hydrophobic o-methylated cyanate was unique in that no signs of moisture plasticization were observed after 500 hours boiling.

Toughness properties of the polycyanurates as measured by notched Izod impact, $G_{IC}$ and peel strength are unusually high for Tg 250-280°C thermosets. Adhesive shear and peel strengths are maintained to 230°C. No loss in thermal or mechanical performance was observed in a flame retardant dicyanate synthesized from a sulfur-linked bisphenol.

New applications in field repairable multilayer PWB's, high signal speed (low Dk) electrical insulation and structural adhesives/composites for supersonic aircraft are forseen for cyanate functional resins.

## 5. ACKNOWLEDGEMENT

The author appreciates many useful insights and suggestions contributed by Mary Potter (thermal analysis) and Joe Farmer (infrared analysis).

## 6. REFERENCES

1. McGowan, D., Circuits Manuf., May 1984, p 78.

2. Guiles, C.L., P C Fab., June 1985, p 47.

3. Shimp, D.A., ACS Meeting New York, April, 1986, PMSE preprints, pp. 107-113.

4. Hudock, F.A. and Ising, S.J., "Cyanate Ester Resins for PWB Laminates", IPC Fall Meeting, San Diego, 9/21/86.

5. Shimp, D.A., 18th International SAMPE Tech. Conf., Seattle, WA, October 1986, pp. 851-862.

6. Oehmke, R.W., US3,694,410, Use of Chelates in Preparing Polycyanurates, Sept. 1972.

7. Shimp, D.A., US4,608,452, Metal Carboxylate/Alkylphenol Curing Catalyst, Aug. 1986.

8. Bargain, M. et al, US3,562,223, Cross-Linked Resins, Feb. 1971.

9. Zahir, S.A. et al., US4,100,140, Crosslinked Polymers Which Contain Imide Groups, July 1978.

10. Ayano, S., "Triazine Chemistry and It's Application Development", Chem. Economy & Eng. Review, March 1978, Vol. 10 No. 3, p 29.

11. Delano, C.B. & Harrison, E.S., Technical Report AFML-TR-73-216, Part II, March 1974.

FIGURE 1

## CHEMICAL STRUCTURE OF ARYL DICYANATES

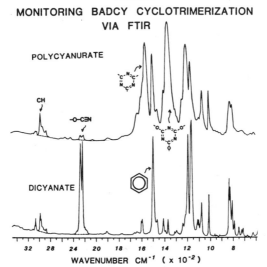

BADCy

M.P. 79 C

BISPHENOL A DICYANATE

METHYLCy

M.P. 106 C

METHYLENE BIS ( 3,5-DIMETHYLPHENYL-4-CYANATE )

THIOCy

M.P. 94 C

4,4'-THIODIPHENYLCYANATE

TABLE 1

## SEMISOLID CYANATE PREPOLYMER RESINS

| Description | RDX 80352 | RDX 80359 | RDX 80371 |
|---|---|---|---|
| Precursor | BADCy | METHYLCy | THIOCy |
| % Trimerization | 31 | 30 | 44 |
| Cyanate Eq. Wt. | 202 | 218 | 240 |
| Melt Viscosity 180 F, cps | 1,000 | 3,100 | 16,700 |
| Sp. Gr. | 1.22 | 1.17 | 1.37 |
| Ionics, ppm | < 5 | < 5 | < 5 |

FIGURE 2

## MONITORING BADCY CYCLOTRIMERIZATION VIA FTIR

POLYCYANURATE

CH

-O-C≡N

DICYANATE

| 32 | 28 | 24 | 20 | 18 | 16 | 14 | 12 | 10 | 8 |

WAVENUMBER CM$^{-1}$ ( x 10$^{-2}$ )

TABLE 2 - METAL CATALYST EFFECTS ON BADCy PROPERTIES

| Reactivity | $Cu^{+2}$ | $Co^{+2}$ | $Zn^{+2}$ | $Mn^{+2}$ | $Fe^{+3}$ | $Al^{+3}$ |
|---|---|---|---|---|---|---|
| Metal Conc., ppm | 360 | 160 | 175 | 435 | 65 | 250 |
| Minutes to Gel | | | | | | |
| 105°C | 60 | 190 | 20 | 20 | 35 | 210 |
| 177°C | 2 | 4 | 1 | 1 | 1.5 | 4 |
| Cured-State[a] | | | | | | |
| Conversion, % | | | | | | |
| 2 phr Nonylphenol | 96.6 | 95.7 | 95.8 | 93.8 | 96.5 | 96.8 |
| 4 phr Nonylphenol | 98.4 | 97.3 | 97.0 | – | – | – |
| HDT, °C | | | | | | |
| Dry | 244 | 243 | 243 | 242 | 239 | 238 |
| Wet[b] | 175 | 193 | 182 | 163 | 143 | 157 |
| Moisture Abs., % | | | | | | |
| 500 Hrs at 100°C | 2.4 | 2.3 | 2.5 | 2.6 | 2.4 | 2.4 |
| Thermal Stability | | | | | | |
| 235°C Air, Hours | 400 | 400 | 375 | 300 | 250 | 200 |

(a)  Cured 1 hour at 150°C + 2 Hours at 250°C.
(b)  Moisture conditioned 64 hours at 92°C + >95% RH.

TABLE 3 - COMPARISON OF CURED RESIN PROPERTIES

| Mechanical | Cyanates BADCy | Cyanates METHYLCy | Cyanates THIOCy | Epoxy TGMDA DDS | Polyimide BMI-MDA | Polyimide BMI-DAB |
|---|---|---|---|---|---|---|
| Flexure at 25°C | | | | | | |
| Strength, ksi | 25.2 | 23.3 | 19.4 | 14.0 | 10.9 | 25.5 |
| Modulus, Msi | 0.45 | 0.42 | 0.43 | 0.55 | 0.50 | 0.53 |
| Strain, % | 7.7 | 8.2 | 5.4 | 2.5 | 2.2 | 5.1 |
| Notched Izod | | | | | | |
| ft.-lbs/in. | 0.65 | 0.8 | 0.8 | 0.4 | 0.35 | 0.5 |
| $G_{IC}$, in-lbs/in$^2$ | 0.8 | 1.0 | 0.9 | 0.4 | 0.4 | 0.5 |
| Thermal | | | | | | |
| HDT, °C | | | | | | |
| Dry | 244 | 242 | 243 | 232 | >270 | 266 |
| Wet[a] | 175 | 234 | 195 | 167 | 262 | 217 |
| TGA | | | | | | |
| Onset, °C (air) | 411 | 403 | 400 | 306 | 369 | 371 |
| Char, % ($N_2$) | 41 | 41 | 46 | 31 | 48 | 29 |
| TMA | | | | | | |
| Tg, °C | 257 | 244 | 270 | 210 | 297 | 263 |
| CLTE, ppm/°C | 64 | 71 | 68 | 67 | 62 | 63 |
| DMA | | | | | | |
| Tg, °C | 289 | 252 | 273 | 246 | 320 | 288 |
| E', GPa | | | | | | |
| at 25°C | 1.62 | 1.61 | 1.41 | 1.97 | 1.65 | 1.90 |
| at 200°C | 1.30 | 1.07 | 1.10 | 1.13 | 1.30 | 1.35 |
| Flammability | | | | | | |
| UL-94, seconds | | | | | | |
| 1st Ignition | >50 | >50 | 1 | >50 | 1 | >50 |
| 2nd Ignition | – | – | 3 | – | 5 | – |

(a)  Moisture conditioned 64 hours at 92°C &  95%RH.

## FIG. 3 - MOISTURE ABSORPTION IN BOILING WATER
### ( 1/8 INCH CASTINGS )

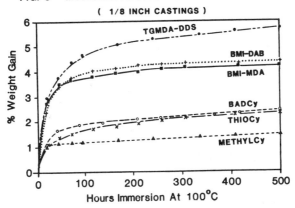

TGMDA-DDS
BMI-DAB
BMI-MDA
BADCy
THIOCy
METHYLCy

% Weight Gain

Hours Immersion At 100°C

## FIG. 4
## THERMOGRAVIMETRIC ANALYSIS IN AIR

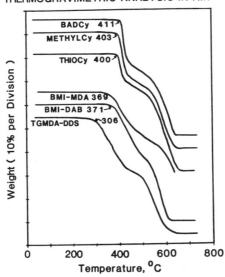

BADCy 411
METHYLCy 403
THIOCy 400
BMI-MDA 369
BMI-DAB 371
TGMDA-DDS 306

Weight ( 10% per Division )

Temperature, °C

## FIG. 5 - THERMAL DEGRADATION RATES IN AIR AT 235°C
### ( 1/8 Inch Castings )

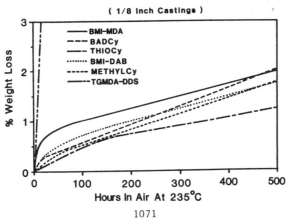

BMI-MDA
BADCy
THIOCy
BMI-DAB
METHYLCy
TGMDA-DDS

% Weight Loss

Hours In Air At 235°C

## FIG. 6
## STIFFNESS VS. TEMPERATURE ( DMA )

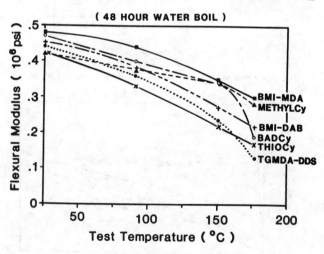

FIG. 7 - FLEXURAL MODULUS (WET) VS. TEMPERATURE

( 48 HOUR WATER BOIL )

CHARACTERIZATION AND ANALYSIS OF PMR-15
POLYIMIDE MATRICES FOR HIGH TEMPERATURE COMPOSITES

E. M. Woo*, J. C. Seferis*, and P. R. Lopez**
Polymeric Composites Laboratory
Department of Chemical Engineering, BF-10
University of Washington
Seattle, WA 98195
and
L. B. Ilcewicz
Boeing Commercial Airplane Company
P. O. Box 3707, 6C-11
Seattle, WA 98124

## Abstract

Neat polyimide matrix (PMR-15) film samples were prepared by two methods: press-forming and autoclave-forming. For each method, three different cure cycles, denoted by the highest temperatures during cure (288°C, 306°C and 329°C), were investigated. The curing mechanism was investigated with thermoanalytical techniques (pressure differential scanning calorimetry and thermogravimetric analysis). Possible structural differences in the samples were correlated to the processing variables and related to the results from dynamic mechanical analysis(DMA). The DMA results showed that the films produced by the three cure cycles either in the press or in the autoclave were quite identical. For the three cure cycles investigated, it was observed that post-curing had the strongest effect on the low temperature cured samples (288°C and 306°C cured samples); while the cure temperature was proportionally related to the resulting sample Tg and stress-free temperature. Finally, the results for composite samples indicate

*Author to whom correspondence should be addressed.
**Present address: Rhone-Poulenc Inc., France.

that the temperature dependence of graphite/PMR-15 moduli was well predicted using micromechanics analysis on neat matrix data.

## 1. INTRODUCTION

The exceptionally high thermo-oxidative stability of wholly aromatic polyimides is, in general, compromised by their difficult processability. An attractive alternative is the in-situ polymerization of monomer reactant resins, e.g., PMR-15, which has been studied extensively (1-10). Past studies have shown that processing dictates both matrix and composite properties (11).

Microcracking and internal stresses are of major concern when this material is used as a matrix for composites. Uneven residual stresses between the matrix and the reinforcing graphite fibers of a composite may cause microcracks in the composite part during its service period. If the microcracks are followed by large-scale dynamic changes in loading and/or temperature, fatigue is prone to occur. This may cause eventual breakdown of the composite.

Another factor contributing to residual stresses is the presence of residual solvents or volatile by-products of cure reactions which may condense at low temperatures but expand at high temperatures and create stress centers from which microcracks might easily propagate. Under these presumptions, it becomes obvious that properties of composite end products are dependent on the cure cycle (processing), during which heat and pressure are applied in different combinations to ensure optimal properties, or simply to adjust the properties of composites for various applications.

The objective of this study was to investigate relationships between property and processing involved with three different cure cycles. Thermal and dynamic mechanical analyses were performed on neat resin samples before and after post cure processing. Finally, temperature-dependent composite moduli were measured and compared to prediction based on micromechanical calculations using the neat resin data.

## 2. EXPERIMENTAL METHODS

### 2.1 Materials and sample preparation

The polyimide resin, PMR-15, was purchased from the Ferro Corporation and kept in an ethanol solution (48% by weight) at -18°C until needed. To obtain a solvent-free prepolymer, the mixture was dried in small quantities in an air-circulated oven at a maximum temperature of 105°C under vacuum.

Film samples were processed in a molding press (manufactured by PHI Co.) and an autoclave (model 466, manufactured by Lipton Inc.).

Three standard cure cycles, designated as 288°C-, 306°C-, and 329°C-cure, were produced by two different processing, press-forming, and autoclaving techniques in as-cured or post-cured states. The 306°C cycle consisted of heating up to 83°C at 1-3°C/min for 60 minutes, followed by heating to 207°C at the same rate for 45 minutes. Procedures prior to this step were under vacuum of 7.6 - 12.7 cm Hg. The vacuum was then released, followed by heating to 235°C at about the same heating rate.

Meanwhile, pressure of 0.021 Pa (150 psi) was applied. The temperature was further increased to 306°C at the same heating rate and under the same pressure for about 185 minutes. The cycle was finished by cooling to 70°C at a cooling rate no greater than 2.8°C/min. The other two cycles, 288°C and 329°C cure cycles, were similar, except that the maximum temperatures used in the cycles were 288°C and 329°C, respectively. Post-curing for each of the cure cycles was carried out by heating to about 320°C for no less than 6 hrs. Both as-cured and post-cured samples were analyzed.

Two different processing methods were used in this study, press-forming and autoclaving. The main differences were the vacuum option and a better pressure control in the latter case. In the former case, the polyimide resin in solution form was first dried in a vacuum oven at 105°C for about 10 hrs. The dried resin was then ground into a fine powder and imidized in a vacuum oven according to the cure cycle. After imidization, the powder was cooled and reground into a finer powder. The powder was placed between a pre-heated press under the pressure of 0.07 to 0.14 Pa, and the standard cure cycle was applied. At the end of the cycle, the film was taken out and put into an oven for a post-cure cycle. For autoclaving process, the procedures were similar, except that an autoclave was used instead of the press and the powder was transformed to a film inside a vacuum bag.

## 2.2  Equipment and Procedures

Apparatus used to characterize the resins were differential scanning calorimetry (DSC, Dupont 9900 thermal analyzer driving 910 and 912 DSC cells), thermogravimetric analysis (TGA, Dupont 951 TGA), and dynamic mechanical analysis (Rheovibron DDV-II).

Dynamic mechanical analysis (DMA) of the samples were performed using a DDV-II Rheovibron viscoelastometer over a temperature range of -120 to 400°C at 11Hz. Selected samples were run only from the ambient temperature up to 400°C. The temperature was increased at an average rate of 1 to 1.5°C/min. The DMA technique was expected to reveal difference in properties of samples processed using different cycles. Thus, structure-property-processing relationships can be interpreted by using the results of dynamic mechanical analysis. Employing the procedures developed by Wedgewood and Seferis (12), the results obtained were compared quantitatively to static modulus values.

In order to get transverse moduli, static tension tests were performed with 3.8cm x 30.5cm [90]$_8$ samples of graphite/PMR-15 (CE6000/PMR-15, fiber volume fraction = 0.63 ) cured with standard 329°C cycle in an autoclave followed by the standard post-cure. Tensile tests were performed over a temperature range between 23°C and 235°C. Static test results were compared to predictions from micromechanics that used DMA neat resin data and fiber properties.

## 3. RESULTS AND DISCUSSION

### 3.1  Thermal analysis

As shown in Figure 1, four thermal transitions were observed. The first endotherm, centered at 150°C, corresponds to the melting stage of the dried prepolymer. The second endotherm, centered at 200°C, can be related to the imidization, characterized by two steps: polycondensation with elimination of methanol and ring-closure with elimination of water. Note that these steps are not well resolved; and there is some controversy about the origins of these peaks. A broad endotherm occurs at 300°C and is not clear regarding its origin. Finally, the crosslinking reactions start to occur at 320°C.

decreases before a jump at about 130°C. This is due to the emission of methanol produced during the first step of imidization. This peak partly overlaps with a second one, centered at 200°C, corresponding to the emission of water produced during the second step of imidization. A third peak occurs at 330°C. At this stage, imidization can be considered to be complete and the by-product may be identified as cyclopentadiene, originating from the crosslinking reaction. Finally, degradation occurs above 400°C. From the TGA data showing dramatic weight loss during the imidization stage, a thorough degassing was employed in making the neat matrix samples.

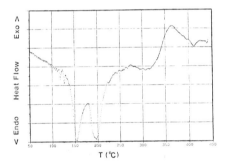

Fig. 1. Typical PDSC scan of prepolymer dried at 90%

Figure 2 shows a typical TGA scan on the partially dried PMR-15 prepolymer sample at 10°C/min. A small loss of weight starts below 100°C, which may be attributed to methanol desorption. The weight loss rate then

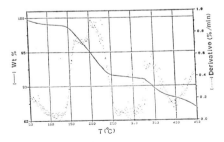

Fig. 2. Typical TGA scan of prepolymer dried at 90%.

### 3.2 Dynamic mechanical analysis

Figure 3 shows the comparison of the dynamic mechanical properties 'of two samples processed in the press according to the 288°C cure cycle, one as-cured and the other post-cured. Storage moduli are nearly the same for both samples below the glass transition temperature. The maximum of the β loss peak of the post-cured sample is shifted to a higher temperature (100°C) compared to that of the as-cured sample (91.5°C). The β loss peak of the post-cured sample is also broader and has higher average intensity than that of the as-cured one.

by the α transitions in Figure 3. The α loss peak of the 288°C post-cured sample has a maximum at 336°C, which is 43°C higher than that of the as-cured sample. This clearly indicates that the post-curing step in the 288°C cycle has a significant effect on the structure of these samples.

The dynamic mechanical properties of all as-cured samples processed according to each of the three cure cycles are summarized in Figure 4.

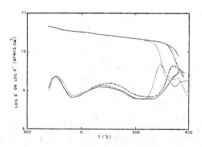

Fig. 4. Dynamic mechanical modulus of as-cured, press-made PMR-15 samples: (-) 329°C cured; (-.-) 306°C cured; and (--) 288°C cured.

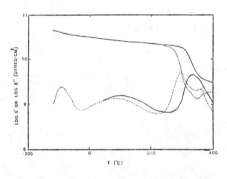

Fig. 3. Dynamic mechanical modulus of press-made 288°C cured samples: (-) post-cured; and (...) as-cured.

The effect of post-curing on the 288°C cured sample is best shown

Before the post-curing step, the 329°C cure cycle produced a material with the highest Tg of 365°C, followed by the 306°C cure cycle (Tg = 343°C). The

288°C cure cycle produced a material with the lowest Tg of 293°C.

Figure 5 summarizes the comparison of the dynamic mechanical properties of all samples that have completed post-curing cycles with maximum post-curing temperatures at about 320°C.

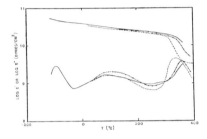

Fig. 5. Dynamic mechanical modulus of post-cured PMR-15 samples: (-) 329°C cured; (-.-) 306°C cured; and (--) 288°C cured.

The dynamic mechanical properties are all similar, except for the temperatures of the glass transition. The temperatures of the maximum of the α loss peaks are 368, 357, and 336°C, respectively, for the samples subject to 329°C, 306°C, and 288°C cure cycles followed by post-curing. Post-curing diminishes, but does not totally eliminate, the difference in properties

of materials processed using different cycles. The comparison of the figure with Fig. 4 indicates that post-curing enhances crosslinking density, as shown by the higher Tg's. The matrix is further stiffened after post-curing treatment. Post-curing, however, does not affect the β or γ transitions.

In summary, for every sample examined there were three peak transitions within the temperature range of -120 to 400°C. In the descending order of temperature at which they occur, these transitions are labeled as α, β and γ peaks which correspond to the α, β, and γ molecular relaxation processes, respectively (13-15). Table I summarizes the temperatures of transitions of the samples subject to a variety of processing conditions.

3.3 Transverse moduli tests

The storage moduli from neat resin tests exhibited a linear decrease with increasing temperature until the glass transition point where a steep drop occurred. Potential engineering applications for composites with a PMR-15 matrix spans a range below the glass transition temperature. Matrix dominated composite properties such as transverse moduli, $E_{22}$, are also expected to be affected by temperature. The question arose as to whether the temperature dependence of matrix dominated composite properties can be predicted using neat resin data. If so, costly static coupon tests could be eliminated in favor of neat resin DMA characterization.

Micromechanical equations can be used to predict $E_{22}$ from matrix and fiber properties (16). Figure 6 shows a comparison between predictions that used the neat resin DMA data from 329°C cured/post-cured samples (see Figure 5) and graphite/PMR-15 unidirectional tape coupon data ($[90]_8$ CE6000/ PMR-15, fiber volume fraction = 0.630). The figure shows that static coupon measurements of $E_{22}$ are very close to the predicted values. Additional composite data (17) also compare favorably with the predictions.

The good comparison between theory and experiment suggests that the modulus measured in neat resin film tests is similar to the in-situ composite property. A good bond must also exist between fiber and matrix. Additionally, DMA tests provided valuable quantitative transition data on matrix properties that can provide insight in further examination of processing- structure- property relationships.

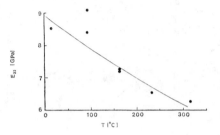

Fig. 6. Comparison of $E_{22}$ experimental data with theoretical prediction: (o) experimental data from Ref. 17; (-) theoretical prediction using $E_{f22}$ = 2.0E6 (from ref.21).

## 4. CONCLUSIONS

A significant contribution of this study was the development of unique methods for making neat PMR-15 matrix films processed in an environment similar to that of composites. Furthermore, DSC and TGA exhibited thermal transitions that were correlated to the reactions involved during the cure process.

Results of dynamic mechanical analysis showed that the three cure cycles in the press and the autoclave yielded quite comparable materials. The cure process probably did not yield much volatile by-products since the starting materials were already imidized. Therefore, the lack of vacuum in the press did not cause any major differences.

Post-curing had more effect on the 288°C cured sample than on either the 306°C or 329°C cured samples. The 288°C as-cured sample was farther from the fully cured state than the other two, as observed in the DSC experiments. Of the three cycles, the 329°C cure cycle gave the highest Tg sample; and the 288°C cure cycle gave the lowest Tg sample.

Storage moduli measured in DMA tests were used in micromechanical calculations to successfully predict temperature dependent composite moduli for PMR-15/CE6000 laminates. This indicated that the neat matrix samples were obtained in a form that was also found in the composite systems.

ACKNOWLEDGEMENTS

The authors would like to acknowledge technical discussions and support by Mahlon Hoover, Harry Thornton and John Quinlivan of the Boeing Company. Financial assistance was provided by Boeing Commercial Airplane Co. to the Polymeric Composites Laboratory at the University of Washington.

REFERENCES
1. Serafini, T. T., P. Delvigs and G.R. Lightsey, J. Appl. Polym. Sci., 16, 905 (1972).

2. Hurwitz, F. I., Polym. Comp., 4, 85 (1983).

3. Delvigs, P., Polym. Comp., 4, 150 (1983).

4. Young, P. R., and A.C. Chang, Proc. of the 30th National SAMPE Symposium, 889 (1985).

5. Hurwitz, F. I., Polym. Comp., 3, 3 (1983).

6. Scola, D. A. and D.J. Parker, Proc. for the 43rd SPE Annual Technical Conference, 399 (1985).

7. Russel, T. P., Polym. Eng. Sci., 24, 5 (1984).

8. Havens, J. R., H. Ishida and J. L. Koenig, Macromolecules, 14, 1327 (1981).

9. Wong, A. C., A. N. Garroway and W.M. Ritchey, Macromolecules, 14, 832 (1981).

10.   Lauver, R. W., J.
Polym. Sci.: Polym. Chem.
Ed., 17, 2529 (1979).

11.   Seferis, J. C. and L.
Nicolais, Eds., "The Role of
the Polymeric Matrix on
Their Processing and
Structural Properties of
Composite Materials," Plenum
Press, New York (1983).

12.   Wedgewood, A. R. and J.
C. Seferis, Polymer, 22, 966
(1981).

13.   Coulehan, R. E., paper
presented at the Sixth
International Congress on
Rheology, Lyon, France Sept.
(1972).

14.   Chartoff, R. P., and T.
W. Chiu, Polym. Eng. Sci.,
20, 244 (1980).

15.   Lim, T., V. Frosini, V.
Zaleckas, D. Morrow, and J.
A. Sauer, Polym. Eng. Sci.,
13, 51 (1973).

16.   Chamis, C.,  NASA TM
83320 (1983).

17.   Cushman, J. B. and S.
F. McClesky, NASA CR 165840
(1982) .

Table I

Molecular Relaxations in Press and Autoclave
Cured PMR-15 Samples

| Sample ID | Processing | $T_\alpha$ (°C) | $T_\beta$ (°C) | $T_\gamma$ (°C) |
|---|---|---|---|---|
| 1 | 329°C cure cycle Press-made As-cured | 365 | 85 | -95 |
| 2 | #2 post-cured | 365 | 83 | -95 |
| 3 | 329°C cure cycle Made in autoclave As-cured | 368 | 103 | -96.5 |
| 4 | #3 post-cured | 367 | 106 | -97 |
| 5 | 306°C cure cycle Press-made As-cured | 343 | 95 | --- |
| 6 | #5 post-cured | 358 | 114 | --- |
| 7 | 288°C cure cycle Press-made As-cured | 293 | 91.5 | -96 |
| 8 | #7 post-cured | 336 | 100 | --- |

EPOXY RESIN CURE FOLLOWED BY
DMTA AND DETA MEASUREMENTS

Raymond E. Wetton
Alan M. Rowe
Margot R. Morton
Polymer Laboratories Ltd.
The Technology Centre, Epinal Way,
Loughborough, U.K.

## Abstract

Dynamic mechanical measurements during isothermal cure of a number of epoxy resin systems show that the loss peak positions due to vitrification move progressively to longer cure times with decreasing measurement frequency. The same is true for dielectric measurements when true molecular relaxation peaks can be followed. These are measured at much higher frequencies than the dynamic mechanical data and therefore occur at correspondingly shorter cure times. The position of these dielectric loss peaks therefore is not a satisfactory measurement of cure. The position is further complicated by high ionic conductivity effects at low measurement frequencies and high cure temepratures.

## 1. INTRODUCTION

The increase in use of epoxy resins for aerospace and similar applications where strength/weight ratios are of paramount importance, requires more definitive methods for characterising the cure process and the state of cure of these resins. This paper applies dynamic mechanical measurements directly to the monitoring of the cure process from the liquid state through to the final cross-linked solid. It also discusses the use of this technique to monitor the state of cure in an epoxy sample, which has to be in a form acceptable by the instrument.

The dielectric relaxation technique studies molecular mobility in much the same way as the DMTA technique. Data is however complicated by ionic and other impurities, which frequencly swamp the main relaxation peaks. The main interest in the dielectric technique is in its potential for in-situ measurements during the cure process. The DMTA cannot at present be usefully employed in the in-situ mode.

In this paper a number of different epoxy systems have been studied by both the dynamic mechanical and dielectric techniques under identical furing conditions. Taking the dynamic mechanical results as being the primary measurement, limitations of the DETA technique are evaluated.

## 2. EXPERIMENTAL

The Polymer Laboratories Dynamic Mechanical Thermal Analyser (DMTA) was utilised in two different modes of clamping. Using the standard head, epoxy resins were applied to a brass shim 0.2mm in thickness and clamped as a single cantilever beam (Fig. 1a). The DMTA was arranged with its drive axis vertical so that the viscous liquid epoxy remained on the brass shim. Rapid isothermal temperature rise was selected on software, so that the isothermal operating temperature was always achieved within 3 minutes. Cure time was recorded from the point of acquisition of the isothermal temperature. The effect of the brass shim could be subtracted off using relationships which we have already published.[1][2] The other mode of clamping employed with the DMTA was to measure the epoxy directly in shear with no added support. This required the use of the DMTA power head providing 14N maximum force. A shear plate was clamped in the DMTA drive and liquid epoxy placed above and below this between clamping studs, to give the shear sandwich geometry, as shown in Fig. 1b.

Dielectric measurements were performed on a fully automatic system, the PL-Dielectric Thermal Analyser (DETA). The same epoxy resin systems which were employed in the DMTA measurements were measured as disc samples, approximately 1mm in thickness between the parallel plate electrodes of the DETA. In order to allow removal of the sample after cure, the electrodes were completely covered in thin aluminium foil prior to insertion of the sample (Fig. 2). The electrodes were locked at the required sample gap and any contraction of the epoxy was taken up by movement of the aluminium foil. The same isothermal temperature control conditions were employed as in the DMTA, with temperature rise again being achieved in 3 minutes or less.

## 3. RESULTS AND DISCUSSION

Data obtained by the shim supported technique is shown in Fig. 3. The measured tan $\delta$ curve is for the shim plus epoxy during cure and the calculated curve is the same data with the effect of the shim subtracted using the relationships

$$(kE')_{\text{composite}} = E'_s b\left(\frac{t_s}{l}\right)^3 + \frac{1}{2}E'_c b\left[\left(\frac{t_s + 2t_c}{l}\right)^3 - \left(\frac{t_s}{l}\right)^3\right]f$$

$$(kE'')_{\text{composite}} = E''_s b\left(\frac{t_s}{l}\right)^3 + \frac{1}{2}E''_c b\left[\left(\frac{t_s + 2t_c}{l}\right)^3 - \left(\frac{t_s}{l}\right)^3\right]f$$

where k is the geometry for the composite, $E'_s$, $t_s$ and $E'_c$, $t_c$ are the moduli and thicknesses of the shim and coating respectively. The sample width, b and the sample length, l are the same for both coating and shim. The empirical factor, f has been found to be close to 0.5 for many practical situations. This data can be compared directly with the results for the same sample measured in shear sandwich geometry, without any additional substrate support. Results in this mode of operation are shown in Fig. 4, with the tan δ at short times clearly reflecting the high damping present in the liquid state. The corresponding modulus curves climb steadily with time. The position of the loss peaks on the log time axes are seen to agree well but the shim support method artificially sharpens the loss curve on the short time side. This is because of inaccuracies in the above correction procedures when using weak coatings, as is the case with the epoxy sample in the early stages of cure. It is apparent that the shim support method gives good sharp loss peaks which are easily identified. This is borne out by the data in Fig. 5 for a one shot epoxy (Fiberdux, Ciba-Geigy Ltd.) at 90°C.

These results from both the shim support method and the direct shear measurement method show clearly that the assessment of cure by dynamic mechanical measurements depend strongly on the measurement frequency. High measurement frequencies sense the freezing out of molecular motions due to cross-linking at an earlier time than lower frequencies. This can be correctly construed in terms of a chemically driven Tg process. The curing systems are being driven into the glassy state by chemical reaction. If we draw an analogy with temperature scans through the Tg of a chemically unchanging system, it is clear that reaction time has a similar effect on molecular mobility for curing systems, as (1/T) has in normal Tg measurements. If Fig. 4 we see that this change in loss peak location is occurring regularly with changing frequency and even at .1Hz is still shifting consistently on a log t basis to longer cure times with decreasing frequency. It is an interesting problem, therefore, as to how cure is best defined by these measurements. Ideally one would like cure to proceed until there is no dispersion in the modulus values with changing frequency and low as the values at all frequencies. The preceding arguments, however, show that on a log time basis this will essentially never occur. Any sensible test for complete cure must either be performed at very low frequency or be extrapolated to a measurement at low frequency.

Dielectric measurements were performed in the DETA cell in the frequency range from 500Hz to 10kHz. A good understanding and interpretation of the results can be seen from a series of measurements on Araldite Rapid, 60/40 mix with cure temepratures 20, 40 and 50°C. The loss factors for these temepratures are shown in Figs. 6, 7 and 8 respectively. In Fig. 6 the loss peaks occurring between 10 and 100 minutes in time are due to the true dielectric relaxation of the epoxy chains. The background fall in tan δ at lower frequencies is due to the freezing out of mobility of ionic impurities. As the temperature is raised, it is seen that the background ionic mobility term increases and eventually dominates the loss peak behaviour. The effect is so strong that the true relaxation peak positions are reversed in sense by the high background at low frequencies and short reaction times.

The best correlation of the loss peak positions from mechanical and dielectric work are obtained by plotting log of cure time versus log of the measurement frequency. The DMTA data always lie on a good straight line running from high frequency, short time cure measurement to low frequency, long time cure measurement. At low temperatures the DETA results follow the same pattern, but as temeprature is raised, the conductivity background perturbs the peak positions and indeed reverses them. (FIGS 9 & 10)

It is seen that the dielectric measurements, because of their higher applied frequency, sense short cure times based on loss peak positions. This clearly will not be a good measurement of cure for the reasons discussed above. It seems, therefore, that any useful dielectric measurements must be based upon the loss factor dropping below some empirically defined value or from the dielectric constant itself showing minimal dispersion.

A dielectric function which minimises the presence of conduction is the dielectric modulus ($1/\epsilon^*$). The loss part of this function is plotted against cure time in Fig. 11. It supresses the conductivity effect enough to resolve relaxation peaks, but these occur at still shorter times than the normal tan $\delta$ function and thus give still worse correlation with the mechanical loss peak positions.

## 4.  REFERENCES

(1) Wetton R.E. in Progress in Polymer Characterization (Ed. Dawkins) Vol. 5 Chapter 5 p.179 Elsevier (1986).

(2) Wetton R.E., Morton M.R. and Gearing J.W.E. in Progress in Advanced Materials (Ed. Bartelds and Schliekelmann) p.293 Elsevier (1985).

## 5.  BIOGRAPHIES

Raymond E. Wetton is Chairman of Polymer Laboratories and Visiting Professor in the Department of Chemistry at Loughborough University U.K.  He holds B.Sc. and M.Sc. degrees from Bristol and a Ph.D from Manchester.  He is author of over 70 publications in the field of Polymer Science.

Alan M. Rowe holds the degrees of B.Sc. and Ph.D from Loughborough University, the former with first class honours.  His Ph.D and subsequent work with B.P. and P.L. has been on structure property relationships in polymers.  He is the author of a number of papers in the field and currently holds the position of Product Manager with Polymer Laboratories.

Margot R. Morton holds the degree of B.Sc. in Chemical Physics from Bristol.  Her work with Polymer Laboratories has been concerned with the dynamic mechanical response of materials and she is the author of a number of papers in the field. She currently holds the position of Product Manager with P.L.

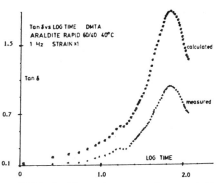

FIG 3  Araldite curing measured on metal shim
Calculated curve is with shim subtracted.

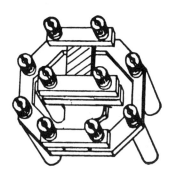

FIG 1a  Epoxy sample on
metal shim, measured in
bending as single cantilever.

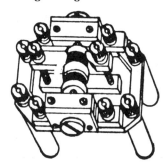

FIG 1b  Epoxy sample measured
in shear sandwich geometery
(sample shaded black).

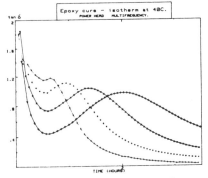

FIG 4  Direct measurement of Epoxy cure in shear
sandwich geometry, without support. Upper curves
tan delta, lower curves modulus change with linear
time.

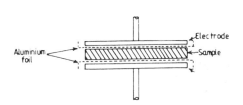

Fig 2

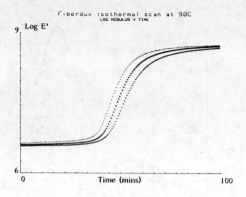

Fiberdux isothermal scan at 90C
LOG MODULUS V TIME

Log E'

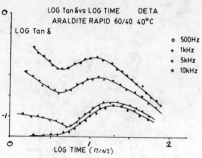

LOG Tan δ vs LOG TIME   DETA
ARALDITE RAPID 60/40 40°C

LOG Tan δ

o  500Hz
+  1kHz
x  5kHz
◇  10kHz

LOG TIME (mins)

FIG 7  Dielectric tan δ for
Araldite Rapid during cure
at 40°C.

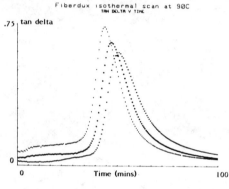

Fiberdux isothermal scan at 90C
TAN DELTA V TIME

tan delta

FIG 5  Measurement of Epoxy cure (Fiberdux) at 90°C
using the shim support method.  Top graph show modulus
change, lower graph tan delta.

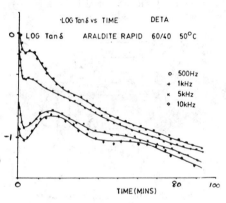

·LOG Tan δ vs TIME     DETA

LOG Tan δ     ARALDITE RAPID  60/40  50°C

o  500Hz
+  1kHz
x  5kHz
◇  10kHz

TIME (MINS)

FIG 8  Dielectric tan δ for
Araldite Rapid during cure
at 50°C.

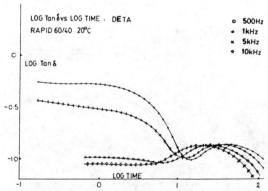

LOG Tan δ vs LOG TIME .  DETA
RAPID 60/40  20°C

LOG Tan δ

o  500Hz
+  1kHz
x  5kHz
◇  10kHz

LOG TIME

FIG 6  Dielectric tan δ for
Araldite Rapid during cure
at 20°C.

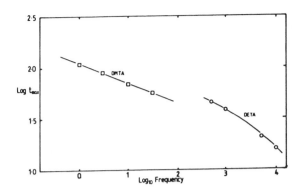

FIG 9 Correlation between DMTA and DETA
loss peak loci, for Araldite Rapid at 20°C.
Log (cure time) for a maximum in tan $\delta$ is
plotted versus log f.

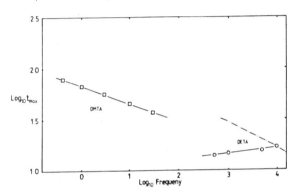

FIG 10 As for Fig. 9 but at 40°C. The
background conductivity has reversed the
DETA loss peak trend with frequency.

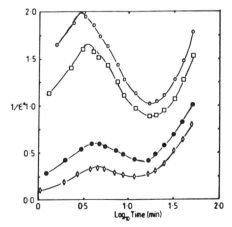

FIG 11 Dielectric loss modulus $[(1/\epsilon^*)\tan \delta]$
versus cure time for the data in Fig. 8. The
peaks in this function follow the correct
trend but occur at even shorter times.

NEW NON-MDA EPOXY RESIN SYSTEMS FOR
RESIN TRANSFER MOLDING (RTM) AND FILAMENT WINDING

Eric B. Stark, Laurie M. Schlaudt,
Benjamin T. A. Chang, and Walt V. Breitigam
Westhollow Research Center
Shell Development Company
P.O. Box 1380
Houston, TX 77251-1380

Abstract

Several resin systems for resin transfer molding (RTM) and wet filament winding have been developed for the manufacture of high performance composites in aerospace applications. All of these resin systems produce ultra-low void content, high-quality composite parts using densely packed fiber forms. The long pot life of these resin systems allows consolidation even during large part fabrication. Among the key features discussed are the good solvent and chemical resistance, good elevated temperature properties, high tensile elongation, and toughness. In addition, a new developmental system providing excellent composite compression strength is introduced and its properties discussed. Non-MDA containing curing agents are available. These curing agents extend the room temperature pot life and improve the safe handling of the EPON$^R$ 9000 resin systems. We also discuss the extensive database sutiable for resin qualification in aerospace programs which has been developed for the EPON$^R$ Resin 9400/EPON CURING AGENT$^R$ 9450 resin system in the advanced composites test center at Westhollow Research Center.

1. INTRODUCTION

With the worldwide decrease in fuel prices over the past two years, the savings realized by reducing the weight of an aerospace structure has been significantly reduced. As a result, the incentive to replace metals with the lighter composite structures has been shifted somewhat away from saving weight to lowering costs incurred in part manufacture (1). Processing techniques other than the costly and labor intensive use of hand lay-up with prepreg are becoming more important. These alternate processing techniques include resin transfer molding (RTM), filament winding, and pultrusion. For these reasons, the EPON[R] 9000 series resin systems were developed. The advantages of using these low cost fabrication techniques have been enumerated previously (2).

The key features of resin systems suitable for these types of processing techniques are low viscosity and long pot life at the processing temperature (usually room temperature), fast gelation at the desired cure temperature, and low volatility and outgassing during cure. These key features were incorporated in the design of the 9000 series resin systems.

The EPON[R] Resin 9400/EPON CURING AGENT[R] 9450 and EPON[R] Resin 9405/EPON CURING AGENT 9470 systems were designed for use in either filament winding or RTM. The EPON CURING AGENT[R] Accelerator 537 may be used with either of these resin systems. The 9400/9450 resin system utilizes a curing agent which contains methylene dianiline (MDA). Because of the safety concerns raised over MDA, an alternate resin system (9405/9470) was developed which contains no MDA and performs comparably to the 9400/9450 system.

This paper will discuss the processing and mechanical characterization of the 9400/9450 and 9405/9470 resin systems and briefly introduce developmental systems aimed at improving the composite compression strength of the 9405/9470 resin system.

## 2. EXPERIMENTAL

A full range of neat resin characterization data was generated for the 9400/9450 and 9405/9470 resin systems with and without 537 accelerator including viscosity, pot life, gel times, heat deflection temperature (HDT), glass transition temperature ($T_g$), compact tension, tensile and flexural modulus, strength, and elongation. In addition, extensive 9400/9450 composite data on filament-wound panels was generated. Composite compression strength was measured on resin transfer molded panels made with 9405/9470 and developmental resin systems, reinforced with 8 harness satin graphite fabric. The experimental procedures utilized for each of the aforementioned analyses are provided below.

### 2.1 Sample Preparation

The neat resin samples for the viscosity, pot life, and gel time experiments were prepared by thoroughly mixing the appropriate quantities of resin and curing agent. The samples used for the remainder of the neat resin tests were formed by casting the systems between glass plates and curing (cure schedules are included with material data). Composite panels, reinforced with graphite fabric or tow, were fabricated using either RTM or wet filament winding.

RTM is the process by which uncured liquid resin is injected into a mold already containing the appropriate reinforcements, cores, etc., then cured, and removed as a finished part. The apparatus utilized for RTM process is shown in Figure I. The reinforcement used for the RTM composite data was an 8 harness satin graphite fabric made with Celion 3K fibers.

Filament winding is the process by which reinforcing fibers are wound following a predetermined pattern onto a mandrel. The fibers may be wound wet, by passing them through a resin bath, or wound dry and impregnated with resin in a secondary step. The composite may be cured on the mandrel, with or without a matching mold, or the uncured material may be removed from the mandrel and placed in a mold or oven for cure. The reinforcement used for the filament winding composite data was Celion 6K fiber tow. Test panels were wet wound onto a hexagonal mandrel, cut into individual panels, placed into a heated press mold, and cured.

## 2.2 Processing Property Procedures

### 2.2.1 Kinematic Viscosity
The kinematic viscosity was determined using a Ubbelohde tube placed in a constant temperature water bath at $25^{\circ}C$. The time required for the material to pass through the capillary is directly related to the kinematic viscosity. The range was determined over many laboratory batches. The viscosity was measured for both the individual components and the mixed resin system.

### 2.2.2 Pot Life
The pot life, as defined in this paper, is the time required to double the initial viscosity when the sample is held at $25^{\circ}C$ (room temperature). A Brookfield viscometer was utilized for this determination on the resin/curing agent mixture.

### 2.2.3 Gel Time
The gel time was determined using a hot plate technique. The resin and curing agent mixture was spread over hot plates heated to $150^{\circ}C$, $180^{\circ}C$, and $210^{\circ}C$ ($302^{\circ}F$, $356^{\circ}F$, and $410^{\circ}F$ respectively). The time required for the resin to stop stringing and become elastic is called the gel time. This determination was done only for the resin/curing agent mixtures.

### 2.2.4 Rheometrics Viscosity
A Rheometrics Dynamic Spectrometer, in the parallel plate meter mode, was used for determining the dynamic cure viscosity for the resin/curing agent mixtures. The

plate gap was 1 mm and the frequency was 11 Hz. The heating
rate selected was 10°C/min.

## 2.3 Neat Resin Mechanical Testing

### 2.3.1 Glass Transition Temperature
The glass transition temperature
($T_g$) was measured using a cured
neat resin specimen in a Rheometrics Dynamic Spectrometer to determine the real and loss shear
moduli and tan delta. The peak in
tan delta is taken as the glass
transition temperature.

### 2.3.2 Water Absorption
The weight gain of a cured neat
resin specimen was measured gravi-
metrically after soaking a 2.5" x
0.5" x 0.125" sample for three
days in 93°C (200°F) water. A temperature controlled water bath was
utilized for this purpose.

### 2.3.3 Mechanical Testing
The heat deflection temperature
(HDT), tensile, flexural, compact
tension, and compression testing
were all performed according to
ASTM test procedures listed below:

| | |
|---|---|
| HDT | ASTM D 648 |
| Tensile | ASTM D 638 |
| Flexural | ASTM D 790 |
| Compact Tension | ASTM E 399-83 |

The testing facility, utilizing
computer controlled, hydraulic
Instron testing machines, is illustrated in Figure II.

## 2.4 Composite Mechanical Testing
The composite mechanical tests
performed on the 9400/9450 system,
and the appropriate ASTM methods,
are listed below:

| | |
|---|---|
| 0°, 90°, ±45° tensile | ASTM D 3039 |
| 0° compression | ASTM D 695 |
| Flexure | ASTM D 790 |
| Short Beam Shear | ASTM D 2344 |

All filament wound test panels
were reinforced with Celion 6K
fiber tows. Tests were performed
under dry and wet conditions at
several temperatures. The environmental conditioning for the
9400/9450 composite samples was
performed in temperature/humidity
chambers. The 9405/9470 and developmental resin systems were only
tested in compression using the
IITRI method. The RTM panels were
carbon-fiber fabric reinforced
with the fabric layers alternated
between warp and fill.

## 3. RESULTS AND DISCUSSION

## 3.1 Processing Characteristics
The typical properties of the
9400/9450 and 9405/9470 mixed
systems are shown in Table I. The
room temperature mix viscosity for
both systems is very low, on the
order of 1000 cp, allowing for excellent flow and wet-out. In addition, the room temperature pot
life is at least 20 hours for the
9400/9450 system, and is greater
than 30 hours for the 9405/9470
system, allowing for greater

working time at room temperature during large part fabrication. On the other hand, the part can be gelled and removed from the tool in short times at elevated temperatures as is shown in the gel times of Table I. The gel times for both the 9400/9450 and 9405/9470 systems may be decreased using the 537 accelerator. Viscosity as a function of temperature for the 9405/9470 resin system at a heating rate of $10^{\circ}$C/min. is shown in Figure III. The 9400/9450 viscosity behavior will be similar under the same conditions. The viscosity for both systems drops below 100 cp at temperatures above $50^{\circ}$C making for easy flow and fiber wet-out even in large, complex parts.

The effect of accelerator on the processing characteristics of these resin systems is presented in Table II. As is shown in the table, the addition of accelerator decreases both the pot life and the gel time at all temperatures. The effect is comparable for both resin systems. By adding as little as 0.25 parts per 100 parts of resin, the gel time can be decreased by about 50% while decreasing the pot life by only about 35%. The use of accelerator may be necessary in the case of high throughput parts and the 537 accelerator can be used to precisely adjust the processing characteristics as needed.

## 3.2 Neat Resin Mechanical Properties

Typical mechanical properties of the unaccelerated neat resin are presented in Table III for both the 9400/9450 and 9405/9470 resin systems. The $T_g$ and HDT of both systems for the stoichiometric concentration of curing agent (28 phr for 9405/9470 and 31.5 phr for 9400/9450) indicate good heat resistance and likely applicability in $180^{\circ}$F hot/wet (H/W) environments. The low water absorption confirms the suitability of even the 37 phr 9470 composition for use in a $180^{\circ}$F H/W environment. The room temperature flex and tensile properties are good with the elongation at break being excellent. The fracture toughness compares favorably to that of typical, untoughened, state-of-the-art TGMDA/DDS epoxy resin systems for prepreg (3).

The effect of accelerator on the HDT and room temperature tensile properties of these resin systems is outlined in Table IV. Mechanical properties are affected minimally except at the highest concentrations of accelerator (4 phr). The HDT appears to be more sensitive, but the effect is significant only in excess of 1 phr. As was detailed in Table II, the processing parameters are affected significantly by adding as little as 0.25 phr. Therefore, the effects on the mechanical properties will typically be minimal. Overall, the neat resin properties show an excellent balance in comparison to current resin systems in $180°F$ aerospace applications with outstanding moisture resistance and ultimate strain.

### 3.3 Composite Mechanical Properties

An extensive composite database has been generated for the 9400/9450 system through its qualification as the filament winding resin for the Beech Starship fuselage. Some representative data from this database are presented in Table V. Reported values are averages of a minimum of 20 test values. Overall, the data generated for the 9400/9450 system indicate that good mechanical integrity is maintained under cold/dry, hot/wet, and chemical environments. Most noteworthy is the retention of tensile and flexural modulus under hot/wet conditions. Limited composite data on 9405/9470 is currently available, but mechanical testing to generate a database for this system is ongoing.

### 3.4 High Compression Strength System

Interest in RTM resin systems with greater compressive strength is growing in the aerospace industry. In response to this, several developmental resin systems are being evaluated. Developmental Resins RSL-1132 and RSL-1133

have shown the most promise in satisfying the high compression strength RTM requirements. The compression strength data on RTM composite panels made with Celion 3K, 8 harness satin graphite fabric are shown in Table VI. As is illustrated there, significant improvements in compressive strength are achieved with both RSL-1132 and RSL-1133 with RSL-1132 showing the greatest improvement. RSL-1132 is a specific blend of EPON Resin 9405 and Experimental Resin RSL-1186. By blending 9405 and RSL-1186, a whole spectrum of performance characteristics can be realized for a specific end use. Table VII provides additional processing and neat resin mechanical property data for EPON Resin 9405/RSL-1186 blends, including RSL-1132. RSL-1132 is currently under evaluation for aerospace applications using the RTM process.

### 4. CONCLUSIONS

The EPON Resin 9405/EPON Curing Agent 9470 resin system offers excellent processing characteristics for utilization in low cost liquid processing techniques such as resin transfer molding and filament winding. In addition, its mechanical properties are comparable to the EPON Resin 9400/EPON Curing Agent 9450 resin system for which a large composite database exists. For aerospace applications using liquid processing, the

9405/9470 resin system offers excellent performance and processing without the presence of MDA. Promising developmental resin systems offer the same outstanding processing behavior combined with improved compression strength.

## 5. REFERENCES

1. Krowlewski, S., and Gutowski, T., Advanced Composites, Vol. 1, No. 3, 49 (1986).

2. Hewitt, R. W., Schlaudt, L. M., and Bonner, D. C., Int. SAMPE Symp. [Proc.] 31, 142 (1985).

3. Johnston, N. J., "Synthesis and Toughness Properties of Resins and Composites", NASA CP-2321, 86 (1984).

## 6. BIOGRAPHIES

Eric B. Stark graduated from the University of Washington with his Ph.D. in Chemical Engineering in 1985. He has authored several papers on the relationship between reaction kinetics and viscosity in high performance epoxy resins and composites. He joined Shell Development Company, Westhollow Research Center in 1985. His work has centered on the characterization and processing of new epoxy resin materials, especially in the area of resin transfer molding.

Laurie M. Schlaudt graduated from Texas A&M University in 1984 with a B. S. degree in Mechanical Engineering. She joined the Resins Department of Shell Development in June 1984. As a member of the Transportations Applications/Futures Group, her responsibilities include material test program management and filament winding technical support.

Dr. Benjamin T. A. Chang is a research engineer with the Material Science and Engineering Department of Shell Development Company, where he has been employed since receiving his Ph.D. in Materials Science from the University of Rochester in 1980. During the past six years he has worked in the area of mechanical behavior of composites, including tensile, compression, fatigue, creep, crashworthiness, hygrothermomechanical resistance and failure analysis.

Walter V. Breitigam obtained his degree in organic chemistry from Ohio State University. He joined Shell Oil Company in Wood River, Illinois as a chemist, developing gasoline and lubricant additives and synthetic lubricants. In 1975 he transferred to Shell Development Company, Westhollow Research Center, where he has been involved in development and technical support of epoxy and vinyl ester resins for industrial uses. Currently, he is project leader in the Resins Department responsible for research and development of advanced composites for the aircraft/aerospace industry.

TABLE I - Processing properties of the unaccelerated
mixed resin/curing agent systems

| | | | |
|---|---|---|---|
| EPON$^R$ Resin 9400 (a) | 100 | | |
| EPON$^R$ Resin 9405 | | 100 | 100 |
| EPON CURING AGENT$^R$ 9450 | 31.5 | | |
| EPON CURING AGENT$^R$ 9470 | | 28 | 37 |
| | | | |
| Viscosity, 25$^o$C, mPa.s (cp) | 850 | 950 | 850 |
| 38$^o$C | | | 248 |
| 65$^o$C | | | 36 |
| 93$^o$C | | | 3 |
| | | | |
| Pot life, 23$^o$C, hrs (b) | 20 | 34 | 41 |
| | | | |
| Gel time, min., 150$^o$C (302$^o$F) (c) | 7.5 | 44 | 36 |
| 180$^o$C (356$^o$F) | 4.2 | 20 | 15 |
| 210$^o$C (410$^o$F) | --- | 8.2 | 6.2 |

(a) Resins and curing agents measured in parts by weight.
(b) Time to 2000 mPa.s (cp) viscosity.
(c) Gel plate.


TABLE II - Effect of EPON CURING AGENT$^R$ Accelerator 537 and curing
agent concentration on processing properties (a)

| Resin, parts | Curing Agent, parts | EPON CURING AGENT Accelerator 537, parts | 210$^o$C Gel Time, sec. | 180$^o$C Gel Time, sec. | 150$^o$C Gel Time, sec. | 25$^o$C Pot Life, hrs. (b) |
|---|---|---|---|---|---|---|
| EPON$^R$ Resin 9405/EPON CURING AGENT$^R$ 9470 | | | | | | |
| 100 | 28 | 0 | 525 | 1200 | 2640 | 4 |
| 100 | 28 | 0.25 | 225 | 615 | 500 | 22.5 |
| 100 | 28 | 0.5 | 130 | 365 | 1050 | 15.8 |
| 100 | 28 | 0.75 | 100 | 260 | 810 | 13.3 |
| 100 | 28 | 1 | 75 | 245 | 660 | 12.0 |
| 100 | 28 | 2 | 88 | 132 | 350 | 7.5 |
| 100 | 28 | 4 | 24 | 75 | 210 | 4.7 |
| | | | | | | |
| 100 | 37 | 0 | 340 | 820 | -- | -- |
| 100 | 37 | 1 | 93 | 225 | 580 | 9.6 |
| 100 | 37 | 4 | 18 | 55 | 175 | 6.0 |
| EPON$^R$ Resin 9400/EPON CURING AGENT$^R$ 9450 | | | | | | |
| 100 | 31.5 | 0 | -- | 210 | 585 | 18.0 |
| 100 | 31.5 | 0.25 | -- | 86 | 185 | 12.0 |
| 100 | 31.5 | 0.5 | -- | 34 | 105 | 8.7 |
| 100 | 31.5 | 1 | -- | 16 | 52 | 6.0 |
| 100 | 31.5 | 2 | -- | 9 | 32 | 3.6 |

(a) Concentrations shown as parts per 100 parts resin.
(b) Time for 25$^o$C kinematic viscosity to reach 2000 mm$^2$/sec (2000 cSt)

| | | | |
|---|---|---|---|
| EPON$^R$ Resin 9400 (b) | 100 | | |
| EPON$^R$ Resin 9405 | | 100 | 100 |
| EPON CURING AGENT$^R$ 9450 | 31.5 | | |
| EPON CURING AGENT$^R$ 9470 | | 28 | 37 |
| | | | |
| $T_g$, dynamic mechanical, $^\circ$C. | 181 | 186 | 147 |
| RT flex strength, MPa (ksi) | 131.0 (19) | 108.9 (15.8) | |
| modulus, GPa (ksi) | 3.0 (430) | 2.8 (411) | |
| ultimate strain, % | | 7.7 | |
| RT tensile strength, MPa (ksi) | 75.8 (11.0) | 77.9 (11.3) | 80.6 (11.7) |
| modulus, GPa (ksi) | 3.1 (450) | 2.8 (413) | 2.9 (427) |
| ultimate strain, % | 4.5 | 6.4 | 8.5 |
| Compact tension, fracture toughness, | | | |
| Kq, MPa-m$^{1/2}$ (psi-in$^{1/2}$) | 0.91 (825) | 0.96 (870) | 0.97 (880) |
| Water absorption, % (c) | 2.25 | 1.17 | 1.31 |

(a) Typical laboratory cure schedules:
   (9405/9470) - 1 hr at 80$^\circ$C, 1 hr at 120$^\circ$C, 1 hr at 150$^\circ$C, 4 hrs
       at 175$^\circ$C.
   (9400/9450) - Tensile specimens: 1 hr at 121$^\circ$C, 1 hr at 177$^\circ$C.
       Flex specimens: 1 hr at 100$^\circ$C, 1 hr at 150$^\circ$C, 1 hr at 177$^\circ$C.
       Dynamic mechanical Tg: 30 mins. at 110$^\circ$C, 1 hr at 150$^\circ$C.
(b) Resins and curing agents measured in parts by weight.
(c) Three day soak of 2.5" x 0.5" x 0.125" sample in 93$^\circ$C. water.


TABLE IV - Effect of EPON CURING AGENT$^R$ Accelerator 537 and curing
agent concentration on mechanical properties (a)

| Resin, parts | Curing Agent, parts | EPON CURING AGENT Accelerator 537, parts | HDT 264 psi, $^\circ$C (b) | R.T. Tensile Strength, MPa (ksi) | R.T. Tensile Modulus, GPa (ksi) | R.T. Tensile Elonga-tion, % |
|---|---|---|---|---|---|---|
| EPON$^R$ Resin 9405/EPON CURING AGENT$^R$ 9470 (c) | | | | | | |
| 100 | 28 | 0 | 160 | 77.9 (11.3) | 2.8 (413) | 6.4 |
| 100 | 28 | 1 | 148,150 | 71.0 (10.3) | 2.7 (396) | 4.7 |
| 100 | 28 | 2 | 141,145 | 75.8 (10.9) | 2.6 (376) | 5.0 |
| 100 | 28 | 4 | 134,135 | 75.1 (10.9) | 2.9 (419) | 4.3 |
| | | | | | | |
| 100 | 37 | 0 | 130 | 80.6 (11.7) | 2.9 (427) | 8.5 |
| 100 | 37 | 1 | 121,122 | 77.9 (11.3) | 2.8 (400) | 5.9 |
| 100 | 37 | 2 | 112,112 | 62.0 (9.0) | 3.0 (430) | 2.6 |
| EPON$^R$ Resin 9400/EPON CURING AGENT$^R$ 9450 (d) | | | | | | |
| 100 | 31.5 | 0 | 166 | 76.5 (11.1) | 2.8 (411) | 5.4 |
| 100 | 31.5 | 1 | 149 | 79.3 (11.5) | 3.0 (436) | 5.4 |
| 100 | 31.5 | 2 | 147 | 77.9 (11.3) | 3.1 (456) | 4.5 |

(a) Concentrations measured in parts per 100 parts resin.
(b) Duplicate determinations.
(c) Lab cure cycle (9405/9470): 1 hour at 80$^\circ$C, 1 hour at 120$^\circ$C,
   1 hour at 150$^\circ$C, 4 hours at 175$^\circ$C.
(d) Lab cure cycle (9400/9450): 1 hour at 80$^\circ$C, 1 hour at 120$^\circ$C,
   1 hour at 150$^\circ$C, 2 hours at 175$^\circ$C.

<u>TABLE V - Composite data for EPON[R] Resin 9400/EPON CURING AGENT[R] 9450</u>
<u>with Celion 6K fiber(a)</u>

| | RT dry | -65°F dry | RT wet (b) | Hot wet (c) | Sol- vent (d) |
|---|---|---|---|---|---|
| 0° Tensile Strength GPa (ksi) | 1.9(283) | 1.9(282) | 1.9(285) | 1.8(259) | ---- |
| 0° Tensile Modulus GPa (msi) | 131(19) | 138(20) | 124(18) | 124(18) | ---- |
| 90° Tensile Strength MPa (ksi) | 35.2(5.1) | ---- | ---- | ---- | ---- |
| 90° Tensile Modulus GPa (msi) | 8.3(1.2) | ---- | ---- | ---- | ---- |
| $\pm$45° Tensile Strength MPa (ksi) | 117(17) | ---- | 110(16) | 117(17) | ---- |
| $\pm$45° Tensile Modulus GPa (msi) | 13.1(1.9) | ---- | 13.1(1.9) | 11.7(1.7) | ---- |
| Compressive Strength GPa (ksi) | 1.3(184) | 1.6(226) | 1.3(182) | 1.1(156) | ---- |
| Compressive Modulus GPa (msi) | 117(17) | 124(18) | 117(17) | 117(17) | ---- |
| Flexural Strength GPa (ksi) | 1.6(231) | 2.0(296) | 1.6(225) | 1.2(169) | 1.6(228) |
| Flexural Modulus GPa (msi) | 110(16) | 110(16) | 110(16) | 110(16) | 110(16) |
| SBS Strength MPa (ksi) | 82.7(12) | 117(17) | 75.8(11) | 55.1(8) | ---- |

(a) Cured 30 min. at 230°F, 1 hr at 300°F; nominally 60% fiber volume
(b) RT wet: conditioned at 160°F, 87% RH to equilibrium, tested at RT
(c) Hot wet: conditioned at 160°F, 87% RH to equilibrium, tested at 180°F, 50% RH
(d) Specimens immersed in 120°F MEK, Jet A fuel, or hydraulic fluid for 7 days, tested at RT

<u>TABLE VI - RTM Composite Panel Compression Strength Data</u>
<u>for Modified 9405/9470 Systems (a)</u>

| System | $T_g$ (°C) | $V_{fiber}$ | $V_{void}$ | Compression Strength Mean MPa (ksi) | cv MPa (ksi) |
|---|---|---|---|---|---|
| 9405/9470 (b) | 137.5 | 56.1% | 0.55% | 544 (78.9) | 48 (6.9) |
| RSL-1132/9470 (c) | 147.0 | 59.6% | 0.77% | 661 (95.9) | 28 (4.0) |
| RSL-1133/9470 (c) | 143.5 | 60.8% | 1.5% | 624 (90.6) | 52 (7.5) |

(a) Panel thickness: 1/8" (125 mils)
    Reinforcement type: 8 harness satin graphite fabric
    Fiber type: Celion 8K
(b) 9405/9470 - Injection temperature: 60°C
                Lab cure schedule: 1.5 hours at 121°C, 1 hour at 149°C, 1 hour at 177°C
(c) RSL-1132 and RSL-1133 - Injection temperature: 80°C
                Lab cure schedule: 1 hour at 80°C, 1 hour at 121°C, 1 hour at 149°C, 1 hour at 177°C

TABLE VII - Processing and mechanical properties of EPON[R] Resin 9405/ EPON CURING AGENT[R] 9470 modified with Epoxy Research Resin RSL-1186(a)

| | | | | | | | | | | |
|---|---|---|---|---|---|---|---|---|---|---|
| EPON[R] Resin 9405 (b) | 100 | 75 | 50 | 25 | | 100 | 75 | 50 | 25 | |
| Epoxy Research Resin RSL-1186 | | 25 | 50 | 75 | 100 | | 25 | 50 | 75 | 100 |
| EPON CURING AGENT 9470 | 28 | 24.4 | 20.7 | 17.1 | 13.4 | 37 | 32.4 | 27.5 | 22.7 | 17.8 |
| Viscosity, 25°C, mPa.s (c) | ----------------------- 800-1000 ----------------------- | | | | | | | | | |
| Pot life, 23°C, hrs (d) | ----------------------- >20 ----------------------- | | | | | | | | | |
| Gel time, min. | | | | | | | | | | |
|   150°C (302°F) | 44 | -- | -- | 35 | 29.2 | 36 | -- | -- | -- | 27.5 |
|   180°C (356°F) | 20 | 16.7 | 15.4 | 13.8 | 10.3 | 15 | -- | -- | -- | 7.9 |
|   210°C (410°F) | 8.2 | 7.4 | 6.8 | 5.0 | 4.3 | 6.2 | -- | -- | -- | 3.0 |
| HDT, 264 psi, °C (e) | 160 | 170 | 186 | 196 | >200 | 130 | 139 | 156 | 182 | 200 |
| R.T. Tensile | | | | | | | | | | |
|   Strength, MPa | 77.9 | 57.2 | 31.0 | 28.3 | -- | 80.6 | 77.2 | 44.1 | -- | 15.2 |
|        (ksi) | 11.3 | 8.3 | 4.5 | 4.1 | -- | 11.7 | 11.2 | 6.4 | -- | 2.2 |
|   Modulus, GPa | 2.8 | 2.8 | 2.9 | 3.1 | -- | 2.9 | 2.7 | 2.8 | -- | 3.1 |
|        (ksi) | 411 | 411 | 424 | 454 | -- | 427 | 386 | 402 | -- | 451 |
|   Ultimate strain, % | 6.4 | 3.1 | 1.2 | 0.9 | -- | 8.5 | 5.2 | 1.8 | -- | 0.5 |
| Compact Tension fracture toughness | | | | | | | | | | |
|   $K_q$, MPa-m$^{1/2}$ | 0.96 | -- | 0.65 | 0.61 | -- | 0.97 | 1.26 | 0.64 | -- | 0.34 |
|     (psi-in$^{1/2}$) | 870 | -- | 593 | 555 | -- | 880 | 1150 | 583 | -- | 311 |

(a) Typical lab cure schedule: 1 hour at 80°C, 1 hour at 150°C, and 4 hours at 175°C.

(b) Resins and curing agent measured in parts by weight.

(c) Time to double initial viscosity.

(d) Gel plate.

(e) Typical lab cure schedule: 1 hour at 80°C, 1 hour at 121°C, and 4 hours at 160°C.

Figure I.  Resin Transfer Molding Apparatus

Figure II.  Instron Test Laboratory

## VISCOSITY CURE SWEEP AT 10 DEG.C/MIN
### 9405/9470 SYSTEM (37 PHR, FN#P41324AA)

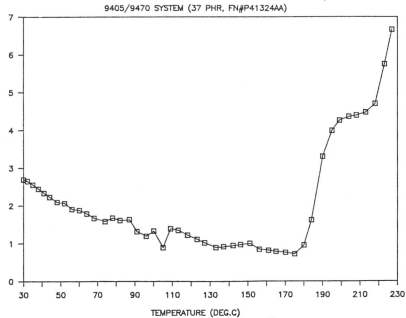

Figure III.  Viscosity Cure Sweep for EPON[R] Resin 9405/
EPON CURING AGENT[R] 9470

# EPOXY RESIN SYSTEMS FOR USE
# IN HIGH PERFORMANCE APPLICATIONS
# AT ELEVATED TEMPERATURES

Ronald S. Bauer, Andrei G. Filippov, Laurie M. Schlaudt, and Walter V. Breitigam

Shell Development Company
Westhollow Research Center
P.O. Box 1380
Houston, Texas 77251-1380

## Abstract

Epoxy resin systems have been developed that have high glass transition temperatures, low water absorption, and consequently, good retention of hot/wet properties. The performance property balance obtained with these new resin systems exceeds that of the state-of-the-art epoxy resin systems currently used in advanced composites.

The resin systems described have potential application in advanced composites for continuous use at elevated temperatures—competing with bismaleimide (BMI) resins in certain temperature ranges. The low water absorption of these new epoxy resin systems provide hot/wet performance comparable to well known toughened bismaleimides, but they can be handled and cured like current epoxy resins.

## 1. INTRODUCTION

In the aircraft/aerospace community there is a pervading impression that epoxy resins as a class of materials have poor hot/wet properties[1]. Because of this perception of epoxy resins, bismaleimides seem to have become the favored matrix resin for applications demanding good hot/wet performance, even though it is acknowledged that bismaleimides tend to be brittle, are prone to microcracking, and have relatively poor processability. It is the purpose of this paper to demonstrate there are specific epoxy resin systems that have hot/wet performance comparable to bismaleimides up to 177°C (350°F).

Unfortunately, the myth about the hot/wet performance of epoxy resins seems to be based entirely on the industry's experience with tetraglycidyl methylenedianiline (TGMDA)/ diaminodiphenyl sulfone (DDS) system. This epoxy resin/curing agent combination has been the basis for almost all epoxy resin formulations used by the industry over the past 15 years for advanced composites. Over the years, the TGMDA/DDS system has been formulated into a variety of matrix resins having an acceptable balance of mechanical properties for many applications along with good processability. However, as can be seen from Table 1, the TGMDA/DDS system can absorb as much as 6.5 percent weight water[2]. This absorbed water[2,3,4] results in a dramatic drop in glass transition and modulus temperature as shown in Tables 1 and 2.

Recently, Shell Chemical Company has introduced two new tetraglycidyl amines[5] and a new diglycidyl ether[6] that were shown to have much improved hot/wet performance at 93°C (200°F) over TGMDA/DDS and yet also have good dry mechanical properties and high glass transition temperatures. These new resins are N,N,N',N'-tetraglycidyl-α,α'- bis(4-aminophenyl)-p-diisopropylbenzene (EP0N HPT™ Resin 1071), N,N,N',N'-tetraglycidly-α',α'-bis(3,5-dimethyl-4-aminophenyl)-p-diisopropylbenzene (EPON HPT™ Resin

1072), and Research Resin RSS-1079 which is a diglycidyl ether based on a stiff backbone bisphenol. The structures and properties of EPON HPT Resin 1071 and EPON HPT Resin 1072 are given in Figure 1 and the properties of Research Resin RSS-1079 are summarized in Table 3. Data has now been obtained that demonstrates these new resins retain their properties in hot/wet environments up to 177°C (350°F) and have hot/wet performance characteristics: 1) superior to TGMDA/DDS, and 2) at least equivalent to those obtained with two commercial toughened bismaleimides up to 177°C.

## 2. DISCUSSION

In Table 4 are summarized the neat resin properties of TGMDA, EPON HPT RESIN 1071, EPON HPT Resin 1072, and Research Resin RSS-1079 cured with a stoichiometric quantity of DDS. As can be seen from the Table, the TGMDA system absorbs 5.7 percent weight water after two weeks immersion in water at 93°C (200°F), where as, the corresponding systems containing the two EPON HPT Resins and Research Resin RSS-1079 absorb only 3.4, 2.6, and 2.8 percent water, respectively. The plasticization of epoxy resins by absorbed water has been well documented and has been shown to result in a loss of modulus at elevated temperatures. In the above systems, for example, when tested in water at 93°C after two weeks immersion in water at 93°C (200°F), the TGMDA system retained only 65 percent of its initial dry, room temperature modulus. The EPON HPT Resin 1071 and 1072, and Research Resin RSS-1079 retained 85, 87, and 93 percent of their initial dry, room temperature moduli.

When cured with a less polar amine than DDS, such as α,α'-bis(4-hydroxyphenyl)-p-diisopropylbenzene (EPON HPT™ Curing Agent 1062) the structure of which is shown in Figure 2, systems are obtained from EPON HPT Resin 1071, EPON HPT Resin 1072, and Research Resin RSS-1079 that absorb even less water. A comparison of the moisture gain values in Tables 4 and 5 show that systems based on EPON HPT Curing Agent 1062 absorb only about half as much water as the corresponding systems based on DDS. Compared

with two commercially available toughened bismaleimides (BMI's), these EPON HPT systems also absorbed, under the same conditions, only about half as much water (see Table 6) as the two BMI's.

The effect of the high levels of absorbed moisture on the hot/wet performance above 93°C (200°F) of a TGMDA/DDS system is apparent from the data summarized in Table 7. Although the flexural moduli for the TGMDA/DDS system measured dry at both 149°C (300°F) and 177°C (350°F) are well above 2.1 GPa (300 ksi), the moduli when measured hot/wet at the same temperatures are only 1.5 GPa (215 msi) and 1.1 GPa (155 ksi), respectively. Compared to the room temperature dry modulus for the TGMDA/DSS system of 3.8 GPa (550 ksi), under hot/wet conditions only 48 percent of the modulus is retained wet at 149°C (300°F) and only 28 percent at 177°C (350°F). Under the same conditions the systems based on EPON HPT Resin 1071, EPON HPT Resin 1072, and Research Resin RSS-1079 retained better than 70 percent and 60 percent, respectively, of their room temperature dry flexural moduli when measured hot/wet (see Figure 4).

In comparison to the BMI's examined, these new EPON HPT Resin systems have lower glass transition temperatures and lower moduli when measured both wet and dry at room temperature (see Table 6). However, as can be seen in Table 7, at temperatures of 149°C (300°F) and 177°C (350°F) the absolute values of the flexural strength and flexural moduli of the epoxy resin systems and the bismaleimides are comparable. On the other hand, the epoxy resin systems retain a greater percentage of their initial room temperature dry moduli under both dry and wet conditions at 149°C (300°F)and 177°C (350°F) as can be seen in Figures 3 and 4, respectively. At 149°C (300°F), for example, the epoxy resin systems retain better than 70 percent of their room temperature dry moduli under hot/wet conditions compared to the BMI's, which retain only about 50 percent of their room temperature dry moduli. Similarly at 177°C (350°F) the BMI's still retain about 50 percent of their room temperature dry moduli and epoxy resin systems between 60 and 80 percent.

Similar trends in retention of hot/wet properties of laminate have been observed with these new high performance matrix resin systems. As shown in Figure 5, for example, laminate prepared from the EPON HPT Resin 1071/EPON HPT Curing Agent 1062 system on Magnamite IM-6 (Hercules Incorporated) retained about 90 percent of its 0° flexural modulus when tested at temperatures of 149°C and 177°C after 96 hours in boiling water. Retention of short beam shear strengths, 60 and 50 percent respectively, were observed under the same test conditions.

## 3. CONCLUSIONS

The evidence presented here should dispel the impression pervading the aircraft/aerospace community, that epoxy resins as a class of materials have poor hot/wet performance characteristics; although TGMDA/DDS does perform poorly under hot/wet conditions. It has been shown that there are now available a family of resin systems that have outstanding hot/wet performance up to 177°C (350°F). These systems have, in fact, hot/wet performance characteristics equivalent to some commercial toughened bismaleimides up to 177°C (350°F). Further, these resins (EPON HPT Resin 1071, EPON HPT Resin 1062, and Research Resin RSS-1079) when cured with a relatively less polar curing agent than DDS such as EPON HPT Curing Agent 1062, not only have performance characteristics equivalent at elevated temperatures to the BMI's mentioned but they also retain a higher level of their room temperature dry properties.

## 4. REFERENCES

1. H. M. Clancy and D.E. Luft, 18th International SAMPE Technical Conference, 135-141, October 7-9, 1986.

2. C. E. Browning, Polym. Eng. Sci., 18, 16(1978).

3. A. Apicella, L. Nicolais, G. Astarita, and E. Drioli, Polymer, 20(9), 1143(1979).

4. E. L. McKague, J. D. Reynold, and J.E. Haskin, J. Appl. Polym. Sci., 22, 1643(1978)

5. R. S. Bauer, 31st International SAMPE Symposium and Exhibition, 1226-1233, Apr. 7-10, 1986

6. R. S. Bauer, 18th International SAMPE Technical Conference, 510-519, Oct. 7-9, 1986.

## 5. BIOGRAPHIES

DR. RONALD S. BAUER obtained his Ph.D. in Organic Chemistry from the University of California at Los Angeles. After graduation, he joined Shell Development Company at Emeryville, California, as a chemist where he worked in the areas of exploratory polymer synthesis, polymer characterization, polymer processing, and polymer process research. During 1971 and 1972 he was an exchange scientist at Shell's Egham Research Laboratories in Surrey, England, where he worked on the development of non-aqueous dispersions of epoxy resins. Currently, he is a Senior Staff Research Chemist at Shell Development Company's Westhollow Research Center in Houston, Texas, and has been involved in the development of a weatherable epoxy resin, high solids epoxy coatings, and new epoxy resins for advanced composites and electrical applications.

Dr. Bauer has publications and patents covering epoxide, olefin, and diene polymerization involving both homogeneous and heterogeneous catalysis. Also, he has written a number of review articles on epoxy resins and has edited two books on the subject. In 1985, he was chairman of the Division of Polymeric Materials: Science and Engineering of the American Chemical Society.

WALTER V. BREITIGAM obtained his degree in organic chemistry from Ohio State University. He joined Shell Oil Company in Wood River, Illinois as a chemist where he worked on the development of gasoline and lubricant additives, and synthetic gas turbine lubricants. In 1975 he transferred to Shell Development Company's Westhollow Research Center in Houston, Texas, where he has been involved in development and technical support activities for structural product applications of epoxy resins for industrial and construction uses. He has published several articles on structural applications of epoxy resins. Currently, he is project leader in the Resins Department responsible for research and development of advanced composites for the aircraft/aerospace industry.

DR. ANDREI G. FILIPPOV joined Shell Development Company after obtaining a Ph.D. in Material Science from the University of Massachusetts in 1981. Currently, he is a Staff Research Engineer in the the Material Science and Engineering Department at the Westhollow Research Center in Houston, Texas. His background is in structure-mechanical properties of polymeric materials. His work since joining Shell has been the study of deformation and fracture behavior of homopolymers, polymeric blends, and composite materials. Other interests include thermodynamics of deformation, stability and fracture of polymeric solids.

LAURIE M. SCHLAUDT graduated from Texas A&M University in 1984 with a B.S. degree in Mechanical Engineering. She joined the Resins Department of Shell Development Company's Westhollow Research Center in Houston, Texas, in June 1984. As a member of the Transportation/Futures Group, she is involved with aerospace applications for epoxy resins. Her current responsibilities include material test program management, database development, and filament winding technical support.

### Table 1. Effect of Water on the Glass Transition Temperatures of TGMDA/DDS Systems

|  | System I[a] | System II[b] | System III[c] |
|---|---|---|---|
| Moisture Gain, %w | 6.5[d] | 5.5[d] | 5.0[e] |
| Glass Transition Temp. | | | |
| Dry, °C | 246 | 175 | 200 |
| Wet, °C | 144 | 112 | 140 |
| °C/%w Water Absorbed | 15.7 | 11.5 | 12.0 |

a) NARMCO 5208 (Reference 4).
b) TGMDA / 32 phr DDS / $BF_3 \cdot H_2NCH_3$ (Reference 3).
c) TGMDA / 50 phr DDS (Reference 2).
d) Immersion in water at 71°C.
e) Immersion in water at 60°C.

Table 2. Effect of Absorbed Moisture on the
Physical Properties of a TGMDA/DDS System

| Components | | |
|---|---|---|
| TGMDA | 100 | 100 |
| BPA Epoxy Novalac[a] | 8.2 | 8.2 |
| DDS | 28 | 51.9 |
| | | |
| Glass Transition Temp. | | |
| Dry, °C | 242 | 262 |
| Wet, °C | 174 | 170 |
| | | |
| Flex Properties (RT/Dry) | | |
| Strength, MPa (ksi) | 117 (17) | 131 (19) |
| Modulus, GPa (ksi) | 4.0 (580) | 3.8 (548) |
| Elongation, % | 3.5 | 4.7 |
| | | |
| Flex Properties (Wet)[b] | | |
| Strength, MPa (ksi) | 83 (12) | 76 (11) |
| Modulus, GPa (ksi) | 3.2 (471) | 2.4 (352) |
| Elongation, % | 3.4 | 3.7 |
| | | |
| Moisture Gain, % | 4.7 | 5.8 |

a) EPI-REZ SU-8 (Interez, Inc.).
b) Tested in water at 93°C after two weeks immersion
at 93°C.

09285-9

Table 3. Neat Resin Properties
of Research Resin RSS-1079

| Typical Properties of Research Resin RSS-1079 | |
|---|---|
| Physical Form | Glassy Solid |
| Melting Point,[1] °C | 80-82 |
| Tg,[2] °C | 51-54 |
| Viscosity,[3] poise at 100°C | 405 |
| 110°C | 95 |
| 120°C | 29 |
| 130°C | 12 |
| 140°C | 5 |
| 150°C | 3 |

1) ASTM D-3461, Mettler, 1°C
2) DSC, 20°C / min
3) Parallel Plate

09285-49

Table 4. Neat Resin Properties EPON HPT™ Resin 1071,
EPON HPT Resin 1072, and Research Resin RSS-1079 Cured[1] with DDS

| Curing Agent | $H_2N-\bigcirc-\overset{\overset{O}{\parallel}}{\underset{\underset{O}{\parallel}}{S}}-\bigcirc-NH_2$ | | |
|---|---|---|---|
| Properties \ Resin | TGMDA[2] | EPON HPT™ Resin | | Research Resin RSS-1079 |
| | | 1071 | 1072 | |
| Tg (Tan δ), °C | 262 | 249 | 258 | 279 |
| Flexural Properties (RT/Dry) | | | | |
| Strength, MPa (ksi) | 138 (20) | 117 (17) | 124 (18) | 124 (18) |
| Modulus, GPa (ksi) | 3.9 (559) | 3.9 (563) | 3.4 (486) | 3.3 (485) |
| Elongation, % | 5.0 | 3.7 | 4.5 | 4.7 |
| Flexural Properties (Hot/Wet)[3] | | | | |
| Strength, MPa (ksi) | 76 (11) | 90 (13) | 83 (12) | 90 (13) |
| Modulus, GPa (ksi) | 2.5 (361) | 3.0 (434) | 3.1 (442) | 3.0 (433) |
| Elongation, % | 4.7 | 4.2 | 3.3 | 4.1 |
| Moisture Gain, %w | 5.7 | 3.6 | 2.7 | 2.8 |

1) Cured with Stoichiometric Amount of Curing Agent, Cure Schedule
   2 hr at 150° and 4 hr at 200°C
2) Formulation Contains 8.2 phr EPI-REZ SU-8 (Interez, Inc.)
3) Tested in Water at 93°C After Two Weeks Immersion at 93°C

Table 5. Comparison of Neat Resin Properties EPON HPT™
Resin 1071, EPON HPT Resin 1072, and Research
Resin RSS-1079 Cured[1] EPON HPT™ Curing Agent 1062

| Curing Agent | $H_2N-\bigcirc-\overset{CH_3}{\underset{CH_3}{C}}-\bigcirc-\overset{CH_3}{\underset{CH_3}{C}}-\bigcirc-NH_2$ (with $CH_3$ substituents) | | |
|---|---|---|---|
| Properties \ Resin | EPON HPT™ Resin | | Research Resin RSS-1079 |
| | 1071 | 1072 | |
| Tg (Tan δ), °C | 239 | 239 | 264 |
| Flexural Properties (RT/Dry) | | | |
| Strength, MPa (ksi) | 131 (19) | 131 (19) | 138 (20) |
| Modulus, GPa (ksi) | 3.4 (497) | 3.4 (491) | 3.2 (462) |
| Elongation, % | 4.5 | 4.7 | 6.4 |
| Flexural Properties (Hot/Wet)[2] | | | |
| Strength, MPa (ksi) | 96 (14) | 90 (13) | 90 (13) |
| Modulus, GPa (ksi) | 3.0 (441) | 3.2 (470) | 2.7 (415) |
| Elongation, % | 3.9 | 3.3 | 4.1 |
| Moisture Gain, %w[3] | 1.6 | 1.4 | 1.3 |

1) Cured with Stoichiometric Amount of Curing Agent, Cure Schedule
   2 hr at 150° and 4 hr at 200°C
2) Tested in Water at 93°C After Two Weeks Immersion at 93°C
3) After Two Weeks Immersion in Water at 93°C

Table 6. Comparison of Neat Resin Properties at Room Temperature of
EPON HPT™ Resin 1071, EPON HPT Resin 1072, and Research Resin RSS-1079
Systems with a TGMDA/DDS System and Commercial Toughened BMI Systems

| Resin System / Properties | DDS[1] TGMDA[1] | EPON HPT™ Curing Agent 1062[2] | | | Commercial Toughened BMI | |
|---|---|---|---|---|---|---|
| | | EPON HPT™ Resin | | Research Resin RSS-1079 | I | II |
| | | 1071 | 1072 | | | |
| Tg (Tan δ), °C | | 250 | 241 | 250 | >310 | >310 |
| **Flexural Properties (RT/Dry)** | | | | | | |
| Strength, MPa (ksi) | 138 (20) | 131 (19) | 90 (13) | 138 (20) | 158 (23) | 158 (23) |
| Modulus, GPa (ksi) | 38 (555) | 3.5 (500) | 3.4 (495) | 3.2 (460) | 3.9 (570) | 4.0 (575) |
| Elongation, % | 3.9 | 4.2 | 2.6 | 5.1 | 4.4 | 4.5 |
| **Flexural Properties (RT/Wet)[3]** | | | | | | |
| Strength, MPa (ksi) | 90 (13) | 145 (21) | 96 (14) | 138 (20) | 138 (20) | 131 (19) |
| Modulus, GPa (ksi) | 3.5 (500) | 3.5 (500) | 3.2 (465) | 3.1 (455) | 3.7 (560) | 3.9 (565) |
| Elongation, % | 2.5 | 4.8 | 3.1 | 4.9 | 4.0 | 3.5 |
| Moisture Gain, %[3] | 3.6 | 0.8 | 1.0 | 1.1 | 2.2 | 2.2 |

1) TGMDA / DDS / EPI-REZ SU-8: 90 / 29.7 / 10, Cure Schedule 2 hr at 150°C and 4 hr at 200°C
2) Cured with Stoichiometric Amount of Curing Agent, Cure Schedule 2 hr at 150°C and 4 hr at 200°C
3) After 48 Hours in Boiling Water

09285-50

Table 7. Comparison of Hot/Wet Neat Resin Properties EPON HPT™
Resins 1071, 1072 and Research Resin 1079 Systems with a
TGMDA/DDS System and Commercial Toughened BMI Systems

| | Temp., °F | Modulus, GPa (ksi) | | Strength, MPa (ksi) | | Elongation, % | |
|---|---|---|---|---|---|---|---|
| | | Dry | Wet[3] | Dry | Wet[3] | Dry | Wet |
| TGMDA/DDS/SU-8[1] | RT | 3.8 (555) | 3.5 (500) | 138 (20) | 90 (13) | 3.9 | 2.5 |
| | 300 | 2.6 (370) | 1.5 (215) | 76 (11) | 41 (6) | 3.2 | 3.5 |
| | 350 | 2.3 (330) | 1.1 (155) | 83 (12) | 34 (5) | 4.2 | 4.0 |
| EPON HPT™ Resin 1071[2] | RT | 3.5 (500) | 3.5 (500) | 131 (19) | 145 (21) | 4.2 | 4.8 |
| | 300 | 2.8 (400) | 2.5 (365) | 96 (14) | 76 (10) | >8 | 5.7 |
| | 350 | 2.5 (365) | 2.2 (320) | 76 (11) | 56 (8.1) | >8 | >8 |
| EPON HPT™ Resin 1072[2] | RT | 3.4 (495) | 3.2 (465) | 90 (13) | 96 (14) | 2.6 | 3.1 |
| | 300 | 2.6 (370) | 2.4 (345) | 103 (15) | 90 (13) | 5.1 | 5.1 |
| | 350 | 2.3 (335) | 2.1 (300) | 76 (11) | 62 (9) | 6.8 | 6.0 |
| Research Resin RSS-1079[2] | RT | 3.2 (460) | 3.1 (455) | 138 (20) | 138 (20) | 5.1 | 4.9 |
| | 300 | 2.6 (380) | 2.5 (355) | 96 (14) | 76 (11) | 6.4 | 5.9 |
| | 350 | 2.5 (355) | 2.3 (330) | 69 (10) | 48 (7) | 7.1 | >8 |
| Toughened BMI (I) | RT | 3.9 (570) | 3.9 (560) | 158 (23) | 179 (26) | 4.4 | 4.0 |
| | 300 | 2.7 (385) | 1.9 (280) | 90 (13) | 69 (10) | 3.5 | 4.1 |
| | 350 | 2.5 (360) | 2.0 (285) | 90 (13) | 83 (12) | 3.8 | 5.2 |
| Toughened BMI (II) | RT | 4.0 (575) | 3.4 (565) | 158 (23) | 131 (19) | 4.5 | 3.5 |
| | 300 | 2.8 (400) | 2.0 (295) | 90 (13) | 48 (7) | 3.5 | 1.9 |
| | 350 | 2.4 (345) | 1.9 (280) | 76 (10) | 62 (9) | 3.1 | 3.7 |

1) TGMDA / DDS / EPI-REZ SU-8: 90 / 29.7 / 10, Cure Schedule 2 hr at 150°C and 4 hr at 200°C
2) Cured with Stoichiometric Amount of EPON HPT™ Curing Agent 1062,
Cure Schedule 2 hr at 150°C and 4 hr at 200°C
3) After 48 hr in Boiling Water

09285-56

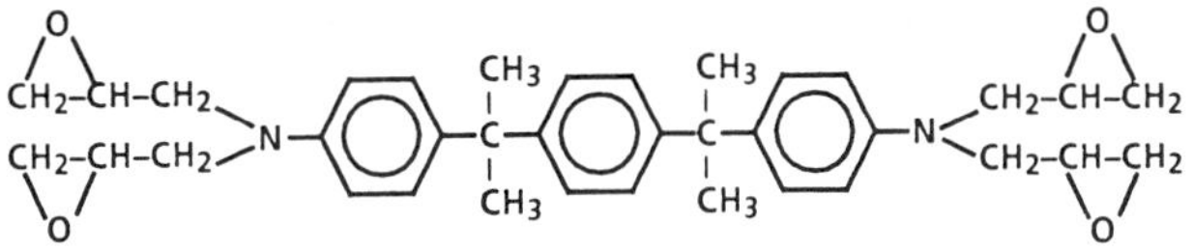

EPON HPT™ Resin 1071

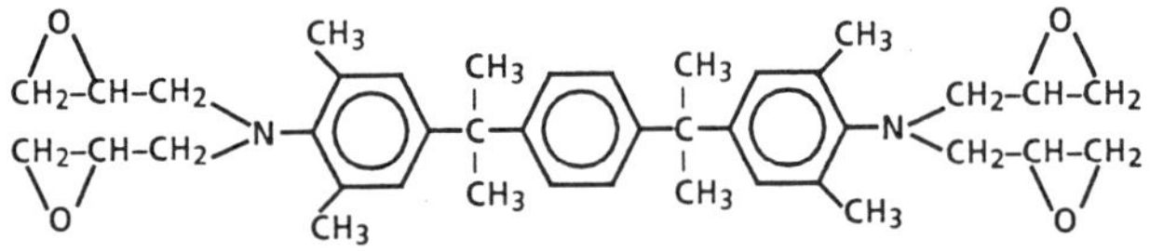

EPON HPT™ Resin 1072

| Typical Neat EPON HPT Resin Properties | | |
|---|---|---|
| Property | EPON HPT Resin 1071 | EPON HPT Resin 1072 |
| Weight per Epoxide (WPE) (Shell Method HC–427C–81 Perchloric Acid Method) | 150–170 | 185–205 |
| Melt Viscosity, Brookfield at 100°C, Pa·s (p) | 18–22 | 30–34 |
| Tg (DSC), °C | 23° | 41° |
| Melting Point, °C (ASTM D3461, Mettler, 1°C/min) | 50° | 65° |

Figure 1. Structures and Typical Neat Resin Properties

09285-26

1111

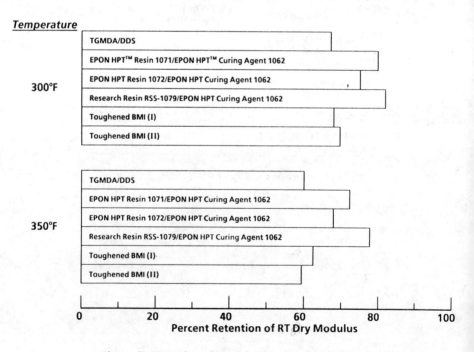

EPON HPT™ Curing Agent 1062

| Typical Properties EPON HPT Curing Agent 1062 | |
|---|---|
| Physical Form | Crystalline Solid |
| Color | Tan to Cream |
| Melting Point | 151°C – 154°C |
| Approximate Equivalent Weight/Active Hydrogen | 100 |

Figure 2.  Structure and Typical Properties
EPON HPT Curing Agent 1062

09285-55

**Temperature**

**300°F**
- TGMDA/DDS
- EPON HPT™ Resin 1071/EPON HPT™ Curing Agent 1062
- EPON HPT Resin 1072/EPON HPT Curing Agent 1062
- Research Resin RSS-1079/EPON HPT Curing Agent 1062
- Toughened BMI (I)
- Toughened BMI (II)

**350°F**
- TGMDA/DDS
- EPON HPT Resin 1071/EPON HPT Curing Agent 1062
- EPON HPT Resin 1072/EPON HPT Curing Agent 1062
- Research Resin RSS-1079/EPON HPT Curing Agent 1062
- Toughened BMI (I)
- Toughened BMI (II)

0    20    40    60    80    100
**Percent Retention of RT Dry Modulus**

Figure 3.  Retention of Modulus (Hot/Dry) for Neat Resin Systems

09285-53

1112

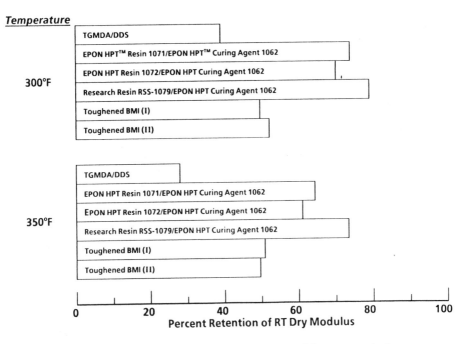

**Temperature**

**300°F**

TGMDA/DDS

EPON HPT™ Resin 1071/EPON HPT™ Curing Agent 1062

EPON HPT Resin 1072/EPON HPT Curing Agent 1062

Research Resin RSS-1079/EPON HPT Curing Agent 1062

Toughened BMI (I)

Toughened BMI (II)

**350°F**

TGMDA/DDS

EPON HPT Resin 1071/EPON HPT Curing Agent 1062

EPON HPT Resin 1072/EPON HPT Curing Agent 1062

Research Resin RSS-1079/EPON HPT Curing Agent 1062

Toughened BMI (I)

Toughened BMI (II)

Percent Retention of RT Dry Modulus

**Figure 4. Retention of Modulus (Hot/Wet) for Neat Resin Systems**

09285-54

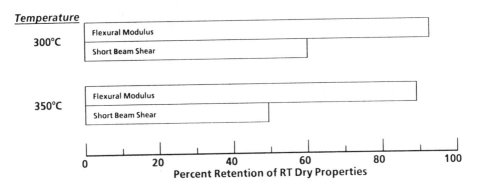

**Temperature**

**300°C**

Flexural Modulus

Short Beam Shear

**350°C**

Flexural Modulus

Short Beam Shear

Percent Retention of RT Dry Properties

**Figure 5. Retention of Flexural Modulus (Hot/Wet)[a] and
Short Beam Shear Strength (Hot/Wet)[a] of an EPON HPT™
Resin 1071/EPON HPT™ Curing Agent 1062 Composite[b]**

*a) After 96 Hours in Boiling Water*
*b) Magnamite IM-6 (Hercules Incorporated)*

09285-57

32nd International SAMPE Symposium
April 6-9, 1987

# DEGREE OF CURE AND VISCOSITY OF HERCULES HBRF-55 RESIN

S.T.Bhi, R. Scott Hansen, and Brian A. Wilson
Aerojet Strategic Propulsion Corporation
Sacramento, California 95813
and
Emilo P. Calius and George S. Springer
Department of Aeronautics and Astronautics
Stanford University, Stanford, California 94305

## Abstract

In this investigation the rate of cure and viscosity were measured for Hercules HBRF-55 resin. The rate of cure was measured by differential scanning calorimetry, while the viscosity was measured by a parallel disc and plate type apparatus. The data were fitted to analytic expressions to make it suitable for use in numerical calculations.

## 1. INTRODUCTION

Recently, it has become evident that manufacturing composites by autoclave curing or by filament winding is greatly eased if the process variables are selected with the use of suitable analytic models. However, models require knowledge of the relevant material properties, including the heat generated by the reaction, the rate of cure, and the viscosity. Therefore, these three parameters were measured for Hercules HBRF-55 resin. This resin system was selected because it is representative of advanced filament winding systems now coming into wider use. A brief characterization of the resin is given in Table 1.

The experimental and analytical procedures employed in the investigation were similar to those used in our previous studies of Hercules 3501-6 and Fiberite 976 resins.[1,2] Therefore, only those aspects of the procedure are given which are needed for interpreting the results.

## 2. EXPERIMENTAL

The heat of reaction and the rate of cure were measured using a DuPont model 910 differential scanning calorimeter. Isothermal scanning tests were performed at 371, 381, 391, 401, 411, 421, and 431°K using resin samples weighting 11.4 to 15 milligrams. Tests were also conducted with the sample heated at a constant rate of 20°K/min (dynamic scanning).

The viscosity was measured with a Rheometrics model RMS-605 parallel disc and plate type spectrometer with a disc radius of 25 mm and gap size 0.5 mm. Data were generated with the disc oscillating at 10 rad/sec. The apparatus provided directly the complex viscosity in poise as a function of time. The complex viscosity was taken to be the same as the shear viscosity. This approximation is reasonable at least at low degrees of cure. It becomes more inappropriate as the gel point is approached.

The viscosities were measured at the constant temperatures of 371, 381, 391, 401, 411, 421, and 431°K. The time required for each test ranged from 38 minutes (at 371°K) to 6 minutes (at 431°K). At every temperature steady state was reached in two to four minutes.

## 3. RESULTS

### 3.1 Heat of Reaction

The ultimate heat of reaction $H_U$ is the total amount of heat generated by taking the cure reaction as far as it will go until the curing reaction is completed. From the results of the dynamic scanning measurements $H_U$ was found to be

$$H_U = 545 \text{ J/g} \tag{1}$$

The isothermal heat of reaction $H_T$ is the total amount of heat generated at constant temperature. $H_T$ determined from the data generated during isothermal scanning together with

## Table 1

### Properties of Hercules HBRF-55 Resin

Hercules HBRF-55 is a mixture of:

> 80 parts/weight of Epon 826 DGEBA resin
> 20 parts/weight of RD2 diluent (equivalent to Celanese 5022)
> 24 parts/weight of Tonox 60/40 curing agent.

Density of cured resin (lb/in$^3$):   0.0437

Cured neat resin data

| | |
|---|---|
| Tensile Modulus (Msi): | 0.44 |
| Shear Modulus (Msi): | 0.16 |
| Poisson's Ratio: | 0.38 |
| Coefficient of Thermal Expansion $\left( \mu \frac{in/in}{°F} \right)$: | 37.8 |
| Heat Distortion Temperature (°F): | 250 |

the analytic procedure given in ref. (2). The results showed that, within the temperature range of the experiments, the isothermal heat of reaction was constant, having the same value as the ultimate heat of reaction

$$H_T/H_U = 1.0 \qquad (2)$$

Note that this result differs from that obtained for Fiberite 976 resin.[2] For the Fiberite 976 resin $H_T$ was found to depend on temperature.

### 3.2 Rate of Cure

The total degree of cure $\alpha$ of the resin is defined by

$$\alpha \equiv \frac{1}{H_U} \int_0^t \left( \frac{dQ}{dt} \right)_T dt \qquad (3)$$

where $(dQ/dt)_T$ is the rate of heat flow during isothermal scanning. From Eq. 3 the rate of cure is

$$\frac{d\alpha}{dt} = \frac{1}{H_U} \left( \frac{dQ}{dt} \right)_T \qquad (4)$$

Since $H_U$ is known and $(dQ/dt)_T$ is available from the results of the DSC isothermal scans, we can numerically integrate Eq. 4 to obtain $\alpha$.

An expression for $d\alpha/dt$ was obtained by writing

$$\frac{d\alpha}{dt} = (K_1 + K_2\alpha^a)(B - \alpha)^b(1 - \alpha)^c \qquad (5)$$

where $K_1$, $K_2$, $B$, $a$, $b$, $c$ are coefficients which must determined from data. To find these coefficients, the DSC data was fitted by curves using a Levenberg-Marquardt algorithm.[2] This procedure resulted in values of $K_1$, $K_2$, $B$, $a$, $b$, $c$ at each isotherm. The exponents $a$, $b$, $c$ and the coefficient B were treated as constants, independent of temperature. The $K_1$ and $K_2$ coefficients depend on the temperature, according to the Arrhenius expression

$$K_i = A_i e^{-\left( \frac{\Delta E_i}{RT} \right)} \qquad i = 1, 2 \qquad (6)$$

where $A_i$ are pre-exponential factors and $\Delta E_i$ are activation energies, $R$ is the universal gas constant, and $T$ is the absolute temperature.

The values of the above parameters are given in Table 2. Results computed by Eq. 5, together with the constants in Table 2, were compared to the data at four typical temperatures. These comparisons, presented in Figure 1, show good agreement between the results of our equation and the data.

It is noted that Pearce and Mijovic also proposed an expression for describing the rate of degree of cure of the HBRF-55 resin system.[3] Their expression is similar, but not identical, to Eq. 5. The constants proposed by Pearce and Mijovic (which are to be used in conjuction with Eq. 5) are included in Table 2.

We also calculated the rate of degree of cure

**Table 2**

The Constants in the Rate of Degree of Cure Expression

|  | Present | Pearce and Mijovic[3] |
|---|---|---|
| $A_1(s)^{-1}$ | 490.6 | $2.5 \times 10^4$ |
| $A_2(s)^{-1}$ | 2052.4 | 3450 |
| $\Delta E_1(\text{J/mol})$ | $4.4714 \times 10^4$ | $5.69 \times 10^4$ |
| $\Delta E_2(\text{J/mol})$ | $4.2848 \times 10^4$ | $4.53 \times 10^4$ |
| $B$ | 1.065 | -- |
| $a$ | 0.840 | $1.43 - 1.41 \times 10^{-3}\ T$ |
| $b$ | 0.54 | $(2-a)$ |
| $c$ | 1.0 | -- |

by Pearce and Mijovic's expression and included the results in Figure 1. As can be seen, the present model agrees somewhat better with the data than the expression of Pearce and Mijovic.

### 3.3 Viscosity

The viscosity data were approximated, as in previous work,[1,2] by an expression of the form

$$\mu = K_\mu\, e^{\kappa\alpha} \qquad (7)$$

where $K_\mu$ and $\kappa$ are unknown coefficients. $K_\mu$ and $\kappa$ were determined by a least-squares fit to the data at each isotherm.[2] The exponent $\kappa$ was treated as a constant independent of temperature. The $K_\mu$ coefficient depends on the temperature according to the following form

$$K_\mu = \mu_\infty\, e^{U/RT} \qquad (8)$$

where $\mu_\infty$ and $U$ are constants, independent of temperature. These constants were again determined by a least-squares fit to the data. The results are

$$\left. \begin{array}{l} \mu_\infty = 9.08\ \text{Poise} \\ U = -1.087 \times 10^4\ \text{J/mol} \\ \kappa = 2.502 \end{array} \right\} \quad \alpha \le 0.6$$

$$(9)$$

Results computed by Eq. 7, together with the constants in Eq. 9, were compared to the test data at four typical temperatures. These comparisons, presented in Figure 2, show that there is reasonable agreement between the results of our equation and the data when the

degree of cure is about 0.6. At higher degrees of cure, the resin hardens very fast and Eq. 7 is no longer applicable.

## 4. CONCLUDING REMARKS

The foregoing results can be used in simulating the cure processe of composites employing Hercules HBRF-55 resin. It is emphasized that the results are meant primarily to assist in engineering studies. They were not meant to provide detailed information on the molecular processes during cure.

## 5. ACKNOWLEDGMENT

This work was supported in part by NASA–Marshall Space Flight Center.

## 6. REFERENCES

1.  Lee, W. I., Loos, A. C., and Springer, G. S. "Heat of Reaction, Degree of Cure, and Viscosity of Hercules 3501-6 Resin," *Journal of Composite Materials, Vol. 16,* 510–520 (1982).

2.  Dusi, M. R., Lee, W. I., Ciriscioli, P. R., and Springer, G. S. "Cure Kinetics and Viscosity of Fiberite 976 Resin," *Journal of Composite Materials, Vol. 21,* in print (1987).

3.  Pearce, E. M. and Mijovic, J. "Final Report to NASA-Ames Research Center on Characterization-Curing-Property Studies of HBRF55A Resin Formulations," Polytechnic Institute of New York, January 31, 1985.

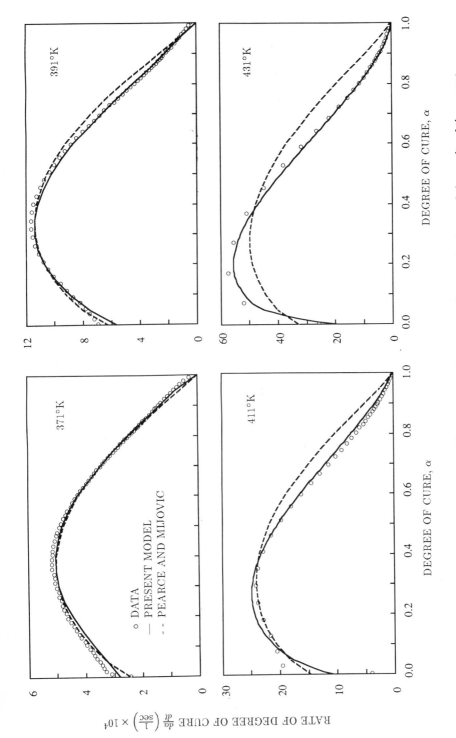

**Figure 1.** Rate of degree of cure as function of degree of cure. Comparisons between the data and the results of the present model and the model of Pearce and Mijovic.

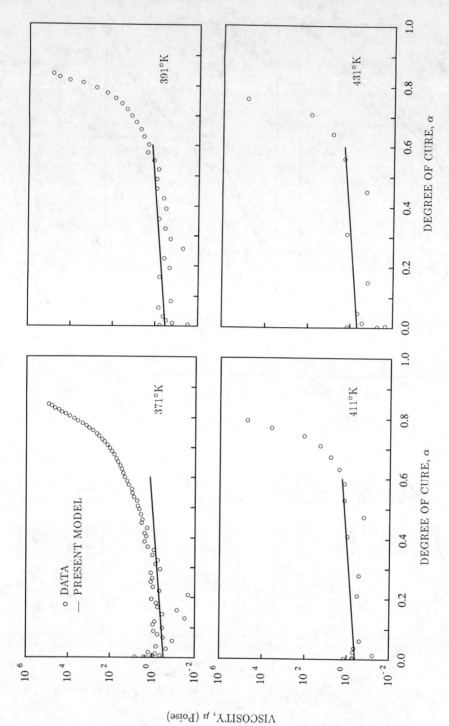

**Figure 2.** Viscosity as function of degree of cure. Comparisons between the data and the results of the present model.

A NOVEL TRI-FUNCTIONAL EPOXY RESIN WITH

LOW VISCOSITY AND HIGH HEAT RESISTANCE

Y.Saito,K.Kamio,A.Morii and T.Adachi

Sumitomo Chemical Co.,Ltd.

Osaka Research Laboratory

3-1-98,Kasugade-Naka

Konohana,Osaka 554 JAPAN

## Abstract

A novel epoxy resin will be introduced and its application for carbon fiber reinforced composite (CFRP) will be discussed. Sumi-epoxy ELM-100 is a tri-functional amino-epoxy resin of which EEW is about 105. It has very low viscosity and excellent heat and moisture resistance. Low viscosity (ca.10Ps/25°C)can be achieved without a troublesome molecular-distillation process.

Cured resin with 4,4'-Diamino diphenylsulfone (DDS) shows high Tg (261°C) which is 20-40°C higher than that of conventional tri-functional amino epoxy resins. Water absorption of the cured resin is least among those resins.

CFRP was prepared using a matrix resin system of ELM-100 and DDS,and its physical pro-perties were observed. The CFRP has very good mechanical properties and heat resistance (especially hot-wet prope-rties).ILSS of the CFRP is more than 13 $Kg/mm^2$ at room temperature and 11.3 $Kg/mm^2$ at room-temperature after 48 hrs boiling in water.

Fracture-strength after 48 hrs boiling in water is 204 $Kg/mm^2$ at room-temperature (ca. 15-35 $Kg/mm^2$ higher than that of other tri-functional amino-epoxy resins) and 113 $Kg/mm^2$ at 121°C (ca.35-90 Kg $/mm^2$ higher than that of above-mentioned other resins). It is concluded that the no-vel resin, ELM-100, represe-nts a significant advance in matrix resins for CFRP where ease of processing, and heat and moisture resistance are required.

# 1. INTRODUCTION

The most widely used structural composit materials are based on epoxy, polyester or vinylester resins combined with carbon or glass fibers. The data presented here is directed toward carbon fiber reinforced epoxy resins because of its relatively good performance characteristics for structural composites. With the increasing use of epoxy resins in reinforced plastics, one of the constantly needed improvements is to make them more easily processable. For example, epoxy resins with lower viscosity and higher reactivity are needed in order to produce void free composite structures easily. The purpose of this paper is to present a higher performance new epoxy resin which meets this need.

# 2. EXPERIMENTAL
## 2.1 MATERIAL

Sumiepoxy ELM-100 is a trifunctional amino-epoxy resin developed recently by Sumitomo Chemical Co.,LTD. It has a very low viscosity at room temperature and a low EEW. Low viscosity can be achieved without a troublesome molecular distillation process. Table-1 lists several conventional diglycidylether of bisphenol A type epoxy resins

and poly-functional amino-epoxy resins. It is easily understood that ELM-100 has the lowest viscosity and EEW among these conventional epoxy resins.

## 2.2 NEAT RESN DATA

Epoxy resins were cured using one of two hardeners; methylhexahydrophthalic anhydride ( MHPA) and 4,4'-Diaminodiphenyl sulfone (DDS). Clear and homogeneous resin was obtained by mixing a epoxy resin, MHPA and ,if necessary, 2-ethyl-4-methyl imidazole (2E4MZ) as an accelerator and stirring the mixture at room temperature for 10 minutes. As the DDS is a crystalline powder with a melting point of 168°C,homogeneous mixture could not be obtained using above mentioned procedure; DDS did disolve at 130°C, and $BF_3$-monoethylamine ($BF_3$MEA ) was used as an accelerator. Cured neat resin sheet were prepared by degassing the resin under vaccum,pouring it into a mould, and curing it at a temperature shown in table 2 and table 3. Properties of uncured and cured neat resin were listed in table 2 and 3. It is revealed from table 2 that ELM-100/MHPA, which is particular useful in applications requiring low viscosity and high thermal resistance such as filament winding use, has extreamly lower viscosity than that of bisphenol A type epoxy resins or polyfunctional amino-epoxy resins.

And its thermal resistance is the same as polyfunctional aminoepoxy resins.

Consequently, curing system of ELM-100/MHPA improves a processability without making a sacrifice of thermal resistance. Table 3 shows that ELM-100 /DDS provides a resin mixture with lower viscosity than that of the other formulations. Furthermore, ELM-100/DDS is superior in thermal and moisture resistance to the other tri-functional glycidylamine epoxy resins. For example, Tg of ELM-100 is 19-39°C higher than that of conventional trifunctional aminoepoxy resins, and water absorption is least among those resins. these characteristics of ELM-100 seem to be desirable for a matrix resin of composit materials, because most works are using "wet" processes such as filament winding, compression moulding, pultrusion, and autoclave method.

2.3 COMPOSIT DATA

Prepreg can be made by either solution method or hot melt( "wet") method. In this study, unidirectional prepregs were prepared from the resin disolved in methylethylketone and carbon fiber (Hercules' carbon fiber AS4-6Kf G-size).

Composit laminates were fabricated by the conventional compression method. Flexural properties (ASTM D-790 ) and short beam shear strength (ASTM D -2344 ) of both dry and wetted laminates were measured at 25° C,121°C,and 177°C. The wetting of the laminates was made by dipping the specimens in boiling water for 48 hrs. Curing conditions and mechanical properties of composite laminates are surmarized in Table 4.

All the dry properties at room temperature are almost the same, but the most important point is the wet strength. CFRP prepared using ELM-100 and DDS has very good mechanical properties and heat resistance, especially hot-wet properties. ILSS is 11.3 $Kg/mm^2$ at room temperature and 4.9 $Kg/mm^2$ at 121°C after 48 hrs boiling in water, which is about 2.6-3.0 $Kg/mm^2$ and 1.1-3.5 $Kg/mm^2$ higher than that of tri-functional aminoepoxy resins respectively.

After 48 hrs boiling in water flexural strength is 204 $Kg/mm^2$ and 26-90 $Kg/mm^2$ higher than that of above mentioned other resins. These properties of ELM-100 is almost the same as those of tetra-functional aminoepoxy resin, ELM-434, which has good heat and water resistance. This illustrates the importance of martix Tg and water absorption in retaining elevated temperature properties. It is concluded that the novel resin, ELM-100, represents a significant advance in matrix resins for CFRP where ease of processing, and heat and moisture resistance are required.

## 3. APPLICATIONS

These resin systems are suitable for selected high performance and conventional applications. More conventional uses would include electrical laminates (printed circuit boards), adhesives, and high temperature resistant filament wound pipe. ELM-100 provides high performance properties and will find application in the more demanding end uses.

Potential applications are being carefully screened to focus efforts on end uses requiring the higher level of performance.

## 4. CONCLUSION

ELM-100 has good handleability and processability because it has low viscosity and high reactivity. Composites made from the resin have almost the same or better properties at room temperature and have much better high temperature properties and hot-wet properties compared with those of conventional tri-functional amino-epoxy resins.

Consequently, the curing systems using ELM-100 are being tailored for specific, selected end uses requiring higher levels of performance.

## BIOGRAPHY

Mr. KUNIMASA KAMIO recieved his B.S. and M.S. degree in chemistry from Tokyo Institute of Technology in 1972 and 1974, respectively. He is a research scientist at Osaka Research Laboratory of Sumitomo Chemical Co., LTD. Currently he is responsible for application research of thermosetting resins in composite materials and electrical uses.

Mr. YASUHISA SAITO recieved his B.S. and M.S. degree in polymer science from Kyoto University in 1970 and 1972, respectively. He is a research associate at Osaka Research Laboratory of Sumitomo Chemical Co., LTD. He is currently responsible for the research work on the synthesis of thermosetting resins.

Mr. AKIRA MORII is a research scientist at Takatsuki Research Laboratory of Sumitomo Chemical Co., LTD. His main research activities have been in the area of mechanical characterization, materials behaviour, processing improvement of composite materials.

Mr. TERUHO ADACHI recieved his

B.S and M.S. degree in hydro-
carbon chemistry from Kyoto
University in 1967 and 1969,
respectively. He is a research
associate at Osaka Research
Laboratory of Sumitomo Chemical Co.,LTD. He is currently
responsible for the research
and development of specialty
polymers.

Table 1. Properties of epoxy resins

|  | Viscosity P(at °C) | EEW(g/eq) |
|---|---|---|
| Bisphenol A Type |  |  |
| Sumi epoxy ELA-128 | 110~140(25) | 184~194 |
| " ELA-134 | semi-solid(25) | 230~270 |
| " ESA-011 | solid(25) | 450~500 |
|  |  |  |
| Amino epoxy resin |  |  |
| Sumi epoxy ELM-100 | 11.2 (25) | 105.6 |
| " ELM-120 | 70~100(40) | 110~130 |
| " ELM-434L | 50~80 (50) | 110~130 |
| " ELM-434 | 70~140(50) | 110~130 |
| " ELM-434HV | 200~250(50) | 110~130 |

Table 2. Cure with MHPA

| | No. | 1 | 2 | 3 | 4 |
|---|---|---|---|---|---|
| ELM-100 | | 100 | | | |
| ELA-128 | | | 100 | | |
| ELM-434L | | | | 100 | |
| ELM-434 | | | | | 100 |
| MHPA | | 145 | 85 | 135 | 130 |
| 2E4MZ | | – | 0.5 | – | – |
| Vicosity (25°C) *1 | | 153 | 503 | 775 | 1000 |
| (after 4hr at 25°C) | | 212 | 1040 | 955 | 1210 |
| (30°C) | | 136 | 445 | 560 | 620 |
| (after 4hr at 30°C) | | 156 | 720 | 655 | 840 |
| Gel.Time at 160°C(min) | | 5.6 | 1.8 | 7.4 | 5.3 |
| Cure conditon | | 160°C/24hrs. | | | |
| Tg (by TMA) °C | | 200 | 156 | 208 | 203 |
| Flexural Strength *2 kg/mm$^2$ | | 11.9 | 12.5 | 10.9 | 9.4 |
| &#x2033; Modulus &#x2033; | | 367 | 298 | 350 | 345 |

  *1 Brook-fild rotally viscometer was used.

  *2 ASTM D - 790


Table 3. Cure with DDS

| | No. | 1 | 2 | 3 | 4 |
|---|---|---|---|---|---|
| ELM-100 | | 100 | | | |
| ELA-128 | | | 100 | | |
| ELM-120 | | | | 100 | |
| ELM-434 | | | | | 100 |
| DDS *1 | | 58 | 33 | 52 | 50 |
| BF$_3$MEA | | 0.58 | 0.33 | 0.52 | 0.50 |
| Gel.Time at 160°C(min) | | 5.3 | 26.7 | 27.0 | 25.5 |
| Viscosity at 90°C(cp) | | 9.2 | – | 900 | 650 |
| Cure condition | | 180°C/3hrs | | | |
| Tg (by TMA) | °C | 261 | 206 | 222 | 250 |
| Tg (by DMA) | °C | 270 | 210 | 231 | 262 |
| Wet Tg (by DMA) *2 | °C | 260 | 203 | 210 | 260 |
| Flexural Strength | kg/mm$^2$ | 18.6 | 14.5 | 18.6 | 16.0 |
| &#x2033; Modulus | &#x2033; | 442 | 316 | 493 | 400 |
| Water Absorption(boil 2hr) | % | 12.4 | 0.93 | 1.45 | 0.96 |
| &#x2033; (boil 24hr) | &#x2033; | 5.23 | 2.83 | 5.83 | 3.98 |
| &#x2033; (boil 48hr) | &#x2033; | 6.57 | 3.09 | 7.46 | 5.18 |

  *1 Stoichiometric amounts of DDS were used.

  *2 Tg after boiling for 48hrs.

Table 4. Composite properties[*3]

| No. | | 1 | 2 | 3 |
|---|---|---|---|---|
| Resin System | | ELM-100 DDS[*1] BF$_3$.PP[*2] | ELM-120 DDS[*1] BF$_3$.PP[*2] | ELM-434 DDS[*1] BF$_3$.PP[*2] |
| Gel.Time at 180°C (min) | | 5 | 8.5 | 9.5 |
| Cure Conditions | | 180° C X 10kg/cm$^2$ X 6hr (Compression mould) | | |
| ILSS (dry) RT | (kg/mm$^2$) | 13.1 | 13.0 | 12.7 |
| 121°C | // | 9.2 | 8.1 | 8.7 |
| 177°C | // | 7.6 | 2.3 | 7.3 |
| ILSS (wet) RT | (kg/mm$^2$) | 11.3 | 8.3 | 10.1 |
| 121°C | | 4.9 | 1.4 | 5.7 |
| Flexural Strength (dry) RT | (kg/mm$^2$) | 210 | 197 | 213 |
| 121°C | // | 181 | 162 | 187 |
| 177°C | // | 141 | 64 | 155 |
| Flexural Strength (wet) RT | (kg/mm$^2$) | 204 | 188 | 200 |
| 121°C | // | 113 | 23 | 124 |
| Flexural Modulus (dry) RT | (T/mm$^2$) | 12.3 | 12.0 | 12.1 |
| 121°C | // | 11.5 | 11.4 | 11.7 |
| 177°C | // | 11.2 | 10.1 | 12.1 |
| Flexural Modulus (wet) RT | (T/mm$^2$) | 11.8 | 11.6 | 12.0 |
| 121°C | // | 10.9 | 4.3 | 11.3 |
| Water Absorption | ( % ) | 1.9 | 2.4 | 1.5 |
| Tg | (°C ) | 260 | 165 | 230 |

*1   0.8 equimolar amounts of DDS was used.

*2   Borontrifluoride piperidine complex.

*3   Normalized to 60% fiber volume.

CARBON FIBER/RESIN MATRIX INTERPHASE:
Effect of Carbon Fiber Surface Treatment & Resin
Toughness on Composite Performance

Cindy Megerdigian
Hysol AIP Division, The Dexter Corporation and Hysol Grafil
Pittsburg, California

Ron Robinson
Courtaulds Research and Hysol Grafil
Coventry, U.K.

Stan Lehmann
Hysol AIP Division, The Dexter Corporation and Hysol Grafil
Pittsburg, California

## ABSTRACT

The degree of adhesion between carbon fiber and resin matrix has been shown to be dependent on the type and level of fiber surface treatment. This study investigates the effect of varying the level of a commercial carbon fiber surface treatment on composite performance of both a brittle and a ductile matrix resin with a 228 GPa (33 MSI) modulus carbon fiber. Composite interlaminar shear, flexure, tensile, compression and shear modulus data are presented.

## 1. INTRODUCTION

The early challenges in composite technology were prepreg quality, processing and test method development to obtain consistently high quality composite performance. Although there remains room for improvement in these areas, recent significant research effort has been directed towards a descriptive model of fiber/matrix adhesion and composite failure mechanism to describe and determine those elements which might improve composite performance. Much has been learned. However, a descriptive model appears far from complete.

Matrix resin chemistry and carbon fiber chemistry are both fairly well understood in the bulk. However, the chemistry and failure mechanisms in the three dimensional interphase between bulk matrix resin, bulk fiber and the role of the interphase in composite performance need further research for adequate definition.

The interphase, 5 to 5000 angstroms thick depending on the system, lies between bulk matrix resin and bulk fiber. It is comprised of distinct layers [1].

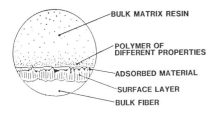

A layer on the surface of the fiber is distinctly different from the fiber bulk and is more highly ordered, i.e. higher modulus. The higher modulus outer layer is less reactive and has a lower surface energy than bulk fiber. The thickness of the high modulus layer is dependent upon the graphitization temperature during carbon fiber manufacture with higher temperature resulting in thicker layers. A layer of materials such as water and carbon dioxide is adsorbed on the high modulus surface layer. A layer consisting of a polymer of different properties may exist. It has been shown to exist when commercial sizings for carbon fibers, which are typically uncured epoxides, use a portion of the curing agent from the matrix resin. The resulting polymeric layer is under stoichiometry in curing agent and exhibits higher

modulus and lower fracture toughness [2,3,4].

The high modulus outer layer of a carbon fiber must be surface treated to obtain improved composite performance. Surface treatment is the last stage of the Hysol Grafil manufacturing process for converting special polyacrylonitrile fiber (SAF) to carbon fiber. The method is oxidative and details are proprietary. Surface treatment penetrates the high modulus exterior layer of the fiber and subsequently reacts polar hydroxyl, carboxyl, ether and other functional groups to the end planes of the polymeric carbon planes [5]. This process is described in detail elsewhere [6]. Significant literature exists which explores the influence of numerous types and levels of carbon fiber surface treatment. Much of this work has been done with surface treatment methods and matrix resins which are not commercially viable for aerospace applications. However, some work has explored practical structural systems and shows a significant variation in composite performance with varying fiber surface treatment [6,7].

This study explores the relative composite performance of a brittle matrix resin and a ductile matrix resin on carbon fiber with varying levels of surface treatment.

## 2. EXPERIMENTAL METHODS

### 2.1 Carbon fibers
Numerous investigators have explored many different surface treatments on carbon fibers on a laboratory scale.

### 2.1.1 Carbon fiber production
The continuous carbon fiber tow was made from Courtaulds special acrylic fiber (SAF). The SAF fiber was converted to carbon fiber on Hysol Grafil commercial carbon fiber production equipment during the course of a specially dedicated trial. The resultant carbon fiber was Hysol Grafil XA continuous tow with a nominal tensile strength of 3.8 GPa (550 KSI) and tensile modulus of 228 GPa (33 MSI). Fiber was collected unsized in order to allow characterization/composite testing to be conducted where necessary on virgin surface.

### 2.1.2 Carbon fiber surface treatment
Surface treatment of Grafil carbon fiber is accomplished by a proprietary treatment. The degree of surface treatment can be varied quite easily. Grafil XA fibers were manufactured with surface treatments varying from none (OX) through 4 times (4X) with 1 times surface treatment being equivalent to the normal level of surface treatment used in commercial production of Grafil XA fiber.

### 2.1.3 Carbon fiber sizing
All fibers used in this study were unsized.

### 2.2 Matrix resins
HG 9101, a complex epoxide based matrix resin formulation with several kinds of resins and curing agents, represents the low toughness (brittle) system with a neat resin compact tension $G_{IC}$ of 0.06 KJ/m$^2$ (0.35 in-lb/in$^2$). HG 9101 was cured in an autoclave for 1 hour at 177°C (350°F) with a 1.7°C (3°F) per minute heatup and 586,200 Pa(85 psi) pressure with vented bag.

HG 9106-3, a complex epoxide based matrix resin formulation with several kinds of resins, curing agents and second phase toughening, represents the high toughness (ductile) system with a neat resin compact tension $G_{IC}$ of 0.70 KJ/m$^2$ (4.0 in-lb/in$^2$). HG 9106-3 was cured in an autoclave for 4 hours at 180°C (355°F) with a 2.8°C (5°F) per minute heatup and 689,400 Pa (100 psi) pressure with vented bag. Neat resin cured properties for HG 9101 and HG 9106-3 are shown in Table 1.

### 2.3 Prepreg manufacture
Prepregs were made by doctoring a thin coat of matrix resin on a sheet of release agent coated paper. The resin coated paper was then placed on a drum winder with heated drum.

Tows of fiber were hoop wound onto the drum. The amount of resin and spacing of fiber tows was calculated to give 33±3% resin and 145±5 g/m$^2$ areal weight. The prepreg was removed from the drum winder and passed between two heated and pressurized rolls to further impregnate and spread the fiber bundles. Single layers were laminated to form the composite. Baseline materials were evaluated from unidirectional prepreg tape of 33±3% resin content and 145±5 g/m$^2$ areal weight manufactured on a commercial unidirectional prepreg machine.

## 2.4 Composite fabrication

All composites were fabricated by the Boeing proprietary string bleed method [8]. This method consists of the layup being surrounded by a 25mm (1 inch) edge breather with connection to a vacuum source. A single glass string is placed from opposite corners of the layup to the edge breather. All lay ups used nylon peel ply and a pressure plate for surface texture.

## 2.5 Composite testing

Most composite testing was done with the Instron Universal tester, Model #1125 and Instron Environmental Chamber, Model #3116 for elevated temperature testing. Testing was conducted on 0° tensile, 0° flexure, 0° interlaminar shear, 0° compression, shear modulus and glass transition. Composites were 62±2% fiber volume. The relative uncertainty for all composite testing was 5 to 8%.

### 2.5.1 Longitudinal tensile

Tensile testing was conducted in accordance with ASTM D638. Specimens were cut from an 8 ply, 0° laminate with 5 ply EA 9680 glass prepreg tabs bonded with EA 9689 supported film. The tensile specimens measured 216 mm x 13 mm (8.5 inch x 0.5 inch) with a 150 mm (6 inch) test area. Specimens were tested in the tension mode at a rate of 1.3 mm (0.05 inches) per minute.

### 2.5.2 Longitudinal flexure

Flexure testing was conducted in accordance with ASTM D790. Specimens were cut from a 13 ply, 0° laminate to a size of 13 mm x 100 mm (0.5 in.x 4 in.). Specimens were tested at a rate of 5.1 mm (0.2 inches)/minute in the 3 point bending mode with loading nose and center supports of 3.2 mm (0.125 inch) diameter at a span of 32 t (average thickness).

### 2.5.3 Interlaminar shear

Interlaminar shear (short beam shear) testing was conducted in accordance with ASTM D2344. Specimens were cut from a 13 ply, 0°

laminate to a size of 13 mm x 6.3 mm (0.5 in x 0.25 in). Specimens were tested at a rate of 1.3 mm (0.05 inches) per minute in a 3 point bending mode with loading nose and supports of 3.2 mm (0.125 inch) diameter at a span of 4 t (average thickness).

### 2.5.4 Longitudinal compression
Compression testing was conducted in accordance with ASTM D695. Specimens were cut from a 7 ply, 0° laminate with 14 ply, 0° laminate carbon fiber tabs bonded by Hysol EA9689 supported film. The compression specimens measured 81 mm x13 mm (3.2 in. by 0.5 in.) with a 4.8 mm (0.188 in) gage length and were tested at 1.3 mm (0.05 in) per minute.

### 2.5.5 Shear modulus
Shear modulus and glass transition data were obtained with a Rheometrics dynamic spectrometer, RDS 7700 Series II, in the solids rectangular torsion mode on 57 mm x 13 mm x 0.66 mm (2.24 in x 0.5 in x 0.026 in) specimens.

## 3.0 RESULTS AND DISCUSSION

Scanning electron photomicrographs of fiber to matrix adhesion and composite performance data were generated with carbon fibers of varying surface treatment for both EA 9101 and EA 9106-3 matrix resins.

### 3.1 Composite performance
Longitudinal flexural strength of both matrix resins at 22°C and 150°C showed increasing performance from OX up to approximately 1X surface treatment and remained the same or decreased slightly at higher surface treatment levels. See Figure 2. Longitudinal interlaminar shear and longitudinal compression showed the same performance versus surface treatment relationships seen with 0° flexural strength data. See Figures 3 and 4 respectively. Longitudinal tensile strength appears to rise very quickly from OX to 0.1X or 0.5X surface treatment level and then diminishes at higher surface treatment levels. See Figures 5 and 6. The very sharp increase in tensile performance is probably due to the partial removal of the exterior high modulus layer on the fiber which is fracture sensitive. Higher levels of surface treatment result in poorer tensile performance because increasing the surface treatment level increases adhesion between the fiber and matrix and the fiber is not fully loaded before failure occurs in the matrix.

Several generalizations can be made concerning the composite data. The increase in composite performance due to fiber surface treatment is less for 150°C testing than 22°C testing. This probably occurs because the matrix resin performance properties become the weak link. The HG 9106-3 ductile resin appears less sensitive to variations in fiber surface treatment than the HG 9101 brittle resin.

## 3.2 Fracture analysis

0° flexure specimens taken from those used to obtain the data shown in Figure 2 were failed and the tensile side was observed with the scanning electron microscope (SEM) to determine the relative fiber pullout and fiber to matrix adhesion. Photo 1(a) of HG 9101 brittle system on fiber with OX surface treatment shows much fiber pullout and no matrix resin adhered to the fiber. The crenulated fiber surface is typical of Hysol Grafil fibers. Photo 1(b), HG 9101 with XA fiber of 1X surface treatment shows much less fiber pullout and significant matrix resin adherence to the fiber. Photo 2(a) shows HG 9106-3 ductile resin with OX surface treatment XA fiber and 2(b) shows HG 9106-3 with 1X surface treatment XA fiber. Both HG 9101 and HG 9106-3 photo micrographs show much improved fiber to matrix adhesion with increased fiber surface treatment.

## 3.3 Shear modulus

Plots of G' and Tg for HG 9101 and HG 9106-3 on XA fiber of varying surface treatments are shown in Figures 7 and 8 respectively. Both G' and Tg clearly show variation with fiber surface treatment. The technical explanation for significant changes in G' at 0.5X surface treatment for HG 9101 and 4X surface treatment for HG 9106-3 are not understood. Other internal data infer that either surface energetics of the fiber or the spacial relationship of fiber functional groups and resin functional groups or both may be significant factors. These factors are also being explored as explanations of the marginal variation of Tg.

## 4.0 CONCLUSIONS

Composite performance of Hysol Grafil XA fiber varys significantly with varying levels of fiber surface treatment. The sensitivity of composite performance appears to be matrix resin dependent with ductile systems being less sensitive. Fiber to matrix adhesion determined with the SEM shows increases with increasing surface treatment. The optimal surface treatment level for XA fiber appears to be 1X to 2X. The optimal blend of composite properties appears to be a compromise between fiber dominated and resin dominated composite properties.

## 5.0 ACKNOWLEDGEMENTS

Among the many that have contributed to this work special thanks goes to

Jim Browne and Rocco Papalia of
Hysol and Greg Askew, Dave Wilford
and Rachael Coulthard of Courtaulds
Research.

## 6.0 REFERENCES

[1] L.T. Drazl, The Role of the
    Polymer Substrate Interphase
    in Structural Adhesion, AFML-
    TR-77129 (1977)

[2] L.T. Drazl, M.J. Rich, M.F.
    Koenig and P.F. Lloyd, J.
    Adhesion 16, 133 (1983)

[3] K. Selby and L.E. Miller, J.
    Mater. Sci. 10, 12 (1975)

[4] S.L. Kim, et al, Poly. Engr.
    and Sci. 18, 1093 (1978)

[5] H.P. Boehm, E. Diehl, W. Heep
    R. Sappok, Agnew. Chem. 3, 669
    (1964)

[6] S. Lehmann, R. Robinson and
    M.K. Tse, Proceedings of
    International SAMPE Symposium
    31, 291 (1986)

[7] S. Lehmann, C. Megerdigian and
    R. Papalia, SAMPE Quarterly
    (Apr. 1985)

[8] Boeing Material Specification
    BMS 8-212, Boeing commercial
    Airplane, Seattle, WA

[9] M.K. Tse, M.E. Hibbs and W.L.
    Bradley, Symposium on Toughened
    Composites, sponsored by ASTM
    D30 Committee and the
    Department of Defense, Houston,
    TX (Mar. 1985)

## 7.0 BIOGRAPHY

Stan Lehmann has worked in various
capacities in research and
development and quality control
since starting with Hysol in 1963.
He currently works on thermoplastic
resin matrices for composite
applications and carbon fiber-resin
matrix interphase. He holds a
Bachelor of Science in chemistry and
Masters in Business Administration
from California State University,
Hayward.

Dr. Ron Robinson is a project leader
at Courtaulds Research, Coventry,
England. He received a Doctor of
Philosophy from Salford University
in 1974 and performed a 3 year post
doctoral study of Fiber/Resin
Interaction at Sheffield University.
His current research is on carbon
fiber surface physical and chemical
characteristics.

Cindy Megerdigian is a Senior
Laboratory Technician for Hysol
Composites R&D. She has four years
experience in composites and polymer
chemistry. She holds a Bachelor of

Science in polymer technology from
California State University, Chico.

All work was performed for Hysol
Grafil Co., a joint venture between
The Dexter Corporation and
Courtaulds Ltd.

TABLE 1

NEAT RESIN CURED PROPERTIES

|  | EA 9101 | EA 9106-3 |
|---|---|---|
| Tensile strength @ 25°C, MPa (Ksi) | 41.0 (5.9) | 90.3 (13.1) |
| Tensile modulus @ 25°C, GPa (Ksi) | 3.1 (450) | 3.5 (501) |
| $G_{IC}$, compact tension KJ/m$^2$ (in-lb/in$^2$) | 0.06 (0.35) | 0.70 (4.0) |
| Glass Transition, Tg, Dry, °C (°F) | 213 (415) | 188 (370) |

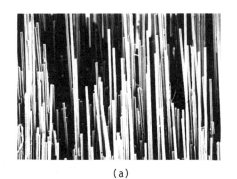
(a)

(b)

Photo 1.  SEM Photomicrographs of HG 9101/Hysol Grafil Unsized XA Fiber
Composite.  (a) 0 Times Surface Treatment at 150 Magnification
and (b) 1 Times Surface Treatment at 750 Magnification.

(a)                                    (b)

Photo 2.  SEM Photomicrographs of HG 9106-3/Hysol Grafil Unsized XA Fiber
         Composite.  (a)  0 Times Surface Treatment at 150 Magnification
         and (b) 1 Times Surface Treatment at 750 Magnification.

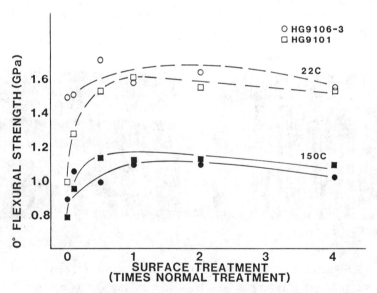

Figure 2.  Longitudinal Flexural Strength Versus Fiber Surface Treatment for
         HG 9101 and HG 9106-3 On Unsized Hysol Grafil XA Fiber.

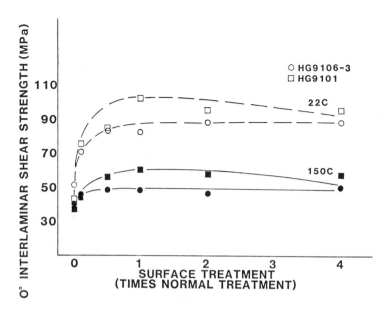

Figure 3. Longitudinal Interlaminar Shear Strength Versus Fiber Surface Treatment for HG 9101 and HG 9106-3 on Unsized Hysol Grafil XA Fiber.

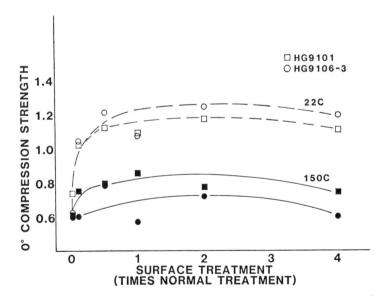

Figure 4. Longitudinal Compressive Strength Versus Surface Fiber Treatment for HG 9101 and HG 9106-3 on Unsized Hysol Grafil XA Fiber.

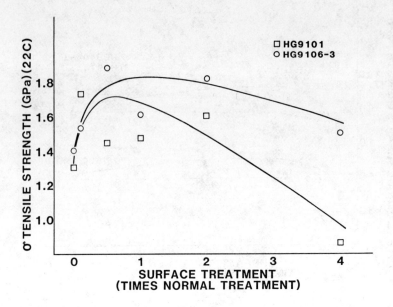

Figure 5. Longitudinal Tensile Strength Versus Fiber Surface Treatment for HG 9101 and HG 9106-3 on Unsized Hysol Grafil XA Fiber.

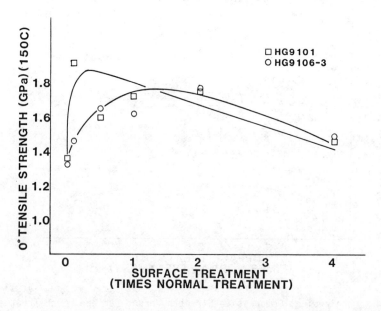

Figure 6. Longitudinal Tensile Strength Versus Fiber Surface Treatment for HG 9101 and HG 9106-3 on Unsized Hysol Grafil XA Fiber.

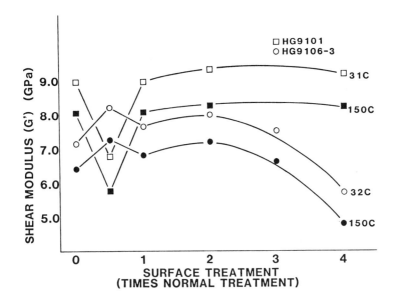

Figure 7.  Composite Shear Modulus Versus Fiber Surface Treatment for
HG 9101 and HG 9106-3 on Unsized Hysol Grafil XA Fiber.

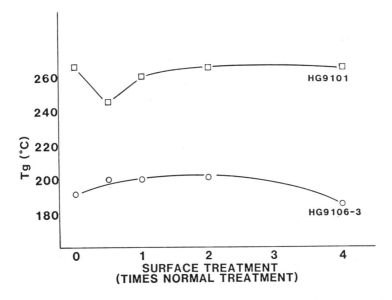

Figure 8.  Glass Transition (Tg) Versus Fiber Surface Treatment for HG 9101
and HG 9106-3 on Unsized Hysol Grafil XA Fiber.

32nd International SAMPE Symposium
April 6-9, 1987

The Automation of the Lay-up and Consolidation of
PEEK/Graphite Fiber Composites

Prof. Jonathan S. Colton
John Baxter, Jack Behlendorf, Tabassum Halim, Bryan Harris,
Gary Kiesler, Kuou-Tung Lu, Susan Sammons, Wanda Savage-Moore

Georgia Institute of Technology
George W. Woodruff School of Mechanical Engineering
Atlanta, GA 30332-0405

## ABSTRACT

The processing of PEEK/graphite fiber composites is characterized by an inordinate amount of hand labor. This greatly increases the cost, time, and inaccuracy involved in production. Automation of the lay-up and consolidation steps are a prerequisite for the widespread application and acceptance of these, as well as all, high performance materials. Automated techniques for the lay-up and consolidation of PEEK/graphite fiber composites have been developed and are presented in this paper. Discrete part lay-up schemes utilize robotic manipulators and specially constructed tables and allow any lay-up configuration to be automatically produced at high production rates. The size of the discrete parts was limited to 4 by 4 meters by design, not by any inherent difficulties. This size can be increased by enlarging the lay-up table. The length of a continuous sheet was limited by the length of the PEEK/graphite fiber prepreg material currently available. Its width was also restricted to 4 meters by designer, but also can be increased to the desired width. Continuous consolidation methods were designed for both discrete and continuous parts. Heating is performed by quartz radiative heaters and consolidation pressure is produced by pressure rollers. These processes were designed taking into consideration the size of the parts being supplied by the lay-up processes. Therefore, the lay-up and consolidation schemes form an integrated system.

## 1. INTRODUCTION

There is a considerable amount of governmental and industrial interest in the development of strong, light-weight, durable materials that can be easily processed. Materials with these properties are of notable interest to the aerospace industry due to the fact that weight reduction and fuel efficiencies are of utmost importance, as is the need for high environmental resistance. The automotive industry has also expressed similar interest in these materials [1].

Unfortunately, the processing of these materials, such as PEEK/graphite fiber composites, is characterized by an inordinate amount of hand labor. This greatly increases the cost, time, and inaccuracies involved in production. As a result, fabricated sheets of composites are subject to delaminations and microcracking. Automating the lay-up and consolidation processes will provide the consistancy required to produce high quality laminates at resonable cost. Thus, the objectives of this paper are as follows:

1) Establish the technical feasibility of automating the lay-up and consolidation of PEEK/graphite fiber.

2) Design an automated lay-up and consolidation process for an arbitrary configuration.

3) Design a process which fabricates a sheet of PEEK/graphite fiber, free of any delaminations or microcracks.

## 2. LAY UP

In order to fabricate sheets of PEEK/graphic fiber composites, layers of continuous carbon fibers, preimpregnated with PEEK (prepreg), must be laid up before consolidation into a final shape. During lay-up, the prepreg material is unwound from rolls, laid up in uniaxial layets, and tacked into place. Each layer can have a different axis of orientation than the layers directly above or below it in order to take advantage of the unidirectional strength and stiffness characteristics of the prepreg material. Tacking of the layers has been successfully accomplished with heat or ultrasonic welders. The completed lay up for a part, consisting of several layers tacked together, is then sent to the consolidation phase of the manufacturing process. While the entire lay-up process can be accomplished at room temperature, it requires considerable control in the placement of the prepreg layers to produce acceptable finished products.

### 2.1 Discrete Lay-Up Process Description

A discrete part lay-up process for composite sheet production must take into account the inherent properties of the composite. The raw material is supplied on rolls which are 0.1397 m or 0.2032 m in width, 0.2286 m in diameter, and 51 m long. The lay-up process must be capable of receiving, storing, and delivering the raw material to the work stations, preferably with a minimum amount of human intervention. Based on these considerations, the following discrete part lay-up process for PEEK/graphite fiber was designed. The proposed facility design (Figure 1) effectively eliminates manual handling of the rolls following delivery to the loading dock. PEEK/graphite fiber rolls are shipped into the loading dock. They are unloaded and placed into gravity feed storage bins at each lay up work station. A lay-up device (Figure 2) removes the rolls from the bins, loads itself and lays up the prepreg onto a lay up table (Figure 3), tacking the layers onto the layers below as it moves along. The table is designed to rotate to allow lay up in any orientation. As PEEK requires no special storage environment, the material is stored at room temperature close to the work station area. Once the number of plies has been fabricated, conveyors move the unconsolidated sheet to a buffer area to await consolidation.

### 2.2 Discrete Lay-Up Table Description

The table is designed to pivot about a centrally positioned shaft which is located underneath the table. The gear drive, thrust bearings, motor design, and cross section are shown in Figure 3. A square table with 4 m dimensions was chosen instead of a circular table with an equivalent diameter because the former represents a 36.3% reduction in the mass, thus reducing the forces needed to start and stop rotation of the table. Holes in the table top to allow a vacuum to be pulled via a pump located at the base of the hollow shaft. This vacuum is used to hold the first layer of prepreg in place during the lay-up process. The vacuum is on for the entire time the sheet is being fabricated.

Once a sheet is completed, the vacuum is shut off, and air pumped through the holes to loosen the sheet for transport to the conveyor.

## 2.3 Discrete Lay-Up Device Description

Perhaps the most complex piece of machinery in this system involves the actual lay-up device, which is shown in Figure 2. This intricate machine serves several purposes: it receives the raw material from the storage bin, aligns the various rolls of prepreg into the desired form, cuts the strips when they reach the required length, tacks the uppermost layer to the one directly beneath it, and removes the fabricated quasi-isotropic sheet from the lay-up table when the sheet is completed.

The lay-up device itself is designed to operate on overhead rails that provide the accurate guidance needed for placement of the PEEK/graphite fiber strips. Also, the lay-up device can rotate 180 degrees about this cart. This rotation allows for placement of the prepreg strips in two directions, increasing the throughput of the process by reducing the travel needed to complete one ply.

The mechanical arms which are used in the storage and retrieval of the prepreg rolls are attached to the lay-up device. Three rolls were determined to be the optimum number based on size considerations of the lay-up table. With three rolls in reserve on secondary arms at any time, as the primary arms become depleted, they exchange places with the secondary arms. The prepreg is then automatically threaded through the lay up machine and the lay-up process continues.

The rollers play an important role in the discrete part lay-up process. Aside from flattening the prepreg before it is tacked to the layer below it, the roller support shaft contains a piston device that provides an even pressure across the length of the roll. This piston can be used as a trigger device to let the lay-up

device know when a strip has been completed. Also contained in the rollers is a counter which keeps track of how much prepreg from a particular roll has been applied. When the counter reaches a predetermined value (for example, 45 m, the primary arm completes its current lay-up with the primary rolls and then exchanges with the secondary arms. This ensures that a roll will not end in the middle of a pass and that all three rolls are replaced at the same time.

A very important characteristic of the lay-up device is its ability to tack one layer of prepreg to the layer directly underneath it. This is accomplished by a series of ultrasound tack welders, located directly behind the pressure rollers. As the lay-up device moves along the guidance rails, the tack welder tacks the new layer to the layer beneath it. Once the weld is completed, the tack welder returns to its position aft of the pressure rollers, and the welding process begins again.

After a sheet has been laid up, it must be removed from the base table and placed on a conveyor which brings it to the consolidation step. The lay-up device's final component is a series of grippers, which are used to move the prefabricated sheet onto the transport system. These grippers have two degrees of freedom: along a vertical axis to reach the table level, and about a wrist joint to pick up the sheet. As the vacuum is reversed to loosen the sheet, then the grippers grip the sheet and move it to the conveyor for transportation to a buffer area to await consolidation. At that point in time, the lay-up device begins fabricating the next sheet.

## 2.4 Continuous Lay-Up Process

A continuous lay-up process was also designed to lay up enough quasi-isotopic sheets of PEEK/graphite fiber composite (Figure 4) to feed an average automobile plant. Considering the 4 m width, this value is 23,225 square meters of fabricated product assuming 24 hours/day production and 80% utilization. A sheet of eight layers was chosen as the basis for the design because two such sheets can be used to produce a standard 16 ply quasi-isotropic laminate. The continuous lay-up process consists of a main web transport system to lay the base layer and auxiliary stations (Figure 5) to lay up each additional cross ply as required. The main web consists of 27 rolls of 0.1397 m wide tape located on a single shaft. This number of rolls is required to provide a finished sheet approximately 4 m wide.

Each auxiliary station carries five rolls. The system includes dual reciprocating stations for each required direction. These track the main web alternately and allow the main web to move continuously while these heads lay tape across. A total of 12 auxiliary stations are required (two reciprocating stations for each of six layers) in addition to the main set and a duplicate main web system (to provide the final zero degree layer) to accomplish the required lay-up.

The material handling system for the process is operator assisted. Manually operated overhead cranes are used to remove shafts containing the depleted rolls from the machines and place them in drop bins, as well as to reload shafts with new material. Each station is equipped with an auxiliary shaft position so that reloading can be accomplished without interrupting the process.

## 2.5 Continuous Lay-Up Equipment Description & Operation

1. ROLL MOUNTING EQUIPMENT: The prepreg rolls are mounted on a roll holder which is then loaded in a top position (Figure 4). A secondary set of rolls is mounted in a bottom position. This set is provided as a spare so that, as the primary rolls run out, the secondary set of rolls can be spliced on with the splicing machine. A new set of rolls will then be loaded into the primary position.

2. LEADER: The leader is a fabric or a cloth liner which is in position throughout the length of the equipment whenever the latter is not in use. The leader required for this process is about 4 m wide and 57 m long. The leader is used to initially guide the tape through the equipment.

3. SPLICING MACHINES: The splicing machines (Figure 4) are used to maintain the continuity of the process. As the lay-up process is continuous, the prepreg rolls must be spliced to the next set of rolls. The main splicing machine is located between the mounted rolls and the accumulator. A small vacuum conveyor is used to feed the tape into the splicing machine. A combination of ultrasonic and optical sensors are provided to successfully operate the splicing machine. The sensors are used to activate the splicing machine indicating when it is to start cutting, to start splicing, to stop, and so forth.

The splicing system used is similar to those in the film industry. When a roll reaches its end, the splicing box is activated by signal sent by a photosensor monitoring tape presence. The ends are cut off to both the old and new tape to ensure a clean edge for bonding. There are two processes that could be used to join the edges. The first uses ultrasonic butt welding. Here, the two pieces of tape are brought together end to end, and an ultrasonic element is passed over them. If its strength is too low, a second process could be used which reinforces the bond. Instead of bonding the two tapes end to end, the tapes are bonded to a strip of PEEK laid across the joint. This would give a stronger weld because of the larger surface area.

4. ACCUMULATORS: Accumulators (Figure 4) are provided for the main tape, as well as for the cross tape, laying heads. The accumulators consist of both fixed and movable rolls which are mounted in a frame. The movable rolls take care of any slack in the tape. This slack is taken up while the

tape end is being held stationary for roll changing and splicing operations. The use of an accumulator allows continuous tape laying.

5. TAPE LAYING TABLE: It is on this table (Figure 4) that the tapes of different orientation are laid up. The table is provided with a intermittent set of idle rollers situated at regular intervals. These reduce friction between the sheet and the table. The idle rollers are provided only at intervals due to the fact that ultrasonic tacking requires a solid substrate.

6. CROSS TAPE LAYING MACHINES: These machines (Figure 4) will lay tape across the '0' orientation (or "main") layer. These machines are mounted on a pivoting mechanism which enables them to lay the tape at any specified angle and moves simultaneously along a short length of the lay-up table. Accumulators maintain the continuity of the process. Its subassemblies are descirbed below.

6a. GRABBER MECHANISM: The grabber mechanism pulls the tape across the width of the table. Sensors feed information through a microprocessor as to whether or not there is tape, when the mechanism should start traversing, when it should stop, etc. To accommodate for the movement of the bottom layer, as the tape laying head is traversing, the grabber mechanism tracks the movement of the main layer.

6b. CUTTING MECHANISM: The cutting and tacking mechanisms work in conjunction with the grabber mechanism, and are mounted on the same block. This block moves vertically via pneumatic cylinders and traverses along the length of the table where cutting and tacking is occurring. The cutting mechanism itself consists of two guillotine cutters.

6c. TACKING MECHANISM: The tacking mechanism forms an integral part of the cutting mechanism. The tacking of prepreg plies is accomplished ultrasonically. They must be tacked to prevent slipping during further processing. Tacking and cutting take place simultaneously; once completed, the head is lifted up.

7. VACUUM CONVEYOR: A vacuum conveyor is provided at the end of the tape laying table and pulls the laid up sheet through the process. It consists of a perforated belt with a vacuum chamber located underneath. Once the leader has been rolled up on the winding roll and the end of the leader is cut from the prepreg tape, the vacuum chamber is activated.

## 2.6 Continuous Lay-Up System Control

The control system for the continuous lay-up process has two main duties: synchronization of the tape laying machines with the main conveyor, and minimization of the gaps between the sheets. To accomplish this, a system of numerically controlled devices, ultrasonic sensors, and a dedicated microprocessor are utilized. A requirement for the process is to produce laminated sheets which have a void content of less than 0.5%. This requires that the side-to-side gap between any two sheets of tape be less than 0.0016 m. It also requires that the two sheets do not overlap, therefore the gap cannot be less than zero. Ideally, this would be accomplished by tracking the edge of the previously laid tape and constantly adjusting it to keep the gap at zero. Unfortunately, because the prepreg tape has very little lateral flexibility, these adjustments would tend to cause buckling, which in turn would create voids between the layers. Therefore, the tape machines are designed to lay the sheets along a straight line.

## 2.7 Design Issues

There are several factors which need to be addressed before the continuous tape laying system can be implemented. The first concern is that the tape length of the prepreg rolls. Presently, the rolls come in 51 m lengths. If this length is used, the rolls would need to be replaced, on the average, once every 2 1/2 minutes. One method to eliminate this problem is to request that the manufacturers of the prepreg tape supply rolls in longer lengths. This should not be too difficult as the rolls are made continuously and cut into manageable lengths for shipping. Increasing the tape width will help reduce void content, because of the reduction in the number of gaps in the final sheet. This also decreases the demand placed on the tape layers to minimize any gaps.

Product demand should also be considered. One of the advantages of the continuous lay-up system is that it can be adjusted easily. If wider sheets are desired, the lay-up machines could easily be widened and an extra conveyor added to carry the finished sheet. If sheets with more than eight layers are needed, the system could either be lengthened, or operated in a parallel/series arrangement with identical systems.

## 3. CONSOLIDATION

When the sheets of PEEK/graphite fiber are received from lay-up, they must be heated to a temperature above their melting point to achieve consolidation. Any heating method could be used which is capable of generating the required heat, but radiant infrared quartz tube heating was selected for the process because it is the most effective means of heating PEEK/graphite fiber sheets [3]. Other methods such as microwave, induction, and laser heating have been explored with limited success [3].

The continuous process under study consists of a tunnel oven with radiant heaters located on its top and bottom (Figure 6) which heat the sheets as they pass. The design questions which need to be answered are as follows:

1. How long does it take to heat eight plies of PEEK/graphite fiber?

2. How long is the oven?

3. What should the heater power, size, and location be?

4. Should several sheets of eight ply be heated separately and consolidated together to form a many ply sheet as required?

## 3.1 Numerical Methods for Heat Transfer Calculations

To address these questions, a computer program was developed using Turbo Pascal running on an IBM-PC. The program uses a one dimensional model to calculate the time required for the temperature of the center layer to reach 400°C [4]. The model was derived by treating each prepreg layer as an individual element. The first step in the calculation process is to determine the heat flow into and out of each element, using standard heat transfer equations. The model assumes that the temperature within each element is constant. The next step in implementing the model is to calculate the temperature of each element after a specified amount of time. If too large a time increment is chosen, the model becomes numerically unstable. A trial and error process is needed to choose a time increment small enough for stability and still yield a reasonable calculation time. After all of the temperatures are calculated the procedure is repeated until the desired inside temperature is achieved.

## 3.2 Heating Unconsolidated Sheets Prior to Rolling

The difference between heating an unconsolidated sheet and a consolidated sheet is that there is an air layer between each of the layers in the unconsolidated sheets. If the air layer is such that the sheets never touch, then the heat flow from an element is controlled by the conduction of heat through air. When the program is executed with a heat input of 50,000 KJ/m$^2$-hr and only conduction through air, it takes 105 minutes for the center layer if an unconsolidated sheet to reach 316°C, with a heat input of 1,000,000 KJ/m$^2$-hr, it takes 66 minutes to reach this temperature. This is far greater than the actual time needed to heat the sheets. More realistic air fractions of 0.1 and 0.2 are used for the remainder of the computer simulations because they produce results close to the actual values of 10-20 minutes.

If one assumes a heat flux of 270,000 KJ/m$^2$-hr and a line speed of 4.88 m/min, the required oven is 76.51 meters long. At 1,000,000 KJ/m$^2$-hr, the oven's length is 32.62 meters. An economic evaluation must be performed to determine if it is more feasible to use higher power heaters and a shorter oven or to buy lower power heaters and lengthen the oven.

The process design calls for manufacturing both continuous and discrete sheets. There is no difference between these two cases in the heating required prior to consolidation. Differences occur in the method by which the continuous or discrete sheets are supported as they travel through the process. In the oven, the two types of sheets can be handled with the same equipment. Wire mesh can carry discrete sheets to the consolidation rollers. Continuous sheets can be driven forward by consolidation and drive rollers.

## 3.3 Consolidation Rolling

Consolidation is the use of elevated temperatures and pressures to melt the resin, force out any entrapped air, and thereby consolidate the different layers of prepreg into a void free sheet. This process determines the quality of the product. Any void formation or residual stress may cause fracturing or weakening of the final material.

After the prepreg reaches thermal equilibrium, the sheets then go through a set of consolidation rollers (Figure 7). The use of rollers to provide the consolidation pressure produces a laminar resin flow in the consoidation direction. This will squeeze out any entrapped air and produce a void free laminate. At the beginning of consolidation, the internal and external temperatures of the sheets are approximately the same. During the transportation

from heater to rollers temperature gradients arise as the material cools. To correct for this, a set of infrared radiant heaters are installed directly before the rollers to maintain the composite's temperature [5]. Twelve inch diameter rollers applying a pressure of approximated about 1.378 MPa can be used to produce the necessary consolidation force [6].

## 3.4 Cooling of Consolidated Sheets

The entire sheet-forming process has been designed to provide 4 m wide sheets of varying lengths. These sheets will then be formed into discrete parts. Part forming requires heating above the melting point of PEEK at 343°C [7]. This causes the crystalline structures formed during the sheet fabrication process to melt. To ensure the proper crystalline morphology in the final part, the cooling rate during forming must be accurately controlled. As a result, control of the consolidation cooling rate for crystallinity control is not needed. The prime constraints on the choice of the cooling rate are the delamination of layers from excessively rapid cooling and the tradeoffs between slower cooling rates and equipment costs. These must be studied further.

## 3.5 Cooling Rate Model

A model of the cooling of the consolidated sheets, supported by more detailed finite element analysis, has also been developed [4] and is used to estimate equipment size and processing time. The model allows for heat sink temperatures, heat transfer coefficients, number of layers and physical properties to be varied to determine their effects. The model is described in detail above.

The model predicts that three different cooling stages occur. The first is a rapid cooling of the outer layers followed by the establishment of a profile of constant cooling rates. This implies that the entire sheet has experienced a similar thermal history and should be thermally uniform throughout. The final stage is an asymptotical approach to the heat sink temperature [4].

The model was used to predict values for the length of cooling equipment. Lengths of approximately 15 m result for the design line speed of 4.88 m/min. Consideration must be made to accommodate higher line speeds as a prime means of reducing costs in the future. The final process should be designed to include extra length or some other means to lower the heat sink temperature as a means of increasing the line speed.

## 3.6 Process Design

A double belt continuous consolidation process has been proposed for the cooling operation (Figure 8). The heated, consolidated sheets are fed into a set of rotating belts. These belts are mounted on adjustable rollers so that the required pressure for consolidation can be sustained. The adjustability can also be used to compensate for varying thickness of feed stock. The overall length of the consolidator will be determined by the number of layers, the maximum cooling rate, and the heat sink temperature, and can be calculated using the cooling model discussed above.

There are a number of points that must be considered in determining the design of the belts. The surface characteristics, heat transfer coefficients with PEEK/graphite fiber, resistance to thermal stresses, and resistance to adhesion to the PEEK/graphite fiber sheet of the belts will each affect the final sheet quality [3,8]. To determine the belt material a number of points must be compared. The cooling model's results clearly show that steel (a material with a high conductivity) provides a rapid cooling rate. However, the low compliance of a steel belt may allow air to become entrapped between the belt and the outer sheet layer [9]. This will lead to irregularities in the cooling profile, possible delamination of the outer layer, and a surface that is not smooth. Finally, the cost of a steel belt versus other materials must be considered. Thus, it is recommended that a more compliant material with a high tensile strength properties and a high heat transfer coefficient with PEEK/graphite fiber be used as the belt material.

The other major question in the design of this process is what media should be used to cool the belts. To estimate the amount of heat removal required, a steady state model approximation was used. Since the sheets are traversing with the belts, the temperature at any specific point is assumed to be constant. Thus the entire length is assumed to be at a constant temperature profile starting at $400^{\circ}$C and reducing to $220^{\circ}$C. At the latter temperature, the sheet has cooled adequately so that the application pressure is no longer needed to prevent delamination.

The average enthalpy of the sheet (429.3 kJ/kg) was used to determine the heat flow out of the sheet, 97,864 kJ/hr/layer. This heat flow is used to determine the heat transfer coefficients required. The results (coefficients between 70 and 700 $W/m-^{\circ}C$) are in the range of coefficients for forced air convection. Forced air cooling could be incorporated easily into the design of the consolidator by forcing air over the internal surfaces of the belts. Only sufficient space for the air handling equipment must be incorporated. If other heat transfer media are desired, a means of providing contact between it and the belts must be devised. Chilled rolls do not provide sufficient contact area to the flat areas of the belt because of their line contact area. Stationary plates containing the media could be used for cooling. They may also be used to assist in maintaining flatness across the belt width. However, the friction between the cooling plates and the belts may lead to large power requirements and inordinate wear.

After leaving the consolidator, the sheet must be cooled from 220°C to near 20°C for handling. It is recommended that this be accomplished by ambient air cooling under no pressure. The continuous sheets can then be cut to the desired length for the optimal layout of the final parts.

## 3.7  Cooling of Discrete Parts

The cooling of discrete parts raises several design concerns not addressed above. Flat discrete sheets can be processed with the same equipment as continuous sheets with the addition of equipment to feed and remove the sheets. A major concern with discrete sheets is the adjustment of the consolidation rollers and the cooling belts for varying sheet thicknesses. Two options are available to accomplish their processing. The first is to run the parts in lots of discrete thicknesses. The second would be to use bar codes to trigger adjustment of the consolidator height and speed for each part. This method raises questions as to the speed at which the sheets could be fed to the cooling belts after their rolling to remove the air. Time is required to make adjustments which could be reflected in the process as over or under cooling. This would lead to voids and delaminations; and therefore, the former method is the most acceptable.

The use of rotating dies instead of laminator belts has been proposed for discrete parts. While this method seems at first glance to be a nice compromise, the heat transfer and logistic problems involved in moving a large number of dies through a heating and cooling sequence need much more study.

## 4.  CONCLUSIONS

Technically feasible automated processes for the lay up and consolidation of flat PEEK/graphite sheets have been proposed. These processes are based on currently available technologies and require limited experimental work to verify their actual feasibility. The lay-up and consolidation processes can produce any arbitrary ply configurations and sheet thicknesses. Two lay-up processes have been proposed for discrete and continuous quasi-isotropic sheets. The consolidation process has been designed to accommodate the products of both lay-up processes.

Recommendations to consider in continuing this work are as follows:

1) Longer rolls of raw material are needed to reduce the excessive number of roll changes and reduce waste.

2) Experimental assessment of the lay-up processes' positioning accuracy.

3) Experiments to verify the feasibility of the consolidation process through the determination of actual heating times and the effectiveness of air removal through rolling.

## 5.  ACKNOWLEDGEMENTS

The assistance of Dr. Walt Cremens, Mr. Jay Shukla and Mr. Tim Greene of the Lockheed-Georgia Company in Marietta, GA is greatly appreciated.

## 6.  REFERENCES

1. "Carbon-Fiber Composites: A Light Weight Alternative," Mechanical Engineering, pp. 24-40, Sept. 1985
2. Thomas, Ralph H., Sr., Ultrasonics in Packaging and Plastics Fabrication, Cahneyz Books, Boston, pp. 93-99.
3. "Manufacturing Fabrication with Aromatic Polymer Composite: APC 2", ICI Provisional Data Sheet APC PD4, August 1984.
4. Blundell, D.J. and F.W. Willmouth, "Crystalline Morphology of the Matrix of PEEK-Carbon Fiber Aromatic Polymer Composites, III. Prediction of Cooling Rates During Fabircation," SAMPE Quarterly, to be published.
5. Cattanach, J.B. and R.C. Harvey, Documentation "Rolling Forming," APC: PEEK-Carbon Fiber Composite

Application Note, ICI.

6. "Making Consolidated Sheets From Aromatic Polymer Composite: APC 2," ICI Provisional Data Sheet, APC PD3, August 1984.

7. Blundell, D.J., J.M. Chalmers, M.W. MacKanzie and W.F. Gaskin, "Crystalline Morphology of the Matrix of PEEK-Carbon Fiber Aromatic Polymer Composites I. Assessment of Crystallinity," SAMPE Quarterly, Vol. 16, No.4, pp. 21-30 (1985)

8. "Adhesive Bonding", ICI Provisional Data Sheet, Pk. Pd. 11, September 1982.

9. Cattanach, J.B. and F.N. Cogswell, "Processing with Aromatic Polymer Composites," ICI Documentation.

## 7. BIOGRAPHY

Jonathan S. Colton received his Ph.D. in Mechanical Engineering from the Massachusetts Institute of Technology in February 1986. His dissertation concerned the development of a theoretical understanding of the nucleation of microcellular foam in thermoplastic polymers.

He joined the George W. Woodruff School of Mechanical Engineering at the Georgia Institute of Technology in November 1985. His research interests, in addition to bubble nucleation theory, include high performance thermoplastic composite processing and control, materials data base applications for design, and biomechanical device design and fabrication.

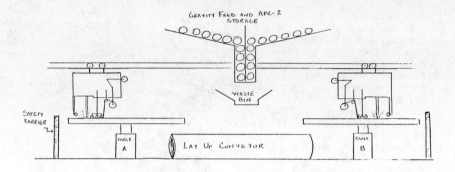

Figure 1:   Facility Design for Discrete Lay-up Process

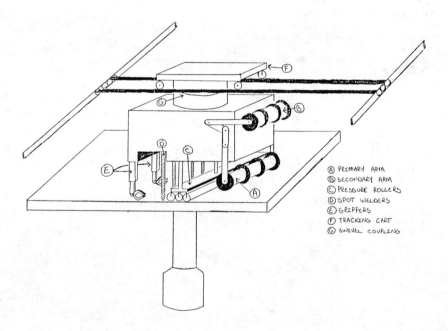

Ⓐ PRIMARY ARM
Ⓑ SECONDARY ARM
Ⓒ PRESSURE ROLLERS
Ⓓ SPOT WELDERS
Ⓔ GRIPPERS
Ⓕ TRACKING CART
Ⓖ SWIVEL COUPLING

Figure 2:   Discrete Lay-up Process Device

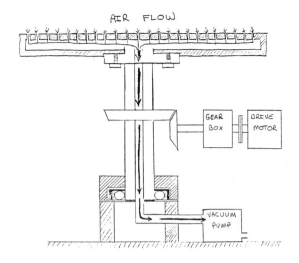

AIR FLOW

Figure 3: Discrete Lay-up Table Cross Section

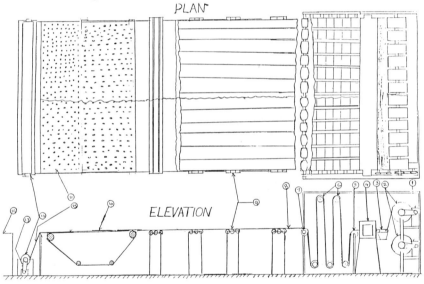

PLAN

ELEVATION

Figure 4: Continuous Lay-up Process Line  (key below)

1. Drive for APC-2 Roll Holders.

2. Primary (top) and Secondary (bottom) APC-2 rolls.

3. Vacuum Conveyor for Splicing Machine.

4. Splicing Machine.

5. Idle Roll.

6. Main Accumulator.

7. Crown Rollers

8. APC-2 main layer (bottom)

9. Idle Rollers of Lay up Table.

10. Main Vacuum Conveyor.

11. Perforated Conveyor belt.

12. Winding Roll for Leader.

13. Drive for Winding roll.

14. Conveyor Table for Consolidation.

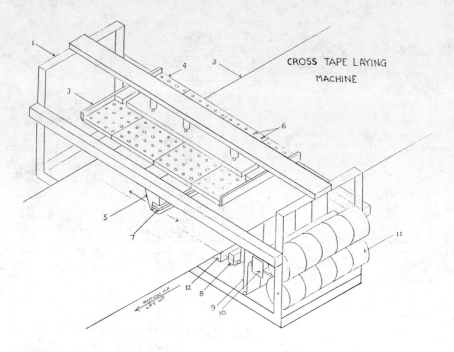

CROSS TAPE LAYING
MACHINE

Figure 5:  Continuous Lay-up Process Auxillary Station   (key below)

1. Tape laying unit, mounted on swivel base on floor.

2. Main conveyor, 12 feet wide, traveling at 17.5 feet/second.

3. Cutting unit, guillotine type.

4. Tacking / cutting unit.

5. Grabbing mechanism.

6. Locations of the ultrasonic sensors.

7. Jaws of grabbing mechanism.

8. Tape clamp.

9. Accumulator.

10. Splicing box.

11. APC-2 Rolls, each 5.5 inches wide.

12. Support table for clamping.

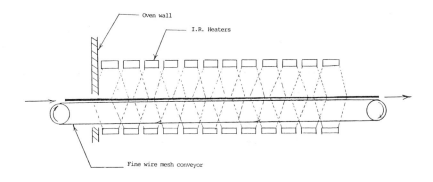

Figure 6: Consolidation Tunnel Oven

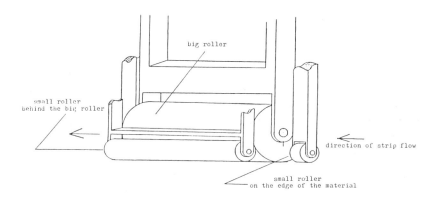

Figure 7: Consolidation Rollers

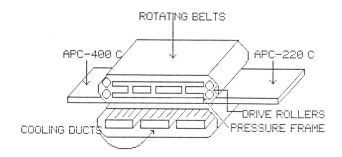

Figure 8: Consolidation Cooling Device

# AUTOMATED FABRICATION OF
# GRAPHITE - EPOXY COMPOSITES

Walter J. Kau
Michael W. Matson
Jonathan P. Russell

**Boeing Military Airplane Company**
**P.O. Box 3707**
**Seattle, WA 98124**

## ABSTRACT

The use of graphite-epoxy composite materials in advanced aircraft structures is discussed. The reasons for their use is given and reasons for automating the fabrication are reviewed and then contrasted with automated processes now in use at the Boeing Co.

The status of Direct Numerical Control at BMAC–Seattle is discussed.

Process monitoring for Quality Control is reviewed.

The evolution of the automated equipment is illustrated by several examples of continuing development and plans for second generation machines.

In conclusion, Boeing has made a commitment to automation of graphite-epoxy fabrication. Boeing's experience and lessons learned are reviewed.

## 1. INTRODUCTION

Graphite-Epoxy (G/E) composites are used in aircraft structures because they offer potential weight savings of up to 30% over metals. The resulting benefits are reduced fuel consumption and operation costs per passenger mile, increased payload, range and performance.

Briefly, Graphite-Epoxy materials are produced in a variety of forms. Unidirectional tapes have all the graphite fiber running in the longitudinal direction. Woven fabrics are available in a variety of styles, most common is a plain weave 0°/90° broad good. A bias weave fabric + 45°/ – 45° is also available for special requirements.

All of the material used at Boeing is preimpregnated with epoxy resin.

A drawback to the use of these materials is their higher cost per pound. The raw material can cost up to 17 times more than aluminum. To reduce the final cost of the product to an acceptable level, automation is being used. Other reasons for automating graphite epoxy fabrication are: configuration control for complex designs, completion of part layup within material out-time restrictions, and to reduce employee exposure to epoxy materials that can cause allergic reactions or dermatitis.

## 2. FABRICATION PROCESSES

The Boeing film "Composites Manufacturing" gives an overview of manufacturing processes used at Boeing's Developmental Center in Seattle for the fabrication of large composite aircraft structures. Manual methods for fabrication are shown. Manual layup involves cutting out the materials, locating plies, hand compacting each piece, bagging and vacuum compacting each ply. It is an extremely labor intensive procedure. Automated machinery that now does many of the same tasks is shown. Machines seen in the film are: Flat and Contour Tape Laminators (FTLM and CTLM), Channel Stiffener Manufacturing System (CSMS), Pultrusion, Autoclave, 5-Axis Router and automated inspection equipment.

The FTLM automatically dispenses, compacts, and cuts either 3-inch or 6-inch wide unidirectional tape. Flat parts for 737 empennage have been produced on this machine.

The CTLM is capable of laying 3-inch unidirectional tape over compound contour surfaces. The A-6 re-wing program is using the CTLM to produce skin panels.

The CSMS uses a tape dispensing head and side-ply formers to construct blade stiffeners.

The Pultrusion Machine produces long lengths of constant cross section parts. This replaces repetitive hand layup of clips, brackets and stiffeners with lengths of stock shapes. The pultrusion is only partially cured so that a length of a particular shape can be drawn from stores, trimmed to rough shape, cured on a lay-up manorel or co-cured to another part and finished.

The 90-foot autoclave is computer controlled so that part heatup rate can be carefully controlled and full process records are maintained.

The 5-Axis Router is used to trim the fully cured detail parts.

Boeing has several pieces of automated equipment for nondestructive test of finished parts. Through Transmission Ultrasonic (TTU) testing is done by passing ultrasonic sound through a part and measuring decibel changes. Streams of water transmit the sound to the part. There is a 10-axis TTU for large panels, a 10-foot TTU for smaller parts and a specialized blade and radius machine. All of these are computer controlled and provide a computer-generated printout showing indications of decibel change in the parts that can be interpreted as voids, porosity or other defects.

Boeing is in the process of converting to using computer-aided design to generate a data base that can be used by Manufacturing. This data base is becoming the sole authority for manufacturing in place of drawings. NC programmers can use the data base to generate part programs for the automated equipment.

### 3. DIRECT NUMERICAL CONTROL

Direct Numerical Control (DNC) is in use throughout the Developmental Center. Machine Controllers on the factory floor are fed the part programs through a Front End Processor from a Post Processor that translates the APT program to machine control language.

### 4. PROCESS MONITORING

Real time Process Monitoring is implemented on several machines. The Pultrusion Machine is a good example. The Process Monitor scans thermocouples, pressure sensors and a tachometer. The information is displayed on the CRT and if preset limits are exceeded a paper printout is generated for Quality Control.

### 5. FUTURE DEVELOPMENT

Boeing is involved in research efforts to improve productivity of the present automated processes. The existing Channel Stringer Manufacturing System has had the side ply formers extended allowing contoured side plies for the full length of the stiffeners.

A second generation CSMS is currently being developed. It will build two channel stiffeners simultaneously. While a ply is being laid on one Layup Mandrel (LM), the ply just laid on the other is being compacted. This system will be combined with an automated handling and positioning system.

The CTLM has been improved by the extension of its gantry rails and addition of a second tape head. This enables work on two parts simultaneously. New commercial CTLM's are being installed further expanding Boeing's capabilities for production. The new CTLM's are much faster, have bidirectional laying capability and are fully programmable in all axes.

Pultrusion is actively being worked with the implementation of I and T stiffeners in addition to the angles that have been in production for a year. Boeing funded development of the process is continuing with programs on full cure process, post-pultrusion forming and low cost tooling being started up.

### 6. LESSONS LEARNED

Boeing's experience with fabrication of composite parts shows that automation is best applied to large structures, difficult layups, and parts made from standard cross sections. Large, complex structures require automation so layup can be accomplished within the work life of the epoxy resins. The laying and cutting of the thousands of pieces of tape could not be accomplished by hand. Automation also assures configuration control. Once a part program is proved then each part thereafter is guaranteed to have all the material plies and ply orientations called for by the Design Engineer. A 3-fold reduction in production costs has been realized on large parts.

### 7. CONCLUSION

To summarize, The Boeing Company has concluded that for selected applications the use of composites for performance benefits is the best design solution and, in order to deliver these composite parts at competitive prices automation is required.

## 8. BIOGRAPHIES

Walter J. Kau is Chief of Aircraft Manufacturing Technology (AMT) for Boeing Military Aircraft Company in the Seattle area.

Michael W. Matson is Manager of the Automation Group within AMT.

Jonathan P. Russell is Lead Engineer responsible for Pultrusion and Composite Machining automation development efforts.

EVALUATING THE THERMAL PROTECTIVE INSULATION PROPERTIES
OF ADVANCED HEAT-RESISTANT FABRICS

Dr. Roger L. Barker and Dr. Young M. Lee
Department of Textile Engineering and Science
North Carolina State University
Raleigh, North Carolina 27695

Abstract

This paper discusses the thermal
insulation properties of high heat
resistant fabrics made with PBI,
aramid and blends of PBI with aramid
or with FR rayon fibers when exposed
to a variety of high intensity
radiant and convective heat
assaults. The thermal protective
performance of various fabric forms
is discussed in light of basic
thermophysical properties. The
contribution of heat transfer
mechanisms is discussed as well as
the effects of moisture and the
degradation behavior of different
materials in high intensity
heating.

1. INTRODUCTION

The most versatile laboratory
instrument available for testing
thermal protective insulation of
fabrics is a Thermal Protective
Performance (TPP) tester. This
device uses a combined flame and
radiant source in time controlled
exposures to measure a thermal
protective index. This research
uses this apparatus to explore
correlations between insulation and
fabric properties. The test
conditions used simulate a variety
of hazardous environments ranging to
highly intense exposures to radiant
and convective heat sources.

2. EXPERIMENTAL

2.1 Fabrics

Fabrics tested and their initial
properties are described in Table 1.
Test fabrics were chosen to include
several of the commercially
available "state-of-the-art" high
performance, thermally stable
materials. Woven, knit and nonwoven
constructions are tested in weights
ranging from 4.5 to 9 oz/yd$^2$.

2.2 Test Methods

2.2.1 Measurement of Thermal
Protective Performance

A Thermal Protective Performance
(TPP) tester (Custom Scientific

## Table 1. Test Materials

| Code | Fabric | Weight (oz/yd$^2$) | Thick- ness (mm) | Air Volume Frac- tion |
|------|--------|--------|--------|--------|
| | **Woven** | | | |
| W1 | PBI | 8.0 | 0.83 | 0.75 |
| W2 | PBI/Kevlar | 7.2 | 0.94 | 0.82 |
| W3 | Kevlar | 7.9 | 0.95 | 0.80 |
| W4 | Nomex | 7.5 | 0.90 | 0.80 |
| W5 | PBI/Rayon | 8.7 | 0.69 | 0.70 |
| W6 | PBI/Kevlar | 5.9 | 0.68 | 0.73 |
| W7 | PBI/Kevlar | 4.1 | 0.61 | 0.84 |
| W8 | Nomex | 4.8 | 0.64 | 0.81 |
| S1 | PBI/Nomex | 6.6 | 0.87 | 0.82 |
| S2 | PBI/Nomex | 4.2 | 0.81 | 0.88 |
| Z | PBI/Kevlar | 7.8 | 0.76 | 0.76 |
| | **Knit** | | | |
| K1 | PBI | 8.7 | 1.49 | 0.87 |
| K2 | PBI/Kevlar | 8.8 | 1.59 | 0.87 |
| K3 | Nomex | 8.8 | 1.47 | 0.85 |
| K4 | PBI/Kevlar | 5.1 | 1.53 | 0.92 |
| N1* | PBI/Kevlar | 4.5 | 2.03 | 0.95 |
| N2** | PBI/Kevlar | 4.5 | 2.03 | 0.95 |
| | **Felt** | | | |
| F1 | PBI | 8.7 | 2.23 | 0.91 |
| F2 | PBI/Kevlar | 8.3 | 1.82 | 0.89 |
| F3 | Nomex | 7.0 | 1.97 | 0.90 |
| F4 | PBI/Kevlar | 5.3 | 1.98 | 0.95 |

*Napped side toward heat source.
**Napped side toward heat sensor.

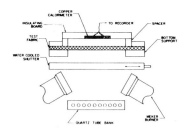

Figure 1. Schematic of TPP Tester.

Instrument Co.) was used throughout this study. The TPP tester consists of two meker burners and a bank of nine electrically heated quartz tubes controlled by powerstats (Figure 1). A pneumatic shuttering mechanism activated by a digital timer allowed control of exposure time to within 0.2 seconds. The heat transferred was measured by an instrumented copper calorimeter located 6.3 mm (1/4 in) behind the test sample. The calorimeter face was blackened and it was mounted in a maronite insulating board. A stainless steel spacer plate provided light transverse friction intended to simulate semi-restrained conditions existent in many clothing assemblies. The TPP tester was adjusted to deliver the desired heat output as indicated by the reading of the copper calorimeter. The balance of radiant to convective heat was determined using a commercial radiometer (Hy-Cal Model No. R-8015-C-15-072).

The rise in temperature of the calibrated copper calorimeter was recorded on a high speed strip chart recorder with a resolution of 0.1 second. The rates of temperature rise or the slope of the temperature vs. time trace is used in conjunction with calorimeter constants to compute the heat flux received. A square wave exposure sequence was used so that results could be related to the values obtained by Stoll [1]. A human tissue tolerance overlay, obtained by integration of the Stoll curve with respect to time, was used to determine tolerance times to second degree burns directly from the recorder trace.

## 3. RESULTS AND DISCUSSION

Table 2 summarizes thermal protective performance data for fabrics tested at nine different heat exposure conditions. Fabrics are rated by the time required for heat transfer to cause a second-degree burn on the reverse side of the fabric. This time, multiplied by the intensity level of the heat exposure, is used as an index of protective capabilities called the Thermal Protective Performance (TPP) rating.

Figure 2 shows the relationship between the intensity of the heat exposure, the percentage of radiant heat in the exposure and the protection afforded by various construction of PBI fabrics.

### 3.1 Effect of Exposure Conditions

The absolute intensity of the heat exposure is the overriding factor determining thermal protective insulation, as measured by protection indices against second degree burn injury (Figure 2). Substantial differences in fabric performance are observed at different exposure conditions. The predicted protection time is lowest when the exposure is to 100% radiant energy. In mixtures of radiant heat, the 50/50 condition is a less severe challenge than either the flame or purely radiant exposures. Flame and 50/50 tests give similar TPP results because of the comparable percentage of radiant energy generated by these sources.

Table 2. Thermal Protective Performance ($cal/cm^2$) of Test Fabrics Tested at 8.4 Watts/$cm^2$

| Fabric | Flame | 50/50 | Radiant |
|--------|-------|-------|---------|
| **Woven** | | | |
| W1 | 16.2 | 17.6 | 11.4 |
| W2 | 14.0 | 16.2 | 12.0 |
| W3 | 13.8 | 15.6 | 12.2 |
| W4 | 16.2 | 16.4 | 10.0 |
| W5 | 13.8 | 13.4 | 9.4 |
| W6 | 13.6 | 13.4 | 12.2 |
| W7 | 11.2 | 11.0 | 8.8 |
| W8 | 12.0 | 11.2 | 8.4 |
| S1 | 15.8 | 15.8 | 12.2 |
| S2 | 11.0 | 10.8 | 7.4 |
| Z | 15.4 | 15.4 | 12.0 |
| **Knit** | | | |
| K1 | 20.4 | 21.6 | 10.8 |
| K2 | 17.4 | 18.6 | 11.8 |
| K3 | 19.2 | 19.2 | 10.8 |
| K4 | 13.4 | 12.8 | 9.0 |
| N1 | 15.4 | 14.2 | 8.6 |
| N2 | 12.6 | 12.6 | 8.6 |
| **Felt** | | | |
| F1 | 26.2 | 28.4 | 20.6 |
| F2 | 23.0 | 26.0 | 19.8 |
| F3 | 17.0 | 19.8 | 14.6 |
| F4 | 20.2 | 20.4 | 14.2 |

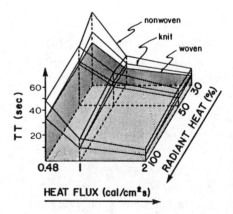

Figure 2. Relationship between predicted protection time and the intensity and percentage of radiant energy in the heat exposure.

The relative severity of purely radiant TPP tests is due to real differences in total energy emitted at the source and the energy detected at the copper sensor.

We believe that another reason for the apparent severity of a radiant exposure is the role played by the fibers protruding from the surface of the test fabrics. It is reasonable to assume that stagnant air layers entrapped by surface fibers play an important part in heat transfer, expecially in convective heat exposures. On the other hand, surface fibers are expected to pay a less important role in purely radiant exposures.

3.2  Effect of Fabric Type and
      Construction

The effect of fabric construction on protective evaluation can be summarized as follows:

1. Effective thermal thickness is the controlling factor in determining thermal protective performance. Bulk density plays a significant role, although the relationships between density and thermal protection are not simply linear. Fabric weight is an important property determining thermal protective insulation: lighter weight fabrics provide less protection than heavier fabrics.

2. Fabric porosity, measured by air fraction, determines heat transfer in intense exposures: air and fiber conduction dominate in dense and heavier weight woven and nonwoven fabrics. Light weight woven and knit fabrics allow convective and radiant heat to penetrate through porous areas in their structure.

3. Ounce for ounce, nonwoven constructions provide better thermal protection than knit or woven structures. We believe that this is largely due to thickness, including the contribution of surface fibers, and especially to the lower bulk density. Nonwovens afford fewer air channels than knits, thereby minimizing radiant and convective transport through the fabric.

Our data show that differences in fabric constructions (e.g., thickness and weight) have greater influence on the TPP of single-layer, thermally stable materials than does difference in fiber content. However, it is significant that relative comparisons of the TPP of fabrics made with different fibers (for example, PBI vs. Kevlar®) depend on the nature of the exposure condition. We suspect that reversals in comparison of TPP based on fiber type is related to the changing role of moisture in different types of heat exposures.

3.3  Moisture Effects in Radiant
      Exposures

Our experiments show that the presence of moisture in the fabric lowers thermal insulation in high intensity exposures to purely radiant heat energy. Calorimetric traces (Figure 3) show that the rate of heat transfer of dry and moist fabrics vary throughout the exposure.

Initially, the greater heat absorbing capacity of moisture lowers the rate of heat transfer through wet fabric (stage 1 in

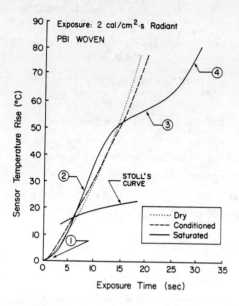

Figure 3. Typical sensor temperature-time trace for preconditioned PBI fabric in high intensity radiant exposure.

Figure 3). The fact that slowing in fabric heating does not translate to give wet fabrics superior thermal protective performance must be due to the effect of vaporized moisture in increasing the efficiency of energy transmission from the fabric. (This is indicated by the rapid rate of heat transfer through wet materials (stage 2 in Figure 3). It is reasonable to expect moisture, evolving from the back surface of heated fabric, to increase the efficiency of the energy transfer in the 1/4-inch air gap located between the fabric and the heat sensor. Therefore, the higher rate of

heat transfer produced by vaporized moisture dominates in tests against purely radiant high intensity thermal energy.

The rate of heat transfer through wet fabric decreases following a period of rapid gain (stage 3 in Figure 3). This is probably due to revaporization of water from the heated surface of the copper calorimeter. In the late stages (stage 4) of this high intensity TPP test, the fabric temperature reaches the level at which PBI degrades (>600°C). Degradation produces hot volatile products which contribute to increase transfer to the heat sensor. This is evidenced by a sharp rise in the calorimetric trace following 25 seconds of exposures.

3.4  Moisture Effects in Combined Radiant/Convective Exposures

Moisture improves TPP in high intensity (2.0 cal/cm$^2$·sec) exposures, provided the heat exposure contains a large component of convective energy. A calorimetric trace (Figure 4) shows that the rate of heat transfer to the heat sensor is considerably slowed by wetting the fabric prior to testing against a heat source containing a high flame component. We believe that this is evidence of the ablative effect of the moisture in which thermal energy is removed from the fabric and carried away from the front surface by the convective action of the flames.

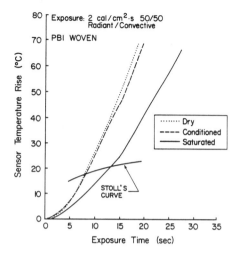

Figure 4. Typical sensor temperature-time trace for preconditioned PBI fabric in combined radiant/convective exposure.

## 3.5 Measurement of Fabric Property Changes in TPP Exposures

An interrupted exposure approach was used to obtain a series of test fabrics exposed for incrementally (3 sec) increasing duration. These samples were then analyzed offline, using various techniques for measuring physical properties. Figure 5 shows how fabric mass and thickness change during a TPP exposure. These data show that intense heating causes drastic changes in the thermophysical properties of heat resistant fabrics. The changes we observe can be summarized as follows:

1. Heating diminishes the mass, thickness and density and increases the heat capacity of these materials. The relationship between the retention of transient thermophysical properties and thermal protective performance is critically driven by the nature of the heat exposure.

2. Two factors are crucially important in determining the response of heat resistant materials in TPP tests: one is moisture in the fabric and its effect on fabric heating and transmission. The other is the retention of effective thermal thickness. Moisture slows fabric heating in intense radiant or convective exposures, although the rate of heating is greater in a convective assault. Degradative losses in thermal thickness can be related to the effects of the exposure environment on reduction in the number of fibers on the surface of the fabric. In this regard, thermal oxidation and decomposition of surface fibers are much more severe in tests that allow direct flame contact with the fabric.

Specific comparisons of the TPP of heat resistant materials made on the basis of different fibers and fabric constructions derive from an understanding gain concerning the effects of moisture. Therefore, woven PBI fabrics provide superior performance in convective tests because of the ablative moisture losses and the somewhat greater thermal stability of fibers (in comparison to aramids). Significantly, moisture in PBI is clearly demonstrated by monitoring fluctuation in fabric mass. The role of surface fibers has been documented by microscopic examination of exposed fabric surfaces (2). Kevlar® fabrics show an increase in thickness after a purely radiant exposure. Moreover, fabrics with lower moisture regain (e.g., Kevlar®) continue to have an ability to retain TPP in radiant tests.

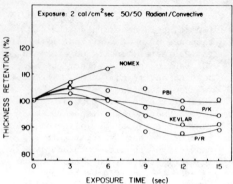

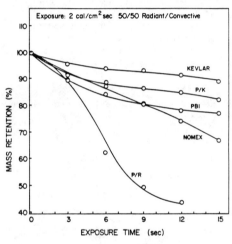

Figure 5. Transient properties retention of fabrics exposed to 8.4 watts/cm², 50/50 radiant/convective heat.

## 4. CONCLUSIONS

This research demonstrates that exposure conditions and fabric properties affect the thermal protective performance and control heat transfer mechanisms through single layer high-performance thermally-stable fabrics.

Comparisons made in this study should be useful for the comparative selection of heat-resistant materials for protective apparel and other applications where insulation against intense heat is important. They also provide foundation for deeper analysis explaining the relationships between fabric properties and thermal insulative performance.

## 5. AUTHORS' NOTE

Care must be taken in deriving conclusions concerning safety benefits from these data. These data describe the properties of fabrics in response to the controlled laboratory exposure and conditions that are specified. They are not presented to predict the actual field conditions where the nature of the thermal exposure can be physically complicated and unqualified. The authors wish to emphasize that it is not their intention to recommend, exclude or predict the suitability of any commercial product for a particular end-use. Neither do they claim that the test samples are unqualifiably representative of given fabric types or that they represent the best fabrics that have been developed for protective applications.

## 6. ACKNOWLEDGEMENTS

This research was funded by Celanese Fibers Operations. We are grateful for advice and material support received from technical personnel at this organization, especially T. E. Schmidt, M. T. Stanhope and N. Byars. We also acknowledge the contribution of Celanese in providing fabric samples for analysis.

## 7. REFERENCES

1. Stoll, A. M. and Chianta, M. A., "Method and Rating System for Evaluation of Thermal Protection," Aerospace Medicine, Vol. 40, No. 11, 1232-1238 (1969).

2. Lee, Young M., "Analysis of Thermal Protective Performance of Heat Resistant Polymeric Fabrics," Ph.D. Dissertation, North Carolina State University, Raleigh, NC, 1985.

## 8. BIOGRAPHIES

Roger Barker is an Associate Professor in the Department of Textile Engineering and Science at North Carolina State University. He holds B.S. and M.S. degrees in Physics from the University of Tennessee and a Ph.D. in Textile and Polymer Science from Clemson University. He has been an Assistant Professor at Cornell University. His industrial experience includes research with DuPont in the area of spunbonded materials. His research interests include measurement of the mechanical and surface properties of textiles, heat transfer through textiles and heat resistant fabrics.

Young Lee is a postdoctoral Fellow in the Department of Chemical and Environmental Engineering, Rensselaer Polytechnic Institute. He holds a Ph.D. in Fiber and Polymer Science from North Carolina State University. He also holds an M.S. degree in Industrial Chemistry and a B.S. degree in Polymer Engineering from Hanyang University, Seoul, Korea. His research interests include heat resistant materials and the permeability of polymeric membranes.

32nd International SAMPE Symposium
April 6-9, 1987

# CERAMIC FIBER PRODUCTS
## AS AN ASBESTOS REPLACEMENT FOR ELECTRICAL AND THERMAL INSULATION

Edward M. Fischer
Ceramic Materials Department
3M Company
3M Center, St. Paul, Minnesota

## ABSTRACT

Health hazards associated with Nextel 312 and 440 ceramic fibers are judged to be minimal based upon current theory and knowledge.

These fibers from 3M are continuous metal oxide fibers suitable for producing textiles such as sleevings, tapes, fabrics, threads, and overbraids.

Products made with these fibers retain strength and flexibility above 1200°C.

Applications for textile products made of Nextel ceramic fibers have been developed for the aerospace, industrial and electrical markets.

## 1. INTRODUCTION

Sizeable research efforts have been made on asbestos substitutes for products such as clutch facings, brake linings, papers, roofing products, filters, gaskets, seals, conveyor belts and blankets.

Textile-grade aramid, fiberglass, fused silica, and leached fiberglass have been found to be suitable replacements for low and intermediate temperature asbestos applications.

The 3M Ceramic Fiber Laboratory has been developing a family of high performance textile-grade products that are candidates for high temperature asbestos replacements.

This paper discusses the health safety aspects, the physical properties and describes some of the non-structural uses of Nextel ceramic fiber products.

## 2. HEALTH SAFETY ASPECTS

Health hazards associated with the inhalation of Nextel 312 and 440 fibers are judged to be minimal based upon current theory and knowledge.

A typical Nextel 312 or 440 fiber may be characterized as having a large diameter (7 to 13 microns) in addition to great length. As a result, these fibers would not be considered to be in the respirable range.

Data compiled at the Fulmer Research Institute indicates that fibers between about five microns and 100 microns long and about two microns or less in diameter are more suspect in causing health problems due to inhalation. Small diameter fibers from materials such as asbestos can be inhaled into the lungs and may cause fibrosis or cancer.

While no health standards exist for employee exposure to Nextel 312 or 440 fibers, a comparison can be made with the current National Institute for Occupational Safety and Health (NIOSH) recommended standard for fiberglass exposure. The OSHA asbestos fiber standard does not allow more than two fibers per cubic centimeter of air based on a count of fibers greater than five microns in length. The NIOSH recommended standard for fibrous glass states that no more than three fibers per cubic centimeter of air having a diameter less than 3.5 microns and a length greater than ten microns shall be present in the workplace air as a time-weighted average concentration for up to a ten hour workshift in a 40 hour work week.

The 7 to 13 micron diameter Nextel 312 and 440 fibers are outside of the diameter range of fibers considered by the recommended standard. In addition, the concentration of airborne Nextel 312 or 440 fibers monitored in typical processing areas have been found to be very low and well within the proposed NIOSH standard.[1]

## 3. PROPERTIES

### 3.1 Physical Properties

Nextel 312 and 440 fibers are continuous metal oxide fibers suitable for producing textiles such as sleevings, tapes, fabrics, threads, and overbraids. Individual fibers are transparent, smooth, and continuous (see Figure 1). The fiber compositions are aluminum oxide, boron oxide, and silicon oxide.

Table 1 lists some of the physical properties of the Nextel 312 and 440 fibers compared to other textile grade continuous fibers.[2]

### 3.2 Temperature Limits

All of the textile-grade continuous fibers have maximum use temperature as indicated in Figure 2.

The difference in strength retention versus temperature of Nextel 312 and 440 fibers is shown in Figure 3 and has been reported separately by Sawko of NASA-Ames.[3]

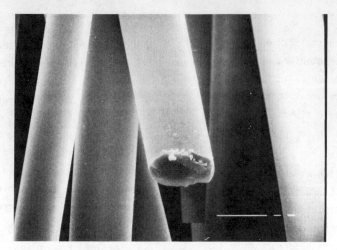

## Figure 1: Individual Nextel Ceramic Fibers
## (2000X)

### Table I: Physical Properties of Continuous Oxide Fibers

| PROPERTY | NEXTEL 440 | NEXTEL 312 | FUSED SILICA | LEACHED SILICA |
|---|---|---|---|---|
| Composition (% by Wgt.) | | | | |
| $Al_2O_3$ | 70 | 62 | | |
| $B_2O_3$ | 2 | 14 | | |
| $SiO_2$ | 28 | 24 | 99.95 | 97.9 |
| $ZrO_2$ | | | | |
| Misc. | | | | 2.1 |
| Density ($gm/cm^3$) | 3.05 | 2.7 | 2.2 | 2.1 |
| Diameter (avg μm) | 11 | 11 | 9 | |
| Tensile Strength (GPa) | 2.14 | 1.72 | | 0.21–0.41 |
| (KSI) | 300 | 250 | 500 | 30–60 |
| Tensile Modulus (GPa) | 200 | 152 | 69 | 72 |
| (MSI) | 28 | 22 | 10 | 10.5 |
| Melt or Liquidus Temp (°C) | >1800 | 1800 | 1700 | >1760 |
| Coefficient of Thermal Expansion $10^{-6}$/°C | 5 | 3.5 | 0.54 | |
| Dielectric Constant (10 GHz) | 5.7 | 5.4 | 3.8 | |
| Loss Tangent (10 GHz) | 0.015 | 0.018 | | |
| Resistivity ($\Omega \times cm$) (20°C) | | | $10^{18}$ | |
| Refractive Index | 1.617 | 1.572 | 1.459 | |

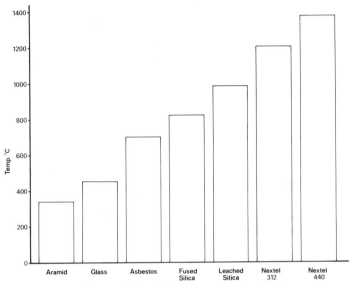

Figure 2: Textile Use Temperature Limits

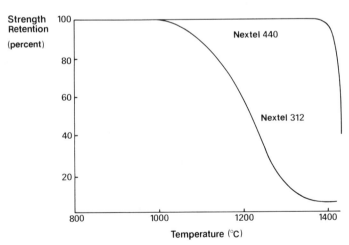

*Samples treated three hours at temperature, tested at room temperature,
3-inch gauge length, warp direction 5HS weave, 25 oz/yd²

Figure 3:    Fabric Breaking Strength versus
             Heat Treatment Temperature*

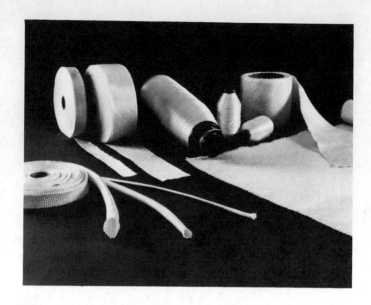

**Figure 4: Nextel 312 Textile Products**

## 4. APPLICATIONS OF NEXTEL CERAMIC FIBER

Two significant properties of Nextel fibers are their textile-grade quality and their high temperature resistance.

Consequently, the textiles produced with the Nextel fibers include fabrics, tapes, sleevings, belts, 3-D woven parts, drop warp tape, gaskets and wire overbraids. Figure 4 shows the variety of textile structures which can be produced. Due to the continuous fiber nature and handleability of Nextel yarns it is possible to design blended textile forms with lower temperature fibers that add additional mechanical properties to the textile product.

Applications for textile products made of Nextel fibers have been developed to fill needs in the aerospace, industrial and electrical markets.

### 4.1 Firewall Blankets (Aerospace)

Blankets for aircraft firewalls offer high temperature protection to engine struts and for cowls on jet engines. A multi-layered firewall blanket containing a fabric of Nextel 312 fibers was developed for this job.

The blanket consisted of quilted Nextel 312 fiber fabric, bulk fiber insulation and fiberglass cloth which was covered with a silicone-coating. To protect the blanket from chemical damage and abrasion, the Nextel 312 fabric side was also shielded with aluminized fiberglass.

Figure 5: Nextel 312 Fabric on Jet Engine Fan Cowls

The blanket significantly reduced maintenance costs because of its durability, ease of installation, and longevity, and if a fire should occur, Nextel 312 fabric has been proven in tests to meet the FAA's 2000°F 15-minute flame penetration requirement.

Boeing is currently using this blanket (see Figure 5) containing Nextel 312 fabric to protect composite fan cowls on their 757-200 and 737-300, and the underside of the engine and air inlet ducts on their 757-200.

## 4.2 Vacuum Furnace Linings (Industrial)

The high temperatures involved in heat treating tool steel, dies, injection molds, and carbon materials cause most linings to fail and need replacement about once every two years. Many heat treating firms are choosing a vacuum furnace lining of Nextel fabric and a bulk insulation. Repair time is cut down significantly. If the fabric is damaged, the bad piece can be cut out and replaced by a new piece. Nextel fabric does not oxidize, so the material will not suffer damage if the furnace is opened before cooling down completely. The Nextel fabric retains its strength and flexibility with high resistance to thermal shock, low shrinkage, and low specific heat.

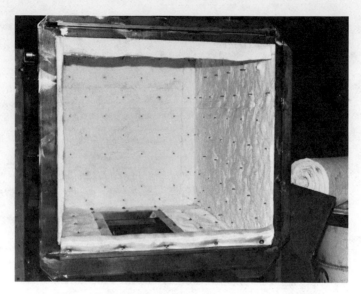

**Figure 6:** **Nextel 312 Fabric Lining in Vacuum Furnace Inner Box**

Figure 6 shows a typical vacuum furnace inner box containing a Nextel fabric lining.

### 4.3 Gaskets (Industrial)

For applications over 550°C where a textile type gasket is needed the Nextel fiber materials can be used. Two examples of such gaskets are woven drop warp and tadpole styles (see Figure 7). These examples are currently being used in industrial and vacuum furnace applications.

### 4.4 Electrical Insulation (Electrical)

Continuous ceramic fibers have been used as an asbestos substitute for braided insulation on thermocouple products and aircraft fire zone wire insulation. Most major thermocouple producers supply type K wire with Nextel fiber insulation because of its high electrical resistance (see Figure 8) at elevated temperatures.

High temperature platinum thermocouples fail prematurely when in contact with silica containing insulation materials. The silica reacts with the platinum destroying the usefulness of the platinum thermocouple. Nextel 440 fiber sleevings have been used successfully as insulation for platinum thermocouple and resistance heating wires. No reaction with the platinum has been observed.

### 4.5 Other Applications

High-temperature-resistant textiles of the 3M fibers are being used in aircraft/aerospace for composite firewalls, manifold line protection,

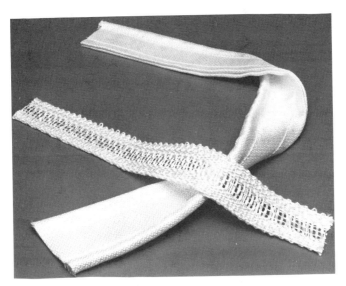

Figure 7: Nextel 312 Woven Drop Warp and
Tadpole Gaskets

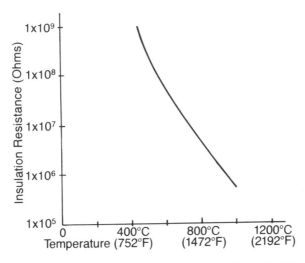

Figure 8: Insulation Resistance of Nextel 312
Braid on Thermocouple Wire (In Air)

and ceramic reinforcement; in industry for flame curtains, furnace zone dividers, tube seals, metal filtration, hot conveyor belts, duct linings, pipe wraps, fire sleeves, flue sleeves and expansion joints; and in the electric field for wire and cable insulation, and heating element and stress relief blankets.

New uses being developed for these fibers include high-temperature toughened ceramic-ceramic composites made by coating fibrous shapes or preforms, such as braided or woven tubes of the fibers, with silicon carbide via chemical vapor deposition techniques as well as high temperature filter bags, and hybrid fabrics.

## 5. CONCLUSION

Nextel ceramic fiber products have been determined to be non-hazardous based on current knowledge and theory. Two distinct advantages of these materials are their textile quality and high temperature resistance. They have been tested and found to be the highest temperature resistant materials of the candidate non-structural asbestos replacements.

## 6. REFERENCES

1. Industrial Hygiene and Toxicology Aspects of 3M Nextel 312 Ceramic Fibers by Mr. Robert S. Larsen and Mr. William McCormick, 3M Company, St. Paul, Minnesota, Proceedings of the National Workshop on Substitutes for Asbestos, November, 1980.

2. H.G. Sowman and D.D. Johnson, "Ceramic Oxide Fibers", Ceramic Engineering and Science Proceedings, Vol. 5, No. 9-10, September-October 1985.

3. P.M. Sawko and H.K. Tran, "Strength and Flexibility Properties of Advanced Ceramic Fabrics", SAMPE Quarterly, Vol. 17, No. 1, October, 1985, pp 7-13.

## 7. BIOGRAPHY

Edward M. Fischer is a Senior Technical Service Technologist in the Ceramic Fiber Products Laboratory at 3M. Nine of his twelve years experience in that laboratory have included major responsibilities in technical service and new product development. This involves the designing and processing of new ceramic fiber textile forms and most recently the development of hybrid textile structures for industrial use. He has been awarded two U.S. Patents entitled "Composite Sewing Thread of Ceramic Fibers" and "High Temperature Oil Boom Cover Blanket".

# KEVLAR® ARAMID PULP, A SHORT FIBER REINFORCEMENT TO REPLACE ASBESTOS

David E. Hoiness
Arnold Frances
E. I. du Pont de Nemours & Company, Inc.
Chestnut Run, Building 701
Wilmington, Delaware   19898

## Abstract

During the mid-1970's, industry was urgently searching for alternatives to asbestos. These alternatives required cost effectiveness, processability on existing manufacturing equipment, in-use performance equivalent to or better than asbestos, and nontoxicity. The potential value of Kevlar* fiber was recognized but the cost barrier, 50 to 100 times as expensive as asbestos, seemed insurmountable. However, the development of a special engineered form, a short fiber with attached fine fibrils, led to the use of Kevlar® as a primary reinforcement in many applications. This paper will discuss the development of aramid pulp to fit an industry need and the laboratory and field testing done to provide realistic cost considered value of this fiber for use in automotive and truck brakes, clutches, gaskets, adhesives, sealants, and short fiber reinforced elastomers.

## 1.  INTRODUCTION

Asbestos is a mineral fiber having many excellent characteristics that led to its use in many applications such as friction products and gaskets. These properties include: hair-like fibers, high tensile strength, flexibility, heat and chemical resistance as well as a stable coefficient of friction with increasing temperature. However, well-documented evidence that asbestos was linked to serious health problems implied future government action which could threaten availability. Consequently, a concerted search for substitutes was undertaken by industry in the mid-1970's.

The favorable cost/performance features of asbestos alone made successful substitution difficult to achieve.  For example, no one substitute material could match the several useful properties available in asbestos, particularly in friction products.

Alternative fibers require cost effectiveness, processability on existing manufacturing equipment, in-use performance equivalent to or better than asbestos, and nontoxicity.  Although the use of asbestos still persists, today many commercial asbestos-free friction products and gaskets satisfy a broad range of applications.

*Du Pont registered trademark

The discovery in the early 1960's of liquid crystalline solutions of extended chain aromatic polyamides led to a new family of high performance organic fibers with a unique combination of engineering properties. These fibers are known generically as aramids. The first para-aramid fiber was introduced by Du Pont in 1971 under the trademark Kevlar®. The potential value of Kevlar® as an asbestos replacement was recognized but the cost barrier, 50 to 100 times as expensive, seemed insurmountable. The critical new technology that helped fill the asbestos replacement need was the discovery that proper application of excessive torsional or tractional strain causes a gross lateral slippage of internal fiber regions leading to a highly fibrillated structure. This special form of Kevlar® is called pulp and is an engineered short fiber with attached fibrils. The fibers' short length and excellent dispersibility provide effective reinforcement when coupled with

high surface area and high length-to-diameter ratio. These characteristics, combined with the inherent toughness, strength and heat stability are the reasons why only very low weights of Kevlar® pulp are required to achieve composite reinforcement.

The remainder of this paper will discuss the development of aramid pulp to fit an industry need and the laboratory and field testing done to provide realistic cost considered value of this fiber for use in automotive and truck brakes, clutches, gaskets, adhesives, sealants, and short fiber reinforced elastomers. In order to provide a basis for understanding the properties of Kevlar® in relationship to other reinforcing fibers, their comparative properties are listed in Table 1.

Consideration of tensile strength indicates that aramid is about two to nine times as strong as asbestos, about 40 percent stronger than high strength

Table 1

Kevlar® Versus Other Materials: Property Comparison

|  | Kevlar® | Asbestos | High Strength Graphite | "E"-Glass | Stainless Steel |
|---|---|---|---|---|---|
| Specific Tensile Strength* lb/in² (MPa) | 10,000,000 (2,500) | 1,000,000- 4,000,000 (270-1,100) | 7,000,000 (1,800) | 3,800,000 (945) | 880,000 (220) |
| Density lb/in³ (gm/cm³) | 0.052 (1.44) | 0.090 (2.50) | 0.063 (1.75) | 0.092 (2.55) | 0.283 (7.83) |
| Elongation at Break (%) | 3-4 | 0.4-1.7 | 1.25 | 3.5 | 2.0 |

*Tensile strength divided by density as tested per resin impregnated strand test--ASTM D2343.

## Table 2

### Thermochemical Properties of Kevlar® Aramid Fiber

| | |
|---|---|
| Thermal Expansion Coefficient | $-2 \times 10^{-6}/°C$ ($-1.1 \times 10^{-6}/°F$) |
| Heat Shrinkage | None |
| Moisture Shrinkage | None |
| Chemical Resistance | High (except hot, concentrated acids and bases) |
| Coefficient of Friction | Stable to 570°F (300°C) |
| Degradation Temperature | 800-1000°F (425-530°C) |
| Hot Wear Resistance | High |
| Moisture Content | 6% at 55% RH, 72°F (22°C) |
| Abrasiveness | Low |

graphite, over twice as strong as "E" glass, and over five times as strong as steel on an equal weight basis. Kevlar® also has the lowest density of these reinforcing materials, as shown in row two. It is less dense than graphite. Its density is approximately half as much as asbestos and glass fiber, and about 1/5 as much as steel. The last row lists the comparative stretch of each material before breaking. Notice that both aramid and glass fiber have a substantial elongation-to-break compared to asbestos, graphite and steel. Table 2 lists some thermomechanical properties of Kevlar® aramid fiber. Note that Kevlar® has excellent dimensional stability. It resists shrinking in heat or water, has a high resistance to chemicals, and its coefficient of friction remains stable in excess of 570°F (300°C). The fiber is inherently flame resistant and will not melt or support combustion, but it will begin to carbonize at about 800°F (427°C).

Equally important, but difficult to quantify, is aramid's inherent toughness. Its unique properties prevent it from exhibiting the brittle characteristics of mineral or manmade inorganic fibers.

## 2. FRICTION PRODUCTS (BRAKES AND CLUTCHES)

Brake manufacturers and users report that brake materials reinforced with aramid in combination with other materials provide several advantages over brakes reinforced with other fibers. These include: longer brake wear with less vehicle maintenance, better friction properties, no rust-bonding performance problems and high reinforcement strength.

Recent on-truck field test results show that brakes with aramid also exhibit less wear on brake drums. Independent testing was conducted by the Transportation Research Center of Ohio, based on the Federal Motor Vehicle Safety Standard 121, modified burnish test. Two thousand stops were made rather than the required five hundred. All tests were done using a GMC tractor and a 45-foot van trailer. The trailer was loaded to 20,000 lbs. per axle. The gross vehicle weight was 82,730 lbs. Each brake wear test required driving about 5,000 miles with a total of 2,000 stops made from 50 mph at a deceleration rate of 10 ft/sec per sec. Brake blocks were measured and weighed before installation and after removal to determine material loss. Brake shoes and drums were changed at the end of each test series. At this time, all cam rollers, springs and anchor pins were replaced and new oil seals were installed. Ten commercial brake formulations were tested. These formulations included: two premium asbestos reinforced, five reinforced with Kevlar® aramid, and three reinforced with glass as the primary fiber.

Although all of the formulations were commercially available and met FMVSS-121 requirements, there were significant differences in wear life and drum abrasion performance. In general, brake formulations reinforced with aramid pulp showed significant wear advantages over other brake compounds even though there was a wide range of performance among the various formulations. All nonasbestos formulations, including glass reinforced brakes, were equivalent to or, in most cases, exceeded the wear performance of the premium asbestos formulations. Glass-reinforced formulations, with one exception, were more abrasive to the drums than those reinforced with Kevlar® aramid or the asbestos formulations. High wear and low drum abrasion translates to low maintenance for truck fleet owners and thus about 25% of the truck fleets that are nonasbestos are using Kevlar® aramid reinforced brakes.

Du Pont maintains a complete friction materials laboratory, which allows the formulation, mixing, molding, machining and mounting to the backing plates of experimental friction pads and blocks. Complete physical testing facilities are also maintained and dynamometer and field testing is done with contracted test sites. This capability allows direct assistance to friction manufacturers worldwide during their change to nonasbestos reinforced formulations.

Similar tests for automotive disc brakes were also contracted with the Transportation Research Center of Ohio. In this case, 14 disc pad formulations were compared and the complete FMVSS-105 test was run. Brake pad formulations included asbestos fiber, metallic fiber and Kevlar® aramid reinforcement formulations. Similar results to those reported on the heavy duty truck test were obtained. The nonasbestos reinforced brakes outwore the

asbestos pads by a factor of 2X-3X. However, the metal fiber reinforced pads proved to be abrasive to the rotors and noisy when compared with the aramid reinforced pads.

European friction manufacturers have taken advantage of the ability of aramid fiber to contribute synergistically to improve wear and reduce noise by modification of the metallic formulations. Reduction of the metal content and the use of alternate high temperature fillers allows the production of superior performance brake pads that retain the high temperature advantages of metal but with reduced noise and abrasiveness. These disc brake pads are now standard OEM equipment on a high percentage of all automobiles manufactured in Europe.

A similar trend is now occurring in Japan among friction brake manufacturers and some 1987 model automobiles will contain nonasbestos disc brake pads. American automotive manufacturers have been somewhat slower to adopt aramid reinforced brake pads although one of the major companies has adopted a similar brake in at least three of its model lines. All friction companies, worldwide, are working to replace asbestos by the utilization of blends of nonasbestos fibers as no single fiber can do the complete job required. Aftermarket disc brake pads and rear drum linings formulated with Kevlar® aramid have been available in the U.S. for several years.

Clutches reinforced with aramid fiber were developed for high performance sports cars in Europe four years ago. These clutches contained as much as 70 volume percent of aramid and thus were very expensive. Primary advantage of aramid in the application is its low density, high strength and good engagement. Smooth engagement of manual transmission clutches is important to European drivers who like to sense the "feel" of the clutch. Field experience with these high performance clutches later led to the reduction of aramid content by judicious blending of fibers and now most European automobile manufacturers have adopted nonasbestos reinforced clutches in many of their model lines. In contrast to brake formulations, most clutches utilize a spun staple aramid yarn. There is at least one exception where in the U.S. a manufacturer of clutches has produced an aramid pulp based clutch that is molded rather than wound. This clutch has also found OEM applications in light trucks and a limited line of automobiles. Finally, automatic transmission clutches based on a paper made from aramid pulp have also found OEM application worldwide and several manufacturers produce nonasbestos paper based structures.

## 3. GASKETS

Asbestos gaskets are made by two methods: Compressed asbestos gaskets (CA), which are calendered from a solvent-based mix, and beater addition asbestos gaskets (BA), which are produced on a paper machine. The compressed gaskets dominate the industrial or service segment of the market, while the beater addition gaskets are used predominately in the automotive industry. Typically asbestos gaskets are formulated with 80-85% asbestos fiber and the remaining percentage composed of the elastomeric binder and filler. Laboratory work at Du Pont during the early 1980's showed that the adjustment of key fiber parameters with the selection of the proper filler(s) could lead to optimization of physical properties and sheet formation. Sheets formed with 5-15 % Kevlar® aramid pulp, 5-20% elastomer and proper fillers performed equal to or better than a compressed asbestos sheet in simulated end-use testing. Sheets with as

little as 5% Kevlar® pulp outperformed a highly densified beater addition asbestos sheet. Nonasbestos gasket sheets reinforced with aramid can also be easily formulated to give a much lower chloride content than the corresponding asbestos sheets.

The highly fibrillated form of Kevlar® pulp with a high surface area was chosen for gaskets because the high surface area enhances chemical/mechanical bonding with the elastomeric binder. In addition, its similarity to cellulosic pulp makes it practical for the beater addition gasket papermaking process. The length of the pulp particle is very important to dispersion in a papermaking process and thus a "short length" version was specially developed for this end-use.

Kevlar® is more expensive than asbestos and this added cost cannot be offset by replacing the remaining asbestos with an inexpensive inorganic filler even though only a relatively small amount of Kevlar® would be used. With proper filler selection, good fiber dispersion, etc., this cost penalty can be minimized making Kevlar® a very viable product. However, depending upon the economic penalty associated with handling asbestos, gaskets reinforced with Kevlar® aramid pulp can even be a very cost effective alternative.

Although aramid pulp can be processed on the same equipment as asbestos, some process modifications are needed. One important consideration in BA gaskets is the difference in the zeta potential between asbestos and aramid. Zeta potential is the work required to bring a unit charge from infinity to the surface around the pulp particle where the ions become free to move in a direction opposite to the particle. Unlike asbestos which has a positive zeta potential and thus a natural affinity for the negatively charged latex binders, Kevlar® is strongly negative (~40mV) as are most fillers and latices. Consequently, the surface charge on the aramid pulp must be neutralized by the addition of cationic inorganic or organic materials and/or pH control.

Laboratory work showed that the tensile strength of both types of gasket sheeting is proportional to the level of Kevlar® pulp. Sealability of the gasket is a measure of its leakage rate. The rate improves as less Kevlar® is used, but in all cases the leakage rate is more than 10X better than the compressed asbestos gasket and 30X better than the beater addition asbestos gasket. Stress retention at high temperature (288°C) is also extremely critical to the performance of a gasket. If the creep is high, the gasket will lose its seal and the flange will have to be retorqued. The high modulus of Kevlar® at elevated temperature gives good creep resistance. Du Pont lab results indicate that from 5-15% Kevlar® pulp (depending on end use) is required to equal the performance of asbestos gaskets. Finally an important gasket parameter is chloride content, especially for the nuclear power industry or when sealing stainless steel. Since the chloride content of Kevlar® pulp is very low, proper filler selection enables gasket manufacturers to compound a gasket with much lower chloride content than with asbestos.

Based on Du Pont laboratory results and particularly commercial experience, Kevlar® aramid pulp is proving to be a very successful replacement for asbestos in both compressed and BA gasketing for most applications.

## 4. SEALANTS AND ADHESIVES

Aramid has also been found to be a substitute for asbestos when used to reinforce sealants, caulkings and adhesives. Kevlar® pulp is very effective in this application and works in a ratio of 10-25:1,

asbestos to Kevlar®. In addition, aramid imparts thermal and chemical resistance. Physically it imparts the networking characteristic that resists flow at low shear, however, at very high shears the fibers orient from a three-dimensional structure to a two-dimensional plane that allows for pumping and flow during processing.

The primary need for a fiber in these applications is to provide rheological control, that is: high viscosity at low shear for slump resistance when applied to vertical surfaces. However, for practical processing in industrial applications, the sealant materials must be pumpable, meterable, and sprayable and thus must thin under high shear forces. Secondarily, the fiber also provides reinforcement during thermal cycling by preventing crack propagation and provides surface texture and provides tear resistance. An example of the viscosity boosting character of Kevlar® pulp is found in Table 3 describing its effect in PVC Plastisol adhesives.

Table 3

Viscosity Increase in Kevlar® Reinforced PVC Adhesives

| Wt. % Kevlar® Pulp | Viscosity (CPS) |
| --- | --- |
| 0 | 400 |
| 0.5 | 40,000 |
| 0.7 | 100,000 |
| 2.5 | Paste |
| 5.0 | Dough |

Commercially, Kevlar® pulp is used in epoxies, industrial asphalt coatings and sealants, hot melt adhesives, thin film foamable caulks and for nonstructural aircraft adhesives.

## 5. REINFORCED ELASTOMERS

The area of reinforcement of elastomers is one of the most recent uses for aramid pulp. Commercial applications for reinforcement have been achieved in the following end uses:

- "V"-belts, where the high modulus of aramid pulp provides stiffness perpendicular to the aramid cord which is the primary strength element. The stiffness improves the wear at the surface that is in contact with the pulley. That is accomplished by proper fiber orientation in the elastomer. This application is not an asbestos replacement.

- Industrial seals in oil well drilling applications where the pressures are so high that the seals creep or extrude. Kevlar® pulp provides the creep resistance under both the high temperature and high pressure of the end use.

- Ablative insulation for missiles and rockets. The nonmelting, high char formation of aramids inhibits thermal erosion while maintaining the network structure necessary for reinforcement. Kevlar® has become the preferred asbestos replacement material in several defense missiles. Development work is in progress with most major missile and rocket programs.

## 6. SUMMARY

Kevlar® aramid fiber and pulp are providing a viable alternative to asbestos fiber for reinforcement in many end-use application areas. Formulations containing Kevlar® are widely used in friction brakes and clutches, gaskets, sealants, caulkings, adhesives, and in reinforced elastomers. Super wear resistance has also been shown in thermoset bulk molding compounds and more recently in thermoplastic resins as well. Continued effort is being applied to identify and provide optimumly engineered aramid fiber forms for new end uses.

## 7. ACKNOWLEDGMENTS

The authors wish to thank their co-workers, C. Slaughter and R. Hayes for their technical assistance, S. McKinney for typing the manuscript, and W. D. Hewett for reviewing this paper.

## 8. REFERENCES

(1) Blades, H., U.S. Patent 3,767,756 (1973).

(2) Kwolek, S. L., Liquid Crystalline Polyamides, 179th Amer. Chem. Soc. Meeting, Houston, TX, March 1980.

(3) Tanner, D., Cooper, J. L., Dhingra, A. K., and Pigliacampi, J. J., Future of Aramid Fiber Composites As A General Engineering Material, Japan Society for Composite Materials, 10th year anniversary issuance, 1985.

(4) Frances, A., Nonasbestos Beater Addition Gaskets Reinforced with Kevlar® Aramid Pulp, TAPPI Nonwovens Symposium, p. 125, 1984.

PHOSPHATE FIBER: A NEW MULTI-PURPOSE REINFORCING FIBER

A. R. Henn and M. M. Crutchfield
Monsanto Company
800 N. Lindbergh Blvd.
St. Louis, MO 63167

## Abstract

Phosphate Fiber (PF) is introduced as a new, multi-purpose fiber that can be used to replace asbestos in many applications. Given its chemical composition and structure, PF is hypothesized to be biodegradable and, therefore, less hazardous than biodurable fibers such as asbestos. The physical and chemical properties of PF are detailed, and the use of PF in various applications is discussed. In particular, the utility of PF in friction materials, paper and gasket materials, plastics reinforcements, and as a thixotropic agent is described. Comparisons to asbestos are noted.

## 1. INTRODUCTION

Phosphate Fiber, hereinafter denoted as PF, is a synthetic, inorganic, crystalline fiber that has been developed by Monsanto Chemical Company[1]. The chemical composition of PF is approximated by the formula $2CaO \cdot Na_2O \cdot 3P_2O_5$ and has the empirical formula of calcium sodium metaphosphate, $CaNa(PO_3)_3$. The crystalline fiber consists of long, parallel, polyphosphate chains of $PO_3^-$ groups that are held together by the ionic bonds existing between the $PO_3^-$ groups and the calcium and sodium cations.

Figure 1 depicts models of the polyphosphate backbone. Since the P-O-P covalent bonds are much stronger than the ionic bonds, energetic milling preferentially cleaves the ionic bonds to give fibers with aspect ratios of 20 or higher.[2]

## 2. MORPHOLOGY

Phosphate Fiber is a soft, white, fluffy, fibrous crystalline material. Figures 2 and 3 are high resolution scanning electron micrographs (SEM) of two different grades of PF. Figure 2 shows our standard fiber product, and

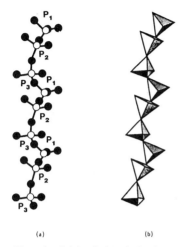

Figure 1. Models of the polyphosphate backbone in PF. a) stick and ball model b) linked tetrahedrons.

figure 3 is a coarser, longer fiber. (Note that the difference in magnification between the two micrographs is about 7x.) The fibrillated nature of this crystalline material is very much evident at these magnifications, as is the naturally wide distribution of fiber diameters and lengths. The wide distribution of particle sizes provides some of the easy processing characteristics of the product because the smaller particles can act as a flow conditioning agent for the larger fibers. The fuzzy and splayed ends of the fibers give a relatively large surface area and porosity, both of which allow for high liquid absorption by the fiber.

Figure 4 compares the fibrillated character of PF with that of chrysotile asbestos. Although qualitatively similar to asbestos, Phosphate Fiber, in actuality, is quite different. This is illustrated by the high resolution transmission electron micrographs of the cross-sectional views of the respective fibers shown in Figure 5. In contrast to asbestos, there are no inherent, cylindrically shaped fibrils present in the individual Phosphate Fibers.

As a consequence of the gross, qualitative similarity with asbestos, PF handles and processes much like asbestos; it is not prickly, and it disperses well in liquids.

3. PROPERTIES OF PHOSPHATE FIBER

3.1 Physical Properties

Table 1 lists some of the important physical properties of PF and the possible advantages associated with those properties[2]. In many instances, PF's physical properties are similar to asbestos or glass fiber. For example, the tensile strength of PF is 2.6 GPa compared to about 2.5 GPa for chrysotile asbestos and 2.7 GPa for fiberglass; other properties of PF that are similar to those of asbestos are the thermal conductivity, the dielectric constant, the index of refraction, the Mohs hardness, and the high oil absorption.

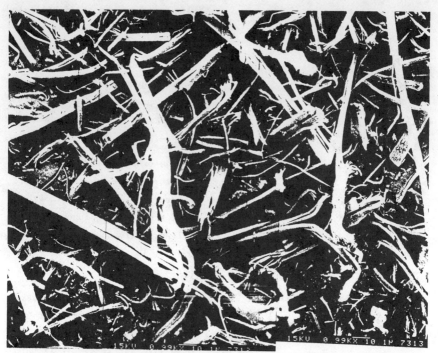

Figure 2.  SEM of standard PF.

Figure 3.  SEM of coarse PF.

**Table 1**
**PHYSICAL PROPERTIES OF STANDARD PHOSPHATE FIBER[2]**

| Property | Value | Advantage |
|---|---|---|
| **Form** | | |
|     Particle Shape | Fibrillar | Handling and processing properties similar to asbestos. |
|     Texture | Soft | Not prickly, no skin irritation. |
| **Dimensional** | | |
|     Density, g/cc | 2.86 | Similar to existing mineral fibers, less than steel. |
|     Bulk Density, g/cc (lb/ft³) | | |
|         Loose | .08-.16 (5-10) | |
|         Packed | .24-.40 (15-25) | |
|     Fiber Diameter, μ | 1-5 | |
|     Aspect Ratio | | |
|         From SEM | 20-30 | No fiber entangling problems. |
|         From Tap Density | 50-80 | Effective as a reinforcing fiber. |
|     BET Surface Area, m²/g | 1-2 | Allows good fiber-to-matrix contact and bonding. |
|     Avg. Aerodynamic Equiv. Diameter[1], μ | 8 | Major portion of fibers not considered respirable. |
| **Mechanical** | | |
|     Tensile Strength, GPa (Kpsi) | 2.55 (370) | Stronger than glass. |
|     Tensile Modulus, GPa (Mpsi) | 124 (18) | Higher modulus than glass. Similar to Aramid fiber. |
| **Thermal** | | |
|     Stability | melts 749°C | Maintains integrity at higher temperature than many other fibers. |
|     Heat Capacity, cal/g-°K | 0.19 | |
|     Linear Expansion, 1/°C | | |
|         Longitudinal | $5 \times 10^{-6}$ | Imparts dimensional stability to composite, similar to glass. |
|         Transverse | $10\text{-}11 \times 10^{-6}$ | |
| **Electrical** | | |
|     Dielectric Constant | 8 (1 KHz) | Similar to other fibers. |
|     Surface Conductivity, 1/ohm, (18% RH) | $3 \times 10^{-9}$ | |
| **Optical** | | |
|     Avg. Refractive Index | 1.572 | |
|     UV/VIS/NIR Reflectance | >80% (300-2400 nm) | Not degraded by UV/VIS/NIR light. |
| **Miscellaneous** | | |
|     Oil Absorption[2], g/100g | 350-380 | Useful as a thickener. |
|     Moisture Pick-up, % | 1.1 (82% RH) | |
|     Mohs Hardness | 3.7-4.3 | Softer than steel and glass fiber. |

Notes:   [1] Calculated
          [2] ASTM D281-31

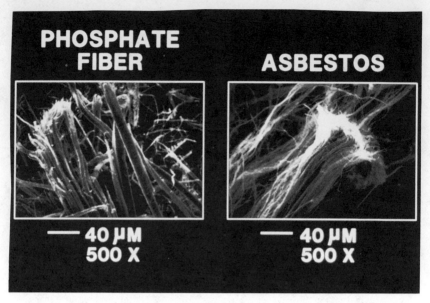

Figure 4. SEM close-ups of the fibrillating nature of PF and asbestos.

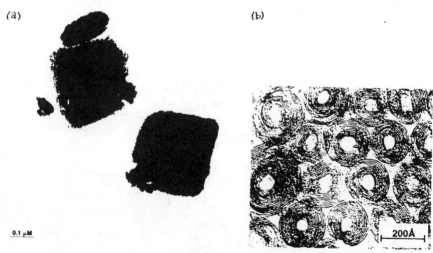

Figure 5. Comparison of the cross sections of a) PF and b) chrysotile asbestos (taken from ref. 10).

When fibers are being evaluated for use in certain applications, a property that should receive more consideration than it does is the softness of the fibers. Having a Mohs Hardness of about 4 (The Mohs hardness of asbestos is 2.5-4.0.[3]), PF is softer than many inorganic fibers, and should therefore cause less equipment wear and fewer handling problems.

Another desireable property of fibers in some applications is a high oil absorption. Through capillary action, standard-grade PF absorbs a large amount of oil, as does asbestos. (The OA values for standard PF and #5 chrysotile are 365 and 193 g/100g, respectively.) PF is thus able to provide a controllable consistency to resin-fiber mixes, and it helps prevent resin extrusion out of molds during pressurized curing.

Phosphate Fiber possesses excellent thermal stability. Figure 6 exhibits the differential thermal analysis (DTA) and thermogravimetric analysis (TGA) curves for $[CaNa(PO_3)_3]_n$ crystal. The only significant thermal transition evident is the sharp melting point of the fiber at 750°C. There is no weight change in the material until around 950°C, when the melt starts to lose weight in the form of $P_2O_5$. The fiber itself can absorb over 1% water at 25°C and 82% relative humidity, but it is not particularly hygroscopic[2].

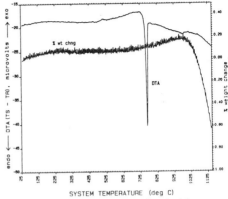

Figure 6. DTA/TGA Curves for $[CaNa(PO_3)_3]_n$.

The average fiber diameter of the standard grade of PF ranges from 1 to 5 microns, with a typical mean of 2 to 3 microns, while the length ranges from about 10 to 100 microns. The average diameter of the coarser grade is about 10-15 microns. Figure 7 is the cumulative particle size distribution of typical standard-grade PF as measured by the Elzone 180XY[2]. The Elzone instrument measures the displaced volume of a particle and reports an equivalent spherical diameter as if the particle were a

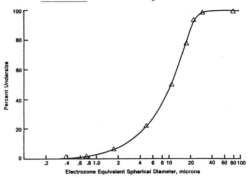

Figure 7. Particle size distribution of standard Phosphate Fiber: cumulative curve for mass-weighted, electrozone equivalent spherical diameter.

sphere. The mass-weighted spherical diameter of the longer fiber in Figure 3 is about 43 microns.

## 3.2 Chemical Properties

PF has good resistance towards most common organic solvents, especially aprotic ones. Table 2 below lists the weight losses suffered by PF when it is contacted with refluxing solvent in a Soxhlet extractor for 21 hours. When water is present at elevated temperatures, the dissolution of the fiber is fairly rapid due to hydrolysis of the polyphosphate chains. At room temperature, however, a continuous leaching experiment indicated that the exposed fiber has a half-life of over 6 months.

Table 2
The Solvent Resistance of PF:
Weight Loss Suffered in Contact
with Refluxing Solvent for
21 hours

| Solvent | Weight Loss % |
|---|---|
| Acetic Acid | 4.2 |
| Methanol | 1.5 |
| Ethanol | 0.84 |
| Ethylene Glycol | 9.8 |
| Isooctane | 0.14 |
| Xylenes | 0.15 |
| Trichloroethylene | 0.11 |

Although relatively inert towards organics, PF readily accepts most common coupling agents[4]. The surface of the fiber has some exposed, acidic P-O-H groups, which can allow bonding to the coupling agent. The pH of a 1 wt.% slurry of PF in water is 3 to 4.

## 4. APPLICATIONS

### 4.1 Friction Materials

One of the most important uses of asbestos is in friction materials for brake pads and linings and clutch facings. Based on its properties, PF should do well as an asbestos replacement in this use. At a loading level of 25 wt.%, PF can provide very good properties for brake pieces similar to those of asbestos-containing brakes. The following simple formulation was used for testing prototypical brake pieces:

```
 25% PF
 50% Barytes
 15% Phenolic Resin
 10% Friction Particles
100% Total
```

The components were dry blended and pressed for one hour at 18.3 MPa (2650 psi) and 171°C and post-cured for 4 hours at 177°C. The resulting composite gave test data that compare favorably with those of a similarly made composite containing 55% #5 chrysotile asbestos (only 20% barytes). The comparative data are listed in Table 3[5].

While not imparting as high a green strength to the preform piece as does aramid fiber, PF provides higher green strength than other asbestos substitutes such as rockwool, basalt, polyacrylonitrile modacrylic, and chopped steel fibers, all of which are a millimeter or more in

Table 3. Properties of Friction Composites Containing PF or Asbestos

| Property | Test | 25% PF | 55% Asbestos |
|---|---|---|---|
| Flexural Strength, MPa (kpsi) | ASTM D790 | 89 (13) | 69 (10) |
| Flexural Modulus, GPa (Mpsi) | ASTM D790 | 16 (2.3) | 15 (2.2) |
| Tensil Strength in Rivet Direction, MPa (kpsi) | -- | 15 (2.2) | 12 (1.7) |
| Swell and Growth, % | SAE J160 | 0.44/0 | 0.75/0 |
| Wear, % | SAE J661a | 11.9 | 10.1 |
| Friction Class | SAE J866a | E E | E E |

length. Moreover, using a different resin or adding 5% water to the mix and letting it air-dry for 24 hours produces preform pieces with green strengths up 0.483 MPa (70 psi). It is believed the reason the addition of water increases green strength is that a small amount of soluble phosphate glass present in the PF product acts as an _in situ_ cement[6].

Figure 8 compares the friction results of the extended SAE J661a fade test for the PF formulation versus the asbestos formulation. Here again the performance properties of the PF formulation rate very well against the standard asbestos piece.

Figure 9 demonstrates the excellent fade and recovery behavior attainable with >30% PF in a totally different, experimental brake pad formulation containing additional proprietary components[6].

An important property of any brake composite is its thermal conductivity. Several brake pieces, including some commercial ones,

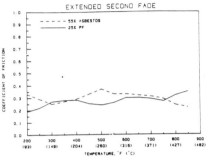

Figure 8. Comparison of extended fade behavior of the 25% PF formulation and the 55% asbestos formulation.

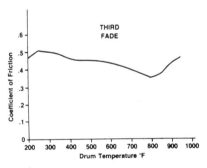

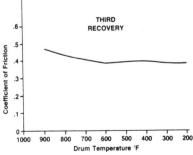

Figure 9. Fade and recovery of a brake piece containing 32% PF.

were measured for their relative thermal conductivities to see where our simple PF formulation stands in this regard. The thermal conductivity of the simple 25% PF formulation turned out to be the same as that of a commercial asbestos brake from Bendix.

## 4.2 Paper Mats

The feasibility of using Monsanto's Phosphate Fiber as an asbestos substitute in the mat backing for continuous sheets of vinyl floor coverings and in gasket materials has been demonstrated[7]. Handsheets (8" X 8") containing PF have been made in the laboratory routinely with very good properties. Furthermore, a continuous mat, 725 feet long, has been produced on a 36" continuous Fourdinier paper machine. Table 4 details the composition and properties of the handsheet and continuous mat. If, when making the handsheet, the pH of the 1.2% fiber-resin aqueous slurry is adjusted to 8.5 with concentrated $NH_4OH$ before the latex is added, the resulting tensile strengths are normally increased by a factor of two.

## 4.3 Gaskets

A fiber application related to paper mats and requiring some of the same processing and physical properties of paper is that of gasket materials. Due to its excellent thermal stability, chemical resistance, and high porosity, PF should function well in gaskets.

An important factor in selecting a fiber for gasket materials is the fiber's "springback". In order to

Table 4. Paper Mat Compositions and Properties

| | Handsheet wt.% | Continuous Mat wt.% |
|---|---|---|
| Composition (nominal) | | |
|     Phosphate Fiber | 76.5 | 73.5 |
|     Dow XD 30192 SBR Latex | 19.1 | 18.4 |
|     Santo-Res® CM (cationic resin) | 4.4 | 4.2 |
|     Albacel Softwood Kraft | -- | 3.8 |
| Properties | | |
|     Caliper Thickness, mm (mils) | 0.81 (32) | 0.79 (31) |
|     Apparent Density, g/cc (lb/ft3) | 0.80 (50) | 0.60 (37) |
|     Sheffield Smothness | | |
|         Wire Side | 275 | 398 |
|         Felt Side | 170 | 359 |
|     Cold Tensile Strength, MPa (psi) | | |
|         Machine | 8.6 (1250) | 6.7 (970) |
|         Cross | -- | 5.1 (740) |
|     Hot Tensile Strength @ 375°F (191°C), MPa (psi) | | |
|         Machine | 3.2 (470) | 2.0 (300) |
|         Cross | -- | 1.6 (225) |
|     Elmendorf Tear (16 ply), g | | |
|         Machine | 260 | 195 |
|         Cross | -- | 200 |

ensure a good seal throughout the service lifetime of a gasket, the gasket material should, after being compressed and released, still be able to "springback," or return, to a sizable percentage of its original dimensions. As a component in gasket materials, PF should be able to provide substantial "springback" for the gasket. Absolute "springback" is defined by us as

$$100(V_f-V_c)/(V_o-V_c),$$

where $V_c$ and $V_f$ are the bulk volumes of the compressed and free (released) fiber, respectively, at the corresponding applied pressure, and $V_o$ is the initial bulk volume at zero applied pressure (very gently tamped). Incremental "springback" is defined similarly, but $V_o$ is replaced by the $V_f$ of the previous pressure.

Table 5 lists the porosity and "springback" of standard PF over a range of pressures up to 345 MPa. The porosity is a measure of the volume percent of matrix and filler needed to fill all fiber pores at a given pressure. The data were acquired on a 5-gram sample of standard PF using a cylindrical die having a diameter of 2.54 cm.

It is noteworthy that the incremental "springback" goes through a maximum at around 13.8 MPa (2000 psi), a realistic pressure for many gasket applications.

4.4 Plastics Reinforcement

Phosphate Fiber, both coupled and uncoupled, has been shown to reinforce a broad range of plastic resins, as seen in Table 6. Although apparently unnecessary in phenolic systems because of PF's acidic surface, application of coupling agents to PF to improve the mechanical properties of composites made from resins other than phenolics has been demonstrated[4]. An effective, yet common, coupling agent for PF is A-1100. Composite pieces containing PF have been prepared by several different methods. However, due

Table 5. Porosity and "Springback" of PF

| Applied Mpa | Pressure Psi | Porosity,% | Absolute Springback, % | Incremantal Springback, % |
|---|---|---|---|---|
| 0 | 0 | 92 | -- | -- |
| 0.6895 | 100 | 78 | 28.9 | 28.9 |
| 1.379 | 200 | 71 | 21.5 | 56.4 |
| 3.447 | 500 | 64 | 19.4 | 71.0 |
| 6.895 | 1,000 | 58 | 17.0 | 74.9 |
| 13.79 | 2,000 | 55 | 15.3 | 83.3 |
| 34.47 | 5,000 | 48 | 11.1 | 62.1 |
| 68.95 | 10,000 | 40 | 7.0 | 52.5 |
| 137.9 | 20,000 | 32 | 4.0 | 45.6 |
| 344.7 | 50,000 | 20 | 1.9 | 31.5 |

Table 6.  Mechanical Properties of Resins Reinforced with PF

| Resin | Wt.% | Tensile Strength, GPa (kpsi) | Tensile Modulus, Gpa (Mpsi) | Flexural Strength, MPa (kpsi) | Flexural Modulus, GPa (Mpsi) |
|---|---|---|---|---|---|
| Nylon 6,6 | 20 | 98 (14.2) | 4.9 (0.71) | 150 (22) | -- |
| Nylon 6 | 15 | 83 (12.0) | -- | -- | 3.8 (0.56) |
| ABS | 20 | 54 ( 7.8) | -- | 90 (13.1) | 5.5 (0.80) |
|  | 25 | 56 ( 8.2) | -- | 98 (14.2) | 6.3 (0.91) |
|  | 30 | 81 (11.8) | -- | 113 (16.4) | 9.1 (1.32) |
| ABS/SAN Alloy | 6 | 42 (6.1) | 2.4 (0.34) | 77 (11.1) | 2.7 (0.39) |
|  | 10 | 52 (7.6) | 3.5 (0.51) | 68 (9.8) | 3.6 (0.52) |
|  | 20 | 46 (6.7) | 4.4 (0.64) | 77 (11.2) | 4.5 (0.65) |
| PBT | 30 | 76 (11.0) | -- | 117 (17.0) | 5.9 (0.85) |
|  | 40 | -- | -- | 124 (18.0) | 7.6 (1.10) |
| Novolak Phenolic | 80 | -- | -- | 207 (30.0) | 30.6 (4.44) |
| Unsat'd Polyester Thermostat (UPT) | 25 | 63 (9.1) | -- | -- | 11.0 (1.60) |
| Epon Epoxy[1] | 15* | -- | -- | -- | 5.7 (0.82) |
| Sulfur[8] | 23* | -- | -- | 45 (6.6) | 10.4 (1.51) |

*Volume %

to the fibrillated nature and high oil absorption of PF, preparation by cold compression molding of highly loaded (>20%) composites is not recommended because of the difficulty in eliminating all voids in the composites.

In addition to good mechanical and isotropic properties, there are other features of PF that make it attractive and useful as a reinforcing fiber in composites. These include imparting excellent surface finishes, causing low equipment abrasion due to fiber softness, and, in general, ease of dispersing and processing.  PF also acts as a reasonably good drip suppressant and char support in flammable resins, and it provides good heat distortion resistance.

Due to its high oil absorption and relatively high surface area, the ability of PF to act as a thickener (viscosifier) in unsaturated polyesters was investigated and is shown in Figure 10.  The efficacy of PF in this role is not as good as Calidria RG244 (asbestos) because its surface area is a factor of 30X less than the asbestos.  However, when PF is combined with fumed silica in a 1:1 or 2:1 ratio, the mixture provides good thickening and thixotropic properties as indicated by its thixotropic index (cf. Figure 10).  The thixotropic

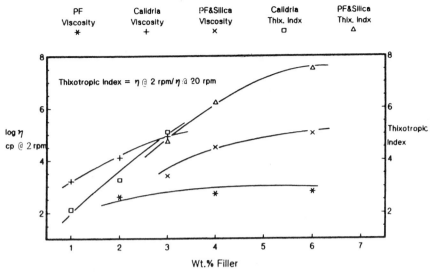

PF          Calidria      PF&Silica      Calidria        PF&Silica
Viscosity    Viscosity      Viscosity     Thix. Indx      Thix. Indx
   *            +              ×             □               △

Figure 10.  The viscosity and thixotropic index of a
filled unsaturated polyester resin at room temperature.

index, defined here as the ratio of the viscosity at 2 rpm to the viscosity at 20 rpm, is very similar for the PF/silica mixture and Calidria RG244. If one is looking to replace asbestos in a particular application in which thixotropy and strength reinforcement are desired, then a mixture of PF and silica may be a suitable substitute.

### 4.5 Miscellaneous Applications

In addition to the applications and uses described above, PF has been evaluated qualitatively for use in paints and in sealants such as caulking. Rope caulk has been made from PF, and the addition of 5% to both oil and latex paints results in a less drippy, faster drying paint that gives a flatter, textured, and more durable finish when dried.

### 5. TOXICITY AND SAFETY

PF is composed of very long condensed polyphosphate chains (Fig. 1). Since biological systems can hydrolyze polyphosphates, it is anticipated that PF should be degraded in biological systems. Biological experiments to test this hypothesis have been completed. For a discussion of the results, the reader is directed to reference 9. In addition, a two-year, chronic fiber inhalation study with rats to assess the potential for carcinogenicity is more than halfway completed.

## 6. REFERENCES

1. Griffith, E. J., U.S. Patent 4,346,028, Asbestiform Crystalline Calcium Sodium or Lithium Phosphate Preparation and Compositions, Aug. 1982.

2. Henn, A. R. and Crutchfield, M. M., Handbook of Reinforcements for Plastics, H. S. Katz and J. V. Milewski, eds., Van Nostrand Reinhold Co., in press, chap. 7.

3. Axelson, J. W., Handbook of Fillers and Reinforcements for Plastics, H. S. Katz and J. V. Milewski, Eds., Van Nostrand Reinhold Co., 1978, p. 423.

4. Monzyk, B. Plastics Compounding, Sept./Oct. 42, (1986).

5. Hinkebein, J. A., "Polyphosphate Fibers: Reinforcement of Friction Materials Composites with Calcium Sodium Metaphosphate", presented at the Society of Automotive Engineers 3rd Annual Colloquium on Brakes, Atlantic City, N.J., Oct. 21-23, 1985.

6. Hinkebein, J. A., "Applications of Phosphate Fibers in Friction Materials", presented at the Society of Automotive Engineers 4th Annual Colloquium on Brakes, Atlantic City, N.J., Oct. 20-22, 1986.

8. Hinkebein, J. A., U.S. Patent 4,484,950, Composites of Sulfur and Crystalline Phosphate Fibers, Nov. 1984.

9. Nair, R. S., "In Vitro and In Vivo Toxicity and Biodegradation Studies with Phosphate Fiber", presented at The Society of Automotive Engineers 4th Annual Colloquium on Brakes, Atlantic City, N.J., Oct. 20-22, 1986.

10. Yada, K., Acta. Crystallogr. 23, 704 (1967).

Marvin M. Crutchfield holds a B.S. Degree in chemistry from Duke University and a Ph.D. degree in physical chemistry from Brown University. He has been employed by the Monsanto Company in St. Louis, Missouri, since 1959 in various R&D capacities. He presently holds the title of Senior Fellow in the Central Technology Organization of the Monsanto Chemical Company where he manages the product research and applications efforts on Monsanto's calcium sodium metaphosphate, a new developmental product more widely known as Phosphate Fiber (PF).

Arthur R. Henn has been a member of the Phosphate Fiber product and applications group since 1985. He joined Monsanto Company in St. Louis in 1982 after obtaining a Ph.D. degree in physical chemistry from Princeton University. He holds a B.S. degree in chemistry from Virginia Polytechnic Institute & State University.

ADVANCED FIBERS FOR
NON-STRUCTURAL APPLICATIONS

Gregory J. LaCasse
Raymark Industries
Manheim, PA

## Abstract

Asbestos replacement represents a large potential market for several of the advanced fibers. Friction materials, packings, seals and gaskets have been making the transition gradually, due to both technical and economic considerations. Man mades, like the aramids, carbons, S-2 Glass® and PBI® are in use where necessary. Fiberglass, cotton, rayon and other synthetics are meeting a large portion of these market requirements.

New Fire blocking applications in both upholstery and apparel are evaluating aramids, oxidized PAN & PBI® in a variety of blends & forms, both woven & non-woven. Ablative & high temperature isolation shielding are rapidly changing from Asbestos. Depending on the environment, candidates could range from carbon & graphite to silica & ceramics, again either in a woven or non-woven form.

Hybrid blends of fibers spun into yarn are finding good acceptance in "semi-structural", "lower tech" applications. Cardable fiberglass and other synthetics have provided desirable characteristics not generally expected from non-continuous fibers.

In several applications the product that has replaced asbestos has proven to be a significant improvement; in others the opposite is true and this is where the opportunity lies for the hybrid blends of advanced fibers.

## INTRODUCTION

One of the largest potential markets for the family of materials generally referred to as "advanced fibers" is in asbestos replacement. For many decades asbestos fiber, yarns & fabrics have been used in literally thousands of applications. The uses ranged from household faucet washers to components in sophisticated military defense systems. The wide spread popularity of asbestos stemmed from; low cost, ready availability, ease of use, always worked, and everyone has been using it since the ancient Egyptians! When the hazards of asbestos exposure gained public attention, fast paced and expensive development programs were initiated to find replacements. In many instances the researchers first had to define or characterize what the asbestos really did in a given application. Frequently this provided some surprises and usually the key requirements of the replacement candidate. It was often found that the "problem" of asbestos replacement provided the "opportunity" to produce technically superior products. This led to significant changes in business strategies and market share distribution.

We will look at a variety of markets where asbestos was the key component and how it has been replaced. No one fiber has been found to replace asbestos in all applications. What once was a relatively simple market with one base product, produced by a small number of companies in large volumes, $10^4$ metric tons per year each, has changed dramatically. Many small companies producing low volume specialty items have emerged and they are providing much new and innovative technology.

Friction Materials

The one industry that consumed more asbestos in all forms than any other is friction materials. When you consider that all the automotive brakes and clutches, both original equipment and after-market, were asbestos a tremendous opportunity existed. Roughly 50% of this market has been converted to non-asbestos products and, if current projections are accurate, the remainder will be converted by 1989. The applications or working environments vary substantially; consequently, several different materials have gained acceptance. E-type fiberglass is generally the predominant replacement fiber due primarily to cost and availability. A variety of aramid containing products have also gained acceptance due to improved processing

and/or durability characteristics.

Carbon fibers have been considered good candidates for friction materials but until recently
not much work had been done due
to their high price and availability.

In critical and extremely demanding applications carbon-phenolic
and carbon-carbon have completely
replaced asbestos. Lower cost
hybrid blends containing carbon
fibers are currently under evaluation for high volume applications. As with many of the other
fibers, carbon is available from
several sources and in this environment each performs differently. Which carbon is better;
PAN based, pitch based, phenolic
based, cellulose based, activated
or non-activated? Time will
tell.

One manufacturer of brakes &
clutches uses approx. 150
formulations to satisfy the requirements of his market. Each
formulation uses different fiber
blends or ratios. Many 2 and 3
component hybrid systems have
been evaluated with good results.
Some of the blends are: E-Glass-
Acrylic (PAN), Pre-ox-PAN, Carbon
(Pitch)-Aramid, E-Glass-Pre-ox-
Aramid, S-2® Glass-Phenolic
fibers, Basalt-E-Glass-Aramid,

and Stainless Steel-Carbon-
Aramid. While friction materials
is a highly competitive business,
the use of these hybrid systems
has resulted in products with
price/performance ratios very
nearly the same as asbestos products. This is a significant
accomplishment considering the
differences in raw material costs.

GASKETS, SEALS, PACKINGS

In this market asbestos was used
in virtually every application for
the same reasons it was used in
friction materials. As asbestos
was replaced with other fibers, an
amazing variety of products
emerged.

Typical applications range from
ambient temperature with no
pressure requirements to 1000°F
at 1000 psi; and asbestos yarns
were used in both environments.

Virtually any product would work
in the first environment so
issues like price, ease of handling and cosmetics were of concern. In the second environment
performance is the primary issue.
In the lower end applications
asbestos yarns have been replaced
with texturized fiberglass or a
hybrid of Fiberglass and PAN made
by two dissimilar processes. One
process employs a core yarn of
fiberglass and a covering layer

of PAN fibers (DREF®). The other is an intimate blend of the two fibers, carded and spun, in the same manner that was used for asbestos. In mid-range applications virtually every fiber has found a niche in this market. Aramids, phenolics, PTFE fibers, Pre-ox and many others have found acceptance in this market; because of the wide selection of products a relatively low volume situation exists. Coupling these associated costs with high raw material costs, products, that are considerably more expensive and generally perform no better than their asbestos predecessors, have been created.

More severe applications, where high temperature and pressure are encountered, require even more expensive products. One of the most popular products is produced from yarns spun from carbon (95% min. assay). Since it is used in an anaerobic environment at temperatures up to 1000°F, it works much better than its asbestos counterpart. One manufacturer relates a 600% life improvement at only a 400% cost penalty. In most circles that would be considered a bargain, especially if the true cost of downtime can be calculated.

In several applications the various sealing products are exposed not only to high temperature and pressure, but also a hostile chemical environment. Care must be taken to evaluate materials under actual service conditions. It has been observed that both temperature and pressure can affect chemical resistance. In applications of this type, products like polyphenylene sulfide (PPS), PBI and S-2 glass® are usually considered. Currently hybrids of S-2 glass®-PBI and carbon-PBI are being evaluated for a broad range of asbestos replacement applications.

Only a small percentage of these products are being produced from asbestos. A mid '86 estimate was 10% to 20%. With the new regulations coming into play on fiber exposure, this percentage should decline rapidly.

## FIRE BLOCKING

Evidence of the spinning and weaving of asbestos textiles for fire barriers has been found in the Hieroglyphics of Egypt. This same product has been used for thousands of years for the same purpose, until recently.

Welding curtains, turn out coats, gloves, fire hose, and many others needed to be redesigned using non-asbestos textiles. Again it was an application by application journey through the maze of new and old products to find replacements.

Welding curtains is one of the more interesting conversion studies. Once there was only asbestos cloth and it did the job. Now we have rubber coated cotton cloth, texturized Fiberglass cloth, Fiberglass hybrid clothes containing PAN or aramids, fiberglass roving cloth, wire inserted fiberglass cloth, silica (leached glass) cloth, and ceramic cloth. These also work. Some are less expensive than asbestos but most are much more expensive. The work site, type of welding, vertical or horizontal curtain placement, economics, and other factors make this array of products necessary.

In the heavy garment area for fire protection, asbestos has been replaced by a narrow range of products. Hybrids of fiberglass with an aramid or PBI cover (DREF® yarn) are being used. Pre-ox and phenolic fibers are also being spun and woven into the appropriate weight fabrics for coats, pants, hoods & gloves.

There has been considerable interest recently in institutional decorative fabrics that will not burn. With the advanced fibers available, the color selection would be limited. A hybrid using a high percentage of cardable fiberglass and a colored synthetic fiber produces an acceptable appearing fabric which does not burn.

Light weight fabrics produced from a hybrid blend of aramid and pre-ox have been evaluated for fire blocking applications in aircraft seats. Durability, flame retardancy, comfort and cost are all key concerns. PBI-aramid and pre-ox candidates have also been evaluated with all submissions showing good results.

Even lighter weight fabrics of the same blends are being considered as clothing for the military and sporting goods markets. Cost vs. durability is an important factor in this application. Yarn spinning technology with hybrids of these fibers has been satisfactory with heavier textiles but serious manufacturing problems have been encountered when "fine" hybrid "advanced" yarns are attempted.

## ABLATIVE PRODUCTS

For many years asbestos-phenolic papers and molding compounds have been used as liners for rocket motors to protect the outer shell from a burn through. The operating conditions of various motor designs are quite different. The operating temperature could vary from one design to another, by 3000°F. The expected life of the part might be microseconds or minutes. The operating environment might be oxidizing or reducing. "Aero heating" may be a consideration. Shelf life of the finished part has become a concern with many non-asbestos materials.

Many applications have made the switch relying on two basic products, silica fabric and carbon fabric. Asbestos is still being used in many of the higher volume applications but extensive evaluations of hybrid systems are underway. Generally non-wovens are preferred over wovens and the hybrids that are receiving the most attention contain aramids or PBI, along with the fiberglass products. There are products that can withstand the hostile environment of these rocket motors but in several instances processability or scrap rate has prevented them from being qualified. As with any development program, the objective is to make a product that processes the same as asbestos with a lower scrap rate, lower specific gravity, better performance and a lower cost. To date it appears that two of the five may be achievable.

## SUMMARY

Replacing asbestos has been a long up-hill struggle for a variety of reasons. Firstly, it worked, and 40 to 50 years ago there were no alternate materials. Systems were built around the characteristics of asbestos. Its strengths were enjoyed and its weaknesses were compensated for. As new materials arrived on the scene with their own strengths and weaknesses, they were expected to fit the asbestos mold, and they didn't. A commitment to the new product and a redesign of the system were needed. Companies who followed this path have done well.

A second consideration had been cost. As the availability and acceptance of replacement materials increased, the price of asbestos fiber dropped. In late 1986 fiber was selling in the $.17 to $.54 per pound range. Replacement fibers were in the $1.00 to $40.00 per pound range. Customers expected similar selling prices. This caused some hard management decisions. As insurance companies, OSHA and lawyers got involved, the directional decisions became easier; asbestos products at any price are too expensive.

In the development of replacement materials, care should be taken to study the fiber's characteristics, both chemical and physical, to hopefully anticipate future health problems. Fracture mechanics, fiber diameter, aspect ratio and reactivity need to be considered.

## BIOGRAPHY

Gregory J. LaCasse is presently Technical and Marketing Development Coordinator for the Raytech™ venture group with Raymark Industries. He joined Raymark in 1977 as a product development engineer and worked exclusively on asbestos replacement programs. Mr. LaCasse received a B.S. in Chemistry from Millersville University and has completed approx. 100 SH of advanced work in Chemistry, Mathematics and Business Administration, including a grant from the National Science Foundation in 1971.

PRINTED CIRCUIT BOARDS REINFORCED BY
TECHNORA® PAPERS AND FABRICS

Kunio Nishimura and Tadashi Hirakawa
Fiber and Textile Research Laboratories,
Teijin Limited
3-4-1, Minohara, Ibaraki, Osaka 567, Japan

## Abstract

Technora® aramid papers and fabrics can be used for reinforcement of flexible or rigid printed circuit boards. Technora® woven fabrics and Technora® papers impregnated with epoxy resins have lower thermal expansion coefficient (TEC) in the X-Y plane than conventional glass/epoxy or paper/phenol substrates, and have also lower hygroscopic expansion coefficient (HEC) in the X-Y plane than PPTA/epoxy laminates. TEC and HEC of Technora® parallel to the fiber axis were found to be negative and this behavior was utilized to make lower coefficient papers and fabrics. The coefficients of the laminates varied depending on the resin volume fraction. Through-holes by drilling were observed using a microscope. Electrical and physical properties of several kinds of laminates were investigated and it was found that the laminates can be used for substrates of advanced printed circuit boards especially for surface mounting of leadless ceramic chip carriers.

## 1. INTRODUCTION

Increasing demands for large scale integration and very high speed processing of information have led to increasing density of chips on printed circuit boards (PCB), and this has accelerated the development of leadless chip carrier packages. Since the leadless ceramic chip carrier (LCCC) has no lead, any mismatch of TEC occuring between LCCC and PCB will result in a stress on the solder joints. Especially, the solder joints will crack due to fatigue after a certain number of thermal cycles[1,2,5].

Current PCB reinforced by E glass fabric is not thermally compatible with LCCC. The TEC of E glass fabric/epoxy laminated in the X-Y plane is of the order of 12-18 ppm/$^{\circ}$C and that of E glass/polyimide laminate is of the order of 13 ppm/$^{\circ}$C[3], whereas the TEC of ceramic chip carriers is around 6 ppm/$^{\circ}$C. Therefore it has been very important to develop an advanced PCB which has a TEC in the X-Y plane as close to 6 ppm/$^{\circ}$C as possible. Trials have been made to obtain a PCB with lower TEC. One measure was to use poly-p-phenylene terephthalamide (PPTA) or quartz fabrics for reinforcement. It has been reported that PPTA/epoxy or PPTA/polyimide laminates yield TEC of about 4-8 ppm/$^{\circ}$C in the X-Y plane and solder joint cracking may be reduced or eliminated[2,3,5]. Also the use of quartz fabric reinforced PCB for this application has been disclosed. Desirable TECs have been reported with quartz fabric reinforced epoxy or polyimide laminates[4].

In spite of the improvement of TEC, it has also been recognized that these two reinforcements have several disadvantages. PPTA is likely to absorb moisture comparably and the laminates also tend to cause microcracking on the surface of resin during thermal cycling[5]. Furthermore, drilling is difficult because of remarkable fibrillation caused by high drill revolution. So, several processes to remove fiber fuzz and the resin smear are required[6]. Quartz fabric laminate, on the other hand, have difficulty in the drilling process because of its remarkable hardness compared with E glass fabric laminates so that drill bits are easy to be worn[4]. Consequently, a demand has grown for substrates having a TEC closely matched with ceramic chip carriers, lower moisture absorption, no microcracking and better drilling performance.

This paper describes the development of improved printed circuit boards reinforced by Technora® aramid papers and fabrics. It is concluded that these new type boards have great advantages to conventional ones.

## 2. EXPERIMENTAL

### 2-1 Preparation of Samples

Four kinds of reinforcements were investigated in this paper. They are listed in Table 1. For PPTA fiber, a high modulus type (Kevlar 49) was used.

Technora® is a new-type high-performance fiber, independently developed by Teijin Limited, Japan. Technora® is made from poly-p-phenylene/3,4'-diphenylethertere-phthalamide,

$$\left(HN-\bigcirc-NHOC-\bigcirc-CO\right)_{m}\left(HN-\bigcirc-O-\bigcirc-NHOC-\bigcirc-CO\right)_{n}$$

whereas the structure of PPTA is

$$\left(HN-\bigcirc-NHOC-\bigcirc-CO\right)_{n} .$$

Technora® has excellent properties such as high tensile strength, high tensile modulus and low moisture content. In addition, this fiber has a negative TEC ( Thermal expansion coefficient) and a negative HEC (Hygroscopic expansion coefficient) parallel to the fiber axis, moreover excellent heat and chemical resistance. Table 2 shows a standard performance of Technora®.

Experiments were made using papers and fabrics made of this new fiber. Technora® papers were newly developed for this study. To compare different types of reinforcements, thickness of each material was equalled roughly and basis weight of Technora® paper, Technora® fabric and PPTA fabric were equalled. All materials except for E glass fabric were cleaned in boiled cyclohexane for 30 min. in order to eliminate the influence of coupling and oiling agents.

The resin system used here was a standard FR-5 type resin consisting of a mixture of brominated DGEBA and phenol novolac type epoxy cured with 4,4'-diamino diphenyl sulphone and $BF_3$-monoethylamine complex. Single-sided copper-clad laminates were prepared by impregnating the reinforcements with the resin system under conditions commonly used for epoxy resins. And the copper was eliminated by etching completely except for the case of evaluation of drilling performance.

## 2-2 TEC Measurement

TEC was measured along the X-Y plane using a Rigaku-Denki type Thermoflex thermal mechanical analyzer (TMA). The specimen was prepared as a size of 4.5 mm width and 20 mm gauge length. Original loading on tray was 5 g and heating rate was $10^{\circ}$C/min. The specimen was heated from room temperature to 200 $^{\circ}$C and then cooled to $50^{\circ}$C, and then heated to $200^{\circ}$C again. This process was controlled by a computer. The thermal expansion rate and the TEC were calculated from the second obtained curve. The thermal expansion rate (dL/L %) for individual temperature was calculated and plotted against temperature, and then the thermal expansion coefficient (TEC; dL/L/$^{\circ}$C ppm) along the curve of 50-100$^{\circ}$C was calculated and used for discussion.

## 2-3 HEC Measurement

HEC was measured along the X-Y plane using a Shinkuriko type TMA-3000 thermal mechanical analyzer with a thermohygrostat. Specimen was prepared as a size of 5.0 mm width and 15 mm gauge length. Original loading was 5 g. The specimen was exposed to hygroscopic conditions of $30^{\circ}$C x 30%RH - $30^{\circ}$C x 70%RH for more than 96 hours. The hygroscopic expansion coefficient (HEC; dL/L/%RH ppm) along the second obtained curve was calculated and used for discussion.

## 2-4 Drilling Performance

Fibrillation on drilling was evaluated using a Schmoll Machinen type MBK-201 drilling machine. Diameters of drill bits were 0.9 - 2.1 mm, feed rates were 1.5 - 3.0 m/min. and the cutting speed was 25,000 - 50,000 rpm. Observation of the drilled holes was made by optical and electron microscopes.

## 2-5 Chemical Resistance

Chemical resistance to acid and alkali was investigated. The specimen was dipped in a 20% $H_2SO_4$ aq. solution or a 10% NaOH aq. solution for $95^{\circ}C \times 48$ hours.

## 3. RESULTS AND DISCUSSION

### 3-1 TEC Measurement

Table 3 shows TEC values of reinforcements, a resin and laminates, where resin volume fractions (VR %) are from 40 to 80 %. It was revealed that TECs of Technora® papers, Technora® fabrics and PPTA fabrics were of negative values. Especially, Technora® fabrics have larger absolute figures than PPTA fabrics. Technora® papers were revealed to have almost the same values as those of PPTA fabrics. On the other hand, E glass fabrics have positive values as already known.

Curves of these thermal expansion rate (dL/L) are shown in Fig. 1 for reinforcements, Fig. 2 for resin and Fig. 3 for laminates of VR 60 - 70% respectively. Three types of reinforcements protect the laminates from expanding in a large scale, whereas in the case of E glass fabric, such a phenomenon is not observed. However, the common phenomena are observed near the $T_g$. As the laminates are heated, the resins tend to expand at a larger rate and the fibers are kept in tension. Near the $T_g$, the resins become soft and can no longer hold the fibers in tension. Therefore the residual stress within the

fibers relaxes and causes the laminates to shrink[7]. So, there are almost the same peak points in the case of aramid reinforced laminates and bending point in the case of E glass fabric reinforced laminates.

The influence of VR upon dL/L are shown in Fig. 4 for Technora® paper/epoxy laminates, Fig. 5 for Technora® fabric/epoxy laminates, Fig. 6 for PPTA fabric/epoxy laminates and Fig. 7 for E glass fabric/epoxy laminates respectively. It can be seen that the peak or the bending points are shifted to higher temperatures as the increase of resin volume fraction. Moreover, the areas which have negative slopes are narrowed gradually as the increase of resin volume fraction in the case of aramid reinforced laminates. It is understood that this is caused by the decline of reinforcement effect as the resin volume fraction increases.

Fig. 8 shows the relation between TEC and the resin volume fraction, where the superiority of Technora® papers and fabrics are clearly expressed, as in the case of PPTA fabrics which have already reported[3,5,6,7,8]. Fig. 9 and Fig. 10 show the influence of in-plane angle of laminates on the TEC in the cases of Technora® papers and fabrics. Fig. 11 shows the relation between the tensile strength and in-plane angle of these two laminates. In spite of a remarkable difference of the tensile strength for each angle, TEC patterns of Technora® fabric reinforced laminates seem to be almost the same as those of Technora® paper reinforced laminates in which single fibers are randomly oriented.

## 3-2 HEC Measurement

Table 4 shows the difference of HECs among four kinds of laminates. It should be noticed that the HECs of Technora® papers and fabrics show larger negative values than PPTA fabrics and needless to say than E glass fabrics. As a result, laminates reinforced by Technora® papers or fabrics have lower HECs than PPTA. This will result in a better copper-clad laminate as it is exposed to high humidity.

## 3-3 Drilling Performance

Fig.12 shows the results of the optical and electron microscopic observations after drilling. They are the results of the samples that were drilled under the same conditions. The diameter of drill bit was 2.1 mm, the feed rate was 3.0 m/min. and the cutting speed was 50,000 rpm. It can be seen that the Technora® paper/epoxy laminate shows less amount of frayed fibers and is almost the same as the case of E glass fabric/epoxy laminate. Moreover, Technora® fabric/epoxy laminate has a less amount of fibrillated fibers compared with the case of PPTA/epoxy laminate. This seems to be ascribed to the difference of fibrillation tendency between Technora® and PPTA.

## 3-4 Chemical Resistance

Table 5 shows the change of physical properties of laminates after the laminates are treated by acid and alkali. It can be seen that the chemical resistance of Technora® paper/epoxy and fabric/epoxy laminates is remarkably better than PPTA fabric/epoxy or E glass fabric/epoxy laminates.

## 3-5 Total Properties of Technora® Laminates

Electrical and physical properties of Technora® paper/epoxy or Technora® fabric/epoxy laminates to be used for printed circuit boards are shown in Table 6. It is found that the laminates can be used as advanced printed circuit boards especially for surface mounting of LCCC.

## 4. CONCLUSIONS

(1) Technora® papers and fabrics have negative values of TEC in the X-Y plane, so that the epoxy laminates reinforced by Technora® have sufficiently lower TEC to be used for printed circuit boards compatible with leadless ceramic chip carriers.

(2) Technora® fabrics have a little bit larger negative values of TEC than those of PPTA fabrics in the X-Y plane.

(3) Technora® papers are composed of completely and uniformly dispersed single fibers so that their laminates have uniform tensile strength and TEC in all directions.

(4) Technora® papers and fabrics have larger negative values of HEC

1204

than PPTA fabrics, so that their laminates have lower positive HEC values than those of PPTA fabrics.

(5) Technora$^{®}$ paper or fabric reinforced laminates grow less fibrils than PPTA reinforced laminate during drilling. Especially, Technora$^{®}$ paper reinforced laminates show excellent drilling performance presumably because of uniform impregnation of resins.

(6) Chemical resistance of Technora$^{®}$ paper or fabric reinforced laminates against acid and alkali is superior to that of PPTA reinforced laminate.

(7) Technora$^{®}$ paper or Technora$^{®}$ fabric reinforced laminates may be used as substrates for advanced printed circuit boards for surface mounting of LCCC.

## REFERENCES

1) G. F. Love, IPC Technical Rev., Dec. 1981, p. 12.

2) W. Woodruff, Electronics, Jan. 27, 1982, p. 119.

3) S. E. Greer, IEEE Trans. on Components, Hybrids and Manufacturing Technology, CHMT-2,1, 140 (1979).

4) L. E. Gates, Jr. and W. G. Reimann, Proc. 2nd Ann. Int'l. Electronics Packaging Conf., p. 605, Nov., 1982.

5) Z. N. Sanjana, J. Valentich and J. R. Marchetti, Proc. of 28th SAMPE Symposium, Anaheim, CA, 1983, SAMPE Ser. 28, p. 1240.

6) Du Pont, Corlam$^{®}$ Laminates and Prepregs, A Processing Guide for High Performance Printed Wiring Boards.

7) Z. N. Sanjana, R. S. Raghava and J. R. Marchetti, ACS Symp. Ser. 242, 379 (1984).

8) E. W. Tokarsky, Proc. of 28th SAMPE Symposium, Anaheim, CA, 1983, SAMPE Ser. 28, p. 1251.

## BIOGRAPHY

Kunio Nishimura and Tadashi Hirakawa are members of Fiber and Textile Research Laboratories at Teijin Limited, Japan, and are involved in the development of advanced materials using aramid fibers such as Technora$^{®}$ and Teijinconex$^{®}$.

Table 1  Four kinds of reinforcements used for investigation

| Reinforcement | Construction | Thickness (mm) | Basis weight (g/m$^2$) |
|---|---|---|---|
| Technora® paper | 1.5 de | 0.098 | 62 |
| Technora® fabric (Plain weave) | 200 de X 200 de 34end/inch X 34end/inch | 0.103 | 62 |
| PPTA fabric (Plain weave) | 195 de X 195 de 34end/inch X 34end/inch | 0.096 | 62 |
| E glass fabric (Plain weave) | 22.5 tex X 22.5 tex 60end/inch X 57end/inch | 0.097 | 101 |

Table 2  Standard performance of Technora®

| | | |
|---|---|---|
| Color | gold | |
| Density | 1.39 | g/cm$^3$ |
| Filament diameter | 12 | µmφ |
| Tensile strength | 310 | kg/mm$^2$ |
| Tensile modulus | 7100 | kg/mm$^2$ |
| Elongation to break | 4.4 | % |
| Thermal decomposing temperature | 500 | °C |
| Heat of combustion | 6800 | cal/g |
| Specific heat | 0.26 | cal/g°C |
| Equilibrium moisture content | 3.0 | % |

Table 3 TEC(In X-Y plane)of reinforcements,resin and laminates(ppm/°C:50-100°C)

| Material | VR(%) | Angle from MD direction(deg) | | | | |
|---|---|---|---|---|---|---|
| | | 0(MD) | 30 | 45 | 60 | 90(CD) |
| Technora® paper | 0 | -3.3 | -3.8 | -4.3 | -4.2 | -4.8 |
| | 42 | -0.7 | -2.1 | -0.9 | -0.2 | -1.0 |
| | 59 | 9.6 | 7.8 | 7.4 | 6.5 | 7.3 |
| | 79 | 17.2 | 13.5 | 16.7 | 15.8 | 17.4 |
| Technora® fabric | 0 | -5.2 | * | * | * | * |
| | 44 | 0.3 | 4.3 | 2.6 | 1.3 | 2.0 |
| | 67 | 6.2 | 8.0 | 7.0 | 7.4 | 7.1 |
| | 75 | 13.2 | 11.4 | 9.5 | 14.0 | 13.7 |
| PPTA fabric | 0 | -4.8 | * | * | * | * |
| | 41 | 0.9 | * | * | * | * |
| | 61 | 3.8 | * | * | * | * |
| | 71 | 9.7 | * | * | * | * |
| E glass fabric | 0 | 4.3 | * | * | * | * |
| | 49 | 8.8 | * | * | * | * |
| | 69 | 13.2 | * | * | * | * |
| | 71 | 14.2 | * | * | * | * |
| Epoxy resin | 100 | 54.8 | | | | |

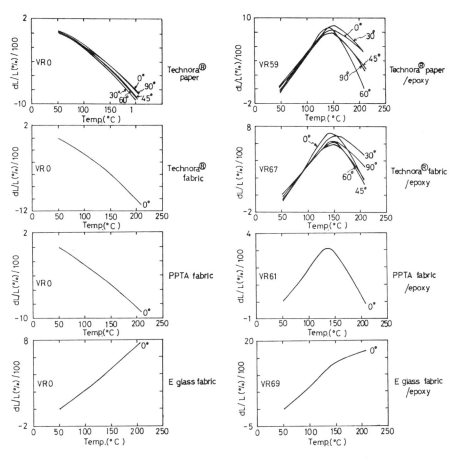

Fig.1  dL/L of reinforcement
versus temperature

Fig.3  dL/L of laminate
versus temperature

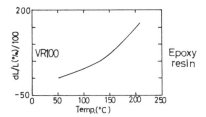

Fig.2  dL/L of epoxy resin
versus temperature

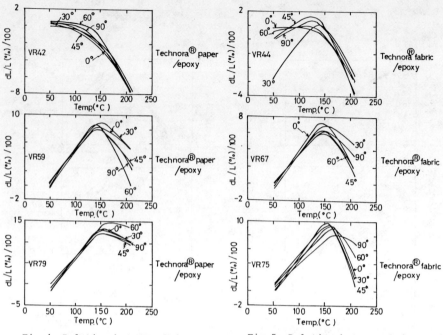

Fig.4 Relation between dL/L
and VR of laminate
versus temperature

Fig.5 Relation between dL/L
and VR of laminate
versus temperature

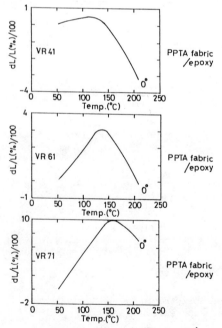

Fig.6 Relation between dL/L
and VR of laminate
versus temperature

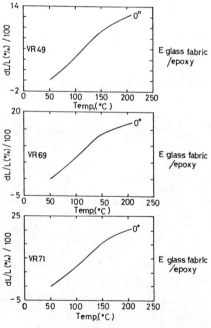

Fig.7 Relation between dL/L
and VR of laminate
versus temperature

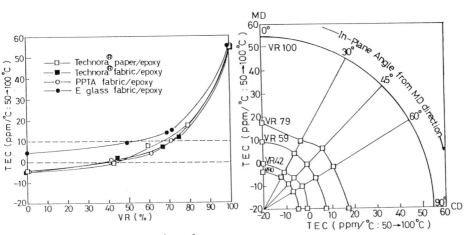

Fig.8  TEC versus resin volume
fraction

Fig.9  TEC of Technora® paper
/epoxy laminate versus
in-plane angle

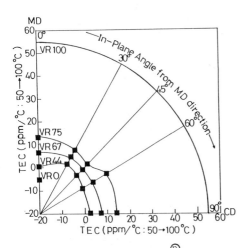

Fig.10  TEC of Technora® fabric
/epoxy laminate versus
in-plane angle

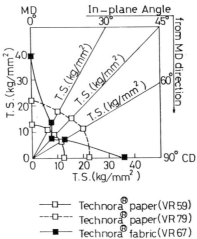

Fig.11  Tensile strength (T.S.)
versus in-plane angle

Table 4  HEC(In X-Y plane)of reinforcements,resin and laminates
(ppm/%RH :30°C X 30%RH - 30°C X 70%RH)

| Material | VR(%) | Angle from MD direction (deg) | | | | |
|---|---|---|---|---|---|---|
| | | 0(MD) | 30 | 45 | 60 | 90 |
| Technora® paper | 0 | -5.7 | * | * | * | * |
| | 59 | 4.7 | 10.2 | 5.8 | 7.1 | 4.0 |
| Technora® fabric | 0 | -9.5 | * | * | * | * |
| | 67 | 7.1 | 6.9 | 6.5 | 4.7 | 5.7 |
| PPTA fabric | 0 | -2.0 | * | * | * | * |
| | 61 | 39.0 | * | * | * | 16.4 |
| E glass fabric | 0 | 1.0 | * | * | * | * |
| | 69 | 6.4 | * | * | * | 8.5 |
| Epoxy resin | 100 | 50.8 | | | | |

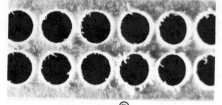 

Technora® paper/epoxy laminate with single-sided copper

Technora® fabric/epoxy laminate with single-sided copper

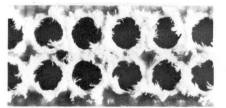 

PPTA fabric/epoxy laminate with single-sided copper

 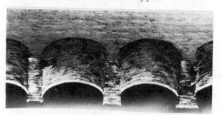

E glass fabric/epoxy laminate with single-sided copper

Fig.12  Optical and electron microscopic
observation of drilled holes

Table 5   The change of physical properties after
chemical treatments

| Laminate | Blank | Tensile strength retention | |
|---|---|---|---|
| | | 20%H$_2$SO$_4$aq.soln. (95°C X 48hrs.) | 10%NaOHaq.soln. (95°C X 48hrs.) |
| Technora®paper/epoxy | 100 | 92.8 | 81.0 |
| Technora®fabric/epoxy | 100 | 95.9 | 67.8 |
| PPTA  fabric/epoxy | 100 | 23.9 | 20.1 |
| E glass  fabric/epoxy | 100 | 2.0 | 43.1 |

Table 6   Properties of Technora® paper/epoxy laminate
and Technora® fabric/epoxy laminate

| Evaluation | Condition | Technora® paper/epoxy | Technora® fabric/epoxy |
|---|---|---|---|
| Solder temperature resistance (20secOK) | Dry C-96/20/65 C-96/20/65+C-96/40/90 | 310< 290 260 | 310< 290 260 |
| Peel strength (kg/cm) | A Solder float(260°Cx20sec) | 1.5 1.3 | 1.5 1.3 |
| Water absorption (wt %) | E-24/50 +D-24/23 | 0.7 | 0.5 |
| Volume resistivity (Ω-cm) | C-96/20/65 C-96/20/65+C-96/40/90 | 6.4X10$^{15}$ 7.3X10$^{14}$ | 2.4X10$^{15}$ 1.5X10$^{15}$ |
| Surface resistance ( Ω ) | C-96/20/65 C-96/20/65+C-96/40/90 | 9.4X10$^{14}$ 1.3X10$^{13}$ | 9.0X10$^{14}$ 6.2X10$^{14}$ |
| Insulation resistance ( Ω ) | C-96/20/65 C-96/20/65+D-2/100 | 3.5X10$^{13}$ 4.5X10$^{10}$ | 1.8X10$^{13}$ 1.7X10$^{10}$ |
| TEC(ppm/°C) VR=60 % | In X-Y plane(50-100°C) | 7.7 | 5.0 |
| HEC(ppm/%RH) VR=60 % | In X-Y plane(30°C) | 6.4 | 6.1 |

BEHAVIOR OF FLAME RETARDANT NYLONS IN ELECTRICAL TESTING
COMPARATIVE TRACKING INDEX INVESTIGATIONS

James J. Duffy
Occidental Chemical Corporation
Technology Center
Grand Island, New York

## Abstract

Materials used in electrical and electronic components have historically met high performance standards. Nylon molding compounds have been very acceptable because of their excellent strength, toughness, high use temperatures, chemical resistance and electrical properties. The additional need to provide flame retardance has led to developments of additive systems based on halogen or phosphorus to provide this extra margin of safety. Further concerns have led to the evaluation of current carrying components by the Comparative Tracking Index (CTI) test. This is directed to insuring safety of components that may be subject to surface contamination or moist conditions, particularly in poorly maintained areas.

This investigation is directed to a comparison of nylon compositions in the CTI test procedure especially as to sample integrity. Loss of sample integrity is felt to be a major factor in providing a quality electronic product subject to electrical stress.

Data has been developed on the effect of additive systems on integrity in non reinforced systems, the effect of glass reinorcement on sample performance, and the comparison of halogen based systems with phosphorus.

Results of these studies give the designer an option in his choice of materials for electronic components. Halogen based systems perform very favorably in overall properties and especially in sample integrity.

## 1. INTRODUCTION

Plastic usage in electrical or electronic equipment continues to grow annually. Concerns for the safety of operation in these applications have led to the imposition of performance standards for the plastic components used.

Criteria are in place which relate to physical, mechanical, electrical, and flammability performance.

In current carrying devices, the primary characteristics of the plastics used are their ability to provide insulation between metal parts and to isolate operating electrical circuits. This study is of the performance characteristics of nylon resin materials which are designed to provide both insulating and reduced flammability characteristics. The primary focus will be on the performance of flame resistant nylon compositions in a test designed to measure insulation failure under moist conditions. This measurement of failure has been described as tracking or arc tracking. When polymeric materials decompose under electrical or even thermal stress, they may produce carbonaceous residues which are conductive. Several tests have been devised to measure this phenomenon when surface tracking or failure of insulating properties are concerned. The most widely accepted test is the measurement of Comparative Tracking Index (CTI). This measures the ability of a material to withstand the formation of a conducting track between two electrodes on the surface of a plastic specimen as a conducting solution is dropped between the electrodes. The voltage at which a prescribed current flow is established between the electrodes is measured

and recorded as the CTI value in volts. Test methods are available for the measurement of this property.(1,2,3) The methodology and definition of CTI is indicated in Figures 1 and 2.

FIGURE 1
Description of Comparative Tracking Index (CTI)

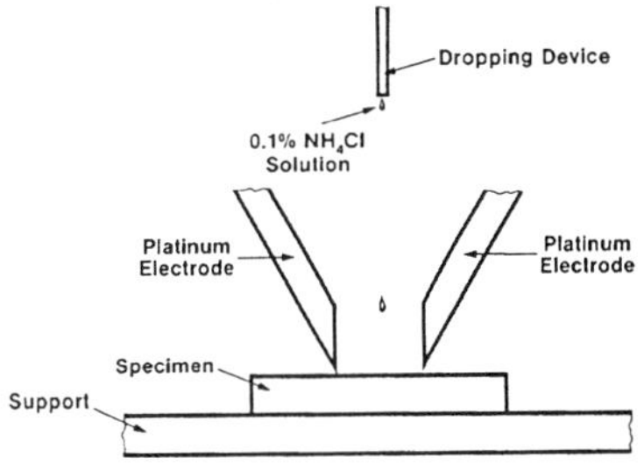

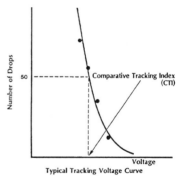

Electrode Arrangement

FIGURE 2
Determination of Comparative Tracking Index (CTI)

Typical Tracking Voltage Curve

In addition to failure in the CTI test by passage of current, two additional concerns have been raised. One is flaming combustion of the sample during test, even if current is not being passed.(4) The second is physical damage or loss of sample materials prior to failure.(5) This loss of material or sample integrity we will define as erosion. Evaluation of each of these phenomenon in conjunction with the actual CTI value itself should provide additional data for evaluation of safety in use when choosing a material. The concern for safety of materials as related to CTI testing is concerned with such applications as cable jacketing (6), coal mines (7), or ship board electronics.(8) The nylon compositions described in the present study are generally small insulating devices such as bobbins or connectors. Failure here through loss of insulating capacity as well as loss of sample integrity should be of serious material selection concern.

Various formulations have been developed to render nylon more flame resistant. Additives are normally used consisting of phosphorus, mainly encapsulated red phosphorus (Red P), thermally stable halogenated additives in combination with metal oxides, or melamine derivatives. Previous work has shown the comparison of the various additive systems in regards to flammability and response in the CTI test.(9,10) These are summarized in Tables 1 through 6. This shows the ability of halogen based formulations to effectively compete with phosphorus based materials which may previously have been chosen for their CTI performance. The present investigations will extend these studies to consideration of sample

integrity during electrical stress, which could be as important a criteria for material selection as a CTI value itself.

TABLE 1
Nylon 6
Typical Flame Retardant Formulations
Non-Reinforced

| Additive | Weight % | Flammability | | Comparative Tracking Index (Volts) | |
|---|---|---|---|---|---|
| | | UL-94 @ 1.6mm | Oxygen Index | $K_c$ | $K_b$ |
| Red Phosphorus | 6-10 | V-2 | 24-25 | 400-500 | 400-500 |
| Dechlorane Plus | 20-25 | V-0 | 28-33 | 225-275 | 225-275 |
| Antimony Oxide | 5-10 | | | | |
| Decabromodiphenyl Oxide | 15-20 | V-0 | 30-33 | 250 | 225 |
| Antimony Oxide | 5-10 | | | | |
| Pyro-Chek 68 PB | 17-20 | V-2, V-0 | 24-27 | 300 | 250 |
| Antimony Oxide | 5-10 | | | | |
| Melamine Cyanurate | 15-25 | V-0 | 30 | 550 | 425 |

TABLE 2
Nylon 66
Typical Flame Retardant Formulations
Non-Reinforced

| Additive | Weight % | Flammability | | Comparative Tracking Index (Volts) | |
|---|---|---|---|---|---|
| | | UL-94 @ 1.6mm | Oxygen Index | $K_c$ | $K_b$ |
| Red Phosphorus | 6-10 | V-0 | 26-28 | 500 | 500 |
| Dechlorane Plus | 18-25 | V-0 | 26-31 | 225 | 150 |
| Antimony Oxide | 5-10 | | | | |
| Decabromodiphenyl Oxide | 15-20 | V-0 | 27-31 | 225 | 150 |
| Antimony Oxide | 5-10 | | | | |
| Pyro-Chek 68 PB | 18-25 | V-2,V-1,V-0 | 26-29 | 275 | 200 |
| Antimony Oxide | 5-10 | | | | |
| Melamine Cyanurate | 15-25 | V-0 | 30 | 550 | 450 |

TABLE 3
Nylon 6
Typical Flame Retardant Formulations
Non-Reinforced
Use of Various Synergists with Dechlorane Plus

| Synergist(s) | Flammability | | Comparative Tracking Index (Volts) | |
|---|---|---|---|---|
| | UL-94 @ 1.6mm | Oxygen Index | $K_c$ | $K_b$ |
| Antimony Oxide | V-0 | 29 | 275 | 275 |
| Ferric Oxide | V-1 | 29 | 325 | 250 |
| Zinc Oxide | NC | 27 | 550 | 400 |
| Antimony Oxide/Zinc Oxide | NC | 25 | 475 | 400 |
| Ferric Oxide/Zinc Oxide | NC | 27 | 450 | 450 |
| Antimony Oxide/Zinc Borate | V-0 | 27 | 425 | 350 |

**TABLE 4**

**Nylon 66**

**Typical Flame Retardant Formulations**

**Non-Reinforced**

**Use of Various Synergists with Dechlorane Plus**

| Synergist(s) | Flammability | | Comparative Tracking Index (Volts) | |
|---|---|---|---|---|
| | UL-94 @ 1.6mm | Oxygen Index | $K_c$ | $K_b$ |
| Antimony Oxide | V-0 | 32 | 225 | 225 |
| Ferric Oxide | V-0 | 30 | 225 | 225 |
| Zinc Oxide | V-1 | 27 | >500 | >500 |
| Antimony Oxide/Zinc Oxide | V-0 | 29 | 550 | 450 |
| Ferric Oxide/Zinc Oxide | V-0 | 27 | 550 | >500 |
| Antimony Oxide/Zinc Borate | V-0 | 30 | 375 | 275 |

**TABLE 5**

**Nylon 6**

**Typical Flame Retardant Formulations**

**25% Glass Reinforced**

| Additive | Flammability UL-94 @ 1.6mm | Comparative Tracking Index (Volts) $K_c$ |
|---|---|---|
| Red Phosphorus | V-0 | 375 |
| Dechlorane Plus Antimony Oxide | V-0 | 225 |
| Brominated Additives Antimony Oxide | V-0 | 225 |
| Melamine Cyanurate | NC | |
| Dechlorane Plus Zinc Oxide | V-1, V-0 | 275 |
| Dechlorane Plus HV Antimony Oxide | V-0 | 400 |

**TABLE 6**

**Nylon 66**

**Typical Flame Retardant Formulations**

**25% Glass Reinforced**

| Additive | Flammability UL-94 @ 1.6mm | Comparative Tracking Index (Volts) $K_c$ |
|---|---|---|
| Red Phosphorus | V-0 | 400 |
| Dechlorane Plus Antimony Oxide | V-0 | 225 |
| Brominated Additives Antimony Oxide | V-0 | 225 |
| Melamine Cyanurate | NC | — |
| Dechlorane Plus Zinc Oxide | V-0 | 250 |
| Dechlorane Plus HV | V-0 | 425 |

## 2. MATERIALS

The nylon polymers and other materials used in the investigations were available from commercial sources and used as received. The flame retardant additives were secured from their suppliers.

Additives used include Dechlorane® Plus (Occidental Chemical Corp.), Dechlorane® Plus HV (Occidental Chemical Corp.), Decabromodiphenyl Oxide (Dow Chemical Corp.), Pyrochek® 68PB (Ferro Corp.), Zinc Oxide (New Jersey Zinc), Firebrake® ZB (U.S. Borax), Antimony Oxide, Thermoguard S (trademark of M&T Chemicals) and red phosphorus, Novared 120 (80% Phosphorus content from Rinkagaku Kogyo Co.).

## 3. EXPERIMENTAL PROCEDURES

### 3.1 Sample Preparation

The experimental samples were prepared by dry blending the ingredients and melt compounding on a Brabender extruder. The compounded extrudate was chopped and injection molded into 3.2 mm. thick samples.

### 3.2 Physical Properties' Determinations

Properties were determined by standard ASTM methods.

### 3.3 Flammability Testing

Testing was accomplished by procedures for UL-94 and Oxygen Index.

### 3.4 Comparative Tracking Index and Erosion

The evaluation for Comparative Tracking Index (CTI) was conducted on a Beckman Insulation Tracking Test Set according to ASTM procedure D-3638 utilizing the electrolyte solutions described in DIN 53,480. The composition of the solutions are as shown in Table 7. These same solutions were used in the erosion studies.

TABLE 7

Electrolyte Solutions Used for the Determination
of Comparative Tracking Index (CTI)
and Erosion Studies

| | In 100 ml. of Distilled Water |
|---|---|
| Solution for $K_c$ | 0.1g NH$_4$Cl |
| Solution for $K_b$ | 0.1g NH$_4$Cl and<br>0.5g Cationic Surfactant |

In order to evaluate the phenomenon we have defined as erosion, the following procedure was followed. The CTI apparatus voltage applied was set at 100 volts and increased in 50 volt intervals to 550 volts. At each voltage, the Kc conductive solution was applied to failure in the test or 50 drops, whichever occurred first. After the test, the sample was examined for deterioration of material. The depth of penetration of the damaged area into the sample was measured with calipers. The volume of the sample destroyed was measured by displacement of a water, wetting agent solution from a micro syringe needed to fill the cavity. The results are reported in mm. of depth and micro liters of volume respectively.

4. RESULTS AND DISCUSSION

Figure 3 is a schematic graphical display of the results of a typical evaluation. The horizontal axis indicates the voltage applied in 50 volt intervals during the experiment. The vertical axis indicates the erosion of the sample either in depth of penetration or the volume of the total excavation. A typical curve is

also illustrated with an indication of the CTI found for that sample and a vertical dashed line indicating the acceptance criteria for the CTI value. In addition to the actual values obtained, a visual inspection of the sample gives valuable evidence of effects such as the morphology of the residue and the relative amounts of carbonaceous residue produced.

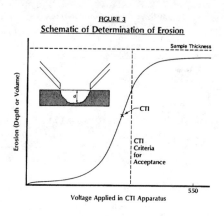

FIGURE 3
Schematic of Determination of Erosion

Figure 4 shows the results of the erosion evaluation of a typical halogenated additive as different synergists are used. Each of the samples gave excellent flammability performance but only the zinc oxide modified material gave high CTI values of >500 volts. In the erosion evaluation, zinc oxide alone gave good retention of integrity and little evidence of accumulation of conductive carbonaceous residue. Both of the other samples showed considerable evidence of carbonaceous residue and loss of material at voltages above their CTI value, approximately 250 volts. The formation of conductive material accounts for both the lowered CTI values and loss of sample due to heating effects during the testing.

FIGURE 5
Erosion of Flame Retardant Nylon 66 in CTI Testing.

Comparison of Halogen Systems with Phosphorus

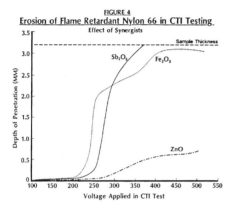

FIGURE 4
Erosion of Flame Retardant Nylon 66 in CTI Testing.

Effect of Synergists

Figure 5 illustrates a comparison of a red phosphorus modified nylon with samples containing halogenated additives. The phosphorus containing sample showed little residue between the electrodes, and little evidence of carbonaceous materials. Much appeared to have been volatilized. Thus, we have a non reinforced nylon composition with quite variable response depending on the additive system used. The relatively high CTI value of the phosphorus system does not guarantee sample integrity.

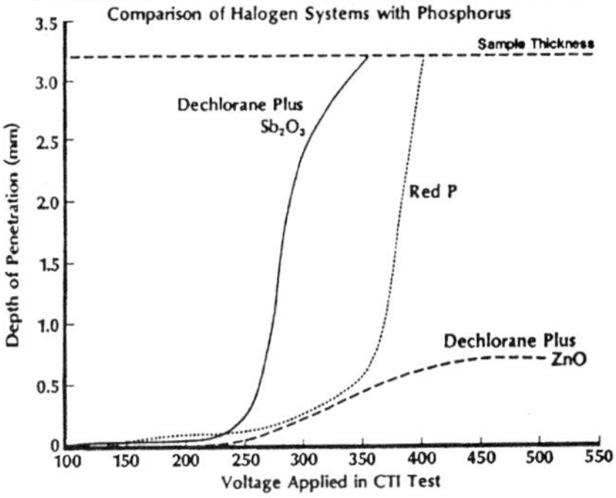

Figures 6 through 9 show the effect of glass reinforcement in two different nylon polymers. In Nylon 66, the glass generally provides a refractory support for the decomposing system yielding lower CTI values in the halogen containing system, especially with zinc oxide. At voltages below the CTI value, considerably more material loss is exhibited for both halogen based systems; likely due to thermal volatilization during inductive heating while at higher voltages, only the non carbon yielding zinc system is adversely affected. Greater evidence of carbonaceous residue is observed here. Neither of the glass systems completely result in a hole being formed in the sample but provides sample integrity. This is true even to voltages much higher than the CTI failure point. In Figure 7 versus Figure 6, the effect of the polymer becomes evident. Nylon 6 is not as good a residue or char former as Nylon 66 due to a different mode of decomposition. Addition of glass adversely affects the CTI as with Nylon 66. Also sample with the antimony oxide yield decreased sample deterioration at high voltages. This is not true for the zinc sample due to a test anomaly. Failure is so fast, no volatilization occurs and little carbon is formed. This illustrates the effect glass can have in actual performance. While it is added to provide rigidity in samples, its effect in electrical testing must be considered. Its refractory nature accentuates the char formation and inductive heating effects yielding a deleterious effect. However, after failure or current passing in the CTI test, it may actually improve the overall sample integrity. Other non melting fibrous additives have similar effects. To achieve a high CTI and high sample integrity, a non reinforced material may be preferred if other component properties can be met.

The effect of glass reinforcement in the phosphorus systems, Figures 8 and 9, is not as dramatic. At no time does the addition of glass lead to a significant improvement in sample integrity. In the Nylon 6 system, the effect is similar to the Dechlorane Plus antimony oxide system. Effects here are obscured as in most every sample, the phosphorus based materials actually burn during testing resulting in vigorous loss of

material.

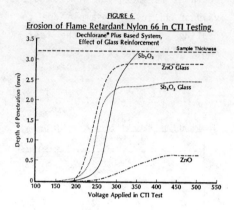

FIGURE 6
Erosion of Flame Retardant Nylon 66 in CTI Testing.
Dechlorane® Plus Based System,
Effect of Glass Reinforcement

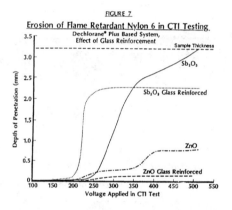

FIGURE 7
Erosion of Flame Retardant Nylon 6 in CTI Testing.
Dechlorane® Plus Based System,
Effect of Glass Reinforcement

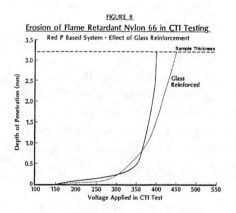

FIGURE 8
Erosion of Flame Retardant Nylon 66 in CTI Testing.
Red P Based System - Effect of Glass Reinforcement

1218

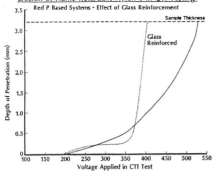

FIGURE 9
Erosion of Flame Retardant Nylon 6 in CTI Testing.
Red P Based Systems - Effect of Glass Reinforcement

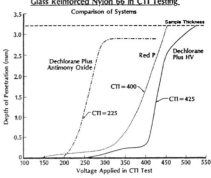

FIGURE 10
Erosion of Flame Retardant
Glass Reinforced Nylon 66 in CTI Testing.
Comparison of Systems

Figure 10 illustrates the effectiveness of a system designed for high CTI glass reinforced nylons based on a halogenated additive. Significantly less loss of sample integrity below the CTI value is obtained versus a normal halogen modified or a red phosphorus modified material. At the same time, it overcomes the deleterious effect on CTI noted in reinforced materials based on halogenated additives while maintaining excellent flame retardant performance.

The information developed and shown in the examples above should alert one to consider the entire performance of a sample in testing. An awareness of the variables involved and their effect on performance is necessary. Such changes as sample composition, synergist, reinforcement, operating voltage, or environmental conditions could serve to dramatically affect performance and choice of material.

5. SUMMARY AND CONCLUSIONS

It was the intent of this study to evaluate the sample integrity of flame retardant nylon compositions whose performance in the CTI electrical test was of interest. This was done by defining and measuring a property of erosion of the sample undergoing testing. It is hoped that this will stimulate the evaluation of a more complete range of performance properties in component parts where safety of performance is a major criteria. The materials evaluated in this study are representative of those which are found in commercial electronic components.

## TABLE 8

## Summary of Flame Retardant-Comparative
## Tracking Index Performance

|  | Flammability | CTI | Erosion |
|---|---|---|---|
| **Nylon 66** | | | |
| Halogenated Additives Antimony Oxide | Excellent | Poor | Poor |
| Red P | Excellent | Good | Poor-Good |
| Dechlorane Plus Zinc Oxide | Excellent | Excellent | Excellent |
| Nitrogen Derivatives | Excellent | Excellent | Excellent |
| **Nylon 66 Glass Reinforced** | | | |
| Halogenated Additives Metal Oxide Synergists | Excellent | Poor | Poor-Good |
| Red P | Excellent | Good | Poor-Good |
| Dechlorane Plus HV | Excellent | Good | Good |
| Nitrogen Derivatives | Not Acceptable | | |
| **Nylon 6** | | | |
| Halogenated Additives Antimony Oxide | Excellent | Poor | Poor-Good |
| Red P | Excellent | Good | Poor-Good |
| Halogenated Additives Metal Oxides | Good-Excellent | Poor-Excellent | Poor-Excellent |
| Nitrogen Derivatives | Excellent | Excellent | Excellent |
| **Glass Reinforced Nylon 6** | | | |
| Halogenated Additives Metal Oxide | Good-Excellent | Poor | Poor-Excellent |
| Red P | Excellent | Good | Poor-Good |
| Dechlorane Plus HV Antimony Oxide | Excellent | Good | Good |
| Nitrogen Derivative | Unacceptable | | |

The investigations of Comparative Tracking Index testing and associated sample erosion has shown wide resulting performance. This is summarized in Table 8 for the general materials of this study. This confirms the notion that consideration of single reported values are insufficient. A more complete investigation of the entire performance response and the variables that affect the results are necessary. Figures 11 and 12 are two examples, at a single voltage value, where consideration of full results could affect a choice of materials.

Results have shown that nylon formulations based on halogenated materials which may be preferable for reasons such as ability to color and tint, ease of processing and years of performance in service, may perform quite comparably to other alternatives available. When one considers the aspect of sample integrity, they are superior and deserving of consideration in design evaluations.

## 6. REFERENCES

1) ASTM D-3638 (1977).

2) International Electrochemical Commission, Publication 112, 2nd Edition (1971).

3) DIN 53,480.

4) Suhr, H., Materialprufung, 19, 208 (1977).

5) Suhr, H., Z. F. Werkstofftechnik, 5, 437 (1974).

6) Paciorek, K.J.L., et al, IEEE Trans. Electr. Insul., Vol. EI-17, 423 (1982).

7) Weller, M.G., IEEE Conf. Publn., 105 (1982).

8) Day, A.C. and Stonard, D.J., IEEE Trans. Electr. Insul., Vol. EI-12, 191 (1977).

9) Duffy, J.J. and Ilardo, C.S., Ninth Europ. Conf. on Flamm. and Fire Retr., May 1985.

10) Ilardo, C.S. and Duffy, J.J., SPI Ann. Tech. Conf., May 1985.

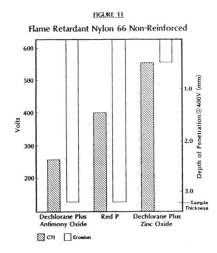

**FIGURE 11**

**Flame Retardant Nylon 66 Non-Reinforced**

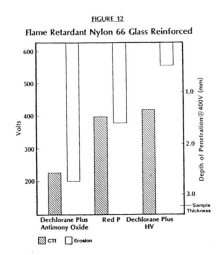

**FIGURE 12**

**Flame Retardant Nylon 66 Glass Reinforced**

## BIOGRAPHY

James J. Duffy, Ph.D., is presently Manager of Environmental Technology at the Occidental Chemical Corporation, Technology Center, Grand Island, New York. Dr. Duffy has held various technical and management positions in technology. His interests include organohalogen and organophosphorus chemistry, polymer additives, polymer processing and properties, and flammability studies in addition to environmental technology.

He holds a B.S. Degree from Canisius College and a Ph.D. in Chemistry from the University of Kentucky.

SILICONE SPRAY CONFORMAL COATING PARAMETER OPTIMIZATION
DAVID TUCKER
PROCESS ENGINEERING, ELECTRO-OPTICS SYSTEMS OF
TEXAS INSTRUMENTS, DALLAS, TEXAS

ABSTRACT

Environmental protection of printed circuit boards used in military systems is achieved by application of a conformal coating. The coating process is characterized by labor intensive masking and demasking of PCB parts such as connectors and test terminals that are required to remain coating free. The common mode of application is a dipping process. Automatic spray conformal coating presents an attractive labor saving alternative to dipping through a decrease in wicking and subsequent ease of masking. Labor savings may reach approximately one third of the present mask and demask requirements. This paper reviews the optimization of spray machine process parameters by determining their effects on the silicone polymer's cure rate, thoroughness and uniformity of circuitry coverage, and overspray; a cosmetic problem involving a 'dusting' effect on the bottom of PCBs incurred while spraying the top. Conveyor speed, oven settings, choice of solvent, spray gun orientation, atomization pressure, and details about the curing reaction are reviewed. Labor saving masking techniques and quality enhance ments are also discussed.

1. INTRODUCTION

In the Electro-Optics Systems, assembled printed circuit boards (PCB's) are conformally coated in an automatic spray application with a silicone resin, Dow Corning's R-4-3117, to provide environmental protection. This is the first automated spray conformal coating system employed at TI. It replaces the former application method: dipping PCB's in a tank and allowing the coating to drip off. Spray coating provides the opportunity for tremendous labor savings and quality enhancements and may soon be adapted by additional groups within Texas Instruments. The primary focus of this article is the determination of optimal settings for critical process parameters. This is achieved by focusing on parameter's effects on the silicone polymer's cure rate, the ability to smoothly and thoroughly cover circuitry, and the effects on overspray; a cosmetic problem involving a 'dusting' effect on the bottom of PCB's incurred while spraying the top.

## 2. CHOICE OF COATING AND COATING APPLICATION SYSTEM

MIL-I-46058C governs the use of conformal coating and allows the use of five types of coatings: acrylic, epoxy, silicone, polyurethane and paraxylene. Silicone was chosen primarily for its excellent thermal stability and repairability. Dielectric strength, moisture resistance and ionic particle entrapment are other essential attributes of this coating. Silicones coating has provided extremely reliable service to functioning systems for over a decade.

The dipping application was accomplished by dipping parts in open tanks of primer and coating. With dip tanks open daily for two to five hours, contamination and pieces of semi-cured coating were easily transferred to parts and appeared as blemishes often requiring touch-up. The extent of these defects was minimized with spray coating as the coating and primer are housed in sealed pressurized tanks. Unlike the spray application, dipped boards dry in a vertical orientation. This provides lumps on ends of integrated circuits and may cause coating streaks on the board. Masking requirements are stricter and more labor intensive for dipping than spraying due to the extreme increase in wicking associated with dipping. An important example of this is the liquid masking required to prevent coating from wicking under connectors. Since spray application of coating incurs minimal wicking, liquid masking is not required and easy to apply caps may be used to mask connectors.

## 3. BASIC FOR PARAMETER OPTIMIZATION

The determination of optimal parameters was driven by the desire for process enhancements as well as the need for refinements necessary or machine operation. Process enhancements include minimizing operating costs and cosmetic defects. Necessities include assuring effective coverage of all exposed circuitry and disallowing the formation of bubbles; a defect caused by thermal expansion of solvent beneath the coating surface resulting in the entrapment of gas bubbles. Minimization of cycle time became a necessity since the projected decrease in labor spurred a reduction from two shifts to one. The shift reduction is extremely vital as the spray machine must coat a run of parts twice, once for each side, within eight hours; the time required to dip parts once. It is in this light that the determination of process parameters was conducted.

## 4. CURE RATE

Dow Corning's R-4-3117 cures via cross linking of the liquid silicone polymer to achieve, when heat cured, an increase in specific gravity from 1.07 to 1.12 and a Shore D Durometer hardness of 45. Although a tack free state may be reached in two to four hours with heat and 12 to 24 hours in air, complete curing requires up to seven days.

Chain linking of the polymer occurs by the following simplified reaction (1) (parenthesis denote part of a polymeric chain): Reaction (1) may be viewed as the summation of reactions (2) and (3). Equation (1) reveals a dependence on water in the

$$
\begin{array}{c}
\text{OCH3} \\
| \\
2(-Si-O-Si-) + H2O \longrightarrow (-Si-O-Si-O-Si-) + 2CH3OH \qquad (1)
\end{array}
$$

$$
\begin{array}{cc}
\text{OCH3} & \text{OH} \\
| & | \\
(-Si-O-Si-) + H2O \longrightarrow (-Si-O-Si-) + CH3OH & (2)
\end{array}
$$

$$
\begin{array}{cc}
\text{OCH3} \quad \text{OH} \\
| \qquad | \\
(-Si-O-Si-) + (-Si-O-Si-) \longrightarrow (-Si-O-Si-O-Si-O-Si-) + CH3OH \quad (3)
\end{array}
$$

form of atmospheric humidity. This helps explain the extensive cure time since water is insolu-bility in recommended coating solvents: toluene, 1,1,1 trichloroethane and xylene. This is why solvent addition is minimized to an amount that will give acceptable coverage.

### 5. EFFECTS OF PARAMETERS ON CURE RATE

Thermal energy has a strong positive effect on cure time via accelerating the endothermic reaction and vaporizing the solvents. Ovens were set at the highest allowable temperatures to maximize these benefits while preventing bubbling. Thermal profiles of PCBs show temperatures of 95 to 105°C (203 to 221°F) well below the 125°C (257°F) maximum curing temperature allowed by MIL-I-46058C.

It is important to minimize the amount of coating applied due to the increased propensity for bubbling with increased thickness. Thinner coating also cures faster due to the lower amount of water insoluble solvent applied and decrease in the distance moisture must transverse to effect the cure. Thus, coating is applied only thick enough to provide effective coverage.

Atomization pressure was found to affect cure primarily through its affect on film thickness. Atomizing coating differs from atomizing conventional spray paint where higher atomization often increases solvent flash off and results in a higher solids quantity applied and subsequent decrease in cure time. Increasing coating atomization does not achieve this since the coating solvent xylene has a low room temperature vapor pressure. Coating and solvent quantities decreased in tests with increasing atomization in a manner proportional to their mixture ratios. The result showed a shorter cure time at the expense of thinner coating and obvious coating waste. For this reason coating atomization was placed only high enough to prevent lumpiness.

Although a considerable amount of methanol evolves as a reaction by-product, its properties do not indicate significant inhibition to curing: low density (0.79 g/cc (49.3 lb/ft3)) to provide buoyancy, low boiling point of 65 °C (149°F) to speed evaporation, and infinite solubility in water.

Conveyer speed affect the cure by allowing various time intervals in the ovens. Five inches per minute was chosen to provide 24 minutes of oven time. This relatively slow conveyor speed aids in producing less tacky boards at the expense of increased cycle time. Several of the above parameters' effects on cure time were empirically verified using a tack free test method similar to Federal Test Method Standard 141B Method 4061. Since the time required to reach a tack free state is indicative of the extent of cure, it was used as the primary test variable. Tack free time is also the most important variable in the manufacturing viewpoint, as it influences cycle time. Tack free time was defined as the time required to reach a quantifiable state of tackiness acceptable to demask. Figure 1 displays effects of critical process parameters on tack free time.

The results tabulated in Figure 1 were determined from 31 tests in which the dependent variable - tack free time - was isolated from other variables by incorporating the other variables' effects. For

| PARAMETER | CHANGE IN PARAMETER | EFFECT ON TACK FREE TIME |
|---|---|---|
| Dry film thickness | Decrease .025 mm (1.0 mil) in 4.0 to 6.0 mil range | 0.5 Hours decrease |
| Relative humidity | 10% increase in 30-70% range | 1.2 Hours decrease |
| PCB temperature | 20 °C (36 °F) increase in 80 to 120 °C (176 to 248 °F) range | 0.6 Hours decrease |
| Conveyor speed | 10 to 5.0 RPM | 1.1 Hours decrease |
| Solvent dilution ratio | Decrease xylene from 1:2 to 1:3 | 0.4 Hours decrease |
| Solvent | Xylene to toluene | 0.3 Hours decrease |

Figure 1. Process parameter's effects on tack free time

example, while testing the effect of dry film thickness after determining the effect of humidity, the results may show no apparent correlation as seen in figure 2a. Accounting for humidity variations: 30% for 4.0 mils test, 40% for 4.5, 5.0 and 5.5 mils tests, and 50% for the 6.0 mils test, by standardizing humidity at 40% will reveal a definite

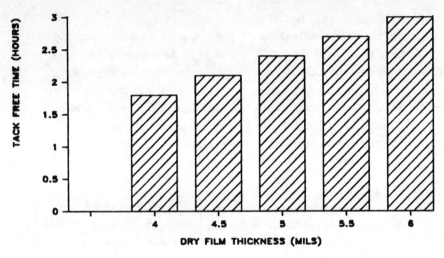

Figure 2(B): Hypothetical test results showing direct relation between tack free time and dry film thickness after isolating dry film variable.

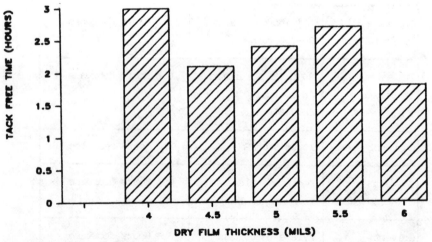

Figure 2(A): Hypothetical test results displaying no apparent relation between tack free time and dry film thickness. Tests were run on different days with different humidities.

correlation as seen in figure 2b. This was achieved by allowing 1.2 hours for a 10% humidity change. The resulting correlation is a 0.5 hour increase in tack free time with a one mil increase in dry film thickness.

6. SOLVENT CHOICE

Although Dow Corning R-4-3117 is supplied with no greater than 30% xylene, additional solvent is added to achieve sufficient coverage; i.e., to flow under components, coating the entire circumference of solder joints. Solvents evaluated include 1,1,1-trichloroethane, toluene and xylene. All solvents impart similar flow to the coating as evidenced by the similarities of surface tensions in Figure 3 and similar coverage of parts observed at a three to one coating to solvent dilution ratio. The critical difference between solvents was their effect on the time required to achieve a tack free state. As seen in Figure 3, trichloroethane has the lowest boiling point, indicating the quickest solvent evaporation. This hastens the removal of tackiness by ridding the liquid solvent and providing a solvent free medium conducive to moisture absorption. But when oven temperatures were considered, xylene and toluene were found to be more effective than trichloroethane in achieving a tack free state by allowing much higher oven temperatures. Trichloroethane expanded to form bubbles in the coating at oven

| SOLVENT | BOILING POINT degrees C (degrees F) | DENSITY gm/cc (lb/ft3) | SOLUBILITY PARAMETER (cal/cc)$^{\frac{1}{2}}$ | SURFACE TENSION dynes/cm |
|---------|------------|---------|-----------|---------|
| 111-Trichloroethane | 74 (133) | 1.34 (83.7) | 8.6* | 25.9* |
| Toluene | 111 (200) | .87 | 8.9 | 28.5 |
| Xylene | 138-144 (248-259) | .86-.88 (53.7-55.0) | 111-8.8 | 28.4-30.1 |
| Coating as supplied | | 1.07 (66.9) | | |

*Estimated from Handbook of Chemistry and Physics, p. C-726

Figure 3: Physical properties of evaluated coating solvents.

temperatures well below bubbling temperatures for toluene or xylene. In addition to oven temperature constraints, trichloroethane also inhibits the achievement of a tack free state by its higher density than coating which prevents buoyancy form aiding evaporation. Thus, trichloroethane provided the slowest tack free time.

The solubility parameter is an important factor due to its indication of water dissolution rate. However, little can be concluded about the relative solubility of water in these solvents since they are nearly identical and differ greatly with the solubility parameters of water: 23.4. With xylene and toluene possessing similar densities and, consequently similar buoyancy effects, the choice of solvent depended on the effect of toluene's lower boiling·point - toluene was empirically shown to provide a slightly shorter tack free time while not increasing bubbling.

7. SPRAY GUN SETTINGS

The height and angle of the six spray guns was extremely critical in providing effective coverage and minimizing overspray. Spray gun orientation affected component coverage in a logical manner: when guns were placed vertically, the top surface of components were covered well, the sides poorly and solder joints under can transistors received almost no coating. If all spray guns were placed at steep angles, the sides of components would receive adequate coating, but the board space between densely packed components as well as adjacent solder joints would receive poor coverage. Thus, it was decided to place four guns at approximately 45 degree angles to the PCB's and each other to effectively cover all sides of components. The remaining two guns were oriented nearly vertically to assure coverage of horizontal sides of components. Coverage was empirically determined to optimal at the minimum allowable gun height of 4.2 inches above parts. A probable explanation involves the high energy level and small particle size found closer to the gun nozzle and the increased spreading associated with these factors. Heights lower than 4.2 inches were deemed unacceptable due to the frequent use of tall components.

Spray gun settings also had strong effects on overspray. This defect appeared on parts with components on the lead side. This caused the lead side to be elevated during second pass and receive a "dusting" or overspray on the prevy coated lead side. Although the primary cause of overspray, either bouncing of liquid particles off the tray or upward motion of particles in turbulence is debatable, it is certain that both contribute. Turbulence was minimized by lowering atomization pressure and assuring that the paths of sprayed coating do not collide. Overspray caused by the bouncing effect was more prevalent at the chosen 45 degree gun angles than if the guns were vertical, but it was not at a detrimental level.

## 8. CONCLUSIONS

Correct parameter settings achieved necessary goals by allowing a one shift operation - 300 boards may be coated in approximately five hours. Bubbles were avoided by lowering oven settings while providing high enough temperatures to significantly reduce cure time. Coating gun parameters were set to provide acceptable coverage while minimizing coating usage and overspray. Toluene was found to be the most effective solvent in reducing tack free time. Humidity was found to be perhaps the most critical process parameter affecting cure rate, yet the only uncontrollable one. The use of proper settings on the conformal coating system will ensure its continued success in producing quality coating in a cost effective manner.

## BIOGRAPHY

David Tucker received his B.S. in Chemical Engineering from Cornell University in 1984. He began work at Texas Instruments in September 1984 as a Diffusion Engineer responsible for furnace operations in the LSI Bipolar Front End at Houston. In 1985 David attained a position in TI's Electro Optics Systems as a Process Engineer responsible for Vapor Phase Soldering in a printed circuit board assembly area. Thereafter he began the task of determining process parameters for the newly installed spray conformal coating.

David is an active member of Toastmaster International, the TI Corporate Track Team and AICHE.

THERMALLY HARDENED INSULATION FOR TWISTED-PAIR COMMUNICATION WIRE

Brian R. Schallhorn, Anthony J. Baba, and Dr. Stewart Share

Harry Diamond Laboratories, Adelphi, MD

## Abstract

Nylon-coated polyethylene is widely used as an insulating coating for twisted wire pairs because of its low cost. Unfortunately, it is susceptible to melting at relatively low levels of thermal radiation. This paper presents thermal response test data for the following wire insulation candidates which are significantly less susceptible to high-intensity thermal radiation than nylon-coated polyethylene: (1) radiation-cross-linked ethylene tetrafluoroethylene, (2) polyalkene under radiation-cross-linked poly-vinylidene fluoride, (3) woven silica glass, (4) woven fiberglass covered with Teflon, (5) Kapton, and (6) clear fluorocarbon. Test samples were exposed at the White Sands Solar Facility to thermal radiation with peak fluxes as high as 3.3 MW/m$^2$ (80 cal/cm$^2$-s) and fluence levels to 8.8 MJ/m$^2$ (210 cal/cm$^2$). The test results indicate that these insulations did provide varying amounts of enhanced hardness to thermal radiation without significantly increasing the volume of insulation required.

## 1. INTRODUCTION

Polyethylene is commonly used to insulate electrical wire because of its low cost. However it can melt, disintegrate, or otherwise breach when exposed to high-intensity thermal radiation, baring the metallic conductor and posing the possibility of short circuits.[1] For example, black polyethylene jackets (0.4-mm-thick) breach at fluences around 1.3 MJ/m$^2$ (30 cal/cm$^2$). Increasing the jacket thickness provides some resistance to thermal damage: 0.7-mm-thick polyethylene jackets did not breach at fluences under 2.7 MJ/m$^2$

(65 cal/cm$^2$). While the increase in the jacket thickness enhanced thermal hardness, it also increased the weight and volume of the wire. Likewise, methods used to thermally harden fiber optic cables could not be used without significantly increasing the jacket thickness.[2,3] These measures also have an associated cost penalty.

To explore the extent to which thermal hardness was compatible with thinness in insulation, the following types of commercially available wire insulation were tested: (1) radiation-cross-linked ethylene tetrafluoroethylene (ETFE), (2) polyalkene under radiation-cross-linked polyvinylidene fluoride (PVF$_2$/PK), (3) a woven silica glass, (4) woven fiberglass covered with Teflon, (5) Kapton, and (6) clear fluororocarbon. This paper describes these insulating materials, their response to high-intensity thermal radiation, and the postexposure behavior in air and wet environments (conducting liquid baths).

## 2. EXPERIMENTAL PROCEDURE

Experimental data on the wire and cable responses to thermal radiation were obtained by exposure of 0.23-m (9-in.) lengths of each wire type to the thermal radiation pulse environment at the White Sands Solar Facility (WSSF). The WSSF uses a 12.2 by 11.0-m (40 by 36-ft) heliostat to reflect solar radiation onto a spherical concentrator which has a focal length of 10.7-m (35-ft). Test samples are irradiated at the focal plane of the concentrator, inside the elevated test room between the heliostat and the concentrator. The thermal flux at the sample position (the focal plane) is periodically measured by WSSF personnel with a Hycal asymptotic calorimeter. The flux at the focal plane is fairly uniform (i.e., within a few percentage points) over a 0.02-m (3/4-in.) radius circle at the center of the irradiated area. At larger distances from the center, the flux drops rapidly: at a 0.025-m (1-in.) radius the flux drops by 10 percent, at a 0.05-m (2-in.) radius it drops by 50 percent, and at a 0.076-m (3-in.) radius it drops by 95 percent.

The pulse shapes used for these exposures had a time period of 5 to 12 s. In all cases, the peak irradiance ($H_p$) occurs at a time ($t_p$) which is 0.1 of the total pulse width, i.e., 0.5 to 1.2 s. The irradiance then slowly falls to zero at $10t_p$. The fluence (Q) incident on the samples is obtained from the time integral of this pulse and is equal to $2.08H_p t_p$. This value is accurate to within ±7 percent.

The integrity of the wire insulation was evaluated by measurement of the resistance between each conductor in the wire pair. For this resistance measurement, each wire was connected across a digital ohmmeter, which applies a 1-V potential and has a maximum range of $16 \times 10^6$ ohms. Resistance measurements were made in air and then repeated with the thermally exposed center portion of the wire immersed in a salt water bath.

### 3. WIRE DESCRIPTIONS

Two types of ETFE insulated wire were tested. In one, a 0.8-mm-diameter bundle of stranded metal (AWG26) was covered by a 0.25-mm-thick layer of red radiation-cross-linked, extruded, modified ETFE. The total outside diameter (OD) of the wire is 1.3-mm. Two wires were twisted together for the thermal tests. In the other type, a 1.6-mm bundle of metal strands (AWG20) was covered by a 0.4-mm-thick layer of the same ETFE, but colored black. The total OD of this wire was 2.4-mm.

Two types of $PVF_2$/PK wire were also tested. In the first type, a 1-mm-diameter metal bundle was covered by a 0.3-mm-thick inner layer of green radiation-cross-linked, extruded PK and a 0.15-mm-thick outer layer of clear radiation-cross-linked, extruded $PVF_2$. The total outside diameter of the wire is 1.9-mm. Again, two wires were twisted together for the thermal tests. In the second type, both insulating layers are 50-percent thinner.

The next type of wire tested consisted of two solid metal wires, each 0.9-mm in diameter and covered with a layer of 0.2-mm-thick Refrasil (woven high-silica glass fiber). Two of these covered wires were paired and covered by another layer of woven Refrasil. The cross section of this composite is about 2.0 x 3.2-mm.

Another type of wire tested had a 1-mm-diameter metallic bundle which was covered by a 0.3-mm-thick inner layer of woven fiberglass and a 0.3-mm-thick outer layer of green Teflon. The total outside diameter of this combination is 2.2-mm.

Also tested was a combination of two 1.4-mm-diameter wires, each of which was covered by 0.1-mm-thick Kapton, and a single uninsulated 1-mm-diameter solid wire; these were combined by being wrapped with a layer of 0.1-mm-thick aluminized Mylar and three layers of 0.033-mm-thick Kapton. The total outside diameter of the wire, including a coat of yellow paint, is 3.6-mm.

The final type of wire tested was a 0.8-mm-diameter bundle of standard metal (AWG26) which was covered with a 0.4-mm-thick layer of uncolored, transparent fluorocarbon.

## 4. RESULTS AND DISCUSSIONS

Most of the organic material charred at levels between 2.1 and 2.9 $MJ/m^2$ (50 to 70 $cal/cm^2$) as shown in Table 1. The only exception was the clear fluorocarbon jacket, which apparently survived unharmed because it does not absorb much of the incident thermal energy; instead, it transmits much of the thermal radiation to the metal conductor, which reflects much of it back out through the transparent jacket, leaving the jacket unharmed. The main disadvantage of this arrangement is that no color coding of the wires is possible. Another disadvantage is the relatively high cost of using fluorocarbons compared to nylon-coated polyethylene.

The woven silica jackets survived the thermal exposures with only superficial discoloration, but the jacket being woven is not watertight. This produces a large reduction in the electrical resistance between the wires: from more than 16 Mohms in air to 400,000 ohms when the wires were immersed in the salt water bath. When the wires were flexed in this bath, values as low as 10,000 ohms were observed. Thus, this type of insulation cannot be recommended for use in wet environments. The Teflon-covered fiberglass is much more suitable for wet environments, since the Teflon appears to seal the fiberglass weave and remains intact even after a thermal exposure of 4.98 $MJ/m^2$ (119 $cal/cm^2$).

The thicker (0.4-mm) colored ETFE appeared to provide undiminished electrical insulation up through 3.3 $MJ/m^2$ (80 $cal/cm^2$) even though it was badly charred and even when it was flexed while in the water bath.

The Kapton also appears to provide undiminished electrical insulation up through 5.6 $MJ/m^2$ (135 cal/cm).

## 5. CONCLUSIONS

Although several types of insulated wire have been shown to be capable of surviving intense thermal radiation exposure, the

TABLE 1.  THERMAL RESPONSE OF TYPES OF WIRE INSULATIONS

| Jacket material | Jacket thickness (mm) | Fluence ($MJ/m^2$) | ($cal/cm^2$) | Postexposure appearance* | Resistance (Mohms) air | water |
|---|---|---|---|---|---|---|
| ETFE** | 0.25 | 2.09 | 50 | C | 16 | 16 |
| | | 2.51 | 60 | C | 16 | 1 |
| | | 3.35 | 80 | BW | 16 | 0.1 |
| | 0.4 | 3.35 | 80 | C | 16 | 16 |
| | | 5.02 | 120 | C | 16 | 0.4 |
| | | 6.90 | 165 | B | 16 | 0 |
| | | 8.79 | 210 | B | 0.34 | 0 |
| | | 8.79 | 210 | B | 9 | 0.6 |
| $PVF_2$/PK** | 0.22 | 3.35 | 80 | C | 16 | 0.5 |
| | | 4.61 | 110 | BW | 0 | 0 |
| | 0.45 | 2.01 | 48 | C | 16 | 16 |
| | | 2.51 | 60 | C | 16 | 16 |
| | | 3.35 | 80 | I | 16 | 4 |
| | | 4.18 | 100 | I | 16 | 0.1 |
| Woven silica | 0.4 | 3.64 | 87 | ND | 16 | 0.4-0.0 |
| | | 4.98 | 119 | SD | 16 | 0.4-0.0 |
| Teflon/ fiberglass | 0.6 | 2.93 | 70 | Teflon C & B | 16 | 16 |
| | | 4.98 | 119 | Teflon D, Fiberglass ND | 16 | 16 |
| Kapton/Al/ Kapton | 0.2† | 2.93 | 70 | Top Kapton D | 16 | 16 |
| | | 3.77 | 90 | Top Kapton D | 16 | 16 |
| | | 5.02 | 120 | Al D | 16 | 16 |
| | | 8.79 | 210 | Inner Kapton B | 16 | 0.6 |
| Clear fluorocarbon | 0.4 | 3.64 | 87 | ND | 16 | 16 |
| | | 7.53 | 180 | ND | 16 | 16 |

* C = charred, BW = bare wire, I = ignited, ND = no damage, B = breached, D = disintegrated, BC = badly charred, SD = slight discoloration.
** Cross-linked.
† Jacket thickness does not include the 0.1-mm-aluminized Mylar which separates the Kapton.

ideal insulation which would be
inexpensive and allow color-coding
has not yet been identified.
Further evaluation will be
conducted as candidates are
uncovered. At the present time we
are looking for an inexpensive,
lightweight, thin, nonflammable,
transparent insulating material
that could be coated with a thin,
sacroficial, colored top skin.

### REFERENCES

1.  Share, S. and Baba, A.J.,
    Response of Polymeric
    Marerials to Thermal
    Radiation, SAMPE Journal
    21(2), 7-11 (1985).

2.  Anderson, J.C., et al., Fiber
    Optic Cable for Applications
    in Thermal Radiation
    Environments, Proceedings of
    33rd International Wire and
    Cable Symposium 1984, pp 258-
    260.

3.  Share, S. and Baba, A.J.,
    US Patent 4,304,462, Thermal
    Hardened Fiber Optic Cables, 8
    December 1981.

DIMENSIONALLY STABLE, HIGH PERFORMANCE, QUATREX EPOXY

COMPOSITES FOR ADVANCED PRINTED CIRCUIT BOARDS - I+

A. Mahammad Ibrahim*, Kirsten A Green**,

Boro B. Djordjevic, and John D. Venables

Martin Marietta Laboratories

1450 South Rolling Road

Baltimore, Maryland 21227

## ABSTRACT

The processing-structure-property relationships of E-glass fabric-reinforced high-performance QUATREX epoxy composites for printed circuit board (PCB) applications are discussed. The trifunctionality of this QUATREX epoxy resin results in a three dimensional, tightly crosslinked network structure with a high glass transition temperature, $T_g$, of 180°C, which is well above that of the existing PCB epoxies. In addition to its high $T_g$ which translates into better dimensional stability, its non-amine cure chemistry (no-DDS, no-MDA, no-Dicy) with reduced toxicity has generated a lot of interest in the PCB industry. In this work, the optimum processing (cure) schedules developed using ultrasonic non-destructive evaluation (NDE) and differential scanning calorimetry (DSC) are addressed as are the thermomechanical, dynamic mechanical, and electrical properties of QUATREX epoxy composites determined using a combination of techniques such as thermomechanical analysis (TMA), thermogravimetric analysis (TGA), dynamic mechanical analysis (DMA), and impedance analysis (IA). The effects of thermal cycling on the stability of the fiber-matrix interface with relevance to microcracking are also addressed.

\* To whom correspondance should be addressed

\** Summer student from the Johns Hopkins University, Dept. of Materials Science and Engineering, Baltimore, Maryland 21218

+ Presented in part at the IPC (The Institute for Interconnecting and Packaging Electronic Circuits) Fall Meeting, San Diego, CA, Sept. 1986.

## 1. INTRODUCTION

There is a growing need to develop dimensionally stable high-performance thermosetting epoxy composites for fabricating advanced PCB's that will be used for very-high-speed integrated circuit (VHSIC) applications. Although the epoxy composites currently used as substrate material for PCB applications have many advantages besides their long history with the PC board industry, they do have some disadvantages, the primary one being their relatively low $T_g$, (< 125°C).[1] The $T_g$ is a major critical material parameter affecting the dimensional stability of the PC board laminate; higher $T_g$ (well above the military thermal cycling range, -55 to +125°C) translates into better dimensional stability. At and around $T_g$, a polymer matrix and its composite experience drastic changes in their thermal and mechanical properties including thermal expansion and hence in their dimensional stability. The $T_g$'s of the currently used epoxies, which are close to the upper thermal cycling temperature, impact on several facets of the PCB performance. Their low $T_g$ significantly increases the amount of thermal stress imposed on solder joints and plated through holes (PTH's) on thermal cycling, leading to failures.[2,3]

The coefficient of thermal expansion (CTE) of the matrix resin is significantly larger than the CTE of the copper. Thus, on thermal cycling, it induces thermal stresses which will lead to PTH failures.[2,3] In general, the CTE of the resin decreases as its $T_g$ increases. This is another reason why the PCB industry needs to develop new high $T_g$ and high-performing epoxy composite laminates. QUATREX epoxy is one such new material currently under serious investigation as a potential substrate material for circuit board applications. QUATREX epoxy resins are based on the triglycidyl ether of tris(hydroxyphenyl) methane isomers and higher oligomers (Fig. 1).[4] The trifunctionality of the resin results in a three dimensional, tightly cross-linked network structure with high $T_g$ and good thermal oxidative stability. In addition to its high $T_g$, its non-amine (no-MDA, no-Dicy, no-DDS) cure chemistry results in reduced toxicity. The actual curing agent used in the QUATREX epoxy system is based on aromatic dihydric phenols. Thus, the cure reactions involved are mostly epoxy-phenol and epoxy-epoxy.

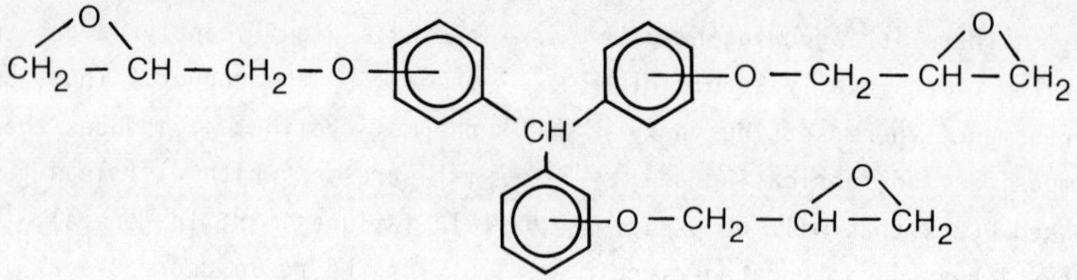

Figure 1. Chemical structure of QUATREX epoxy resins (curing agent is based on aromatic phenols)

This work focuses on the processing-structure-property relationships for E-glass fabric-reinforced high-performance QUATREX epoxy composites for PCB applications. In particular, the optimum processing (cure) schedules developed using ultrasonic non-destructive evaluation (NDE) and differential scanning calorimetry (DSC) will be addressed as will be the thermomechanical, dynamic mechanical, and electrical properties of QUATREX epoxy composites determined using a combination of techniques such as thermomechanical analysis (TMA), thermogravimetric analysis (TGA), dynamic mechanical analysis (DMA), and impedence analysis (IA). The effects of thermal cycling on the stability of the fiber-matrix interface with relevance to microcracking will also be addressed.

## 2. EXPERIMENTAL

QUATREX 5010 epoxy prepregs reinforced with woven E-glass fabric were made and supplied by Dow Chemical Company, Freeport, Texas.

Nondestructive evaluation (NDE) was performed on the QUATREX epoxy composites, using an ultrasonic through-transmission c-scan system. The transmitted and received ultrasonic signals were coupled to the panel by Low-Noise Ultrasonic Liquid-Jet Probes (LUP™).[5] An inspection frequency of 2.5 MHz was found to be sensitive to variations in resin content as well as to porosity and delaminations.

The thermal and thermomechanical properties of the uncured prepregs and cured laminates were studied by DSC, TGA, and TMA, using a Mettler TA3000 thermal analysis system. The dynamic mechanical properties of the laminates were determined using a Dupont 982 Dynamic Mechanical Analyzer attached to a

Dupont 1090 Thermal Analysis System. These experiments were performed in a nitrogen atmosphere at a heating rate of 5°C/min, in the temperature range -150 to +260°C. The data obtained are reported in the form of traditional tan $\delta$ quantity.

## 3. RESULTS AND DISCUSSION

### 3.1 Processing of QUATREX Epoxy Composites

QUATREX epoxy composites reinforced with E-glass fabric were processed by compression molding techniques using match-die molds. NDE and DSC were used to develop optimum processing pressure and time, respectively, for fabricating composites with maximum resin uniformity and maximum possible $T_g$. The procedure developed is provided below:

o  Lay up the prepreg sheets in the mold
o  Place the mold in a (microprocessor-controlled Tetrahedron) press; apply 50-psi contact pressure
o  Heat to 600°F at 5°F/min
o  Apply 300-psi pressure and cure for $1\frac{1}{2}$ hr
o  Reduce pressure to 50 psi and cool to room temperature at 5°F/min.

Processing pressures of 150, 300, and 500 psi were tried and the resultant laminates were examined by ultrasonic and visual inspection techniques. The laminate obtained using 300 psi pressure had minimum porosity and very good resin uniformity in all parts of the panels. Visual inspection of the 500-psi panel showed a lot of resin squeezed out at the edges and the ultrasonic scans also showed less resin uniformity. For the 150-psi panel, the pressure was insufficient to compact the plies together. Ultrasonic tests also showed less resin uniformity in this panel than in the 300-psi panel. Based on the NDE test data and visual inspection, the pressure selected for fabricating panels with good resin uniformity was 300 psi.

Using 300-psi pressure, we cured the laminates for different times, i.e., 1/4, 1/2, 1, 1 1/2 , and 2 hr. It is well established that the $T_g$'s strongly depend on the extent of cure in the epoxy matrices and composites.[6,7] The $T_g$'s of the resultant laminates are shown in Fig. 2. As can be seen, in order to obtain a maximum possible $T_g$ of 180°C under the above processing conditions, these laminates must be cured for at least $1\frac{1}{2}$ hr.

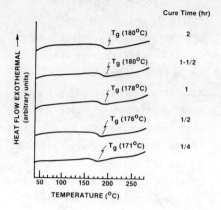

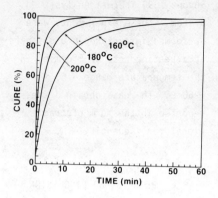

Figure 2. The effect of cure time on $T_g$ for QUATREX epoxy/E-Glass composites @ 10°C/min.

Figure 3. DSC isothermal cure predictions for QUATREX epoxy/E-glass prepregs using a dynamic scan @ 10°C/min

In order to further understand the cure chemistry and kinetics, a dynamic DSC experiment was carried out for the uncured QUATREX epoxy/E-glass prepreg. This experiment showed a cure exotherm starting from 80°C to 250°C with a peak temperature around 190°C. This dynamic DSC data as a function of temperature was used to theoretically predict the degree of cure as a function of time at isothermal cure temperatures.[8] The predicted isothermal data at 160°C, 180°C (cure temperature), and 200°C are shown in Fig. 3. It is clear that at 180°C, as much as 90% degree of cure was achieved within the first 15 min, giving rise to a high $T_g$ of 171°C.

The laminates were further post-cured in the DSC cell by raising them to 280°C from room temperature at a heating rate of 10°C/min. The DSC thermograms of these post-cured samples are shown in Fig. 4. All the post-cured samples showed a same $T_g$ around 185-186°C despite the fact that they were initially cured for different times, i.e., 1/4 - 2 hr. This suggests that the maximum attainable $T_g$ using QUATREX epoxy composites is ~ 185°C, as measured by DSC techniques at 10°C/min. These series of DSC experiments clearly show that QUATREX epoxy is a high-performance epoxy with a high $T_g$ of 180-185°C, which is well above the existing PCB epoxies.

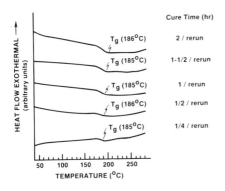

Figure 4. The effect of post-cure on $T_g$ for QUATREX epoxy/E-glass composites using DSC @ 10°/min

## 3.2 Properties of QUATREX Epoxy Composites

The ultimate thermal, mechanical, and electrical properties of the laminates depend strongly on their resin contents. Using our method developed earlier by TGA-DTG techniques[9], the resin contents in these QUATREX epoxy/E-glass laminates were first evaluated; typical TGA and DTG thermograms obtained are shown in Fig. 5. Since glass would not decompose below 800°C, any weight loss in the laminate below this temperature is due only to the decomposition of the QUATREX epoxy matrix. Thus, the two decomposition peaks in the DTG correspond to the QUATREX epoxy only and the total weight loss in the DTG, which is 44.6% by weight, corresponds to the QUATREX resin content in the final cured laminates.

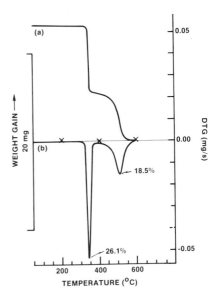

Figure 5. Dynamic (a)TGA and (b)DTG thermograms of QUATREX epoxy/E-glass composite in air @ 10°C/min.

In order to ensure that the two peaks seen in the DTG come from the decomposition of QUATREX epoxy matrix only, the TGA and DTG experiments were carried out for the neat QUATREX epoxy casting. Figure 6 shows the DTG's obtained for the neat QUATREX matrix casting and for the QUATREX epoxy/E-glass laminate. Each of them has two decomposition peaks at more or less the same temperature locations, thus confirming the TGA results. The average resin content of the laminates was ~ 45% by weight.

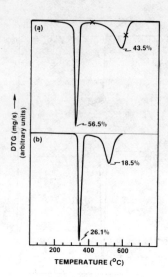

Figure 6. Dynamic DTG's of (a) neat QUATREX epoxy resin matrix and (b) QUATREX epoxy/E-glass composite. The total loss by weight is 100% for (a) and 44.6% for (b).

### 3.2.1 Thermomechanical Properties.

The CTE's of the QUATREX epoxy/E-glass composites were measured in the x-y and z directions, using TMA at a heating rate of 5°C/min at temperature ranges of -50 to +140°C, -50 to 250°C, 25 to 140°C, and 25 to 250°C. The x-y and z CTE values obtained are shown in Table 1. Because the x- and y-direction CTE's are almost equal, they are reported as x-y direction CTE. An average x-y CTE of 15.7 ppm/°C was obtained in the temperature range -50 to +140°C for the laminates, which was expectedly high because of the E-glass reinforcement. It is a well established fact that the fabric

Table 1

CTE* of QUATREX Epoxy/E-glass Composites

| Temp. Range (°C) | CTE (ppm/°C) x-y | z |
|---|---|---|
| -50 to 140 | 15.7 | 67.0 |
| 25 to 140 | 15.7 | 69.4 |

---

* Determined using TMA @ 5°C/min; average of 3 measurements

largely controls the thermal expansion in the in-plane (x-y) direction, whereas the resin controls in the out-of-plane (z) direction.

Following thermal cycling in the TMA instrument from -50 to +140°C, the average CTE was still around 16.3 ppm/°C. This minimal change in the CTE after a few thermal cycles shows the dimensional stability of the QUATREX epoxy composites in the in-plane direction.

The out-of-plane (z) CTE, which is dominated by the resin matrix, was around 67 ppm/°C, which is typical of the existing epoxy/glass laminates. In order to determine the effect of $T_g$ on the CTE, the x-y and z-direction CTE's were also determined from -50 to 250°C (through the $T_g$); these are compared with those obtained from -50 to 140°C in Table 2.

## Table 2

Effect of $T_g$ on CTE's of QUATREX
Epoxy/E-glass Composites

| Temp. Range (°C) | CTE (ppm/°C) | |
|---|---|---|
| | Below $T_g$ -50 to 140 | Thru $T_g$ -50 to 250 |
| x-y | 15.7 | 12.5 |
| z | 67.0 | 103.0 |
| Temp. Range (°C) | 25 to 140 | 25 to 250 |
| x-y | 15.7 | 11.5 |
| z | 69.4 | 125.4 |

---
* Determined using TMA @ 5°C/min;
average of 3 measurements

property, our results indicate the influence of $T_g$ in slightly decreasing the CTE in the fabric direction as well.

Similar results were obtained when the CTE's from 25 to 140°C and 25 to 250°C were determined. However, when taken through the $T_g$ from 25 to 250°C, the out-of-plane CTE was nearly double that of the 25 to 140°C range. This nearly two-fold increase in the CTE of the laminate on passing through the $T_g$

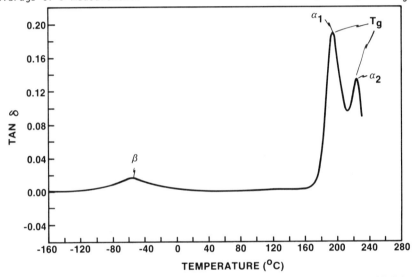

Figure 7. Typical tan δ as a function of temperature data for QUATREX epoxy/E-glass composites @ 5°C/min.

As expected, the average out-of-plane CTE increased drastically from 67 to 103 ppm/°C when passing through the $T_g$ of the resin matrix. On the other hand, the average in-plane CTE decreased from 15.7 to 12.5 ppm/°C on passing through the $T_g$. Even though the in-plane CTE is a fabric-dominated

clearly demonstrated the need for high-performance materials with $T_g$'s well above the required operating temperature ranges.

3.2.2 Dynamic Mechanical Properties. The dynamic mechanical properties, such as E' (tensile storage modulus), E'' (tensile loss modulus), and tan δ (E''/E'), were determined using the Dupont DMA. Figure 7 shows tan δ as a function of temperature for the QUATREX epoxy/E-glass composites obtained at 5°C/min. As observed in other high-performance epoxy systems currently used in the aerospace industry,[6,7,10] this QUATREX epoxy also showed two major transitions, a low-temperature β transition and a high-temperature α transition. The β transition, which appears at -50°C, may be due to the segmental motion of the groups that resulted from the epoxide-phenol and phenol-phenol reaction.[6,7,10] The α transition is due to the extensive movement of the entire epoxy network structure resulting from the glass transition. For this QUATREX epoxy system, two high-temperature transitions, $\alpha_1$ and $\alpha_2$, were obtained, suggesting the possibility of possible $T_g$'s at 193 and 230°C, respectively. Although unconfirmed in this work, these two possible $T_g$'s may be arising from the epoxy isomers or their higher oligomers based on the triglycidyl ether of tris(hydroxyphenyl) methane resins. The origin of these two high-temperature transitions is being investigated using further neat QUATREX epoxy castings.

3.2.3 Electrical Properties. The dielectric constants were deter-mined using an impedance analyzer at room temperature for two QUATREX epoxy/E-glass composites cured for 15 min and 1 1/2 hr (standard cure). The results are shown in Table 3.

Table 3

Dielectric Constants* of QUATREX Epoxy/E-glass Composites

| Frequency (kHz) | Dielectric Constant | |
| --- | --- | --- |
| | 15-min Cured Sample | 1-1/2-hr Cured Sample |
| 1 | 4.64 | 5.08 |
| 10 | 4.61 | 5.03 |
| 100 | 4.50 | 4.94 |
| 1,000 | 4.35 | 4.79 |
| 10,000 | 4.13 | 4.57 |

----------------
* Determined using an impedance analyzer at R.T.

The dielectric constant decreased steadily with increasing frequency for both the laminates in the 1 kHz - 10MHz range studied. With the standard 1-1/2-hr cure, the dielectric constant at 1 kHz is 5.08, which is not unexpected due to the E-glass reinforcement used. But for the 15-min cure sample, the dielectric constant is almost 9 - 10% lower. Although the thermal properties suggest an almost 90% cure within 15 min, giving rise to a $T_g$ as high as 171°C (compared to maximum attainable $T_g$ of 180°C at that cure temperature), the changes in the dielectric constant are quite significant. Since the propagation delay in VHSIC devices is proportional

1246

to the dielectric constant of the PCB, this dielectric data suggest the importance of the "proper cure" in epoxy composite laminates in order to realize the proper dielectric constant.

## 3.3  Structure of QUATREX Epoxy Composites

The QUATREX epoxy/E-glass composite laminates were subjected to 300, 500, 800, and 1000 thermal cycles from -55 to +125°C in accordance with military specifications. These thermally cycled laminates and a control uncycled laminate were examined using optical and scanning electron microscopy (SEM) to determine the resin microcracking in these QUATREX epoxy/E-glass composites. Both the optical and SEM investigations have shown no microcracking whatsoever in these QUATREX composites even after 1000 thermal cycles, thus showing the stability of their fiber-matrix interface.

## 4.  CONCLUSIONS

Processing-structure-property relationships pertinent to PCB applications were developed for the new high-performance QUATREX 5010 epoxy composites reinforced with woven E-glass fabrics.

A combination of NDE and DSC techniques were used to develop optimum processing pressure and cure time, respectively, for fabricating composites with maximum resin uniformity and a maximum possible $T_g$ of 180°C. DSC experiments on the uncured QUATREX epoxy/E-glass prepregs indicated that a degree of cure as large as 90% was achieved within the first 15 min, giving rise to a high $T_g$ of 171°C compared to 180°C for 1-$\frac{1}{2}$- and 2-hr cured laminates. When the laminates initially cured for 1/4, 1/2, 1, 1-$\frac{1}{2}$, and 2 hr were post-cured, the same $T_g$, ~ 180-186°C, was obtained, suggesting that the maximum attainable $T_g$ using QUATREX epoxy is 185°C as measured by DSC.

Using our method developed earlier by TGA-DTG techniques, the average resin content in the QUATREX epoxy/E-glass composites was found to be about ~ 45% by weight.

The thermal expansion properties obtained using TMA showed average CTE's (in the temperature range -50 to 140°C) of 15.7 ppm/°C in the in-plane (x-y) direction and about 67 ppm/°C in the out-of-plane (z) directions for the QUATREX epoxy composites. The high value obtained for the x-y in-plane CTE is due to the E-glass reinforcement used. However, even after a few thermal cycles from -50 to 140°C, this CTE remained almost constant, showing the dimensional stability of the

QUATREX epoxy composites in the in-plane direction.

The thermal expansion properties of these composites were significantly affected on passing through the glass transition; the average out-of-plane CTE increased drastically from 67 to 103 ppm/°C, whereas the in-plane CTE decreased from 15.7 to 12.5 ppm/°C. These significant changes in the dimensions of the laminate on passing through the $T_g$ clearly demonstrated the need for high-performance materials with $T_g$'s well above the required operating temperature ranges.

The DMA showed two major transitions for these QUATREX epoxy composites, a low-temperature (-50°C) β transition and a high-temperature α transition, which were also seen in other high-performance epoxy systems currently used by the aerospace industry. The two high-temperature transitions, $\alpha_1$ and $\alpha_2$, at 193°C and 230°C respectively, suggest the possibilities of two glass transitions that might have arisen from the epoxy isomers or their higher oligomers based on the triglycidyl ether of tris(hydroxyphenyl) methane resins.

These high-performance composites showed no microcracking even after 1000 thermal cycles in the temperature range -55 to +125°C, indicating the stability of the QUATREX epoxy matrix-glass fabric interface.

## 5. ACKNOWLEDGEMENT

The authors thank Mr. Dale Aldridge of Dow Chemical Company, Freeport, Texas, for supplying QUATREX epoxy prepregs and Messers Gail Love and William Pattison of Martin Marietta Orlando Aerospace, Orlando, Florida, for helping with the thermal cycling experiments.

## 6. REFERENCES

1. Z.N. Sanjana, J. Valentich, and J.R. Marchetti, IPC-TP-488, 1983.

2. G.F. Love, IPC Technical Review, 12, 1981.

3. J.K. Lake and R.N. Wild, Natl. SAMPE Symp. Exhib., 28, 1406 (1983).

4. D. Aldridge, Dow Chemical Company, Private Communications.

5. B.B. Djordjevic, J.D. Venables, M. Kroll, and L.J. Matienzo, Nondestructive Evaluation (NDE) and Processing of Advanced Composite Materials, Martin Marietta Independent Research and Development (IRAD) Report, Feb 1985.

6. A.M. Ibrahim and J.C. Seferis, in Interrelations Between Processing, Structure and Properties of Polymeric Materials, J.C. Seferis, and P.S. Theocaris (eds.), Elsevier, Amsterdam, 1984, pp. 325-341.

7. A.M. Ibrahim and J.C. Seferis, Natl. SAMPE Symp. Exhib., $\underline{28}$, 581 (1983).

8. Mettler TA 3000 Thermal Analysis System, DSC Kinetic Analysis Program.

9. A.M. Ibrahim, T.K. Shah, L.J. Matienzo, and J.D. Venables, Int. SAMPE Symp. Exhib. $\underline{31}$, 669 (1986).

10. A.M. Ibrahim and J.C. Seferis, Polymer Composites, $\underline{6}$, 47 (1985).

PERFORMANCE OF CERAMIC FIBER COMPOSITE UNDER
STATIC AND CYCLIC LOADING

Dr. Srinivasa L. Iyer          and          Mr. Rana Gupta
Professor of Civil Engineering                Graduate Student

South Dakota School of Mines and Technology
Rapid City, South Dakota   57701

## Abstract

The usage of ceramic fibers in composites is relatively new and its performance and properties need to be studied for the proper design of parts. This study deals with the mechanical properties and strength of ceramic fibers with F650 (Hexcel) resin system. Unidirectional and cloth fabric ceramic reinforcements are used to make the tension specimens. These specimens were subjected to static and cyclic (tension-tension) loading using the MTS closed loop system machine. Photo stress method was used to evaluate the deterioration of composites under cyclic loading.

## 1. INTRODUCTION

The usage of ceramic fibers in composites is relatively new and its performance under loading both static and cyclic is of great importance to its proper use in the design. This study deals with ceramic fibers (Nextel produced by 3 M Company) generally used for aircraft firewalls and F650 resin developed by Hexcel. Very little information is available for this composite, subjected to cyclic loading with and without notches. This study compares the increase in strain due to cyclic loading on notched and unnotched specimens. Photostress method and strain gages were used to measure the strains on (Nextel/F650 resin composites) unidirectional and mat specimens.

## 2. OBJECTIVE AND SCOPE

The main objective of this study is to determine the degree of damage (increase in strain) caused around a hole of a ceramic/resin composite (Nextel/F650) with fibers in the uniaxial direction. For comparisons, tests were conducted on unnotched specimens with fibers in the uniaxial direction and on mat. The specimens were tested for ultimate tensile strength under static loading. Companion specimens were tested for cyclic loading. The strains were measured using strain gages and reflection polariscope.

## 3. PREPARATION OF SPECIMENS

The test specimens were made using prepreg tape supplied by Hexcel and the properties of fibers and resin are shown in Table 1. Standard lay up recommended by the manufacturer was used and a nine ply laminate was used for all specimens. Figure 1 shows basic geometry of specimens and the actual dimensions of the specimens are shown in Table 2. The curing cycle used for all the specimens is shown in Figure 2. Aluminum tabs were used at the ends of specimens for proper gripping of specimens and these tabs were bonded to the specimens using film adhesive supplied by HYSOL.

## 4. TESTING PROCEDURE

MTS closed loop system testing machine was used to test all specimens subjected to static and cyclic loading. A sinusoidal loading was used for cyclic loading at a rate of 900 cycles per minute. The maximum load was 60 percent of the ultimate load whereas the minimum load was 30 percent of the ultimate load. Electrical strain gages and reflection polariscope were used to measure the strain under various loading conditions. Polariscope, Model 031A was used to generate a polarized light and analyzed the reflected light from the birefringant photoelastic sheet bonded to the stressed

specimen. Photoelastic sheet (PS-2B) was bonded to the test specimen using PC-1 adhesive. Points were selected near the free edges and near the hole of the test specimens to monitor the strain difference under different loads. A null-balance compensator, Model 232, was used to measure the fringes accurately. Electrical strain gages were used to measure strain in the axial and transverse directions for unnotched specimens. The properties of photoelastic sheets and strain gages are shown in Table 3. The polariscope, photoelastic sheets and strain gages were supplied by Micro-Measurements Group.

## 5. TEST RESULTS

The unnotched unidirectional fiber specimens, #1 was subjected to static tension test and failed at the ultimate tensile stress of 63,850 psi and the strain at failure was 0.63 percent. Specimen #2 was subjected to (tension-tension) cyclic loading with load varying from 1100 to 2200 lbs. (30 to 60 percent of ultimate load) and strain readings were recorded at frequent intervals and the results are shown in Table 4. From the test data it can be seen that very little damage was done to the specimen after one million cycles of loading. The residual strain after the test was 96 micro. in/in. (less than three percent). Figures 3 and 4 show the load-strain diagrams before and after one million cycles of loading and very little change is observed in the diagram. Figure 5 and 6 show no appreciable increase in strain in the longitudinal and transverse direction due to cyclic loading. This indicates no damage was done on the specimen due to cyclic loading. The mechanical properties such as modulus of elasticity and Piosson's Ratio do not change during and after the cyclic loading.

The notched unidirectional fiber specimen #3 failed at a load of 3400 lbs and at about 2000 lbs some failure happened near the edge of the notch. The failure initiated at 2000 lbs near the notch and hence 60 percent of this load (2000 lbs) was used for maximum load for cyclic loading on specimen #4. Strain gage located away from the notch in the longitudinal direction was used to monitor the maximum and minimum load applied on the specimen during cyclic loading (600 to 1200 lbs). The test results are shown in Table 5. The strain near the edge of the hole was 2692 micro. in/in for a load of 1200 lbs and it was comparable with the strain measured on the unnotched specimen at this load level. At frequent intervals strain readings were taken at the mean load (900 lbs) at various points on the specimen. From the strain difference at the edge of the notch, it can be noticed that there was a considerable increase in strain (from 1935 to 2491 micro. in/in). Figure 7 shows the increase in strain at all points with number of cycles. The strain near the notch shows the maximum increase compared to the other points. The discontinuity of fiber near the notch develops an interlaminar shear in this area and shows the continuous

damage to the composite due to cyclic loading.

The unnotched mat specimen #5 was subjected to static tension and failed at the ultimate tensile stress of 22,680 psi. The unnotched mat specimen #6 was subjected to (tension-tension) cyclic loading with load varying from 300 to 600 lbs (30 to 60 percent of the ultimate load). The increase in strain due to one million cycles of load was not significant (less than 5 percent) in the longitudinal direction and the test results are shown in Table 6. From the test data it can be seen that very little damage was done to the specimen after one million cycles. This agrees with the unnotched unidirectional specimen test results as discussed earlier. No temperature increase was noticed during the cyclic loading on all the specimens.

6. CONCLUSIONS

On the basis of limited tests conducted on ceramic epoxy composites, the following initial conclusions were drawn:

1. The damage on the unnotched specimens due to cyclic loading (30 to 60 percent of ultimate load) was very minimal.

2. The damage on the notched specimen due to cyclic loading (30 to 60 percent of ultimate load) was significant (an increase of strain of 28 percent in the longitudinal direction).

Further tests on more number of specimens are recommended

to confirm the above conclusions

7. ACKNOWLEDGEMENTS

The authors would like to thank HEXCEL for supplying the ceramic/epoxy prepreg and HYSOL division of THE DEXTER Corporation for supplying the film adhesive. These studies would not have been possible without the generous help of the Civil Engineering Department, South Dakota School of Mines and Technology for providing equipment and facilities.

8. REFERENCES

1. Hexcel Technical Bulletin for Resin Systems for Advanced Composites - F650, 1985.

2. Iyer, Srinivasa L., and Haugan, Michael D., "Cyclic Loading and Stress Concentration on Composites," 31$^{st}$ International SAMPE Symposium, Las Vegas, Nevada, 1986.

3. Majidi, Azav P., and Chen, Tsu-Wei, "Impact Tolerance of Braided Alumina Fiber Reinforced Aluminum Composites," 31$^{st}$ International SAMPE Symposium, Las Vegas, Nevada, 1986.

4. Tsai, Stephen W., "Composites Design 1986," Think Composites: Dayton, Ohio.

9. BIOGRAPHY

Dr. Srinivasa L. Iyer received B.S. in Civil Engineering, 1956; M.S. in Structural Engineering 1960 from University of Kerala, India and Ph.D., 1974 from South Dakota School of Mines and Technology, Rapid City, South Dakota, U.S.A. He joined the faculty at SDSM&T in 1974 and is currently active in Research, Teaching and Consulting as a full professor in Civil Engineering Department. He has formed his own corporation of "Stress Steel Co. Inc." He has received one patent on polymer shrinkage device and has applied for two more patents in epoxy graphite implant and prestressed steel bridges/beams. He has published several papers in the stress analysis and composite areas. He is a member in ASTM, ASCE, NSPE, Sigma Xi and SAMPE.

Mr. Rana Gupta was born on September 4, 1959 in Calcutta, India. He graduated with his Bachelor of Technology (Honors) in Civil Engineering from I.I.T. Kharagpur on May 1981. He is currently a graduate student in Civil Engineering at the South Dakota School of Mines and Technology working actively in Advanced Composites.

Table 1 -- Properties of Ceramic - Epoxy Prepreg

Mat Prepreg: XC568/F650 BMI

Unidirectional Prepreg: Nextel 480/F650

Fiber Composition: Alumina-Boric-Silica   Nextel 3M

Resin: Polymide Resin  F650 - Hexcel

Resin Content: 33 - 35 percent by volume


* Prepregs supplied by Hexcel


Table 2 -- Identification and Dimensions of Test Specimens

| Id. No. | Length (in.) | Width (in.) | Thickness (in.) | No. of Plys | Orentation of Fibers | Notched or Unnotched |
|---------|--------|-------|-----------|------|------------|------------|
| 1 | 8 | 0.972 | 0.058 | 9 | [0] | Unnotched |
| 2 | 8 | 0.977 | 0.058 | 9 | [0] | Unnotched |
| 3 | 8 | 0.980 | 0.058 | 9 | [0] | Notched |
| 4 | 8 | 0.985 | 0.058 | 9 | [0] | Notched |
| 5 | 8 | 0.938 | 0.047 | 9 | Mat | Unnotched |
| 6 | 8 | 0.942 | 0.047 | 9 | Mat | Unnotched |


Table 3 -- Properties of Photoelastic Sheet & Strain Gages

| Type of Sheet | Elongation | Sheet Thickness | Strain Optical Property | Fringe Value | Bonding Adhesive |
|---------|-----------|--------|-----------|------|---------|
| PS2B | 3 percent | 0.08 | 0.1300 | 1090 | PC 1 |

| Where Used | Strain Gage Designation | Resistance (Ohms) | Gage Factor |
|---------|------------------|-----------|---------|
| Tape | CEA-06-250UN-120 | 120 | 2.05 |
| Mat | CEA-06-1250UT-120 | 120 | 2.04 |

Table 4 -- Cyclic Loading on Unnotched Unidirectional
Specimen #2

Load Range 1100 to 2200 lbs at 15 Hz - Sinusoidal

| Loading Before Cycling (lbs) | Longitudinal Strain (micro in./in.) | | Transverse Strain (micro in./in.) | |
|---|---|---|---|---|
| 0 | 0 | | 0 | |
| 1000 | 1864 | | -418 | |
| 1100 | 2056 | | -459 | |
| 1650 | 3064 | | -689 | |
| 2000 | 3682 | | -845 | |
| 2200 | 4036 | | -938 | |

| Number of Cycles | Max. | Min. | Max. | Min. |
|---|---|---|---|---|
| 0 | 4006 | 2095 | -477 | -963 |
| 100,000 | 3993 | 2154 | -480 | -970 |
| 200,000 | 4011 | 2178 | -465 | -953 |
| 350,000 | 4017 | 2185 | -470 | -955 |
| 750,000 | 4025 | 2168 | -481 | -972 |
| 1,000,000 | 4029 | 2162 | -458 | -947 |

| Loading After Cycling (lbs) | | | | |
|---|---|---|---|---|
| 2200 | 4130 | | -960 | |
| 2000 | 3768 | | -864 | |
| 1650 | 3143 | | -693 | |
| 1100 | 2124 | | -436 | |
| 1000 | 1960 | | -390 | |
| 16 | 96 | | +55 | |

Table 5 -- Cyclic Loading on Notched Unidirectional
Specimen #4
Load Range 600 - 1200 lbs at 15 Hz - Sinusoidal

| Loading Before Cycling | Longitudinal Strain (micro in./in.) | $(\epsilon_y - \epsilon_x)$ micro in./in. | | | | | |
|---|---|---|---|---|---|---|---|
| | | Pt1 | Pt2 | Pt3 | Pt4 | Pt5 | Pt6 |
| 0 | 0 | 0 | 0 | 0 | 0 | 0 | 0 |
| 600 | 1045 | 1945 | 1290 | 1401 | 1401 | 1290 | 1245 |
| 700 | 1567 | 1757 | 1812 | 1924 | 1924 | 1812 | 1757 |
| 1200 | 2090 | 2391 | 2469 | 2692 | 2692 | 8469 | 2391 |

| Number of Cycles | Max | Min | Static Mean | | | | | | |
|---|---|---|---|---|---|---|---|---|---|
| 0 | 2095 | 1012 | 1564 | 1868 | 1868 | 1935 | 1935 | 1868 | 1868 |
| 100,000 | 2080 | 1054 | 1549 | 1780 | 1802 | 1935 | 1935 | 1802 | 1780 |
| 200,000 | 2093 | 1067 | 1568 | 1824 | 1846 | 2024 | 2024 | 1846 | 1824 |
| 300,000 | 2097 | 1070 | 1569 | 1868 | 1891 | 2091 | 2091 | 1891 | 1868 |
| 800,000 | 2106 | 1090 | 1572 | 2113 | 2202 | 2335 | 2335 | 2202 | 2113 |
| 900,000 | 2095 | 1084 | 1580 | 2157 | 2224 | 2469 | 2469 | 2224 | 2157 |
| 1,000,000 | 2097 | 1089 | 1587 | 2202 | 2268 | 2491 | 2491 | 2268 | 2202 |

| Loading After Cycling | | | | | | | |
|---|---|---|---|---|---|---|---|
| 1200 | 2095 | 2400 | 2470 | 2710 | 2710 | 2470 | 2400 |
| 900 | 1569 | 1750 | 1815 | 1942 | 1942 | 1815 | 1758 |
| 600 | 1050 | 1247 | 1292 | 1420 | 1420 | 1292 | 1247 |
| 0 | 4 | 0 | 0 | 0 | 0 | 0 | 0 |

Table 6 -- Cyclic Loading on Unnotched Mat Specimen #6

Load Range 300 to 600 lbs - 10 Hz - Sinusoidal

| Number of Cycles | Longitudinal Strain (micro in./in.) | | Transverse Strain (micro in./in.) | |
|---|---|---|---|---|
| | Max | Min | Max | Min |
| 0 | 2691 | 1465 | -4 | -27 |
| 50,000 | 2696 | 1485 | 35 | 14 |
| 100,000 | 2770 | 1502 | 44 | 22 |
| 150,000 | 2728 | 1527 | 54 | 33 |
| 200,000 | 2745 | 1543 | 68 | 46 |
| 250,000 | 2763 | 1560 | 74 | 53 |
| 300,000 | 2784 | 1585 | 81 | 58 |
| 500,000 | 2801 | 1600 | 90 | 67 |
| 700,000 | 2809 | 1606 | 77 | 55 |
| 1,000,000 | 2818 | 1619 | 67 | 50 |

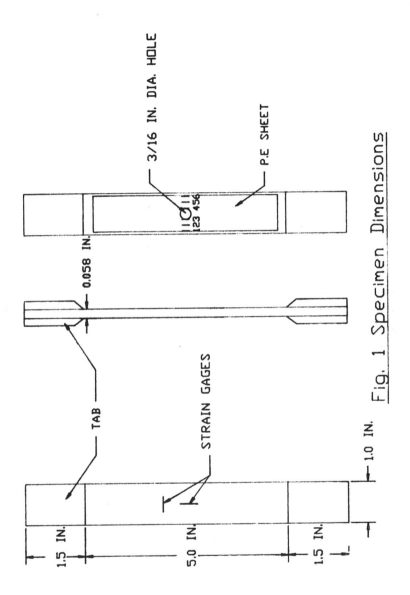

3/16 IN. DIA. HOLE

P.E SHEET

123 456

0.058 IN.

TAB

STRAIN GAGES

1.5 IN.

5.0 IN.

1.5 IN.

1.0 IN.

Fig. 1 Specimen Dimensions

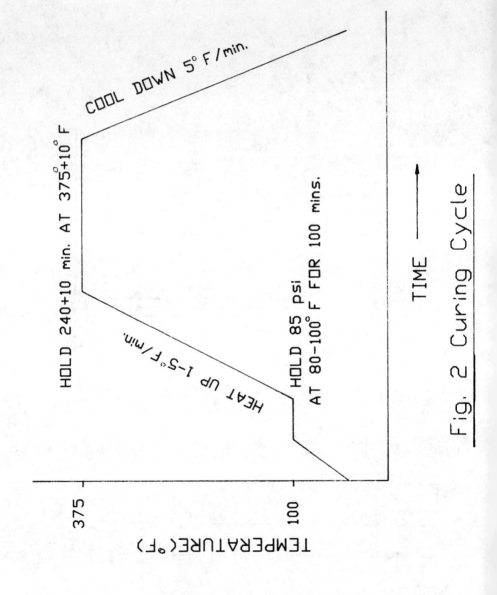

Fig. 2 Curing Cycle

COOL DOWN 5° F / min.

HOLD 240+10 min. AT 375$^{o}$+10$^{o}$ F

HEAT UP 1-5°F/min.

HOLD 85 psi AT 80-100° F FOR 100 mins.

TEMPERATURE(°F)

375

100

TIME

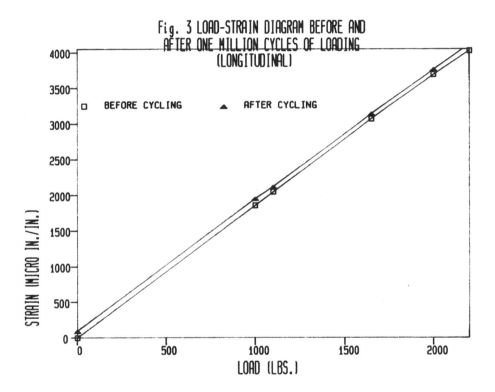

Fig. 3 LOAD-STRAIN DIAGRAM BEFORE AND AFTER ONE MILLION CYCLES OF LOADING (LONGITUDINAL)

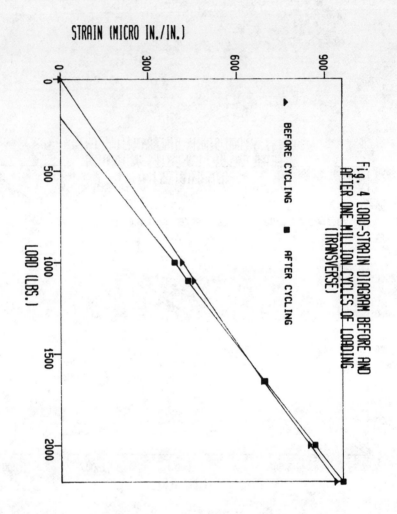

Fig. 4 LOAD-STRAIN DIAGRAM BEFORE AND
AFTER ONE MILLION CYCLES OF LOADING
(TRANSVERSE)

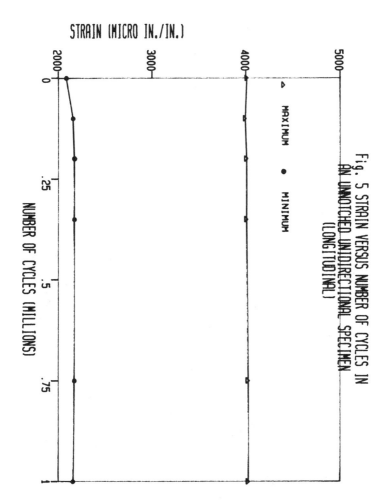

Fig. 5 STRAIN VERSUS NUMBER OF CYCLES IN AN UNNOTCHED UNIDIRECTIONAL SPECIMEN (LONGITUDINAL)

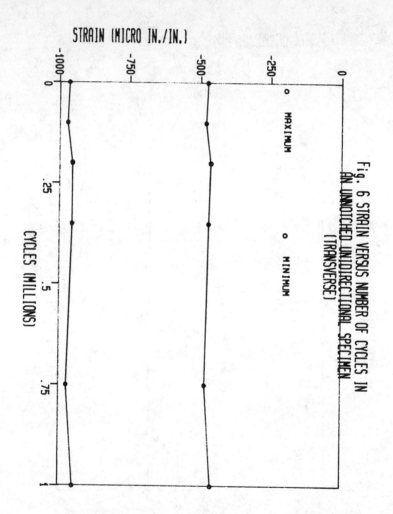

Fig. 6 STRAIN VERSUS NUMBER OF CYCLES IN AN UNNOTCHED UNIDIRECTIONAL SPECIMEN (TRANSVERSE)

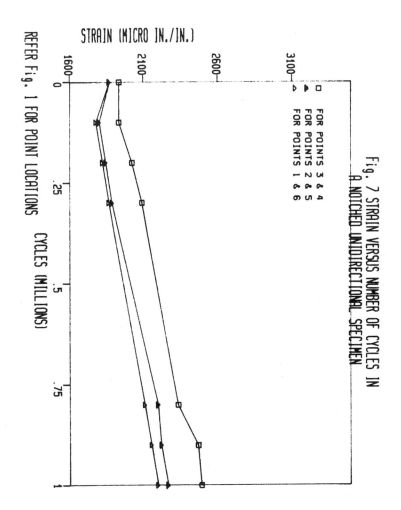

STRAIN (MICRO IN./IN.)

Fig. 7 STRAIN VERSUS NUMBER OF CYCLES IN
A NOTCHED UNIDIRECTIONAL SPECIMEN

□ FOR POINTS 3 & 4
▶ FOR POINTS 2 & 5
△ FOR POINTS 1 & 6

REFER Fig. 1 FOR POINT LOCATIONS    CYCLES (MILLIONS)

# LASER CURING OF EPOXIES AND EPOXY-MATRIX COMPOSITES

L.L. Clements, Materials Engineering Department
H.S. Lakarraju, Physics Department
R. Kellman, Chemistry Department
San Jose State University
San Jose, CA 95192
and
D.R. Osment and B.A. Wilson
Advanced Manufacturing Technology
Aerojet Strategic Propulsion Company
P.O. Box 13618
Sacramento, CA 95853

Aerojet Strategic Propulsion Co.
Release # PRA-SA-ASPC/1 Jan. 8, 1987

## Abstract

This paper describes work on the non-thermal curing of epoxy and epoxy-matrix composites using laser energy. The only energy truly needed to cure an epoxy is that required to open the epoxy ring and produce crosslinking, but in a thermal cure process much energy -- and time -- goes into heating the air, the oven walls, supports, and so forth. Thus, direct energy input to the epoxy would produce a great savings of both time and energy. In this study various means of directly producing cure were considered. The most obvious is to rupture the epoxy ring directly by vibration and distortion. The approach of this study, however, was to activate the curing agent using a U.V. laser of appropriate wavelength. Under some circumstances the approach led to rapid partial cure, but the technique currently suffers from a lack of control and must be optimized. While the laser is unable to penetrate through a layer of carbon fiber and produce complete cure of a composite, it may be quite suitable for producing a layer-by-layer "skin" of gelled material in a composite. Also, for such applications as magnetic media it may be able to quickly gel in place oriented particles in an epoxy carrier.

## 1. INTRODUCTION

Traditional thermal cure of epoxy-matrix composites imposes a tremendous energy penalty since much of the thermal energy input is used to heat the oven walls and air, and the composite mandrel, supports, and/or fixturing. The thermal cure process wastes energy and is unnecessarily time consuming and expensive. Thermal cure may also lead to problems such as excessive resin movement or bleed-out due to the long times at temperature before cure occurs. These problems are also seen in the cure of epoxies used for other applications, such as adhesives and coatings, or as carriers for magnetic particles on disks. However, such difficulties are avoidable, at least in principle, since the only needed energy is that required to activate the resin cure process and effect cure. This should be an efficient and fairly rapid process. For this reason, direct energy input into the epoxy to activate the cure process is of great interest -- if such a process can be controlled, is non-damaging, and can effect complete and appropriate cure of the epoxy.

Thus, cure of epoxies by direct energy input would have significant advantages, such as saving both energy and time, and producing cure exactly when and where needed. But the means of achieving this is not easily found, or it would have long since become the standard method of cure. This paper reports the results to date of one approach to the problem -- curing of epoxies and epoxy-matrix composites using a laser of appropriate frequency. The work reported here is neither complete nor conclusive, but it has produced some interesting results that may eventually lead to a successful non-thermal cure technique.

## 2. THEORY

An uncured epoxy resin is by definition a resin containing the oxirane ("epoxide") ring, $-\overset{\overset{\text{O}}{\diagup\diagdown}}{C}-C-$. Cure of such a resin proceeds by rupturing this ring and then bonding ("crosslinking") the epoxy chain to other components of the system. Although epoxy resins can "homo-" (self) polymerize, most epoxies are cured with a "curing agent" that assists in opening the epoxy ring and in crosslinking the epoxy molecules together. Some cure processes, however, proceed with the aid of an "accelerator", which speeds the opening of the epoxy ring and thus the overall cure reaction. Other components may also be involved in the cure process, such as "flexibilizers" (which are merely flexible epoxy resins), and "diluents" (which are very low molecular weight mono- or di-functional epoxies). For all epoxy cure reactions, however, the key to achieving crosslinking is to activate the reaction that ruptures

the epoxy ring and produces crosslinking. Since the crosslinking process is exothermic, this is the only energy necessary to achieve cure. But, of course, the correct reaction must be activated.

## 3. MATERIALS

This work was initiated as part of a study of non-thermal curing of large filament-wound structures made from carbon-fiber/epoxy. Thus, two epoxy resin systems used in filament winding were studied. The first was an anhydride-cured system designated SRF-102, and the second was an amine-cured system designated SRF-205. The components of these systems are listed in Table 1. Details on the structures of the components can be found in References 1 and 2.

The normal thermal cure for both of these systems is 16 hours at 120 °C (250 °F).

Table 1
Components of Epoxy Systems

---------------------------------
SRF-102
---------------------------------
epoxy resin: ERL-4221
(3,4-epoxy cyclohexylmethyl-3,4-epoxy cyclohexyl carboxylate)

curing agent: NMA
(nadic methyl anhydride)

accelerator: EMI-24
(2-ethyl-4-methylimidazole)
---------------------------------
---------------------------------
SRF-205
---------------------------------
epoxy resin: EPON 828
(digylcidyl ether of bisphenol A)

epoxy resin: ERL-4206
(vinyl cyclohexene dioxide)

curing agent: MDA
(4,4'-methylene dianiline)
---------------------------------
---------------------------------

In the Appendix are given Fourier-transform infrared (FTIR) and U.V./visible absorption spectra of these components, as well as FTIR spectra and differential scanning calorimetry (DSC) scans and typical data for the mixed resin systems. It should be noted that although the general features remain unchanged, the details of cure -- as determined by DSC -- change with each new batch of components and to some extent with each batch of resin that we mix. Thus, for such matters as characterizing degree of cure, each batch of resin had to be characterized individually.

In addition to the neat resins, composites of carbon fiber in these resins were also studied. The fiber used was IM-6 by Hercules.

## 4. APPROACH AND RESULTS

The most obvious way to achieve non-thermal cure of an epoxy resin would be to provide energy that directly ruptures the epoxy ring. The ruptured ring would then be available to bond with available species. One way of doing this would be to use a laser to sequentially distort and then rupture this ring, working with the vibrational frequencies of the epoxide ring which are in the infrared range. There are three such vibrational bands; the "11 μm band" occurs between 950 and 810 $cm^{-1}$, and corresponds to assymetric epoxy ring stretching. Other bands occur at "8 μm" (1250 $cm^{-1}$) and at "12 μm" (840-750 $cm^{-1}$). The mechanism of such rupture would be multi-photon dissociation (MPD), which has been used to produce vibrational dissociation of polyatomic gas molecules such as $SF_6$, $CF_3$, and $BCl_3$. Figure 1 shows schematically a plot of potential energy versus distance between atoms in a molecule. In order to dissociate the molecule, the overall energy of the molecular system must be raised from the level at the bottom of this potential energy "well" to or above the level it would reach at complete dissociation and

separation -- here taken as zero energy. Multi-photon dissociation may occur if the molecule absorbs many photons from the laser, each one of which moves it one more (quantum) step up the potential energy well. Eventually, if it absorbs enough photons, it will dissociate.

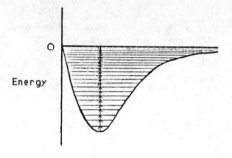

Figure 1: Schematic of multi-photon dissociation process

As Figure 1 shows, the quantum steps become closer together as the overall energy of the molecular system is increased toward dissociation. Thus, this process requires a tuneable laser so that the radiation frequency can be adjusted to these step differences. Unfortunately, early work showed this process to require a much higher energy density than could be provided by continuously tuneable lasers. A $CO_2$ laser, however, would have sufficient energy density for MPD. This laser is not truly tuneable, but can be adjusted in wavelength between about 9.0 and 11.4 μm by changing the isotopic $CO_2$ gas mixture. Such a powerful laser would have to be carefully regulated to avoid incinerating the epoxy, but does have the potential for efficiently opening the epoxy ring and producing cure. To date, we have not pursued this approach, and it will not be further discussed in this paper. Nonetheless, we do feel that there is merit to MPD, and hope to pursue it in the future.

An alternative to achieving cure by direct rupture of the epoxy ring is to produce cure by activating the curing agent, which will then attack and rupture the epoxy ring. FTIR, visible light, and U.V. absorption studies of the components of two epoxy systems showed that MDA (4,4'-methylenedianiline), the amine curing agent of SRF-205, has a strong absorption in the U.V. wavelength range of about 200 to 340 nm. This is shown in Figure 2. We had available a pulsed nitrogen laser having a wavelength of 337 nm. Although this was far from an ideal frequency, experiments with this laser produced color and viscosity changes indicative of some degree of cure in SRF-205. (This nitrogen laser has 10 nanosecond pulses which repeat at 20 Hz, and about 500 kW peak power.) This experiment was repeated with various formulations with and without MDA, and the reaction was found to occur only when MDA was present. The formulations containing MDA were also found to fluoresce in the green during exposure.

(We also found that we had to stir SRF-205 for 3 hours before laser exposure, as is done commonly in production with the system, or we got a great deal of sputtering in the laser. This sputtering was presumably due to undissolved particles of MDA.) In addition, the nitrogen laser was found to produce cure in a U.V.-activated system, but to produce no reaction in the aliphatic epoxy system SRF-102, whose components had no absorptions in the 337 nm region.

Examination of the U.V. absorption spectra of the SRF-205 components, and especially of MDA, led us to conclude that appropriate wavelengths to produce efficient cure could be achieved by use of an "excimer" laser. Table 2 shows the wavelengths achievable with such lasers. Since the MDA absorbs most strongly from about 220 to 320 nm, KrF, XeBr, and XeCl (and perhaps KrCl) all have the potential for efficiently activating the cure reaction. For our experiments we used a XeCl laser (308 nm) at XMR Corporation in Santa Clara.

Table 2
Available Excimer Laser Wavelengths

| Excimer | Wavelength (nm) |
| --- | --- |
| ArF | 193 |
| KrCl | 222 |
| KrF | 248 |
| . XeBr | 282 |
| XeCl | 308 |
| XeF | 351 |

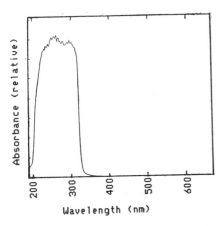

Figure 2: U.V. absorption of MDA

The reaction, however, was slow, inefficient, and incomplete, which was not surprising since the MDA absorption peak has dropped to almost zero by 337 nm. In fact, the occurrence of any reaction was more surprising than its slowness.

The experimental set-up used for these excimer laser experiments was to use a mirror to divert the horizontal beam downward through a beam shaper, and onto the specimen. Thus, in all cases the beam impinged upon the specimen at 90° to the surface, and was rectangular in shape. The specimen was carried in an aluminum foil "boat", which was placed on a metal "lab jack". A "house vacuum" was used to carry away (some of) the gases generated.

In our initial XeCl experiments we used pulse rates between 10 and 300 Hz, a pulse energy of about 275 mJ, and exposure times of 1 second to 5 minutes. The specimens were placed in the center of round aluminum foil pans. The rectangular beam was about 5 by 10 mm in size. In most cases this impinged as a "spot" in the center of the specimen area, although in some cases the beam was scanned over the specimen. We tested epoxy specimens and a few fiber and fiber/epoxy specimens. The primary problem was one of control -- numerous specimens burned up, while most others showed no reaction at all. The specimens were also found to liberate a lot of volatiles -- far more than in a thermal cure -- and for those specimens which burned, the flames seem to originate above the surface of the specimen in these volatile gases. The most promising result was the only specimen which was tested at 300 Hz, which appeared to be curing when the volatile gases ignited and charred the surface of the specimen.

Our analysis of these results led us to conclude that we were concentrating too much power in too small an area of the specimen. We were also concerned that the relationship between the specimen size and shape and the laser spot size or scan area was somewhat uncontrolled. We thus performed a second set of experiments, where we adjusted the beam to a rectangular shape of 9 mm by 15 mm, we placed all specimens in rectangular boats 9mm by 25 mm in size, and we reduced the pulse energy. In all cases the specimens filled the boats to about 2-mm thickness, and the beam impinged upon the center of the specimen and was not scanned. We used frequencies of 100 to 500 Hz, a pulse energy of 180 to 270 mJ, and exposure times of 1/5 to 15 seconds. We tested both epoxy and fiber/epoxy specimens. Significant viscosity increases were observed in several of these. The degrees of cure of these specimens were analyzed by comparing the subsequent curing reaction enthalpies to that of

uncured material, using DSC.

Table 3 and Figure 3 summarize the findings of this set of experiments on epoxy specimens. At some conditions the specimens either burned or were charred as volatile gases ignited. This occurred with all specimens exposed at 500 Hz, as well as with exposure times of 10 seconds and longer at 300 Hz, and for some of the specimens at 200 Hz. (At 500 Hz, the specimens caught fire within 0.2 seconds!) As can be seen, the most promising results were achieved at 200 Hz, where 32.2, 26.8, and 25.6% cure were obtained (without burning) after 10, 12.5, and 15 seconds exposure at a pulse energy of 180 mJ, and 25.2% cure was obtained after 10 seconds exposure at 220 mJ. However, for this condition as well, control was difficult: one specimen which was exposed for 15 seconds and one exposed for 20 seconds at 180 mJ burned rather than cured. Furthermore the reaction with specimens which contained carbon fiber seemed to be harder to control than that for the epoxy alone: almost all of these specimens either burned or were charred by gas ignition.

Although these results are far from conclusive, significant quick cure of the epoxy has been achieved under some circumstances. The approach has promise, but must be controlled. It seems likely that the burning and charring problem -- particularly with the "composite" specimens -- is still one of concentrating too much power in too small an area. The beam shaping and broadening must be improved, and it would appear desirable, if possible, to further decrease the pulse energy.

It is theorized that an excimer laser of different wavelength might be more efficient and controllable. The KrF excimer laser has a wavelength of 248 nm, which is at the center of the absorption band of MDA. EPON 828 also has a bimodal absorption band between about 200 and 300 nm, so the KrF excimer would also potentially

activate the epoxy species as well
as the curing agent. (This excimer
also might be able to produce cure
in the SRF-102 system; all of the
components of that system have
absorption bands which trail into
the 248 nm region.)

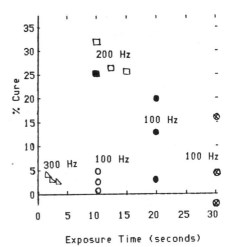

Figure 3: Percent cure versus
exposure time for XeCl excimer
laser experiments.
Key:
○ 100 Hz, 180 mJ pulses
● 100 Hz, 220 mJ pulses
⊗ 100 Hz, 270 mJ pulses
▢ 200 Hz, 180 mJ pulses
◼ 200 Hz, 220 mJ pulses
◿ 300 Hz, 180 mJ pulses

Table 3
Second Set of XeCl Excimer Laser
Experiments on Epoxy Specimens

| Freq. (Hz) | Pulse Energy (mJ) | Time (s) | Viscosity Change * | %Cure (from DSC) |
|---|---|---|---|---|
| 100 | 180 | 10 | 0 | 4.7 |
| 100 | 180 | 10 | 0 | 0.2 |
| 100 | 180 | 10 | 0 | 2.2 |
| 100 | 220 | 20 | 2 | 19.8 |
| 100 | 220 | 20 | 0 | 2.6 |
| 100 | 220 | 20 | 1 | 12.9 |
| 100 | 270 | 30 | 2 | 4.5 |
| 100 | 270 | 30 | 2 | -2.0 |
| 100 | 270 | 30 | 2 | 16.0 |
| 200 | 180 | 10 | 3 | 32.2 |
| 200 | 180 | 12.5 | 3 | 26.8 |
| 200 | 180 | 15 | – | Burned |
| 200 | 180 | 15 | 3 | 25.6 |
| 200 | 180 | 20 | – | Burned |
| 200 | 220 | 10 | 2 | 25.2 |
| 300 | 180 | 1.7 | 0 | 4.3 |
| 300 | 180 | 2.5 | 0 | 2.7 |
| 300 | 180 | 3.3 | 0 | 2.5 |
| 300 | 180 | 5 | (3) ** | (34 ) |
| 300 | 180 | 10 | – | Burned |
| 300 | 180 | 20 | – | Burned |
| 500 | 180 | 0.2 | – | Burned |
| 500 | 180 | 0.2 | – | Burned |
| 500 | 180 | 0.2 | – | Burned |
| 500 | 180 | 0.4 | – | Burned |
| 500 | 180 | 0.4 | – | Burned |
| 500 | 180 | 0.4 | – | Burned |
| 500 | 180 | 0.6 | – | Burned |
| 500 | 180 | 0.6 | – | Burned |
| 500 | 180 | 0.6 | – | Burned |

* Viscosity change is relative,
ranging from  0 (= no change)
to 3 (= significant increase).

** This specimen charred due to
ignition of volatile gases.

5.   ANALYSIS OF LASER CURE REACTION

Although we have demonstrated that
under some circumstances we can
produce rapid non-thermal cure of
an epoxy system using a laser, the
question still remains as to
whether the resulting reaction is
comparable to that produced during
thermal cure. Our work to date has
produced some conflicting findings
on this. It is encouraging that
the DSC scans for the laser-exposed
specimens are comparable to those
for specimens which have been
partially cured using thermal
means. The FTIR diffuse
reflectance results, however, show
not only the changes expected
following partial thermal cure
(such as a decrease in the epoxy
ring vibrational absorption at
about 910 cm$^{-1}$), but also show
a greater-than-expected increase in
the hydroxyl group vibration at
about 3400 cm$^{-1}$. This may
indicate that part of the apparent
"cure" is actually reaction of the

MDA with environmental oxygen. In addition, there are far more gases coming off of the laser-exposed specimens than thermal specimens. This may be because excessive laser power is vaporizing some of the components, but it also could be because the reaction is producing something different from that produced during thermal cure. Clearly the nature of the reaction produced by laser exposure must be further studied.

## 6. APPLICATIONS

Even with a polarized beam from the nitrogen laser, we found the transmittance of SRF-205 "prepreg" and of the bare IM-6 fiber bundle to be very poor -- less than 7%. We have no reason to believe that any other U.V. laser will be more efficient than the nitrogen laser in penetrating the fiber, and we thus have concluded that the excimer laser cannot be used to efficiently penetrate more than the surface of a fiber/epoxy bundle. Thus, this U.V. laser technique will not be able to achieve cure of the composite by itself. Furthermore, these two epoxy systems during cure become increasingly opaque to U.V. radiation. Thus, it may not be possible to completely cure even a neat epoxy sample using this technique.

Does this mean that this approach, even if optimized, has no use? Although it may not be able to yield complete cure of a composite or of anything but thin sections of epoxy, it may nonetheless be useful for some applications. For example, other non-thermal (and some thermal) cure techniques are currently unworkable because the epoxy viscosity drops very rapidly in the initial stages and the epoxy literally drips off of the part. This U.V. technique may be able to solve this problem by producing a gelled "skin". For example, such a gelled layer could be produced rapidly during filament winding by directing the laser at the interface between the previous top layer of the part being filament

wound and the next layer being wound on. Then delamination due to winding upon a partially cured layer would be avoided, and an "onion skin" of gelled material would be produced throughout the composite part. Even if broken by winding pressure in places, this skin would then greatly restrict resin run-off during the rapid viscosity decrease produced during other non-thermal cure techniques. In another application, a similar approach could also quickly lock in place (in an orientation produced by an applied magnetic field, for example) the magnetic particles in the epoxy applied to a computer disk.

An interesting side benefit to the use of the laser on composites is that even brief excimer laser exposure seems to greatly improve the penetration of the resin into the fiber. The reasons for the behavior, however, are not clear.

It may be legitimately asked whether a laser is the correct tool for this approach. If laser light of a certain frequency produces such cure, why couldn't a cheaper, more efficient U.V. lamp produce the same result? Indeed, it may be able to do so. The disadvantage of the U.V. lamp, however, is that it is not monochromatic and thus may initiate reactions other than the one(s) desired. Also, it is not as easily focused to the desired location. We feel that the correct approach at this time is to perform experiments with the monochromatic source of the laser, and then once this approach is optimized and understood, to see if it can be duplicated by a non-monochromatic source.

## 7. CONCLUSIONS

We have demonstrated the possibility that gel or cure of an epoxy resin might be produced by use of a U.V. laser of appropriate wavelength. We have produced rapid partial cure, but the technique currently suffers from a lack of control. We also must study further whether the cure produced is the equivalent of thermal cure. We have, however, determined that this technique will not be able to penetrate through a layer of carbon fiber, but that it might be suitable to produce a layer-by-layer "skin" of gelled material in a composite. It may also be useful in such applications as magnetic media, where it could be used to quickly gel in place oriented particles in an epoxy carrier.

## 8. ACKNOWLEDGEMENTS

The sponsorship by Aerojet Strategic Propulsion Company of this work is gratefully acknowledged. The authors wish to acknowledge and thank San Jose State University students Richard Huang, Larry McNeil, and Rod Pugh for their extremely able experimental assistance with this work.

## 9. REFERENCES

1. H. Lee and K. Neville, Handbook of Epoxy Resins, McGraw-Hill, 1967.

2. L.S. Penn and T.T. Chiao, "Epoxy Resins," in Handbook of Composites, G. Lubin, ed., Van Nostrand Reinhold, 1982, pp.57-88.

## 10. BIOGRAPHIES

LINDA L. CLEMENTS is a Professor in the Materials Engineering Department at San Jose State University, where she has been since 1981. Her specialties are composite materials and polymers, with recent emphasis on thermoplasic matrix composites, non-thermal cure techniques, failure mechanisms, and test technique development. Prior to joining San Jose State she was a Research Scientist on contract to NASA/Ames Research Center, and an Engineer and Program Manager in the Fiber Composites and Mechanics Project at Lawrence Livermore National Laboratory. She received both her B.S. and Ph.D. degrees in Materials Science and Engineering from Stanford University, and her M.S.E. in Metallurgy and Materials Science from the University of Pennsylvania.

HARINARAYANA S. (SARMA) LAKKARAJU is an Associate Professor in the Physics Department at San Jose State University. Sarma joined the SJSU faculty in 1981. Prior to that he spent two years at Texas A&M, having received his Ph.D. in Physics from State University of New York in Buffalo in 1979. His specialties are laser spectroscopy, non-linear optics, and applications to surfaces and to condensed matter.

RAYMOND KELLMAN, Associate Professor, has been with the Chemistry Deparment at San Jose State, where his specialty is polymer chemistry, since 1982. Ray also spent five years on the faculty of the University of Texas, San Antonio, worked for five years as a Research Associate and Instructor in polymers at the University of Arizona, and spent three years as a Research Polymer Chemist for Uniroyal, Inc. He received his Ph.D. in Organic Chemistry from the University of Colorado in Boulder in 1968, and spent one postdoctoral year at the University of Wisconsin in Madison. His specialties are polymer synthesis, mechanisms, and characterization, and the synthesis and properties of composite matrices.

BRIAN A. WILSON is Manager of Advanced Manufacturing Technology for Aerojet Strategic Propulsion Company. He has over 25 years experience in the aerospace industry, having worked for Bendix, Rockwell, Brunswick, and Aerojet in a variety of mechanical and materials disciplines. During the past eight years, he has been responsible for design, material research, and advanced development in strategic and tactical solid rocket motors. He is International Secretary of SAMPE, and a past chairman of the Sacramento Chapter. He received his B.Sc. in Civil and

Mechanical Engineering from St. Andrews University in Scotland, and his M.S. in Mechanics and Materials from California State University at Northridge. He is currently finishing the dissertation for his Ph.D. from the University of Nebraska.

DONALD R. OSMENT is a Manufacturing Engineer in Advanced Manufacturing Technology at Aerojet Strategic Propulsion Company. He has over six years research experience in organic matrix composite materials. Before joining Aerojet in 1985, he worked at Morton-Thiolol's Wasatch Division in filament winding research and development. He attended the University of Arizona majoring in chemistry and engineering. He is currently investigating cost saving measures for composite rocket motor case manufacturing. This includes high temperature composite reliability, new materials investigation, and advanced cure coupled with improved processing techniques.

## 11. APPENDIX

Appended here is resin information which we feel may be of use to others. The first plots given are FTIR transmittance for uncured SRF-102 and SRF-205 and for their components. Next are shown U.V./visible absorption spectra of the components of the two resin systems. Finally we give typical DSC scans for the two resin systems and typical DSC data obtained from such scans.

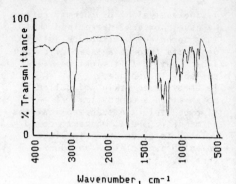

Figure 5:  FTIR spectrum of ERL-4221 (epoxy in SRF-102).

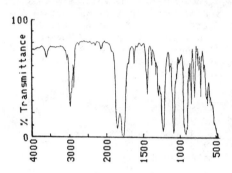

Figure 6:  FTIR spectrum of NMA (curing agent in SRF-102).

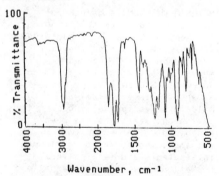

Figure 4:  FTIR spectrum of uncured SRF-102.

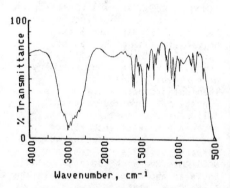

Figure 7:  FTIR spectrum of EMI-24 (accelerator in SRF-102).

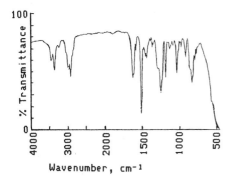

Figure 8: FTIR spectrum of uncured SRF-205, stirred 3 hours.

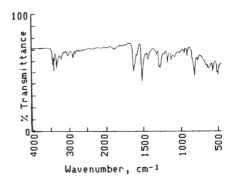

Figure 11: FTIR spectrum of MDA (curing agent in SRF-205).

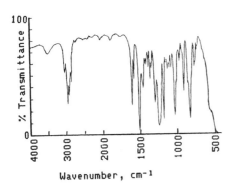

Figure 9: FTIR spectrum of EPON 828 (epoxy in SRF-205).

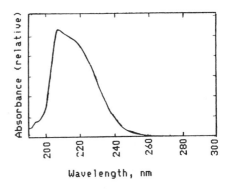

Figure 12: U.V. spectrum of ERL-4221 (epoxy in SRF-102).

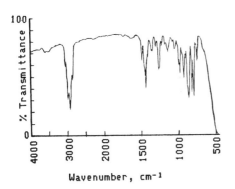

Figure 10: FTIR spectrum of ERL-4206 (epoxy in SRF-205).

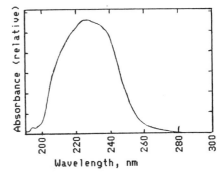

Figure 13: U.V. spectrum of NMA (curing agent in SRF-102).

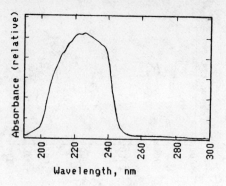

Figure 14: U.V. spectrum of EMI-24 (accelerator in SRF-102).

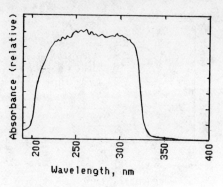

Figure 17: U.V. spectrum of MDA (curing agent in SRF-205). Note different scale from previous plots.

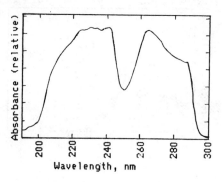

Figure 15: U.V. spectrum of EPON 828 (epoxy in SRF-205).

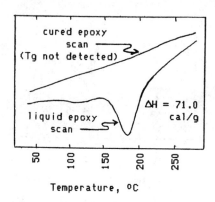

Figure 18: Typical DSC scan for SRF-102, at 5 °C/min.

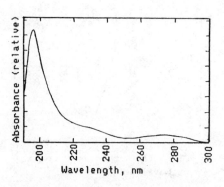

Figure 16: U.V. spectrum of ERL-4206 (epoxy in SRF-205).

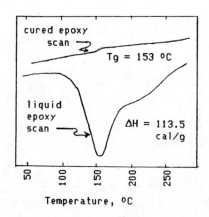

Figure 19: Typical DSC scan for SRF-205, at 5 °C/min.

Table 4
Typical DSC Data for
SRF-102 and SRF-205

| | SRF-102 | SRF-205 |
|---|---|---|
| Typical heat of reaction (cal/g) | 70* | 100* |
| Tg of cured resin (°C) | no obvious Tg! | 153 °C |
| Temp. of maximum exotherm at heating rate of 5 °C/min. | 183 °C | 152 °C |
| Type of DSC curve (at 5 °C/min.) | small peak at 110 °C; main peak at 183 °C. | main peak at 152 °C; small shoulder at 210 °C. |

* These values vary somewhat for each batch of resin mixed.

FABRICATION AND CHARACTERIZATION OF AMINE TERMINATED
POLY(ARYLENE ETHER SULFONE) MODIFIED
EPOXY-CARBON FIBER COMPOSITES
James A. Cecere, James S. Senger, James E. McGrath*
Department of Chemistry and Polymer Materials and Interfaces Laboratory
Virginia Polytechnic Institute and State University
Blacksburg, VA
and
Paul A. Steiner, Raymond S. Wong, Yesh Sacheva
Hysol Aerospace and Industrial Products Division
The Dexter Corporation
Pittsburg, CA


*To whom correspondence should be addressed

ABSTRACT

Multifunctional epoxy resin
networks were chemically modified
with thermoplastic amine terminated
poly(arylene ether sulfones) of
controlled molecular weights.  This
system was then examined as both
neat resin and as a matrix resin
for carbon fiber composites.  The
neat resin displayed a significant
increase in both $K_{1C}$ and $G_{1C}$
values.  This was attributed to the
altered morphology, which could be
varied from particles of polysulfone in an epoxy matrix to that
of a quasi-continuous polysulfone
phase.  In general, the composite
samples displayed increased
mechanical property values over the
control as evaluated by a wide
range of tests.  However, the
increase in toughness for the
composite was not as significant as
the neat resin results, possibly
due to a need to optimize the
preferred morphology in the
composite system.

1. INTRODUCTION

Composite based materials are
already a mainstay in the aerospace industry.  As stronger and
tougher systems are developed,
opportunities to replace conventional type materials (such as
metals) in structural parts are
increasing.  However, the maximum
properties of a composite system
are currently limited by the matrix
resin.  As a result, much work is
being done to either develop new
resin systems with optimum properties or to modify existing systems
to overcome their deficiencies.

The two main classes of polymeric
matrix resins utilized are thermosets and thermoplastics.  The
former exhibits many desirable
properties as solvent resistance,
low creep, good adhesion and high
Tg.  However, they are generally
brittle.  Examples include epoxy
resins, bismaleimides and unsaturated polyesters.  By contrast,
thermoplastics are characterized as
relatively tough, moldable, easier
to process, and extrudable.  Yet
they are often susceptible to
swelling and premature failure by
solvents and may also require high
processing temperatures.

In our laboratories, we are
attempting to combine the most
desirable properties of both types

of systems into a multicomponent resin. Specifically, we are utilizing thermoplastic based materials to modify thermosets. The idea of using physically blended thermoplastic modifiers as toughening agents is not a new one. For example, tough ductile thermoplastics such as polyether sulfones[1] and non-reactive polyether imides[2] have been reported to be effective toughening agents. However, these studies were based on physically blended modifiers which were difficult to control. As a result, the morphologies of such blends are extremely dependent upon processing conditions. We have taken this idea one step further by using bisphenol-A based polysulfones, (PSF), which are functionally terminated with either hydroxyl[3] or amine end groups[4]. These materials are able to chemically react into the network, allowing them to be uniformly dispersed and at the same time, making them less susceptible to chemical attack.

The toughening mechanism is based upon the concept of microphase separation. The thermoplastic modifier is designed to be initially soluble in the resin. However, since the solubility parameter of the PSF is different from that of the epoxy/diamine network, it will phase separate into discreet particles (on the order of 1-2μ) during the curing process. These particles remain ductile and actually plastically yield during fracture, thus absorbing energy. These particles may also act as an energy dispersing medium for the matrix in the vicinity of the PSF particle. An added advantage in using thermoplastics as modifiers is that they do not sacrifice the overall modulus of the system. This is in contrast to rubber based materials, such as butadiene-acrylonitrile copolymers with amine or carboxyl endgroups[5-7], which are liquids at room temperature. They generally sacrifice modulus to achieve toughness. This point is sometimes neglected, especially since many investigators report $G_{1C}$ values,

which increase with lower modulus.

Our initial work involved using amine terminated poly(arylene ether sulfones) and poly(arylene ether ketones) to toughen epoxy systems based upon EPON 828 and diaminodiphenyl sulfone (DDS)[7]. Since these materials were synthesized in our laboratories[8,9], the effect of molecular weight and percent incorporation on "toughenability" was studied and found to be fairly significant in some cases. In all compositions, only a minimum decrease in flexural modulus was observed. We also have shown that if the molecular weight and modifier composition is high enough, a partial phase inversion will take place. This new morphology is characterized by the observation that the PSF appears to be semi-continuous and has a definite honeycomb appearance. The fracture surface of the failed test specimen is roughened The amount of new surface area formed is thus dramatically increased, which may explain the high toughness values that were obtained.[8]

This present study represents a joint project of Virginia Tech and Dexter-Hysol. An initial study was made to determine if amino functional PSF oligomers could also improve graphite fiber based composites. The resin utilized was supplied by Dexter-Hysol. It was a multifunctional based epoxy resin cured with    -diaminodiphenyl sulfone (   -DDS). This system was modified using 13,000 Mn PSF at a weight % loading level of 15% and 21,000 Mn PSF at 15% and 30%. Various composite and neat resin tests were conducted to determine the feasibility of these materials as toughening agents in composites.

## 2. EXPERIMENTAL

(2.1) <u>Preparation of Amino Terminated Poly(Arylene Ether Sulfone)</u> Two molecular weights of amine terminated polysulfone (PSF) were synthesized according to a procedure developed in our lab.[9] The resulting polymers were purified by precipitation and thoroughly dried.

They were titrated with a HBr/Glacial acetic acid solution to determine their number average molecular weights ($\overline{Mn}$). This value was compared to that obtained from gel permeation chromatography using polystyrene standards. In general, the agreement was satisfactory.

(2.2) Reaction of PSF With The Epoxy Resin-Curing Agent System
The resin system utilized in this system was a mixture of various epoxy based materials which is termed a multifunctional epoxy resin. The concentration of polysulfone modifier used was determined on a weight basis. The amount of curing agent required to yield a 2:1 molar ratio of epoxy to amine groups was calculated taking into consider-ation the number of PSF end groups. The resin and the PSF were initially mixed and heated with stirring to ~90°C, whereupon the PSF completely dissolved and formed a homogeneous solution. The curing agent ( -DDS) was then added and the mixture was heated to ~140°C and stirred until a homogeneous solution was again achieved. The resin was then cooled to room temperature and used subsequently for cure studies.

(2.3) Neat Resin Testing
Samples of neat and modified resin were cured under comparable conditions in order to obtain data such as water uptake, Tan δ and loss modulus (DMTA), $K_{1C}$, $G_{1C}$, and lap shear strength. These samples were also used to study the influence of system variables on morphology. Samples were cured in an autoclave according to the cure schedule shown in Table 1. Water uptake studies were determined by boiling samples for 48 hours and observing a mass change. Test specimens were then analyzed using a Polymer Labs DMTA and compared with samples which were run dry.

Compact tension samples were used to obtain $G_{1C}$ and $K_{1C}$ data. These specimens, after casting and curing, were milled to exact size and then precracked. This was accomplished by first cutting a small slot with a reciprocating

table saw, then precracking with a clean razor blade, which had been cooled with dry ice. The samples were tested on a United Test machine and stressed to failure. The $K_{1C}$ data was calculated according to the following equation:

$$K_{1C} = \frac{P}{\sqrt{W} \cdot B} \cdot f(a/W) \qquad (1)$$

where P=load, W=width, B=thickness, and a=length of precrack. The energy release rate, $G_{1C}$, was then calculated by

$$G_{1C} = \frac{(K_{1C})^2(1-\nu^2)}{E} \qquad (2)$$

where $K_{1C}$=fracture toughness, $\nu$=Poissons ratio, and E=modulus.

The modulus data were obtained from the DMTA and Poisson's ratio, which was assumed to be 0.25. Tensile lap shear data was generated using aluminum panels bonded with both neat and modified resin. The surface of the aluminum utilized was prepared using standard degreasing and etching techniques to remove any undesirable oxides and form the best bonds. In order to obtain uniform thickness of the adhesive layer, glass beads (~5μ) were mixed into each resin system to be tested. The resins were spread on one aluminum panel edge and the second panel was overlapped by ½". The panels were then layed up and cured in an autoclave ac-cording to the schedule in Table 1. The cured panels were cut into 1" wide strips and tested in tension at various temperatures.

Scanning electron microscopy photographs were taken using a JEOL JSM-35C instrument. The samples utilized were the same used for DMTA data generation. The samples were broken and SEM photographs were taken of the fracture surface.

(2.4) Composite Fabrication
The composite data were generated from prepreg which was fabricated as 6" wide tape, using a hot melt

process on a California Graphite Corporation prepreg machine. The fiber utilized was Hysol Grafil Apollo XAS fiber with an average tensile strength of 650 Ksi (4.48 GPa) and a modulus of 33 Msi (227 GPa). These fibers received standard oxidation surface treatments followed by the application of an epoxy size ("A" size) which was approximately 1% by weight. The prepreg had an approximate areal weight of 145 $g/m^2$ and a total weight of 220 $g/m^2$.

Various sample geometries were fabricated in order to generate a wide range of mechanical property values. The sample geometries and the specific mechanical properties obtained from each are listed in Table 2. The samples were layed up, autoclaved according to the cycle listed in Table 1, cut to size on a diamond cut-off wheel and milled to exact size.

### 3. RESULTS AND DISCUSSION

#### (3.1) Neat Resin Specimens
The thermal properties of the systems were determined with a DMTA at a frequency of 1Hz and a heating rate of 4°C/min for dry samples and 10°C/min for wet samples. The data obtained are listed in Table 3. As seen in the dry samples, the upper Tg is effected slightly by the incorporation of PSF. In the case of 30% loading of 21K PSF, two transitions are seen, with the first occurring at 180°C. This corresponds exactly with the Tg of the PSF which was previously determined by DSC. This slight lowering of the upper Tg suggests some phase mixing is taking place. At 15% loading, only one transition is seen, indicating that the amount and Mn of the PSF incorporated did not produce a detectable second transition.

The effect of water appears to be dramatic in this system. In the control sample, two distinct transitions are seen, one at 142°C and the second at 182°C. This may be due to the fact that the various components in the epoxy resin have different susceptibilities to water. There is a slight change in the location of the peaks as PSF is added. At the 30% level, a shoulder is seen at ~175°C. This may correspond to the PSF peak observed in the dry samples. This suggests that the water is not in the PSF phase, but in the epoxy phase.

The water uptake results are provided in Table 4. As shown, there is a decrease in the water uptake as the amount of PSF incorporated in the resin system increases. This may be attributed to the fact that PSF itself is reasonably hydrophobic. As it is added to the system, one can expect a decrease in the number of pendant hydroxy amine groups (on a weight basis) of the type depicted in Scheme 2. Since these groups, which are the result of the ring opening epoxide reaction, are the most probable sites for hydrogen bonding, their elimination should decrease the water uptake. This is supported by the DMTA data, since there is little change in the peak positions of the PSF in the wet samples with increasing PSF concentration. This supports the assumption that the water is mostly in the epoxy matrix phase.

The neat resin mechanical properties are listed in Table 5. There appears to be a dramatic effect upon the toughness of the resin system as the amount of PSF incorporated is increased. There is a greater than two-fold increase in the stress intensity factor ($K_{1C}$) and almost a five-fold increase in the energy release rate ($G_{1C}$). This type of toughening confirms and expands what was previously shown in the EPON 828-DDS[3,4,8] system, indicating that the results were not unique to that chemistry.

The morphology observed with the SEM was as expected, based on our earlier work. At 15% loading of 21K PSF, discreet particles of PSF are seen as 1-2μ spheres uniformly distrib-uted throughout the epoxy matrix (Figure 1). They appear to be drawn out and cracks are seen to initiate through each. However, at 30% loading, 21K PSF, the morphology changes dramatically (Figure 2). As was seen in the EPON 828/DDS system, a honeycomb-like morphology developed. Clearly, a semi-continuous phase of PSF has developed. This would explain the high toughness, since a dramatic increase in surface area occurs as the sample is fractured. It should also be noted that the fracture surface is rougher and the size of each hexagonal unit is about 3μ in diameter. A much more detailed study is required to further elucidate this interesting and possibly technologically important morphology.

Finally, the tensile lap shear data is plotted as a function of temperature in Figure 3. The PSF appears to increase the adhesive character of the resin system, most likely as a result of enhanced toughness. This effect is seen up to 350°F (177°C), where the strength of the system diminishes. This is to be expected, since it is approaching the Tg of the PSF and, indeed the epoxy network system.

(3.2) Composite Results

The data obtained from the 90° flexure samples are given in Table 6. The PSF appears to increase both the ultimate strength and the flexural modulus. This is a significant achievement, since many tougheners (e.g., rubber-based materials) usually have the draw-back of lowering the modulus. Though the observed increase in this system is not large, it is real.

The compression samples were tested dry and in a hot/wet condition by first boiling for 48 hours and then testing at 180°F (82°C) after holding for about 2 min. The results of these tests are tabulated in Table 7. In the dry samples, there appears to be an overall decrease in the compressive strength, but it is minimized as a function of the molecular weight and weight % of the PSF. The changes in the hot/wet samples are not as clear cut as the dry case. For example, an actual increase in strength at the 15%-21K loading was

noted. This may again be an indication that since the PSF is decreasing the water uptake, the mechanical properties are affected in a less dramatic way.

The data obtained from the double cantilever beam samples are given in Table 8. The $G_{1C}$ data were calculated by measuring the area under the curve of load vs. cross head deflection. The $G_{1C}$ is then calculated by the formula:

$$G_{1C} = \frac{\Delta A}{\Delta l \cdot B} \qquad (3)$$

where $\Delta A$=change in area under load-deflection curve, $\Delta l$=change in crack length and B=sample width. As expected, the sample modified with 30%-21K PSF gave the greatest increase over that of the control. However, the magnitude is not as great as seen in the neat specimens. One possible explanation for this may have to do with the morphology of the system. This will be discussed later.

It does appear that the weight % of modifier has a much greater effect than the molecular weight over the limited range examined. This too is somewhat surprising, since preliminary work did show a toughness dependence on molecular weight. However, in the earlier studies much lower molecular weights were also studied.

The last set of data generated, ±45° tensile, are shown in Table 9. All of the data generated were obtained using strain gauges (Micro-measurements Division, Measurements Group Inc., Raleigh, NC, type CEA-06-250UW-350), which were attached using a cyanoacrylate adhesive. These specimens were also tested both dry and hot/wet. It appears that most of the results show a slight decrease compared to that of the control. However, there is an increase in the ulti-mate strength as the amount of PSF is increased. This is seen in both sets of data. SEM photomicro-graphs were also taken of some of the composite specimens. Figure 4 is the edge-on view of the 15%-21K PSF modified ±45° tensile sample. As seen, the fibers are well covered with resin and there appears to be little fiber pullout. This indicates that the PSF doesn't diminish the resin-fiber adhesive properties. A similar type photograph of the ±45° tensile specimen modified with 30%-21K PSF is shown in Figure 5. Again there appears to be little fiber pullout, and the fracture surface is relatively even between the fiber and the resin, indicating good adhesion.

This photograph provides a possible explanation of why the composite $G_{1c}$ value increases are not as large as indicated by neat resin testing. The spacing between the fibers is apparently only approximately 1μ, which is much less than the size of the "hexagonal" units of the modifier in the neat resin castings. This may indicate that there is not enough room for the tough honeycomb morphology to develop. Alternatively, it may be hypothesized that it is the fibers themselves that do not allow the honeycomb morphology to develop, possibly by changing the way the PSF precipitates during curing. This is inconsistent with Figure 6, which is a photomicrograph of a different region of the same specimen. Here, the fiber density is much less and it can be plainly seen that the honeycomb will develop provided the geometry is appropriate. It should also be noted that the resin-fiber adhesion remains high. Further work is in progress to better understand where the PSF exists in these systems. This may provide further insight into the toughening phenomena and suggest methods for additional improvements.

## 4. CONCLUSIONS

Amine terminated poly(arylene ether sulfone) thermoplastic oligomers are good modifiers for epoxy based graphite composite systems. The characteristics of high solubility, ease of fabrication and thermal stability are positive in terms of processing. The modified systems show a number of improved properties. Significantly, the modified networks show a slight increase over unmodified systems under hot/wet conditions. Finally, the PSF increases the adhesive properties of the system over the entire usable temperature range.

## 5. ACKNOWLEDGEMENTS

We would like to thank those at Dexter-Hysol for their support, expertise and materials utilized in this study, especially G. Gong and P. Naye. We would also like to thank J. C. Hedrick for his help in synthesizing the functionalized polysulfone. Portions of this work were funded by a grant from the Polymer Materials Branch of the NASA Langley Research Center.

## 6. REFERENCES

1. Bucknall, C.B., Patridge, I.K. Polymer 24, 639 (1983).
2. Diamant, J., Moulton, R.J. 29th Natl. SAMPE Symp. 29, 422 (1984).
3. Hedrick, J.L., Yilgor, I., Wilkes, G.L., McGrath, J.E. Polymer Bulletin 13, 201 (1985).
4. Hedrick, J.L., Jurek, M.J., Yilgor, I., McGrath, J.E. ACS Polym. Prep. 26(2), 283 (1985).
5. Rowe, E.H., Siebert, A.R., Drake, R.S. Mod. Plast. 47, 110 (1970).
6. Sultan, J.M., McGarry, F. Polym. Eng. Sci. 13, 29 (1973).
7. Drake, R.S., Egan, D.R., Murphy, W.T., in "Epoxy Resin Chemistry II", Bauer, R.S., ed. ACS Symposium Series, 221, chapt. 1, (1983).
8. Cecere, J.A., Hedrick, J.L., McGrath, J.E. 31st Natl. SAMPE Symp. 31, 580, (1986).
9. Jurek, M.J., McGrath, J.E. Polym. Prep. 27(1), 315 (1986); M. J. Jurek, Ph.D. Thesis, VPI&SU, 1987 and forthcoming publications.

## 7. BIOGRAPHIES

James A. Cecere, a native of Queens, N.Y., received his B.S. in Chemistry from Rochester Institute of Technology in 1983. He entered the Ph.D. program in Material Engineering Science the following fall at Virginia Polytechnic Institute and State University. His research has primarily been in the field of impact modification of thermosetting resins.

James S. Senger received his B.S. in Chemical Engineering in 1983 and his M.S. in Chemistry in 1986 from Virginia Polytechnic Institute and State University. He is currently enrolled in the Materials Engineering Science Program at VPI. His research to date has been focused on the synthesis and characterization of epoxy resins.

James E. McGrath was born in Easton, New York and received his B.S. in Chemistry from Siena College in 1956. After being employed ITT Rayonier in Whippany, NJ and the research division of the Goodyear Tire and Rubber Co., he obtained an M.S. degree in Chemistry from the University of Akron in 1964 and Ph.D. in Polymer Science from the same university in 1967. Professor McGrath joined the Union Carbide Corporation in August of that year. Professor McGrath joined VPI & SU in September 1975. He was named a tenured Associate Professor in 1977 and was promoted to Full Professor in 1979. He is a consultant to industry and government. He is also a Co-Director of the Polymer Materials and Interfaces Laboratory. Professor McGrath has coauthored what is considered a definitive book on Block Copolymers (Academic Press, 1977), and has edited books on Anionic Polymerization, Ring Opening Polymerization and Advances in Polymer Synthesis (with B. M. Culbertson). He has over 100 contributions in the literature, and also 21 U.S. patents.

$$H_2N\text{-}R\text{-}O\text{-}R'\text{-}\!\!\left[\!O\text{-}R\text{-}O\text{-}R'\right]\!\!\text{-}O\text{-}R\text{-}NH_2$$

$$\underset{1}{}$$

Where $R = $

$R' = $

$$\underset{2}{}$$

## TABLE 1

### AUTOCLAVE CURING CYCLE

1. Pressurize to 0.59MPa (85psi)
2. Heat up to 120°C (248°F) at 2°C/min
3. Hold at 120°C (248°F) for 60 min
4. Heat up to 179°C (355°F) at 2°C/min
5. Hold at 179°C (355°F) for 240 min
6. Cool down at 2°C/min

## TABLE 2

### COMPOSITE SAMPLES UTILIZED

| SAMPLE | # PLIES | SIZE (in) | GEOMETRY | DATA OBTAINED |
|---|---|---|---|---|
| DOUBLE CANTILEVER BEAM | 24 | 1x9 | [0] | $G_{1C}$ |
| COMPRESSION | 8 | 3.5x0.5 | [0] | $\sigma_{ult}$ |
| TENSILE | 16 | 1x9 | [±45] | $\sigma_{ult}, \nu, E_{11}, G_{12}$ |
| FLEXURE | 16 | 1x5 | [90] | $\sigma_{ult}, E$ |

## TABLE 3

### TAN$\delta$ TRANSITIONS OBSERVED AT 1Hz

| SAMPLE | TRANSITIONS (°C) | | |
|---|---|---|---|
| CONTROL | 194 | | |
| 15% - 21K | 193 | | |
| 30% - 21K | 180 | 191 | |
| CONTROL/WET | 142 | 182 | |
| 15% - 21K/WET | 142 | 179 | |
| 30% - 21K/WET | 148 | 173 | 175* |

*APPEARS AS SHOULDER ON 173° PEAK

## TABLE 4

### WATER UPTAKE RESULTS

| SAMPLE | % WT. INCREASE |
|---|---|
| CONTROL | 4.47 |
| 15% - 21K | 3.92 |
| 30% - 21K | 3.16 |

## TABLE 5

### NEAT RESIN MECHANICAL PROPERTIES

| SAMPLE | $K_{1C}$ N/m$^{3/2}$ (psi/$\sqrt{in}$) | | $G_{1C}$ J/m$^2$ (lb/in) | |
|---|---|---|---|---|
| CONTROL | 0.58 | (496) | 103 | (0.59) |
| 15% - 21K | 1.06 | (961) | 318 | (1.82) |
| 30% - 21K | 1.32 | (1202) | 513 | (2.93) |

## TABLE 6

### 90° FLEXURE SAMPLES

| SAMPLE | $\sigma_{ult}$ MN/m$^2$ (ksi) | | MODULUS MN/m$^2$ (msi) | |
|---|---|---|---|---|
| CONTROL | 121 | (17.5) | 8960 | (1.30) |
| 30% - 13K | 138 | (20.1) | 9370 | (1.36) |
| 15% - 21K | 124 | (18.0) | 9780 | (1.42) |
| 30% - 21K | 128 | (18.6) | 9510 | (1.38) |

## TABLE 7

### COMPRESSION RESULTS

| SAMPLE | $\sigma_{ult}$ MN/m$^2$ (ksi) | |
|---|---|---|
| CONTROL | 1430 | (208) |
| 30% - 13K | 890 | (128) |
| 15% - 21K | 1010 | (146) |
| 30% - 21K | 1190 | (173) |
| CONTROL/WET* | 930 | (133) |
| 30% - 13K/WET* | 820 | (119) |
| 15% - 21K/WET* | 1000 | (145) |
| 30% - 21K/WET* | 640 | (93) |

*MEASURED AT 82°C (180°F)

## TABLE 8

### DOUBLE CANTILEVER BEAM RESULTS

| SAMPLE | $G_{1C}$ J/m$^2$ (lb/in) | |
|---|---|---|
| CONTROL | 203 | (1.16) |
| 30% - 13K | 236 | (1.35) |
| 15% - 21K | 198 | (1.13) |
| 30% - 21K | 257 | (1.47) |

## TABLE 9

±45 TENSILE SPECIMEN RESULTS

| SAMPLE | $\sigma_{ult}$ MN/m$^2$ (ksi) | | | $E_{11}$ GN/m$^2$ (msi) | | $G_{12}$ GN/m$^2$ (msi) | |
|---|---|---|---|---|---|---|---|
| CONTROL | 223 | (32.3) | .762 | ------ | ------ | ----- | ------ |
| 15% - 21K | 220 | (32.0) | .766 | 18.2 | (2.64) | 5.15 | (0.75) |
| 30% - 21K | 267 | (38.8) | .762 | 15.5 | (2.25) | 4.41 | (0.64) |
| CONTROL/WET* | 200 | (29.0) | .783 | 14.9 | (2.17) | 4.27 | (0.62) |
| 15% - 21K/WET* | 208 | (30.2) | .764 | 18.4 | (2.67) | 4.42 | (0.64) |
| 30% - 21K/WET* | 213 | (30.9) | .746 | 12.9 | (1.87) | 3.69 | (0.54) |

*MEASURED AT 82°C (180°F)

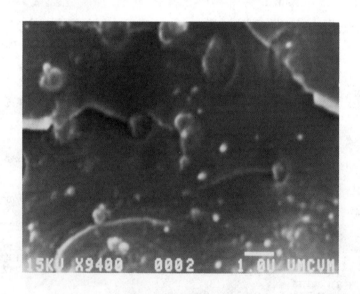

FIGURE 1   15%-21K PSF
Particles of PSF in Epoxy Resin Casting

1284

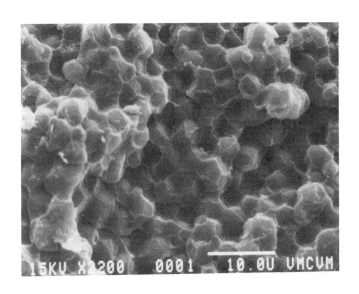

FIGURE 2   30%-21K PSF
Honsycomb Morphology Develops

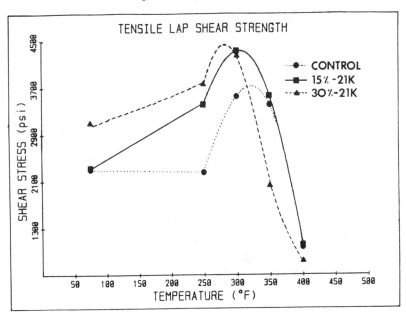

FIGURE 3   Tensile Shear Strength
of Aluminum Lap Joints vs. Temperature

1285

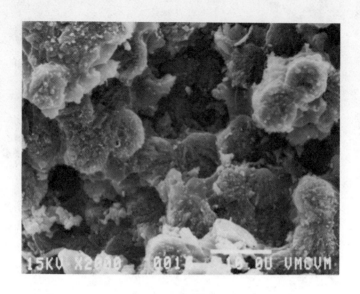

FIGURE 4 ±45° Tensile Specimen 15%-21K PSF
Edge on View

FIGURE 5 ±45° Tensile Specimen 30%-21K PSF
Edge on View

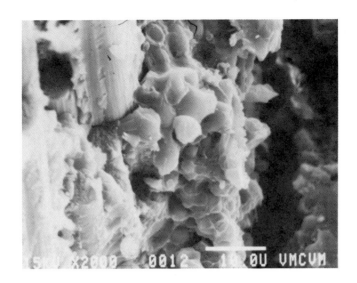

FIGURE 6   ±45° Tensile Specimen 30%-21K PSF
Note the Hexagonal Particles that Developed

THERMAL ANALYSIS FOR PULTRUSION PROCESS MODELLING
Walter X. Zukas and Noel Tessier
Organic Materials Laboratory
U.S. Army Materials Technology Laboratory
Watertown, Massachusetts 02172

### ABSTRACT

An anhydride cured epoxy resin formulation was characterized by thermal analysis techniques in order to obtain material properties for a pultrusion process model. Differential scanning calorimetry (DSC) was used to obtain rate of reaction and extent of reaction information from which the kinetic parameters, including reaction order, heat of reaction, activation energy, and pre-exponential factor, were derived. The times to gelation were determined by torsional braid analysis (TBA) and with a hot pot gel timer and compared with the extent of reaction by DSC. The effects of an epoxy viscosity diluent on the cure behavior and properties was also followed by DSC and TBA. The kinetic parameters were inputed into the model and predictions for the extent of cure and the temperature profile in the pultrusion die were made. These were compared to the

results of processing studies of the formulations on graphite where thermocouples were passed through the die with the composite. Reasonable agreement between the actual and predicted temperature profiles was observed.

## 1. INTRODUCTION

The pultrusion process consists of drawing a resin-saturated fiber bundle through a heated die where the resin cure takes place. What leaves the die is a rigid part of constant cross-sectional area. The cure of epoxy formulations is exothermic and temperatures which can lead to the degradation of the material can be reached, especially in thick parts. A pultrusion process model was developed, based on a control-volume finite difference method applied to continuity and energy balances (1), to predict the cure behavior and temperature profiles for optimization of the process. The energy balance requires an expression for the rate of reaction as a function of extent of reaction and temperature to calculate the rate of heat evolution due to chemical reaction. This was accomplished through the use of thermal analysis techniques.

Prime (2) provides an excellent review of the application of thermal analysis to thermosetting systems. More specifically, Werner and Kusibab (3) applied thermal analysis techniques to pultrusion and the evaluation of pultrusion resins. They established correlations between laboratory DSC measurements and pultruder runs. The kinetics of the cure of epoxy resins is quite complex and a number of different kinetic models are available (2). Thermal analysis by DSC has provided the experimental information for a number of these models. Fava (4) applied a number of DSC procedures to an anhydride cured epoxy resin formulation. Acitelli et al. (5) compared DSC results to other analysis techniques for the amine cure of an epoxy formulation. Levy (6) applied DSC to an epoxy system in order to develop an nth order reaction model for the cure behavior. An approach similar to Goldfarb and Adams (7) was used in this study to provide the kinetic parameters for the pultrusion process model. This approach originally applied by Friedman (8) to thermogravimetric analysis.

## 2. EXPERIMENTAL

The resin formulations used in this study consisted of Epon 826, methyltetrahydrophthalic anhydride (MTHPA), diglycidyl ether of 1,4 butanediol (BDE), as a viscosity diluent, and benzyldimethylamine (BDMA), as an accelerator. All materials were used as received from the manufacturer. The resin formulations addressed in this study will be referred to in parts by weight 826/MTHPA/BDE/BDMA, respectively.

Differential scanning calorimetry data was obtained on a Perkin-Elmer DSC-2 equipped with an intercooler. Two methods of analysis were applied. Samples were scanned at five heating rates: 80, 40, 20, 10, and 5°C/min under nitrogen or were scanned at 320°C/min to an isothermal curing temperature ranging from 116 to 171°C and held at that temperature until cure was complete.

Torsional braid analysis was carried out on an automated TBA system from Plastic Analysis Instruments, Inc. A resin formulation would be coated on a glass braid and the dynamic mechanical properties as a function of time at an isothermal cure temperature would be monitored. Gelation was associated with a peak in the log decrement trace. Subsequent dynamic scans of the cured material revealed the glass transition temperatures of the fully cured resin formulation. Gel times were also obtained from measurements made on a hot pot gel timer.

Processing studies were done on pultruded bar stock 12.7mm (1/2") thick by 25.4mm wide. The materials for the pultrusion runs were Fortafil 5, 40K, graphite fiber tows at 60% volume loading and the 2000/1832/200/20 (826/MTHPA/BDE/BDMA) resin. The die utilized was a standard pultrusion die 762 mm long, equipped with three-zone heating control. Line speed and die temperature varied from 152 to 305 mm/min (6 to 12 inches/min) and 160 to 171°C. All heating zones were controlled to the same die temperature. The material was run at the given conditions until steady-state was reached (approximately 20 min) and a thermocouple was fed into the die with the fiber bundles. The temperature was monitored as a function of time.

## 3. KINETIC ANALYSIS

The kinetic analysis carried out in this study closely parallels previous work (7,8) and is outlined here. The chemical reaction can be expressed by

$$d\alpha/dt = kf(\alpha) \qquad (1)$$

where the rate of conversion, $d\alpha/dt$, is expressed as the product of a rate constant, $k$, and some function of reactants, $f(\alpha)$. The rate constant can be further expressed in the Arrhenius form

$$k = A \exp(-E/RT) \qquad (2)$$

where $E$ is an activation energy, $R$ the gas constant, $T$ the absolute temperature, and $A$ the pre-exponential factor. A typical form of the concentration function for systems which follow nth order kinetics can be expressed by

$$f(\alpha) = (1-\alpha)^n \qquad (3)$$

where $\alpha$ is the fractional amount of reactants consumed (extent of reaction) and $n$ is the reaction order.

The first type of experiments carried out were under different heating rates so that the same $\alpha$ is obtained at different temperatures. If $f(\alpha)$ is assumed independent of temperature, equations 1 and 2 can be combined and rewritten

$$\ln d\alpha/dt = -E/RT + \ln Af(\alpha) \qquad (4)$$

At constant $\alpha$, $\ln Af(\alpha)$ is a constant and a plot of $\ln d\alpha/dt$ vs. $1/T$ for different heating rates should yield a straight line with a slope equal to $-E/R$ and an intercept of $\ln Af(\alpha)$. Repeating this process at different values of $a$ can verify that the activation energy is constant during the course of reaction. If not, it indicates that equation 1 with 2 is not a valid representation of the reaction.

This analysis also yields $Af(\alpha)$ as a function of $\alpha$. If a plot of $\ln Af(\alpha)$ vs. $\ln (1-\alpha)$ results in a straight line then the slope of that line gives the reaction order, n, as well as the pre-exponential factor, A.

The second type of experiments carried out involve holding the temperature constant. Equations 1 and 2 can be rearranged and integrated to yield

$$\int_0^\alpha d\alpha/Af(\alpha) = \exp(-E/RT) \int_0^t dt \quad (5)$$

Substituting the function $g(\alpha)$ for the left hand side, equation 5 becomes

$$g(\alpha) = t \exp(-E/RT) \quad (6)$$

or rearranging and taking logarithms of both sides,

$$\ln t = E/RT + \ln g(\alpha) \quad (7)$$

which relates reaction time to extent of reaction for each temperature. Plots of $\ln t$ vs. $1/T$ should yield straight lines with slope $E/R$ and an intercept of $g(\alpha)$ for each extent of reaction. Interpretation of $g(\alpha)$ will be addressed later in this paper.

4. RESULTS AND DISCUSSION

The degree of conversion and rate of reaction were determined from the DSC traces as follows. Assuming a constant total heat of reaction, $\Delta H_{rxn}$, for the polymerization, the extent of reaction, $\alpha_t$, is assumed proportional to the heat of

reaction, $\Delta H_t$, generated up to time t and $d\alpha/dt$ is assumed proportional to the rate of heat generation, $dH/dt$, yielding

$$\alpha_t = \Delta H_t / \Delta H_{rxn} \quad (8)$$

and

$$d\alpha/dt = (1/\Delta H_{rxn})(dH/dt) \quad (9)$$

The values of $\alpha$ vs. T for a series of dynamic DSC scans on a sample of 2000/1832/200/20 are shown in Figure 1 where a value for $\Delta H_{rxn}$ of 73 cal/g from the 5°C/min run was used for all the curves. The plateau value for the extent of reaction at the end of cure is lower for the faster scan rates.

Goldfarb and Adams (7) observed no decrease in total heat of reaction for similar heating rates, using similar instrumentation, for an acetylene terminated resin, so instrumental error does not seem to be a factor. Fava (4) had observed a similar decrease in heat of reaction with heating rate and attributed it to thermal decomposition.

A sample of 100/80/0/1 similarly showed a decrease in total heat of reaction with heating rate. This decrease thus appears to be due to either a change in reaction mechanism or thermal decomposition of the anhydride cured system at the higher temperatures reached with the faster heating rates. The exotherm peak positions and total heat of reaction as functions of heating rate are shown in Table 1 for both of these samples. The addition of BDE was observed to have little or no effect on the peak position or total heat of reaction. It appears that the procedure of using different scanning rates to obtain the kinetic parameters of the reaction is not valid for these formulations. Total heat of reaction is not a constant and an analysis based on these numbers would not be correct.

Isothermal experiments were carried out and rate of reaction, extent of reaction, and total heat of reaction values were calculated. Table 2 shows the total heat of reaction values as a function of isothermal cure temperature for a sample of 2000/1832/200/20. The values show some scatter, but are essentially constant with the exception of the cure at 171°C. At this temperature the significantly lower value may be reflecting a change in reaction mechanism or an onset of thermal decomposition. Reaction at this temperature is very fast and the initial part of the exotherm may also have been missed during the heat up portion of the curve. The assumption of constant heat of reaction appears to be valid for most of the isothermal cure curves obtained. A total heat of reaction of approximately 90 cal/g was observed. This is significantly higher than the dynamic cured samples at 5°C/min and appears to support the

possibility of a change in reaction at higher temperatures. Rescans of the isothermally cured resins showed no residual amount of cure.

Values for E and $g(\alpha)$ were obtained by a least linear squares fit of the data through equation 7 and are shown in Table 3. The values for E decrease somewhat with extent of reaction, with larger differences observed at the highest conversions. The times to extents of reaction greater than 80% are subject to larger experimental error due to baseline determination which may account for some of the variation observed for E above 80% conversion. A varying value for E would indicate that the form of equation 1 may not be valid for the system considered, but a mathematical expression for the kinetics is still required for the process model. Since the variation in E is not that great the analysis was continued.

The values of $g(\alpha)$ represent the integral of $d\alpha/Af(\alpha)$ from 0 to the $\alpha$ under consideration. In order to evaluate values for A, nth order kinetics were assumed for $f(\alpha)$ of the form of equation 3. The order, n, was varied until constant values of k were obtained as a function of $\alpha$ at a fixed temperature. The best fit was obtained when a value of 1 was used for n. Table 4 contains the values of A obtained and of k at 160°C. Figure 2 shows a plot of rate of reaction vs. conversion at 160°C and the predicted rates from the above values of E, A, and n. Although varying values of E and A are used, a relatively good fit of the experimental data is obtained.

Isothermal TBA was carried out on a sample of 2000/1832/200/20 and the times to gelation were fit to equation 7. Gelation should theoretically occur at the same extent of reaction, independent of temperature. A plot of the linear least squares fits of the times to 20, 50, and 80% conversion by DSC and the times to gelation by TBA are shown in Figure 3. An activation energy of 17.85 kcal/mole was obtained for gelation with the times to gelation essentially occurring at an extent of reaction of 50% by DSC. These values can be compared to the results from the hot pot gel timer indicated by points on the same plot. Gelation as measured by this technique occurs later and lines up with the 80% conversion curve by DSC.

TBA was also used to determine the effects of BDE on the mechanical properties of the cured resin. Samples were cured at 160°C with 0, 10, and 15 parts BDE per 100 parts 826 with the corresponding stoichiometric amount of MTHPA and 1 part BDMA per 100 parts 826. The fully cured glass transition temperatures corresponded to 107, 100, and 95°C, respectively. As the amount of viscosity diluent was increased the glass transition temperature of the fully cured resin decreased.

The kinetic parameters of E, A, n, and total heat of reaction were used in the pultrusion process model for the 12.7 mm pultruded bar. Plots of predicted center-line temperature and experimentally measured center-line temperature vs. distance (time) into the die are shown in Figure 4 for a cure of 2000/1832/200/20 at approximately 160°C and a line speed of 152 mm/min. The dimensionless die temperature profile is indicated by the surface temperature trace in Figure 4. The predicted values somewhat underestimate the temperature observed experimentally. This variation could be due to either an experimental or a modelling error. Experimentally, the thermocouples may not be properly centered in the bar and thus indicate a temperature profile closer to that for the surface. On the other hand, the correct heat transfer coefficient for the composite may not have been used in the model and the effects of this coefficient on the temperature profile is currently being investigated.

The corresponding predicted values for extent of reaction vs. distance (time) into the die for the center-line and surface are shown in Figure 5. The surface is predicted to leave the die at about 70% conversion and the center at about 65%. The actual pultruder run was thought to produce an undercured part which subsequent postcure resulted in a significant improvement in physical properties. The model thus appears to give reasonable values for the extent of reaction in the process.

Further work is underway to refine some of the other constants used in the pultrusion process model, as well as developing a better understanding of the apparent change in reaction mechanism at higher temperatures. Although this may indicate that a different form of the rate of reaction should be used, this preliminary study yielded reasonable temperature and extent of reaction predictions from the derived kinetic parameters.

## 5. CONCLUSIONS

Dynamic DSC kinetic analysis proved to be inconclusive due to a changing total heat of reaction with heating rate. It is thought that the change may be indicating a change in reaction mechanism or thermal decomposition. Kinetic parameters were obtained through analysis of isothermal cures by DSC. A changing activation energy with cure indicates a possible mechanism change, but the resulting kinetic expression appeared to give good estimates of rate of reaction as a function of cure. A reaction order of one was found to best fit the data. Both TBA and hot pot gel timer results indicate gelation as an isoconversion process when compared to the DSC results, but at different extents of reaction. Preliminary results from the pultrusion process model indi-

cate correlation with actual pultruder runs.

## 6. REFERENCES

1. S.M. Walsh and M. Charmchi, unpublished work.

2. R.B. Prime, in "Thermal Characterization of Polymeric Materials", Chap. 5, E.A. Turi, ed, Academic Press, New York(1981).

3. R.I. Werner and Z. Kusibab, 38th Annual Conference, Reinforced Plastics/Composites Institute, The Society of the Plastics Industry, Inc., Feb., 1983.

4. R.A. Fava, Polymer, 9, 137 (1968).

5. M.A. Acitelli, R.B. Prime, and E. Sacher, Polymer, 12, 333(1971).

6. N. Levy, Preprints of the ACS Division of Organic Coatings and Plastics Chemistry, 45, 485(1981).

7. I.J. Goldfarb and W.W. Adams, Preprints of the ACS Division of Organic Coatings and Plastics Chemistry, 45, 133(1981).

8. H.L. Friedman, J. Polymer Sci., 6C, 183(1965).

| TABLE 1 Dynamic DSC Results | | | |
| --- | --- | --- | --- |
| Formulation | Rate (°C/min) | Peak T (°C) | ΔH (cal/g) |
| 100/80/0/1 | 5 | 144 | 73.0 |
| | 10 | 157 | 68.4 |
| | 20 | 169 | 65.2 |
| | 40 | 184 | 58.9 |
| | 80 | 200 | 49.2 |
| 200/183.2/20/2 | 5 | 146 | 73.4 |
| | 10 | 159 | 72.0 |
| | 20 | 172 | 66.9 |
| | 40 | 187 | 58.4 |
| | 80 | 204 | 50.6 |

TABLE 3
Isothermal DSC Results for
2000/1832/200/20

| $\alpha$ | E(kcal/mole) | $g(\alpha)(\min^{-1})$ |
| --- | --- | --- |
| 0.05 | 19.88 | 1.06E-11 |
| 0.10 | 19.82 | 2.23E-11 |
| 0.20 | 19.45 | 7.02E-11 |
| 0.30 | 19.02 | 1.78E-10 |
| 0.40 | 18.67 | 3.73E-10 |
| 0.50 | 18.34 | 7.26E-10 |
| 0.60 | 17.99 | 1.42E-09 |
| 0.70 | 17.50 | 3.24E-09 |
| 0.80 | 16.78 | 1.02E-08 |
| 0.90 | 14.85 | 1.55E-07 |
| 0.95 | 12.26 | 5.23E-06 |

TABLE 2
Isothermal DSC Results for
2000/1832/200/20

| Isothermal T (°C) | ΔH (cal/g) |
| --- | --- |
| 115.6 | 95.7 |
| 126.7 | 89.8 |
| 137.8 | 90.6 |
| 148.9 | 88.3 |
| 160.0 | 84.9 |
| 171.1 | 66.6 |

TABLE 4
Isothermal DSC Results for
2000/1832/200/20 at 160°C for n=1

| $\alpha$ | $A(\min^{-1})$ | $k(\min^{-1})$ |
| --- | --- | --- |
| 0.10 | 4.73E+09 | 0.470 |
| 0.20 | 3.18E+09 | 0.492 |
| 0.30 | 2.00E+09 | 0.506 |
| 0.40 | 1.37E+09 | 0.520 |
| 0.50 | 9.59E+08 | 0.533 |
| 0.60 | 6.46E+08 | 0.544 |
| 0.70 | 3.75E+08 | 0.550 |
| 0.80 | 1.59E+08 | 0.544 |
| 0.90 | 1.50E+07 | 0.483 |

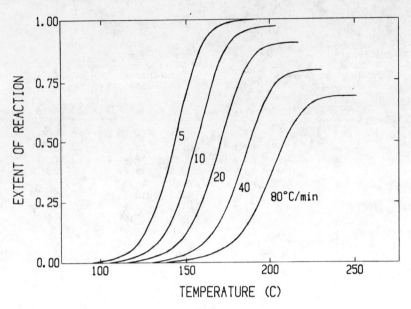

Figure 1. Extent of reaction vs. temperature for dynamic DSC scans on samples of 2000/1832/200/20 resin for different heating rates.

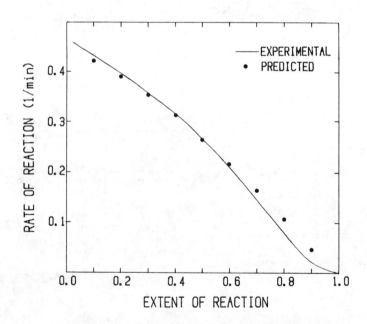

Figure 2. Comparison of rate of reaction of 2000/1832/200/20 at an isothermal cure temperature of 160°C to the rate predicted with the derived kinetic parameters.

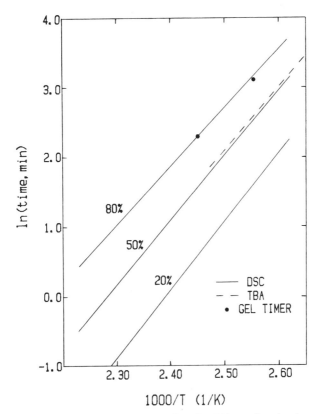

Figure 3. Arrhenius plot comparison of DSC, TBA, and gel timer experiments.

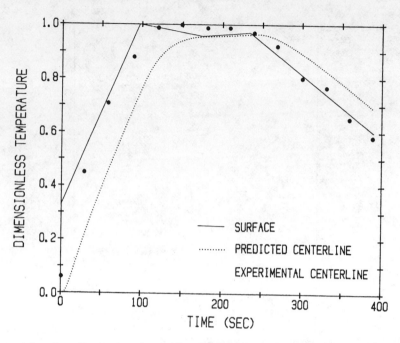

Figure 4. Experimental and predicted temperature profiles for a pultrusion run at 160°C at 152 mm/min of 2000/1832/200/20 from the process model.

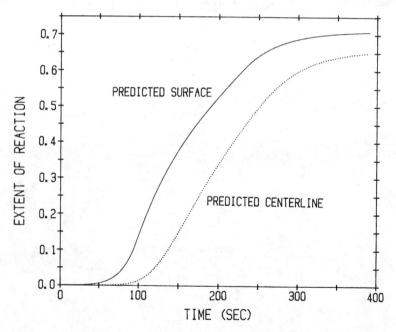

Figure 5. Process model predicted extent of reaction profiles for a pultrusion run at 160°C at 152 mm/min of 2000/1832/200/20.

A SIMPLE, LOW-COST, TENSIONING DEVICE
FOR PULTRUDING COMPOSITE MATERIALS
Everett S. Arnold and Noel J. Tessier
Army Materials Technology Laboratory
Watertown, Massachusetts 02172-0001

## Abstract

In order to obtain optimum properties from a pultruded composite material, it is important to insure parallel alignment of the fibers through the resin bath and die without rupturing filaments. In this paper, the design and construction of a simple, low-cost, tensioning device for composite materials is discussed.

Experiments were performed with the tensioning device using graphite tow and epoxy resin to produce .95cm x 2.5 cm (3/8" x 1") and 1.2cm x 2.5cm (1/2" x 1") rectangular pultruded stock.

The results of trial runs using the new tensioning device indicated an improvement in productivity of the cross-sections. Fewer snags and ruptured fibers were observed during processing.

The pultrusions made with the simple, low-cost, tensioning device showed more uniform cross-sections than a conventional device. Photomicrographs of sections cut from these pultrusions also showed more parallel alignment of the fibers and a better surface finish. The results from physical tests also showed that the pultrusions produced with the new device exhibited better mechanical properties than were previously obtained.

## 1.  INTRODUCTION

The Army has many potential applications requiring very high stiffness to weight materials. Several of these candidate components lend themselves to pultrusion processing. In an effort to optimize this process, it was determined that controllable fiber tension was desired. For the cross sections of interest this would require the processing of 400-600 ends of a 40,000 filament graphite tow.

A review of past efforts to tension large numbers of fiber tows showed problems associated with multiple end tension devices, for example, nip rolls. The major problems were poor tension control across the web as well as fiber breakage.

However, in order to control the tension on each of 600 fiber tows it would be necessary to have a low cost system. The object of this work was to investigate low cost, individual fiber tension devices to control fiber tension for optimum material properties.

2. DEVELOPMENT & FABRICATION

In this investigation, Army requirements dictated the use of high modulus graphite fibers. Great Lakes Fortafil 5 40K piddle packages were selected to minimize the number of ends required to produce larger cross sections and to keep the cost of prototypes as low as possible.

Originally, the 40K graphite tows were taken from plastic bags, run over a rod directly overhead, through gromets at the front of the creel, through a carding rack, condensing rack, into the resin bath, through preforms, and into the die. Depending upon the position of the tows in relation to the inside or outside (in contact with plastic bag) they were either slack or under tension (Figure 1). We felt that to improve the properties of the pultruded piece that some control of tension was necessary. Some of the standard tension devices were tried with the following observations and conclusions.

2.1 Multi Rod Arrangement: This did give some control of the tension but also had the tendency to chafe or rupture filaments (Figure 2). The larger the diameter of the rods the less rupture of filaments. We felt that this arrangement was not satisfactory.

2.2 Three Post Tensioning: Tows would travel around the three posts and a weighted disc could be placed on each post and weights added to increase or taken off to decrease tension. It was found that filaments would rupture and build up on the posts thereby decreasing the control and amount of pre-set tension.

2.3 Single Disc Compensator and Twin Disc Compensator Devices: Tension is pre-set by means of a spring or springs. The problem of filament rupture and build up under the disc(s) gave the same problem as the three post device.

2.4 P.E. Tubing: Tubes of .95cm (3/8") O.D. and .64cm (1/4") I.D. were incorporated into an adjustable wooden frame. Tube length of 45.7cm (18") was selected to enable the tubes to be bent to see what effect, if any, this would have on the tension. It was found that just running the tows through the straight tubes evened the tension out dramatically. Weights ranging from 2 grams up to 60 grams were then hung on the tows and starting with the straight tube the amount of sag the tows showed below $C_L$ was measured. The frame was adjusted to provide radii of 6.4cm (2 1/2"), 8.26cm (3 1/4") and 10.8cm (4 1/4") for increased tension. The results of static loading of the bent tubes showed basically a decrease in the tow sag as the bend in the tubes was increased (Figure 8). Some variation was observed and it was suggested that we try to pull all the tows from the center of the package to minimize contact with the plastic bags. This was accomplished by mounting funnels over the center of each package (Figure 5). This simple device further decreased the tension variations observed especially using the straight tubes.

Once it was decided to go with the tube device the wooden frame was replaced by aluminum frames built and mounted on the front of the creels (Figure 3). These figures also show the arrangment used to bend the tubes to increase the tension. To do this the frame length is decreased and a rod inserted, if needed (Figure 4). The tows were then threaded through a carding rack which is used to start condensing the width of the tows in the creel to the die width (Figure 6). Between the carding rack and the resin bath an eyeboard was incorporated to further narrow the width and also maintain the alignment of the tows (Figure 6). In recent runs we have found that, depending on the width of the tows in the carding rack, the eyeboard is not necessary. For example, when running 600 tows of 40K graphite the width in the carding

rack is 86cm (34") and because of the limited space we have between the creels and resin bath the eyeboard aids in condensing the tows to approximately 20.3cm (8") wide. This is further narrowed in the guide just prior to the resin bath. Maintaining the alignment or orientation of the filaments through the resin bath and die was greatly increased by this relatively simple tensioning device. Preforms were used between the resin bath and the die to further condense the tows to the die dimensions. For the 2.5cm x 1.3cm (1" x 1/2") piece the first preform dimensions were 2.6cm wide x 2.6cm high (1 1/32" wide x 1 1/32 high) and the second was 2.5cm wide x 1.9cm high (1" wide x 3/4" high). With no tension on the tows extreme ballooning (Figure 7) was observed especially at the first preform. This ballooning tended to further misalign the tows prior to entering the die and in many cases ruptured fibers and stopped the process. This condition was virtually eliminated when the tows were run under tension.

### 3. PROCESSING

Two dies were used during this work, one produced a 2.54cm x 1.3cm (1" x 1/2") cross section and the other produced a 2.54cm x .95cm (1" x 3/8"). Both dies were 76cm (30") long and had three heating zones along their length. The temperature was maintained at $160^{o}C$ ($320^{o}F$) in all zones. The line speed used for comparing tension vs. no tension runs varied 15cm (6") per minute to 38cm (15") per minute. The samples to be evaluated were post cured at 177°C (350°F) for two hours and slowly cooled down to room temperature. Resin formulation used is listed in Table 1.

### 4. DISCUSSION OF RESULTS

The primary concerns of this work were the effect of fiber tension on ease of processing and mechanical properties.

The amount of fiber tension had a large effect on the processing of the material studied. As fiber tension increased, the amount of resin picked up in the resin bath decreased dramatically. With no fiber tension, the cross section of the wetted fiber bundles exiting the resin tank were twelve times larger than the finshed cross section. With the high modulus graphite fibers, the squeezing out of this excess resin becomes very difficult. Ballooning at the preforms as shown in Figure 7 occurred at slow speeds 15cm (6") per minute. This effect results in fiber rupture and misalignment as well as limiting line speed. With the straight tube/no bend radius tension devices, the wetted fiber bundles exiting the resin bath were only three times the area of the finished cross section. Little ballooning occurred at slow line speed and line speeds of 38cm (15") per minute were achieved. Finally,

at the 11.4cm (4.5") bend radius for the tension device the wetted fiber bundles were only twice the area of the finished cross section. The viscosity of the resin for these runs was between 450 and 500 centipoise.

The effect of fiber tension on mechanical properties was determined from flexural testing. The poorest properties were obtained with no tension on the fibers. The highest properties were obtained from the straight tube/no bend radius runs. This indicates that there is an optimum level of tension that can be used to maximize properties. This is a result of the fact that at no tension the fibers in the final composite are misaligned and damaged due to problems with resin squeeze out. As tension is added, alignment of fibers improved and no damage occurred during resin squeeze out. Finally, at higher tension, fiber damage occurrs due to the increased surface contact with process hardware.

### 5. CONCLUSIONS

The simple low cost tensioning device developed in this work proved to be cost effective as well as improve properties. A 47% increase in flex strength was realized as well as a 21% increase in flex modulus. This was a result of less fiber damage during processing and improved fiber alignment.

## 6. ACKNOWLEDGEMENTS

We would like to thank Dr. Stephen Petrie for his support. The authors greatly appreciate the time and effort that was spent on this paper by Miss Susan M. Reagan and Miss Susan A. Mehigan.

## Table 1

### Resin Formulation

| | |
|---|---|
| Epon 826 Epoxy (Shell) | 2000 pts |
| Methyltetrahydrophthalic Anhydride (Anhydrides & Chemicals) | 1832 pts |
| RD-2 Reactive Dilient (Ciba-Geigy) | 200 pts |
| Carnauba Wax LT-1 (Ceura) | 36 pts |
| Benzyldimethylamine (Anhydrides & Chemicals) | 20 pts |
| Percipitated Silicon Zeothix 265 | 80 pts |

## Table 2

| Tensioning | Sample Thickness | Flex Strength Pa (psi) | Flex Modulus Pa (msi) |
|---|---|---|---|
| None | 1/2" | $9.27 \times 10^8 \pm 2.1 \times 10^7$ (134,400 $\pm$ 3,100) | $18.2 \times 10^{10} \pm 1.2 \times 10^9$ (26.4 $\pm$ .17) |
| None | 3/8" | $8.37 \times 10^8 \pm 8.6 \times 10^7$ (121,400 $\pm$ 12,500 psi) | $16.1 \times 10^{10} \pm 1.4 \times 10^{10}$ (23.3 $\pm$ 2.0) |
| Straight Tube | 3/8" | $12.31 \times 10^8 \pm 2.5 \times 10^7$ (178,500 $\pm$ 3600) | $19.4 \times 10^{10} \pm 2.1 \times 10^9$ (28.1 $\pm$ 0.3) |
| Bent Tube 4.5" Bend Radius | 1/2" | $9.90 \times 10^8 \pm 1.9 \times 10^7$ (143,500 $\pm$ 2,800) | $18.3 \times 10^{10} \pm 1.4 \times 10^9$ (26.6 $\pm$ 0.2) |

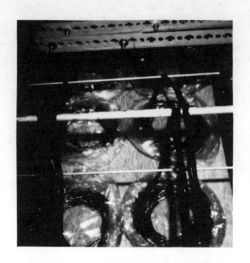

FIGURE 1

FIGURE 2

FIGURE 3

FIGURE 4

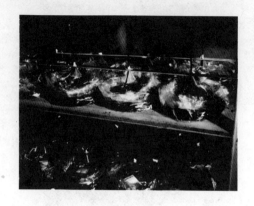

FIGURE 5

FIGURE 6

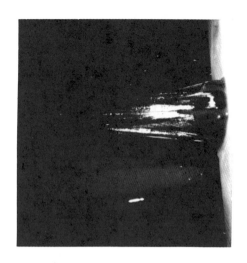

FIGURE 7

# STATIC FIBER TENSION

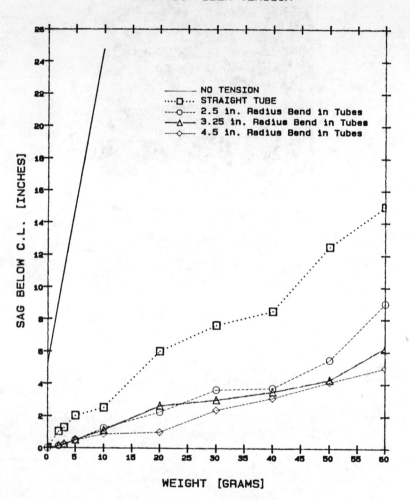

FIGURE 8

# PULTRUDED TYPE
# THERMOPLASTIC COMPOSITE STRUCTURES

William H. Beever and James E. O'Connor
Phillips 66 Company
Advanced Composites Division
Bartlesville, OK 74004

## ABSTRACT

With the advent of continuous fiber
reinforced thermoplastic composites,
investigation into the use of these
materials in processes which
previously utilized thermoset
composites has begun. As a part of
our thermoplastic composites program
we are developing fabrication methods
for structures using continuous fiber
reinforced composite materials based
primarily on polyphenylene sulfide.
Polyphenylene sulfide is a
semicrystalline polymer exhibiting
excellent corrosion resistance, good
thermal stability, inherent flame
resistance and good physical
properties. We have now developed a
proprietary process that continuously
fabricates a pultruded type
thermoplastic composite structure.
The mechanical properties of these
carbon or glass reinforced structures
are comparable to those of pultruded
thermoset structures. Several
products of differing cross-sections
have been produced ranging from a
simple rod or bar to a more complex
hat or channel. This paper presents a
description of polyphenylene sulfide
and its characteristics. Mechanical
properties of the glass fiber-PPS and
carbon fiber-PPS pultruded type
structures are compared with those of
typical thermoset pultruded
structures. The chemical resistance
of the unidirectional reinforced-PPS
pultruded type structures is discussed.
The important aspect of postforming
the PPS thermoplastic structures is
likewise discussed. Pultruded type
structures of carbon fiber with PAS-2
(high temperature polyarylene sulfide),
PEEK and nylon 6 as the matrix
materials are presented.

Keywords: Polyphenylene sulfide,
thermoplastic polymer composites,
pultruded type structures, pultrusion,
composite materials.

## 1.0 INTRODUCTION

The pultrusion process since its
inception in the 1950's has been
primarily associated with thermoset
polymers such as polyester, vinylester
and epoxy. The process is now well
established and experiencing good
growth rates within the reinforced
plastics industry. It remains the only
composites manufacturing process
which affords profiles, shapes and
structures continuously and in very
long lengths.

Recent research efforts have
introduced modifications such as
pulforming (1) and rolltrusion (2) to
the basic process to afford structures
of non-uniform cross-sections. In
order to give additional transverse
strength to pultruded products
nonwoven biaxial reinforcement
materials have been developed (3).
Research on faster curing
methacrylate and epoxy resins for use
in the pultrusion process has also been
reported (4,5).

With the introduction of continuous oriented fiber reinforced engineering thermoplastic materials an opportunity to develop a new class of pultruded type composite structures exists (6,7). These thermoplastic materials are now available employing polyphenylene sulfide (PPS), polyetheretherketone (PEEK), polyamid-imide (TORLON), polysulfone (PSO) and polyetherimide (ULTEM) as matrix polymers. In a finished product these composites offer advantages over thermoset composites in areas such as increased toughness and impact resistance, improved repairability, increased moisture resistance and the important ability to postform or reshape the structure.

In our continuing efforts in thermoplastic composites, the fabrication of pultruded type structural products has been investigated. Although emphasis has been on the polymer polyphenylene sulfide other thermoplastics such as PEEK, nylon and a new polyarylene sulfide PAS-2 have also been fabricated into pultruded type structures.

## 2.0 POLYMER AND COMPOSITES

Polyphenylene sulfide (PPS) is the simplest member of a family of polyarylene sulfide polymers. It consists of a polymer backbone of alternating aromatic and sulfur moieties (Figure 1). It is a semicrystalline polymer exhibiting a glass transition temperature ($85°C$) and a crystalline melting point ($285°C$). The polymer is characterized by excellent corrosion resistance, good thermal properties, inherent flame resistance and good physical properties (8,9). Nominal properties of the polymer are listed in Table I. In order to capitalize on these properties PPS-composites containing long random fibers (10), woven fibers (11), and continuous oriented fibers (12,13) have been developed. Several methods are currently being developed to fabricate the fiber reinforced polyphenylene sulfide composite materials into structural parts.

## 3.0 PULTRUDED TYPE PRODUCTS

Fiber reinforced composite products having constant cross section profiles of continuous lengths are generally fabricated by a pultrusion process. These types of products have now been developed using polyphenylene sulfide as the matrix material. In addition pultruded type rods of carbon fiber reinforced PEEK, nylon and a new polyarylene sulfide, PAS-2 have been fabricated.

### 3.1 Simple Shapes

We have defined simple shapes as those containing only unidirectional reinforcement. These type structures will have maximum strength along the fiber direction and much reduced strength in the transverse direction.

A large number of "simple shapes" of various sizes have been fabricated in continuous lengths. Figure 2 shows rods ranging in diameter from 0.125" to 0.750" produced using either glass or carbon reinforcements. Figure 3 shows fiber reinforced PPS half-circle and bar shapes. Carbon fiber reinforced tube structures (0.187" outside diameter x 0.100" inside diameter) and similar sized rods are shown in Figure 4. These rods, bars and tubes have excellent surface appearance and dimensional tolerances. Void contents are low indicating good consolidation. Nominal properties of the polyphenylene sulfide pultruded type composite shapes are listed in Table II. The mechanical properties were measured on annealed specimens using standard ASTM procedures. The properties of the PPS pultruded type composites are comparable to similar properties measured on compression molded unidirectional PPS composite laminates (14).

Comparison of properties of the PPS-glass composites with those of typical thermoset-glass composites shows the expected improvements in impact strength and moisture resistance of the thermoplastic composite (Table III). The remaining mechanical properties are comparable.

Figure 5 shows 0.125" pultruded type carbon fiber composite rods using PEEK, nylon and PAS-2 as the matrix materials. As with the PPS composite structures excellent surface appearance was obtained. This is the first reported fabrication of these materials into a pultruded type structure.

### 3.2 Complex Shapes

Pultruded type structures of a more complex nature generally require strengths in all directions. Thus these shapes generally have in addition to unidirectional reinforcements, reinforcements in the form of woven fabric or random mat to impart transverse properties.

Figures 6, 7 and 8 show several complex structures fabricated continuously from glass or carbon fiber reinforced polyphenylene sulfide. These structures include a channel, hat, angle, z-bar, wedge, rectangular lobe and 1/2" diameter overbraided rod. Again excellent compaction and surface appearance of these structures were obtained. The amount and type of reinforcements included within the complex cross-section will determine the properties.

### 4.0 CHEMICAL RESISTANCE

Chemical resistance of the long random fiber reinforced polyphenylene sulfide composites was found to be excellent upon exposure to a number of common solvents and water (15,16). This corresponds well with the excellent chemical resistance of the polyphenylene sulfide matrix itself (8,9). Figures 9-12 show the effect of 5 common solvents and hot/wet conditions on unidirectional carbon fiber-PPS and glass fiber-PPS pultruded type bars. The bars were immersed in the solvents for 1 week at 90°C. At room temperature little effect on properties is observed with the PPS composites thus the use of the higher

test temperature. The hot/wet tests were run after immersion of the bars in water for two weeks at 71°C. Even after the extended exposure at elevated temperature, the pultruded type polyphenylene sulfide bars exhibited very good retention of mechanical properties.

### 5.0 POSTFORMABILITY

The thermoplastic nature of the polyphenylene sulfide polymer which enables it to be remelted and reformed results in one of the most intriguing advantages of thermoplastic pultruded type composite structures. This advantage is the ability to postform the initial structure into a new one simply by applying heat and pressure. Figure 13 shows the reshaping (twisting) of a glass-PPS and carbon-PPS bar afforded by heating the entire bar during postforming whereas Figure 14 shows a variety of postforming operations in which a localized area of several glass fiber-PPS rods was heated and reshaped. This postformability of thermoplastic composite structures will open up various application areas previously unavailable to thermoset pultruded composites.

### 6.0 CONCLUSIONS

Various pultruded type polyphenylene sulfide composite structures have been fabricated. Simple shapes such as bars, rods and tubes containing only unidirectional reinforcements as well as more complex shapes such as channels, hat-sections, z-bar sections, wedges, rectangular lobes and overbraided rods that contain off-axis reinforcements as well as unidirectional reinforcements have been produced. Good consolidation of the structures as well as good surface appearance was obtained. Mechanical properties compared well with those of pultruded thermoset composites with improved toughness and reduced moisture absorbance realized. The excellent chemical resistance of the PPS matrix was carried over to the pultruded type composite structures.

As a result of the thermoplastic nature of the PPS matrix, applications of heat and pressure allowed the pultruded type PPS composite structures to be postformed or reshaped. Pultruded type carbon fiber composite structures using PEEK, nylon and PAS-2 as matrix materials were fabricated for the first time.

## 7.0 ACKNOWLEDGEMENT

The authors wish to thank Messrs. A. Y. Lou, J. R. Bohannan, F. J. Burwell, E. L. Martin, V. H. Rhodes, Jr., L. M. Selby, D. L. Straw and J. R. Wareham of Phillips Petroleum Company Research Center for their valuable assistance in materials processing, specimen preparation and property testing.

## 8.0 REFERENCES

1. G. Ewald, "Details on a New Way to Make Curved RP Shapes", Modern Plastics, 74 (May 1981).

2. J. R. Plumer, M. A. Yates and S. B. Dirscoll, "Rolltrusion of Composite Structures", SPI Reinforced Plastics/Composites Symposium Proceedings, 38, 20-B (1983).

3. R. Birsa and P. Taft, "A New Materials Approach for Providing Transverse Strength in Pultruded Shapes", SPI Reinforced Plastics/Composites Symposium Proceedings, 39, 1A (1984).

4. J. A. Kershaw, "New Epoxy Resin Systems for Pultrusion", SPI Reinforced Plastics/Composites Symposium Proceedings, 38, 6C (1983).

5. J. A. Kershaw, T. J. Tulig and G. I. Yamashita, "Pultrusion Processing of Epoxies into Advanced Composite Structures", COGSME Fabricating Composites Conference Proceedings, EM86-711 (1986).

6. D. E. Beck, "New Processes and Prospects in Pultrusion", SPI Reinforced Plastics/Composites Symposium Proceedings, 38, 6B (1983).

7. W. B. Goldsworthy, "Thermoplastic Composites: The New Structurals", Plastics World, 56 (Aug. 1984).

8. H. W. Hill and D. G. Brady, "Polymers Containing Sulfur", in Kirk-Othmer Encyclopedia of Chemical Technology, 18, 3rd Edition, p. 793 (1982).

9. D. G. Brady and H. W. Hill, "How Some High Performance Resins Stack Up in Chemical Resistance", Modern Plastics 51 (5), p. 60 (1974).

10. D. G. Brady, et al, "Long Fiber Reinforced Polyphenylene Sulfide - A New Dimension in Toughness", SPE ANTEC Proceedings, 42, 690 (1984).

11. T. P. Murtha, et al, "High Performance Composites Based on Polyphenylene Sulfide", National SAMPE Proceedings, 29, 870 (1984).

12. J. E. O'Connor, C. C. Ma and A. Y. Lou, "Polyphenylene Sulfide: Resin, Prepreg and High Performance Composites", SPI Reinforced Plastics/Composites Symposium Proceedings, 39, 11E (1984).

13. J. E. O'Connor, A. Y. Lou and W. H. Beever, "Polyphenylene Sulfide - A Thermoplastic Polymer Matrix for High Performance Composites", Fifth International Conference on Composite Materials Proceedings, 5, 963 (1985).

14. W. H. Beever, C. L. Ryan, J. E. O'Connor and A. Y. Lou,

"Ryton PPS Carbon Fiber Reinforced Composites: The How, When and Why of Molding", Toughened Composites ASTM STP937, Norman J. Johnston Ed., American Society for Testing and Materials, Philadelphia, PA, to be published in 1986.

15. A. Y. Lou and T. P. Murtha, "Chemical Resistance of Glass Fiber Reinforced PPS Stampable Composites", 17th International SAMPE Technical Conference Proceedings, Oct. 22-24, 1985.

16. A. Y. Lou and T. P. Murtha, "Environmental Effects on Glass Fiber Reinforced PPS Stampable Composites", presented at the ASM International Conference and Exposition of Fatigue, Corrosion Cracking, Fracture Mechanics, and Failure Analysis at Salt Lake City, Utah, December 2-6, 1985.

## 9.0 BIOGRAPHIES

William H. Beever is responsible for development of processing and production technology for high performance thermoplastic composite shapes, profiles and structures. He is a member of the Advanced Composites Division of Phillips 66 Company. Since receiving his Ph.D. from Colorado State University in 1979 he has worked for Phillips Petroleum Company in the advanced composite and engineering materials area.

James E. O'Connor is Manager of the Thermoplastic Composites Structures and Shapes Section in the Advanced Composites Division at Phillips 66 Company. He is responsible for development of the technology, products and markets for high performance thermoplastic composite structures, shapes and profiles. He received his Ph.D. from the University of Cincinnati in 1968 and has been involved with the advanced composites effort at Phillips Petroleum Company since 1980.

TABLE I

Nominal Properties of Polyphenylene Sulfide

| | |
|---|---|
| Density, g/cc | 1.36 |
| Tensile Strength, Ksi | 12 |
| Elongation, % | 4 |
| Flexural Strength, Ksi | 22 |
| Flexural Modulus, Msi | 0.5 |
| Izod Impact, ft lb/in | |
| Notched | 0.4 |
| Unnotched | 8.0 |
| Oxygen Index, % | 44 |

TABLE II

Properties of Unidirectional Fiber Reinforced
Pultruded Type Polyphenylene Sulfide Composite Structures

| Property | ASTM Test | Reinforcement Type | |
|---|---|---|---|
| | | Glass | Carbon |
| Density, g/cc | D792 | 1.92 | 1.52 |
| Fiber Content, Vol% | D3171 | 55 | 55 |
| Tensile Strength, Ksi | D3039 | 115 | 200 |
| Tensile Modulus, Msi | D3039 | 6.0 | 17.0 |
| Elongation, % | D3039 | 2.0 | 1.1 |
| Flexural Strength, Ksi | D790 | 140 | 170 |
| Flexural Modulus, Msi | D790 | 5.0 | 14.0 |
| Compressive Strength, Ksi | D3410 | 100 | 85 |
| Compressive Modulus, Msi | D3410 | 6.0 | 15.0 |
| Izod Impact, ft lb/in | D256 | | |
| Notched | | 58.2 | 39.3 |
| Unnotched | | 64.6 | 55.4 |
| Water Absorption (24 h), % | D570 | 0.07 | 0.15 |

TABLE III

Comparison of Properties of Unidirectional Glass Fiber Reinforced
Pultruded Type Polyphenylene Sulfide and Polyester* Structures

| Property | Polymer Matrix Type | |
|---|---|---|
| | Polyphenylene Sulfide | Polyester |
| Density, g/cc | 1.92 | 2.05 |
| Fiber Content, Wt% | 70-72 | 70-75 |
| Tensile Strength, Ksi | 115 | 100-120 |
| Tensile Modulus, Msi | 6.0 | 6.0-6.5 |
| Flexural Strength, Ksi | 140 | 100-120 |
| Flexural Modulus, Msi | 5.0 | 6.0-6.5 |
| Compression Strength, Ksi | 100 | 60-70 |
| Notch Izod Impact, ft lb/in | 58 | 35-40 |
| Water Absorption (24 h), % | 0.07 | 0.25 |

* Values compiled from Creative Pultrusion Co. "Design Guide" and
Morrison Molded Fiberglass Co. "Engineering Manual".

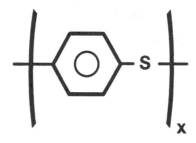

Figure 1:  Polyphenylene Sulfide

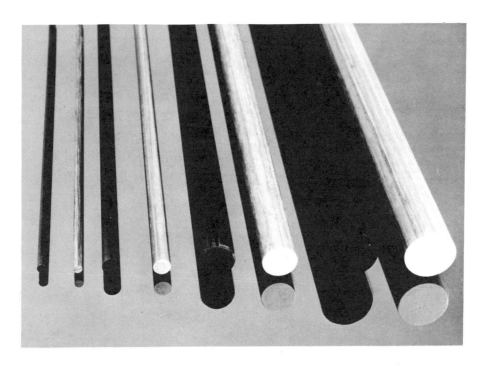

Figure 2:  Carbon Fiber-PPS and Glass Fiber-PPS Pultruded Type
Solid Rods (0.125", 0.250", 0.500" and 0.750" Diameter).

Figure 3:     Glass Fiber-PPS and Carbon Fiber-PPS Pultruded
              Type Bar and Half-Circle Structures.

Figure 4:     Carbon Fiber-PPS Pultruded Type Solid Rod (0.187"
              Diameter) and Tube (0.187" OD x 0.100" ID) Structures.

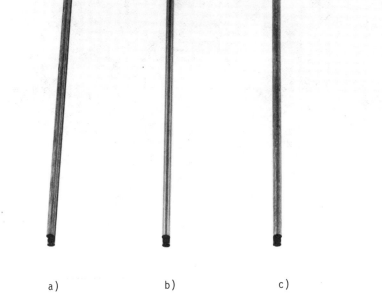

a)            b)            c)

Figure 5:     Carbon Fiber Reinforced Pultruded Type 0.125"
Diameter Rods of a) PEEK, b) Nylon, c) PAS-2

Figure 6:     Glass Fiber-PPS and Carbon Fiber-PPS Pultruded
Type Channel, Hat, Angle and Z-Bar Structures.

Figure 7:    Glass Fiber-PPS and Carbon Fiber PPS Pultruded Type
             Wedge and Rectangular Lobe Structures.

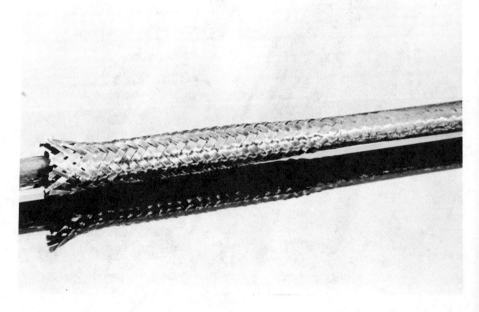

Figure 8:    Overbraided Carbon Fiber-PPS .500" Diameter
             Pultruded Type Rod.

FIGURE 9

# EFFECT OF SOLVENTS & WATER ON TENSILE PROPERTIES
## OF PULTRUDED TYPE PPS/CARBON BARS

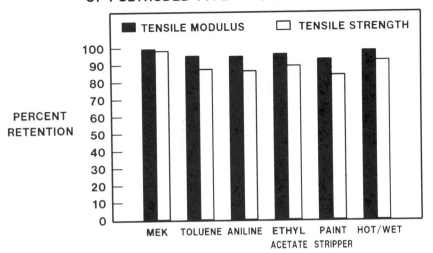

SOLVENTS (IMMERSED 1 WEEK AT 90°C)
HOT/WET (IMMERSED 2 WEEKS AT 71°C)

FIGURE 10

# EFFECT OF SOLVENTS & WATER ON FLEXURAL PROPERTIES
## OF PULTRUDED TYPE PPS/CARBON BARS

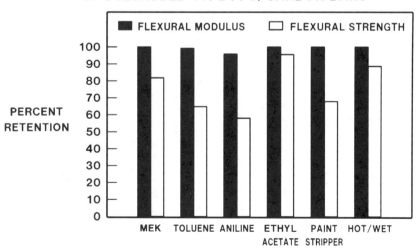

SOLVENTS (IMMERSED 1 WEEK AT 90°C)
HOT/WET (IMMERSED 2 WEEKS AT 71°C)

# FIGURE 11
## EFFECT OF SOLVENTS & WATER ON TENSILE PROPERTIES OF PULTRUDED TYPE PPS/GLASS BARS

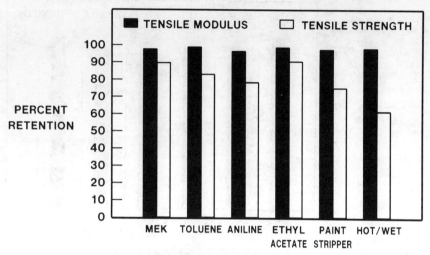

SOLVENTS (IMMERSED 1 WEEK AT 90°C)
HOT/WET (IMMERSED 2 WEEKS AT 71°C)

# FIGURE 12
## EFFECT OF SOLVENTS & WATER ON FLEXURAL PROPERTIES OF PULTRUDED TYPE PPS/GLASS BARS

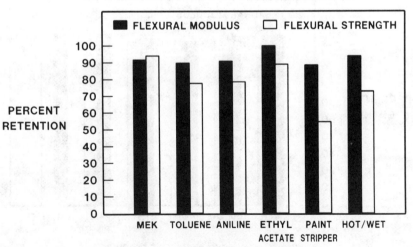

SOLVENTS (IMMERSED 1 WEEK AT 90°C)
HOT/WET (IMMERSED 2 WEEKS AT 71°C)

1320

Figure 13:     Postformed Glass Fiber-PPS and Carbon Fiber-PPS
               Bar Structures.

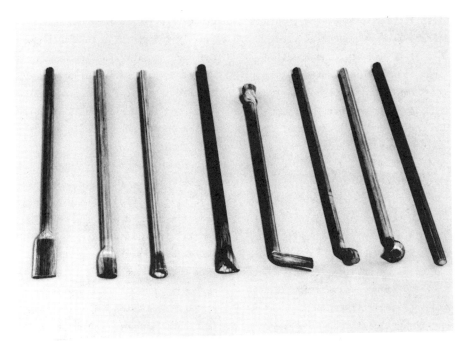

Figure 14:     Postformed Glass Fiber-PPS 0.250" Diameter
               Rod Structures.

YARN SELECTION GUIDE FOR
ASBESTOS REPLACEMENT APPLICATIONS
Frank Scardino
IFAI Professor of Industrial Fabrics
Philadelphia College of Textiles and Science
Philadelphia, Pennsylvania

## Abstract

In many asbestos replacement applications, it is often necessary to
produce fabric constructions which
are equivalent to standard asbestos
fabrics in thickness and cover. To
achieve equivalent thickness and
cover in standard constructions,
fabrics must be made from yarns of
equivalent diameter and cross section. This guide is offered as an
aid in selecting yarns of equivalent diameter and cross section for
the one-on-one replacement of asbestos yarns in standard woven,
knitted, and braided fabric constructions.

## 1. INTRODUCTION

After a century of commercial success in a variety of industrial applications, asbestos has been classified as an unsafe material. Thus
far, experience has shown that no
single fiber will replace asbestos
in all applications, one-on-one, on
a cost/performance basis. Consequently, several alternate fibers,
fiber blends, yarn structures, fabric constructions and finishes are
being used in asbestos replacement
applications.

In analyzing textile asbestos products (1), it is found that a carrier
fiber (e.g. cotton, rayon, etc.) is
usually incorporated in a blend with
the much shorter and finer asbestos
fiber. Asbestos rich blends are
usually mixed in a woolen picker,
processed into roving on a modified
woolen card, and wet twisted into
spun yarn on a modified cotton frame
with no draft, or a tension draft
at most, during twisting. The carrier fiber is used to help provide
cohesion in card webs and rovings
and in wet twisting.

While most experts agree that the
primary function of the carrier fiber is to convert the very-short and
very-fine asbestos fiber into spun
yarn, some experts contend that the
carrier fiber also helps the asbestos perform well especially in dy-

namic, high-frictional/high-abrasion applications such as brake linings, clutch facings, etc. Perhaps the possibility of a positive synergistic effect by a blend of fibers should be considered in developing a replacement system rather than a one-on-one replacement fiber for asbestos (2,3).

## 2. REPLACEMENT MATERIALS

The leading fiber contenders in the asbestos replacement race are compared in Table I for a number of critical properties. While glass fiber is well positioned from a cost/performance point of view, each fiber has unique advantages in various aspects of performance, processability, durability, comfort or aesthetics. Accordingly, it would appear that fiber blends will be even more important in the next generation of asbestos replacement products.

Yarn properties depend upon the physical properties of the constituent fibers and upon the yarn structural form. Most of the candidate fibers listed in Table I are available in various spun and filament yarn structural forms. Intimate fiber blending (as opposed to hybrid combining) requires a spun yarn structure made from staple fibers.

Yarn structural features depend mainly upon the constituent fiber geometry and the intrinsic characteristics of the processing system. A comparison of the structural features in various types of spun and filament yarns and of the major factors affecting yarn structural features is presented in Table II. As suggested in Table II, yarn structural features (fiber packing density, fiber segment length between points of entanglement, and mobility of fiber segment lengths) are dependent upon constituent fiber geometric properties (fiber length, fineness, and crimp) and upon the effect of the yarn processing system characteristics (fiber orientation in relation to the yarn axis, and the extent of fiber entanglement).

Referring to Table II, it can be seen that, based on the fiber geometric properties and inherent characteristics of the processing systems, spun yarns have substantial fiber density, short fiber segment lengths between points of entanglement, and minimal mobility of the fiber segments. This is the basis for the retention of the intrinsic fiber contiguity and thus the excellent dimensional stability of spun yarn structures under low levels of stress. All yarns, except stretch or elastomeric, tend to have excellent dimensional stability when loaded in the direction of the yarn axis. However, spun yarns also have good dimensional stability when loaded or deformed cross-sectionally or normal to the yarn axis. This means that spun yarns retain their natural bulkiness, good covering power, excellent hand and outstanding insulating quality under various extensional, compressional, and bending deformations to a much greater extent than would be expected in air-textured filament yarns, for example (5).

In many asbestos replacement applications, it is often necessary to produce fabric constructions which are equivalent to standard asbestos fabrics in thickness and cover. To achieve equivalent thickness and cover in standard constructions, fabrics must be made from yarns of equivalent diameter and cross section. This guide is offered as an aid in selecting yarns of equivalent diameter and cross section for the one-on-one replacement of asbestos yarns in standard woven, knitted, and braided fabric constructions.

### 3. ASBESTOS YARN REPLACEMENT GUIDE (6)

The major factors to be considered in selecting an asbestos replacement yarn are, as follows: (1) the linear density of the asbestos yarn; (2) the replacement fiber density $(g/cm^3)$ versus the asbestos fiber density $(g/cm^3)$; and, (3) the replacement fiber packing fraction versus the asbestos fiber packing fraction. The fiber packing fraction is the ratio of the fiber specific volume $(cm^3/g)$ to the yarn specific volume $(cm^3/g)$.

The yarn specific volume or bulk density is a measure of how the fibers are packed together in the yarn cross section. Accordingly, the specific volume of textile yarns is determined by the volume occupied by the fibers and the air spaces incorporated into the body of the yarn. The yarn specific volume is always greater than the fiber specific volume, except in the case of monofilament yarns wherein the fiber and yarn specific volumes are equal.

Formulas for approximating the linear density of equivalent replacement yarns are, as follows:

## Indirect Yarn Numbering System

$$\begin{array}{c}\text{Yds/lb of}\\\text{Replacement}\\\text{Yarn}\end{array} = \begin{array}{c}\text{Yds/lb of}\\\text{Asbestos}\\\text{Yarn}\end{array} \times \frac{\begin{array}{c}\text{Asbestos}\\\text{Fiber}\\\text{Density}\end{array}}{\begin{array}{c}\text{Replacement}\\\text{Fiber}\\\text{Density}\end{array}} \times \frac{\begin{array}{c}\text{Asbestos}\\\text{Fiber Pkg}\\\text{Fraction}\end{array}}{\begin{array}{c}\text{Replacement}\\\text{Fiber Pkg}\\\text{Fraction}\end{array}} \quad (1)$$

## Direct Yarn Numbering Systems

$$\begin{array}{c}\text{Denier of}\\\text{Replacement}\\\text{Yarn}\end{array} = \begin{array}{c}\text{Denier of}\\\text{Asbestos}\\\text{Yarn}\end{array} \times \frac{\begin{array}{c}\text{Replacement}\\\text{Fiber}\\\text{Density}\end{array}}{\begin{array}{c}\text{Asbestos}\\\text{Fiber}\\\text{Density}\end{array}} \times \frac{\begin{array}{c}\text{Replacement}\\\text{Fiber Pkg}\\\text{Fraction}\end{array}}{\begin{array}{c}\text{Asbestos}\\\text{Fiber Pkg}\\\text{Fraction}\end{array}} \quad (2)$$

$$\begin{array}{c}\text{Tex of}\\\text{Replacement}\\\text{Yarn}\end{array} = \begin{array}{c}\text{Tex of}\\\text{Asbestos}\\\text{Yarn}\end{array} \times \frac{\begin{array}{c}\text{Replacement}\\\text{Fiber}\\\text{Density}\end{array}}{\begin{array}{c}\text{Asbestos}\\\text{Fiber}\\\text{Density}\end{array}} \times \frac{\begin{array}{c}\text{Replacement}\\\text{Fiber Pkg}\\\text{Fraction}\end{array}}{\begin{array}{c}\text{Asbestos}\\\text{Fiber Pkg}\\\text{Fraction}\end{array}} \quad (3)$$

Where: Fiber Density is in $g/cm^3$, and

$$\text{Fiber Packing Fraction} = \frac{\text{Fiber Specific Volume } (cm^3/g)}{\text{Yarn Specific Volume } (cm^3/g)}$$

Denier No. is weight (g) of 9000 meters of yarn.
Tex No. is weight (g) of 1000 meters of yarn.

Density values $(g/cm^3)$ for fibers of current interest in asbestos replacement applications are listed in Table I. If blends of various fibers are used in asbestos yarns or in asbestos replacement yarns, an average fiber density should be calculated based upon the proportion of blend components in the yarn. For example, the average fiber density $(g/cm^3)$ in an 80% asbestos/20% cotton yarn would be (2.50 x 0.8) plus (1.52 x 0.2) or 2.30, approximately, assuming that the blend proportions in the yarn were correct.

Fiber packing fractions for a var-

iety of asbestos yarn constructions are listed in Table III. These fractions are approximations based upon measuring the linear density (g/cm) and cross sectional area $(cm^2)$ of the various asbestos yarn constructions. Variations in processing factors (eg., blend components and levels; processing systems; twist factors; ply constructions; wet or dry twisting; tension and compression in spinning, plying, winding, warping, weaving, knitting, braiding, finishing, etc.) can significantly affect the yarn cross-sectional area, specific volume and packing fraction.

If an unusual asbestos yarn construction or linear fiber assemblage must be replaced (e.g. asbestos sliver or roping around a filament core), the average diameter (cm) of the construction in its finished form must be carefully measured. The cross sectional area $(cm^2)$ of the assemblage can be estimated based upon the average diameter (cm). The specific density $(g/cm^3)$ is determined by dividing the linear density (g/cm) by the cross sectional area $(cm^2)$. The specific volume $(cm^3/g)$ is the reciprocal of the specific density $(g/cm^3)$.

The fiber packing fractions for a variery of replacement yarn constructions are listed in Table IV. Once again, these fractions are based upon conventional or standard processing and constructions and, consequently, are first approximations at best. As with asbestos yarn constrictions, variations in fiber geometrey (length, fineness, shape, crimp, surface, etc.) and in

processing can significantly affect the packing fractions and, accordingly, the specific volumes can be measured as described in the previous paragraph.

Examples of applications of the proposed formula for selecting equivalent yarns in asbestos replacement are shown in Table V. Once again, it is pointed out that the formula is offered only as a first approximation. With experience, modifications of the formula can be made to fit particular needs.

In general, the same approach can be used in the selection of nonwoven fabric constructions for equivalent thickness, cover, and insulating value in asbestos replacement applications. Fabric areal density (oz. /sq.yd.) can be substituted for yarn linear density (denier or tex). Fiber packing fractions and insulating values would depend on fiber orientation, method of fiber bonding or entanglement, and fabric finishing technique.

TABLE I

COMPARISON OF ASBESTOS REPLACEMENT FIBERS (2)

| | Limiting Oxygen Index % | Service Temperature °F | Filament Diameter Micron | Specific Gravity g/cm³ | Resistance to Acids | Resistance to Alkali | Cost $/Pound |
|---|---|---|---|---|---|---|---|
| Asbestos | 100 | 1200 | < 1 | 2.5-3.4 | Good | Excellent | $1-$2 |
| Glass Fiber | 100 | 900 | 3.5,6.0,9.0 | 2.54 | Good | Good[1] | $1-$3 |
| Ceramic Fiber | 100 | 2200 | 12+ | 2.70 | Excellent | Excellent | $50 |
| HM Aramid | 29 | 400 | 12+ | 1.44 | Fair | Fair | $ 8 |
| Aramid | 29 | 500 | 12+ | 1.4 | Poor | Good | $ 8 |
| PBI | 38 | 750 | 12+ | 1.4 | Excellent | Good | $38 |
| Acrylic | 18 | 350 | 12+ | 1.4 | Good | Fair | $1-$2 |
| OPF[2] | 60 | 500 | 10+ | 1.4 | Good | Poor | $6-$8 |

[1]utilizing alkali resistant (AR) glass

[2]oxidized polyacrylinitrile fiber

## TABLE II

COMPARISON OF THE STRUCTURAL FEATURES IN VARIOUS TYPES OF YARN
AND SOME OF THE FACTORS AFFECTING YARN STRUCTURAL FEATURES (4)

| | Fiber Geometric Properties + Processing Characteristics | | | | | = Yarn Structural Features | | |
|---|---|---|---|---|---|---|---|---|
| | Length | Fineness | Crimp | Orientation | Entanglement | Density[a] | Segment[b] | Mobility[c] |
| **Spun Yarns** | | | | | | | | |
| Carded cotton | short | fine | slight | medium | high | medium | short | minimal |
| Combed cotton | medium | fine | slight | high | v. high | high | v. short | minimal |
| Woolen (carded) | medium | coarse | high | low | medium | low | medium | slight |
| Worsted (combed) | long | coarse | medium | high | v. high | high | v. short | minimal |
| **Filament Yarns** | | | | | | | | |
| Untwisted | continuous | fine | none | v. high | minimal | medium | v. long | great |
| With twist | | medium or | none | high | medium | high | long | slight |
| Air textured | | coarse | high | low | high | low | medium | large |
| Stretch textured | | | v. high | v. low | minimal | v. low | v. long | tremendous |

[a]Fiber or filament packing density in the yarn cross section.

[b]Average fiber or filament segment length between points of entanglement.

[c]Freedom of movement or mobility of fiber segments held between points of entanglement.

1328

TABLE III

FIBER PACKING FRACTIONS IN
ASBESTOS YARN CONSTRUCTIONS (6)

| Yarn Construction | Fiber Packing Fraction |
|---|---|
| Spun | |
| Soft | 0.12 |
| Medium | 0.25 |
| Hard | 0.40 |
| Thread | 0.40 |
| Cord | |
| Coarse (1/4 in) | 0.4 |
| Medium (1/8 in) | 0.5 |
| Fine (1/16 in) | 0.6 |
| Rope | |
| Coarse (1.5 in) | 0.3 |
| Medium (1 in) | 0.4 |
| Fine (0.5 in) | 0.5 |

TABLE IV

FIBER PACKING FRACTIONS IN YARN CONSTRUCTIONS OF
INTEREST FOR ASBESTOS REPLACEMENT APPLICATIONS (6)

| Yarn Construction | Fiber Packing Fraction |
|---|---|
| Monofilament | 1.00 |
| Multifilament | |
| Untwisted | 0.25 |
| Slightly Twisted | 0.30 |
| Regularly Twisted | 0.60 |
| Hard Twisted | 0.90 |
| Air Textured | 0.33 |
| Spun | |
| Soft Twisted | 0.33 |
| Hard Twisted | 0.60 |

## TABLE V

### EXAMPLES OF APPLICATIONS OF PROPOSED FORMULA

Replacement of 1020 Asbestos Yarn
with Soft-Twisted Glass Spun Yarn

$$\text{Yds/lb Replacement Yarn} = \text{Yds/lb Asbestos Yarn} \times \frac{\text{Asbestos Fiber Density}}{\text{Replacement Fiber Density}} \times \frac{\text{Asbestos Fiber Pkg Density}}{\text{Replacement Fiber Pkg Density}}$$

$$\text{Yds/lb Glass Yarn} = 500 \times \frac{2.50}{2.54} \times \frac{0.25}{0.33} = 373 \text{ yds/lb}$$
(11969 denier)
(1330 tex)

Replace 1020 Asbestos Yarn with
Reg. Twisted Aramid Filament Yarn

$$\text{Yds/lb of Replacement Yarn} = \text{Yds/lb of Asbestos Yarn} \times \frac{\text{Asbestos Fiber Density}}{\text{Replacement Fiber Density}} \times \frac{\text{Asbestos Fiber Pkg Fraction}}{\text{Replacement Fiber Pkg Fraction}}$$

$$\text{Yds/lb of Aramid Yarn} = 500 \times \frac{2.50}{1.44} \times \frac{0.25}{0.60} = 362 \text{ yds/lb}$$
(12,333 denier)
(1370 tex)

## 4. ACKNOWLEDGEMENTS

The development of this yarn selection guide was sponsored by Owens-Corning Fiberglas Corporation. Fiber packing fractions for asbestos and replacement yarn constructions were measured by Mark Braune and Ping Fang, PCT&S graduates, during the summer of 1984. Amatex supplied a variety of the yarn and fabric samples used in preparation of the guide.

## 5. REFERENCES

1. Handbook of Asbestos Textiles
   3rd Edition
   Asbestos Textile Institute
   LCCC No. 66-30393, 1967, 102 p.

2. Hinton, R.B.
   New Cardable Glass Fiber for
   Asbestos Replacement
   Intimate Blend Fabrics
   Textile Technology Forum
   '83 Proceedings
   Industrial Fabrics Association
   International
   St. Paul, Minnesota, October
   31, 1983, p. 135

3. Ko, Williams, Pastore, and
   Scardino
   Cardable Glass Fibers for
   Composites
   Textile Technology Forum
   '86 Proceedings
   Industrial Fabrics Association
   International
   St. Paul, Minnesota, October
   19, 1986

4. Goswami, Martindale, and
   Scardino
   Textile Yarns: Technology,
   Structure and Applications
   John Wiley and Sons, 1977, p.84

5. Scardino, F.
   The Effect of Yarn Structure on
   Fabric Aesthetics
   Proceeding of the Third Japan-
   Australia Joint Symposium On
   Objective Measurement
   The Textile Machinery Society
   of Japan
   Osaka, Japan, 1985, p. 717

6. Scardino, F.
   Yarn Selection Guide for Asbes-
   tos Replacement Applications
   Textile Technology Forum
   '86 Proceedings
   Industrial Fabrics Association
   International
   St. Paul, Minnesota, October
   19, 1986

ASBESTOS SUBSTITUTION
- AN OVERVIEW
G F Heron
T&N Materials Research Ltd
Rochdale, Lancashire, U.K.

## Abstract

The paper reviews the main applications of asbestos fibre - in fibre
cement, friction materials, sealing
materials, textiles and reinforced
plastics. It highlights the properties of importance in each
application. The main substitute
fibres used in these applications
are identified and significant
differences in processing and
properties of the substitute
product, compared to asbestos
discussed. General indications
of the extent of market penetration
and the price premium commanded
by substitutes are given.

The paper concludes that whilst
no single fibre has the balance
of properties and cost to be
directly substituted for asbestos
in most of its uses, five fibres
have emerged as the main replacements. First generation substitute products containing one or
more of these fibres, are now
commercially available and in use
for most of the important application areas, though there are
notable exceptions.

In general, substitute fibres are
used in smaller proportion than
the asbestos they replace, the
balance being made up by other
fibres and fillers. Very substantial reformulation has therefore
been necessary, to achieve acceptable properties and to allow
manufacture on existing plant,
designed for asbestos processing.

Many substitutes are superior in
some respects to the asbestos
products they replace but almost
invariably their costs are higher.

## 1. INTRODUCTION

Note Except where otherwise stated
Asbestos in this paper refers to
white chrysotile asbestos which
is by far the most important
commercial grade.

The combination of mechanical,
thermal, frictional, chemical and
electrical properties of asbestos

have generated an enormous range of applications.

No other fibre has matched this variety and balance of properties (and if one did, its biological activity might be similar to that of asbestos!).

Equally important, many conversion processes have been devised specifically around the properties of chrysotile. The tough fibrillar structure confers a number of important processing and property benefits. It requires and can withstand high shear mixing to progressively "open" and separate the fibre bundles into finer and finer bundles, which are flexible and "hairy" (see Fig 1).

This leads to maximum mechanical interlocking of fibrils with one another to give excellent "web forming", in the sheet making processes used to manufacture asbestos cement, asbestos paper, gaskets and other materials. It also confers good textile processability. These webs are highly absorptive, trapping and holding fillers and additives to allow maximum versatility in formulation. The absoptivity of asbestos is used in filtration and even ultra-filtration of bacteria in, for example, beer and wine making.

The "hairy" fibre bundles not only interlock with themselves but with matrices of polymer or cement, to give good reinforcement.

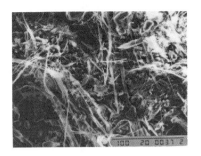

CHRYSOTILE ASBESTOS

ARAMID PULP

100 μm

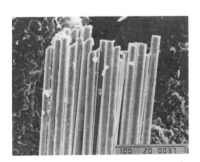

E-GLASS FILAMENT

FIG 1    SEM COMPARISON OF THREE FIBRE TYPES

The hydroxylated fibre surface is relatively polar and wets well with most liquids and solvents, including water. It also bonds well to many of the resins and rubbers with which it is frequently combined. This unusual and formidable range of properties and characteristics, combined with low cost, has generated a world market of about 4 million tonnes/year. The total output is broken down by producing countries in Table 1 which shows that USSR produces over half the world's output and its output is increasing. That of the rest of the world (except China) is falling. It is estimated that, excluding USSR, world output fell from about 2.6 million to about 1.7 million tonnes between 1980 and 1985.

Canada's production almost halved in this period.

Not all the reduction can be attributed to concern about health. Part must be due to recession in key markets, such as construction. More detail of United States trade in asbestos is given in Table 2 and Table 3 which lists the 1984 U.S. consumption, by end-use and fibre types.

This paper will review the main end-uses of asbestos, identify the most important substitute fibres for each application and estimate the extent of substitution in western markets.

Health aspects of substitutes

Though the paper is concerned principally with the technical aspects of substitution and not

Table 1

WORLD PRODUCTION OF ASBESTOS 1980-85 (tonnes)

|  | 1980 | 1981 | 1982 | 1983 | 1984 | 1985 (E) |
|---|---|---|---|---|---|---|
| USSR* | 2,000,000 | 2,000,000 | 2,000,000 | 2,200,000 | 2,250,000 | 2,300,000 |
| CANADA | 1,323,053 | 1,121,845 | 834,249 | 857,404 | 836,655 | 744,000 |
| ZIMBABWE | 250,900 | 247,600 | 194,400 | 153,000 | 160,000 | 175,000 |
| S AFRICA | 277,734 | 235,943 | 211,860 | 221,000 | 167,390 | 160,000 |
| CHINA | 131,700 | 140,000 | 140,000 | 160,000 | 160,000 | 160,000 |
| BRAZIL | 170,403 | 138,417 | 144,832 | 149,612 | 136,080 | 148,300 |
| ITALY | 157,794 | 137,086 | 116,410 | 139,000 | 147,280 | 135,000 |
| USA | 80,079 | 75,618 | 63,515 | 69,900 | 57,420 | 61,000 |
| GREECE | - | 30 | 17,500 | 32,000 | 45,000 | 55,000 |
| OTHERS | 200,270 | 145,919 | 145,551 | 108,132 | 107,315 | 109,600 |
|  | 4,591,933 | 4,242,458 | 3,868,317 | 4,090,048 | 4,067,140 | 4,047,900 |

*Estimate

Source: "ASBESTOS"

## Table 2

### US TRADE IN ASBESTOS 1980-85 (tonnes)

|  | 1980 | 1981 | 1982 | 1983 | 1984 | 1985* |
|---|---|---|---|---|---|---|
| Production (sales) | 80,079 | 75,618 | 63,515 | 69,906 | 57,422 | 61,000 |
| Imports | 327,296 | 337,618 | 241,737 | 196,387 | 209,963 | 152,000 |
| Exports & re-exports | 48,671 | 64,419 | 58,771 | 54,634 | 39,919 | 6,000 |
| Consumption | 358,700 | 348,800 | 246,500 | 217,000 | 226,000 | 207,000 |

*Estimate

Source: US Bureau of Mines

## Table 3

### US ASBESTOS CONSUMPTION BY END USE AND TYPE (tonnes)

|  | Chryso-tile | Croci-dolite | Amo-site | Total asbestos |
|---|---|---|---|---|
| **1984** |  |  |  |  |
| Asbestos-cement pipe | 31,300 | 5,700 | – | 37,000 |
| Asbestos-cement sheet | 12,100 | – | – | 12,100 |
| Coatings & compounds | 21,600 | – | – | 21,600 |
| Flooring products | 46,400 | – | – | 46,400 |
| Friction products | 48,400 | – | – | 48,400 |
| Insulation |  |  |  |  |
| Thermal | 500 | – | 100 | 600 |
| Electrical | 2,200 | – | – | 2,200 |
| Packing & gaskets | 13,200 | – | – | 13,200 |
| Paper | 1,700 | – | – | 1,700 |
| Plastics | 1,100 | – | – | 1,100 |
| Proofing products | 7,100 | – | – | 7,100 |
| Textiles | 1,500 | – | – | 1,500 |
| Other | 32,800 | – | 300 | 33,100 |
| **TOTAL** | 219,900 | 5,700 | 400 | 226,000 |
| **1983** | 210,200 | 6,200 | 600 | 217,000 |

Source: US Bureau of Mines

with their possible effects on health, these effects cannot be ignored.

The main factors determining the harmful effects of inhaled fibres are reasonably well understood. The fibre must be respirable i.e. have diameter below about 3-5 microns. Length is also important. Very short fibres below 1 micron in length may not be retained when breathed and if retained are more likely to be removed by the lung's defence mechanisms. Very long fibres (greater than a few hundred microns) are more likely to be trapped, harmlessly, in the outer airways. So fibres below 3 microns in diameter and in the 1-100 micron length range would be particularly liable to be harmful. Durability is also important. Fibres which can withstand the physical and chemical processes (transport, digestion and dissolution) which might otherwise remove them, may persist long enough to be harmful.

Against these criteria some substitute fibres can be confidently expected to be harmless. E-glass fibres have controlled diameters of 6-20 microns and are not respirable. Steel fibres are also very coarse. However aramids (see Fig 1), particularly in pulp form contain fibres in the respirable range and are physically and chemically durable. It is too soon to know whether they are significantly harmful, but they may be

and we should treat them accordingly.

## 2. FIBRE CEMENT

Asbestos cement, world-wide, uses far more asbestos fibre than any other product range, though this is less true in the USA where flooring and friction materials use comparable amounts (see Table 3). Western world output of asbestos cement in the mid seventies was about 17.5 million tonnes/year, consuming about 1.75 million tonnes /year of asbestos. Since then, there has been intensive development of substitutes, particularly in Europe, and with minor but locally important exceptions a clear picture has emerged.

The asbestos product range comprises flat and corrugated sheets, roof slates and tiles (shingles), profiles and moulded shapes, pipe for sewage and drainage and pressure pipe for potable water. Several different manufacturing processes are used, all taking a dilute slurry containing asbestos fibre, with ordinary Portland Cement (OPC) and relatively few additives. The slurry is dewatered through a sieve, and transferred to a felt prior to wrapping on a forming bowl to give a sheet of the required thickness. It is then dried by air-curing at room temperature over a period of 2-4 weeks or by autoclaving at 180°C in steam.

Substitutes for the simple flat sheets were first available. followed by corrugated sheets and then the more complex shapes and profiles. Pipes have proved very difficult to substitute and there is still no fully acceptable alternative to asbestos pressure pipe. The key properties of asbestos for cement reinforcement, which must be broadly matched by a substitute, are shown in Table 4.

Table 4

Properties of Asbestos Fibre
Important in Fibre Cement
Construction Materials

1. High Stiffness - Modulus 165 GPa
2. Very fine fibres - fibrils are sub-micron in diameter
3. Does not burn or support combustion
4. Resistant to alkali attack
5. Non-biodegradable
6. Not moisture sensitive - dimensionally stable
7. Easy to process - tough and an effective web former
8. Retains 'fines' in wet processes
9. Low cost

Throughout Europe and Japan the fibres most frequently used as substitutes are shown in Table 5.

Whichever substitute fibre is used it is necessary to reformulate the slurry to include a range of additives not required with asbestos, to at least partly match the requirements listed in Table 4. Choice of prime reinforcement fibre varies, too, with the process used.

Inclusion of some cellulose as web-former and to retain particles in the web is almost universal. Also cellulose is increasingly being used as the prime reinforcement, particularly in autoclaved sheets. For non-autoclaved, air-cured sheets, thermoplastic fibres such as polyvinyl alcohol and polyacrylonitrile (which can't withstand autoclaving), provide the prime reinforcement. In all the cement sheets, the amount of substitute fibre is substantially less than the asbestos it replaces. The balance is made up by inorganic fillers, such as PFA (Pulverised Fly Ash) which is pozzolanic.

Use of alkali-resistant glass has been limited by concern about its long-term durability and of carbon fibre - even the lower cost, low modulus types - by price.

Some new processes are being developed in which sheets of reinforcement are impregnated with cement slurry. Examples are the Surrey University/Montedison "Netcem" process, using fibrillated polypropylene netting - and the Natural Fibre Concrete AB (Sweden) process using cellulose fibre mats.

In summary, first generation fibre cement products to replace all A.C. products except pressure pipe are now commercially available. They sell at a price premium of about 10% over the asbestos product and have a different balance of properties, with the main benefit

## Table 5

| Type | Composition | Source/Manufacturer/Trade Name |
|---|---|---|
| Natural organic fibres | cellulose | Scandinavia, Canada, Australia and New Zealand |
| Synthetic organic fibres | High modulus Polyvinyl Alcohol | Unitika (Mewlon); Kuraray (Kuralon) |
| | Polyacylonitrile | Hoechst, (Dolanit) + others |
| | Low modulus carbon fibre | Kureha Ashland |
| Glass fibre | Alkali resistant glass | Pilkington "Cemfil" |

that they are much tougher and better able to withstand impact damage.

No figures are available for western world output of non-asbestos fibre cement sheets, but in UK for example, in 1985, of the 350-400,000 tonnes of fibre cement sheets (flat, corrugated and slates) produced, over 50% (about 200,000 tonnes) was non-asbestos. This proportion will have increased in 1986, with the advance in sales of non-asbestos corrugated sheet.

## 3. FRICTION MATERIALS

A list of properties of asbestos which are particularly useful in friction materials is given in Table 6.

Some of the many different fibres and fillers which have been evaluated as substitutes, are listed, with their volume cost, in Table 7. None has all the required characteristics and blends of two or more are almost always used. The materials adopted have varied with the producer, the product type, and the demands of the process.

### 3.1 Disc Pads (Pucks)

Compromising cost and performance, steel fibres have been widely adopted as the main reinforcement. For many years, both in USA and Europe, so-called "semi-metallic" disc brake pads containing high volumes of steel fibre, roughly equivalent to the asbestos replaced, have been available. Though commercially successful, their very high thermal conductivity can lead to overheating of the hydraulic brake fluid. More recent formulations replace some of the steel fibre content by organic fibres, including aramid or by inorganic fillers. These later formulations have lower thermal conductivity and a better overall balance of friction and wear properties over a range of duties. Their life is typically 1.5 to twice as long as asbestos pads, at only slightly higher cost.

### 3.2 Drum Brakes

Because of its radiused shape, the manufacturing processes for drum brake linings are more complex

Table 6

PROPERTIES OF ASBESTOS USEFUL IN FRICTION MATERIALS

1. Heat resistant - stable to 400°C, retains useful properties above 400°C

2. Decomposes above 650°C to other solid silicates

3. Decomposition products have higher friction than asbestos

4. Low thermal conductivity

5. Strong

6. Flexible

7. Non-springy - low rate of fibre recovery

8. Resistant to breakage under shear

9. High surface area

10. Absorbent to resins and rubbers molten or in solution

11. Surface readily wetted by and adheres to resins and rubbers

12. Readily available in a range of fibre lengths in quantity at relatively low cost

than for disc brakes and so matching the processing characteristics of asbestos is even more important, if investment in expensive new plant is to be avoided. In general, tough fibres have been chosen, to ease processing.

In Europe, for car brakes, steel fibre with some added organic fibre (for example aramid) has been the main reinforcement. E-Glass fibre is brittle and has found only limited use. Though it offers good strength, reasonable heat-resistance and low cost, its wear products are abrasive and can damage the brake drum. At high temperatures, under heavy duty it can melt and cause frictional fade.

E-glass is much more widely used in commercial vehicle brake linings and in the yarns for wound clutch facings. In both these applications it competes with and is frequently used in combination with aramid fibre.

A summary of the main European substitute materials, and the price premium they command is given in Table 8. The extent of market penetration in Europe varies with product from country to country, but the following generalisations can be made. In the replacement or aftermarket there is virtually no penetration by substitutes. In original equipment (O.E.) sales, substitution has usually been linked to introduction of new models. Virtually 100% conversion of car disc brakes, drum linings and clutch facings has been achieved in Germany and Scandinavia. In UK the conversion has been slower, but still substantial (30%-70%). In contrast, there has been very little substitution in France and Italy.

Table 7

Asbestos substitute fibres and
fillers and their volume costs

| Material | Sp. Gr. | Volume Cost (P) |
|---|---|---|
| Chrysotile asbestos (short-medium grades) | 2.60 | 50-100 |
| Wollastonite | 2.90 | 116 |
| Vermiculite | 2.63 | 53 |
| Mica | 2.48 | 75 |
| Basalt Fibre | 2.95 | 103 |
| Rockwool | 2.50 | |
| Manilla (cellulose) | 1.20 | 182 |
| Processed Mineral Wool | 2.70 | 193 |
| Ceramic Fibre | 2.66 | 204 |
| Polyacrylonitrile (P.A.N.) | 1.18 | 226 |
| Polyester | 1.38 | 269 |
| Chopped E-Glass Fibre | 2.54 | 315 |
| P.V.A. Fibre | 1.33 | 326 |
| Steel Fibre | 7.86 | 448 |
| Low Mod. Carbon | 1.63 | 815 |
| Oxy P.A.N. | 1.43 | 786 |
| Aramid Pulp | 1.44 | 1318 |
| Aramid Staple | 1.44 | 1735 |

Commercial vehicle linings have
proved technically more difficult
to substitute and the rate of
conversion is much lower than for
car linings.

In USA, as in Europe, there has
been virtually no substitution
in the aftermarket, but substantial
conversion in O.E. particularly
of disc brake pads.

As in Europe, steel fibre is dominant in disc pads, but with
significant use of aramid and glass
fibre. Similarly, with passenger
car linings. Clutch facings and
commercial vehicles linings both
use aramid and/or glass fibre.

Both in Europe and USA, many of
the substitute friction products
have superior properties, particularly life but also high friction
stability and temperature resistance, compared to asbestos based
products.

A European aramid fibre producer,
ENKA, recently gave estimates of
the present and prospective aramid
market in Western Europe. Their
estimates for friction materials,
gaskets, packings and textile products are given, in Table 9.

### 4. SEALING MATERIALS

Asbestos is used in a number of
sealing materials, of which CAF
sheet packing, Beater sheet and
textile packings (dynamic and static) are the most important.

#### 4.1 Sheeting Materials

Both CAF and Beater sheet contain
a high proportion of asbestos fibre,
bonded with natural or synthetic
rubbers and containing fillers.
Both are used as gaskets to seal
flanges in applications ranging
from chemical plant to automotive
carburettors and cylinder heads.

CAF is produced by calendering
a dough of fibre/filler/rubber
in solvent, on a 2-bowl "IT" calen-
der, to form a dense resilient
sheet on the heated back bowl.

Beater sheet is formed on a papermaking machine from a slurry of

## Table 8

FRICTION MATERIALS IN EUROPE
SUBSTITUTE FIBRES USED AND TYPICAL PRICE PREMIUMS

| PRODUCT | MAIN SUBSTITUTE FIBRE USED | PRICE PREMIUM |
|---|---|---|
| Car Disc Pad | Steel and some aramid | 10-30% |
| Car Brake Lining | Steel and some aramid | 25-50% |
| Commercial Vehicle Lining | Aramid/Glass | 70-100% |
| Clutch Facings | Aramid/Glass | About 50% |

## Table 9

# ARAMID MARKET
## WESTERN EUROPE

IN TONS p.a.

- NON-WOVENS
- PROTECTIVE CLOTH
- PACKINGS
- GASKETS
- FRICTION MATERIAL

1800 + to
1600
1400
1200
1000
800
600
400
200
0

| | 1984 | 1985 | 1986 | 1987 | 1988 |
|---|---|---|---|---|---|
| | 380 | 585 | 830 | 1075 | 1210 |

w.lub

2.27 **Enka Symposium '86**

Enka

1341

fibre, fillers and rubber latex. Their uses overlap but CAF, having higher density and better sealing properties, is used for the higher duty applications. CAF mixing and calendering plant is designed to handle the tough, fibrillar, non-springy chrysotile fibres. Not surprisingly, all but one of the major manufacturers of CAF have used aramid fibre, alone or with other synthetics, in their first generation substitute products. In contrast, TBA Industrial Products in UK has pioneered the development of E-glass fibre reinforced sheeting.

The product (Permanite AF 2000 series) has a number of advantages over aramid based material - not least the lower cost of glass fibre and the fact that the fibres are non-respirable.

Aramid-based material sells at a price 2-3 times that of CAF sheet packing.

The substantial price-premium on substitute gaskets and the fact that asbestos is perceived to be "locked into" CAF has meant that sales of substitutes have been relatively slow in Western Europe. They account for less than 10% of the sales of CAF sheet packing, in a market of about 10000 tpa. In the slightly larger USA market (12000 tpa), up to 30% penetration has been achieved by non-asbestos sheeting.

Beater sheet has tended to revert to cellulose, in replacement of asbestos for many lower temperature, lower sealing pressure applications. For higher duty, aramid fibre is being increasingly used, particularly in USA.

## 4.2 Textile Packings

Asbestos yarns, braided or plaited into packings, with added lubricants, have been used for about a hundred years in dynamic sealing applications. First use was in steam engines at higher temperatures than cellulose packings allowed. Throughout the 20th Century, asbestos usage has grown and diversified (for example in pump shaft seals) in parallel with the growth of the chemical process industries. Cellulosics, meanwhile, have continued to be cost-effective in lower duty applications and are regaining some ground, where asbestos is now unacceptable and has been over-specified.

In recent years a number of speciality packings, developed for severe duty have extended the range beyond asbestos capability. PTFE, offers exceptional chemical resistance, very low friction and wear and good heat resistance: carbon and graphite yarns have even higher heat resistance and good chemical resistance. Despite their high price, these materials are cost-effective in resisting aggressive environments and extending the periods between plant downtimes for maintenance. Now aramid fibre packings are making significant in-roads in this medium to high duty range and had secured about 10% of the packings market in W. Europe in 1985. Their combination of chemical, heat and abrasion resistance is proving very competitive and usage is expected to double in the next five years.

Static Packings in braided or rope form have mainly been used for heat insulation, in applications where limited resilience is required and relatively poor flex or abrasion resistance could be tolerated. Since these are relatively dusty asbestos products they were amongst the first to be substituted. At moderate temperatures of 200-600°C E-glass, being non-respirable, proved acceptable and highly cost-effective. For higher temperatures, up to about 1200°C, the so-called "ceramic" fibres (staple glassy alumino -silicate fibres) such as Fibrefrax and Kaowool) are used. Both E-glass and ceramic fibres are skin-irritants and can be uncomfortable to handle and the ceramics may contain respirable fibres. For highest temperatures, up to 1600°C, zirconia or alumina fibres (ICI Saffil) are used.

### 6. TEXTILES

The variety of forms and uses of asbestos textiles is so great that generalising is particularly diff-icult. However one very important category of use is in heat protection and insulation. Since this was associated with some of the dustiest processes, it was one of the earliest to be substituted.

Asbestos yarns, with their flexible staple fibres, are self-insulating by virtue of entrapped air. They decompose only slowly above 400°C, to a solid product, only melt above 1400°C and were very widely used. They work well and "fail safe" and in protective clothing applications are reasonably comfortable to wear. Conventional E-glass yarns and cloths were obvious potential substitutes, not least because of their easy textile processability, reasonable heat resistance and comparable cost. They did however suffer serious disadvantages. The smooth continuous filaments (see Fig 1) form denser yarns with higher thermal conductivity than asbestos. They soften above 400°C and eventually melt - potentially catastrophically - for example if splashed with molten metal in protecting welding operations.

A very important development was that of air-texturing these yarns to increase their bulk and their insulation value and generally make them much more like asbestos. Air textured fabrics such as TBA's Fortaglas are now widely adopted as replacements for asbestos in insulation and heat protection.

Further development of special treatments to improve their resistance to molten metal splash ("WELD-STOP") and to increase their abrasion resistance ("GRAFLEX") has extended the range of use.

Where abrasion resistance is paramount, aramid (Nomex and Kevlar) fabrics have proved especially effective, providing good heat resistance, very good insulation and the wearer-comfort typical of synthetic organic fibres.

For highest heat resistance, broadly the same materials are used as for dry packings (i.e. ceramic fibres, silica fibres and alumina or zirconia). Some of these are textile processed only with great difficult and all are expensive.

## 7. REINFORCED PLASTICS

Asbestos is used to reinforce both thermoplastics and thermosetting resins.

### 7.1 Reinforced Thermoplastics

Chrysotile was a general purpose fibre reinforcement, offering a good combination of mechanical, thermal, chemical and electrical properties, at relatively low cost. Substitution has not been very difficult and is now well advanced, due to the very wide range of formulation options available to the plastics compounder. Chopped (typically 6 mm-12 mm long) E glass strand is the main fibre substitute, offering similar mechanical properties at similar cost. A range of fillers, for example talc, chalk or mica, provide specific properties at lower cost and are used alone or in combination with fibres. The principal problem in replacing asbestos has been in fire resistant grades. Due to its peculiar fibrillar structure and chemical composition, chrysotile readily conferred non-burning or self-extinguishing and, particularly important, non-drip characteristics on thermoplastics. Glass fibre can't readily achieve this but in combination with fillers, alternatives are now available.

### 6.2 Reinforced Thermosets

Asbestos reinforced phenolics (ARP) have been a very important product group for many years.

ARP are normally produced by pre-impregnation of an asbestos paper, felt or cloth then moulding or laminating under heat and pressure.

### (i) Heat and fire protection and insulation

Low pressure moulded felt laminates are used to make automotive heat shields for underbody and engine component protection and insulation.

Glass fibre felts are now replacing asbestos. A small but vitally important market has been in missile blast shields and rocket throats, where the missile body itself or the launching pad are protected from the plasma-temperatures of burning rocket fuel, by the unique ablation

characteristics of asbestos/
phenolics.

These have proved to be amongst
the most difficult products to
substitute and asbestos/phenolic
pre-pregs such as TBA's Durestos,
are still widely used. Carbon
fibre is the nearest equivalent,
but at much higher cost.

(ii)Wearing and bearing applications

High pressure ARP laminates are
used for a range of engineering
applications. In particular where
a combination of high compressive
strength or low coefficient of
thermal expansion, with low friction
and low wear rate are required.
Important applications are in dry-
running and lubricated bearings
and pump rotor vanes. In a typical
product range, such as the Fero-
bestos range from TENMAT, in UK,
all the materials whether for low,
medium or high duty were based
on asbestos felt or cloth. The
first generation substitutes in
the FEROFORM range, use six different fibres alone or in
combination, i.e. two cellulosics,
two synthetic organics (polyester
and aramid), E glass and carbon
fibre. Not surprisingly, for these
very demanding applications, in
which the perceived health risk
is low, the asbestos products are
still widely used and there is
less than 20% penetration by substitutes.

8. CONCLUSIONS

8.1 World production of asbestos
is still about 4 million tons p.a.
Over half of this comes from USSR
where production is apparently
still rising. The rest of the
world's producers (except China)
have seen a sharp fall in output
in recent years.

8.2 First generation and in some
cases second generation substitute
products are now commercially
available and being used in almost
all the main product groups, though
there are important exceptions.

8.3 No single fibre replaces
asbestos in all its applications.

8.4 The main substitute fibres
in first generation products are
Polyvinyl alcohol and cellulose
(in fibre cement); steel fibre
(in friction materials); aramid
fibre (in friction materials, sealing materials and textiles); and
E-glass fibre (in friction
materials, sealing materials and
textiles).

8.5 Polyacrylonitrile is partly
replacing polyvinyl alcohol in
second generation fibre cement
products.

8.6 In many products the substitute fibre is used in lower proportion - in the case of aramid fibre,
in very much lower proportion -
than asbestos, because of its high
cost.

8.7 This means that direct substitution has not usually been

possible and substantial reformulation has been necessary. More than one fibre has frequently been used in combination with one or more fillers, not only to match the asbestos product properties, but to achieve satisfactory processing.

8.8 Most substitute products are being made on existing plant designed and optimised for the particular processing characteristics of asbestos. This has necessitated the addition of special processing aids (e.g. web-forming fibres such as cellulose).

8.9 Most substitutes command a price premium. This can be as low as 10% or as high as 200% or more.

8.10 Many substitutes now have superior properties to the asbestos product they replace.

## ACKNOWLEDGEMENTS

The author thanks Turner and Newall for approval to publish the paper. He gratefully acknowledges assistance from colleagues in T&N who provided data, particularly Dr J D Crabtree, Mr C J Hill and Mr G A Ross. He is indebted to Mrs Desley Ashworth for her preparation of the typescript.

## BIOGRAPHY

Gordon F. Heron is a Fellow of the Royal Society of Chemistry. He is a Director of Turner and Newall Materials Research Ltd, Rochdale, U.K. with special responsibility for Composite Materials Development. His earlier career, all with Companies of the Turner and Newall Group, has involved research, development and technical management in friction materials, asbestos textiles, sealing materials and reinforced plastics, latterly as Research and Engineering Director of TBA Industrial Products Ltd.

# AN OVERVIEW OF FIRE BLOCKING FABRICS

Neil Saville
Research and Development Director
Universal Carbon Fibres Limited, Gomersal Mills, England

## Abstract

The paper reviews the manufacture of fire blocking aircraft seats and the problems encountered in trying to pass FAR 25 853 (C) appendix F. Factors considered are fibres, fabric and foam properties, seam problems and passenger comfort. The paper raises a number of question marks over the repeatability of test method.

## 1 TYPES OF FIBRE COMBINATION FOR BLOCKING LAYER FABRICS

There are many combinations of fire resistant fibres possible in production of fire blocking fabrics. Each fibre has its own role and advantages Panox/Kevlar for instance has a high quality char and the Kevlar particularly adds strength to that char and assists Panox by improving abrasion resistance. The quality of the char when using Panox and Kevlar is such, that there is little or no shrinkage of the fabric under the flame. This can be, and often is, a major advantage, but if no shrinkage occurs then the tightness of the weave has to be such, that molten foam does not escape through the pores. Panox/aramid blends have a higher abrasion resistance than Panox/Kevlar and are far easier to cut, the aramid being for example Nomex or Conex. Aramid fibres offer high abrasion but low cutting resistance, and also contract 30% under flame. The contraction of aramids is a very powerful one and therefore this technique can be used to close the pores of the fabric whilst under flame. Panox/Kevlar/aramid blockers can in effect take advantage of two factors: the Kevlar adding strength of char whilst the aramid assists in the closing of the pores and giving exceptional abrasion resistance, whilst the Panox acts as the major donor of carbon char and also allows the price to be at a lower level than using Kevlar or aramid only. Kevlar/aramid structures are quite commonly used, and again have higher abrasion than the Panox rich fabrics, but in Europe the price levels of fabrics made from these two fibres tend to preclude them from wide usage. This is even more marked when using PBI Kevlar. PBI has a far higher abrasion resistance than Panox whilst having a similar blocking effect; it is a very valuable fibre and combined with Kevlar allows the use of lower weight fabrics. Low weight is a major advantage to most carriers in the Aircraft Industry who multiply the weight of the fabric by the factor, usually between £20 and £60 per kilo per year, and set that at the side of the actual cost of the fabric. Quite often the dearer fabric wins out, purely on having a lower weight. Panox/aramid wool fabrics have been used as fire blockers often in felt form.

The aramid is used for the scrim of the non woven whilst the Panox and wool form the needle bat for the insulation layer. This system could also be used with Panox/Kevlar and was suggested and encouraged by I.W.S. The problem with wool is the loss of weight under flame forcing the fabric to be at least 30% heavier than a Panox/aramid or Panox/Kevlar. Wool rich fabrics tend to be ruled out on weight consideration and the mere reduction in cost of pound or two pounds per metre is insufficient to over-ride the extra fuel cost of carrying the blocker. It is suggested that on a Jumbo Jet of British Airways, a cost of £14,000 per year is typical for the cost of flying the blocker alone and a range of costs figures are given in appendix 1.

## 2 METHODS OF BLOCKING

The blocking of an aircraft seat by present conventional techniques tends to be split into three sections. A standard F R Foam covered by a blocking layer fabric, further covered with the dress or outer cover in the specified design of the carrier. This is perhaps the most common way. Blocking foams are sufficiently advanced to be preferred on certain airlines, particularly on the short flight routes and the commuter jet. The problem that arises with the fire blocked foams is not one of blocking but of comfort. It would be admitted by most people that up to the inception of fire blocked foams, the foams used on the seat backs were of a lower density allowing the person to sink into the chair. Fire blocking foams are commonly of a higher density and a greater hardness, keeping the person held away from the chair or held at the first point of contact. There is the difficulty on seat bases of high density and hardness, or to put it in common parlance the numb bum syndrome, where people tend to get pins and needles from the pressure applied by the seat. There is a tendency because of high cost, to glue a one inch layer of the stiffer, non burn foam around the softer core. This is satisfactory for the centre of the seat but the edges are at least twice as stiff as the middle. This tends to cause problems of comfort and produce an unusual stress on the glue line, which leaves a question mark over the long term durability of such a system. Perhaps the cheapest, most comfortable and advanced system is the use of the standard F R Foams as they currently exist together with a fire blocking fabric which combines the dress cover and the blocker as one woven fabric. The combined blocker gives a considerable cost saving. Usually the cost of making up the seat cover is greater than the cost of the fabric from which the seat cover is made. The combined blocker saves manufacture of one of the seat covers and it also simplifies the fastening of the outer cover to the chair. There is no doubt that having an intermediate layer between the foam and the outer dress cover does cause a number of anchorage problems. Each system of blocking aircraft seats has its own advantages and disadvantages. Standard F R Foams have been designed over many years to produce the maximum comfort with the minimum weight, at the same time an attempt has been made to reduce the flammability, although the F R is fairly limited; the highly F R Foam systems which are currently available leave much to be desired in terms of choice of comfort. People would wish to indicate that the present fire blocking foams are satisfactory comfort wise but it has to be said that the range of densities and comfort possibilities are severely restricted when compared with the standard materials. The blocking layers perform their function but produce the disadvantage of the extra making up, together with the problems of fixing it to the outer cover. Blocking fabrics also produce an intermediate layer which has a stiffening effect, which itself can become uncomfortable. This is particularly true of hard woven fabrics. It has to be appreciated that, when sitting on foam cushions, the thickness of the cushion itself can be reduced by 60% at least. This means that the fabrics surrounding the cushion becomes slack. If the materials have insufficient elasticity to take up this slack, spare material on the seat becomes folded and these folds over a period of two or three hours during a flight, can become semi-permanently set.

Once a seat fabric creases, the creasing is re-emphasised by the next passenger who sits in exactly the same place. The creases already preformed then tend to be established as a permanent feature, giving a rod-like feel to the topside of the seat. Creasing can be counteracted by the honeycomb type of weave. As the cushion is depressed the fabric takes up the slack by becoming thicker, thus preventing the creasing problem. The actual extra making up and extra velcro fastenings required for the intermediate blockers do create problems but there are also advantages. The foam itself tends to adhere to the under side of the normal seat fabric and the two materials tend to abrade against each other. The extra layer prevents this. On the whole the fabrics are stronger than the foams so it is the foam that suffers and gradually it crumbles away by the abrasion of the top fabric. With the introduction of an intermediate layer, expecially if this intermediate layer has stretch characteristics, the blocker acts as a lubricant between the top surface of the foam and the underside of the outer fabric. Wear life is increased by factors of 30 to 40%, so that the increased cost is in some way mitigated by the improvement in wear life of all the components. The combined blocker, with its obvious advantages of avoiding the extra making up, can have a number of other advantages. The reason for its acceptance, is due to the pressure from the airlines for a weight reduction. It has become very clear that it is more than twice the cost of the fabric to fly the blocker and therefore the primary cost of the blocker itself is only one factor. The pressure is on the fabric manufacturers to reduce the total weight of the seat covering material. Weight reduction can be achieved from a blocking point of view fairly readily and on a typical 470 gm/m square outer cover and a 250 gm/m square blocker, the weight reduction can be between 15% and 20%. The points that have to be watched, of course, are that in reducing the weight of the outer cover, one is likely to dramatically reduce the abrasion resistance; the abrasion resistance of textile fabrics tend to increase by the square of the increase in the set of the fabric, ie. any reduction in the end and picks per inch inserted, without further adjustment of the fabric, would result in a dramatic fall off in durability. Reduction in abrasion resistance is hard to justify when you consider the combined blockers are likely to cost a least 50% more than the two separate fabric costs added together. To make a cost effective combined outer dress cover and blocking fabric one has to introduce the concept of blocking into the dress cover as well as into the back of the double cloth where blocking yarns are of traditional type ie. Panox/Kevlar, Panox/aramid. It is not essential to produce a double or treble cloth

in order to block the foam, but certainly if the exotic fibres are used entirely in the outer dress cover, then the cost, particularly the cost of dyeing and colouring of the fibres would soar to an uneconomical level. It certainly is advantageous to use for instance 80% wool/20% aramid in the top cover whilst using sparingly more expensive Panox/aramid and Panox/Kevlar yarn in the back fabric. If no attempt is made to use aramids in the dress cover, then more of the backing yarn has to be used to block the fire and as these yarns are approximately twice to three times the cost of the outer cover, it is therefore an unwise move. If wool alone is used in the outer cover, then we are back to the abrasion resistance problems, and also have a problem of stitching and maintaining the two fabric chars into one single layer. It is aesthetically unacceptable to bring the black or strongly yellow yarns to the face of the fabric in order to stitch, so you are left with the only alternative - to bring the outer cover yarns down into the back of the fabric in order to tie the two fabrics together. If the stitching yarns have no char quality, then the two fabrics will separate and the char will be less effective in blocking the flame, and also insulating and absorbing the hot foam and particularly in the exclusion of air from the burning cushion. This air flow into the cushion can be quite critical in the extinguishing of the flame after the burner is switched off. Combined fire blockers can now be made from brightly coloured 100% synthetic blends whose properties include high abrasion, non-drip, non-stain and easy care.

## 3 TYPES OF FABRIC STRUCTURES USED

In reviewing this section, the most common weave is a tight flat woven twill, usually 2/2 twill. This structure will give reasonable durability and allows tight packing of the yarns. Tightly woven fabrics allow little of the molten foam to drip from the cushion and they also exclude the movement of air into the cushion, whilst the test is in progress. The major disadvantage of tight woven non stretch fabrics, of course, is that just like the bagging of trousers at the knees, the continual seating or distortion of the fabric, caused by the squirming of the passenger during the flight, causes the fabric to permanently bag. All textiles for whatever use suffer from bagging, unless either the yarn has a stretch component or the fabric structure is allowed to stretch. Distortion occurs when the fabric has no room for manoeuvre and the yarns are stretched. Once the yarn is stretched, there is no way the fabric can be put back into its original configuration. Bagged fabric not only leads to creasing and discomfort but the appearance of the seat tends to make the interior of the aircraft look second hand from a very early stage. Stretch woven honeycomb fabrics have the disadvantage of having open pores, but the structure allows the fabric to stretch, particularly on the diagonal, to a degree of 25-30%. As the yarn has not stretched, but only the fabric, on the removal of the pressure the fabric itself returns to normal ie. it gets thicker and thinner with pressure changes, recovering as the pressure is removed. Abrasion resistance is much more of a problem with the honeycomb structure, ie. if the martindale abrasion tester is used. In practise because fabrics move, rather than are stationary abraded, as they are in a laboratory, the sophisticated Furniture Industry's Research Association simulated wear test is much more satisfactory. BS 6261 is an abrasion or wear life test developed by the Furniture Industries Research Association and the CAA. This test checks the interaction of materials and like the Squirming Herman Test is much more practical in its assessment of fabric durability. Double Jersey knits have been used in fire blocking and there are indications that some progress is being made, but I suspect the major problem is one of bagging in that the internal fiction within the fabric is such that, once distorted, the yarns find great difficulty in returning to their normal position. There is also a secondary problem with knitted fabrics in that the air passage through the fabric is twice that of the same weight woven. Certainly we have much experience of this problem with knitted trousers, particularly for men. This material was fashionable in the early sixties when the drafty trouser syndrome was a well known problem that was never overcome. During the second stage of the burner test the quantity of air allowed into the cushion is a vital factor. One has to bare in mind that a very high level of heat has been pumped into the seat cushion from the flame thrower. This heat causes a rising air current and draws fresh air into the cushion from anywhere that it can obtain it. The high air permeability of knitted fabrics requires the contraction and char quality of the knitted fabric to be better than that of its woven counterpart. The one thing in favour of knitted fabrics is, of course, their comfort and ease of application. When trying to fit them to cushions certainly initial appearance is in their favour but long term use tends to be against them. Regarding the combined blocker, much is still unknown. Obviously it will react at least as well as the single outer cover does at the moment, but I suspect that the abrasion against the foam will be similar to the present seat situation, ie. the foam will probably wear out significantly sooner than when a separate blocking layer is used. From a comfort and appearance point of view, the combined blocker has most of the answers. It may be necessary, on certain parts of the chair to still use the single blocking fabric as it may be hard to justify having the expensive combined blocker on the underside of the seat. Non-woven fabrics are invariably comfortable, in that their surfaces are soft, they also absorb more readily than most other materials. Non woven are also excellent insulators and therefore for any given weight loss than most other materials, but usually require a centre scrim for stability. Non-woven fabrics suffer a number of problems in all textile applications; one is abrasion resistance and it is invariably difficult to stop fibre pilling and fluffing on the surface of the fabric. A futher problem with non-woven fabrics is that although most materials have a centre woven scrim, they are far more subject to distortion and bagging, resulting in a relatively poor appearance after only a short time in use. Amongst the early fabrics for fire blocking, there were a number of laminated materials. Particularly common was to aluminised surfaces to reflect the heat and also to contain the melting foam. These worked very well indeed, showing excellent blocking characteristics and giving very low weight loss figures, but the problems of manufacture and the discomfort to the passenger caused these materials, on the whole to be unacceptable.

## 4 COATED FIRE BLOCKERS

There are a number of commercially available coated fabrics. Most of their base materials are Panox/Kevlar, Panox/aramid but require additional blocking power especially when confronted with low density or relatively poor quality foams. Coatings take up a number of forms. High density fire blocking materials can be used or materials which simply fill the holes in the fabric to give a high air impedance. These fabrics can be either foam coated or air free coated. There is also a special class of intumescent coatings which are based on expandable graphite. These materials expand up to 100 fold on heating and as well as absorbing the melting foam and impeding the air, they give a high insulation layer of up to 6mm thick on the inside of the fabric. There is quite a deal of prejudice against coated fabrics on the basis that they may be removed during wear. This is of course a possibility and if the coating wears, then the performance will obviously be diminished. We have supplied to a number of companies, coated materials but we do take the precaution of testing the fabric on the Furniture Industry Research Association simulated wear test of a 100,000 cycles before applying the FAR 25853 flame test to the composite. We feel that this wear testing before flame testing gives reasonable degree of assurance that certainly, the coatings which we would apply could expect to last the full life of the fabric.

## 5 COMMON PROBLEMS IN FIRE BLOCKING

If nylon is blended with wool in the dress cover of the aircraft seat, as is quite common, the introduction of the nylon does prevent char formation and instead of the top dress cover acting in any way to block the flame, the char literally is blown away by the thrust of the burner. It is probably worth commenting that not only is wool/nylon poor value for money in blocking terms, in that usually a higher weight of blocking material of higher value is usually required on the back, to block the flame. Wool/nylon whilst wearing longer when any test is applied ie. martindale or stolflex or taber, the pilling performance is significantly impaired and although a wool/nylon seat cover will last longer, it usually has a poor appearance from a very early stage due to the pilling problem. With fancy jacquard patterns manufacturers are often tempted to use as one of the colours, a yarn with a high percentage of meltable synthetic fibre as one of its main constituents. The result is that instead of burning or charring, the top cover will split from one side to the other following the pick or warp line. The outer cover using such a yarn will then offer in blocking performance terms, no assistance and can be very misleading when a statement of 90% wool/10% nylon is put as the fabric composition. One can then be lead to believe that a 90/10 wool nylon fabric containing 100% wool and 100% nylon yarns is similar to a fully integrated blended 90/10. Using the same blocker an intimate blend may pass with flying colours, whilst the fabric containing high percentage of nylon in one yarn may fail miserably. One can imagine situations where ostensibly the same fabric of the same composition and pattern is put into two separate aircraft. One which would pass and one which would fail and this use of percentages of none intimate blended yarns can be a hazardous business. Shrinkage of the outer cushion material can be disasterous, particularly with the use of leather in seats. Leather itself shrinks dramatically, causing the base cushion to arch its back, the flame thrower then inpinges on the under side of the seat, burning out and cracking the base so that the interior of the cushion literally pours out onto the floor. Leather materials are often used in executive jets and these require usually a special heavy non-woven fabric to absorb and protect the foam from these difficult conditions. A secondary problem does arise when the base cushion shrinks. It reveals the under edge of the back cushion causing weight losses of up to 50% if conventional blocking fabrics are used. Strength of char is also a major factor in obtaining a good pass, in that if the fabric and the blocker split open it allows the flame to penetrate and scour out material from the interior of the cushion, so that splitting should be prevented at all costs. For any given weight of fabric acting as a blocker, the finer the yarn from which it is woven and more the ends and picks per inch the greater will be the efficiency of the blocker, ie. if the pores of the fabric are large, then invariably the performance of the blocker will be poorer, unless some other mechanism within the fabric comes into play. As I said before stretch fabrics have a major advantage in terms of comfort and appearance, but they also have one major disadvantage in that under the weight of melting foam during the first 30 seconds of the test, the fabric stretches and produces a funnel effect at the front edge of the cushion, and as the material is de-natured and charred to carbon a permanent funnel forms near to the front centre edge of the cushion. Once a funnel is formed the melting foam runs to this spot and should the bottom of the stalactite, as it were break away, then the liquid foam pours onto the floor immediately below the cushion, forming a secondary fire which once the burner is turned off, continues to flame for a considerable time thus accentuating weight losses. Manufacturers of blocking layers will all recognise how much easier it is to pass with the first class and business class seats, in that usually the weight of fabric and the density of the foam is higher. This allows a greater weight loss, not as a percentage but in total. Economy seats with their poorer quality foam and low weight can be a problem; to lose significantly less than 200 gms is difficult as there is a fixed amount of energy pushed into the cushion and therefore the blocking of economy seats, ironic as it may seem, has to be of a higher order than first class or business class. It is well recognised, that the breaking open of seams is critical, or shall we say the non breaking open of seams is an important factor in the sucess of the test and from this point of view Kevlar has been recommended as essential to the stitching of the blocking fabrics. The main seam is usually a polyester sewing thread or polyester cotton, primarily because there is the lack of availability of an other non-contracting sewing thread. None the less the use of any sewing thread which has significant power to contract, can distort the cushion allowing the flame to reach the far side of the seat back allowing an excess of the outer cover to be burnt off and a poor result to ensue. The use of exceptionally high tenacity fibre in the blocking layer fabrics does produce major maker up problems for conventional seat manufacturers and special steps have had to be taken, and certainly on the shop floor there is a resistance to these materials.

## 6 COMMON CAUSES OF PERFORMANCE DIFFERENCES BETWEEN TEST RESULTS OF THE SAME MATERIAL COMBINATION

I have noticed from time to time that when the test rigs are being assessed for temperature, followed by wattage or vice versa, that first of all temperature may be adjusted to the right range. That rig is then removed and the wattage test is put into place. Adjustments have to be made and it is often difficult to be in specification with both temperature and wattage because conditions appear to drift. It seems to me the wattage is more vital than the temperature, because it is really the wattage which destroys the cushion. I have certainly seen instances when the temperature has been adjusted into the correct range and the wattage has been 0.5 watts/sq cm too high. The result of a test can depend on the degree of patience of the test house, as to whether or not adjustments will continue to be made until both figures are within spec. This difficulty of adjustment does in my opinion lead to some significant differences between one test and a similar test carried out at a different time.

A large number of repeat tests have been carried out on similar combinations of fabrics and foams which were to specification and I am led to the belief that the temperature and humidity of the test room in which the burner is placed, is a critical factor which is ignored in the original F.A.A. specification. I have a theory that the melting foam and its ability to run free from the buring cushion, particularly after the burner has been turned off, ie. during the final five minutes of the test, can be the case that the burners are placed in outbuildings of relatively poor quality. These outbuildings during winter can be approaching freezing point. The melting foam reaches congealing point far more rapidly than it would during the summer period when these tin buildings can often be far higher than the normal ambient temperature. Variability can be quite serious as a pass achieved in winter may be a fail in summer. If this is the case, then claims can be made against the supplier of the material, or simple ignorance may intervene and aircraft may be flying about with combinations which in effect on re-test would fail. From my own point of view, I think the fact that the temperature and humidity of the test room where the burner is placed is not standardized, is a major oversight and stops the burner from being used as a major scientific piece of equipment to guide the researcher. There is a second effect of not controlling the temperature of humidity of the test room in which the burner operates and that is that some materials have little or no moisture regain. Up to 16%, ambient conditions vary with the time of the year and you will appreciate a wool outer cover with a high moisture regain could

lose up to 5% in weight simply by taking it out of the condition atmosphere without subjecting it to heat or flame, but simply by re-adjusting its moisture level down to the new relative humidity. If the regain was consistent for all fibres then the test would be consistent for all materials but as I have said before there is an enormous difference in moisture regain between the aramids on the one hand and the wools and PBI on the other and therefore fibres such as wool have a significant penalty* which is nothing to do with the degree of burn or loss of char. Different laboratories use different fuels and any specific laboratory has to make adjustments to the air intake, in order to reach the correct temperature and heat flux. What concerns me is the effect of changing the air speed. If a low air speed is used the flame hits the chair frame and sweeps upwards. If on the other hand the air speed is higher, there is a tendancy for the flame to split as it strikes the chair, 30% going straight on burning the base and 70% eating into the side of the seat. The indications are that higher weight losses occur with higher air speed with any given temperature and wattage.

## 7  CONCLUSIONS

Whilst the new regulations have brought us many problems, they have certainly brought us a great deal of business which has had the side effect of giving confidence to the fire resistant fibre, yarn and fabric industry and we have invested most of our profits this year into establishing new growth areas. Understanding what happens when materials are burnt and how best to combine materials, not only with good fire blocking properties, but with the use of low calorific value materials so that the burning or injury to people in fires can be significantly reduced has been advanced by the investigation of fire blocking seats during this last year. We have developed a shirt and blouse material which can be any colour, as light as 120 gms/sq mt which will withstand the thrust of a blow lamp for at least two minutes and has no tendency to melt or stick to the wearer and similarily for uniform fabrics and outer clothing of all types. On the back of the development of fire blocking layers we have been able to produce an F R thread as strong as the normal poly-cotton, with good char properties sufficient to prevent a seam from breaking open under flame. The novelty of this new thread is that whilst Kevlar has excellent F R properties it is always yellow. This new fibre mixture, on the other hand, allows us to achieve any colour desired, and whilst it has not got quite the char quality of Kevlar it will last as long as most of the fire blocking materials under flame. You will appreciate the most significant property of any material has been proved to be its price. We have fought hard against this but we now recognise that fire blocking capacity is expected to be just an additional property. A customer expects the normal abrasion, colour, drape, washability and durability of fabrics he normally uses and there will be no concession to the manufacturer of fire resistant materials.

We as manufacturers of oxydized acrylic fibre, sat on the fence for three years and said that the customers will recognise the qualities of our material and will beat a path to our door. As insolvency approached we became absolutely convinced that Henry Ford was wrong. Every colour but black will be asked for and unless you satisfy the customer there is little or no future. We are now convinced that the customer is always right.

* For Panox read Oxidised Polyacrylic Nitrile. Panox is the trade name of R K Textiles Limited.

"By kind permission of Ken Bouch, British Airways."

## OPERATIONAL COST OF FIRE BLOCKING PASSENGER SEATS IN ALL BRITISH AIRWAYS AIRCRAFT

| A/C type | No. Seats | No a/c per a/c | Wt inc per a/c kg | Cost of wt inc £/kg/yr | Cost per a/c £ | Cost per fleet £ |
|---|---|---|---|---|---|---|
| 1-11 (500) | 99Y | 21 | 59 | 37 | 2183 | 45843 |
| 1-11 (400) | 81Y | 5 | 49 | 37 | 1813 | 9065 |
| HS74B | 44Y | 5 | 26 | 18 | 468 | 2340 |
| B737 | 114Y | 35 | 68 | 26 | 1768 | 61880 |
| B737 (Airtours) | 130Y | 8 | 78 | 26 | 2028 | 16224 |
| B757 | 189Y | 15 | 113 | 17 | 192? | 28815 |
| B757 | 195Y | 9 | 117 | 17 | 1983 | 17847 |
| L1011-1 | 352Y | 3 | 211 | 14 | 2954 | 8862 |
| L1011 (Airtours) | 393Y | 4 | 236 | 14 | 3304 | 13216 |
| L1011-200/50 | 18F/62SC/139Y | 10 | 165 | 56 | 9240 | 92400 |
| L1011-500 | 12F/26SC/188Y | 2 | 157 | 39 | 6123 | 12246 |
| B747 | 18F/90SC/262Y | 28 | 267 | 57 | 15219 | 426132 |
| B747 (Combi) | 22F/92SC/220Y | 2 | 250 | 57 | 14250 | 28500 |
| Concorde | 100F | 7 | 100 | 65 | 6500 | 45500 |

YEARLY TOTAL COST FOR ALL AIRCRAFT    £808,870 to carry extra weight
£1,529,043 to replace fire
blocking

Weight increase for  fireblocking

First Class seat      1.1 kg per seat place
Super Club seat       1.0 kg per seat place
Economy and Club seat 0.6 kg per seat place
Concorde seat         1.0 kg per seat place

These are typical figures for each type of seat using Firotex 46538 and Panotex P503 on Economy and Super Club seats; Tex Tech XD192:26R on First Class seats; Tex Tech XD192:26R and Panotex P503 on Concorde seats.

Replacement costs assume fireblocking is replaced at every other seat refurbish, i.e. every 4 years. It is dry cleaned at other refurbish.

BIOGRAPHY

Trained at a local technical college; spent the first seven years in woollen carding and spinning. Studied on a part-time basis obtaining, the ATI as the age of 25 and then three years later completed the examinations for membership of the Institute of Production Engineers. Other areas of study from Leeds University extra mural courses were psychology, religion and philosophy and also fine art.

For the last 28 years I have been involved in all aspects of worsted processing with the exception of 1 year when I taught textiles full-time. Most of my time has been spent in industrial research and development and Universal Carbon Fibres is one of the projects which started as an investigation into fire resistant products and finished up as a separate company.

THE DESIGN OF FIBER REINFORCED COMPOSITE
SPRINGS FOR AN IMPACT LINE PRINTER

John W. Huffman
Hewlett-Packard Company, Boise, Idaho
Ronald F. Gibson
University of Idaho, Moscow, Idaho

## Abstract

This study describes the design of
fiber reinforced composite flexural
springs that functionally replace
stainless steel springs in an impact
line printer. A theoretical model
of the printing mechanism is
presented and used to illustrate the
process of designing the traditional
stainless steel flexural springs.
Using the procedure for homogeneous
steel springs as a framework, the
differences in design for fiber
reinforced composite springs are
shown and a computer program is used
to solve the equations of classical
lamination theory. Predicted
dynamic properties for composite
spring design agree well with
measured values and preliminary
fatigue testing shows adequate life.

## 1. INTRODUCTION

One very versatile means of
mechanically printing text and
graphics on paper is by the use of a
dot matrix type of character cell.
A printer which operates in this way
will form dots by impacting a small
wire or hammer behind an inked
ribbon onto paper that is backed by
a hard platen. Because of hammer
size limitations, a horizontal
linear array of hammers (assembled
into a "printbar") is moved side-to-
side in order to print a horizontal
line of connected dots. In this
way, each hammer prints only a small
portion of every horizontal line of

dots, and the total distance that
the printbar must move is
correspondingly small. To form
letters and graphics, the paper is
incremented vertically by one dot
row height during the time that the
printbar changes direction in its
back and forth cycle. Although
various velocity profiles for the
printbar motion can be used, the
motion is typically sinusoidal.

One common way for this motion to be
achieved is to design a spring-mass
system of the desired
natural frequency, where the
printbar is the mass supported on
stiff flexural springs. Present
successful designs for the printbar
"springs" call for a high strength,
precision machined and ground
stainless steel. This material is
necessary in order to achieve the
fatigue life and system natural
frequency tolerance needed for
energy efficient operation and
balance.

Although stainless steel springs
have performed very well in several
print mechanism designs, there are
some distinct disadvantages. The
most serious is that small material
flaws may be enough to allow a
catastrophic failure of the spring
in service, resulting in costly
repairs. Also, because of the many
process steps needed to produce high
quality steel springs, the part cost
is high, and the ability to rework
the material is very limited.

As a result of these difficulties, the use of continuous fiber reinforced composite materials as a direct functional replacement was considered. Although such a "direct replacement design" obviously does not allow the designer to take full advantage of the capabilities of composite materials, it is a way of gaining experience with composites at relatively low risk. The material chosen for this analysis was "Scotchply" fiberglass reinforced epoxy type 1002, manufactured by 3M Company.

The printing mechanism used in this work was a Hewlett-Packard 2564A line printer which prints at a speed of 600 lines of text per minute. Since any new composite spring design must directly replace the existing stainless steel springs, certain restrictions are needed to maintain compatibility. With this in mind, Section 2 presents the dynamic model of the printbar/spring system used to design steel springs.

There are certain differences in designing with anisotropic materials like fiber reinforced plastics, particularly in the estimation of overall laminate properties for use in design equations. Section 3 explains a way to estimate the flexural modulus of elasticity and Poisson's ratio for a symmetric assembly of laminae using a desktop computer program to solve the equations of classical lamination theory. The system model developed in Section 2 along with the calculated material properties for a composite spring then enable a suitable combination to be chosen for a workable design.

Section 4 presents a summary of experimental testing and concluding remarks. Although the experimental verification of the design procedures presented is adequate in many cases for illustrating the feasibility of a spring design, further work needs to be done to insure that possible new and unusual modes of failure are not overlooked.

## 2. STEEL SPRING DESIGN

The process of designing steel or composite springs falls into three distinct phases. First, the relevant printing process parameters are defined and the printbar excursion and natural frequency of operation are calculated. Second, a model of the print mechanism is used in order to allow the determination of the spring dimensions for the correct natural frequency and to predict bending stresses and fatigue life. And third, actual parts are made and tested to make sure that the overall print mechanism balance and flexural fatigue life are acceptable.

Because the print mechanism used in this study was a Hewlett-Packard 2564A line printer, several important printing parameters were established by definition. Basically, the printbar must be driven with a sinusoidal velocity profile horizontally back and forth at a frequency of 35.5 Hz and a peak to peak excursion of 6.354 mm (0.25 in). Since the mass of the printbar approaches 2 kg., a counterweight mass is provided and driven at the same frequency, but 180 degrees out of phase to help cancel the forces generated by the moving printbar. Each of these masses is supported on flexural springs designed to give a simple spring-mass system with a fundamental natural frequency of 35.5 Hz. The printbar/counterweight assembly is mounted in an aluminum die cast housing and is driven by a dc motor/crankshaft/connecting arm assembly similar in many ways to a typical internal combustion engine piston/crank arrangement.

Figure 1 shows a simplified cross section of the printbar and counterweight supported on springs, with the print mechanism casting represented as ground. In actual practice, the attachment points of the cantilever-type springs to the printbar or counterweight masses and to the print mechanism casting are not infinitely stiff and must be corrected for in the spring design process.

In formulating a useful dynamic model of the print mechanism assembly, the primary consideration is the ability to predict the fundamental natural frequencies of the two spring-mass systems. For optimum balance, the natural frequency of the printbar should be close to that of the counterweight, and both in turn will be close to the forcing frequency, 35.5 Hz. By driving the system at or near resonance in this way, the energy input is minimized.

If we assume that the two masses will be driven 180 degrees out of phase and that we can control their displacements, then the forces generated during turnaround will be minimized. Therefore, each system can be analyzed independently. Using a single-degree of freedom model to predict the system natural frequency:

$$f_n = \frac{1}{2\pi} \sqrt{\frac{k_{eff}}{m_{eff}}} \qquad (2.1)$$

where $k_{eff}$ is a spring stiffness which is corrected for mounting losses, and $m_{eff}$ is the total effective mass including contributions from the mass of the springs.

Referring to Fig. 2, we see a free body diagram of a typical spring in its deformed state, with the following definitions:

x,y — the axes that describe the undeformed shape of the spring
W — the applied force
a — the distance of W from the y-axis
L — the unsupported or "active" length
h — the thickness of the spring
Ra,Rb — forces at the clamped ends
Ma,Mb — moments at the clamped ends
$\theta$a,$\theta$b — slopes at the clamped ends

This general configuration is described in detail by Roark and Young [1], and after some simplification the equation for the deflection, y(x) is

$$y(x) = \frac{W}{2EI} \left[ \frac{Lx^2}{2} - \frac{L^3}{6} - \frac{x^3}{3} \right] \qquad (2.2)$$

where $I = bh^3/12$ is the moment of inertia of the cross sectional area of the spring with respect to the neutral axis and E is Young's modulus for the material.

Since the flexural spring rate is the force per unit deflection, if we divide the applied force W, by the deflection, y(o) from Eq. (2.2), we get the stiffness of the equivalent beam element. Since the spring is more like a plate than a beam, from Den Hartog's analysis of the bending of flat plates [2], the stiffness of a plate will be higher than that predicted by beam theory by a factor of $1/(1-v^2)$, where $v$ is Poisson's ratio for the material of the plate. Thus, the spring stiffness is

$$k = \frac{12EI}{L^3(1-v^2)} \qquad (2.3)$$

There are several aspects of the spring design which contribute to the inaccuracy of Eq. (2.3). As mentioned earlier, the stiffness at the mounting points of the springs is not infinite, and there will be a rotation associated with the end moments. However, experience has shown that the effective stiffness, $k_{eff}$, can be approximated by a constant, C, multiplied by the theoretical spring stiffness, k.

$$k_{eff} = C k \qquad (2.4)$$

With the spring supports, casting, and associated mounting hardware for this print mechanism, $k_{eff}$ for the printbar is approximately 0.760(k) and $k_{eff}$ for the counterweight is approximately 0.736(k).

In contrast to the development of Eq. (2.4) for $k_{eff}$, the effective mass of the system, $m_{eff}$ is simply the actual mass of the printbar or counterweight, associated screws, clamps, cables, transducer elements and so on, added together, plus a contribution from the mass of the springs themselves. In general then

$$m_{eff} = m + m_{eq} \qquad (2.5)$$

where m is the combined mass of the printbar or counterweight not including the springs, and $m_{eq}$ is a fraction of the spring mass. The quantity $m_{eq}$ can be easily calculated from energy methods as described by Thompson [3], and for this configuration is found to be

$$m_{eq} = (2) \frac{13}{35} m_f \qquad (2.6)$$

where $m_f$ is the mass of the spring, and the quantity 2 is added since two springs are always used to support each mass.

Combining the equations for $f_n$, $m_{eff}$ and $k_{eff}$, we get an expression for the natural frequency in terms of the spring dimensions and material properties

$$f_n = \frac{1}{2\pi} \sqrt{\frac{2CEbh^3}{(m+m_{eq})L^3(1-v^2)}} \qquad (2.7)$$

where a factor of 2 has been introduced in the stiffness to account for the two parallel springs that support each mass.

Once the thickness of the spring is fixed, the maximum stress at the root can be calculated and used to estimate fatigue life. Using the familiar equation for bending $\sigma = My/I$, and substituting appropriate values for this spring configuration, the maximum fiber stress is

$$\sigma = \frac{3y_a \, h \, E}{L^2} \qquad (2.8)$$

where $y_a$ is the deflection of the end of the spring.

The stress value that we obtain from Eq. (2.8) will always overestimate the actual stress since $\theta a$ and $\theta b$ are not equal to zero. Strain gage measurements taken near the root of the springs were very dependent on the geometry of the clamp used to hold the spring as well as the tightening torque of the screws holding the assembly together. In addition, finite element modeling of the clamp area indicated that the clamp geometry could cause as much as a 15 percent difference in stress on opposite sides of the spring at the root. As a general guideline, however, it is convenient to estimate the maximum stress as that calculated in Eq. (2.8) multiplied by the factor C, measured earlier.

3. COMPOSITE SPRING DESIGN

As with the process of designing steel springs, the retrofit design of composite springs will use the same print mechanism model developed in Section 2, but there will be differences in how the spring dimensions in Eq. (2.7) and spring stresses in Eq. (2.8) are calculated.

The first major difference encountered in designing with a conventional material such as steel and an inhomogeneous, anisotropic material like a fiber reinforced composite, is in the estimation of the flexural modulus, $E_f$. For an isotropic material like steel, Young's modulus is the same regardless of position, and Young's modulus is the same as the flexural modulus. For a layered composite, on the other hand, the overall modulus may be different for any given frame of reference and so will be unique to each different sequence of ply orientations. In addition, the flexural modulus is not the same as Young's modulus.

There is also a difference in Poisson's ratio, $v$, for the laminate. Although each lamina will have a particular value for $v$, the laminate will have a different overall value.

Assuming that $E_f$ and $\upsilon$ are known, there is still no simple way to compute the maximum stress as in Eq. (2.8). This is because the maximum stress will not vary linearly through the cross section of the spring, but will depend on the particular stacking sequence of laminae selected. Also complicating the analysis is the fact that the laminated spring can only be fabricated in certain increments of thickness, corresponding to the thickness of an individual lamina.

In order to provide a way of analyzing potential laminated plate configurations for use as springs, a desktop computer program "Composite" was written [4]. The program solves the equations of classical lamination theory. A complete description of the equations is presented by Jones [5] and it is his nomenclature that is used in this work.

In general, the resultant forces, N, and moments, M, on a laminate are related to the middle surface strains, $\epsilon^{\circ}$, and the curvatures $\kappa$, by the following set of equations

$$\left\{ \begin{array}{c} N \\ \hline M \end{array} \right\} = \left[ \begin{array}{c|c} A & B \\ \hline B & D \end{array} \right] \left\{ \begin{array}{c} \epsilon^{\circ} \\ \hline \kappa \end{array} \right\} \quad (3.1)$$

The $A_{ij}$ terms are known as the extensional stiffnesses, the $B_{ij}$ terms are known as the coupling stiffnesses and the $D_{ij}$ terms are known as the bending stiffnesses.

Inversion of Eq. (3.1) allows us to solve for a set of laminate strains and curvatures, given an initial set of applied forces and moments per unit length. Thus,

$$\left\{ \begin{array}{c} \epsilon^{\circ} \\ \hline \kappa \end{array} \right\} = \left[ \begin{array}{c|c} a & b \\ \hline c & d \end{array} \right] \left\{ \begin{array}{c} N \\ \hline M \end{array} \right\} \quad (3.2)$$

where the $a_{ij}$, $b_{ij}$, $c_{ij}$ and $d_{ij}$ form the inverted coefficient matrix. The stresses in the $n^{th}$ lamina are then

$$\{\sigma\}^n = [\bar{Q}]^n (\{\epsilon^{\circ}\} + z\{\kappa\}) \quad (3.3)$$

where the $\bar{Q}_{ij}^n$ are the lamina stiffnesses, and z is the distance from the laminate middle surface [5].

The laminate analysis equations can also be used to find effective engineering constants such as the flexural modulus and Poisson's ratio for use in the spring design equations of Section 2. Since the effective flexural modulus and Poisson's ratio for the laminate depend upon the stacking sequence and number of plies, as well as the ply properties, both must be calculated for each individual laminate. First we use the "Composite" program to calculate the stiffness matrix for a symmetric laminate. From Tsai [6] we know that for a symmetric laminate, $B_{ij} = 0$, and that

$$E_f = \frac{12}{h^3 d_{11}} \quad (3.4)$$

$$\upsilon = - \frac{d_{21}}{d_{11}} \quad (3.5)$$

where the [d] matrix is the inverse of the bending stiffness matrix [D].

Figures 3 and 4 show the results for Scotchply 1002 crossply and spring orientation laminates. The spring orientation material consists of a typical stacking sequence of $[0/90/0/0/0/...]_s$ where 10 to 15 percent of the plies under the surface are oriented at 90 degrees to the direction of bending. Crossply laminates having an even number of plies are not symmetric and are not presented as design alternatives.

The final expression for the natural frequency was given by Eq. (2.7).

After substituting the estimated values for the flexural modulus and Poisson's ratio for the composite into this equation, and assuming

1364

that the effective mass of the spring is small in comparison with the printbar mass, the following relationship emerges

$$\frac{E_f h^3}{1-\upsilon^2} = A \qquad (3.6)$$

where,

$$A = \frac{2 f_n^2 \pi^2 L^3 m}{Cb} \qquad (3.7)$$

Note that the parameters on the right hand side of Eq. (3.7) are fixed for a particular printbar. Thus, A is fixed and the parameters $E_f$, $\upsilon$ and h are selected so as to satisfy Eq. (3.6). These equations outline a design procedure for composite springs. First, assume a trial thickness, h, and stacking sequence and estimate $E_f$ and $\upsilon$ using the "Composite" program. Next, calculate the constant A according to Eq. (3.6), and compare with the required value of A from Eq. (3.7). If a close match is not obtained, then choose a new thickness or stacking sequence and try again. Once an acceptable fit is made according to Eq. (3.6), then use "Composite" to solve for the lamina stresses and strains in the composite spring, and make a judgement as to the resulting fatigue life. Fatigue life tests can then be used to verify the design.

## 4. EXPERIMENTAL TESTING

The ability to estimate effective material properties and predict the system natural frequency was experimentally tested with two samples of fiberglass reinforced epoxy material fabricated into printbar springs and two samples fabricated into counterweight springs. The procedure outlined in Section 3 for both sets of springs yielded value of $f_n = 33.7$ Hz for the printbar, and $f_n = 37.2$ Hz for the counterweight. The descriptions of the spring systems are given in Table 1.

The experimental procedure for measuring the natural frequency of the printbar spring or counterweight spring system involved the use of an HP 5423A structural dynamics analyzer. First, the printbar, springs and print mechanism casting were assembled and mounted on an aluminum fixture which was in turn bolted to a granite flat table for stability. An accelerometer was placed on the printbar and energy supplied to the system by impact with a rubber nosed hammer. The output from the accelerometer was used by the structural dynamics analyzer to present the autospectrum response. From this data, the fundamental natural frequency of the system can be calculated by a single-degree of freedom curve fitting routine available in the analyzer, or alternately by visual examination of the curve.

The actual measured fundamental natural frequency for the printbar was 33.4 hz, and for the counterweight was 37.3 Hz. With this and other similar tests the agreement between the predicted values of the fundamental natural frequencies and the actual measured natural frequencies was close enough for preliminary design specification. These results suggest that the design procedure of Section 3 can be effectively used to predict ply combinations that will ultimately yield a spring of the right overall stiffness for the desired operating frequency. The next step in the design process is to estimate the fatigue life of the material and judge if it is adequate by virtue of testing and/or theory.

The fatigue testing performed with composite springs consisted of basically a survival test of four samples of [0/90/0/90/0/90] Scotchply 1002 crossply material. The four springs were assembled into a print mechanism of slightly different configuration, and earlier vintage, than previously described. As far as the spring material itself is concerned, the design and analysis procedures of Section 3 still apply, with the fundamental natural frequency of the spring/printbar system being approximately 18.5 Hz.

The print mechanism was run continuously for over 270 million cycles. The maximum strain amplitude was estimated to be 0.0023 in the 90 degree plies, which had a failure strain of 0.0029. Periodic measurements were taken of the system natural frequency and damping, and recorded along with the number of cycles on the springs. Damping measurements were taken with the structural dynamics analyzer, but using two different methods. Initially, the log-decay technique was used on the time record response of the accelerometer mounted on the printbar, and for additional comparison purposes, a damping value was calculated from the autospectrum response data by the bandwidth method. The purpose of monitoring damping values during the life of the fatigue test was to look for any increase that could signal structural damage to the flexural material.

After 270 million cycles the fatigue cycling was discontinued, with no failures of any of the composite springs. In this sense the crossply material showed adequate fatigue resistance in a closely simulated printing mechanism. The values of the natural frequency and damping through the life of the test are presented in Figs. 5 and 6, respectively.

From Fig. 5 we can see that the natural frequency started to decrease after about 100 million cycles and dropped through the remainder of the test. From Fig. 5 the damping values (bandwidth method) remained roughly constant. Examination of the spring material at the end of the cycle showed virtually no visible changes other than a slightly abraded surface of the spring under the clamping area at each end of the part.

Obviously, this preliminary type of fatigue testing is not rigorous enough to provide really meaningful data for application to other ply configurations. In addition, before actual field application, longer term viscoelastic and environmental effects need to be investigated. However, it does indicate the potential serviceability of materials of this type, and further testing is planned.

5. CONCLUSIONS

The following list summarizes the important findings and conclusions of this investigation:

A.   The printbar/spring and counterweight/spring system can accurately be represented individually by a simple single-degree of freedom model. Consideration must be taken into account for the mounting stiffness of the spring supports and the effective mass of the printbar or counterweight.

B.   The same general design process used for steel springs can be extended to composite springs by using laminate analysis to estimate the effective flexural modulus and Poisson's ratio for the composite.

C.   Preliminary experimental testing of Scotchply 1002 composite for direct replacement of steel as a spring material showed not only the ability to specify a ply configuration likely to have the required stiffness, but also showed excellent fatigue life.

Details of the investigation are given in the thesis by Huffman [7].

6. REFERENCES

[1]   Roark, R.J. and Young, W.C., Formulas for Stress and Strain, McGraw-Hill, Inc., N.Y., 1975, p. 96.

[2]   Den Hartog, J.P., Advanced Strength of Materials, McGraw-Hill, Inc., N.Y., 1953, p. 113.

[3]   Thompson, W.T., Theory of Vibration with Applications, Prentice Hall, Inc., Englewood Cliffs, N.J., 1972, pp. 17-20.

[4] Huffman, J.W., "Composite Materials Analysis Program," (229004), Hewlett-Packard Co., 1000 NE Circle Blvd., Corvallis, OR.

[5] Jones, R.M., Mechanics of Composite Materials, Scripta Book Co., Washington D.C., 1975, pp. 147-56.

[6] Tsai, S.W. and Hahn, H.T., Introduction to Composite Materials, Technomic Pub. Co., 1980, pp. 176-77.

[7] Huffman, J.W., The Design of Fiber Reinforced Composite Springs for An Impact Line Printer, M.S. Thesis, University of Idaho, 1985.

7. BIOGRAPHIES

Born in Panama, John W. Huffman studied Mechanical Engineering at Oklahoma State University (BS 1975), and the University of Idaho (MS 1986). As a Development Engineer with Hewlett-Packard since 1979, John works on the mechanical and acoustic design of high speed dot matrix and laser printers. Before coming to Hewlett-Packard, he worked at Halliburton Services designing well logging instrumentation. John and his wife Helen live in Meridian, Idaho and enjoy raising Arabian horses, camping and astronomy.

Ronald F. Gibson received his BS degree in Mechanical Engineering from the University of Florida in 1965, and was employed as a Development Engineer by the Nuclear Division of Union Carbide Corporation. He received a MSME degree from the University of Tennessee in 1971, and a PhD degree in Mechanics from the University of Minnesota in 1975. Since that time, he has served on the faculties of Iowa State University, the University of Idaho, and the University of Florida. He is currently a Professor of Mechanical Engineering at the University of Idaho, where he teaches courses in applied mechanics and materials, and conducts research in dynamic behavior of advanced materials.

Table 1

Description of Composite Spring System

| Property | Printbar | Counterweight |
|---|---|---|
| Spring thickness, h(m) | 0.00381 | 0.00381 |
| Spring width, b(m) | 0.051 | 0.051 |
| Spring length, L(m) | 0.118 | 0.089 |
| Spring material | E-glass/epoxy* | E-glass/epoxy* |
| Number of plies | 15 | 15 |
| Stacking sequence | $[0/90/0_5/\bar{0}]_s$ | $[0/90/0_5/\bar{0}]_s$ |
| Mass, m(kg) | 1.8408 | 3.3948 |
| Calculated flexural modulus, $E_f$(GPa) | 31.0 | 31.0 |
| Calculated Poisson's Ratio, $\nu$ | 0.134 | 0.134 |
| Calculated natural frequency, $f_n$(Hz) | 33.7 | 37.2 |
| Measured natural frequency, $f_n$(Hz) | 33.4 | 37.3 |

* Scotchply 1002, 3M Company, St. Paul, Minnesota

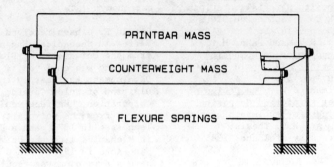

Figure 1. Simplified spring-mass
printer configuration

PRINTBAR MASS

COUNTERWEIGHT MASS

FLEXURE SPRINGS ⟶

Figure 2. Free body diagram of spring

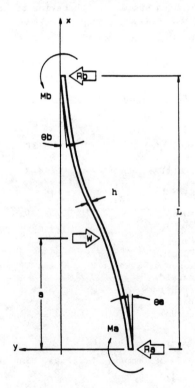

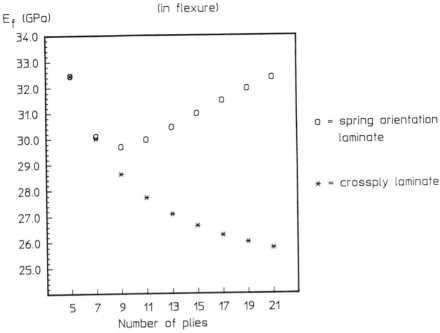

Figure 3. Effective Young's modulus (in flexure)

o = spring orientation laminate

* = crossply laminate

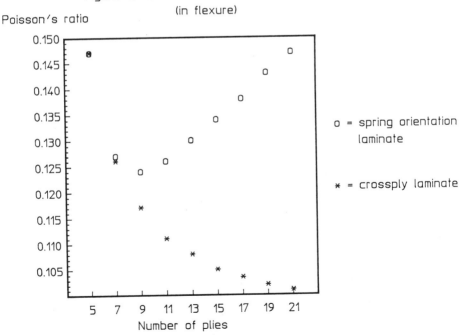

Figure 4. Effective Poisson's ratio (in flexure)

o = spring orientation laminate

* = crossply laminate

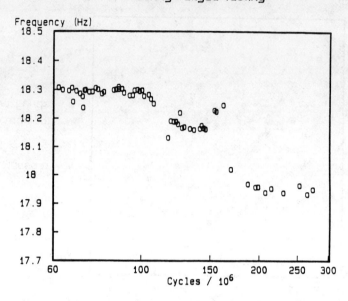

Figure 5. Printbar natural frequency during fatigue testing

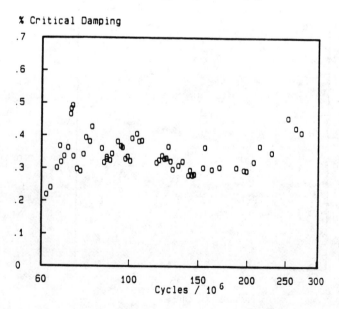

Figure 6. Printbar system damping during fatigue testing

ADHESION AND CHARACTERIZATION OF
PREPREG PROCESSABLE POLYIMIDES

Steven B. Driscoll and Thomas C. Walton
Department of Plastics Engineering
University of Lowell
Lowell, Massachusetts

ABSTRACT

Polyimides have gained wide acceptance for use in many aerospace
composite, electrical, and industrial
applications. The intent of this
work is to share with the reader
practical knowledge of how some of
the currently available commercial
systems perform.
Several prepreg processable polyimide
systems were evaluated for adhesive
properties and characterized with
the use of SEM, TGA, DSC, TMA,
Dynamic Spectroscopy, and Force vs.
Time Electronic Impact Analyses for
comparison.
The chemistry and nature of these
resin systems is reviewed, including
several BMIs (new hot melts examined),
Amide-Imides (AI) and a Thermoplastic
Polyimide (TPI). PMR-15 and a high
temperature epoxy resin are included
for comparison of high temperature
properties.
Of major consideration is the
processing characteristics and
thermal-oxidative stability of each
hot melt system or, (for those
systems which lack low stable melt
viscosities), solutions in N-methyl
Pyrrolidone, DMF or methanol.

1. INTRODUCTION

Although polyimides (PI) have outstanding high-temperature properties,
they have been slow to move from
the laboratory to the marketplace.
Processing and mechanical-
property balance balance problems
related to the same molecular
structures[1,2] that gave them
their high thermal stability have
plagued this class of polymers.
However, in the last couple of
years, PIs have undergone a
renaissance into a wide range of
materials, which have struck winning
balances in their processing and
physical properties. This work
attempts to differentiate the
sucessful materials from those that
will never be more than laboratory
curiosities. It is not sufficient
to produce an occasional high
quality molded part through the
highly skilled efforts of scientists
and technicians. Production parts
are made on a routine basis by
average factory employees, therefore
the process must be well understood
and under control.

## COMPARISON FOR THIS STUDY

The polyimides chosen for comparison of thermal, adhesive, mechanical and processing properties in this work are as follows:

### 1.1 ADDITION TYPE POLYIMIDE (API) (PMR-15) [3]

This type PI involves an initial condensation reaction, which yields a low molecular weight PI that melts and flows easily at low temperature allowing release of condensation volatiles. Next, an addition type free radical polymerization occurs, which builds the molecular weight usually to a highly cross-linked cured matrix. However, shrinkage occurs during cross-linking, the cured matrix is more brittle and has lower thermal oxidative stability than the typical Condensation PI(CPI), (eg. Dupont's Avimid N-old NR150B2-CPI-TPI), due to the lower content of resonance stabilized (aromatic) structures or more non-aromatic hydrogen.

● PMR-15 (NASA-Lewis) Most sucessful 2nd generation API-after P13-N

● Use of Pure, well Characterized, monomer reactants which simplify material and process control

● Low boiling solvents(eg. Methanol) can be used to reduce the problems associated with volatile removal during cure

● Monomeric prepreg has better tack and drape than original P13-N system

● Needs high temperatures and pressures to process ($315^{o}$C and 1.03MPa (150 psi))

### 1.2.0 API BISMALEIMIDES (BMIs)

By far the fastest growing area of new PI matrix materials is that of BMIs. New products and applications appear with astonishing frequency and new gains in performance are continually demonstrated.

### 1.2.1 KERIMID 601 (BMI) [4]

● Mfgd. by Rhone-Poulenc (France)

● Introduced around 1970, forerunner of modern BMIs

● Processes like epoxy ($175^{o}$C, autoclave cure, good tack and drape, post cured at 200°C to high $T_g$)

● Has tendency to microcrack

● DMA, DMF, and NMP solvents are hard to remove, which leads to blistering, delamination, and reduced elevated temperature performance

● Rhone Poulenc now has new melt processable BMI resins (Based on low M.W. Kerimid 601 plus catalyst)

### 1.2.2 COMPIMIDE FAMILY OF BMIs [5,6,7,8] (Boots-Technochemie)

● Major supplier of toughened BMIs and BMI building block raw materials (453 and 896 studied here)

### 1.2.3 MATRIMID 5292 BMI [9] (Ciba Geigy)

● Two component hot melt which is easy to process

● Currently devoting major research into a new generation of toughened BMIs.

### 1.2.4 BTL FAMILY OF BMIs (formerly Reichhold Chemicals)

● BTL can custom tailor resins for a balance of toughness and thermal-oxidative stability

### 1.2.5 UNIVERSITY OF DAYTON RESEARCH INSTITUTE (UDRI) BMIs [10]

● UDRI is looking for improved processability, thermal and mechanical performance, a low dielectric constant and no susceptability to moisture

In general, however, the BMIs do not provide the same thermal-oxidative stability as other polyimides. New laboratory innovations, provided they can be scaled up to production level, are expected to show notable improvement. BMI formulations are preferred to epoxies when high temperature resistance, good hot/wet environmental stability, and improved

Fire, Smoke and Toxicity (FST) properties must be combined in one material. For example, in aviation interiors, FST requirements favor BMI resins for floor and wall sandwich panels.

Their low Coefficients of thermal expansion (CTE), high dimensional stability, and resistance to chemical attack, radiation and flame have made them widely used in aerospace composites and molded multilayer printed-circuit boards[11]. Except for the new hot melt processable BMIs, they are limited to flat planes or simple laminates due to prepreg-solvent retention problems.

## 1.3.0 THERMOPLASTIC POLYIMIDES(TPIs)

Most TPIs are linear, fully reacted polymers which have true thermoplastic characteristics. The final processing step is simply melt, flow and solidification. Ideally no chemistry is involved, only physical change. This allows bubbles, porosity, delaminations, and other flaws to be eliminated.

## 1.3.1 AMOCO'S TORLON AI AND 4000 SERIES AMIDE IMIDE (TPIs)[12]

● Non- reactive and reactive TPI's respectively (4000 series achieves high temperature properties only after long, slow post cure)

## 1.3.2 MATRIMID 5218 TPI[13]
(Ciba Geigy)

● Soluble thermoplastic polyimide for solution coatings

## 2. EXPERMENTAL

### 2.1 Tensile Shear Strength

Adhesive joint strength of materials shown in Table 1 were compared at 25 and 250°C using the overlap shear test as outlined by ASTM D1002-72 for aluminum to aluminum (Alloy 2024, Temper T-3).

The aluminum specimens were degreased with MEK, then acid etched in a 2.44% sodium dichromate/ 24.4% sulfuric acid bath at 70°C for 10 minutes, sprayed with deionized water then air dried. (SEM photos

before and after etching in Figure 1 shows a typical treated (pitted) aluminum surface from the etching solutions treatment which demonstrates the importance of proper control of etching time, temperature and degree of mixing for maintaining even surface pitting from strip to strip.

### 2.2 Thermal-Oxidative Stability

A comparison of the thermal-oxidative stabilities of different polyimide resins was accomplished using a Dupont 1090 Thermal Analysis System in the TGA mode. Each 10-20 mg cured sample was scanned from 25 to 620°C at a rate of 10°C/min in 100 ml/min air.

### 2.3 Differiential Scanning Calorimetry (DSC)

Neat resin DSC studies were carried out using the Dupont 1090 Thermal Analysis System in the DSC mode. All samples had heat-up rates of 10°C/min in the open configuration exposed to 50 ml/min air. Sample sizes ranged from 10-20 milligrams.

### 2.4 Thermal Mechanical Properties (TMA)

Thermal Mechanical Properties were measured and compared using the Dupont Thermal Analysis System in the TMA mode. All cast cured neat resin samples were 6.35mm (¼ inch) cubes heated from 25 to 320°C at 10°C/min. Proceedures for checking $T_g$ using dual run, back to back TMA scans, (of the same sample), which has a tendency to Post Cure the sample was avoided.

### 2.5 Dynamic Impact Testing

Impact testing was performed on 6.35mm (¼ inch ± 0.001) thick by 62.2mm (2.45 inches) OD cast samples using ASTM D3763 and the new state-of-the-art Rheometrics RDT 5000, 50 lb. (electronic) dynamic dart tester. All samples were centered and clamped over a 38.1mm (1.5 inches) I.D. support collar. The dart speed was 3386.mm/sec (8000. inches/min) or 7.6 miles per hour and used for all samples tested since this is the Izod impact test speed.

## 2.6 Dynamic Mechanical Spectroscopy

A 0.6 to 0.8 millimeter resin gap was used on 25 millimeter parallel plates for dynamic mechanical spectroscopy measurements[14,17] done on a Rheometrics RDSII in which Eta* (complex viscosity) vs. temperature scans of the uncured resins and G' (storage modulus) vs. temperature measurements on cured rectangular specimens (Figure 7) were carried out for comparison of rheology during cure and modulus vs. temperature for the hot melt BMI type resins. ASTM standard D4065-82 was followed for all runs . The cured rectangular specimens used in G' vs. temperature testing were 2"Lx ½"Wx 1/10"THK.

## 3. RESULTS AND DISCUSSION

### 3.1 Adhesive Properties

A comparison of the adhesive properties as shown in Table 3 show a wide range of bond strengths.
Performing best at 25°C were the Amide -Imide/NMP solutions followed by the high temperature Epoxy and the polyimides show up lower in the scale.
All the solvent base systems (except the highly volatile solvent used in PMR-15), failed miserably at 250°C were PMR-15, Compimide 896, Matrimid 5292, BTL 94-606, and the other hot melt BMI's held their strengths quite well.
Although Amoco's 4000 TF/NMP performed best at 25°C, it melted and lost all strength at 250°C.
The PMR-15 system pre- processed under vacuum performed to a higher degree than that not pre-processed under low pressure conditions.

### 3.2 Thermal-Oxidative Stability

Table 4 gives a summary of how the cured resins compare by TGA 5% and 10% weight loss temperatures.
performing at the top, as shown in Figure 2, as expected is PMR-15.
very close to this performance was Matrimid 5218 (non-processed, eg. no solvent used). Compimide 896, BTL 94-606, and Matrimid 5292 also performed particularly well.
Most of the NMP and DMF solvent base systems performed poorly at the 5% loss point, but the Amoco/ NMP resins held-up fairly well at their 10% loss point. The Matrimid 5218/NMP processed system did very poorly even though the neat resin TGA took the top of the list. Similar performance was noted for Kerimid 601 with and without NMP but at lower temperatures.

### 3.3 Differential Scanning Calorimetry (DSC)

Figure 4 shows some typical endotherm-exotherm behavior for the various hot melt BMI resins. In Table 4 PMR-15 as expected showed a high exotherm temperature $T_{max}$) with Compimide 453, BTL 94-606, Matrimid 5292, and Compimide 896 showing $T_{max}$ between 250° and 275°C. BTL 94-606 has a high exotherm indicating that a gradual cure was the right choice as was recommended by the vendor.
The Matrimid 5292 peak shows quite a difference from that bimodal curve reported in the literature.[9,14] This is due to the prolonged (2½ hrs @ 100°C) aging of the hot melt sample during processing vs. the fresh hot melt sample used for Ciba Geigy's DSC scan. The aged resin shows the disappearance of the first peak's associated reaction (eg. ENE -addition of allyl group to the Maleimide double bond).

### 3.4 Thermal-Mechanical Analysis (TMA)

A typical TMA scan is shown in Figure 3. Coefficients of thermal expansion (CTE) for the resins tested are typical for Polyimides (eg. 40-50 microinches/inch°C). No signs were found in these resins tested of an extremely low CTE Polyimide, as was developed by Hitachi Ltd.[16]
Second scans of each sample by the DMA were avoided because of the post curing effect on the sample by the first scan.

### 3.5 Dynamic Impact Testing (RDT 5000)

Table 6 summarizes the impact data from plots like that shown in Figure 5 for those resins that are amendable to easy casting.
This is one of the best impact tests in which a pile of data can be collected from a single sample

(especially when samples are limited) The Matrimid 5292 and Compimide 896 performed well, whereas the other samples didn't perform better than the high temperature epoxy. This method, run at the Izod velocity of 7.6 miles per hour is much more practical than the Izod single data point test. The only draw-back is it's lack of notch sensitivity, but this problem can be overcome by incorporating the proper sample holder for rectangular lever type notched samples rather than flat plates.

It is interesting to note that a 1/8 inch thick Polycarbonate sample[14], tested for comparison, deformed but did not break in brittle failure like the other samples and absorbed an order of magnitude greater energy during deformation after impact.

This test method will surely be one to watch data from in the future.

3.6 Dynamic Mechanical Spectroscopy

Figure 6 and 7 demonstrate the usefulness of RDS II data in optimizing the hot melt processing temperature or post cure schedule by examining either the viscosity or the elastic modulus(G') vs temperature for the raw uncured resin or step cured rectangular specimens respectively.

This technique helps designers get the most desired properties from their resin through a deeper understanding of the cure rheology. A more exhaustive explanation of this technique[17] and analysis of these resins[15] is given in the Literature.

## 4. CONCLUSIONS

The preceeding work reviewed the processability and performance of comercially available polyimide resin systems. The following conclusions were drawn from this work:

• ASTM D1002-72 Overlap Shear Test results are highly dependent on the method of substrate preparation, (eg. the substrates original roughness, time, temperature, degree of mixing and the nature of the etching solution used), which is not completely specified in the ASTM test.(fig.1)

• NMP and DMF solvent processing affects the high temperature strength of the polymer matrix (Table 3)

• Several processable hot melt BMIs were found to have excellent thermal oxidative stability when compared to the highest temperature PMDA/Epoxy (Table 4)

• Several of these new BMIs have very good hot melt prepreg processability with good tack and drape. High melt viscosity resin hampers processing because of restricted flow and difficulty in metering resin from prepreg to the desired resin content. (Table 5)

• The Falling Dart Dynamic Electronic Impact Tester (RDT 5000) gives excellent comparative impact data for dynamic toughness testing. Use of this data is highly recommended in addition to $G_{IC}$ to get a better overall picture of toughness. (Table 6)

• Dynamic Mechanical Spectroscopy has sucessfully been used to compare hot melt processing characteristics and cured polymer mechanical properties over a wide range of temperatures. (figs. 6 and 7)

## ACKNOWLEDGEMENTS

The research and development staffs at Amicon Inc., Rheometrics, Inc. and Chomerics Inc. are greatfully acknowledged for their experienced advice in interpreting; TGA, DSC, TMA, RDS, RDT and Scanning Electron Micrograph data.
The authors are grateful for the inspiration and suggestions provided by Dr. Justin C. Bolger, without his help this work would not have been possible.

## REFERENCES

1) J.C. Bolger, "Adhesives in Manufacturing", edited by G.L. Schneberger Marcel Dekker, Inc., New York, 1983, Pg. 135

2) R.D. Deanin, "Polymer Structure Properties and Applications", Cahners Publishing Co., Boston, 1972, Pgs.-456,457

3) T.T. Serafini, "Handbook of Composites", edited by George Lubin, Van Nostrand Reinhold Co., New York, (1982), Pgs. 89-114

4) C.L. Leung, et al., 28th National SAMPE SYMP., 28, p. 818, (1983)

5) C.L. Segal, H.D. Stenzenberger, et al., 17th SAMPE Technical Conf., 17, p. 147, (1985)

6) H.D. Stenzenberger, et al., 30th National SAMPE SYMP., 30, p. 1568, (1985)

7) H.D. Stenzenberger, et al., 31th International SAMPE SYMP., 31 , p. 920. (1986)

8) H.D. Stenzenberger, et al., 18th International SAMPE Technical Conf., 18, p. 500, (1986)

9) M. Chaudhari, et al., SAMPE JOURNAL,21(4), pgs. 17-21, July / August (1985)

10) J.A. Harvey, R.P. Chartoff et al., 18th International SAMPE Technical Conf., 18, p. 705, (1986)

11) C. Guiles, PC FAB, 28 (6), pgs. 47-59, June, (1985)

12) B. Cole, 30th National SAMPE SYMP., 30, p. 799, (1985)

13) C.A. Cobuzzi and M.A. Chaudhari, 17th National SAMPE Technical Conf., 17, p. 318, (1985)

14) T.C. Walton, "Adhesion and Characterization of Currently Available Prepreg Processable Polyimides", Master's Thesis, University of Lowell, Lowell, Massachusetts, pgs. 42-87, December, (1986)

15) ibid, pgs. (28-32)

16) S. Numata, et al., "Proceedings of the Second International Conf. on Polyimides", Society of Plastics Engineers, Gen. Chairman: Dr. M.R. Gupta, IBM East Fishkill, N.Y., p. 492, (1985)

17) S.B. Driscoll, "Polymeric Material Systems,Properties and Performance", Published by S.B. Driscoll, Dept. of Plastics Engineering, University of Lowell, Lowell, Massachusetts, (1981), pgs. II-116 to II-139

## BIOGRAPHY

STEVEN B. DRISCOLL
University of Lowell
Lowell, Massachusetts
Professor of Plastics Engineering at the University and Chemical / Marketing Consultant for Rheometrics, Inc. of Piscataway, N.J.
He received his M.S. in Plastics Engineering from the University of Lowell.


THOMAS C. WALTON
Emerson & Cuming Inc.
Woburn, Massachusetts
He is currently a New Product Development Engineer with the Polymer and Electronic Material Group. He received his B.S. Degree in Chemical Engineering from Notheastern University, Boston, and is a candidate for the degree of M.S. in Plastics Engineering at the University of Lowell.

TABLE 1

## RESIN INFORMATION

| RESIN TRADE NAME | MANUFACTURER | COMPOSITION | CURED SPECIFIC GRAVITY | % SOLIDS USED IN PROCESSING | COST [c] $/LB. |
|---|---|---|---|---|---|
| Kerimid 601 In NMP[a] | Rhone-Poulenec | (7% MDA) | 1.95 | 61% | $16.73 |
| Bisphenol A Bismaleimide (BPA-BMI) | Univ. of Dayton | AROMATIC ETHER BMI | 1.27 | 100% Hot Melt | NA |
| Compimide 896 | Boots – Technochemie | Proprietary With Reactive Liquid Toughening Agent | 1.30 | 100% Hot Melt | $36.29 [d] |
| Compimide 453 | Boots – Technochemie | Proprietary With 33% CTBN – Rubber In Eutectic BMI Mixture (Comp. 353) | 1.19 | 100% Hot Melt | $31.75 [d] |
| IM-AD 94-606 | BTL Specialty Resins | Proprietary Modified BMI | | 100% Hot Melt | $25-30 |
| Matrimid 5292 | CIBA – GEIGY | (1 Mole) (.87 Mole) | 1.23 | 100% Hot Melt | $25 |
| PMR-15 (CPI-2237) | FERRO | $H_2N$—◯—$CH_2$—◯—$NH_2$ + MDA  3.087 M  BTDE 2.087 M  NE 2 M | 1.323 | 50% Solution In Methanol | $70 |
| Epoxy/PMDA (IN MEK) | Emerson & Cuming | Proprietary | – | 75% | $13.90 |
| Eccobond 104 (2 Part Epoxy | Emerson & Cuming | Proprietary | 1.36 | 100% | $13.90 |
| AI-10 In NMP[a] (Amide – Imide) | Amoco | (Fully Polymerized) (BYPRODUCT OF TRIMELLITIC ANHYDRIDE AND AROMATIC DIAMINE) | – | 40% | $12.38 |
| 4000 TF In NMP[a] (Amide – Imide) | Amoco (Torlon) | (With Reative Oligomers) (BYPRODUCT OF TRIMELLITIC ANHYDRIDE AND AROMATIC DIAMINE) | – | 27% | $5.90 |
| IM-AD 94-396 In DMF[b] | BTL Specialty Results | Proprietary MDA Based Modified BMI | – | 67% | $18.66 |
| Matrimid 5218 In NMP | CIBA GEIGY | DAPI/BTDA Thermoplastic | 1.2 | 25% | $100 |

(a) N-Methy-2-Pyrrolidone
(b) Dimethyl Formamide
(c) 12/86, 1000 lb. order, FOB warehouse, Prices expected to become very competitive, (cost: per dry lb. basis)
(d) Prices, FOB destination via sea route

TABLE 2

CURE SCHEDULES FOR LAP SHEAR[a] AND TGA SPECIMENS

| RESIN TRADE NAME | OPEN SOLVENT REMOVAL METHOD | CURE SCHEDULE[b,c] |
|---|---|---|
| Kerimid 601 (In NMP) | 3 Hrs. @ 40°C/ 2 Hrs. @ 100°C | 2 Hrs. @ 180°C/40 Hrs. @ 200°C |
| BPA-BMI | None | 2 Hrs. @ 200°C/5 Hrs. @ 250°C |
| Compimide 896 | None | 2 Hrs. @ 190°C/5 Hrs. @ 210°C/ 5 Hrs. @ 250°C |
| Compimide 453 | None | 2 Hrs. @ 160°C/8 Hrs. @ 180°C/ 8 Hrs. @ 210°C/40 Hrs. @ 200°C |
| IM-AD 94-606 | None | 2 Hrs. @ 190°C/10 Hrs. @ 210°C |
| Matrimid 5292 | None | 1 Hr. @ 180°C/2 Hrs. @ 200°C/ 6 hrs. @ 250°C |
| PMR-15 (CPI-2237) (In Methanol) | 2 Hrs. @ 60°C @ 380 mmHg | 2 Hrs. @ 80°C@ 152 mmHg/ 1 Hr. @ 200°C @ 152 mmHg/ 1 Hr. @ 230°C @ 760 mmHg/ 3 Hrs. @ 315°C @ 760 mmHg |
| Epoxy/PMDA (In MEK) | 3 Hrs. @ 80°C/ 1 Hr. @ 100°C | 3 Hrs. @ 160°C/12 Hrs. @ 230°C |
| Eccobond 104 (2 Part Epoxy) | None | 1 Hr. @ 200°C/12 Hrs. @ 225°C |
| AI-10 (In NMP) | 1 Hr. @ 25°C/ 3 Hrs. @ 40°C/ 1 Hr. @ 100°C | 3 Hrs. @ 180°C/40 Hrs. @ 200°C |
| 4000 TF (In NMP) | 1 Hr. @ 25°C/ 2 Hrs. @ 100°C | 2 Hrs. @ 180°C/40 Hrs. @ 200°C |
| IM-AD-94-396 (In DMF) | 1 Hr. @ 25°C/ 2 Hrs. @ 100°C | 1 Hr. @ 180°C/40 Hrs. @ 200°C |
| Matrimid 5218 | 1 Hr. @ 100°C | 1 Hr. @ 180°C/ 2 1/2 Hrs. @ 200°C |

(a)  ASTM D1002
(b)  Ambient pressure at 760 mmHg unless otherwise specified
(c)  (2) 0.127 mm, (0.005inD) bond line shims in all specimens clamped with two No. 50 Acadia binder clamps after solvent removal steps

TABLE 3

ASTM D1002 LAP SHEAR  DATA, ALUMINUM TO ALUMINUM

| RESIN TRADE NAME | ROOM TEMP. AVG. LAP SHEAR STRENGTH MPa (psi) | (PSEUDO) ELONGATION % of 0.5 INCH GAGE LENGTH | MODE OF FAILURE (R.T.) | CONSISTENCY OF CURED BROKEN RESIN SURFACE (R.T.) | LAP SHEAR STRENGTH AT 250°C MPa(psi) |
|---|---|---|---|---|---|
| 4000 TF (In NMP) | 24.9 (3610) | 6.3 | Cohesive | ca. 15% Voids Visible | 0 (0) (Melted) |
| AI-10 (In NMP) | 12.0 (1740) | 5.1 | Cohesive | ca. 25% Voids Visible | 4.69 (680.) |
| Eccobond 104 (2 Part Epoxy) | 11.8 (1710) | 3.6 | Cohesive | Smooth | 9.96 (1445.) |
| PMR-15 (CPI-2237) (In MEOH) | 9.86 (1430) | 3.9 | Cohesive | ca. 10% Voids Visible | 10.34 (1500.) |
| Matrimid 5292 | 8.69 (1260) | 3.9 | Adhesive | Smooth | 7.03 (1020.) |
| Compimide 896 | 8.69 (1260) | 3.1 | Cohesive | Smooth | 10.2 (1480.) |
| Kerimid 601 (In NMP) | 8.34 (1210) | 3.1 | Adhesive | Smooth | 4.28 (620.) |
| Compimide 453 | 8.07 (1170) | 3.0 | Adhesive | Smooth | 5.24 (760.) |
| BTL 94-606 | 7.17 (1040) | 2.6 | Adhesive | Smooth ca. 5% Voids Visible | 8.27 (1200) |
| Epoxy/PMDA (In MEK) | 5.59 (810) | 2.0 | Cohesive | ca. 40% Voids Visible | 2.62 (380.) |
| BTL 94-396 | 5.52 (800) | 1.9 | Adhesive | Smooth | 1.36 (197.) |
| BPA-BMI | 5.24 (760) | 1.7 | Cohesive | ca. 40% Voids Visible | 3.93 (570.) |
| Matrimid 5218 | — — | — | — | — | — |

(a)   2.4% Sodium Dichromate/24.4% Sulfuric Acid Etched
Specimens (10 minutes @ 70°C)

TABLE 4

## THERMAL-OXIDATIVE STABILITY DATA BY TGA[a]

### (ALL SAMPLES PRECURED AS SHOWN IN TABLE 2 )

| HIGH TEMP. RESIN TRADE NAME | TEMPERATURE FOR 5% WT. LOSS °C (RANK) | TEMPERATURE FOR 10% WT. LOSS (°C) | Tmax[b] ( °C) |
|---|---|---|---|
| PMR-15 (IN MEOH) (CPI-2237) | 466 (2) | 514 | 238, ≥350 |
| Matrimid 5218 (Neat) | 482 (1) | 513 | — |
| Amoco AI-10 (In NMP) | 380 (7) | 478 | — |
| Compimide 896 | 439 (3) | 460 | 252 |
| Amoco 4000 TF (In NMP) | 345 (10) | 456 | — |
| Compimide 453 | 436 (3) | 449 | 275 |
| BTL 94-606 | 432 (4) | 446 | 261 |
| Matrimid 5292 | 428 (5) | 438 | 256 |
| BPA-BMI | 409 (6) | 426 | 217 |
| Kerimid 601 (Neat) | 410 (6) | 424 | 210 |
| Matrimid 5218 (In NMP) | 310 (11) | 418 | — |
| Eccobond 104 (2 Part Epoxy) | 377 (8) | 389 | 236 |
| Epoxy/PMDA | 370 (9) | 380 | — |
| Kerimid 601 (In NMP) | 347 (10) | 380 | — |
| BTL 94-396 (In DMF) | 295 (12) | 344 | — |

(a)  All samples:  using DuPont 1090 TGA; @ 10 °C/Min. in 100 ml/Min Air
(b)  Tmax, DSC Peak Exotherm Temp. (+ 15°C) @ 10°C/Min

## TABLE 5

### PREPREG [a] PROCESSING CHARACTERISTICS

| PREPREG [a] PROCESSING RANGE | | | | | | | | | |
|---|---|---|---|---|---|---|---|---|---|
| 1 | 2 | 3 | 4 | 5 | 6 | 7 | 8 | 9 | 10 |
| VERY STICKY, TACKY, MAXIMUM DRAPE | VERY TACKY TO ITSELF, GOOD DRAPE | VERY TACKY TO ITSELF, DRAPE OK | SOFT, TACKY TO ITSELF | SOFT, FLEXIBLE, NOT TACKY TO ITSELF | SLIGHT FLEX, BRITTLE | NO TACK, STIFF, MINOR FLEX | | STIFF, FRAGILE | FRAGILE, VERY BOARDY |

| RESIN TRADE NAME | % SOLIDS | PROCESSING METHOD | PREPREG QUALITY [b] (AFTER 48 HOURS) | PREPREG QUALITY [b] (AFTER 18 DAYS) |
|---|---|---|---|---|
| Compimide 453 | 100% | Hot Melt[c] Through Calendar Rolls | 1 | 1 |
| BTL IM-AD 94-606 | 100% | Hot Melt[c] Through Calendar Rolls | 1 | 1 |
| Epoxy/PMDA | 75% In MEK | Through Calendar Rolls | 2 | 9 |
| Matrimid 5292 | 100% | Hot Melt[c] Through Calendar Rolls | 3 | 6 |
| PMR-15 (CPI-2237) | 50% In Methanol | Through Calendar Rolls | 4 | 7 |
| Kerimid 601 | 61% In NMP | Through Calendar Rolls | 5 | 7 |
| BTL IM-AD 94-396 | 67% In DMF | Through Calendar Rolls | 5 | 5 |
| AI-10 | 40% In NMP | Through Calendar Rolls | 7 | 8 |
| TF-4000 | 27% In NMP | Through Calendar Rolls | 7 | 8 |
| Compimide 896 | 100% | High Viscosity Hot Melt[c] Through Calendar Rolls | 9 | 9 |
| BPA-BMI | 100% | Very High Visc. Hot Melt[c] Through Calendar Rolls | 10 | 10 |

(a) A light weight scrim glass fabric (CS 104) with CS440 finish was impregnated with 100-200% resin content
(b) Stored and measured in 30-40% relative humidity air at $25^{\circ}$C
(c) Hot melts when used ranged in temp. between $100-130^{\circ}$C

TABLE 6

RDT (Rheometrics Dart Tester)[a]

DYNAMIC IMPACT PROPERTIES

| RESIN TRADE NAME | PEAK FORCE KNewtons ($LB_F$) | | SLOPE[c] (LB/INCH) | ABSORBED ENERGY[b] DURING IMPACT | |
|---|---|---|---|---|---|
| | | | | JOULES | ($IN \cdot LB_F$) |
| Matrimid 5292 | 4.63 | (1040.) | 27,850 | 4.12 | (36.5) |
| Eccobond 104 (Epoxy) | 2.31 | (520.) | 15,330 | 2.34 | (20.7) |
| Compimide 896 | 2.00 | (450.) | 27,500. | 2.19 | (19.4) |
| Compimide 453 | 1.82 | (410.) | 20,000 | 1.56 | (13.8) |
| BTL 94-606 | 1.96 | (440.) | 27,990 | 1.33 | (11.8) |
| Merlon 40 (For Comparison) Mobay Polycarbonate (1/8" Thick) | 6.05 | (1360.) | 3,040 | 62.7 | (555.) |

(a) All samples: 6.35 mm (1/4 inch) thick over 38.1 mm (1.5") ID backing ring and tested at 3386. mm/sec (7.6 miles/hr.) 50 lb. dart speed. (ASTM D3763)

(b) From area under the curve to full penetration (all impacts except Merlon 40F (which was intact after impact) were brittle fractures)

(c) From straightest section of curve before peak.

FIGURE 1

SCANNING ELECTRON MICROGRAPHS SHOWING THE PITTED SURFACE FROM THE ETCHING PROCESS ON 2024(T3) ALUMINUM USED IN ADHESION STUDY

(DIPPED IN 24.4% $H_2SO_4$, (2.44% $Na_2Cr_2O_7 \cdot 2H_2O$) SOLUTION @70°C FOR 10 MINUTES)

BEFORE                                    AFTER

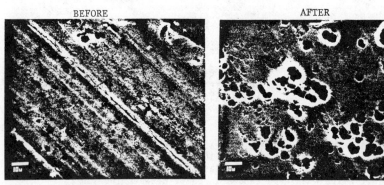

SEM  1000 X                                SEM  1000 X

## FIGURE 2
### TYPICAL TGA CURVE FOR A COMPARISON OF THERMAL OXIDATIVE STABILITY FOR CURED POLYIMIDES (SEE TABLE 4) (SAMPLE: PMR-15 FERRO)

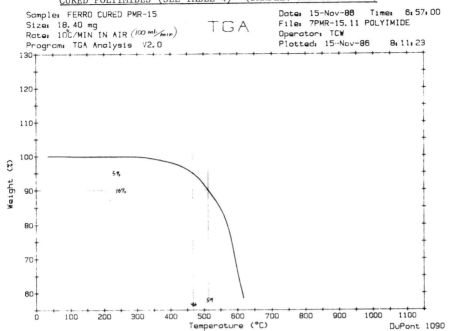

Sample: FERRO CURED PMR-15
Size: 18.40 mg
Rate: 10°C/MIN IN AIR $(100\,ml/min)$
Program: TGA Analysis V2.0

TGA

Date: 15-Nov-88  Time: 6:57:00
File: 7PMR-15.11 POLYIMIDE
Operator: TCW
Plotted: 15-Nov-86   8:11:23

## FIGURE 3
### TYPICAL TMA CURVE FOR COMPARISON OF COEFFICIENT OF THERMAL EXPANSION AND $T_g$ OF CURED POLYIMIDES (SAMPLE: KERIMID 601)

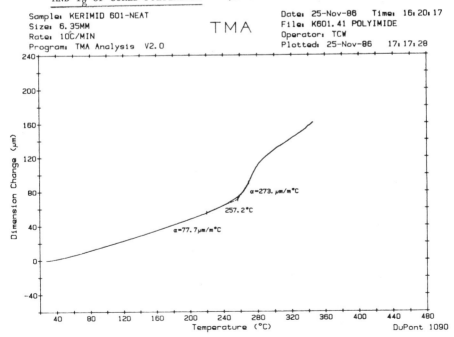

Sample: KERIMID 601-NEAT
Size: 6.35MM
Rate: 10°C/MIN
Program: TMA Analysis V2.0

TMA

Date: 25-Nov-86  Time: 16:20:17
File: K601.41 POLYIMIDE
Operator: TCW
Plotted: 25-Nov-86   17:17:28

## FIGURE 4

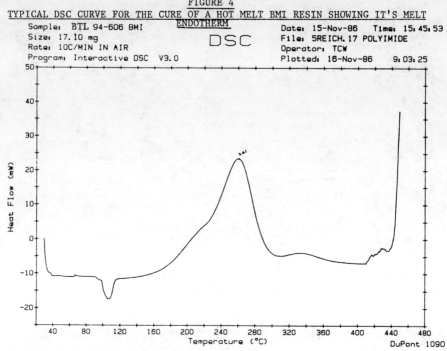

Sample: BTL 94-606 BMI
Size: 17.10 mg
Rate: 10C/MIN IN AIR
Program: Interactive DSC V3.0

DSC

Date: 15-Nov-86    Time: 15:45:53
File: 5REICH.17 POLYIMIDE
Operator: TCW
Plotted: 16-Nov-86    9:03:25

DuPont 1090

## FIGURE 5

TYPICAL FALLING DART TIP-FORCE vs. DISTANCE THROUGH SPECIMEN FOR A 50 1b. DART PIERCING A ¼ INCH THK SAMPLE AFTER ACCELERATING TO 3.386 m/sec.
(SEE TABLE 6)

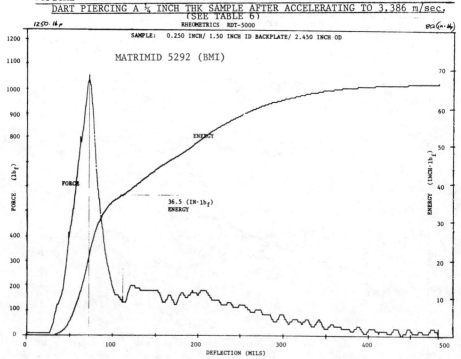

MATRIMID 5292 (BMI)

RHEOMETRICS RDT-5000

SAMPLE:   0.250 INCH/ 1.50 INCH ID BACKPLATE/ 2.450 INCH OD

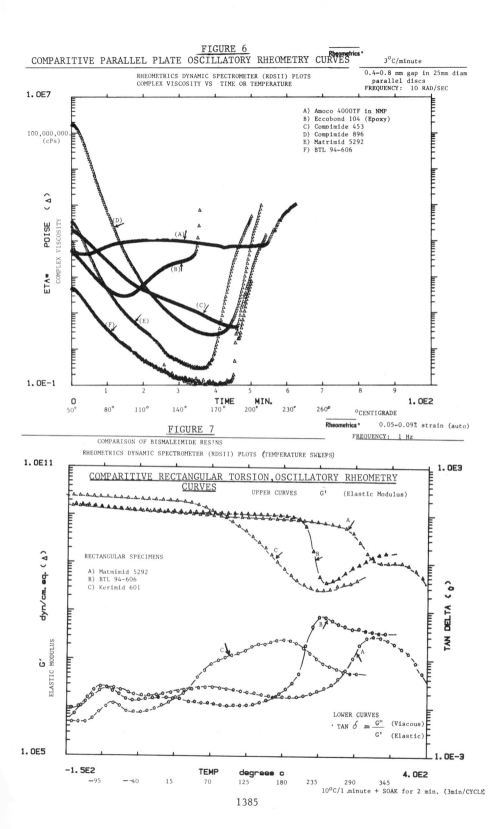

### FIGURE 6
COMPARITIVE PARALLEL PLATE OSCILLATORY RHEOMETRY CURVES

RHEOMETRICS DYNAMIC SPECTROMETER (RDSII) PLOTS
COMPLEX VISCOSITY VS  TIME OR TEMPERATURE

3°C/minute

0.4-0.8 mm gap in 25mm diam
parallel discs
FREQUENCY:  10 RAD/SEC

A) Amoco 4000TF in NMP
B) Eccobond 104 (Epoxy)
C) Compimide 453
D) Compimide 896
E) Matrimid 5292
F) BTL 94-606

1.0E7

100,000,000.
(cPs)

ETA* POISE ⟨Δ⟩
COMPLEX VISCOSITY

1.0E-1

0      1    2    3    4    5    6    7    8    9    1.0E2
50°   80°  110°  140°  170°  200°  230°  260°   °CENTIGRADE

TIME  MIN.

0.05-0.09% strain (auto)
FREQUENCY:  1 Hz

### FIGURE 7
COMPARISON OF BISMALEIMIDE RESINS

RHEOMETRICS DYNAMIC SPECTROMETER (RDSII) PLOTS (TEMPERATURE SWEEPS)

1.0E11

COMPARITIVE RECTANGULAR TORSION, OSCILLATORY RHEOMETRY
CURVES

UPPER CURVES    G'  (Elastic Modulus)

RECTANGULAR SPECIMENS

A) Matmimid 5292
B) BTL 94-606
C) Kerimid 601

dyn/cm. sq. ⟨Δ⟩
G'
ELASTIC MODULUS

TAN DELTA ⟨○⟩

LOWER CURVES
TAN $\delta = \dfrac{G''}{G'}$ (Viscous)
(Elastic)

1.0E5

1.0E3

1.0E-3

-1.5E2              TEMP   degrees c        4.0E2
-95    -40    15    70    125   180   235   290   345

$10°C/1$ minute + SOAK for 2 min. (3min/CYCLE)

ADVANCED BISMALEIMIDE COMPOSITES, PT II[1]
CONVERSION OF HIGH PERFORMANCE STRUCTURAL
FIBER PROPERTIES INTO BISMALEIMIDE
COMPOSITE PROPERTIES
by J. D. Boyd and T. F. Biermann

BASF Structural Materials Inc.
Narmco Materials, Anaheim, California

## Abstract

Carbon fiber composites are being evaluated for aircraft primary structural applications under increasing environmentally severe use conditions. For these applications improved performance over epoxy matrix resins is required. With the discovery of new curing reactions, bismaleimide resins (BMI's) are emerging as a versatile resin class for application under conditions too severe for epoxies. This paper describes advanced BMI based composites designed for both toughness (evaluated by post impact compression) and use up to 450°F (depending on the evaluation method). When possible, performance comparisons are made with well known commercial systems. Data are also presented on composites of BMI resins with intermediate modulus carbon fibers looking at modulus and strength translation and damage tolerance.

For certain applications, fibers other than carbon are required. The application of BMI resins to unidirectional composites of HVR Nicalon™ (silicon-carbide), Nextel 480, S-2 and Astroquartz II is presented.

## 1. INTRODUCTION

BASF Structural Materials has an extensive effort to fully develop the properties of BMI composites. From this effort two commercially available products, 5250-2 and 5250-3, have been developed. These composites have a higher service temperature potential and greater damage tolerance than the best currently used high temperature epoxy systems. While both BMI resins have good damage tolerance, the majority of the data shows 5250-3 composites to be somewhat tougher than 5250-2 composites. Further, the 5250 products do not form volatiles during curing, and they can be easily fabricated into complex parts using the same techniques as those used for epoxies.

## 2. DAMAGE TOLERANCE AND TEMPERATURE PERFORMANCE OF 5250 COMPOSITES

For airframe applications a composite must meet the specified requirements of a wide variety of tests. The two most difficult to meet simultaneously are use temperature and damage tolerance. The plastic matrix plays the major role in both of these tests, and the designer of the composite matrix quickly finds that these two criteria oppose each other. Hence, these are the best tests for evaluation of matrix resins.

### 2.1 Damage Tolerance
The most widely used damage tolerance test is compression after impact (CAI) residual strength and strain[2]. Figure 1 shows CAI strain data for the 5250 composites and the well known MY-720 based epoxy, 5208. 5250-2/-3 systems exhibit significantly better residual strain than 5208. The data further show 5250-3 is more damage tolerant than 5250-2.

Although there is always a desire for improved damage tolerance, MY-720 based composites like 5208 are used successfully in primary structures such as the emponage of fighter aircraft. Since the 5250 composites have greater damage tolerance than 5208, they could also be used in such primary structures.

### 2.2 Temperature Performance
Compressive strength is often the limiting design parameter. Since, composites lose strength when heated, especially saturated with moisture, hot/wet compression strength was used to establish the use temperature.

Hot/wet compression strength for 5250-2, 5250-3, and 5208 on Celion 12K is compared in Figure 1. At room temperature and 250°F there is no difference in compression strength. However, at 350°F, there is a major difference. The BMIs retain >60% of the RT strength while the epoxy is past its use temperature. This shows that 5250 systems can be used at 350°F while the use temperature of 5208 is significantly lower.

### 2.3 Cure Conditions of 5250
All 5250 BMI composite data presented in this paper were generated on material cured with an initial autoclave cycle of 6 hrs at 350° followed by a free standing

post cure of 6 hours at 470°F. This cycle was found to offer a good balance of hot/wet and toughness properties.

### 3. NEW FIBERS OFFER HIGHER STRENGTH/MODULUS 5250 COMPOSITES

Intermediate modulus (IM) fibers offer the advantage of lighter composite structures. This advantage can only be realized if the 5250 resin can efficiently translate the increased fiber properties. Therefore, 5250-2 and 5250-3 were tested on most of the commercially available IM fibers for:

- Translation of fiber properties
- Damage tolerance

The modulus of the IM fibers examined ranged from 40 to 44 MSI as shown in Table 1. AS-4 was used as a lower modulus (35 MSI) control. The results of these tests showed that the IM fibers gave higher composite strength and stiffness properties, but with some decrease in damage tolerance.

### 3.1  0° Tensile Properties of 5250/IM Fibers Composites

The 0° tensile composite test data is given in Table 2. BMS-8-294[3] testing methods were used, and all values were normalized to 60% fiber volume.

The data in the table shows the IM fibers composites are all higher in modulus than the AS-4 control. Depending on resin and fiber, modulus increases ranged 10 - 20% higher than AS-4. The coupon strengths of the IM composites were also higher. Again, depending on resin and fiber the range was from -3 to 19% higher than AS-4.

The % translation of strand modulus to coupon modulus is also given in Table 2. All the values fall in the narrow range between 87-95%. The strength translation, however, showed a wider range from 79 to 97%.

Since BMI resins have relatively low $G_{IC}$, 0° tensile translation of 5250 composites were compared to a toughened resin composite with nearly double the $G_{IC}$, 5245C[1]. The data in Table 2 show 5250 resins translate fiber properties as well as the toughened matrix, 5245C control. This indicates that resin toughness is not an important factor in translation of fiber properties, and the 5250 BMI resins can translate fiber properties as required for high performance applications.

### 3.2  Compression Modulus of 5250 IM Fibers Composites

As expected, composites made from IM fibers gave higher compression modulus than the AS-4 composite. The modulus of quasiisotropic panels (25% in zero direction) is given in Figure 2[4].

The AS-4 tested lowest at 6.7 to 6.8 MSI. The IM fibers all fell into a higher range from 7.9 to 8.4 MSI.

Although a higher compression modulus was expected for the IM fibers, it was somewhat unexpected that the magnitude was the same as for tensile modulus. The 5250 IM fiber tensile modulus was 10 - 20% higher than the AS-4 control, and the compression modulus was 16 - 22% higher.

### 3.3 Damage Tolerance of 5250 IM Fibers Composites

For damage tolerance assessment on IM fibers, the CAI test was again used. Quasiisotropic panels (32 ply) were impacted at 1000 in-lb/in, then tested in compression by Boeing Technology Services[4]. This impact level was selected because the NASA CAI test, uses approximately the same impact level (960 in-lb/in).[2] The fiber and CAI strength, strain, and damage are reported in Table 3.

Impact damage area was measured by C-scan. In every case the damage area was greater for the IM fiber composites than the AS-4 control.

None of the IM fibers were found to offer improved damage tolerance over the lower modulus AS-4 fiber. The CAI strengths fell into a range between 26 to 33 ksi. From this

data alone, no single IM fiber stands out as a particularly superior performer. However, the data does indicate that 5250-3 resin is better than 5250-2 in damage tolerance. The damage tolerance of the 5250-3 tested about 10% higher than the 5250-2 on most fibers. This means in damage critical areas 5250-3 might be preferred.

An interesting aspect of the data is the relationship between damage area and residual strength (Figure 3). Disregarding the particular fiber or resin, there is a general trend to lower strength with increased damage area. Even though the relationship is far from linear, reduction in the damage caused by impact can be seen as a key in designing damage tolerant composite materials.

The reduced impact resistance of the IM fibers has been noted before and is reportedly caused by their lower surface functionality[5]. Typically IM fibers have 40% less polar functionality than lower modulus fibers. This is theorized to give poorer adhesion between the fiber and supporting matrix and consequently less delamination resistance and less impact resistance. It is interesting that the theorized poor adhesion did not significantly affect the 0° properties.

### 3.4  Conclusions on IM Fiber Tests

The intermediate modulus fibers offer significant improvement in tensile and compressive properties over the lower modulus fibers like AS-4. However, the impact resistance of the IM fibers is not as good – they suffer greater damage by impact and the CAI strength and strain are lower. The amount lower (typically 10% in strength at 1000 in-lb/in impact) may be at an acceptable level when the increased modulus is considered.

### 4.  NON-CARBON FIBERS

Certain applications and cost considerations may dictate fibers other than carbon fiber. Since some of these fibers are relatively new, a comparison of composite properties on a single resin system, 5250-2, will help to define the best fiber for a particular application.

Several of the commercially available non-carbon fibers were tensile tested. The numbers are reported in relative terms. 0° tensile strength and modulus were divided by the density to give specific values and then normalized to the lowest value. This data is shown in Figure 4.

Of the fibers tested, the carbon fibers have by far the highest specific specific strength and modulus. For non-carbon fibers, the highest tensile modulus is HVR Nicalon™. Nextel 480 has a lower specific modulus with the S-2 and Astroquartz II even lower still. The HVR Nicalon™ has the further outstanding property of compression strength and modulus at 270 ksi and 16.7 MSI, respectively (60% FV 5250-2 composite).

In specific strength, Astroquartz II came out on top followed in order by S-2 glass, HVR Nicalon™, and Nextel 480.

Since there are such large differences in the strengths and moduli among the various fibers, the particular fiber selected will depend upon application requirements.

### 5.  EFFECT OF MOISTURE ABSORPTION ON 5250 PERFORMANCE

Absorbed moisture reduces the elevated temperature performance of composites especially the compression strength. The new technique of dynamic mechanical analysis (DMA) was used to evaluate the effect of absorbed moisture on elevated temperature composite properties of 5250. In the DMA test, a thin specimen is heated at a constant rate while the flexural storage modulus is continually recorded. The advantage of this technique over direct measurement of hot/wet compression strength is speed and simplicity[6].

Since DMA could be considered as a simplified replacement of compression testing, a comparison of % retention of wet DMA flexural storage modulus and % retention of wet compression strength is given in Figure 5 for 5250-3 composites. In both cases the specimens were saturated by water boiling (5 days). The DMA was ramped at 10°C/min.

The compression curve was obtained by measuring strength at several points in the temperature range. To reduce specimen dry out, the compression test was complete within 5 minutes of placement in the environmental chamber.

The results show a close but not identical correlation between DMA and hot/wet compression. This suggests DMA is useful for comparative purposes, but not currently useful to define the exact % retention of compression strength.

Since DMA is a good comparative tool, it was used to compare the performance of composites with varying levels of moisture. A useful number from the DMA flexural storage modulus curve is the temperature of modulus change

$(T_{\Delta G'})$. The $T_{\Delta G'}$ was taken as the intersection of the tangent line to the initial linear modulus portion and the final linear modulus portion reflected back to the DMA flexural storage modulus/ temperature curve.[7] The $T_{\Delta G'}$ values were measured for both 5250-2/T-300 and 5208/T-300 fabric composites.

To condition with moisture, 5250-2/ T-300 and 5208/T-300 fabric specimens were boiled in water for varying lengths of time. Figure 6 shows the moisture uptake with time for these thin specimens (0.08 in. thick). Both composites take up moisture at about the same rate. However, the epoxy reaches equilibrium at a higher moisture content than the BMI.

The variously moisturized specimens were analyzed by DMA, and the $T_{\Delta G'}$ was determined for each moisture level. Figure 7 gives the $T_{\Delta G'}$ vs moisture content. As expected, the $T_{\Delta G'}$ decreases with increased moisture level. Again as expected, 5250-2 has a higher $T_{\Delta G'}$ than 5208 at every moisture level.

The most probable reason for reduced composite performance with moisture absorption is the effect of moisture on the supporting matrix. At elevated temperatures the water may be acting as a plasticizer, reducing matrix resin modulus, and thereby, composite properties.

## 6. CONCLUSIONS

The Narmco 5250 series BMI based composites have superior properties which make them candidate materials for advanced, high temperature applications. 5250 resins were found to exhibit high translation of high performance fiber properties on all fibers tested. IM fiber composites have increased modulus over the extensively used lower modulus fibers. However, this gain in modulus is at the expense of damage tolerance.

Although 5250 was the vehicle for these tests, all BMI matrices and for that matter other matrix resins are expected to behave similarly with the various fibers.

The greatest deficiency uncovered by these tests is the decreased damage tolerance of BMI composites made from IM carbon fibers vs regular modulus carbon fibers. Therefore, not only should the matrix resin designer develop more damage tolerant resins, but also the fiber suppliers should focus on understanding and improving the damage tolerance of IM carbon fiber composites.

## 7. ACKNOWLEDGEMENTS

The assistance of Philip Ferrel, Judith Anderson, and Albert Kuo is gratefully acknowledged for doing the majority of the laboratory work.

## 8. REFERENCES

1. For the first part of this series see: J. D. Boyd and T. F. Biermann, 31st International SAMPE Symposium and Exhibition, April 7-10, 1986, pp. 977-986.

2. NASA Publication 1092 revised as NASA 1142.

3. Boeing Material Specification -8-294B.

4. Boeing Material Specification -8-276B.

5. Manders P. W. and Harris W. C., SAMPE Journal, 22(6), 1986, pp. 47-51.

6. Sichina W. J. and Gill P. S., 31st International SAMPE Symposium and Exhibitor, April 7-10, 1986, pp. 1104-1112.

7. Boll D. J., Bascom W. D. and Motiee B., Composites Science and Technology, 24, 1985 pp. 253-273.

BIOGRAPHY

Jack Boyd is currently a Senior
Chemist at Narmco.  Previous to
Narmco he worked as an R&D Chemist
with Occidental Research
Corporation.  Before joining
Occidental, he received his Ph.D.
in Organic Chemistry from the
University of Utah in 1977 and then
served as a Post doctoral Fellow at
the University of California, Los
Angeles.

Ted Biermann is the Research and
Development Manager at Narmco.  He
trannsferred to Narmco in late 1982
from Celanese Fibers Operations
where he held various technical
management positions in R&D and in
plant locations.  He joined
Celanese after receiving his Ph.D.
in Organic Chemistry from Purdue
University in 1970.

## Table 1

### Strand Properties of Carbon Fibers Used

| Fiber | Strength (KSI) | Modulus (MSI) |
|-------|----------------|---------------|
| AS-4 | 530 | 35.0 |
| G40-600 | 655 | 42.7 |
| IM-7 | 739 | 40.9 |
| Hi Tex 46-8B | 721 | 44.3 |
| Apollo 43-600 | 573 | 43.3 |

## Table 2

### Translation of Intermediate Fiber Properties into 0° Tensile Coupon Properties

0° Tensile Testing[1,2]

| Fiber | Resin | Coupon Strength (KSI) | Percent Translation | Coupon Modulus (MSI) | Percent Translation |
|-------|-------|-----------------------|---------------------|----------------------|---------------------|
| AS-4 | 5250-3 | 310 | 97% | 20 | 95% |
| IM-7 | 5245C | 380 | 86 | 22 | 90 |
| | 5250-2 | 360 | 82 | 22 | 90 |
| | 5250-3 | 370 | 84 | 22 | 90 |
| G40-600 | 5250-2 | 330 | 84 | 24 | 93 |
| Apollo 43-600 | 5245C | 310 | 90 | 24 | 92 |
| | 5250-2 | 300 | 87 | 24 | 92 |
| Hi Tex 46-8B | 5245C | 330 | 76 | 24 | 91 |
| | 5250-2 | 340 | 79 | 23 | 87 |
| | 5250-3 | 350 | 81 | 24 | 91 |

1.  All modulus and strength values were normalized to 60% fiber volume
2.  Boeing testing methods BMS-8-294B; secant modulus P1-P6

Table 3

Comparative Compression After Impact Testing of

Intermediate Modulus Fibers[1,2]

| Fiber | Resin | Strength (KSI) | Strain Percent | Damage[3] Area (in[2]) |
|---|---|---|---|---|
| AS-4 | 5250-3 | 33 | .50% | 1.2 |
|  | 5250-2 | 31 | .43 | 2.0 |
| Hi Tex 46-8B | 5250-3 | 33 | .42 | 1.4 |
|  | 5250-2 | 26 | .32 | 2.2 |
| IM-7 | 5250-3 | 31 | .40 | 1.8 |
|  | 5250-2 | 28 | .37 | 1.9 |
| Apollo 43-600 | 5250-3 | 28 | .35 | 1.9 |
|  | 5250-2 | 29 | .37 | 2.5 |
| G40-600 | 5250-3 | 29 | .41 | 2.0 |

1.  32 ply quasi-isotropic panels, AFW of 145 $g/m^2$ nominal, fiber volume of 61%, average

2.  Impact level was 1000 in-lb/in, 3 specimens per test

3.  C-scan damage area

# FIGURE 1
## COMPARATIVE COMPOSITE DATA
### 5250 AND 5208

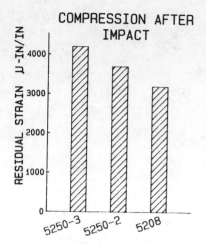

COMPRESSION AFTER IMPACT

AS-4 FIBER, NASA 1142
QUASIISOTROPIC PANELS
960 IN-LBS/IN IMPACT

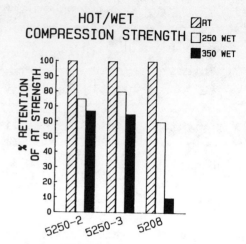

HOT/WET COMPRESSION STRENGTH

CELION 12K FIBER
16 PLY QUASIISOTROPIC SPECIMENS
MOISTURE SATURATED

# FIGURE 2
## COMPRESSION MODULUS OF
### QUASIISOTROPIC PANELS

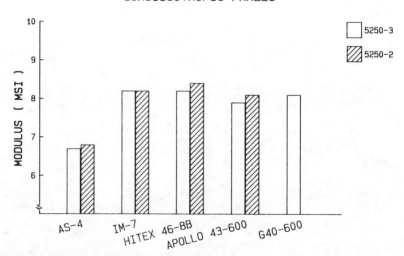

BOEING TESTING SERVICES USING BMS-8-276
FIBER VOLUME APPROXIMATELY 61%

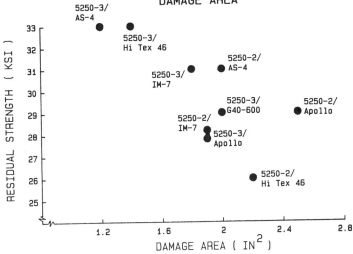

FIGURE 3
RESIDUAL CAI STRENGTH vs.
DAMAGE AREA

SEE TABLE 3 FOR DATA

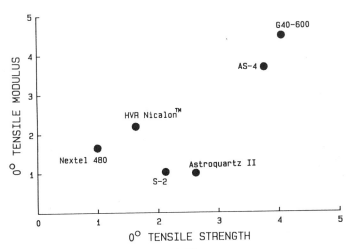

FIGURE 4
SPECIFIC TENSILE STRENGTH AND
MODULUS OF 5250-2/NON-CARBON FIBER COMPOSITES

0° TENSILE STRENGTH AND MODULUS WERE DIVIDED
BY THE DENSITY AND NORMALIZED

## FIGURE 5
## COMPARISON OF METHODS TO MEASURE HOT/WET PERFORMANCE OF 5250-3 COMPOSITES

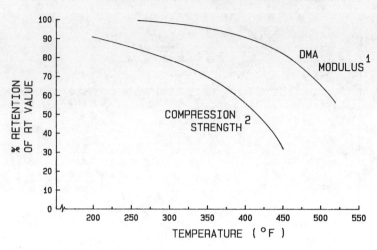

1. 5250-3/T300   MOISTURE SATURATED
2. 5250-3/AS-4   QUASIISOTROPIC SPECIMENS,
   MOISTURE SATURATED

## FIGURE 6
## MOISTURE UPTAKE WITH WATER BOIL

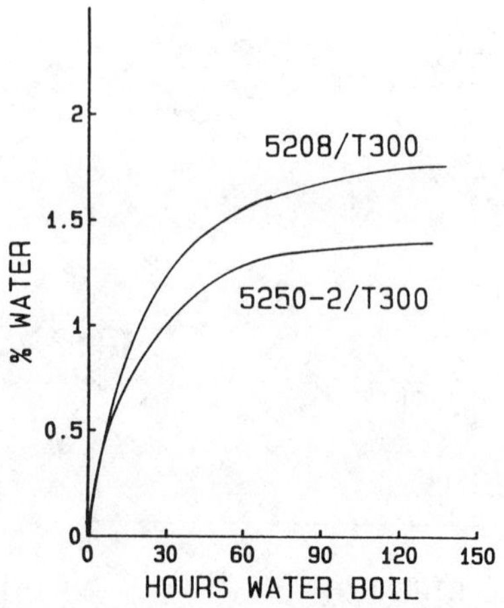

## FIGURE 7
## $T_{\Delta G'}$ DECLINES WITH ABSORBED MOISTURE

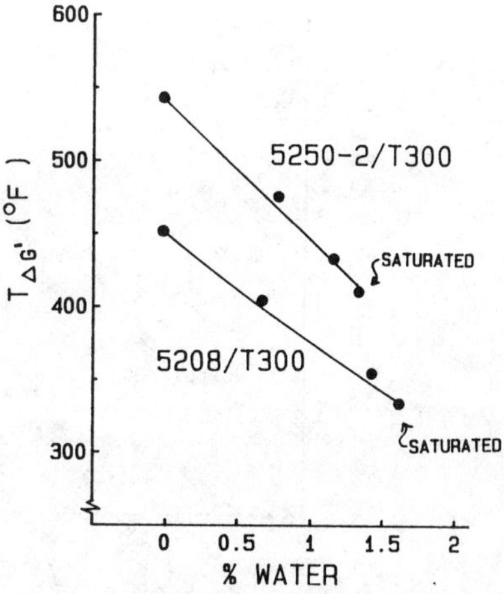

ARAMID POLYMERS, THEIR PROPERTIES AND APPLICATION
by
PAUL R. LANGSTON
E. I. du Pont de Nemours & Co., Inc.

Abstract

Two aramid polymers, introduced in Advanced Composites in the 60s and 70s, have dramatically influenced how aircraft are being built. Certainly from a materials standpoint, they would be second only to carbon fibers in their impact.

Nomex$^R$ mechanical paper introduced by DuPont in the 60s and developed into honeycomb by Ciba Geigy and Hexcel, has pushed the cored skin design concept. Designers of secondary structures and interiors now weigh merits of the honeycomb core concepts versus ribs and stringers on an ever increasing number of applications. Honeycomb core parts on the Boeing 747 where over one acre of Nomex$^R$ honeycomb is used on each aircraft has been successful and acknowledged even by the most skeptical engineers after 15 years of grueling service. Data from this 747 experience has helped establish design criteria for other commercial aircraft, helicopters and general aviation aircraft. New general aviation trend setters for composite aircraft such as the Avtek 400 and Beech Starship have adopted Nomex$^R$ Honeycomb core design not only for secondary structures, but also primary structures including empennage, wing skin, and fuselage.

Kevlar$^R$, the second of the aramid family introduced in the early 70s, has provided alternatives to fiberglass and carbon. In comparison to fiberglass it offers a 40% density advantage plus significant improvement in tensile strength and modulus. Versus carbon it offers a 15-20% density advantage with significant advantages in toughness. A hybrid of Kevlar$^R$/carbon skins has proven a good compromise for maximizing desired composite properties. Hybrids of Kevlar$^R$/carbon cover 21-23% of the wetted surface of the highly successful Boeing 767.

The physical properties and design concepts that aramids offer will be detailed along with examples of their application in this paper.

HIGH PERFORMANCE THERMOPLASTICS: A REVIEW OF NEAT RESIN AND
COMPOSITE PROPERTIES

Norman J. Johnston and Paul M. Hergenrother
NASA Langley Research Center
Hampton, Virginia 23665-5225

ABSTRACT

A review was made of the principal
thermoplastics used to fabricate
high performance composites. Included are neat resin tensile and
fracture toughness properties,
glass transition temperatures
(Tg), crystalline melt temperatures (Tm) and approximate processing conditions. Mechanical properties of carbon fiber composites
made from many of these thermoplastics are given (flexural,
longitudinal tensile, transverse
tensile and in-plane shear properties as well as short beam shear
and compressive strengths and
interlaminar fracture toughness).
Attractive features and problems
involved in the use of thermoplastics as matrices for high performance composites are discussed.

INTRODUCTION

The purpose of this paper is to
give a timely review of the
properties of the principal candidate thermoplastic (TP) resins
that either have been studied as
composite matrices or are currently undergoing detailed experi-
mental/developmental evaluation as
matrices. This review is
especially appropriate at this
time because of the heavy emphasis
being placed on the development
and application of TPs as matrices
for fiber reinforced composites on
advanced Air Force weapons systems
such as the Advanced Tactical
Fighter (ATF).

The reader should note that the
data included in this paper were
obtained from a large number of
sources, mostly materials
suppliers. Variations in test
methods from source to source are
inevitable. There is no guarantee
that ASTM procedures were
employed. No attempt was made to
normalize composites data to
constant fiber volume fraction.
Mechanical properties of a parti-

cular composite matrix will vary with fiber type and within a fiber type the date that fiber was produced. Problems with resin reproducibility, consistency, and processability also cause property variations as new materials mature in their development from experimental to commercial products. Consequently, the reader should bear these caveats in mind in using this data. Care and sound judgment should be exercised in making comparisons between materials.

## GENERAL FEATURES OF THERMOPLASTIC MATRICES

Three of the most attractive features offered by TPs as composite matrices are listed in Table I. The majority of applications involving aircraft structures demand that components have superior damage tolerance and delamination resistance. Most high performance TPs offer outstanding interlaminar fracture toughness and acceptable post-impact compression response.

They also offer an even bigger payoff: the potential of low cost manufacturing. By taking advantage of the inherent chemical nature of TP molecules to undergo thermally-induced flow, shaped articles can be fabricated at elevated temperatures by relative-

ly fast processing methods. Consequently, time is profitably exchanged for temperature. Prodigious and ambitious programs are underway throughout the industry to develop cost-effective processing technology for TP materials. This effort is partly catalyzed by progressive Air Force contractual activities.

Other attractive benefits of TP composites include indefinite prepreg life, ability in certain processes to recycle a flawed part, and the ease with which TP resins can be quality controlled during their manufacture. The latter is due, in part, because of their less complex formulations and their inherent stability at ambient conditions.

The major problems facing TP composites are outlined in Table II. TP prepreg of the quality generally associated with standard 350°F cure epoxies is difficult to fabricate. Either hot/melt or solution methods have to be employed. Both have their problems. In the former, temperatures generally above 650°F have to be employed in order to achieve a melt viscosity sufficiently low to wet out 6,000-12,000 filament carbon fiber tow. Solution prepregging to produce uni-tape is not an advanced state-of-the-art process in industry because of the heavy emphasis in the past with solventless epoxy coatings and the strict air quality standards now in force. Further, many solvents, especially aprotic organic liquids, are difficult to remove during prepreg B-staging or composite fabrication. Solvents also can offer fire, explosion and health hazards. These problems can lead to non-uniform prepreg that contains misaligned and wavy fibers, resin-rich and resin-poor areas, gaps, and poor fiber wetting. TP prepreg generally has very little drape, unless powder impregnated, and no tack, unless tackifiers, solvents, or plasticizers are added.

The processing of TP prepreg into shaped composite structure affords many problems which are listed in detail in Table II. One or more of these difficulties is always present regardless of the fabrication technique employed. Edge buckling, fiber misalignment, porosity, and part uniformity are also continuous problems in thermoforming TP composites. Finally, the time-dependent properties of most TP composites need to be fully characterized so that creep and fatigue problems will not become a Pandora's box.

PRINCIPAL HIGH PERFORMANCE
THERMOPLASTICS

Sixteen principal thermoplastics considered as candidates to fabricate high performance composites are shown in Table III along with their supplier, Tg, Tm (for semi-crystalline polymers), and approximate maximum processing temperature. The first five are polyarylene ether or sulfide polymers, three of which are semicrystalline. The next three are amide or amideimide compositions, followed by four polyimides. Three polysulfones and one polyester complete the list. All except the J-polymer are heavily aromatic in character. The dominent chemical structure of all polymers except the five latest compositions (APC-HTX, PXM 8505, PAS-2, Torlon AIX638, and Avimid K-III) are known. In these eleven systems, the chemical flexibilizing groups between phenyl rings in the backbone other than the propagating linkages include isopropylidene, carbonyl, oxygen, sulfur, and sulfone.

It is noteable that the average dry Tg value of the polyimides and polyamideimides is higher than the average Tg from any other polymer class. Three of the five latest compositions are polyarylene ethers or sulfides and have much higher Tg values than their predecessors (e.g., APC-HTX vs. PEEK, PAS-2 vs. PPS). The Tg values were increased in order to afford improved wet 350°F properties for ATF applications. Most of the polyarylene ethers and polyimides are candidate 350°F/AFT resins. J-Polymer's Tg is too low. Torlon is too difficult to process and the three polysulfones are too moisture and solvent sensitive.

Processing temperatures for all 16 polymers are extremely high and range from 329° to 420°C. Processing pressures range from 100 to 300 psi for conventional press moldings while processing times vary from 1 to 13 hours. Most TPs use short cycles; Avimid K-III requires a much longer cure because chain growth occurs with the evolution of condensation volatiles (and residual solvent).

## NEAT RESIN PROPERTIES

Table IV lists the tensile and fracture toughness properties of 14 of the principal TPs shown in Table III. Values for APC-HTX were not available and J-Polymer was omitted because of its low Tg. Although tensile strengths are extremely notch sensitive, most of the newer polyarylene ethers, polysulfides and polyimides have very respectable values in the range from 14 to 17 Ksi.

Tensile moduli should be above 450 Ksi to achieve acceptable composite compression strengths. The polysulfones fall way below this value; PXM 8505 and Torlon AIX638 appear to be marginal; most of the polyimides are well above it. The high value for $PISO_2$ stands out and helps explain why some of the $PISO_2$ composite strengths are outstanding. The Xydar SRT-300 is a liquid crystalline polyester whose tensile properties are outstanding, as would be expected for an ordered molecule. In this case, the high value might not translate into high composite compression strength; resin compression modulus might be a better indicator.

Fracture energies ($G_{Ic}$) between 5 and 10 in-lb/in$^2$ should give composites with acceptable interlaminar fracture toughness. Values above 10 in-lb/in$^2$, while excellent, are overkill and should not be obtained at the expense of some other property such as hot/wet strength or modulus. All the values listed in Table IV are either acceptable or excellent. Notched Izod impact strengths have not been correlated with composite interlaminar fracture toughness.

## COMPOSITE PROPERTIES

Of the original 16 TPs listed in Table III, composite properties were obtained for 10 and are given in Tables V and VI. Flexure strengths above 250 Ksi are excellent; only Udel P-1700 and PPS fall substantially below this value, the latter probably because of either poor fiber-resin adhesion or fiber damage. Flexure moduli listed in Table V are very good; values should generally be in the 16-19 Msi range.

Short beam shear strengths generally should be above 14 Ksi, where most 350°F cure epoxies fall. The values listed in Table V are acceptable but not outstanding and suggest premature failures on the compressive side of the specimen. The lower than desired compressive strengths tend to substantiate this. Room temperature compressive strengths for 350°F cure epoxies average well above 210 Ksi. Compressive strengths for all the TPs except one are below 165 Ksi. Yet the resin moduli for these same materials, in all cases except Udel P-1700, are sufficiently high (450 to 667 Ksi; Table IV) that one would predict their composite compression strengths to be above 200 Ksi.

This suggests that poor fiber alignment and poor fiber-resin interfacial adhesion may be playing a destructive role in compression response. The low transverse strengths in all cases except PEEK-APC-2, which is known to have excellent fiber-resin interfacial adhesion, indicates that the remaining TPs have an interface problem. Transverse tensile strengths for epoxies are generally above 8-10 Ksi and they exhibit excellent interfacial adhesion.

It also seems possible that in some cases fiber damage due to harsh prepreg fabrication might be the culprit. The generally poor longitudinal tensile strengths, which are a fiber dominated property, seem to suggest poor fiber alignment and/or fiber damage. Values above 280-300 Ksi are obtained with 350°F cure epoxies.

Interlaminar fracture toughness ($G_{Ic}$) values for all the TP composites are excellent as expected, based on neat resin fracture toughness. Values between 4 and 6 in-lb/in$^2$ are good and should afford acceptable post-impact compression strengths.

## THERMOSETS VERSUS THERMOPLASTICS

Table VII summarizes the trade-off in properties between thermosets and TPs as composite matrices.

350°F Cure epoxies and bismale-imides can be considered thermo-sets for purposes of this compari-son. In this paper, most of the listed properties have been discussed in terms of the advan-tages and deficiencies of TPs; it is felt TPs have to overcome certain disadvantages. On the other hand, it should be noted that the thermosets have very few disadvantages and, even then, most of those are acceptable and can be dealt with. The one key element that dominates the trade-off list and can tip the balance to TPs is that of fabrication costs. However, the potential for low cost manufacturing of TPs remains to be demonstrated.

## CAVEATS

Utilization of TP composites will raise some of the same performance concerns inherent in toughened thermosets, except the problems could be exacerbated by a higher anticipated use temperature. These concerns include fatigue and creep response and load-rate sensitivity, especially at elevat-ed temperatures and at high stress levels. No long-term environment-al exposure experience even with small coupons has been obtained, especially in the presence of corrosive fluids under load. The effects of large built-in residual stresses generated by higher

processing temperatures need to be understood, especially in larger structures containing cut-outs, cavities and built-up areas. Very little flight experience even with TP secondary structure exists to gauge either durability or maintenance requirements, let alone damage tolerance. The newer TP materials are generally more costly than current generation epoxies and bismaleimides. Extensive use of composites will be tempered by costs and cost/performance trade-offs. The total performance characterization of these promising "new-improved" composite matrices (toughened thermosets as well as TPs) is incomplete and new, untried materials will be introduced with caution.

## GENERAL REFERENCES

1. Engineering Thermoplastics. J. M. Margolis, editor, Marcel Dekker, Inc., 1985.
2. N. J. Johnston and T. L. St. Clair, 18th International SAMPE Technical Conference, 18, 53 (1986); SAMPE J., 23(1), 12 (1987).
3. N. J. Johnston, NASA CP-2321, 1984, pp. 75-95.
4. D. L. Hunston, R. J. Moulton, N. J. Johnston and W. D. Bascom, ASTM STP-937, 1987 (in print).

5. N. J. Johnston, T. K. O'Brien, D. H. Morris, and R. A. Simonds, 28th National SAMPE Symposium and Exhibition, 28, 502 (1983).

6. D. L. Hunston, Composites Technology Review, 6, 176 (1984).

7. "The Place for Thermoplastic Composites in Structural Components," National Materials Advisory Board, National Research Council, NMAB-435, 1987.

8. Product literature and research data, D. R. Nethero, Amoco Performance Products, Inc., Advanced Composite Systems, Ridgefield, CT 06877.

9. Product literature and research data, H. H. Gibbs, DuPont Company, Composites Division, Wilmington, DE 19898.

10. Product literature and research data, R. S. Shue, Phillips Petroleum Company, Thermoplastic Composites, Bartlesville, OK 74004.

11. Product literature and research data, M. Harvey, ICI Americas Inc., Advanced Materials Group, Wilmington, DE 19897.

12. Product literature, Dartco Manufacturing Inc., P. O. Box 5867, Augusta, GA 30906.

13. Product literature, General Electric Company, Plastics Group, One Plastics Avenue, Pittsfield, MA 01201.

14. Handbook of Composites, G. Lubin, editor, Van Nostrand Reinhold Company, NY, 1985, pp. 243 ff.

## BIOGRAPHIES

Norman J. Johnston, Senior Scientist, Materials Division, NASA Langley Research Center, received his Ph.D. in Organic Chemistry from the University of Virginia in 1963. Prior to joining NASA in 1966, he was an advanced development chemist working with polyesters and polyimides for the General Electric Company in Schenectady, NY, and from 1963-1966, an Assistant Professor of Chemistry at Virginia Polytechnic Institute and State University in Blacksburg, VA. At Langley, he has worked with pyrrones and polyimides and more recently on high performance toughened composite materials.

Paul M. Hergenrother, Senior Polymer Scientist in the Materials Division, NASA Langley Research Center, received a B.S. degree from Geneva College and took graduate work at the University of Pittsburgh and Carnegie Mellon University. He held various technical and management positions at Mellon Institute, Koppers Co.,

Whittaker Corp. (R&D Div.) and
Boeing Aerospace Co. and was an
Adjunct Research Professor with
Virginia Polytechnic Institute and
State University. His work has
been primarily in the area of high
performance polymers.

## TABLE I

### ATTRACTIVE FEATURES OFFERED BY THERMOPLASTICS AS COMPOSITE MATRICES

● POTENTIAL OF LOW COST MANUFACTURING
- O   INDEFINITE PREPREG STABILITY
- O   THERMOFORMING OF FLAT SHEET STOCK
- O   REPROCESSING TO CORRECT FLAWS
- O   FAST PROCESSING CYCLE

● HIGH TOUGHNESS (DAMAGE TOLERANCE)

● EASY QUALITY CONTROL

## TABLE II

### PROBLEMS TO RESOLVE WITH THERMOPLASTIC COMPOSITE MATRICES

● QUALITY PREPREG
- O   DIFFICULT TO MAKE
- O   FIBER WETTING
- O   NON-UNIFORMITY
- O   NO TACK
- O   NO DRAPE (UNLESS POWDER IMPREGNATED)
- O   PROPER SIZING

● PROCESSING PROBLEMS
- O   LAY-UP WITH BOARDY PREPREG
- O   COST OF TOOLING
- O   BAGGING MATERIALS
- O   HIGH TEMPERATURES
- O   MODERATE TO HIGH PRESSURES

● UNKNOWN FATIGUE PERFORMANCE
● UNKNOWN CREEP BEHAVIOR
● SOLVENT SENSITIVITY (EXCEPT SEMI-CRYSTALLINE POLYMERS)
● CONTROL OF MORPHOLOGY WITH SEMI-CRYSTALLINE POLYMERS

## TABLE III

## PRINCIPAL HIGH PERFORMANCE THERMOPLASTICS*

| POLYMER | SUPPLIER | $T_G$, °C ($T_M$, °C) | PROCESSING TEMP., °C |
|---|---|---|---|
| POLYETHERETHERKETONE (PEEK) | ICI | 143 (343) | ~400 |
| POLYARYLENE KETONE (APC-HTX) | ICI | 205 (386) | ~420 |
| POLYARYLENE KETONE (PXM 8505) | AMOCO | 265 | -- |
| POLYPHENYLENE SULFIDE (PPS, RYTON) | PHILLIPS PET. | 90 (290) | ~343 |
| POLYARYLENE SULFIDE (PAS-2) | PHILLIPS PET. | 215 | ~329 |
| POLYARYLAMIDE (J-POLYMER) | DuPONT | -120 (279) | ~343 |
| POLYAMIDEIMIDE (TORLON) | AMOCO | 275 | ~400 |
| POLYAMIDEIMIDE (TORLON AIX638) | AMOCO | 243 | ~350 |
| POLYETHERIMIDE (ULTEM 1000) | G. E. | 217 | ~343 |
| POLYIMIDE (AVIMID K-III) | DuPONT | 251 | ~343-360 |
| POLYIMIDE (LARC-TPI) | MITSUI TOATSU | 264 (275, 325**) | ~343 |
| POLYIMIDESULFONE (PISO₂) | HIGH TECH SERVICES | 273 | ~343 |
| POLYSULFONE (UDEL P-1700) | AMOCO | 190 | ~300 |
| POLYARYLSULFONE (RADEL A400) | AMOCO | 220 | ~330 |
| POLYARYLETHERSULFONE (VICTREX 4100G) | ICI | 230 | ~300 |
| POLYESTER (XYDAR SRT-500) | DARTCO | 350 (421) | ~400 |

*DIFFERENT GRADES GENERALLY AVAILABLE; DATA FROM VARIOUS SOURCES

**SEMICRYSTALLINE POWDER FORM ONLY

## TABLE IV

## NEAT RESIN MECHANICAL AND FRACTURE TOUGHNESS PROPERTIES OF HIGH PERFORMANCE THERMOPLASTICS*

| POLYMER | TENSILE PROPERTIES AT 25°C | | | FRACTURE ENERGY ($G_{Ic}$), IN-LB/IN$^2$ | NOTCHED IZOD, FT. LB/IN |
| --- | --- | --- | --- | --- | --- |
| | STRENGTH (YIELD), KSI | MODULUS, KSI | STRAIN (BREAK), % | | |
| POLYETHERETHERKETONE (PEEK) | 14.5 | ~450 | > 40 | > 23 | 1.6 |
| POLYARYLENE KETONE (PXM 8505) | 12.7 | 360 | 13 | -- | -- |
| POLYPHENYLENESULFIDE (PPS) | 12.0 | 630 | 5 | -- | 3.0 |
| POLYARYLENE SULFIDE (PAS-2) | 14.5 | 470** | 7.3 | -- | 0.8 |
| POLYAMIDEIMIDE (TORLON) | 9.2 | 667 | 1.4 | 19.4 | 2.7 |
| POLYAMIDEIMIDE (TORLON PIX638) | 13.0 | 400 | 30 | 20 | -- |
| POLYETHERIMIDE (ULTEM 1000) | 15.2 | 430 | 60 | 19 | 1.0 |
| POLYIMIDE (AVIMID K-III) | 14.8 | 546 | 14 | 11 | -- |
| POLYIMIDE (LARC-TPI) | 17.3 | 540 | 4.8 | -- | -- |
| POLYIMIDESULFONE (PISO2) | 9.1 | 719 | 1.3 | 8 | -- |
| POLYSULFONE (UDEL P-1700) | 10.2 | 360 | > 50 | 14 | 1.2 |
| POLYARYLSULFONE (RADEL A400) | 10.4 | 310 | 60 | 20 | 12 |
| POLYETHERSULFONE (VICTREX PES 4100G) | 12.2 | 380** | > 40 | 11 | 1.6 |
| POLYESTER (XYDAR SRT-300) | 20.0 | 2400 | 4.9 | 6.9 | 2.4 |

*TAKEN MOSTLY FROM PRODUCT LITERATURE
**FLEXURE MODULUS

TABLE V

MECHANICAL PROPERTIES OF HIGH PERFORMANCE THERMOPLASTIC/
UNIDIRECTIONAL CARBON FIBER COMPOSITES AT 25°C*

| POLYMER | FIBER | FLEXURAL ST., KSI | FLEXURAL MOD., MSI | SHORT BEAM SHEAR ST., KSI | COMPRESSIVE ST., KSI | $G_{Ic}$, IN-LB/IN$^2$ |
|---------|-------|-------------------|--------------------|---------------------------|----------------------|------------------------|
| PEEK (APC-2) | AS-4 | 273 | 19.4 | 15.2 | 150-160 | 10.7 |
| APC-HTX | AS-4 | 257 | 19.0 | 12.0 | 164 | 12.7 |
| PPS | AS-4 | 187 | 17.6 | -- | 95 | 5.1 |
| PAS-2 | AS-4 | 241 | 16.0 | -- | 130 | -- |
| TORLON C | C-6000 | 300 | 18.5 | 16.0 | 200 | 10.0 |
| ULTEM 1000 | AS-4 | -- | -- | -- | -- | 6.1 |
| AVIMID K-III | IM-6 | 230 | 18.0 | 13.9 | 144 | 9.7 |
| LARC-TPI | AS-4 | 285 | 14.1 | 13.8 | -- | 4.8 |
| PISO2 | AS-4 | 300 | 18.8 | 18.4 | -- | 7.0 |
| UDEL P1700 | AS | 214 | 14.0 | 14.3 | 151 | 7.7 |

*FIBER VOLUME VARIES; DATA FROM VARIOUS SOURCES

## TABLE VI

## MECHANICAL PROPERTIES OF HIGH PERFORMANCE THERMOPLASTIC/ UNIDIRECTIONAL CARBON FIBER COMPOSITES AT 25°C*

| POLYMER | FIBER | LONGITUDINAL TENSILE | | TRANSVERSE TENSILE | | IN-PLANE SHEAR | |
|---|---|---|---|---|---|---|---|
| | | STRENGTH, Ksi | MODULUS, Msi | STRENGTH, Ksi | MODULUS, Msi | STRENGTH, Ksi | MODULUS, Msi |
| PEEK (APC-2) | AS-4 | 309-356 | 19.4-20.5 | 11.9 | 1.3-1.5 | 44-47 | 0.74-0.91 |
| APC-HTX | AS-4 | -- | 19.7 | -- | -- | 39.0 | 2.18 |
| PPS | AS-4 | 238 | 19.6 | 4.6 | 1.3 | -- | -- |
| PAS-2 | AS-4 | 194 | 21.0 | 5.5 | -- | -- | -- |
| TORLON C | C-6000 | 200 | 20.6 | -- | -- | -- | -- |
| ULTEM 1000 | T-300 | -- | 19.7 | -- | 1.2 | -- | 0.71 |
| AVIMID K-III | AS-4 | -- | -- | 7.0 | 1.5 | 17.3 | 0.6 |
| | IM-6 | -- | -- | 6.8 | 1.2 | 32.8 | -- |
| LARC-TPI | AS-4 | -- | -- | -- | -- | 16.2 | 0.8 |
| UDEL P1700 | AS | 193 | 18.7 | 8.6 | 1.0 | -- | -- |

*FIBER VOLUME VARIES, DATA FROM VARIOUS SOURCES

TABLE VII

TRADE-OFFS OF THERMOSETS VERSUS THERMOPLASTICS AS COMPOSITE MATRICES

| PROPERTY | THERMOSETS | THERMOPLASTICS |
|---|---|---|
| FORMULATIONS | COMPLEX | SIMPLE |
| MELT VISCOSITY | VERY LOW | HIGH |
| FIBER IMPREGNATION | EASY | DIFFICULT |
| PREPREG TACK | GOOD | NONE |
| PREPREG DRAPE | GOOD | NONE TO FAIR |
| PREPREG STABILITY | POOR | EXCELLENT |
| PROCESSING CYCLE | LONG | SHORT TO LONG |
| PROCESSING TEMPERATURE/PRESSURE | LOW TO MODERATE | HIGH |
| FABRICATION COST | HIGH | LOW (POTENTIALLY) |
| MECHANICAL PROPERTIES -54 TO 93°C, HOT/WET | FAIR TO GOOD | FAIR TO GOOD |
| ENVIRONMENTAL DURABILITY | GOOD | UNKNOWN |
| SOLVENT RESISTANCE | EXCELLENT | POOR TO GOOD |
| DAMAGE TOLERANCE | POOR TO EXCELLENT | FAIR TO GOOD |
| DATA BASE | VERY LARGE | SMALL |

32nd International SAMPE Symposium
April 6-9, 1987

PROPERTIES AND FORMABILITY OF A NOVEL ADVANCED THERMOPLASTIC

COMPOSITE SHEET PRODUCT

R. K. Okine, D. H. Edison and N. K. Little
E. I. Du Pont de Nemours and Company
Wilmington, Delaware 19898

ABSTRACT

A thermoformable thermoplastic
composite sheet product is
described. The sheet consists of
aligned discontinuous reinforcing
fibers in a thermoplastic resin.
It is drawable to 50% or more
axial draw ratio and has
mechanical properties close to
those of straight continuous fiber
composite sheets. Development and
preservation of fiber alignment
enables the achievement of high
fiber volume fraction (FVF) and
mechanical properties. In general,
tensile modulus and strength are
90% or better of continuous fiber
composites at the same FVF.
Fatigue and creep properties are
also comparable. This paper
details the mechanical and
formability characteristics of the
sheets and presents various
applications that show the
processing advantages of this
advanced composite sheet concept
over straight continuous fiber or
woven fabric sheets.

--------------------------------

1. INTRODUCTION

Some of the most important
potential advantages offered by
advanced thermoplastic composites
over thermosets are process
automation and lower cost to
convert intermediate material
forms into final end-use parts.
The belief is that advantage can
be taken of the processability of
the thermoplastic component both
to eliminate the long thermoset
curing step and also offer a
system where reprocessing is
feasible. These potentials have
yet to be realized. On the other
hand there are activities at most
major composites manufacturers to
develop the technologies required
to make these advantages a
reality.

One of the most attractive processing techniques for converting
thermoplastic composites to parts
is sheet stamping. Cold sheet
stamping is a widely used low cost
and fast rate process in the sheet
metal forming industry. It has
been applied in various forms to
unreinforced plastics, usually in
high temperature processes.
Examples of this are thermoforming, solid-phase high pressure
forming and more recently hot
stamping, such as used in the
processing of random glass fibers
in polypropylene [1]). Various

attempts have been made to apply these techniques to advanced thermoplastic composite forming. The thermoplastic, which controls to a large extent the forming conditions, can be heated and cooled rapidly. Since there is no need for expensive and time consuming autoclaving, as in thermoset processing, forming cycle times are considerably reduced. If processing can be done in existing metal forming equipment with minimal modifications, these composites will find acceptance more easily as replacements for sheet metal.

One of the biggest obstacles faced in using advanced thermoplastic composites in such processes is the virtual absence of extensibility of the reinforcing fibers. However, the property requirements for advanced composites dictate continuous fiber-like properties. The forming processes currently under discussion in the literature primarily deal with the use of available straight or woven continuous fiber thermoplastic composites in the form of sheet or tape prelam. [2,3]. There is also mention of random fiber thermoplastic composite formable sheets such as Phillip's Ryton®, PPG's Azdel® and others [4]. Forming of straight continuous or woven fiber composites into fairly complex multi-cavity parts results in fiber wrinkling and distortions. Hence, the parts that have been produced by these techniques have tended to be generally of very gentle curvature. The random fiber systems have excellent formability but lack the high directional properties desirable in advanced composites.

The development program reported in this paper was undertaken to overcome the shortcomings stated above. The goal of the program was to develop a system of material and forming technology that enables the production of complex composite articles with continuous fiber-like properties (within 90%), in a cycle time of the order of one minute. The heart of this technology is a formable sheet consisting of aligned staple (discontinuous) fibers that permit extensive deformations while maintaining nominally continuous fiber properties. This sheet is referred to as an ordered staple formable sheet.

## 2. SHEET DESCRIPTION

The ordered staple formable sheets consist of aligned staple (discontinuous) fibers in a thermoplastic matrix. The fibers are typically either Kevlar®, carbon or glass with average fiber length in the 1 to 6 inch long range. In general, over 85% of the fibers are aligned within $\pm 5°$ of the axial direction. The general characteristics of the sheets are given in Table 1. Thermoplastic matrix resins can include amorphous, semi-crystalline or crystalline materials. The formable sheet comes in various forms that include unidirectional and angle plied laminated sheets (see Figure 1). Figure 2 is a schematic depicting the uniaxial drawing behavior of the sheet. Drawing is achieved by the sliding of the short fibers past one another. If the deformation is uniform, at least a 50% axial draw is achievable and continuous fiber-like properties are maintained.

## 3. MECHANICAL PROPERTIES

### 3.1 Tensile Modulus/Strength

The goal of the sheet development program was the preservation of straight continuous fiber-like properties in the sheets both before and after forming into 3-dimensional parts. The maintenance of fiber orientation or order during the sheet making process ensures good fiber packing and hence good quality at high fiber volume fractions (FVF). For advanced composites, a FVF of at least 50% is desirable. These are not practical in void-free random fiber systems which are generally limited to below 40% FVF. Table 2 gives representative values of the degree of fiber alignment in the ordered staple sheets along with those of continuous fiber tape and Highform®, which is produced by Richard Klinger, Ltd. in the U.K. and consists of 1/8 inch long aligned carbon fibers in poly-ethersulfone (PES) [5,6]. As the table shows 85% of the fibers in the ordered staple sheets lie within +5° of the 0° direction. This degree of alignment falls between the 1/8 inch Highform® product and a continuous fiber tape.

Selected mechanical properties of the ordered staple sheets are listed in Table 3. For the tensile properties, a more convenient way of looking at the data is shown in Figures 3 and 4 where the modulus and tensile properties relative to those of the reinforcing fibers are plotted versus the fiber volume fraction. The plots show that tensile modulus and strength of the ordered staple sheets are within 90% of the corresponding straight continuous fiber composite controls. Within the limits of the fiber lengths of interest (1/2 - 6 inch) the tensile modulus and strength are not a strong function of the staple length. The data for Highform® indicates that this is true for staple length down to 1/8 of an inch for carbon fibers. This observation for modulus is in agreement with theoretical predictions from the Halpin-Tsai equations. As Figure 5 shows, the short fiber composite modulus should be over 90% of the infinite (continuous) fiber modulus when the fiber length-to-diameter ratio is greater than 1000. For carbon fibers this is equivalent to 1/4 inch long fibers. The data for Highform®, which consists of 1/8 inch long fibers indicate that the reinforcing efficiency of the fibers is good enough to produce 95% of the straight continuous fiber composite modulus with a fiber l/d ratio of about 500.

## 3.2 Creep

The limited data obtained so far for creep and fatigue indicates similar behavior for discontinuous and continuous fiber composite. Figure 6 is a plot of the creep behavior of Kevlar®/J-2 polymer composites for both the ordered staple and continuous fiber composites. A gage length of 20 inches was used in the tests. As the plot shows the behavior of the two sheet structures for 25% and 50% of the ultimate tensile stress loading are very similar. The limited data available in the literature indicate that the creep and fatigue behavior of aligned short fiber composites are approximately the same as those of continuous fiber composites [7].

## 4. FORMABILITY

The most important characteristic of the ordered staple formable sheet is its drawability. To study the forming behavior of the sheets, two deformation modes were used. These were high temperature tensile drawing and 3-dimensional forming using a hemisphere.

### 4.1 High Temperature Tensile Drawing

Tensile drawing provides the simplest deformation mode, and gives information on the basic drawing behavior of the sheets. The set-up for the tensile drawing tests is as shown in Figure 7. This consisted of a heated chamber for heating the test section of test samples. As will become clear shortly, the clamped sections of the test samples were kept cold and outside the heat chamber. The heated section of the test samples was either 8 or 15 inches and these were drawn in an Instron or MTS test machine, respectively. The drawing load, strain rate and temperature were recorded.

Figure 8 shows a typical plot of the tensile drawing stress versus strain for high temperature tensile drawing of an ordered staple sheet. The three points of interest in such plots are the shape of the curve before and after the peak stress and the value of the peak stress.

The declining slope of the curve after the peak stress is the result of necking or non-uniform deformation in the test sample. This is similar to the tensile deformation instability that occurs in most materials. The drawing stress is due to the shearing stresses resulting from the sliding of the staple fibers past each other. A horizontal or a gentler sloping curve means that the deformation is more uniform along the length of the test specimen. In general, we have found that the sheets with a lower fiber volume fraction (FVF) have more uniform deformation than those with high FVFs. Lower deformation rate seems to also produce a gentler slope.

The peak stress value in the plot is of particular interest in forming because this stress value has to be overcome before any appreciable deformation can take place in the sheet. This stress can be likened to the yield stress in the deformation behavior of ductile solid materials. If it is assumed that the peak stress indeed corresponds to the point during high temperature deformation of the ordered staple sheets where yielding occurs, then the behavior of the sheet before this point should be elastic. The straightness of the stress/strain curve before the peak stress, such as in Figure 8, suggests that this is the case. This means that it should be possible to use elasticity arguments, such as the micromechanics of short fiber composites, to predict the value of the peak stress. This will then provide an insight into the pre-yield behavior of the sheets and how they might be advantageously controlled.

### 4.1.1 Stress Analysis

If it is assumed that the ordered staple sheet behaves elastically until the peak yield stress, the yield or ultimate flow stress can be represented by the equation for the tensile strength of ordered staple solid composite [8]

$$\sigma_p = \tau_y \frac{\ell}{d} V_f + \sigma_{mu} V_m \qquad (1)$$

where

$\sigma_p$ = Peak drawing stress

$\tau_y$ = Matrix shear yield stress

$\sigma_{mu}$ = Matrix tensile strength or flow stress

$\ell/d$ = Fiber length-to-diameter ratio

$V_f$ = Fiber volume fraction

$V_m$ = Matrix volume fraction

Equation (1) assumes that the sheet failure does not result in fiber breakage but in the shear yielding of the matrix. This assumption can also be checked by ensuring that the critical fiber length required for fiber breakage is longer than the staple length. The critical fiber length is given by the relationship

$$\frac{\ell_c}{d} = \frac{\sigma_{fu}}{2\tau_y} \qquad (2)$$

where

$\ell_c$ = Critical fiber length

$\sigma_{fu}$ = Fiber tensile strength

For a PETG matrix resin, equation (2) predicts a critical fiber length for Kevlar® of the order of 60 inches. This is based on fiber diameter and tensile strength of 11.9 microns (0.47 mils) and 525 KPSI, respectively. The matrix shear yield stress at 220°C and .02 in/min draw speed is estimated at 2.13 psi (see Figure 11). This critical fiber length is an order of magnitude longer than the actual maximum staple length of 6 inches in the ordered staple sheet structures tested. Using the Tresca yield criteria that $\sigma_{mu} = 2\tau_y$ , equation (1) can be re-written as

$$\sigma_p = \sigma_{mu}\left(\frac{\ell}{d}\frac{V_f}{2} + V_m\right) \qquad (3)$$

$V_m$ in general will be much smaller than the first term in the bracket in equation (3) and thus can be ignored. Hence equation (3) can then be written as

$$\sigma_p = \left(\frac{\ell}{d}\right)\left(\frac{\sigma_{mu}}{2}\right) V_f = \frac{\ell}{d}\tau_y V_f \qquad (4)$$

Equation (4) shows that the peak drawing stress is proportional to the FVF, the fiber l/d ratio, and the matrix shear flow stress. For a 6-inch length Kevlar® fiber the l/d ratio is of the order of 12,000 while the matrix flow stress, at the temperatures of interest, is of the order of 4 (see Figure 11). The main effect of the matrix flow stress $\sigma_{mu}$ (or $\tau_y$) is to determine the temperature and strain rate dependence of equation (4) since the material flow stress behavior is viscoelastic.

4.1.2 Experimental Verification

To determine how well this equation estimates the peak drawing stress, the data plotted in Figure 9 were generated. In the figure, the ratio of the peak drawing stress and the fiber length-to-diameter ratio is plotted against fiber volume fraction. At the low draw rate of $2 \times 10^{-4}$/sec., the plot is linear as is expected from equation (4). At the higher strain rate of $2 \times 10^{-2}$/sec., the curve levels off beyond 25% FVF. Non-linear viscoelastic effects could account for this deviation due to the increased shear rate as a result of increased apparent draw rate and high fiber volume fraction that acts to increase the effective shear rate in the resin. The shear rate and temperature

dependence of the peak drawing stress is a result of the shear flow stress behavior of the resin. Figure 10 is a plot of peak stress values versus deformation rate of Kevlar® ordered staple fibers in PETG resin. The figure shows that the peak stress per fiber volume fraction at 255°C is reduced from about 32,000 psi to 1,300 psi (a 25X drop) over a 2-decade drop in the draw rate. Even though this low draw rate might not be a practical rate to use in actual production, the figure does point out the effect of the deformation rate on the draw stress. This result is in qualitative agreement with the expected dependence on the draw rate of the viscoelastic behavior of the matrix as suggested by equation (4). Figure 11 is a plot of an estimate of the shear yield (or flow) stress behavior of PETG versus strain rate. The plot was obtained from capillary rheometry data for shear viscosity determination. The shape of the plots agree with the behavior of the composite as depicted in Figure 10 and predicted by equation (4). Using the value of the shear flow stress of the matrix at the lowest pressing rate (0.02/sec), the peak stress per FVF can be estimated by equation (3) to be 26,000 psi at 220°C, which although is 38% off the measured 42,000 psi, puts us in the right ball park for estimating the peak stress level.

The results for J-1 Polymer matrix using the same fiber are similar to those for PETG. For tensile drawing rates of 10 and 0.02 in/min. (for an 8-inch gage length) at 305°C, the peak stress per FVF were measured to be 30,556 and 4,032 psi, respectively. The predicted value using equation 4 and an estimated shear yield stress value of 2.16 psi from capillary rheometry data was 27,168 psi for the 10 in/min. draw rate.

### 4.1.3 Implications for Sheet Forming

The preferred deformation mode for the ordered staple formable sheets during forming is tensile. This deformation mode allows better control of fiber placement and reduces fiber wrinkling. This contrasts with the squeezing type deformation used for forming Azdel®. To enable fiber drawing in the ordered staple sheet structure the edges of the sheet must be clamped. The clamping stress must be high enough to overcome the peak tensile drawing stress before any significant drawing will take place. Unfortunately, since the clamping load is transmitted to the fibers through the shearing of the resin, the same mechanism that allows fiber sliding in the bulk of the sheet is at play in the clamped region. The resin at the interface of the clamp and the sheet acts as a lubricant. The clamp stress achievable is, therefore, determined mainly by the shear characteristics of the resin whereas the drawing stress is determined by the composite as a whole. If the load due to the drawing stress of the sheet is larger than that due to the shear yield stress at the interface of the clamps, the sheet will slide from under the clamps. For a 6-inch long fiber, it has been observed that the clamped region has to be kept at a much lower temperature than the bulk of the sheet to prevent slippage in the clamps. Using such a temperature distribution to ensure sheet clamping is not the desirable way to go. If equation (4) is assumed to describe the mechanics of the drawing of these sheets, then a way of reducing the peak drawing stress without affecting the resin shear stress at the clamp interface is to reduce the fiber l/d ratio, for a given FVF.

### 4.2 Forming of Hemisphere

A 6-inch diameter hemisphere was formed from the formable sheets to analyze the 3-dimensional deformation behavior. The forming tests were done in a 100-ton variable speed hydraulic stamping press using a set of matched metal dies. Figure 12 is a schematic of the set-up for forming the hemisphere. A photograph is shown in Figure 13.

A typical forming cycle involves heating the sheet to the resin melt temperature in an oven outside the forming press, then quickly transferring it into the dies, which are kept at the same or a lower temperature. With the sheet edges clamped, the press is then closed to form the sheet. After forming the sheet is then cooled under pressure inside the dies to below the resin glass transition temperature. This procedure is typical for thermoplastic composites forming. For forming J-2 polymer composite, the sheet and the dies were heated to 315°C. For PET the temperature was 285°C. Cooling was done under an equivalent of 1400 psi over the projected area of the hemisphere. Due to the high drawing stresses in the ordered staple sheets as stated in the previous section, the edges of the sheets were clamped cold using a set of hydraulic grips. This provided a normal stress of 296 psi on a set of diamond-shaped serrated-faced clamps. Figures 14a, b and c show typical hemispheres formed from carbon/J-2 polymer, Kevlar®/J-2 polymer and glass/PET. The figures show that the sheets can be deep drawn to over 50% areal increase. In contrast, a straight continuous or woven fiber sheet requires the pulling in of fiber into the cavity region resulting in fiber wrinkling and distortion around the periphery of the cavity and in the flange area (see Figure 15). This wrinkling problem is expected to be worse in a more complex multi-cavity part consisting of say three or four hemispheres. Figure 16 depicts how such parts would look. This fiber distortion issue has not yet been addressed by investigators trying to form complex shapes using straight continuous or woven fiber composites. It is not clear what level of fiber wrinkling can be tolerated in formed parts, and where they exist, what their effects on mechanical properties are. However, as use of stamped thermoplastic composites parts moves in the structural arena, these issues will have to be faced and solved. The main advantage of the ordered staple formable sheets lies in helping to solve these problems.

### 4.3 Other Formed Shapes

Various 3-dimensional shapes besides the hemisphere have been formed to demonstrate the versatility of the ordered staple sheet structure. Photographs of some of these are shown in Figures 17 (rib shape) and 18 (beaded rib panel).

### 5. CONCLUSIONS

The forming characteristics and capabilities of the ordered staple formable thermoplastic composite sheets have been described. A model proposed to describe the tensile drawing behavior up to the yield point seems to agree with experimental data. The work done to date on the sheets demonstrate advantages over straight continuous or woven fiber composite sheets. These sheets enable the forming of complex 3-dimensional advanced composite articles without any appreciable loss of mechanical properties. The complexity of wrinkle-free shapes that can be formed with controlled fiber placement are higher than obtainable from straight continuous or woven fiber sheets that are currently available on the market.

### 6. REFERENCES

1. S. Sikes, "Stampable Plastic Composites Gaining Ground", Machine Design, May 22, 1986, p. 68.

REFERENCES (cont.)

2. T. P. Kueterman, "Advanced Manufacturing of Thermoplastic Composites", Proceedings, Advanced Composites Conference, ASM & ESD, December 2, 1985, Dearborn, MI, p. 147.

3. L. Post and W. H. M. van Dreumel, "Continuous Fiber Reinforced Thermoplastics", Progress in Advanced Materials and Processes: Durability, Reliability and Quality Control, G. Bartelds and R. J. Schliekelmann, editors, Elsevier Science Publishers, 1985, p. 201.

4. R. H. Wehrenberg II, "Stampable Reinforced Thermoplastic Sheet", Materials Engineering, Vol. 99, June, 1984, p. 35.

5. U. K. Patent Nos. 1,249,291 and 1,389,539.

6. K. D. Potter, "The Influence of Accurate Stretch Data for Reinforcements on the Production of Complex Structural Moldings: Parts 1 and 2", Composites, July, 1979, p. 161.

7. K. Schulte, G. Horstenkamp and K. Friedrich, "Mechanical Properties of Aligned Short Carbon Fiber Reinforced PES and PI Laminates". European Sapce Agency translated report of DFVLR-FB-84-24

8. B. D. Agarwal and L. J. Broutman, ANALYSIS AND PERFORMANCE OF FIBER COMPOSITES, John Wiley and Sons, Inc., 1980, p. 81.

## 7. ACKNOWLEDGMENT

The authors would like to acknowledge the valuable contributions made by several people at the Du Pont Company in the course of this work, particularly, those of H. G. Lauterbach, J. F. Pratte, S. J. Medwin, W. C. Walker, C. M. Cavanaugh, T. N. Rielly, T. E. Armiger, N. E. Hopkins, K. M. Robinson, R. C. Groeber, J. Serio, M. W. Kroeber and K. Haight. The help from Prof. Tim Gutowski of MIT is also acknowledged.

Richard K. Okine is a Senior Engineer in the Composites Division (Textile Fibers Department) at E. I. Du Pont Company. Prior to joining this Division in September 1985, he was a Research Engineer in the Engineering Research and Development Division (Engineering Department). He is currently working on developing the forming technology for Du Pont's formable advanced composites sheets. He joined Du Pont in 1981 after receiving his S.B, S.M. and Sc.D. degrees in Mechanical Engineering, all from MIT.

David H. Edison received an A.B. from Oberlin College in 1955 and a Ph.D. in Chemistry from the University of Rochester in 1959. Since that time he has had a variety of technical assignments in the Textile Fibers Department of E. I. Du Pont Company prior to joining the Composites Division in November of 1985.

Nancy K. Little is an engineer with E. I. Du Pont Company at the Engineering Development Laboratory. She is currently developing processing equipment for the Composites Division. She came to Du Pont in 1981 after receiving her B.S. in mechanical engineering from Drexel University.

TABLE 1

ORDERED STAPLE SHEET CHARACTERISTICS

| FIBER | KEVLAR® | | | CARBON | | GLASS | |
|---|---|---|---|---|---|---|---|
| TYPE | 29 | 49 | 49 | AS4 | AS4 | OCF T-30 | P353B |
| AVERAGE LENGTH (INCH) | 5.0 | 2.0 | 3.8 | 1.3 | 3.2 | 1.6 | 3.4 |

EXAMPLES OF RESINS INCLUDED IN EXPERIMENTAL PROGRAM

J-2 Polymer: An amorphous polyamide copolymer based upon PACM - bis(para-aminocyclohexyl)methane.

J-1 Polymer: A semi-crystalline polyamide homopolymer also based on PACM).

PEEK - Polyetheretherketone

PET - Polyethylene terephthalate

PETG - Glycol-modified PET (Copolyester)

TABLE 2

EXAMPLES OF FIBER ORIENTATION IN SHEETS

| | ORDERED STAPLE SHEETS | | HIGHFORM | CONTINUOUS FIBER TAPE |
|---|---|---|---|---|
| FIBER | CARBON | | CARBON | CARBON |
| RESIN | J-2 POLYM | PETG | PES | J-1 POLYM |
| FVF | 55 | 40 | 49 | 50 |
| PERCENT WITHIN | | | | |
| ± 5 DEG | 92 | 85 | 74 | 93 |
| ±10 DEG | 96 | 95 | 93 | 98 |
| ORIENTATION STANDARD DEVIATION | 6.13 | 6.78 | 7.13 | 3.85 |

TABLE 3

TYPICAL  MECHANICAL PROPERTIES
ORDERED STAPLE VS. CONTINUOUS FIBER SHEETS

| SHEET TYPE | AVERAGE FIBER LENGTH (") | FVF % | MODULUS (MPSI) | STRENGTH (KPSI) | SBS (KPSI) | COMP STR. (KPSI) |
|---|---|---|---|---|---|---|
| | | ORDERED STAPLE | | | | |
| AS4-CARBON/ J-2 POLYMER | 3.2 | 55 | 18.8 | 246 | 13.7 | - |
| KEVLAR®-49/ J-2 POLYMER | 3.8 | 58 | 8.9 | 154 | 7.2 | 35.6 |
| KEVLAR®-29/ J-2 POLYMER | 5.0 | 50 | 7.2 | 132 | - | - |
| PES/CARBON (HIGHFORM) | 0.125 | 49 | 14.4 / 13.0 | 131 / 145 | - | - |
| | | STRAIGHT CONTINUOUS | | | | |
| KEVLAR®-49/ J-2 POLYMER | - | 60 / 54 | 10 / 10.6 | 157 / 147 | 8-10 | 37-40 |
| AS4-CARBON/ J-2 POLYMER | - / - | 60 / 55 | - / 18.2 | - / 243 | 14-15 / - | 151 / - |

Figure 1:  Ordered Staple Product in Sheet Form.

Before Drawing                    After Drawing

Figure 2:  ORDERED STAPLE SHEET CONCEPT
           SHOWING FIBER SLIDING

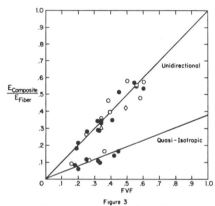

Figure 3
**Tensile Modulus Versus Fiber Volume Fraction:**
Ordered Staple ( O Kevlar®-29, Δ Kevlar®-49, □ AS4 Carbon );
Straight Continuous Fiber ( ● Kevlar®-29 & Kevlar®-49, ● AS4 Carbon),
Highform ( ◊ XAS Carbon)

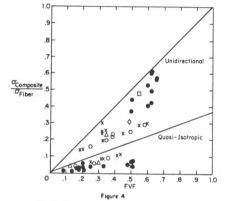

Figure 4
**Tensile Strength Versus Fiber Volume Fraction:**
Ordered Staple ( O Kevlar®-29, Δ Kevlar®-49, □ AS4 Carbon);
Straight Continuous Fiber ( X Kevlar®-29 & Kevlar®-49, □ AS4 Carbon),
Literature Data ( ● Various Fibers); Highform ( ◊ XAS Carbon)

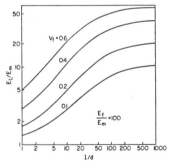

Figure 5:  Variation of Longi-
tudinal Modulus of Short Fiber
Composites Against Fiber Aspect
Ratio (l/d) for Different
Volume Fractions ($E_f/E_m$ =100) [8]

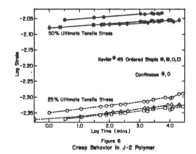

Figure 6
Creep Behavior in J-2 Polymer

1423

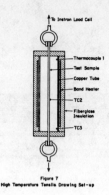

Figure 7
High Temperature Tensile Drawing Set-up

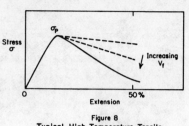

Figure 8
Typical High Temperature Tensile
Drawing Stress/Strain Behavior

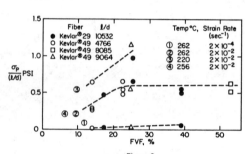

Figure 9
High Temperature Tensile Drawing Stress
Versus Fiber Volume Fraction

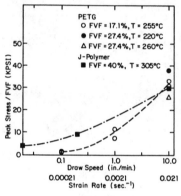

Figure 10
Peak Tensile Stress Versus Draw Rate
of 6-Inch Kevlar® Ordered Staple

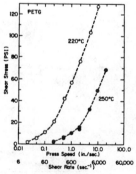
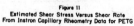

Figure 11
Estimated Shear Stress Versus Shear Rate
From Instron Capillary Rheometry Data for PETG

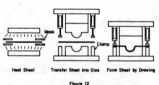

Figure 12
Schematic of Hemisphere Forming Sequence

Figure 13:    Forming Press and Die Set-up.

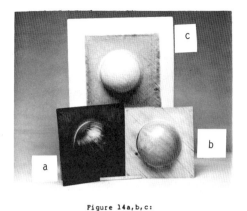

Figure 14a,b,c:

Typical Formed Hemispheres of Ordered
Staple Carbon/J-2 Polymer,
Kevlar®/J-2 Polymer and Glass/PET.

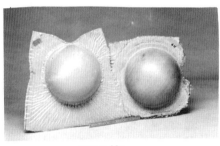

Figure 15:

Formed Hemisphere of Straight Continuous
Kevlar®/PETG.

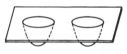

Ordered
Staple
Sheet

Wrinkles

Straight
Continuous
or Woven
Fiber
Sheet

Figure 16
Schematic of Formed Double Hemisphere

Figure 17:    Formed and Trimmed Rib Stiffener
of Kevlar®/PETG.

Figure 18:    Formed and Trimmed Beaded Rib of
Kevlar® and Carbon/J-2 Polymer.

# THE EFFECT OF FILLERS ON THE STABILITY OF EPOXY ADHESIVES

David Elmendorf and Edward J. Hughes
Kearfott, A Division of The Singer Company
1150 McBride Avenue
Little Falls, NJ 07424

## Abstract

Torsion shear creep measurements have been used to characterize the dimensional stability of filled and unfilled epoxy adhesive systems. The effect of filler type, geometry, and concentration on the creep properties was measured at $65^\circ$ and $79^\circ$C. Data are presented which show the effect of each of these variables.

## 1. INTRODUCTION

Epoxy adhesives are commonly used to join components in precision inertial guidance systems such as gyroscopes and accelerometers. Any instability in these joints can be detrimental to the system performance; therefore, epoxies with a high degree of dimensional stability are required. There is an abundance of data in the literature on mechanical properties of bonded adhesives such as lap shear strength, flexural strength, peel strength, etc., but dimensional stability information is usually not available.

Earlier work at Singer-Kearfott [1,2] was devoted to the development of techniques for accurately measuring the tensile and shear properties of adhesively bonded joints. As a result, testing procedures now exist for measuring adhesive deformations at extension sensitivities of $2.5 \times 10^{-5}$ mm ($10^{-6}$ inch). These techniques have recently been adapted to measuring the creep properties of adhesives to characterize their dimensional stability. Creep measurements were used to evaluate the role of fillers (geometry, type, and amount) on dimensional stability.

This work represents a new approach to the selection of adhesives for applications requiring a high degree of dimensional stability (a characteristic not usually associated with epoxy adhesives). Heretofore selection was usually made on the basis of property values commonly available in the literature, but which were not necessarily an indication of the stability of the particular epoxy system.

## 2. EXPERIMENTAL

Araldite 6005 was used as the matrix to determine the effects of fillers on dimensional stability. Five types of filler materials were incorporated in the 6005 matrix (see Table 1).

Napkin ring adherends (outside diameter 7.62 cm, wall thickness 0.32 cm) made from Ti-6Al-4V were bonded together end-to-end with the adhesive under test. Prior to bonding they were subjected to standard cleaning procedures. In all cases the manufacturers' recommended cures were used with bond line thicknesses in the range 0.10 to 0.15 mm. Full details of bonding procedures and torsion shear testing are described elsewhere[3,4]. The torsion shear apparatus is shown schematically in Figure 1. Creep tests were performed by loading the specimens to the required stress level by rotating one adherend about an axis coincident with the longitudinal axis of the specimen while the other adherend remained fixed. The load was applied in an Instron tensile testing machine and the deformation was meas-

ured with a capacitance type extensometer which has a measuring sensitivity of $2.5 \times 10^{-5}$ mm. In most cases, creep was monitored for a period of no more than three hours. Experience has shown that over longer periods of time, unexpected and unpredictable changes in the extensometer output may occur as a result of electronic or temperature drifts.

Creep measurements were made as a function of stress at $65^{\circ}$ and $79^{\circ}$C. A typical family of creep curves for an epoxy tested at different stress levels is shown in Figure 2.

## 3. RESULTS AND DISCUSSION

### 3.1 Effect of Filler Content on Dimensional Stability

Earlier work with unfilled epoxies showed that Araldite 6005 was the most stable of the systems tested so it was chosen as the matrix to study the effect of filler content. Lithium aluminum silicate was chosen as the filler for the initial phase of the investigation. Different amounts of filler were incorporated into the matrix and the creep results are shown in Figure 3. At 30 weight percent (wt %) filler, no apparent improvement in stability was observed. However, at greater filler levels, the stability improved dramatically.

## 3.2 Effect of Filler Geometry and Type

In addition to lithium aluminum silicate, the effect of two additional types of filler (aluminum oxide and silicone carbide) on the stability of Araldite 6005 were investigated. Two geometric forms of each filler were used.

### 3.2.1 Aluminum Oxide

Aluminum oxide in both tabular and powder form was added to Araldite 6005 as a filler material. Initial data (Figure 4 and Table 2) show that the powdered $Al_2O_3$ was considerably more effective than the tabular $Al_2O_3$. This can be seen by comparing the normalized creep strains (Table 2) and creep curves for specimens filled with 40 weight percent powdered $Al_2O_3$ and tabular $Al_2O_3$. Since both forms of the filler should be comparable in reducing internal stresses due to shrinkage, coefficient of expansion, peak exotherms, etc. (all of which influence stability), it seems to suggest that the difference may lie in the different effects of the two forms on the rheology of Araldite 6005. As a result, the amorphous powder form may be more effective in restricting movement of the molecular chains under the influence of an external stress.

### 3.2.2 Silicon Carbide

The third filler material incorporated into the Araldite 6005 resin was silicon carbide in both whisker and powder form. A comparison of the two types of fillers at $65^\circ C$ is shown in Figure 5. The effect seems to be the opposite of that observed in the Araldite $6005/Al_2O_3$ system. The whisker reinforcement appeared to be much more effective than the amorphous powder in inhibiting creep. The normalized values of the creep strains obtained with silicon carbide are included in Table 2 and they indicate that for a given volume fraction of filler, silicon carbide whiskers are the most effective of the fillers investigated in improving the dimensional stability of the Araldite 6005.

In this investigation three different types of powdered fillers were used to fill the Araldite 6005 system. Normalized creep strains (after 7000 seconds of creep) for each of the powder-filled Araldite systems are listed in Table 3 which also includes both volume and weight percent of each filler. From Table 3, it appears that those epoxies with similar volumes of filler had similar creep behavior as evidenced by the normalized strains. This suggests that in terms of creep, the type of powder filler may not be a factor and stability is determined only by the relative amounts of filler and epoxy.

## 4. SUMMARY

This work has demonstrated that it is feasible to classify the dimensional stability of adhesively bonded joints through the use of creep tests. As a result, in those applications where stability is a requirement, it offers design engineers a means of adhesive selection based on quantitative measurements rather than mechanical properties not necessarily related to stability. Some of the more specific findings of this study can be summarized as follows:

o For fillers in powdered form, the type of powder did not appear to play a role in determining the creep behavior. The volume fraction of filler powder was the controlling factor.

o Geometry of the filler particles influenced creep behavior of the filled epoxy adhesives.

o For a given volume fraction of filler, whisker form silicon carbide was the most effective in reducing creep.

## ACKNOWLEDGEMENTS

The authors would like to thank Bud Dunham who performed much of the testing described in this paper and Elizabeth Gears who mixed the epoxies prior to bonding.

## REFERENCES

1. F.C. Bossler, M.C. Franzblau, and J.L. Rutherford, J. Sci. Instr. (J. of Phys. E.), (1968), 1, 829-33.

2. J.L. Rutherford, F.C.Bossler, and E.J. Hughes, Rev. Sci. Inst., (1968), 39, 666-71.

3. E.J. Hughes, W. Althoff, and R.B. Krieger, Adhesive Bonding of Aluminum, Edited by E.W. Thrall and R.W. Shannon, Marcel Dekker, New York, 1985, 141-174.

4. E.J. Hughes and J.L. Rutherford, "Study of Micromechanical Properties of Adhesive Bonded Joints," Final Rept., The Singer Co., Contract No. DAA21-67-C-0500, Picatinny Arsenal, Dover, NJ, 1968.

BIOGRAPHIES

David Elmendorf, Materials Engineer, has been with Singer-Kearfott for over two years and has worked extensively with microstrain behavior of adhesives. He graduated from Lehigh University in 1984 with a B.S. in Metallurgy and Materials Science Engineering and is currently pursuing an M.S. in Engineering Management at New Jersey Institute of Technology. Mr. Elmendorf is a member of the Metallurgical Society of A.I.M.E.

Dr. Edward J. Hughes, Principal Scientist, has been with Singer-Kearfott for twenty years where he has conducted extensive research of metals and adhesives. Prior to that he was a Senior Scientist at the Materials Research Corporation, Orangeburg, NY. Dr. Hughes received his B.Sc. (Honors) and Ph.D. degrees in Physical Metallurgy at Birmingham University in England. He is a member of S.A.M.P.E. and ASM.

Table 1 - Filler Materials

| Composition | Shape | Size (microns) |
|---|---|---|
| alumina | tabular | 18 |
| alumina | powder | 60 |
| silicon carbide | whiskers | 0.6 Dia. |
| silicon carbide | powder | 70 |
| lithium aluminum silicate (mixture of $Li_2O$, $Al_2O_3$, $SiO_2$) | powder | 20 |

Table 2 – Creep of Filled 6005 at 65$^\circ$C

| Type Filler | Amount Filler | | Normalized Creep Strain at 7000 Seconds ($\times 10^{-9}$/psi/sec) |
|---|---|---|---|
| | Wt % | Vol % | |
| Tabular Al$_2$O$_3$ | 72 | 55 | 0.21 |
| Tabular Al$_2$O$_3$ | 40 | 30 | 0.53 |
| Powder Al$_2$O$_3$ | 40 | 30 | 0.28 |
| Powder SiC | 50 | 35 | 0.33 |
| Whisker SiC | 33 | 23 | 0.19 |
| Powder LiAlSiO$_3$ | 60 | 46 | 0.18 |
| Powder LiAlSiO$_3$ | 50 | 38 | 0.28 |

Table 3 – Effect of Powdered Fillers on Creep of Araldite 6005 at 65$^\circ$C

| Type Filler | Amount Filler | | Normalized Creep Strain at 7000 Seconds ($\times 10^{-9}$/psi/sec) |
|---|---|---|---|
| | Wt % | Vol % | |
| LiAlSiO$_3$ | 50 | 38 | 0.28 |
| Al$_2$O$_3$ | 50 | 30 | 0.23 |
| SiC | 50 | 35 | 0.34 |

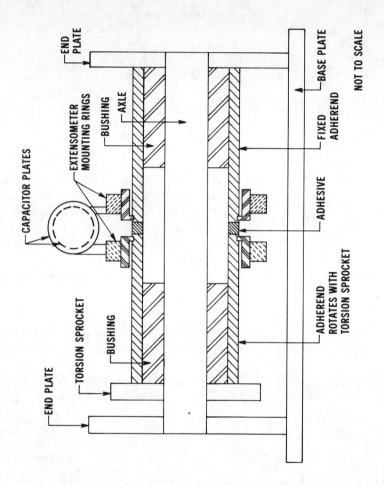

Figure 1 – Schematic of Torsion Shear Apparatus

END PLATE

CAPACITOR PLATES

EXTENSOMETER MOUNTING RINGS

BUSHING

AXLE

TORSION SPROCKET

BUSHING

END PLATE

BASE PLATE

FIXED ADHEREND

ADHESIVE

ADHEREND ROTATES WITH TORSION SPROCKET

NOT TO SCALE

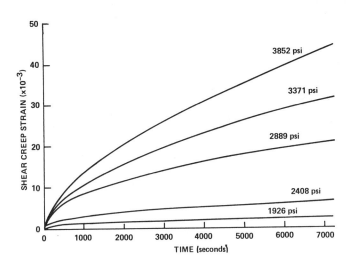

Figure 2 – Typical Torsion Shear Creep Curves for an Epoxy Adhesive at 79°C

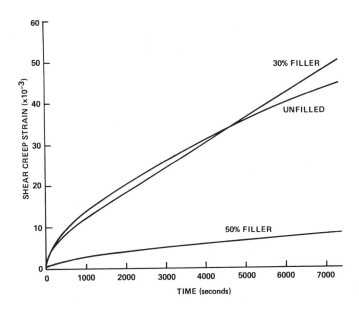

Figure 3 – Effect of Filler Content on Creep of Araldite 6005 at 79°C and 3850 psi

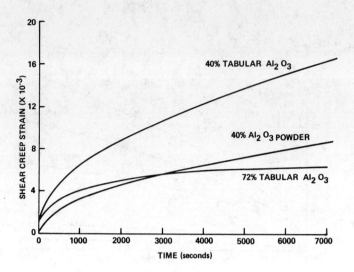

Figure 4 – Effect of Alumina Filler Geometry on Creep of Araldite 6005 at 65°C and 4300 psi

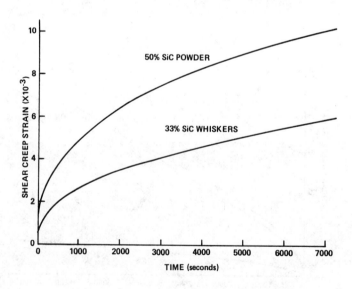

Figure 5 – Effect of Silicon Carbide Filler Geometry on Creep of Araldite 6005 at 65°C and 4300 psi

32nd International SAMPE Symposium
April 6-9, 1987

RESPONSES OF PCS-TYPE SiC/Al COMPOSITE TO

THERMAL- AND/OR NEUTRON EXPOSURES

N.Igata, A. Kohyama and H.Tezuka

Y. Imai*, H. Teranishi* and T. Ishikawa*

Department of Materials Science,

The University of Tokyo

7-3-1 Hongo, Bunkyo-ku, Tokyo 113, Japan

*): Nippon Carbon Corporation

2-6-1 Hachicho-bori, Chuo-ku, Tokyo 104, Japan

## ABSTRACT

Characteristics of polycarbosilane
(PCS) type SiC/Al preform wires and
their hot-pressed sheets were
investigated. The objectives of
this investigation are to clarify
responses of preform wires and hot-
pressed sheets to thermal- and
neutron exposures. Mechanical
properties were measured by tensile
test and three point bending test.
Microstructures were inspected by
means of SEM and TEM. The low
induced activation characteristic
of SiC/Al composite is anadvantage
as fusion reactor material. As one
of the important characteristics
for the application to fusion
reactors, the effects of neutron
irradiation were studied using
fission neutrons from JOYO and JMTR
and with 14 MeV neutrons from RTNS-
II. The SiC/Al composite and SiC
fiber showed excellent stability
under neutron irradiations. Increments of tensile strength and
Young's modulus were clearly observed by irradiation in FBR (JOYO).
The microstructure evolution during
electron irradiation in HVEM proved
suppression effect of interfacial
SiC formation by irradiation.

## 1. INTRODUCTION

SiC/Al composites have been inten-
sively studied in Japan as the
important part of the National R &
D project on advanced composite
materials[1]. One of the emphasis
of this project is to develop high
quality and low cost SiC/Al
composite utilizing the continuous
and multifilament-yarn polycarbosilane (PCS) type SiC fibers

(Nicalon) obtained from a spun polymer precursor[2].

Besides application in aeronautical industries, SiC/Al and C/Al FRMs are thought to be potential candidate materials for nuclear fusion reactors because of their unique properties of interest, especially of their low activation characteristics.[3,4]

SiC/Al preform wire has been produced in a test plant continuously without any kind of prior surface treatment to SiC fibers[5]. This work was carried out on the basis of the recent extensive advancement in fabrication processes for Nicalon fiber and SiC/Al preform wire and of our intensive works on fundamental radiation damage study for fusion reactor materials. One of the objectives of this investigation is to evaluate the effects of thermal exposure, which is the essential factor to affect the mechanical properties of final products. The others are to clarify the mechanism of neutron damage, and to develop SiC/Al composite for application to fusion reactor[6,7].

## 2.EXPERIMENTAL

SiC/Al preform wires including 500 PCS-type SiC fibers were produced from the test plant at the R & D Laboratory, Nippon Carbon Co. Ltd, by the liquid metal infiltration method. Industrial grade pure aluminum, Al050, was selected as a matrix material because of its excellent compatibility with PCS-type SiC fibers[2]. Preform wires with the fiber volume fraction about 40 % were tensile tested and three point bending tested for evaluating mechanical properties. In order to evaluate the fiber, matrix and interfacial strengths, separately, the aluminum matrix was chemically removed, and SiC fibers were extracted. Then they were tensile tested by monofilament tensile test. The surface morphologies of as received and as extracted fibers and the fracture surfaces were inspected by scanning electron microscopy (SEM). Microstructures of the composite were precisely inspected by means of transmission electron microscopy (TEM) with EDX capability.

Isochronal heat treatments were done at 823, 873 and 923 K for 64 minutes in air to clarify the effect of heat treatment on mechanical properties of SiC/Al preform wire. Microstructure change at elevated temperatures was in-situ observed under HVEM. Condition used in HVEM experiment was,

Electron energy: 1 MeV

Max. dose: 11 dpa

Test Temp.: 300, 370, 475, 760 K

From flux differences of electrons, thermal- and radiation- effects have been separately evaluated. Neutron irradiations were carried out using neutrons from Japan Material Test Reactor (JMTR) at 423 K to $7 \times 10^{23}$ n/m$^2$, neutrons from JOYO at 723 K to $1 \times 10^{24}$ n/m$^2$ and $1 \times 10^{25}$

n/m². 14 MeV neutrons from RTNS-II were also utilized at ambient temperature to the fluence of $5 \times 10^{20}$ n/m². Post irradiation experiments were performed at Reserch Institute of Iron, Steel and Other Metals, Tohoku Univ., Oarai Branch, using the testing facilities of the same types as those for cold samples. The length of neutron irradiated preform wires was 12.3 mm. Post irradiation mechanical tests of preform wires were carried out by three point bending test, only.

## 3.RESULTS AND DISCUSSION

### 3.1Effects of Heat Treatments

The PCS-type SiC/Al preform wire used in this work is subjected to elevated temperature exposure during forming processes for final products. Therefore, fundamental understandings of thermal exposures, i.e. heat treatments, are definitly required. The excellent thermal stability of PCS-type SiC/Al upto 723 K in air has been

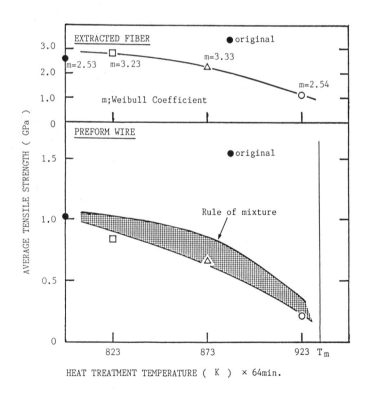

Fig. 1 : Effect of heat treatment ( T x 64 min. ) on tensile strength of SiC/Al preform wires and SiC mono-filaments extracted from them.

reported.[7] Here, we had three kinds of isochronally heat treated (at 823, 873 and 923 K for 64 minutes) preform wires. Those preform wires and SiC monofilaments extracted from them were tensile tested at room temperature. Figure 1 shows the results of the tensile tests. Both of the tensile strengths of SiC mono-filaments and preform wires decrease with increasing heat treatment temperature. The line named "Rule of mixture" is indicating the calculated values by a simple rule of mixture using tensile strength of extracted SiC mono-filaments and that of the matrix (Al050). The difference between the two solid lines; experimental and calculated, would be representing the degradation of interfacial strength. At 873 K, the degradation of interfacial strength is the dominant factor to reduce the strength of preform wire, but at 923 K, the major part of the degradation comes from the degradation of SiC fibers. The absolute values of Weibull coefficient, in this case, are not important, because they are reflected the effect of very short gauge length.

SEM fractographs of SiC/Al preform wires with different heat treatment are shown in Fig.2. As produced preform wires (without heat treatment) and preform wires with 823 K heat treatment have similar rugged surfaces with some isolated pullouts indicating the soundness of

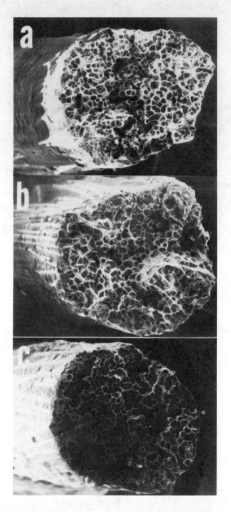

100µm

Fig. 2 : SEM fractographs of tensile tested preform wires after isochronal heat treatments as follows

    a ; 823 K x 64 min.
    b ; 873 K x 64 min.
    c ; 923 K x 64 min.

interfacial bonding. In the case of 873 K heat treatment, Fig.2 (b), the rugged surface includes some even surface with a little pull-outs, indicating the degradation of crack arrest property at matrix or fiber-matrix interface. Figure 2 (c) shows a fracture surface of the 923 K heat treated preform wire. The even surface without pull-outs is typical for very brittle materials. This is come from the low critical stress for crack initiation and the poor crack arrest property. Many pore like openings at the fracture surface and cracks along interface could be produced during tensile tests due to low strength of interfacial phase ( or layer). And the fracture at the interface would become a trigger for the fracture of the preform wires. Figure 3 is a high magnification image of the fracture surface in 923 K heat treated preform wire. The cracks, observed in Fig. 3, seem to propagate across the interface and to cause brittle fracture. Figure 4 shows the microstructure of the as produced

preform wire, indicating that there is no visible interfacial phase and no dislocation pile up at the interface. The selected area electron diffraction analyses reveal that SiC fiber has amorphous structures.

## 3.2 Correlation Between Tensile and Three Point Bending Tests

Because of the limitation of irradiation volume, we adopted a specimen length of 12.3 mm and the post irradiation mechanical test had been selected to be three point loading flexural test, i.e. three point bending test. The mechanical properties of preform wires from three point bending test were compared with those from tensile test and a linear relation between two test methods was obtained[8]. Further three point bending test was found to be more sensitive than tensile test to detect interfacial micro pores. That is, in case of high strength preform wire, no

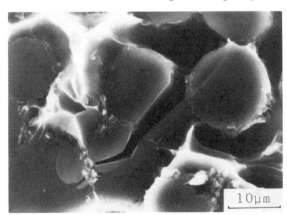

Fig. 3 : High magnification SEM fractograph of the same preform wire as shown in Fig. 2-c

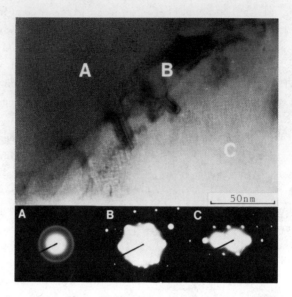

Fig. 4 : TEM micrograph showing microstructure of SiC/Al preform
wire. Diffraction patterns A, B and C correspond to positions
marked as A, B and C.

shrinkage cavities were observed
after tensile test, but some micro
cavities were observed after three
point bending test. As the micro-
cavities increased, strength of
preform wire decreased. This behavior seems to indicate that the
micro cavities degrade the strength
of preform wire.

### 3.3 Neutron Radiation Effects

In order to study the effects of
neutron irradiation to SiC/Al composite materials, preform wires
were irradiated with fission neutrons from JOYO and JMTR and with
14 MeV neutrons from RTNS-II[3].
The SiC fibers extracted from
SiC/Al preform wires showed irradiation induced hardening by fast

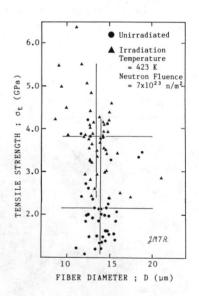

Fig. 5 : Tensile strength of extracted SiC mono-filaments from
SiC/Al preform wire irradiated in
JMTR as the dependence on fiber
diameter.

1440

neutron irradiation in JOYO at 723 K and by mixed spectrum neutrons in JMTR at 423 K (Fig. 5), although the extent of hardening is different. The origin of the difference is not clear yet, but the difference in irradiation temperature might be one of the important factors. Figure 6 shows the fluence dependences of Young's modulus and ductility for fibers extracted from preform wires. Young's modulus has a tendency to increase with increasing neutron fluence. But, the ductilities have different fluence dependences for JOYO and JMTR irradiation, i.e. the former decreases and the latter increases with increasing total fluence. These neutron fluence dependences of tensile strength and Young's modulus indicate the same trends as reported for Nicalon fiber by K.Okamura et al.[9] and G.Hopkins et al.[10].

The calculated rule of mixture value using the strength of extracted fiber and that of aluminum matrix is shown as solid line in Fig. 7, together with the bending test results. The coincidence of the both values suggests that there is little degradation of interfacial strength by neutron irradiation. But on the contrary, mixed spectrum neutron irradiation in JMTR at 423 K only showed strengthening of extracted fibers but slight increase in strength of preform wires. To clarify the degrading factor of preform wire strength in the JMTR irradiation, Weibull's analysis for extracted SiC fibers was performed. Figure 8 indicates that neutron irradiation produces no significant difference for JMTR and JOYO which implies that formation of initiation sites for fracture of SiC fiber does not

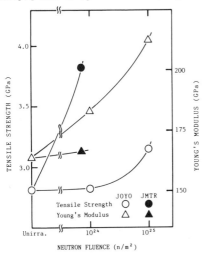

Fig. 6 : Fluence dependence of tensile strength and Young's modulus of SiC mono-filaments

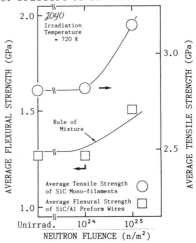

Fig. 7 : Fluence dependence of flexural strength for preform wire and tensile strength for extracted SiC mono-filaments.

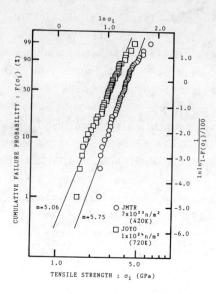

Fig. 8 : Chart of Weibull's analysis for SiC mono-filaments extracted from preform wires irradiated in JOYO and JMTR

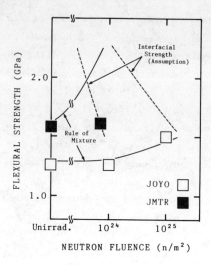

Fig. 9 : Fluence dependence of flexural strength of preform wires irradiated in JOYO and JMTR. The solid lines indicate the calculated rule of mixture values from the results of extracted SiC fibers. The broken lines are the schematic drawings of the assumption for the fluence dependence of SiC-Al interfacial strength.

seriously take place in JMTR and in JOYO irradiations. Therefore, these results are believed to be influenced through the neutron fluence dependence of interfacial strength and those of matrix and fiber, as schematically illustrated in Fig. 9. The dotted lines in the figure are the assumptions about fluence dependences of interfacial strength assuming that the degradation rate on neutron fluence was the larger for the lower irradiation temperature. In the case of JOYO irradiation, interfacial strength was assumed to be larger than fiber strength, thus strength of preform wire coincided well with the rule of mixture value. On the contrary, in the case of JMTR irradiation,

interfacial strength was assumed to become lower than fiber strengths at higher neutron exposure, thus strength of preform wire reflected the SiC-Al interfacial strength and showed no increase by neutron irradiation. 14 MeV neutron irradiation to the fluence of $5 \times 10^{21}$ n/m$^2$ was performed to preform wires at ambient temperature. There could be observed no effects on mechanical properties. This result is consistent with the results by K.Okamura et al.[9], where JMTR neutrons reduce crystalline size of SiC, but 14 MeV neutrons from RTNS-

II show little effect on crystalline size of SiC and radiation
hardening was observed in JMTR
irradiation, but was not observed
in RTNS-II irradiation. At present,
one of our concerns is the degradation process of mechanical properties of preform wire and the extent
of the degradation at higher neutron fluence.

### 3.4 IN-SITU Observation in HVEM

In-situ observations of microstructure changes have been carried out
under a high voltage electron
microscope (HVEM) from room temperature to 760 K. As for electron
damage study, we used 2 um diameter
electron beam and as for image
recording, we used weak and uniform
beam. Figure 10 indicates an example of damaged microstructure at

760 K for 93 min. (11 dpa). In
aluminum grains, well developed
dislocation structures are shown,
but in SiC, no detectable change
can be seen. The most significant
difference in the area with thermal
effect only and that with thermal
and electron damage is the formation of whisker like SiC crystal at
SiC/Al interface from SiC side to
Al, mainly along Al cell boundary.
That is significant SiC crystal
formation takes place with thermal
exposure at 760 K for 93 min., but
simultaneous electron damage sup-
presses such SiC crystal formation
at interface. The suppression
effect of SiC crystal formation has
been also observed by lower temperature electron irradiations
prior to the thermal exposure at
760 K for 93 min. [11]. Another
characteristic feature of damage

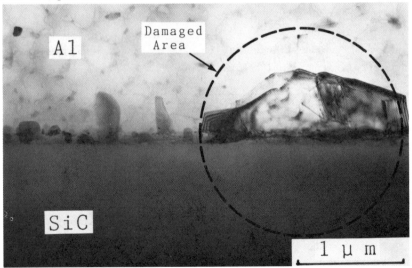

Fig. 10 : HVEM in-situ observation of SiC-Al interface indicating
microstructural difference between damaged (inside of the dotted
circle) and non-damaged area.

microstructure is the suppression of void formation in Al between 300 and 475 K, where void formation in Al has been reported with the same HVEM and with the same experimental condition[12]. The difference of Al in the two cases is the grain size and interstitial impurity contents. In case of SiC/Al, very fine sub-grain structure and relatively high impurity content (supposed to be induced from SiC fibers) produced too high dislocation density to form visible voids even at 11 dpa. This behavior seems to suggest the excellent resistivity to radiation damage even in JOYO and also the thermal degradation of mechanical properties above 823 K.

### 3.5 Comparison of Thermal- and Radiation- Effects

As mentioned the above, there are significant difference of the dependences of tensile strength and Young's modulus on thermal treatment and on neutron irradiation. Figure 11 is a schematic illustration of those dependences. The dependences on heat treatment temperature for SiC fibers have been previously reported and they were related with the crystallization of SiC from amorphous state[4]. The increase in Young's modulus due to structural change to crystalline state agrees well with physics based model. Our present interpretation is that neutron irradiation produces very fine crystalline particles in amrphous SiC which introduce hardening instead of reducing tensile strength. There are no evidence as to the formation of very fine crystals in the amorphous SiC, but sub-cascades induced by neutron damage could be heterogenious nucleation sites for SiC. That means, if we can distribute a great number of nucleation sites, the same order as those produced by neutron irradiation, the improvement of tensile strength and

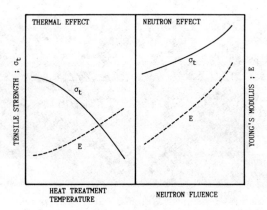

Fig. 11 : Schematic illustration of thermal and neutron effects on tensile strength and Young's modulus of preform wire.

Young's modulus, simulteneously, will be easily achieved. The production of high modulus and high strength type SiC/Al composite materials brings great advantages especially for applications in aeronautical and space technologies.

## 4. CONCLUSIONS

(1) The strength of PCS-type SiC/Al preform wires was discussed relating with strengths of matrix, fiber and interface and with microstructures.
(2) The effect of thermal treatment was studied and degradation mechanisms based on crystallization of amorphous SiC was proposed.
(3) The effects of neutron irradiation were studied using fission neutrons from JOYO and JMTR and with 14 MeV neutrons from RTNS-II. The SiC/Al composite and SiC fiber showed excellent stability under neutron irradiations.
(4) Increments of tensile strength and Young's modulus were observed by irradiation in FBR (JOYO). A heterogenious nucleation of SiC crystal due to sub-cascade damage was discussed.
(5) In-situ experiment under 1 MV HVEM suggests that the suppression of SiC crystal at interface with simultaneous exposures to 760 K and to electron damage is an important mechanism as to the excellent resistivity of PCS-SiC/Al for neutron damage. Also SiC crystal formation at SiC/Al interface at

760 K would be an origin of thermal degradation of mechanical properties at above 823 K.

## ACKNOWLEDGEMENTS

The authors would like to express their appreciation to Professor K. Okamura (Tohoku University) for his assistance in mono-filament tensile test and for discussions and to Professor H.Takahashi and Dr. S.Ohnuki (Hokkaido Univ.) for their assistance and discussions in HVEM experiment. They acknowledge gratefully contributions by J.Tanaka, K.Umemura, Y.Nagata and M.Takeda (Nippon Carbon Co.).

## REFERENCES

[1] T.Hayashi: Proceedings of the 5th International Conference on Composite Materials, TMS-AIME, (1985)1641-1654.
[2] S.Yajima and K.Okamura, J.Hayashi and M.Omori: J. Am. Ceram. Soc., 59(1976)324-327.
[3] R.L.Engelstad, E.G.Lovell and S.J.Covey: J. Nucl. Mater., 122&123 (1984) 816-820.
[4] G.R.Hopkins and R.J.Price: Nucl. Engn. and Design/Fusion, 2 (1985)111-143.
[5] A.Kohyama, N.Igata, Y.Imai, H.Teranishi and T.Ishikawa: Proceedings of the 5th International Conference on Composite Materials, TMS-AIME, (1985)609-621.
[6] A.Kohyama:"Plasma-wall interaction data needs critical to a

burning core experiments (BCX)",
US-DOE CONF-8506177, vol.1
(1985)347-369.

[7] A.Kohyama, H.Tezuka, N.Igata,
Y.Imai, H.Teranishi and T.Ishikawa:
J. Nucl. Mater., (1986)-in press-.

[8] H.Tezuka, A.Kohyama and
N.Igata: Abstracts for Annual
Meeting of JIM, 98(1985)305.

[9] K.Okamura, T.Matsuzawa,
M.Sato, Y.Higashiguchi and
S.Morozumi: J.Nucl.Mater.,133&134
(1985)705-708.

[10] R.J.Price, G.R.Hopkins and
G.B.Engle: Proceedings of The 17th
Biennial Conference on Carbon,
Lexington, Ky., June, 1985.

[11] A.Kohyama, S.Ohnuki and
H.Takahashi: to be published.

[12] S.Ohnuki, T.Kinoshita and
H.Takahashi: to be published.

# PITCH-BASED CARBON FIBERS
## FOR COST EFFICIENT COMPOSITES

Richard P. Krock, D. Chris Boyer, Melvin D. Kiser
Ashland Carbon Fibers
A Division of Ashland Petroleum Company

## Abstract

The combination of desirable properties characteristic to pitch carbon fibers is enabling designers to manufacture strong, lightweight, dimensionally stable, thermally and electrically conductive, durable composites. The isotropic nature of pitch carbon fibers in a random orientation imparts uniform composite properties which is imperative where tight tolerances are concerned. Affordable carbon fibers are now available from Ashland Carbon Fibers Division in a wide variety of forms for those applications where low cost composites must be designed to replace conventional metal parts. Applications of various forms of carbon fiber such as chopped, milled, and nonwovens will be discussed.

## 1. INTRODUCTION

Ashland Petroleum Company manufactures a unique carbonizeable pitch known as A-240, from a specially selected heavy feedstock from our crude oil refining operations. A-240 is primarily used as a binder and densifier for carbon and graphite electrodes which are consumed in steel and aluminum smelting processes. Many additional uses of A-240 pitch have been explored, and one of the most promising was determined to be conversion into a low cost carbon fiber. Over the past ten years, intensive research and development efforts resulted in novel methods of forming fibers and heat treatment processes that allow substantial reductions in manufacturing expenses to be realized compared with conventional processes that convert polyacrilonitrile precursor fibers into carbon fibers. In 1984 Ashland Carbon Fibers was formed as a division of Ashland Petroleum Company and is currently manufacturing and marketing low price, multipurpose carbon fibers under the brand name CARBOFLEX[R] for a diverse range of industrial, automotive, aerospace and defense applications.

## 2. CHARACTERISTICS AND PROPERTIES OF CARBON FIBERS

The outstanding properties of carbon fibers make them a prime choice material for today's composite designers. As indicated in Table 1, carbon fibers are strong, stiff, lightweight, abrasion resistant, and self-lubricating, all of which combined provide excellent mechanical properties. They conduct heat, have a low coefficient of thermal expansion, retain their mechanical properties at elevated temperatures, and resist oxidation at temperatures up to 500°C. Carbon fibers are good electrical conductors but are nonmagnetic and x-ray transparent. In addition, they are highly chemical resistant, are biologically inactive, and will only absorb very low levels of moisture, which are extremely important characteristics for long composite life. Highlights of Ashland's pitch-based carbon fiber properties listed in Table 2 indicate that they are as strong as steel yet one-fifth the weight, which makes them a good reinforcement for secondary structural composites. Where high strength and high modulus carbon fibers are priced too high for general composite applications outside of aerospace and sporting goods, multipurpose pitch-based carbon fibers, having the

aforementioned properties, are now available for cost efficient component designs. Ashland Carbon Fibers' quality assurance methods will be briefly described prior to discussing composite applications.

## 3. QUALITY ASSURANCE METHODS

Innovative methods have been developed at Ashland Carbon Fibers' Quality Assurance Laboratory for analyzing pitch feedstock, and fiber strengths in order to appropriately determine quality of raw material and products from Ashland's unique manufacturing technology.

### 3.1 Characterization of Petroleum Pitches

Physical properties measured include softening point, glass transition temperature, viscosity at high temperatures, specific gravity, and specific heat. Chemical properties analyzed include coking value, ash content, elemental analysis, average molecular weight, spectral analysis, and solubility in various solvents. In addition to the normal characterization techniques, pyrolysis gas chromatography is employed in order to simultaneously investigate physical and chemical properties.

### 3.2 Fiber Strength

The tensile strength of Ashland carbon fibers is calculated from the amount of stress applied in a longitudinal direction required to cause a single 19mm fiber to break or fail. In order to compare fibers of various diameters and lengths, this value is normally expressed in force per unit area (KPa). But tensile strength measured for a 19mm fiber length does not accurately predict the reinforcing characteristics of the same fiber after it has been reduced to a length of 13mm or to 2mm. It is possible to obtain an approximately linear progression by plotting the logarithm of tensile strength vs. the logarithm of the testing length. For example, use of 1600 micron long pitch carbon fibers would provide a fibrous reinforcement with 1.38 Gpa (200,000 psi) tensile strength. This relationship is typical for many fibers and has been measured at the shorter lengths for Ashland's fiber.

4. ISOTROPIC COMPOSITES

In certain situations, isotropic composites are desirable. Influencing factors which determine whether or not a composite would be isotropic are materials and fiber orientation. Reducing the pitch carbon fiber length to very short ranges on the order of 100 to 1600 microns allows the fibers to achieve a more random orientation in the finished composite. Long fibers have a tendency to become oriented by gates, edges and portals within an injection mold. Short fibers usually travel through less affected and retain their random orientation. Isotropic properties become important for electrical applications where uniform conductivity in all directions is essential, and for certain dimensionally restricted composites where expansion and contraction in all directions must be uniform. Isotropic pitch fibers can provide enhancement of both electrical and dimensional characteristics in all directions. Because of the ultra-short nature of milled carbon fibers, dispersion into the composite becomes enhanced. Adequate fiber dispersion is necessary to achieve isotropic properties in the finished composite. Fibers chopped from a tow have a tendency at times to remain clustered together during low shear blending operations.

5. ASPECT RATIO AND LOADINGS

Aspect ratio (length divided by diameter) and fiber loading in a composite are two very important parameters which the designer must consider. Depending upon the type of composite being made and the processing technique being utilized, aspect ratio and fiber loading play different roles. Evaluation of selected strength and modulus properties for injection molding compounds (Table 3) show that fiber loading is more of a determinant than is aspect ratio. At equal loadings in nylon 6/6 injection molding compound, properties for 200 micron and 400 micron length carbon fiber reinforcements are very similar. This is primarily due to the intense shearing action experienced during the compounding, pelletizing and injection molding operations which tend to reduce fiber length. However, aspect ratio does play an important role in strength properties for compression molded composites.

Certain properties are more sensitive to fiber loadings than others and points of diminishing returns are reached at higher loadings for other properties. Depicted in Table 4 is the sensitivity of heat deflection temperature under load at 264 psi due to carbon fiber loading. An increase of 100% was achieved with just a 10% carbon fiber loading. While not all properties are this sensitive, it is important to note those which are for maximum utilization of composite ingredients.

Other properties are influenced both by aspect ratio and loading.[1] For instance, data depicted in Table 5 indicates that mold shrinkage can be reduced substantially by increasing fiber length. Doubling the fiber length reduces the mold shrinkage by more than a factor of two. Similarly doubling the loading cuts the mold shrinkage by approximately half. Electrical resistivity is another property which is very sensitive to both aspect ratio and fiber loading as depicted in Figure 3. This data represents analysis of epoxy composites where 100, 400 and 1600 micron length carbon fibers were employed at various loadings to compare the effectiveness on reducing electrical resistivity. Clearly longer fibers perform better for this property and higher loadings can be used to achieve the same results when utilizing shorter fibers as dictated by other more constraining properties.

6. HYBRID COMPOSITES

Because a balance of physical, mechanical, electrical, thermal and economical properties are necessary in composites, no one fiber is ideally suited for all applications. It can be deceiving to make direct comparison of properties to determine the best fiber because no one comparison can take into account the balance of properties, the processing and cost considerations simultaneously. Composites generally consist of four categories of materials: 1) resins, 2) fibers, 3) fillers, and 4) additives and modifiers. Ashland is pursuing applications in composite friction materials where carbon fibers provide enhanced friction coefficients, longer wear and distribution of heat across a friction surface, all of which combine to provide a low fade, safe brake lining without the use of asbestos[2]. As many as ten items from a single materials category are selected sometimes for a given friction composite to obtain the best balance of uniform friction, fade and wear throughout the life of the composite. The same philosophy can be applied to reinforced plastic composites in order to obtain the proper balance of properties at the lowest cost. Hybrids of pitch-based carbon fiber, aramid, and glass fibers can be employed to make strong, durable, yet lightweight, thermally and electrically conductive composites. Individual properties can be analyzed such as previously presented for aspect ratio and loading, and tailor-made composites can be designed to meet finished product specifications at the lowest cost. Pitch carbon fiber can impart electrical and thermal conductivity, mold shrinkage, dimensional stability, and light-weight properties in an isotropic manner to a composite whereas glass can add strength and low cost properties, while aramid fibers provide stiffness, toughness and lightweight properties all in a single composite. Many composite properties can be tailored with the use of various hybrid fiber reinforcements.

## 7. SPECIAL FORMS OF CARBON FIBERS AVAILABLE

Our discussion has focused upon composites containing chopped or milled carbon fibers. A number of low priced nonwovens such as mat, stitched mat, paper and needled mat are available for compression molded composites. Also, hybrid yarns are available for thermoplastic coating and weaving to fabrics for certain applications.

### 7.1 Nonwoven Mat

Nonwoven mat is a three-dimensional material comprised of semi-continuous carbon fibers having an areal weight of from 136 to 238 grams per square meter (6 to 10 oz./sq. yd.) In certain advanced structural composites, high modulus unidirectional carbon fibers are shaped and placed into a mold and structural thermoset resins are injected to complete the composite. By nature of their high modulus, a bending radius is found at each location where the carbon fibers change directions. As a result of this bending radius, a resin rich area can be found which serves as a site for initiation of cracks in the matrix. Nonwoven, lower modulus, multipurpose pitch-based carbon fibers can be utilized to fill and reinforce these problem areas; thereby, preserving the structural integrity of the composite[3].

### 7.2 Stitched Mat

Stitching the carbon fiber mat with fiberglass, Kevlar or conventional stitching thread is performed to increase the pull strength of the nonwoven mat or to form hybrid nonwovens incorporating fiberglass or some other type scrim as a backing.

### 7.3 Paper

Carbon fiber paper, ranging in weights from 25 to 204 grams per square meter (0.7 to 5.0 oz./sq. yd.) can be made in wet systems from chopped carbon fibers.

Binders can be used that are compatible with the matrix system and carrier fibers such as Kevlar[R] or glass can be added to increase wet and dry strength when necessary. Various thermoplastic fibers can be added to impart thermoforming qualities to the paper. Situated at the surface of a composite, carbon fiber paper creates an electrically conductive layer for static dissipation. Surface resistivities of 40 ohms per square were measured on a composite made from Ashland Chemical's Arimax[R] reaction injection molded compound and CARBOFLEX paper. Effective EMI and RFI shielding can be accomplished in composite housings for protecting sensitive electronic equipment by alternating layers of paper and fiberglass mat so that waves transmitted through surface layers can be further attenuated. Composite walls have been developed for heating animal stalls where the carbon fiber paper was connected to an electrical supply for resistance heating.

### 7.4 Needled Mat

Consolidation of the nonwoven mat and needling together of multiple layers of nonwoven mat can be accomplished by punching a certain percentage of the fibers into the "Z" direction. Areal weights ranging from 204 to 1530 grams per square meter (6 to 45 oz./sq. yd.) are available in thicknesses from 6mm to 51mm (1/4" to 2"). The pull strength of the needled mat is five times greater than that of the nonwoven mat. Although the primary use of the needled mat is high temperature furnace insulation in both rigid and flexible forms, the needled mat can be impregnated with phenolic or epoxy resins and hand-laid up as a sound or thermal insulation in aircraft composites. The needled mat can also be preimpregnated with Ashland's new AEROCARB[R] high carbonizing impregnant material for a low cost carbon/carbon fiber composite. Selected properties of AEROCARB are shown in Table 6. As a companion product to CARBOFLEX, AEROCARB can most readily be used in a hot isostatic press (HIP) process for densifying carbon/carbon fiber composites. Its coking value (by ASTM D2416) of 74% versus 35 to 55% for phenolics enables carbon/carbon composite manufacturers to meet finished product density specifications in fewer impregnating/carbonization cycles.

### 8. COMBINATION OF FIBER FORMS

In addition to the aforementioned applications of short pitch-based carbon fibers, one under evaluation at this time is its use as an eliminator of interlaminar shear between layers of woven reinforcing high strength fibers in advanced composites. By incorporating these short milled fibers in a liquid resin and coating sheets of woven prepregs with the mixture, reduction of interlaminar shear is believed to be possible.

## 9. CONCLUSION

Many areas of applications are envisioned for carbon fibers which are indicated in Table 6. Some of the most prospective of these uses are in composites, friction materials, packings and gaskets, ablatives, filters and coalescers, and static dissipant yarns and materials[4].

## 10. ACKNOWLEDGEMENT

Ashland Carbon Fibers Division thanks the program chairman for inviting us to deliver this paper and extends their appreciation to SAMPE for providing us the opportunity to participate in the 32nd International SAMPE Symposium.

## 11. BIOGRAPHIES

Richard P. Krock: Manager of Commercial Development of Ashland Carbon Fibers Division, he received his Chemical Engineering degree from the University of Cincinnati, and an MBA from Marshall University. He is involved in the composites development program at Ashland Carbon Fibers.

D. Chris Boyer: Sales Manager for Ashland Carbon Fibers Division, he earned his degree in Oceanography from the University of South Carolina and is participating in the sales and development of carbon fibers in composites for Ashland Carbon Fibers.

Melvin D. Kiser: Supervisor of the Quality Assurance Laboratory, he received his bachelors degree in chemistry from Marshall University and is presently responsible for the day-to-day operations of the Quality Assurance Laboratory and for providing analytical support for technical service activities.

## 12. REFERENCES

(1) Krock, R.P., et al, "Versatility of Short Pitch-Based Carbon Fibers in Cost Efficient Composites", 42nd Annual SPI RP/CI Conference, Cincinnati, Ohio, February 5, 1986.

(2) Hettinger, W.P., et al, "CARBOFLEX and AEROCARB - Ashland's New Low Cost Carbon Fiber and Carbonizing Products for Future Brake Applications", SAE Earthmoving Industry Conference, Peoria, Illinois, April 8, 1986.

(3) Krock, R.P., et al, "CARBOFLEX Nonwoven Carbon Fibers, A Cost Efficient Reinforcement for Lightweight Composites", TAPPI Annual Nonwovens Conference, Atlanta, Georgia, April 24, 1986.

(4) Hettinger, W.P., "New Low Cost Pitch-Based Carbon Fibers and Carbonizing Materials", Clemson University Symposium on New Directions in Carbon and Graphite and Other High Performance Fibers and Fabrics, Clemson, South Carolina, February 4, 1986.

# TABLE 1

## OUTSTANDING CHARACTERISTICS OF CARBON FIBER

| Property | Characteristic |
|---|---|
| Mechanical | Strength |
| | Low Density |
| | Stiffness-rigidity |
| | Fatigue resistant |
| | Abrasion resistant |
| | Tribological |
| Thermal | Heat conductive |
| | Dimensional stability |
| | Thermal stability |
| | Oxidation resistant |
| Electrical | Electrically conductive |
| | Non-magnetic |
| | X-ray transparent |
| Biological | Biologically inert |
| Chemical | Chemically inert |
| | Chemically resistant |

## TABLE 2

### SELECTED PROPERTIES OF CARBOFLEX PITCH-BASED MULTI-PURPOSE CARBON FIBER

| Typical Fiber Properties (Filaments) | Units |
|---|---|
| Density | 1.6 gm/cc |
| Electrical Resistivity | 0.004 ohm-cm |
| Diameter | 12 micron |
| Tensile Strength | 80,000 psi |
| Tensile Modulus | 6.5 million psi |

## TABLE 3

### COMPARISON OF INJECTION MOLDING COMPOUND PROPERTIES INFLUENCED BY FIBER LOADING NYLON 6/6 AND PITCH CARBON FIBER

| | | |
|---|---|---|
| Average Fiber Length, Microns | 200 | 400 |
| Loading, Wt.% | 20 | 20 |
| Deflection Temperature UL @ 264 psi, °C | 229 | 230 |
| Tensile Strength, KPa | 82 | 79 |
| Flexural Modulus, GPa | 5.1 | 5.2 |

## TABLE 4

### SENSITIVITY OF THERMAL PROPERTIES
### TO FIBER LOADING
### NYLON 6/6 AND PITCH CARBON FIBER MOLDING COMPOUND

|  | Unfilled Nylon 6/6 | Nylon 6/6 and 10% Wt. 200 Micron Carbon Fiber |
|---|---|---|
| Heat Deflection Temperature Under Load @ 264 psi, °C | 90 | 217 |

## TABLE 5

### COMPARISON OF INJECTION MOLDING COMPOUND PROPERTIES
### INFLUENCED BY FIBER ASPECT RATIO AND LOADING
### NYLON 6/6 AND PITCH CARBON FIBERS

| Average Fiber Length, Microns | 100 | 200 | |
|---|---|---|---|
| Loading, Wt.% | 20 | 20 | 40 |
| Mold Shrinkage, % | 1.9 | 0.86 | 0.39 |

## TABLE 6

## SELECTED PROPERTIES OF AEROCARB IMPREGNANT

| Typical Properties | Units |
|---|---|
| Density | 1.27 gm/cc |
| Softening Point | 204°C |
| Glass Transition Temp. | 171°C |
| Carbon Assay: | |
|    Coking Value | 74 Wt. % |
|    Total Carbon | 94 Wt. % |
| Quinoline Insolubles | 0.4 Wt. % |
| Toluene Insolubles | 17 Wt. % |
| Viscosity Poise: | Temp., °C |
| 1 | 342 |
| 5 | 301 |
| 10 | 288 |
| 20 | 277 |
| 60 | 262 |
| 100 | 256 |
| 140 | 252 |

## TABLE 7

### CARBON FIBER APPLICATIONS

- ° Brakes – Aircraft, Industrial, Automotive, Racing; Carbon Fiber and Carbon/Carbon Fiber Composites.

- ° Friction Materials, Clutch Plates

- ° Packings, Gaskets, High Temperature Insulation Reinforcement

- ° Plastics–Carbon Fibers Composites, with Thermoplastics and Thermosets; Injection, Reaction Injection and Compression Molding; Reinforcement, Shielding, Conductance, Fatigue Enhancement

- ° Electrostatic Dissipant – Garments, Fabrics, Yarns, Carpets; Composites – Circuit Boards, Electrical Equipment, Operating Room Furniture

- ° Electromagnetic Shielding – Communications Equipment, Computers, Typewriters, Telephones

- ° Ablatives & Ablative Coatings, Paints

- ° Battery Electrodes

- ° Filter Material, High Temperature, Chemical Inertness Capability

- ° Cement Reinforcement

- ° Conductive Paints

- ° Heating Tapes, Strips, Walls

- ° Fire Retardants, Fire Protection Equipment, Clothing, Fillers

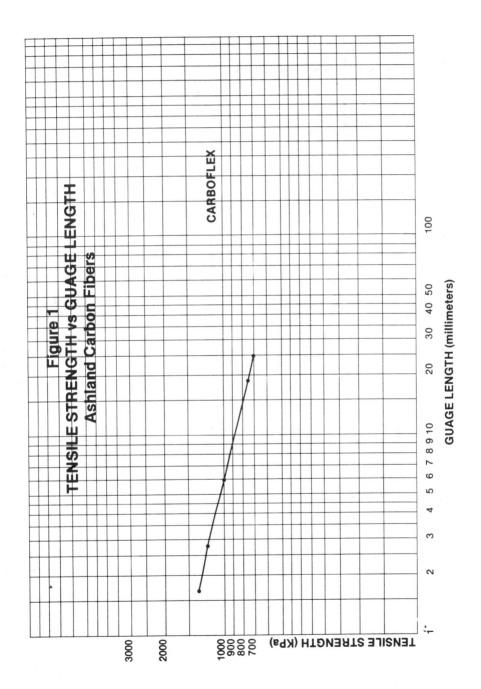

Figure 1
TENSILE STRENGTH vs GUAGE LENGTH
Ashland Carbon Fibers

CARBOFLEX

GUAGE LENGTH (millimeters)

TENSILE STRENGTH (kPa)

Figure 2

## Performance Evaluation at Various Lengths and Loadings
## Carboflex in Epoxy Composites

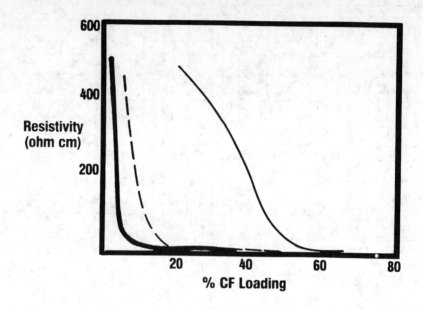

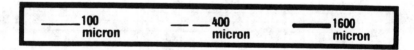

IMPACT PERFORMANCE OF VARIOUS FIBER
REINFORCED COMPOSITES AS A FUNCTION OF TEMPERATURE

Richard S. Zimmerman and Donald F. Adams
Composite Materials Research Group
University of Wyoming
Laramie, Wyoming 82071

## Abstract

Spectra 900 and Spectra 1000 are high strength, high modulus polyethylene fibers recently developed by Allied Fibers. They have demonstrated an excellent capability for energy absorption when combined with polymer matrix materials. Many impact applications operate under a varying temperature environment, which requires retention of impact performance at both subambient and elevated temperatures. Kevlar 49 and E-glass composites have, until recently, been the choice when selecting an impact resistant material system.

The present study compares Spectra 900, Spectra 1000, Kevlar 49, and E-glass fiber composites incorporating the same polymer matrix, at three test temperatures, viz., -50°C, 23°C, and 100°C. The relative impact performance of these four materials over this temperature range offers alternatives in selecting a material system for impact resistant applications.

The polymer matrix selected was Shell EPON 826, an epoxy, cured with Millikin Chemical's Millamine 5260 curing agent. This polymer matrix has a strain to failure capability, when cured with the Millamine 5260, that is compatible with the four fibers used in this study. Both Spectra fibers offer a 40 percent lower density than Kevlar 49, and a 150 percent lower density than E-glass fibers, which can be important in weight-critical applications.

Some limited impact testing was also performed on fabric laminates using a Dow Derakane 411-45 vinyl ester resin as the matrix material, to compare matrix effects and construction effects on impact performance at different temperatures.

Peak force, energy to peak force, and total energy absorbed during a flat plate penetrating impact will be presented. All materials were tested at approximately 2.3 mm plate thickness, on a 127 mm square, clamped-edge anvil base, using a 12.7 mm diameter spherical tup.

## 1. INTRODUCTION

Considerable work has been devoted over the last 20 years to the impact performance of various fiber-reinforced composites. Glass and Kevlar 49 fibers have been studied extensively, and have been found to be quite effective in energy absorption. Recently, drop weight impact tests on fabric and cross-ply composites incorporating two new high strength, high modulus polyethylene fibers, viz., Spectra 900 and Spectra 1000 produced by Allied Fibers (1), have shown energy absorbing properties higher than those of both E-glass and Kevlar 49 fiber-reinforced composites.

The purposes of the present study were to determine the relative impact performance of Spectra 900, Spectra 1000, Kevlar 49, and E-glass composites as a function of temperature, to determine any effects of the use of fabric rather than a cross-ply configuration, and also to compare, on a limited basis, the effects on impact of an epoxy versus a vinyl ester matrix material.

## 2. SPECIMEN CONFIGURATIONS

All four fiber types were prepregged and then fabricated into laminates at the University of Wyoming, in the configurations shown in Table 1. The matrices used were Shell EPON 826, a 120°C cure epoxy, and Dow Derakane 411-45, a room temperature cure plus elevated temperature post cure vinyl ester. The epoxy cure cycle used was 690 kPa (100 psi) pressure, at 50°C for one hour, 85°C for one hour, and 120°C for two hours, with no postcure. The vinyl ester was cured at 690 kPa (100

TABLE 1
LAMINATE CONFIGURATIONS

| Material System | Laminate Type | Fiber Volume (percent) | Laminate Thickness (mm) |
|---|---|---|---|
| Spectra 900/826 | $[0/90]_{2s}$ | 65 | 2.5 |
| Spectra 1000/826 | $[0/90]_{2s}$ | 60 | 2.3 |
| Kevlar 49/826 | $[0/90]_{2s}$ | 66 | 2.0 |
| E-Glass/826 | $[0/90]_{2s}$ | 61 | 2.5 |
| Spectra 1000/826 | Plain Weave | 57 | 2.4 |
| Spectra 1000/411-45 | Plain Weave | 57 | 2.4 |
| Spectra 900/826 | Plain Weave | 53 | 2.5 |
| Spectra 900/411-45 | Plain Weave | 47 | 2.7 |

psi) at room temperature for two hours, then postcured at 100°C for four hours.

The cross-ply laminates were eight plies thick, fiber volumes being in the vicinity of 63 percent (see Table 1). The fabric laminates were eight plies thick also, with the fiber volume being near 55 percent in all cases.

### 3. TEST METHODS

All instrumented drop weight impact testing was performed using the test apparatus described in References (2,3), and shown in Figures 1a and 1b. The impact velocity was measured for each individual test, typically being in the range of 4.4 m/s (14.5 ft/sec.). The 150 mm square specimen, nominally 2.3 mm thick, was clamped on all four edges to round-nose rails arranged in a 127 mm square pattern. The falling crosshead mass was 16 kg (35 lbm), with the included impacting tup being a spherical-nosed rod 12.7 mm in diameter.

All plates were fully penetrated, with a complete load versus time plot being obtained for each test. Data were acquired using a Nicolet Explorer III digital oscilloscope, then transferred to a Hewlett Packard 21 MX-E minicomputer for preprocessing. The data were then sent to a Control Data Corporation Cyber 760 main frame computer for data reduction and plotting of the load versus time and energy versus time curves.

Nonambient temperature drop weight impact testing was conducted using a temperature chamber designed and built to fit into the drop weight impact apparatus. Liquid nitrogen injection was used to achieve subambient temperatures, and electric resistance heaters to obtain the elevated temperatures. Thermocouples at the top and bottom surfaces of the specimen were used to monitor the specimen temperature. A small sliding panel above the test specimen was opened immediately prior to impact, to expose the specimen to the falling tup. The operating temperature range of this chamber is approximately -100°C to 180°C.

### 4. EXPERIMENTAL RESULTS

Drop weight impact test results are summarized in Table 2. The average values given typically represent five individual tests. Also included in Table 2 are cross-ply and fabric data taken from References (2,3) for Kevlar 49 and E-glass fibers in Hercules 3501-6 epoxy. Results have been normalized with respect to plate thickness and density. Energy values are plotted in Figures 2 and 3 for ease of comparison. Figure 2 shows the specific energy at peak force absorbed by the four composites in the EPON 826 epoxy at the three test temperatures. As will be noted, there is a significant effect of temperature on the Spectra 900 and Spectra 1000 composites, and very little or no

Figure 1a.  Overall Photograph of Instrumented Drop Weight Impact
Apparatus Developed at the University of Wyoming.

Figure 1b.  Close-up of Temperature Conditioning Box, Sliding Cover,
and Impacted Composite Laminate Still Clamped to the Base
of the Instrumented Drop Weight Impact Apparatus.

TABLE 2

AVERAGE DROP WEIGHT IMPACT TEST DATA

| Material System | Specimen Thickness (mm) | Specimen Density (g/cm$^3$) | Test Temp. (°C) | Specific Peak Force* ($\frac{N \cdot m^2}{g}$) | Specific Energy at Peak Force* ($\frac{J \cdot m^2}{kg}$) | Specific Total Energy* ($\frac{J \cdot m^2}{kg}$) |
|---|---|---|---|---|---|---|
| Spectra 900/826 cross-ply | 2.5 | 1.03 | -50 | 3.1 | 6.9 | 17.5 |
|  | 2.7 |  | 23 | 3.9 | 13.9 | 23.8 |
|  | 2.1 |  | 100 | 4.3 | 27.4 | 38.6 |
| Spectra 1000/826 cross-ply | 2.3 | 1.03 | -50 | 4.9 | 36.7 | 41.4 |
|  | 2.3 |  | 23 | 5.2 | 52.3 | 63.1 |
|  | 2.2 |  | 100 | 5.1 | 61.4 | 64.9 |
| Kevlar 49/826 cross-ply | 1.9 | 1.30 | -50 | 1.2 | 3.8 | 7.8 |
|  | 2.0 |  | 23 | 1.5 | 6.0 | 9.9 |
|  | 2.1 |  | 100 | 1.9 | 7.7 | 10.8 |
| E-glass/826 cross-ply | 2.4 | 2.00 | -50 | 2.2 | 10.3 | 10.7 |
|  | 2.7 |  | 23 | 1.9 | 6.3 | 11.3 |
|  | 2.4 |  | 100 | 1.8 | 7.2 | 9.2 |
| Spectra 900/826 plain weave (4) | 2.6 | 1.05 | -50 | 2.6 | 11.9 | 19.2 |
|  | 2.6 |  | 23 | 2.3 | 12.7 | 19.1 |
|  | 2.4 |  | 65 | 2.6 | 17.8 | 24.0 |
| Spectra 900/411-45 plain weave (4) | 2.6 | 1.07 | -50 | 2.5 | 14.9 | 19.7 |
|  | 2.6 |  | 23 | 2.4 | 13.8 | 18.0 |
|  | 2.7 |  | 65 | 2.4 | 13.1 | 21.0 |
|  | 2.8 |  | 100 | 2.4 | 13.9 | 19.1 |
| Spectra 1000/826 plain weave (4) | 2.4 | 1.05 | 23 | 2.8 | 18.9 | 27.2 |
| Spectra 1000/411-45 plain weave (4) | 2.4 | 1.05 | 23 | 2.9 | 20.9 | 28.6 |
| Kevlar 49/3501-6 cross-ply (2,3) | 2.7 | 1.27 | 23 | 0.9 | 4.3 | 6.4 |
| E-glass/3501-6 cross-ply (2,3) | 2.5 | 1.78 | 23 | 1.6 | 8.0 | 10.5 |

*Normalized by dividing by specimen thickness and laminate density

temperature effect on the Kevlar 49 and E-glass composites.

The Spectra 1000/826 composite exhibited a very high energy absorption capability at all test temperatures. Corresponding values obtained in previous testing of the Spectra 1000 fiber in the EPON 826 and Derakane 411-45 matrices were 20 to 30 percent lower than measured in the present testing (4). The previous testing was done on slightly thicker plates (2.6 mm vs. 2.3 mm) with lower fiber volumes (45 percent vs. 60 percent), which undoubtedly had some effect on the

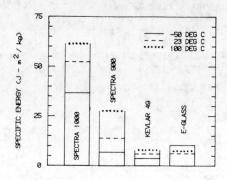

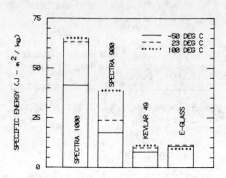

Figure 2. Normalized Energy at Peak Force for Four Cross-Ply Laminates at Three Test Temperatures.

Figure 3. Normalized Total Energy For Four Cross-Ply Laminates at Three Test Temperatures.

results.

The Spectra 900/826 cross-ply drop weight impact results were one-half to one-fourth those of the Spectra 1000/826, yet still higher than the Kevlar 49/826 or E-glass/826 values, except at the -50°C test temperature for the E-glass. The results for three of the four materials showed that the least energy was absorbed at the lowest temperature. The E-glass/826 and Kevlar 49/826 composites exhibited little temperature dependence. Specific energies absorbed by the Kevlar 49/826 and E-glass/826 composites up to peak force were consistently lower than the Spectra/826 composites at all but the one condition mentioned above. Specific total energy (plotted in Figure 3) indicated similar trends, with the Spectra 1000/826 composite energy absorption being much higher than for the other three material

systems. The Spectra 900/826 composite energy absorption was also consistently higher than that of the Kevlar 49/826 and E-glass/826 composites at all test temperatures.

Failure modes were very similar for all four material systems. Figure 4 is a typical view of the entrance side of an impacted plate. The circular hole made by the penetrating tup will be noted, along with the broken fibers that sprung back after the tup was removed, closing the hole somewhat. The exit side of a typical impacted plate is shown in Figure 5. The bottom ply was disrupted the full width of the plate, with broken fibers extending into the hole area from all directions.

The Kevlar 49/826 and E-glass/826 composites exhibited no permanent "dishing in" in the center of the plate due to the tup forcing

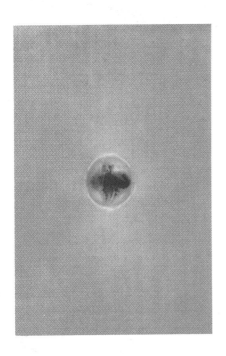

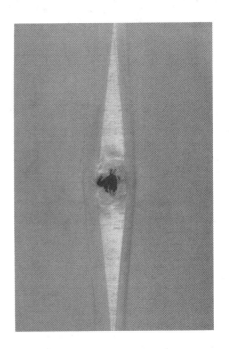

Figure 4. Photograph of Entrance Side of Hole in Penetrated Spectra 900/EPON 826 Cross-Ply Laminate.

Figure 5. Photograph of Exit Side of Hole in the Penetrated Spectra 900/EPON 826 Cross-Ply Laminate of Figure 4.

through and penetrating. Both of the Spectra/826 composites exhibited some "dishing in" in the center hole areas, which undoubtedly assisted in the energy absorbing capability of both materials. A permanent depression was evident in failed specimens of both types of Spectra/826 composites, due to the greater flexibility of the Spectra plates in bending.

No effect on failure mode due to temperature was seen in any of the penetrated plates. All failures were very similar at the three test temperatures used in this program, for all four material systems. Damage zones extending away from the circular holes were largest for the E-glass/826 composites and smallest for the Spectra/ 826 composites. Slight internal delaminations could be seen extending approximately 15 mm from the edges of the holes into the E-glass/826 plates. The Kevlar 49/826 delamination zone extended only 4 to 5 mm into the plate from the sides of the hole, and the Spectra 900/826 and 1000/826 composites had a 2 to 3 mm visible delamination zone surrounding the tup hole area.

Some of the plain weave fabric laminates were drop weight impact tested at different temperatures.

Spectra 900 plain weave fabric was combined with EPON 826 epoxy and Derakane 411-45 vinyl ester resins. Figures 6 and 7 are plots of specific energies, comparing the Spectra 900/826 cross-ply laminate with Spectra 900 fabric laminates incorporating the two polymer matrices mentioned above. The cross-ply laminate exhibited a dramatic temperature effect, with the specific energy more than doubling when the test temperature was increased from -50°C to 100°C. The two fabric laminates exhibited very little influence of temperature between -50°C and 100°C. It will be noted that the highest test temperature for the Spectra 900/826 plain weave fabric composite was only 65°C, compared to 100°C for the other two laminate types.

There were essentially no differences in impact performance between the two resins used in the fabric-reinforced composites tested at the three temperatures. The EPON 826 epoxy and Derakane 411-45 vinyl ester performed equally well in impact when combined with the Spectra 900 plain weave fabric. The three laminate types exhibited almost equal impact energies at room temperature, with the cross-ply laminate absorbing less energy at the low test temperature and somewhat more energy at the elevated test temperature.

Drop weight impact energies with matrix materials and fiber types as variables are plotted in Figures 8 and 9, respectively. Only room temperature data are compared here since no nonambient temperature data were available for many of the materials. Specific energies are shown, the energy values being normalized with respect to plate thickness and material density. Spectra 1000

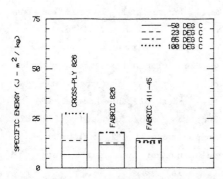

Figure 6. Normalized Energy at Peak Force for Spectra 900 Laminates at Three Test Temperatures.

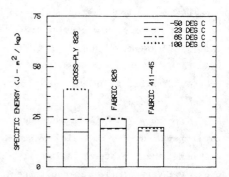

Figure 7. Normalized Total Energy for Spectra 900 Laminates at Three Test Temperatures.

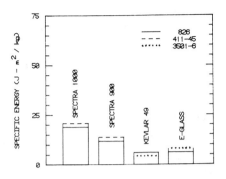

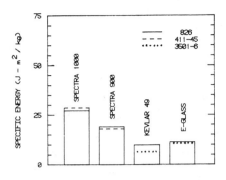

Figure 8. Normalized Energy at Peak Force for Four Fiber-Reinforced Laminates Incorporating Different Matrices at Room Temperature.

Figure 9. Normalized Total Energy for Four Fiber-Reinforced Laminates Incorporating Different Matrices at Room Temperature.

exhibited the highest specific energy absorption of the four fiber types, for both of the resin systems. The Spectra 900 exhibited specific energies slightly lower than those of the corresponding Spectra 1000 composites. No significant differences were noted between the two resin systems in terms of their energy dissipation capability.

Kevlar 49 had the lowest specific energy absorption of the four fiber types tested, E-glass being only slightly better. No difference was noted in energy dissipation capability between the two epoxy matrices used with the Kevlar 49 and E-glass. The EPON 826 is a 120°C cure epoxy with a tensile strain capability of 5 to 6 percent (4). The 3501-6 epoxy is a 177°C cure epoxy with a tensile

strain capability of 1 to 2 percent, i.e., about one-third that of the EPON 826 epoxy (5). The Derakane 411-45 vinyl ester was cured at room temperature, with a 100°C post cure, and exhibits tensile strain capability of 5 to 6 percent, i.e., about the same as the EPON 826 epoxy (6). Strain capability of the matrix resin appears to have had little effect on the impact energy absorption capability of any of the composites of this study.

## 5. DISCUSSION

The present investigation showed that the Spectra 900 and Spectra 1000 composites consistently performed better in impact energy absorption than either Kevlar 49 or E-glass composites over the temperature range from -50°C to 100°C. The Spectra fiber composites exhibited a more pronounced temperature dependence than either the Kevlar 49 or E-glass composites, however. Impact energies were typically lowest at the -50°C test temperature, as might be expected. Elevated temperatures up to the 100°C used in this study resulted in a large improvement for the Spectra fiber composites, but had almost no effect on the Kevlar 49 or E-glass fiber-reinforced composites.

Impact energy absorption capability was not dependent on the matrix materials used in this study. Comparisons with other data available for similar composites with another epoxy of a more brittle nature (Hercules 3501-6) also supported the conclusion that matrix system variations had minor effects on impact energy absorption.

Some effect of laminate construction on impact energy absorption was noted. Cross-ply laminates appeared to be affected by temperature, while fabric laminates were not, at least over the temperature range used in the present study.

The excellent impact properties and low density of the Spectra polyethylene fiber composites make them very attractive for impact applications previously dominated by E-glass and Kevlar 49 composites.

## 6. ACKNOWLEDGEMENTS

The Spectra 900 and 1000 polyethylene fibers were provided by Allied Fibers, Petersburg, Virginia. The EPON 826 epoxy and Millamine 5260 catalyst were provided by Shell Development, Houston, Texas.

The authors also gratefully acknowledge the assistance of various undergraduate student members of the Composite Materials Research Group at the University of Wyoming in performing the composite plate fabrication and testing.

## 7. REFERENCES

1. "Spectra Fiber Properties," Allied Fibers, Petersburg, Virginia, 1986.

2. J.D. Winkel and D.F. Adams, "Instrumented Drop Weight Impact Testing of Composite Materials," Report UWME-DR-301-108-0, Department of Mechanical Engineering, University of Wyoming, December 1983.

3. J.D. Winkel and D.F. Adams, "Instrumented Drop Weight Impacted Testing of Cross-Ply and Fabric Composites," Composites, Vol. 16, No. 4, October, 1985, pp. 268-278.

4. Unpublished data generated for Allied Fibers, 1986.

5. D.F. Adams, R.S. Zimmerman, and E.M. Odom, "Polymer Matrix and Graphite Fiber Interface Study," Report No. UWME-DR-501-102-1, Department of Mechanical Engineering, University of Wyoming, June 1985.

6. K. Hawthorne, R. Stavinoha, and L. Craigie, "High Bond-Super Tough-CR Resin," Proceedings of the 32nd Annual Technical Conference, Reinforced Plastics/Composites Institute, The Society of the Plastics Industry, Inc., 1977, Section 5-E, pp. 1-3.

## 8. BIOGRAPHIES

Richard S. Zimmerman is presently a Staff Engineer in the Composite Materials Research Group at the University of Wyoming. He is responsible for fabrication and characterization of a wide variety of composite materials. He graduated in 1979 with a B.S.M.E. from the University of Wyoming. From 1979 to 1981 he worked at Hercules Aerospace in Salt Lake City, Utah, in a composite materials design and analysis group.

Dr. Donald F. Adams is presently Professor of Mechanical Engineering at the University of Wyoming and head of the Composite Materials Research Group, where he has been since 1972. Prior experience was with the Rand Corporation, five years, the Aeronutronic Division of Philco-Ford Corporation, four years, and Northrop Aircraft, three years. He has over 20 years of full-time experience in composite materials research and development.

SENSOR DEVELOPMENT FOR PMR15 CURE MONITORING AND CONTROL

David R. Day
David D. Shepard
Micromet Instruments, Inc.
21 Erie Street
Cambridge, MA 02139

## Abstract

The design of dielectric sensors used in the cure monitoring of PMR15 is critical in optimizing the quality of the data obtained. Limitations of traditional high conductivity sensors and how they are overcome with the newly designed P3 sensor are discussed. Curing of PMR15 using traditional high conductivity sensors and the new P3 sensors are compared. The effects of pressure and vacuum on monitoring PMR15 cure using the new P3 sensor are examined.

## 1. INTRODUCTION

As the use of PMR15 grows in the structural composite industry, there is an increasing need for an easy to use, disposable cure monitoring sensor. In recent years, several dielectric cure monitoring sensors have been developed (1-3) and used with PMR15. (3) However, the commonly available sensors cannot withstand the high temperatures or are not sensitive to the wide range of ionic conductivity required for monitoring PMR15 cure, or they are made of expensive ceramic. In addition, graphite fibers, small particles, and bubble formation can potentially interfere with measurements and further complicate the testing procedure. This paper discusses these issues and the results obtained in the development of a new dielectric sensor designed especially for the PMR15 environment.

## 2. EQUIPMENT

Dielectric properties were monitored using a Micromet Instruments, Inc. Eumetric System II Microdielectrometer, Eumetric System II Software, and an IBM personal computer. The PMR15 curing cycle was monitored and compared using both a "high conductivity sensor" (similar in design to traditional comb electrodes) and the newly developed sensors (P3) discussed in this paper. Heat and pressure were applied to the samples when necessary using a Tetrahedron compression molding press and vacuum was applied using standard vacuum bag techniques.

## 3. DISCUSSION AND RESULTS

Loss factor and ionic conductivity levels generally vary inversely with the resin viscosity, making dielectrics an ideal method for cure monitoring. However, the PMR15 cure cycle is difficult to monitor with a single sensor due to temperature and ionic conductivity excursions. The traditional high conductivity sensor follows the 95C viscosity minimum region very well (exhibiting a loss factor maximum) but then loses signal during the subsequent viscosity maximum due to the very low conduction level. During the second viscosity minimum (260-300C), the high conductivity sensors have frequently shown an inflection but not the expected maximum in loss factor. Evidence presented in this paper shows that the absence of this maximum is due to internal leakage currents at high temperatures within the standard high conductivity sensor.

To improve upon these limitations a new high conductivity sensor, subsequently referred to as a Porous Parallel Plate (P3) sensor, was developed. It consists of a copper electrode on a Kapton backing with high temperature glass cloth placed on top of this electrode, then a fine copper mesh, and finally a top layer of high temperature glass felt. The copper mesh acts as the second plate of a classical parallel plate cell while the intermediate glass layer acts as a porous electrode separator. The upper glass layer acts as a fine filter paper to strain out and prevent small graphite particles from shorting the sensor. This new geometry results in a sensor that is about 100 times more sensitive than the high conductivity sensors, placing it in a range where it is sensitive enough for the high viscosity regions, while not being totally overwhelmed in the low viscosity/high ionic conduction regions. The P3 sensor is also made with a high temperature adhesive so that internal leakage currents are below noise level. These improvements enable monitoring of the entire cure including the second viscosity minimum (loss factor maximum).

The data plots are presented in terms of loss factor or ionic conductivity. The relation between the two is quite simple for both the high conductivity sensor and P3 sensor at low frequencies:

$$\text{Loss Factor} = E'' = \frac{s}{w*Eo} \quad (1)$$

where:

$s$= ionic conductivity
$w$= $2*PI*$frequency of measurement
$Eo$= $8.85*10^{-14}$  Farads/cm

Ionic conduction is a useful quantity since the frequency dependence is removed and only a single curve need be interpreted. The lowest ionic conduction level that can be measured by a standard high conductivity sensor is about $10^{-9}$ (ohm-cm)$^{-1}$. The new P3 sensor is about 100 times more sensitive which extends the range to $10^{-11}$ (ohm-cm)$^{-1}$.

### 3.1 Temperature Influence

Both a traditional high conductivity sensor and a new P3 sensor were heated to 315C to observe the stray signal measured on a bare sensor. Figure 1 shows a high conductivity sensor output that was no more than the lower noise limit (Log Conductivity = -9 to -10) until a temperature of 175C was attained. Above this temperature, the background

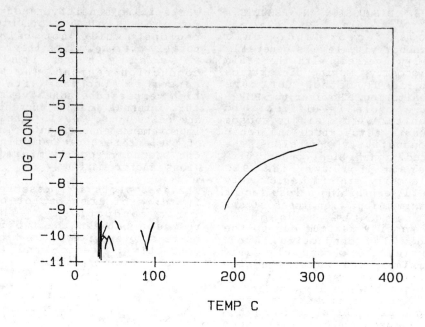

**Figure 1.** Output from bare high conductivity sensor during heating to 315C.

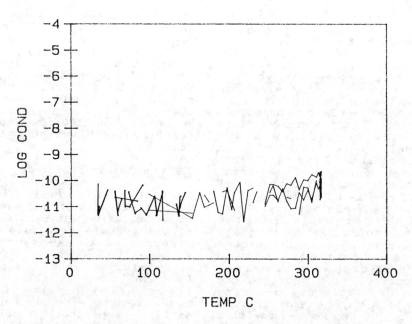

**Figure 2.** Output from bare P3 sensor during heating to 315C.

signal from the sensor increased to about $10^{-7}$ at 315C. Although this is still a fairly low signal level, it does indicate a loss of lower sensitivity limit with increasing temperature.

The P3 sensor data in figure 2 show that the design and change in materials cause the signal to remain at the lower noise limit for the entire temperature range.

## 3.2 Comparison Of Actual Cures

Cures of PMR15 are shown in Figure 3 for a high conductivity sensor and in figure 4 with the P3 sensor. Several glass cloth layers were used with the high conductivity sensor to prevent contact of the graphite fibers with the electrodes. Since very small graphite particles can penetrate glass cloth layers, the P3 sensor incorporates a superior built-in high temperature glass felt material for filtering. Both cures were carried out under 200 PSI of pressure.

The loss factor data collected using a high conductivity sensor broke up and essentially disappeared between 175C and 230C due to the imidization and solidification of PMR15 in this range. The loss factor was very low and was at or below the sensitivity limit of a high conductivity sensor. During the final ramp to 315C the signal reappeared and leveled out at an equivalent conduction of about $4*10^{-6}$ a value not much greater than the bare sensor output at that temperature.

The P3 sensor performed significantly better throughout the entire cure as shown in figure 4. The vicosity minimum (loss factor maximum) in the 175C to 230C range was easily tracked. During the final 315C ramp and hold, the loss factor increased, went through a maximum, and then leveled out. This last maximum, attributable to the high temperature cross-linking reaction, had been expected

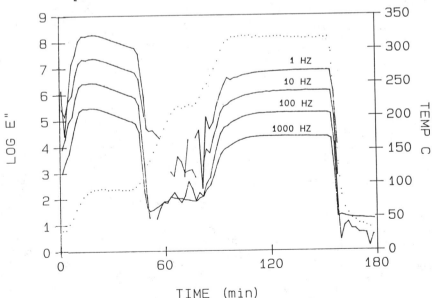

**Figure 3.** Experimental cure data from high conductivity sensor during cure of PMR15.

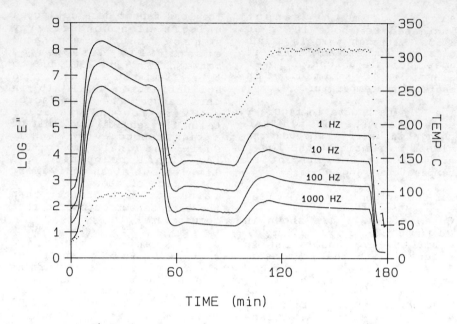

**Figure 4.** Experimental cure data from P3 sensor during cure of PMR15.

but never observed previously with the high conductivity sensor.

### 3.3 Influence Of Pressure

Large quantities of volatile products are generated during the early stages of PMR15 cure. If the volatiles are not allowed to leave, voids will be formed and trapped during cure. Press cures of PMR15 often have little or no pressure applied during the early stages of cure to allow volatiles to escape. With no pressure applied and the extreme degree of bubbling which occurs in PMR15, it is almost inevitable that bubbles will form on the sensor and disrupt the measurement. A high conductivity sensor is especially prone to this problem due to the large separation between sensing electrodes. The P3 sensor is less affected by bubbling but is not immune. The following figures demonstrate how pressure (and thus bubbling) influences the P3 measurement.

Figures 5, 6, and 7 are cures with identical P3 sensors under pressures of 200, 15, and 0 PSI respectively. After a hold at 215C the pressure was increased to 200 PSI in all cases. Pressure was applied before the start of the 0 PSI cure to get initial contact of the material to the sensor.

Figure 5 shows the 200 PSI cure which appears noise free and quite similar to the 15 PSI cure (figure 6). However, close examination shows the conduction levels of the 15 PSI cure was about a factor of three less than the 200 PSI cure. At about 150 minutes, when the pressure was increased from 15 to 200 PSI, the conductivity level became identical to that of the 200 PSI cure. These data suggest that 15 PSI is enough pressure to prevent excess foaming but that some bubbles are present resulting in some loss of conductivity. Figure 7 shows the 0 PSI cure and demonstrates the result of poor sensor contact. After a

1476

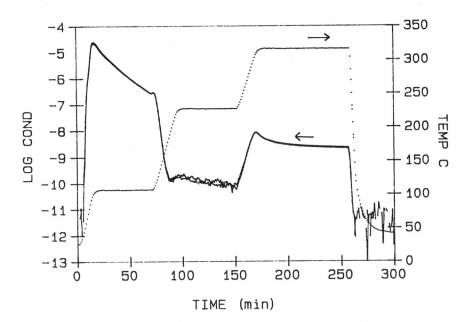

**Figure 5.** Log conductivity data from P3 sensor during cure of PMR15 under 200 PSI.

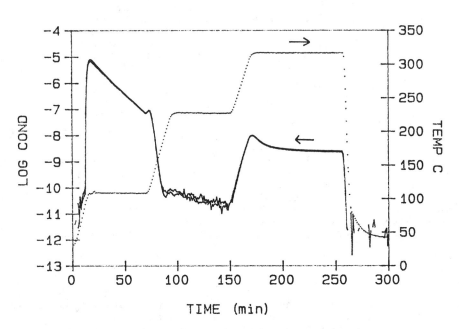

**Figure 6.** Log conductivity data from P3 sensor during cure of PMR15 under 15 PSI. Pressure was raised to 200 PSI after 150 minutes.

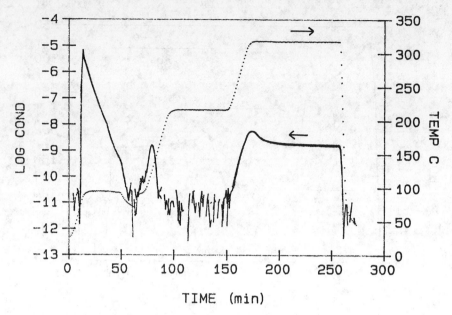

**Figure 7.** Log conductivity data from P3 sensor during cure of PMR15 under 0 PSI. Pressure was raised to 200 PSI after 150 minutes.

brief reapplication of pressure at 80 minutes, some signal was measured but it was extremely noisy. When the pressure was raised to 200 PSI at 150 minutes, the conductivity increased and smoothed out to become identical to that of Figures 5 and 6.

### 3.4 Influence Of Vacuum

A cure of PMR15 was carried out in a vacuum bag with vacuum applied from the beginning of cure. After 90 minutes (during the 215C hold) the cure was stopped, the sample removed from the vacuum bag, and the cure was continued in a press at 200 PSI. The data from both the vacuum bag and the press were merged into a single plot shown in figure 8. Note that the tail in the data just before 90 minutes is due to the press heating up during the restart of the cure in the press.

The data in Figure 8 are quite smooth suggesting that 15 PSI in combination with fast volatile removal result in good sensor contact. It is important to note that the data from the restart of the cure in the press at 90 minutes (where 200 PSI was applied) merge exactly with the data from the end of the vacuum bag part of the experiment. This is good evidence for little or no bubble formation at the sensor during vacuum bag cure.

### 4. CONCLUSIONS

These studies have shown that there are three primary difficulties that may be encountered when monitoring dielectric properties during PMR15 cure. The first is leakage current at high temperatures in a high conductivity sensor. The second is that a high conductivity sensor is unable

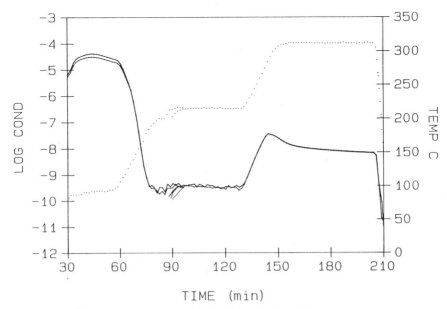

**Figure 8.** Log conductivity data from P3 sensor during cure of PMR15 under vacuum. Pressure was raised to 200 PSI after 90 minutes.

to follow the solidified region of the PMR15 cure where conductivity is low. The last cause for difficulty lies in the nature of PMR15 itself. Graphite particles and the generation of volatiles interfere with the measurement by shorting electrodes or by decreasing the area of the electrode that is in contact with the resin. Through careful design and choice of materials, the P3 sensor minimizes all of the difficulties. P3 sensors should prove extremely useful for closed loop control through dielectric feedback during curing of PMR15.

## 5. REFERENCES

1) S.D. Senturia, et al., SAMPE J., 19, 22 (1983).

2) M.L. Bromberg, et al., Proc. 2nd Conf. Advanced Composites, Dearborn, MI, 307 (1986).

3) Kranbuehl, D., Delos, S., Yi, E., SPE Tech Papers, 31, 311 (1985).

## 6. BIOGRAPHIES

David R. Day received his Ph.D. in Polymer Science from Case Western Reserve University in 1980. He was Research Fellow in Electrical Engineering at the MIT Center for Materials Research until 1983. Since that time he has been employed at Micromet Instruments, Inc as Vice President for Technology.

David D. Shepard was graduated from The Pennsylvania State University with a B.S. in Chemistry in 1982. He was employed as an Analytical Chemist for Reichhold Chemicals, Inc until 1986. Recently he joined Micromet Instruments, Inc as an Applications Engineer.

NOMEX® AND KEVLAR® ARAMID FIBERS FOR FLAME PROTECTION
Wallace P. Behnke and Michael A. Blaustein
Industrial Applications Research
Industrial Products Division
E. I. du Pont de Nemours & Company, Inc.

## Abstract

Nomex* and Kevlar* aramid fibers
were developed by Du Pont to serve
different industrial markets, but
we are now discovering desirable
synergism when these fibers are
combined. Nomex® fiber has
excellent toughness properties and
outstanding long term high temperature durability which is useful
for applications such as hot gas
filtration. Kevlar® fiber has very
high tenacity and modulus, even at
elevated temperatures, which is the
performance needed as a reinforcing
material. The properties of these
two fibers can be combined to
develop structures with excellent
flame barrier performance.
Examples of these are the unique
combination of Nomex® and Kevlar®
fibers in the fireblocking
structure used in aircraft seats to
prevent ignition and smoke

*Du Pont Registered Trademark

generation of the cushion foam.
Intimate staple blends of Nomex®
and Kevlar® fibers give fabrics
with low shrinkage and with
resistance to breaking open when
exposed to a flash fire. Tests
have been developed to evaluate
these structures under realistic
exposure conditions to assure the
user of the protective performance
of these materials. Du Pont has
developed the technology to provide
combinations of Nomex® and Kevlar®
fibers to meet a wide variety of
thermal requirements.

## 1. INTRODUCTION

Du Pont determined during the early
60's that there was need for
high temperature resistant fibers
for industrial uses and initiated
fundamental studies on the
relationships between molecular
structure and the properties of
fibers. The aramid family of

fibers resulted from these studies. Two uniquely different fibers were identified and developed.

## 1.1 Nomex® Aramid Fiber

Nomex® aramid fiber, commercialized in 1967, was developed as a fiber with excellent textile properties. It has the strength and recovery properties of polyester, and the toughness of nylon. It can be dyed to a wide range of bright colors. In addition, the molecular structure gives fabrics of Nomex® fibers resistance to ignition and flame propagation when exposed to an igniting source, and in addition excellent durability to long term exposure to high temperatures. These physical and high temperature resistant properties are widely used in industrial applications such as hot gas particulate filtration. Filter bags of Nomex® fiber will withstand the mechanical stress of a jet pulse bag house for 4 years in a noncorrosive atmosphere at temperatures up to about 400°F. The textile and flame resistant properties of Nomex® fiber have been used extensively in protective clothing against flash fire exposures. In addition, an all aramid paper, made of short straight fibers and filmy fibrids, is used as high temperature insulation in electric motors and transformers.

## 1.2. Kevlar® Aramid Fiber

Kevlar® fiber, the other member of the Du Pont aramid class of fibers, was commercialized in 1972. It has very high strength and modulus when compared to other organic fibers which are useful for rubber and composite reinforcement, and its retention of strength at elevated temperatures is excellent. The very high strength provides mechanical protection against physical threats such as projectiles, knives and shattered glass particles. In addition, Kevlar® does not ignite when exposed to a flame, and fibers and fabrics of Kevlar® do not shrink, but actually grow slightly in length while degrading to a carbon-like structure. These properties give Kevlar® fiber utility as reinforcement in structures exposed to high temperature and where flame resistance is required. Kevlar® pulp has unique properties for an organic material which make it an ideal substitute for asbestos in composites requiring high temperature resistance such as brake shoes and clutches.

## 1.3 Kevlar® Fiber in Fabric Blends

We are becoming increasingly aware of the advantages of combining Nomex® and Kevlar® fibers in structures to take advantage of the fiber interaction to improve the performance of the blend compared to that of the individual fibers. Fabrics of Nomex® fiber shrink about 40-50% and pull apart when exposed to flame temperatures because the high temperature strength of the fiber is less than the that of the shrinkage forces. A small amount of Kevlar® fiber, for example only 5%, in an intimate blend with Nomex® fiber, reduces shrinkage to only 20-25% and the breaking open of the fabric is completely eliminated. As the amount of Kevlar® fiber is increased, the flame shrinkage of the fabric is reduced to essentially 0% at the 50/50 blend level. As expected, the flame strength of the fabrics increases proportionately to the Kevlar® fiber concentration.

Du Pont has recently introduced a blend of Nomex® and flame retardant rayon which also contains 5% Kevlar® fiber. Substituting rayon for a portion of the Nomex® fiber improves comfort because of the moisture properties of the cellulosic fiber and improves aesthetics by permitting use of a permanent press treatment on the fabric. However, the high temperature strength of the rayon is less than that of Nomex® fiber, and use of the Kevlar® fiber insures high temperature strength

and fabric integrity during a flash fire exposure.

## 2. PERFORMANCE EVALUATION

A critical and especially important requirement in developing and commercializing materials for protective clothing is the evaluation and the prediction of the performance actually provided to the user. Protective clothing requires both the properties of general work clothing and protection from the chance thermal exposure. Wear performance and aesthetics requirements are well understood and can be evaluated with available techniques. Further wear testing or obtaining customer experience does not entail any undue liability. The relative flame resistance of the materials is easily determined with recognized tests, but prediction of thermal protection requires special considerations.

### 2.1 Thermal Protective Performance Test

The Thermal Protective Performance (TPP) test has been widely used to measure the relative performance of different materials when exposed to the energy of a flash fire. This test is especially useful as a guide for developing new materials, for setting specifications, and for maintaining quality control of protective clothing. However, translation of the results of this test to actual protection provided by the protective garment is difficult. The TPP test is necessarily limited in its representation of the condition of fabric in actual garments; therefore, successful prediction of garment performance from TPP data relies on experience. Use of the TPP test to predict the performance of a new material that responds differently from those in the experience base to thermal exposure is very difficult, if not impossible.

## 2.2 Instrumented Manikin Tests

An interesting conflict occurred which demonstrates the difficulty of predicting garment performance from TPP data. Early TPP evaluations of blends of Nomex® fiber with PFR rayon indicated that these fabrics would have a performance level which was inferior to that of Nomex® III, fabrics with contain 95% Nomex® fiber and 5% Kevlar® fiber. However, thorough evaluation of the cellulosic blend fabric included instrumented manikin tests using actual exposures to flash fires; the manikin tests indicated that the blend of Nomex® and rayon had protective performance fully equal to that of Nomex®. This experience indicated the lack of an exact

correlation between the TPP and full scale garment tests results and emphasized the need to utilize full scale tests for garment qualification. "Thermo-man", a fully instrumented manikin, was originally developed for the military in the early 1970's and was used to evaluate and qualify military protective clothing. Du Pont installed a "Thermo-man" test facility in about 1975 to measure the hazard of flammable fabrics. In the late 70's propane torches were installed to enable measurement of flame protection provided by a garment. However, this system did not adequately simulate the intensity or flame coverage experienced in a flash fire. More realistic exposures were achieved in outdoor tests first at a Conoco Refinery at Ponca City, OK, and later at the Dover, Delaware Fire School in order to determine in as realistic way as possible the real protective performance of the test garments. However, these tests suffered from the variability of outdoor testing. Convinced that full scale manikin tests are necessary in the evaluation and qualification of protective materials, we undertook and have recently completed an up-grade to our "Thermo-man" facility which provides realistic exposure, control provided by lab conditions, and improved protection evaluation techniques.

## 3. "THERMO-MAN" FACILITY UPGRADE

There are two basic requirements in a full scale manikin test: use of a realistic exposure and measurement of garment performance. Garment performance is determined from the heat received by 122 sensors distributed uniformly over the manikin surface; the heat received is related to the thermal protection and flame resistance of the garment.

### 3.1 Test Exposure

Measurements of the military full-scale manikin test exposure, and evaluation of garments used by workers in actual flash fire exposures indicated that the intensity of these events ranged from 2.0 to 2.8 $cal/cm^2$ sec. This is achieved in our lab with 8 torches mounted on quadrants at two levels around "Thermo-man". These torches were modified to provide a large volume of flame at a temperature of about 820°C (1500°F) when supplied with full flow of propane gas. The torches were adjusted to give a relatively uniform coverage over the total area of the manikin of 2.0 to 3.0 $cal/cm^2$ sec with a variation of about +/- .2 $cal/cm^2$ sec. Intensity can be controlled by gas pressure. For safety, each torch is equipped with a pilot light, which is ignited prior to the exposure.

### 3.2 Test Procedure

The test protocol consists of dressing the manikin in the test garment, exposing it to the controlled flash fire and recording sensor response, and then calculating the effect received heat would have had on human tissue. The procedure is facilitated by computer software that controls the test, provides safety checks, and calculates the results of the test exposure. Temperature data from each of the manikin sensors is recorded every second for the duration of the exposure and for an extended period of time after (usually 60 seconds) in order to record all of the heat that is transmitted to the manikin as a result of the exposure.

Using the temperature of the sensors vs. time as input, the incident heat flux at each sensor is determined for the duration of the data collection time. This heat flux is then used to calculate the temperature that human tissue would reach at ten depths under similar heat exposure. The temperature/time histories at two depths, 100 um and 2000 um below the surface, are used to predict class "C" and class "D" burns. The total body area represented by the sensors reaching each of these predicted burn levels is a measure of the protection provided by the test garment.

The upgrading of the "Thermo-man" test has provided Du Pont with the ability to determine the performance of materials in a full-scale garment test using realistic exposures representative of industrial flash fires.

3.3 "Thermo-Man" Test Results

Table I gives the results of "Thermo-man" data on a number of different fabrics containing Nomex® and Kevlar® fibers. The performance is dependent on the weight of the fabric and the way in which the fabric is dyed and finished. Lighter weight fabrics are less protective than heavier ones, and the reduced shrinkage achieved with fabrics of 100% Kevlar® fiber further increases protection, as indicated by lower predicted manikin burn level. These data also demonstrate that fabrics containing PFR rayon in a blend with aramids give the same protection as those of 100% aramid fibers.

Table I

Thermal Protection of Fabrics
of Aramid Fibers
"Thermo-Man" Prediction

| Fabric | | Manikin Burn (%) | | |
|---|---|---|---|---|
| Type | Weight | "C" | "D" | Total |
| Nomex®/ Kevlar® (95/5) | 4.5 | 17.2 | 7.4 | 24.6 |
| Nomex®/ Kevlar® (95/5) | 6.5 | 8.6 | 7.4 | 16.0 |
| Nomex®/ Rayon/ Kevlar® (60/35/5) | 6.5 | 8.6 | 8.2 | 16.6 |
| Kevlar® (100%) | 7.0 | 5.4 | 4.9 | 10.2 |

4. FIREBLOCKING PERFORMANCE

Another area in which we have found synergism between Nomex® and Kevlar® fibers is in fireblocking layers for aircraft seat cushions. About 15 years ago, the Federal Aviation Administration identified the need for reduced flammability and smoke generation in aircraft interiors in order to increase the escape time in the event of a fire. In the late 70's, covering the seat cushion material with a flame resistant layer was identified as a practical method of reducing smoke generation and flame spread. The addition of another layer of material to the seat cushion structure resulted in some restrictions on that material. These included use of the lightest weight possible to accomplish the fireblocking protection, flame resistance, durability under the mechanical torture from passengers, and compatibility with the manufacturing of the seats.

## 4.1 Performance Evaluation

As with clothing, the textile and inherent durability properties of candidate materials could be accessed using standard evaluation techniques. Ease of construction could be determined by construction of seat cushions utilizing the test materials. Mechanical seat wearlife testers were in use by the aircraft manufacturers and could be used to evaluate new materials. Flame resistance could be evaluated and weight/cost considerations accessed. However, a test to predict the performance of candidate materials in an aircraft fire was needed. The approach adopted by FAA was to test the candidate materials in a composite cushion structure using realistic exposure conditions, as opposed to testing individual components in the laboratory.

The FAA Technical Center developed a test method using a modified oil furnace burner to produce the intensity of heat estimated to be present in a burning aircraft. A heat flux of 2.84 +/- 0.14 cal/cm$^2$ sec. (10.5 +/- 0.5 BTU/Ft$^2$ sec.)

and a flame temperature of 1040 +/- 60°C (1900 +/- 100°F) were selected as the thermal conditions to be used. Fireblocked cushions, consisting of foam cushioning encapsulated in the fireblocking layer and dressed with an outer upholstery fabric, were tested. The flames from the oil burner impinge upon the test specimen for 2 minutes to determine the ability of the fireblocking layer to isolate and protect the cushioning material from ignition and degradation, which would result in flame propagation or smoke generation. Loss in weight and char length were chosen as the performance criteria, since these measurements could be used to indicate fuel contribution and flame spread. The performance standard chosen requires that the average weight loss for 3 specimens must not exceed 10%, and that any charring from the exposure cannot exceed 17 inches in 2 out of 3 specimens.

## 5. SPUNLACE ARAMID SHEET STRUCTURES

Spunlaced aramid nonwovens recently developed by Du Pont using Nomex® and Kevlar® fiber appeared to be ideal to meet the weight and performance criteria for effective fireblocking materials. These structures have shown very high utilization of fiber in a random physical structure. The spunlace process is capable of making stable sheet structures in which the individual fibers are very uniformly distributed. This structure has proven to be very efficient for filtering particulates from hot gases and in providing protection from flash fire exposure in limited use over-garments.

5.1 Multiple Layers

Early scouting tests showed that multiple light-weight layers in a structure were more effective in fireblocking that 1 or 2 heavier weight layers (Table II). The interface between the layers, even though each layer is extremely light, appears to improve the ability of the structure to insulate and contain the melted seat cushion material.

Table II

Effect of Multiple Layers
on Fireblocking Performance
of Spunlace Aramid Blends

| No. Layers in 10 oz/yd$^2$ Quilt Structure | Weight Loss (%) in FAA Test |
|---|---|
| 2 | 12.6 |
| 3 | 10.5 |
| 4 | 7.8 |

5.2 Kevlar® Fiber Content and Placement

Fireblocking efficiency is also affected by content of Kevlar®

fiber and the manner in which Kevlar® fiber is used to reinforce the Nomex® fiber. Increasing the concentration of Kevlar® fiber in each layer of a 4 layer quilt reduces the weight loss in the fireblocking thermal test (Table III). This is the result of decreased shrinkage in the quilted structure. Improved performance can also be achieved by substituting 100% Kevlar® fiber layers for one or more of the blend layers (Table IV). Finally, a surprising fireblocking improvement was achieved when the individual spunlace layers were cross-direction stretched to increase dimensional stability. A five layer spunlace aramid quilt showed a weight loss reduction from 8.0 to 6.0% as a result of stretching.

Table III

Effect of Fiber Content on
Fireblocking Performance

| Kevlar® Content in Spunlace Quilts of Nomex® and Kevlar® | Weight Loss (%) in FAA Test |
|---|---|
| 15 | 9.6 |
| 30 | 7.4 |
| 100 | 6.1 |

## Table IV

### Effect of Number of Layers of Spunlaced Structures of Kevlar® on Fireblocking Performance

| Number of Layers of Kevlar® in 4 Layer Quilt | Weight Loss (%) in FAA Test |
|---|---|
| 1 | 9.7 |
| 2 | 8.4 |
| 3 | 7.8 |

### 5.3 Optimized Structure

The final structures selected for commercialization consisted of 3 or 4 layers of spunlaced aramid quilted with a woven shell fabric of Nomex® fiber spun yarn. At least one of the spunlaced layers is 100% Kevlar® aramid; the balance is 70/30 Nomex®/Kevlar®. These combinations allow the properties of the two fibers to work together to provide superior fireblocking performance when used with any of a variety of dress covers. The stability and high strength of Kevlar® fiber at flame temperatures provides a stable fiber network around which the Nomex® fiber can soften, foam, and develop a "seal". This insulates the cushion foam from excessive heat, and also isolates and contains any degraded material, to give the low flame spread and weight loss needed for effective fireblocking performance.

### 6. SUMMARY AND CONCLUSIONS

Although developed for entirely different markets and with distinctly different balances of properties, Nomex® and Kevlar® fibers show synergism when combined to provide protection from flame exposure. Realistic test procedures have been developed and utilized to evaluate and qualify structures designed to have superior barrier and insulative performance in protective clothing and in aircraft seat cushion fireblocking. As more challenging thermal requirements are identified, we are confident that the combination of Nomex® and Kevlar® fibers will provide superior performance in meeting those needs.

### 7. BIOGRAPHY

Wallace P. Behnke has a Bachelor of Science degree from Northwestern University, and has worked for Du Pont for 36 years. He is a Research Fellow and a member of Industrial Applications Research of the Industrial Products Division. He has conducted Research and Development of Nomex® fiber and paper products for over 21 years. His area of special expertise is in the development and evaluation of materials for thermal protective clothing.

Michael A. Blaustein received his Bachelor of Arts degree from Columbia University and his Doctorate in Chemistry from Yale

University. Since joining Du Pont
in 1982 as a member of Industrial
Applications Research, he has
worked in the areas of electrical
insulation, mechanical papers, and
thermal protective clothing and
structures.

# SYSTEM SURVIVABILITY IN NUCLEAR AND SPACE ENVIRONMENTS

by Dr. Norman J. Rudie
IRT Corporation
101 S. Kraemer Blvd., Suite 132
Placentia, California 92670

## ABSTRACT

Space systems must operate in the hostile natural environment of space. In the event of a war, these systems may also be exposed to the radiation environments created by the explosions of nuclear warheads. The effects of these environments on a space system and hardening techniques are discussed in the paper.

## INTRODUCTION

The space and man-made nuclear environments are many and varied. The primary space environments are electrons, protons, and the complete spectrum of heavy ion cosmic rays. The primary man-made space nuclear weapon environments are x-rays, gamma rays, neutrons, beta particles, and electromagnetic pulse (EMP).

Systems operating within the earth's atmosphere are not exposed to the natural space environments, but may be exposed to the man-made nuclear environments accompanying a nuclear war. Atmospheric/surface explosions add to the nuclear driven environment. Blast, thermal, dust, debris, cratering, and ground shock appear in the wake (see Figure 1).

| Environment | Exoatmospheric Burst | Atmospheric Burst | Surface Burst |
|---|---|---|---|
| Electrons | x | | |
| X-Rays | x | x | |
| Fission Products | x | x | x |
| Neutrons | x | x | x |
| Gammas | x | x | x |
| EMP | x(1) | x | x(2) |
| Thermal | x | x | x |
| Blast | | x | x |
| Turbulence | | x | x |
| Ground Shock | | | x |
| Ejecta, Dust | | | x |
| Crater | | | x |
| Fallout | | | x |

(1) HEMP
(2) SREMP

RT-28859

**Figure 1. Nuclear explosion environments as a function of burst altitude**

A Nuclear Hardness and Survivability (NH&S) program consists of a combination of tests and analyses designed to verify that the system will withstand exposure to the criteria explosion and/or natural radiation environments. The NH&S program is characterized by extensive analyses to evaluate parts selection, circuit designs, system mechanization, shielding, and other hardening techniques. The analysis is supported by testing to provide design data, validate the analysis results, and confirm system survivability.

## NUCLEAR HARDNESS AND SURVIVABILITY (NH&S) PROGRAM

A NH&S program flow diagram is presented in Figure 2. There are many parallel tasks performed during the program. All candidate design alternatives are evaluated and traded off. Several factors are considered during a tradeoff:

4. Technical risk of hardening technique,

5. Hardness margin of safety and,

6. Hardness Assurance and Hardness Maintenance costs.

For ease of understanding, systems hardening will be discussed by the type of effect. We will begin with Transient Radiation Effects in Electronics (TREE).

## TREE

TREE refers to dose rate, total dose, and neutron effects in electronics.

### Dose Rate (G)

X-rays and gamma rays, for example, interact with the electrons of matter. The number of electrons driven out of their orbits is defined in terms of the radiation absorbed dose (rad). One rad(Si) corresponds to the excitation of $4 \times 10^{13}$ electrons/cm$^3$. Since the x-rays arrive in a pulse, the rad will also have a rate; i.e., rad(Si)/s. The rate is symbolized by G.

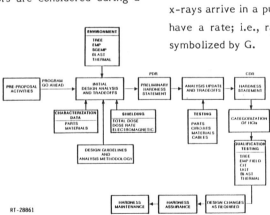

**Figure 2. NH&S Program Flow Diagram**

1. Physical and functional design constraints,

2. Hardness design verifiability,

3. Design to cost (DTC) and life cycle costs (LCC),

Electrons and holes (absence of an electron) freed within the depletion region of a PN junction or within a diffusion length of the junction will be swept out by the built-in E-field and appear as a current in the external circuit (see Figure 3). This current is called a photocurrent, $I_p$. $I_p$ flows in the leakage current direction and is in the range of $10^{-10}$ to $10^{-12}$ A-s/rad(Si). Power diodes have the largest photocurrent response.

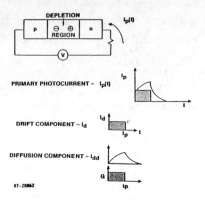

**Figure 3. Primary photocurrent**

This PN junction photocurrent is what causes the dose rate effects in all devices. These photocurrents will drive transistors into saturation, cause logic to upset, and drive the outputs of linear ICs to the rail. The transient recovery times of logic devices is typically hundreds of nanoseconds, whereas linear ICs recover in tens of microseconds.

Some devices experience permanent damage in high dose rate $10^{10}$ rad(Si)/s if they are not protected. Junction isolated integrated circuits (JIICs) may also latchup; i.e., an SCR action is initiated by the dose rate event causing the device to latch up in a non-operating mode. Some latchup is destructive, whereas other JIICs can be recovered by cycling power. Dose rate hardening techniques are summarized at the end of this section.

## Total Dose (D)

Total dose effects are related radiation (X-ray, gamma ray, beta, electron, neutron) induced changes in the oxide layer on a semiconductor device. Radiation drives electrons out of the oxide (insulation) layer leaving behind a positive charge and causes states at the silicon/oxide interface. The amount of charge/states is related to the total radiation absorbed dose accumulated.

Total dose effects appear as a change in the electrical parameters of semiconductor devices. These effects are illustrated in Figure 4 for a 2N3866A bipolar transistor. Gain decreases, and $V_{CE(sat)}$ and the leakage currents increase.

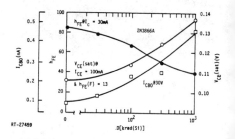

**Figure 4. Total dose effects in a 2N3866A transistor biased at $V_{CB} = 10$ V during irradiation**

MOS devices are the most sensitive to total dose. Unhardened NMOS devices may fail functionally for doses as low as 1 krad(Si). Whereas, unhardened PMOS and CMOS are approximately an order of magnitude harder. Hardened parts are also available.

### Neutron Fluence (N)

Neutrons interact with the nuclei of matter causing lattice displacement damage. The resulting carrier lifetime reduction degrades the electrical parameters of semiconductor devices. This is illustrated in Figure 5 for a 2N3251 bipolar transistor. Gain decreases, and leakage current and $V_{CE(sat)}$ increases.

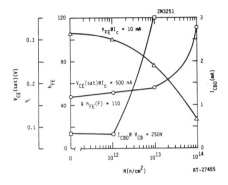

**Figure 5.  Permanent neutron damage in a 2N3251 transistor**

SCRs and linear ICs are relatively sensitive to neutron effects. Some fail in the range of $10^{12}$ to $10^{13}$ n/cm$^2$. Low frequency transistors (e.g., power) are also more sensitive than high frequency transistors.

### TREE Hardening

Hardening is an iterative process of test and analysis (see Figure 2). Design engineers require radiation effects data during the early design phase of a contract. Design guidelines provide that initial information. The initial hardness tradeoff study considers all attributes and variables of the TREE environments which may influence unit performance. Logic upset that does not impair unit performance need only be qualitatively addressed. Permanent parameter degradation of parts which have no effect on unit performance can be handled similarly. The analysis concentrates on permanent change in $V_Z$, $h_{FE}$, $V_{CE(sat)}$,

$V_{os}$, $I_{os}$, etc., and the possibility of burnout and the effects of circuit upsets.

Tests are performed to verify and support the analysis of the radiation response of a unit and to determine the inherent vulnerability of the unit. Verification of analysis with empirical data will provide complete understanding of the operation of the circuits and unit in the radiation environments. Analysis verification is essential since no facility can simulate all the environments and/or their effects. Analysis is used to predict the response to criteria environments.

Each component of the TREE radiation environment can be reduced with shielding. The weight and volume penalty is, however, severe for reducing the gamma ray and neutron environments. On the order of two inches of lead is required to reduce the gamma ray dose by an order-of-magnitude. About three inches of polyethylene are required to reduce the fission neutron fluence by an order-of-magnitude.

Space radiation, beta particles, and x-rays can be reduced with shielding. These environments are applicable to exoatmospheric systems such as space-crafts and ICBMs. X-rays cause the dose rate and the other environments contribute to the total dose.

## ELECTROMAGNETIC PULSE (EMP)

### Coupling to Systems

EMP interacts with an electronic system in much the same way as radio waves excite an antenna. Once a system is excited, currents and voltages can be conducted into the system and effect internal electronics. Entire systems, in fact, behave as antennas to EMP. This is quite different from the way that gamma and neutron radiation interact with individual piece parts in systems. This dependency on system configuration means that a standard military inventory radio or computer that finds itself in a B-52 bomber will likely encounter different EMP stresses than the same unit housed in a ground-based $C^3$ shelter, even though both have the same external EMP environment.

In general, however, exposure to an EMP wave will excite large currents on the skin and cabling of a system. The peak values of these currents may be thousands of amperes. Interior electronics can experience deleterious effects if even small fractions of this external current is able to leak into the system. For this reason, extensive shielding and hardening are usually required of survivable systems.

The nature of the currents that are induced on the system are a natural function of the body resonances possible for the configuration. For this reason, many times these currents exist as damped sinusoids, such as the example seen in Figure 6.

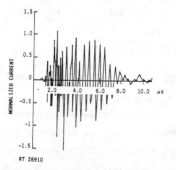

**Figure 6. EMP induced current transient**

There are three fundamental categories of EMP penetration into mod-

erately well-shielded systems. These types of penetrations are (1) conductive penetrations, (2) aperture leakage, and (3) diffusion. Antennas and cable runs form the majority of conductive paths of penetration into the interior of systems. These antenna-like penetrations may be intentional, such as a whip antenna, or unintentional, like an external cable run that conducts EMP energy. Apertures can also pose significant penetration problems. Examples of these may be air vents, displays, hatches, doors, etc. Diffusion merely recognizes that EMP fields may seep through a solid metal wall.

Antennas and conductive penetrations often represent dominant penetration mechanisms. An example of high altitude burst EMP coupling to an HF 7.5 m monopole antenna over a perfectly conducting ground plane is presented in Figure 7. It is obvious, even from cursory examination, that the voltage and current signals induced on the monopole are high enough to burn out most electronics that may be attached to the antenna.

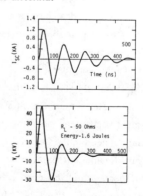

**Figure 7. EMP response of 7.5 m monopole (Ref 3)**

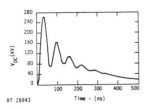

RT 26943

**Figure 7. (Continued)**

## System Response

EMP effects in a system are classed as operational upset or functional damage. Operational upset refers to the case where the system responds to an EMP induced transient as if it were normal information and an erroneous output results. Permanent parametric changes or catastrophic failures are referred to as functional damage.

The allowable degradation in a piece of equipment is dependent on the role which that equipment plays in the success of a mission. The transient loss of information on an audio line, for example, may have negligible effect on the communication channel. Whereas, the operational upset of a missile computer may cause it to miss its intended target.

Electronic circuits are generally the most sensitive components of a system to operational upset. The susceptibility of a circuit depends on the time signature of the EMP excitation and the terminals on which this excitation appears.

The energy required to cause functional damage is one to two orders of magnitude above that required to cause operational upset.

Failure energy threshold data for a number of device types are depicted in

Figure 8. Note that microwave diodes are very sensitive to pulsed power burnout.

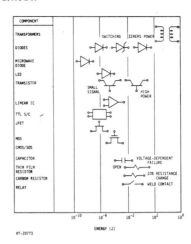

RT-20773

**Figure 8. Parts permanent damage threshold energy for 100 ns square pulse (Ref 5)**

### EMP Hardening

There are a number of techniques for reducing the EMP vulnerability of a system. The principle techniques are shielding, filtering, component selection, circuit layout, circuit design, and function hardening. Each technique incurs cost, size, and weight penalties. The severity of the penalty is a function of how each technique is used and individual system operational requirements.

Shielding provides an effective method of reducing the EMP environment at a sensitive component. A continuous (Al) shield around a system can provide complete protection from EMP. For many aircraft system, 25 to 60 mils is adequate. Continuous shielding (Faraday cage) is, however, seldom achieved in practice. Aperture, joints, exposed cables, etc., are nearly always present.

Cables exposed to the direct EMP must be shielded. When picking a shield material, one must consider durability in a vibration environment, corrosion resistance, contact resistance, tensile strength, and weight as well as shielding effectiveness (SE). Tinned copper and nickel plated copper are good materials.

A cable shield must be circumferentially terminated at both ends in conductive backshells. Care must be taken in choosing the connector with this feature. Usually, good RFI connectors are also good EMP connectors.

Additional protection may be required at the input of a functional unit. For example, the EMP signal picked up by an antenna may be unacceptably high for a transmitter or receiver. Both active and passive protection techniques are available.

### EMP Testing

There are a number of reasons for testing. These include: (1) verification of analysis, (2) extension of analysis, (3) surprises, (4) quality assurance, (5) surveillance, and (6) assessment. In general, each of these categories is appropriate to different phases of the system's life (cradle to grave).

Verification of analysis testing is applicable to the conceptual and early design phases. Electromagnetic (EM) coupling may be employed to confirm currents coupled onto the skin and/or cabling of a system. EM testing may also be used to determine scattered fields, aperture local fields, fields behind apertures, and points-of-entry in general.

Testing at the component and circuit levels can be used to verify or extend an analysis. Current injection testing is a good tool for determining the upset response and burnout thresholds of a circuit.

EM coupling analyses are confined to relatively simple geometries. Testing is usually required to complete a system EMP qualification. Testing will also uncover design weaknesses (surprises) resulting from improper modeling.

## SYSTEM GENERATED ELECTROMAGNETIC PULSE (SGEMP)

SGEMP is an all-encompassing term referring to the electromagnetic effects in a system resulting from the interaction of x-rays or gamma rays with the system. The three basic categories of SGEMP are: (1) structural replacement currents ($I_R$s) and fields, (2) cavity fields and cable coupling, and (3) cable direct injection currents ($I_I$s).

### Coupling
### Replacement Currents and Fields

X-rays and gamma rays (photons) interact with the electrons of matter. Some of the energized electrons on the outermost surface of a material will escape the material. The electron yield is a function of the material and the photon energy. Back emission yields are presented in Figure 9 as a function of backbody spectrum.

The motion of electrons away from a system act like ideal current sources exciting the electrical characteristics of the structure. The resulting current flowing over the structure is called a replacement current ($I_R$).

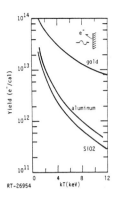

**Figure 9.** Electron emission versus blackbody spectrum

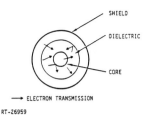

→ ELECTRON TRANSMISSION

RT-26959

**Figure 10.** Injection current phenomena

## Cavity Fields and Coupling to Cables

The emission of electrons from the walls of a cavity creates an internal electromagnetic pulse (IEMP). The emission of an electron leaves behind a positive charge. This charge displacement sets up a restoring E-field. The motion of electrons in the cavity constitutes a current which in turn sets up an H-field. Since the generating x-rays are emitted in a pulse, the cavity fields are also a pulse.

Both the E-and H-field can couple energy to cables threading through the cavity. H-field pickup can be minimized by minimizing the cable loop area. E-field coupling can be minimized by routing the cable close to a ground plane.

### Injection Current ($I_I$)

The current flowing in the load of a cable, exposed to x- or gamma-radiation is called an injection current. Injection currents result from the radiation driven charge motion between the shield and core of a cable (see Figure 10).

$I_I$ is a strong function of the cable geometry and materials, and the radiation environment. Small gaps between the shield and dielectric of a cable can enhance the $I_I$ by an order-of-magnitude or more.

### SGEMP Hardening

SGEMP hardening techniques can be classified as (1) environment reduction (ER), (2) coupling reduction (CR), (3) interface protection (IP), and (4) susceptibility reduction (SR). Environment reduction refers to either reducing the photon intensity incident on a surface or reducing the electron emission efficiency of the surface. High-Z materials can be used to reduce the x-ray intensity. A severe weight penalty must be paid to make a significant reduction in a gamma ray environment (see TREE Hardening). In general, the lower the Z of a material, the lower the electron emission efficiency. Thus, an emitting surface can be painted with a low Z material to reduce its electron emission efficiency.

Coupling reduction refers to (1) reducing the injection current $I_I$ in a cable (2) reducing the replacement current $I_R$ flowing onto a cable, and (3) reducing the cavity field coupling to a cable. Injection currents can be minimized by making a cable of low-Z materials. The core, shield, and dielectric materials should have similar atomic numbers (Zs) and densities. Gaps between the shield and the dielectric accentuate the $I_I$. Gaps should be minimized.

Electromagnetic shielding, grounding, and cable routing can be used to reduce the $I_R$ and cavity field coupling to cables.

Interface protection and susceptibility reduction techniques are also used. These are illustrated in Table 1 for functional upset and permanent damage protection.

### Table 1. Electronics Hardening Techniques (Ref 5)

| Hardening Technique | Illustrative Example | Operating Principle | Significant Design Penalties |
|---|---|---|---|
| Filter (discrete, filter pin, incidental) | BOX WALL | Filters pass signals in selected frequency band and stop other signal transmissions. | Weight, cost, reliability |
| Limiters (Zeners, MOVs, spark gaps, resistors) | LIMITER | Limits voltage and thus energy applied to interface electronics | Weight, cost, reliability |
| Hybrids | | Combination of filters and limiters | Weight, cost, reliability |
| Fiber Optic | | Eliminates SGEMP signals from the cabling | Weight, cost, power, reliability |
| Hardened IC Circuits | | High noise immunity logic, burnout resistant ICs | Cost, schedule, power (?) |
| Clocked Flip-Flop | FLIP FLOP / CLOCK PULSE | Minimizes upset probability by minimizing the times at which upset can occur | Cost, power, reliability |
| Signal Averaging | DATA / AVERAGE | EMP/SGEMP transient does not change averaged signal | Cost, power, reliability |
| Store and Forward | A DATA G DATA / REPLY PATH | Automatic data check and retransmission if required. | Weight, power, cost, reliability |
| Circumvention | | Technique for recovering from EMP/SGEMP-induced logic upset. | Weight, cost, schedule, reliability |

RT-28863

### Thermal Radiation

The nature and degree of damage to an exposed material is a function of thermal pulse and the properties of the exposed material. Relevant material properties are: thickness, thermal diffusivity, color, transparency, and surface condition. Color, surface condition, and transparency affect the amount of thermal energy absorbed. Thickness and thermal diffusivity determine the thermal gradient in the material.

Thermal hardening is achieved through: (1) increased material thickness, (2) material selection, (3) fabric sleeving, (4) coatings, (5) smoke producing materials, (6) shutters, and (7) deployment.

### Blast

Blast damage is primarily due to the deforming action of the blast overpressure and the translational motion of the target. Target translation is due to the dynamic pressure. Blast damage can also occur from blast energized debris. The deforming action (crushing, distortion, buckling, etc.) results from the sudden application of a differential pressure to the target. For a closed target, such as a radio, the differential pressure is due to the differential pressure between the outside and inside of the target. For open targets, the differential pressure is due to the finite time it takes the overpressure to envelop the target.

The two target characteristics of primary importance for determining blast loading are overall geometry and mass distribution. The influence of geometry can be understood by considering two targets of equal surface area: one target a cube and the other target a sphere. The sphere being more streamlined will tolerate more blast than will the cube with its flat faces. The influence of the mass distribution can be understood by considering the center of gravity. A target with a high center of gravity is more likely to experience translational motion than one with a low center of gravity.

The ground surface condition will also effect the response of a surface target. For soft ground, such as sand, the surface tends to resist sliding and the target may experience tipping, rolling, and/or lofting. The same target is more likely to slide on a hard surface.

Blast hardening is achieved through: (1) structural reinforcement, (2) rattle space, (3) transparent armor, (4) spare parts, (5) roll bars, (6) outriggers, (7) tie-downs, and (8) deployment.

has worked in the Radiation Effects Department at Autonetics, at Rockwell International and IRT. He joined IRT in 1980 where he now works.

His experience covers satellite, missile, aircraft, and ground based systems. Since 1970, he has been teaching radiation effects courses at private, military, and university facilities. He has taught at California State University at Fullerton and the University of Alabama at Huntsville, Alabama. He is a prolific author having written many articles for IEEE, Defense Science and Electronics Magazines, and is the author of the of the enclopedia 19-volume text entitled "Principles and Techniques of Radiation Hardening," 3rd Edition.

## REFERENCE

1. N. J. Rudie, Principles and Techniques of Radiation Hardening, 3nd Edition, Western Periodicals Co., 13000 Raymer St., North Hollywood, CA. 1980.

2. System Survivability in the Nuclear Age, IRT Corporation, 1985.

3. SC-M-71-0346, "Electromagnetic Pulse Handbook for Missiles and Aircraft in Flight," Sandia, September 1972.

4. "Radiation Effects and Systems Hardening Short Course," Section 4.0, IEEE, NS, July 1980.

5. "DNA SGEMP Users Manual," December 1981.

## ABOUT THE AUTHOR

Norman J. Rudie holds a Ph.D. degree in electrical engineering from Montana State University, Bozeman, Montana. Since graduating in 1967, he

DEGREE OF IMPREGNATION OF PREPREGS -
EFFECTS ON POROSITY
by B. Thorfinnson and T. F. Biermann
BASF Structural Materials Inc.
Narmco Materials, Anaheim, California

## Abstract

Previous work has demonstrated that
the degree of impregnation of an
uncured prepreg is one of the key
factors affecting porosity in thick
laminates.  In typical autoclave
cures, a partially impregnated
prepreg is much more likely to
produce a void free composite part
than a fully impregnated prepreg.
Although the degree of impregnation
of a prepreg has been shown to be a
key variable, it is a variable that
is rarely measured.  This paper
discusses prepreg degree of
impregnation with respect to
porosity and introduces a technique
to measure this important variable.

## 1.  INTRODUCTION

In today's environment of
optimizing mechanical properties
and reducing costs, it is very
important to understand and prevent
potential quality problems.  One of
the most common quality problems in
the processing of laminates is
porosity.  Porosity (or voids) are
internal defects which result in
reduced mechanical properties,
rejected parts, and extra costs in
terms of time and money.  Although,
the problems are often solved
through a combination of experience
and trial and error, these
solutions usually result in
nonoptimized lay-ups and cures.
Clearly, a better understanding of
this important problem is needed.

This paper deals with the building
of void free laminates using
typical autoclave cure cycles but
without the need for debulking.
Here, debulking refers to the
intermittent application of vacuum
throughout the lay-up.  It is a
very common method of reducing or
eliminating porosity problems in
thick laminates.  Debulking is
expensive, time consuming, and does
not necessarily guarantee a good
part.

## 1.1 Correlation Between Prepreg Process and Laminate Porosity

As discussed in a previous paper (1), a strong correlation between porosity and the prepreg process was identified; the prepreg process being defined as the process used to impregnate fibers with resin to make prepreg. By simply adjusting the prepreg process it was possible to produce a prepreg that consistently resulted in void free laminates or a prepreg that consistently resulted in porosity. One prepreg was called TLP (Thick Laminate Prepreg) because void free thick laminates were produced from this material without any debulking. The other product was called VPP (Void Producing Prepreg) because this material produced voids in thick laminates if the laminates were not debulked.

## 1.2 Correlation Between Prepreg Impregnation and Laminate Porosity

Further work identified the relationship between porosity and the prepreg process. The major difference between TLP and VPP was determined to be the degree of impregnation (DI) of the prepreg. As would be expected, changes in the prepreg process can affect the degree of impregnation of the prepeg. Less obvious, however, is the fact that the prepreg with the higher degree of impregnation would result in voids, whereas the prepreg with the lower degree of impregnation would result in a void

free laminate. VPP (void producing) is essentially a fully impregnated prepreg and TLP (non-void producing) is only partially impregnated. Since the fibers in TLP are not fully wet out with resin, the TLP is actually porous. In other words, it takes a porous prepreg to make a nonporous laminate.

## 1.3 Theory Behind Correlations

The concept that describes how TLP works is illustrated in Figure 1. Whenever a laminate is laid up in an air environment, there will be some air entrapped in between the plies. As described above, TLP is a porous prepreg and therefore the entrapped air (and/or other volatiles) can escape through the pores when vacuum is applied to the laminate. In other words, the porous prepreg allows the vacuum to be evenly felt throughout the entire laminate. At this point the TLP laminate consists of resin, fiber, and vacuum (or evacuated spaces). With the application of heat and pressure, the resin flows into the evacuated areas and cures resulting in a void free laminate.

The concept that describes why VPP does not work is illustrated in Figure 2. Since VPP is a nonporous prepreg, the entrapped air tends to remain entrapped even after vacuum has been applied. The nonporous prepreg "seals off" immediately and does not allow the vacuum to be

# FIGURE 1 – TLP PANEL THROUGH A TYPICAL AUTOCLAVE CURE

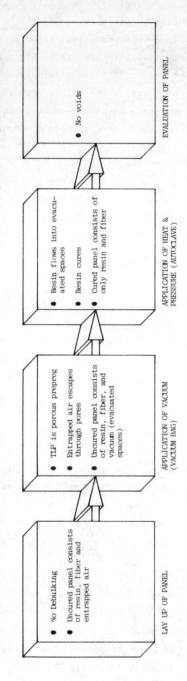

| LAY UP OF PANEL | APPLICATION OF VACUUM (VACUUM BAG) | APPLICATION OF HEAT & PRESSURE (AUTOCLAVE) | EVALUATION OF PANEL |

- No Debulking
- Uncured panel consists of resin, fiber and entrapped air

- TLP is porous prepreg
- Entrapped air escapes through pores
- Uncured panel consists of resin, fiber, and vacuum (evacuated spaces)

- Resin flows into evacuated spaces
- Resin cures
- Cured panel consists of only resin and fiber

- No voids

# FIGURE 2 – VPP PANEL THROUGH A TYPICAL AUTOCLAVE CURE

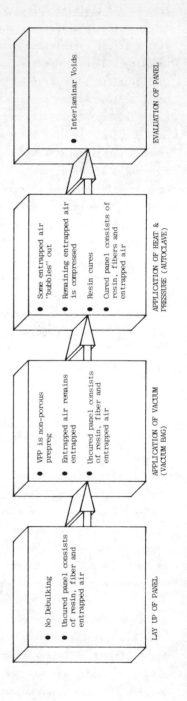

| LAY UP OF PANEL | APPLICATION OF VACUUM (VACUUM BAG) | APPLICATION OF HEAT & PRESSURE (AUTOCLAVE) | EVALUATION OF PANEL |

- No Debulking
- Uncured panel consists of resin, fiber and entrapped air

- VPP is non-porous prepreg
- Entrapped air remains entrapped
- Uncured panel consists of resin, fiber and entrapped air

- Some entrapped air "bubbles" out
- Remaining entrapped air is compressed
- Resin cures
- Cured panel consists of resin, fibers and entrapped air

- Interlaminar Voids

distributed throughout the laminate. At this point, the VPP laminate still consists of resin, fiber, and entrapped air. Even though the remaining entrapped air is somewhat compressed by the resin pressure, the VPP prepreg still results in a laminate with voids.

The concept behind TLP and VPP is based on many experiments, some of which were described in a previous paper (1). These materials have behaved consistently despite experimental variations in resin content, areal fiber weight, tack, drape, viscosity, flow, chemistry, and moisture.

1.4  Demonstration of Theory

A good demonstration of this concept is the "half and half" panel, the cross-section of which is shown in Figure 3. The half and half panel was a 12" X 48" X 200 ply unidirectional (lengthwise) panel, where 100 plies were TLP and 100 plies were VPP. All 200 plies were laid up without any debulking. The TLP and VPP used in this panel were made from the same resin batch, same fiber lot, same resin content, and same areal fiber weight. Since the materials were laid-up into the same panel, the bagging and curing conditions were obviously the same. The only difference between the TLP and VPP was the prepreg process used to produce them and therefore the DI

of each. The difference in the way the two materials perform in a cured laminate is dramatic (Figure 3).

2.  PREPREG IMPREGNATION

Despite its importance, DI is barely mentioned in the literature. In most studies it would appear that DI was either not recognized as a significant variable or that the DI of the prepregs was assumed to be 100%. This is very understandable considering that variations in DI are not always readily apparent. It is therefore worthwhile to define in detail what is meant by DI and two related terms: Impregnation Index (II) and Prepreg Permeability.

2.1  Degree of Impregnation

A dictionary definition of DI might be "level of saturation or permeation." For the purposes of this paper, however, a more mathematical definition is required. Figure 4 represents cross sections of prepreg with 3 hypothetical levels of DI at 2 very different resin contents. It is obvious that Figures 4a and 4d represent 100% DI - the fibers are surrounded by resin and there are no pores in the prepreg. It is also obvious that if the resin and fiber are completely separated then that must represent 0% impregnation as is shown in Figures 4c and 4f.

# Half N' Half Panel Demonstrates Effect of Different Prepreg Processes

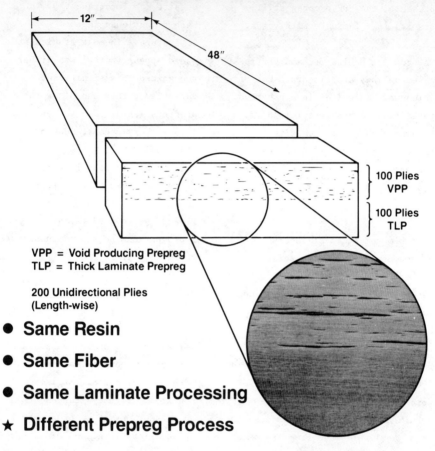

**VPP** = Void Producing Prepreg
**TLP** = Thick Laminate Prepreg

**200 Unidirectional Plies (Length-wise)**

- **Same Resin**
- **Same Fiber**
- **Same Laminate Processing**
- ★ **Different Prepreg Process**

Figure 3.

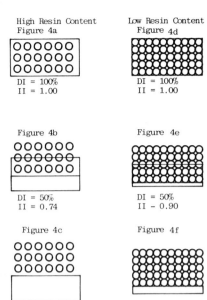

High Resin Content
Figure 4a

DI = 100%
II = 1.00

Low Resin Content
Figure 4d

DI = 100%
II = 1.00

Figure 4b

DI = 50%
II = 0.74

Figure 4e

DI = 50%
II - 0.90

Figure 4c

DI = 0%
II = 0.58

Figure 4f

DI = 0%
II = 0.82

Figure 4. Hypothetical Cross
Sections of Prepreg at
Different Levels of Impregnation

And although it may be a little
less obvious, most would agree that
Figures 4b and 4e represent 50% DI.

Based on these diagrams the
following equation can be written:

$$DI = [(IV - PV)/IV]\ 100\%$$

where:  DI is the Degree of
        Impregnation in percentage
        IV is the Interstitial
        Volume in cc/gram prepreg
        PV is the Pore Volume of
        the prepreg in cc/gram
        prepreg.

In other words, DI is equal to the
percent of interstitial volume that
is not pore volume, or DI is equal
to the percent of interstitial
volume filled with resin. In this
situation the interstitial volume
is basically defined as the spacing
between the fibers. It can be seen
from Figure 4 that a 100% DI the
interstitial volume is equal to the
resin volume. If the interstitial
volume is assumed to be independent
of DI, the interstitial volume at
0% DI and 50% DI must also be equal
to the resin volume.

The assumption that the
interstitial volume is independent
of DI is made to simplify the use
of the equation. In reality, the
interstitial volume of a prepreg is
actually a function of DI and
several other variables. For the
purposes of this paper, however, an
assumption that the interstitial
volume to be independent of DI is
acceptable. To develop equations
to predict variations in
interstitial volume would greatly
and unnecessarily complicate this
work.

Figure 4 is, of course, a greatly
simplified illustration of DI and
is not meant to represent TLP or
VPP. Figure 5 is a SEM photograph
of the cross section of a VPP
prepreg and Figure 6 is a SEM
photograph of the cross section of

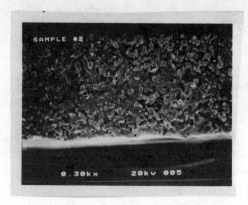

Figure 5. SEM Photograph of
Uncured VPP Cross Section Shows
High Degree of Impregnation

Figure 6. SEM Photograph of
Uncured TLP Cross Section
Shows Low Degree of Impregnation

a TLP prepreg. These photographs
give a much better idea of true DI.

## 2.2 Impregnation Index

In terms of the elimination of
entrapped lay-up air, the
Impregnation Index (II) is a more
useful term than DI. It is more

directly related to prepreg
permeability (discussed below) and
it does not require any assumptions
about the interstitial volume. The
impregnation index is defined by
the following equation:

$$II = (BV - PV)/BV$$

where: II is the Impregnation
Index as a ratio
BV is the Bulk Volume of
the prepreg in cc/gram
prepreg
PV is the Pore Volume of
the prepreg in cc/gram
prepreg

The bulk volume is the total volume
(resin, fibers, pores) per gram
prepreg. Therefore, II is equal to
one minus the volume fraction of
prepreg pores. Obviously, the
percent of prepreg porosity could
be used instead of II, but this
terminology was purposely avoided
to prevent confusion with percent
of laminate porosity.

## 2.3 Permeability

In terms of laminate porosity,
permeability is the property of the
prepreg that is of most interest.
At this point in time, DI and II
(and pore distribution - discussed
below) are used to estimate
permeability. Efforts to directly
measure permeability are ongoing.
Permeability can be defined as the
volume of air flow per unit time
per length of flow per cross-

sectional area. The permeability of a prepreg can be measured horizontally (in the direction of the fibers) or vertically (perpendicular to the plane of the prepreg).

### 3. MEASUREMENT OF PREPREG IMPREGNATION

Given the obvious importance of this variable, a quantitative method is needed to determine prepreg DI and the related properties, II and permeability. Experiments done to date, indicate that the mercury porosimeter is an excellent tool for evaluating these properties. The test procedure that has been developed shows excellent reproducibility and correlates well to laminate porosity results. With the information from the mercury porosimeter, DI and II can easily be calculated. The information obtained also gives a good indication of the prepreg permeability.

Mercury Porosimeter - Procedure

The mercury porosimeter used in this study is built by Micromeritics and is called Pore Sizer 9310. The test is run on a small sample (12" X 0.5") of prepreg. The sample is carefully weighed, rolled into a spiral configuration, and placed in a small glass chamber. The chamber is then thoroughly evacuated until a vacuum of about 30 micrometers of Hg is achieved. Upon reaching the desired vacuum, the mercury is allowed to flow into the chamber and surround the sample. Mercury is nonwetting so it does not penetrate into the prepreg at this time. The volume of the chamber is known and the volume of the mercury is determined by weight. From this information one can calculate the bulk volume of the prepreg sample. The bulk volume includes the volume of the prepreg pores. After the mercury has surrounded the sample, pressure is applied to force the mercury into the prepreg pores. The pressure required to force the mercury into the pores is inversely proportional to the pore size (2). The volume of the mercury penetrating the prepreg is measured and recorded with respect to pressure. This is called the incremental intrusion volume. Pressure is increased until 30,000 psia is reached. At this pressure the pores should have been penetrated or compressed. The volume of mercury that had intruded into the prepreg at this point is called the total intrusion volume. Pressure is then decreased back to atmospheric. The volume of mercury that had extruded out of the prepreg at this point is called the total extrusion volume.

The test can provide a tremendous amount of information including:

Bulk prepreg volume (BV)
Incremental intrusion volume
Incremental extrusion volume
Total intrusion volume (pore
  volume) (PV)
Total extrusion volume
Density
Degree of impregnation (DI)
Impregnation index (II)
% porosity in the prepreg
Distribution of pore sizes
  in the prepreg
Resin content by density
Resin volume (RV)

## 4. DISCUSSION

In order to determine the relevance of the prepreg DI measurements, a porosity test laminate is needed. In this study a quick and easy test laminate, called the "Long panel," is used to evaluate a material's ability to produce void free laminates. The Long panel is made from 22 unidirectional plies, each 1 foot wide and 4 feet long, laid up without any intermittent debulking. A unidirectional panel is used because the nesting action of the unidirectional fibers increases the likelihood of entrapping air. Previous work has indicated that air, volatiles, and resin tend to flow in the direction of the fibers (3). The length of the Long panel makes the removal of the entrapped air more difficult. For thes reasons the Long panel is a very sensitive test of a material's ability to produce void free laminates.

Table 1 shows that the DI and II values obtained via the mercury porosimeter correlate well to porosity results of the Long panels. Also shown in Table 1 is a material with an intermediate level of DI and II. The first Long panel built from this material was found to have an high level of porosity. A second long panel was built using a better vacuum source for a longer period of time. This panel was void free. Based on porosimeter and long panel results, there appears to be a logical relationship between the degree of impregnation and the time and quality of vacuum needed to produce a good part. Further work needs to be done to better define this relationship.

Table 1. Mercury Porosimeter
Results Correlate Well to
Laminate Porosity Results

| | DI | II | LONG PANEL EVALUATION |
|---|---|---|---|
| TLP | 60% | 0.86 | VOID FREE |
| INTERMEDIATE | 82% | 0.93 | DEPENDENT ON QUALITY & TIME OF VACUUM |
| VPP | 93% | 0.97 | INTERLAMINAR VOIDS |

## 5. ACKNOWLEDGEMENTS

The authors greatfully acknowledge the assistance and support of their colleagues at BASF Structural Materials Inc., in particular Lloyd Grant and Dave Ross.

## 6. REFERENCES

1. Thorfinnson, B., Biermann, T. F. "Production of Void Free Composite Parts Without Debulking" 31st International SAMPE Symposium, April 7-10, 1986, pp. 480-490.

2. Rootare, H. M. "A Review of Mercury Porosimetry" Advanced Experimental Techniques in Powder Metallurgy, Plenum Press, 1970, pp. 225-252.

3. Brand, R. A., Brown, G. G., McKague, L. E. "Processing Science of Epoxy Resin Composites" Final report, January 1984, AFWAL-TR-83-4124.

## 7. BIOGRAPHY

Brad Thorfinnson has been employed by Narmco for four years as a Process Development Engineer in the R&D department. Prior to working for Narmco, he spent two years as a development engineer in Celanese Water Soluble Polymers Group. He received his B.S. in Chemical Engineering from South Dakota School of Mines and Technology in 1981.

Ted Biermann is the R&D Manager at Narmco. He transferred to Narmco in late 1982 from Celanese Fibers Operations were he has held various technical management positions in R&D and plant locations. He joined Celanese after receiving his Ph.D. in Organic Chemistry from Purdue University in 1970.

# COMPOSITE STRUCTURES IN HOMEBUILT SPORT AIRCRAFT

Andrew C. Marshall
Northern California Chapter

## Abstract

The use of composite materials in aircraft built in home workshops has resulted in a major improvement in appearance, styling and performance. This paper examines some of the reasons behind this shift in the small airplane industry. A brief summary of construction materials along with the performance and appearance of a few of these modern homebuilt airplanes are presented.

Homebuilt, or "amateur built" aircraft have been part of the aviation scene ever since the proprietors of a bicycle shop in Ohio first demonstrated powered flight. The aircraft industry which evolved from this remarkable event has undergone such a transformation during the enusing 84 years, that we can barely see a resemblance between the first aircraft to fly and the large commercial and military vehicles that now ply our airways.

During this entire period, however, individuals have continued to design and construct their own aircraft. The materials used to fashion these hand-built flying machines have followed the path lighted first by the pioneers, and later by the commercial aircraft builder. In recent years, however, the amateurs have been much more ingenious and resourceful than the airplane factories,

and today more new 1 to 4 seat aircraft are built and flown each year by the home workshop builders than by all the airplane factories combined! This unexpected phenomenon first occurred about 1982, and was due as much to the rapid escalation in the price of factory built airplanes, as to any other cause. Figure 1 illustrates this price escalation, along with the corresponding drop in production.

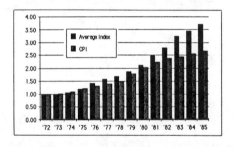

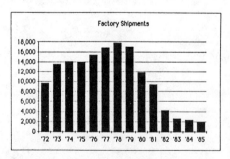

Another cause driving this switch was a surge in both airplane design technology and in composite materials technology. Neither can be termed break-throughs, as all the bits and pieces of information and materials had been either on hand and ready, or under development and well understood for many years. The most visible surge started in 1972 with the introduction of the BD-5, shown in Figure 2. This aircraft was promoted heavily and received instant acceptance. The promise of 200 mph on an engine of 50 to 70 hp (the company folded before the final choice of engines was developed and delivered), together with the beautiful lines of the craft, resulted in some 3,000 kit sales, with another 6,000 cash deposits on kits that were never delivered. Although the first prototype BD-5 was a composite structure, the version finally offered for sale was entirely based on aluminum, used in a completely conventional manner. Only about 80 have ever been finished and flown, a fact which has probably saved many lives, since their operational accident rate has been very high.

A primary problem with the BD-5 was working the aluminum alloy skins into the smooth, sweeping lines of the fuselage and fairings. Such shapes require either an extremely high degree of craftsmanship, or the use of expensive tooling on a stretch press. Since most homebuilders have neither, these parts could only be provided by a properly equipped factory.

This difficulty began to be solved when Burt Rutan, a young engineer working on the BD-5 project, set off on his own in 1974, to develop his own designs for sale to home builders. His second attempt, the "Vari-eze", shown in Figure 3, was able to capture the visual excitement of the BD-5, along with the promise of easy construction. ("Vari-eze" was supposed to mean Very Easy to build!).

The easy construction promised by Rutan results from his patented "Moldless Construction" technique, which produces beautifully shaped structural sandwich structures by the simple expedient of covering a carefully shaped foam core with layers of structural

Fig. 3. Dick Rutan, Jeana Yeager & Mike Melville with the "Long-Eze" that was built prior to the Voyager's construction.

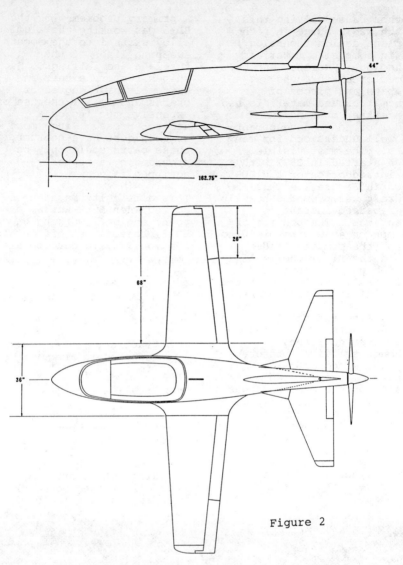

Figure 2

fibers and epoxy resin. Critical shapes, such as wings or control surfaces, are accurately sculptured by using a resistance-heated nichrome wire which is slowly moved along a template attached to each end of a block of polystyrene foam. Numbered stations along the templates permit the operator at each end of the wire to manually advance the wire in concert, such that the desired airfoil contour is produced. Three dimensional contours are made by carving and sanding the foam to the desired shape. Internal vertical load-carrying members, such as spar webs, are provided by making a cut through the foam, laying in the required layers of material, and then gluing the cut-off foam back in place, as shown in Figure 4.

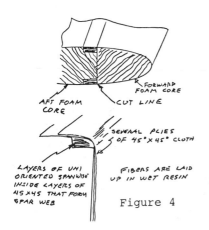

FORWARD
FOAM CORE

AFT FOAM
CORE

CUT LINE

SEVERAL PLIES
OF 45°X45° CLOTH

LAYERS OF UNI
ORIENTED SPANWISE
INSIDE LAYERS OF
45×45 THAT FORM
SPAR WEB

FIBERS ARE LAID
UP IN WET RESIN

Figure 4

The use of polystyrene foam is required wherever hot-wire contouring is used, since most other foams, particularly the urethanes, give off extremely toxic gases when contacted by the hot wire. This foam is quickly dissolved by the styrene used in polyester or vinylester resins, so that epoxy resin must be used to impregnate and attach the layers of structural fiber in contact with the foam.

The new composite technology was an instant success, and in the years between 1976, when the Vari-eze was first shown, and 1985, when Rutan closed down his company, more than 4,000 sets of plans were sold. There were two versions, nearly identical, which were offered. The second, the "Long-Eze", had more wing area and better handling qualities, but was very close to the Vari-eze in appearance. These airplanes were the first to offer both high performance (200 mph, 1200 mile range) along with a reasonable building time. As a result, about 2,000 are actually flying, as compared to some 50 BD-5 models. This is even more impressive, when it is noted that the BD-5 builders received kits containing prefabricated parts, while the Eze builders

bought only plans, and fabricated most of their structures from scratch.

The obvious simplicity of Rutan's composite construction method, along with the freedom to build in any desired shape, resulted in its acceptance by not only the Eze builders, but by builders and designers of many other homebuilt aircraft. The "Quickie", shown in Figure 5, is a Rutan design which was sold to another company, Quickie Aircraft, for their own promotion and sale. It is a bare minimum airplane, operating on only 18 hp, and reaching about 130 mph. That company also bought the US rights to a Canadian design of similar appearance and construction, but having 2 seats and better performance. Its name, obviously, is the "Quickie II". It is shown in Figure 6.

Another look-alike which uses nearly the same construction methods, is the Dragonfly, shown in Figure 7. This is actually a quite different airplane from the Quickie II, although the appearance and general arrangement are very similar.

Two other composite homebuilt aircraft receiving current acclaim are the Glasair and the Lancair. These aircraft are substantially different than the composite craft mentioned previously, in that they are both sold as completely premolded kits. In effect, you buy a full-scale model airplane, gluing the premolded sections together in order to complete the structure. Although it sounds very simple, the actual labor involved in completing these kits to test-flight can run to 2,000 man-hours or more. Both aircraft are of conventional

Figure 5

Figure 6

Figure 7

Figure 8

Figure 9

Two other composite homebuilt aircraft receiving current acclaim are the Glasair and the Lancair. These aircraft are substantially different than the composite craft mentioned previously, in that they are both sold as completely premolded kits. In effect, you buy a full-scale model airplane, gluing the premolded sections together in order to complete the structure. Although it sounds very simple, the actual labor involved in completing these kits to test-flight can run to 2,000 man-hours or more. Both aircraft are of conventional configuration, tail-feathers to the rear, rather than the Canard type represented by the Eze's Quickies and Dragonfly. The Glasair is shown in Figure 8, Lancair is shown in Figure 9.

The Glasair is at the top of the class in performance, with a top speed of over 300 mph on 300 hp, in their most recent model. An earlier model gives 245 mph on just 180 hp!

The Lancair is smaller and lighter than the Glasair, but still yields 225 mph on just 118 hp. It features the use of Nomex honeycomb, rather than the foam core of its competitors, and also uses carbon fiber spar caps.

The fact that construction of these aircraft, by definition, is done by an individual working in a home workshop dictate the use of wet resins and dry fabric layups in projects where the materials are not premolded. As a result, all of the earlier designs mentioned end up with a higher structural weight at the same strength than similar structures on the factory molded kits. This is nearly inevitable, since the resin content must remian on the high side to prevent dry areas of either air entrainment or un-wetout areas of fiber. Typically, estimates of fiber volumes will run from 50% at best, down to less than 30% in some cases, for these wet laid-up structures. Prototype aircraft built by Rutan or his technicians all appear to be at or above 50%, but few homebuilders seem to be able to demonstrate the finesse required to reach the low structural weights achieved by the Rutan group. As a consequence, many Eze's, Quickies and Dragonfly's are from 20 to 50 pounds heavier than the plans say they should be.

1515

The obvious solution to this need for weight reduction would be to switch to prepregs. The use of prepregs, however, requires low temperature storage, elevated temperature cure and vacuum bags on female molds. Although such an array is unlikely to be found in a homebuilders shop, it is quite feasible for a factory environment, and can lower cost as well as weight.

The largest and most famous of the homebuilt aircraft is the Voyager, which circled the earth on a single tank of gas. (Actually, 17 tanks occupying almost the entire airframe.) This aircraft was built and certified as a "Homebuilt Aircraft", and except for some rather timely and expert assistance, was entirely built by the two pilots, Dick Rutan and Jeana Yeager. The help they received varied from many entirely unskilled volunteers, on up to autoclave and shop time donated by Hercules in the fabrication of the wing spars, and engineering assistance from a legion of technically qualified enthusiasts. The skill and determination of this group has resulted in establishment of a very significant milestone in aviation history.

Even this rather sophistocated craft, however, is almost a typical homebuilt when compared to current commercially produced airplanes.

For example:

- Only one set of parts was produced for each section, and tooling development was carried only far enough to produce those parts.

- Critical tools, such as the autoclave mold for the main spars, were made from surplus steel I-beams and channels, ground and finished by volunteer machinists, and sold back to the scrap-yard when the spars were completed.

- Female molds for sandwich wing skins were fabricated by forming sheet metal to ribs cut from particle board, and bonding on a back-up structure consisting of the remaining piece of particle board. By using heat-resistant adhesives, the tools were usable for a cycle or two at 250 degrees F.

- The oven used for cure of the bag-molded sandwich skins was made from a junked home gas furnace. (And would never pass an OSHA inspection!)

The list is endless, but is really representative of the expeditious methods used in the field of amateur built aircraft.

The variety of composite materials in the homebuilts is huge. Many foam cores are used, as well as balsa, plywood and honeycomb. Fibers include E-glass, in both heavy and very light weave styles, as well as roving used right off the spool. Kevlar appears in a great many models, as does S-glass and carbon fiber. The trend, however, is to finer fabrics and better resins. Glasair began the use of 7781 weave, rather than the coarser 7725 and 7715 used by Rutan. They also use a vinylester resin,

Dow 411-45, which greatly reduces layup time because it wets out so fast.

Lancair, although not the first kit to use carbon fiber in the spar caps, uses heat and pressure to make an even lighter and more efficient structure than Dragonfly, which uses dry Orcoweb carbon fiber and wet, room temperature cure epoxy in a wet layup.

All of the airplane models mentioned above represent only a few of the current crop of homebuilts, not even covering a majority of the composite models. The single factor about all of the new homebuilts which stands out above all else, is performance. The new composite developments make them easier to build and, therefore, more popular, but the selling feature is their performance. A brief comparison makes the point, as shown is Figure 11.

Fig. 10. Voyager, as it appeared on its first cross-country trip to Oshkosh, WI.

| MODEL | SEATS | HP | MAX SPEED | UNREFUELED RANGE | FUEL BURN, GPH |
|---|---|---|---|---|---|
| Cessna 152 | 2 | 118 | 125 | 300 | 6 |
| Piper Tomahawk | 2 | 118 | 130 | 300 | 6 |
| Beech Bonanza | 4 | 285 | 190 | 600 | 14 |

These 3 factory-built airplanes can be compared to the following homebuilt aircraft, using some of the same engines.

| MODEL | SEATS | HP | MAX SPEED | UNREFUELED RANGE | FUEL BURN, GPH |
|---|---|---|---|---|---|
| QUICKIE | 1 | 18 | 130 | 300 | 2 |
| BD-5 | 1 | 70 | 220 | 900 | 4 |
| Long-Eze | 2 | 118 | 210 | 1200 | 6 |
| Lancer | 2 | 118 | 225 | 900 | 6 |
| Glasair II | 2 | 180 | 245 | 1200 | 10 |
| Glasair III | 2 | 285 | 300 | 1000 | 14 |
| Voyager | 2 | 250 | 120 | 28000 | 5 |

Figure 11

A comparison of three factory built airplanes to several home built airplanes, including Voyager. Note that the best of the factory built airplanes simply do not measure up.

While it is not at all clear that the use of composite structures is the primary cause for the present success of the homebuilt market, it is certainly an obvious fact that superior performance can only be achieved there. As long as the prices are so high for comparatively poor performance of factory built small airplanes, it is unlikely the commercial builders will recover.

## Biography of the Author

Andrew C. Marshall was born in 1921 in Stockton, Calif. Received his degree in Mechanical Engineering from the University of California in Berkeley in 1943. After spending the war as an officer in the U.S. Naval Reserve, he began his professional career as a Materials and Process Engineer at United Airlines, in South San Francisco, CA. He moved to the Hexcel Corp. in 1950, initially as Chief Engineer, and later was promoted to Vice President, Applications Engineering. After retiring from Hexcel in 1978, he founded his own consulting firm in the field of composite materials. He is currently employed by the Orcon Corp. Union City, California, as Vice President, Marketing.

FIBER REINFORCED CERAMIC COMPOSITES

Helen H. Moeller & William G. Long
Babcock & Wilcox
Lynchburg, Virginia
and
Anthony J. Caputo and Richard A. Lowden
Oak Ridge National Laboratory
Oak Ridge, Tennessee

Abstract

Ceramic fiber ceramic composites
have been fabricated by a novel
chemical vapor infiltration
process. Silicon carbide fiber
reinforced silicon carbide tubes
were fabricated and tested. This
material exhibits non-brittle
fracture and excellent strength
at room-temperature, 1000°C and
1200°C.

1. INTRODUCTION

Advanced ceramics such as silicon
carbide and silicon nitride are
receiving a great deal of attention in the technical press for
high temperature applications.
In a review on advanced ceramics
prepared for the National Bureau
of Standards,[1] the benefits of
advanced ceramic materials were
described in terms of productivity improvement, leverage on
U.S. competitiveness in world
markets, and reduced U.S. dependence on foreign sources of supply
for critical materials. From a
materials performance standpoint,
only the productivity improvement
is important. The other two
benefits are largely political.

The productivity improvement
generally relates to the higher
temperatures available in energy
systems and heat engines. Monolithic ceramics have excellent
high temperature properties, but
suffer from brittle behavior.

The most significant advanced
materials technology now under
development is ceramic composite
materials. Both whisker reinforced composites[2] and continuous fiber reinforced composites[3] represent a new
approach to reliability in high
temperature applications. The
brittle failure of monolithic
ceramics under stress essentially
compromises their reliability and
highlights the need for the
development of fiber reinforced
ceramic composites.

Babcock & Wilcox (B&W), in a
cooperative program with Oak Ridge
National Laboratory (ORNL), has
been developing a silicon carbide
fiber reinforced silicon carbide
composite system using chemical
vapor infiltration (CVI) techniques for the past two years.
This paper presents the results
generated to date on this project.

2. EXPERIMENTAL PROCEDURE

Cylindrical preforms of silicon
carbide fiber, Nicalon, manu-

factured by Nippon Carbon Company, were first formed by filament winding. These preforms were infiltrated by gases at high temperature to deposit a silicon carbide matrix and to form a silicon carbide fiber-silicon carbide matrix composite. Pyrolytic carbon in thicknesses ranging from 0.3 to 1.5 microns was applied by chemical vapor deposition to several preforms prior to infiltration of the matrix.

In this program, we have utilized preforms approximately 1.50 in. I.D. by 5 in. long to demonstrate the feasibility of fabricating larger composite hardware.

In a joint effort the preforms which were wound on mandrels in the Babcock & Wilcox R&D Laboratory in Lynchburg, Virginia were later converted into composites at Oak Ridge by B&W and ORNL personnel using a novel chemical vapor infiltration process. Composite tubes with a 0.25 inch wall thickness were produced in less than three days. Figure 1 shows a composite tube with a section removed by cutting with a diamond saw.

C-rings were then cut from this tube for evaluation of mechanical properties at room temperature and elevated temperatures to 1200 °C. A schematic of a C-ring is shown in Figure 2, with the C-ring under compressive load. Mechanical testing was performed on a computer controlled MTS test machine equipped with a furnace capable of generating temperatures to 1600°C. Strengths were determined after a soak of 1 hour at temperature to assure uniform temperature distribution in the test specimen.

In addition to mechanical testing, density measurements were made on each individual C-ring. Polished sections were prepared to examine fiber-matrix distribution in the composite and to study the fiber-matrix interface.

## 3. DISCUSSION

### Applications

Ceramic composites will find wide applications in areas where the inherent refractory properties of ceramics can be taken advantage of, combined with the toughness realized in a fiber reinforced material. Ceramic composites development is considered by many an enabling technology, one which will allow the use of ceramics at temperatures beyond the temperature limits for metals, but without the familiar brittle fracture of conventional ceramics.

Ceramic composites will provide the design engineer with a high temperature materials system with the advantages of a conventional ceramic material, but with substantially increased reliability. Once reliability, or the absence of catastrophic brittle fracture, is demonstrated, ceramic composites will be introduced as an engineered material into critical applications. Typical systems which will require ceramic composites include:

o Heat Exchangers - Recuperators must function in the exhaust ducts of industrial furnaces at temperatures to 1300°C. Thermal cycling and thermal shock can cause cracking and subsequent failure in conventional ceramics. Tubes in radiant heat furnaces represent one potential market.

o Turbine Engines - The drive toward increased efficiencies available with higher temperature operation dictate ceramics, and the reliability issue will require ceramic composites. Extremes in

1520

thermal cycling and thermal shock, together with compatibility with metallic components, will be the main considerations. Static components will be examined initially, and rotating hardware will be evaluated once service history is available on static components.

o Space Power Systems - The temperatures being discussed for systems in the Strategic Defense Initiative (SDI) will require ceramic composite materials. SDI and turbine engines should drive the development of ceramic composites for the next 10-15 years.

The development and introduction of ceramic composites should follow the same historical pattern as organic matrix composites. Military and high performance industrial applications will be the primary focus, but as confidence in these materials is achieved, substitution for metals and traditional ceramics will be a cost-performance tradeoff.

All of the composites discussed in this paper contain silicon carbide fiber in a silicon carbide matrix. The fiber volume in these composites range from 35-60 percent.

Density

The density of the composite tubes was typically 80-85 percent of theoretical. We believe this range of densities is sufficient and may represent the optimum level for composite behavior. Research at the Societe Europeenne de Propulsion (SEP)[4] indicated that densities above approximately 85 percent of theoretical produced composites with brittle properties at room temperature. We have reproduced these results in one tubular specimen with a density of approximately 90 percent of theoretical.

The fiber distribution and porosity of a specimen are shown in Figure 3. Each fiber appears to be coated with silicon carbide matrix, indicating the penetrating efficiency of the CVI method. This coating efficiency is important in elevated temperature service, as the matrix tends to protect the fiber from oxidation.

Mechanical Strength

Strengths generated from a C-ring test can be compared to a standard three-point flexural strength performed on rectangular specimens.

Our mechanical test results will be discussed in two categories; those without a pyrolytic carbon coating on the fibers and those specimens with a carbon coating deposited as the first step prior to introduction of the matrix.

For composites without a carbon coating, a maximum fracture stress value of 101,000 psi was obtained at room temperature. In addition to this excellent strength value, good composite behavior was exhibited during failure.

Figure 4 illustrates the benefits of fiber reinforcement in ceramic composites. The maximum stress for this specimen was 88,865 psi in this room temperature test. Of particular interest is the load carrying capability of the composite following initial indication of fracture. This composite behavior, a measure of fracture toughness, is attributed to fiber pullout during testing.

This successful demonstration of composite behavior is indicative of good fiber integrity as well as the appropriate level of fiber-matrix bonding to absorb energy during the fracture process. Figure 4 also shows the catastrophic failure of a unreinforced alpha silicon carbide C-ring specimen. The failure strain of the composite is substantially higher, indicating that toughness of the composite is much greater than the unreinforced material.

All specimens tested at elevated temperature showed good composite behavior. In C-ring tests conducted at 1000°C, the maximum fracture stress achieved was 60,000 psi. In limited testing at 1200°C, values between 24,000 and 29,000 psi were obtained. Figure 5 shows a fracture surface of a C-ring specimen tested at 1000°C, with fiber pullout evidence of the non-brittle failure generated in these composite materials. All tests, room and elevated temperature, were conducted in an oxidizing environment.

Fracture usually initiated at the tensile surface of the C-ring, with the crack propagating radially to approximately the midpoint of the C-ring, and then circumferentially around the tube causing delamination parallel to the direction of fiber winding.

For composites with a pyrolytic carbon coating applied to the fibers prior to matrix formation, mechanical strengths were increased slightly, with a maximum of 117,000 psi achieved in C-ring room temperature testing.

Figure 6 shows the fiber-matrix interface on a fracture surface of SiC-SiC composite with pyrolytic carbon coating. This composite had been tested at room temperature. Note the thin layer separating the fiber from the matrix. This specimen failed at a fracture stress of 97,000 psi.

Figure 7 shows the carbon layer still intact at the interface. This specimen was tested at 1000°C, following a 1 hour soak at temperature. The carbon coating will be vulnerable in an oxidizing environment, particularly following cracking of the matrix. However, these results indicate that the matrix protected the carbon coating at the fiber-matrix interface for a limited time at 1000°C. A maximum fracture stress of 71,000 psi was obtained in testing at 1000°C, as seen in Figure 8. This specimen also shows non-brittle failure.

The advantages of a fiber coating will require further study. For some applications, a fiber coating may be dictated to obtain maximum strength. The whole area of optimization of the fiber-matrix interface will be examined over the next several years in the development of high temperature ceramic composites.

### 4. SUMMARY

Fiber reinforced ceramic composites were fabricated by infiltrating ceramic fiber preforms with gaseous precursors which form a silicon carbide matrix. Pyrolytic carbon coatings were introduced on some of the fiber preforms. These silicon carbide fiber reinforced silicon carbide composites have exhibited strengths of greater than 100,000 psi at room temperature and 70,000 psi at 1000°C.

The ceramic composites did not exhibit the typical brittle failure properties of monolithic ceramic materials. Following the initial indication of fracture, the ceramic fiber provided sufficient toughness to delay fracture which will be a requisite for critical applications in heat exchangers, turbine engine components, and SDI hardware.

### 5. REFERENCES

1. "Technology and Economic Assessment of Advanced Ceramic Materials," National Bureau of Standards Report No. GCR 84-470-1, August 1984.

2. Becher, P. F., Tiegs, T. N.,
   Ogle, J. C., Warwick, W. H.,
   "Toughening of Ceramics by
   Whisker Reinforcement,"
   Proceedings of 4th Interna-
   tional Symposium on the
   Fracture Mechanics of
   Ceramics," Blacksburg, VA,
   June 19-21, 1985 (in press).

3. Brennan, J. J. and Prewo, K.
   M., "Silicon Carbide Fiber
   Reinforced Glass-Ceramic
   Matrix Composites Exhibiting
   High Strength and Toughness,"
   J. Mater. Sci. 17 p. 2371-82
   (1982).

4. Naslain, R. et al., "An
   Analysis of Properties of Some
   Ceramic-Ceramic Composite
   Materials Obtained by CVI-
   Densification of 2D C-C
   Preforms," Proceedings of the
   Fifth International Conference
   on Composite Materials, The
   Metallurgical Society, pp
   499-514, 1985.

## ACKNOWLEDGEMENT

This research was co-sponsored by
Babcock & Wilcox and the U. S.
Department of Energy, AR&TD
Fossil Energy Materials Program,
under contract DE-AC05-84OR21400
with Martin Marietta Energy
Systems, Inc.

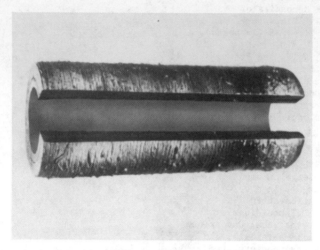

FIGURE 1.   SiC-SiC COMPOSITE TUBE WITH
LONGITUDINAL SECTION REMOVED

FIGURE 2.   SCHEMATIC OF C-RING TEST SPECIMEN

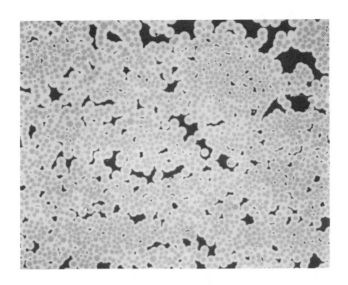

FIGURE 3. POLISHED SECTION OF SILICON CARBIDE FIBER
REINFORCED SILICON CARBIDE COMPOSITE

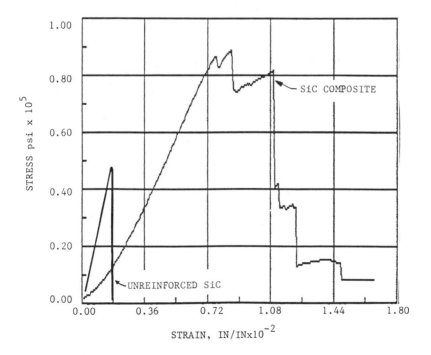

FIGURE 4. STRESS-STRAIN CURVE FOR REINFORCED AND
UNREINFORCED SILICON CARBIDE AT
ROOM TEMPERATURE

FIGURE 5.  FRACTURE SURFACE OF SiC-SiC COMPOSITE
TESTED AT 1000°C SHOWING FIBER PULLOUT

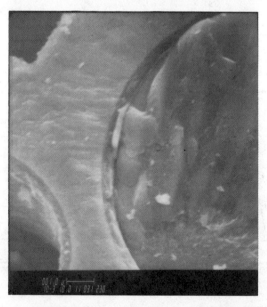

FIGURE 6.  FIBER-MATRIX INTERFACE OF CARBON COATED FIBER
COMPOSITE FOLLOWING ROOM TEMPERATURE TESTS

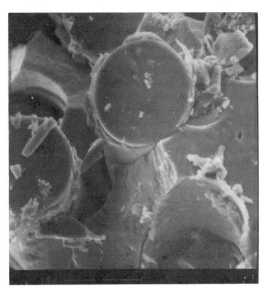

FIGURE 7.   FIBER-MATRIX INTERFACE OF CARBON COATED FIBER
COMPOSITE FOLLOWING TESTING AT 1000°C

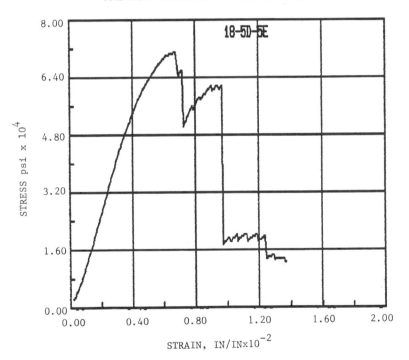

FIGURE 8.   STRESS-STRAIN CURVE FOR SiC-SiC COMPOSITES WITH
CARBON COATING ON FIBERS, TESTED AT 1000°C

32nd International SAMPE Symposium
April 6-9, 1987

DAMAGE TOLERANCE/FRACTURE CONTROL APPROACH TO GRAPHITE/EPOXY
FILAMENT WOUND CASE (FWC) FOR SPACE SHUTTLE MOTORS

S. W. Beckwith
M. E. Morgan

Hercules Aerospace Company
Materials Technology Department
Magna, Utah 84044

J. R. Kapp
G. P. Anderson
Morton-Thiokol Incorporated
Brigham City, Utah 84302

## Abstract

A damage tolerance program was conducted in support of the Space Shuttle Filament Wound Case (FWC) composite segments in order to develop a Fracture Control Document to guide the disposition of segments which incur processing/handling damage or defects. An extensive effort was made to provide analytical and experimental data to assess the overall damage tolerance of the AS4-12K graphite composite case segments. The effort included an empirical series of tests on various regions of the case through the use of tag end joint coupons and quarter-scale (36-inch) pressure vessels to simulate the membrane and transition regions.

Tests included delaminations, surface cuts, and impact damage to the subscale analogs. Each analog was subjected to various NDE techniques, acoustic emission (AE) monitoring, and multiple load cycles prior to ultimate load-to-failure. A series of AE pressure cycles in the 160-840 psig range was followed by a proof cycle (1110 psig) and four MEOP cycles (988 psig) prior to full-scale burst.

The results of these experimental tests and current state-of-the-art finite element analyses were used to formulate an accept/reject criteria for the full-scale (146-inch) FWC segments. The planned use of a full-scale damage tolerance segment (DT-001) for verification of the current criteria and analysis methods is also discussed.

## 1. INTRODUCTION

Filament wound composites have been used in structural applications for several decades, particularly in solid rocket motor cases. The Space Shuttle Filament Wound Case (FWC) is one application where graphite/epoxy composites are being developed as a replacement for the current D6AC steel case for some missions.[1] The present design consists of four FWC segments with a nominal 146-in. diameter: forward section, two interchangeable center sections, and an aft section with external buckling rings. Each section has a center region defined as the "membrane" or cylindrical section, a transition (build-up region), and a joint area on either side. The membrane region acts as a normal pressure vessel in that it carries the tension-tension loads normally exerted by the internal pressure of a closed-end vessel. The joint region carries the axial line loads which act through the steel (D6AC) ring and pin (Inconel 718) connections between segments. Figures 1 and 2 show the basic concept and design details of the various regions, respectively. The primary advantage of the FWC is the nominal 40% weight savings due to the lightweight advanced graphite composites used in fabricating the structure.

As with the steel case solid rocket motors (SRM), handling and processing damage must be a consideration in the FWC segments as well. The fracture control aspects of the shuttle components is inherent in the overall design and evaluation of the steel case and the FWC both from a potential economic loss, and, more important, because the case is man-rated. Damage tolerance is therefore a very important area to the FWC program. This area has been studied extensively in the aircraft industry[2,3] as well as the solid rocket motor case industry[4-6].

In fact, some of the more recent work[5,6] was aimed specifically at damage tolerance in thick-wall FWC composite designs. It should also be noted that the FWC design shown in Figure 2 possesses some inherent damage tolerance features not normally used in pressure vessels. These features are the use of the double helical layer ($\pm$ 33.5°)$_2$ on the outer surface and a graphite/epoxy cloth (0°/90°) layer on the inner surface. The inner cloth layer is non-structural and protects the first hoop layer from internal damage. The stress ratio of the design is around 0.3 which means the helical fiber stresses are 30% of the hoop fiber stresses by design. Consequently, the lower stress loading on the outer helical layers will not result in significant loss of performance should they be damaged. External paint and

thermal insulation also provide damage protection but were not considered in the initial fracture control design.

The intent of this paper is to discuss the damage tolerance work performed to date on the membrane and transition regions using subscale analog pressure vessels and supporting finite element analysis. Work on the joint region is also covered. The overall Fracture Control Plan[7] which outlines the analysis and experimental approach, depended very heavily on subscale and full-scale analog tests to derive an empirical data base for FWC damage tolerance due to the infancy of composite material analysis capability for impact, surface cut, and delamination failure criteria. It is believed that the appropriate analytical techniques will be available in the future, but at the present time the empirical approach seemed to have the most chance for success.

2. DAMAGE TOLERANCE PROGRAM

The overall damage tolerance program relied on 2-D finite element analysis and empirical tests using joint tag end analogs and quarter-scale (36-inch) units to evaluate delaminations, surface cuts, and impact damage. The program is defined in more detail in the following sections.

2.1 Approach

The damage tolerance program was designed to provide sufficient experimental and analytical data to support the accept/reject criteria required by the FWC Fracture Control Document[8,9]. Since only limited analytical tools are currently available for predicting failure and fracture behavior in composite materials, the emphasis was placed on developing an empirical data base which is analyzed and extended to full-scale geometries and loading conditions. Each type of damage or defect was inflicted in subscale test analogs that were loaded to failure. For membrane and transition region damage, quarter-scale (36-inch) FWC units were damaged and subsequently tested. For the joint region damage, standard tag end coupons were tested in a two-hole and ten-hole specimen configuration. The test results were evaluated using 2-D finite element techniques and then scaled (where appropriate) to the full-scale (146-inch) article using the most conservative approaches available.

As a final confirmation, to ensure that a conservative approach has been taken, a full-scale unit (DT-001) with all the acceptable upper bound damage and defects, will be tested to demonstrate damage tolerance to a safety factor of 1.4 as required by design.

## 2.2  Fracture Control Document

The current Fracture Control Document[9] summarizes all the pertinent empirical data and supporting analysis related to the full-scale FWC articles. The primary function of the document is to define the disposition procedures for discrepancy reports (DR) which are written for damaged full-scale articles. It contains two major components in addition to the supporting empirical test data and finite element analyses. These components are: (1) an accept/reject criteria and (2) procedures for dispositioning the discrepancies which fall outside these criteria. The document also includes the metal (D6AC) rings and the attach pins (Inconel 718). These areas are not discussed in the present paper which concentrates on the FWC composite segments.

## 2.3  FWC Segment Materials

The FWC segments are manufactured using a filament winding process coupled with hand layup of 0-degree broadgoods in the joint region. Hercules AS4-12K graphite fiber is used in the full-scale and quarter-scale test units. The basic winding resin is HBRF-55A. It has a nominal tensile modulus of 460 ksi, tensile strength of 12 ksi, and an elongation of 6-8 percent at room temperature. The prepreg broadgoods used in the joint regions use HBRF-3501 resin, which has been widely employed in the aircraft industry, with AS4-12K graphite fiber.

The full-scale construction is shown in Figure 2. The membrane region consists of hoop (90°) and helical ($\pm$ 33.5°) layers. The stress ratio of the units is 0.30 such that the hoop layers carry significantly more load than tactical motor cases which are designed to stress ratios closer to a 0.65-0.85 range. A double helical layer is placed on the outside of the unit to take advantage of this feature from a damage tolerance standpoint. Earlier in the Development Program the helical design consisted of 29° layers carried throughout the full length of the segments. The final version is the 33.5° design with several cut-helical layers which end at various points in the transition region. Quarter-scale tests defined later take advantage of these designs to model the membrane and transition regions.

## 2.4  Damage Types

Three types of damage or defects were selected on the basis of their potential during FWC processing, handling, transportation, and operational environments. The types selected were also identified for each region (membrane, transition, joint) of the segments: delaminations, surface cuts, and impact damage. Porosity was not considered since it is inherent in all the units (full-scale and quarter-scale) as well as tag end joint samples. Since the A-basis strength allowables were determined from the quarter-scale and tag end joint samples, the effects of nominal porosity were included in the allowables which

1530

were cycled through a proof test, four (4) MEOP load cycles, and a final burst test.

Delaminations were induced during the filament winding process in the quarter-scale units and the joint tag ends. A double layer of Teflon tape cut into a rectangular pattern with sharp corners or round circles (near pin holes) was used to induce the delamination. Some preliminary tests were conducted on joint specimens which had delaminations induced artificially with a wood chisel to provide the starting energy.

Surface cuts were made only after nondestructive evaluation (NDE) was performed to insure that delaminations were not present. The cuts were made using a diamond abrasive cutter with a kerf of 0.06-inch. For the quarter-scale units, the cuts were made in an axial or circumferential orientation, at the maximum helical fiber bridging (crossover) region, to a depth of 0.020-0.035 inch. A variety of cut lengths and depths were induced in the tag end joint coupons.

Based on previous experience with impact damage events observed both in-plant and from the field, impact damage velocities were limited to below 50 ft/s. The quarter-scale units were supported internally to provide a rigid backing to magnify FWC damage. The impact damage was induced using a drop-tower device in which the energy-displacement-time history was recorded. Blunt impact damage was induced using a 1-inch diameter steel hemispherical rod. Sharp impact damage was induced with the corner of a steel block oriented 45° to its normal axis.

## 2.5 Delaminations

To test the effects of delaminations, quarter-scale units with wound-in delaminations and joint tag end specimens machined to a two-hole or ten-hole

configuration were used. The basic 36-inch quarter-scale unit, designated DV-36-XX, is shown in Figure 3. These units were used to simulate the membrane region (DV-36-29) and the transition region (DV-36-30 and DV-36-31). The strength allowables in these vehicles were derived after low level acoustic emission cycles, proof test to 1110 psig, four (4) MEOP cycles to 988 psig, and subsequent burst test. Figures 4 and 5 show the delamination regions of these respective units as well as the strain gage locations around the delamination zones. While a delamination itself may not produce failure, it may produce significant stress redistribution in regions where shear, bending, or compressive loads are present. Under such conditions delamination growth may occur and result in further stress redistribution. Consequently, these units underwent intensive instrumentation and NDE during all phases of the test series.

Full-scale tag end coupon tensile test were conducted using two-hole (shearout) and ten-hole (tensile) geometries with the delaminations shown in Figure 6. The externally induced delaminations were produced between the fifth broadgood stack and the helical layer immediately above it using a chisel and hammer. The extent of the delaminations was monitored by visual inspection and confirmed by pulse echo ultrasonic inspection. Offset elongated bushings were used in the two-hole specimen test fixture to allow pin rotation and produce a through-the-thickness stress gradient similar to that present in the FWC full-scale joint. The wound-in Teflon envelope delaminations shown in Figure 6 were not tested at the time of this presentation.

The empirical testing of strength was supported by analyses of delaminations in each of the three regions. Axisymmetric finite element analyses were performed to determine the energy release rate during a small extension of the delamination under a given loading condition. Delamination growth was predicted by comparing these values to the critical energy release rate determined from interlaminar Mode II fracture toughness tests. In addition, the redistribution of stress as a result of the delaminations was addressed. By using such analyses to support the empirical testing, an accept/reject criteria has been defined for the membrane, joint, and transition regions of the composite.

## 2.6 Surface Cuts

The same approach used for evaluating the effects of delaminations, namely quarter-scale units and tag end joint coupons, was used to assess the effects of surface cuts on performance. Surface cuts in the 36-inch units were made to a length of 10-inch in the axial as well as the hoop direction. Both the membrane region (4 units) and the transition region (2 units) were simulated. Surface cuts were inflicted in the outer helical layer and ground rules established early in the program that the hoop layer would not be cut.

The same type of joint two-hole and ten-hole samples were used as shown in Figure 7 to simulate surface cuts in the joint region. Cuts were made between pin holes, pin hole-to-free edge, and through-the-thickness (within the hole and at the free edge).

Since fracture mechanics are only beginning to be understood for crossply failure in composites, especially for shallow, part-through cracks, nonclassical approaches to failure predictions were used. Three analyses were used to determine the effect of cuts in the full-scale FWC case both in the membrane and transition regions. The first analysis, a plane stress model of the cut region was used to determine cut length effect.

This model defined the minimum lengths which need to be detected through visual inspection. The second model, based on linear elastic fracture mechanics, was used to predict when delaminations form near a crack in the membrane and transition regions (an event noted both experimentally and analytically prior to catastrophic failure). The third model was used to determine catastrophic failure in the presence of the crack and subsequent delamination through finite element analysis.

For joint analysis, the same failure criteria were used as in the baseline failure analysis. It is an extension of the average stress criteria. The loaded holes with flaws adjacent to these holes was modeled and the stress distribution was determined. Based on this stress distribution, strength reduction with the presence of a flaw was calculated and subsequently tested.

## 2.7 Impact Damage

To test for impact damage, a conservative approach with respect to testing was employed. Rigid back support was used to allow no energy absorption through global deformation due to bending. Thus upper bound energies were developed in these series of tests. Quarter-scale units were used for the membrane region (2 units) and the transition region (2 units). Impacting was done as described earlier with a dead weight drop test and two head shapes (blunt 1-inch sphere and a sharp corner impactor).

The impact levels were selected in the 25-100 ft-lb range for the quarter-scale units and a 125-300 ft-lb range for the joint region tag end coupons. Damage to the two-hole and ten-hole joint coupons was inflicted between the pin holes based on the analysis conducted to support the baseline design.

As a part of understanding the scaling behavior, a static analysis of the impact event was conducted in order to obtain conservative scaling parameters. This analysis was conducted to determine incipient hoop ply failure in the quarter-scale and full-scale cases. Since incipient hoop ply failure is more severe in the quarter-scale units than in the full scale (which has more hoop plies), this analysis is considered very conservative. The second analysis of the membrane and transition regions was conducted to predict failure with a given amount of inflicted damage. A maximum stress criterion was used to predict catastrophic failure. For joint strength predictions at a given level of damage, the average stress criterion was used. The stresses around the damage zone adjacent to the holes were determined and residual strengths with this damage were predicted.

## 2.8 NDE and Test Support

Because of the nature of this damage tolerance program, the use of various NDE techniques and extensive test instrumentation were employed for both the quarter-scale units and the joint tag end coupons. The primary NDE techniques were visual examination, ultrasonic through transmission, ultrasonic pulse echo, and acoustic emission (AE). Pulse echo was used to monitor changes in the damage zone throughout the test phases.

Instrumentation on all quarter-scale and joint tag end coupons was also extensive. Strain gages were used to measure axial and hoop composite strains in far-field and near-damage zone regions. Long wire gages and LVDT's were used to confirm overall case growth and verify axial growth, a critical parameter in the FWC segment design. High speed film coverage was used on the final burst pressure cycle in order to identify the failure mode and location.

## 3. DISCUSSION OF RESULTS

The results of this test and analysis effort to evaluate the damage tolerance of the AS4-12K graphite composite FWC segments were an accept/reject criteria for dispositioning discrepancy reports (DR's) involving delaminations, surface cuts, and impact damage. This represents the first time that such a criteria or dispositioning procedure has been derived from tests and analysis for a composite solid rocket motor case and put into actual practice through a Fracture Control Document[9]. The results of these tests and analyses are covered in the next few sections.

### 3.1 Delaminations

The DV-36-29 unit was used to evaluate membrane delaminations. The primary failure occurred in the membrane region at a circumferential edge of the delamination as determined from high speed film. The primary failure mode is believed to have been hoop failure. No unusual behavior was seen in the strains monitored around the delaminations. Table I shows a comparison of membrane axial and hoop strains recorded in DV-36-29 with previous units tested from the same design and with the same fiber lots. No significant difference was observed in membrane strains due to the presence of the delamination. In addition, no significant effect was observed in the burst pressure due to the delamination. Acoustic emission results from DV-36-29 indicated that the case was of good quality. No unusual AE data was seen in the areas of delamination, indicating that the delamination did not grow during testing. In addition, pulse echo inspection performed in the delamination region of the case following the proof and MEOP pressure cycles indicated no delamination growth. Based on the hydrotest of DV-36-29, it may be concluded that a 9.0 in. axial length membrane delamination has no effect on the performance of a quarter-scale unit.

The effect of transition region delaminations on the FWC strength allowable was derived from the hydroburst of quarter-scale units DV-36-30 and DV-36-31. These two units were of the newer design which had a cut helical layer in the transition region as in the full-scale units and therefore served as a better representation of the transition region than the earlier (29°) quarter-scale design.

Unit DV-36-30 burst at 1635 psig corresponding to a maximum hoop fiber stress of 503.4 ksi by finite element analysis. The primary failure as determined from high speed film occurred in the aft end transition or peak hoop region near a circumferential edge of the delamination and is believed to have been a hoop failure. The burst pressure for DV-36-31 was 1586 psig corresponding to a maximum hoop fiber stress of 514.6 ksi by finite element analysis. The primary failure as determined from high speed film occurred in the forward transition or peak hoop region. It is noted that the delamination was at the aft end of the case. The primary failure mode is believed to have been a hoop failure.

The delamination size in the membrane region was chosen such that no growth was predicted. This same philosophy was applied to the transition region with an additional unit selected such that delaminations were expected to grow. In each instance, no strength loss due to delaminations was shown either analytically or experimentally. The results of these tests may be found in Table II. All tests were within one standard deviation of the undamaged quarter-scale units.

Based on the hydrotest of DV-36-30 and DV-36-31, it may be concluded that a 4.5 in. axial length transition region

delamination has no effect on the performance of a quarter-scale unit.

Test results for the tag end delamination tests are presented in Table III. No significant effect on strength is seen in either the two-hole or ten-hole tests due to the presence of the externally-induced delaminations. Testing of the two-hole and ten-hole specimens with wound-in delaminations was not completed at the time of this presentation and is therefore unavailable for comparison.

## 3.2 Surface Cuts

Table IV summarizes the results of the quarter-scale surface cut testing. From observing the high speed test films, it was clear that all but one of the units failed at the inflicted surface cut. There is no clear reason why DV-36-35 did not fail at the inflected surface cut in the transition region. The actual failure occurred in the membrane region. The burst pressure was consistent with other undamaged cases, so it is concluded that the cut had no effect.

Some small delamination initiation and growth was detected in units DV-36-22 and DV-36-23. In both cases, these delaminations where found near the cut region. The relative sizes of these delaminations, when compared with the accept/reject criteria, indicate that they had no significant effect on the burst pressures. Local material around the inflicted cut was possibly weakened during the cutting operation, producing the delaminations when the hydroproof pressure was applied. These delaminations did not propagate beyond the cut region. Therefore, their effect is considered small in comparison to the surface cuts.

Predictions of residual strength were made prior to inflicting the impact damage. It was felt that the analyses were conservative so damage greater than that analyzed was inflicted on the quarter-scale units to provide a data base which would bound damage. Unfortunately only one of each type of damage was tested and statistical variation cannot be addressed. The impact damage present was such that the failure strength would fall outside the normal population in quarter-scale units. Based on these data it can be concluded that the analytical techniques used are conservative. Not enough data are available to quantify this degree of conservatism, however. It should be noted that in all instances hoop fibers were damaged during the impact. In fact, impact damage from the blunt impact on DV-36-34 actually extends through the first hoop and to the next hoop layer. Thus, if the same techniques are used to predict strength degradation (plane stress, maximum stress criterion) in the full-scale units, the results should be conservative. The FWC Damage Tolerance Working Group felt that a damage verification unit (DT-001) would provide an additional confirmation and scaleup.

Fortunately, no scaling of joint tag end test data is needed for full-scale tests, since the strength degradation for a given flaw size can be used directly. If these data are used directly the approach should be conservative since failure around a single hole will not constitute catastrophic failure. Table VIII summarizes the analytical/ experimental correlations for the two-hole joint impact specimens. No significant strength reduction was predicted for any of the five specimens and no strength reduction was determined experimentally.

## 4. FULL-SCALE DAMAGE TOLERANCE VERIFICATION UNIT (DT-001)

It was established by the FWC Damage Tolerance Working Group that DT-001 should include the delaminations, surface cuts, and impacts which represent the final accept/reject criteria. The final accept/reject criterion, however, is that the case strength may not be degraded below a factor of safety of 1.4. If a particular segment is judged to have a pre-damage factor of safety higher than 1.4, then flaws or damage will be allowed which are larger than the specified criteria (provided the degraded factor of safety is at least 1.4). This unit-specific factor of safety will be determined based on case strains at proof pressure and the strain-to-failure data of the fiber lots used in the DT-001 case manufacture.

Since the unit-specific factor of safety for DT-001 will not be known until after it is proof-tested, the severity of the final cuts and impacts cannot be defined until after the proof test. Therefore, DT-001 will be put into the test bay twice: once for the proof test (with the cuts, impact damage, and delaminations allowed by the acceptance criteria), then again (after final cuts and impacts are inflicted) for the criteria verification test. This is the only way felt to determine and then verify the real accept/reject criteria.

### 4.1 Configuration

DT-001 will be an aft unit except that it will have only one of the buckling rings (the one nearest the aft end of the membrane). The aft segment design was chosen because the membrane stresses are higher than forward or center segments at a given pressure. Since the hydrotest hardware is limited to 1500 psig, the aft segment is most likely to be able to achieve a burst. A burst is desirable because the effects of damage on failure mode could be seen. Because of the burst potential, the reusable bladder will not be used for the criteria verification test. A bladder will be vulcanized into DT-001 after the post-damage NDE inspection. A reusable bladder will be used for the initial hydroproof test. One buckling ring is needed to verify the effects of flaws and damage in the buckling ring region. More than one buckling ring is not needed since the rings act independently and is undesirable because it would reduce the testable membrane area.

## 4.2  DT-001 Test Procedure

1. Proof Test
   a. Low pressure cycles for acoustic emission data
   b. One cycle to proof pressure

2. Criteria Verification Test
   a. Low pressure cycles for acoustic emission data
   b. One cycle to proof pressure
   c. Four cycles to MEOP
   d. One high-rate pressurization to burst or 1475-1500 psig, whichever comes first.

## 4.3  DT-001 Delaminations, Cuts, and Impacts

Delaminations will be wound in to DT-001 in the same manner as they were wound in to DV-36-29, DV-36-30, and DV-36-31. A sealed envelope of temperature-resistant, non-porous Teflon material will be placed atop a finished DT-001 layer before the next layer is wound over it. Cuts will be machined with a circular saw blade, as was done with the DV-36 units, or with an equivalent method (small-bit router, etc.). Impacts will be performed with an instrumented impact head, using a pulley device, pendulum device, or etc.

The following symbols were used to numbering the delaminations, cuts, and impacts:

M = membrane
J = joint
T = transition
B = buckling ring region
D = delamination
C = cut
I = impact

For example, MD-1 is Membrane Delamination #1. For an overview of the current damage locations, Figure 9 represents the DT-001 configuration.

## 5.  CONCLUSIONS

The results of the quarter-scale tests and the joint tag end coupon tests were combined with analysis to arrive at the current accept/reject criteria outlined in the present FWC Fracture Control Document[9]. The criteria allows for dispositioning discrepancy reports (DR's) for composite FWC segments with delaminations, surface cuts, and impact damage. The empirical data base on the AS4-12K graphite composite pressure vessel and joint areas provides the first step in developing an acceptable criteria. Future tests will be conducted on a full-scale verification unit (DT-001) and additional quarter-scale units in order to extend the data base to full-scale units and verify the analytical predictions.

## 6.  ACKNOWLEDGEMENTS

This work was conducted as part of the Space Shuttle Filament Wound Case (FWC) Program, Contract Number 114010 with Morton-Thiokol Incorporated. The technical guidance provided by Dick Brolliar and John Knadler (NASA MSFC), and Dr. Ward Hill (Hercules Aerospace) during the course of the program is appreciated. NDE support was provided by Jim Kordig and Don Gardiner and their staff. The analysis support was provided by Dan Adams, Rob Cilensek, and Dr. Robert Riddle during the program.

The support of these people and others in test support areas in appreciated.

7. REFERENCES

1. Solid Rocket Motor (SRM) Filament Wound Case Feasibility Study, Final Report H250-12-1-012, Volume 1, NASA Contract NAS 8-34356, Hercules Aerospace Division, February 1982

2. Chow, E. and Kam, C. Y., "Damage Tolerance and Durability Design of Composite Structures for Commercial Aircraft," 13th National SAMPE Technical Conference, October 1981, p. 13.

3. Caravasos, N. and Donnelly, J. P., "Damage Tolerance and Repairability of Advanced Composite Structures," 27th National SAMPE Symposium, May 1982, p. 865.

4. Steck, K. P., Kawabata, C. M., Beckwith, S. W., and Bau, H., "Damage Assessment in Kevlar/Epoxy Composite Pressure Vessels Due to External Flaws," 1982 JANNAF Structures and Mechanical Behavior Subcommittee Meeting, CPIA Publication 374, April 1983.

5. Beckwith, S. W. and Bau, H., "Fracture Mechanics Evaluation of Filament Wound Case Materials Subjected to Operational Environments," AIAA/SAE/ASME 18th Joint Propulsion Conference, June 1982, AIAA Paper 82-1070.

6. Riddle, R. A. and Beckwith, S. W., "Fracture Toughness Characteristics of Thick-Wall Graphite/Epoxy Structural Composites," 29th National SAMPE Symposium, April 1984, p. 304.

7. Pearce, T. M., Cairns, D. S., Riddle, R. A. and Kordig, J. W., Fracture Control Plan, WDI-FWC-11, Revision 1, Hercules Aerospace Division, Contract Number 114010, March 1984.

8. Cairns, D. S., Kordig, J. W., Adams, D. S., and Cilensek, R. F., Fracture Control Document, WDI-FWC-11, Revision 0, Hercules Aerospace Division, Contract Number 114010, June 1984.

9. Adams, D. S., Colvin, G. E., Dodson, C. L., Itchkawich, T. J., Kordig, J. W., Messick, M. J., Goodro, J. B., and Beckwith, S. W., Fracture Control Document, WDI-FWC-11, Revision 2, Hercules Aerospace Division, Contract Number 114010, Volumes I and II, January 1985.

Table I

Comparison Between DV-36-29 and Previous
29° Design Units

| Unit* | Membrane Axial Proof Strain (%) | Membrane Hoop Proof Strain (%) | Burst Pressure (psig) |
|---|---|---|---|
| DV-36-6 | 0.2033 | 1.0030 | 1720 |
| DV-36-18 | 0.1939 | 0.9736 | 1725 |
| DV-36-19 | 0.1953 | 0.9807 | 1779 |
| DV-36-29 (Membrane delamination) | 0.2070 | 0.9975 | 1749 |

\* Fiber lot 630-3A units with and without delaminations tested to proof pressure (1120 psig) and final burst

Table II. Delamination Test Results

| Quarter-Scale Unit | Delamination Size | Predicted Growth | Actual Growth | % Strength of baseline |
|---|---|---|---|---|
| DV-36-29 | Membrane Region, 9.0 in. axial length 27.0 in. circumferential length | None | None | 100.3 |
| DV-36-30 | Transition Region, 3.0 in. axial length 57.8 in. circumferential length | None* | None | 99.1** |
| DV-36-31 | Transition Region, 4.5 in. axial length 57.8 in. circumferential length | None* | None | 98.1 |

\* Negligible growth predicted at location of cut helical.
\*\* No baseline available for this design; value was based on DV-36-29 baseline and ratioed to account for the difference in fiber lot strength.

Table III. Joint Delaminations

| Test Type | Delamination Type | Unit | Number of Specimens | Average Bearing Strength (ksi) | % Strength of Baseline |
|---|---|---|---|---|---|
| 2 Hole | externally induced | DA-001 | 3 | 107.9 | 101.8 |
| | wound-in | DA-005 | 3 | TBD | TBD |
| 10 Hole | externally induced | DA-001 | 3 | 65.2 | 97.0 |
| | wound-in | DA-005 | 3 | TBD | TBD |

1538

Table IV. Quarter Scale Surface Cut Test Results

| Unit | Description | Burst Pressure (Psig) | Acoustic Emission | Pulse/Echo Results | Comments* |
|------|-------------|------------------------|-------------------|---------------------|-----------|
| DV-36-22 | Membrane, Axial, 10.0" x 0.020" | 1532 | Some activity near cut | No delam growth, New delam (0.7 in$^2$) near cut but at a greater depth | Failed at cut |
| DV-36-23 | Membrane, Hoop, 10.0" x 0.020" | 1596 | No unusual activity | Small new delam initiated and grew near cut | Failed at cut |
| DV-36-24 | Membrane, Axial, 10.0" x 0.035" | 1436 | No unusual activity | No change | Failed at cut |
| DV-36-32 | Membrane, Axial, 10.0" x 0.020" | 1494 | No unusual activity | No change | Failed at cut |
| DV-36-33 | Transition, Axial, 10.0" x 0.020" | 1659 | Some activity near cut | No change | Failed at cut, unusually high burst |
| DV-36-35 | Transition, Hoop, 10.0" x 0.020" | 1470 | No unusual activity | No change | Failed away from cut |

* Failure locations were determined from the high speed hydroburst test films.

Table V. Analytical Quarter Scale Surface Cut Predictions

| Unit | Description | Uncut/Cut Ratio LPT | FE | Uncut Average Test Burst (Psi) | Predicted Cut Burst (Psi) LPT | FE | Predicted Strength % of Baseline | Actual Strength % of Baseline |
|------|-------------|---------------------|-----|-------------------------------|-------------------------------|-----|----------------------------------|-------------------------------|
| DV-36-22 | Membrane, Axial 10.0" x .020" | 1.153* | 1.117 | 1744 | 1512* | 1561 | 90 | 88 |
| DV-36-23 | Membrane, Hoop 10.0" x .020" | 1.144 | 1.117 | 1744 | 1524 | 1561 | 90 | 92 |
| DV-36-24 | Membrane, Axial 10.0" x .035" | 1.153* | 1.215 | 1744 | 1512* | 1435 | 82 | 82 |
| DV-36-32 | Membrane, Axial 10.0" x .020" | 1.110 | 1.097 | 1626 | 1465 | 1482 | 91 | 92 |
| DV-36-33 | Transition, Axial 10.0" x .020" | 1.110 | 1.097 | 1626 | 1465 | 1482 | 91 | 102 |
| DV-36-35 | Transition, Hoop 10.0" x .020" | 1.100 | 1.097 | 1626 | 1477 | 1482 | 91 | 90 |

* LPT assumed full helicals were cut at the crossover location;
thus, there is no difference between the .020" and .035" depths.

Table VI. Cut Tag End Coupon Results for Joint Region

| Test Type | Cut Type | Predicted Strength % of Baseline | Actual Strength % of Baseline |
|-----------|----------|----------------------------------|-------------------------------|
| 10-hole | Surface, 2.97" x 0.10" | 91 | 100 |
| 10-hole | Edge, 1.62" x 0.10" | 94 | 98 |
| 2-hole | Surface, 2.25" x 0.10" | 91 | 105 |
| 2-hole | Edge, 1.62" x 0.10" | 94 | 105 |

Table VII. Quarter Scale Impact Units

| Unit | Damage Region | Energy (ft-lbs) | Burst Pressure (psig) | Predicted Strength % of Baseline | Strength % of Baseline |
|------|--------|--------|--------|--------|--------|
| DV-36-34 (Membrane) | 1" x 0.10" | 100 blunt | 1371 | 91 | 84 |
| DV-36-37 (Membrane) | 1" x 0.10" | 25 sharp | 1440 | 91 | 89 |
| DV-36-38 (Transition) | 1" x ~ 0.15" | 75 blunt | 1540 | 93 | 100.4 |
| DV-36-36 (Transition) | 1" x ~ 0.15" | 35 sharp | 1630 | 93 | 95 |

Table VIII. Joint Impact Analytical/Experimental Correlations

| Specimen | Impact Energy (ft-lb) | Damage Region Diameter (in) | Depth (in) | Strength (% of Baseline) Predicted | Actual |
|------|--------|--------|--------|--------|--------|
| DA01-22-9 | 181 (blunt) | 0.50 | 0.06 | 100 | 105 |
| DA01-24-9 | 127 (blunt) | 0.25 | 0.06 | 100 | 103 |
| DA01-23-8 | 150 (sharp) | 0.30 | 0.13 | 99 | 98 |
| DA01-23-9 | 295 (sharp) | 0.75 | 0.13 | 97 | 109 |
| DA01-23-12 | 158 (sharp) | 0.50 | 0.13 | 98 | 108 |

Figure 1. Space Shuttle Filament Wound Case (FWC)

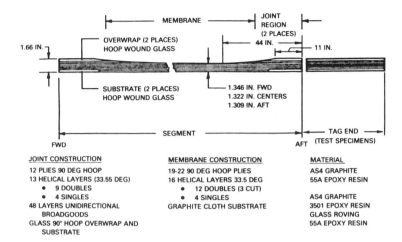

Figure 2. Typical FWC composite segment construction

JOINT CONSTRUCTION

12 PLIES 90 DEG HOOP
13 HELICAL LAYERS (33.55 DEG)
  • 9 DOUBLES
  • 4 SINGLES
48 LAYERS UNIDIRECTIONAL
    BROADGOODS
GLASS 90° HOOP OVERWRAP AND
    SUBSTRATE

MEMBRANE CONSTRUCTION

19-22 90 DEG HOOP PLIES
16 HELICAL LAYERS 33.5 DEG
  • 12 DOUBLES (3 CUT)
  • 4 SINGLES
GRAPHITE CLOTH SUBSTRATE

MATERIAL

AS4 GRAPHITE
55A EPOXY RESIN

AS4 GRAPHITE
3501 EPOXY RESIN
GLASS ROVING
55A EPOXY RESIN

Figure 3. DV-36-XX quarter-scale unit ready for test

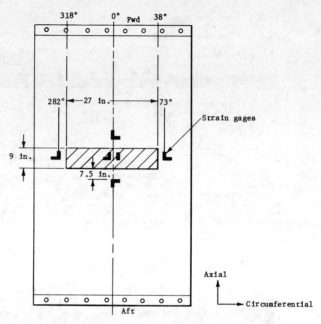

Figure 4. Membrane delamination in DV-36-29

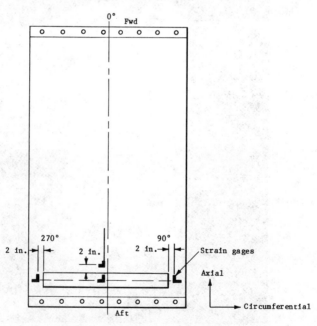

Figure 5. Transition delamination in DV-36-30
and DV-36-31

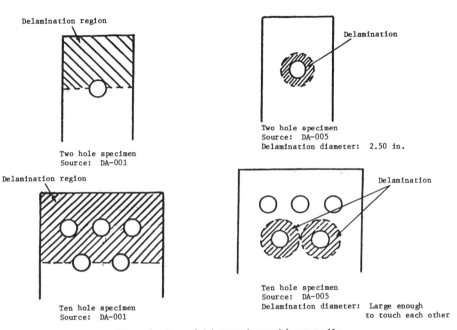

Figure 6.  Tag end joint specimens with externally
induced and wound-in delaminations

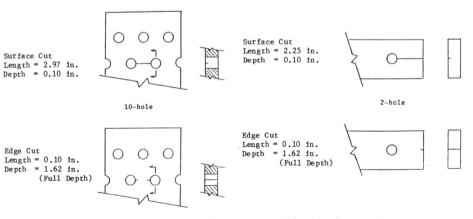

Figure 7.  Tag end joint specimens with external
surface cuts

$G_{Ic}$ AND $G_{IIc}$ CHARACTERISTICS OF
FILAMENT WOUND CARBON/EPOXY COMPOSITES

T. J. Todaro
C. L. Dodson
S. W. Beckwith

Hercules Aerospace Company
Materials Technology Department
Magna, Utah 84044

## Abstract

Critical strain energy release
rates were determined for mode I
($G_{Ic}$) and mode II ($G_{IIc}$) in
filament wound composites made
with Hercules AS4-12K carbon fiber
and HBRF-55A epoxy resin. Mode I
tests were conducted on thickwall
composites (greater than one inch)
using a double cantilever beam
specimen. Finite element analysis
of the specimen evaluated the
effects of mesh size and crack
length. Values of $G_{Ic}$ ranged
from 0.45 to 2.84 in-lb/in$^2$.

Tests for mode II were
conducted on similar materials and
the values of $G_{IIc}$ were
calculated using three (3)
methods: crack closure finite
element, semi-direct compliance,
and direct compliance.
Experimental data were derived
from end-notched flexure (ENF)
experiments. Samples were also
tested at several temperatures
between 30 and 300°F. The ENF
samples were 0.6 inches thick.

## 1. INTRODUCTION

Growth of a delamination in a
composite case can be
characterized by the loading
conditions which cause the
growth. Mode I (opening or
tensile mode), mode II (sliding or
shear mode), and combinations of
mode I and mode II are common.
$G_c$, the critical strain energy
release rate, is a measure of the
fracture toughness, the amount of
energy it takes to make the crack
grow. Currently, tests are used
to characterize unidirectional
material. But in structural
applications (such as the Space
Shuttle Filament Wound Case or
FWC), where the effects of
processing or laminate
configuration may be significant,
tests to characterize "in-situ"
material must be developed.
Double cantilever beam (DCB)
samples were tested to determine
the mode I critical strain energy
release rate, $G_{Ic}$, for
thick-walled carbon/epoxy filament
wound composite material under
ambient conditions. To determine
$G_{IIc}$, the mode II critical
strain energy release rate,
end-notched flexure (ENF) tests
were performed under ambient
conditions on samples from two
different units. Temperature
tests were carried out on samples
from one of the units to determine
$G_{IIc}$ as a function of
temperature. To determine $G_{Ic}$,
the double cantilever beam results
were combined with finite element
analysis results using the crack
closure method. The end-notched
flexure results were reduced using
the crack closure method, as well

as two closed form solutions: semi-direct compliance method,[1] and a direct compliance method,[2] to determine $G_{IIc}$. The results of these three methods compared and are discussed.

## 2. TEST SPECIMEN

### 2.1 Mode I

The double cantilever beam sample is shown in Figure 1. The DCB samples were taken from full scale filament wound cases (FWC), having a thickness of 1.3 inches. The composite consists of approximately 30 layers of alternating helical (33.5°) plies and hoop (90°) plies. Because the case is so thick, a 0.125 inch thick notch with a 90° included angle can be machined at the center of the sample to create two arms of equal size. Holes for pins were drilled in each arm to apply the tensile load. An extensometer was placed on the axial face of the specimen, straddling the notch, to measure the crack opening displacement (Figure 2).

### 2.2 Mode II

A schematic of the ENF test specimen is shown in Figure 3. Each specimen was machined from full-scale FWC cases as shown in Figure 4. The delamination was introduced, using a 0.01 inch thick razor blade, at the neutral axis of the specimen to a length approximately half-way between the outer and center supports. The neutral axis location was chosen because bending stresses are zero and shear stresses are maximum. Thus, delamination growth will occur under in-plane shear (mode II) loading. A fluorescent dye solution was applied at the edge of the delamination with an eye dropper to mark the location of the crack tip prior to testing.

## 3. TEST METHOD

### 3.1 Mode I Tests

The double cantilever beam sample was tested in uniaxial tension to provide opening mode test conditions. The Instron test machine was operated in displacement control with a ramp of 0.05 in/min. As the cross-heads moved apart, the applied load increased; however, once the crack started to propagate, the load dropped. After the crack stopped propagating, the load began to increase again until it reached a critical value (not necessarily the same as the first critical load), and the crack began to propagate again. This repeated stop-start behavior is the method of crack propagation in this material and for this specimen configuration. Data from this test consisted of load vs. crack opening displacement plots. Figure 5 is a plot of load vs. displacement for one of the DCB samples, and shows the stop-start behavior of the crack.

### 3.2 Mode II Tests

Before inducing the delaminations, the specimens were loaded in a three-point-bend fixture to 400 lbs to determine the specimen compliance without a crack. The delaminations were introduced at the neutral axis of each specimen, and the fluorescent dye was applied. Each specimen was again loaded in the three-point-bend fixture. The loading was increased at a constant cross-head speed (0.05 in/min) until the failure load was reached. Failure occurred when the load dropped, and the crack extended rapidly toward the center loading pin. A load vs. crosshead displacement curve was obtained for each specimen.

The specimen crack lengths were measured after the test by prying open each specimen along the delamination and inspecting it under a black light. The presence of fluorescent dye made it possible to trace the original crack length, which was then used in the GIIc calculations.

The temperature tests were conducted at 31, 120, 180, 240 and 300°F. The samples were "soaked" at the test temperature for an hour prior to the test, then the three-point bend test was carried out in an environmental chamber at the desired temperature.

## 4. ANALYSIS

### 4.1 Mode I Evaluation

The double cantilever beam specimen was analyzed using the Hercules in-house linear-elastic finite element program with quadrilateral elements containing two additional internal degrees of freedom. Samples were taken from two cases with a slightly different layup. The two cases were modeled and analyzed separately. Each specimen type was modeled layer by layer near the crack, but ply properties were averaged farther out. (Table 1 contains the nominal lamina properties used in the analyses.)

Five mesh sizes were used: 0.025, 0.050, 0.080, 0.100, and 0.125 inches. Figure 6 shows a mesh of refinement of 0.100 in. Four crack lengths were modeled: 1.0, 1.25, 1.50, and 1.75 inches. Crack length is defined as the axial distance from the loading pins to the crack front. A crack length of 1.0 inch means that the crack front is located at the machined notch tip. Only six of the specimen's actual 14-inch length were modeled because the crack was extended only three-quarters of an inch from the notch tip in the analyses.

To determine GIc, test results (crack length, crack opening displacement, and load readings) were combined with the results of the finite element analysis using the crack closure method. GI, the strain energy release rate, was calculated from the nodal forces and displacements at the crack tip using equation (1).

$$GI = \frac{FR \ (\Delta \ Yab)}{2b \ \Delta a} \qquad (1)$$

where, FR = radial force constraining the nodes when Pcrit is applied to the model
$\Delta$ Yab = relative radial displacement
t = specimen width
$\Delta$a = length of the element

Two stress analyses are required the obtain this information. In the first run, the crack is closed (i.e. the elements are held together and are not allowed to move apart) (Figure 7a). The radial force (Frad) required to hold the nodes together is taken from this run. In the second run, under the same load, the nodes are untied and allowed to move apart (Figure 7b). From this run the relative radial displacement ($\Delta$ Yab) of the two nodes is found. This process was repeated for the different mesh sizes and crack lengths.

To find values of GIc, the critical strain energy release rate, for each test specimen, it was necessary to multiply GI by a scaling factor. Because this analysis is linear-elastic, Frad and $\Delta$ Yab vary linearly with the applied nodal load. Because GIc is a function of the product of these two variables, GI varies with the square of the nodal load and can be found using equation (2).

$$G_{Ic} = (P_{crit}/P_o)^2 \ G_I \qquad (2)$$

where Po is the load used in the stress analysis, and Pcrit is the load at which the crack begins to propagate in the test sample.

The peak stress before the first drop in load is the $P_{crit}$ used in the calculations of $G_{Ic}$. It is the only point where both the load and the crack length are known. As shown in Figure 5, the general trend is for the critical load to increase as the crack grows. Although the critical load associated with other crack advances cannot be pinpointed, it would be conservative to use the initial peak load in calculations of $G_{Ic}$ for longer crack lengths, as was done. However, it would have been more precise to use the crack lengths and corresponding loads in the finite element analysis and $G_{Ic}$ calculations had this data been available.

## 4.2 Mode II Evaluation

$G_{IIc}$ was determined using three different methods, illustrated schematically in Figure 8. The first method was the well established crack closure method described above. The second was a closed form solution for laminates developed by Russell(1). The third, a modification of the Russell and Street solution, was a closed form expression based on the change in experimental specimen compliance with crack length(2).

### 4.2.1 Finite Element (Crack Closure) Method

Finite element analyses were carried out on models using two different layer thicknesses (as-built and design). Schematics of these models are shown in Figure 9. Figure 9 also shows a schematic for a model using smeared layers (a hoop and a helical were combined into one layer). Both design thicknesses, rather that as-built thicknesses, and the technique of smearing layers are used in the analyses of the full-scale cases. Since the $G_{IIc}$ allowables generated from these tests are used in the full-scale analyses, this study was conducted to determine the effects of layer thicknesses and

smearing on finite element results.

$G_{IIc}$, critical strain energy release rate, was calculated from the nodal loads and displacements at the crack tip in much the same way that $G_{Ic}$ was calculated; however, instead of using the radial forces and displacements, the axial forces and displacements are used:

$$G_{IIc} = \frac{FA \ (\Delta \ X_{ab})}{2b \ \Delta a} \qquad (3)$$

where, $FA$ = axial force contraining the nodes when $P_{crit}$ is applied to the model

$\Delta X_{ab}$ = relative axial displacement between nodes a & b

$b$ = specimen width

$\Delta a$ = length of the element

### 4.2.2 Semi-Direct Compliance Method

This closed form solution for laminates was developed by Russell.(1) The specimen flexural compliance (c) is calculated using:

$$C = \frac{d_{11}^{(1)}}{12b} [2L3 + a3 \ (4\alpha-1)] \qquad (4)$$

where, $d_{11}^{(1)}$ = laminate flexural flexural compliance, calculated from laminated plate theory (superscripts refer to sections of sample shown in Figure 10)

$b$ = specimen width

$L$ = 1/2 span length

$a$ = crack length

$\alpha = \dfrac{d_{11}^{(2)} \ d_{11}^{(3)}}{4d_{11}^{(1)} [d_{11}^{(2)} + d_{11}^{(3)}]}$

Refer to Figure 10 for an illustration of these parameters on a test specimen. The expression for specimen compliance without the delamination is obtained by setting a = 0 in equation (4).

The expression for $G_{IIc}$ requires a knowledge of the applied load at the onset of delamination growth, the crack length, and the flexural compliance with a crack.

$$G_{IIc} = \frac{P2}{2b}\left[\frac{dC}{da}\right] \quad (5)$$

$$= \frac{3a2}{2b}\frac{\left[P2\ C\ (4\alpha-1)\right]}{\left[2L3 + (4\alpha-1)a3\right]} \quad (6)$$

where, P = critical failure load

### 4.2.3 Semi-Direct Compliance Method

This direct compliance method, developed by Adams(2), requires a knowledge of the critical load at delamination growth and the specimen compliance with and without a delamination.

The expression for GIIc

$$= \frac{3\ P^2}{2\ ab}\left[C_{cracked} - C_{uncracked}\right] \quad (7)$$

All of the terms in equation (7) were measured directly from the specimen and the load-deflection curve.

## 5. RESULTS AND DISCUSSION

### 5.1 Mode I

Although the actual ply thicknesses differed from the nominal case designs values, the finite element models of the nominal test specimen configurations predicted the actual crack opening displacements accurately for one of the units. The displacement for the TFS-1 specimen under a 100-lb load was predicted to be 0.0081 inches. The actual displacements fell within 9%. For the DA-003 specimen, however, the samples were stiffer than what the model predicted. The predicted displacement was 0.0080 inches, and the actual displacements ranged from 0.0058 inches to 0.0078 inches. It is believed that the matrix of the TFS-1 case was damaged during a high pressure hydroproof test, causing the compliance of the TFS-1 samples to increase, and coincidently, fall closer to the predicted values.

Tables 2 and 3 contain the results of the calculations of $G_I$ for TFS-1 and DA-003 material, respectively, for a load of Po equal to 100 lb. These calculations show that for both sources, $G_I$ increases with mesh size and with crack length. These calculations also show that the effect of mesh size on GI decreases as the crack length increases. At the notch tip, there is a difference of 30%, approximately, between the GI calculated for a mesh size of .025 inches and GI calculated for a mesh size of .125 inches. One-quarter of an inch from the notch tip, the difference in GI decreases to approximately 4%, and for three-quarters of an inch away, the difference is about 3%.

Using equation (2), the $G_{Ic}$ matrix ($G_{Ic}$ as a function of mesh size and crack length) for each case was calculated using the corresponding average initial peak loads. Figure 11 is a plot of GIc as a function of crack length. According to the finite element analysis, the structural responses of the two layups are not very different: the GI values for TFS-1 differ from the DA-003 values by an average of approximately 1%. The GIc values do differ, however, by an average of 43%. The reason is that the DA-003 samples failed at a much higher load than the the TFS-1 samples did. The average initial failure load was 151.5 lb for the ten TFS-1 samples ($C_V = 12\%$), and 202.5 lb ($C_V = 12\%$) for the four DA-003 samples. The difference in critical loads is due to the hydrotest of the TFS-1 unit.

For the DA-03 samples, the value of GIc at crack initiation, 0.78 in-lb/in2 (136 J/m2), agrees with $G_{Ic}$ values reported for both unidirectional samples and crossply laminates (at initiation).(3-5) The initiation

GIc value for the TFS-1 samples, 0.45 in-lb/in$^2$ (79 J/m$^2$), is low because of the damaged matrix.

Figure 11 also shows that GIc is dependent on crack length; GIc increases as the crack propagates. Russell and Street found similar results when testing DCB samples, and attributed the increase to fiber bridging(3,4). The TFS-1 samples reached maximum values of 1.50 to 1.59 in-lb/in2, and the DA-03 samples reached a maximum of 2.67 to 2.84 in-lb/in2, depending on mesh size. While these values of GIc agree with Russell and Street's results, they achieved a constant GIc value at crack lengths as low as 0.4 inches. Although these results show GIc starting to level off, the critical strain energy release rate is still increasing at a crack length of 1.75 inches.

GIc as a material property, however, should be independent of crack length. This implies that the GIc determined in these tests is not a true material property, but is a related to the structure of the specimen. Thus, this GIc is more of a "structural" property, unique to the particular structures tested, the TFS-1 and the DA-003 layups, and to the location of the crack in the layup. While using full thickness material, directly from a case or tag end, has not provided a material property, the structural property found instead may be more indicative of the behavior of the actual full-scale case.

## 5.2 Mode II

### 5.2.1 Ambient Results

Figure 12 shows a typical fracture surface of a failed specimen. Although the photo is not clear, the fluorescent dye was visible along the entire initial crack length. The photo also illustrates the difference between fracture surfaces that are propagated in shear (mode II) and

in opening (mode I). The lighter (or dull) surface is where the delamination propagated in shear during the test. The crack stopped just under the loading pin location, where the compressive forces are large enough to suppress further crack growth.

It was assumed that crack growth began at the point where the loading curve became nonlinear for two reasons. First, the linear finite element analysis used models only the linear behavior of the specimen. Second, some damage at the crack tip occurred during the nonlinear portion of the curve, before ultimate failure at the load drop.

The compliances for specimens without a delamination, calculated using equation (4), were much stiffer than the measured experimental and finite element compliances. A reason for this discrepancy is the shear deformation of the specimen in the 1-3 plane not accounted for in equation (3).

Figure 13 shows the comparison between the measured and calculated specimen compliance with the delamination. Again, the closed-form compliances were lower than the experimental compliances. The finite element analysis assumes a frictionless delamination surface, which is not the case upon studying a specimen's fracture surface after testing. Therefore, the finite element model, although accurate for a specimen without a crack, becomes too compliant for the specimen with a crack.

Figure 14 shows the change in experimental compliance, dC, from no delamination to the measured delamination, a, for each specimen. The Adams(2) solution uses this value for calculating GIIc. Figure 15 is a plot of the failure load squared, P$^2$, as a function of crack length for each specimen. All three methods use this value for calculating GIIc. The measured experimental specimen compliances (C) with a delamination, shown in Figure 12, are used in the Russell(1) solution for $G_{IIc}$. Any variance in these experimental values, C, dC and $P^2$, will influence the variability in $G_{IIc}$ calculated from any of the three methods.

Table 4 shows an example of the effects of different models on finite element results. There was no difference between using nominal and as-built layer thicknesses in the finite element model. Smearing the layers increases the specimen compliance but decreases $G_{IIc}$, because the element length ($\Delta a$) has increased from 0.02 inches to 0.10 inches (see equation (3)). In general, smearing the layers decreased the finite element values for $G_{IIc}$ by about 9-10%.

Table 5 shows the values for $G_{IIc}$ calculated by the three different methods. In general, the finite element results were higher than the other two methods, because the finite element compliances were larger than the experimental compliances.

5.2.2  Temperature Results

Table 6 shows the values for $G_{IIc}$ calculated by the three data reduction methods; Adams, Russell and Crack Closure. The degraded material properties used in the finite element crack closure method (for 180, 240 and 300°F) were calculated by assuming the matrix properties only are affected by temperature, and are contained in Table 1. In addition, the finite element grid used smeared nominal layer thicknesses and a 0.1 in. grid spacing.

Figure 16 shows the effect of temperature on $G_{IIc}$ using the three methods. Although neat resin fracture toughness increases with increasing temperature, these results show that $G_{IIc}$ decreases with increasing temperature. This trend agrees with the results obtained by Russell and Street(4).

Russell and Street show that $G_{IIc}$ decreases with increasing temperature for AS1/3501-6 composite, but their values for $G_{IIc}$ are lower than the values determined in this test for AS4/55A. There are some possible explanations for this discrepancy. First, the materials are different, and $G_{IIc}$ is a material property. HBRF-55A is a slightly tougher resin than 3501-6. Second, the specimens tested by Russell and Street(4) were unidirectional with a 0.001 in. thick layer of teflon inserted between the center plies. Our specimens were machined from tag end unit DA-010, had a hoop/helical layup, and had a razor blade-induced delamination in the center.

6.  CONCLUSIONS

6.1  Mode I

$G_{Ic}$ for full thickness TFS-1 and DA-003 material was found to be dependent on crack length, and was thus determined to be a structural rather than material property.

These results indicate that for TFS-1 and DA-003 material, it becomes increasingly difficult to propagate a crack as the crack grows. The F. E. analysis was not extensive enough to determine whether $G_{Ic}$ varies linearly with crack length, or whether $G_{Ic}$ levels off at some maximum.

The finite element analysis showed that $G_I$ varied with the mesh size as well as with crack length. It was found that $G_{Ic}$ for DA-003 varied from .78 to 2.84 in-lb/in2 and that $G_{Ic}$ for TFS-1 membrane varied from 0.45 to 1.59 in-lb/in2 depending on the two variables. It was also determined that the effect of mesh size decreased as the crack length increased.

## 6.2    Mode II

### 6.2.2    Ambient Tests

Compliances (without a delamination) calculated for the axial specimens using equation (2) were much stiffer than the experimental and finite element compliances.

With the delamination, the closed form axial compliances were still stiffer than the experimental compliances. The finite element results are more compliant than the experiment.

There was no difference between using nominal and as-built layer thicknesses in the finite element model. In general, smearing the layers decreased the finite element values for GIIc by about 9-10%.

Using the three methods: Russell(1), Adams(2) and crack closure, the average values for GIIc ranged from 6.1 to 14.6 in-lb/in$^2$ for DA-010 and 5.3 to 9.4 in-lb/in$^2$ for DC-001. The range of these values includes some scatter in the experimental data. Note that in all cases the values are higher than the 3.0 value used as the initial allowable.

### 6.2.2    Temperature Tests

These tests show that GIIc decreases with increasing temperature for FWC membrane material. The values for GIIc were calculated using three different methods for comparison purposes. Since full-scale analysis uses the crack closure method for predicting delamination growth, the GIIc allowables calculated with the crack closure method should be used.

## 7.    REFERENCES

1.    Russell, A. J. "On the Measurement of Mode II Interlaminar Fracture Energies," Defence Research Establishment Pacific, Victoria, BC, 1982

2.    Adams, D. S., "RI85B10203: Analytical Damage Assessment, Final Report," Jan., 1986.

3.    Russell, A. J., and Street, K. N., "Factors Affecting the Interlaminar Fracture Energy of Graphite/Epoxy Laminates," "Progress in Science and Engineering of Composites," T. Hayashi, K. Kawata, and S. Umekawa, ED., ICC-1V Tokyo, 1982, pp. 279-286.

4.    Russell, A. J., and Street, K. N., "Moisture and Temperature Effects on Mixed Mode Delamination Fracture of Unidirectional Graphite/Epoxy," Defence Research Establishment Pacific, Victoria, BC, 1983.

5.    Bascom, W. D., Bullman, G. W., Hunston, D. L., Jenson, R. M., "The Width-Tapered Double Cantilever Beam for Interlaminar Fracture Testing," Proceedings of the 29th National SAMPE Symposium, Reno, NV, April 3-5, 1984, pp. 970-978.

## 8.    BIOGRAPHIES

Tammy J. Todaro is an Engineer in the Materials Characterization Group at Hercules Aerospace Company, Magna, Utah. She received a BS degree in Engineering Science and Mechanics from Virginia Polytechnic Institute and State University (1984). She has been engaged in several programs related to composite materials characterization, fracture analysis, and damage assessment of composite structures.

Carroll L. Dodson is an Engineer in the Materials Characterization group at Hercules Aerospace Company, Magna, Utah. She received a BS degree in Aeronautical and Astronautical Engineering from Massachussetts Institute of Technology (1983). She has been involved in several programs with the experimental determination of composite fracture properties and the finite element analysis of composite structures.

Scott W. Beckwith is Manager
of the Materials Characterization
group at Hercules Aerospace
Company, Magna, Utah. He has been
engaged in research and development
programs pertaining to solid
rocket propellants and composite
materials. He has been conducting
several programs on composite
damage assessment and fracture
mechanics. He received a BS
degree from Texas A&M University
(1964) a MS degree from the
California Institute of Technology
(1965) and a PhD degree from Texas
A&M University (1974).

## TABLE 1. LAMINA PROPERTIES OF FWC MATERIAL

| Property | Temperature (°F) | | | | | | |
|---|---|---|---|---|---|---|---|
| | 31 | 77[a] | | 120 | 180 | 240 | 300 |
| | | AS4/ HBRF 55A | AS4/ HBRF 3501-6 | | | | |
| $E_{11}$ (msi) | 17.70 | 17.70 | 17.30 | 17.70 | 17.70 | 17.70 | 17.70 |
| $E_{22}$ (msi) | 0.300 | 0.300 | 1.173 | 0.300 | 0.230 | 0.190 | 0.149 |
| $G_{12}$ (msi) | 0.670 | 0.670 | 0.575 | 0.670 | 0.515 | 0.424 | 0.334 |
| $G_{23}$ (msi) | 0.114 | 0.114 | 0.426 | 0.114 | 0.114 | 0.114 | 0.114 |
| $v_{12}$ | 0.267 | 0.267 | 0.303 | 0.267 | 0.267 | 0.267 | 0.267 |
| $v_{23}$ | 0.309 | 0.309 | 0.375 | 0.309 | 0.309 | 0.309 | 0.309 |

a. Properties for both materials are shown for 77°F only. The other temperatures show only the properties for AS4/HBRF-55A

## TABLE 2. AVERAGE $G_I$ VALUES FOR TFS-1 MEMBRANE TESTS

| Mesh Size | $G_I$ Crack Length (in.) | | | |
|---|---|---|---|---|
| | 1.0 | 1.25 | 1.50 | 1.75 |
| 0.025 | 0.192 | 0.434 | 0.534 | 0.644 |
| 0.050 | 0.231 | 0.442 | 0.542 | 0.652 |
| 0.077 | 0.247 | 0.440 | 0.566 | 0.670 |
| 0.100 | 0.250 | 0.468 | 0.549 | 0.684 |
| 0.125 | 0.267 | 0.452 | 0.553 | 0.666 |

## TABLE 3. AVERAGE $G_I$ VALUES FOR DA-003 MEMBRANE TESTS

| Mesh Size | $G_I$ Crack Length (in.) | | | |
|---|---|---|---|---|
| | 1.0 | 1.25 | 1.50 | 1.75 |
| 0.025 | 0.187 | 0.430 | 0.534 | 0.645 |
| 0.050 | 0.219 | 0.435 | 0.539 | 0.651 |
| 0.080 | 0.242 | 0.437 | 0.538 | 0.683 |
| 0.100 | 0.257 | 0.465 | 0.548 | 0.685 |
| 0.125 | 0.271 | 0.449 | 0.553 | 0.666 |

**TABLE 6. TESTS RESULTS**

| Temp | Specimen | Experimental Compliance (x 10⁻⁵ in.) w/o crack | Experimental Compliance (x 10⁻⁵ in.) w/crack | Failure Load (lb) | Crack Length (in.) | $G_{IIc}$ (in-lb/in²) Adams | $G_{IIc}$ (in-lb/in²) Russell | $G_{IIc}$ (in-lb/in²) Crack Closure |
|------|----------|------|------|------|------|------|------|------|
| 31°F | 5B | 2.460 | 3.127 | 3000 | 2.0 | 20.01 | 15.13 | 13.26 |
|  | 23A | 2.460 | 3.152 | 2500 | 2.3 | 12.01 | 13.65 | 12.69 |
| 77°F | DA11-4B | -- | 2.050 | 2680 | 2.1 | -- | -- | 10.83 |
|  | DA11-1B | -- | 2.130 | 2400 | 2.1 | -- | -- | 8.69 |
|  | DA11-4A | -- | 2.080 | 2480 | 2.3 | -- | -- | 11.49 |
| 120°F | 1B | 2.600 | 3.333 | 2400 | 2.1 | 13.19 | 11.35 | 9.50 |
|  | 3A | 2.560 | 3.317 | 2300 | 2.0 | 13.35 | 9.44 | 7.79 |
|  | 22A | 2.460 | 3.317 | 2300 | 2.4 | 11.93 | 12.98 | 11.82 |
| 180°F | 8B | 2.360 | 3.250 | 2000 | 2.2 | 10.47 | 8.69 | 7.83 |
|  | 13B | 2.460 | 3.152 | 2300 | 2.2 | 10.77 | 11.15 | 10.35 |
| 240°F | 12A | 2.800 | 3.822 | 1800 | 2.2 | 9.74 | 8.28 | 6.88 |
|  | 12B | 2.700 | 3.774 | 1855 | 2.1 | 11.55 | 7.68 | 6.55 |
|  | 22B | 2.660 | 3.333 | 2250 | 2.1 | 10.65 | 11.30 | 9.64 |
| 300°F | 1A | 2.800 | 6.944 | 900 | 2.4 | 8.83 | 4.16 | 2.32 |
|  | 3B | 2.700 | 5.750 | 1000 | 2.2 | 8.97 | 3.84 | 2.35 |
|  | 10A | 2.660 | 4.167 | 1200 | 2.1 | 6.78 | 3.55 | 3.03 |

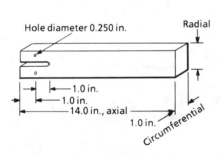

FIGURE 1. MODE I DOUBLE CANTILEVER BEAM TEST SAMPLE

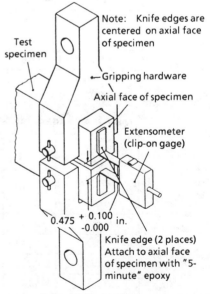

FIGURE 2. INSTRUMENTATION FOR MODE I INTERLAMINAR FRACTURE TOUGHNESS TEST

1554

**TABLE 4. EFFECTS OF DIFFERENT MODELS ON FINITE ELEMENT RESULTS**
(EXAMPLE: DC-01 AXIAL - 1)

|  | Nominal Thicknesses | Smeared Nominal Thicknesses | As-Built Thicknesses | Smeared As-Built Thicknesses |
|---|---|---|---|---|
| Compliance w/o delamination ($\times 10^{-5}$ in.) | 2.08 | 2.42 | 2.08 | 2.32 |
| Compliance w/delamination ($\times 10^{-5}$ in.) | 3.01 | 3.24 | 2.99 | 3.11 |
| $G_{IIc}$ (in-lb/in$^2$) | 9.8 | 8.9 | 9.6 | 8.6 |

**TABLE 5. CALCULATED VALUES FOR $G_{IIc}$ USING DIFFERENT METHODS**

| Specimen | Failure Load (lb) | $G_{IIc}$ (in-lb/in$^2$) | | |
|---|---|---|---|---|
|  |  | Closed-Form (Russell) | Closed-Form (Adams) | Finite Element[a] |
| • DA-10 axial |  |  |  |  |
| 10B | 2320 | 9.6 | 5.6 | 11.7 |
| 11B | 2960 | 13.1 | 6.8 | 17.2 |
| 14B | 2760 | 11.8 | 6.0 | 14.9 |
|  | Avg | 11.5 | 6.1 | 14.6 |
|  | $C_v$ (%) | 15.7 | 9.8 | 19.2 |
| • DC-01 axial |  |  |  |  |
| 1 | 2000 | 8.1 | 4.7 | 9.6 |
| 2 | 2000 | 8.6 | 5.8 | 9.6 |
| 3 | 2040 | 8.5 | 5.5 | 9.0 |
|  | Avg | 8.4 | 5.3 | 9.4 |
|  | $C_v$ (%) | 3.6 | 11.3 | 3.2 |

a. As-built

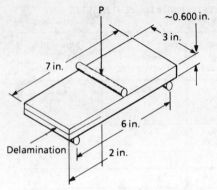

FIGURE 3. END-NOTCHED FLEXURE TEST
SPECIMEN CONFIGURATION

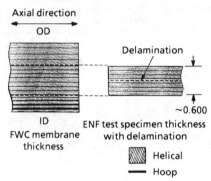

Axial direction

FIGURE 4. END-NOTCHED FLEXURE
TEST SPECIMEN THICKNESS AND
DELAMINATION LOCATION

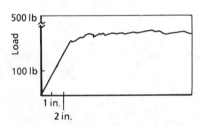

DISPLACEMENT 1 IN. CHART = 0.005 IN.

FIGURE 5. LOAD-DISPLACEMENT PLOT
FOR DA-003-163-FDC-2

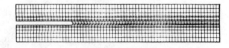

FIGURE 6. FINITE ELEMENT MESH WITH
0.10 IN. LONG ELEMENTS NEAR NOTCH TIP

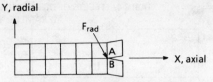

FIGURE 7a. CRACK CLOSURE METHOD:
RADIAL FORCE REQUIRED TO HOLD NODES
A AND B TOGETHER.

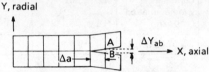

FIGURE 7b. CRACK CLOSURE METHOD:
RELATIVE RADIAL DISPLACEMENT OF NODES
A AND B WHEN CRACK IS ALLOWED TO
OPEN ONE ELEMENT, Δa

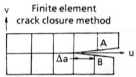

$v$    Finite element
crack closure method

$T_{ab}$    tangential nodal forces required
to hold nodes A and B together

$(u_a - u_b)$    displacement between
nodes A and B

$$G_{II_c} = \frac{1}{2\Delta a} \, T_{ab} \, (u_a - u_b)$$

Compliance methods
Direct method

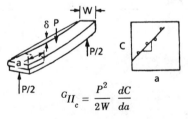

$$G_{II_c} = \frac{P^2}{2W} \, \frac{dC}{da}$$

Semi-direct method

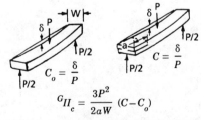

$$C_o = \frac{\delta}{P} \qquad C = \frac{\delta}{P}$$

$$G_{II_c} = \frac{3P^2}{2aW} \, (C - C_o)$$

FIGURE 8. SCHEMATIC OF THREE MODE II
DATA REDUCTION METHODS

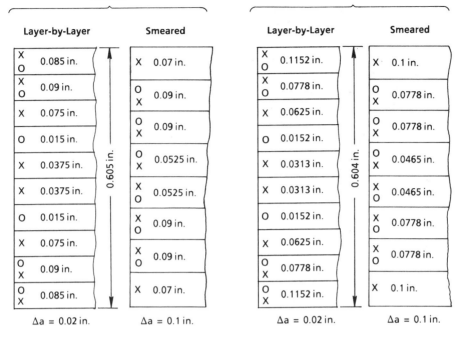

|  | As-Built | | | Nominal | | | |
|---|---|---|---|---|---|---|---|

**As-Built**

**Layer-by-Layer** / **Smeared**

| Layer-by-Layer | | Smeared | |
|---|---|---|---|
| X O | 0.085 in. | X | 0.07 in. |
| X O | 0.09 in. | O X | 0.09 in. |
| X | 0.075 in. | O X | 0.09 in. |
| O | 0.015 in. | O X | 0.0525 in. |
| X | 0.0375 in. | X O | 0.0525 in. |
| X | 0.0375 in. | X O | 0.09 in. |
| O | 0.015 in. | X O | 0.09 in. |
| X | 0.075 in. | X | 0.07 in. |
| O X | 0.09 in. | | |
| O X | 0.085 in. | | |

0.605 in.

Δa = 0.02 in.          Δa = 0.1 in.

**Nominal**

**Layer-by-Layer** / **Smeared**

| Layer-by-Layer | | Smeared | |
|---|---|---|---|
| X O | 0.1152 in. | X | 0.1 in. |
| X O | 0.0778 in. | O X | 0.0778 in. |
| X | 0.0625 in. | O X | 0.0778 in. |
| O | 0.0152 in. | O X | 0.0465 in. |
| X | 0.0313 in. | X O | 0.0465 in. |
| X | 0.0313 in. | X O | 0.0778 in. |
| O | 0.0152 in. | X O | 0.0778 in. |
| X | 0.0625 in. | X | 0.1 in. |
| O X | 0.0778 in. | | |
| O X | 0.1152 in. | | |

0.604 in.

Δa = 0.02 in.          Δa = 0.1 in.

HOOP        O
HELICAL     X

**Figure 9. Finite Element Model Thicknesses**

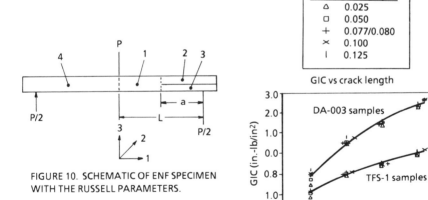

FIGURE 10. SCHEMATIC OF ENF SPECIMEN WITH THE RUSSELL PARAMETERS.

| Mesh size (in.) | |
|---|---|
| △ | 0.025 |
| □ | 0.050 |
| + | 0.077/0.080 |
| × | 0.100 |
| \| | 0.125 |

GIC vs crack length

DA-003 samples

TFS-1 samples

FIGURE 11. $G_{IC}$ AS A FUNCTION OF CRACK LENGTH AND MESH SIZE FOR TFS-1 AND DA-003 SAMPLES

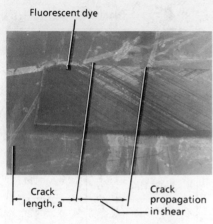

FIGURE 12. FIGURE SURFACE OF AN
END-NOTCHED FLEXURE SPECIMEN

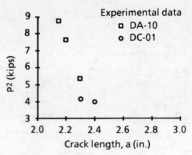

FIGURE 15. FAILURE LOAD SQUARED
VERSUS CRACK LENGTH

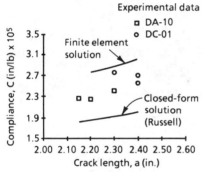

FIGURE 13. COMPLIANCE VERSUS
CRACK LENGTH

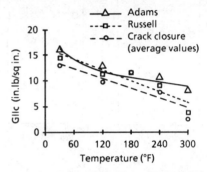

FIGURE 16. COMPARISON OF $G_{IIc}$ CALCULATED
USING THE THREE METHODS

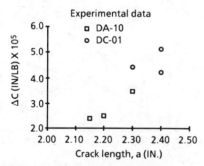

FIGURE 14.   CHANGE IN EXPERIMENTAL
COMPLIANCE FROM NO DELAMINATION TO
THE MEASURED DELAMINATION (a)

# Design, Fabrication and Performance
## of Fiber FP/Metal Matrix Composite Connecting Rods

F. Folgar, J.E. Widrig and J.W. Hunt
E. I. du Pont de Nemours & Co. (Inc.)
Textile Fibers Department
Wilmington, Delaware

## Abstract

Aluminum and magnesium castings reinforced with Fiber FP, Du Pont's alumina fiber, are being developed for high-performance automotive applications. The connecting rod has been selected as the model to assess the value of these reinforced castings, and to develop design and fabrication technology. While they are lighter and may offer better compressive strength, stiffness and fatigue resistance than conventional engine materials, their design still represents a major technical challenge. A comparative study was undertaken to predict the structural behavior of connecting rods using two- and three-dimensional finite element stress analysis models, and to determine the most cost effective modeling and analysis approach. Results from all the analyses matched closely and correlated well with static and fatigue testing of FP/Aluminum and FP/Magnesium cast connecting rods.

Useful alloy composition, casting conditions and machining techniques were found. Computerized industrial tomography was identified as a reliable method of determining casting quality.

## 1. INTRODUCTION

Overall performance of internal combustion engines is affected by the high inertial forces generated by the moving parts. Consequently, there is growing interest in use of lightweight materials for reciprocating components of these engines. For example, in the case of in-line 4-cylinder engines, a 50% reduction in reciprocating weight may be required to reduce secondary harmonic vibrations.(1)* Reciprocating components thus appear to be potentially attractive applications for aluminum or magnesium castings which have been reinforced with Fiber FP, Du Pont's alumina fiber. Aluminum castings containing 50-55% by volume of Fiber FP have less than half the density of steel and their fiber-direction modulus greater than that of steel; in addition, they exhibit very high compressive strength and fatigue resistance.

Before FP/metal matrix composites could be considered by design engineers, an analysis approach was required which predicted the structural behavior of the components, and prototypes had to be fabricated and tested to verify the stress analysis. We chose to work with the connecting rod, because it is one of the most severely stressed components in the engine. In this paper, two- and three-dimensional finite element analyses of connecting rods are discussed to determine the most cost effective modeling technique. Prototype rods have been fabricated using a vacuum infiltration process and tested both under static loading conditions to verify the linear elastic stress analysis, and under fatigue conditions to estimate their durability in an engine.

In addition, preliminary guidelines for the machining of these reinforced castings have been developed, and a reliable non-destructive test method, Computerized Industrial Tomography (CIT), has been identified which will assess the quality of these reinforced castings and will eventually predict their performance.

## 2. DESIGN

### 2.1 Kinematic Analysis

A kinematic model was developed to analyze the motion of the piston-connecting rod-crankshaft assembly; it permitted calculation of total forces

---

*Numbers in parentheses designate references at end of paper.

on the rod as a function of specific engine conditions (2). As an example, forces on connecting rods in a 4-cycle, 4-cylinder engine were calculated to assess the effects of lighter-weight rods under the same engine conditions. Table I lists key engine parameters, including firing pressure, rpm and piston weight.

Table II shows the maximum axial forces on the rods as a function of engine speed. It clearly shows that for rods running in the same engine at 6000 rpm, the 33.5% lighter FP/aluminum rod will experience a 16.4% lower maximum axial load at the crank end than will a steel rod. Loading conditions can also be determined from this analysis, i.e., the maximum forces under tension and compression that a connecting rod of a given material will experience at a specified engine speed. Once the forces are calculated, a finite element stress analysis can be performed to determine the structural design adequacy of a specific rod.

## 2.2 Stress Analysis

A comparative study was undertaken to predict the structural behavior of connecting rods and to determine the most cost-effective modeling approach. Both two- and three-dimensional models were generated using PATRAN, and then analyzed using NASTRAN or PATCHES (3). This study focussed on evaluation of alternative modeling approaches to the design of a part with complex geometry, multiple loads, and anisotropic material properties, and on comparison of modeling results with experimental data.

Figure 1 shows the section view and fiber orientation pattern of a typical split crankshaft end connecting rod with the cap attached to the main body with two bolts. The rod is selectively reinforced, i.e., the fibers are oriented to carry the principle stresses parallel to the fiber axes. For example, fibers in the shank and bolt-boss areas are axially oriented to carry large compressive forces. Fibers wound as a continuous belt around the entire rod transfer stresses from one end to the other, while fibers in the inner ring at the crank end provide the stiffness needed to minimize deformation of the inside bearing surface.

The required fiber orientation distribution was established from previous experience with steel connecting rods. For example, the piston-pin end assembly can be closely simulated by the end of an eyebar, where the distribution of pressure along the upper half of the inner edge of the hole can be approximated by a cosine function. The stress distribution on the eyebar is a two-dimensional problem for which a series solution in polar coordinates has already been derived (4). Figure 2 shows the typical stress distribution across the thickness of a critical area. These results indicated that concentric rings of reinforcing fibers were needed: an inner ring to carry large tangential tensile stresses parallel to the fibers and an outer ring to transfer the stresses to the shank area.

A detailed 3-D NASTRAN model of the Fiber FP/Aluminum connecting rod was generated (Figure 3). The orthotropic material properties of 50-55% by volume Fiber FP/Aluminum composites used in the stress analysis are listed in Table III. In actual operation, loading of the connecting rod is quite complex. Each section sees a different combination of bearing contact, internal and body forces, plus a bolt pretension force in the crankshaft end. The loading distributions used in these analyses were determined experimentally (5), and are shown in Figure 4.

Several models were attempted to find intersections usable for defining element boundaries at the crankshaft end because of its circular wall, fiber layup pattern, and cylindrical bolt hole. Fiber orientation was input throughout the model by specifying the first parametric direction of each hyperpatch.

The end-cap on the crankshaft was modeled independently, then joined to the main body via multipoint constraint equations. Several iterations were required in the tension load case to determine which nodes along the inner surface had to be released to allow the separation indicated by the NASTRAN results (Figure 5). By taking advantage of the connecting rod symmetry, only a quarter-section had to be modeled. Despite this, over 1200 eight-node hexa-elements were used in the NASTRAN model. Tension and compression loads were applied as pressure forces on the bearing surfaces, using the data patch capability. The compression load was distributed over $63°$, rather than the nominal $60°$ arc, due to hyperpatch boundary locations. Bolt pretension forces were applied directly to the bolt heads.

A partial 3-D model, including only the piston end of the connecting rod, was generated relatively easily (Figure 6). Symmetry boundaries similar to those used in the full-model analysis were employed. In addition, the NASTRAN analysis of the full connecting rod indicated an area of small axial displacement in the shank. Axial constraint of the nodes in that area completed specification of boundary conditions. A higher nodal density was used in the partial model to help determine minimum meshing requirements. Material property, orientation, and loading were specified as described previously.

An even simpler approach was to use the partial 2-D model also shown in Figure 6. Symmetry and boundary conditions similar to those used in the partial 3-D model were used. Variable thicknesses were assigned to the four-node plate elements to approximate the connecting rod's 3-D geometry. Material orthotrophy was specified by aligning patch first parametric direction with fiber orientation. Load distribution was specified using data line interpolation capability.

Results from all the analyses matched fairly closely. For example, all of the analyses predicted a highly stressed, highly deformed piston-end bearing surface where a similar connecting rod failed in the pull test. To obtain an overall picture of stresses, the two-dimensional analysis was the least expensive, although it is not possible to directly model the oil and bolt holes with the 2-D model. The assigned thickness of the plate element at that location was reduced to simulate the effect of the holes. Modeling to the detail described in the NASTRAN 3-dimensional model is very expensive and is recommended only for optimizing a design, and not for determining feasibility of a prototype.

## 3. PERFORMANCE

### 3.1 Static

Three connecting rods, similar to the ones shown in Figure 7, were tested under axial tensile load to verify the corresponding finite element analysis and expected mode of failure. Maximum axial forces as a function of engine rpm are given in Table IV. Results from a 2-D NASTRAN analysis are presented in Figure 8 as stress contours at the crank end under a tensile load of 8,900 N (2000 lbs).

All the rods failed at the location of maximum tensile stress predicted by the NASTRAN stress analysis. Cracks initiated at the inner wall of the piston end, 90° from the longitudinal axis of the rod. Figure 7 also shows the fracture mode of the rods. The average experimental breaking load for the three rods was 20,460 N (4,600 lbs.). When this load value is applied to the 2-D NASTRAN model the calculated stress at the point of failure is 519 MPa (75.2 ksi), which is very close to the 552 MPa (80 ksi) ultimate tensile strength of the rod material in the fiber direction.

The rods were further tested to investigate the strength of the crank end. The broken end of each rod was held with grips, and the load was applied through a pin-clevis fixture at the crank end. The crank end failed at an average load of 53,380 N (12,000 lbs.) (Figure 8.) The 2-D finite element stress analysis gives a stress of 519 MPa (75.3 ksi) at the point of failure which, again, is very close to the 552 MPa (80 ksi) ultimate tensile strength. Failure of the crank end also started at the inner wall, 90° from the axis of the rod, as predicted by the NASTRAN analysis.

### 3.2 Fatigue

Performance testing of FP/Aluminum connecting rods under fatigue conditions is underway. Results presented here are preliminary and come from a joint research program between Chrysler Corporation and Du Pont Company. The Chrysler rod is shown in Figure 9 and results from the kinematic analysis are presented in Table V.

Du Pont design was developed to survive 28 hours at a constant maximum engine speed of 6000 rpm, which is equivalent to 10 million cycles under a cyclic loading of 26,690 N (±6000 lbs.). Results from the stress analysis at critical areas of the rod are given in Table VI as a function of engine rpm, using a bolt preload of 53,380 N (12000 lbs) per bolt.

The first rod survived 10 million testing machine cycles at 23,570 N (±5300 lbs.) at approximately 30 Hz. Then cyclic loading was increased by 23%, to 28,910 N (±6500 lbs.), in order to reduce the testing time. Rods have still survived 10 million cycles under these new loading conditions. In the case of one that failed at the crankshaft end, near the bolt bosses, after 8 million cycles (Figure 10), careful analysis of the mode of failure revealed that cracks propagated along areas with poor bonding between the fibers and the aluminum matrix; SEM micrographs of the rod are shown in Figure 11. These findings are consistent with the stress analysis, which indicates that the stress levels in the area of failure are very low.

All rods experienced a loss of bolt torque during testing as a result of yielding of the aluminum matrix. This means that fiber reinforcement is necessary all the way up to the surface in contact with the bolt head and nut. Losing bolt torque translates into higher critical tensile stresses at the crank end which may produce eventual rod failure. In addition, bolt and nut assembly is recommended over a threaded hole design because the thread strength of the composite material is less than 40.7 - 54.2 N.m (30-40 ft.-lb.) torque depending upon the relative orientation of the fibers and the bolt thread.

## 4. FABRICATION TECHNOLOGY

### 4.1 Casting

Fiber FP reinforced aluminum and magnesium connecting rods were fabricated in a laboratory facility by vacuum infiltration of molten aluminum—2% lithium alloy into a mold containing a rod preform. Various prepregging techniques, including filament winding, were utilized to make Fiber FP rod preforms. Figure 12 shows the preform, with fibers placed in the desired orientation and proper volume loading, and the steel casting molds. Figure 13 schematically shows the step-by-step vacuum infiltration casting process. Before casting, the assembly is placed in an oven to burn off the wax used as a binder to hold the fibers together during the preform-making step. The mold is infiltrated with molten metal, directionally solidified, and the composite rod removed from the mold.

Addition of lithium to aluminum provides wetting of Fiber FP and allows the casting of composites by the vacuum infiltration technique;

however, control of the lithium concentration is required to prevent excessive fiber reaction which causes strength degradation of the fiber and a poor fiber/matrix bonding(6). The best results were obtained by using aluminum/1.8-2.2% by weight lithium and 700°C molten metal temperature. In addition, preheating the mold assembly at 700°C prior to the metal infiltration is necessary.

Figure 11 shows scanning electron micrographs of fracture surfaces of FP/Aluminum casting. Figure 11a is an example of a good bonding case between Fiber FP and the aluminum matrix, as indicated by the ductile nature of the fracture. In contrast, Figure 11b shows a clean fiber surface with no signs of the aluminum matrix sticking to the fibers.

Despite the careful handling of the Al/Li alloy required during casting, the vacuum infiltration technique has attractive features. It is a low cost and versatile process suitable for research and development of FP/Al-Li composite prototypes and unique for the magnesium metal matrix system. Magnesium and commercially available magnesium alloys naturally wet Fiber FP, thus allowing infiltration and developing a strong fiber/metal matrix bond with no fiber degradation(7).

## 4.2 Machining

Major roadblocks to the commercialization of Fiber FP/metal connecting rods have been their difficult machinability and lack of effective quality assurance techniques. Although methods are being developed to fabricate connecting rods to near net shape, machining will continue to be required to produce close dimensional tolerances and surface finish. Cleveland Twist Drill Company, a division of Acme-Cleveland, Cleveland, Ohio, has developed for Du Pont guidelines for cost effective machining of Fiber FP/metal matrix composites(8). They have found that relatively high productivity rates can be achieved with the proper selection of speed, feed rate, tool material and tool geometry.

In conventional processes such as drilling, turning and milling, best results were obtained at a speed of 30.5 surface meters per minute (100 surface feet per minute). Drilling is optimum when solid carbide high helix drills are used at a feed of 0.305 mm/revolution (.012 inches per revolution). Turning and milling are best performed with C-2 grade carbide or ceramic coated carbide inserts, and with the highest feed rate that produces acceptable surface finish. For turning, Cleveland Twist Drill Co. recommended using large nose radius or round inserts to improve surface finish at high feed rates and no oil coolant. Coolant applied to the tool resulted in accelerated wear because an abrasive slurry was generated at the cutting edge. Hence, compressed air directed to the tool to remove chips is preferred.

Less expensive solid carbide drills outperformed polycrystalline and natural diamond tool materials.

## 4.3 Non-Destructive Testing

Non-destructive test procedures are needed to monitor the quality of fiber reinforced castings and to predict their service performance. Ultrasonics and computerized industrial tomography (CIT) techniques have been considered as means to characterize defects in Fiber FP/metal components(9).

Ultrasonic scanning techniques appear adequate for quality control of flat shapes. We have investigated two ultrasonic methods, threshold and intensity level. The intensity level ultrasonic scans use shades of grey or pseudocolors to represent attenuation levels that give more information about the overall quality of the sample than do threshold ultrasonic scans. Unfortunately, both techniques have significant limitations which are compounded when specimens of irregular geometry are evaluated. As a result, the ultrasonic scans did not give reliable information on the quality of connecting rods due both to reflection and refraction of the beam at the curved surfaces and to the varying thickness of the rod.

To date, Computerized Industrial Tomography (CIT) is the most reliable (albeit expensive) non-destructive test method for flaw detection in complex shape components; Figure 14 shows the system involved in a typical tomography scan. The CIT system is able to reproduce the 3-dimensional nature of the test part by taking successive 2-dimensional slices which are then synthesized by a computer into a 3-dimensional picture. Figure 15a is a CIT scan of a Fiber FP reinforced aluminum connecting rod taken through the middle plane of the rod thickness along the length of the rod. The slice plane is 1mm thick and the defects, which are regions of low density, are shown in the scan as dark areas. To find the size of the flaw, the cross-section shown in Figure 15b was taken through the shank of the rod. When the rod was physically cross-sectioned, the observed porosity (voids and non-infiltrated fibers) in that area was clear evidence of the reliability of the CIT system. Therefore, CIT scanning systems offer several advantages over ultrasonic inspection methods and are particularly useful when only selected areas of a complicated shape part need to be analyzed.

## 5. CONCLUSIONS

We have developed high quality Fiber FP metal matrix composites that show potential to meet the demanding stiffness, strength and fatigue-endurance requirements for materials suitable for automotive engine applications.

We have designed connecting rods with the aid of a kinematic model and stress analysis using the finite element method. All analysis results were consistent with experimental data, and closely predicted the stress level at failure under static loading conditions. The 2-dimensional analysis was found to be the most cost-effective approach to determine the feasibility of a connecting rod design.

We have fabricated connecting rod prototypes that have survived more than 10 million cycles under loading conditions in excess of the maximum 6000 rpm engine speed of a typical 4-cylinder passenger car. These rods have the potential to replace steel at much lower weight and offer benefits in improved fuel economy and engine performance.

We have developed guidelines for the cost effective machining of Fiber FP reinforced metal matrix composites, and have demonstrated the reliability of Computerized Industrial Tomography (CIT) as a non-destructive analytical technique.

## 6. ACKNOWLEDGEMENTS

We would like to thank Chrysler Corporation for sharing with us their preliminary fatigue testing results of Fiber FP/aluminum rods.

## 7. REFERENCES

1. Taylor, C.F., "The Internal-Combustion Engine in Theory and Practice", the MIT Press, 1968, Vol. II, Chapter 8.

2. Folgar, F.; Krueger, W.H. and Goree, J.G., "Fiber FP/Metal Matrix Composites in Reciprocating Engines", 8th Annual Conference on Composites and Advanced Materials, The American Ceramic Society, Cocoa Beach, Florida, 1984.

3. Folgar, F.; Perez, E.H.; Hunt, J. and McCabe, D.D., "Finite Element Analysis of Fiber FP/Metal Matrix Composite Connecting Rods", 1986 PATRAN Users Conference, Newport Beach, California, June 3-5, 1986.

4. Timoshenko, S.P. and Goodier, J.N., "Theory of Elasticity", McGraw-Hill Book Company, 3rd Ed., 1970, (page 138).

5. Webster, Jr., W.D.; Coffell, R. and Alfaro, D., "A Three Dimensional Finite Element Analysis of a High Speed Diesel Engine Connecting Rod", Society of Automotive Engineers, Paper No. 831322.

6. Champion, A.R.; Krueger, W.H.; Hartmann, H.S. and Dhingra, A.K., "Fiber FP Reinforced Metal Matrix Composites", Second International Conference on Composite Materials, Toronto, Canada, 1978.

7. Dhingra, A.K. and Krueger, W.H., "New Engineering Material—Magnesium Castings Reinforced with Continuous Alumina Fiber FP", 36th Annual World Conference on Magnesium, Oslo, Norway, 1979.

8. McGinty, M.J.; Preuss, C.W., "Machining Ceramic Fiber Metal Matrix Composites", ASM's International Conference on High Productivity Machining, Materials and Processing, New Orleans, Louisiana, 1985.

9. Widrig, J.E.; McCabe, D.D. and Conner, R.L., "Non-Destructive Evaluation of Fiber FP Reinforced Metal Matrix Composites", ASTM Symposium on Testing Technology of Metal Matrix Composites, Nashville, Tennessee, November 18-20, 1985.

## Table I

### Kinematic Analysis
### 4-Cylinder Engine Parameters

| | |
|---|---|
| Connecting Rod Length | 157.00 mm |
| Crankshaft Radius | 52.00 mm |
| Crank to C.G. Distance | 42.50 mm |
| Engine RPM | 4000-6000 rpm |
| Maximum Firing Pressure | 53.70 ATM |
| Maximum Compression Pressure | 10.00 ATM |
| Piston Diameter | 87.5 mm |
| Piston Assembly Mass | 565.00 g |
| Connecting Rod Mass | |
|   Steel | 662.00 g |
|   FP/Aluminum | 440.00 g |

## Table IV

### Maximum Axial Forces On The
### 4-Cycle Engine Connecting Rods
Newtons (Lbs.)
*(Shown in Figure 7)*

| RPM | Crank-End | | Piston-Pin End | |
|---|---|---|---|---|
| | Tension | Compression | Tension | Compression |
| 4000 | 4,270 (960) | 14,680 (3300) | 2,670 (600) | 16,460 (3700) |
| 6000 | 9,790 (2200) | 9,340 (2100) | 6,000 (1350) | 13,340 (3000) |

Maximum Static Loading Conditions:
Crank End: ± 21,350 N (± 4800 lbs.)
Piston-Pin End: ± 9,340 N (± 2100 lbs.)

## Table II

### Maximum Axial Forces On A
### 4-Cylinder Engine Connecting Rod
Newtons (Lbs.)

| RPM | Material* | Crank-End | | Piston-Pin End | |
|---|---|---|---|---|---|
| | | Tension | Compression | Tension | Compression |
| 5500 | Steel | 25,440 (5720) | 17,260 (3880) | 12,990 (2920) | 19,080 (4290) |
| | FP/Aluminum | 21,260 (4780) | 13,880 (3120) | 12,990 (2920) | 19,080 (4290) |
| 6000 | Steel | 30,250 (6800) | 20,550 (4620) | 15,440 (3470) | 16,640 (3740) |
| | FP/Aluminum | 25,310 (5690) | 16,590 (3730) | 15,440 (3470) | 16,640 (3740) |

*Connecting Rod Weight
Steel: 662 g
FP/Aluminum: 440 g

## Table V

### Maximum Axial Forces
### On The Connecting Rod
Newtons (Lbs.)

| RPM | Crank-End | | Wrist-Pin End | |
|---|---|---|---|---|
| | Tension | Compression | Tension | Compression |
| 4800 | 16,190 (3640) | 15,880 (3570) | 9,920 (2230) | 2,220 (4990) |
| 5500 | 21,260 (4780) | 13,880 (3120) | 12,990 (2920) | 19,080 (4290) |
| 6000 | 25,310 (5690) | 16,590 (3730) | 15,440 (3470) | 16,640 (3740) |

Design Loading Conditions:
RPM: 6000
Crank End: ± 26,690 N (6000 lbs.)
Piston-Pin End: ± 17,790 N (4000 lbs.)

## Table III

### Material Properties
*50-55% by Volume Fiber FP Reinforced Aluminum*

| PROPERTY | DESIGNATION | N/m² | psi |
|---|---|---|---|
| Tensile Modulus, 0° | $E_1$ | 207 x 10⁹ | 30 x 10⁶ |
| Tensile Modulus, 90° | $E_2, E_3$ | 144 x 10⁹ | 20.9 x 10⁶ |
| Shear Modulus | $G_{12}, G_{23}, G_{31}$ | 48 x 10⁹ | 7 x 10⁶ |
| Poisson's Ratio, 0° | $v_{12}$ | 0.244 | 0.244 |
| Poisson's Ratio, 90° | $v_{13}$ | 0.17 | 0.17 |
| Poisson's Ratio, (Matrix) | $v_{23}$ | 0.3 | 0.3 |
| Tensile Strength, 0° | | 552 x 10⁶ | 80 x 10³ |
| Tensile Strength, 90° | | 172 x 10⁶ | 25 x 10³ |
| Compressive Strength, 0° | | 2760 x 10⁶ | 400 x 10³ |
| Compressive Strength, 90° | | 345 x 10⁶ | 50 x 10³ |
| Shear Strength | | 83 x 10⁶ | 12 x 10³ |
| Endurance Limit | | 414 x 10⁶ | 60 x 10³ |

## Table VI

### Critical Stresses
### On The Connecting Rod

| RPM | Crank-End | | Piston-Pin End | |
|---|---|---|---|---|
| | Tension Force | Tensile Stress | Tension Force | Tensile Stress |
| 4800 | 16,190 (3640) | 168 (24,300) | 9,920 (2230) | 89 (12,850) |
| 5500 | 21,260 (4780) | 220 (31,900) | 12,990 (2920) | 116 (16,800) |
| 6000 | 25,310 (5690) | 262 (38,000) | 15,440 (3470) | 138 (20,000) |

Force: Newtons (Lbs.)
Stress: MPa (psi)

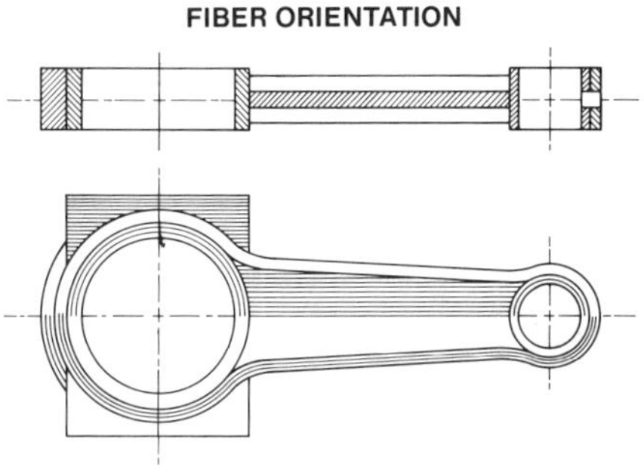

## FIBER ORIENTATION

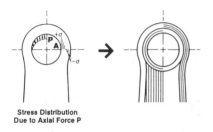

Figure 1

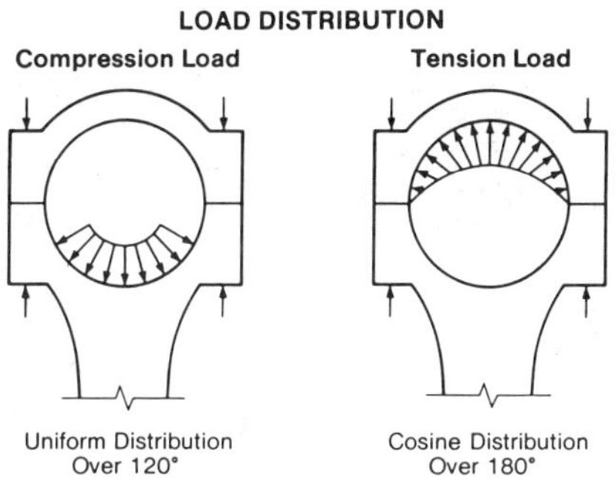

## LOAD DISTRIBUTION

**Compression Load**    **Tension Load**

Uniform Distribution
Over 120°

Cosine Distribution
Over 180°

Figure 4

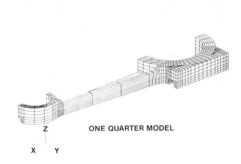

Stress Distribution
Due to Axial Force P

Figure 2

Z

X    Y

ONE QUARTER MODEL

**FIBER FP COMPOSITE CONNECTING ROD**

Figure 3

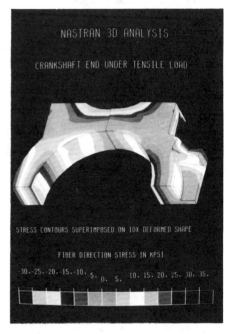

NASTRAN 3D ANALYSIS

CRANKSHAFT END UNDER TENSILE LOAD

STRESS CONTOURS SUPERIMPOSED ON 10X DEFORMED SHAPE

FIBER DIRECTION STRESS IN KPSI

-30. -25. -20. -15. -10. -5. 0. 5. 10. 15. 20. 25. 30. 35.

Figure 5

1565

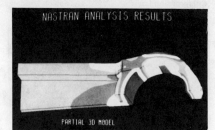

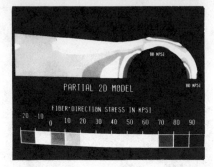

*Figure 6*

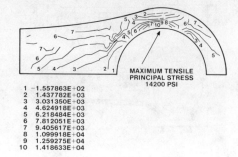

MAXIMUM TENSILE
PRINCIPAL STRESS
14200 PSI

| | |
|---|---|
| 1 | −1.557863E+02 |
| 2 | 1.437782E+03 |
| 3 | 3.031350E+03 |
| 4 | 4.624918E+03 |
| 5 | 6.218484E+03 |
| 6 | 7.812051E+03 |
| 7 | 9.405617E+03 |
| 8 | 1.099918E+04 |
| 9 | 1.259275E+04 |
| 10 | 1.418633E+04 |

*Figure 8*

*Figure 9*

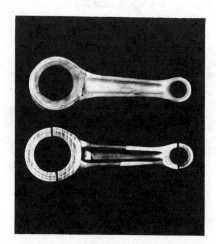

*Figure 7*

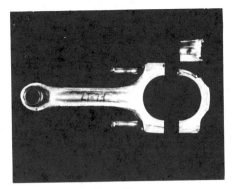

Figure 10

Figure 11b

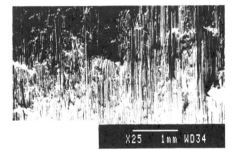

Figure 11

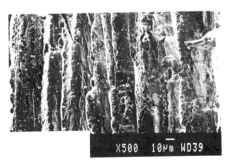

Figure 11a

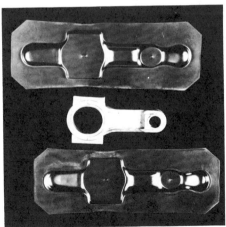

Figure 12

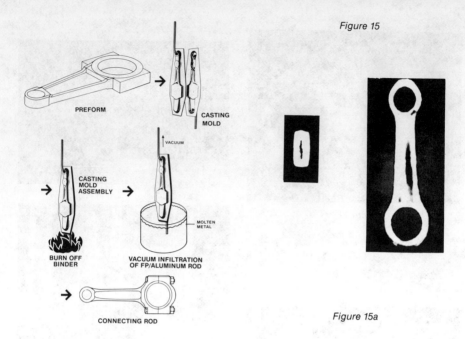

Figure 15

Figure 13

Figure 15a

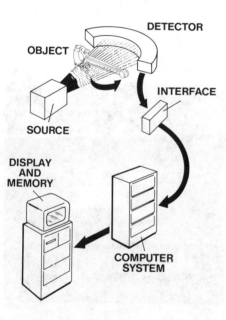

DETECTOR

OBJECT

INTERFACE

SOURCE

DISPLAY
AND
MEMORY

COMPUTER
SYSTEM

Figure 14

Figure 15b

PROTECTIVE COATINGS FOR COMPOSITE
TUBES IN SPACE APPLICATIONS

Harry W. Dursch and Carl. L. Hendricks
Boeing Aerospace Company
Seattle, Washington 98124

## Abstract

Protective coatings for graphite/epoxy (Gr/Ep) tubular structures for a manned Space Station truss structure were evaluated. The success of the composite tube truss structure depends on its stability to long-term exposure to the low Earth orbit (LEO) environment with particular emphasis placed on atomic oxygen. Concepts for protectively coating Gr/Ep tubes include use of inorganic coated metal foils and electroplating. These coatings were applied to Gr/Ep tubes and then subjected to simulated LEO environment to evaluate survivability of coatings and coated tubes. Evaluation included: atomic oxygen resistance, changes in optical properties and adhesion, abrasion resistance, surface preparation required, coating uniformity, and formation of microcracks in the Gr/Ep tubes caused by thermal cycling. Program results demonstrated that both phosphoric and chromic acid anodized Al foil provided excellent adhesion to Gr/Ep tubes and exhibited stable optical properties when subjected to simulated LEO environment. $SiO_2/Al$ coatings sputtered onto Al foils also resulted in an excellent protective coating. Electroplated Ni exhibited unacceptable adhesion loss to Gr/Ep tubes during atomic oxygen exposure.

This work was funded by NASA Langley Research Center, Contract NAS1-16854, titled "Development of Composite Tube Protective Coatings".

## 1.0 INTRODUCTION

Gr/Ep tubular struts comprise the baseline design for the large "dual keel" rectangular truss structure of the manned Space Station. These tubular struts will be approximately 2-in-diameter and up to 23-ft-long. There are many requirements for these tubes including high stiffness, dimensional stability, close dimensional tolerance and long life in LEO. The success of the composite tube truss structure depends on its ability to endure long-term exposure to various LEO environmental factors such as atomic oxygen, thermal cycling, charged particle radiation, ultraviolet radiation, micrometeriods and space debris. Atomic oxygen environment at LEO is especially severe. The recombination of atomic oxygen absorbed on unprotected Gr/Ep surfaces causes substantial erosion of the composite. Combined effects must also be taken into account. For example, micrometeroid penetration of protective coatings would allow a mechanism for atomic oxygen degradation of the Gr/Ep tube.

This paper describes the development and evaluation of LEO protective coatings applied to Gr/Ep tubular struts. Candidate coatings included anodized Al foil, sputtered $SiO_2$/sputtered Al/Al foil, bare Al foil and electroplated Ni. Adhesive systems and fabrication techniques were evaluated for bonding foils to the tubes and surface treatments were evaluated for promotion of electroplating adherence. Evaluation of coatings and coated tubes included resistance to atomic oxygen (using Boeing's large scale plasma atomic oxygen screening facility), thermal cycling, abrasion resistance, bond surface preparation, formation of microcracks in tubes, changes in optical properties and adhesion after testing, and coating uniformity.

Four 2-in-diameter by 8-ft-long Gr/Ep tubes, fitted with representative space-erectable truss structure end fittings, were fabricated with the selected protective coating to verify full scale fabrication techniques.

## 2.0 GR/EP COMPOSITE SELECTION AND PROCESSING

The composite material selected for tube fabrication was P75S/934 Gr/Ep manufactured by Fiberite Corporation. It was supplied as a unidirectional prepreg tape with a per ply thickness of 0.005-in. This material meets the primary design requirements of high composite stiffness (longitudinal tensile modulus > 45 Msi), relatively large data base, and commercial availability. P75S is a high modulus graphite fiber manufactured by Amoco Performance Products (formally Union Carbide) and 934 is an epoxy resin manufactured by Fiberite.

Composite tube ply orientation selected was $(0_2,\pm20,0_2)_s$. This selection was based on analysis of P75S/934 composite ply orientations using a Boeing-developed computer program called INCAP. Analysis methods contained in INCAP are based on classical lamination theory; the base material properties of P75S/934 were developed from results of industry test data. Three basic ply stacking techniques were analyzed: $(0_2,\pm\theta,0_2)_s$, $(\theta,0,-\theta,0,\theta,0,)_s$ and

$(\pm\theta,0_2,\mp\theta)_s$. Table 1 shows the results of this analysis. The orientation $(0_2,\pm20,0_2)_s$ was selected based on possessing a minimum modulus of 40 Msi and adequate crushing strength.

Construction technique selected for fabrication of the required composite tubes was convolute wrapping. Prepreg piles were wrapped onto the mandrel using a rolling table. Before the initial ply is wrapped, the mandrel has had several coats of releasing agents applied to ensure release of the cured composite tube. The wrapped mandrel was vacuum bagged, thermocoupled and then cured at 350°F for 2 hours.

## 3.0 PROTECTIVE COATING EVALUATION

Several concepts for protectively coating Gr/Ep tubes were evaluated including; anodized Al foil, Al foil sputter-coated with Al and $SiO_2$, electroplated Ni (with and without $SiO_x$ coatings) and inorganic sol gel solutions. Except for the large area $SiO_2$ coatings and $SiO_x$ depositions, all the above coatings (along with the Gr/Ep tubes) were deposited or fabricated in Boeing facilities. The coatings were required to meet targeted optical values of an AM-O solar absorptance = 0.20 to 0.35 and a thermal emittance = 0.15 to 0.25. Low absorptance values reduce the maximum tube temperatures when exposed to direct or albedo radiation and low values of emittance reduce minimum temperatures when exposed to deep space. A low specular reflectance was also a design goal. This would provide astronauts with a non-mirror like surface when conducting EVA. Figure 1 shows the predicted temperature range that a Gr/Ep tube, wrapped with Al foil possessing the required optical values, would undergo in LEO orbit. The maximum temperature is on the front side and is predicted to be +65F and the minimum temperature would be -55F on the backside of the tube.

Screen testing consisted of establishing a coating's ability to achieve the desired optical properties and possessing processing parameters that are compatible with the Gr/Ep tubes. Results of screen testing narrowed the initial list of coatings down to the following five: chromic and phosphoric acid anodized Al foil, sputtered $SiO_2$/sputtered Al/Al foil and electroplated Ni with and without an $SiO_x$ coating.

Various thicknesses and tempers of Al foil were evaluated. All foil evaluated was Al alloy 1145. The 0.002-in-thick, 1145-H19 (fully strain-hardened) Al foil was selected as the lightest weight Al foil that could be consistently wrapped onto the 2-in-diameter tubes without tearing or pinholes caused by handling. Thicker Al foil can be bonded to the tubes to improve the resistance to impact damage, if it is required.

## 3.1 Anodized Al Foil

Anodizing of Al foil was performed initially using Boeing specifications. After anodizing, optical properties of the foil were determined. This established what could be achieved using Boeing specifications. Follow-up samples were then fabricated using modified anodizing parameters in an attempt to achieve the targeted optical values. Because anodizing was performed in production tanks, it was impossible to modify various acid solution/water percentages. Parameters that were varied included immersion time in the acid solution and/or ramp time to the desired voltage. The foils were not sealed because of expected adverse effects on the foil to Gr/Ep bonding strengths (ref. 1).

## 3.2 Sputtered $SiO_2$/Sputtered Al/Al Foil

Several iterations of sputtered $SiO_2$ and sputtered Al were deposited onto Al foil to determine the thicknesses required to obtain required optical values. Emittance was tailored by controlling thickness of the deposited $SiO_2$, and absorptance was tailored by controlling thickness of sputtered Al. It was found that, using Al foil as a substrate, sputtered Al layers of less than 1000-angstroms exhibited little if any grain growth, therefore no change in reflectance. Sputtered Al layers greater than 1000-angstroms developed increasing grain structure and, as a result, reflectance decreases (increased absorptance) as the grain structure increases. The optimized thickness proved to be 1-micron of $SiO_2$ and 3000-angstroms of Al.

Flexibility of the 1-micron layer of $SiO_2$ on Al foil was a concern because the coated foils would be wrapped around 2-in-diameter tubes. However, testing showed that no crazing of the $SiO_2$ took place unless the foil was actually creased by folding it in half.

## 3.3 Electroplated Nickel with and without $SiO_x$

Electroplating was selected as potential coating because of its low cost application methods, good corrosion resistance, good uniformity and ability to coat irregular-shaped surface such as tube end fittings. The exterior surfaces of the Gr/Ep tubes were sanded prior to plating to improve adhesion. These tubes were immersed in an electroless copper solution to provide the conductive surface required for electroplating. Because of expected degradation during atomic oxygen testing, $SiO_x$ coatings were deposited onto Ni. The $SiO_x$ also increased the emittance of Ni to within the targeted range.

## 4.0 COATING EVALUATION TEST RESULTS

The five selected coatings were bonded to 1-ft-long tubular sections of P75S/934 Gr/Ep tubes. These coatings were then subjected to LEO environmental testing that included atomic oxygen testing, thermal cycling, adhesion testing and abrasion resistance. Criteria used for evaluating coatings was change in optical properties and change in coating adherance.

### 4.1 Thermal Cycling

Coated tubes were initially thermal cycled in an LEO environment by placing them in a vacuum chamber with AM-0 solar simulation capabilities. Solar simulation was generated with xenon lamps providing a 35-in-diameter uniform beam. The heat sink of space was simulated using $LN_2$ shrouds. The tubes were subjected to 50 thermal cycles with each cycle consisting of 57-mins of solar and 37-mins without solar radiation. This cycle closely matches that of the Space Station at LEO. Using this testing technique, each coated tube was allowed to seek its own temperature versus time profile. Several of the tubes had thermocouples bonded to their surfaces. These profiles were used to verify analytical predictions of the temperatures during thermal cycling of the tubes. After completion of the cycling the tubes were optically evaluated, checked for formation of microcracks and evaluated for coating and foil

adhesion. There were no detectable changes in any properties including formation of microcracks. Because of the unexpected lack of microcracking, further thermal cycling was undertaken. The tubes that had been exposed to the previous 50 cycles were subjected to an additional 500 56-min, +120F to -150F thermal cycles. This testing was performed in a thermal cycling chamber and not under vacuum as the initial 50 cycles were. After completion of 500 cycles, the tubes were re-examined for microcracks using 200X photomicrographs and X-ray analysis. Again, no microcracks were found in any specimens.

### 4.2 Atomic Oxygen Testing

The coated tubes were exposed in the Boeing built large-scale plasma atomic oxygen materials screening (PAMOS) test facility. Tube specimens were exposed in the PAMOS facility for 11-hours and then removed for evaluation. The specimens were then placed back in the chamber for an additional 22-hours of exposure. This would be an equivalent of 10 months at a 305-km orbit for Kapton-H. The tubes were placed parallel to the flow to minimize turbulence. Edges and interior surfaces of the tubes were left exposed. Anodized and $SiO_2/Al$ coated specimens had minimal changes in optical and adhesion properties. All samples exhibited loss of Gr/Ep on the unprotected surfaces. Edges that were once flush with Al foil at the specimen ends, had recessed 1/16-in during the 33-hour exposure. The downstream edges degraded at similar rates as the upstream edges. Electroplated Ni exhibited total adhesion loss to the Gr/Ep. The $SiO_x$ coating did prevent degradation of optical properties that took place on uncoated tubes but this coating did not improve adhesion of Ni to the composite. During atomic oxygen testing, one of the tubes was pre-punctured to produce a 0.015-in-diameter pin hole through the coating and foil to simulate potential damage caused by micrometeoroids. Figure 2 shows a photomicrograph of a cut made through the pinhole after 22-hours of exposure in the PAMOS facility. The photo shows that 2 of the 12 plies were eroded away. Because the Al foil is inert to effects of atomic oxygen, the diameter of the pinhole through the foils remains constant. This limits the flux of atomic oxygen to the composite. Therefore, it is expected that while continued exposure would erode the Gr/Ep at a constant mass loss, because of increasing surface area, the rate of penetration would be expected to decrease. No structural testing was performed to determine the effect of erosion on mechanical properties of the tube section.

## 4.3 Adhesion Testing

Anodized and unanodized foil bonded to Gr/Ep specimens were subjected to 80 72-min, +250F to -250F thermal cycles to determine bond strengths before and after cycling. The Al foil had primer sprayed to the backside prior to bonding with 0.005-in-thick epoxy sheet adhesive to the Gr/Ep substrate. Testing of control specimens showed that while unanodized foil (backside of $SiO_2$ coated foil) was able to be peeled off the composite (average peel strength of 4-in-lb/in), the peel strength of anodized foil exceeded the tensile strength of the 0.002-in foil. Peel testing of the cycled specimens showed no decrease in peel strengths.

## 4.4. Abrasion Resistance

Abraiding the tubes by rubbing tubes with like coatings together caused the $SiO_2$/Al/Al foil tube to become darkened along the line of contact. There was no change in any of the anodized foil tubes even after being aggressively rubbed together.

## 5.0 SUMMARY AND CONCLUSIONS

Both phosphoric and chromic acid anodized Al foil proved to possess very good durability to LEO environment (no UV testing was performed) and also possessed excellent adhesion to Gr/Ep tubes. Chromic acid anodizing can be easily tailored to meet a variety of optical values by varying anodizing parameters. Phosphoric acid anodizing was not as versatile. Anodized Al foil possesses an additional benefit of being produced in large volume without a major R&D effort.

Sputtered $SiO_2$ and sputtered Al onto Al foil also possessed environmental stability similar to anodized foils although the bond strength to the composite was not as high. During abrasion testing the coating showed signs of optical degradation, but this would be a small percent of total area. A major disadvantage is the need to have large area vacuum coaters to deposit these coatings onto Al foil.

While electroplated Ni has the potential of providing conformal coatings to tubes and any irregular shaped surfaces, such as end fittings, adhesion loss during exposure to LEO environment needs to be improved. $SiO_x$ coatings demonstrated the capability of improving the durability of the electroplated Ni and also improved the ability to tailor optical properties of the Ni.

No microcracks were found in any of the P75S/934 tubes after undergoing 50 94-min, +175F to -180F thermal cycles and 500 56-min, +120F to -150F themal cycles. The use of low angle off-axis plies required to meet stiffness requirements of the Space Station truss structure minimizes microcracking.

As part of this effort, four 8-ft-long Gr/Ep tubes, wrapped with chromic acid anodized 0.002-in Al foil were fabricated and latched together using a typical space-erectable structural end fitting. The foil surface was textured during tube fabrication to increase diffuse reflectance as shown in Figure 3. The hub and stud assembly shown in Figure 4 represents a corner of an interlocking network of the Gr/Ep struts and aluminum hubs that can easily be erected by a single astronaut without tools. Threaded aluminum inserts are bonded to the inside of each Gr/Ep tube. The latching mechanism is then screwed into the tube and a locking ring is tightened to hold the devise in place. The four tubes with the latching mechanism in place are shown in Figure 5. Strut attachment to the hub is accomplished by latching the strut and hub assemblies together and then sliding the locking collar foward and rotating to secure the strut to the hub. Figure 6 shows the four 8-ft-long tubes latched together.

Chromic acid anodized 0.002-in Al foil was selected as the best coating for protecting the tubes from LEO environment due to:

o   Environmental durability in LEO including retention of foil to Gr/Ep bond strength and retention of optical properties during LEO exposure.

o   Excellent adhesion to Gr/Ep.

o   Optical tailorability.

o   Ease of manufacture and low cost.

o   Excellent handling properties.

## 6.0 REFERENCES

1.  Wernick, S. and Pinner, R., "Surface Treatment of Aluminum", Vol. 2, 1972, pp. 725.

| Layup | $E_x$ (Msi) | $E_y$ (Msi) | $G_{xy}$ (Msi) | $\alpha_x$ ($\mu$in/in) | $\alpha_y$ ($\mu$in/in) | $F_L^{tu}$ (ksi) | $F_T^{tu}$ (ksi) |
|---|---|---|---|---|---|---|---|
| $(0_2, \pm10, 0_2)_s$ | 45 | 0.8 | 1.1 | -0.8 | 17.5 | 105 | 2.4 |
| $(0_2, \pm15, 0_2)_s$ | 42.5 | 0.9 | 1.6 | -1.0 | 17 | 100 | 2.5 |
| $(0_2, \pm20, 0_2)_s$ | 40 | 1.0 | 2.2 | -1.1 | 14 | 95 | 2.7 |
| $(0_2, \pm30, 0_2)_s$ | 35 | 1.7 | 3.5 | -1.2 | 8 | 80 | 4.0 |
| $(10, 0, -10, 0, 10, 0)s$ | 42.5 | 0.8 | 1.4 | -0.9 | 17 | 95 | 2.5 |
| $(20, 0, -20, 0, 20, 0)s$ | 35.5 | 1.0 | 2.9 | -1.3 | 13.5 | 80 | 2.7 |
| $(30,0, -30, 0, 30, 0)s$ | 28 | 1.8 | 4.5 | -1.3 | 7 | 65 | 4.0 |
| $(\pm10, 0_2, \mp10)_s$ | 42.5 | 0.8 | 1.6 | -1.0 | 16.5 | 102 | 3.0 |
| $(\pm20, 0_2, \mp 20)_s$ | 32 | 1.0 | 3.8 | -1.6 | 12.5 | 74 | 4.4 |
| $(\pm30, 0_2, \mp 30)_s$ | 22 | 1.8 | 6.2 | -1.6 | 6 | 52 | 8.6 |

*Table 1. Composite Matrix Properties*

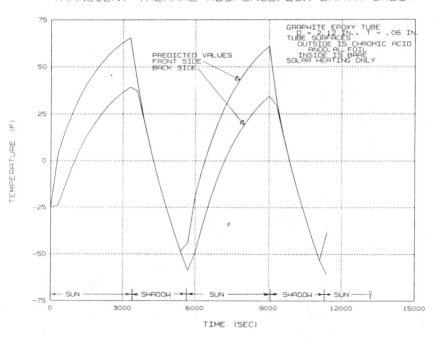

*Figure 1. Predicted LEO Temperature Range*

1.4X magnification

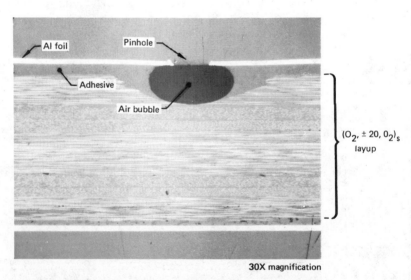

30X magnification

Figure 2. Effects of Pin Holes During Atomic Oxygen Exposure

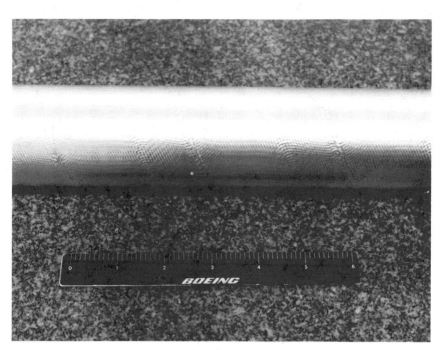

Figure 3. Chromic Acid Anodized Foil With Textured Surface

Figure 4. Space-Erectable End Fitting

Figure 5. Four 8-ft-Long Gr/Ep Tubes Wrapped With Aluminum Foil

Figure 6. Four 8-ft-Long Gr/Ep Tubes Latched Together